13659

KT-152-343

303218696

MICROBIOLOGY

Fundamentals and Applications

E £32·00
SM

MICROBIOLOGY

Fundamentals and Applications

Ronald M. Atlas

University of Louisville

LINACRE COLLEGE WITHDRAWN

LINACRE COLLEGE LIBRARY
UNIVERSITY OF OXFORD

Macmillan Publishing Company
New York

Collier Macmillan Publishers
London

Copyright © 1984, Macmillan Publishing Company, a division of Macmillan, Inc.
Printed in the United States of America.

All rights reserved. No part of this book may be reproduced or transmitted in any form or by any means, electronic or mechanical, including photocopying, recording, or any information storage and retrieval system, without permission in writing from the Publisher.

Macmillan Publishing Company
866 Third Avenue, New York, New York 10022

Collier Macmillan Canada, Inc.

Library of Congress Cataloging in Publication Data

Atlas, Ronald M.
  Microbiology: fundamentals and applications.

  Includes bibliographies and index.
  1. Microbiology.   I. Title.
QR41.2.A84 1984   576   83-13603
ISBN 0-02-304550-7

Printing: 2 3 4 5 6 7 8   Year: 4 5 6 7 8 9 0 1 2

ISBN   0-02-304550-7

Preface

Microbiology plays an essential role in our daily lives and an understanding of microbiology is needed by all educated citizens. One cannot read a daily newspaper or weekly news magazine without finding articles relating to microbiology, whether it be about human health or environmental quality. This textbook is intended for students who require a knowledge of microbiology for their future careers, as well as for students who wish a knowledge of microbiology for use in their daily lives. *Microbiology: Fundamentals and Applications* examines the broad vistas of microbiology and provides the student with a basis for understanding the varied aspects of this discipline, thus meeting the needs of students even if they are experiencing their first exposure to college-level biology. At the same time, sufficient material and depth is provided to stimulate and satisfy the more advanced student.

The book is intended to be comprehensive and to meet the varied teaching needs of different institutions. Instructors will be able to assign some of the basic chapters as review to ensure that all students in the course begin on an equal plane. Undoubtedly, there is more material in this text than can be reasonably handled in a one-semester course. Instructors may pick from the various chapters to meet their own course needs. The book is intended to be flexible, and instructors may wish to change the order of the chapters for use in their courses. Most instructors are creative in the way they teach, and a good text should augment, not restrict, the instructor's approach to teaching microbiology. One suggestion is to use the individual sections of Chapter 1 on the history of microbiology to supplement the discussions of the specific areas examined later in the book. Students who find that topics of interest to them are not covered in the lecture material will be able to supplement their coursework by reading additional chapters.

Clearly there are many good textbooks presently on the market for introductory microbiology and bacteriology courses. One must question at the outset: "Why write another book, and what are the unique features of this new textbook?" First, microbiology is a relatively new science, which continues to expand at an exceptionally rapid rate. New information continuously replaces accepted dogma, and textbooks must keep up with this outpouring of new information, making revisions or entirely new textbooks necessary every few years to update information. Second, as educational and career goals of students change, textbooks also must change to present information so that students see the relevance of what they are learning. A textbook must act as a communication device, easing the learning process and making the wealth of information available to the student on a basis that he or she can assimilate.

It has been my experience that the student population of introductory microbiology classes has been changing. Today, in addition to the traditional biology and microbiology majors, including premed and predental students, there are many students preparing for careers in the various paramedical professions who take introductory microbiology courses as part of their program. Many of these students, however, do not take the same number or kinds of advanced biology and chemistry courses as the traditional students who in the past largely populated introductory bacteriology courses. Many of today's students are ill prepared to deal with the advanced biochemical approaches used in some textbooks. Additionally, many older students are returning to the classroom. The chemistry and biology taught even a decade ago often does not provide sufficient background for considering our modern molecular understanding of biological systems. This book, therefore, has several chapters—Chapter 3, "Introduction to Biochemistry," for instance—which aim to remove such deficiencies students may have when they enter a course in microbiology. These chapters remove obstacles to those students wishing to take courses in microbiology, but who are hesitant to do so because of inadequate backgrounds. Depending on the nature of a particular course, the instructor may wish to make these chapters optional reading for the students or require their coverage to ensure a uniform background of information and understanding from which he or she may begin their lectures.

The book is organized from the basic to the practical. Following an examination of the historical development of microbiology and the basic techniques of the microbiologist, the fundamental aspects of microbiology are examined in sections on structure, metabolism, genetics, and taxonomy. These sections build from the subcellular to the whole-organism level and estáblish the fundamental aspects of microbiology needed to understand the practical or applied aspects of microbiology covered in later chapters. I have tried to achieve a balance between the fundamental and applied aspects of microbiology. Of the 23 chapters, 12 are in basic areas and 11 are in applied areas of microbiology. Within each chapter I have also related the development of a basic understanding of microbial structure and function to the applications and ramifications of that information.

Within the basic chapters, microbial metabolism is thoroughly discussed, including the impact of the discovery of chemiosmosis on our understanding of ATP generation. In Chapter 5 metabolism is considered from the viewpoint of bioenergetics and the different strategies for generating ATP. Chapter 6 then examines the flow of carbon through metabolism to convert starting substrates into the macromolecules comprising the structural units considered in Chapter 4. Many of the recent developments in the molecular level

aspects of genetics are examined in Chapters 7 and 8. This is undoubtedly the area of microbiology in which the greatest advances have occurred in the last decade and the one that therefore contains many new and exciting revelations, such as the split genes of eukaryotes. Genetic engineering, a topic of much discussion in the lay press and an area of microbiology that holds great promise for future improvements in the quality of life, is thoroughly covered in this text. The basic aspects of genetic engineering are discussed, along with other processes that give rise to genetic variation, in Chapter 9, and the applications of genetic engineering are then viewed within the applied contexts of medicine, agriculture, foods, and ecology.

The book includes coverage of both the eukaryotic and prokaryotic microorganisms, and although the emphasis is undoubtedly placed on bacteria, eukaryotic organisms and viruses are also thoroughly covered. By comparing the similarities and differences among viruses, bacteria, and eukaryotic microorganisms, the student can develop an understanding of some of the applied aspects of microbiology, such as the use of antimicrobial agents in treating infectious diseases. Though many textbooks and most introductory microbiology laboratory courses use the taxonomic approach, it has been my experience, which I am sure others have shared, that students do not find this an exciting or relevant approach to the introduction of microbiology. Students are generally much more interested in the practical aspects of microbiology, such as disease processes, and I have therefore included some of these applied aspects in the discussion of microbial taxonomy in Chapters 11 and 12. Separate treatment is given to the identification of medically important microorganisms in Chapter 16, on clinical microbiology.

Section seven, "Medical Microbiology" (Chapters 13–18) should be of great interest to students taking introductory microbiology because we all can relate to being sick. The medical microbiology section includes chapters on the interactions of microorganisms and humans, and immunology, providing the necessary introduction to pathogenicity and host defense mechanisms required to understand human resistance to disease and why these defense systems occasionally fail. "Prevention and Treatment of Human Diseases Caused by Microorganisms," Chapter 15, covers the use of antimicrobial agents for treating and preventing diseases. This chapter will be particularly useful for prepharmacy, as well as students who want to know why a physician prescribes a particular treatment. Chapter 16 deals with clinical microbiology, a topic omitted from most introductory microbiology books, but one that is quite relevant to the career goals of many students taking introductory microbiology courses. The discussion of infectious diseases (Chapters 17 and 18) is divided according to portal of entry. I find this a particularly useful approach because it emphasizes not only the disease, but the modes of transmission, and hence the means of prevention. Many diseases are covered in this chapter. Perhaps some will think the treatment too encyclopedic, but it is always the disease I forget to cover that is the one of particular interest to a student that semester because he or she has had a personal experience with it. I leave it to the instructor to choose which diseases to cover and which to omit.

Section eight deals with food and industrial microbiology. Many students are preparing for eventual careers in industry, and this section will be of particular interest and value to them. Even if course time is inadequate to include these topics within the lectures, many students will find these chapters useful and relevant supplementary reading to the main body of the introductory microbiology course. And of course, in addition to the increasing number of dieticians studying microbiology, all students everywhere are interested in food. The homemaker applies daily the principles of food preservation, and because most food-borne outbreaks of disease occur as a result of improper handling of foods, it is essential to know how to prevent microbial spoilage of the food we eat. Also, many of the products we buy are produced by microorganisms. Recombinant DNA technology promises to revolutionize industrial microbiology, and this area is a growing source of employment for microbiologists.

The environmental microbiology section covers various topics that are relevant to the quality of life as we know it. The discussion of the basic aspects of microbial ecology emphasizes those aspects of microbial metabolism and the interactions among biological populations involving microorganisms that support life on earth. The inclusion of a chapter on agricultural microbiology is relatively novel and should prove quite relevant to students in the agricultural sciences. The concluding chapter on environmental quality and pollution abatement should prove of great interest and importance to students of sanitary engineering, as well as to those of us who have a real concern about our environment. Pollution is a major problem of the world today created by humans, but one that microorganisms have been

studying for a long time and one that microorganisms are helping us solve.

Several pedagogic aids have been included in the text to help students using this text. The book is elaborately illustrated, and each figure is accompanied by a clear and thorough figure legend. Many of the illustrations also include a discussion of the discovery process to help students develop an appreciation of how microbiological research leads to our current state of knowledge. Each chapter ends with a postlude that summarizes and places the information discussed in the chapter in proper perspective. A list of questions is included at the end of each chapter for the student's use in determining if the information has been assimilated. Also, each chapter contains a list of suggested readings. This list is meant neither to be complete, nor of great depth, but to supplement the text for more advanced courses and to sustain the interest of the student who finds a particular topic relevant to his or her purpose for having enrolled in an introductory microbiology course. An extensive glossary has also been included to help the student understand the information of the book. The extensive indexing of the book permits students to cross-reference information, allowing the instructor to cover a topic in lecture and let the students use the textbook as a reference work to supplement the lecture material. The extensive use of subheadings also is meant to help cross referencing.

Many individuals contributed to the writing of this book, and their efforts indeed are appreciated. Deborah Harper labored hard at typing the manuscript. Burt Monroe Jr., chairman of my department, supported me throughout this effort. Many of my colleagues, to whom I am indebted, contributed micrographs. Carl May of Biological Photo Service found many of the photographs that help illustrate this text. I am especially grateful to the following reviewers for their comments and helpful suggestions:

Robert Bender, *University of Michigan*

John A. Breznak, *Michigan State University*

Alfred E. Brown, *Auburn University*

Warren Cook, *Georgia State University*

Arnold Demain, *Massachusetts Institute of Technology*

Kenneth W. Dobra, *University of Louisville*

Douglas Eveleigh, *Rutgers University*

Scott Graham, *University of Louisville*

James Hauxhurst, *University of Wisconsin, Fox Valley*

Alice Helm, *University of Illinois*

John Holt, *Iowa State University*

Roger Lambert, *University of Louisville*

Hubert Lechevalier, *Rutgers University*

James MacMillan, *Rutgers University*

Glenda Michaels, *Western State College, Gunnison, Colorado*

Thomas Montville, *USDA, Philadelphia*

William O'Dell, *University of Nebraska, Omaha*

Raymond Otero, *Eastern Kentucky University*

Harry E. Peery, *Tompkins Cortland Community College*

Morrison Rogosa, *National Institutes of Health*

Antonio Romano, *University of Connecticut*

Chester Roskey, *Framingham State College*

Ramon Seidler, *Oregon State University*

John Smith, *Oklahoma State University*

George Somkuti, *USDA, Philadelphia*

John Wallace, *University of Louisville*

I particularly want to thank the editors at Macmillan, Kate Aker and Greg Payne, for helping develop this book. The many hours that they spent and their constant support made the writing of this text as painless as possible, allowing me to focus on the task of writing. Many other people at Macmillan also deserve thanks for their efforts in bringing this book to fruition: Harvey Kane made sure that no problems developed on the homefront; John Snyder coordinated the various efforts in New York; and Bob Hunter and Bob Pirrung oversaw the production of the work.

Finally I want to thank my family who not only tolerated my obsession with writing this book, but who contributed to its writing. My wife Michel spent many hours researching parts of this book and editing numerous drafts; my children, Matt and Debbie, also made themselves an integral part of the work (see Figure 13.16).

R.M.A.

Contents

x

Contents

Detailed Contents

Microbial anatomy,   98

SECTION TWO

chapter 4
Organization and structure of microorganisms, 100

chapter 8
Genetic variation: the introduction and maintenance of heterogeneity within the gene pool   245

Microbial growth,   284

SECTION FIVE

chapter 9
Growth and reproduction of microorganisms,   287

Microbial taxonomy,   360

SECTION SIX

xvi

Detailed contents

Medical microbiology, 456

SECTION SEVEN

chapter 13
Interactions between microorganisms and humans, 459

chapter 17
Human diseases caused by microorganisms: the respiratory, gastrointestinal, and genitourinary tracts as portals of entry, 586

chapter 18
Human diseases caused by microorganisms: the skin as a portal of entry and diseases of superficial body tissues, 626

Food and industrial microbiology 662

SECTION EIGHT

chapter 19
Food microbiology, 665

chapter 20
Industrial microbiology, 704

Environmental microbiology, 754

SECTION NINE

chapter 21
Microbial ecology, 757

chapter 22
Agricultural microbiology, 812

Appendices

SECTION TEN

MICROBIOLOGY

Fundamentals and Applications

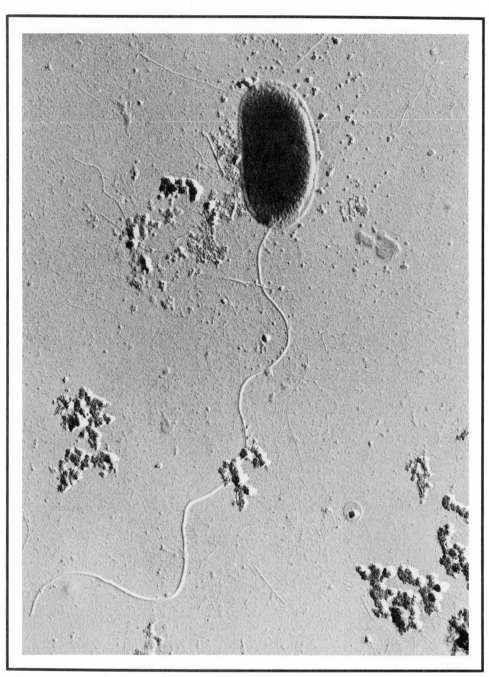

Electron micrograph of a Pseudomonas *species. (Courtesy Peter Hirsch, University of Kiel.)*

Background to the study of microbiology

SECTION ONE

Introduction to microbiology

1

Scope of microbiology

Microbiology is the science that studies microscopic organisms, including **viruses** (virology), **bacteria** (bacteriology), **fungi** (mycology), **algae** (phycology), and **protozoa** (protozoology) (Figure 1.1). These organisms are very small, and a microscope is generally required to view individual microorganisms. Despite their microscopic size and although their structural organization is simpler than that of higher plants and animals, microorganisms do exhibit great diversity of form and metabolic complexity. Higher plants and animals have an organizational structure in which cells form differentiated tissues that have specialized functions, while true tissue differentiation is lacking in microorganisms, and many microorganisms are in fact unicellular. For the most part, each microbial cell must contain the total capacity for complete self-sufficiency and reproduction.

The taxonomic position of microorganisms in the living world

When living organisms were initially classified into systematic units, the existence of microorganisms was unknown. Early classification systems recognized the plant and animal kingdoms, with the primary criterion for separation being that animals were motile and plants were stationary. Classification systems, as expounded by Aristotle and other ancient philosophers, were developed centuries before the beginnings of evolution theory. Microorganisms were considered to be little plants and animals when they were discovered in the late seventeenth century. One of the earliest classification systems that attempted to reflect evolutionary relationships was developed by a German zoologist, Ernst Heinrich Haeckel, who was a disciple of Darwin and his theories. Haeckel proposed a three-kingdom classification system that divided living organisms into animals, plants, and protists (Figure 1.2). The **Protista** were defined by a lack of tissue differentiation; therefore, in this system all microorganisms were classified as protists and, conversely, all protists as microorganisms. This system has been superseded by other theoretical systems but still retains its pragmatic appeal for microbiologists because Haeckel's kingdom Protista encompasses all the organisms traditionally studied by microbiologists.

Today many biologists use the **five-kingdom classification system** proposed by Robert H. Whittaker in 1969. According to Whittaker's proposed classification, evolutionary divergence occurs along three principal modes of nutrition: synthetic, to evolve plants; adsorptive, to evolve fungi; and ingestive, to evolve animals. In this system microorganisms exclusively comprise three of the five kingdoms: in addition to the kingdoms of Plants

Viruses
0.01–0.2 μm

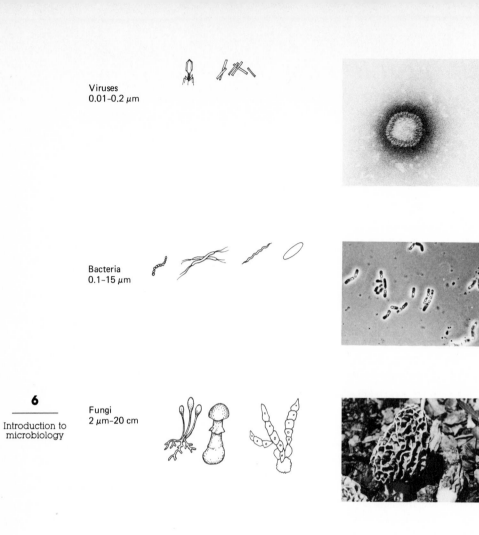

Bacteria
0.1–15 μm

Fungi
2 μm–20 cm

Algae
1 μm–many meters

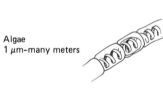

Protozoa
2–200 μm

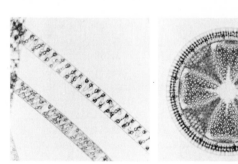

Figure 1.1 (facing page)

The term **microorganism** encompasses five major groups of organisms; the viruses, bacteria, fungi, algae, and protozoa. These organisms exhibit diversity of form and size but are similar in that a microscope is generally required for the observation of individual microbes. The micrographs shown in this figure show the following: **Viruses**—respiratory syncytial virus (150,000×) (courtesy Eugenie C. Ford, Georgetown Medical Center); Influenza A virus (200,000×) (courtesy Erskine Palmer, Centers for Disease Control, Atlanta); **Bacteria**—Bacillus thuringinesis (1500×) (courtesy B. N. Herbert, Shell Research, England) and Saprospira grandis (700×) (courtesy Hans Reichenbach, Institute for Biotechnical Research, Germany); **Fungi**—Morchella esculenta (courtesy Orson Miller, Virginia Polytechnic Institute and State University); and Baker's yeast, Saccharomyces cerevisiae (courtesy Robert Apkarian, University of Louisville); **Algae**—Spirogyra elongans; a marine diatom, Actinoptychus (courtesy Varley Wiedman); **Protozoa**—Paramecium sp. and Amoeba sp. (courtesy Eugene McArdle, Northeastern Illinois University).

Figure 1.2

Haeckel's classification system, the phylogeny of living beings, was expressed in a formal tree-like diagram in 1866. This system, based almost entirely on morphological observations, was an extension of Charles Darwin's theories on evolution and natural selection. The Protista as defined in this system are essentially synonymous with the microbial world, encompassing the bacteria, fungi, algae, and protozoa, all of which lack the organizational complexity that characterizes plants and animals.

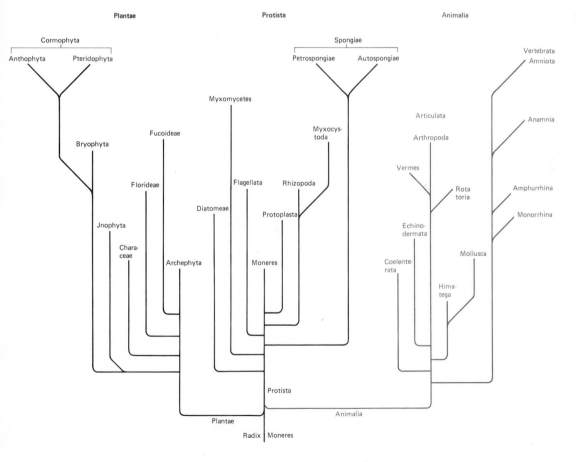

and Animals, there are the kingdoms **Monera**, which includes all bacteria; the **Protista**, which includes the algae and protozoa; and the **Fungi**, which is composed of the fungi (Figure 1.3). Whittaker's system assumes that the Monera (bacteria) represent the most primitive form of life and that evolutionary processes led to the development of the Protista (algae and protozoa), which then underwent further evolution to form the remaining three kingdoms.

Determining the validity of classification systems that aim at reflecting the process of evolutionary separation of species has been greatly bolstered by the development of molecular biology during the 1970s, which has provided the necessary tools for assessing the phylogenetic relationships among organisms at the molecular level. Closely related organisms should have similar genetic compositions. Basing his classification of living organisms on molecular-level genetic analyses, Carl Woese in 1980 proposed a new classification system with three primary king-doms: **Archaebacteria**, **Eubacteria**, and **Eukaryotes** (Figure 1.4). Both eubacteria and archaebacteria have **prokaryotic cells**, which are simply organized, lack a nucleus, and are structurally quite different from **eukaryotes**, which have a nucleus and various other organelles. The archaebacteria and eubacteria, however, are genealogically as distantly related to each other as each is to eukaryotes. Woese proposed that the **archaebacteria**, **eubacteria**, and an **urkaryote**, which was the original eukaryotic cell (German *ur* = proto-type), all evolved from a common simple ancestor, the *progenote*. According to Woese's theory, eukaryotes evolved after the urkaryote became a "host" for bacterial endosymbionts that developed into mitochondria and chloroplasts. This system is receiving a great deal of attention from microbiologists but has yet to gain wide support from general biologists. The progression of systems for classifying organisms has gone from the proposal that the biological world contains two kingdoms, plants and animals, made before the

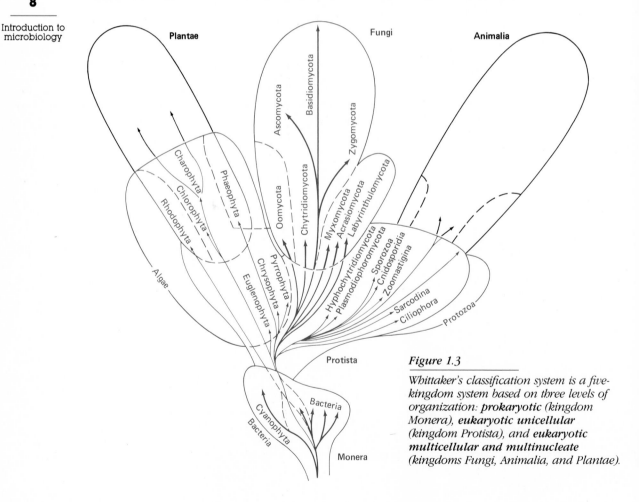

Figure 1.3

*Whittaker's classification system is a five-kingdom system based on three levels of organization: **prokaryotic** (kingdom Monera), **eukaryotic unicellular** (kingdom Protista), and **eukaryotic multicellular and multinucleate** (kingdoms Fungi, Animalia, and Plantae).*

Figure 1.4

The three-kingdom classification system Woese proposed is based on the discovery that the archaebacteria are fundamentally different from all other organisms. This theory was developed by using the modern techniques of molecular biology and in particular the examination of the macromolecules comprising key structures of the organism, especially the RNA of the ribosomes. Unlike previous classification systems, the use of analysis of conserved gene products permits a direct assessment of genetic and thus evolutionary relatedness.

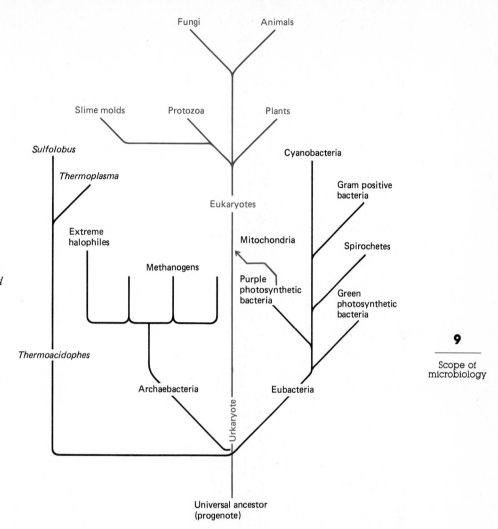

existence of microorganisms was suspected, to one in which microorganisms represent the majority of the primary kingdoms.

The **viruses** are not included in any of these classification systems and are treated as nonliving entities. The reproductive capacity of viruses is integrally linked to host organisms in which the viruses multiply, and therefore viruses are normally classified along with their hosts as either animal, plant, fungal, algal, protozoan, or bacterial viruses.

Microbiology as a basic science

The nature of microorganisms makes them suitable for use in many basic scientific studies. Consequently, the field of microbiology is often ex-tended into areas such as basic molecular biology and biochemistry. Our fundamental understanding of molecular genetics and the relationship of DNA and enzymes (genetics and metabolism) has been developed using microorganisms. Microorganisms reproduce rapidly, and large numbers of organisms with the same genetic composition can easily be grown for genetic and biochemical studies. Compared to higher multicellular organisms, unicellular microorganisms are simple, both genetically and biochemically; thus, microbial systems are easier to deal with and understand than those of higher, more complex organisms. Besides, no one cares if bacteria are sacrificed for the advancement of science. Well, almost no one cares: Japanese microbiologists, feeling a subtle sense of guilt for killing microorganisms as part of their studies, have established a memorial to microorganisms in which they buried a Buddhist prayer scroll and the ashen remains of *Bacillus subtilis*.

Microbiology as an applied science

Microbiology is also an applied science, and the many practical aspects of microbiology affect our daily lives and influence the overall quality of life (Figure 1.5). Applications of microbiology are important in medicine, industry, agriculture, and ecology. The health of humans and planet earth depends on microorganisms, making microbiology a relevant and pragmatic science.

Microbiology includes the study of pathogenic microorganisms that cause diseases in plants and animals, including humans. Microbiology also includes the study of the prevention and treatment of diseases caused by microorganisms. Medical microbiology (the study of medical science relating to microorganisms) represents an important applied area in the study of microbiology. An un-

derstanding of the causative agents of disease has allowed us to develop preventive and treatment methods that have reduced **morbidity** and **mortality** arising from certain diseases, resulting in an increase in life expectancy. Since microorganisms are involved in causing infectious diseases, the field of microbiology has logically been extended to include the response of the diseased plant or animal. As such, the disciplines of immunology (the study of the immune response of higher animals) and plant pathology are included in the study of microbiology.

Microbiology is an integral part of many industrial processes, and thus microorganisms have great economic importance. Quality control in many industries is concerned with preventing microbial contamination of products that could lead to the spoilage of the products, reducing their economic value. The proper handling of food products is based on an understanding of microbiology. Various products, including many pharmaceuticals and food products such as beer, wine, and spirits, are produced by microbial fermentations.

Microorganisms also play an important role in maintaining environmental quality. Microbial ecology includes the study of biogeochemical cycling reactions, which are important in maintaining air, water, and soil quality. The maintenance of soil fertility, the role of microorganisms in causing plant diseases, and the interactions of microorganisms with pesticides and fertilizers are important topics considered in agricultural microbiology. The ability of microorganisms to degrade waste materials extends microbiology into the field of sanitary engineering.

It is thus apparent that microbiology encompasses a very broad field. Some microbiologists are concerned with the basic sciences and the development of a fundamental understanding of living systems; others are concerned with the application of basic scientific knowledge. Microbiology overlaps with a number of other scientific disciplines, including biochemistry, genetics, zoology, botany, ecology, pharmacology, medicine, food science, agricultural science, industrial science, and environmental science. The broad scope of microbiology attests to the diversity of the microorganisms themselves, their ubiquitous distribution in nature, and the importance of microorganisms in virtually all aspects of life; the unity of microbiology rests with its central subject matter of the organisms which are considered to be microbes. The vistas and challenges for the student of microbiology are virtually limitless.

Figure 1.5

The science of microbiology has many applied aspects that have impact on our daily lives and affect the quality of modern life. Microbiology and medical practice are interrelated in many ways: the control of microbial growth is used to prevent many infectious diseases; some substances derived from microorganisms (vaccines) are used to prevent and eliminate diseases; and microbial products (antibiotics) are used routinely to treat many diseases that were once deadly. Many food products, including beer, wine, spirits, cheese, and yogurt, are produced by using microorganisms. Sewage and other waste products are decomposed by microorganisms, which detoxify many substances and transform wastes into the substances necessary to support life on earth.

Historical perspective of microbiology

Historical background of microbiology

Microbiology as a science did not exist before the latter part of the nineteenth century, but man has undoubtedly been a practitioner of microbiology since prehistoric times. Early man could see the macroscopic structures formed by some microorganisms and made use of some microbes, such as the fruiting bodies of certain fungi (mushrooms), as food sources. Early man also had to contend with disease. Without any scientific understanding of the reasons for their effectiveness, early man used various herbs and other plants to treat disease. Some of the effective substances used in these treatment methods were potent poisons that had antimicrobial properties. In the third century B.C., Theophrastus, successor to Aristotle as head of the Lyceum, wrote several volumes on plants, which included a discussion of the medicinal properties of various herbs; most of the remedies discussed by Theophrastus originated in the Orient and were imported to Greece.

People produced and drank beer and wine before they understood the role of microorganisms in fermentation (Figure 1.6). Salt was used to preserve foods long before the involvement of microorganisms in food spoilage was understood. Long voyages of discovery, such as those to the Americas, depended on the ability to control microbial food spoilage by drying and salting foods. Francis Bacon, the eminent English philosopher of the early seventeenth century, had a theory that freezing could retard the spoilage of chicken, a commonplace practice for preserving foods today. However, being a better philosopher than scientist, Bacon caught cold while testing his theory and died.

In the mid-fourteenth century more than 25 million people died in medieval Europe during a severe outbreak of plague (Figure 1.7). Lacking any true understanding of microbiology, the populace turned to such unscientific and unrewarding practices as flagellation, beating themselves to drive out the force causing "black death." A more relevant method of dealing with plague was that of quarantining individuals suspected of having the disease; although this practice dates back to Biblical times, the term "quarantine" comes from the Italian *quarantenaria*, which originated at the height of the fourteenth century plague epidemic when sea voyagers coming into Sicily had to wait 40 days before entering the city. Plague still occurs today but is effectively controlled through the use of antibiotics. Two centuries after the major outbreaks of plague in Europe, Girolamo Fracastoro of Verona, a contemporary of Copernicus, published a work on contagious diseases and

Figure 1.6

The long history of man's use of fermentative techniques is illustrated in this painted relief discovered on the wall of a Fifth Dynasty Egyptian tomb dating from about 2400 B.C. The top and middle panels show the production of bread. In the middle panel, leavening of the dough and baking of the "beer-bread" are shown. The bottom panel shows mash being strained into the fermenting vat, which rests on a stand that looks like a coiled rope. After being left to ferment for a few days, the finished beer was then poured into pottery jars, which were quickly capped and sealed with clay to be placed in storage. These early Egyptian brewers originally used yeasts in the air or on the skin or husk of fruits and cereals to carry out these fermentations but later developed an almost pure yeast. Their ancient breweries were able to produce different types of beer, some of which may have had an alcohol content as high as 12 percent. (Courtesy M. J. Vinkesteyn, National Museum of Antiquities, Leiden, The Netherlands.)

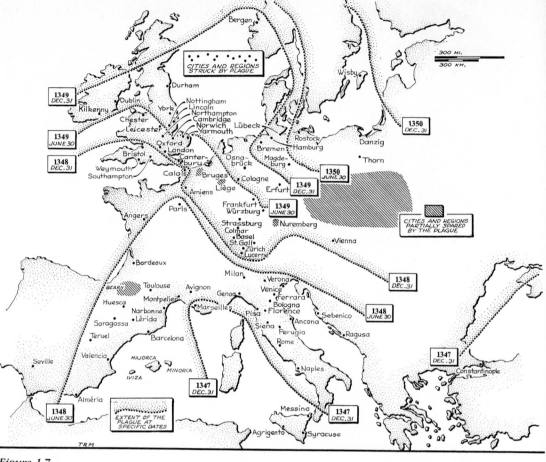

CITIES AND REGIONS
STRUCK BY PLAGUE

1349
DEC. 31

1349
JUNE 30

1348
DEC. 31

1350
DEC. 31

1350
JUNE 30

1349
DEC. 31

1349
JUNE 30

CITIES AND REGIONS
PARTIALLY SPARED
BY THE PLAGUE

1348
DEC. 31

1348
JUNE 30

1347
DEC. 31

1347
DEC. 31

1347
DEC. 31

1348
JUNE 30

EXTENT OF THE
PLAGUE AT
SPECIFIC DATES

300 MI.
300 KM.

Figure 1.7

This map shows the spread of plague, the "black death," in the mid-fourteenth century. Plague was apparently introduced by seaborne rats from Black Sea areas. The spread of the disease generally followed trade routes from the Near East to Europe, reaching Scandinavia by 1350 and spreading perhaps as far as Iceland and Greenland, whereas areas outside the major trade routes remained virtually unaffected. (Reprinted by permission of Macmillan Publishing Company, New York, from D. Kagan, S. Ozment, and F. M. Turner, 1979, The Western Heritage.*)*

their treatment. This work of 1546, entitled *De Contagione*, was largely philosophical and did not recognize the true nature of microorganisms. Nevertheless, Fracastoro did hypothesize that some diseases were caused by the passage of "germs" from one thing to another. He described three processes for the **transmission of contagion**: direct contact, transmission via **fomes** (inanimate objects, such as clothing, which may pick up and preserve germs), and transmission from a distance. Fracastoro had the insight to recognize the similarity between **contagion** (disease processes) and **putrefaction** (e.g., the turning of wine into vinegar) centuries before the discoveries of Pasteur and Koch. He further recognized that germs that cause disease exhibit specificity, which re-

sults in different diseases occurring in different hosts and different processes of transmission for different germs.

The advent of the microscope

The **microscope** appears to have originated in the mid-sixteenth century and probably was invented by the Dutch spectacle maker Hans Janssen or the Italian astronomer Galileo. It was not until the mid-seventeenth century, when the development of the microscope permitted the visualization of microorganisms, that the great diversity of the microbial world began to be recognized. Robert Hooke's *Micrographia or Some Physiological De-*

Figure 1.8

Various species of fungi can clearly be identified in the drawings of Robert Hooke (1635–1703). This 1664 drawing shows the growth of blue mold on leather. Although macroscopic fungal structures had been described much earlier, Hooke's represent some of the earliest descriptions of many of the common fungi. (Reprinted by permission of the Royal Society, London, from R. Hooke, 1665, Micrographia.*)*

Figure 1.9

Antonie van Leeuwenhoek (1632–1723), here seen holding one of his microscopes, opened the doors to the hidden world of microbes when he described bacteria. Although only an amateur scientist, Leeuwenhoek's keen interest in optics and his diligence allowed him to make this important discovery. (Reprinted by permission of Russell and Russell from C. E. Dobell, 1932, Antony van Leeuwenhoek and His "Little Animals."*)*

scriptions of *Minute Bodies Made by Magnifying Glasses with Observations and Inquiries Thereupon,* published in 1665, contains drawings of the fruiting bodies of fungi (Figure 1.8). Hooke was an English experimental philosopher who also made significant contributions to physics, architecture, and clockmaking, in addition to microscopic observations of biological specimens.

It was the amateur Dutch microscope maker Antonie van Leeuwenhoek (Figure 1.9) who first recorded observations of bacteria, yeasts, and protozoa. Leeuwenhoek was a cloth maker and tailor by trade; he was also a surveyor and the official winetaster of Delft, Holland. His interest in microscopes was probably related to the use of magnifying glasses by drapers to examine fabrics. Leeuwenhoek also was a close friend of Christian Huygens, the noted Dutch physicist, mathematician, and astronomer, who undoubtedly encouraged his interest in science and lenses. Leeuwenhoek's findings were transmitted in a series of letters, from 1674 to 1723, to the Royal Society in London, through which his observations were widely disseminated; during part of this period, Leeuwenhoek sent his communications to Robert Hooke, who was at that time sec-

retary of the Royal Society. The letters contained detailed drawings, some of which clearly showed microorganisms (Figure 1.10). Unfortunately, while Leeuwenhoek shared his observations with the scientific community, he never revealed the details of his methods of microscopy, nor how he constructed his hundreds of microscopes.

The microscope continues to be the basic tool of the microbiologist, through which the organisms being studied can be observed. The early microscopes used by Hooke and Leeuwenhoek permitted only a fuzzy view of the microorganisms they were observing (Figure 1.11). Many advances in the art of microscope making and the science of microscopy were needed before observations could be made on the fine structure of microorganisms. Major advances were made in microscopy in the late nineteenth century, a period of great advancement in microbiology. Ernst Abbe, a German physicist, developed microscope lenses in the early 1880s, which corrected for chromatic and spherical aberrations, the major distortions inherent in microscopes developed to that time. Abbe also developed the oil immersion lens, which allowed for improved resolution in light microscopy. Microscopic visualization of

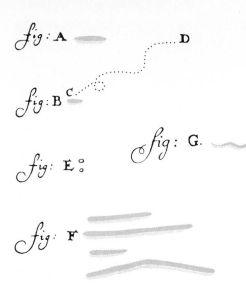

bacteria was greatly improved by the introduction by Paul Ehrlich in 1881 of vital staining of bacteria with methylene blue and the development of a differential staining method for bacteria by Hans Christian Gram in 1884. The Gram staining technique is basic to the taxonomic description of bacteria even today. Modern microscopes and the current art of microscopy will be discussed in Chapter 2, and throughout this book micrographs can be seen that illustrate the types of information that can be obtained about microorganisms by using different microscopic techniques.

Pasteur and the beginning of food and industrial microbiology

A great debate of the eighteenth and early nineteenth centuries concerned the **theory of spontaneous generation**. Several major advances in the development of microbiology were tied to disproving this theory. In order to disprove spontaneous generation, it was necessary to begin to develop a true understanding of microbiology. There had frequently been observed a relationship between the growth of infusoria (microorganisms) in organic infusions and the onset of chemical changes in the broth known as fermen-

Figure 1.10

Leeuwenhoek's sketches of bacteria from the human mouth illustrate several common types of bacteria, including rods and cocci. (A) A motile Bacillus: *(B)* Selenomonas sputigena; *(C) and (D) show the path of B's motion; (E) Micrococci; (F)* Leptothrix buccalis; *and (G) a spirochete, probably* Spirochaeta buccalis. *(Reprinted by permission of the Royal Society, London, from* Letters *39, September 17, 1683.)*

Figure 1.11

Some early microscopes used in the late seventeenth century. (A) This microscope, used by Leeuwenhoek, consists of a single biconvex lens (1) inserted into a small hole on the left side of the leather base. The specimen was placed on the small pointed wire attached to the screw, which moved the specimen back and forth to bring it into the focus of the fixed glass, rather than focusing the lens on a fixed object, as is done on a modern microscope. Leeuwenhoek apparently kept the method for using his very tiny "magnifying-glasses" to himself. (Reprinted by permission of Russell and Russell from C. E. Dobell, 1932, Antony van Leeuwenhoek and His "Little Animals." *) (B) Hooke's compound microscope of the same period more closely resembles the appearance of a modern microscope. (Courtesy American Optical.)*

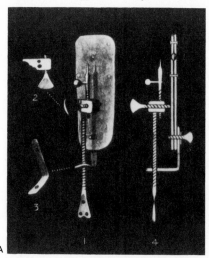

Figure 1.12

Louis Pasteur (1822–1895), seen working in his laboratory in this 1885 woodcut, began as a chemist but soon became a pioneer microbiologist. Pasteur's work encompassed both pure research and many areas of applied science that produced many important practical discoveries. Among his many accomplishments, Pasteur discredited the theory of spontaneous generation, introduced vaccination to treat rabies, and solved industrial problems related to the production and spoilage of foods. (Courtesy Bettmann Archive.)

tation and putrefaction. The experiments of Lazzarro Spallanzani, an eighteenth-century priest who had an exceptionally inquiring mind and willingness to challenge the conventional wisdom of his time, demonstrated that the putrefaction of organic substances is caused by microorganisms that multiply by reproductive divisions and that do not arise by spontaneous generation. He demonstrated that heating destroyed these organisms and that in sealed flasks spoilage was prevented indefinitely. Nineteenth-century advocates of spontaneous generation, though, claimed that the elimination of oxygen compromised these experiments. Noted chemists, such as Justus von Liebig, Jons Jakob Berzelius, and Friedrich Wöhler, held that changes in organic chemicals, such as the putrefaction of proteins and the transformation of sugar into alcohol, occurred by strictly chemical processes without the intervention of living organisms. This view was opposed in the 1830s by Charles Cagniard de Latour of France and by Theodor Schwann and Friedrich Kützing of Germany, who separately proposed and conducted experiments to demonstrate that the products of fermentation, ethanol and carbon dioxide, were produced by microscopic forms of life. Schwann used a flame, and Cagniard de Latour and Kützing used cotton plugs to prevent microorganisms from entering the heat-sterilized broth. Each of these experiments, aimed at showing that living forms were responsible for fermentation, was criticized by the chemists because it destroyed or eliminated some essential component in air that was needed for the spontaneous generation of the fermentation products.

It took several more decades before Louis Pasteur (Figure 1.12) settled the dispute, definitively discrediting the theory of spontaneous generation and establishing that living microorganisms are responsible for the chemical changes that occur during fermentation. Pasteur's report, *On the Organized Bodies Which Exist in the Atmosphere; an Examination of the Doctrine of Spontaneous Generation* (1861), effectively ended the controversy concerning spontaneous generation. In this experiment Pasteur demonstrated that liquids subjected to boiling remained sterile as long as microorganisms in the air were not allowed to contaminate the liquid. By using a **swan-necked flask** (Figure 1.13), Pasteur was able to leave a vessel containing a fermentable substrate open to the air and show that fermentation did not occur. The shape of the flask prevented airborne microorganisms from entering the liquid, and the fact that air could enter the flask overcame the main argument that chemists had leveled against earlier studies using sealed flasks, namely, that oxygen was essential for the chemical reactions involved in the formation of alcohol.

Pasteur was trained as a chemist, and this training had a marked influence on his approach to scientific questions. As a chemist, Pasteur was able to separate a racemic mixture into its two optically active components, explaining the riddles concerning the optical activities of liquids and why light passing through a solution was sometimes bent to the left or right. He followed the same investigative approach throughout his long scientific career; he identified the problem, sought out all the available information on the topic, formed a hypothesis, and devised experiments to test the validity of his theory. The problem of

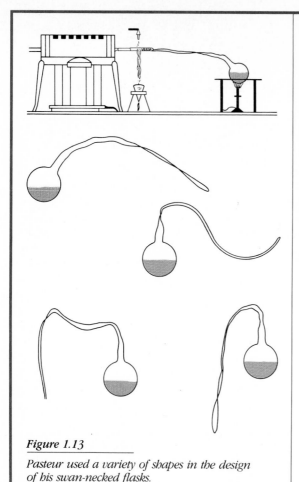

Discovery process

Settling the question of spontaneous generation was one of Pasteur's great accomplishments. Pasteur began this classic experiment in 1861 by placing yeast water, sugared yeast water, urine, sugar beet juice, and pepper water into ordinary flasks. He then reshaped the necks of the flasks under a flame so that there were several curves in each neck, hence the term *swan-necked flasks*. Next, Pasteur boiled the liquids until they steamed through the necks. By using curved ends, Pasteur could leave the flasks open to the air, thus overcoming a major criticism of previous experiments aimed at disproving spontaneous generation where air, an "essential life force," had been excluded. Dust and microbes settled out in the curved neck of the flask and thus, while exposed to air, the broth did not become contaminated with microorganisms. Contrary to the opinion of those who believed in spontaneous generation, no change or alteration appeared in the liquid. These flasks, later sealed, may be seen at the Pasteur Institute in Paris.

Figure 1.13

Pasteur used a variety of shapes in the design of his swan-necked flasks.

applied versus basic or pure research never bothered this aggressive, ambitious, argumentative, and highly patriotic Frenchman. Much of his work stemmed from the requests of local manufacturers to help solve the practical problems of their industrial processes; he loyally responded, attempting to solve these problems in order to improve the French economy and demonstrate French superiority. He was concerned with problems such as why French beer was inferior to German beer. The answer to this practical question eventually led him to the basic discovery of the existence of anaerobic life: life in the absence of air.

In 1854 Pasteur was appointed Dean of the Faculty of Science at the University of Lille. Following his appointment, one of the first problems Pasteur attacked was at the request of a local industrialist and concerned the souring of alcohol produced from beets; Pasteur's agreement to help

solve this problem of the wine industry placed him directly on the main path of a scientific career which would lead him from chemistry to microbiology. By comparing, with the aid of a microscope, samples taken from productive and nonproductive vats, Pasteur observed budding yeast cells in the productive vats and rod-shaped organisms in the nonproductive ones. He was able to demonstrate that these organisms determined the course of the chemical processes, that the yeasts were responsible for the production of alcohol, and that the rod-shaped bacteria produced the lactic acid that caused the wine to sour, thus demonstrating that different microorganisms are responsible for different fermentations.

The work of Pasteur led to an understanding of the role of microorganisms in food spoilage and the use of heat to destroy microorganisms in food products. In 1857 Pasteur demonstrated that the souring of milk also was caused by the action of microorganisms, and about 1860 he showed that heating could be used to kill microorganisms in wine and beer. This process of **pasteurization**, as it has come to be known, was introduced around 1867. Commercial pasteurization of milk began in Germany in 1880 and in the United States in 1890. Canning as a method of preserving foods has its roots in the works of Spallanzani, Schwann, and others who steadfastly held that putrefaction was due to living organisms. The process of canning was patented in 1810 by a French candy maker, Nicolas Appert, and by 1820 the commercial production of canned foods had begun in the United States.

Early work on heat sterilization was complicated by the occasional presence of endospore-forming bacteria. The English physicist John Tyndall, trying to confirm the results of Pasteur's experiments refuting spontaneous generation, determined that the variability of results was due to the capacity of bacteria to exist in two forms—a heat-labile form that was killed by exposure to elevated temperatures and a heat-resistant form that could survive at such high temperatures. He found that intermittent heating could eliminate viable microorganisms and sterilize solutions. Repeated heating on successive days, a process known as **tyndallization**, successfully sterilizes even solutions containing endospore-forming bacteria. The bacterial endospores of *Bacillus subtilis* and their relationship to heat resistance in sterilization processes were described by Ferdinand Cohn in 1877. In 1881 several disciples of Cohn, Robert Koch, Georg Gaffky, and Friedrich Loeffler, published *Observations on the Effectiveness of Hot Steam for Disinfection*, in which they showed that it was necessary to achieve a temperature of 160°C for at least one hour in order to achieve sterility. The first laboratory autoclaves, which resembled large pressure cookers, were designed by Charles Chamberland, a colleague of Pasteur. The autoclave, which permits the heating of steam under pressure to achieve the temperatures above 100°C needed to kill bacterial endospores, is an essential instrument in every bacteriology laboratory.

Koch and the beginning of medical microbiology

The studies of Robert Koch (Figure 1.14) greatly advanced the field of **medical microbiology**. Koch, a German country physician, began his scientific studies isolated from any contact with the scientific community, working alone with primitive tools and materials. As a result of his medical practice,

Koch was well aware of the diseases of humans and other animals. From 1873 to 1876, Koch conducted experiments to show that the spores of anthrax bacilli isolated from pure cultures could infect animals. Koch demonstrated for the first time that germs grown outside the body could cause disease, and that specific microorganisms caused specific diseases. Recognizing that to be of real use his findings would have to be published, Koch contacted Ferdinand Cohn, the esteemed director of the Botanical Institute at Breslau. Cohn quickly saw the significance of Koch's studies and arranged for their publication. This was the beginning of Koch's illustrious career. He went on to determine the causative organisms for several other diseases, including tuberculosis and cholera.

In describing the cause (**etiology**) of tuberculosis, Koch set forth the basic principles for establishing a cause-and-effect relationship between a given microorganism and a specific disease. Koch's four postulates for identifying the etiologic agent of a disease are

1. The organism should be present in all animals suffering from the disease and absent from all healthy animals.
2. The organism must be grown in pure culture outside the diseased animal host.
3. When such a culture is inoculated into a healthy susceptible host, the animal must develop the symptoms of the disease.
4. The organism must be reisolated from the experimentally infected animal and shown to be identical with the original isolate.

These postulates, which are applicable to plant as well as animal diseases, still form the basis for determining that a particular disease is caused by a given microorganism. For example, the search for the cause of Legionnaire's disease in 1976 followed Koch's 1890 postulates, resulting in the eventual identification of the bacterial etiologic agent. After many attempts, the bacterium *Legionella pneumophila* was isolated from patients with this disease; grown in the laboratory; inoculated into test animals, which then caused the onset of disease symptomology; and reisolated from the experimentally infected animals. Some modifications to Koch's postulates are required in cases when the disease is caused by opportunistic pathogens that are normally associated with healthy animals, when the experimental host is immune to the particular disease, when the disease process involves cooperation between multiple organisms, and when the causative agent

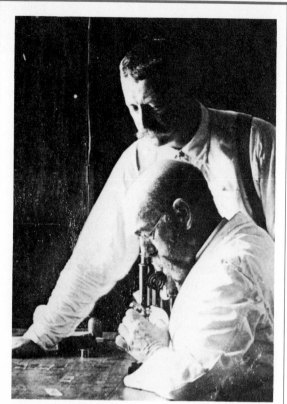

Figure 1.14

Robert Koch (1843–1910), seen here viewing a specimen while a disciple looks on, pioneered studies in medical microbiology and developed many of the basic methods essential for the study of microbiology. Koch's postulates for establishing the etiology of infectious diseases and the methodological techniques he developed are still used today in scientific investigations. Many of Koch's students also made significant contributions to the development of the field of microbiology. (Courtesy Bettmann Archive.)

Discovery process

"To obtain a complete proof of a causal relationship, rather than mere coexistence of a disease and a parasite, a complete sequence of proofs is necessary. This can only be accomplished by removing the parasites from the host, freeing them of all tissue elements to which a disease-inducing effect could be ascribed, and by introducing these isolated parasites into a healthy animal with the resulting reproduction of the disease with all its characteristic features. An example will clarify this type of approach. When one examines the blood of an animal that has died of anthrax one consistently observes countless colorless, non-motile, rod-like structures.... When minute amounts of blood containing such rods were injected into normal animals, these consistently died of anthrax, and their blood in turn contained rods. This demonstration did not prove that the injection of the rods transmitted the disease because all other elements of the blood were also injected. To prove that the bacilli, rather than other components of blood produce anthrax, the bacilli must be isolated from the blood and injected alone. This isolation can be achieved by serial cultivation.... The serial transfers can be continued for 3 or as many as 50 passages and in this manner the other blood components can be eliminated with certainty. Such pure bacilli produce fatal anthrax soon after injection into a healthy animal, and the course of the disease is the same as if produced with fresh anthrax blood or as in naturally occurring anthrax. These facts proved that anthrax bacilli are the unique cause of the disease." (Robert Koch, 1884, *The Etiology of Tuberculosis*.)

cannot be grown in pure culture outside of host cells. But in general, the philosophy of Koch's postulates for identifying the causes of infectious diseases remains intact.

Many basic microbiological methods developed by Koch also are still in use today with only minor modifications. It was Koch who, between 1881 and 1883, developed simple methods for the isolation and maintenance of pure cultures of microorganisms on chemically defined solid media. Prior to that time, microorganisms were grown in liquid broths. At first, Koch cultured bacteria on solid fruits and vegetables, such as slices of boiled potato, but many bacteria cannot grow on such substrates. Koch developed a way of solidifying liquid broths that could support the growth of a greater variety of microorganisms, initially using gelatin and later **agar** (an algal extract) as the **solidifying agent**. The suggestion for using agar originated with the New Jersey-born wife of one of the investigators at Koch's Institute, Mrs. Hesse, who had seen her mother using agar to make jellies. The use of these solidified media permitted the isolation and unequivocal identification of disease-causing microorganisms. The isolation and growth of microorganisms in pure culture has dominated most microbiological studies since the necessary techniques were developed by Koch. One of the modifications to Koch's original method was made by Richard J. Petri, who in 1887 described the use of a new type of culture dish for growing bacteria on semisolid media. The basic design has become known as the **petri dish** and is used in virtually all microbiological laboratories in essentially the same design described by Petri.

Chemotherapy

Responding to the growing awareness that microorganisms are associated with disease processes, Joseph Lister, an English Quaker and physician, revolutionized surgical practice in 1867 by introducing **antiseptic** principles (Figure 1.15). The discovery in the early 1850s of anaesthesia and its administration to patients made surgery much easier but of course did nothing to reduce the incidence of post-surgical disease, which often was as high as 90 percent, especially in military hospitals. Lister knew that in the 1840s, Ignaz Semmelweis, a Hungarian physician who worked in maternity wards in Vienna, had shown that physicians who went from one patient to another without washing their hands were responsible for transmitting childbed fever (puerperal fever). He was also aware that Pasteur had demonstrated that microorganisms are present in the air. Lister used carbolic acid, phenol, as an antiseptic during surgery. He first used bandages soaked in carbolic acid to dress wounds from compound fractures in order to diminish the likelihood of infection. Later, he used a carbolic acid spray in addition to direct application of this compound during surgical procedures. He eventually discarded the practice of spraying after 17 years of trials as unnecessary, but he retained the use of direct application.

Various chemical formulations for preventing microbial growth and infection were described by Koch and his disciples, including Paul Ehrlich, who, like Pasteur, had been trained as a chemist. From 1880 to 1896, Ehrlich worked in Koch's laboratory, and in 1896 he became director of the first of his own institutes, which he dedicated to finding "substances which have their origin in the chemist's retort," that is, substances produced by chemical synthesis, to cure infectious diseases. Ehrlich's research between 1880 and 1910 established the early basis for modern **chemotherapy**. He established the correct formula for atoxyl, an arsenical, which was being considered for use in treating sleeping sickness, and developed almost a thousand new derivatives of this compound. Compound 606, salvarsan, proved to be effective in treating syphilis. Sahachiro Hata, a Japanese expert on spirochetes, came to Ehrlich's laboratory on the recommendation of Shibasaburo Kitasato, whom Ehrlich knew from their years together in Koch's laboratory. Hata tested the atoxyl derivatives, using syphilitic rabbits. He found compound 914, neosalvarsan, to be a reliable drug for curing syphilis and relapsing fever, both diseases caused by spirochetes. The use of neosalvarsan in 1912 represents the first widespread use of synthetic drugs; these drugs became known as **"magic bullets"** and were portrayed as being able to find and kill disease-causing germs.

Figure 1.15

Joseph Lister (1827–1912) recognized the importance of preventing the contamination of wounds in order to curtail the development of infection. Here Lister is shown spraying carbolic acid over a patient undergoing an operation (ca. 1867). (Courtesy Bettmann Archive.)

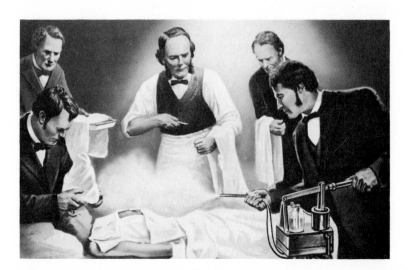

Figure 1.16

Sir Alexander Fleming (1881–1955), shown here working on penicillin development in his laboratory, had the insight to recognize the significance of the inhibition of bacterial growth in the vicinity of a fungal contaminant when most other scientists probably would have simply discarded the contaminated plates. (Courtesy Central Office of Information, London.)

A major breakthrough in chemotherapy occurred in 1929, when the Scottish bacteriologist Alexander Fleming, working in a London teaching hospital, reported on the antibacterial action of cultures of a *Penicillium* species (Figure 1.16). Fleming observed that the mold *Penicillium notatum* killed his cultures of the bacterium *Staphylococcus aureus* when the fungus accidentally contaminated the culture dishes. It is likely that the fungal contaminant of Fleming's cultures, which was to bring medical practice into the modern era of drug therapy, blew into his laboratory from the floor below, where an Irish mycologist was working with strains of *Penicillium*. Such a serendipitous event can change history, but in science it takes a special individual like Fleming to recognize the significance of the observation. As Pasteur said, "Chance favors the prepared mind." After growing the fungus in a liquid medium and separating the fluid from the cells, Fleming discovered that the cell-free liquid was an inhibitor of many bacterial species. His publication on the active ingredient, which he called penicillin, was the first report of the production of an antibiotic. However, Fleming did not isolate pure penicillin, nor did he demonstrate its chemotherapeutic effects. It was ten years after Fleming's initial report that Howard Florey and Ernst Chain successfully isolated and purified penicillin. Other scientists established the therapeutic value of penicillin, and this antibiotic remains as a cornerstone of the modern medical treatment of many infectious diseases.

Major advances in the development of chemotherapeutic agents continued to be made in the 1930s. Gerhard Domagk, a German physician employed as a chemist by a dye works company, the forerunner of modern pharmaceutical companies, developed prontosil, the first sulfa drug, effective against streptococcal infections. At the Pasteur Institute a husband-and-wife research team, the Trefouëls, discovered the active constituent of prontosil, sulfanilamide, the first real wonder drug, so called because of its amazing ability to cure serious diseases. In the early 1940s the Russian immigrant soil microbiologist, Selman Waksman, and coworkers at Rutgers University in New Jersey found that various bacteria of the actinomycetes group produced antibacterial agents. Streptomycin, produced by *Streptomyces griseus*, became the best known of the new antibiotic wonder drugs. The antibiotics produced by actinomycetes generally have a broader spectrum of action than penicillin and thus can be used for treating a number of diseases against which penicillin is ineffective. Most antibiotics in current use are produced by actinomycetes. The importance of penicillin and the subsequently discovered antibiotics in treating diseases of microbial origin cannot be overestimated.

Immunology

Whereas chemotherapy is effective in treating disease, **immunization** is used for preventing infectious diseases. The practice of immunization was used in the Far East for centuries before it was first introduced into England in 1718 by Lady Mary Montagu, whose husband had been the British Ambassador to Turkey. Lady Mary used her considerable influence at the court of King George I to gain publicity for the increased use of immunization. She even arranged for testing of her idea, though she had no explanation for how or why it worked, on prisoners and orphans, then a common practice. Despite her efforts, immunization was not accepted by the scientists and physicians of the time as a useful practice for preventing disease. It was not until the report by Edward Jenner to the Royal Society in London 80 years later that credence was given to the practice of immunization. Jenner's 1798 report on the value of vaccination with cowpox as a means of protecting against smallpox established the basis for the immunological prevention of disease (Figure 1.17). Jenner was a middle-class English country doctor, whose interest in science, like that of Leeuwenhoek, was typical of his class: scholarly but amateur. There was no hurry, no great impetus to discovery in his work, just the careful, methodical observation of outbreaks of disease. The work begun with Jenner's discovery of the effectiveness of **vaccination** for preventing smallpox culminated in the 1970s with the eradication of smallpox from the face of the earth.

Much work was needed beyond Jenner's report to achieve the elimination of smallpox and to use vaccines for preventing many other diseases. Pasteur greatly furthered the development of **vaccines** when in 1880 he reported that attenuated microorganisms (modified to reduce virulence) could be used to develop effective vaccines against chicken cholera. The production of attenuated vaccines that were effective against fowl cholera depended on prolonging the time between transfers of the cultures, a fact accidentally discovered through an error by Charles Chamberland, who used an old culture during one of the experiments he was conducting with Pasteur. Following his work on chicken cholera, Pasteur directed his attention to the study of anthrax; this irritated Koch, who considered this disease his exclusive domain. There was an intense rivalry between Pasteur and Koch due to personal egotistical feelings and intense nationalistic pride. Because he enjoyed being the center of attention and controversy, Pasteur staged a very dramatic public demonstration to test the effectiveness of his anthrax vaccine. Witnesses to the demonstration were amazed to see that the twenty-four sheep, one goat, and six cows that had received the attenuated vaccine were in good health, but that all the animals that had not been vaccinated were dead of anthrax.

Even considering the impressiveness of this display, Pasteur's greatest success in developing vaccines was yet to come. In 1885 Pasteur was able to announce to the French Academy of Sciences that he had developed a vaccine for preventing rabies. Although he did not understand the nature of the causative organism, Pasteur developed a vaccine that worked. Pasteur's motto was "seek the microbe," but the microorganism responsible for rabies is a virus, which could not be seen under the microscopes of the 1880s. He nevertheless was able to weaken the rabies virus

Figure 1.17

Edward Jenner (1749–1823) vaccinated James Phipps about 1800 with cowpox material, resulting in the development of resistance to smallpox infection by the boy and establishing the scientific credibility of vaccination to prevent disease. (Courtesy Culver Pictures.)

by drying the spinal cords of infected rabbits and allowing oxygen to penetrate the cords. Thirteen inoculations of successively more virulent pieces of rabbit spinal cord were injected over a period of two weeks during the summer of 1885 into Joseph Meister, a nine-year-old boy who had been bitten by a rabid dog (Figure 1.18); this treatment met with a highly publicized, personal success for Pasteur. The development of the rabies vaccine culminated Pasteur's distinguished career. Donations sent to Pasteur as a consequence of his discoveries were used to erect l'Institute Pasteur, the first aim of which was to provide the proper facilities for the production of vaccines.

While Pasteur used attenuated microorganisms for his vaccines, Daniel Salmon and Theobold Smith (1886) demonstrated that it was not necessary to use a live attenuated microbial strain in order to achieve artificial immunity. They developed the first killed vaccine for the control of hog cholera, a significant advance in developing vaccines for preventing disease.

The development of vaccines demonstrated that the immune response could prevent infections but did not elucidate the mechanisms of immunity, that is, the underlying basis of the immune response. Eli Metchnikoff sought to establish the principle that the activity of phagocytes, white blood cells that engulf foreign particles, was responsible for the protection of animals against infectious diseases and for the development of acquired immunity (Figure 1.19); he was the first

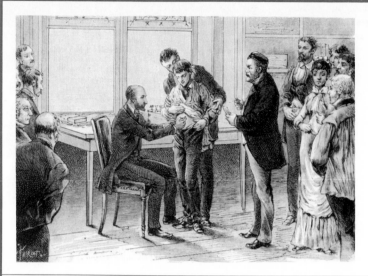

Figure 1.18

The vaccination of a child against rabies conducted under the direction of Louis Pasteur. With the successful development of a vaccine for preventing rabies, crowds flocked to Pasteur's laboratory. (Courtesy Bettmann Archive.)

Discovery process

"Since the death of the child was almost certain, I decided in spite of my deep concern to try on Joseph Meister the method which had served me so well with dogs.... I decided to give a total of 13 inoculations in ten days. Fewer inoculations would have been sufficient, but one will understand that I was extremely cautious in this first case. Joseph Meister escaped not only the rabies that he might have received from his bites, but also the rabies which I inoculated into him." (Louis Pasteur, 1885, *Compt. Rend. Acad. Sci.*)

to describe the phenomenon of **phagocytosis** (1884). Metchnikoff was a moody, temperamental Russian who did his most creative scientific work in a manic state, applying detailed, detached observation to his studies. He meticulously reported his microscopic observations of how a microbe

Figure 1.19

Eli Metchnikoff (1845–1916), here shown at work in his laboratory, first proposed the cellular theory of immunity, opposing the accepted humoral theory, which said that soluble substances in body fluids were responsible for immunity. (Courtesy Bettmann Archive.)

intruding into an organism is dealt with by that organism; he began by watching starfish larvae stuck with thorns and then the transparent water-flea *Daphnia* diseased with a microbial infection. Metchnikoff saw mobile cells, which he called phagocytes, migrate to the area of infection and destroy the infecting microbes by digesting them. This pioneering work established the role of cellular components of the blood in destroying disease-causing microorganisms and the basis for the field of cellular immunology.

Koch and his disciples also made significant contributions to the field of immunology and our understanding of the complexities of the immune response. Workers in Koch's laboratory were dominated by Koch's strong personality. Koch's desire for secrecy fostered competitiveness and limited cooperation between the scientists working in his laboratory. Koch's assistants, Emil von Behring and Shibasaburo Kitasato, however, jointly discovered that substances in blood sera were able to neutralize foreign materials. They found that **antitoxin**, a substance from the serum of infected animals, could neutralize bacterial toxins and cure some diseases. Their studies on diphtheria and tetanus established the existence of antibodies, which are proteins made in response to antigens that circulate in blood and react with the antigen molecules. **Antigen-antibody reactions** are extremely important in establishing immunity against disease and form the cornerstone of the field of **serology**. Kitasato and von Behring published a joint paper on their results for tetanus, but von Behring, having learned Koch's zealotry for individual recognition, only a week later published his experiments on diphtheria toxin, which he had kept secret from Kitasato. Von Behring engineered an agreement with a dye works company and the Ministry of Culture to commercially produce diphtheria antitoxin and was able to assume the rights to Ehrlich's discoveries for the effective production of antitoxin as well as his own. He was one of the few scientists of his time to gain substantial financial rewards for his discoveries. While still in Koch's laboratory, Paul Ehrlich also worked on developing an understanding of the immune response. In 1891 he published a paper in which he differentiated active immunity, which is immunity acquired as a result of a prior interaction with pathogens or their products, from passive immunity, which is immunity that results from the injection of blood serum containing antibodies. The development of active immunity provides long-term protection against disease, whereas passive immunity, such as that transferred from mother to fetus, provides only temporary **prophylaxis** against infectious microorganisms.

Koch was not to be outdone by his assistants. Although his research was far from complete, the popular media of the day interpreted Koch's talk before the Tenth International Medical Congress in 1890 as stating that he had found a method for developing resistance against tuberculosis. Koch had been under extreme pressure from the German government to produce a breakthrough in tuberculosis research prior to the Medical Congress that was held in Berlin to demonstrate Germanic intellectual supremacy. In this case his lifelong habit of secrecy to ensure his own priority of discovery prevented a thorough, impartial examination of the treatment; so powerful was the force of Koch's reputation that even scientists of international renown did not challenge the lack of specific detail in Koch's presentation and allowed the premature and indiscriminate use of tuberculin on patients. Koch originally did his testing of the tuberculin vaccine on guinea pigs, but humans proved to be far more sensitive to the vaccine, incurring serious side effects within hours of vaccination. The indiscriminate use of tuberculin by physicians and the fatalities the vaccine caused turned the tide of public opinion against Koch. When data accumulated on 2,000 cases showed tuberculin to be ineffective, its use as a therapeutic agent was abandoned. The development of tuberculin for preventing tuberculosis represents one of Koch's few failures.

The work in Koch's laboratory, together with the discoveries of Metchnikoff, established the foundations for understanding the basis for the humoral immune response, which is antibody-mediated, and the cellular immune response, which involves cell components. These two mechanisms of the immune response system represent major lines of defense of animals against infectious agents.

The field of immunology was furthered by the experiments of Jules Bordet, who discovered in 1895 that a component of blood serum, called **complement**, could bring about the destruction of infecting bacteria. When complement reacts with antibody-coated (sensitized) bacteria, it causes the cells to lyse (rupture), killing the infecting bacteria. Important advances in the use of immunology for diagnosing disease were made by August von Wasserman (1906) for the detection of syphilis and by George and Gladys Dick (1924) for the diagnosis of scarlet fever. These diagnostic procedures make use of the reactions between antigen, antibody, and complement molecules, which can be visualized *in vitro*. For example, the Wasserman test for syphilis is designed to determine if complement molecules are fixed by reaction with the antigens of the bacterium that causes syphilis and the antibodies that are made in response to a syphilitic infection; the test is designed to determine whether there are free complement molecules in a patient's blood serum that can bring about lysis of cells when mixed with those cells in a test system, or whether the patient's complement is not free because it has been used up in fighting a syphilitic infection. Significant advances in our basic understanding of the defense mechanisms of animals against microbial infection are still being made in the latter half of the twentieth century. In 1960 Alick Issacs found that cells infected by a virus produce a substance called **interferon** that inhibits viral replication. The discovery of interferon is a significant advance in determining how animals recover from viral diseases. Interferon does not show great specificity in the viruses against which it is active and thus may be quite useful in treating diseases caused by viruses. The potential use of interferon for treating viral and other diseases of man, including cancer, is being actively investigated and may prove to be a breakthrough, comparable to the discovery of penicillin, in treating these diseases.

Virology

The word **virus** was used by the ancient Romans to mean "poison," "venom," and/or "secretion,"—all unpleasant-sounding meanings. When the field of bacteriology began to develop, the medical use of the term virus implied any microscopic etiologic agent of disease. It was not until the invention of bacterial filters in 1884 by Charles Chamberland, Pasteur's coworker (who also invented the autoclave) that the field of virology really began to develop. In 1882, when Dmitrii Ivanowski reported that the agent responsible for tobacco mosaic disease could pass through a bacteriological filter, it became apparent that the microbial world contained even smaller members, the viruses, than had been previously recognized.

While the observation of most viruses awaited a further advance in microscopy, the development of the electron microscope, the field of virology continued to progress. In 1898 Friedrich Loeffler and Paul Frosch reported that foot-and-mouth disease was caused by an agent that passed through a bacteriological filter, and they suggested that the causal agents of a large number of other infectious diseases, including smallpox, cowpox, and measles, might be caused by similar filterable agents (viruses). Their discoveries opened up the field of animal virology. Also in 1898, Martinus Beijerinck, unaware of Ivanowski's work, ascribed tobacco mosaic disease to a "contagious living liquid." In 1911 Peyton Rous demonstrated that a cell-free filtrate could cause malignant growths in animals. The work of Beijerinck and Rous established the basis for tumor virology. The work of Walter Reed and others on yellow fever in the early 1900s showed that this disease was caused by a filterable agent (virus) that could be transmitted by a mosquito carrier (vector); this work demonstrated that viruses could infect more than one animal species and that viral diseases could be transmitted to man by biting arthropods. Bacteriophage, viruses that infect bacteria, were discovered separately by Frederick Twort and Felix d'Herelle, and consequently, it was known by 1915 that viruses could infect even the smallest organisms that had been seen to that date.

In the mid-1920s F. Parker and R. N. Nye successfully cultivated viruses by using tissue culture techniques. Further advances were made by others during the following three decades, including the culture of viruses by Ernest Goodpasture and colleagues, using chick embryos (1931), and the establishment of the HeLa cell line (isolated from a cervical carcinoma of Henrietta Lacks) by G. O.

Gay and coworkers (1952). The ability to grow viruses in culture facilitates the study of these organisms and permits the performance of various experiments with viruses.

Although it was not until the 1940s that electron microscopy was introduced, permitting microbiologists to observe viruses in detail and discover that even these smallest microbes exhibit a diversity of form, Wendell Stanley, in 1935, successfully crystallized tobacco mosaic virus (TMV) and showed that it was largely composed of protein. TMV was later shown, in 1937, to contain 6 percent RNA (ribonucleic acid) by Frederick Bawden and Norman Pirie. Other viruses subsequently have been shown to contain RNA or DNA in addition to protein. The only essential components of a virus are a protective protein coat and a nucleic acid, either DNA or RNA, containing the genetic information of the virus.

The latter half of the twentieth century has been marked by great advances in our understanding of molecular biology. With respect to viral genetics, Heinz Fraenkel-Conrat and Robley Williams reported, in 1955, on the reconstitution of tobacco mosaic virus from its inactive protein and nucleic acid components. That same year Seymour Benzer published a paper on the fine structure of the genetic region in bacteriophage. In 1969 Max Delbrueck, Alfred Hershey, and Salvator Luria shared the Nobel Prize for their studies on viral genetics. These researchers had performed studies on the reproduction of viruses with the finding that the viral nucleic acid coded for the reproductive sequence of the viral components within a host cell. Viruses are well suited for use in genetic studies because of their relatively small genomes.

Bacterial classification

Today, genetics pervades all areas of microbiology. Even classical taxonomy, which relies on classifying organisms largely on their appearance, is giving way to genetic analyses to classify microorganisms. The current taxonomic status of microorganisms represents a long history of developments in biological classification systems. The first comprehensive **classification system** proposed for all living organisms was put forth by the Swedish botanist Carl Linnaeus in 1743. In this system the organisms (infusoria) observed by Antonie van Leeuwenhoek were all placed in the genus *Chaos*. The Linnaean system weighted features in differentiating one organism from another; that is, it gave more importance to certain characteristics than to others. Although the actual classification of organisms proposed by Linnaeus did not adequately treat microorganisms, the Linnaean classification system has had a great influence on traditional microbiological classification systems. This system established the basis for using dichotomous keys for identifying organisms, that is, for distinguishing organisms according to a hierarchical series of tests, with separations between groups made according to a specified order of importance.

An alternative classification system, proposed by Michael Adanson in 1763, did not give differential weighting to particular features for determining an organism's taxonomic position. In this system of classification, organisms are grouped on the basis of their overall similarity rather than separating them based on an individual key test. Adanson's taxonomic system had little influence on bacterial classification systems until the early 1960s, when it was incorporated into the computerized identification systems of Peter Sneath and others.

Early monographs on the classifications of bacteria called infusoria, the term used at that time to describe microbes occurring in aqueous infusions of vegetable matter and meat broths, were published by Otto Muller (1786) and Christian Ehrenberg (1838). In 1857 Karl Nageli proposed that the bacteria should be placed in a class of their own, Schizomycetes (fission fungi). Ferdinand Cohn made major contributions to the development of bacterial taxonomy; between 1872 and 1876 he published a new system of classification for bacteria based largely on microscopic observations. The methodological developments introduced by Robert Koch and others had a marked impact on bacterial classification systems. Gram's staining method was incorporated as a key feature of bacterial classification systems by W. Migula in 1894. K. B. Lehman and R. O. Neumann (1896) proposed a different classification system that also utilized staining reactions and endospore formation as diagnostic features. It is clear that the foundations of classical bacterial classification systems as we know them today developed step by step during the late nineteenth century and were tied to methodological advances in the art of culturing and characterizing microorganisms.

Many of the advances in microbial systematics were incorporated by Robert Buchanan, who, starting in 1915, published a series of articles on the nomenclature and classification of the bacteria. Buchanan proposed a code of nomenclature for bacteria that was different from the botanical and zoological codes and developed a classification system, using a wide range of features, including morphological, biochemical, and pathogenic characteristics, in order to organize the bacteria into families, tribes, and genera. His efforts led to the formation by the American Society of Microbiology of the Committee on Characterization and Classification of Bacterial Types. This committee, under the chairmanship of David Bergey, subsequently published the first edition of *Bergey's Manual of Determinative Bacteriology* in 1923. In 1936 Bergey established a trust to continue the work. Buchanan later chaired the *Bergey's Manual* trust for 17 years. The manual is now in its eighth edition and is still viewed as the authoritative work on bacterial classification.

In the eighth edition of *Bergey's Manual*, the bacteria are divided into 19 major groups, which are further divided into systematic units of orders, families, tribes, genera, and species. As with other living organisms, only the genus and species names are used in bacteriology to designate a given bacterium, e.g., *Escherichia* (genus) *coli* (species); the genus and species names are both italicized or underlined; the first letter of the genus name is capitalized, the first letter of the species name is not. The periodic revisions of *Bergey's Manual* have incorporated numerous changes in the descriptions of bacteria based on up-to-date developments in bacterial classification. Features based on new methodological developments, including genetic analyses, have found their way into bacterial classification systems. Today, modern versions of both the classical Linnaean dichotomous approach and the Adansonian approach to taxonomy are used for the classification of bacteria and other microorganisms. Modern taxonomic descriptions of microorganisms, which periodically appear in journal articles and revisions of *Bergey's Manual*, increasingly rely on molecular analyses to determine similarities among microorganisms and the proper taxonomic positions of bacteria. Starting with the ninth edition, *Bergey's Manual* will be published in parts, describing the taxonomy of particular groups of microorganisms, rather than as one comprehensive volume on bacterial classification.

Phycology

Many algae form macroscopic structures that are visible to the naked eye, and references to algae are found in early Eastern and Western literature. It was not until the mid-eighteenth century, though, that microscopic methods were used to examine algae. Once the reproductive phases of algae were recognized, life history studies could proceed, and taxonomic systems of classification could be developed. The elucidation of algal sexual reproductive cycles was pioneered by Gustave Thuret, a wealthy Parisian amateur scientist, in studies conducted from 1840 to 1854 with *Fucus*. Nathaniel Pringsheim, working with *Vaucheria* during the same period, described the growth of algae and the development of sexual stages, allowing algal classification to be based on reproductive systems rather than just on superficial resemblances. During the early nineteenth century, many **phycologists** (algologists) published works that advanced the taxonomic classification of algae. At the end of the nineteenth century, phycologists employed their accumulated knowledge of algal morphology and reproduction to revise the classification schemes for the taxonomy of algae. Many of today's views on the systematics of the algae, which are discussed in Chapter 12, date from the late nineteenth and early twentieth centuries.

Mycology

Like algae, some fungi form macroscopic structures, and fungi were used and studied from early times. The ancient Romans knew which fungi were epicurean delicacies, which were lethal, and which had hallucinogenic effects. Hooke's *Micrographia* (1665) included illustrations of microscopic fungi. Yeasts are recognizable in the drawings of Leeuwenhoek. The first book solely about fungi was the *Theatrum Fungorum* by Johannes Franciscus Van Starbeeck in 1675, using the drawings Charles de l'Escluse, also known as Clusius, prepared in 1601. The fungi *Rhizopus, Mucor*, and *Penicillium* are identifiable in the 1679 drawings of Marcello Malpighi.

The science of **mycology**, the study of fungi, however, probably owes its origins to Pier Antonio Micheli, an Italian botanist who, in 1729, published *Nova Plantarum Genera*, which included his studies on fungi. Almost half of the plants Micheli described were fungi, and many of the generic names still used today were first presented in this study. Micheli's most important contribution was the observation of the production of spores and the demonstration that the spores reproduce plants similar to their parent. Heinrich Anton deBary made major contributions to the field of mycology in his studies on plant pathology. He elucidated the life cycles of many rusts and defined the broad groups of Phycomy-

cetes and Myxomycetes. In 1885 deBary proved that the blight that had caused the great Irish potato famine of the preceding decade was caused by a fungus; this was the first demonstration that a specific microorganism is the causative agent of disease.

A practical system of classification of the **Fungi Imperfecti**, those fungi that do not exhibit sexual reproductive phases according to spore groups, was developed by Pier Andrea Saccardo, who also collaborated on the 25-volume *Sylloge Fungorum* (1882–1925), which critically compiled most of the literature on fungal systematics published prior to 1920. In the early twentieth century, A. H. R. Buller also published a major monograph on fungal systematics. The first major compilation of yeast systematics was published in 1896 by Emil Hansen; this taxonomic system was greatly revised by A. Guilliermond between 1920 and 1928. The systematics proposed by Hansen and Guilliermond included the use of physiological, sexual, and phylogenetic relationships, as well as morphological observations, to determine classification. These characteristics, supplemented with direct analyses of fungal genetic information, form the basis of today's classification of the fungi as discussed in Chapter 12.

Protozoology

Protozoa were among the first microscopic organisms ever observed. The protozoa *Vorticella, Volvox*, and *Euglena* clearly are shown in the 1677 sketches of Leeuwenhoek. *Paramecium*, as well as other protozoa, were described in 1678 by Christian Huygens. Leeuwenhoek continued to report his drawings of protozoa and in 1681 described what appears to be *Giardia intestinalis*, thus discovering parasitic protozoa. Louis Joblot, a professor of mathematics with an interest in optics that led him to microscopy, published the first treatise on protozoa in 1718. G. A. Goldfuss introduced the term Protozoa in 1817, and a chapter about this group of organisms appeared in a book on comparative anatomy of invertebrates in 1848 by Karl T. E. von Siebold. The term *protozoa* is derived from the Greek "protos," meaning first, and "zoon," meaning animal. In 1838 Christian Ehrenberg published a major monograph on the protozoa, describing more than 500 species, including descriptions of their digestive and reproductive systems. Felix Dujardin, a French professor of zoology, published a classification system for the protozoa, using primarily morphological features in 1841; Dujardin was an accurate observer, and his classification system was sounder than that of Ehrenberg.

Medical protozoology began in the mid-1800s. Pasteur, in 1870, reported that a protozoan was responsible for a disease of silkworms that had devastated the French silk industry during the 1800s. Also in 1870, T. R. Lewis observed *Amoeba* in the stools of individuals suffering from choleric symptoms and, shortly thereafter, F. Losch described *Entamoeba histolytica* as a causative agent for dysentery in man. Transmission of this disease by ingestion of *Entamoeba histolytica* was shown by E. L. Walker and A. W. Sellards in 1913.

The discovery that disease-causing microorganisms could be transmitted by animal vectors, which was made by Theobold Smith, represented a major advance in medical protozoology and our understanding of the mechanisms of disease transmission. Smith and coworkers (1893) proved that Texas cattle fever was caused by a protozoan and that transmission of the disease involved a tick vector. This proof opened the way for discovering that a number of other diseases are transmitted by arthropod vectors. Alphonse Lav-

eran and Camillo Golgi (1881–1886) were able to show that malaria was caused by a parasitic protozoan. Ronald Ross, in 1897–1898, found that the malarian parasite is transmitted in birds by a mosquito vector. The mode of transmission of the protozoan that causes malaria in man was discovered by Battista Grassi, who disputed with Ross the priority of discovery of vector transmission of this disease. Joseph Dutton (1902) found that a trypanosome protozoan causes African sleeping sickness and is transmitted by the tsetse fly. William B. Leishman and C. Donovan (1903) discovered that Kala azar disease is caused by a protozoan that subsequently was named *Leishmania donovani*; many microbial species names are derived from the names of the individuals who studied them.

The beginning of microbial ecology and physiology

The works of Sergei Winogradsky (Figure 1.20) and Martinus Beijerinck (Figure 1.21) largely established the basis for **microbial physiology** (the activities of microorganisms, including their me-

Figure 1.20

Sergei Winogradsky (1856–1953) established the concept of microbial chemoautotrophy and made major contributions to the development of the enrichment culture technique and soil microbiology. (From the collection of H. A. Lechevalier, Waksman Institute of Microbiology.)

Figure 1.21

Martinis Beijerinck (1851–1931) was the first to demonstrate the existence of a virus by extracting the juice of a diseased tobacco plant, filtering it, and allowing the clear filtrate to transmit the disease to healthy plants. (From the collection of H. A. Lechevalier, Waksman Institute of Microbiology.)

tabolism) and **microbial ecology** (the interactions of microorganisms with their biotic and abiotic surroundings). Beijerinck, a Dutchman, received his initial training in chemistry at the Delft Polytechnical School, to which he was later appointed professor in 1895. He was difficult to get along with and berated students for even their smallest errors. Beijerinck was a keen observer with a strong general scientific background and liked performing neat experiments with simple tools. Winogradsky was a Russian scientist, but he performed many of his studies in France and Switzerland, including work at the University of Strasbourg while it was headed by the mycologist deBary, and at the Pasteur Institute. Whereas Pasteur had concentrated on the microbial use of organic compounds (**heterotrophic growth**—requiring organic compounds), Winogradsky and Beijerinck made significant discoveries concerning microbial transformations of inorganic compounds and the **autotrophic** (self-feeding—not requiring preformed organic compounds) means of metabolism. In 1877 Theophile Schloesing and Achille Müntz had proposed that nitrification, the biochemical transformations of inorganic nitro-

Figure 1.22

Professor A. J. Kluyver (1888–1956) toward the end of his life. His greatest contribution was perhaps his comparative approach, which stressed the unifying metabolic features found within the diverse microbial world. (From the collection of H. A. Lechevalier, Waksman Institute of Microbiology.)

gen compounds, is a microbial process. Winogradsky was able to isolate and describe the autotrophic nitrifying bacteria and show that these bacteria are responsible for transforming ammonium ions to nitrate ions in soil. He showed that microorganisms can derive energy from inorganic chemical reactions and their carbon from carbon dioxide, and by doing so he defined chemoautotrophic metabolism. Winogradsky also described the microbial oxidation of sulfur, hydrogen sulfide, and ferrous iron and anaerobic nitrogen-fixing bacteria.

During the same period as Winogradsky's work on the metabolism of inorganic substances, 1888 to 1901, Beijerinck reported on symbiotic and nonsymbiotic aerobic nitrogen fixation by bacteria, the process by which atmospheric nitrogen is combined with other elements, making this essential nutrient available to plants, animals, and other microorganisms. Beijerinck also was able to isolate sulfate-reducing bacteria, which are important in the cycling of sulfur compounds in soil and sediment. All of these reactions form the basis of important transformations and movements of elements in soil ecosystems and determine the fertility of soil. Another particularly significant advance made by Beijerinck was the development of the technique of enrichment culture, which permits the isolation of a bacterium with a particular metabolic activity by adjusting incubation conditions.

The works of Beijerinck and Winogradsky were primarily concerned with soil processes, but the microbial transformations that they discovered formed the basis for understanding biogeochemical cycling reactions and the critical role of microorganisms in transforming elements on a global scale. These microbially mediated cycling reactions are essential for maintaining environmental quality and are necessary for supporting life on earth.

During the early twentieth century major advances were made in the understanding of biochemistry and microbial metabolism. Albert Kluyver (Figure 1.22) followed in the footsteps of Beijerinck, both as director of the Delft school and in the biochemical direction of his study, continuing the great tradition in the small town of Delft of microbiological study that had been begun by Leeuwenhoek 300 years earlier. Kluyver examined the unity and diversity in metabolism of microorganisms, emphasizing the unifying features of microbial metabolism, correctly recognizing the nature of intermediary metabolism, and establishing that hydrogen transfer (oxidation–

reduction reactions) is a basic feature of all metabolic processes. The flow of carbon and energy through a series of metabolic transformations is an essential feature of living microorganisms. Kluyver was instrumental in developing an understanding of the role of central metabolic pathways in microbial metabolism. C. B. van Niel, who was a student of Kluyver, continued the tradition of advancing microbial physiology. He made important contributions to our understanding of photosynthesis, recognizing the similarity between H_2S and H_2O in the photosynthetic processes of the anaerobic photosynthetic sulfur bacteria and in higher plants.

Important later discoveries, which emphasized the unity of intermediary metabolism, included the elucidation of the citric acid cycle, for which Sir Hans Krebs received the Nobel Prize in 1953, and the elucidation of the chemical steps in carbon dioxide fixation during photosynthesis, for which Melvin Calvin received the Nobel Prize in 1961. In 1978 the Nobel Prize was awarded to Peter Mitchell for the development of chemiosmotic theory to explain how the biochemical reactions occurring at membranes generate energy in the form of ATP. While these Nobel Prizes were awarded in chemistry, they represent fundamental advances in our understanding of microbial metabolism.

Microbial genetics

Microbial genetics is probably the most rapidly expanding field in microbiology today. The beginnings of our understanding of microbial genetics did not start until the middle of the twentieth century. In 1941 George Beadle and Edward Tatum published their studies on the genetic control of biochemical reactions in the fungus *Neurospora*. Tatum also showed in 1945 that exposure to X rays increased the mutation rate in the bacterium *Escherichia coli*. In 1944 Oswald Avery, Colin MacLeod, and Maclyn McCarty published their studies on the nature of the substance that induces transformation of pneumococcal types and concluded that a nucleic acid of the deoxyribose type is the fundamental unit of the transforming principle of *Streptococcus pneumoniae*. Joshua Lederberg and coworkers, between 1946 and 1956, studied genetic exchange processes in bacteria and made the first reports on conjugation and transduction. In 1958 George Beadle, Edward Tatum, and Joshua Lederberg shared the Nobel Prize for their studies on microbial genetics. Their pi-

oneering studies led to a flourish of many works aimed at using genetic recombination processes to map the genomes of microorganisms.

A major breakthrough occurred in 1953, when James Watson and Francis Crick proposed the double-helical structure of DNA (Figure 1.23). The revelation of DNA truly opened the field of molecular genetics for major new investigations. The establishment of the structure of the molecule housing the genetic information of living organisms permitted the unravelling of the ways in which genetic information is stored and expressed. As with any new model, new information can lead to refinements, and it now appears that there are regions of the DNA that form a Z-structure that does not conform to the α-helix model proposed by Watson and Crick. The functions of the Z-regions of the DNA are yet to be fully understood, but they probably are involved in controlling expression of the DNA. The basic structural model proposed by Watson and Crick does provide the basis for understanding how the vast amount of genetic information of a cell can be encoded within a macromolecule and used to direct the synthesis of other macromolecules involved in cellular functions.

With the establishment of the structural and biochemical nature of the genetic material of the cell, the time was right for a unification of concepts in microbial genetics and biochemical metabolism. In 1960 Francois Jacob and Jacques Monod proposed the **operon theory**, which explains how the genetic information controls protein synthesis, that is, the nature of the control regions of the DNA molecule that act as switches, turning on and off the synthesis of enzymes. Tsutomu Watanabe (1961) and Naomi Datta (1962) reported on episome-mediated transfer of drug resistance, identifying the role of plasmids in genetic exchange, an important discovery for our understanding of why some microorganisms are resistant to antibiotics. Crick and coworkers correctly proposed that three nucleic acid bases in DNA code for one amino acid. Several research groups determined which triplet base sequences specify which amino acids, and in 1968 the Nobel Prize was awarded to Robert Holley, H. G. Khorana, and Marshall Nirenberg for their work on breaking the genetic code. These researchers had examined different facets of the translation of the genetic code into protein structure. A number of other investigators also contributed to the basic understanding of the genetic code. Frederick Sanger and Walter Gilbert shared the 1980 Nobel Prize for their studies on sequencing the bases in

Figure 1.23

James Watson (left), age 23, and Francis Crick, age 34, posed with their DNA model in the Cavendish Laboratory at Cambridge University, England, in 1953, when they announced their discovery of the molecular structure of deoxyribonucleic acid. They shared the Nobel Prize for Medicine in 1962 with Maurice Wilkins. (Photograph by Barr-Brown Camera Press, London, from Central Office of Information, London.)

Discovery process

This particular scientific breakthrough was not accomplished through the research we often think of as involving many test tubes and petri plates, guinea pigs, and bubbling chemical retorts, but rather was the product of a great deal of thought, discussion, examination of evidence already available, and intuition, as exemplified by the decision to build two-chain models simply because of the "repeated finding of twoness in biological systems." Watson and Crick knew they had to rely on the simple laws of structural chemistry. "The essential trick was to ask which atoms like to sit next to each other. In place of pencil and paper, the main working tools were a set of molecular models superficially resembling the toys of preschool children." So after getting the tautometric forms of guanine and thymine serendipitously corrected by a visiting scientist, who pointed out that the structures in the organic chemistry books were wrong, Watson and Crick began "shifting the bases in and out of various pairing possibilities." In the next step their tools were a plumb line and a measuring stick to determine the relative positions of all the atoms in a single nucleotide. By assuming a helical symmetry, the locations of the atoms in one nucleotide would automatically generate the other position.

DNA; determining the correspondence of the nucleic acid base sequence in DNA and the amino acid sequence in proteins was a major accomplishment.

Genetic studies during the early 1960s established the basis for understanding how genetic information is stored in the DNA molecule and how that information is transcribed and translated into proteins that act as enzymes in determining the metabolic activities of microorganisms. By the early 1970s the genetics of bacteria was sufficiently understood to perform experiments that could "create" new organisms. In 1978 the Nobel Prize was awarded to Werner Arber, Hamilton O. Smith, and Daniel Nathans for the development of restriction endonuclease enzymes that can be used to study genetic organization (genetic mapping) and to manipulate DNA for genetic engineering. Along with Sanger and Gilbert, the 1980 Nobel Prize was shared by Paul Berg for his work on constructing **recombinant DNA molecules**. Pioneering efforts in genetic engineering also were made in the early 1970s by Harold Boyer and Stanley Cohen, who used plasmids and recombinant DNA methodology to clone genes, opening up an entirely new area of applied microbiology and spurring the formation of many genetic engineering companies. **Genetic engineering**, using **recombinant DNA technology**, holds the promise of solving many problems of mankind, including the ability to produce drugs for treating currently incurable diseases, to produce sufficient food to feed the world's population, and to solve problems of environmental pollution.

In just over one hundred years, microbiology has developed as a major scientific discipline. Microbiology grew out of several other scientific disciplines, which undoubtedly contributes to its broad multidisciplinary nature.

Without question the two greatest historic figures in microbiology were Louis Pasteur and Robert Koch. They set the stage for the future development of this field of science, emphasizing the relationships between germs and disease (including host resistance), and microbes and fermentation (including metabolism and biochemistry). Louis Pasteur made major contributions to our understanding of fermentations; beginning as a chemist, Pasteur emerged as an outstanding microbiologist. Pasteur studied both the lactic acid and alcoholic fermentations, describing the chemical changes that occur during fermentation and the growth of yeast when air is present (aerobic conditions) and when it is absent (anaerobic conditions). Robert Koch extended his training as a physician to tackle the problems of treating and preventing infectious diseases. He and his disciples discovered the etiologic agents of several diseases and developed methods for their cure and prevention. The pure culture techniques developed by Koch have dominated microbiological research up to the present day.

Pasteur's unique contribution was his conceptual ability, his ability to analyze and to bring together the pieces of the puzzle that form the grand scheme of microbiology. Koch's genius lay in his ability to devise the methods and techniques that allowed the science to progress. Both men were excited by the process of scientific discovery and by solving the problems they had set for themselves. Their inquiring minds, sense of adventure, egos, and competitiveness caused them to strive ahead at a furious pace. They set a pattern for many great microbiologists who followed. The race to be first in making microbiological discoveries of basic and practical importance continues today.

Both Koch and Pasteur trained students to carry on their work and to publish extensively to make available the knowledge they had acquired. The science of microbiology is still young; advances in microbiology parallel progress in other fields of science (Table 1.1). The development of appropriate methodologies for advancements in one field often depends on developments in other areas of scientific endeavor, and this is apparent in the historical development of microbiology. The science of microbiology has evolved from many fields—from the contributions of numerous amateur scientific observers who, like Leeuwenhoek, took pleasure in searching for the very small and discovering the wonders of the microbial world, to professional scientists who, like Pasteur and Koch, brought their training in other fields to the development of the new discipline of microbiology. Scientists continue to enter the field of microbiology from other disciplines. Francis Crick, for example, the codiscoverer of the nature of the genetic molecule DNA, was trained as a physicist, and it was his background in X-ray crystallography that provided the necessary expertise for unravelling the arrangement of DNA as a double helix.

Currently, studies in molecular genetics are producing new knowledge, enhancing our fundamental understanding of microorganisms and other living systems. Industrial and environmental problems provide many opportunities for the practical application of basic microbiological principles. Improvements in scientific communication have allowed microbiologists to capitalize rapidly on scientific advances, hastening the rate of development in the field of microbiology; today, there are numerous journals and publications through which worldwide distribution of microbiological information is made possible (Table 1.2). The field of microbiology promises to continue its rapid development for many more years. It is an exciting and challenging field of science for students and professionals alike.

table 1.1

Time-line showing development of microbiology relative to other sciences

1200–1220	1221–1240	1241–1260	1261–1280	1281–1299
			Bacon describes laws of reflection and refraction of light and prototype magnifying glass Marco Polo goes to China	Invention of glass mirror Invention of spectacles
1300–1320	**1321–1340**	**1341–1360**	**1361–1380**	**1381–1399**
		Plague in Europe	Quarantine used in Europe against plague	
1400–1420	**1421–1440**	**1441–1460**	**1461–1480**	**1481–1499**
		Gutenberg begins printing		Da Vinci observes capillary action Mercury used to treat venereal disease
1500–1520	**1521–1540**	**1541–1560**	**1561–1580**	**1581–1599**
Sewers begun in England		Copernicus' treatise on heliocentric solar system Fracastoro's theory of contagious disease	Manufacture of pencils begun in England	**Janssen invents compound microscope**
1600–1620	**1621–1640**	**1641–1660**	**1661–1680**	**1681–1699**
Kepler calculates paths of planets Invention of telescope Galileo studies gravitation	Harvey describes theory of blood circulation Descartes develops analytical geometry		Malpighi describes plant and animal anatomy Boyle's law of gases **Hooke's Micrographia** Newton's theory of gravitation **Leeuwenhoek observes bacteria and protozoa**	
1700–1720	**1721–1740**	**1741–1760**	**1761–1780**	**1781–1799**
Fahrenheit constructs mercury thermometer **Montague introduces smallpox vaccination**	**Micheli publishes work on fungi**	Osmosis described Benjamin Franklin shows lightning to be electricity	Watt perfects steam engine Hydrogen, oxygen, and nitrogen discovered Linnaeus describes classification system for living organisms **Spallanzani disputes theory of spontaneous generation**	Lavoisier produces table of chemical elements **Jenner introduces vaccination**

1800–1820	1821–1840	1841–1860	1861–1880	1881–1899
Dalton's atomic theory	Faraday's electric motor	Liebig studies biochemistry	Maxwell's electromagnetic theory of light	Koch develops methods for pure culture of bacteria
Discovery of uv rays	Ohm's law of current	Joule defines 1st law of thermodynamics	Mendel studies genetics	Koch's postulates
Invention of battery	**Schwann, Kutzing, and Cagniard de Latour report that yeasts cause fermentation**	Clausius outlines 2nd law of thermodynamics	**Lister begins practice of antiseptic surgery**	Pasteur inoculates against rabies
Wave theory of light		Morse invents telegraph	Nobel produces dynamite	Discovery that mosquitoes carry malaria
Avogadro proposes molecular theory of matter		Bunsen invents gas burner	Periodic table of elements	Hertz demonstrates radio waves
Lamarck emphasizes effects of environment on species		**Pasteur studies fermentation**	Bell patents telephone	Roentgen discovers X rays
Appert develops canning		Darwin's theory of natural selection	Edison patents gramophone	Marconi invents radio
			Zeiss constructs modern microscope	Discovery of electron, 1st atomic particle
			Koch studies bacteria-causing disease	**Metchnikoff describes phagocytosis**
			Cohn discovers endospores of bacteria	**Petri's culture dish**
			Development of autoclave—use of steam for sterilization	**Beijerinck studies nitrogen fixation**
			Abbe improves microscope lenses	**Winogradsky studies autotrophy**
			Gram stain developed	Discovery of antitoxins
				Field of virology begins with discovery of plant and animal viruses
				Ford builds cars
				Complement discovered
				Smith shows protozoan transmitted by tick vector

1900–1920	1921–1940	1941–1960	1961–1980
Planck's quantum theory	Birdseye deep-freezes food	Discovery of neutron	**Discovery of interferon**
Curies discover radium	Development of centrifuge	**Discovery that actinomycetes produce antibiotics**	Discovery of quasars
Wright brothers make first airplane flight	**Fleming discovers penicillin**	**Beginning of molecular genetics**	Man lands on moon
Einstein's theory of relativity	Invention of Geiger counter	**Avery shows bacterial transformation**	Fertilization of human egg in test tube
Bohr's theory of atomic structure	**Invention of electron microscope**	**Watson and Crick describe DNA double helix**	**Episomes discovered**
Ehrlich studies chemotherapy	**Stanley crystallizes a virus**	Sputnik	**Discoveries on gene regulation of enzyme production**
Discovery of bacteriophage	Splitting of atom	Use of antibiotics in medicine begins	**Breaking of the genetic code**
Discovery that viruses can cause malignancy	Theory of gene developed	Antipolio vaccine	**DNA recombination studies lead to genetic engineering**
	Invention of phase contrast microscope	Invention of transistor	
	Beadle and Tatum isolate mutants		
	First edition of *Bergey's Manual of Determinative Bacteriology*		
	Kluyver and Van Niel work on comparative microbial metabolism		

33

Postlude

table 1.2

Some journals concerned primarily with microbiological studies

Annales de Microbiologie. France. Institut Pasteur. 1887–.

Prominent journal publishing research papers in microbiology and immunology. Initial volumes from time of Pasteur and Koch.

Antimicrobial Agents and Chemotherapy. U.S.A. American Society for Microbiology. 1972–.

Research papers covering all aspects of knowledge relating to antimicrobial agents and chemotherapy, including cancer chemotherapy.

Antiviral Research. The Netherlands. Elsevier. 1982–.

Articles concerning effective controls of viral infections of humans and other organisms, including treatment and prevention (vaccines).

Antonie van Leeuwenhoek Journal of Microbiology. The Netherlands. Stichting Antonie van Leeuwenhoek. 1935–.

Publishes research papers of investigations of limited scope devoted to microbiology and related fields of science and technology.

Applied and Environmental Microbiology. U.S.A. American Society for Microbiology. 1953–. (Formerly *Applied Microbiology.*)

This research journal is dedicated to applied studies in microbiology, including studies on clinical microbiology, virology, immunology, food microbiology, toxicology, environmental microbiology, ecology, taxonomy, metabolism, and fermentation products.

Archives of Microbiology. U.S.A. Springer-Verlag. 1930–.

Subtitled: *Journal for the Investigation of Microorganisms.* Original papers and review articles covering the entire field of microbiology.

British Phycological Journal. United Kingdom. British Phycological Society. 1953–.

Publishes papers on all aspects of the study of algae, including original research reports, review articles on topics of phycological interest, field studies, and techniques relevant to algae.

Canadian Journal of Microbiology. Canada. National Research Council of Canada. 1954–.

Comprehensive, general microbiology journal, part of the Canadian Journal of Research series; includes research papers in microbial ecology, infection and immunity, physiology, virology, molecular biology, and genetics.

Comparative Immunology, Microbiology and Infectious Disease. U.S.A. Pergamon Press. 1978–.

Research papers related to medical microbiology, including works relating to immunological responses to infection, pathogenic microorganisms, and infectious disease processes.

CRC Critical Reviews in Microbiology. U.S.A. CRC Press. 1971–.

Prominent scientific authorities present in-depth critical reviews, generally four per issue, centered on a single topic.

Current Microbiology. U.S.A. Springer-Verlag International. 1978–.

Publishes brief papers that concisely yet thoroughly deal with significant facts and ideas in all areas of microbiology. This journal is predicated on the idea that important benefits can be gained by substantially reducing the length of time required for scientific publication.

Diagnostic Microbiology and Infectious Disease. U.S.A. Elsevier Biomedical. 1983–.

Publishes articles, critical reviews, and commentaries oriented to the diagnostic microbiology laboratory and its role in the identification and treatment of infectious disease in the areas of bacteriology, immunology, immunoserology, infectious disease, mycology, parasitology, and virology.

Infection and Immunity. U.S.A. American Society for Microbiology. 1970–.

Devoted to the advancement and dissemination of fundamental knowledge concerning pathogenic microorganisms and infections and with ecology, epidemiology, and host factors as well as immunology.

International Journal of Systematic Bacteriology. U.S.A. International Association of Microbiological Societies. 1951–.

Devoted to the advancement of the systematics of bacteria, yeasts, and yeast-like organisms.

Journal of Applied Bacteriology. United Kingdom. Society for Applied Bacteriology. 1954–. (Formerly Proceedings of the Society for Applied Bacteriology.)

Review articles, original papers, and short communications on microbiology and its applications to agriculture and other industries.

Journal of Bacteriology. U.S.A. American Society for Microbiology. 1916–.

The leading American journal in the field is concerned with general, fundamental reports on bacteria, their morphology and ultrastructure, genetics and molecular biology, physiology, and metabolism and enzymology.

Journal of Clinical Microbiology. U.S.A. American Society for Microbiology. 1975–.

Disseminates new knowledge concerning the applied microbiological aspects of human and animal infections and infestations, particularly regarding their etiologic agents, diagnosis, and epidemiology.

Journal of General and Applied Microbiology. Japan. Microbiology Research Foundation. (English vol. 19.) 1973–.

Research papers relating to applied aspects of microbiology and the underlying fundamentals.

Journal of General Microbiology. United Kingdom. Society for General Microbiology. 1947–.

Fundamental studies on bacteria, microfungi, microscopic algae, etc., discusses development and structure of microorganisms, their physiology and growth, biochemistry, genetics, and ecology.

Journal of General Virology. United Kingdom. Society for General Microbiology. 1967–.

Original works in general virology, including studies of bacterial, plant, and animal viruses, their structure, genetics, systematics, and interactions with host cells, as well as investigations of pathogenesis.

Journal of Immunology. U.S.A. American Association of Immunologists. 1916–.

Covers all areas of immunology, including cellular, tumor, viral, and microbial immunology, immunochemistry, immunogenetics, and transplantation.

Journal of Infectious Diseases. U.S.A. University of Chicago Press. 1904–.

Presents laboratory and clinical findings concerning infectious diseases, medical microbiology, and immunology, with emphasis on research.

Journal of Medical Virology. U.S.A. Alan R. Liss, Inc. 1977–.

Original reports of major research developments in epidemiology, structure, and composition of viruses, diagnosis, pathology, and treatment.

Journal of Microbiological Methods. U.S.A. Elsevier. 1983–.

Original papers and reviews on new and revised methods in all areas of microbiology.

Journal of Phycology. Canada. Phycological Society of America. 1965–.

Covers all aspects of marine and freshwater algae, including taxonomy, biochemistry, physiology, morphology, cytology, molecular biology, and ecology.

Journal of Protozoology. U.S.A. Society of Protozoologists. 1954–.

Original papers on both descriptive and experimental studies on protozoa. This journal provides a common medium for the taxonomist, physiologist, parasitologist, and biochemist working with protozoa.

Journal of Virology. U.S.A. American Society for Microbiology. 1967–.

Contains reports of original research in all areas of basic virology about the viruses of bacteria, plants, and animals.

Medical Microbiology and Immunology. Federal Republic of Germany. Springer-Verlag. 1886–

Founded by Robert Koch, publishes original papers in virology, bacteriology, immunology, and epidemiology, with pathogenic mechanisms as underlying causes of origin, treatment, and prophylaxis the core of its contents.

Microbial Ecology. U.S.A. Springer-Verlag. 1974–.

An international journal presenting research papers concerning those branches of ecology in which microorganisms are involved.

Microbiological Reviews. U.S.A. American Society for Microbiology. 1937–. (Formerly *Bacteriological Reviews*.)

Authoritative and critical review papers on the current state of research in all areas of microbiology, often including historical analyses.

Microbiology and Immunology. Japan. Japanese Society for Bacteriology, Society of Japanese Virologists, and Japanese Society for Immunology. 1957–. (Formerly *Japanese Journal of Microbiology*.)

Publishes original reports of research concerning significant findings in bacteriology, immunology, virology, and related fields.

Microbios. United Kingdom. Faculty Press. 1969–.

Transworld biomedical research journal devoted to fundamental studies of viruses, fungi, microscopic algae, and protozoa, with emphasis on chemical microbiology.

Mikrobiologiya. Soviet Union. Akademiya Nauk S.S.R. (English translation) 1973–.

Research papers on microbiological studies conducted primarily in the Soviet Union. Original papers published in Russian, with volumes published in English after several years' delay.

Mycologia. U.S.A. Mycological Society of America. 1909–.

Original research papers in all areas of mycology, including works on taxonomy, morphology, physiology, and ecology of fungi.

Phycologia. United Kingdom. International Phycological Society. 1961–.

Review articles, reports, and notes concerning material of phycological interest.

Virology. U.S.A. Academic Press. 1955–.

Articles on the biological, biochemical, and biophysical aspects of viral research, stressing contributions of a fundamental rather than applied nature.

Zeitschrift für Allgemeine Mikrobiologie. German Democratic Republic. Akademie-Verlag. 1960–.

Research papers relating to morphology, physiology, genetics, and ecology of microorganisms.

Study Questions

1. What is a microorganism?

2. What is the current taxonomic position of bacteria?

3. What major contributions to microbiology were made by each of the following microbiologists?
  a. Pasteur
  b. Koch
  c. Winogradsky
  d. Beijerinck
  e. Metchnikoff
  f. Fleming
  g. Ehrlich
  h. Kluyver

4. Why were Leeuwenhoek's observations the critical first step in the development of microbiology?

5. What was the theory of spontaneous generation? How was it disproved, and what is the significance of having disproven this theory?

6. What are Koch's postulates? What are their significance to medical microbiology? Discuss a recent use of Koch's postulates to determine the cause of a disease outbreak.

7. What major discoveries were made in microbiology during each of the following time periods? For each of these time periods, in what areas of microbiology were the most significant advances made?
  a. 1700–1800   **d.** 1901–1950
  b. 1801–1850   **e.** 1951–2000
  c. 1851–1900

8. What is the significance of the discovery made by Watson and Crick?

9. What recent achievements in microbiology have been recognized by the awarding of the Nobel Prize, and to whom were the prizes awarded? What was the significance of the the contribution that was recognized?

10. What attributes make a great microbiologist?

Adelberg, E. A. (ed.). 1966. *Papers on Bacterial Genetics*. Little, Brown and Co., Boston.

Ainsworth, G. C. 1976. *Introduction to the History of Mycology*. Cambridge University Press, London.

Allen, G. E. 1978. *Life and Science in the Twentieth Century*. Cambridge University Press, London.

Brock, T. D. (ed.). 1975. *Milestones in Microbiology*. American Society for Microbiology, Washington, D.C.

Bulloch, W. 1938. *The History of Bacteriology*. Oxford University Press, London. (1979. Dover Publications, Inc., New York.)

Christensen, C. M. 1965. *The Molds and Man*. University of Minnesota Press, Minneapolis.

Clark, P. F. 1961. *Pioneer Microbiologists of America*. University of Wisconsin Press, Madison.

Cole, F. J. 1926. *The History of Protozoology*. University of London Press, Ltd., London.

De Kruif, P. 1926. *Microbe Hunters*. Harcourt, Brace and Co., New York. (1966. Harcourt Brace Jovanovich, Inc., New York.)

Dobell, C. (ed.). 1932. *Antony van Leeuwenhoek and his "Little Animals."* Constable and Co., Ltd., London. (1960. Dover Publications, Inc., New York.)

Eron, C. 1981. *The Virus That Ate Cannibals: Six Great Medical Detective Stories*. Macmillan Publishing Co., Inc., New York.

Grainger, T. H. 1958. *A Guide to the History of Bacteriology*. Ronald Press Co., New York.

Hooke, R. 1665. *Micrographia*. Royal Society, London. (1961. Dover Publications, Inc., New York.)

Judson, H. F. 1979. *The Eighth Day of Creation: Makers of the Revolution in Biology*. Simon & Schuster, Inc., New York.

Lechevalier, H. A., and M. Solotorovsky. 1965. *Three Centuries of Microbiology*. McGraw-Hill Book Co., New York. (1974. Dover Publications, Inc., New York.)

Lederberg, J. (ed.). 1951. *Papers in Microbial Genetics: Bacteria and Bacterial Viruses*. University of Wisconsin Press, Madison.

Peters, J. A. (ed.). 1959. *Classic Papers in Genetics*. Prentice-Hall, Inc., Englewood Cliffs, New Jersey.

Reid, R. 1975. *Microbes and Men*. Saturday Review Press, New York.

Tunevall, G. (ed.). 1969. *Periodicals Relevant to Microbiology and Immunology*. John Wiley & Sons, Inc., New York.

Waksman, S. A. 1954. *My Life with the Microbes*. Simon and Schuster, New York.

Waterson, A. P., and L. Wilkinson. 1978. *An Introduction to the History of Virology*. Cambridge University Press, New York.

Watson, J. D. 1968. *The Double Helix*. Atheneum Publishers, New York.

Whittaker, R. H. 1969. New concepts in kingdoms of organisms. *Science* 163: 150–160.

Woese, C. R. 1981. Archaebacteria. *Scientific American* 244(6): 98–122.

Microbiological methods: pure culture and microscopic techniques

2

The major advances in our understanding of the microbial world have been tied closely to the development of appropriate methods for growing and viewing microorganisms. The microscope and inoculating loop are essential tools of the microbiologist. In beginning microbiology laboratory courses, students first must learn the methods for using a microscope before they can observe microorganisms, and the techniques for growing microorganisms in pure culture before they can perform additional experiments. The development of proper aseptic technique is critical for performing microbiological experiments, both for avoiding contamination of pure and valuable microbial cultures and for avoiding contamination of the experimenter and others in the laboratory with potentially pathogenic microorganisms. In this chapter we shall consider some aspects of the microscopic methods that permit visualization of microorganisms and the ways by which microorganisms can be isolated, safely transferred, and grown in pure culture. This examination of methods for handling and viewing microorganisms, though, cannot replace the learning experience of actually culturing and viewing microorganisms in the laboratory.

Visualization of microorganisms

Through the microscope many microorganisms appear beautiful, diverse and graceful in form, and surprisingly colorful. Although some of this colorful world can be seen with the naked eye, it is the microscope that truly opens the vistas of the microbial world (Plates 1–4).

Light microscope

The **microscope** is the basic tool employed for the visualization of the microbial world. It is used to magnify (enlarge) the size of the apparent image of an object and makes it possible to see the detailed structure of microorganisms. A simple magnifying glass with a convex-convex lens permits some enlargement in the apparent size of the object and is useful in examining the details of macroscopic structures of microorganisms such as mushrooms but does not normally permit the visualization of microorganisms that are the size of bacteria. The normal light microscope that is used for viewing bacteria and other cellular organisms is a compound microscope that has a light source, a condenser lens that focuses the light on the specimen, and two sets of lenses that contribute to the magnification of the image (Figure 2.1). Through the refraction or bending of light rays by the system of microscope lenses, an image of the specimen (object) is formed that is

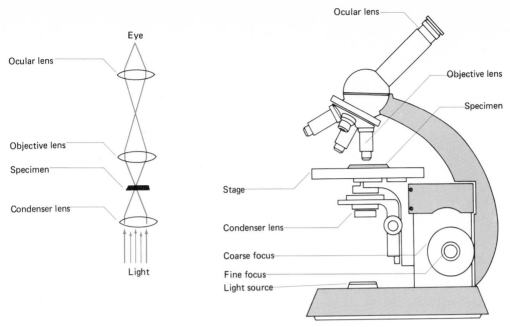

Figure 2.1

A light microscope and the path of light as it is modified by passage through the lenses.

larger than the object itself, permitting examination of the detailed structure of the specimen.

Magnification

The **magnifying capability** of a compound microscope is the product of the individual magnifying powers of the ocular lens (the one nearest the eye) and the objective lens (the lens nearest the specimen). A typical microscope used in bacteriology has objective lenses with powers of 10×, 40×, and 100× and an ocular lens of 10× and thus is capable of magnifying the image of a specimen 100, 400, and 1,000 times (Figure 2.2). If the microscope is parfocal, the various lenses are adjusted so that once the specimen is focused with one lens, it remains in focus when switching to other objective lenses. This permits focusing with a low-power objective and changing to higher-power objectives without refocusing. At a magnification of 1,000×, bacteria and larger microorganisms can be visualized, but viruses and much of the fine structural detail of bacteria cannot be seen (Figure 2.3).

Optical defects and corrections

It is critical that the magnified image not be distorted and that the enlarged image retain the essential detail of the specimen. Seeing a large blur is of no use! The detail of the specimen should be resolved without distortion in the enlarged

Figure 2.2

A bright-field microscope. (Courtesy Leitz Advanced Clinical Laboratory Microscope.)

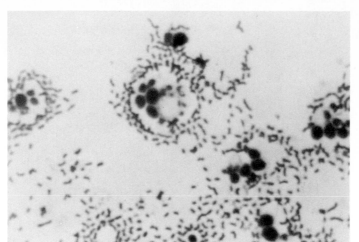

Figure 2.3

This photomicrograph was produced by using bright-field microscopy and shows bacteria and yeasts. By using bright-field microscopy, the characteristic sizes, shapes, and arrangements of the microbial cells can be observed. In this micrograph the bacteria are the small rods, and the yeasts are the larger oval-shaped cells.

microscopic image, and it is important that in addition to magnifying power, the microscope be designed to avoid distorting the image.

There are several types of distortions, including spherical and chromatic aberrations and curvature of field, which occur in light microscopy as a result of the optical properties of lenses. Distortions of the image occur when the microscope's system of lenses fails to focus the image of the specimen in a single plane. Light passing through glass lenses is refracted, or bent, and in theory, parallel light entering a convex-convex lens will be focused at a single point, the focal point of the lens. Modern microscopes contain complex lens systems that are designed to correct for various distortions that occur when light is refracted by passage through microscope lenses, including the various objective lenses that are designed to compensate for the deviations inherent in simple ground-glass convex-convex lenses (Table 2.1). These objective lenses are actually a system containing several lenses of various shapes, dimensions, and types of glass, including crown glass, flint glass, and fluorite (Figure 2.4). The design of these compound objectives attempts to counteract defects caused by one surface with equal and opposite defects in other surfaces, so that the whole objective acts as a single lens to produce a high-quality image. It is essential for the production of an undistorted image that all light passing through a given point focus at the same point; any deviation from this results in a distorted image.

Chromatic aberration Chromatic aberration results from the fact that light of varying wavelengths is refracted slightly differently when passing through a convex-convex lens (Figure 2.5). A

table 2.1

Properties of various types of objectives for light microscopy

Objective lens	Description
Achromatic	Chromatic aberration corrected for two colors and spherical aberration for one color. These are relatively inexpensive objective lenses that are used on many microscopes intended for routine observations of microorganisms, including many of the microscopes used in introductory microbiology laboratory courses.
Planachromatic	A flat-field achromatic lens with little curvature of field.
Fluorite	Chromatic aberration corrected for two colors and spherical aberration for two colors. The mineral fluorite is included in the glass of the lens. Particularly useful for phase microscopy.
Apochromatic	Chromatic aberration corrected for three colors and spherical aberration for two colors. Produces very-high-quality images revealing true colors of specimen without distortion of shape. Excellent for photomicrography.
Planapochromatic	A flat-field apochromatic lens with minimal curvature of field.

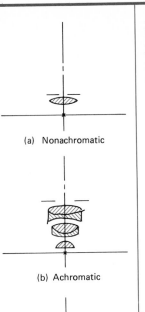

(a) Nonachromatic

(b) Achromatic

(c) Apochromatic

Figure 2.4

The construction of a compound objective lens showing the complexity of lenses corrected for spherical and chromatic aberrations.

Discovery process

Eighteenth-century microscopes were primitive instruments, limited in their usefulness by the chromatic and spherical aberration of their lenses. Chromatic aberration, which derives from the fact that a simple double-convex lens does not focus light rays of various wavelengths in the same plane, results in a blurred, multicolored image. The severity of blurring of the image due to chromatic aberration led many scientists in the late 1700s to believe that the microscope was useful only in creating artifacts. Chromatic aberration was the cause of a brisk argument between Isaac Newton, an Englishman who pessimistically thought that the aberration was an inherent property of light, and Leonhard Euler, a Swiss mathematician who reasoned that the aberration could be corrected if lenses were made of two different materials. Euler thought that glass and water might be suitable, but this idea did not materialize. In 1759 an English artisan, John Dollond, successfully fashioned an achromatic objective for a telescope by combining two lenses made of glass with different indices of refraction; the different refractions of the two glasses canceled the aberration. This practical realization by Dollond made possible microscopy as we know it today. Despite the advances in the construction of microscope objectives, the manufacture of microscopes remained an art rather than a science until Carl Zeiss of Jena, Germany, formed a partnership with a young physicist, Ernst Abbe. From 1833 to 1895, Abbe solved the basic problems of light microscopy, particularly spherical aberration, which occurs because rays of light passing through the edges of a lens have a focal point different from those passing through the center. Abbe constructed objectives in which a concave lens was added to the basic convex lens system in order to diverge the peripheral rays of light slightly to form an almost flat image. Schott's Optical Glass Works, another firm in Jena, furnished the glasses needed by the Zeiss factory. Jena became the mecca of microscopy, and it was there, in 1883, that the first lenses that corrected for both chromatic and spherical aberration were made. By 1900 the light microscope largely had reached the state of technical development used in the production of light microscopes today.

microscopic lens should not act as a prism, and therefore it is essential that microscope lenses be corrected for this type of distortion. Chromatic aberration is corrected in achromatic and apochromatic lenses by using multilayered lenses constructed with several different types of glass. Chromatic aberration can also be reduced by using monochromatic light, that is, light of a single wavelength.

Spherical aberration and related defects
Spherical aberration occurs when light passing through the thinner portions of a convex-convex lens does not focus in exactly the same plane as light passing through the wider central portion of the lens (Figure 2.6), evidenced by a loss of contrast in the image. This type of defect in the

image is largely corrected for in apochromatic lenses by using concave lenses to compensate for the differential refraction and to focus the light in a single plane. The light rays must be focused

Figure 2.5

Distortion due to chromatic aberration. Red light focuses more distantly from the lens than blue light.

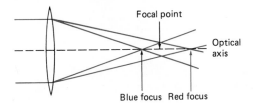

Focal point

Optical axis

Blue focus   Red focus

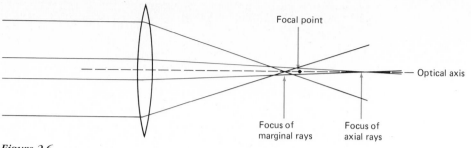

Figure 2.6

Distortion due to spherical aberration. The axial rays focus more distantly from the lens than the marginal rays.

in a single plane for the images in the field of view to be clear, and so this is a serious defect in many microscope lenses.

Another defect that may arise from spherical aberration, called **distortion**, occurs when the image formed by the peripheral light rays lies in a different plane from the image formed by the axial light rays. The magnification of the image, therefore, varies with distance from the axis of the lens, causing pincushion or barrel distortion of the image (Figure 2.7). Astigmatism is related to distortion and occurs when the lens has unequal magnification at various azimuths. Both astigmatism and distortion alter the image so that it does not accurately correspond to the shape of the specimen.

Curvature of field **Curvature of field** is yet another defect and results in a curved image of a flat field due to marginal portions of the microscope field coming into focus at a different plane than the central portion of the field. This defect is present in even the finest **apochromatic objectives** and is generally reduced by using compensating eyepieces. These ocular lenses correct for the residual defects in the image formed by the objective lens that result in curvature of field. **Planapochromatic objective lenses** greatly reduce the problem of curvature of field. Flat-field, planachromatic objectives that focus the majority of light within a single plane are also now available; these lenses are used on many student microscopes.

Resolution

In addition to magnifying the image of a specimen so that it can be viewed without distortion, the usefulness of a microscope is dependent on its **resolving power**, that is, the degree to which the detail present in the specimen is retained in the magnified image. **Resolution** is defined as the closest spacing between two points at which they still can clearly be seen as separate entities (Figure 2.8), and the resolving power of a microscope, therefore, is the distance between two structural entities of a specimen at which they can still be seen as individual structures in the magnified image. The resolving power of a microscope is dependent on the wavelength of light (λ) and on a property of the lens known as the

Figure 2.7

Barrel and pincushion distortions are illustrated in these three diagrams: (A) shows negligible distortion (B) pincushion distortion and (C) barrel distortion.

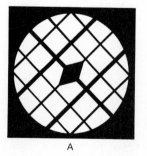

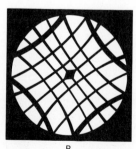

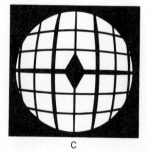

A

B

C

Plate 1

(A) Phase contrast micrograph of a Beggiotoa *sp. (500×). (B) Phase contrast micrograph of the cyanobacterium* Spirulina *(590×). (C) Gram stain of* Acinetobacter, *a Gram negative coccobacillus (3600×). (D) Gram stain of* Bacillus subtilis, *a Gram positive rod (4600×). (A,B from BPS—Paul Johnson; C,D BPS—Leon Lebeau.)*

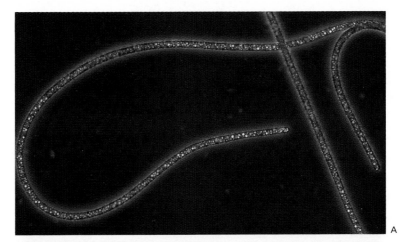

A

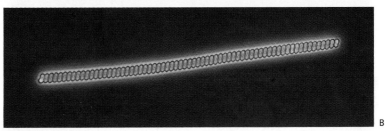

B

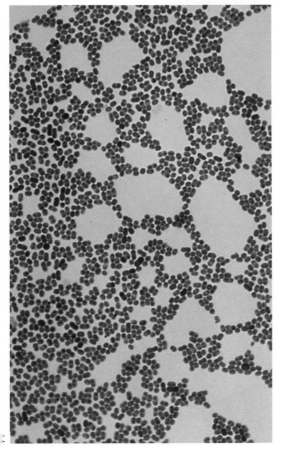

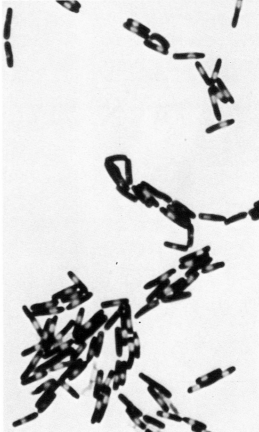

D

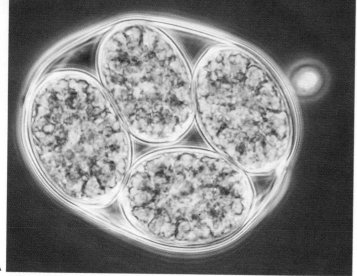

Plate 2

(A) Phase contrast micrograph of the green alga Oocystis (1100×).
(B) Bright-field micrograph of the red alga Polysiphonia *releasing
carpospores (210×). (C) Dark-field micrograph of the marine
brown alga* Ectocarpus *(160×). (D) Dark-field micrograph of the
diatom* Tabellaria *(180×). (A–D from BPS—J. Robert Waaland.)*

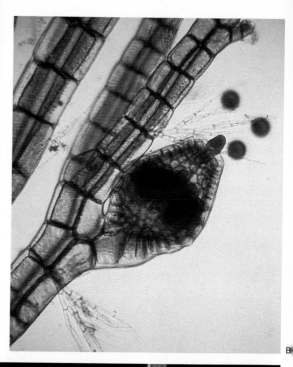

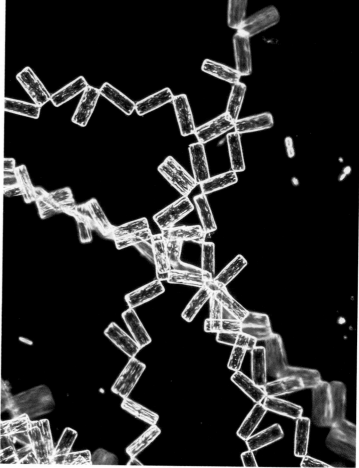

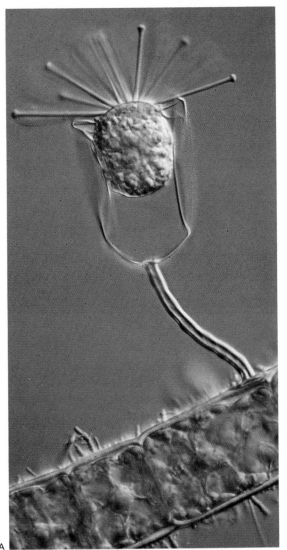

A

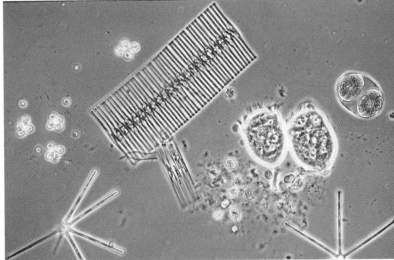

B

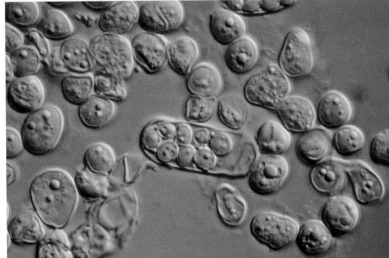

C

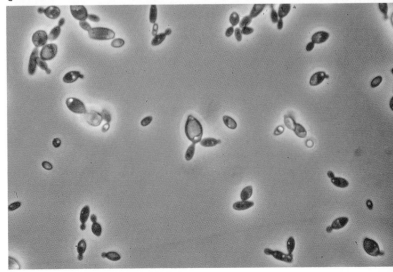

D

Plate 3

(A) Nomarski differential interference micrograph of the protozoan Paracineta *on the alga* Spongomorphora *(700×). (B) Phase contrast micrograph showing the great variety of diatoms in freshwater (235×). (C) Nomarski differential interference micrograph of the yeast* Schizosaccharomyces *containing ascospores (1500×). (D) Phase contrast micrograph of the yeast* Brettanomyces bruxellensis *showing budding mode of reproduction (900×). (A from BPS—Paul Johnson; B,C from BPS—J. Robert Waaland; D courtesy C. P. Kurtzman.)*

A

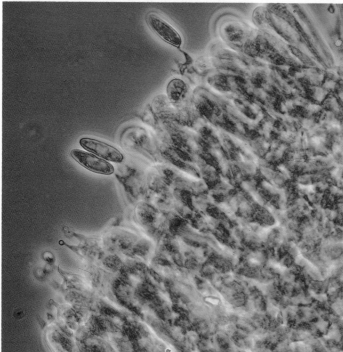

B

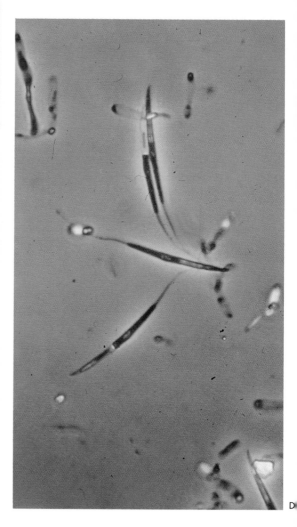

C

D

Plate 4

(A) The edible and good-tasting mushroom Pleurotus ostreatus. *(B) The deadly poisonous mushroom* Amanita virosa. *(C) Phase contrast micrograph showing basidia on gills of the fungus* Gomphidius glutinosus *(18,000×). (D) Phase contrast micrograph of the needle-shaped ascospores with whip-like tails of the yeast* Nematospora coryli *(1100×). (A,B,C courtesy O. K. Miller; D courtesy C. P. Kurtzman.)*

Figure 2.8

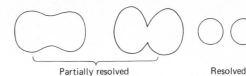

Unresolved   Partially resolved   Resolved

numerical aperture (NA). The limit of resolution of a microscope is approximately equal to $0.61\lambda/$NA, which for a light microscope is approximately 200 nanometers (nm). The lower the wavelength of light and the higher the numerical aperture of the lens, the better the resolving power of the microscope. It is thus apparent that the resolving power of a light microscope is restricted by the obtainable numerical apertures of the lens systems and the wavelengths of the visible light spectrum.

Effects of wavelength on resolution Because blue light has a shorter wavelength than red light, greater resolution can be achieved by using a blue light source to illuminate the specimen. Ultraviolet (*uv*) light, which has a still shorter wavelength, would be preferable to visible light for increasing resolution, but because ultraviolet light will not penetrate glass lenses well, and because viewing uv light directly results in permanent eye damage and blindness, it normally is not possible to capitalize on the improved resolving power that could be achieved by using this shorter wavelength of light. Specialized uv microscopes with approximately twice the resolving power of the normal light microscope can be constructed using quartz lenses and a screen or photographic plate for indirectly viewing the image of the spec-

imen. The advent of the electron microscope, which utilizes electrons, having even shorter wavelength electromagnetic waves, has made such uv light microscopes obsolete.

Numerical aperture Along with using blue light for illumination, using a lens with a **high numerical aperture permits high resolution**. The numerical aperture of a lens is dependent on the refractive index (*N*) of the medium filling the space between the specimen and the front of the objective lens and on the angle (*u*) of the most oblique rays of light that can enter the objective lens (Figure 2.9). The formula for calculating the numerical aperture is

$$NA = N \times \sin\frac{u}{2}$$

Air has a refractive index of 1, which limits resolution that can be achieved, but the numerical aperture can be increased by placing immersion oil between the specimen and the objective lens, improving the resolving power of the microscope. Immersion oil has a refractive index of 1.5, which considerably increases the numerical aperture and thus improves the resolving power of the microscope. The observation of fungi, algae, and protozoa can be achieved with dry objectives, that is, where air occupies the space between the

Figure 2.9

The numerical aperture is improved by the use of immersion oil to replace the air between the objective lens and the specimen, as shown by the wider cone of light obtained using an oil immersion compared with a dry objective. u = angle of peripheral light entering lens.

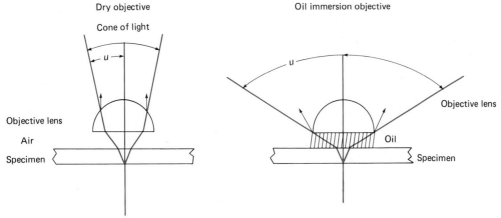

Dry objective   Oil immersion objective

Cone of light

u

Objective lens

Air

Specimen

u

Objective lens

Oil

Specimen

(1) Place a loopful of the culture on a clean slide

(2) Spread in a thin film over the slide

(3) Air dry

(4) Fix by passing the slide rapidly through the bunsen flame

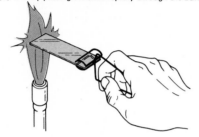

(5) Stain (e.g., with crystal violet)

(6) Wash off stain with water

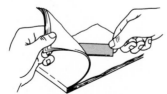

(7) Blot off excess water

(8) View under microscope

Figure 2.10

In a simple staining procedure, microorganisms are affixed to a glass slide and stained with an appropriate dye to increase the contrast between the cells and the background so that they can easily be seen using a light microscope.

specimen and the objective, but the viewing of bacteria in sufficient detail to determine the shape and arrangement of cells normally requires the use of an **oil immersion lens**. As a rule of thumb, the useful magnification of a microscope is 1,000 times the numerical aperture being used, and so it is possible with an oil immersion lens with a numerical aperture of 1.4 to achieve a useful magnification of approximately 1,400×. At higher magnifications the quality of the image deteriorates, and the magnification is therefore considered to be empty. A high-numerical-aperture lens has a short focal length, and therefore there is a short working distance between the lens and the object; that is, the lens and the specimen are very close to one another. Another consequence of the short focal length is a very shallow depth of field, and only a very thin section can be in focus at any one time. Because of the short working distance and shallow depth of field of an oil immersion lenses, many students at first have great difficulty trying to focus the microscope on a specimen of bacteria without breaking the slide and scratching the objective lens, but with a little practice this problem is easily overcome.

Contrast and staining

Another factor that must be considered in light microscopy is contrast, because without adequate contrast it is impossible to discern a structure from the surrounding background. Microorganisms are largely composed of water, as is the medium in which they are normally suspended, and simply viewing microorganisms with a light microscope without performing procedures to increase contrast can be likened to trying to see a white object on a white background.

Staining is used to increase the contrast between the specimen and the background, but, unfortunately, staining generally precludes the observation of living microorganisms. In most staining procedures (Figure 2.10) a suspension of microorganisms is transferred to a glass microscope slide and allowed to dry. The slide is then quickly passed through a flame to heat fix the cells to the slide, that is, to affix the cells to the slide so that they cannot be washed off. A stain is then added to the slide, allowed to stand for a period of time, rinsed to remove excess stain, and viewed.

Figure 2.11

The interaction of a cell with negative and positive stain reagents. Because the outer layer of a cell is negatively charged, a positive stain is attracted to the cell, whereas a negative stain is repelled.

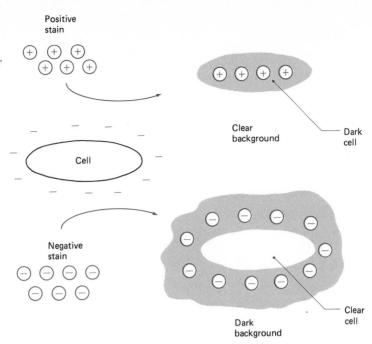

Visualization of microorganisms

Simple staining procedures In a **simple staining procedure**, a single stain reagent is used, and the procedure does not attempt to produce different staining reactions for different structures or types of microorganisms. Simple staining of microorganisms may be positive where the stain is attracted to the microbial cells, and so the cells appear dark or colored on a light or clear background; or they may be negative, in which case the background is stained, and the microorganisms appear bright on a dark background. Both of these staining procedures depend on the fact that bacteria and other cellular microorganisms have a negative charge associated with the outer surface of the cell, largely because of the PO_4^{3-} groups of the cell membrane (Figure 2.11). In **positive staining procedures** an acidic (cationic) stain, which has a positively charged chromophore (colored portion of the stain molecule), is attracted to the negatively charged outer surface of the microbial cell. A stain such as methylene blue has a colored blue portion of the molecule, which is attracted to the cell surface, resulting in positive staining of the microorganism (Figure 2.12). In **negative staining procedures**, a basic (anionic) stain is used, which has a negatively

Figure 2.12

Positively stained bacteria. This photomicrograph shows a mixture of E. coli *and* Staphylococcus aureus *(50×). (From BPS—Leon J. LeBeau, University of Illinois Medical Center, Chicago.)*

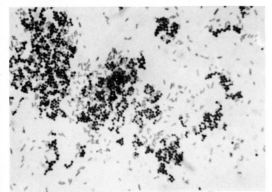

Figure 2.13

The bacterium Thiocapsa floridiana *following negative staining with india ink. Note that the capsule surrounding this bacterium can be readily seen. (From BPS—Stanley C. Holt, University of Massachusetts.)*

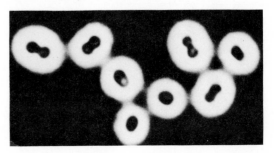

charged chromophore that is repelled by the negatively charged microorganism, resulting in negative or indirect staining of the microbial cell. Nigrosin and india ink are frequently used for negative staining of microbial cells, and this type of staining is particularly useful for viewing some structures such as the capsules that surround some bacterial cells (Figure 2.13).

Differential staining procedures In **differential staining procedures** specific types of microorga-nisms and/or particular structures of a microorganism exhibit different staining reactions that readily can be distinguished. The **Gram stain procedure** undoubtedly is the most widely used differential staining procedure in bacteriology (Figure 2.14). This staining procedure includes primary staining with crystal violet, which stains all bacterial cells blue-purple, application of Gram's iodine—a **mordant** that increases the affinity of the primary stain for the bacterial cells, rinsing with acetone-alcohol or other decolorization agent, and

Discovery process

The development of the Gram stain procedure by the Danish physician Hans Christian Gram remains one of the most important methodological contributions in bacteriology. At the time Gram published the description of his staining method, most bacteriologists, such as Loeffler at the Koch Institute in Berlin, were concerned with simply seeing difficult-to-detect bacteria and differentiating infecting bacteria from mammalian nuclei. Gram worked in the laboratory of Friedländer at the morgue of the City Hospital of Berlin. Gram's paper describing the staining technique, *The Differential Staining of Schizomycetes in Sections and in Smear Preparations*, was published in 1884. In this paper Gram described primary staining with aniline-gentian violet, treatment with iodine-potassium iodide (Gram's iodine mordant), and decolorization with absolute alcohol followed by further decolorization with clove oil. "Bacteria are stained intense blue while the background tissues are light yellow...." Gram observed that not all bacteria were stained in this procedure and suggested that counterstaining was possible. It was the detailed reporting of which bacteria were stained and which were not that was crucial to the recognition of the value of this staining procedure. "I. The following forms of schizomycetes retain the aniline-gentian violet after treatment with iodine followed by alcohol: (a) cocci of croupous pneumonia (19 cases) ... (k) tubercule bacilli (5 cases).... II. The following schizomycetes are decolorized by alcohol subsequent to treatment with iodine: (a) encapsulated cocci from croupous pneumonia...." Gram undoubtedly was disappointed in his procedure. He wanted to stain all bacteria and not the surrounding mammalian tissues. It is not known who thought of using the Gram stain procedure to differentiate bacteria, as routinely employed in bacteriology laboratories today, but a textbook published by Flügge in 1886, only two years after Gram's initial report, stated that "The method of Gram is mainly useful for the differential staining of bacteria in tissues and for the diagnostic differentiation of species." Gram died in 1935 without further developing his staining procedure but with the hope stated in the conclusion to his paper that "the method would be useful to other workers." Frequently in science, the discoverer of a method does not recognize the full potential of a discovery, and it remains for later scientists, working in an era of different concerns and enlightened by later discoveries, to realize the significance of the original finding.

Figure 2.14

The Gram stain procedure is widely used to differentiate major groups of bacteria.

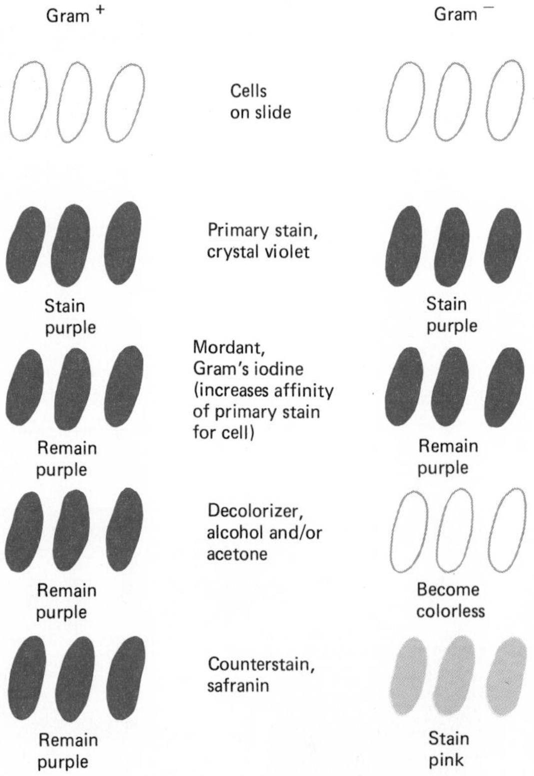

Gram +		Gram −
Stain purple	Cells on slide	Stain purple
Remain purple	Primary stain, crystal violet	Remain purple
Remain purple	Mordant, Gram's iodine (increases affinity of primary stain for cell)	Become colorless
Remain purple	Decolorizer, alcohol and/or acetone	Stain pink
	Counterstain, safranin	

Figure 2.15

Photomicrograph of the stained bacterium, Bacillus subtilis, *showing endospores (the light areas within cells) and free endospores (the lightly stained bodies in the background). (Courtesy R. A. Smucker, Chesapeake Bay Laboratory, University of Maryland.)*

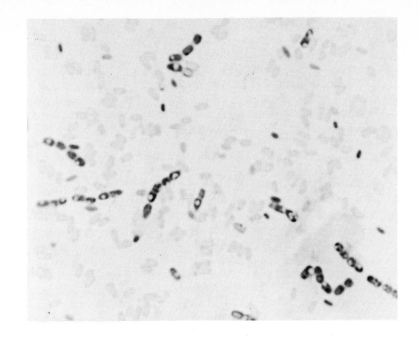

application of a counterstain, which stains those bacteria that were decolorized in the previous step so that they can be easily visualized. The decolorization step is the critical step that differentiates bacterial species based on their cell-wall structure. About half of the species of bacteria are Gram positive and appear blue-purple at the end of this procedure, and the remaining bacterial species are Gram negative and appear red-pink following Gram staining. The Gram stain procedure has great diagnostic value as a result of its ability to differentiate between different bacterial species and therefore is a key feature employed in many bacterial classification and identification systems.

Another differential staining procedure frequently used in bacteriology is **acid-fast staining**. The acid-fast stain procedure is especially useful in identifying members of the bacterial genus *Mycobacterium* and is important in identifying the causative organism of tuberculosis, *Mycobacterium tuberculosis*. The bacterial endospore, which is produced by members of relatively few bacterial genera but is very important because of its resistance to high temperatures, can be visualized by using a differential staining procedure. In this procedure the bacterial endospore is stained one color and the rest of the bacterial cell another color, permitting differentiation of the endospore from the vegetative cell (Figure 2.15).

Fluorescence microscopy

Other stains used in microbiology include fluorescent dyes that when illuminated by light of one wavelength give off light at a different wavelength; for example, a dye that when illuminated with blue light emits green or orange light. Microscopy that uses such fluorescent dyes is known as **fluorescence microscopy**, and the microscope used for viewing such specimens is referred to as a fluorescence microscope. If the excitation light is transmitted to the specimen through the objective lens, the system is referred to as "epifluorescence." If the excitation light is transmitted from below the specimen, it is termed "transmitted fluorescence." The principles of magnifying power and resolution are no different for a fluorescence microscope than for the normal light microscope. The wavelength of the light used to excite the dye may be in the ultraviolet range, but the emitted light that is viewed must be in the visible range. Fluorescent dyes can be conjugated with antibodies for immunofluorescent microscopy, which provides great specificity in staining procedures, because of the specificity of immunological reactions. **Immunofluorescence staining** is one of the applications that make fluorescence microscopy an important method for viewing microorganisms.

Dark-field microscopy

There are several alternative microscope designs that enhance the contrast between the specimen and the background without the use of staining, thus permitting the visualization of living specimens. In the simplest of these microscopes, the **dark-field microscope**, the normal condenser of the light microscope is replaced with a dark-field condenser that does not permit light to be transmitted directly through the specimen and into the objective lens (Figure 2.16). The dark-field condenser focuses light on the specimen at an oblique angle, such that light that does not reflect off an object does not enter the objective lens. Thus, only light that reflects off the specimen will be seen, and in the absence of a specimen the entire field will appear dark. Bacteria viewed with a dark-field microscope appear very bright on a dark (black) background (Figure 2.17). The contrast between the specimen and the background is sufficient to permit the visualization of even small bacteria and large viruses, but it is not generally possible with dark-field microscopy to distinguish the internal structures of the microorganisms being viewed.

Phase contrast microscopy

The **phase contrast microscope** is a useful microscope for visualizing living microorganisms since it permits the viewing of microbial structures without the necessity of staining. Light passing

through a cell of higher refractive index (greater ability to change the direction of a ray of light) than the surrounding medium is slowed down relative to the light that passes directly through the less dense background medium. The greater the refractive index of the cell or cellular structure, the greater the retardation of the light wave (Figure 2.18). Thus, when light passes through a microorganism there is a slight alteration in the phase of the light wave. The phase contrast microscope takes advantage of the differences in refractive indices between structures of a microbial cell, translating differences in phase into changes in light intensity that are visible to the eye. The conversion of differences in the phase of the light waves to differences in light intensity is based on interference between light waves that are out of phase with one another (Figure 2.19). The interaction of light waves that are in phase produces addition, which increases the intensity of the light, but the interaction of light waves that are 90° out of phase results in destruction of the propagating light wave, that is, a decrease in intensity. Thus, **interference of light waves** accomplishes the task of creating visible light intensity differences that are necessary for visualizing microorganisms.

In order to permit the observation of unstained microorganisms, the phase contrast microscope is designed to separate the direct background light from the light passing through the object, causing these two different light waves to be approximately 90° out of phase with one another so that they destructively interact and cause changes in light intensity (Figure 2.20). To accomplish this,

Figure 2.16

Diagram of a dark-field microscope showing the path of light passing through the background and light striking the specimen.

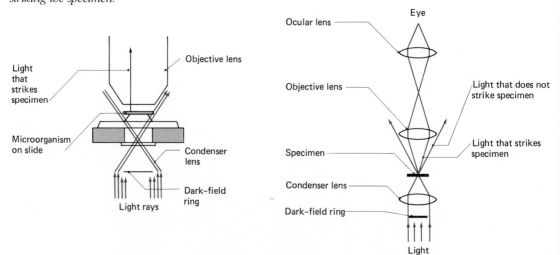

Figure 2.17

(A) This photomicrograph shows the appearance of the cyanobacterium Gleocapsa *(1470×) using dark-field microscopy. In this type of microscopy the cells appear bright against a dark background. Dark-field microscopy is useful for visualizing bacterial cells without the need for staining to enhance contrast. For a comparison of how these cells of* Gleocapsa *appear in other types of microscopy, see Figures 2.21 and 2.23. (From BPS—J. Robert Waaland, University of Washington.) (B) Micrograph of a colony of the green alga* Volvox *as it appears in dark-field microscopy. (C) Micrograph of a colony of the green alga* Volvox *as it appears in light microscopy is shown for comparison. (B and C Courtesy Gary B. Collins, USEPA, Cincinnati.)*

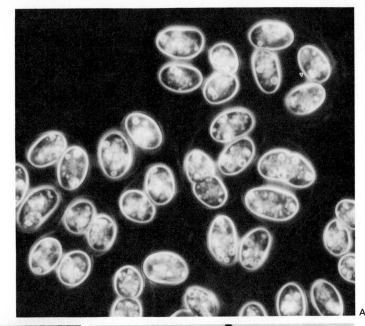

A

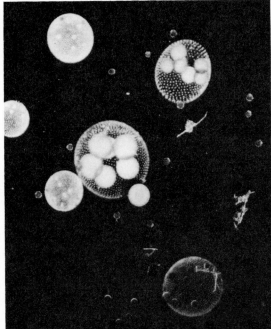

B

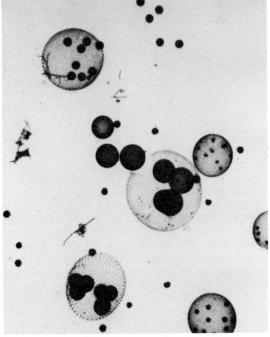

C

Figure 2.18

The retardation of phase of light waves as they pass through a transparent living cell mounted in saline is shown. As compared to the waves that do not pass through the cell (A), the waves passing through the full thickness of the cytoplasm (B) have been retarded by $\frac{1}{4}\lambda$, and those passing through a more highly refractive infusion (C) have been retarded by $\frac{1}{2}\lambda$.

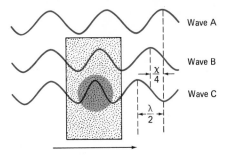

Wave A

Wave B

$\frac{\lambda}{4}$

Wave C

$\frac{\lambda}{2}$

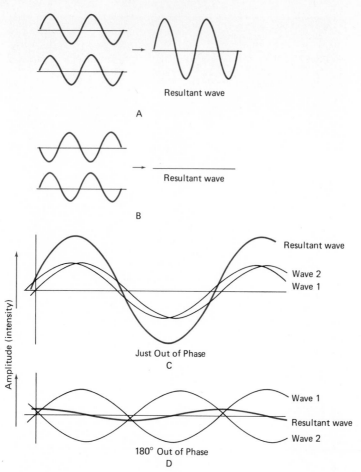

Figure 2.19

Additive and destructive interference of light waves. (A) Shows two waves in phase, combining to produce a wave with twice the amplitude. (B) When the two waves are one-half wavelength out of phase, they combine to cancel each other out. (C) The superimposition of two waves of equal frequency and amplitude that are almost in phase results in a wave of almost twice the amplitude of either component. (D) The superimposition of two waves of equal frequency and amplitude, almost 180° out of phase, results in a wave whose amplitude is nearly zero. Note that in all these cases the frequency is unchanged. The ability to alter the amplitude of a light wave by combining two different light waves forms the basis for phase contrast microscopy, allowing the visualization of different intensities of light passing through a specimen and light passing through the background.

the phase contrast microscope has an annulus, a ring-shaped disk, on the substage condenser that is opaque except for a thin ring through which light can pass. Thus, the phase contrast condenser produces a ring of light that illuminates the specimen and enters the objective lens. An objective lens of a phase contrast microscope is constructed with a ring-shaped phase plate at its back focal plane that corresponds geometrically with the ring of direct light from the condenser; that is, the annulus ring is aligned with the ring of the phase plate. Normally, the ring-shaped area of the phase plate is thinner than the rest of the plate, so that the phase of the light wave passing through the ringed area is altered by 90° or one-quarter of a wavelength. Light that passes directly through

Figure 2.20

the object passes through the ring-shaped area of the phase plate and is accordingly advanced by 90°. Light that is diffracted by the specimen passes through the thicker portion of the phase plate and is retarded by 90° relative to the light that is transmitted directly. When recombined, light waves that are out of phase interfere with each other, producing alterations in the amplitude of the light waves, which are seen as differences in light intensity.

If a specimen has the same refractive index as the surrounding medium, the direct light and the diffracted light will be 90° out of phase, causing destructive interference and producing a uniform decrease in intensity; that is, a uniform dark field would be seen. However, if the specimen has a higher refractive index than the surrounding medium, the light diffracted by the specimen will be retarded relative to the light passing directly through the surrounding background. The degree of retardation depends on the difference in magnitude between the **refractive indices** of the specimen and its background and the thickness of the specimen. The greater the refractive index and thickness of the specimen, the greater the retardation. Similarly, differences in thickness and/ or refractive index between structures within the specimen will produce differences in the retardation of the diffracted light, that is, phase changes that are convertible into visible differences in light intensity. Even difficult-to-stain structures often are conspicuous under a phase contrast microscope because small phase changes result in interference, giving rise to high-contrast images (Figure 2.21). Thus, with the phase contrast microscope, living organisms can be clearly observed in great detail without staining them, permitting the study of their movements in the medium in which they are growing.

Interference microscopy

Besides the phase contrast microscope, there are several other types of microscopes that rely on light interference for producing a visible image. Both phase contrast and **interference microscopes** make use of the fact that light travels as waves and that the addition of light waves that are

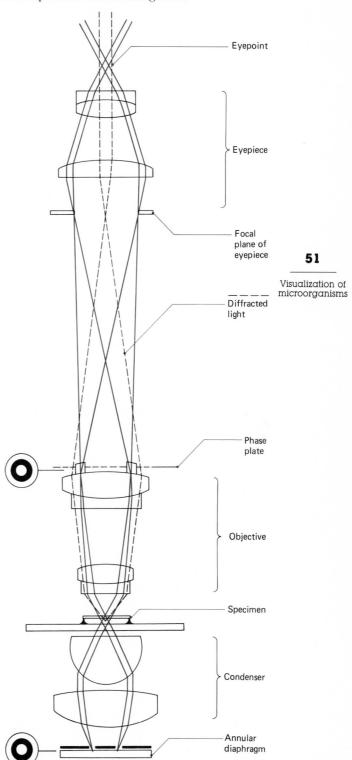

Eyepoint

Eyepiece

Focal plane of eyepiece

Diffracted light

Phase plate

Objective

Specimen

Condenser

Annular diaphragm

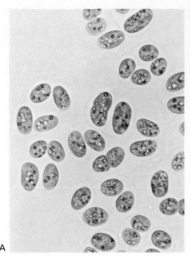
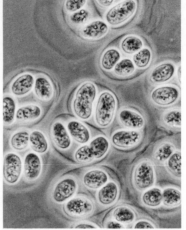

A   B

Figure 2.21

These photomicrographs of the cyanobacterium Gleocapsa *compare its appearance by using (A) bright-field and (B) phase contrast microscopy. The cell structures are easier to see in the micrograph taken with the phase contrast microscope. (From BPS—J. Robert Waaland, University of Washington.)*

out of phase with each other produces interference that alters the amplitude of the light wave. Compared to the phase contrast microscope, interference microscopes can have higher numerical apertures, can vary the degree of contrast to suit the subject, and can use colors to increase contrast. Like phase contrast microscopes, interference microscopes permit the visualization of living cells and, as in the phase contrast microscope, phase changes that occur when light passes through the specimen are translated into changes in light intensity.

Nomarski differential interference contrast microscopy

Of the various types of interference microscopes, microbiologists most frequently employ **Nomarski differential interference contrast** (NDIC) microscopy (Figure 2.22), which produces high-contrast images of unstained, transparent specimens in what appears to be three dimensions (Figure 2.23). The Nomarski interference microscope has three special features: a polarizing filter, an interference contrast condenser, and a prism-analyzer plate. The prisms permit the complete separation of direct and diffracted light beams. Unlike the phase contrast microscope, where a single beam of light is transmitted through the specimen, the Nomarski interference microscope condenser produces a second beam of light that passes through the specimen. In the Nomarski interference microscope, polarized light is split into two beams at right angles to each other that travel parallel and very close together to each other through the specimen. The two beams of light are then combined and pass through an analyzer. When the two light rays that are differentially dif-

Figure 2.22

Diagram of a Nomarski interference microscope, showing the path of light.

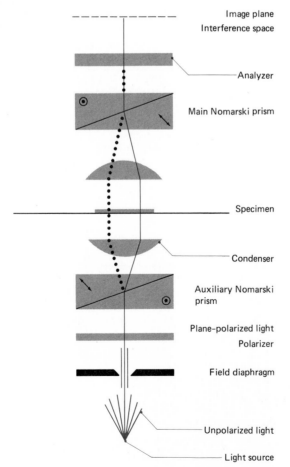

Image plane
Interference space

Analyzer

Main Nomarski prism

Specimen

Condenser

Auxiliary Nomarski prism

Plane-polarized light
Polarizer

Field diaphragm

Unpolarized light

Light source

fracted by the specimen are recombined, they produce an interference pattern.

The pseudo-three-dimensional image of NDIC is produced because the two beams of light travel very close to each other through the specimen, producing a stereoscopic effect. The degree of topographic relief is a function of the refractive index differences at the boundary surfaces of the specimen. Contrast in NDIC depends on the rate of change of refractive index across a specimen, and consequently, especially good contrast is produced at the edges of the specimen, where there is a large refractive index differential. Structures such as cell walls are very well defined when viewed with interference microscopes. Different structures of microorganisms appear in different colors that are related to the phase changes in the light passing through each individual structure, and images seen through the interference microscope are normally brilliantly colored. Interference microscopes are very useful for qualitative observations of unstained cells because they produce images with high contrast and striking topographic relief.

Electron microscopy

The advent of the electron microscope marked a significant improvement over light microscopes for the visualization of microorganisms. Many of the fine structures of microorganisms have been elucidated by using electron microscopy. The electron microscope permits better resolution, and therefore, higher useful magnifications than can be achieved with light microscopy. There are two basic types of electron microscopes: the **transmission electron microscope (TEM)** and the **scanning electron microscope (SEM)**.

Transmission electron microscope
The design of the **transmission electron microscope** is similar to that of the compound light microscope, except that an electron beam is substituted for the light source and a series of electromagnets is substituted for the glass lenses (Figure 2.24). The source of the electrons is a hot tungsten filament in an electron gun. Electrons are drawn from the filament and accelerated as a fine electron beam past an anode by a high voltage that is established between the filament and the anode. The electron beam is focused on the specimen with an electromagnetic condenser lens by varying the current to the lens. Instead of placing the specimen on a glass slide, as is done in light microscopy, the specimen normally is placed

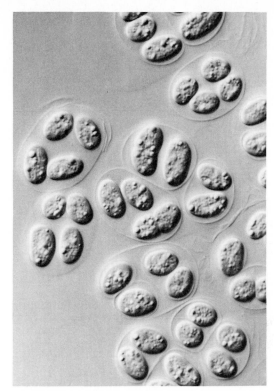

Figure 2.23

Photomicrograph of the cyanobacterium Gleocapsa, *using Nomarski interference microscope (1260×). (From BPS—J. Robert Waaland, University of Washington.)*

on a copper mesh grid within the evacuated column of the electron microscope. The transmission electron microscope has an objective lens, as does the light microscope, but instead of an ocular lens, the TEM has a projector lens that projects the image onto a fluorescent viewing screen or film plate, which is necessary because the electron beam cannot be viewed directly.

In an electron microscope, air is removed from the path of the electron beam to prevent collisions with gas molecules that would scatter the electron beam and make it impossible to resolve a high-magnification image. Additionally, the removal of air reduces the heat associated with the electron beam, reducing destruction to biological specimens in the path of the electron beam and to the filament that otherwise would rapidly deteriorate. Many of the technical advances in electron microscopy have involved improvements in the vacuum system.

The transmission electron microscope permits much greater resolution and thus much higher useful magnifications than the light microscope because the wavelength of an electron beam, gen-

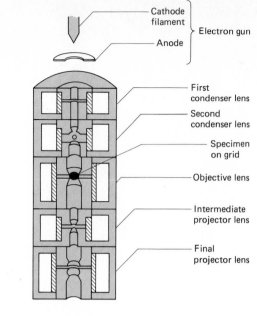

Cathode filament — Electron gun
Anode —

First condenser lens
Second condenser lens
Specimen on grid
Objective lens
Intermediate projector lens
Final projector lens

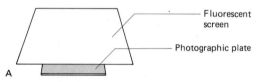

Fluorescent screen
Photographic plate

A

B

Figure 2.24

The transmission electron microscope (TEM) allows the visualization of the fine detail of the microbial cell. (A) Diagram of a TEM and (B) photograph of a high-resolution TEM. (Courtesy JEOL, Peabody, Massachusetts.

erated at a high accelerating voltage, is much shorter than for light in the visible range of the electromagnetic spectrum. The actual wavelength of the electron beam depends on the accelerating voltage of the microscope. At 60,000 volts, a typical accelerating voltage used in a transmission electron microscope, the wavelength of the electron beam is approximately 0.005 nm, permitting a theoretical resolution of approximately 0.2 nm, which is about a thousand times better than can be achieved when using light microscopy. The useful magnification for an electron microscope, consequently, is in excess of $100,000 \times$, which permits the visualization of all microorganisms, including viruses (Figure 2.25). In fact, the resolving power of the electron microscope permits visualization of molecules and atoms. Contrast in electron microscopy occurs as a result of some electrons colliding with and being scattered by the specimen, while other electrons penetrate without collision. The amount of electron scattering is proportional to the electronic density (mass thickness) of the specimen, that is, the number of atoms per unit area and their atomic densities. Because the atoms composing biological cells have

relatively low electron densities, biological specimens produce little scattering of the electron beam and, hence, low contrast. Staining is used to improve the contrast between the specimen and the background, but instead of the dyes used in light microscopy, the stains used for electron microscopy contain electron-dense heavy metal salts. The heavy metal stains scatter the electron beam and the stained areas thus appear dark. Particular metals exhibit selective affinities for different biological structures, permitting differential staining and visualization of the detailed ultrastructure of microorganisms.

There are several problems in viewing biological specimens, including microorganisms, with the transmission electron microscope. There is a great potential for creating artifacts that could be mistakenly viewed as real structures in electron micrographs. An artifact is the appearance of something in an image or micrograph that is due to causes within the optical system and not a true representation of the features of the specimen on view. This is a problem common to all microscopes but is particularly troublesome in electron microscopy because of the high magnifications

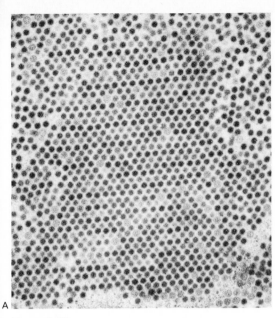
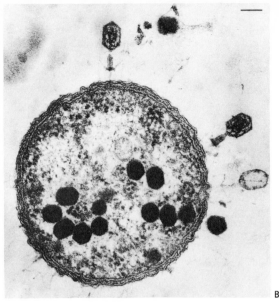

A   B

Figure 2.25

To view viruses, an electron microscope is needed. (A) Here an electron micrograph shows an intracellular crystalline array of adenoviruses viewed with a TEM (26,400×). (From BPS—S. Dales and S. L. Wilton, University of Western Ontario.) (B) This electron micrograph shows phage infecting a bacterium. Some viruses are outside of the cell, and new viruses are being made within the bacterium (111,000×). (Courtesy Lee Simon, Rutgers, The State University.)

that are used, the need to dehydrate the specimen, and the fact that the specimen must be placed in a high-vacuum chamber. Improper adjustment of the electron beam, excessive magnification, and improper sample preparation can all cause the formation of artifacts.

Preparation of specimens for TEM   Before they can be viewed with a transmission electron microscope, biological specimens, including microorganisms, must be prepared in order to maximize the amount of detail that can be visualized and to avoid the formation of artifacts (Figure 2.26). Biological specimens containing water cannot simply be placed under high vacuum because the water would boil, destroying the integrity of the organisms. Therefore, before viewing a microorganism with a transmission electron microscope, it is necessary to preserve (fix) and dehydrate the specimen. The fixation and dehydration process must be carried out carefully in several stages, because during the fixation process it is possible to shrink or otherwise distort the microorganisms. Additionally, microorganisms are too large (thick) to view with a transmission electron microscope and see the maximum amount of detail that is possible with the high resolving power of

the TEM. Therefore, it is normally necessary to slice the microorganisms into **thin sections** in order to view their ultrastructures (Figure 2.27). The thin sectioning of microorganisms is achieved by using a microtome with a diamond or glass knife. The microorganisms are normally embedded in a plastic resin to facilitate handling during thin sectioning. Living microorganisms cannot be viewed in a vacuum. These represent technical difficulties that require great care in preparing specimens for TEM and great caution in interpreting micrographs obtained with transmission electron microscopy.

There are several special preparation procedures that are used in electron microscopy for revealing detailed structures of microorganisms. **Freeze etching**, for example, is used to reveal various biochemically defined layers of a microorganism, including organelle structures (Figure 2.28). In this procedure, a frozen specimen is fractured by striking it with a knife blade, a carbon replica is made, and the biological specimen is oxidized (etched) away. The carbon replica is then viewed with an electron microscope. The freeze etching method reveals great detail of both internal and external surface structures (Figure 2.29) and also eliminates some problems with

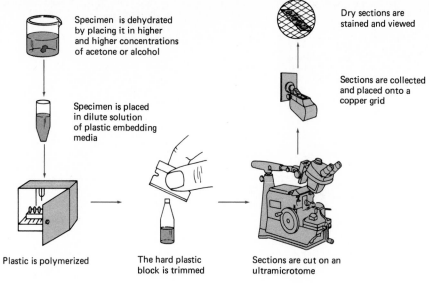

Specimen is dehydrated by placing it in higher and higher concentrations of acetone or alcohol

Specimen is placed in dilute solution of plastic embedding media

Plastic is polymerized

The hard plastic block is trimmed

Sections are cut on an ultramicrotome

Dry sections are stained and viewed

Sections are collected and placed onto a copper grid

Figure 2.26

Preparation of a specimen for viewing by transmission electron microscopy. The elaborate preparation requires far more time than the actual visualization of the specimen.

Figure 2.27

The detailed structure of a bacterium is seen in this electron micrograph of a thin section of Pseudomonas aeruginosa *(61,800×) viewed with a transmission electron microscope. (From BPS—John J. Cardamone, Jr., University of Pittsburgh.)*

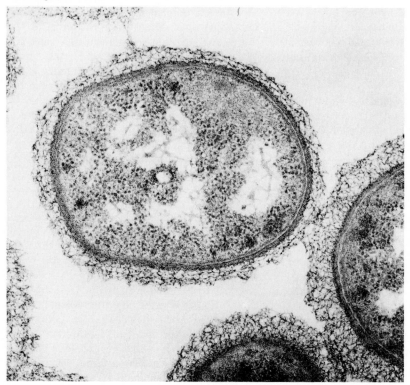

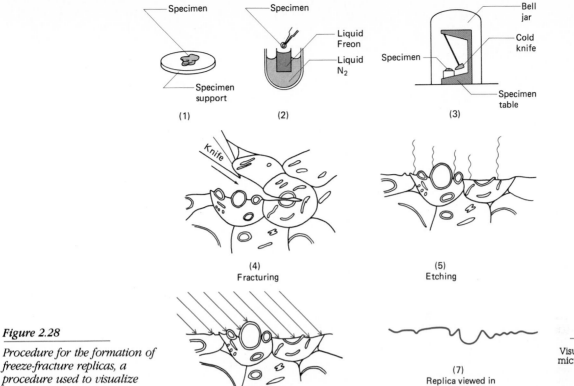

(1)   (2)   (3)

(4)
Fracturing

(5)
Etching

Figure 2.28

Procedure for the formation of freeze-fracture replicas, a procedure used to visualize surface structures in conjunction with transmission electron microscopy.

(6)
Shadowing and replicating

(7)
Replica viewed in electron microscope

Figure 2.29

Electron micrograph of the endospore-forming bacterium Clostridium botulinum *(type A) following freeze-etching as viewed under a TEM (49,000×). Note the prominence and topographical relief of the terminal endospore. (From BPS—T. J. Beveridge, University of Guelph.)*

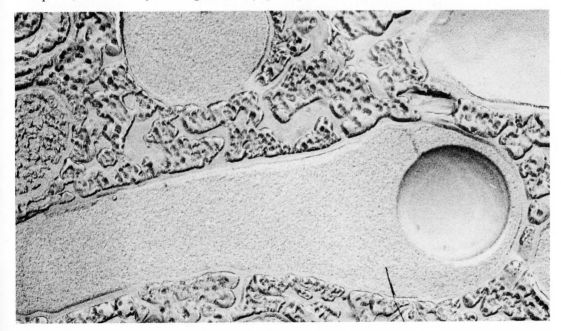

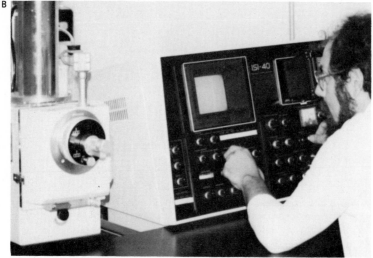

A

Electron gun { Cathode filament / Anode

Condenser lenses

Beam deflector coils

Scan generator

Secondary electrons

Specimen

Detector

Amplifier

CRT display

B

Figure 2.30

The scanning electron microscope (SEM) is used to view surface structures and their three-dimensional spatial relationships. (A) Diagram of SEM. (B) Photograph of SEM.

artifacts that arise through chemical fixation and sectioning of biological specimens.

Scanning electron microscope

The **scanning electron microscope** is primarily used for viewing surface details rather than the internal structures of microorganisms that are revealed by using the transmission electron microscope. The operational principles and design of the SEM are quite different from those of the TEM (Figure 2.30). In scanning electron microscopy, an accelerated electron beam is focused on the specimen by using a condenser lens. The lenses of the scanning electron microscope are designed so that an extremely small electron beam is produced. The beam is rapidly scanned across the surface of the specimen, forming a raster (a set of straight lines as is used in a television set).

The primary electron beam knocks electrons out of the specimen surface, and the secondary electrons produced in this process are transmitted to a collector, amplified, and used to generate an image on a cathode-ray tube (CRT) screen. Some of the primary electrons also will be reflected or backscattered from the specimen surface, but the number of backscattered electrons will be far fewer than the number of secondary electrons emitted from the specimen surface. Because the number of low-energy secondary electrons reaching the collector will be far greater than the number of backscattered electrons, a more intense signal will be developed by the secondary electrons than by the backscattered electrons. A cathode-ray tube is scanned across its screen by another beam of electrons and the intensity of the electron beam used to illuminate the cathode-ray tube is controlled by the number of electrons collected by the detector. The detector consists of a scintillator that emits flashes of light when struck by electrons and the emitted light is converted to an electrical current that is used to control the brightness of the spot on the CRT screen.

The secondary electrons emitted from each point on the specimen are characteristic of the surface at that point. The intensity of the image seen on the CRT screen, thus, reflects the composition and topography of the specimen surface. Contrast in the SEM is primarily determined by surface topography, which controls the number of secondary electrons reaching the detector. The image shown on the CRT screen shows a shadowing effect that gives a three-dimensional appearance to the image, and the topography of the specimen surface is easily seen with the SEM (Figure 2.31).

Magnification in a scanning electron microscope is not achieved by using lenses, as is the case in both light microscopy and transmission electron microscopy. Magnification in the SEM is achieved by having different lengths of scan for the specimen and the CRT display and is determined by the ratio of the length of the scan across the specimen surface to the length of the scan of the CRT. If the electron beam scans a raster of 100 nm × 100 nm on the specimen and the image is displayed on a CRT screen of 100 mm × 100 mm, the magnification will be 100,000×. The primary beam that scans across the specimen is synchronized with the raster in the CRT display so that the image on the CRT screen is an accurate reproduction of the scanned image.

As with other forms of microscopy, resolution is essential for achieving useful magnification. Resolution of the SEM depends on the spot size of the screen and the size (diameter) of the primary electron beam. Most scanning electron microscopes are capable of achieving resolution in the 1–10 nm range, and thus, useful magnification of the SEM is of the range 10,000–100,000×.

Preparation of specimens for SEM As with the transmission electron microscope, the electron beam in the scanning electron microscope must be transmitted through a vacuum, and therefore, biological specimens must be fixed and dehydrated. The goal of specimen preparation is to retain the natural shape of the living organism when it is viewed in the scanning electron microscope. Artifacts may be easily created when biological specimens are dehydrated during preparation for viewing in the SEM. Particularly, destructive effects result from the surface tension

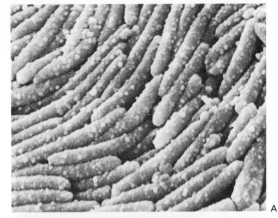
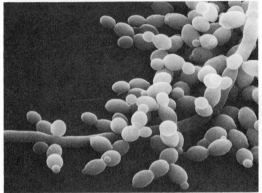
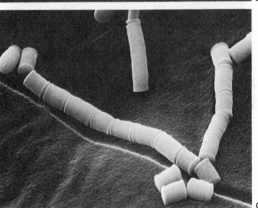

Figure 2.31

Scanning electron micrographs of microorganisms showing their three-dimensional appearance. (A) Note the bumps on the outer membranes of these rod-shaped bacteria; (B) Candida albicans yeast and hyphal phases (1450×); and (C) Geotrichum candidum, a fungus associated with food spoilage forming arthrospores (1500×). (A from BPS— Z. Skobe, Forsythe Dental Center; B and C from BPS—Garry T. Cole, University of Texas, Austin.)

properties of the air–water interface when specimens are simply air dried.

Critical-point drying is normally utilized to minimize artifact creation, as this technique avoids distortions due to the effects of surface tension. This procedure takes advantage of the fact that there is a critical point in terms of temperature and pressure, for a solvent in a closed container where the density of the liquid is equal to the density of the vapor. At the critical point, the liquid in a specimen and the gaseous phase over it become indistinguishable, permitting conversion of the liquid to a gas without any surface tension damage to the specimen. When all the gas is removed, the specimen is dry. In critical-point drying, the tissue water is replaced by a liquid with a low critical point, such as liquid carbon dioxide. The specimen is first placed in ethanol or acetone to remove water, and the dehydrating solvent is then replaced in a sealed chamber by using pressurized liquid CO_2 as the transitional liquid. The temperature is raised above 32°C, the critical point of CO_2, and the fluid vaporizes without any surface tension. The pressure and gas are released slowly, leaving a dry, undistorted specimen.

Once the specimen is dehydrated, it is coated with a layer of gold or gold-palladium to improve its characteristics as a target for the electron beam of the scanning electron microscope. Coating is normally accomplished with a sputter coater to ensure even coverage with a thin film of metal. Proper coating of the specimen minimizes surface charging, a serious problem in scanning electron microscopy that can produce an intense emission of electrons that precludes achieving contrast and viewing surface detail. Surface charging occurs because biological specimens have nonconducting surfaces. Coating with a metal produces a conductive surface that permits dissipation of the electrons without surface charging. Following metallic coating, the specimens normally are mounted and viewed. Unlike the transmission electron microscope, one need not worry about thin sectioning because only surface structure is effectively viewed. It is possible, though, to expose and then view subsurface layers. This can be accomplished by cryofracturing the specimen. In this technique the specimen is frozen at very low temperature, usually in liquid nitrogen, and then fractured with a sharp blade. The specimen fractures along planes that correspond to the internal surfaces of the organism. The organism can then be coated with a suitable metal and viewed with the scanning electron microscope.

Pure culture methods

The ability to examine and study the characteristics of microorganisms, including obtaining organisms for microscopic visualization, depends in large part on being able to grow the organisms in pure culture. As will be discussed in subsequent chapters, there are many different types of microorganisms, and for each type of microorganism there are minimal requirements, tolerance limits, and optimal conditions for growth. In order to grow, microorganisms require a suitable environment, including a growth medium that can support their nutritional needs. Many bacterial species can be grown in the laboratory on a **defined medium**, which normally includes an organic carbon growth substrate, such as glucose or protein; mineral nutrients, including a source of nitrogen and phosphorus; and water. There are numerous different types of media that are used for growing bacteria and fungi in pure culture. Even viruses, which can only grow inside of host cells, can be grown *in vitro* (within glass), using tissue culture, and within embryonated eggs. In addition to a defined growth medium, the culture of microorganisms requires careful control of various environmental factors, including temperature, which normally is maintained within narrow limits by using a temperature-controlled incubator. By understanding the growth requirements of a given microbial species, it is possible to establish the necessary conditions *in vitro* to support the optimal growth of that microorganism.

The growth of a microorganism in pure culture further mandates that all other microbial species be eliminated. Since microorganisms are ubiquitously distributed in nature, obtaining and maintaining pure cultures requires the elimination of other microorganisms from the growth medium (**sterilization**), the separation of the microorganism being cultured from a mixture of microbes (**isolation**), the movement of the microorganism from one place to another without contamination (**aseptic transfer**), and the maintenance of the pure culture (**preservation**).

Sterilization

Sterilization procedures eliminate all viable microorganisms from a specified region. Culture dishes, test tubes, flasks, pipettes, transfer loops, and media must be free of viable microorganisms before they can be used for establishing pure cultures of microorganisms. The culture vessels must be sealed or capped with sterile plugs to prevent contamination. There are various ways of sterilizing liquids, containers, and instruments used in pure culture procedures; these include exposure to elevated temperatures or radiation levels to kill microorganisms, and filtration to remove microorganisms from solution. Sterilization by filtration is accomplished by passage through a 0.2-μm or smaller pore size bacteriological filter; most bacteria are trapped on the filter, but viruses and some very small bacteria pass through the filter.

Media preparation for the microbiology laboratory involves the use of an autoclave for sterilization, which permits exposure to high temperatures for a specified period of time. Generally, a temperature of 121°C (achieved by using steam at 15 lb/in.^{2}) for 15 minutes is used to heat sterilize bacteriological media. Much of the time spent in preparation for the bacteriology laboratory involves cooking the media for growing bacteria, that is, mixing and sterilizing the growth media in suitable sterile culture vessels.

Aseptic transfer technique

Many of the petri dishes and tissue culture plates that are used for growing pure cultures of microorganisms now are plastic and come presterilized from the manufacturer, and filling these vessels with a sterile medium requires the use of aseptic technique. **Aseptic technique** is required when filling presterilized glassware with microbiological media. Pouring media requires flaming the mouth of the vessel to kill surface contaminants (Figure 2.32). Aseptic technique is also required to transfer the pure culture from one vessel to another. By aseptic technique the microbiologist means taking all prudent precautions to prevent contamination of the culture. Proper aseptic transfer technique also protects the microbiologist from contamination with the culture, which should always be treated as a potential pathogen. Aseptic technique involves avoiding any contact of the pure culture, sterile medium, and sterile surfaces of the growth vessel with contaminating microorganisms. To accomplish this task (1) the

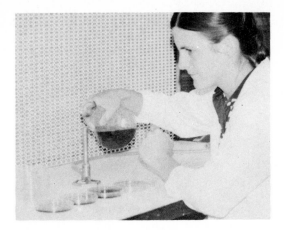

Figure 2.32

A laboratory technician flaming the mouth of a flask during aseptic pouring of medium into a petri dish. It is critical that microorganisms from the environment not contaminate the medium as it is being poured; flaming the mouth of the flask eliminates a source of potential contaminating microorganisms.

work area is cleansed with an antiseptic to reduce the numbers of potential contaminants; (2) the transfer instruments are sterilized—for example, the transfer loop is sterilized by heating with a bunsen burner before and after transferring (Figure 2.33); and (3) the work is accomplished quickly and efficiently to minimize the time of exposure during which contamination of the culture or laboratory worker can occur.

Figure 2.33

A laboratory technician is shown here flaming an inoculating loop. To prevent contamination and spread of bacteria, the loop is flamed before and after transferring bacterial cultures.

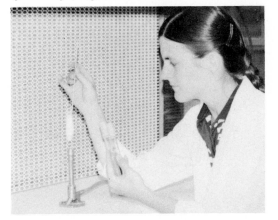

The normal steps for **transferring a culture** from one vessel to another are (1) flame the transfer loop; (2) open and flame the mouths of the culture tubes; (3) pick up some of the culture growth and transfer it to the fresh medium; (4) flame the mouths of the culture vessels and reseal them; and (5) reflame the inoculating loop (Figure 2.34). Essentially the same technique is used for inoculating petri dishes, except the dish is not flamed, and for transferring microorganisms from a culture vessel to a microscope slide.

Isolation

Several different methods are used for the **isolation of pure cultures** of microorganisms. These methods involve separating microorganisms on a solid medium into individual cells that are then allowed to reproduce to form a clone, which is a pure culture of a single microbial type. Isolation is achieved by the physical separation of the

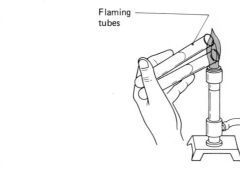

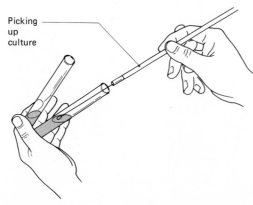

Figure 2.34

Steps in the aseptic transfer of bacteria.

Flaming loop

Flaming tubes

Picking up culture

Figure 2.35

Streaking for the isolation of pure cultures, showing two different streaking patterns. In this procedure a culture is diluted by drawing a loopful of the organism across a medium until only single cells are deposited at a given location. The growth of each isolated cell results in the formation of a discrete colony.

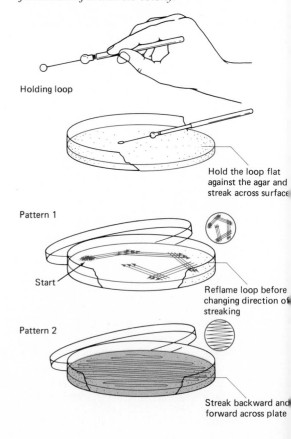

Holding loop

Hold the loop flat against the agar and streak across surface

Pattern 1

Start

Reflame loop before changing direction of streaking

Pattern 2

Streak backward and forward across plate

microorganisms, but the success of an isolation method also involves the maintenance of the viability and growth of a pure culture of the microorganism. Care must be taken to ensure that the microorganisms are not killed during the isolation procedure, which can easily occur by exposing the microorganisms to conditions they cannot tolerate, such as air in the case of obligately anaerobic microorganisms that are sensitive to oxygen. The success of an isolation method also depends on being able to grow the microorganism, that is, to define the growth medium and to establish the appropriate incubation conditions that permit the growth of the microorganism.

Streak plate

In the **streak plate technique** a loopful of bacterial cells is streaked across the surface of an agar solidified plate of a nutrient medium. Several different streaking patterns can be used to achieve separation of individual bacterial cells on the agar surface (Figure 2.35). The plates are then incubated under favorable conditions to permit the growth of the microorganisms. The key to this method is that, by streaking, a dilution gradient is established across the face of the plate, so that while confluent growth occurs on part of the plate where the bacterial cells are not sufficiently separated, isolated colonies do develop in another region of the plate. The colonies are macroscopic and easily seen with the naked eye. Each well-isolated colony arises from a single bacterium and represents a clone of a pure culture. The isolated colonies can than be picked, using a sterile inoculating loop, and restreaked onto a fresh medium to ensure purity. A new colony is then picked and transferred to an agar slant or other suitable medium for maintenance of the pure culture.

Spread plate

In the **spread plate method** a drop of a suspension of microorganisms is placed on the center of an agar plate and spread over the surface of the agar by using a sterile glass rod (Figure 2.36). The glass rod is normally sterilized by dipping in alcohol and flaming to burn off the alcohol. By spreading the suspension over the plate, a dilution gradient is established so that individual microorganisms are separated from the other organisms in the suspension and deposited at a discrete location. In order to accomplish this, it is often necessary to dilute the suspension before application to the agar plate to prevent overcrowding and the formation of confluent growth rather than the desired development of isolated colonies. After incubation isolated colonies are picked and streaked onto a fresh medium to ensure purity.

Figure 2.36

The spread plate technique for enumerating microorganisms. (1) Aseptically apply a known volume of a suspension to a suitable solid medium; (2) sterilize a spreading rod by dipping in alcohol and flaming; (3) use a sterile rod to spread suspension over surface of the medium; (4) incubate; (5) count colonies and calculate number of microorganisms in the original suspension.

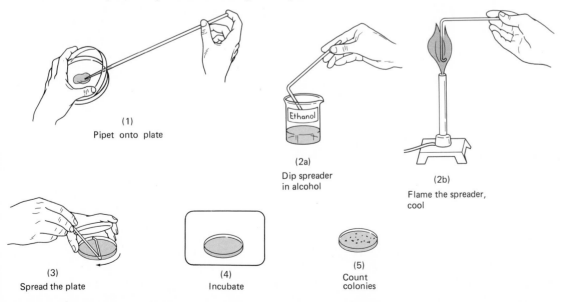

(1)
Pipet onto plate

(2a)
Dip spreader in alcohol

(2b)
Flame the spreader, cool

(3)
Spread the plate

(4)
Incubate

(5)
Count colonies

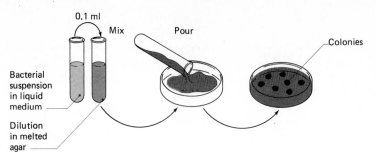

Colonies

Figure 2.37

The pour plate technique for enumerating microorganisms. A known volume of a microbial suspension is mixed with a liquefied agar medium and poured into a petri plate. After incubation the numbers of colonies that develop are counted, and the concentration of microorganisms in the original suspension is calculated.

Pour plate

In the **pour plate technique**, suspensions of microorganisms are added to melted agar tubes that have been cooled to approximately 42–45°C (Figure 2.37). The bacteria and agar medium are mixed well and the suspensions are poured into sterile petri dishes by using aseptic technique. The agar is allowed to solidify, trapping the bacteria at separate discrete positions within the matrix of the medium. While the medium holds bacteria in place, it is soft enough to permit growth of bacteria and the formation of discrete isolated colonies both within the fluid and on the surface of the agar. As with the other isolation methods, individual colonies are then picked and streaked onto another plate for purification. In addition to its use in isolating pure cultures, the pour plate technique is used for the quantification of numbers of viable bacteria. The facts that agar solidifies below 42°C and that many bacteria survive at these temperatures ensure the success of this isolation technique. But because in some cases significant numbers of bacteria can be killed by the heat shocking, this method cannot always be used.

Preservation

After isolating a microorganism in pure culture, it is necessary to maintain the viable culture, free of contamination, for some period of time. There are several methods available for maintaining and **preserving pure cultures**. The organisms may simply be subcultured periodically onto or into a fresh medium to permit continued growth and ensure viability of a stock culture. Proper aseptic technique must be used each time the organism is transferred, and there is always a risk of contamination. Furthermore, repeated subculturing is extremely time-consuming, making it difficult to successfully maintain large numbers of pure cultures for indefinite periods of time. Additionally, genetic changes are likely to occur when cultures are repeatedly transferred.

Therefore, a variety of methods have been developed for preserving pure cultures of microorganisms in a relatively dormant state, so that they do not require frequent subculturing into a fresh medium to maintain viability. These methods include refrigeration at 0–5°C for short storage times, freezing in liquid nitrogen at −196°C for prolonged storage, and **lyophilization (freeze-drying)** to dehydrate the cells. By sufficiently lowering the temperature or removing water, microbial growth is precluded but viability is maintained, and the microorganisms can be preserved for extended periods. The choice of the preservation method depends on the nature of the culture and the facilities that are available. Often, valuable cultures are deposited in centralized **type culture collections**, such as the American Type Culture Collection in Rockville, Maryland, where they are preserved. It is especially important that all new microbial species be deposited in such culture collections to ensure their indefinite preservation and to make them available for scientific study.

Postlude

The development of the various microscopes for viewing microorganisms, each having its strengths and weaknesses (Table 2.2), has gone hand in hand with the unravelling of the mysteries of the microbial world. The critical concern of the microbiologist is that the image produced in the microscope be large enough to

table 2.2
_____

Comparison of various types of microscopes

Type of microscope	Maximum useful magnification	Resolution	Comments
Bright-field	1500×	200 nm	Extensively used for the visualization of microorganisms; usually necessary to stain specimens for viewing.
Dark-field	1500×	200 nm	Used for viewing live microorganisms, particularly those with characteristic morphology; staining not required; specimen appears bright on dark background.
Ultraviolet	2500×	100 nm	Improved resolution over normal light microscope; largely replaced by electron microscopes.
Fluorescence	1500×	200 nm	Uses fluorescent staining; useful in many diagnostic procedures for identifying microorganisms.
Phase contrast	1500×	200 nm	Used to examine structures of living microorganisms; does not require staining.
Interference	1500×	200 nm	Used to examine structures of microorganisms, particularly those near edges; produces sharp, multicolored image with three-dimensional appearance.
Transmission electron microscope, TEM	500,000–1,000,000×	1 nm	Used to view ultrastructure of microorganisms, including viruses; much greater resolving power and useful magnification than can be achieved with light microscopy.
Scanning electron microscope, SEM	10,000–100,000	10 nm	Used for showing detailed surface structures of microorganisms; produces three-dimensional image.

permit visualization of the organisms being studied and that the detail of the specimen be resolved in the microscopic image, while at the same time the image is magnified, it is not distorted, nor are artifacts produced. Obviously, it is critical to know that what you see through the microscope is real. The ability to visualize the fine ultrastructure of microorganisms, such as is accomplished with an electron microscope, must be matched with the critical intellectual ability of the microbiologist to interpret properly the significance of what is seen. The advances in microbiology are tied to developments in methodology, but it is the creativity of the microbiologist in using these tools that is essential for scientific discovery.

The pure culture and microscopic methods discussed in this chapter have opened the way for conducting studies in microbiology. Great microbiological discoveries, such as those made by Koch and Pasteur, depend on these basic techniques, as do routine microbiological procedures, such as are conducted in clinical laboratories for the diagnosis of infectious diseases and by the student in the introductory microbiology laboratory. The ability to grow a microorganism in pure culture and visualize even the smallest microbes is basic to the development of our understanding of the fundamentals of microbiology. Future advances in microbiology will depend on the application of these methods in original and novel experiments and on the development of new techniques for examining microorganisms. Miniaturization and computers are now being used to develop improved microbiological methods that will be used to enhance basic research and the application of microbiology for improving the quality of life. The development of new methodologies permits the more detailed examination of microorganisms, revealing the finer aspects of their structure and function, and technological developments permit the use of this fundamental information for major applications.

1. What is resolution, and why is it important in microscopy?

2. What factors influence resolution? Why do we consider an electron microscope superior to a light microscope? What organisms and structures can be seen with a light microscope? What organisms and structures can be seen with a transmission electron microscope?

3. What is meant by the term *useful magnification*?

4. Why do we stain microorganisms before viewing them with a microscope?

5. Name five types of microscopes and discuss the advantages and disadvantages of each.

6. What are some optical defects of light microscopes, and how are they corrected?

7. How does a SEM differ from a TEM? What are the different applications for each of these electron microscopes?

8. What is the difference between a simple and a differential stain?

9. What is a pure culture? Why do we place such importance on obtaining and maintaining pure cultures?

10. What is aseptic transfer technique? Why must you master this technique to work in a microbiology laboratory?

11. Discuss three methods for isolating pure cultures of microorganisms.

Microbiological
methods: pure
culture and
microscopic
techniques

**Suggested
Supplementary
Readings**

Berlyn, G. P., and J. P. M. Miksche. 1976. *Botanical Microtechnique and Cytochemistry.* The Iowa State University Press, Ames.

Burrells, W. 1978. *Microscope Techniques: A Comprehensive Handbook for General and Applied Microscopy.* Halsted Press, New York.

Collins, C. H., and P. M. Lyne. 1976. *Microbiological Methods.* Butterworth, Woburn, Massachusetts.

Gerhardt, P. (ed.). 1981. *Manual of Methods for General Bacteriology.* American Society for Microbiology, Washington, D.C.

Goodhew, P. J., and L. E. Cartwright. 1975. *Electron Microscopy and Analysis.* Crane, Rusack and Co., Inc., New York.

Gray, P. 1973. *Encyclopedia of Microscopy and Micro-Technique.* Van Nostrand Reinhold Co., New York.

Hayat, M. A. 1978. *Introduction to Biological Scanning Electron Microscopy.* University Park Press, Baltimore.

James, J. 1976. *Light Microscopic Techniques in Biology and Medicine.* Kluwer Boston, Inc., Higham, Massachusetts.

Laskin, A., and H. A. Lechevalier. 1977–1981. *Handbook of Microbiology* (4 volumes). CRC Press, Inc., Boca Raton, Florida.

Meek, G. A. 1976. *Practical Electron Microscopy for Biologists.* John Wiley & Sons, London.

Norris, J. R., and D. W. Ribbons (eds.). 1969–. *Methods in Microbiology.* Academic Press, New York.

Rash, J., and C. S. Hudson. 1982. *Electron Microscopy Methods and Applications*. Praeger Publishers, New York.

Rochow, T. G., and E. G. Rochow. 1979. *An Introduction to Microscopy: by Means of Light, Electrons, X-Rays or Ultrasound*. Plenum Publishing Co., New York.

Ross, K. F. A. 1967. *Phase Contrast and Interference Microscopy for Cell Biologists*. St. Martin's Press, Inc., New York.

Sirockin, G., and S. Cullimore. 1969. *Practical Microbiology*. McGraw-Hill Publishing Co., Ltd., London.

Wischnitzer, S. 1981. *Introduction to Electron Microscopy*. Pergamon Press, New York.

Introduction to biochemistry

3

Biochemistry is an integral part of the biological sciences, and the information covered in this chapter is fundamental to the study of microbiology. All living systems share in common the fact that they are carbon based; that is, **carbon-containing organic molecules** form the essential components of all living organisms. Carbon atoms are able to establish strong bonds with hydrogen, oxygen, nitrogen, and other carbon atoms; because these four elements comprise 99 percent of the mass of living organisms, this makes the element carbon well adapted for uniting the atoms of living systems into stable macromolecules. Microorganisms are composed of a large variety of organic molecules representing four major classes of biochemicals: **carbohydrates**, **lipids**, **proteins**, and **nucleic acids**. These classes of biochemicals form the backbone structures of the majority of the macromolecules that are used to construct the structural units of microorganisms

and to house the genetic information. Biochemical reactions involving these same classes of molecules are essential for the metabolic activities of living systems, and they establish the basis for microbial growth and reproduction. The dynamics of living organisms reflect their perpetual metabolism, and the continuous flow of biochemicals and their stored energy through the system is an essential aspect of growing microorganisms. The study of microbiology requires an appreciation of the biochemical macromolecules that comprise the structures of microorganisms and the biochemical reactions involved in the metabolic activities of microorganisms. Examining the molecular level aspects of microorganisms is necessary for the development of a basic understanding of microbial structure and function. The material in this chapter will serve as a review and/ or introduction to biochemistry, providing the necessary background for studying microbiology.

Nature of organic molecules

Molecules are specific combinations of chemical elements, formed when atoms establish chemical bonds by sharing or exchanging electrons in their outer orbitals. The **electrons** in an **atom** are specifically arranged in a defined order, with electrons located at specified energy levels or shells (Figure 3.1). The **electron shell**, or **orbital** of electrons, represents a volume occupied by an elec-

tron cloud at some distance from the central nucleus of the atom. The maximum number of electrons that can occupy a particular energy level increases with distance from the nucleus; for example, the first orbital holds a maximum of two electrons and the second orbital can hold eight electrons. The electrons in the outermost valence shell normally enter into **chemical bonds**, with

H

First electron shell

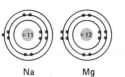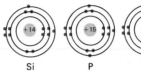

C   N   O

Second electron shell

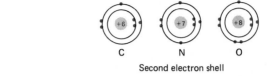

Na   Mg   Si   P   S   Cl

Third electron shell

Figure 3.1

The electrons of an atom are arranged in orbitals. Each orbital represents a different energy level.

stable chemical bonds occurring when atoms fill their valence shells. The number of bonds that an element can form depends on the number of electrons required by that atom to fill its outer electron shell; carbon, the central atom in organic molecules, for example, can establish four bonds because it has four electrons in its outer electron shell that can hold a maximum of eight electrons. Carbon-containing molecules—with the exception of carbon dioxide, carbon monoxide, and a few others which contain carbon but do not also contain hydrogen—are **organic molecules**; all others are inorganic molecules. In addition to carbon, hydrogen, oxygen, and nitrogen, the elements phosphorus, sulfur, sodium, potassium, magnesium, calcium, iron, and chlorine also occur in relatively high abundance within the biochemicals of microorganisms. To fill their outer electron shells, nitrogen requires three electrons, oxygen requires two electrons, and hydrogen re-

quires one electron and, therefore, nitrogen can establish three chemical bonds, oxygen two bonds, and hydrogen one bond. Both inorganic and organic molecules are essential components of microorganisms, and although life is based on organic carbon, it is likewise totally dependent on inorganic water.

Types of chemical bonds

Two basic types of chemical bonding can hold molecules together. **Ionic bonds** are based on charge differences between chemical elements that develop when atoms donate or acquire electrons to fill their outer electron shells. For example, the salt sodium chloride represents a molecule formed by an ionic bond between the elements sodium and chlorine (Figure 3.2). The sodium atom donates an electron, forming a positively

Figure 3.2

The formation of ionic bonds involves the loss and gain of electrons. In the formation of NaCl, the chlorine atom gains an electron to fill its outer electron shell, and the sodium atom loses an electron so that all the remaining electron shells are filled. After the formation of the ionic bond, the sodium has a positive charge and the chloride a negative charge.

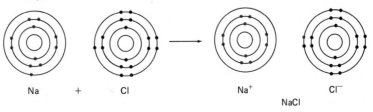

Na   +   Cl   ⟶   Na⁺   Cl⁻

NaCl

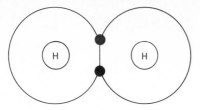

Figure 3.3

*The formation of covalent bonds involves the
sharing of electrons. In molecular hydrogen (H_2)
each atom shares only its first orbital electron with
the other, effectively completing the first orbital of
each atom.*

charged sodium ion (**cation**); the chlorine ac-
cepts the electron, forming a negatively charged
chloride ion (**anion**); the positively charged sodium ions and negatively charged chloride ions
are held together by electrostatic forces. Ionic
bonds are relatively weak, and in aqueous solution the ionic bond is readily broken with the
dissociation of ions. Consequently, ionic bonds
are not strong enough to hold together the
macromolecules of living systems.

In contrast to the ionic bond, where an electron is completely transferred from one element
to another, **covalent bonds** form when elements
share electrons (Figure 3.3). The number of covalent bonds an element is capable of forming depend on its electronic structure. Each covalent
bond involves the sharing of a pair of electrons,
with each atom contributing one electron. In some
cases atoms share two pairs of electrons, giving
rise to a double bond. Once formed, covalent
bonds require a relatively high amount of energy

table 3.1

Relative strengths of some chemical bonds

Bond	Relative strength
Covalent (energy of interaction 30–100 kcal/ mole)	
H—H	1.0
C—C	0.8
C—H	1.0
C—O	0.8
C—N	0.7
Ionic (energy of interaction 10–20 kcal/mole)	
Na···Cl	0.3
Hydrogen (energy of interaction 2–10 kcal/ mole)	
H—O···O	0.1
H—N···N	0.1

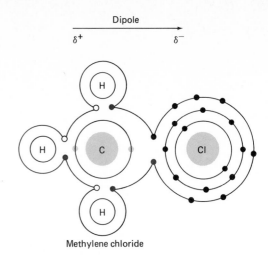

Methylene chloride

Figure 3.4

*The covalent bonds of methylene chloride show the
formation of dipole moment due to the uneven
sharing of electrons. The shared electron draws
closer to the chlorine (Cl) than to the carbon (C),
thus establishing a charge separation.*

to break them; they provide the stability needed
to establish the macromolecules required for the
existence of microbes and other living organisms.

Covalent bonds exhibit differences in relative
strength, depending on which elements are sharing electrons (Table 3.1). There also are differences in the distribution of electrons between
atoms forming covalent bonds. If two atoms of
the same element share electrons to form a molecule, the electrons are evenly distributed between the two atoms. This occurs in a molecule
such as hydrogen (H_2). However, when a covalent
bond forms between the atoms of two different
elements the electrons are shared unevenly, with
the electrons exhibiting a greater affinity for one
of the atoms and being drawn closer to that atom
(Figure 3.4). This gives rise to **polarity** in which
one of the atoms has a greater positive charge
and the other atom a greater negative charge. The
molecule is said to have a **dipole** resulting from
the separation of charge, with the polarity of the
covalent bond depending on the specific elemental atoms involved in establishing the bond. The
polarity of covalent bonds establishes the basis
for charge interactions and the formation of additional weak bonds between organic molecules.

Water, for example, is an excellent polar solvent because it has a large dipole moment, that
is, a large charge separation due to the unequal
sharing of electrons between the oxygen and hydrogen atoms (Figure 3.5). The internal separa-

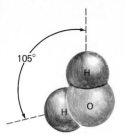

105°

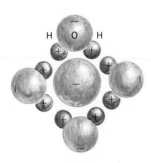

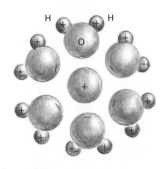

Figure 3.5

The spatial arrangement of atoms in a water molecule results in a dipole moment due to unequal distribution of electrons between hydrogen and water. As a result, water is a good polar solvent, because water can surround both positively and negatively charged ions.

tion of charge in the water molecule permits intermolecular associations with other polar molecules. The positive regions of solute molecules associate with the negative oxygen atom of water, and the negative regions of the solute associate with the positive hydrogen atoms of water. These electrostatic interactions are responsible for separating polar solute molecules, enhancing their dissolution in aqueous solution. The solvent properties of water are essential for the occurrence of biochemical reactions within microbial cells. The interactions between polar groups also lead to the exclusion of nonpolar molecules, or

nonpolar portions of large molecules, creating hydrophobic interactions between nonpolar molecules. The separation of the polar (***hydrophilic*** means water-loving) and nonpolar (***hydrophobic*** means afraid of water) ends of phospholipids, discussed later in this chapter, establishes the basis for the selective permeability of biological membranes, an essential function of living systems.

Weak intermolecular bonds may occur between neighboring molecules. A hydrogen atom that is covalently bonded to oxygen or nitrogen within a molecule may be simultaneously attracted to a nitrogen or oxygen atom of a neighboring molecule, forming a weak link known as a **hydrogen bond** between the two molecules (Figure 3.6). The strength of the hydrogen bond is only about one-tenth that of the covalent bond, but such weak bonds have very important functions in living systems. Hydrogen bonds are essential for maintaining a double helical structure of DNA. Hydrogen bonding accounts for the fluid nature of water and its ability to act as a good polar solvent. The capillary action of water, its ability to move upward through narrow pores, is dependent on the hydrogen bonding of the water molecules to the solid surfaces of the porous material, such as glass. In proteins, besides the strong covalent bonds that link the carbon skeleton of the molecule, intramolecular hydrogen bonds between the functional groups of different amino acids act to stabilize and determine the configuration of these large molecules. It is the fact that the weak hydrogen bonds can be readily formed and broken that allows the flexibility needed for these various biochemical functions.

Figure 3.6

Hydrogen bonding occurs between molecules as a result of charge interactions stemming from the dipole moments of the molecules.

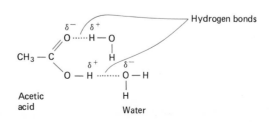

Functional groups and structure of organic molecules

Organic molecules have vastly differing properties that are determined by their chemical composition, size, and the arrangement of the atoms within the molecule, that is, how the specific elements of the molecule are bonded together. Organic molecules consisting of only carbon and hydrogen, hydrocarbons, are relatively nonreactive, and although such compounds are important in fossil fuels, they do not comprise the major biochemicals of microorganisms. Rather, the major biochemical constituents of microorganisms include oxygen, nitrogen, sulfur, and other elements in addition to carbon and hydrogen. As a rule, the organic molecules that enter into biochemical reactions of microorganisms possess **functional groups**, and the major classes of organic molecules comprising living systems are based on the presence of particular functional groups (Table 3.2). The substitution of a functional group for one of the hydrogens on the carbon skeleton of an organic molecule renders the molecule more reactive, establishing the basis for microbial metabolism.

An important property of a functional group is its ability to react with a different functional group of another molecule. The interactions between functional groups involve the formation of weak bonds. The nature of these weak bonds ensures that only complementary functional groups can react to form a bond, establishing a specificity of chemical reactions. The reactions between organic acids and bases, which form the basis for many fundamental biochemical reactions of microorganisms, involve such specific interactions between functional groups and the formation of weak ionic bonds between complementary groups. The ability of biochemicals to dissociate and form acids, which in water solution produce hydrogen ions (H^+), and bases, which in water solution produce hydroxyl ions (OH^-), is important in many of the biochemical reactions of microorganisms. The concentration of hydrogen ions determines the pH of a solution, which is defined as $-\log[H^+]$. Pure water contains 10^{-7} mole of hydrogen ions per liter and thus has a

table 3.2.

Common functional groups of organic molecules and some representative classes of molecules in which they occur

Group	Name	Class of molecules
R—OH	Hydroxyl group	Alcohols, carbohydrates
OH | R=O	Carboxyl group	Carboxylic acids, fatty acids, amino acids
R | R=O	Keto group	Ketones, sugars
H | R—C=O	Aldehyde group	Aldehydes, sugars
H | R—C—H | H	Methyl group	Hydrocarbons, lipids
H | R—N—H	Amino group	Amines, amino acids
OH | R—P=O | OH	Phosphate group	Organic phosphates, phospholipids, nucleotides
R—SH	Sulfhydryl group	Mercaptans, a few amino acids

pH of 7.0. At pH 7.0 the concentrations of H^+ and OH^- are equal; water therefore has a neutral pH. pH values below 7 are acidic, and those above 7 are basic. The pH affects the degree of dissociation of acids and bases, which has a marked influence on the properties of biochemical molecules. For example, the enzymatic activities of microorganisms are determined in large part by the degree of dissociation of the functional amino and carboxylic acid groups of the enzyme molecules.

The location of a functional group within an organic molecule can be quite important in determining the biochemical properties of that molecule. In order to keep track of the location of functional groups, a convention has been adopted for numbering the carbon atoms in an organic molecule. The specific numbering systems vary for different classes of organic molecules. An example of the numbering of a simple hydrocarbon and a carboxylic acid is shown in Figure 3.7. In some molecules, in addition to the formal numbering system, the term α-carbon is used to describe the carbon atom to which the functional group that typifies that class of compounds is directly attached, and the succeeding carbons are also designated by continuing the Greek alphabet. By establishing formal conventions the exact arrangement of atoms within the biochemical can be specified, permitting discussion of the properties and function of that specific molecule in the biochemistry of microorganisms.

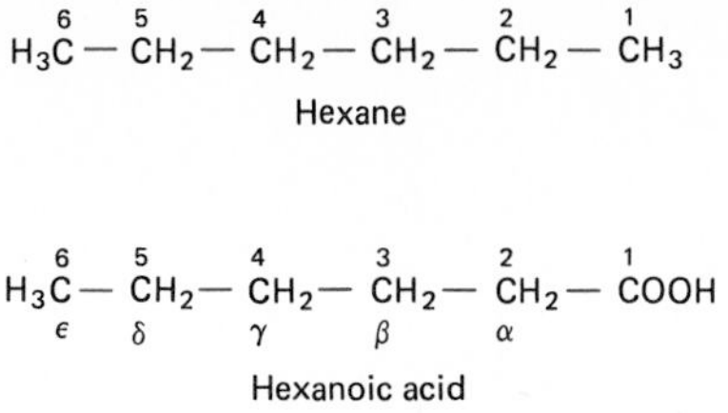

Isomerism

Compounds can have the same chemical composition, that is, the same molecular formula, but still have very different chemical and physical properties, depending on the arrangement of the molecule. Such molecules with identical molecular formulas, but different spatial arrangements, are called **isomers**. The carbon atoms of isomers can be bonded at different positions, producing structures with different spatial arrangements (**structural isomers**) (Figure 3.8). Additionally, because organic molecules are three-dimensional, compounds that have the same structural formulas can still have different spatial arrangements. The spatial arrangements of the atoms in the molecule are defined by the configuration and conformation of the molecule. **Configuration** describes the relative arrangement in space of the atoms that make up the molecule, and **conformation** describes the different spatial arrange-

Figure 3.7

An example of the numbering of a carbon chain of a hydrocarbon and a carboxylic acid and the use of Greek letters to designate position relative to functional group.

ments of the molecule that result from rotation about single covalent bonds (Figure 3.9).

Stereoisomers are compounds with the same molecular formula and the same skeletal structure but different three-dimensional arrangements of the atoms in space. Because a carbon atom can form bonds with four different groups, creating an asymmetric center, many organic compounds occur as stereoisomers. Organic molecules can have two different forms (configurations) about an asymmetric carbon atom that represent mirror images of the molecule (Figure 3.10). The mirror images of the molecule are not superimposable and represent truly different configurations. The different configurations of the stereoisomers can have optical activity, and when a plane of polarized light shines through the two different isomers, it will be rotated in opposite directions. The direction of rotation of the light by the optical isomers is termed dextrorotatory (+) if it is bent to the right, or levorotatory (−) if it is bent to the left. The configurations of all optically active compounds can be related to the

Figure 3.8

Structural isomers have identical molecular formulas, but the substituents are located at different positions. In this example, methylbutyric acid, the only difference is the position of the methyl group. Nevertheless, these structural isomers have different physicochemical properties.

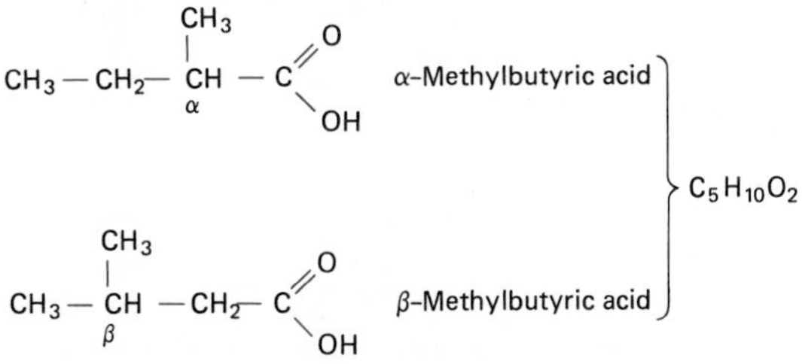

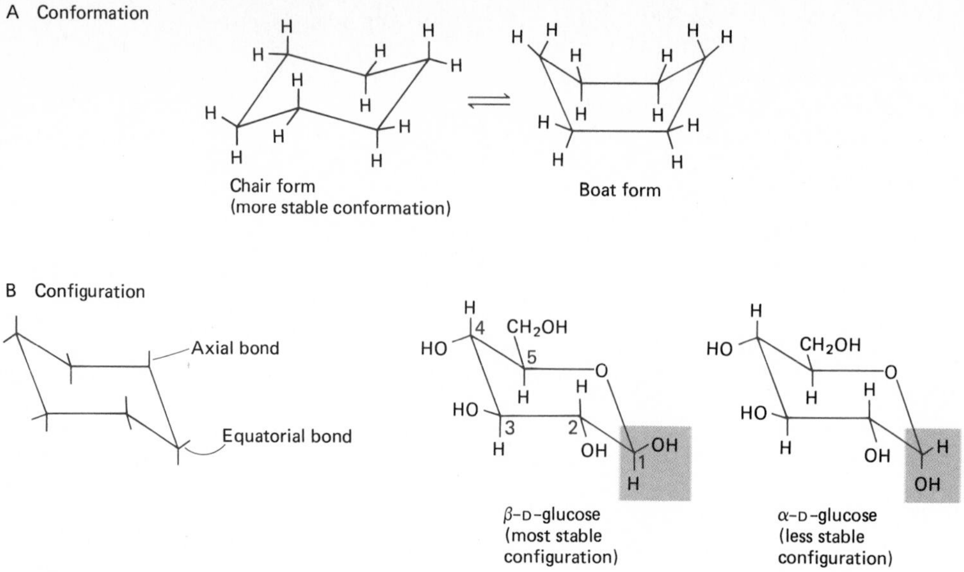

A   Conformation

Chair form
(more stable conformation)

Boat form

B   Configuration

Axial bond

Equatorial bond

H
4 CH₂OH
HO
5
H
O
HO
H
3   2
H   OH   OH
1
H

β-D-glucose
(most stable
configuration)

H
HO   CH₂OH
O
H   H
HO
H   OH   H
OH

α-D-glucose
(less stable
configuration)

Figure 3.9

74

Introduction to
biochemistry

Organic molecules can vary in their configuration and conformation, as shown in this
illustration. The stabilities of the differing configurations and conformations vary, and in each
case there is a preferred (most stable) form. (A) The chair and boat conformations of
cyclohexane. (B) The two possible configurations of the chair conformation of D-glucose.

Figure 3.10

An asymmetric carbon allows molecules to exist in
two different forms that are mirror images.

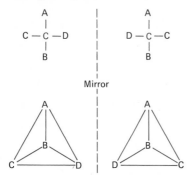

Figure 3.11

The D and L forms of a glyceraldehyde molecule
and its relation to L-alanine.

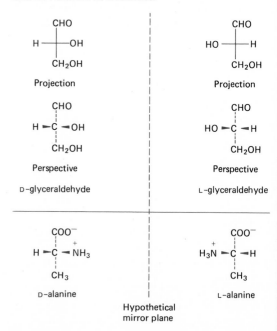

arbitrarily chosen reference molecule glyceraldehyde, the smallest sugar that has an asymmetric
carbon atom. The two stereoisomers of glyceraldehyde are designated D and L (Figure 3.11). All
stereoisomers that are related to L-glyceraldehyde
are designated L, and those related to D-glyceral-
dehyde are designated D, regardless of the direction of rotation of polarized light given by the
isomer. The designations D and L therefore denote the absolute configuration of a molecule.

Many biochemicals found in microorganisms contain either D or L forms of the molecule but not both. Proteins exclusively contain L-amino acids, although D-amino acids occur as specific constituents of other macromolecules. The importance of stereoisomerism rests with the fact that the different stereoisomers of a molecule exhibit different reactivities in chemical reactions. Generally, only one of the optical isomers is capable of participating in a given biochemical reaction.

Macromolecules comprising microorganisms

In the biochemical molecules of living systems, elements are linked together by covalent bonds to form relatively high-molecular-weight macromolecules with differing functional groups and spatial arrangements. The major classes of macromolecules that make up the structural and functional units of microorganisms are the carbohydrates, lipids, proteins, and nucleic acids (Table 3.3).

Carbohydrates

Carbohydrates have the basic chemical formula $C_n(H_2O)_n$ and include the simple sugars, such as glucose, and the macromolecules formed from such simple monosaccharides. By definition, carbohydrates are either polyhydroxy-aldehydes or polyhydroxy-ketones or molecules that yield polyhydroxy-aldehydes or polyhydroxy-ketones upon hydrolysis. Those carbohydrates that cannot be hydrolyzed are called monosaccharides (Figure 3.12). Individual monosaccharides may be linked to form larger polymeric carbohydrate mole-

cules. A disaccharide contains two monosaccharide units, an oligosaccharide contains from two to ten monosaccharide units, and a polysaccharide contains more than ten monosaccharide units.

In aqueous solution the five- and six-membered monosaccharides can form ring structures, and the structures of these carbohydrates frequently are illustrated in the ring form as op-

Figure 3.12

Structural formulas of common monosaccharides.

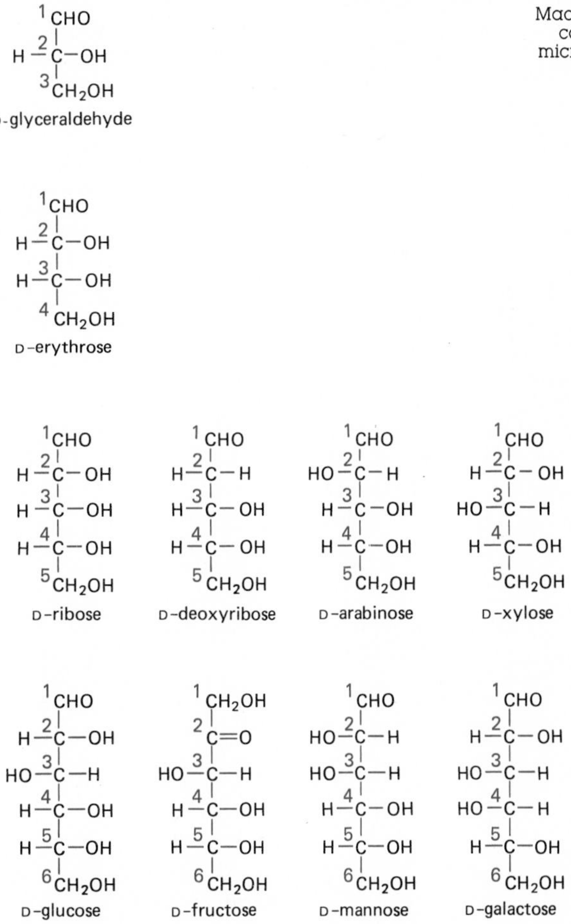

table 3.3

Molecular components of a cell of the bacterium
Escherichia coli

Biochemical molecule	Percent of total weight	Approximate number of each kind
Water	70	1
Proteins	15	3000
Nucleic acids		
DNA	1	1[*]
RNA	6	1000
Carbohydrates	5	50
Lipids	2	40
Other small organics	2	500
Inorganic Ions	1	12

[*]Many strains of E. coli have additional plasmids containing different types of DNA.

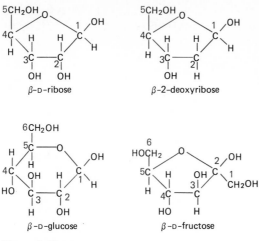

β-D-ribose

β-2-deoxyribose

β-D-glucose

β-D-fructose

Figure 3.13

The ring structures of some representative carbohydrates.

posed to the simpler straight-chain representation (Figure 3.13). As a result of ring formation, monosaccharides act as if they have an extra asymmetric carbon, and therefore, in addition to being designated D or L, these compounds are designated α or β.

The bond formed between monosaccharides to form disaccharides and larger polysaccharides is called a **glycosidic bond**. The glycosidic bond forms between the aldehyde group of one of the monosaccharide units and one of the alcohol groups of the other monosaccharide unit, with the elimination of water (Figure 3.14). When a glycosidic bond forms, the functional aldehyde or ketone group of one of the monosaccharide units is generally left free and therefore remains reactive. The disaccharides sucrose and trehalose are exceptions in which the reactive groups of both monosaccharides are linked. The glycosidic bond is specified by the position numbers of the carbons that are linked by the bond and by the three-dimensional orientation of the bond. Glycosidic bonds are termed α or β, depending on the three-dimensional orientation of the bond. The same monosaccharides can form various disaccharides and polysaccharides by forming glycosidic bonds with different orientations. For example, maltose is formed by an α-1,4 linkage of two glucose units, and cellobiose is formed by a β-1,4 linkage of two glucose units. Similarly, starch represents a polymer of glucose units linked by α-1,4 glycosidic bonds, and cellulose represents a polymeric chain of glucose-monosaccharides that are linked with

Figure 3.14

The formation of a glycosidic bond showing the α and β orientations.

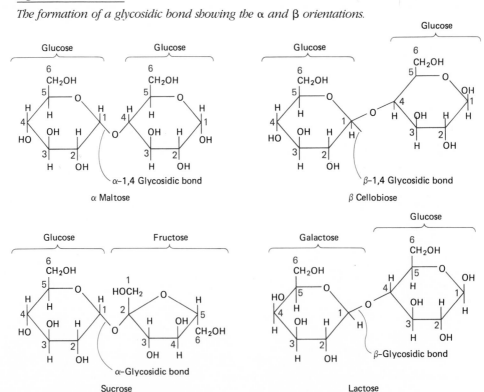

Figure 3.15

Several polymers of glucose—cellulose, starch, and glycogen.

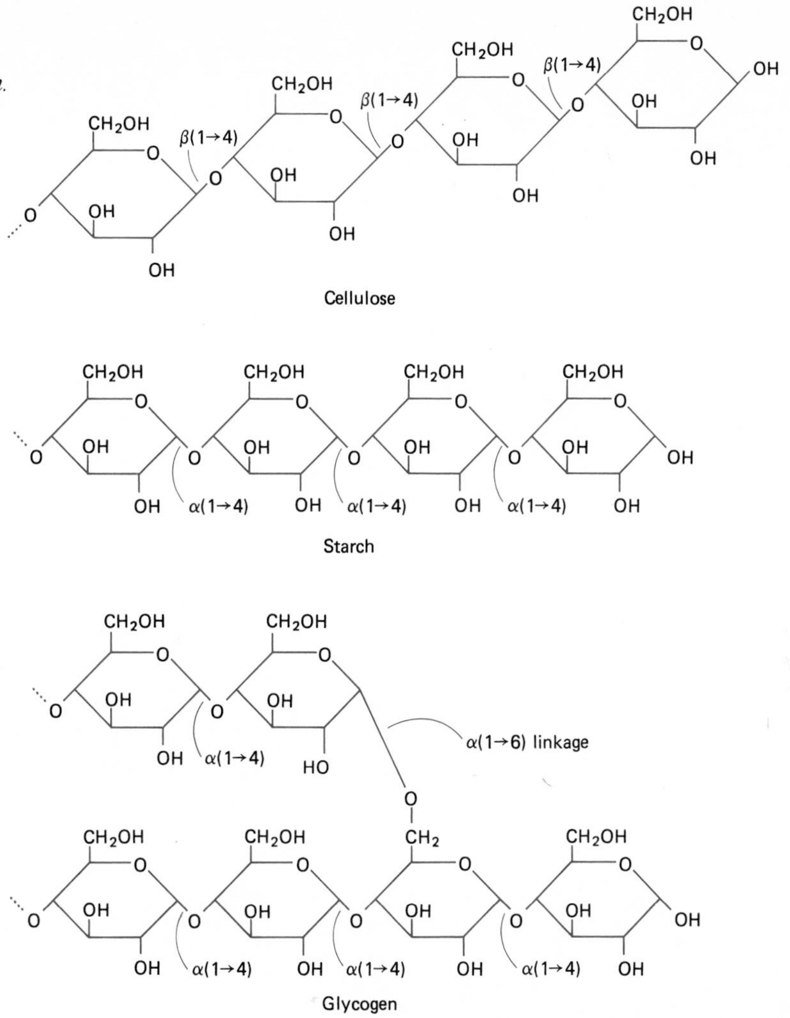

Cellulose

Starch

α(1→6) linkage

Glycogen

β-1,4 glycosidic linkages (Figure 3.15). Starch occurs in two forms: amylose, which is an unbranched polymer of glucose with α-1,4 glycosidic bonds; and amylopectin, which has branches formed by α-1,6 glycosidic bonds, in addition to the normal α-1,4 bonding found in amylose. The branches normally occur every 20–30 glucose units. Glycogen resembles amylopectin but has even greater branching.

Carbohydrate molecules also may establish bonds with other molecules to form carbohydrate derivatives. For example, phosphate may bond with glucose to form glucose 6-phosphate (Figure 3.16). The 6 refers to the position of the phosphate on the glucose. The bond formed between the monosaccharide and phosphate is an ester linkage. **Phosphate ester linkages** are important in linking the monosaccharide units that form the

Figure 3.16

The structural formulas of glucose 6-phosphate and fructose 6-phosphate.

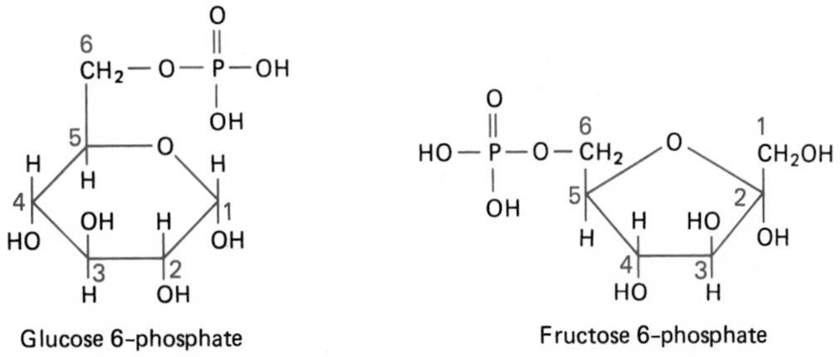

Glucose 6-phosphate

Fructose 6-phosphate

N-acetylglucosamine

N-acetylmuramic acid

Figure 3.17

The structural formulas of N-acetylglucosamine and N-acetylmuramic acid.

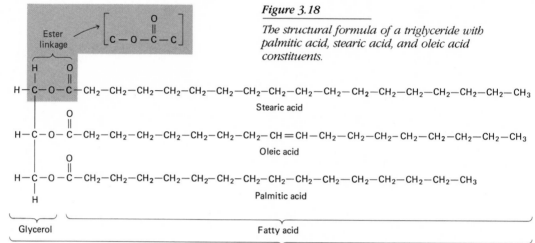

Figure 3.18

The structural formula of a triglyceride with palmitic acid, stearic acid, and oleic acid constituents.

Ester linkage

Stearic acid

Oleic acid

Palmitic acid

Glycerol

Fatty acid

Lipid (triglyceride)

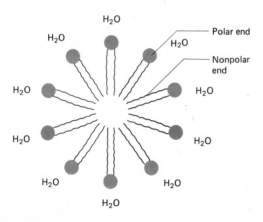

H_2O

H_2O

H_2O

H_2O

H_2O

H_2O

H_2O

H_2O

H_2O

H_2O

H_2O

Polar end

Nonpolar end

Figure 3.19

Because part of the molecule is more polar than the other nonpolar part, lipids form clusters, called micelles, in water. The nonpolar portions point inward and the polar sections toward the water.

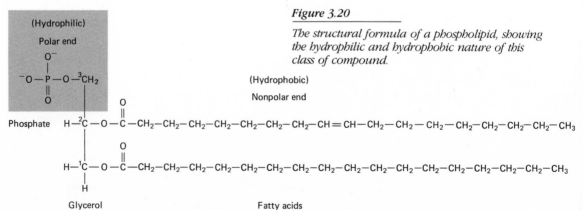

(Hydrophilic)

Polar end

Phosphate

Glycerol

(Hydrophobic)

Nonpolar end

Fatty acids

Figure 3.20

The structural formula of a phospholipid, showing the hydrophilic and hydrophobic nature of this class of compound.

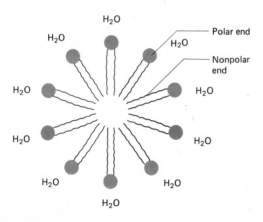

78

Introduction to biochemistry

backbone of the molecules DNA and RNA, discussed later in this chapter in the section on nucleic acids. In addition to bonding with phosphate, carbohydrates may form derivatives containing nitrogen molecules, such as amino-sugars. Two such carbohydrate derivatives are *N*-acetylglucosamine and *N*-acetylmuramic acid, which occur in the cell walls of bacteria (Figure 3.17). Carbohydrates also may bond with peptides and protein molecules to form peptidoglycans and glycoproteins. In a peptidoglycan molecule a polysaccharide forms the backbone of the molecule and peptides form the side-chains, whereas a glycoprotein is a protein macromolecule with oligosaccharide or polysaccharide side-chains; that is, the protein forms the backbone of the molecule and the carbohydrate the minor side-chains. Carbohydrates may also bind with lipids to form **lipopolysaccharides**.

Lipids

Lipids are water-insoluble biochemicals that are soluble in nonpolar solvents, such as chloroform. Lipids are the major constituent of biological membranes, and the solubility properties of lipid molecules enable them to function as the components of these semipermeable barriers. There are two major classes of lipids: **complex lipids**, which normally are composed of fatty acids bonded to an alcohol, and **simple lipids**, such as steroids. The most common complex lipids are the fats, which are combinations of fatty acids bonded to glycerol. The **triglycerides**, for example, are complex lipids formed by the bonding of fatty acids with glycerol (Figure 3.18). The linkage of the fatty acid to the alcohol is by an ester bond that is formed between the carboxyl group of the acid and the alcohol group of the glycerol molecule. In triglycerides three fatty acids, which may be the same or different, are linked to the three carbons of the glycerol molecule. The most common fatty acids found in complex cellular lipids are palmitic acid, a 16-carbon saturated fatty acid; stearic acid, an 18-carbon saturated fatty acid; and oleic acid, an 18-carbon unsaturated fatty acid having one double bond. The nonpolar nature of these relatively long-chain fatty acids causes lipids to exhibit hydrophobic reactions with water. The salts of C_{16}–C_{18} fatty acids are soaps and form micelles in water that are stabilized by hydrophobic interactions; in micelles the negatively charged carboxyl groups of the fatty acids all point outward toward the water, and the nonpolar hydrocarbon portions are buried within the micelle (Figure 3.19). Similarly, in biological membranes the negatively charged portion of the lipid interfaces with water, and the nonpolar portion is buried within the membrane structure.

The principal lipids of biological membranes are phospholipids. These phospholipids are derivatives of glycerol in which fatty acids are bonded to only two of the carbons of the glycerol molecule (Figure 3.20). The third carbon is linked by an ester bond to phosphate. The negatively charged phosphate acts as a hydrophilic head of the lipid molecule and is attracted to water, and the hydrocarbon portion of the fatty acid acts as a hydrophobic tail. The phosphate may further be linked by a second ester bond to another alcohol, forming more complex phospholipid molecules. Lipid molecules may combine with carbohydrates to form glycolipids or with proteins to form lipoproteins. Combinations of lipids with other **moieties** is important in some structures such as the cell walls of Gram negative bacteria (see discussion in Chapter 4).

Most of the simple lipids are unsaturated hydrocarbon molecules. The steroid molecules are four-ringed compounds that can form a variety of derivatives. The sterol cholesterol (Figure 3.21) is an important steroid derivative in animal tissues, and the accumulation of cholesterols is important in cardiovascular disease; other steroid derivatives in mammalian systems include the bile acids, adrenocortical hormones, and both male and female sex hormones. Different sterols occur in plants, fungi, algae, and protozoa, but bacteria generally lack sterols in their membranes. The presence of sterols in the membranes of fungi establishes the site of action of several antifungal drugs used in treating fungal infections of humans.

Figure 3.21

The structural formula of the sterol cholesterol.

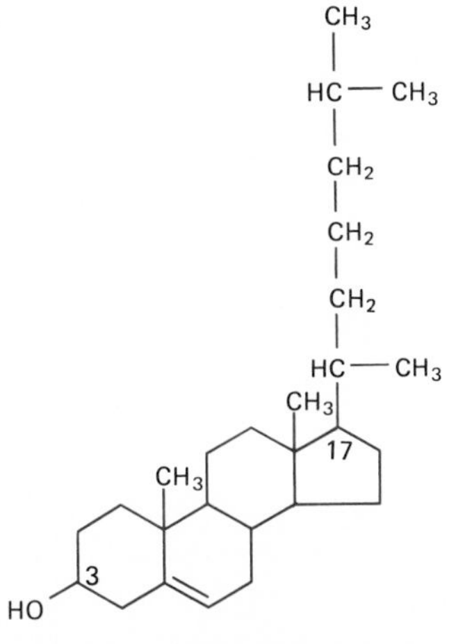

Cholesterol

Proteins

Proteins are large macromolecules of considerably higher molecular weight than lipids; the molecular weights of proteins range from 6000 to the millions. **Proteins are composed of long chains** of amino acids linked by peptide bonds and are extremely important molecules in biological systems, often composing 50 percent of the cell's dry weight. The amino acid subunits of proteins possess two functional groups: an amino group and a carboxylic acid group. Only 20 amino acids nor-

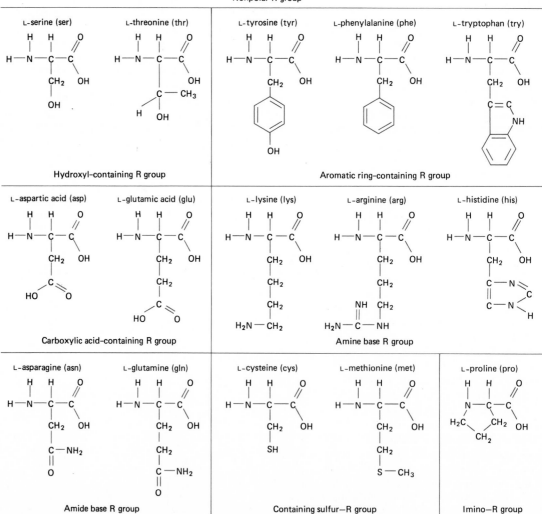

Figure 3.22

The structural formulas of 20 common amino acids. Each is an L-α-amino acid. The structures differ in the other constituents.

mally are found in the protein macromolecules. In some cases these 20 amino acids are modified to form derivatives such as hydroxyproline. All 20 of the essential amino acids have an L configuration, and in each of these amino acids both the amino group and the carboxylic acid group are linked to the same central α-carbon atom (Figure 3.22). Each of the amino acids also has an additional chemical group bonded to the α-carbon atom that is designated as an "R" group. These R groups distinguish the amino acids and establish the chemical properties of the protein molecule.

The bonding of the carboxyl group of one amino acid to the amino group of another amino acid establishes the **peptide linkage** that characterizes the protein molecule (Figure 3.23). The formation of the peptide bond can be considered as a condensation reaction between the amino group of one amino acid and the carboxyl group of another amino acid with the elimination of water. A protein may be composed of one or more polypeptide chains of amino acids. Two amino acids linked together by a peptide bond are termed a dipeptide; three amino acids so linked, a tripeptide; and many amino acids linked by peptide bonds are called a **polypeptide**. In the middle of the polypeptide chain, the carboxyl and amino groups are linked to each other. When a polypeptide forms, one end of the molecule will have a free carboxylic acid group, and the other end of the molecule will have a free amino group (Figure 3.24). The free amino end is termed the **amino** (NH_2) or **N-terminal** and the free carboxyl end, the **carboxyl** (COO) or **C-terminal**. These two ends of a peptide chain are biochemically distinguishable and impart an ordered direction to the protein molecule. The ability of biochem-

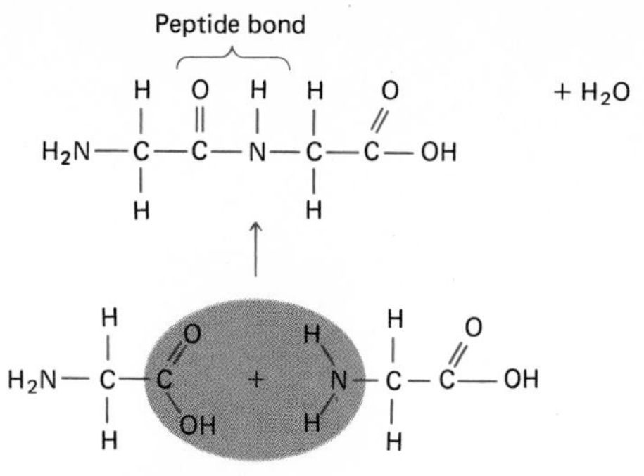

Figure 3.23

The peptide bond links amino acids into polymeric units, binding together the units of proteins.

ical molecules to exhibit directionality is important in their functioning in biological systems.

The **protein molecule** also has a three-dimensional structure that determines its biological properties (Figure 3.25). The sequence of amino acids in the polypeptide forms the primary structure of a protein. It is the number and order of specific amino acids within the peptide chains of the protein—that is, the **primary structure**—that establish the properties of the protein macromolecule. The long polypeptide chain of a protein molecule is twisted, often forming a helix. The specific conformational pattern of a protein, stabilized by hydrogen bonding, is known as the **secondary structure**. Protein molecules also exhibit folding of the polypeptide chains that establishes **tertiary structure**. In globular proteins the polypeptide chains are tightly folded into a tight spherical or globular shape. Folding of the polypeptide molecule is also stabilized by interactions

Figure 3.24

A polypeptide has a free amino end and a free carboxyl end.

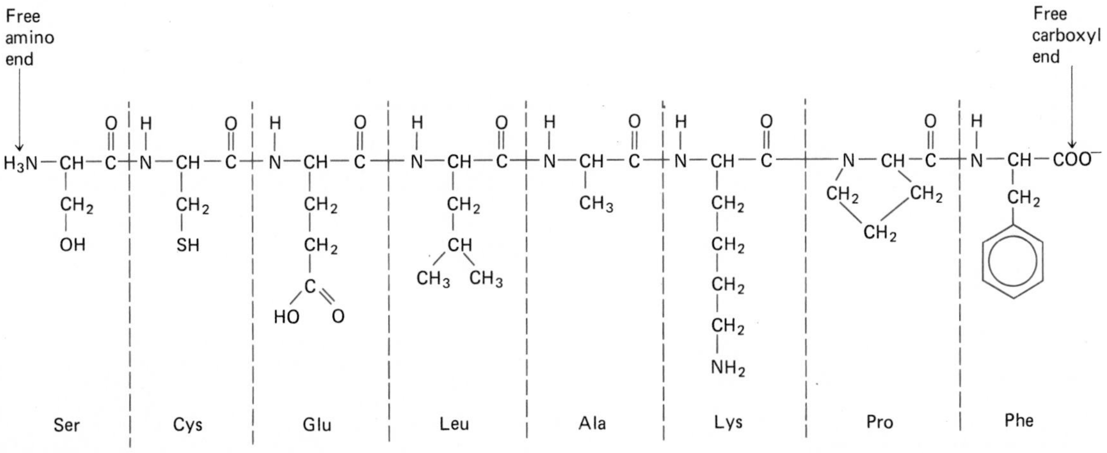

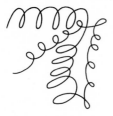

Primary

Secondary

Tertiary

Quaternary

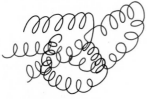

Figure 3.25

*Proteins have primary, secondary, tertiary, and
quaternary structures.*

Figure 3.26

*(A) The immunoglobulin G (IgG) molecule is
composed of four polypeptide chains held together
by disulfide bridges; (B) disulfide bonds also can
establish the structure of a single peptide chain.*

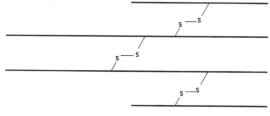

A

B

of the sulfhydryl groups of the sulfur-containing
amino acids. The interactions of sulfhydryl groups
of the amino acid cysteine form disulfide bridges
that contribute to the protein molecule's tertiary
structure. Some proteins are composed of more
than one polypeptide molecule (Figure 3.26). The
quaternary structure of a protein reflects the interactions of two or more polypeptide chains and
describes how the peptide chains are arranged or
clustered in space.

Ultimately it is the primary structure that determines the higher-order structures that describe the three-dimensional nature of the protein molecule. The distinctive characteristics of
the R groups of the amino acids in the polypeptide chain impose constraints on the shape of the
protein molecule. Hydrophobic groups tend to
associate with each other at the interior of folded
chains, and hydrophilic groups tend to assume
positions where they can form weak bonds with
other polar groups. These weak bond interactions
specify the folding patterns of the protein and the
interactions between different polypeptide chains
in determining the configuration of the protein
macromolecule. It is the orientation in space of
the various amino acid groups of a protein, established by the configuration of the protein, that
is essential for that protein to function properly
in a biological system.

**One of the most important functions of pro-
teins is their role as enzymes.** Although not all
proteins are enzymes, **all enzymes are proteins.**
Enzymes are responsible for catalyzing biochemical reactions in living systems. Every microbial
cell must possess many enzymes, thousands in
fact, in order to carry out the essential metabolic
activities involved in its growth and reproduction.
The order of the amino acids at the substrate
binding and active sites of the enzyme are especially critical in determining the specificity of action of that enzyme, that is, which specific biochemical reactions the enzyme is capable of
catalyzing. Enzymes with different three-dimensional conformations at their active sites catalyze
different metabolic reactions. The high specificity
exhibited by enzymes is based on their biochemical composition, which establishes their proper
three-dimensional configuration, and even slight
changes, such as a change in the order of relatively few amino acids in a long polypeptide chain,
can alter the activity of an enzyme.

Active enzyme molecules must have the proper
three-dimensional conformation to exhibit catalytic activity. **Denaturation of the protein molecule**, that is, disruption of the three-dimensional
shape, eliminates the activity of an enzyme mol-

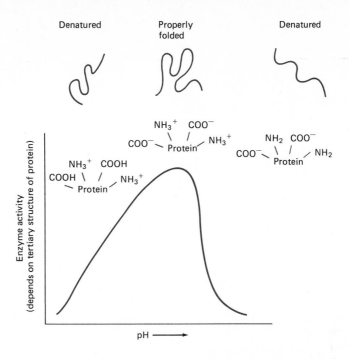

Figure 3.27

The effects of pH on the dissociation of the amino acids in a protein and its effect on tertiary structure is shown in this graph. Each protein is properly folded at a particular pH value.

ecule. Enzymes are globular proteins, and the tight folding pattern of the protein determines the active sites for the catalytic activities demonstrated by these macromolecules. The critical three-dimensional configuration of a protein can be easily disrupted without breaking the primary polypeptide chain. Changes in the ionic strength of a solution, such as occur when salt is added to a solution, alter the weak bond interactions stabilizing the secondary and tertiary structure of the protein. Similarly, changes in pH alter the weak bond interactions that maintain the structure of the protein macromolecule (Figure 3.27). As a result the protein unfolds and loses its biological activity. In most cases such denaturation of proteins is irreversible, but in some cases it has been found that a denatured protein can spontaneously refold, establishing its proper three-dimensional shape and once again its capability of exhibiting biological activity. At high temperatures, changes in the shapes of protein molecules also occur and, hence, enzymes are inactivated at elevated temperatures. We use this fact to heat-kill microorganisms.

Nucleic acids

Nucleic acids, like proteins, are large macromolecules, but instead of being composed of amino acids, the subunits of nucleic acid macromolecules are **nucleotides** linked together to form a polymer. The nucleotide subunits of nucleic acids

are themselves composed of smaller units. A nucleotide is composed of three covalently linked individual units: a nitrogenous base (a heterocyclic ring structure containing nitrogen as well as carbon, sometimes referred to as a nucleic acid base or simply as the base), a five-carbon monosaccharide (either ribose or deoxyribose), and a phosphate group (Figure 3.28). The nitrogenous base is linked to the 1-carbon of the monosac-

Figure 3.28

The structural formula of a nucleotide, the basic unit of the informational macromolecules of living systems.

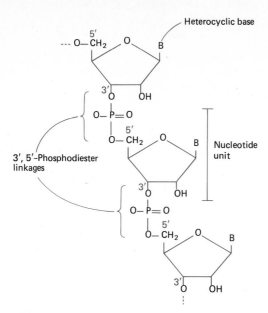

Figure 3.29

Nucleotides are linked by phosphate diester bonds to form dimers and polymeric units.

charide by an *N*-glycosidic bond. The nitrogenous base plus the monosaccharide, without the phosphate group, is called a nucleoside. The phosphate group forms a diester bond between the 3-carbon of one monosaccharide and the 5-carbon of another monosaccharide unit and it is this bond that links the molecule together (Figure 3.29).

There are two types of nucleic acids that occur in biological systems: deoxyribonucleic acid (DNA) and ribonucleic acid (RNA) (Figure 3.30). In most organisms DNA stores the genetic information of the organism. There are several types of RNA molecules that serve different functions, all of which are involved in converting the genetic information into proteins in order to carry out the biological functions of the organism. In DNA the monosaccharide that occurs is deoxyribose; ribose is the monosaccharide found in RNA. **Four nitrogenous bases occur in DNA: adenine, guanine, cytosine, and thymine**. Adenine (A) and guanine (G) are substituted purine bases that are two-ringed structures. Cytosine (C) and thymine (T) are substituted pyrimidine bases and have only one ring. The pyrimidine base **thymine does not occur in RNA**. Rather, another pyrimidine, **uracil (U), occurs in the RNA molecule**. The bases adenine, guanine, and cytosine also occur in RNA. By having thymine in DNA and uracil in RNA, these two macromolecules are differentiated so that they can perform separate functions within the cell; if

they contained the same nucleotides, errors could occur in the storage of genetic information, in the passage of that information from one generation to the next, and in the use of genetic information to specify the functional properties of the organism.

In both RNA and DNA the nucleotides are linked by 3′,5′-phosphate diester linkages (Figure 3.31). Consequently, at one end of the nucleic acid molecule, there is no phosphate diester bond to the 3-carbon of the monosaccharide, and thus there is a free hydroxyl group at the 3-carbon position (**3′-OH free end**); at the other end of the molecule, the 5-carbon is not involved in forming a phosphate diester linkage, and there is a free hydroxyl group at the 5-carbon position (**5′-OH free end**). The fact that the ends of the nucleic acid macromolecule differ is extremely important because this permits directional recognition at the biochemical level in the same sense that we can recognize left and right. This is essential for the nucleic acids to perform their principal function in biological systems, which is to store and transmit the genetic information of the cell.

The DNA molecule stores the genetic information in all biological systems, except for some viruses; this informational macromolecule normally occurs as a double helical molecule (Figure 3.32). The **DNA double helix** is composed of two primary polynucleotide chains held together by **hydrogen bonding between complementary nucleotide bases** (Figure 3.33). Within the double helical DNA, adenine always pairs with thymine, and guanine always pairs with cytosine. There are two hydrogen bonds established between the adenine–thymine base pairs and three hydrogen bonds established between the guanine–cytosine base pairs. The sequence of nucleotides within the DNA molecule codes for the sequence of bases in RNA and ultimately the sequence of amino acids in proteins. Reading the sequence in the appropriate order is essential for converting stored genetic information into the functional activities of the organism.

The directional nature of nucleic acid molecules is critical for establishing the necessary direction of reading the genetic information. Within the double helical DNA macromolecule the two polynucleotide chains run in opposite directions; that is, one chain runs from the 3′-OH to the 5′-OH free end and the complementary chain runs in an antiparallel direction from the 5′-OH to the 3′-OH free end. A consequence of the antiparallel nature of the DNA molecule is that different information is stored in each of the chains. This

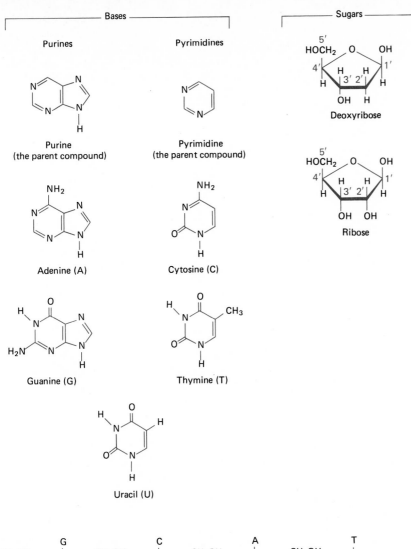

Bases

Purines

Purine
(the parent compound)

Adenine (A)

Guanine (G)

Pyrimidines

Pyrimidine
(the parent compound)

Cytosine (C)

Thymine (T)

Uracil (U)

Sugars

5'
Deoxyribose

5'
Ribose

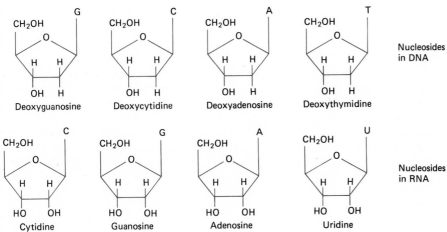

G
Deoxyguanosine

C
Deoxycytidine

A
Deoxyadenosine

T
Deoxythymidine

Nucleosides
in DNA

C
Cytidine

G
Guanosine

A
Adenosine

U
Uridine

Nucleosides
in RNA

Figure 3.30

The structural formulas of bases and sugars in DNA and RNA.

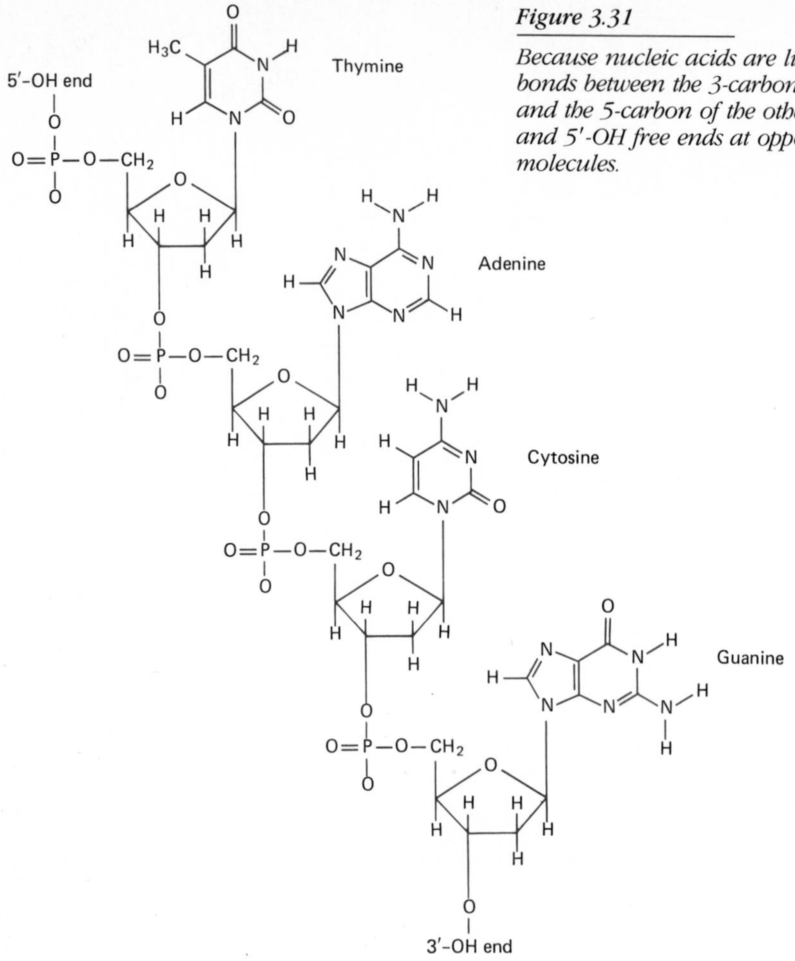

Figure 3.31

Because nucleic acids are linked by phosphodiester bonds between the 3-carbon of one nucleic acid base and the 5-carbon of the other, nucleic acids have 3'-OH and 5'-OH free ends at opposite ends of polymeric molecules.

further means that for a given region of stored information, there must be some mechanism for designating which of the complementary chains is running in the appropriate manner for extracting the appropriate information coding for a particular function. As will be discussed later in the genetics section, there are indeed recognition sequences encoded within the DNA molecule that designate which chain to read, where to begin, and where to end. The genetic code, based on only the few letters (nucleotides) in its "alphabet," provides the necessary biochemical basis for encoding the genetic information of the great diversity of living organisms.

In addition to their involvement in the nucleic acids DNA and RNA, several nucleotides form derivatives that have extremely important functions within microorganisms (Figure 3.34). **Adenosine**

triphosphate (ATP) is a molecule produced in cells, used to store chemical energy until it is needed to drive various energy-requiring biochemical reactions. The production and utilization of ATP is essential to the bioenergetics of the cell, and all the metabolic pathways of microorganisms are involved in either producing or consuming ATP. **Nicotinamide adenine dinucleotide (NAD)** is another nucleotide derivative that plays a critical role in the metabolism of microorganisms. NAD and related compounds act as coenzymes and are involved in oxidation–reduction reactions. Much of the metabolism of microorganisms involves oxidation–reduction reactions requiring the use of coenzymes, accounting for the importance of these nucleotide molecules in biochemical reactions.

Figure 3.32

The double helix is the fundamental structure of DNA in which a cell's genetic information is stored.

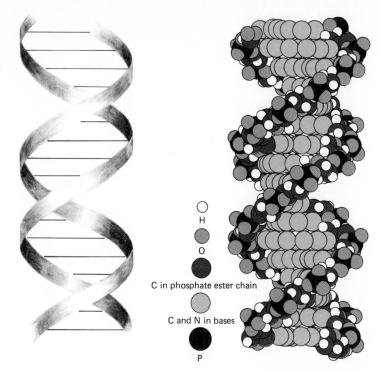

○ H

◐ O

● C in phosphate ester chain

◯ C and N in bases

● P

Figure 3.33

The two strands of DNA are held together by hydrogen bonding between complementary base pairs.

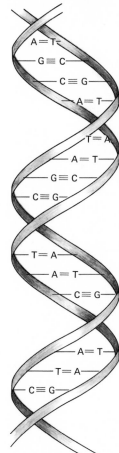

CH₃   O---H—NH

To strand of helix

Thymine (T)   Adenine (A)

T = A

To strand of helix

Cytosine (C)   Guanine (G)

C ≡ G

To strand of helix

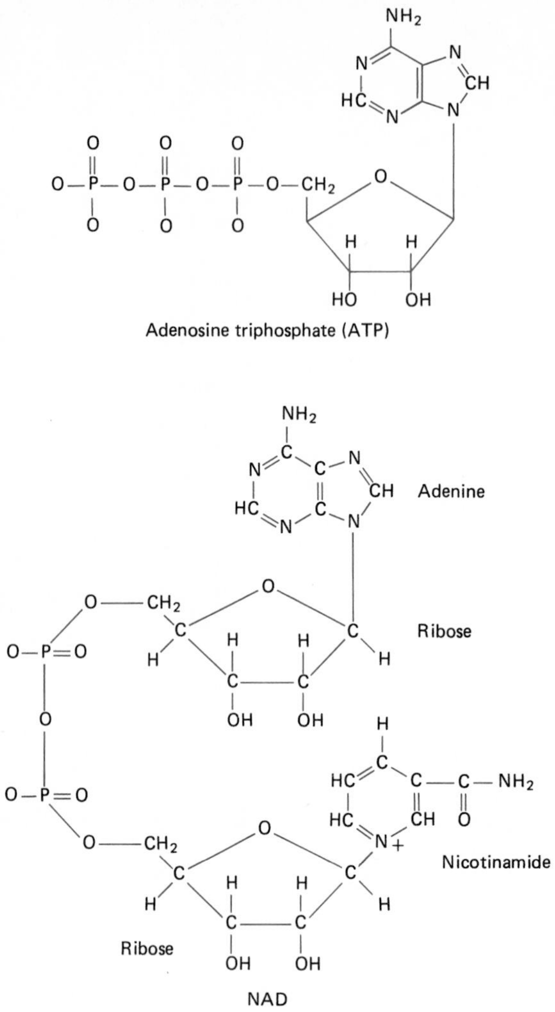

Adenosine triphosphate (ATP)

Adenine

Ribose

Nicotinamide

Ribose

NAD

Figure 3.34

The structural formulas of ATP, the key compound involved in biological energy transfers, and NAD, a key coenzyme involved in hydrogen transfers within a cell.

Discovery process

Concentrations of ATP can be determined by using the firefly assay, so called because the luciferin and luciferase it requires are derived from the tail light assembly of the firefly. Light is produced when reduced luciferin reacts with oxygen to form oxidized luciferin in the presence of luciferase enzyme, magnesium ions, and ATP. The light emitted can be detected and quantitated by a photomultiplier. When ATP is the limiting reactant, the quantity of light is proportional to the concentration of ATP. This assay works regardless of the source of the ATP, demonstrating the versatility of ATP as the universal currency of energy in biological systems.

Biochemical reactions in living systems

Having examined some of the classes of macro-molecules that occur within microorganisms, let us now consider how these molecules react in microorganisms. It is these biochemical reactions that form the basis for the dynamic aspects of living microorganisms.

Bioenergetics

Energy changes and biochemical reactions

Chemical reactions involve the forming and breaking of chemical bonds. In a chemical reaction the bonds holding the atoms of a molecule together may dissociate to form products, and individual molecules may associate to form new bonds. Chemical reactions are governed by the laws of **thermodynamics**, which consider the initial and final state of a system. In biological systems, **thermodynamic principles prescribe the flow of energy through the system**, but they do not consider the rates or mechanisms of the reactions. Biochemical reactions cannot violate the basic laws of thermodynamics. According to the **first law of thermodynamics, energy is conserved**; that is, chemical reactions do not create or destroy energy. It is important to keep this in mind when considering the transformations of chemicals and the flow of energy through a biological system. Whereas energy must be conserved, it can also be transformed, and thus chemical, heat, light, electrical, and mechanical energy can be interchanged. The chemical bonds of a molecule store energy (chemical energy); breaking chemical bonds requires energy; and the formation of new chemical bonds releases energy. In a chemical reaction there will be a net balance among the energy required to break chemical bonds, the energy released by the new bonds that are formed, and the energy—such as heat energy—that is exchanged with the surroundings. The change in the stored energy between the amount contained in the bonds of the reactants and those of the products of a chemical reaction is described by the ΔH (**enthalpy**) of the reaction, the change in heat content of the molecules (Figure 3.35). Reactions that absorb heat have a positive ΔH and are termed **endothermic reactions**; chemical reactions which release heat have a negative ΔH and are termed **exothermic reactions**.

Free energy and biochemical reactions

Of more importance than the change in enthalpy in predicting whether a biological reaction can occur is the change in **free energy**, ΔG, which describes the change in the energy of the system that is available for doing work. The concept of free energy takes into consideration the degree of order as well as the stored energy. The change in free energy of a reaction (ΔG) relates to the change of the heat of reaction (ΔH), the temperature (T, absolute temperature in degrees Kelvin), and the change in the state of order or **entropy** (ΔS) between products and reactants:

$$\Delta G = \Delta H - T\Delta S.$$

According to the **second law of thermodynamics,**

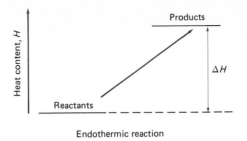

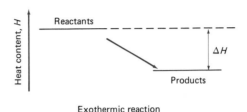

Figure 3.35

Energy diagrams showing ΔH *for exothermic and endothermic reactions.*

all processes proceed in the direction that increases the total entropy of the system and the surroundings, that is, in the direction of maximum randomness or disorder. Order can increase within a biological system only if disorder increases elsewhere in the universe. Whether a chemical reaction occurs depends on the relative state of order of the products and reactants as well as on the energy stored within the reactant and product molecules. Reactions that release free energy have a negative ΔG and are termed **exergonic**; reactions that require the addition of free energy from another source have a positive ΔG and are termed **endergonic**. Viewed in another way, some chemical reactions require free energy to drive them uphill (endergonic reactions), and other chemical reactions release free energy as they run downhill (exergonic reactions) (Figure 3.36). Some chemical reactions are driven primarily by enthalpy (ΔH), and others are driven by entropy (ΔS). When ΔG is negative, reactions occur spontaneously, the reaction favors the formation of products, and there is a loss of free energy.

If a chemical reaction is allowed to proceed to completion, it will reach a point of **equilibrium** where there is no net change in the concentrations of products and reactants. Each reaction attempts to reach equilibrium, and therefore, the net flow of the reaction shifts, either toward the products or toward the reactants, depending on the concentrations of the products and the reac-

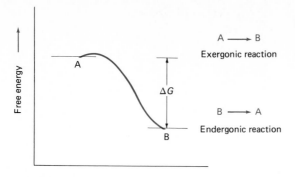

Figure 3.36

This graph shows the changes in free energy during exergonic and endergonic reactions.

tants and on the temperature. Adding or removing reactants or products alters the direction of the reaction.

The **equilibrium constant** (K_{eq}) at a given temperature is defined as the product of the concentrations of the molecules formed in the reaction divided by the product of the concentrations of the reactants. For the reaction $A + B \rightleftharpoons C + D$ the equilibrium constant is given by the equation

$$K_{eq} = \frac{[C][D]}{[A][B]}.$$

At equilibrium the free energy will be at a minimum, and no further change occurs. Regardless of the actual ΔG of a reaction, it is possible to calculate a standard free energy (ΔG^0) for a reaction. The ΔG^0 is a thermodynamic constant for a given chemical reaction. At equilibrium the ΔG^0 of a reaction is given by the equation

$$\Delta G^0 = -RT \ln K_{eq}$$

where R is the gas constant (1.99 cal/mole/deg), T is the absolute temperature, and $\ln K_{eq}$ is the natural logarithm of the equilibrium constant. Stated another way, the standard free energy of a reaction is the difference between the standard free energies of the reactants and the standard free energies of the products. Thus, for any given biochemical reaction the ΔG^0 of the reaction indicates whether the formation of products or reactants is favored.

ATP and bioenergetics of biochemical reactions

The ΔG^0 for an overall reaction will be the same regardless of the number of steps required to go from reactants to products. Consequently, it is possible to couple individual reactions to achieve a favorable overall ΔG^0 that will allow the com-

plete process to occur. **Endergonic and exergonic reactions can be coupled** so that the energy released by the exergonic reaction is used to drive the endergonic reaction. In particular, the formation of ATP, an endergonic reaction, can be driven by various exergonic metabolic reactions, and the hydrolysis of ATP can be coupled with endergonic reactions, allowing such reactions to proceed in biological systems. ATP is called an energy-rich or high-energy compound because it and those like it contain a chemical bond that yields a large release of energy upon hydrolysis. **The hydrolysis of ATP to form ADP (adenosine-diphosphate) and inorganic phosphate (P_i) releases approximately 7.3 kcal (kilocalories) per mole of ATP hydrolyzed** (Figure 3.37).

The same amount of energy (7.3 kcal) is required for the formation of ATP from $ADP + P_i$. Many of the metabolic reactions of microorganisms that break down organic compounds are involved in generating ATP. ATP can also be generated by some microorganisms by using light energy to drive the reaction of $ADP + P_i$ to form the product ATP. Once formed, energy can be stored in ATP and moved through the microbial cell. Energy can then be used to drive energy-requiring reactions, including those involved in biosynthesis. ATP serves almost universally as the energy source for driving energy-requiring reactions in biological systems; most endergonic biochemical reactions are coupled with the exergonic hydrolysis of ATP. As such, ATP can be termed the "universal currency" of energy in biological systems, although some endergonic biochemical reactions specifically require coupling with other exergonic reactions. The usefulness of ATP as a currency of energy within cells rests with the fact that its hydrolysis yields enough free energy to drive other biochemical reactions, yet its formation does not require so much energy that sufficient amounts of ATP could not be synthesized to meet the energy requirements of living systems.

Reaction kinetics and enzymes

Although the ΔG^0 of a reaction describes whether a reaction can occur spontaneously, it does not describe the rate of the chemical reaction. The **kinetics** of a chemical reaction depends on the temperature and the relative concentrations of reactants and products. As a rule, the more collisions between reactant molecules, the greater the rate of product formation. Because at elevated

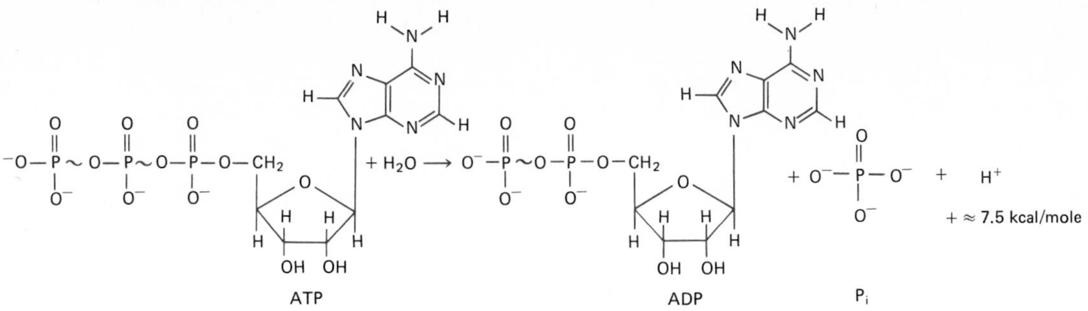

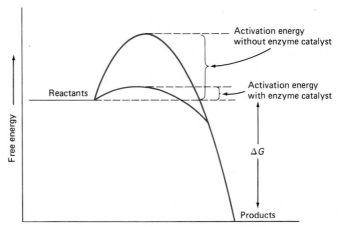

Figure 3.37

The hydrolysis of ATP to ADP is used to drive endergonic reactions.

Note: the chemical equation shows ATP + H₂O → ADP + Pᵢ + ≈ 7.5 kcal/mole, with labels ATP, ADP, and Pᵢ.

temperatures there is increased motion of molecules, reaction rates can be increased by heating. However, as discussed earlier, the macromolecules of microorganisms are disrupted at elevated temperatures and therefore, while chemical reactions can be greatly accelerated in a test tube using a bunsen burner, they cannot similarly be increased in biological systems by increasing the temperature of the reaction. Reaction rates also can be speeded up by increasing the concentrations of reactants. According to the law of mass action, all other things being equal, the rate of reaction is proportional to the concentrations of the reactants, and therefore, high relative concentrations of reactants can drive the rapid formation of products as the reaction moves toward a state of equilibrium. Although changes in concentrations of reactants do affect the rates of chemical reactions in living systems, such changes do not explain the high rates of biochemical reactions, whose rates are higher than can be accounted for by simply increasing the concentrations of reactants.

Enzymes

The rapid rates of biochemical reactions in living systems are possible because of the role of enzymes as biological catalysts. For a chemical reaction to occur, it must achieve the energy of activation, the amount of energy required in a collision between two molecules to bring about the reaction (Figure 3.38). The **energy of activation** must be overcome even if the ΔG^0 of the reaction is negative; that is, the energy must be elevated to a higher-energy transition state before it can start running downhill. At the relatively low temperatures of biological systems few collisions between molecules have the energy required to overcome the energy of activation and to react; therefore, biological reactions are dependent on catalysts. A catalyst is a substance that lowers the energy of activation required to bring about a reaction, and therefore, at a given temperature a catalyzed reaction proceeds more rapidly than an uncatalyzed reaction.

Enzymes are biological catalysts that lower the energy of activation for biochemical reactions and

Figure 3.38

This graph illustrates the energy of activation required for a chemical reaction to start and that an enzyme effectively lowers this energy of activation.

(Graph: y-axis labeled "Free energy"; curves from "Reactants" to "Products" with labels "Activation energy without enzyme catalyst", "Activation energy with enzyme catalyst", and "ΔG".)

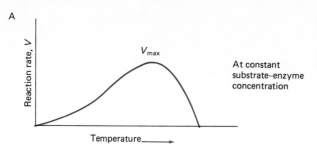

A

At constant
substrate–enzyme
concentration

B

At constant
temperature

Figure 3.39

These graphs show the effect of temperature (A) and substrate concentration (B) on the rate of an enzymatic reaction.

permit essential metabolic steps to proceed rapidly even at the relatively low temperatures at which living systems exist. Enzymes combine with the substrate molecules to form an enzyme–substrate complex that is later broken to form products and return the enzyme to its original state. This sequence of reactions regenerates the enzyme molecules, and therefore, the enzyme is not consumed in the overall reaction of substrate to products and can be used over and over. The rate of an enzymatically catalyzed reaction is dependent on temperature, the concentration of en-

Figure 3.40

This Lineweaver Burk plot of enzyme kinetics shows the reciprocals of velocity and substrate concentration from which the values of K_m and V_{max} are obtainable.

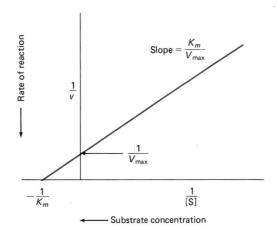

Slope $= \dfrac{K_m}{V_{max}}$

$\dfrac{1}{v}$

$\dfrac{1}{V_{max}}$

$-\dfrac{1}{K_m}$

$\dfrac{1}{[S]}$

Rate of reaction

Substrate concentration

zyme, the concentration of substrate, and the affinity of the enzyme for the substrate (Figure 3.39). The kinetics of enzymatic reactions exhibit a phenomenon of **saturation**, in which raising the concentration of substrate does not continue to increase the rate of the reaction. The maximal rate of the reaction is termed V_{max}, and the substrate concentration at $\frac{1}{2}V_{max}$ is termed the K_m. The K_m is a measure of the affinity of the enzyme for the particular substrate. The **Michaelis-Menten equation** describes the relation of V_{max} and K_m and the kinetics of enzyme reactions:

$$\frac{1}{v} = \frac{K_m}{V_{max}} \times \frac{1}{[S]} + \frac{1}{V_{max}}$$

where [S] is the substrate concentration and v is the velocity (rate) of reaction. Fortunately, this relationship can be described graphically, permitting the determination of V_{max} and K_m by determining the rates of reaction at a limited number of low substrate concentrations (Figure 3.40).

The degree of **substrate specificity** exhibited by enzymes reflects the fact that the enzyme and the substrate must fit together in a specific way for the enzyme to lower the activation energy. The fit has been likened to a lock and key, although the binding of the substrate molecule to the enzyme may alter the three-dimensional configuration of the enzyme, inducing the proper fit (Figure 3.41). The enzyme has a specific active site at which the substrate binds. When all the active sites of an organism's enzyme molecules are occupied, saturation occurs and the reaction proceeds at the maximal rate. The binding of the enzyme to the substrate involves the formation of weak bonds that are sufficient to place a strain on the substrate molecule, lowering the activation energy and permitting the reaction to proceed. In the case of multiple reactive molecules, which bind at the active sites of an enzyme, the enzyme effectively positions the reactants in space so that they are brought together with the right orientation to bring about the biochemical reaction. The precision of fit and structural complementarity between enzyme and substrate molecule permits the establishment of exactly the right spatial orientation so that the numerous biochemical reactions of a microorganism can occur with great rapidity.

The specificity of the configuration needed to form the enzyme–substrate complex, however, renders enzymes quite susceptible to inhibition. The activities of an enzyme can be completely inhibited by denaturing the enzyme, that is, by radically distorting its three-dimensional shape, such as occurs at high temperatures. The activities of an enzyme can also be inhibited by the binding of substances other than the specified substrate to the enzyme (Figure 3.42). If an inhibitory substance exhibits an affinity for the active site of the enzyme because of its similarity to the normal substrate, it can act as a **competitive inhibitor**, competing with the specified substrate for the active site of the enzyme. A competitive inhibitor does not lower the V_{max} of the reaction, but the apparent K_m of the reaction increases. **Competitive inhibition** can be overcome by increasing the substrate concentration. In contrast, increasing substrate concentration does not overcome **noncompetitive inhibition**. **Noncompetitive inhibitors** do not compete with the specified substrate, but they may bind irreversibly at the active site or may distort the enzyme by binding at a site on the enzyme other than the active site, modifying the enzyme molecule so that the V_{max} is decreased.

The ability to modify the three-dimensional shape of the enzyme molecule provides a basis for microorganisms to regulate the rates of their enzymatic activities. Substances may bind to the enzyme at sites removed from the active site and still alter the properties of the enzyme molecule. Additional binding sites of enzyme molecules that are involved in regulating the activities of enzymes are called allosteric effector sites (Figure 3.43). **Allosteric effector sites** may be involved either in the inhibition or the activation of an enzyme. Just as with the reaction between enzyme and substrate, the affinity of a allosteric inhibitor for an allosteric effector site normally shows

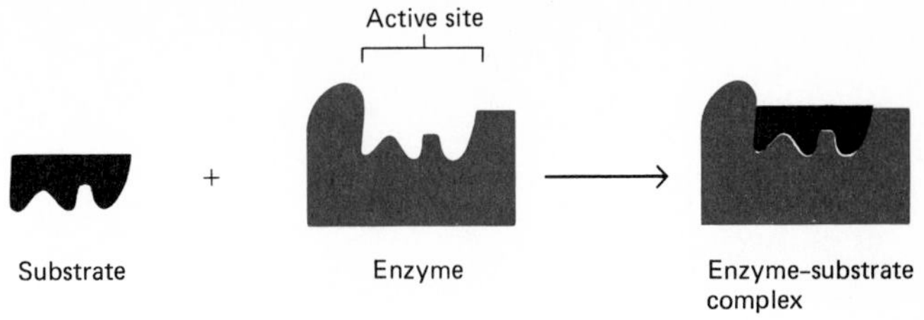

Active site

Substrate + Enzyme → Enzyme–substrate complex

Figure 3.41

The fit between enzyme and substrate has been likened to a lock and key. Actually, the interaction of substrate with enzyme modifies the three-dimensional structure of the enzyme. The precision of fit is responsible for the high degree of specificity of enzymes for particular substrates.

a great deal of specificity. In some cases, the end product of an enzyme reaction sequence may act as an **allosteric inhibitor** by reversibly binding with an enzyme within that sequence. Such a system is self-regulating because there is a feedback mechanism through which the reaction is shut off when excessive product is produced. This type of regulation by allosteric inhibition is called **feedback** or **end product inhibition**. In other cases, the reversible binding of a substance to an allosteric site may activate the enzyme, increasing its activity (**feedback activation**). Both feedback inhibition and feedback activation are important processes that regulate the activities of enzymes and thus the rates of metabolic reactions carried out by microorganisms.

Cofactors and enzymatic reactions

Many enzymatically catalyzed reactions require **cofactors** that are involved in the reaction, often donating or accepting a chemical moiety. Often the cofactor or coenzyme is a vitamin. **Coenzymes** frequently accept a chemical moiety produced in one enzymatic reaction, hold that moiety for a short period of time, and transfer it to another biochemical reaction in which they donate that moiety. Coenzyme A (CoA), for example, is a universal carrier of acyl groups, and many biochemical reactions involve the transfer of a two carbon acetyl group from acetyl CoA. Derivatives of vitamin B_6 (pyridoxine) are involved in transamination reactions, which are reactions in which an amino group is transferred from a given amino acid to α-keto acid to form another amino acid. In these reactions the vitamin-derived cofactor temporarily holds the amino group during the

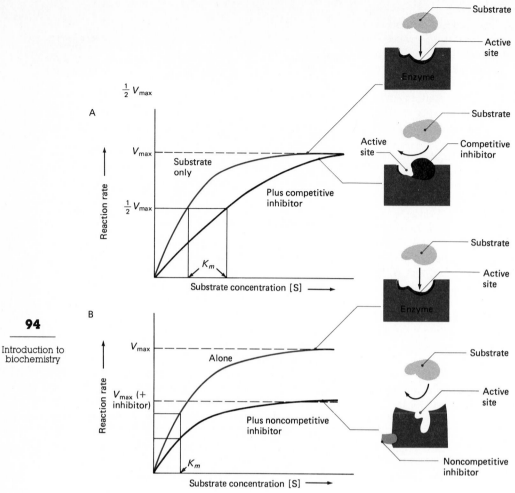

Figure 3.42

Enzymes are subject to various types of inhibition. (A) Competitive inhibition occurs when an inhibitory substance competes with the natural substrate for the active and/or binding sites. (B) Noncompetitive inhibition occurs when an inhibitory substance reacts with the enzyme at a site other than the active site, decreasing the activity of the enzyme.

transfer process, and the coenzymes act as if they were a substrate of the enzyme.

Coenzymes also are important in biochemical oxidation–reduction reactions. Oxidation represents a loss of electrons, whereas reduction is the gain of electrons. Oxidation–reduction reactions always occur together and electrons gained by an oxidizing agent balance those lost by a reducing agent. Metabolic reactions in which a substrate is oxidized must be accompanied by a metabolic reaction in which a substrate is reduced, and thus, the transfer of electrons during metabolic reaction sequences must be balanced. In many enzymatically catalyzed reactions where a substrate is oxidized, the electron is transferred to a coen-

zyme, an organic nonprotein molecule, and the coenzyme becomes reduced in this process. An example of such reaction is the conversion of the coenzyme NAD to its reduced form NADH (Figure 3.44). The reverse is true for an enzymatic reaction in which the substrate is reduced. In such reactions a reduced coenzyme can donate an electron, thus becoming oxidized, such as occurs when the reduced coenzyme NADPH (nicotinamide adenine dinucleotide phosphate) is oxidized to NADP in a reaction that is coupled with the reduction of a substrate. Biosynthetic reactions involve a net conversion of relatively oxidized carbon-containing molecules, such as carbon dioxide, to relatively reduced carbon-

Figure 3.43

The activities of enzymes can be increased or decreased by the binding of substances other than the substrate to allosteric effector sites.

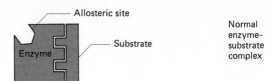

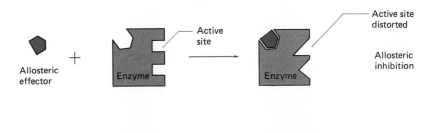

Figure 3.44

The reduction of NAD^+ to $NADH + H^+$ is a critical reaction that often is coupled with the oxidation of substrates within a cell. This reaction can be written several ways; the abbreviated form $NAD \rightarrow NADH$ is employed throughout this book.

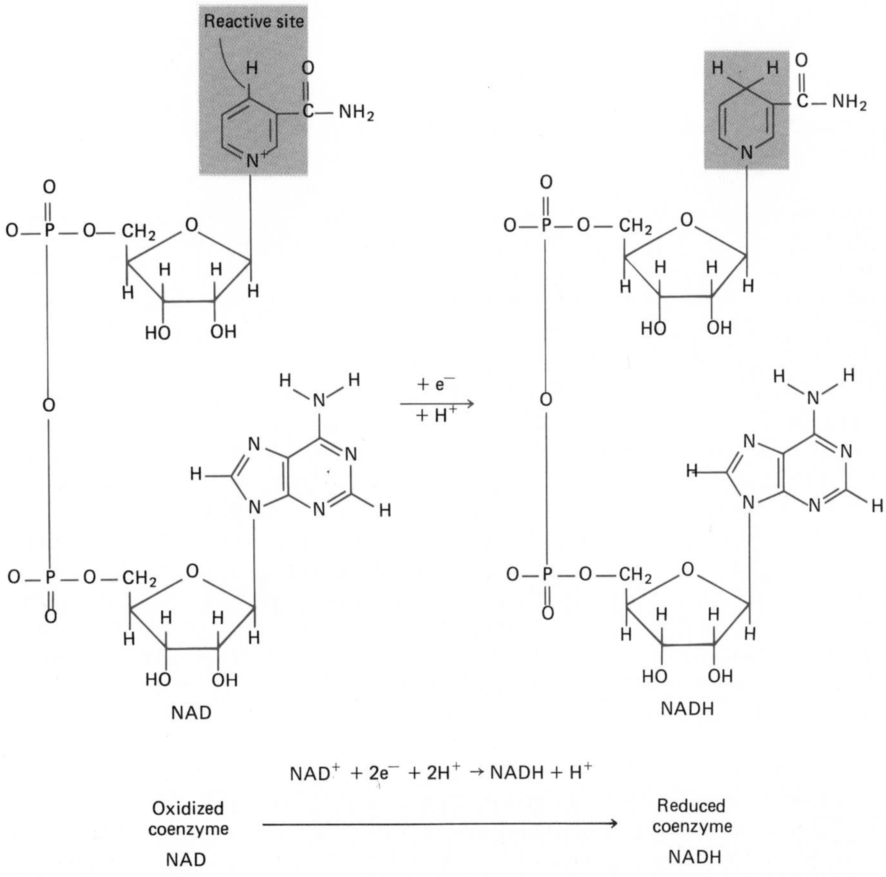

$$NAD^+ + 2e^- + 2H^+ \rightarrow NADH + H^+$$

Oxidized coenzyme		Reduced coenzyme
NAD		NADH

containing molecules, such as carbohydrates ($C_nH_{2n}O_n$). In order to carry out biosynthetic reactions, a cell must generate reduced coenzymes that can be coupled with the reductive biosynthetic reactions. The ability of microorganisms to grow and reproduce depends on the generation of reducing power (reduced coenzymes) and the use of these compounds in microbial biosynthesis.

This introduction to biochemistry has emphasized some of the important properties of the molecules on which life is based. The biochemistry of microorganisms is fundamental to their structure and function. In subsequent chapters we shall see that the properties of microorganisms, what they look like, what they do, and where they live are a direct reflection of their biochemical composition. The carbon-based macromolecules of living systems, with their abilities to form both strong covalent bonds and weak bonds, provide both the strength (stability) and flexibility (dynamics) needed by living microorganisms. The three-dimensional configurations of macromolecules, determined both by their primary structure (covalent bonds) and by higher-order interactions (weak bonds), are especially important in establishing the specificity of roles played by different biochemical molecules in microorganisms. The importance of the three-dimensional configuration of biochemicals can be seen in the case of enzymes, where any distortion of the configuration results in a change in the activity of the enzyme molecule, and in the DNA, which owes its stability to its three-dimensional double helix configuration.

Looking ahead to our examination of the role of biochemistry in developing a fundamental understanding of microbiology:

In Chapter 4, when we examine the functional anatomy of microorganisms, we will see that each structure has a unique biochemical composition. The ability of each structural unit of a microorganism to perform its proper function depends on putting the right biochemicals together in the right place. The cytoplasmic membrane of a microorganism, for example, is composed largely of phospholipid, which, as we have discussed, possesses the necessary hydrophilic and hydrophobic groups to orient the membrane so that it can act as a semipermeable barrier and regulate the flow of material into and out of the microorganism.

In Chapters 5 and 6 we will examine the metabolism of living cells, examining how they transform substrates into the macromolecules composing the organism. Because microorganisms require energy to maintain their structural organization and carry out metabolism needed for growth and reproduction, the synthesis and use of ATP, which we have termed the energy currency of the cell, will occupy a central place in our discussion of microbial metabolism. Many of the specific metabolic pathways that we will examine center around different strategies that microorganisms have evolved for generating ATP and reduced coenzymes that they need for the biosynthesis of new microbial biomass. Throughout the consideration of microbial metabolism, we will keep in mind the importance of enzymes in allowing metabolic reactions to occur. By lowering the activation energies of biochemical reactions, enzymes permit biochemical reactions to occur rapidly at temperatures where living systems can maintain their structural integrity and organization.

In Chapters 7 and 8 we will examine how the genetic information of microorganisms is encoded in nucleic acid molecules and how the information stored in the DNA molecule directs and controls the synthesis of enzymes. In this discussion we will consider the mechanisms for maintaining the integrity of the DNA molecule, which is necessary for the fidelity of hereditary processes, and the mechanisms that establish variability, which is a necessary prerequisite for evolutionary separation of organisms.

The brief introduction to biochemistry given in this chapter should provide the background necessary to understand the nature of the macromolecules of which microorganisms are made and the biochemical reactions carried out in the metabolism of microorganisms, providing the basis for developing an appreciation of the basic structure and function of microorganisms examined in these subsequent chapters.

1. What types of chemical bonds occur in macromolecules? What are the relative strengths of each?

2. What is a the difference between a D and an L form of a molecule, and what is the significance of this distinction in biological systems?

3. What classes of macromolecules comprise microorganisms?

4. What is a protein? What is meant by the primary, secondary, tertiary, and quartenary structure of a protein?

5. What is an enzyme? Why are enzymes essential for microbial metabolism?

6. Why is the configuration of an enzyme critical to its function? What is meant by denaturation? How can proteins be denatured?

7. What are the similarities and differences between DNA and RNA?

8. What is the central role of ATP in microbial metabolism? What properties of this molecule make it well suited as an energy carrier in the cell?

9. What is an oxidation–reduction reaction? What role does the coenzyme NAD play in biological oxidation–reduction reactions?

10. What effect do the following have on the rates of an enzymatic reaction?
   a. Raising the temperature from 10° to 20°C
   b. Raising the temperature from 90° to 100°C
   c. Increasing the substrate concentration
   d. Adding a noncompetitive inhibitor

Davidson, E. A. 1967. *Carbohydrate Chemistry*. Holt, Rinehart and Winston, New York.

Davidson, J. N. 1972. *The Biochemistry of Nucleic Acids*. Academic Press, New York.

Dickerson, R. E., and I. Geis. 1969. *The Structure and Action of Proteins*. Benjamin/Cummings Publishing Co., Menlo Park, California.

Fasman, G. D. (ed.). 1976. *Handbook of Biochemistry and Molecular Biology*. CRC Press, Inc., Boca Raton, Florida.

Koshland, D. E. 1973. Protein shape and biological control. *Scientific American* 229(4): 52–64.

Lehninger, A. L. 1975. *Biochemistry*. Worth Publishers, Inc., New York.

Lehninger, A. L. 1982. *Principles of Biochemistry*. Worth Publishers, Inc., New York.

Stryer, L. 1981. *Biochemistry*. W. H. Freeman and Co., San Francisco.

Watson, J. D. 1976. *Molecular Biology of the Gene*. Benjamin/Cummings Publishing Co., Menlo Park, California.

West, E. S., W. R. Todd, H. S. Mason, and J. T. VanBruggen. 1966. *Textbook of Biochemistry*. Macmillan Publishing Co., Inc., New York.

Whittaker, R. H. 1969. New concepts of kingdoms of organisms. *Science* 163: 150–160.

Widom, J. M., and S. J. Edelstein. 1981. *Chemistry*. W. H. Freeman and Co., San Francisco.

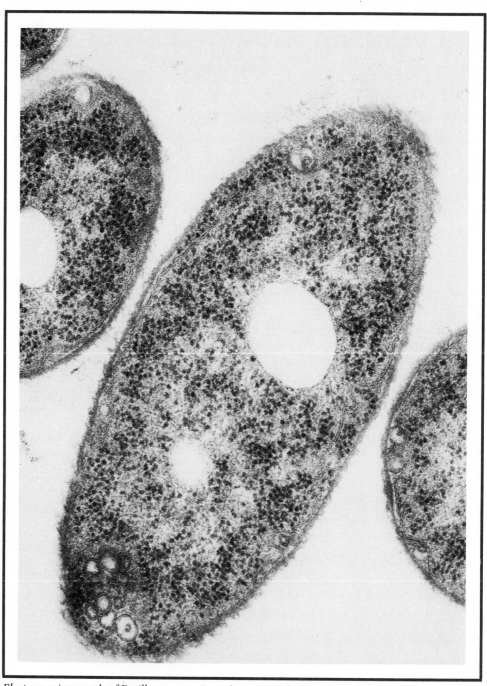

Electron micrograph of Bacillus megaterium. *(Courtesy John Hanson, Ohio State University.)*

Microbial anatomy

SECTION TWO

chapter 4
Organization and structure
of microorganisms

Organization and structure
of microorganisms

Organization of living systems

A fundamental characteristic of living systems is that they are highly organized. **The cell is the basic unit of organization of biological organisms** and provides the spatial framework within which the biochemical reactions of the living system can occur. A cell is separated from its surroundings by a membrane that acts as a boundary layer, allowing the organism to maintain its integrity and identity. The specific organizational pattern of the cell, including its metabolic capabilities and structural components, is determined by the genetic information contained within that cell. This maintenance of cellular organization requires energy. When an organism no longer is able to sustain its cellular organization, it dies and the organism and its cellular components become disorganized.

Another characteristic of living systems is the ability to reproduce. **Cells have the capacity to reproduce their own organizational pattern independently.** Cells come from the reproduction of preexisting cells. Many microorganisms are unicellular and, therefore, the reproduction of a single cell is synonymous with the reproduction of the entire organism. Other organisms are multicellular, consisting of aggregates of cells that are themselves organized. In some multicellular organisms, the cells are differentiated into specialized organizational units of tissues and organs. Although many microorganisms form multicellular aggregations, they lack the high degree of specialized organization of cells into differentiated tissues exhibited by plants and animals. To achieve higher levels of organization, cells exhibit differentiation, whereby different cells of the same organism can express different genetic information and hence perform specialized functions.

Compartmentalization of biological systems permits the establishment of an effective surface area-to-volume ratio. Cells depend on their ability to exchange materials with their surroundings in order to meet their physiological needs. If a cell is too large, it does not possess sufficient surface area to permit adequate exchange across its limiting boundary to furnish its required nutrients and remove its waste products. Even within the cell there is a separation of function. Compartmentalization within cells permits the concentration of particular biochemicals in a specific cellular region where specialized biochemical reactions can occur. In some cases the cell is protected by such division of cellular function; for example, confining certain enzymes within membrane-bound organelles protects the rest of the cell from digestion by these enzymes. A high degree of internal organization enables cells to achieve larger sizes than could otherwise be achieved and still maintain their required functions.

There are two basic types of cells: prokaryotic and eukaryotic (Figure 4.1). Eukaryotic cells are

Figure 4.1

A comparison of structural organization reveals that the eukaryotic cell has far more internal organization than the prokaryotic cell; many of the organelles found in eukaryotic cells do not occur in prokaryotic cells.

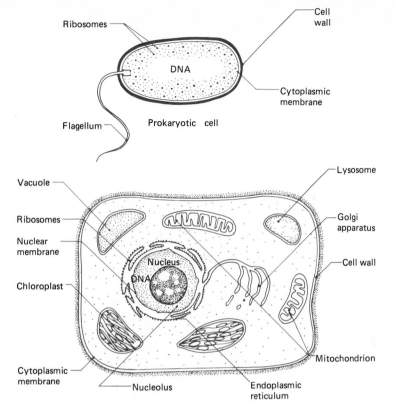

Prokaryotic cell

Eukaryotic cell

characterized by a greater degree of internal structural organization than is found within prokaryotic cells. The larger size of eukaryotic cells, compared to prokaryotic cells, can be attributed to the ability to form and manage subsystems within eukaryotic cells. Eukaryotic cells contain membrane-bound organelles that effectively compartmentalize the specialized cellular functions. Of prime importance is the separation of the genome, containing the cell's essential genetic information, from the rest of the cellular components in eukaryotic cells. The primary genetic information of a eukaryotic cell is contained within a membrane-bound organelle, the nucleus, and is separated from the rest of the cell by a membrane barrier. The DNA in prokaryotic cells is not separated by a membrane barrier from the rest of the cell constituents, as is the DNA within the nucleus of eukaryotic cells. By definition, **all eukaryotic cells have a nucleus, and prokaryotic cells never possess this organelle**. Additionally, there are several other cellular functions, including many of the metabolic reactions involved in the generation of ATP, that occur within membrane-bound organelles in eukaryotic cells and are not so separated within prokaryotic cells. These differences between prokaryotic and eukaryotic cells represent a fundamental division of living systems. Presumably, as cells evolved into more advanced forms, greater separation of function was needed to efficiently carry out the functions of the cell.

Microorganisms can be classified based on their cellular organization (Table 4.1). **Viruses and viroids are acellular** and are not separated from their surroundings by a membrane. The lack of a cellular membrane precludes the ability of the virus or viroid to carry out organized biochemical reactions outside a host cell. As such, viruses and viroids are obligate intracellular parasites and are only capable of acting as living systems within the cells of suitable host organisms. Within the confines of a host cell there is a boundary membrane, provided by the host cell, that permits the genetic material of the virus to program the reproduction of its own macromolecules and structures. Outside such host cells, viruses and viroids are incapable of reproduction and act as nonliving entities. Because viruses are so intimately tied to host cells, they can be thought of as extracellular genetic elements or degenerate portions of the cells that act as hosts for viral reproduction.

table 4.1

Organizational structure of the major groups of microorganisms

Microbial group	Structural organization	Essential macromolecules
Viroids	Acellular—no protective coat around nucleic acid	RNA
Viruses	Acellular—protein coat surrounds nucleic acid	DNA or RNA, protein
Bacteria	Prokaryotic cell	Nucleic acids (DNA + RNA), protein, lipid, carbohydrate
Fungi	Eukaryotic cell	Nucleic acids (DNA + RNA), protein, lipid, carbohydrate
Algae	Eukaryotic cell	Nucleic acids (DNA + RNA), protein, lipid, carbohydrate
Protozoa	Eukaryotic cell	Nucleic acids (DNA + RNA), protein, lipid, carbohydrate

In contrast to the viruses, all other organisms are composed of either prokaryotic or eukaryotic cells. Only the bacteria have prokaryotic cells, and **all organisms possessing prokaryotic cells are classified as bacteria. Plants, animals, protozoa, algae, and fungi all have eukaryotic cells.** The difference between acellular, prokaryotic, and eukaryotic organisms has great practical importance as well as great fundamental interest. For example, the ability to use antibiotics to treat the bacterial diseases of humans is dependent on the ability to target the mode of action of the antibiotic selectively against prokaryotic cells (bacteria) without killing eukaryotic human cells at the same time; it is more difficult to target antibiotics against disease-causing fungi and protozoa, because these organisms, like humans, have eukaryotic cells. Similarly, because viruses reproduce only within host cells, it has been much more difficult to find substances that will selectively inhibit viruses, as those chemicals that inhibit viral reproduction generally do so by killing the host cells, eliminating their therapeutic value. The fundamental differences in cellular organization mean that eukaryotic and prokaryotic microorganisms must possess different structures and strategies for carrying out essentially the same physiological and reproductive functions of generating ATP, synthesizing macromolecules and assembling them into cellular components, replicating the genetic information, and passing the information and organizational pattern to the next generation.

Cytoplasmic membrane

The **cytoplasmic membrane**, the boundary layer of the cell, is a **differentially permeable barrier**. The movement of molecules across the cytoplasmic membrane is selectively restricted; that is, transport of materials into and out of the cell is regulated by the cytoplasmic membrane. Small molecules such as water move across the membrane quite readily; larger molecules such as glucose can be translocated across the membrane; and very large molecules such as cellulose cannot pass through the membrane barrier. The membranes themselves mediate the selective transport process, and the biochemical structure of the membrane determines which molecules can enter and leave the cell.

The structure of the cell membrane

The cytoplasmic membrane is composed largely of **phospholipid**. The fact that **phospholipid molecules have polar (hydrophilic) and nonpolar (hydrophobic) ends** establishes the basis for the orientation of the membrane around the cell. The normal cytoplasmic membrane has a bilipid structure; that is, there are two layers of phospholipid (Figure 4.2). The hydrophobic ends of the phospholipids orient toward each other and form the internal matrix of the membrane; the hydrophilic ends point away from each other, with one layer of polar ends pointing away from the cell and the other layer of polar ends pointing to

Figure 4.2

The typical cytoplasmic membrane of both prokaryotic and eukaryotic cells is a bilipid membrane, as illustrated here showing the orientations of the hydrophilic and hydrophobic ends of the phospholipids that make up this structure.

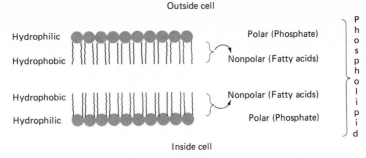

Figure 4.3

The railroad track-like appearance of the cytoplasmic membrane of Bacillus subtilus *is seen in this electron micrograph; bar = 100 nm, W = cell wall outside of membrane, CM = cytoplasmic membrane. (From BPS—T. J. Beveridge, University of Guelph.)*

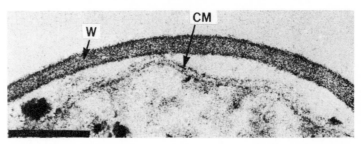

the cell's cytoplasm—the fluid contents within the cell. When viewed with the transmission electron microscope, the cytoplasmic membrane has a railroad track appearance; the dark rail-like portions of the membrane correspond to the electron-dense hydrophobic portions of the phospholipid molecule (Figure 4.3). The specific biochemical structures of a phospholipid molecule differ between organisms, establishing variations in the properties of the cytoplasmic membranes between organisms. Typically, the hydrophobic portions of the molecule consist of fatty acids containing 16–18 carbon atoms. There are some differences in lipid composition between prokaryotic and eukaryotic cytoplasmic membranes (Table 4.2). Sterols, such as cholesterol, occur in eukaryotic cell membranes, but generally are absent from those of prokaryotic cells. As a result, the cytoplasmic membrane of eukaryotic cells is somewhat more rigid than the cytoplasmic membrane of the prokaryotic cell. In addition to the lipid constituents of the cytoplasmic membrane, there are other biochemicals, including proteins, that are integrated into or associated with the basic membrane structure. The protein molecules establish pores or channels through the membrane. The R groups of the amino acids in the proteins that form these channels give the pores some degree of selectivity, and even molecules small enough to fit through the pores can be excluded by their biochemical interactions with the protein substituents. The distribution of proteins establishes a definite "sidedness" to the membrane so that the membrane has a distin-

table 4.2

Differences in membrane lipid composition between prokaryotic and eukaryotic cells

Lipid	Eukaryotic cell	Prokaryotic cell	
	Human erythrocyte	*Escherichia coli*	*Bacillus megaterium*
Cholesterol	25	0	0
Phosphatidyl ethanolamine	20	100	45
Phosphatidyl serine	11	0	0
Phosphatidyl choline	23	0	0
Phosphatidyl inositol	2	0	0
Phosphatidyl glycerol	0	0	45
Sphingomyelin	18	0	0
Lysyl phosphatidyl glycerol	0	0	10
Other	2	0	0

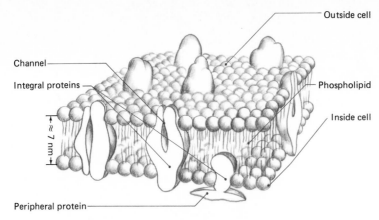

Channel

Integral proteins

≈ 7 nm

Peripheral protein

Outside cell

Phospholipid

Inside cell

Figure 4.4

The fluid mosaic model of membrane structure accounts for the fact that proteins as well as phospholipids comprise an integral part of membranes and that the structure is dynamic as opposed to static in nature.

guishable inside and outside; different functions are associated with the different sides of the membrane.

The fluid mosaic model

Several models have been proposed to explain the relationship between membrane structure and function. The currently accepted **fluid mosaic model** of membrane structure allows for movement of the proteins within the phospholipid matrix of the membrane (Figure 4.4). According to this model, the membrane is a bilipid layer with the proteins associated with the membrane distributed in a mosaic pattern, both on the surfaces and the interior of the membrane (Figure 4.5). Some of the proteins are confined to the membrane surfaces (extrinsic proteins), and others are partially or totally buried within the membrane matrix (intrinsic proteins). Hydrophilic amino acids, those with charged R groups, predominate in the portions of the protein macromolecules that project into the aqueous phase, whereas hydrophobic amino acids, those with nonpolar R

Figure 4.5

This electron micrograph of a freeze-fractured and etched Aquaspirillum putridiconchylium *shows visible particles (proteins) arrayed in a mosaic pattern protruding from the membrane that were exposed by freeze-fracturing; width of cell = 0.6 μm. (From BPS—T. J. Beveridge, University of Guelph.)*

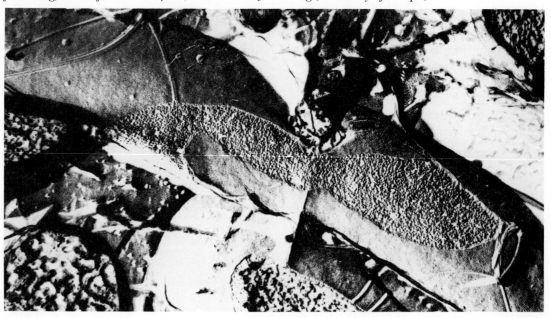

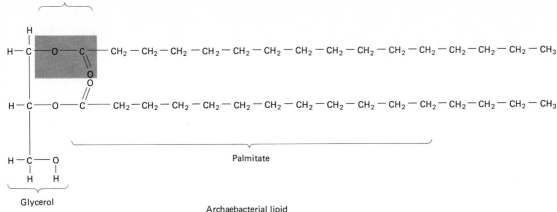

Figure 4.6

The membrane lipids of archaebacteria are different from the lipids found in other organisms. The lipids of both eubacteria and eukaryotes are glycerol esters of straight-chain fatty acids. Archaebacterial lipids are diethers in which a glycerol unit is connected by an ether link to phytanols, which are branched chains in which carbon atoms at regular intervals carry a methyl group. Moreover, glycerol has two optical isomers, distinguished by the configuration of the molecule around the central carbon atom; the optical isomers rotate polarized light in opposite directions. The configuration around the central atom of glycerol found in archaebacterial lipids is the mirror image of the configuration found in both eubacterial and eukaryotic lipids. (Based on C. R. Woese, 1981, Scientific American *244(2):120.)*

groups, predominate within the portions of the protein that are buried within the bilipid membrane. The structure of the membrane is not viewed as static. Lipids held together by weak bonds can move laterally through the fluid membrane matrix. Proteins can also move laterally, but to a lesser extent than the phospholipid molecules. The fluid mosaic model accounts for the biochemical and microscopical analyses of the membrane that have revealed the integral and dynamic relation between protein and phospholipid molecules in the membrane matrix and also explains the permeability properties of the cytoplasmic membrane.

Cell membranes of archaebacteria

The archaebacteria do not have a normal bilipid cell membrane. Rather, the phosphate groups in their membrane lipids are linked by ether bonds instead of the normal ester linkages (Figure 4.6). Many of the archaebacteria live in extreme environments where unusual physiologically specialized membranes are needed for survival. As an example, *Sulfolobus*—a bacterium that lives at high temperatures in acidic environments—has a cytoplasmic membrane that is a monolayer, consisting of a long-chain branched hydrocarbon, instead of the typical bilipid cytoplasmic membrane. Similar unusual membrane structures occur in other archaebacteria living in extreme habitats, including *Thermoplasma*, which lives at high temperatures, and *Halobacterium*, which lives in habitats with high salt concentrations. The membranes of these bacteria are conformed so that they have the necessary orientation of hydrophobic and hydrophilic groups to function as a semipermeable barrier. The biochemistry of these membranes makes them very resistant to conditions that would disrupt the structure and interrupt the function of a normal bilipid layer, enabling them to remain as a semipermeable barrier in extreme habitats. The distinct membrane structure of these bacteria is one of several features that distinguish the archaebacteria from other prokaryotes.

Figure 4.7

As illustrated in this figure, there are several ways for substances to cross a cell membrane and enter a cell. In cytosis the substance is transported into the cell without actually passing through the membrane. Diffusion across a membrane occurs when substances can pass through the pores of the membrane and when there is a favorable concentration gradient; this type of transport represents the downhill flow of a substance along a concentration gradient. In contrast to diffusion, active transport can occur along an unfavorable concentration gradient but requires the input of energy. In some cases the required energy is established by hydrogen ion pumping (chemiosmosis); in other cases the hydrolysis of ATP is used to drive the pumping of substances across the membrane. A special form of active transport, group translocation, occurs exclusively in prokaryotes: in group transport the substrate is chemically modified during transport; i.e., the substance is phosphorylated during transport across the membrane with the energy derived from phosphoenolpyruvate.

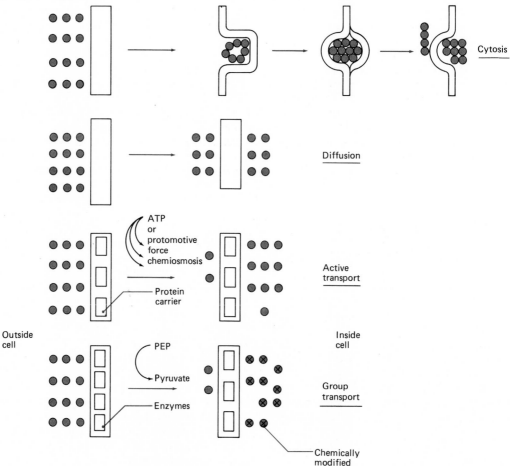

The functions of the cell membrane

In both prokaryotic and eukaryotic cells, **the primary function of the cytoplasmic membrane is to regulate the flow of material into and out of the cell**. There are several different ways by which molecules can enter or leave a cell, including diffusion, active transport, group translocation, and cytosis (Figure 4.7). It is the properties of the membrane that regulate these different **transport mechanisms**, establishing which materials can be translocated across the membrane. Organisms use different transport mechanisms for moving various biochemicals into and out of the cell and have different capabilities for transporting molecules across the cytoplasmic membrane.

Diffusion and osmosis

In accordance with the second law of thermodynamics, molecules of a substance will move from regions of higher concentration to those of lower concentration, thus increasing the degree of randomness and the entropy of the system. The movement of molecules to this equilibrium state of lowest free energy is known as diffusion. The increase in entropy drives diffusion, but differences in pressure and temperature can counteract the effects of concentration gradients; the critical factor is that by **diffusion** the system achieves a state of equilibrium at minimal free energy. Obviously, the macromolecules of living systems are not randomly distributed and concentration gradients exist across the cytoplasmic membrane. The cytoplasmic membrane restricts free diffusion of molecules, and the cell uses energy to maintain biochemical concentration gradients with its surroundings.

When there is a difference in the concentrations of a solute on either side of a membrane, water will move across the membrane from the region of higher to the region of lower concentration of solute until the concentrations of the solute are equalized on both sides of the membrane, or until a pressure force prevents further flow of the water. By doing so, the free energy of the system decreases. The process by which the water crosses the membrane in response to the concentration gradient of the solute is known as **osmosis**. Osmosis may be viewed as a form of passive diffusion in which a small water molecule can cross a membrane, but the larger solute molecules cannot freely move across this barrier. This process exerts a pressure known as the osmotic pressure on the membrane, which represents the force that must be exerted to maintain the concentration differences between solutions on opposite sides of the membrane.

Now, let us examine the consequences of osmosis for a microbial cell. In a medium where the solute concentration inside the cell is equal to the solute concentration outside the cell, water will flow equally in both directions across the membrane (Figure 4.8). However, in a medium where the solute concentration is higher (**hypertonic**) outside the cell than inside the cell, water will tend to flow out of the cell and the cell will tend to shrink. The reverse is true if the cell is in a medium where the solute concentration is lower (**hypotonic**) outside the membrane than inside the cell, in which case water will tend to flow into the cell, causing the cell to expand and—if unrestricted—the cytoplasmic membrane to burst. Microorganisms usually find themselves in this latter situation because the various biochemicals of the cell normally are in considerably higher concentrations than the solutes in the dilute aqueous medium in which most microorganisms exist. In order to survive, microorganisms have developed various strategies—to be discussed later in this chapter—for preventing such **osmotic shock**.

Passive and facilitated diffusion

In addition to water, various small solute molecules can diffuse passively across the membrane. **Passive diffusion** occurs when a solute molecule moves from an area of higher concentration on one side of the membrane to an area of lower concentration on the other side of the membrane. The rates of passive diffusion are determined by the concentration gradient; the greater the gradient, the more rapid the movement. The rates of simple diffusion across the cytoplasmic membrane are not rapid enough for many of the exchanges of biochemicals with the surroundings that must be accomplished by a cell, and microorganisms have developed additional membrane transport mechanisms. Some molecules can move across a membrane at higher rates than would occur by simple diffusion in a process referred to as facilitated diffusion (Figure 4.9).

Facilitated diffusion involves the proteins associated with the cytoplasmic membrane. Some of the membrane proteins—**permeases**—selectively increase the permeability of the membrane for specific solutes. The mechanisms of permease action are not fully understood. These membrane proteins may form channels through which facilitated diffusion occurs or the proteins may act as carriers. Carrier proteins may pick up a molecule

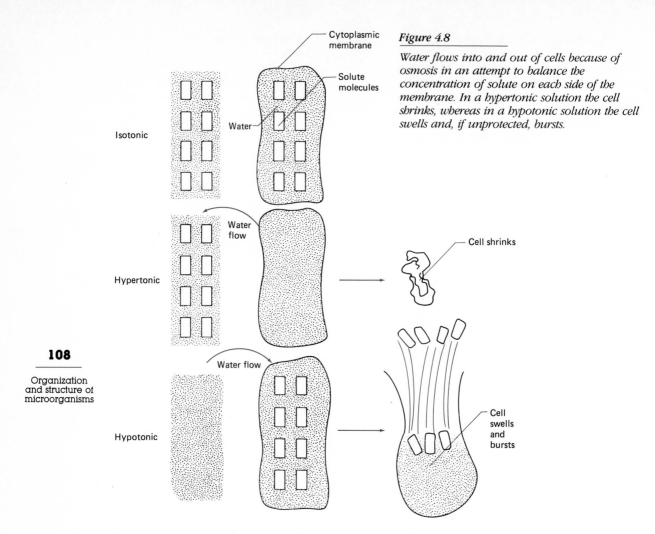

Cytoplasmic membrane

Solute molecules

Isotonic

Water

Water flow

Hypertonic

Cell shrinks

Water flow

Hypotonic

Cell swells and bursts

Figure 4.8

Water flows into and out of cells because of osmosis in an attempt to balance the concentration of solute on each side of the membrane. In a hypertonic solution the cell shrinks, whereas in a hypotonic solution the cell swells and, if unprotected, bursts.

Figure 4.9

A comparison of the relative rates of passive and facilitated diffusion across a membrane shows that the rate of simple diffusion increases linearly with increasing solute concentration, whereas the rate of facilitated diffusion is far more rapid and exhibits saturation kinetics.

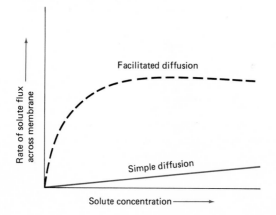

Facilitated diffusion

Simple diffusion

Rate of solute flux across membrane

Solute concentration

on one side of the membrane and transport it across the membrane to the other side. Such carrier-mediated transport is commonly found in yeasts, perhaps because sugar substrates are rarely limiting in the solutions surrounding these organisms. It is likely that the binding of a solute to a permease alters the three-dimensional properties of the permease protein, and this change in conformation moves or carries the substance across the membrane. Regardless of the mechanism, it should be emphasized that both passive and facilitated diffusion involve movement of molecules down a **concentration gradient**; by doing so diffusion lowers the free energy of the system and therefore does not require cellular input of energy to drive the process. The concentration gradient depends on the amount of free solute. In many cases when a substance moves into a cell, it binds with other substances, forming complexes, or is metabolically transformed. These processes prevent the buildup in concentration of the transported substance, allowing diffusion to continue along a favorable concentration gradient.

Active transport

Substances may also move across the membrane against a concentration gradient by **active transport**. Active transport can be likened to a pump that requires energy to move water uphill. The cell must actively work to move substances up a free-energy gradient and therefore, unlike diffusion, active transport requires that the cell expend energy to move biochemicals across the membrane against a concentration gradient. In active transport, membrane proteins act as carriers to move substances across the membrane.

The movement of the carrier–substrate complex across the membrane can be coupled with the hydrolysis of ATP to supply the energy needed to drive the active transport process (Figure 4.10). Such ATP-driven transport occurs in eukaryotic cells and in the transport of some amino acids and proteins into bacterial cells. For example, the uptake of glutamate by some bacteria is dependent on the hydrolysis of ATP as an energy source.

The hydrolysis of ATP does not always supply the energy needed for active transport across the cytoplasmic membrane. Active transport of a substance may also be driven by **chemiosmosis**, the electromechanical force established when there is a hydrogen ion gradient across the cytoplasmic membrane; it is this mechanism that appears to provide the energy for active transport of many

Figure 4.10

Active transport requires the temporary binding of the substance with a carrier protein during transfer across the membrane and energy activation of the membrane from chemiosmosis or the hydrolysis of ATP.

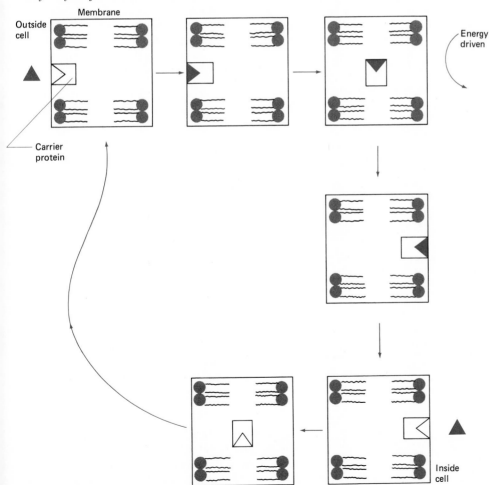

substances, e.g., proline, into bacterial cells. Chemiosmosis will be discussed further in Chapter 5 with respect to ATP generation. Suffice it here to say that some microorganisms possess the metabolic capability to pump hydrogen ions out of the cell, establishing a hydrogen ion concentration gradient and the associated electromechanical gradient that can be used to do work when the hydrogen ions move back into the cell by diffusion. Many substrates are transported into microbial cells by active transport. The active transport capabilities of a cell determine in large part what substrates can move into a cell and thus help define the metabolic potential of that cell.

Group translocation. In the active transport processes that have been discussed so far, the substance is not altered by transport across the membrane. Although temporary binding with a protein may occur during passage through the membrane, the transported substance appears on both sides of the membrane in the same biochemical form. In contrast, **during group trans-**

Figure 4.11

The group transport system, showing the reactions involved in the phosphotransferase system for the transport of sugars into a bacterial cell. In reaction 1, cytoplasmic enzyme I transfers P from P-enolpyruvate to HPr, a small cytoplasmic protein, approximate MW = 9000. In reaction 2, HPr transfers the high-energy P to various sugar-specific enzymes, in conjunction with other sugar-specific, membrane-bound enzymes (enzyme II complex). P is transferred to specific sugars, and the phosphorylated sugar is transported to the inner surface of the membrane.

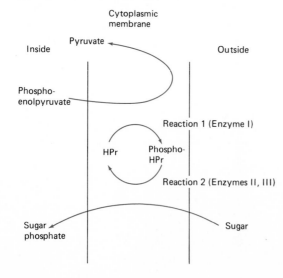

location processes the substance is chemically altered during passage through the membrane (Figure 4.11). Group translocation appears to be restricted to prokaryotic cells. The evolution of this transport system by bacteria allows them to use their energy resources efficiently by coupling transport with metabolism. By altering the chemical form of the substrate during transport a more favorable concentration gradient is established, an important fact because bacteria generally live in aqueous environments with very low concentrations of carbohydrate substrates.

In the **phosphotransferase system (PTS)** a sugar is phosphorylated as it crosses the cytoplasmic membrane. Thus, when glucose is transported into a bacterial cell by this system, it occurs as glucose outside the cell but as glucose 6-phosphate within the cell. In the phosphoenolpyruvate–phosphotransferase system the phosphate for the phosphorylation of the carbohydrate comes from phosphoenolpyruvate (PEP), which is a high-energy phosphate compound. There are three separate enzyme systems in the phosphotransferase system. The initial step in this process, catalyzed by enzyme I, involves the transfer of phosphate from PEP to a heat-stable protein molecule (HPr). This reaction occurs in the cytoplasm or at the inner surface of the membrane. Within the membrane, the phospho-HPr molecule then transfers phosphate to a carbohydrate. There are two additional enzymes, II and III, involved in the transfer of phosphate from the phospho-HPr molecule to the sugar. Different substrate-specific enzymes II and III are needed for the transport and initial phosphorylation of different carbohydrates. Carbohydrate substrates, such as glucose, are normally phosphorylated as part of their metabolism to generate ATP and cell constituents. Thus, during transport by the group translocation process, the initial transformation involved in the metabolism of the substrate is accomplished, providing an energetically efficient mechanism for bringing such compounds into the cell. Group translocation is not restricted to carbohydrates and also occurs in the transport of some fatty acids. The transport of substrates by this mechanism does not occur in all bacteria but is an important process in those bacteria that have the necessary enzymes associated with their membranes to phosphorylate substrates during their translocation across the membrane.

Secretion of extracellular enzymes
Many polymeric substrates, such as cellulose, are too large to transport through the pores of the cytoplasmic membrane. One strategy employed by microorganisms to use such substrates is the

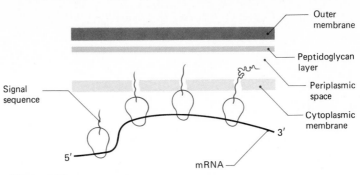

Signal sequence

Outer membrane

Peptidoglycan layer

Periplasmic space

Cytoplasmic membrane

3′

5′

mRNA

Figure 4.12

The utilization of high-molecular-weight compounds requires the secretion of exoenzymes that function outside the cell to break the substance down into molecules small enough to transport across the membrane. Prokaryotic proteins destined for locations other than the cytosol are synthesized by ribosomes bound to the cytoplasmic membrane. A signal sequence on the peptide chain directs the ribosome to the plasma membrane and enables the protein to be translocated across the membrane.

secretion of **extracellular enzymes (exoenzymes)** that degrade the polymeric substance into smaller units that can then be transported across the cell membrane and metabolized (Figure 4.12). Besides their role in converting impermeable substances into useable substrates, exoenzymes are involved in destroying substances that are harmful to the cell. The secretion of extracellular enzymes represents an interesting regulatory mechanism through which the cell recognizes which proteins to export. Many enzymes that are designed to be secreted contain a leader segment at the amino-terminal end of the molecule that acts as a signal to initiate the secretion process. This signal sequence contains about 20 predominantly hydrophobic amino acids. These hydrophobic amino acids react with the membrane, initiating the translocation of the protein across the cell membrane barrier. During transport the leader sequence of the protein is cleaved by an enzyme within the membrane matrix, so that the exoenzyme released is smaller than when it was synthesized. In many cases the secretion of the exoenzyme is initiated before the synthesis of the protein is completed, and secretion continues at the same time that protein synthesis is proceeding. The secretion of exoenzymes is often essential for the survival of the cell because these enzymes can be used both for obtaining nutrition and protection. The biochemical composition of the synthesized exoproteins and their interactions with the components of the cytoplasmic membrane provide the mechanism for the selective secretion of these extracellular enzymes.

Endocytosis and exocytosis

Up till now we have discussed mechanisms for moving substances through the membrane. Some cells possess another mechanism, **cytosis,** by which substances may enter or leave a cell. Cytosis does not involve transport through the membrane but rather involves the engulfment of a substance by the membrane to form a vesicle. **Endocytosis** refers to the movement of materials into the cell by this mechanism; **exocytosis** denotes movement out of the cell. Both endo- and exocytosis are important in the movement of large substances into and out of the cell that could not be accomplished by transport through a membrane. Both endo- and exocytosis are energy-requiring processes in which ATP must be hydrolyzed to drive the movement of materials into or out of the cell. **Phagocytosis** is a specific example of this mechanism in which one cell engulfs a smaller cell or other particle (Figure 4.13). This transport mechanism is particularly important in some protozoa that feed on bacteria and in the ability of certain white blood cells to engulf and digest bacteria as part of the immune response of animals to infections. Interestingly, such endocytosis has not been observed in prokaryotic cells, and this transport mechanism appears to be restricted to eukaryotic cells.

Additional functions of the bacterial cell membrane

Because the prokaryotic cell lacks internal membrane-bound organelles, several functions associated with such organelles in eukaryotic orga-

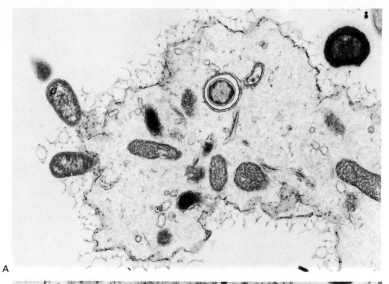

A

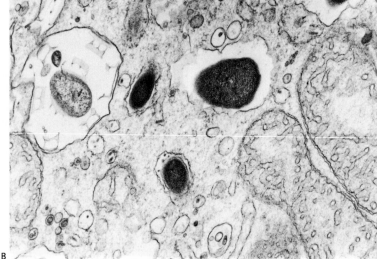

B

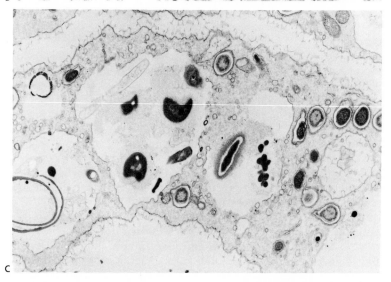

C

Figure 4.13

Phagocytosis of rod-shaped bacteria by an ameboid protozoan. (A) In the early stages of bacterial engulfment, one bacterium is attached to the surface scales while the other is already in the process of being surrounded by the cytoplasmic membrane of the ameba (12,600×). (B) This phagosome enclosed a bacterium shortly after engulfment; there are still scales on the vacuolar membranes. A later stage shows advanced digestion and no scales on the vacuolar membrane (27,700×). (C) Here, digestive vacuoles containing bacteria and other food particles are in advanced stages of digestion (8,800×). (Courtesy O. Roger Anderson, Columbia University.)

nisms are associated with the cytoplasmic membrane in prokaryotes. One such process, **oxidative phosphorylation**, important in the generation of ATP during respiration, will be discussed in Chapter 5. The proteins associated with oxidative phosphorylation occur at the cytoplasmic membrane of prokaryotes; these proteins establish channels through which hydrogen ions are transported to form a concentration gradient that drives the formation of ATP by chemiosmosis. In eukaryotic cells, oxidative phosphorylation occurs within a specialized organelle—the mitochondrion—and oxidative photophosphorylation—another process for generating ATP—occurs within the chloroplast. The involvement of membranes in these ATP-generating processes is essential, as it provides the necessary semipermeable boundary across which an electochemical gradient can be established. By performing work to pump hydrogen ions across the membrane, the return flow can be efficiently coupled with the synthesis of ATP; the ATP formed at the cyto-plasmic membrane can then be an energy source within the cell.

Mesosomes

As an extension of membrane function, prokaryotic organisms form extensively invaginated portions of the cytoplasmic membrane, called **mesosomes**, which do not occur in eukaryotic cells (Figure 4.14). Elucidating the role of mesosomes has proved difficult, probably because anatomically similar-appearing structures can be functionally different. Among their proposed functions, mesosomes may play a role in cell divisional processes, various metabolic processes—including the generation of ATP—and secretion of enzymes from the cell. Mesosomes effectively provide additional membrane surface within bacterial cells; in eukaryotic microorganisms organelles that are bounded by membranes provide the surface area needed for the metabolic reactions that occur in association with membranes.

Figure 4.14

Mesosomes are formed by the invagination of the bacterial cytoplasmic membrane. Note the mesosome structure and its attachment to the membrane. (Courtesy M. I. Higgins, Temple University, reprinted by permission of the American Society for Microbiology, from M. I. Higgins and L. Daneo-Moore, 1972, Journal of Bacteriology *109: 1226.)*

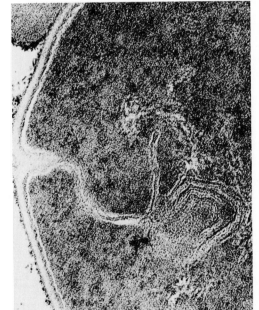

Structures outside the cytoplasmic membrane

Cell wall

A cell wall surrounds the cytoplasmic membrane of almost all prokaryotic cells and many eukaryotic microorganisms (Figure 4.15). The cell wall is a very important structure because it protects the cell against osmotic shock. Without a cell wall the cells of most microorganisms would burst from the osmotic pressure exerted on their cytoplasmic membranes because these organisms normally exist in dilute aqueous environments. Microbial cells that lack a cell wall must possess other mechanisms for protecting the cell against osmotic shock—such as a contractile vacuole to pump water out of the cell—or the organisms are restricted to living in environments where the solute concentration outside of the cell is high enough to prevent the osmotic pressure from causing the cell to rupture.

The cell wall is generally a relatively porous structure that does not restrict the flow of small molecules to or from the cytoplasmic membrane, although very large polymers are unable to pass across the cell wall. Some enzymes that are secreted by the cell can be trapped in the **periplasmic space**, the region between the cell wall and the cytoplasmic membrane; the **periplasm**, therefore, can be a region of intense enzyme activity. Cell wall structures are normally rigid. This rigidity of the prokaryotic cell wall is responsible for maintaining the specific shape of a bacterium. Bacteria owe their shapes, which are characteristic of particular species, to their cell wall structure. Often, bacteria occur as spheres called cocci, cylinders called rods, or spiral shapes, although many diverse forms typify different bacterial species; the diversity of bacterial form can be seen in the micrographs throughout this book.

The bacterial cell wall

The bacterial cell wall is a biochemically unique structure. The cell wall of all but a few bacteria contains **murein,** which is also known as **peptidoglycan** or **mucopeptide.** This peptidoglycan layer is not found in any eukaryotic organism. Murein is composed of a backbone of alternately repeating units of the amino sugars N-acetylglucosamine and N-acetylmuramic acid (Figure 4.16). These repeating amino sugars form the glycan portion of the molecule. Attached to some of the N-acetylmuramic acid units is a short peptide chain consisting of four amino acids. Some of the amino acids occurring in the peptide portion of the molecule are relatively unique in biological systems, and both L- and D-amino acids occur in murein, whereas proteins contain only L-amino acids. Diaminopimelic acid occurs as one of the amino acids in the tetrapeptide sequence of some bacteria; this amino acid does not occur in proteins and is not found anywhere else in biological peptides. The specific amino acids comprising this peptide portion of the peptidoglycan molecule show limited variation between different bacterial species. All the tetrapeptide sequences include L-alanine, D-glutamic acid (which may be hydroxylated in some organisms), either L-lysine or diaminopimelic acid, and D-alanine. The only variation that occurs is the substitution of lysine and diaminopimelic acid, with lysine occurring in most Gram positive bacteria and diaminopimelic acid occurring in all Gram negative bacteria.

The tetrapeptide chains are interlinked by a peptide bridge between the carboxyl group of an amino acid in one tetrapeptide chain and the amino group of an amino acid in another tetrapeptide chain. The greatest variation in the peptidoglycan occurs in these cross-linkages. The cross-linkage can occur between tetrapeptides in different chains, as well as directly between adjacent tetrapeptides so that the peptidoglycan forms a rigid, multilayered sheet. In the Gram positive bacterium *Staphyloccoccus aureus* the cross-linkage involves a pentapeptide. In many Gram negative bacteria cross-linkage often involves the direct bonding of diaminopimelic acid of one chain to the terminal D-alanine of another chain. This cross-linked pep-

Figure 4.15

The bacterial cell wall occurs outside the cytoplasmic membrane, protecting the cell from various types of damage, such as bursting of the cell due to osmotic pressure. The periplasmic space between the cell wall and cytoplasmic membrane is an area of active enzymatic activity.

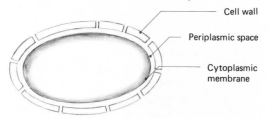

Cell wall

Periplasmic space

Cytoplasmic membrane

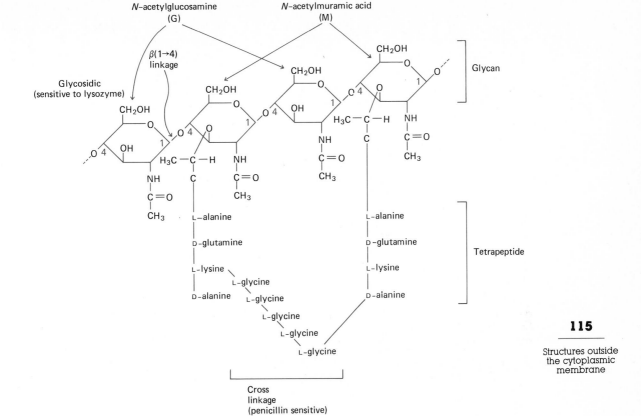

Figure 4.16

Murein or peptidoglycan is the backbone biochemical of the bacterial cell wall; it is composed of repeating alternating units of N-acetylglucosamine and N-acetylmuramic acid and has cross-linked, short peptide chains, some of which have unusual amino acids.

tidoglycan layer can be likened to a girder-supported bridge, and in fact the cross-linkages are referred to as bridges (Figure 4.17). Without the cross-linkage of the peptide chains, the murein layer would not be rigid and would not protect the cell against osmotic shock. When cross-linkage is disrupted, the cell wall is defective and cannot adequately protect the bacterial cell against osmotic shock.

For this reason the antibiotic penicillin is effective in controlling bacterial infections. Penicillin prevents the formation of cross-linkages between the peptides, resulting in the production of defective cell walls and the death of growing bacteria. Penicillin, however, does not destroy cross-linkages of the peptidoglycan layer and thus has no effect on bacteria that are not growing. In contrast to penicillin, the enzyme lysozyme will degrade a preformed peptidoglycan molecule. Lysozyme acts by breaking the glycan rather than the peptide portion of the peptidoglycan layer.

This enzyme is produced by various organisms that consume bacterial cells, aiding in the digestion of the bacteria; lysozyme also occurs as part of various normal body secretions, such as tears, providing protection against would-be bacterial invaders. By using lysozyme, it is possible to remove all or part of the cell wall structure. If the bacterial cell remains intact following the partial removal of the cell wall, it is referred to as a spheroplast; if the wall has been completely removed, such an intact bacterial cell is called a protoplast. Both protoplasts and spheroplasts can exist in a supporting medium of high solute concentration in which the osmotic pressure is not high enough to lyse the cell.

The Gram negative and Gram positive bacterial cell walls As discussed in Chapter 2, bacteria can be differentiated by using the Gram stain reaction. The difference in staining reaction reflects differences in cell wall structure. Regardless

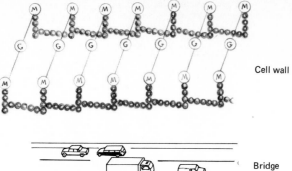

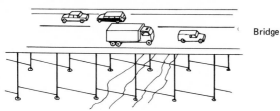

Cell wall

Bridge

Figure 4.17

This diagram illustrates the similarity of the biochemical structure of murein to a girder-supported bridge. In both cases the cross-links provide the needed structural support, and if the cross-links are not present, as occurs in the cell walls of bacteria treated with penicillin, there is structural failure.

of the staining reaction, which sometimes gives anomalous results depending on the condition and age of the bacteria, the difference between Gram negative and Gram positive bacteria is defined by the inherently different structure of the cell wall in these two groups of bacteria (Figure 4.18). Both Gram positive and Gram negative cell-wall structures contain a peptidoglycan layer, but they have different amounts of peptidoglycan and further differ in which biochemicals, in addition to murein, form the cell wall complex.

The **Gram positive cell wall** has a peptidoglycan layer that is relatively thick and comprises approximately 90 percent of the cell wall (Figure

4.19). The murein layer acts to fulfill the primary protective function of the cell wall. It is this thick murein layer that is believed to trap the primary stain within the periplasmic space during the Gram stain procedure, preventing the primary stain from being washed out when the cells are treated with a decolorizing agent. In addition to peptidoglycan, Gram positive cell walls also generally have another component, **teichoic acids**, which are acidic polysaccharides. Teichoic acids contain a carbohydrate such as glucose, phosphate, and an alcohol, which may be either glycerol or ribitol. The teichoic acids may be involved in binding substances at the cell surface or may serve as a

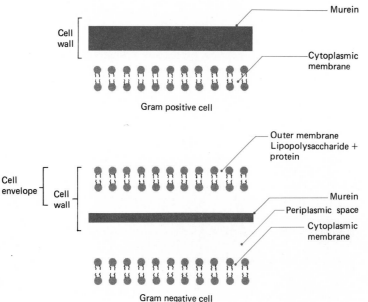

Figure 4.18

Gram positive and Gram negative bacteria differ in the structure of their cell walls. The Gram positive cell wall has a relatively thick murein layer. The Gram negative cell wall has a thin murein layer but also has an outer membrane and additional lipopolysaccharides and proteins not present in Gram positive cell walls.

way of storing phosphorus, among other possibilities.

In contrast to the relatively simple Gram positive bacterial cell wall, which is composed mostly of murein, the **Gram negative cell wall** is biochemically far more complex (Figure 4.20). The peptidoglycan layer of the Gram negative cell wall is very thin and often comprises only 10 percent of the cell wall. Teichoic acids generally do not occur in Gram negative bacteria, but the cell wall of Gram negative bacteria does contain lipids, polysaccharides, and proteins that form an outer layer of the wall. The Gram negative bacterial cell wall effectively forms a complex with an outer membrane layer, which, in addition to the normal membrane phospholipids, contains **lipopolysaccharides (LPS)**. The phospholipids and lipopolysaccharides of this outer cell wall layer are sometimes referred to as the **cell envelope**.

Effectively, Gram negative bacteria possess two membranes—the normal cytoplasmic membrane and the additional membrane of the cell envelope layer surrounding the cell wall—that regulate the flow of material into and away from the cell. Functionally, the **outer membrane** of the Gram negative bacterial cell is a coarse molecular sieve that allows the diffusion of both hydrophilic and hydrophobic molecules up to a molecular weight of about 800, whereas the cytoplasmic membrane excludes almost all hydrophilic substances except water. Despite its permeability to small molecules, the outer membrane is less permeable than the cytoplasmic membrane to hydrophobic or amphipathic molecules such as phospholipids that have polar and nonpolar ends. For this reason Gram negative bacteria are less sensitive to antibiotics because antibiotics with these properties

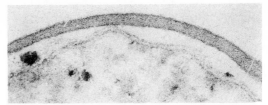

Figure 4.19

Electron micrograph of the cell wall of Bacillus subtilis, *a typical Gram positive cell wall with a thick murein layer (136,200×). (From BPS—T. J. Beveridge, University of Guelph.)*

do not reach the cytoplasm of the cell where such antibiotics act. Additionally, the LPS of the Gram negative outer wall is known as endotoxin and causes disease symptoms such as those that characterize travelers' diarrhea. Both because of the endotoxin and the increased resistance to antibiotics resulting from the biochemical nature of their cell walls, Gram negative organisms are more prominent than Gram positive bacteria in human infection.

Bacteria lacking peptidoglycan walls Not all bacteria contain murein in their cell walls. The archaebacteria do not contain peptidoglycan in their cell walls, but rather the walls of these organisms are composed of proteinaceous subunits (Figure 4.21). Included among the bacteria that do not contain peptidoglycan in their cell walls are the methanogenic bacteria, *Sulfolobus*, and *Halobacterium*. One bacterial genus, *Mycoplasma*, lacks a cell wall entirely. This particular bacterial genus is also unusual in that it incorporates sterols, which it normally acquires from

Figure 4.20

Electron micrograph of the cell wall of Escherichia coli, *a typical Gram negative cell wall with a thin murein layer and an outer membrane (132,000×). (From BPS—T. J. Beveridge, University of Guelph.)*

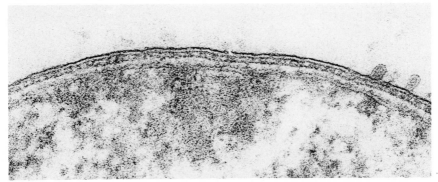

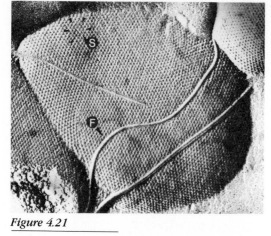

Figure 4.21

Electron micrograph of the freeze etched cell wall of the methanogen Methanogenium marisuigri *(60,700×); S = periodically ordered surface layer (C-fold rotational symmetry), F = flagella. (Courtesy F. Mayer, University of Gottingen.)*

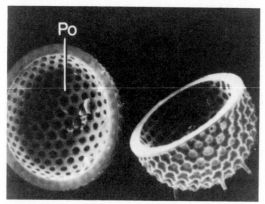

Figure 4.22

Electron micrograph of diatom frustule of Xanthopyxis *(780×); Po = pore. (Reprinted by permission of Springer-Verlag, New York, from R. G. Kessel and C. Y. Shih, 1976,* Scanning Electron Microscopy for Biology.*)*

a eukaryotic host cell, as a component of its cytoplasmic membrane. Because they lack a cell wall, members of the genus *Mycoplasma* will not be inhibited by penicillin because they lack the biochemical that this antibiotic affects.

Cell walls of eukaryotic organisms

Many eukaryotic microorganisms have a cell wall structure, the primary function of which is to protect the cell against osmotic shock. There is a great diversity of biochemical structures found in the cell walls of eukaryotic microorganisms. Many al-

gae have a cell wall comprised primarily of cellulose, but various other polysaccharides, such as pectins, xylans, mannans, and alginic acid, are found as major components of some algae. Some algae have a cell wall structure containing calcium or silicon, sometimes called the test or frustule. The diatoms, for example, have a frustule that is a cell wall composed of silicon dioxide, protein, and polysaccharide. The frustule has two overlapping halves and distinctive markings that give these organisms their characteristically symmetrical and beautiful shapes (Figure 4.22). The coral algae deposit calcium carbonate in their wall structures, forming the basis for coral reefs. These structures protect the cell against physical damage rather than against osmotic shock. The cell walls of these organisms are preserved long after the organisms die. Not all algae, though, have a cell wall, and the genus *Euglena*, for example, lacks a true cell wall structure.

In fungi the cell walls are normally composed of microfibrils of cellulose or chitin or a combination of these biochemicals. Most fungi, including yeasts, have a cell wall with distinctive biochemical characteristics. The composition of the fungal cell wall can change in response to environmental conditions. The cells of some fungi, the slime molds, lack a cell wall and are pleomorphic; that is, they assume different forms rather than having a rigid shape. However, during the normal slime mold life cycle, the cells go through a stage in which they do have rigid cell walls. The biochemical composition of fungal cell walls appears to reflect taxonomic relationships and is a useful criterion in fungal classification systems.

The protozoa usually do not have a true cell wall surrounding the membrane, and many protozoa have developed alternative mechanisms for osmoregulation. Many protozoa do have a thin pellicle surrounding the cell that maintains the shape of the organisms. If the pellicle of a ciliate protozoan such as *Paramecium* is removed, the cell becomes spherical. Some protozoa form an outer wall or shell, composed of calcium carbonate in the foraminifera, and silicon dioxide or strontium sulfate in the radiolaria. These shells are not a basis for osmoregulation, and, in fact, many foraminifera extend their cytoplasm beyond the shell.

Bacterial capsule and slime layers

The cell wall is not always the outermost layer of a cell. Some bacteria form a **capsule** external to the cell wall (Figure 4.23). The capsule is com-

Figure 4.23

Photomicrograph showing the thick capsule of Klebsiella aerogenes *(27,500×). (From BPS—Stanley C. Holt, University of Massachusetts.)*

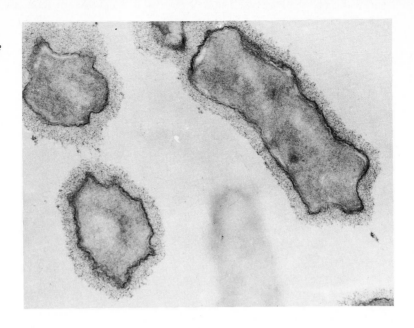

posed of polysaccharides and/or proteins, but the specific biochemical composition of the capsule varies between species of bacteria. The capsule is especially important in protecting bacterial cells against phagocytosis. The presence of a capsule can be a major factor in determining the pathogenicity of a bacterium, that is, the ability of a bacterium to cause disease in the organism that it infects. In some cases a bacterial species will have two variants, one that forms a capsule and is a virulent pathogen, and a nonencapsulated form that is avirulent and does not cause disease. The reason for this is that the nonencapsulated bacteria are subject to phagocytosis by blood cells involved in the immune response of the infected host organism. On the other hand, phagocytizing blood cells involved in the immune response are unable or less able to adhere to, engulf, and digest those bacteria that have a capsule.

Slime layers are similar to capsules but are not as tightly bound to the cell. These external layers may protect the cell against dehydration and a loss of nutrients. In some cases they appear to act as a trap that—because of the viscosity of the slime—restricts the diffusion of substrates away from the cell. The slime layer, or **glycocalyx**, may act to bind cells together, forming multicellular aggregates. Additionally, the glycocalyx of some bacteria are involved in attachment to solid surfaces (Figure 4.24). The glycocalyx is a mass of tangled fibers of polysaccharides or branching sugar molecules surrounding an individual cell or a colony of cells. Some bacteria in aquatic habitats, for example, appear to be held to rocks

Figure 4.24

The bacterial glycocalyx is involved in the attachment of bacteria to solid surfaces; in this electron micrograph the stabilized glycocalyx surrounds enteropathogenic E. coli *cells. (Courtesy. William Costerton, University of Calgary.)*

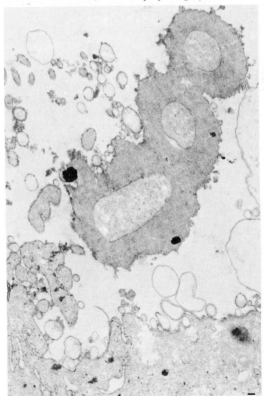

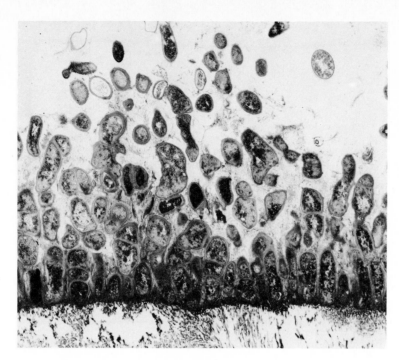

Figure 4.25

The slime layer of Streptococcus *species permits the attachment of bacterial cells to tooth surfaces and the formation of dental plaque. This electron micrograph, showing bacteria in plaque and actinomycetes in the slime (plaque) layer, exemplifies the normal microbiota associated with human tooth enamel (7,800×). (From BPS—Max Listgarten, School of Dental Medicine, University of Pennsylvania.)*

through the slime layers they secrete. Bacteria occurring in the oral cavity on the surfaces of teeth form a polysaccharide slime, dental plaque, which enables them to adhere to the tooth (Figure 4.25). This adherence to the tooth surface is important in the formation of dental caries.

Figure 4.26

Proteus mirabilis, negatively stained with PTA and surrounded by pili (36,000×). (Courtesy J. F. M. Hoeniger, University of Toronto, reprinted by permission of the Society for General Microbiology, from J. F. M. Hoeniger, 1965, Journal of General Microbiology *40:29–42.)*

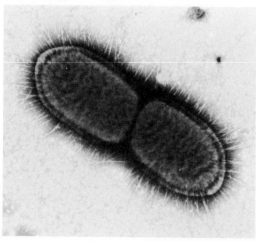

Pili

In addition to slime layers, **pili**, short hair-like projections composed primarily of protein subunits emanating from the surface of some bacteria, appear to be involved in attachment processes (Figure 4.26). There appear to be several different types of pili that may be associated with the bacterial surface, each serving a different function, but attachment is central to these different functions. The F or **sex pilus** is involved in bacterial mating and is found exclusively on the donor cells (Figure 4.27). The F pili are phosphate–carbohydrate–protein complexes with a single peptide subunit, pilin. Mating pairs cannot form in the absence of an F pilus or if the bridge established by the F pilus between the donor and recipient cell is interrupted. The precise role of pili in conjugal cell–cell interactions, however, remains controversial. The cylindrical nature of the F pilus could permit the passage of single-stranded DNA, but once a mating pair is established through pilus–wall contact, the cells rapidly establish wall–wall contact, and the DNA transfer therefore can occur directly through the contacting walls. It does appear that the contact of the F pilus of the donor cell with the wall of the recipient cell establishes the mating pair and initiates the transfer of DNA. Pili also act as the receptor sites for phage, providing these bacterial viruses with a site of attachment to the bacterial cell (Figure 4.28).

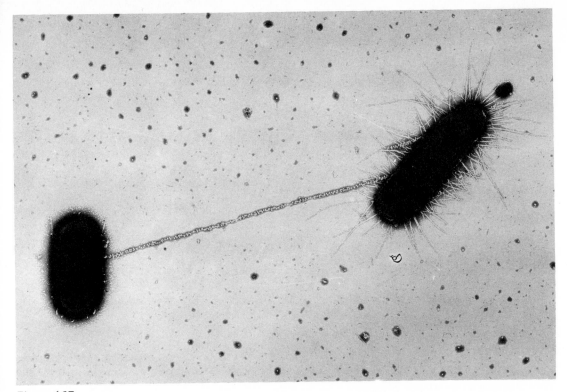

Figure 4.27

Electron micrograph showing F pilus acting as a conjugation bridge. The bacterial cell covered with numerous appendages is a genetic donor connected to a recipient cell (without appendages) by the F pilus. The F pilus is necessary for the genetic donor to transfer bacterial genes to the recipient. In this micrograph the F pilus has been labeled along its length by the use of phage that specifically attach to the F pilus, permitting its recognition. The numerous other appendages on the donor are called type I pili and have no role in conjugation. (Courtesy Charles C. Brinton, Jr., and Judith Carnahan, University of Pittsburgh.)

Figure 4.28

Electron micrograph showing phage attached to bacterial pili. The phage recognize and bind to specific receptor sites on the pilus. (Courtesy Lucien Caro and Edouard Boy de la Tour, University of Geneva.)

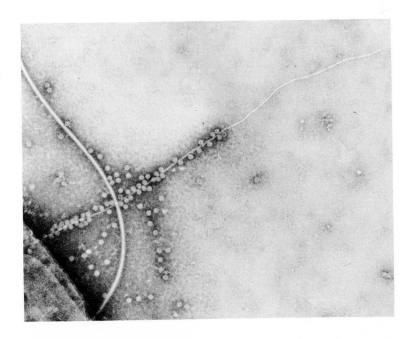

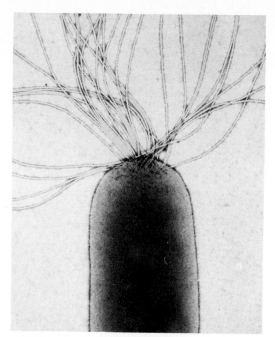

Organization and structure of microorganisms

Figure 4.29

A tuft of flagella appear at the polar end of the negatively stained bacterium Aquaspirillum graniferum *(49,700×). (From BPS—T. J. Beveridge, University of Guelph.)*

The phage attach to the pili and subsequently transfer their genetic information to the bacterial cell. Pili further have been implicated in the ability of bacteria to recognize specific receptor sites on host cell membranes. The pili appear to allow the bacteria to attach to and colonize the host

Figure 4.30

Electon micrograph of a marine vibrio showing a single polar flagellum (25,200×). (From BPS— Paul W. Johnson, University of Rhode Island.)

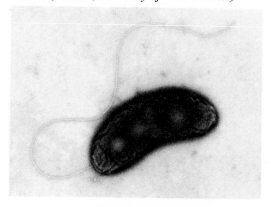

cells, sometimes leading to disease in the host organism. In all of these processes the pili act as a point of specific contact and attachment between the bacterial cell and another surface.

Flagella and cilia

Many microorganisms are motile, moving from place to place to obtain nutrition, grow, and reproduce. Microorganisms have evolved various mechanisms for locomotion. Protozoa, like *Amoeba*, move by extending their cytoplasm in a particular direction, forming "false feet" as they continuously change shape. Other microorganisms, such as the gliding bacteria, move without any obvious organelles of locomotion; mechanisms that have been proposed for how these bacteria move include secretion of a slime layer, reduction of the surface tension at one end of the cell, and production of fibrous extensions that cause unidirectional and submicroscopic contraction waves at the cell surface.

More commonly, microorganisms are motile by means of **flagella** or **cilia** structures that, like pili, project out from the cell surface. In contrast to pili, which are short projections, flagella are relatively long projections extending outward from the cytoplasmic membrane (Figure 4.29). Although flagella serve the same function of locomotion in prokaryotes and eukaryotes, the flagella of bacteria and those of eukaryotic cells are markedly different in structure.

Bacterial flagella

Two basically different types of arrangements of flagella occur in bacteria. In some bacteria such as *Pseudomonas* the flagella emanate from an end of the bacterial cell; such flagella are known as **polar flagella** because they originate from the pole of the cell (Figure 4.30). A bacterial cell may have one or more polar flagella. Bacteria with polar flagella tend to swim rapidly with a corkscrew motion. In contrast to polar flagella, some bacteria, such as those in the genus *Proteus*, have **peritrichous flagella** that surround the cell (Figure 4.31). The specific number of flagella vary, but there are always multiple peritrichous flagella emanating from lateral points around the cell. The arrangement of the flagella is characteristic of a bacterial genus and is an important diagnostic characteristic used in classifying bacteria.

The bacterial flagellum consists of a single filament composed of protein subunits. The protein of the filament is known as flagellin. The shape of the filament varies, depending on the amino

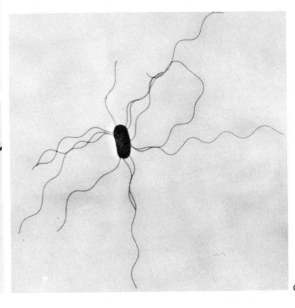

Figure 4.31

(A) Electron micrograph of Escherichia coli *showing peritrichous flagella. (Courtesy Julius Adler, University of Wisconsin.) (B) Light micrograph of* Bordetella bronchiseptica—*stained by Leifson flagella staining procedure—showing peritrichous flagella. (Courtesy Centers for Disease Control, Atlanta.) (C) Negatively stained peritrichously flagellated* Proteus sp. *(Courtesy Lee D. Simon, Rutgers, The State University.)*

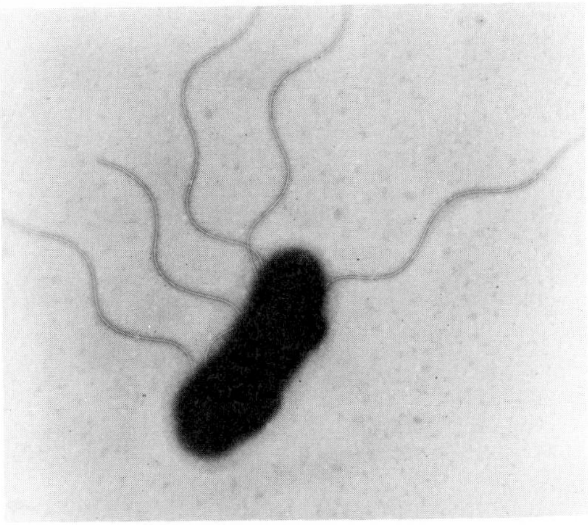

A

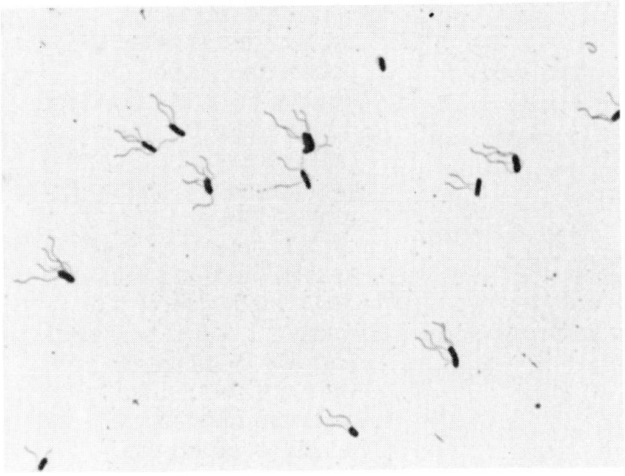

B

C

acid composition of the flagellin. The bacterial flagellum is attached to the cell by a hook and basal body that has a set of rings that attach to the cytoplasmic membrane and a rod that passes through the rings to anchor the flagellum to the cell (Figure 4.32). In Gram negative bacteria the basal body has a second set of rings that attaches to the outer lipid layer of the cell wall, and in

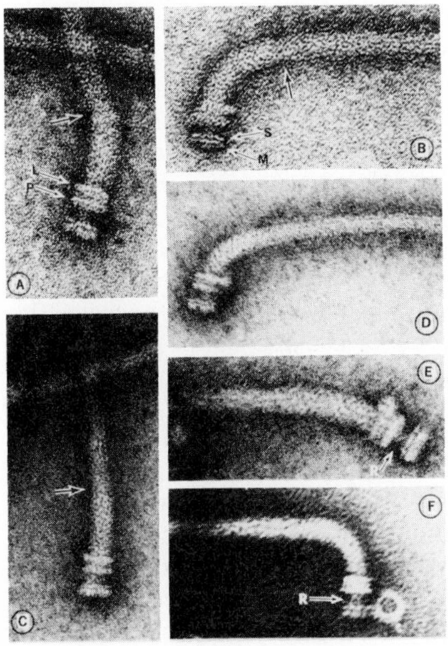

Figure 4.32

High-resolution electron micrographs showing the attachment of flagella to E. coli. *(A) Both the top and the bottom edges of the L and P rings are seen and give a circular appearance to the rings (160,050×). In this and following figures, the arrows mark the junction between the hook and the filament. (B) The top edge of the S ring and the top and bottom edges of the M ring can be seen, giving a circular appearance (146,850×). (C) L, P, and M rings appear about twice as thick as in (A), probably because the top and bottom edges of each ring are not resolved (135,300×). (D) The L ring appears extended on here and on (E) because a fragment of the outer membrane remained attached (130,350×). (F) Here, the stain did not penetrate the top rings. R marks the rod connecting the top and bottom rings. The detached ring has associated with the bottom ring (130,350×). (Courtesy Julius Adler, University of Wisconsin, reprinted by permission of the American Society for Microbiology, from M. L. DePamphilis and J. Adler, 1971,* Journal of Bacteriology *105:384.)*

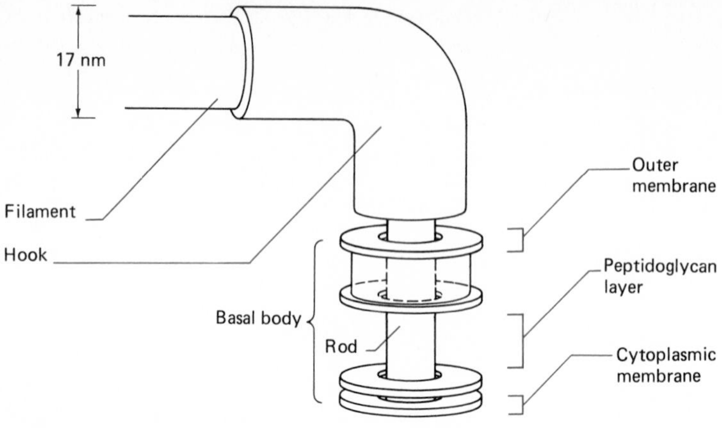

17 nm

Filament

Hook

Basal body {

Rod

Outer
membrane

Peptidoglycan
layer

Cytoplasmic
membrane

Figure 4.33

*Drawing showing how the bacterial flagellum is anchored at the cell.
The flexible hook serves as a universal joint coupling the rod to the
filament. Torque is generated between the M ring, which is rigidly
mounted on the rod and freely rotates in the cytoplasmic membrane,
and the S ring, which is rigidly mounted on the cell wall. Torque is
generated by the translocation through the cytoplasmic membrane and
the M ring of ions that interact with charges on the surface of the S
ring. The energy for rotation is derived from the transport of hydrogen
ions across the cytoplasmic membrane, chemiosmosis, with transport of
approximately 256 hydrogen ions required per revolution. The
additional pair of outer rings (P and L) in this Gram negative
bacterium apparently serves as bearings for the rod's passage through
the cell's more complex wall and minimize friction and leakage; the L
and P rings do not occur in Gram positive bacteria. The rings are
about 0.2 μm in diameter.*

Discovery process

Just how bacteria move
was revealed by
photographing the
movements of bacteria
using high-speed motion
picture cameras. By
tethering *E. coli* to a
microscope slide, Howard
C. Berg was able to
watch as the flagellum
made the body of the
bacteria rotate about in a
counterclockwise
direction. Norbert Pfennig
photographed
Chromatium okenii at
600–1000 frames per
second and showed that
it took about one-
hundredth of a second for
the flagellar fragments to
complete one turn, during
which time the cell body
moved forward a fraction
of its own length. The
films of these experiments
dramatically show the
rotary motion of bacterial
flagella. When combined
with high-resolution
electron microscopy,
which revealed the
structure of the flagellum,
the mechanism of
bacterial movement was
shown to be similar to the
working of an electric
motor.

124

Organization
and structure of
microorganisms

Gram positive bacteria there is only one set of
rings that attaches to the cytoplasmic membrane
(Figure 4.33). The hook structure attaches the filament of the bacterial flagellum to the rod of the
basal body. The structure of the bacterial flagellum allows it to spin like a propeller and thereby
propel the bacterial cell (Figure 4.34). Effectively,
the structure allows the flagellum to spin like the
shaft of an electric motor. The fact that flagella

indeed rotate was neatly demonstrated in experiments in which bacteria were tethered to glass
microscope slides. Through the use of immunologic reactions, the flagella of bacteria can be attached to glass slides, leaving the body of the cell
free to move. Microscopic observations coupled
with high-speed photography clearly show the rotary motion as the bacteria attempt to swim. The
rotation of the flagella also can be visualized directly by coating the flagella with microscopic latex beads. The flagella are seen rotating as a line
extending from the cell, occasionally stopping and
changing direction.

The bacterial flagellum provides the bacterium
with a mechanism for swimming toward or away
from chemical stimuli, a behavior known as
chemotaxis (Figure 4.35). Chemosensors in the
cell envelope can detect certain chemicals and
signal the flagella to respond. Bacteria move to-

Figure 4.34

*Drawing showing the propeller-like rotation of
bacterial flagellum that allows bacteria to swim.*

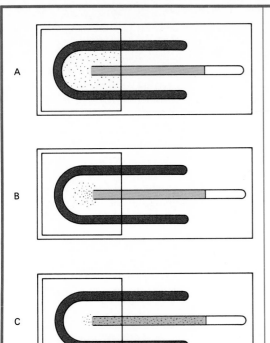

Figure 4.35

A demonstration of a chemotactic response to an attractant.

Discovery process

That bacteria respond to their chemical surroundings by chemotaxis was an important finding, showing that even the "simplest" organisms show approach-avoidance behavior. Chemotactic behavior is readily demonstrated and measured by placing the tip of a thin capillary tube containing an attractant solution in a suspension of motile *E. coli* bacteria. The suspension is placed on a slide, in a chamber created by a U-tube and a coverslip. (A) At first, the bacteria are distributed at random throughout the suspension; (B) after 20 minutes they have congregated at the mouth of the capillary; and (C) after about an hour many cells have moved up into the capillary tube. If the capillary had contained a repellant, few bacteria, if any, would have entered the tube. Using this technique, it is possible to show which chemicals attract bacteria and which substances bacteria choose to avoid.

Figure 4.36

Illustration of patterns of movement of bacteria during chemotaxis. When responding to chemotactic stimuli, bacteria move in characteristic straight lines (runs) and turning movements (twiddles). The runs and twiddles result from changes in direction of the rotation of the flagella. Movement toward a substance is characterized by long runs and few twiddles, whereas when moving away from a substance the bacterial cell exhibits short runs and numerous turns.

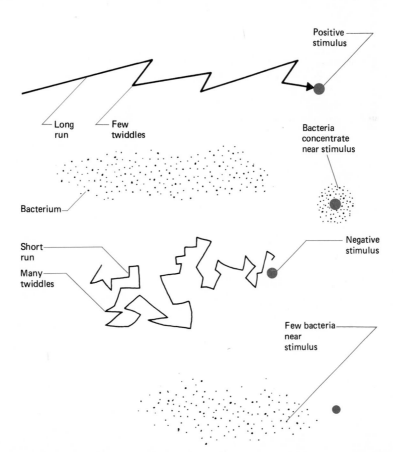

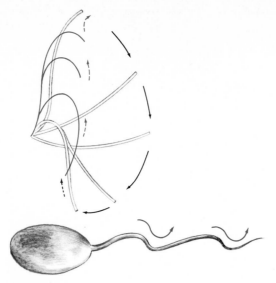

Figure 4.37

The movement of the eukaryotic flagellum occurs with a wave-like motion that propels the cell.

ward certain chemicals, known as attractants, and away from others, known as repellents. When bacteria move, they periodically change direction rather than reaching their destination by swimming in one straight line (Figure 4.36). The straight-line movements of bacteria are known as runs, and the turns are called tumbles or twiddles. The relative proportion of **running** and **twiddling** determines the overall directional movement of the bacterium. When a bacterium is moving toward a positive chemotactic stimulus (attractant), it exhibits relatively long straight runs and relatively little tumbling, whereas when it moves away from a negative chemotactic stimulus (repellent), it tumbles much more. The basis for this movement rests with the ability of the bacterium to reverse the direction of rotation of the flagellum, depending on whether it is moving toward or away from a chemical stimulus. At least in bacteria with peritrichous flagella, the counterclockwise rota-

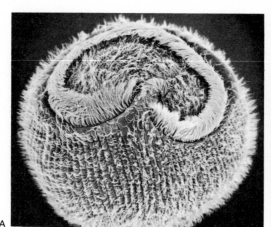

A

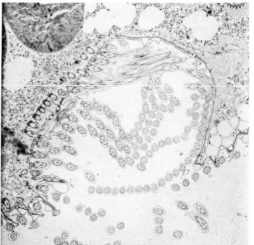

B

Figure 4.38

Micrographs of ciliated protozoans, illustrating the cilia with their characteristic "9 + 2" arrangement of microtubules. (A) A SEM of Stentor coeruleus *(320×). (From BPS—Garry T. Cole, University of Texas, Austin.) (B)* Tetrahymena rostrata *showing cross section of cilia. (Courtesy Eugene W. McArdle, Northeastern Illinois University.) (C)* Paramecium multinucleatum *(29,000×); Pe = pellicle, Cl = longitudinal section of cilium, Mi = mitichondrion, BB = basal body or kinetosome, Tr = trichocysts (Reprinted by permission of Springer-Verlag, New York, from R. G. Kessel and C. Y. Shih, 1976,* Scanning Electron Microscopy for Biology.*)*

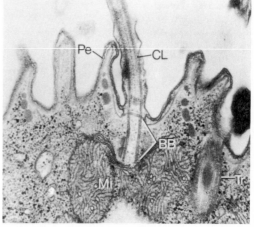

C

tion of the flagella results in a run and the clockwise rotation in a twiddle.

Flagella and cilia of eukaryotic cells

Unlike the bacterial flagella that rotate, cilia and flagella of eukaryotic cells undulate in a wave-like motion to propel the cell (Figure 4.37). Eukaryotic flagella emanate from the polar region of the cell, whereas cilia, which are somewhat shorter than flagella, surround the cell. The flagella and cilia of eukaryotic microorganisms are important taxonomic characteristics. Two of the four major groups of protozoa are defined based on whether they have cilia or flagella; the Ciliophora are grouped taxonomically because of the presence of cilia (Figure 4.38), and the Mastigophora are grouped based on the presence of flagella (Figure 4.39). Both cilia and flagella are generally involved in cell locomotion, but cilia may also be involved in moving materials past the cell surface while the organism or cell remains stationary.

In contrast to the rather simple structure of the bacterial flagellum, the flagella and cilia of eukaryotic microorganisms are far more complex and thick (Figure 4.40). Both the flagella and cilia of eukaryotic cells have the same basic structure and biochemical composition. The eukaryotic flagel-

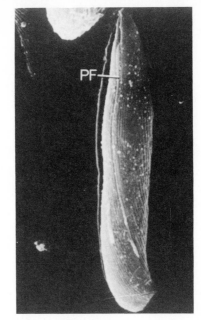

Figure 4.39

Micrograph showing a flagellate protozoa illustrating the flagellum of Peranema *(1,500×): PF = posterior flagellum. (Reprinted by permission of Springer-Verlag, New York, from R. G. Kessel and C. Y. Shih, 1976,* Scanning Electron Microscopy for Biology.)

Figure 4.40

Drawing showing the structure of the eukaryotic flagellum. Nine pairs of microtubules occur like the spokes of a bicycle wheel around the central pair.

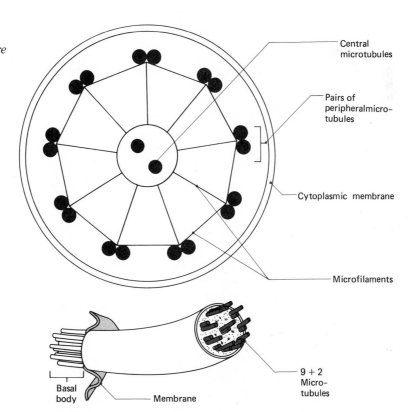

Central microtubules

Pairs of peripheralmicrotubules

Cytoplasmic membrane

Microfilaments

9 + 2 Microtubules

Basal body

Membrane

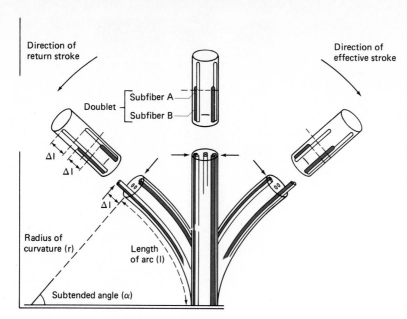

Figure 4.41

The flagella and cilia of eukaryotic cells bend because of the sliding of microtubules past each other. The microtubules on the concave side of the bend (right) have slid upward. Because the tubules are all interconnected, their changes in register can be accommodated only if the stalk of the flagellum or cilium bends.

Direction of return stroke

Direction of effective stroke

Doublet

Subfiber A

Subfiber B

Δl

Δl

Δl

Radius of curvature (r)

Length of arc (l)

Subtended angle (α)

lum consists of a series of microtubules—hollow cylinders composed of proteins—surrounded by a membrane. The tubular proteins are known as tubulin. The normal arrangement of microtubules in eukaryotic flagella and cilia is termed the "9 + 2" system because it consists of nine peripheral pairs of microtubules surrounding two single central microtubules. The nine pairs of microtubules form a circle surrounding the central microtubules. The peripheral microtubule doublets are linked to the central microtubules by radial spokes of protein microfilaments; the peripheral microtubule doublets are also similarly linked

to each other to form a circular network surrounding the spokes. This movement of the microtubules requires energy in the form of ATP. The peripheral spokes of the microtubular network contain the protein dynein, which has ATPase activity and thus is involved in coupling ATP hydrolysis to movement of the flagella or cilia. The movement of flagella or cilia appears to be based on a sliding microtubule mechanism in which the peripheral doublet microtubules slide past each other, resulting in bending of the flagella or cilia (Figure 4.41).

Structures inside the cell

Storage of genetic information

The genetic information, encoded within DNA macromolecules, determines the properties and structural characteristics of both eukaryotic and prokaryotic cells. As already indicated, the way in which the genetic information is stored is the prime distinction between prokaryotic and eukaryotic cells. In the eukaryotic cell the genome is separated from the rest of the cell and contained within the nucleus, whereas in the prokaryotic cell the DNA is not segregated from the rest of the cell constituents by a membrane barrier. Additionally, it is now clear that there are funda-

mental differences in the way that the genetic information is stored within the genomes of eukaryotic and prokaryotic cells.

Bacterial chromosome

The region occupied by the DNA is a prominent feature within prokaryotic cells (Figure 4.42). Most of the genetic information of the bacterial cell is contained within a **bacterial chromosome** composed of a single DNA macromolecule. The area occupied by the DNA is sometimes referred to as the nucleoid region, although the DNA is not contained within a separate membrane-bound organelle. The DNA in the **nucleoid region** occurs as a single large circular macromolecule of a DNA double helix (Figure 4.43). This DNA macromolecule is highly folded, but the nature of the forces maintaining the highly condensed form of the bacterial chromosome is not fully understood. In eukaryotic cells, basic proteins—called **histones**—are involved in the coiling of the DNA, but proteins do not form an integral part of the bacterial chromosome. The bacterial chromosome exhibits coiling, but this does not appear to be equivalent to the specific winding patterns observed in eukaryotic organisms.

The only proteins normally associated with the bacterial chromosome are those that are involved in DNA replication, transcription of the information in the DNA molecule to RNA, and regulation of gene expression. Although bacteria lack the basic histone proteins involved in the coiling of the DNA in eukaryotic chromosomes, a histone-like protein has been reported in at least one bacterium. The organism from which this histone-like protein was isolated lives in hot acid springs, and the basic protein appears to stabilize the DNA against heat denaturation.

Plasmids

In addition to the bacterial chromosome, bacteria may contain one or more small circular macromolecules of DNA—known as **plasmids**. All bacterial cells contain a bacterial chromosome, but not all bacteria contain plasmids. Plasmids contain a limited amount of specific genetic information that supplements the essential genetic information contained in the bacterial chromosome. This supplemental information can be quite important, establishing mating capabilities, resistance to antibiotics, and tolerance of toxic metals. Such supplemental genetic capability can permit the survival of the bacterium under conditions that are normally unfavorable for growth and survival. Pathogenic bacteria containing plasmids that code for multiple drug resistance have become a particular problem in treating some infectious diseases of humans because such bacteria are resistant to many antibiotics and can continue to grow in the body despite antibiotic treatment. Plasmids also are quite useful and are employed in genetic engineering as carriers of genetic information from a variety of sources. Because they are relatively small, plasmids are relatively easy to manipulate; they can be isolated, genetic information from other sources can be spliced into them, and they can be implanted into viable bacterial cells, permitting expression of the genetic information they contain. Genetic engineering appears to have many industrial applications that will be discussed later.

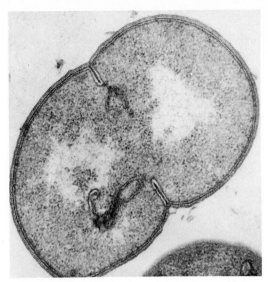

Figure 4.42

In this electron micrograph of dividing Sporosarcina ureae, *the nucleoid region is seen as light areas within the cytoplasm (63,500 ×). (From BPS—T. J. Beveridge, University of Guelph.)*

Figure 4.43

The mass of DNA contained within the circular loop of the bacterial chromosome is shown in this micrograph. (Courtesy Ruth Kavenoff, reprinted by permission of Springer-Verlag from Kavenoff and Ryder, 1976, Chromosoma 55: 23.)

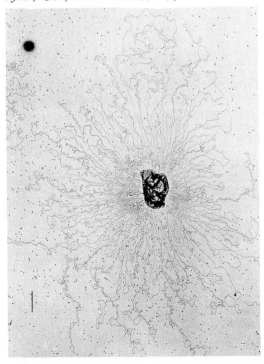

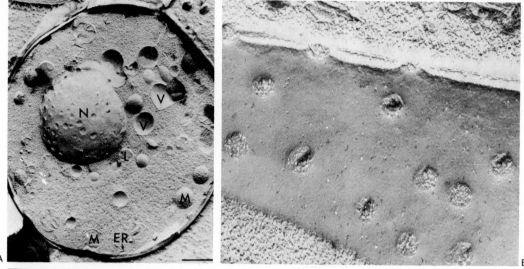

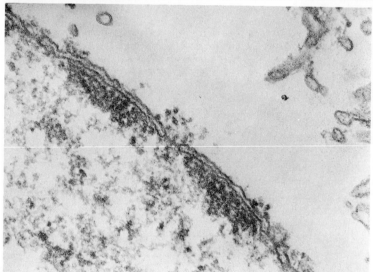

Figure 4.44

*The double membrane
structure and pores of the
nucleus are visible in these
electron micrographs. (A) An
electron micrograph of a
freeze-etched preparation of*
Saccharomyces cerevisiae,
*showing the surface of a
nucleus (7,740×). Note the
large number of pores in the
membrane of the nucleus (N),
making it look rather like a
golf ball. ER = endoplasmic
reticulum; L = lipid granules;
M = mitochondrion; and V
= vacuole. (Courtesy Sam
Conti, University of
Massachusetts, reprinted by
permission of American
Society for Microbiology, from*
E. Guth, T. Hashimoto, and S. F. Conti, 1972, Journal of Bacteriology, *109: 869-880). (B) High-
magnification freeze-etched preparation of nuclear pores (66,000×). (From BPS—D. Branton, Harvard
University.) (C) Thin section of nuclear envelope (69,600×), showing pore through which granular material
is escaping. (From BPS—W. Rosenberg, Iona College.)*

Nucleus and chromosomes
of eukaryotic cells

The genome of eukaryotic cells is contained within
the cell's **nucleus**, which is separated from the
rest of the cell by an inner and an outer membrane (Figure 4.44). In contrast to the cytoplasmic
membrane, the **nuclear membrane** is a double
layer with a distinct space between the two membranes. Further, the nuclear membrane has pores
that permit the exchange of relatively large molecules between the nucleus and the cytoplasm of
the cell. This nuclear membrane nevertheless is

selective and controls which molecules pass into
and out of the nucleus.

Within the nucleus the genetic information is
stored within **chromosomes**, which are composed of chromatin consisting of DNA and protein. The chromosomes are only visible with the
light microscope when the cell is undergoing division, and the DNA is in a highly condensed form
(Figure 4.45); at other times the chromosomes
are not condensed and are not visible as distinct
thread-like structures when the nucleus is viewed
with the light microscope. All the genetic infor-

mation resides in the DNA, although the protein component of chromatin is more abundant than the DNA. Unlike the bacterial chromosome where the DNA forms a circular macromolecule, the chromosomes of eukaryotic cells contain linear DNA macromolecules arranged as a double helix. The chromatin proteins consist primarily of five histone proteins that are basic proteins that bind to the DNA by ionic interactions. These weak bonds determine the three-dimensional configuration of the chromatin.

The DNA coils around the histones to form subunits of the chromatin known as **nucleosomes** (Figure 4.46). Each nucleosome is composed of about 200 nucleotides of DNA coiled around the histones. The resulting structures appear as spherical particles or beads on a string when viewed by electron microscopy (Figure 4.47). The nucleosomes, which establish the structural configuration of eukaryotic chromosomes, are a fundamental unit of the eukaryotic genetic material but are absent in the bacterial chromosome. This highly structured arrangement of the DNA within the nucleus is critical because the genetic information coding for the synthesis of specific proteins is split in eukaryotes and must be extensively processed before it can be properly expressed. The processing of the genetic information to form a readable message (see discussion in Chapter 7) occurs within the nucleus, which provides the necessary separation from the rest of the cell for this process to occur. In prokaryotic cells, where there is no nucleus, the information in the bacterial chromosome is not split, but rather the information coding for specific proteins oc-

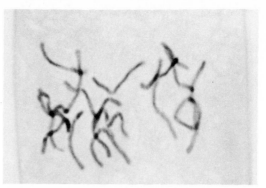

Figure 4.45

Photomicrograph showing appearance of chromosomes of the green alga Oedogonium *(2,000 ×). The chromosomes are only visible when the chromatin is condensed during a reproductive phase. The chromosomes of many eukaryotic microorganisms are more difficult to see than those of higher plant and animal cells. As a result there is much controversy about the number and morphologies of the chromosomes of many microbes. (Courtesy Larry R. Hoffman, University of Illinois, reprinted by permission from L. R. Hoffman, 1967,* American Journal of Botany *54: 273.)*

curs as a contiguous sequence of nucleotides within the DNA macromolecule. Hence, the genetic information of prokaryotes does not have to be extensively processed before it can be expressed.

An exception to the usual chromosomal arrangement occurs in dinoflagellates. These orga-

Figure 4.46

This illustration of a nucleosome shows how histones establish the configuration of DNA in eukaryotic cells. It was originally proposed in 1974 that the structure of the 100-angstrom-wide chromatin fiber showed successive 200-pair stretches of DNA wound on a series of beads—octamers composed of two molecules each of the four histones H2A, H2B, H3, and H4. At the time neither the actual shape of the histone complex nor the path of the DNA was yet known. (Based on R. D. Kornberg, and A. Klug, 1981, The nucleosome, Scientific American *244(2): 55.)*

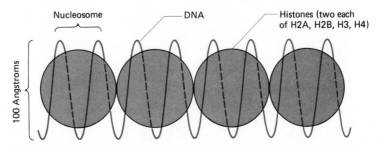

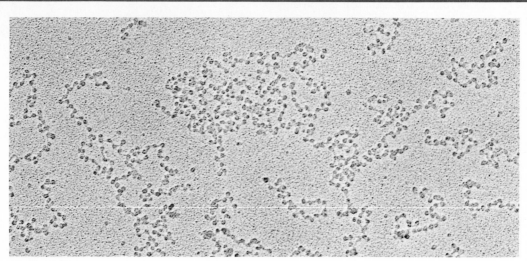

Figure 4.47

In this electron micrograph of chromatin fibers, the nucleosomes look like beads on a string. The chromatin was stretched in the preparation process, increasing the distance between nucleosomes and making it easier to distinguish them. (Courtesy Aaron Klug, Medical Research Council, Cambridge, England.)

Discovery process

The existence of the nucleosome has been established by two avenues of research: electron microscopy and studies on the enzymatic digestion of chromosomal DNA. By improving the methods for preparing chromatin fibers for electron microscopy to the point where clear and regular substructural patterns could be seen, Ada L. and Donald E. Olins at the Oak Ridge National Laboratory, C. L. F. Woodcock of the University of Massachusetts, Pierre Chambon at the Laboratoire de Genetique Moleculaire des Eucaryotes in Strasbourg, and Jack D. Griffith of Stanford University showed the chromatin fibers appearing as linear arrays of spherical particles, about 100 Å in diameter, connected by thin strands of apparently naked DNA. Micrococcalnuclease digestion of chromatin and analysis of the DNA fragments by gel electrophoresis experiments showed that the cleavage of chromatin at sites spaced at regular intervals along the DNA is a property of chromatin structure. The nucleosome was isolated, free of DNA, following dissociation of the ionic bonds between histones and DNA in solutions of high salt concentrations. Combined cross-linking and nuclease-digestion studies established the association of a histome octamer with a 200-nucleotide-pair unit of DNA. Critical to the establishment of the ordered association of histones with chromosomal DNA was the work done by Aaron Klug and colleagues at the Medical Research Council Laboratory of Molecular Biology in Cambridge, England. These studies determined the three-dimensional structure of chromosomes by using high-resolution electron microscopy and X-ray crystallographic analysis of the nuclease digests of chromosomes. Klug's work, which related the cellular and molecular levels of organization of chromatin, exemplifies his approach of chemically dissecting out parts of complex structures for detailed analysis by X-ray diffraction, the results of which are correlated with information about macromolecular organization of the intact assembly from electron microscopy. The elucidation of the histone-DNA organization of chromosomes was among several accomplishments cited in awarding the 1982 Nobel Prize in Chemistry to Aaron Klug.

nisms are eukaryotic, and the DNA is contained within their nuclei, but the DNA is not associated with histones and does not exhibit the supercoiling of the DNA typical of the chromosomes of other eukaryotic organisms. The DNA within the nucleus of dinoflagellates resembles the nucleoid region of prokaryotic cells, but except for this feature the structure of dinoflagellates conforms with that of eukaryotic cells. Dinoflagellates may represent an evolutionary link between bacteria and eukaryotic algae.

Ribosomes

The expression of the genetic information of a cell requires that the information stored in the DNA macromolecules be used to direct the synthesis of functional proteins. The **ribosomes** are the sites where protein synthesis occurs within the cell. A typical prokaryotic cell may have 10,000 or more ribosomes, and eukaryotic cells contain considerably more. Ribosomes are composed of **ribosomal ribonucleic acid (rRNA)** and protein. In the bacterium *Escherichia coli* about two-thirds of the ribosome is rRNA, and the remainder is protein. During protein synthesis the information stored in the DNA is transferred to an RNA molecule (mRNA) that acts as a messenger carrying the transcribed information to the ribosomes located in the cell's cytoplasm, where the information is translated to direct the synthesis of the protein.

There are significant differences between the ribosomes that occur in prokaryotic cells and those that are found in the cytoplasm of eukaryotic cells. The ribosomes of eukaryotic cells are larger and contain different-sized rRNA molecules than the ribosomes of prokaryotic cells. The ribosome is composed of several different rRNA molecules that make up the two structural subunits of the ribosome. The "size" of a ribosome is measured in **Svedberg units (S)**, which are units of movement in density gradient ultracentrifugation and depend on both the mass and shape of the ribosome. Svedberg units are in principle nonadditive; when the subunits combine to form the functional ribosome, one cannot simply add the sizes of the subunits to determine the size of the intact ribosome. Ribosomes are only functional when the two subunits are associated, and the establishment of functional ribosomes depends on the presence of magnesium ions for binding of the subunits.

The prokaryotic cell has 70S ribosomes composed of 50S and 30S subunits (Figure 4.48). The 30S subunit contains about 21 proteins and a 16S rRNA molecule, having approximately 1,540 nucleotides; the 50S subunit is composed of approximately 34 proteins, a 23S rRNA, having approximately 2900 nucleotides, and a small 5S rRNA species, having only about 120 nucleotides. **The**

eukaryotic cell has 80S ribosomes in its cytoplasm, composed of 60S and 40S subunits. The 40S subunit contains 18S rRNA, and the larger 60S subunit has 25–28S rRNA and 5S rRNA. In eukaryotic cells the ribosomal subunits are synthesized within the nucleus in a region known as the **nucleolus** and are transported through the pores of the nuclear membrane to the cytoplasm where the assembly of the 80S ribosome occurs. In addition to their 80S ribosomes, eukaryotic cells also have 70S ribosomes within their mitochondria and chloroplasts. The 70S ribosomes of these organelles appear to be very similar to the ribosomes of prokaryotic cells.

The sequence of nucleotide bases within the rRNA molecules of the ribosome appears to provide a basis for assessing the phylogenetic relatedness of microorganisms. The argument for placing the archaebacteria in a separate kingdom is based largely on the analysis of the nucleotide sequences in their 16S rRNA, which shows that although the rRNA of most of the bacteria examined formed one coherent similarity group, the methanogens and other bacteria—such as *Halobacterium*, *Sulfolobus*, and *Thermoplasma*, which are considered to be archaebacteria—formed a second coherent similarity group. While this type of analysis must be further tested to establish its validity for assessing phylogenetic relationships, it does appear to be an appropriate way of assessing genetic relatedness. The usefulness of using 16S rRNA analyses is based on the supposition that rRNA molecules have been stable over long periods, and the pragmatic ease with which 16S rRNA can be isolated and analyzed.

The differences in the structural composition of prokaryotic and eukaryotic ribosomes forms an important basis for using antibiotics in the treatment of animal and plant diseases caused by bacteria. Protein synthesis that occurs at the ribosomes is essential for cells to carry out life-supporting metabolism, and any disruption of the ribosomal conformation can disrupt this essential process. Many antibiotics, such as erythromycin

Figure 4.48

A basic difference between prokaryotic and eukaryotic cells is the nature of the ribosomes found in the cytoplasm. These differences form the basis for the specificity of action of some antibiotics that inhibit protein synthesis.

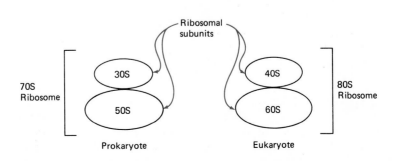

and streptomycin, are effective because they bind to and alter the shape of 70S ribosomes. Such antibiotics are useful therapeutically because they selectively attach to 70S ribosomes and hence disrupt protein synthesis in bacteria but do not exhibit any affinity for 80S ribosomes and therefore do not disrupt protein synthesis in eukaryotic human cells. Here we can see the practical application of a fundamental difference in the cellular structure of eukaryotes and prokaryotes.

Reserve materials

When conditions are favorable, microorganisms often store various biochemicals, which they have synthesized or accumulated, within the cell to act as nutrient reserves that can be used in times of need. Many microorganisms accumulate granules of **polyphosphate**, which are reserves of inorganic phosphate that can be used in the synthesis of ATP. Polyphosphate granules can be seen in light microscopy after staining and are sometimes termed **volutin** or **metachromatic granules**. Microorganisms may also accumulate reserves of organic carbon molecules that can be metabolized at a later time for the generation of ATP and cell constituents. In eukaryotic organisms such reserve materials normally accumulate in membrane-bound vacuoles. In bacteria the reserve materials accumulate as cytoplasmic inclusions not separated by a boundary membrane from the rest of the cytoplasm. Rather, the separation of the reserve inclusions is generally based on differential solubility.

Figure 4.49

Electron micrograph of a nitrogen-starved Alcaligenes eutrophus *(132,000×), showing accumulation of PHB, the white globular areas within the cell. (Courtesy Richard Bartha, Rutgers, The State University.)*

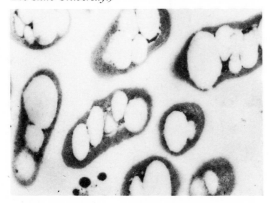

The most common reserve material that accumulates in bacteria is **poly-β-hydroxybutyric acid (PHB)**, a lipid-like molecule (Figure 4.49). Microorganisms may accumulate other lipid molecules that can be differentially stained and viewed by light microscopy. Some microorganisms, including bacteria, accumulate the polysaccharide reserve materials glycogen or starch; these latter polysaccharide reserve materials most commonly occur in eukaryotic microorganisms. Different microorganisms store different reserve materials, and this is sometimes used as a criterion for their classification.

Spores

Another feature often employed to distinguish microbial species is the production of characteristic **spores**. Spores are important in the taxonomic classification of eukaryotic microorganisms, but spore production is also the key feature used to define several bacterial genera. Numerous different types of spores are produced by microorganisms, especially eukaryotes, and the various spores serve different functions. Spores typically are involved in the reproduction, dispersal, or survival of the organism. Those spores involved in reproduction are metabolically quite active, and those involved in dispersal or survival of the microorganism often are metabolically dormant. Spores involved in the dispersal of microorganisms usually are quite resistant to desiccation, and the production of such spores is an important adaptive feature that permits the survival of microorganisms for long periods of time during transport in the air. The spread of many fungi depends on the successful transport of fungal spores from one place to another. Unfortunately, many of these fungi are plant pathogens and cause great agricultural damage as a result of their ability to effectively move from field to field.

Bacterial endospore

Of the many types of microbial spores one specific type—the **bacterial endospore**—has special importance (Figure 4.50). The endospore is a complex seven-layered structure containing murein within its complex spore coat and **calcium dipicolinate** within its core (Figure 4.51). The endospore is highly refractory and resistant to desiccation, retaining its viability over extended periods of time under conditions that do not permit growth of the organism. The major importance of the bacterial endospore rests with its resistance to high temperatures. **Endospores can survive ex-**

A 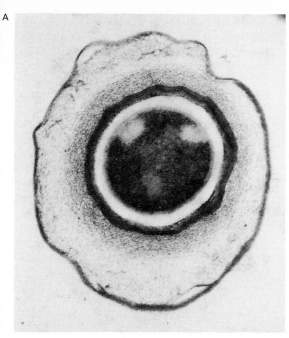 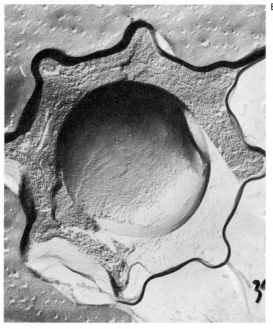 B

Figure 4.50

Electron micrographs of bacterial endospores. (A) Bacillus sphaericus *(59,300×) in thin section; (B) freeze-etched preparation of a* Bacillus polymyxa *spore. (From BPS—Stanley C. Holt, University of Massachusetts.)*

posure to high temperatures for extended periods, whereas normal bacterial vegetative cells are killed by brief exposures to such high temperatures. The mechanism of heat resistance must protect the cell's macromolecules from denaturation and disruption of their critical three-dimensional conformation. It appears that the presence of calcium dipicolinate is involved in conferring heat resistance on the endospore. Cells growing in a medium lacking calcium and mutant strains that cannot form calcium dipicolinate produce endospores that are not particularly resistant to elevated temperatures.

Few bacterial genera are capable of forming endospores; the most important endospore formers are members of the genera *Bacillus* and *Clostridium*. Both *Bacillus* and *Clostridium* are defined as Gram positive rods that form endospores; *Bacillus* is aerobic, growing in the presence of oxygen, and *Clostridium* is obligately anaerobic, growing only in the absence of oxygen. That members of these genera form endospores presents special problems for the food industry, which employs processes that rely on heat to sterilize products. Some endospores can withstand boiling for more than one hour. Endospores are formed when conditions are unfavorable for continued growth of the bacterium. Once formed,

endospores can retain viability for millennia. Under favorable conditions the endospore can germinate and give rise to an active vegetative cell of the bacterium. In order to ensure that endospores are killed, it is necessary to heat liquids to

Figure 4.51

This drawing shows the complex multilayered structure of the bacterial endospore.

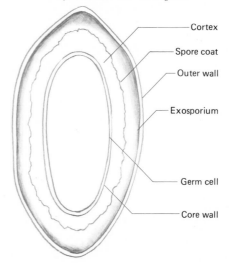

- Cortex
- Spore coat
- Outer wall
- Exosporium
- Germ cell
- Core wall

a temperature greater than 120°C and hold that temperature for at least 15 minutes; dry materials require several hours at this temperature to ensure sterilization.

Membrane structures within the cell

Internal membranes of prokaryotic cells

Although prokaryotic cells are generally characterized by a lack of internal membrane-bound organelles, certain specialized groups of bacteria do contain extensive internal membranes. Such groups of bacteria include some nitrifying bacteria and the photosynthetic bacteria. In these nitrifying and photosynthetic bacteria, the cells literally may appear to be filled with membranes (Figure 4.52). In the photosynthetic bacteria these membranes are the anatomical sites of the light reactions of photosynthesis that occur in the process of oxidative photophosphorylation and result in the conversion of light energy to chemical energy within ATP (see Chapter 5 for a discussion of this process). In the nitrifying bacteria the internal membranes similarly are involved in the generation of ATP, only in this case through the oxidation of inorganic nitrogen compounds. Several other bacteria, including hydrocarbon oxidizers, form similar extensive internal-membrane networks that provide the necessary structure for synthesizing relatively large quantities of ATP. With the development of our understanding of the role of chemiosmosis in ATP generation, it is clear that membranes are essential for forming the separation of the hydrogen ion gradient used in the synthesis of ATP.

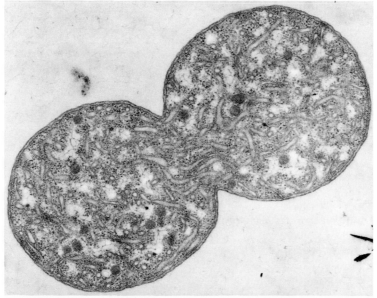

A

B

Figure 4.52

Some specialized bacteria have extensive internal membrane networks, as shown in these electron micrographs. (A) This electron micrograph of the nitrifying bacterium Nitrococcus mobilis *shows the extensive internal membranes (82,000 ×). (Courtesy Stan Watson, Woods Hole Oceanographic Institution, Woods Hole, Massachusetts.) (B) This cyanobacterium,* Synechoococcus lividans, *has been freeze-etched to reveal its internal membrane structure. (From BPS—Stanley C. Holt, University of Massachusetts.)*

Figure 4.53

This micrograph shows the gas vacuoles of Nostoc muscorum, *a cyanobacterium that has been freeze etched (57,600×). (From BPS—J. Robert Waaland, University of Washington.)*

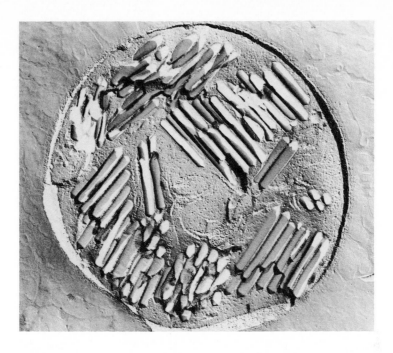

The specialized **photosynthetic membrane** arrangements of photosynthetic bacteria may appear as spherical vesicles or as flattened sheet-like layers. Such photosynthetic membranes occur in the purple and green anaerobic photosynthetic bacteria and in the cyanobacteria (formerly known as the blue-green algae). In these bacteria the photosynthetic membranes occur within the cytoplasm of the cell and are not isolated from the rest of the cytoplasm by a boundary membrane. The photosynthetic membranes of the purple photosynthetic bacteria are contiguous with the cytoplasmic membrane, filling much of the cytoplasm. In the green photosynthetic bacteria the photosynthetic membranes are cylindrically shaped vesicles; these vesicles are attached to the cytoplasmic membrane and have an unusual single-membrane-layered appearance. The cyanobacteria often form extensive multilayered photosynthetic membrane structures known as thylakoids. Unlike the photosynthetic membranes of eukaryotic cells that occur within the chloroplast, the bacterial photosynthetic membranes do not form the stacked structures known as grana.

In addition to these membranous networks, some bacteria form "membrane-bound" **gas vacuoles** (Figure 4.53). Actually, the boundary layers of these vacuoles are not true membranes but rather are composed exclusively of protein. The "membrane" layer appears to be only one protein molecule thick. The protein composing the boundary layer of these vacuoles has both hydrophilic and hydrophobic properties. The formation of gas vacuoles by aquatic bacteria provides a mechanism for adjusting the buoyancy of the cell and thus the height of the bacterium in the water column. Many aquatic cyanobacteria, for example, use their gas vacuoles to move up and down in the water column, depending on light irradiation levels to achieve optimal conditions for carrying out their photosynthetic metabolism. The presence of gas vacuoles thus provides the necessary structural units for phototaxis in aquatic bacteria.

Some bacteria also contain membrane-bound iron granules that permit them to exhibit **magnetotaxis** (Figure 4.54). Bacteria can use these granules to navigate along the earth's magnetic field. Some bacteria move predominantly north, and others move south. For some anaerobic bacteria, magnetotaxis allows them to orient the cells pointed downward into the sediment.

Membranous organelles in eukaryotic cells

In marked contrast to the prokaryotic cell, the eukaryotic cell is filled with membranous organelles. We have already discussed the importance of the nucleus in separating the genome of the eukaryotic cell from the rest of the cell's contents. The extensive internal membrane systems of eukaryotic microorganisms permit the efficient segregation of function, adding versatility to the metabolic functioning of the eukaryotic cell and increasing the need to coordinate and manage the functions of the cell's subunit organelles. Many

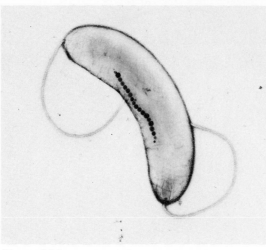 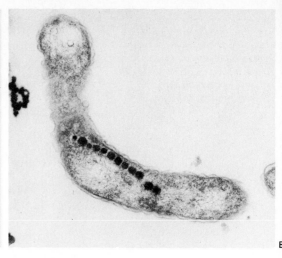

A   B

Figure 4.54

An interesting and relatively new finding is that some bacteria have internal magnetic structures that permit them to navigate. (A) In this electron micrograph, iron granules are prominent within a magnetotactic bacterium, which was found in a water treatment plant in Durham, New Hampshire (17,900×). (Courtesy R. Blakemore and N. Blakemore, University of New Hampshire). (B) This electron micrograph of Aquaspirillum magnetotacticum *(37,200×), shown in thin section, illustrates the membrane around the magnetic iron particles. (Courtesy R. Blakemore and D. Maratea, University of New Hampshire.)*

of the organelles of the eukaryotic cell are linked together so that they can function in a coordinated manner.

Mitochondria   The **mitochondria** are the sites of extensive ATP synthesis in eukaryotic cells and, as one would predict, an organelle carrying out such a function should have a large membrane surface area. The mitochondria have an outer unit membrane that acts as the boundary between this organelle and the cell cytoplasm, and an independent interior membrane that exhibits extensive folding (Figure 4.55). The convolutions of the inner membrane that extend into the interior of the mitochondrion are called **cristae** (Figure 4.56). This inner membrane has a higher proportion of protein associated with it than the outer mitochondrial membrane. Many of these proteins are involved in energy-transferring metabolic reactions. The metabolic pathways of the Krebs cycle and of oxidative phosphorylation occur within the mitochondria (see Chapter 5 for a discussion of these processes). The inner matrix space between the invaginated inner membrane of the mitochondrion is the site of Krebs cycle reactions; the membrane matrix itself is the location of the electron transport chain involved in oxidative phosphorylation. As a result of electron transport through a series of carriers embedded asymmetrically within the membrane, a gradient of hy-

drogen ions across the inner membrane of the mitochondrion is used to drive the synthesis of ATP in eukaryotic cells (see Figure 5.12).

In addition to considering the relationship between structure and function of the mitochondrion, several aspects of the structure of the mitochondrion are of interest in evolutionary theory. These organelles of eukaryotic cells show a marked resemblance to the prokaryotic cell; they are approximately the same size as a bacterial cell, a fact that probably explains why one never finds mitochondria within prokaryotic cells. Within the mitochondrion there are 70S ribosomes and a circular strand of DNA that is arranged in a manner similar to the bacterial chromosome. The mitochondrial membranes appear to lack the sterols found in the cytoplasmic membranes of eukaryotic cells. Further, the 16S rRNA of the 70S mitochondrial ribosomes shows a high degree of similarity in its nucleotide sequence with the 16S rRNA from prokaryotic cells. These similarities have led some to propose that mitochondria evolved from prokaryotic cellular organisms. Based on these facts it would appear that the mitochondria have descended from prokaryotic cells that became entrapped in larger eukaryotic cells, eventually forming a stable relationship and evolving into the present mitochondrial structure. However, the DNA contained in mitochondria is not sufficient to direct the synthesis of the entire

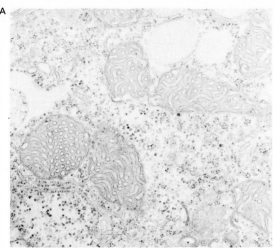

A

B

Figure 4.55

(A) Electron micrograph of Tetrahymena rostrata, *showing the mitochondria. As shown here, the inner membranes of the mitochondrion of many microorganisms appear as tubular structures. (Courtesy Eugene W. McArdle, Northwestern Illinois University.) (B) Cross section of a growing hyphae of the fungus* Trichoderma reesei; M = *mitochondrion with distinct cristae, N = nucleus, NM = nuclear membrane, NP = nuclear pore, RER = rough endoplasmic reticulum, V = vacuole, i = invagination common to vacuoles, SV = smooth surface vesicles, PS = polysaccharide layer on the outside surface of the cell wall. (Courtesy Bijan Ghosh, Rutgers, The State University.)*

structure and it has been recently found that the genetic code is different for mitochondrial DNA than for DNA from other sources, fueling the debate over the relatedness of the mitochondria of eukaryotes to the prokaryotic cell.

Chloroplasts **Chloroplasts** are quite similar in many ways to mitochondria, as they are composed of extensively invaginated membranes, contain 70S ribosomes, and have a circular DNA macromolecule—as do the mitochondria. Chloroplasts are one form of **plastid**, which are large

cytoplasmic organelles occurring within the cytoplasm of photosynthetic eukaryotic organisms (Figure 4.57). Among the microorganisms, chloroplasts occur exclusively in the algae, where they are the sites of ATP synthesis. Like the mitochondrion, the chloroplast contains an outer membrane that separates the organelle from the cytoplasm and an inner membrane, but the chloroplast also has an additional complex internal membranous system—the **thylakoids.** The interior compartment of the chloroplast, defined by the inner membrane, is known as the **stroma,** and

Figure 4.56

Drawing of the structure of a mitochondrion, showing its two distinct membranes and the extensive folding of the internal membrane.

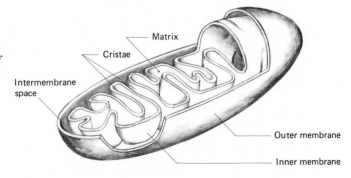

Matrix

Cristae

Intermembrane space

Outer membrane

Inner membrane

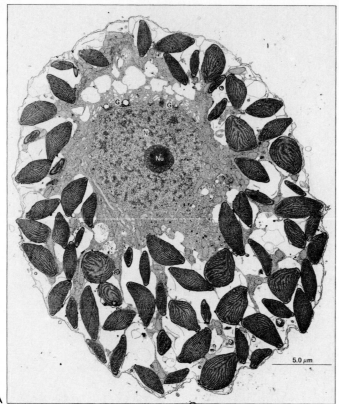

A

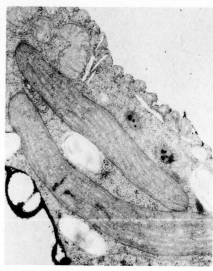

B

Figure 4.57

Electron micrographs showing chloroplasts, the sites where light energy is converted to ATP in algae. (A) Vacuolaria virescens *(6,300 ×) is shown in longtitudinal section. Note that the chloroplasts are in large numbers between the cell membrane and the dense cytoplasmic layer surrounding the nucleus (N). Nu = nucleolus, G = Golgi body. (Courtesy Peter Heywood, Brown University, reprinted by permission from P. Heywood, 1977,* Journal of Phycology, *13: 69). (B) This* Euglena gracilis *shows the extensive internal photosynthetic membranes of the chloroplast (9,900 ×). (From BPS— Stanley C. Holt, University of Washington.)*

it is here that the enzymes involved in the dark reactions of photosynthesis and the fixation of carbon dioxide occur. The thylakoids contain chlorophylls and the proteins responsible for the conversion of light energy to chemical energy in the form of ATP in the process of oxidative photophosphorylation. The photosynthetic pigments associated with the thylakoid membranes, including the chlorophylls, are responsible for trapping light energy and initiating the process of oxidative photophosphorylation. The establishment of a hydrogen ion gradient across the thylakoid membranes drives the synthesis of ATP during oxidative photophosphorylation (see Figure 5.23); this is analogous to the synthesis of ATP in the mitochondria, except that the flow of hydrogen ions is inward in the case of the chloroplast and outward in the case of mitochondria.

Within the chloroplast, the sac-like membranous vesicles (**thylakoids**) may be stacked to form **grana**, which normally are densely packed piles of individual thylakoids. The chloroplast, therefore, may be viewed as containing stacks or col-umns of grana in the matrix space of the stroma (Figure 4.58). This is a somewhat oversimplified view, as there are great variations in the structures of chloroplasts in different algae. In general, the subunits of the chloroplast structure are less organized than the highly specialized organization characteristic of higher plants, and in the brown algae, for example, no grana occur and the thylakoid membranes are not stacked. The algae also show major variations with respect to the auxiliary photosynthetic pigments associated with these membranes. The auxiliary pigments within the chloroplast confer characteristic colors on the algae and determine what wavelengths of light can be used for initiating oxidative photophosphorylation. The biochemical nature of the light-absorbing pigments is a basic taxonomic characteristic used in defining the major algal groups that are referred to by their common colors, such as the green, red, and brown algae.

Endoplasmic reticulum

In addition to the chloroplasts and mitochondria, which are in-

Figure 4.58

Drawing of a chloroplast, showing its double membrane structure and the stacks of membranes (grana) within the chloroplast.

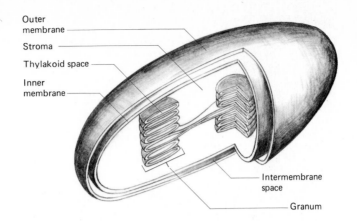

Outer membrane
Stroma
Thylakoid space
Inner membrane
Intermembrane space
Granum

volved in energy conversions, eukaryotic cells contain an extensive membranous network known as the **endoplasmic reticulum**. The appearance of the endoplasmic reticulum varies greatly between different eukaryotic cells but always forms a system of fluid-filled sacs enclosed by the membrane network. This large membranous network appears to serve several functions within the eukaryotic cell. There is evidence that the endoplasmic reticulum forms a continuum with the outer nuclear membrane and may provide a communication network for coordinating the metabolic activities of the cell. The endoplasmic reticulum shows two distinct appearances when examined by electron microscopy: in one case the endoplasmic reticulum (ER) appears rough and has attached ribosomes, and in the other case the ER appears smooth and is not associated with ribosomes (Figure 4.59). The ribosomes of prokaryotic cells do not attach to an analogous membrane structure.

The attachment of ribosomes to the endoplasmic reticulum allows for coordinated activity, and the channels of the endoplasmic reticulum provide a system through which the synthesized protein can be transported. In fact, it appears that many of the proteins synthesized by ribosomes attached to the endoplasmic reticulum are destined to be transported out of the cell or for incorporation into membranes. Proteins synthesized on free ribosomes not associated with the endoplasmic reticulum are not transported through the channels of this membranous network and appear to be destined for use within the cytoplasm of the cell. Additionally, many enzymes are associated with both the smooth and rough endoplasmic reticulum, and this large membrane network appears to provide a large surface for enzymatic activities, including the synthesis of lipids for membrane structures. The endoplasmic

reticulum appears to be the source for some of the membranes of the other organelles of eukaryotic cells.

The Golgi apparatus The **Golgi apparatus** is closely associated with the rough endoplasmic reticulum, and appears to interact with the rough ER to establish an integrated function. Normally, four to eight Golgi bodies, which are flattened membranous sacs, are stacked to form the Golgi apparatus (Figure 4.60). The Golgi apparatus is sometimes referred to as the Golgi complex and the individual stacks of membranes as dictysomes. The proteins formed at the ribosomes of the rough endoplasmic reticulum are transferred to the Golgi apparatus by a process of vesicular budding, called **blebbing** (Figure 4.61). Golgi bodies are the sites of various synthetic activities through which polysaccharides and lipids can be added to proteins to form lipoproteins, glycoproteins, and various polysaccharide derivatives; these complex biochemicals are essential for the synthesis of various cell constituents. Membrane sacs from the endoplasmic reticulum carry protein and lipids to the Golgi apparatus, where repackaging into secretory vesicles occurs. The secretory vesicles, which are formed by the Golgi apparatus, then move to the cytoplasmic membrane where they release their contents through exocytosis. Such a process is important for the construction of cellular structures that occur externally to the cytoplasmic membrane, such as the cell wall. Taken together, the outer nuclear membrane, the endoplasmic reticulum, the Golgi bodies, and the secretory vesicles form a sequence of coordinated activities that move and process biochemicals associated with protein synthesis through the cytoplasm of the cell to the exterior of the cytoplasmic membrane.

Figure 4.59

(A) Drawing of rough and smooth endoplasmic reticulum. The terminology is derived from the fact that when viewed by electron microscopy, the endoplasmic reticulum with attached ribosomes appears bumpy or rough, whereas the endoplasmic reticulum lacking attached ribosomes appears smooth. (B) Electron micrograph of ribosomes arranged along the rough endoplasmic reticulum of the protozoan Tetrahymena rostrata. *(Courtesy Eugene W. McArdle, Northeastern Illinois University.) (C) Electron micrograph of a thin section of the fungus* Trichoderma reesei *showing large numbers of ribosomes (R) attached to the endoplasmic reticulum membrane and the cisternal space (c) of the endoplasmic reticulum. (Courtesy Bijan Ghosh, Rutgers, The State University.)*

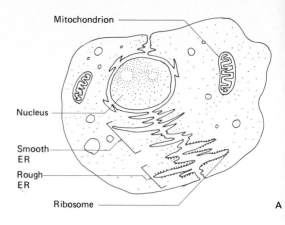

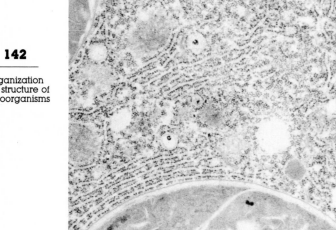

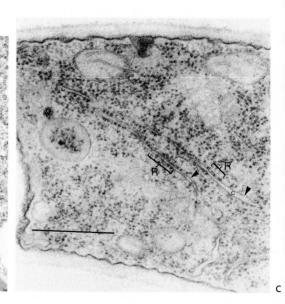

Figure 4.60

Electron micrograph of Golgi apparatus of Tetrahymena rostrata, *showing stacked Golgi bodies (52,200×). (Courtesy Eugene W. McArdle, Northeastern Illinois University.)*

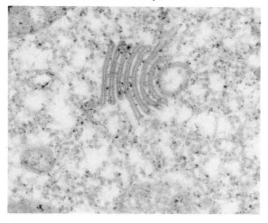

Lysosomes **Lysosomes** are specialized membrane-bound organelles that are probably produced in the Golgi apparatus. They contain various hydrolytic enzymes, and various digestive activities of the cell occur within the lysosome. These hydrolytic enzymes must be packaged rapidly into these membranous organelles to prevent their release into the cytoplasm, and the coordination between the endoplasmic reticulum and the Golgi apparatus establishes the mechanism by which this containment can be achieved. The lysosome membrane is impermeable to the outward movement of these hydrolytic enzymes and also is resistant to their action. This segregation of certain enzymes within the lysosome is necessary to prevent the degradation of essential biochemical components of the eukaryotic cell because the enzymes within the lysosomes often are capable of digesting many of the cell's structural

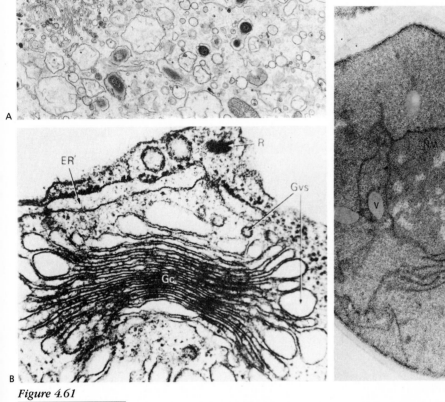

Figure 4.61

(A) Golgi apparatus forming vesicles in an Amoeba. *(Courtesy O. Roger Anderson, Columbia University.) (B)
The Golgi apparatus of the green alga* Dunaliella, *showing the formation of vesicles. GC = Golgi cisternae,
Gvs = smooth vesicles formed from the Golgi, ER = rough endoplasmic reticulum, and R = base of
flagellum. (Reprinted by permission of the Society of Protozoologists from B. P. Eyden, 1975,* Journal of
Protozoology *22: 339.) (C) Cross section of the yeast* Saccharomyces cerevesiae; *G = a single-layered Golgi,
N = nucleus, NM = nuclear membrane, NP = nuclear pore, V = vacuoles. (Courtesy Bijan Ghosh,
Rutgers, The State University.)*

components. Indeed, one of the functions of the enzymes within the lysosomes is to digest prokaryotic cells that have been ingested by phagocytosis.

Microbodies Microbodies have been found in eukaryotic cells and appear to be similar in function to lysosomes in that they isolate specialized enzyme functions within the cell. Microbodies are smaller than lysosomes and appear to isolate metabolic reactions that involve hydrogen peroxide. Microbodies contain catalase, which immediately breaks down the hydrogen peroxide to oxygen

and water. The **peroxisome** is one type of microbody that has been found in eukaryotic microorganisms. Peroxisomes are involved in the oxidation of amino acids that produces hydrogen peroxide. If the peroxides formed in these reactions were not contained or destroyed, they could oxidize a number of biochemicals within the cell, resulting in the death of the cell.

Vacuoles Various types of membrane-bound vacuoles, serving differing purposes, occur within eukaryotic microorganisms (Figure 4.62). Whereas reserve materials are not separated from the cy-

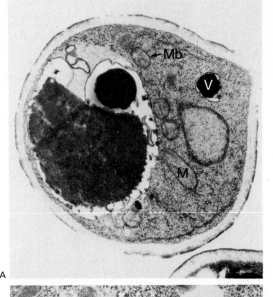

Figure 4.62

Electron micrographs showing various vacuoles found in eukaryotic cells. (A) Cladosporium resinae *(30,000×): M = mitochondrion, Mb = microbody, V = storage vacuole. The large vacular inclusion bodies have not been identified. (Courtesy Richard A. Smucker, Chesapeake Bay Laboratory, University of Maryland.) (B) A food vacuole of the ciliate protozoan* Tetrahymena rostrata *filled with membranes of bacteria that have been ingested. (Courtesy Eugene McArdle, Northeastern Illinois University.) (C) A freeze-etched preparation of the yeast* Saccharomyces cerevisiae *in a late stage of meiosis, showing prominent storage vacuoles (V) and a dumbbell-shaped nucleus (N); ERV = endoplasmic reticulum vesicles. (Courtesy Sam Conti, University of Massachusetts, reprinted by permission of the American Society for Microbiology, from E. Guth, T. Hashimoto, and S. F. Conti, 1972,* Journal of Bacteriology *109: 874)*

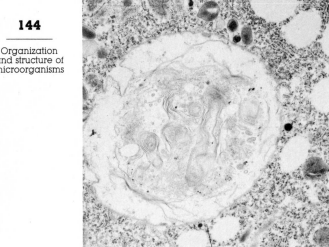

toplasm by a membrane barrier in prokaryotic cells, one type of vacuole, the **storage vacuole**, is involved in maintaining accumulated reserve materials segregated from the cytoplasm within eukaryotic cells. For example, yeast vacuoles can store polyphosphate, amino acids, and uric acid as reserve materials. Other organisms store other forms of organic carbon, nitrogen, and phosphate reserves for times of need. Other vacuoles are involved in the movement of materials out of the cell. These vacuoles can unite with the cytoplasmic membrane during endo- and exocytosis. In some cases a vacuole formed when the cell engulfs a food source fuses with lysosomes, es-

tablishing a digestive vacuole that permits digestion of the contents. A specialized type of vacuole, the **contractile vacuole**, occurs in some protozoa and acts to pump water and wastes out of the cell, affording protection for the organism against osmotic shock and a mechanism for excluding sometimes toxic materials from the cell (Figure 4.63). As water enters the cell because of differences in water potential, it fills the contractile vacuole; the contractile vacuole then suddenly contracts, forcing the water out of the cell (Figure 4.64). This is an interesting adaptation for protecting the cell against osmotic shock in organisms lacking a rigid cell wall structure.

Figure 4.63

(A, B) Electron micrographs showing the contractile vacuole system of Paramecium aurelia. (Courtesy Eugene McArdle, Northeastern Illinois University, reprinted by permission of Elsevier Publishing Company, Amsterdam, from Paramecium: A Current Survey, *1976.) (A) The nephridial canals (nc) connect directly through openings with surrounding tubules; (B) the discharge channel (dc) is surrounded by microtubules that run in two directions; the contractile vacuole opens to the outside of the cell. (C) Electron micrograph of a contractile vacuole pore in* Stentor coeruleus *(1,200 ×); CVP = contractile vacuole pore, MB = membranellar band of cilia, FF = frontal field of cilia, (Reprinted by permission of Springer-Verlag, New York, from R. G. Kessel and C. Y. Shih, 1976,* Scanning Electron Microscopy for Biology.)

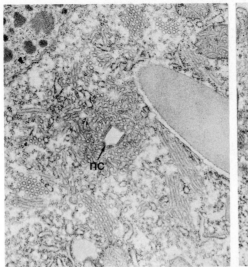

A

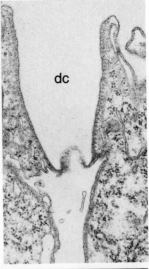

B

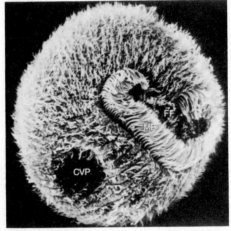

C

Figure 4.64

Diagram showing the action of contractile vacuole.

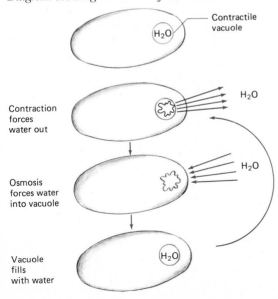

Contractile
vacuole

H_2O

Contraction
forces
water out

H_2O

Osmosis
forces water
into vacuole

H_2O

Vacuole
fills
with water

H_2O

Cytoskeleton

In addition to the numerous membrane-bound organelles that occur in eukaryotic but not in prokaryotic cells, the eukaryotic cell has a **cytoskeletal network**, consisting of microtubules and microfilaments, which is implicated in the ability of the cell to move and to maintain its shape. This cytoskeletal network appears to link the various components of the cytoplasm into a unified structure, the cytoplast, providing the rigidity needed to hold the various structures in their appropriate locations (Figure 4.65). The microtubular-microfilament arrangement of the cytoskeleton runs throughout the eukaryotic cell, connecting membrane-bound organelles with the cytoplasmic membrane. This cytoskeletal structure appears to be involved in both the support and movement of membrane-bound structures, including the cy-

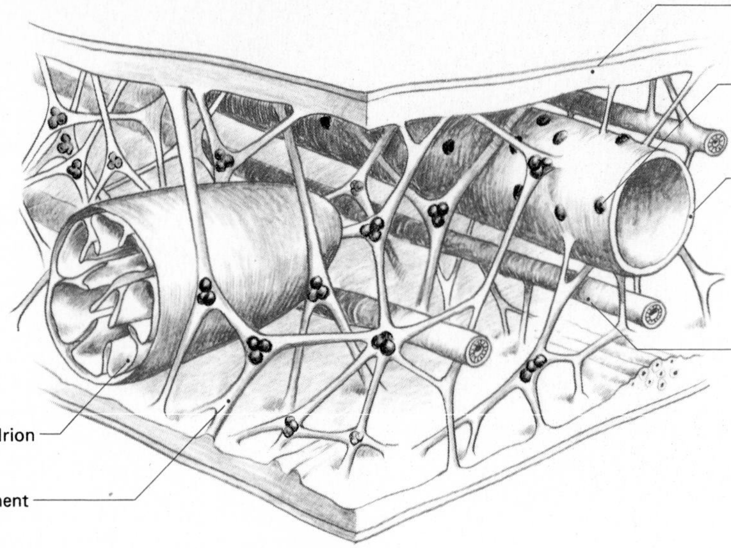

Cell membrane

Ribosome

Endoplasmic reticulum

Microtubule

Mitochondrion

Microfilament

Figure 4.65

Drawing showing the complex cytoskeleton of the eukaryotic cell.

toplasmic membrane and the various organelles of the eukaryotic cell. The involvement of microtubules and microfilaments, composed of various proteins, in eukaryotic cilia and flagella has already been discussed. The movement of the microtubules also permits the extension of the cytoplasmic membrane and cytoplasm, providing the basis for locomotion through the formation of

false-feet (pseudopodia), as occurs in some protozoa, such as *Amoeba*. Similarly, the cytoskeleton appears to be involved in the ability of a cell to carry out cytosis as a means of moving materials into and out of the cell. The lack of a cytoskeleton in prokaryotic cells may explain why bacteria have not been found to be capable of phagocytosis.

Structure of viroids and viruses

Viroids

Unlike prokaryotic and eukaryotic cells, viroids and viruses are acellular. **Viroids** are, in fact, simply composed of RNA. They have no other structures, and their RNA genomes are quite small. Inside a suitable host cell, however, the RNA is capable of initiating its own replication. The presence of viroids sometimes manifests itself as disease symptoms in the host organism, and certain plant and animal diseases have been identified as caused by viroids. It is also suspected that some diseases of humans are caused by viroids, but this has yet to be confirmed. In essence, a viroid is simply a macromolecule that can be preserved and transmitted to cells where it is reproduced. Such macromolecules contain the information needed for directing their own replication but are totally dependent on the metabolic activities of a

host cell for accomplishing this task. It is not yet clear how these molecules survive outside of host cells, and how they are transported. That such macromolecules can be transmitted and cause infectious diseases of higher organisms is a relatively new finding, the ramifications of which have yet to be fully appreciated.

Viruses

Viruses are more highly structured than viroids, having an outer protective layer surrounding their genetic material. A central genetic nucleic acid core surrounded by a protein coat normally constitutes the entire structure of the virus (Figure 4.66). Unlike prokaryotic and eukaryotic cells, viruses only contain one type of nucleic acid, either RNA or DNA. The genome of the virus may consist of double-stranded DNA, single-stranded DNA, single-stranded RNA, or double-stranded RNA. The ability of nucleic acid molecules other than double-stranded DNA to store the genetic information of the organism is unique among biological systems.

The viral coat structure surrounding the nucleic acid, known as the **capsid**, is composed of protein subunits called **capsomers**. Whereas the capsid of some viruses is composed of a single type of protein, many other viruses have more complex capsids containing several different proteins. There are two basic types of capsids, helical

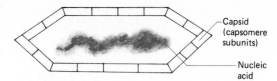

Figure 4.66

Drawing of a virus showing that the essential components are a nucleic acid core within a protein coat.

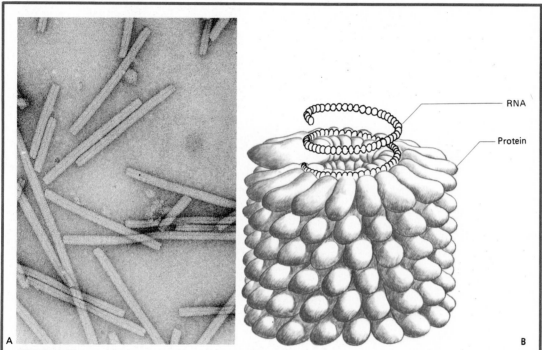

Figure 4.67

In tobacco mosaic virus (TMV) the subunits of the capsid are helically coiled around the virus's RNA nucleic acid. (A) An electron micrograph of TMV (Courtesy Lee Simon, Rutgers, The State University.) and (B) a drawing showing the detailed structure of TMV.

Discovery process

The elucidation of the structure of tobacco mosaic virus was accomplished in large part by work at the Medical Research Council's Laboratory of Molecular Biology in Cambridge, England, under the direction of Aaron Klug. It took 20 years to determine the complete details of the structure and assembly of TMV. Klug developed electron microscopic methods for the visualization of three-dimensional structures; his development of crystallographic electron microscopy permitted the elucidation of nucleic acid-protein complexes that are essential to the construction of viruses such as TMV. Understanding structure is an important part of learning how biological entities function. Specifically, Klug's work showed how the components of TMV are put together; starting with one disk of protein surrounding a strand of RNA, the initial complex serves as a center for the assembly of a stack of 100 such disks composed of a total of 2,200 identical protein molecules that make up the complete virus.

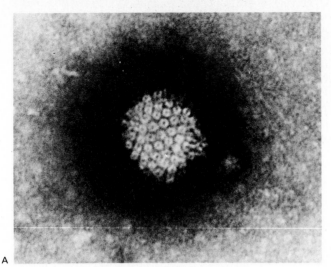

Figure 4.68

(A) Electron micrograph of Herpesvirus, an isometric virus, the capsid of which has been negatively stained. There is no envelope surrounding this virus. (From BPS— B. Roizman, University of Chicago.) (B) Infantile gastroenteritis virus Rotavirus (35,650×). (From BPS—Erskine Palmer, Centers for Disease Control, Atlanta.)

148

Organization
and structure of
microorganisms

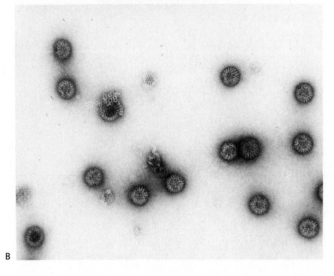

Figure 4.69

Electron micrographs of T-even phage. (A) A negatively stained bacteriophage, T₂ from an E. coli lysate, showing capsid structure and tail fibers. (From BPS—T. J. Beveridge, University of Guelph.) (B) A T-even phage with a contracted sheath. (Courtesy F. Williams, U.S. Environmental Protection Agency, Cincinnati.)

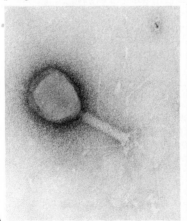

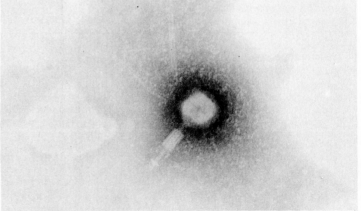

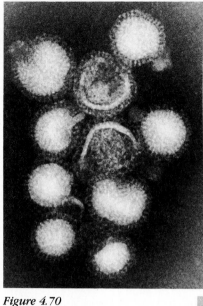

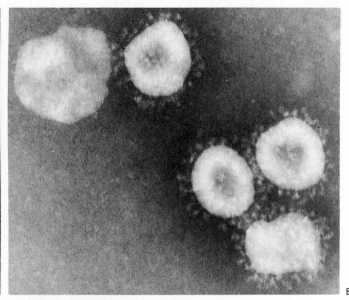

A

B

Figure 4.70

*Electron micrographs showing
enveloped viruses. (A) Electron
micrograph of the influenza A$_2$
Hong Kong virus (120,700×)
showing protruding spikes of the
envelope structure. (From BPS—
F. A. Murphy, Centers for Disease
Control, Atlanta.) (B) A
Coronavirus that causes respiratory
tract infections. (From BPS—F. A.
Murphy, Centers for Disease
Control, Atlanta) (C) A respiratory
syncytial virus (93,000×); the
pleomorphic forms are the usual
structures associated with virus
infectivity. (Courtesy Eugenie C.
Ford, Georgetown University
Medical Center, Washington, D.C.)*

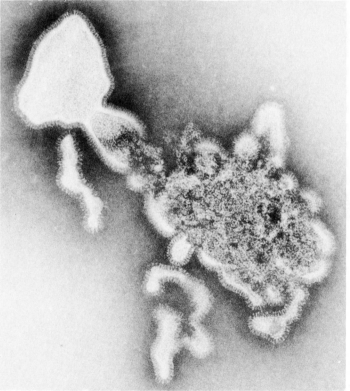

C

and isometric. The capsid of a virus, such as tobacco mosaic virus, forms a helical coil around
the nucleic acid (Figure 4.67), and some isometric viruses form a geometric pyramid-like structure known as an icosahedron (Figure 4.68). In
large isometric viruses the capsomeres may bind
into pentamerous and hexamerous arrangements
to form a somewhat different icosahedron struc-

ture. In some viruses, such as the T-even bacteriophage, the capsid structures are relatively complex. T-even phage have a head and tail structure
that have quite different structural appearances
(Figure 4.69).

In addition to the nucleic acid and capsid components, some viruses have an **envelope** that surrounds the virus particle. Like cellular mem-

branes, these envelopes possess lipid bilayers and proteins with specialized functions. Some of the envelope proteins are glycosylated and the hydrophilic carbohydrate ends of such proteins may protrude from the viral particle. Such glycoproteins often occur as spikes on the outer surface of the virus (Figure 4.70). Some of these proteins are involved in the binding of the virus to a host cell, and others cause cell lysis. In addition to the glycoproteins, other proteins of the viral envelope form a matrix layer that attaches the envelope to the capsid. The proteins of the envelope are specified by the viral genome, but the carbohydrate moieties of the glycoproteins and the lipid components of the viral envelope are obtained from the host cell. When the virus leaves the host cell, it picks up a portion of the nuclear or cytoplasmic membrane and that piece of host cell membrane can surround the viral capsid, forming the lipid portion of the envelope. The presence of host cell membranes surrounding a viral particle can help the virus evade the normal host defense mechanisms designed to recognize and destroy foreign substances.

Even microorganisms, the smallest forms of life, are highly organized living systems. They are composed of a large variety of biochemical constituents and contain a large number of specialized structures. The organizational patterns of viruses, bacteria, and other microorganisms are quite different. The structure of the cell allows for a separation of the organism (living system) from the surrounding environment, with the cytoplasmic membrane acting as a semipermeable barrier that controls the flow of materials into and out of the cell. Viruses, which lack a cytoplasmic membrane, are obligately dependent on host cells to provide a suitable environment for their reproduction. The membrane is critical in biological systems, as it permits the

table 4.3

Comparison of eukaryotic and prokaryotic cell structure

Structure	Prokaryotic cell	Eukaryotic cell
Cytoplasmic membrane	+	+
Sterols in membranes	−	+
Nucleus	−	+
DNA arranged as true chromosomes with associated histone proteins	+	+
Ribosomes	70S	80S
Ribosomal subunits	50S + 30S	60S + 40S
Cell wall	Contains murein	Several types, none with murein
Internal membranous organelles	±*	+
Chloroplasts	−	+
Mitochondria	−	+
Endoplasmic reticulum	−	+
Golgi apparatus	−	+
Vacuoles	±	+
Flagella	+	+
9 + 2 microtubular arrangement	−	+
Cytoskeleton	−	+

* ± denotes general absence, although membranous structures occur in some species of prokaryotes.

inside of the cell to remain structured and functional, while the entropy of the surroundings increases. Transport of materials across this boundary layer must be carefully regulated to prevent the random diffusion of cellular constituents; the bilipid membrane, with its hydrophilic and hydrophobic portions, is well adapted for this function. The cell's boundary is an active layer and the proteins associated with the cytoplasmic membrane play an active role in the transport of materials into and out of the cell. With the recognition of chemiosmosis as a major process in driving the synthesis of ATP, the importance of the membrane in the bioenergetics of the cell, as well as transport, becomes apparent. It now appears that the establishment of a hydrogen ion gradient across a membrane is essential for the generation of the majority of the ATP produced in many organisms.

The protection of the membrane against rupture due to osmotic shock is critical for the survival of microorganisms. The bacteria have evolved a distinctive complex wall structure containing murein that acts as a rigid outer layer to protect the cell. Other external structures, such as the bacterial capsule, also act to protect the cell against physical and chemical forces that could disrupt the cell's organizational structure. Also, some cells possess organelles of locomotion that permit them to move and find more favorable environments away from these disruptive forces for growth and reproduction. Even bacteria can recognize and respond to their chemical surroundings, moving toward or away from chemical stimuli by chemotaxis.

The differences between prokaryotic and eukaryotic cells represent a true split in the architectural strategies of organizing living systems. These different strategies give rise to the formation of distinct structures in prokaryotic and eukaryotic organisms, with some structures occurring exclusively in prokaryotes and others occurring exclusively in eukaryotes (Table 4.3). Although the genetic information of both prokaryotic and eukaryotic cells is encoded within DNA, the storage of the information is inherently different. The genetic information of eukaryotes often is stored as split sections, whereas in prokaryotic cells functional units occur as contiguous segments of the genome. The nucleus of the eukaryotic cell allows for the processing of the genetic information before it is translated into proteins within a protected organelle. In bacteria such processing cannot occur because the DNA is not segregated within a membrane-bound organelle. The membranous organelle substructures of eukaryotic organisms, including the nucleus, establish a compartmentalization of function and permit a high degree of specialized activities to occur within the integrated framework of the eukaryotic cell. The eukaryotic cell also has an extensive support and communication network that permits the coordination of complex cellular activities.

As we use newly developed microscopic and biochemical techniques to delve more deeply into the structures of microorganisms, we develop new insight into the coordination between composition, form, and function. The recognition of the archaebacteria represents the sort of unexpected finding that can result from such detailed analyses, altering our views on evolution and the phylogenetic relationships between organisms. The differences in structural elements of acellular, prokaryotic, and eukaryotic organisms also form the fundamental basis for many applied aspects of microbiology, including the use of particular antibiotics to control various diseases of humans, other animals, and plants, which will be examined in later chapters.

Study Questions

1. What are the fundamental differences between prokaryotic and eukaryotic cells? Which groups of microorganisms are prokaryotic, and which are eukaryotic?

2. What structures occur in eukaryotic cells that are not found in prokaryotic cells? What structures are found in prokaryotic cells and not in eukaryotic cells?

3. How does a virus differ from a bacterium?

4. What are the differences between bacteria and fungi in how the genetic information is stored?

5. What is osmotic pressure, and what strategies have microorganisms evolved for protection against this force?

6. Describe the differences in cell wall structure between Gram negative and Gram positive bacteria.

7. What are the similarities between a mitochondrion and a bacterial cell?

8. What are the structural differences between archaebacteria and eubacteria?

9. What is a bacterial endospore, and what is the significance of this structure?

10. Discuss how materials move into and out of cells. What are the different transport mechanisms in prokaryotic and eukaryotic cells? How is the structure of the cytoplasmic membrane related to transport processes?

Suggested Supplementary Readings

Adler, J. 1976. The sensing of chemicals by bacteria. *Scientific American* 234(4): 40–47.

Berg, H. C. 1975. How bacteria swim. *Scientific American* 233(2): 36–44.

Blakemore, R. P., and R. B. Frankel. 1981. Magnetic navigation in bacteria. *Scientific American* 245(6): 58–67.

Costerton, J. W., G. G. Geesey, and K.-J. Cheng. 1978. How bacteria stick. *Scientific American* 238(1): 86–95.

Dills, S. D., A. Apperson, M. R. Schmidt, and M. H. Saier, Jr. 1980. Carbohydrate transport in bacteria. *Microbiological Reviews* 44: 385–418.

Dyson, R. D. 1978. *Cell Biology: A Molecular Approach*. Allyn & Bacon, Inc., Boston.

Finean, J. B., R. Coleman, and R. H. Mitchell. 1978. *Membranes and Their Cellular Functions*. Blackwell Scientific Publications, Oxford, England.

Giese, A. C. 1979. *Cell Physiology*. Holt, Rinehart and Winston, Inc., New York.

Jain, M. K., and R. C. Wagner. 1980. *Introduction to Biological Membranes*. Wiley-Interscience, New York.

Jensen, W. A., and R. B. Park. 1967. *Cell Ultrastructure*. Wadsworth Publishing Co., Belmont, California.

Karp, G. 1979. *Cell Biology*. McGraw-Hill Book Co., Inc., New York.

Kornberg, R. D., and A. Klug. 1981. The nucleosome. *Scientific American* 244(2): 55–72.

Lake, J. A. 1981. The ribosome. *Scientific American* 245(2): 84–97.

Loewy, A. G., and P. Siekevitz. 1970. *Cell Structure and Function*. Holt, Rinehart and Winston, Inc., New York.

Luria, S. E., J. E. Darnell, Jr., D. Baltimore, and A. Campbell. 1978. *General Virology*. John Wiley & Sons, New York.

Molecules to Living Cells: Readings from Scientific American. 1980. W. H. Freeman and Co., San Francisco.

Rosen, B. P. (ed.). 1978. *Bacterial Transport*. Marcel Dekker, Inc., New York.

Salton, M. R. J., and P. Owen. 1976. Bacterial membrane structure. *Annual Reviews of Microbiology* 30: 451–482.

Satir, P. 1974. How cilia move. *Scientific American* 231(4): 44–63.

Silverman, M., and M. I. Simon. 1977. Bacterial flagella. *Annual Reviews of Microbiology* 31: 397–420.

Walsby, A. E. 1977. The gas vacuoles of blue-green algae. *Scientific American* 237(2): 90–97.

Williams, R. C., and H. W. Fisher. 1974. *An Electron Micrographic Atlas of Viruses*. Charles C. Thomas, Publisher, Springfield, Illinois.

Woese, C. R. 1981. Archaebacteria. *Scientific American* 244(6): 98–122.

Wolfe, S. L. 1981. *Biology of the Cell*. Wadsworth Publishing Co., Belmont, California.

153

Suggested
supplementary
readings

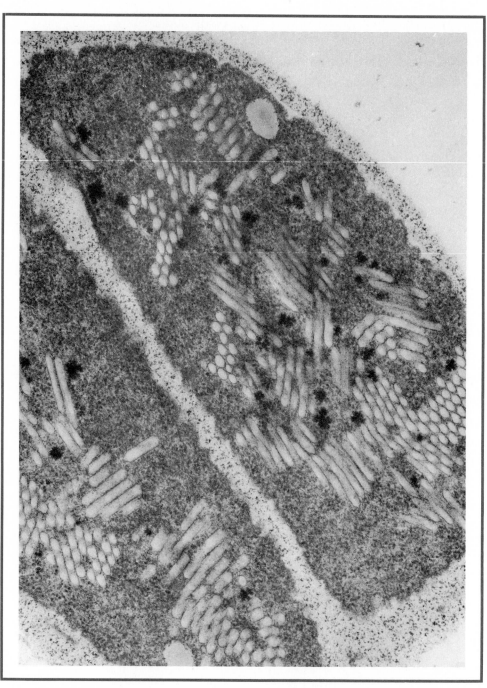

Electron micrograph of the methanogenic bacterium, Methanosarcina.
(Courtesy Jack Pangborn and Robert Mah, University of California, Los Angeles.)

Microbial metabolism

SECTION THREE

chapter 5
Microbial energetics—
the generation of ATP

chapter 6
Carbon flow:
biosynthesis of macromolecules

Microbial energetics— the generation of ATP

5

The formation and maintenance of the highly organized structural system of a microorganism is dependent on a complex integrated network of biochemical reactions, which collectively constitute the **metabolism** of the organism. It is the function of microbial metabolism to extract energy and reducing power from the microbe's surroundings and to use this energy and reducing power to synthesize the macromolecular constituents of the microbe's structure. The **energy**, which during metabolism is transferred to and stored within the molecule **ATP**, is needed to drive the endergonic biochemical reactions of biosynthesis. The **reducing power**, generated during metabolism by coupling the reduction of coenzymes with oxidation reactions, is needed for the conversion of oxidized forms of carbon—such as carbon dioxide—to their reduced oxidation states

that are found in the microorganism's macromolecules. The process of microbial metabolism can be viewed from two perspectives: the flow of energy through the cell, which is needed for biosynthesis and to maintain the organizational structure of a living system; and the flow of carbon through the cell, which is needed for incorporation into the macromolecules of the cell and for the biosynthesis of the structural components of the microorganism. In this chapter we will concentrate on the bioenergetics of microorganisms, and in the next chapter we will consider the flow of carbon through the metabolic systems of microorganisms. Although these topics are treated separately, it must be kept in mind that microbial metabolism is a unified, carefully regulated, and highly integrated system.

Autotrophic and heterotrophic metabolism

Microorganisms have developed several strategies of metabolism for meeting their common needs of synthesizing ATP, reducing coenzymes, and transforming carbon-containing molecules into the macromolecules that constitute the microorganism. Two distinct modes of microbial metabolism have evolved for accomplishing these tasks: **autotrophy** and **heterotrophy**. **Autotrophic** (literally self-feeding) microorganisms do not require preformed organic matter to generate ATP

or as a source of carbon for the biosynthesis of the macromolecules of the cell. Rather, autotrophic microorganisms are able to generate ATP either from the oxidation of inorganic compounds or through the conversion of light energy to chemical energy, and the carbon for the macromolecules of these microorganisms originates from inorganic carbon dioxide. **Heterotrophic** microorganisms, on the other hand, require preformed organic matter both for the generation of ATP and for the synthesis of cellular macromolecules. Microorganisms—such as viruses—that lack the metabolic capabilities of generating sufficient ATP for driving their essential endergonic biochemical reactions and/or are unable to synthesize their own macromolecules cannot exist independently as living systems and have developed obligate relationships with other organisms on whose metabolic activities they are totally dependent to sustain their growth and reproduction.

The central role of ATP in intermediary metabolism

All the metabolic reactions of microorganisms involved in the generation and use of ATP, reducing power, and macromolecular precursors are enzymatic, and, therefore, microbial metabolism occurs rapidly at temperatures that are not disruptive to the three-dimensional configurations of the organism's macromolecules. Thousands of enzymes, involved in the metabolic reactions of each cell, are required for microbial growth and reproduction. These enzymatically mediated metabolic reactions proceed via a series of small discrete steps that establish a metabolic pathway. The sequential steps between the starting substrate molecule(s) and the end product(s) constitute the intermediary metabolism of the cell.

Intermediary metabolism can be likened to moving between the first and second floors of a building. Most of us lack the energy necessary to jump the 10 feet to reach the next floor, and the energy released by jumping down from a height of greater than one story can be quite disruptive to the organization of the human body. In order to move up and down efficiently between floors of a building, therefore, we normally use a staircase by which we are able to alter energy levels in small discrete steps. Likewise, the microbial cell does not convert a substrate into a product in one large step but rather carries out a series of smaller intermediary metabolic reactions. The small steps of intermediary metabolism are nec-

essary for several reasons: first, although the hydrolysis of ATP is exergonic, the amount of energy released by the hydrolysis of the terminal high-energy phosphate bond ATP is limited and thus can be coupled to drive only those biochemical reactions that require less than 7.3 kcal/mole; and second, if a cell carried out reactions that were very highly exergonic, enough free energy would be released to disrupt the organization of the microbial system.

A central focus of the intermediary metabolism of a microorganism concerns the flow of energy through the cell, that is, the synthesis and use of ATP. A growing cell of the bacterium *Escherichia coli*, as a typical example, must synthesize approximately 2.5 million molecules of ATP per second to support its energy needs. The coupling of the free energy released from the hydrolysis of ATP with thermodynamically unfavorable reactions shifts the ratio of products to reactants in such endergonic reactions by a huge factor, of the order of 10^8. Thus, a thermodynamically unfavorable metabolic pathway can be driven by coupling with the hydrolysis of ATP to achieve a favorable ΔG for the reaction. The reverse process must occur for the generation of ATP, and many of the metabolic pathways of microorganisms are involved with coupling thermodynamically favorable reactions with the endergonic conversion of $ADP + P_i$ to ATP. This cycling of ADP and ATP within the cell is fundamental to the bioenergetics of microorganisms.

Before we examine some of the metabolic pathways involved in the generation of ATP, we should consider some of the properties of ATP that make this biochemical well suited for its central role in bioenergetics. When ATP is hydrolyzed, the electrostatic repulsion between the negatively charged phosphate groups is reduced, and it is this fact that accounts for the relatively large release of free energy associated with this reaction. The ADP and P_i, formed when ATP is hydrolyzed, are more stable than ATP because they can better distribute electrons through the molecule and are therefore resonance-stabilized. ATP

table 5.1

Free energies of hydrolysis of some phosphorylated compounds

Compound	ΔG^0 (kcal/mole)
Phosphoenolpyruvate	-14.8
Carbamoyl phosphate	-12.3
Acetyl phosphate	-10.3
Creatine phosphate	-10.3
Pyrophosphate	-8.0
ATP to ADP	-7.3
Glucose 1-phosphate	-5.0
Glucose 6-phosphate	-3.3
Glycerol 3-phosphate	-2.2

is particularly useful because of its intermediate position in terms of stored energy, making it possible for microorganisms to generate as well as to use this molecule as a currency of free energy (Table 5.1). The cell must continuously form and consume ATP, and although more work could be accomplished if the cell used a biochemical with a higher free energy of hydrolysis than ATP for the storage of its energy, it would be inefficient to do so, considering the number of individual endergonic reactions of the cell that must be driven by coupling with an exergonic reaction. It is important that various other phosphorylated compounds have more stored energy so that their hydrolysis can be used to drive the formation of ATP and that others have less, so that ATP can play a pivotal role in energy transfers within the cell.

Heterotrophic generation of ATP

Let us now examine some of the common pathways and variations that microorganisms use in the generation of ATP and in doing so establish the central corridors (**metabolic pathways**) of energy and carbon flow through the microbial system. **The heterotrophic generation of ATP involves the conversion of an organic substrate molecule to end products via a metabolic pathway that releases sufficient free energy so that it can be coupled with the synthesis of ATP.** This process involves the breakdown of an organic molecule to smaller molecules and is called **catabolism.** The biochemical reactions in such a pathway that liberate sufficient energy to drive the conversion of ADP to ATP are oxidation reactions, and for such reactions to occur they must be coupled with a simultaneous reduction reaction, often the reduction of the coenzyme NAD to its reduced form, NADH. To sustain its metabolism, the cell must reoxidize the reduced coenzyme in subsequent biochemical reactions; the reoxidation of NADH ensures the continuous supply of NAD required for use as an oxidizing agent in metabolic pathways aimed at generating ATP. Thus, the heterotrophic generation of ATP is integrally tied to the cell's ability to balance its oxidation–reduction reactions.

Fermentation vs respiration

Heterotrophic microorganisms exhibit two basic strategies, **fermentation** and **respiration**, for oxidizing organic compounds to release the free energy that can drive the formation of ATP while still maintaining the required balance between oxidation and reduction reactions. **In fermentation pathways the organic substrate acts as the internal electron donor (reducing agent) and a product of that substrate acts as an internal electron acceptor (oxidizing agent).** (Oxidation means the loss of electrons, and reduction means that the compound gains electrons.) There is no net change in the oxidation state of the products relative to the starting substrate molecule in fermentation pathways; the oxidized products are exactly counterbalanced by reduced products, and thus the required oxidation–reduction balance is achieved. Such a metabolic pathway can occur in the absence of air because there is no requirement for oxygen or other electron acceptor to balance a change in the oxidation state of the organic molecule.

In contrast to fermentation, **a respiration pathway requires an external electron acceptor**; that is, some molecule other than a product derived from the organic substrate must act as the oxidizing agent for reoxidizing the reduced coenzyme that is formed during the oxidation of the organic substrate molecule. The most common external electron acceptor in respiration pathways is molecular oxygen. The electron acceptor is necessary to terminate the flow of electrons in the metabolic pathway and is thus termed the terminal electron acceptor; by reducing this terminal electron acceptor, the cell is able to balance the change in the oxidation state of the metabolic products relative to the starting substrate. When molecular oxygen serves as the terminal electron acceptor, the pathway is known as **aerobic respiration;** when another molecule, such as nitrate or sulfate, serves as the terminal electron acceptor, the metabolic pathway is called **anaerobic respiration** (literally meaning respiration not requiring the presence of air).

Considerably different energy yields can be achieved by microorganisms using fermentation and respiration pathways. Fermentation yields far less ATP per substrate molecule than respiration. This is because the organic substrate molecule must serve as both the internal electron donor and electron acceptor during a fermentation pathway, and thus, the substrate cannot be completely oxidized to carbon dioxide. The ΔG^0 for the complete oxidation of glucose to carbon dioxide and water is 688 kcal/mole, compared to a ΔG^0 of only 58 kcal/mole when glucose is partially oxidized to the fermentation-product lactic acid. As a result, not as much energy can be released from the substrate molecule to drive the synthesis of ATP as can be obtained during a respiration pathway. Because more ATP can be generated per molecule of substrate by a respiration pathway than by a fermentation pathway, fewer substrate molecules must be metabolized during respiration than during fermentation to achieve equivalent growth, that is, to support the metabolic requirements of an equivalent number of cells. From both the viewpoints of bioenergetics and conservation of available organic nutrient resources, respiration is more favorable than fermentation, and organisms that have the metabolic capability to carry out both types of metabolism will generally use the energetically more favorable respiration pathway when conditions permit and will rely on fermentation only when there is no available external electron acceptor that can be used by the organism.

Respiratory metabolism

There are three distinct phases to a complete respiration pathway: first, a pathway during which the organic substrate is converted to small organic molecules that can enter the Krebs cycle; then, the Krebs cycle, during which organic carbon is oxidized to inorganic carbon dioxide and reduced coenzyme is generated; and finally, oxidative phosphorylation, during which reduced coenzyme molecules are reoxidized, electrons are transported through a series of membrane-bound carriers establishing a hydrogen ion gradient across a membrane, the terminal electron acceptor is reduced, and ATP is synthesized (Figure 5.1). In the case of carbohydrates, the substrate molecule is initially broken down to pyruvate in a metabolic pathway called **glycolysis**. The pyruvate formed during glycolysis can be converted to acetyl coenzyme A (acetyl CoA) that then can enter the Krebs cycle. Glycolysis is a central pathway in the metabolism of carbohydrates by microorganisms. Different metabolic pathways are involved in the conversion of other classes of biochemicals into small organic molecules that can enter the Krebs cycle during respiration. For example, lip-

Figure 5.1

The metabolic pathways of a cell form a complex, interrelated network. This drawing shows a simplified view of the interrelationships of several major pathways: glycolysis, the Krebs cycle, and oxidative phosphorylation. Using these pathways, cells can generate needed cellular energy and carbon constituents from various classes of substrates, including carbohydrates, lipids, and proteins.

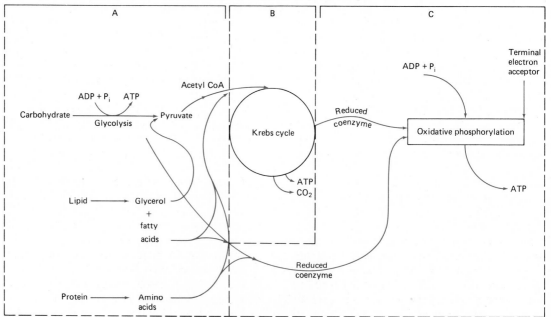

ids are converted to fatty acids that are broken down by β-oxidation to acetyl CoA, which can enter the Krebs cycle; proteins are broken down into amino acids that are deaminated to form various small organic acids that can enter the Krebs cycle.

The overall respiration pathway results in the formation of carbon dioxide from the organic substrate molecule, and in aerobic respiration water is also produced as a result of the reduction of oxygen, whereas in anaerobic respiration molecular nitrogen, hydrogen sulfide, or other reduced compounds are produced in addition to carbon dioxide, depending on the specific terminal electron acceptor. The classic equation for respiration of glucose is

$$C_6H_{12}O_6 + 6 O_2 \rightarrow 6 CO_2 + 6 H_2O$$

For simplicity, only the bioenergetics concerned with carbohydrates will be considered in this section; the metabolism of lipids and proteins relative to carbon flow will be considered in the next chapter.

Glycolysis

There are several different **glycolytic pathways** by which a carbohydrate can be converted to pyruvate. In the most common of these pathways—the **Embden-Meyerhof pathway** of glycolysis (Figure 5.2), one molecule of glucose is converted to two molecules of pyruvate and this is accompanied by the formation of two reduced coenzyme (NADH) molecules and the release of sufficient free energy to permit a net synthesis of two ATP molecules (Table 5.2).

During the initial three reactions of the Embden-Meyerhof pathway of glycolysis, glucose is converted to fructose 1,6-diphosphate. This beginning sequence of reactions of the glycolytic pathway involves endergonic reactions that require energy to drive them, and the pathway starts with the hydrolysis of ATP rather than the synthesis of ATP. In eukaryotes, this series of reactions requires two ATP molecules, one for the conversion of glucose to glucose 6-phosphate and one for the conversion of fructose 6-phosphate to fructose 1,6-diphosphate. As discussed in Chapter 4, bacteria initiate the glycolytic pathway during the transport of the substrate across the membrane in the group transport system with phosphorylation of the carbohydrate coupled with the conversion of phosphoenolpyruvate to pyruvate. Subsequent to the formation of fructose 1,6-diphosphate, the energetic balance of the reactions of this pathway is exergonic and, thus, further utilization of ATP is not required; in fact, ATP is synthesized in two of the steps.

The conversion of fructose 6-phosphate to fructose 1,6-phosphate is catalyzed by the enzyme **phosphofructokinase**. This is an important regulatory step in the pathway and phosphofructokinase is a key enzyme in determining the overall rate of the conversion of glucose to pyruvate. Even though ATP is required for the conversion of fructose 6-phosphate to fructose 1,6-diphosphate, the enzyme phosphofructokinase is inhibited by excess ATP because ATP is an **allosteric inhibitor** of the enzyme. If the cell has a sufficient supply of ATP, the inhibition of this enzyme stops the glycolytic pathway near its beginning, preventing further ATP synthesis. When ATP is depleted, the cell has a relatively high concentration of AMP—the monophosphate formed by the hydrolysis of ADP. AMP is an allosteric activator for the enzyme phosphofructokinase and, thus, when the cell really needs to generate ATP, the key rate-limiting reaction of glycloysis is stimulated, leading to increased synthesis of ATP. The allosteric control of phosphofructokinase is responsible for the paradoxical observation that in the presence of oxygen less carbohydrate substrate disappears during the growth of many microorganisms than during the growth of the same organism in the absence of air. This phenomenon, known as the **Pasteur effect**, happens because during aerobic respiration a high level of ATP accumulates within the cell and inhibits phosphofructokinase, greatly slowing the rate of substrate conversion. In the absence of oxygen, when the cell is using fermentative metabolism, less ATP is produced and glycolysis proceeds without inhibition.

The strategy of these first steps of glycolysis is to form a compound that can be broken down into phosphorylated three-carbon units. The fructose 1,6-diphosphate molecule contains six carbon atoms; it is split into two three-carbon pieces by the action of the enzyme aldolase. One of the three-carbon molecules formed is 3-phosphoglyceraldehyde; the other three-carbon molecule, dihydroxyacetone phosphate, may be isomerized to 3-phosphoglyceraldehyde. The equilibrium constant for the conversion of dihydroxyacetone phosphate and 3-phosphoglyceraldehyde favors the formation of dihydroxyacetone phosphate that is not in the direct glycolytic pathway, but the constant removal of 3-phosphoglyceraldehyde, which is in the direct glycolytic pathway, shifts the balance of reactants and products. Thus, for each six-carbon glucose molecule, two molecules of 3-phosphoglyceraldehyde are formed. It is im-

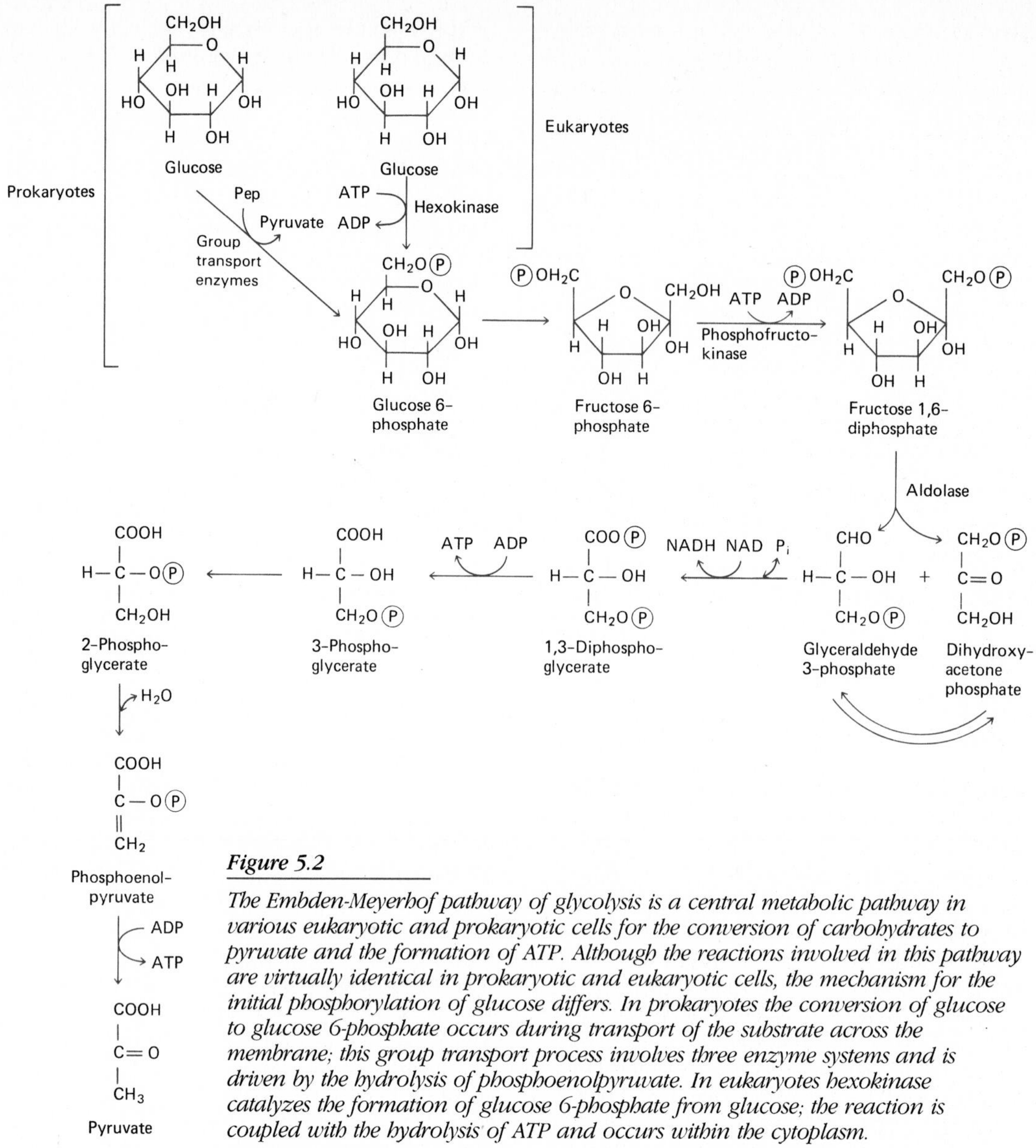

Figure 5.2

The Embden-Meyerhof pathway of glycolysis is a central metabolic pathway in various eukaryotic and prokaryotic cells for the conversion of carbohydrates to pyruvate and the formation of ATP. Although the reactions involved in this pathway are virtually identical in prokaryotic and eukaryotic cells, the mechanism for the initial phosphorylation of glucose differs. In prokaryotes the conversion of glucose to glucose 6-phosphate occurs during transport of the substrate across the membrane; this group transport process involves three enzyme systems and is driven by the hydrolysis of phosphoenolpyruvate. In eukaryotes hexokinase catalyzes the formation of glucose 6-phosphate from glucose; the reaction is coupled with the hydrolysis of ATP and occurs within the cytoplasm.

162

Microbial ener-
getics—the gen-
eration of ATP

portant to keep in mind that two phosphorylated three-carbon molecules are formed for each six-carbon carbohydrate substrate molecule in order to keep track of the net yield of ATP and reduced coenzyme (NADH) molecules formed during the overall pathway.

After forming two molecules of 3-phosphoglyceraldehyde, the next portion of the glycolytic pathway is concerned with using the energy stored in this compound to drive the synthesis of ATP. The 3-phosphoglyceraldehyde molecule is oxidized to 1,3-diphosphoglyceric acid in a substrate-level phosphorylation reaction, so designated because inorganic phosphate (P_i) is incorporated into the molecule during this exergonic reaction. The oxidative conversion of 3-phosphoglyceraldehyde to 1,3-diphosphoglyceric acid is coupled with the conversion of oxidized NAD to the reduced coenzyme NADH. (Because there are two molecules of 3-phosphoglyceraldehyde generated from each glucose molecule, there is a net production of two NADH molecules per molecule of glucose.) The 1,3-diphosphoglyceric acid is converted to 3-phosphoglyceric acid, an exergonic reaction that can be coupled with the synthesis of ATP. Because this reaction

table 5.2

Free energies of reactions of glycolysis

Reaction	Enzyme	ΔG^{0*}	ΔG
Glucose + ATP → glucose 6-phosphate +ADP + P_i	Hexokinase	−4.0	−8.0
Glucose 6-phosphate → fructose 6-phosphate	Phosphoglucose isomerase	+0.4	−0.6
Fructose 6-phosphate + ATP → fructose 1,6-diphosphate + ADP + P_i	Phosphofructokinase	−3.4	−5.3
Fructose 1,6-diphosphate → dihydroxyacetone phosphate + glyceraldehyde 3-phosphate	Aldolase	+5.7	−0.3
Dihydroxyacetone phosphate → glyceraldehyde 3-phosphate	Triose phosphate isomerase	+1.8	+0.6
Glyceraldehyde 3-phosphate + P_i + NAD → 1,3-diphosphoglycerate + NADH	Glyceraldehyde 3-phosphate dehydrogenase	+1.5	−0.4
1,3-Diphosphoglycerate + ADP → 3-phosphoglycerate + ATP	Phosphoglycerate kinase	−4.5	+0.3
3-Phosphoglycerate → 2-phosphoglycerate	Phosphoglyceromutase	+1.1	+0.2
2-Phosphoglycerate → phosphoenolpyruvate	Enolase	+0.4	−0.8
Phosphoenolpyruvate + ADP → pyruvate + ATP	Pyruvate kinase	−7.5	−4.0

*The ΔG^0 is based on the standard free energy; the ΔG represents the approximate actual free-energy change based on measured concentrations of reactants under typical conditions.

occurs for each of the two three-carbon molecules generated from glucose, two molecules of ATP are generated per glucose molecule, and therefore, the synthesis and hydrolysis of ATP balance at this point in the metabolic pathway; that is, the net production of ATP equals zero at this stage of the metabolic pathway because two ATP molecules had been hydrolyzed in earlier steps.

The 3-phosphoglyceric acid is further converted to phosphoenolpyruvate and finally to pyruvate. The conversion of phosphoenolpyruvate to pyruvate can be coupled with the synthesis of ATP. Thus, this glycolytic pathway results in the conversion of the six-carbon molecule glucose to two molecules of the three-carbon molecule pyruvate, with the net production of two molecules of reduced coenzyme, NADH, and the net synthesis of two ATP molecules (Figure 5.3). The overall equation for glycolysis by the Embden-Meyerhof pathway can be written as

Glucose + 2 ADP + 2 P_i + 2 NAD →
2 pyruvate + 2 NADH + 2 ATP

Glucose is not the only carbohydrate that can be converted to pyruvate by glycolysis. Many other carbohydrates can be converted to intermediate products of the glycolytic pathway, and the synthesis of ATP can be driven by the reactions of this central metabolic pathway. Even large carbohydrate polymers, such as cellulose, starch, and glycogen, can be cleaved by extracellular enzymes to yield glucose phosphate units that are translocated across the cytoplasmic membrane and enter the glycolytic pathway (Figure 5.4). For some of these polymers a substrate-level phosphorylation reaction can occur to initiate the glycolytic pathway, and in such cases the hydrolysis of ATP is not required to drive the initial phosphorylation reaction of the glycolytic pathway. The net synthesis of ATP from such a polymer is three rather than two molecules of ATP per glucose subunit of the polymer following the Embden-Meyerhof pathway of glycolysis. Whereas the specific yield of ATP may vary, the Embden-Meyerhof pathway always produces two pyruvate and two NADH molecules per six-carbon carbohydrate unit.

The Embden-Meyerhof pathway is only one of several glycolytic pathways possessed by different microorganisms. The **Entner-Doudoroff pathway** of glycolysis, for example, results in the net production of only one ATP molecule per molecule

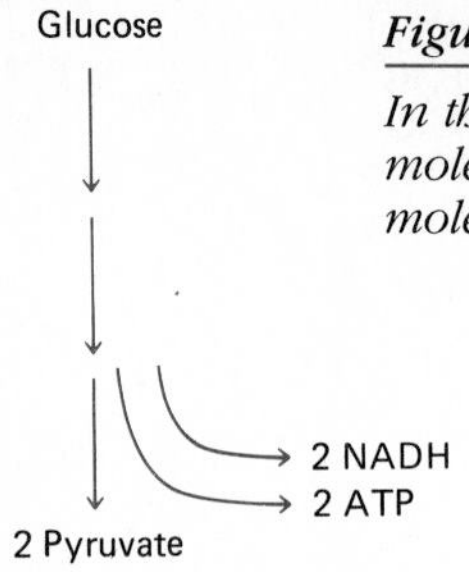

Glucose

→ 2 NADH
→ 2 ATP

2 Pyruvate

Figure 5.3

In the Embden-Meyerhof pathway a molecule of glucose is converted to two molecules of pyruvate with the net production of two molecules of ATP and two molecules of reduced coenzyme NADH.

Discovery process

The delineation of this basic metabolic pathway—the Embden-Meyerhof pathway—was central to the development of the field of biochemistry. The process of discovery began accidentally in 1897 when Hans and Eduard Buchner, while trying to manufacture cell-free extracts of yeast for therapeutic use, attempted to preserve the extracts without using antiseptics, such as phenol, and tried sucrose. They were surprised to find that the sucrose was rapidly fermented to alcohol by the yeast juice, thus demonstrating for the first time that fermentation could occur outside living cells. This contradicted the accepted dogma of the time, held even by Pasteur, that fermentation was inextricably tied to living cells, and allowed the study of metabolism to proceed along biochemical paths. Beginning in 1905, Arthur Harden and William Young studied what happened when yeast juice was added to a glucose solution. They found that fermentation began almost immediately, that its rate decreased unless organic phosphate was added, and that the added inorganic phosphate disappeared during the fermentation. They inferred from this that the inorganic phosphate was incorporated to form a sugar phosphate; later, it was shown that fructose 1,6-diphosphate and other phosphorylated carbohydrates are key intermediates in the metabolism of carbohydrates. They also found that the activity of the yeast juice depended on the presence of two kinds of substances: a heat-labile, nondialyzable component and a heat-stable, dialyzable fraction. Later studies of muscle extracts showed that many of the reactions of lactic fermentation were the same as those of alcoholic fermentation, thus revealing the underlying unity in biochemistry. Contributions made by Gustav Embden, Otto Meyerhof, Carl Neuberg, Jacob Parnus, Otto Warburg, Gerty Cori, and Carl Cori led to the complete elucidation of the glycolytic pathway by 1940.

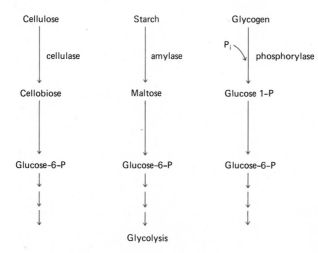

Figure 5.4

Large polymers such as cellulose, starch, and glycogen are hydrolyzed and phosphorylated to products that can be further metabolized by glycolysis.

of glucose substrate metabolized (Figure 5.5). The net equation for the Entner-Doudoroff pathway of glycolysis can be written

Glucose + 2 NAD + ADP + P_i →
2 pyruvate + 2 NADH + ATP

The Entner-Doudoroff pathway is utilized by various *Pseudomonas* species and other bacteria. This pathway provides an important shunt for producing reduced coenzyme and three-carbon building blocks when these are in greater need than ATP for biosynthesis.

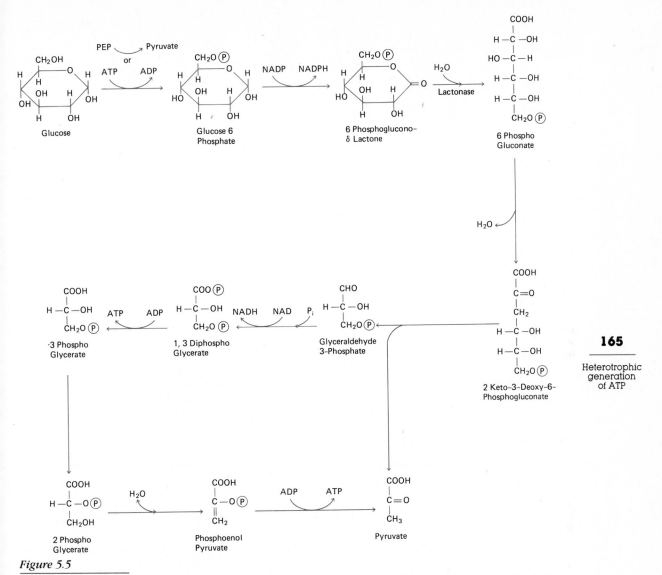

Figure 5.5

The Entner-Doudoroff pathway is one of several variations of glycolysis. Compared to the Embden-Myerhof pathway, less ATP is generated when this metabolic pathway is used.

In another glycolytic pathway, the **phosphoketolase pathway**, only one molecule of pyruvate and one molecule of ATP are generated (Figure 5.6). The remaining carbon atoms of glucose are converted to ethanol and carbon dioxide in this pathway. The overall equation for the phosphoketolase pathway for the glycolysis of glucose can be written as

Glucose + NADP + ADP + $P_i \rightarrow$
pyruvate + ethanol + CO_2 + ATP + NADPH

A key intermediate in this pathway is the formation of ribulose 5-phosphate, a five-carbon phosphorylated carbohydrate. The phosphoketolase pathway also is utilized for the metabolism of pentose carbohydrates such as ribose. When pentose is the substrate, however, the pathway is somewhat different (Figure 5.7). In the case of a pentose molecule, two ATP molecules are generated for each pentose molecule metabolized. The phosphoketolase pathway for the glycolysis of pentose also results in the formation of three molecules of reduced coenzyme NADH. The overall equation for the phosphoketolase pathway of glycolysis for ribose can be written as

Ribose + 2 ADP + 2 $P_i \rightarrow$
lactate + acetate + CO_2 + 2 ATP

CH$_2$-O-(P) ... NADP NADPH ... CH$_2$-O-(P) ... H$_2$O ...

Glucose-6-Phosphate   6-Phosphogluconolactone   6-Phosphogluconic acid   Ribulose-5-Phosphate

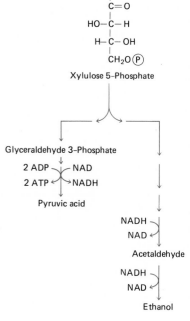

Figure 5.6

The phosphoketolase pathway is another glycolytic pathway carried out by some microorganisms.

166

Microbial ener-
getics—the gen-
eration of ATP

Figure 5.7

The phosphoketolase pathway of glycolysis is used for the metabolism of pentoses as well as hexoses. This pathway provides an essential link between the metabolism of C5 and C6 carbohydrates.

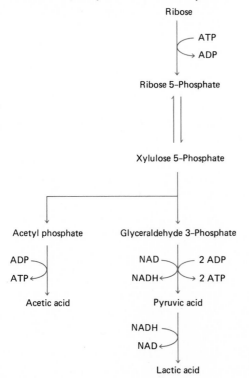

Krebs cycle

The pyruvate generated via one of the glycolytic pathways can feed into the **Krebs cycle**, the next stage of respiratory metabolism (Figure 5.8). (The Krebs cycle is also known as the **tricarboxylic acid cycle** or the **citric acid cycle**.) Each of the pyruvate molecules generated during glycolysis reacts with coenzyme A (CoA) to form acetyl CoA with the release of carbon dioxide. This reaction is coupled with the conversion of the coenzyme NAD to reduced NADH. The acetyl group then enters the tricarboxylic acid cycle by reacting with oxaloacetic acid to form citric acid. During the Krebs cycle there are two reactions that liberate carbon dioxide—the conversion of isocitrate to α-ketoglutarate, and the next step that converts the α-ketoglutarate, through an intermediate of succinyl CoA, to succinate. Reduced coenzyme NADH is generated during three reactions of the

Krebs cycle. Additionally, another coenzyme, flavin adenine dinucleotide (FAD), is reduced to FADH$_2$ during the conversion of succinate to fumarate. Only one of the exergonic reactions of this cycle, the conversion of α-ketoglutarate to succinate, is

Figure 5.8

The Krebs cycle is a central metabolic pathway to respiratory metabolism and provides a critical link in the metabolism of the different classes of macromolecules. The metabolism of pyruvate through the complete Krebs cycle results in the generation of ATP and reduced coenzymes and the formation of CO$_2$. When the pathway is completed, the intermediate carboxylic acids are regenerated and continue to cycle through the reactions.

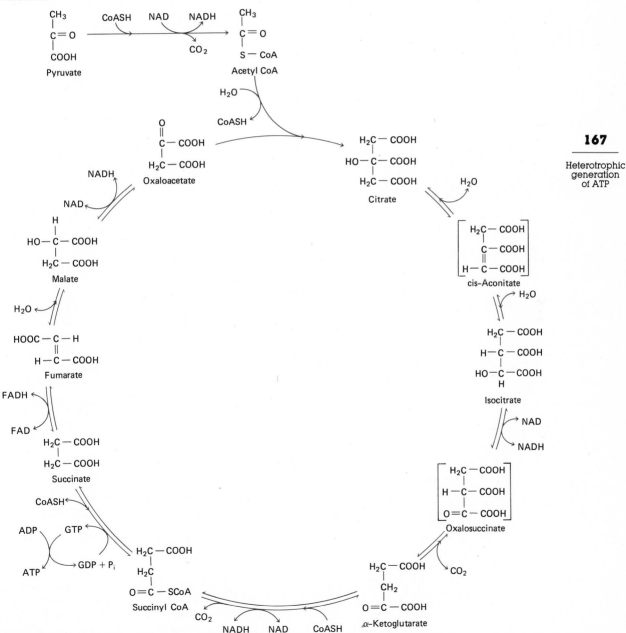

directly coupled with the generation of a high-energy phosphate-containing compound. In this reaction, instead of the normal generation of ATP, guanosine triphosphate (GTP) is synthesized from GDP + P_i. The GTP can be converted to ATP, and for accounting purposes, the GTP generated in this reaction will be treated as equivalent to ATP in determining the net synthesis of ATP during respiration. The Krebs cycle is completed with the regeneration of oxaloacetate, and thus, the intermediary carboxylic acids of the Krebs cycle can be continuously re-formed.

The net reaction of the Krebs cycle, starting with the pyruvate generated from glucose can be written as

2 Pyruvate + 2 ADP + 2 FAD + 8 NAD →
6 CO_2 + 2 ATP + 2 $FADH_2$ + 8 NADH

At the end of the Krebs cycle, the cell has managed to convert all of the substrate carbon of the glucose molecule to carbon dioxide; there also has been a net synthesis of four ATP molecules—the production of ten reduced coenzyme molecules as NADH and the generation of two reduced coenzyme molecules as $FADH_2$ (Figure 5.9). Several of the enzymes mediating the reactions of the Krebs cycle are inhibited by ATP. In particular, ATP is an allosteric inhibitor of the enzyme citrate synthetase that catalyzes the first step of the Krebs cycle—the synthesis of citrate from acetyl CoA and oxaloacetate. This is an important control point, and as the need for ATP decreases, the rate of the Krebs cycle slows through the allosteric inhibition of the enzyme catalyzing the first step of the pathway.

In addition to its role within the overall respiratory generation of ATP, the Krebs cycle occupies a central place in the flow of carbon through the cell, which will be discussed in the next chap-

ter. As a result of its function of supplying small biochemical molecules for biosynthetic pathways, the Krebs cycle rarely acts as a complete cycle. Because some of the intermediates are siphoned out of the cycle, some of the intermediary metabolites of this pathway must be resynthesized to maintain an active Krebs cycle. In many microorganisms only part of the substrate is completely oxidized for driving the synthesis of ATP, and the remainder is used for biosynthesis. Similarly, the reduced coenzymes generated in this pathway can be used for generating ATP or for the synthesis of the reduced coenzyme NADPH for use in biosynthesis.

Oxidative phosphorylation

When we consider the respiration pathway from the standpoint of the synthesis of ATP, the reduced coenzyme molecules—generated both during glycolysis and the Krebs cycle—can be reoxidized during **oxidative phosphorylation** with the generation of additional ATP. During oxidative phosphorylation, electrons from NADH and $FADH_2$ are transferred through a series of steps, known as the **electron transport chain**. This transfer of electrons involves a series of oxidation–reduction reactions of membrane-bound carrier molecules and the eventual reduction of a terminal electron acceptor (Figure 5.10).

Although different microorganisms can have different carriers in their electron transport chains, **cytochrome molecules** play a key role in all electron transport chains. These molecules contain a central iron atom, which can be cycled between the oxidized ferric state (Fe^{3+}) and the reduced ferrous state (Fe^{2+}). With the exception of the final transfer reaction, the same series of electron transfer reactions can occur during aerobic and anaerobic respiration (Figure 5.11). The terminal electron acceptor during aerobic respiration is oxygen that is reduced to form water when it acts as terminal electron acceptor in this process. In the absence of molecular oxygen, nitrate can serve as terminal electron acceptor with the production of molecular nitrogen and water; in the absence of both oxygen and nitrate, sulfate can serve as terminal electron acceptor, becoming reduced to hydrogen sulfide and water in the process.

As a consequence of the electron transport chain, the energy contained in the reduced coenzyme molecule is dissipated and used to drive the synthesis of ATP. The reduced coenzyme NADH contains more stored chemical energy than the reduced coenzyme $FADH_2$. The oxidation of NADH

Figure 5.9

This simplified drawing of the Krebs cycle summarizes the major inputs and products of this pathway.

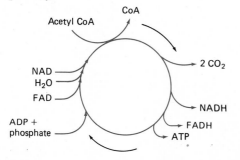

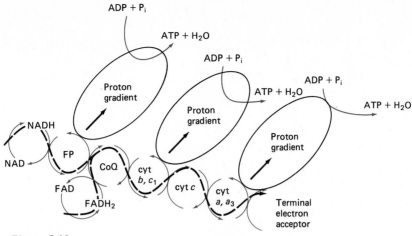

Figure 5.10

The electron transport chain is a membrane-bound series of reactions that results in the reoxidation of reduced coenzymes. The transport of electrons through the cytochrome chain of this pathway results in the pumping of hydrogen ions across the membrane, and the return flow of hydrogen ions resulting from this proton gradient drives the generation of ATP.

liberates 54 kcal/mole, compared to only 42 kcal/mole for the oxidation of FADH$_2$. For each NADH molecule three ATP molecules can be synthesized during oxidative phosphorylation, compared to only two ATP molecules for each FADH$_2$. The 10 NADH molecules generated during glycolysis and the Krebs cycle, therefore, can be converted to 30 ATP molecules during oxidative phosphoryla-

Figure 5.11

The electron chain terminates with reduction of a terminal electron acceptor. The requirement for such a terminal electron acceptor differentiates respiration from fermentation pathways. In aerobic respiration oxygen is the terminal electron acceptor; in the absence of oxygen some microorganisms can carry out anaerobic respiration by using the reduction of nitrate or sulfate to terminate the reaction sequence of this pathway.

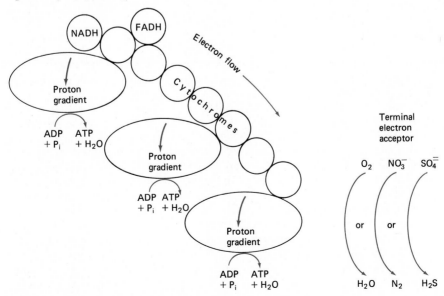

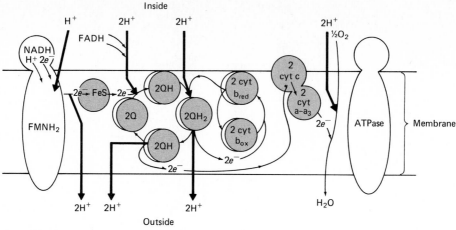

Figure 5.12

The generation of the hydrogen ion gradient across a membrane, which is needed to generate ATP by chemiosmosis, is depicted in this drawing.

tion; the two FADH$_2$ molecules generated during the Krebs cycle can generate four ATP molecules. This synthesis of ATP is in addition to the ATP formed during glycolysis and the Krebs cycle. Thus, the overall reaction for the respiratory metabolism of glucose (using the Embden-Meyerhof pathway of glycolysis) can be written as

Glucose + 6 O$_2$ + 38 ADP + 38 P$_i$ →
6 CO$_2$ + 6 H$_2$O + 38 ATP

Whereas the generation of ATP during oxidative phosphorylation depends on the electron transport chain, it is not directly coupled to the specific biochemical reactions that occur during electron transport. Rather, the generation of ATP during this process is driven by **chemiosmosis**. The transfer of electrons from the reduced coenzyme to the terminal electron acceptor establishes a hydrogen ion gradient across the membrane (Figure 5.12). The intermediate electron carriers of the electron transport chain are asymmetrically distributed through a membrane—in the case of eukaryotic cells, the mitochondrial membrane; and in the case of prokaryotic cells, the cytoplasmic membrane. These membranes contain the enzyme ATPase, which can catalyze the synthesis of ATP from ADP + P$_i$. The movement of hydrogen ions across the membrane, as a result of electron transport, establishes an electrochemical gradient. The diffusive counterflow of hydrogen ions back across the membrane returns through a proton channel established by the membrane-bound ATPase enzyme and drives the synthesis of ATP. The electrochemical gradient

created by the extrusion of hydrogen ions across a membrane establishes the chemiosmotic mechanism for generating ATP.

Fermentation

Unlike respiration, **fermentation** does not involve oxidative phosphorylation for generating ATP. Rather, the synthesis of ATP in fermentation is largely restricted to the amount formed during glycolysis. Because they do not require oxygen, all fermentation pathways are anaerobic and microorganisms that generate their energy by fermentation are carrying out anaerobic metabolism, regardless of whether or not the organism is growing in the presence of molecular oxygen. Some microorganisms, known as **facultative anaerobes**, are capable of carrying out both respiration and fermentation pathways. Other microorganisms that are restricted to fermentative pathways are metabolically **obligate anaerobes**. The use of the terms *obligate anaerobe* and *obligate aerobe* is often confusing. Some fermentative microorganisms are inhibited by oxygen and grow only in the absence of air, and in such cases the designation obligate anaerobe can be used without ambiguity. However, many microorganisms that are metabolically obligate anaerobes, because they are restricted to fermentative metabolism, can grow in the presence of air even though they do not use molecular oxygen. Similarly, as noted earlier, some organisms that are referred to as obligate aerobes, because they are restricted

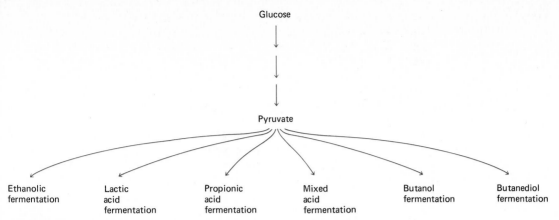

Figure 5.13

Various fermentation pathways, each branching from pyruvate, are carried out by different microorganisms. Compared to respiration, these fermentation pathways are energetically less favorable. In each complete fermentation there is a balance of oxidation–reduction reactions so that there is no net production of reduced coenzyme. Each pathway results in the formation of characteristic products, many of which are of industrial importance.

to respiratory metabolism, can grow in the absence of air by using terminal electron acceptors other than oxygen.

In the fermentation of a carbohydrate, the initial metabolic steps are identical to those in respiration, and the metabolic pathway for carbohydrate fermentation begins with glycolysis. If the microorganism follows the Embden-Meyerhof glycolytic pathway, it generates two pyruvate molecules, two reduced coenzyme NADH molecules, and a net synthesis of two ATP molecules for each molecule of glucose that goes through glycolysis. In general, the two ATP molecules generated during glycolysis represent the total energy conversion of the fermentation pathway, as measured by the number ATP molecules synthesized. The remainder of the fermentation pathway is really concerned with reoxidizing the coenzyme. In respiration the reoxidation of the coenzymes occurs in the electron transport chain and requires an external electron acceptor, whereas in fermentation the reoxidation of NADH to NAD depends on the reduction of the pyruvate molecules formed during glycolysis to balance the oxidation–reduction reactions. Different microorganisms have developed different pathways for utilizing the pyruvate for reoxidizing the reduced coenzyme with the different terminal sequences of the various fermentation pathways, resulting in the formation of various end products (Figure 5.13). The different fermentation pathways generally are named for the characteristic end products that are formed.

For example, the **ethanolic fermentation pathway** derives its name from the fact that ethanol is one of the end products. In this fermentation pathway, pyruvate is converted to ethanol and carbon dioxide (Figure 5.14). This reaction is coupled with the conversion of NADH to NAD. The

Figure 5.14

The ethanolic fermentation pathway results in the formation of ethanol and CO_2. The fermentation of carbohydrates to these end products forms the basis of the beer, wine, and spirits industries.

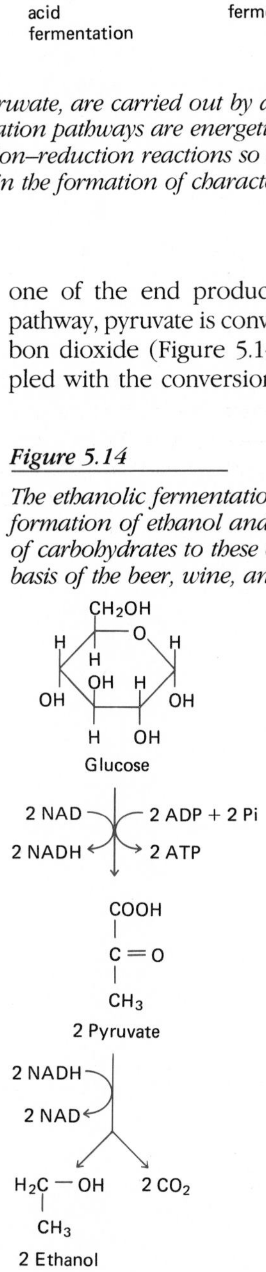

equation for the ethanolic fermentation when glucose is used as a substrate can be written as

$$Glucose + 2\ ADP + 2\ P_i \rightarrow$$
$$2\ ethanol + 2\ CO_2 + 2\ ATP$$

The complete fermentation pathway begins with the substrate, includes glycolysis, and terminates with the formation of end products. There is no net change in the oxidative state of the coenzymes during the overall fermentation pathway, and thus, the coenzyme does not appear in the overall equation for the fermentation. The ethanolic fermentation is carried out by many yeasts, such as *Saccharomyces cerevisiae*, but by relatively few bacteria. This fermentation pathway is quite important in food and industrial microbiology and is used to produce beer, wine, and spirits. Besides

its importance in alcoholic beverages, the ethanol produced by *Saccharomyces cerevisiae* in this fermentation is used as a fuel, in gasohol. In addition to using this fermentation pathway for ethanol formation, *Saccharomyces cerevisiae* (Baker's yeast) is used in the production of bread; the carbon dioxide released by the ethanolic fermentation causes bread to rise. All of these uses of the ethanolic fermentation pathway have great economic importance (see Chapter 20).

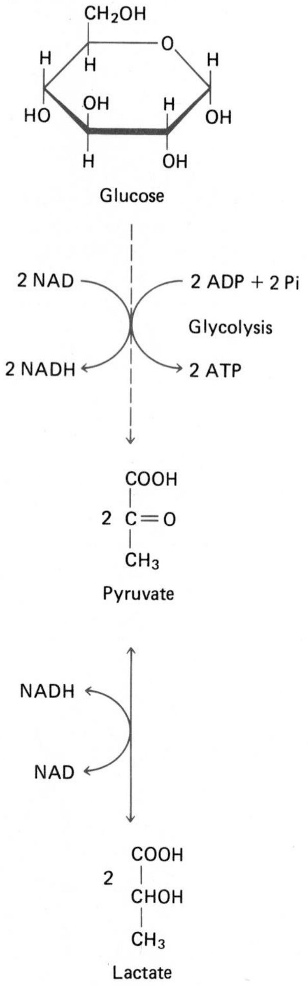

Another fermentation pathway of great economic importance is the **lactic acid fermentation pathway**. This fermentation pathway is carried out by bacteria, which by virtue of their metabolic reactions are classified as the lactic acid bacteria. Two important genera of lactic acid bacteria are *Streptococcus* (Gram positive cocci that tend to form chains) and *Lactobacillus* (a Gram positive rod that tends to form chains). In the lactic acid fermentation pathway, pyruvate is reduced to lactic acid with the coupled reoxidation of NADH to NAD (Figure 5.15). The overall lactic acid fermentation pathway can be written as

$$Glucose + 2\ ADP + 2\ P_i \rightarrow$$
$$2\ lactic\ acid + 2\ ATP$$

When the Embden-Meyerhof scheme of glycolysis is used in the lactic acid fermentation pathway, the overall pathway is a **homolactic fermentation** because the only end product formed is lactic acid. The homolactic fermentation is carried out by *Streptococcus*, *Pediococcus*, and various *Lactobacillus* species. If, however, the phosphoketolase pathway of glycolysis is used, the pathway is **heterolactic** because ethanol and carbon dioxide are produced in addition to lactic acid (Figure 5.16). The ethanol and carbon dioxide come from the glycolytic portion of the pathway. The overall reaction for the heterolactic fermentation can be written as

$$Glucose + ADP + P_i \rightarrow$$
$$lactic\ acid + ethanol + CO_2 + ATP$$

The heterolactic fermentation pathway produces only one molecule of ATP per molecule of glucose substrate metabolized. This fermentative pathway is carried out by *Leuconostoc* and various *Lactobacillus* species.

Both the homolactic and heterolactic acid fermentations have important practical applications. The heterolactic fermentation pathway carried out by *Leuconostoc* species is responsible for the production of sauerkraut. The homolactic acid fermentation pathway is quite important in the dairy industry; it is the pathway that is responsible for the souring of milk and is used in the production

Figure 5.15

The homolactic acid fermentation pathway results in the production of lactic acid.

of numerous types of cheese, yogurt, and various other dairy products. Lactic acid produced by streptococci living on tooth surfaces in the oral cavity is held against the tooth by dental plaque and gradually eats through the enamel of the tooth, creating a cavity. Even though they can grow in the oral cavity (mouth), *Streptococcus* species are metabolically obligate anaerobes. *Lactobacillus* species occur in the digestive tract of humans and aid in the digestion of milk. *Lactobacillus* species are the initial colonizers of the intestinal tract. Adults often lack the ability to digest the carbohydrates in milk and suffer disease symptoms if they consume milk. *Lactobacillus acidophilus* is added to various commercial milk products (acidophilus milk) to aid individuals who are unable to digest milk products adequately. The enzymes produced by *L. acidophilus* convert milk sugars to products that do not accumulate and cause gastrointestinal problems.

Another fermentation pathway of some interest is the **propionic acid fermentation pathway**. This metabolic sequence is carried out by the propionic acid bacteria, which are especially interesting because they have the ability to carry out a fermentation pathway beginning with lactic acid as the substrate (Figure 5.17). The bacterial genus *Propionibacterium* is defined as Gram positive rods that produce propionic acid from the metabolism of carbohydrates. The ability to utilize the end product of another fermentation pathway is quite unusual, and this ability permits species of *Propionibacterium* to carry out a late fermentation during the production of cheese. The lactic acid bacteria convert the initial substrates in the milk to lactic acid. The propionic acid bacteria subsequently convert the lactic acid to propionic acid and carbon dioxide. Only after the cheese curd has formed, through the action of lactic acid bacteria, do the propionic acid bacteria begin their fermentation. The release of carbon dioxide during this late fermentation forms gas bubbles in the semisolid cheese curd, which we recognize as Swiss cheese holes. The propionic acid formed during this fermentation also gives Swiss cheese its characteristic flavor.

In contrast to the propionic acid fermentation, which is restricted to a few bacterial species, the **mixed acid fermentation pathway** is relatively common. The mixed acid fermentation, so named because of the mixture of end products that are formed, is carried out by members of the family Enterobacteriaceae, a large family of bacteria that includes *Escherichia coli* as well as hundreds of other bacterial species (Figure 5.18). In this metabolic pathway the pyruvate formed during glycolysis is converted to a variety of products, including ethanol, acetate, formate, molecular hydrogen, and carbon dioxide. The proportions of the products vary, depending on the bacterial species. During formation of these various products, the reduced NADH is reoxidized to NAD. Also, during the formation of acetic acid, sufficient energy is released to drive the synthesis of additional ATP.

Some bacteria are capable of carrying out multiple metabolic pathways. Such is the case for members of the bacterial genus *Klebsiella*, which in addition to the mixed acid fermentation can carry out a **butanediol fermentation pathway**. Part of the substrate molecule is converted to end products via the mixed acid fermentation pathway, and part of the substrate forms the end products of the butanediol pathway. In the butanediol fermentation pathway, carbon dioxide is released, and NADH is reoxidized to NAD, but no additional ATP is generated (Figure 5.19). An inter-

Figure 5.16

The heterolactic fermentation pathway results in the production of several products: lactic acid, ethanol, and CO_2.

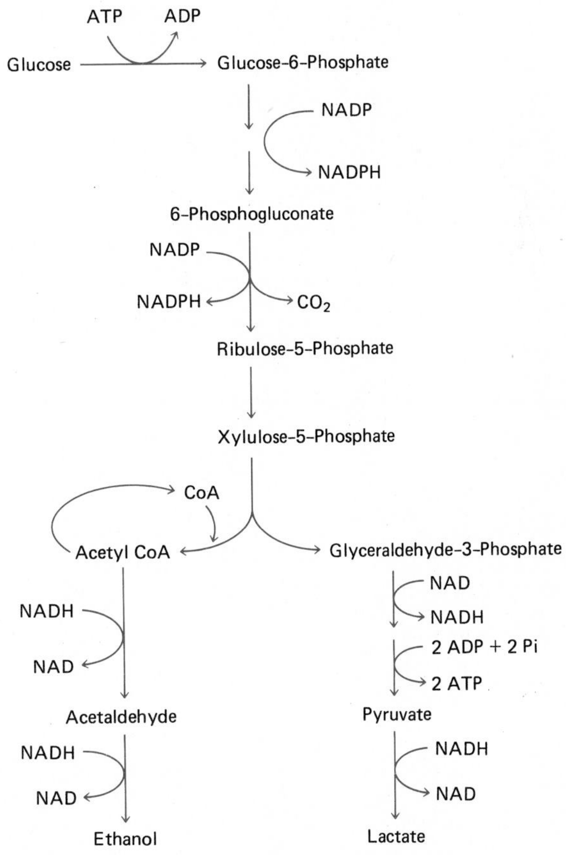

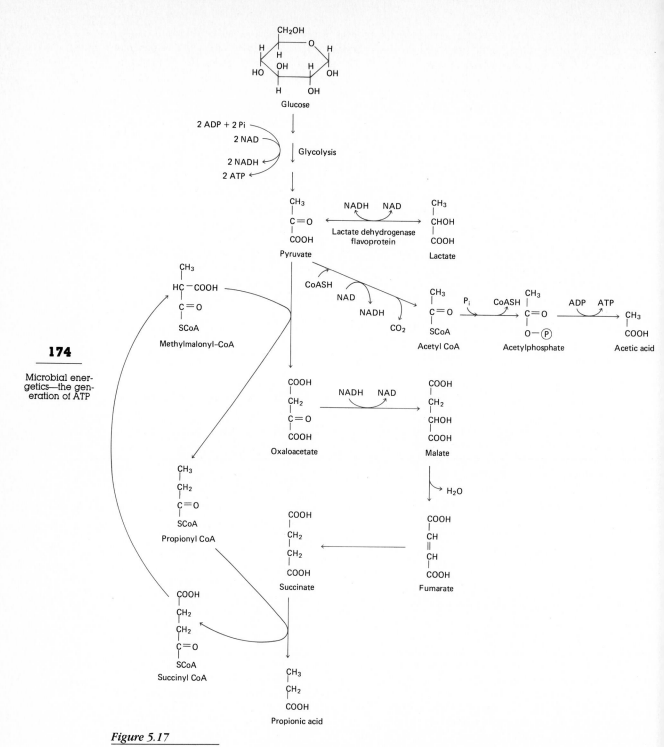

174

Microbial ener-
getics—the gen-
eration of ATP

Figure 5.17

The propionic acid fermentation pathway can use lactate as well as carbohydrates as the initial substrate.

Figure 5.18

The mixed acid fermentation pathway, which is carried out by enteric bacteria such as E. coli, *results in the production of carbon dioxide, hydrogen gas, acetic acid, lactic acid, and ethanol.*

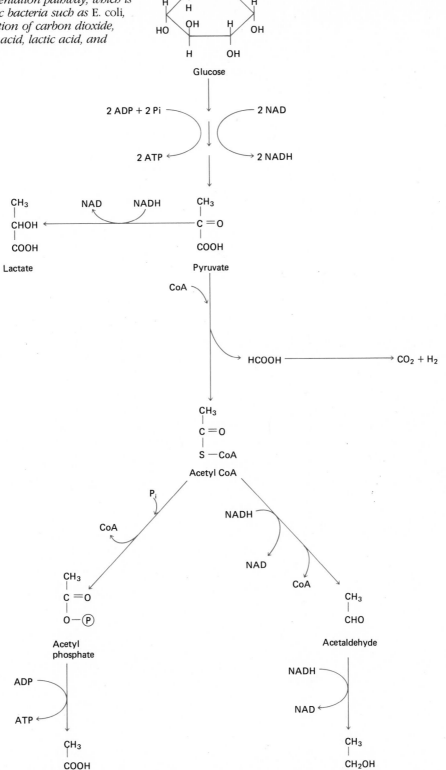

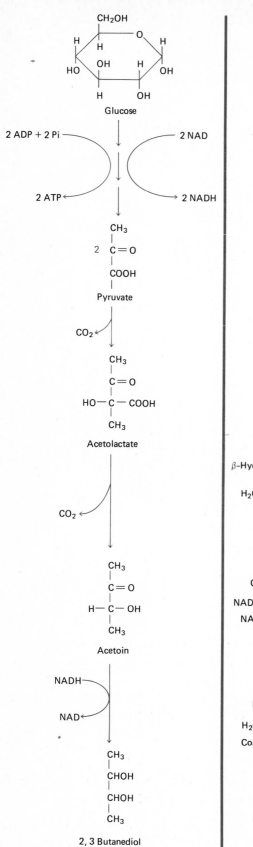

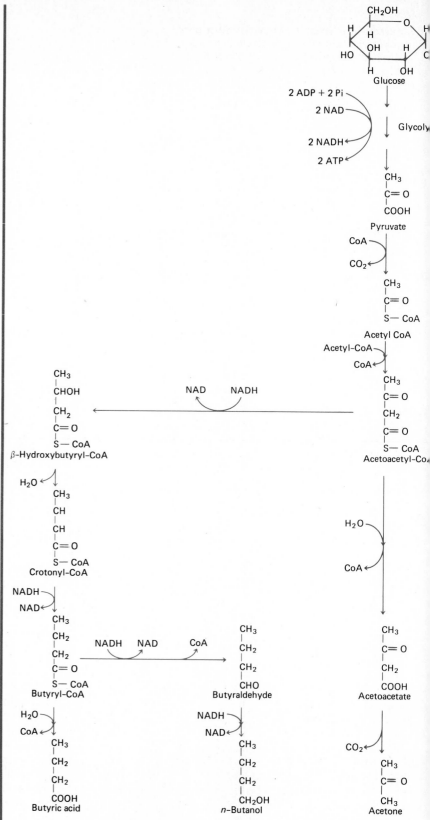

Figure 5.19 *(Facing page, left)*
The butanediol fermentation pathway results in the production of the neutral product 2,3-butanediol. The production of acetoin, an intermediary metabolite, is diagnostic of this pathway.

Figure 5.20 *(Facing page, right)*
The butanol fermentation pathway, which is carried out by different species of Clostridium, *can result in the formation of several different products, including acetone, butanol, and butyric acid.*

mediary product in the butanediol fermentation pathway is acetoin (acetyl methyl carbinol). One of the classic diagnostic tests used for separating *E. coli* from *Enterobacter aerogenes* is the Vogues-Proskauer test, which detects the presence of acetoin. *E. coli* does not carry out a butanediol pathway, whereas *Ent. aerogenes* does. Another test used for separating *Ent. aerogenes* from *E. coli* is the methyl red test, which detects very low pH resulting from high amounts of acid production. Because *Ent. aerogenes* channels part of its substrate into the neutral fermentation end product, butanediol, it does not produce as much acid and thus does not lower the pH as much as *E. coli*, which channels all of its substrate into the mixed acid fermentation pathway. Thus, *E. coli* shows a positive methyl red test and a negative Vogues-Proskauer test, whereas *Ent. aerogenes* yields exactly the opposite reactions in these two test procedures.

In yet another pathway, members of the genus *Clostridium* carry out a **butanol fermentation**. During the butanol fermentation pathway, NADH is reoxidized to NAD, but no additional ATP is generated beyond the amount synthesized during glycolysis. Different *Clostridium* species form a variety of end products via this fermentation pathway, with pyruvate being converted either to acetone and carbon dioxide, propanol and carbon dioxide, to butyrate, or to butanol. This pathway is sometimes referred to as the butyric acid fermentation (Figure 5.20). Many of the fermentation products formed by *Clostridium* species are good organic solvents that have commercial applications. As such, *Clostridium* species have been used in industrial microbiology for the production of organic solvents. The butanol fermentation pathway was particularly important when it was first discovered and used in munitions production during World War I.

In addition to various pathways for the fermentative metabolism of carbohydrates, amino acids can be metabolized by a **mixed amino acid fermentation pathway**. This fermentation is important when there is extensive protein degradation and involves one amino acid serving as an electron donor and another amino acid acting as an electron acceptor to achieve the necessary oxidation–reduction balance. The mixed acid fermentation results in the deamination and decarboxylation of the amino acids. For example, a mixture of alanine and glycine can yield the end products acetate, carbon dioxide, and ammonia. The mixed amino acid fermentation contributes to the pleasant odors of some wines and cheeses but also is responsible in part for the malodorous smell of a gangrenous wound.

Autotrophic generation of ATP

In contrast to both fermentation and respiration pathways, the metabolism of autotrophic microorganisms does not require preformed organic matter to drive the conversion of ADP + P$_i$ to ATP. Instead, **photoautotrophic (photosynthetic) microorganisms use light energy, and chemoautotrophic (chemolithotrophic) microorganisms use the energy derived from the oxidation of inorganic compounds to supply the energy needed for the synthesis of ATP.**

Chemolithotrophy

Chemolithotrophic bacteria are very important because their metabolic activities form critical links in the biogeochemical cycling of various elements. **Chemolithotrophic microorganisms couple the oxidation of an inorganic compound with the reduction of a suitable coenzyme** (Figure 5.21). The transfer of electrons from the reduced coenzyme molecules, through an electron transport

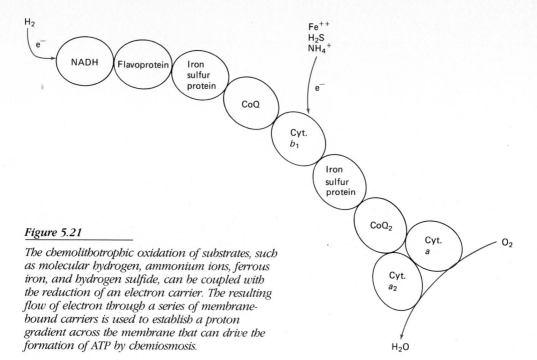

178

Microbial ener-
getics—the gen-
eration of ATP

Figure 5.21

The chemolithotrophic oxidation of substrates, such as molecular hydrogen, ammonium ions, ferrous iron, and hydrogen sulfide, can be coupled with the reduction of an electron carrier. The resulting flow of electron through a series of membrane-bound carriers is used to establish a proton gradient across the membrane that can drive the formation of ATP by chemiosmosis.

Figure 5.22

(A) Micrograph of Thiospira *sp. from upwelling groundwater near Delft. (B) Micrograph of* Thiovulum *cells, showing the deposition of sulfur by this chemolithotroph. (Courtesy J. W. M. laRiviere, International Institute for Hydraulic and Environmental Engineering, Delft.)*

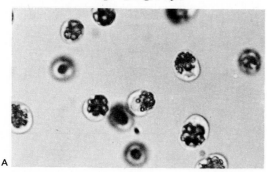

A

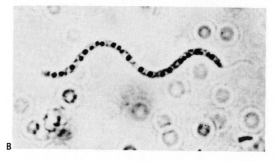

B

chain of membrane-bound carriers, establishes a hydrogen ion gradient that can drive the synthesis of ATP. The normal terminal electron acceptor for chemolithotrophs is oxygen, although in the case of sulfur compounds, nitrate may also serve as a terminal electron acceptor. The chemiosmotic mechanism for the synthesis of ATP is equivalent to the mechanism for ATP generation during oxidative phosphorylation discussed earlier in this chapter.

Some microorganisms produce the enzyme hydrogenase and are able to metabolize molecular hydrogen, forming water and the reduced coenzyme NADH from the oxidation of hydrogen. Other elements that can be oxidized by chemolithotrophs are iron, sulfur, and nitrogen. The oxidation of reduced compounds containing either iron, sulfur, or nitrogen does not produce a sufficient change in oxidation state to drive the reduction of NAD to NADH. Therefore, it appears that the electrons derived from the oxidation of reduced iron, sulfur, and nitrogen compounds enter the electron transport chain via one of the cytochrome carrier molecules that are at a lower energy level than the reduced coenzyme NADH. The amount of ATP that can be synthesized by this mechanism is thus lower than the three molecules of ATP per molecule of NADH that can be generated by hydrogen-oxidizing chemolithotrophs. Consequently, in order to generate the amount of ATP required for growth and reproduction, these chemolithotrophic microorganisms must oxidize very large amounts of reduced nitrogen-, sulfur-, or iron-containing compounds, and chemolithotrophic microorganisms, therefore, normally exhibit high turnover rates of inorganic substrates. This accounts for their importance in large-scale biogeochemical cycling reactions.

Sulfur oxidation

The most common reduced sulfur compounds, which are oxidized by chemolithotrophs, are hydrogen sulfide, elemental sulfur, and thiosulfate. The electrons derived from the oxidation of these substances are fed into an electron transport chain. The chemolithotrophic activities of sulfur-oxidizing microorganisms have received considerable attention as a result of the finding that a highly productive submarine area off the Galapagos Islands is supported by the productivity of chemolithotrophs growing on reduced sulfur released from thermal vents in the ocean floor. It is quite unusual to find an ecological system driven by chemolithotrophic metabolism; most ecosystems are dependent on the primary productivity of photoautotrophs, either higher plants, algae, or photosynthetic bacteria.

When hydrogen sulfide is utilized as the substrate for chemolithotrophic metabolism, elemental sulfur is often deposited by the bacteria, with some such bacteria depositing sulfur granules within the cell and others depositing elemental sulfur extracellularly (Figure 5.22). Some sulfur-oxidizing bacterium, such as *Thiobacillus thiooxidans*, can oxidize large amounts of reduced sulfur compounds with the formation of sulfate. The sulfur-oxidizing activities of this bacterium are important because of their involvement in the formation of acid mine drainage (Chapter 21) and mineral recovery processes (Chapter 20). *T. ferrooxidans*, which oxidizes both reduced sulfur and reduced iron for generating ATP, often is found in acid mine drainage streams where it has an available source of reduced sulfur and reduced iron that it can utilize for the chemolithotrophic generation of ATP. Various other bacteria also have the ability to oxidize ferrous iron to ferric iron, often forming a deposit of insoluble ferric hydroxide.

Nitrification

The nitrifying bacteria oxidize either ammonium or nitrite ions. Bacteria, such as *Nitrosomonas*, oxidize ammonia to nitrite; other bacteria, such as *Nitrobacter*, oxidize nitrite to nitrate. Again, because the chemolithotrophic oxidation of reduced nitrogen compounds yields relatively little energy, chemolithotrophic bacteria carry out extensive transformations of nitrogen in soil and aquatic habitats in order to synthesize their required ATP. The activities of these bacteria are quite important in soil because the alteration of the oxidation state radically changes the mobility of these nitrogen compounds in the soil column and the nitrifying bacteria have a marked influence on soil fertility (see Chapter 21).

Photosynthesis (oxidative photophosphorylation)

Photoautotrophic microorganisms (algae and photosynthetic bacteria) are able to convert light energy to chemical energy in a process known as **oxidative photophosphorylation**. When a chlorophyll molecule absorbs light energy, it becomes energetically excited and emits an electron that is transferred through an electron transport chain, analogous to the electron transport system of oxidative phosphorylation. As in the case of oxidative phosphorylation, discussed earlier in this chapter, the electron transfer reactions of oxidative photophosphorylation involve a series of membrane-bound carriers, collectively known as a **photosystem**. The transfer of electrons through the carriers of the electron transport chain establishes a hydrogen ion gradient across the membrane. The photosynthetic membranes contain ATPase, an enzyme involved in the synthesis of ATP. According to the chemiosmotic mechanism, the extrusion of hydrogen ions to the inside of the thylakoid membrane establishes an electrochemical gradient that supplies sufficient energy to drive the synthesis of ATP (Figure 5.23). In chloroplasts the flow of hydrogen ions is opposite that in the mitochondria; the ATPase is located on the outer side of the thylakoid membrane, whereas the ATPase spheres occur on the inner surface of the inner mitochondrial membrane.

Although the basic chemiosmotic mechanism driving the synthesis of ATP during oxidative photophosphorylation is the same in all photoautotrophs, there is a major difference between the algae and the anaerobic photosynthetic bacteria in the **pathways of electron transfer (photosystems)** that they employ to transfer the energy absorbed from light irradiation to the synthesis of ATP. In the case of the anaerobic green and purple photosynthetic sulfur bacteria, there is only one photosystem, known as photosystem I or cyclic oxidative photophosphorylation. The cyanobacteria and the algae have two photosystems, re-

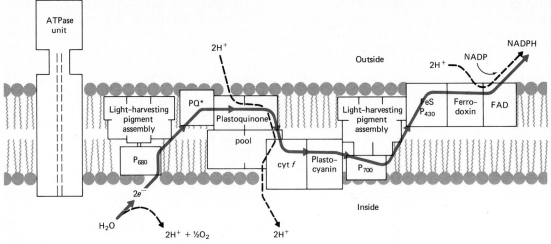

Figure 5.23

In photosynthetic organisms, light activation of a photoreceptor molecule (e.g., chlorophyll) initiates the flow of electrons through an electron transport chain. The proton gradient across the membrane is used to drive the formation of ATP by chemiosmosis in this process of oxidative photophosphorylation.

180

Microbial ener-
getics—the gen-
eration of ATP

ferred to as **photosystem** I and **photosystem** II, or **cyclic** and **noncyclic oxidative photophosphorylation**.

The Z pathway of oxidative photophosphorylation

Photosystems I and II are normally linked in a **Z pathway** of oxidative photophosphorylation (Figure 5.24). The combined photosystems require two separate photo-acts, that is, the absorption of light energy at two different photo-activation centers. The light energy actually is often absorbed by accessory pigments that transfer the energy to the primary photosynthetic pigment of the photo-activation center. In the algae and cyanobacteria the primary photosynthetic pigment is chlorophyll *a*. In photosystem I chlorophyll *a* absorbs light at a wavelength of 700 nm and therefore is designated as a P_{700} molecule. In photosystem II the photoreceptor is designated P_{680}, representing a molecule of chlorophyll *a* that has been modified to absorb light at a wavelength of 680 nm. When the P_{680} chlorophyll reaction center absorbs light from pigments in the photosynthetic unit, it becomes energetically excited and emits an electron, initiating a noncyclic pathway in which the electron is transferred through a series of membrane-bound electron carriers.

In the combined Z pathway the electrons from photosystem II are transferred through a series of membrane-bound carriers to the P_{700} chlorophyll reaction center molecule of photosystem I.

The transfer of an electron from an excited P_{680} chorophyll molecule of photosystem II to the P_{700} chlorophyll molecule of photosystem I establishes a sufficient hydrogen ion gradient across the membrane to drive the synthesis of one ATP molecule. The electron transport chain is continued when a molecule of P_{700} chlorophyll absorbs light energy, initiating the electron transfer sequence of photosystem I. The electron that is transferred from photosystem II is used to balance the oxidation state of the P_{700} chlorophyll molecule. The electron transferred through photosystem I normally is used to reduce the coenzyme NADP to NADPH, providing an essential source of reducing power for biosynthetic metabolic reactions.

The unidirectional flow of electrons to reduce NADP to NADPH leaves a charged P_{680} chlorophyll molecule that must be reduced to balance the oxidation–reduction state of this pathway. In order to accomplish the needed balance of oxidation–reduction molecules, a water molecule is split, forming oxygen and transferring electrons to the P_{680} chlorophyll of photosystem II. The oxygen produced in the pathway is essential for supporting the aerobic respiratory metabolism of heterotrophic microorganisms and animals. It also appears that a second ATP molecule may be generated as a consequence of the electron transport from water to the P_{680} chlorophyll molecule and the hydrogen ion gradient is established across the membrane as a result of this process.

Anoxygenic photosynthesis

Although the movement of the electrons through the entire Z pathway is normally noncyclic, with the electron flow from the electron donor, H_2O, to the electron acceptor, NADP, electrons may also flow cyclically through photosystem I (Figure 5.25). When this occurs, reduced coenzyme NADPH is not generated, but ATP is synthesized; that is, pho-

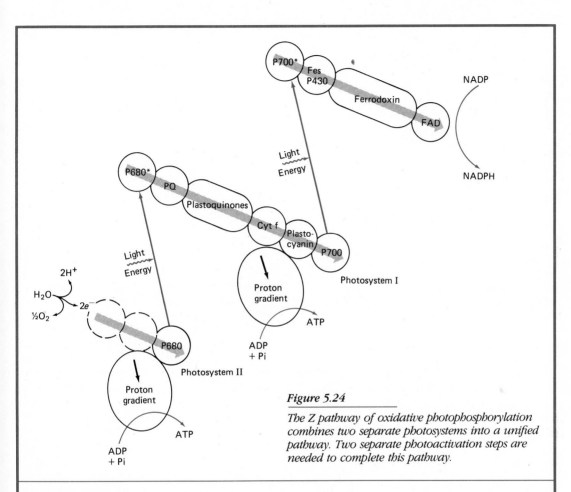

Figure 5.24

The Z pathway of oxidative photophosphorylation combines two separate photosystems into a unified pathway. Two separate photoactivation steps are needed to complete this pathway.

Discovery process

Daniel Arnon and coworkers in 1954 at the University of California first demonstrated complete photosynthesis by isolated chloroplasts, showing that isolated chloroplasts could apply absorbed light purely to the formation of ATP. They discovered this by depriving illuminated chloroplasts of carbon dioxide and pyridine nucleotide (coenzyme), while giving them large doses of ADP and inorganic phosphate. With no raw material (substrate) for making carbohydrates or NADPH, the chloroplasts used light energy to force the third phosphate on ADP to form ATP, which accumulated in substantial quantities. By 1958 Arnon and associates had constructed a theoretical model to fit the experimental facts of photosynthetic phosphorylation based on another photochemical reaction worked out by Gilbert Lewis and David Lipkin, also at the University of California. They proposed that in the primary photochemical act a photon of light strikes a chlorophyll molecule, exciting one of the electrons to an energy level sufficient to remove it from the molecule. After losing this electron, the molecule can now act as an electron acceptor. If the chlorophyll molecule took back the electron directly, it would not be useful in generating ATP, but if the excited electrons are transferred to carriers, the electrons are forced to return to chlorophyll in a series of graded steps; such a chain of electron transfers can lead to the generation of ATP and in some cases to the production of reduced coenzyme. Now we know that the series of electron transfers initiated by the light excitation of chlorophyll leads to the establishment of an electrochemical gradient across the photosynthetic membrane and the chemiosmotic generation of ATP.

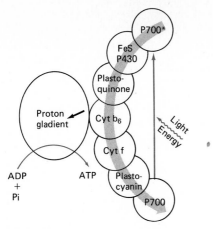

Figure 5.25

The cyclic version of photosystem I, using P_{700}, results in the generation of ATP, but reduced coenzyme NADPH is not produced.

Microbial ener-
getics—the gen-
eration of ATP

tosystem I can act as a cyclic oxidative photophosphorylation system, providing the organism with an increased yield of ATP. At low light intensities the cyanobacteria can carry out **anoxygenic photosynthesis** during which photosystem I follows a cyclic oxidative photophosphorylation pathway. In anoxygenic photosynthesis (the process derives its name from the fact that oxygen is not produced), photosystem II is inoperative, and thus, water is not split to yield oxygen. While carrying out anoxygenic photosynthesis, the cyanobacteria derive their reducing power from the coupled oxidation of hydrogen sulfide with coen-

Figure 5.26

Cyclic oxidative photophosphorylation for anaerobic photosynthetic bacteria uses P_{870}. This pathway generates the proton gradient needed to drive the formation of ATP and does not produce reduced coenzyme.

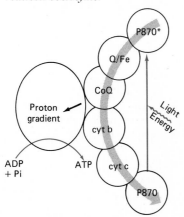

zyme reduction. When cyanobacteria utilize hydrogen sulfide as a reducing agent in this process, they form elemental sulfur granules that deposit outside the cells.

Anaerobic bacterial oxidative photophosphorylation

The process of anaerobic photosynthesis carried out by the photoautotrophic bacteria, other than the cyanobacteria, is similar in many ways to cyanobacterial anoxygenic photosynthesis. The **anaerobic photosynthetic bacteria** only carry out the reactions of photosystem I (Figure 5.26). In the green and purple anaerobic photosynthetic bacteria the primary photosynthetic pigment is **bacterial chlorophyll**, which absorbs light energy of longer wavelengths than does chlorophyll *a*. The primary photoreaction center in the anaerobic bacteria is labelled P_{870} (P_{870} refers to bacterial chlorophyll that absorbs light at infrared wavelengths). As is the case for the cyanobacteria and algae, the incoming light energy may initially be absorbed by accessory pigments and the absorbed energy then transferred to the primary photosynthetic pigments.

The excitation of the bacterial chlorophyll P_{870} causes it to emit an electron that is transported through an electron transport chain. Instead of using an external electron donor to reduce the oxidized chlorophyll molecule, the electron is transferred back to the chlorophyll molecule, forming a cyclical pathway. In this system the chlorophyll molecule acts as an internal electron donor and acceptor for photosystem I. In the course of returning the excited electron to the oxidized chlorophyll molecule, there is a sufficient energy gradient to permit the synthesis of one ATP molecule. Because the electron is returned to the chlorophyll molecule, completing a cycle, and because ATP is synthesized as a consequence of light having excited the chlorophyll molecule, the pathway is referred to as cyclic oxidative photophosphorylation.

Cyclic oxidative photophosphorylation generates ATP without generating reduced coenzyme. The anaerobic photosynthetic bacteria, however, require the reduced coenzyme NADPH for biosynthetic reactions. In order to generate NADPH, these organisms utilize reduced compounds, such as H_2S, as an electron donor. The flow of electrons from the external electron donor to the reduced coenzyme may pass through photosystem I in a noncyclic pathway (Figure 5.27). Insufficient energy, however, is released from hydrogen sulfide, or other sulfur-containing compounds, to directly drive the formation of NADPH. It is possible that the electrochemical potential established by chemiosmosis across a membrane can provide sufficient energy to drive the formation of reduced coenzyme. It is also possible that the formation of NADPH requires the consumption of ATP to drive this reaction. Thus, some of the energy generated as ATP in cyclic oxidative phosphorylation may be utilized to drive the formation of the reduced coenzyme NADPH.

The purple membrane of Halobacterium

Halobacterium has an interesting **purple membrane** that permits the conversion of light energy to chemical energy by a mechanism that is different from the ones already discussed for photoautotrophic microorganisms. The purple membrane contains bacteriorhodopsin, a protein that has a chemical structure similar to the rhodopsin pigment of the human eye. In the presence of oxygen, halobacteria use aerobic respiration to generate ATP, including oxidative phosphorylation for ATP synthesis. In the absence of oxygen, these bacteria turn to a form of oxidative photophosphorylation for synthesizing ATP. When illuminated by light, bacteriorhodopsin pumps protons to the outside of the membrane, establishing a hydrogen ion gradient. The counterflow of hydrogen ions drives the synthesis of ATP by chemiosmosis. The ATPase enzyme used for converting ADP + P_i to ATP is contained in a separate red membrane fraction. The same ATP synthesizing system used for oxidative photophosphorylation is used for oxidative phosphorylation. The halobacterial membrane system for using light energy to establish a hydrogen ion gradient that can drive the synthesis of ATP provides firm evidence for the essential role of chemiosmosis in the synthesis of ATP.

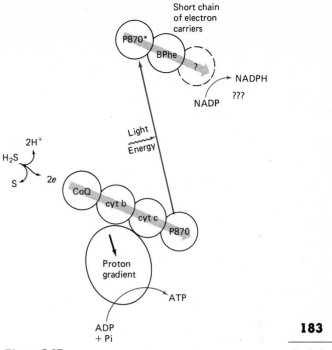

Figure 5.27

The mechanism for anaerobic photosynthetic generation of the reduced coenzyme NADPH has not been fully elucidated. In this drawing a possible mechanism is illustrated, using noncyclic photosystem I hydrogen sulfide as the electron donor. The energetics for this reaction sequence suggest that NADH rather than NADPH may be generated via this pathway and that some of the ATP formed by oxidative photophosphorylation may be used to convert NADH to NADPH needed by the cell for biosynthetic reactions.

Postlude

It is clear that microorganisms exhibit a variety of strategies for converting chemical and light energy into the energy stored within ATP, the central currency of energy of the cell. The synthesis of ATP can be achieved autotrophically—either through the oxidation of inorganic substrates or through the conversion of light energy to chemical energy—or may be generated heterotrophically through the utilization of organic substrates. The amount of ATP that can be synthesized in these processes varies greatly and microorganisms accordingly show great variation in the efficiency with which they can synthesize sufficient ATP to meet their energy requirements. Chemolithotrophic metabolism, for example, is relatively inefficient energetically, and thus, chemolithotrophic microorganisms must metabolize a large number of inorganic substrate molecules to synthesize sufficient ATP for their metabolic needs during growth and reproduction.

Similarly, respiration is energetically far more favorable than fermentation. To take a representative number, many microorganisms are capable of generating 38 ATP molecules per molecule of glucose by respiration, compared to only 2 molecules of ATP per molecule of glucose by fermentation. As a consequence of the "inefficiency" of the fermentation pathway, more substrate must be utilized during fermentation than during respiration to achieve similar amounts of growth. Although fermentation does not yield as much ATP per molecule of substrate as does respiration, the end products of fermentation are of more practical interest than are those of respiration. Respiration normally results in the formation of carbon dioxide and water. The various fermentation pathways carried out by different organisms produce numerous end products of economic value. For example, the ethanolic and lactic acid fermentation pathways are of particular importance in the food industry because they are directly involved in the production of bread, beer, wine, and cheese, among other foods and beverages.

Although microorganisms show great metabolic versatility for generating ATP and reducing power, several central pathways play key roles in the metabolism of microbial cells. The metabolic pathways employed in ATP generation involve various intermediary metabolites, linked together in a series of small steps to form unified biochemical pathways. These metabolic pathways are carefully regulated so that when the cell has sufficient ATP, pathways that result in the synthesis of ATP are shut off through feedback inhibition by an allosteric inhibitor of a critical enzyme in the pathway. For example, phosphofructokinase, a key enzyme in the glycolysis of carbohydrates, is inhibited by ATP, and therefore, when the cell has enough ATP, further synthesis of ATP is restricted.

Considering the variety of ways in which a microorganism can drive the synthesis of ATP, heterotrophic metabolism of an organic substrate may occur by either respiration or fermentation, and autotrophic metabolism may involve the oxidation of inorganic compounds or the conversion of light energy. In respiration an external electron acceptor is required to complete the metabolic pathway; oxygen serves as the terminal electron acceptor in aerobic respiration, and nitrate or sulfate acts as a possible terminal electron acceptor in anaerobic respiration. In fermentation the organic substrate molecules serve as internal electron donors, and products of their metabolism serve as acceptors. Chemolithotrophic microorganisms are able to oxidize inorganic compounds to generate ATP. Photoautotrophic microorganisms are able to trap light energy, in some cases involving one cyclic photosystem and in others involving two different photosystems linked in a Z pathway.

With the exception of fermentation, all of these pathways, which are aimed at the synthesis of ATP, involve an electron transport through a series of membrane-bound carriers and the establishment of a hydrogen ion gradient across a membrane. For example, in photosynthetic microorganisms the flow of electrons—initiated when a chlorophyll molecule is energetically excited by absorbing light energy—establishes an electrochemical gradient across a membrane during the process of oxidative photophosphorylation. Similarly, a hydrogen ion gradient across a membrane is formed during oxidative phosphorylation in both respiration and chemolithotrophic metabolism. The counterflow of hydrogen ions, channeled through membrane-bound ATPase, drives the synthesis of ATP. This process of chemiosmosis is central to many of the ATP-synthesizing pathways of microorganisms.

As a consequence of their metabolic activities, microorganisms are capable of channeling energy into the synthesis of ATP. They are also capable of channeling reducing power into the synthesis of the reduced coenzymes NADH and NADPH. The reducing power and the ATP generated by microbial metabolism are used by microorganisms for their growth and reproduction. The particular pathways that a microorganism can use for generating ATP and reducing power depend on the enzymes that the organism possesses. The additional end products of metabolism that are produced also are a function of the cell's enzymatic potentialities, which are genetically determined. Regardless of the mode of metabolism the strategies are the same: synthesize ATP, reduce coenzyme (NADPH) and small precursor molecules to serve as the building blocks of macromolecules, and then use the energy, reducing power, and precursor molecules to synthesize the macromolecular constituents of the organism.

1. What is autotrophic metabolism? Discuss the different ways autotrophs generate ATP. Consider both photoautotrophs and chemolithotrophs.

2. What is heterotrophic metabolism?

3. What is the difference between fermentation and respiration?

4. What individual pathways are involved in respiratory metabolism?

5. What is a terminal electron acceptor? What is the difference between aerobic and anaerobic respiration?

6. Name five different fermentation pathways. For each pathway, what are the metabolic end products, and what organisms characteristically carry out the pathway?

7. What is the difference between oxygenic and anoxygenic photosynthesis?

8. Based only on their metabolism, should the blue-greens be considered bacteria or algae?

9. What is chemiosmosis, and how does it explain the generation of ATP both in oxidative phosphorylation and photophosphorylation?

10. Where do the following processes occur?
   a. oxidative phosphorylation in bacteria
   b. oxidative phosphorylation in fungi
   c. oxidative photophosphorylation in algae
   d. oxidative photophosphorylation in cyanobacteria
   e. glycolysis in bacteria
   f. the Krebs cycle in bacteria

Becker, W. M. 1977. *Energy and the Living Cell*. Harper & Row Publishers, Inc., New York.

Bronk, J. R. 1973. *Chemical Biology: An Introduction to Biochemistry*. Macmillan Publishing Co., Inc., New York.

Chance, B. 1977. Electron transfer: pathways, mechanisms, and controls. *Annual Reviews of Biochemistry* 46: 967–980.

Dawes, I. W., and I. W. Sutherland. 1976. *Microbial Physiology*. Blackwell Scientific Publications, Oxford, England.

Govindjee. 1975. *Bioenergetics of Photosynthesis*. Academic Press, New York.

Govindjee, and R. Govindjee. 1974. The absorption of light in photosynthesis. *Scientific American* 231(6): 68–82.

Hinkle, P. C., and R. E. McCarthy. 1978. How cells make ATP. *Scientific American* 238(3): 104–123.

Lehninger, A. L. 1971. *Bioenergetics*. Benjamin/Cummings Publishing Co., Menlo Park, California.

Lehninger, A. L. 1982. *Principles of Biochemistry*. Worth Publishers, Inc., New York.

Mandelstam, J., K. McQuillen, and I. W. Dawes. 1982. *Biochemistry of Bacterial Growth*. Blackwell Scientific Publications, Oxford, England.

Mitchell, R. 1979. Keilin's respiratory chain concept and its chemiosmotic consequences. *Science* 206: 1148–1159.

Parson, W. W. 1974. Bacterial photosynthesis. *Annual Reviews of Microbiology* 28: 41–59.

Stryer, L. 1981. *Biochemistry*. W. H. Freeman Co., San Francisco.

Carbon flow: biosynthesis of macromolecules

6

In addition to requiring energy in the form of ATP, microorganisms must carry out metabolic activities that transform the starting substrate molecule into the many different macromolecules of the cell, including those that make up the cell's structural units. An autotrophic microorganism must transform inorganic carbon dioxide into all its necessary macromolecules, including—among others—proteins for enzymes, lipids for membranes, carbohydrates for various structures such as cell walls, and nucleic acids for the storage and expression of genetic information. Similarly, a heterotrophic microorganism growing on a single carbon source, such as glucose, must carry out metabolic activities that transform that substrate into all of the essential macromolecules of the microorganism.

The basic strategy of the cell, in terms of **carbon flow**, is to form relatively small biochemical molecules that can act as central building blocks for establishing the carbon skeletons of a variety of large macromolecules. When a heterotrophic microorganism starts with an organic substrate, like glucose, it carries out a catabolic pathway to break that molecule down into smaller compounds that then act as precursors for the biosynthesis of macromolecules (**catabolism** means

degradative process). Then the microorganism uses an anabolic pathway in which small molecules are transformed into larger molecules (**anabolism** means biosynthetic process).

As a general rule, the macromolecules of microorganisms are in a more reduced oxidation state than the starting substrate molecules, and therefore, biosynthetic pathways require reducing power. Part of the metabolic activities of a microorganism are aimed at generating the reduced coenzyme NADPH required for biosynthesis. There is an interesting and critical difference in the coenzymes used in catabolic and anabolic pathways; in catabolism NAD serves as an oxidizing agent when it is reduced to NADH, whereas in anabolism the coenzyme NADPH serves as the reducing agent and is converted to its oxidized form, NADP. Biosynthesis also requires ATP because anabolic pathways are endergonic and therefore must be coupled with exergonic reactions, generally the hydrolysis of ATP, for biosynthetic reactions to occur. Various ways of coupling exergonic reactions with the synthesis of ATP from ADP and P_i have been discussed in Chapter 5; in this chapter we shall see how the ATP that is generated by the cell is used to drive biosynthetic reactions.

Amphibolic pathways

In addition to ATP and NADPH, biosynthetic pathways require small precursors for the molecules being synthesized, that is, the substrates that can be transformed into macromolecules in the anabolic pathway. Many of the intermediary metabolites of the catabolic pathways serve as these substrates for biosynthesis. These compounds can serve in the generation of ATP and in the biosynthesis of macromolecules and often are siphoned from catabolic, ATP-generating, pathways into anabolic, ATP-utilizing pathways. In fact, there are several **central metabolic pathways** that play key roles in both catabolism and biosynthesis, and the flow of carbon can be reversed through these so-called **amphibolic (dual-purpose) pathways**. Glycolysis and the Krebs cycle, which were discussed in Chapter 5 for their roles in ATP generation, also serve as anabolic pathways and thus play key roles in the flow of carbon through the cell. These pathways are central to the flow of carbon among all four major classes of microbial biochemicals, namely among carbohydrates, lipids, proteins, and nucleic acids; the pathways permit the necessary metabolic connections between the degradation of one class of compound, serving as substrate, and the biosynthesis of another class of compound composing a necessary macromolecular product (Figure 6.1).

Because the same intermediary metabolites in these amphibolic pathways serve both for catabolism and anabolism, there must be mechanisms to regulate the direction of carbon flow, that is, some major differences between the anabolic and catabolic functions of these pathways that determine whether the flow of carbon is from large molecules to small ones or whether larger molecules are being synthesized from smaller molecules. Indeed, while the intermediates are the same, the catabolic and anabolic functions of amphibolic pathways are distinct. A key difference is the separation of coenzymes that are involved in catabolic and anabolic pathways; NAD is involved in catabolic pathways, and the related coenzyme NADP is involved in anabolic pathways. The use of different coenzymes enables the cell to differentiate the direction of carbon flow through the same intermediary metabolites, making it possible for reductive biosynthesis to occur at the same

Figure 6.1

The central metabolic pathways are interconnected so that carbon flows through the cell via a unified network of metabolic reactions. This network connects all the major classes of macromolecules. Small carbon molecules formed from the breakdown of one macromolecular class can be used for the biosynthesis of compounds of the same or other classes of macromolecules.

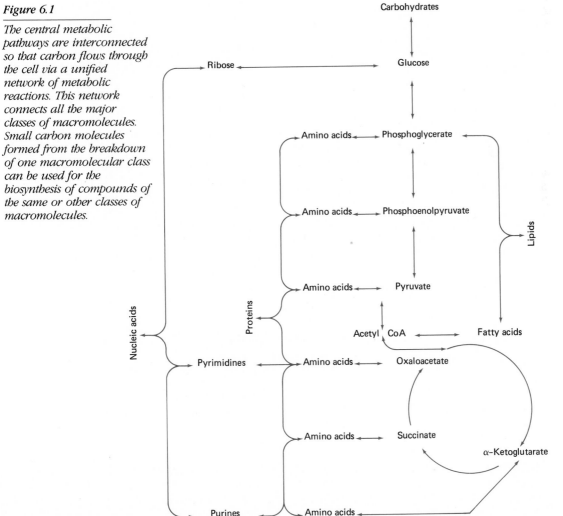

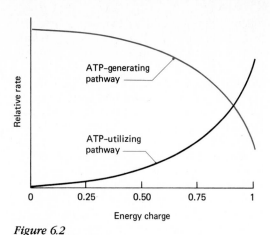

Figure 6.2

The effects of energy charge on rates of ATP-generating (catabolic) and ATP-utilizing (anabolic) pathways. By having ATP and cyclic AMP serve as allosteric effectors for key enzymes, the flow of carbon through identical intermediates of biosynthetic and catabolic pathways is controlled. When more ATP is needed, the energy charge is low and the rate of ATP-generating pathways is accelerated. When the cell has sufficient ATP, the rates of biosynthetic pathways is relatively high and the rate of catabolic pathways relatively low.

molecules. In particular, feedback inhibition, whereby an end product of a metabolic pathway acts as an allosteric inhibitor of an enzyme catalyzing a reaction near the beginning of that pathway, often plays a critical role in the regulation of a cell's metabolism.

A major factor controlling the direction of carbon flow through amphibolic pathways is the energy status of the cell, a measure of which is known as the **energy charge**. The energy charge describes the relative proportions of ATP, ADP, and AMP in the cell and is defined as

$$\text{Energy charge} = \frac{[\text{ATP}] + \frac{1}{2}[\text{ADP}]}{[\text{ATP}] + [\text{ADP}] + [\text{AMP}]}$$

As a rule, ATP-generating pathways are inhibited by high energy charge, whereas ATP-utilizing pathways are stimulated by high energy charge (Figure 6.2). The reason for this is that ATP generally acts as an allosteric inhibitor of a key enzyme in an energy-generating pathway, as we saw for the enzyme phosphofructokinase, which catalyzes the conversion of fructose 6-phosphate to fructose 1,6-diphosphate, when we considered the regulation of the rates of glycolysis in Chapter 5. In contrast, AMP is an allosteric inhibitor of fructose 1,6-diphosphatase, which catalyzes the reverse reaction that converts fructose 1,6-diphosphate to fructose 6-phosphate in the energy-requiring gluconeogenic (glucose-synthesizing) pathway. Thus, when the catabolic ATP-generating pathway is active, the reverse anabolic ATP-requiring pathway is inhibited and vice versa, so that the flow of carbon through an amphibolic pathway at any point in time is unidirectional, regulated by differences in enzymes, their allosteric inhibitors, and the energy charge of the cell.

time that the cell is generating ATP through glycolysis. Additionally, different enzymes are involved in key metabolic reactions that regulate the overall rates of these pathways. These enzymes generally have allosteric effectors that inhibit or accelerate the rates of the metabolic reactions catalyzed by the enzymes, and in this manner, the cell is able to fine-tune the flow of carbon, balancing catabolic activities involved in generating ATP and anabolic activities involved in the synthesis of the microorganism's essential macro-

The conversion of inorganic carbon dioxide to organic carbon

The formation of organic compounds from carbon dioxide is fundamental to the central metabolic pathways of the cell that involve interconversions of organic compounds. Autotrophic microorganisms use the organic molecules generated from the reduction of CO_2 as precursors for the rest of the macromolecules that they require; these organic molecules synthesized by au-

totrophs, both microbial autotrophs and plants, form the substrate organic molecules required for the metabolism of heterotrophic organisms. Because organic carbon is in a more reduced oxidation state than carbon dioxide, the autotrophic conversion of carbon dioxide to organic carbon requires reducing power in the form of reduced coenzyme NADPH, and because the pathway of carbon dioxide fixation is endergonic, carbon dioxide fixation also requires ATP. We have already seen how photoautotrophic microorganisms generate NADPH and ATP as part of their photophosphorylation systems and how chemoautotrophs generate reduced coenzyme and ATP through the oxidation of inorganic compounds. We shall now examine how such autotrophic microorganisms use NADPH and ATP to convert inorganic carbon dioxide to organic carbohydrate molecules.

The Calvin cycle

The **fixation of CO_2 by autotrophic microorganisms** occurs in a metabolic pathway known as the **Calvin cycle** (Figure 6.3). The Calvin cycle is a complex series of reactions that really represents three slightly different but fully integrated metabolic sequences. It effectively takes three turns of the Calvin cycle to synthesize one molecule of glyceraldehyde 3-phosphate, the net organic product of this metabolic pathway. Because glyceraldehyde 3-phosphate contains three carbon atoms, the Calvin cycle is sometimes referred to as a C_3 pathway. The formation of one molecule of glyceraldehyde 3-phosphate requires three molecules of carbon dioxide, nine molecules of ATP, and six molecules of NADPH. It is thus apparent that the conversion of carbon dioxide to glyceraldehyde 3-phosphate requires a great deal of energy and reducing power. In photoautotrophs the ATP and NADPH (energy and reducing power) to drive the Calvin cycle come from the light reactions of photosynthesis and in chemolithotrophs from the oxidation of inorganic compounds. The Calvin cycle itself is known as a dark reaction because, although it requires ATP and NADPH, it does not require any light reactions (photo-acts) as an integral part of the metabolic reactions of this cycle. As long as there is an adequate supply of ATP and NADPH, the Calvin cycle continues in the absence of light.

The initial metabolic step in the Calvin cycle involves the reaction of carbon dioxide with ribulose 1,5-diphosphate to form a six-carbon compound that immediately splits to form two molecules of 3-phosphoglycerate (PGA). The reaction of CO_2 with ribulose 1,5-diphosphate is catalyzed by the enzyme ribulose 1,5-diphosphate carboxylase and is highly exergonic with a ΔG^0 of -12.4 kcal/mole. This initial reaction occurs three separate times so that three molecules of carbon dioxide enter the Calvin cycle by reacting with three molecules of ribulose 1,5-diphosphate and

there is a formation of six molecules of 3-phosphoglycerate. Five of the six molecules of 3-phosphoglycerate go through a series of reactions that regenerate the original three molecules of ribulose 5-phosphate. The remaining 3-phosphoglycerate molecule is reduced to form the net product of the cycle, the glyceraldehyde 3-phosphate molecule. Because the ribulose 1,5-diphosphate is regenerated by these reactions, the metabolic pathway forms a cycle, and the only net carbon flow is the consumption of three carbon dioxide molecules and the production of one molecule of glyceraldehyde 3-phosphate (Figure 6.4).

The glyceraldehyde 3-phosphate molecules that are formed during the Calvin cycle can further react to form glucose and polysaccharides of glucose, such as starch and cellulose. The conversion of glyceraldehyde 3-phosphate to glucose occurs by a gluconeogenic pathway, which effectively is a reversal of the Embden-Meyerhof pathway (Figure 6.5). It takes six turns of the Calvin cycle to form a six-carbon carbohydrate, such as glucose. The net input of energy, as ATP, and reducing power, as NADPH, required for the conversion of carbon dioxide to glucose is 18 ATP and 12 NADPH molecules. The overall conversion of carbon dioxide to glucose is highly endergonic and requires 114 kcal/mole. In algae and cyanobacteria, in order to meet the ATP and NADPH requirements of this process, eight photo-acts, four each in photosytems I and II, are needed. Because a mole of photons is equivalent to approximately 47 kcal, the efficiency of photosynthesis is about $114/(47 \times 8)$ or approximately 30 percent.

The key enzyme that determines the rates of the Calvin cycle is **ribulose 1,5-diphosphate carboxylase**, and thus, the key metabolic step is the reaction of carbon dioxide with ribulose 1,5-diphosphate. This enzyme occurs on the surfaces of the thylakoid membranes of photosynthetic microorganisms; in the chloroplasts of some eukaryotic microorganisms it constitutes approximately 15 percent of the total protein. Ribulose 1,5-diphosphate carboxylase is subject to allosteric control, providing the mechanism for regulating the rates of carbon dioxide fixation via the Calvin cycle. The rates of reaction of CO_2 + ribulose 1,5-diphosphate to form PGA are increased by NADPH, an allosteric activator of ribulose 1,5-diphosphate carboxylase, and thus, when the light reactions are generating large amounts of NADPH, the dark reactions of photosynthesis are stimulated. When reducing power is not available, the Calvin cycle ceases to function. Ribulose 1,5-diphosphate carboxylase also has higher activities at alkaline than at neutral pH

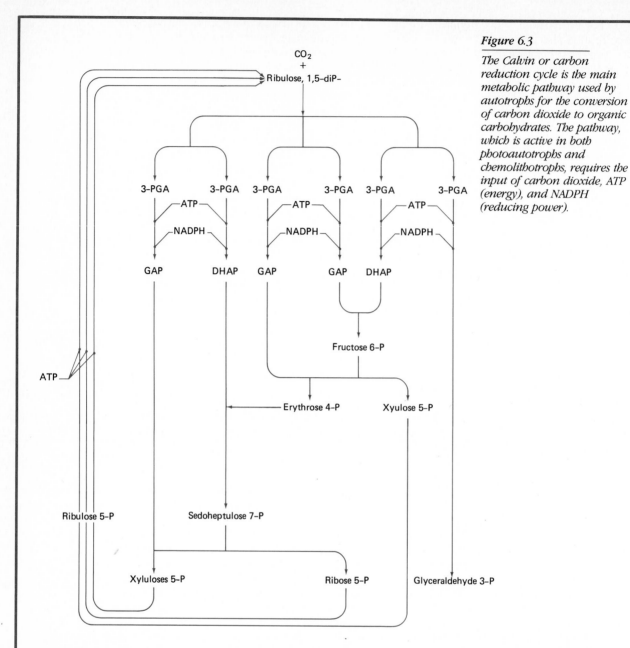

Figure 6.3

The Calvin or carbon reduction cycle is the main metabolic pathway used by autotrophs for the conversion of carbon dioxide to organic carbohydrates. The pathway, which is active in both photoautotrophs and chemolithotrophs, requires the input of carbon dioxide, ATP (energy), and NADPH (reducing power).

Discovery process

In 1945 Melvin Calvin and his coworkers began the work that determined the dark reaction of photosynthesis and that also won him a Nobel Prize in 1961. The unicellular green alga *Chlorella* was their test organism because it is easy to culture in a highly reproducible way. In order to elucidate the pathway by which CO_2 becomes fixed into carbohydrate, Calvin and colleagues centered their experimental strategy on the use of radioactive ^{14}C to trace the fate of the CO_2. Radioactive $^{14}CO_2$ was injected into an illuminated suspension of algae that had been carrying out photosynthesis with normal CO_2. The algae were killed at selected intervals by pouring the suspension into alcohol, which stopped the enzymatic reactions. The radioactive compounds in the algae were then separated and identified by using two-dimensional paper chromatography. The resulting paper chromatogram was pressed against photographic film to produce an autoradiogram in which black spots indicated the locations of the labeled products arising from CO_2 fixation; the next 10 years were devoted to positively identifying these products, thus establishing the CO_2-reduction pathway and all the intermediary metabolites in what later became known as the Calvin cycle.

Figure 6.4

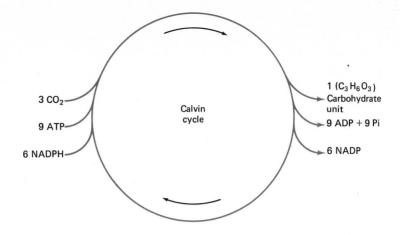

values, and the pumping of hydrogen ions across the photosynthetic membrane that occurs during the light reactions of photosynthesis raises the pH in the stroma of chloroplasts where the Calvin cycle occurs within eukaryotic photosynthetic organisms. Similar hydrogen ion pumping in photosynthetic and chemolithotrophic prokaryotes likewise increases the rates of carbon dioxide fixation by raising the pH so that ribulose 1,5-diphosphate carboxylase functions optimally. By having the regulation of the Calvin cycle occur at the first metabolic step, the cell efficiently conserves its metabolic energy and reducing power.

The C_4 pathway for the fixation of carbon dioxide

While the Calvin cycle is the primary pathway for carbon dioxide fixation in most autotrophic microorganisms, there are other metabolic pathways in which carbon dioxide is converted to organic molecules. In fact, carbon dioxide fixation is a universal reaction of living systems, including heterotrophic organisms; this fact has been employed in designing extraterrestrial life detection systems, and when the *Viking* spacecraft landed on Mars, one of three critical life detection experiments measured the rates of conversion of carbon dioxide to organic compounds. A common pathway for the fixation of CO_2 in both heterotrophs and autotrophs is the **C_4 pathway**, so designated because the product of the pathway, oxaloacetate, is a four-carbon molecule. In this metabolic pathway, pyruvate or phosphoenolpyruvate—metabolites of the glycolytic pathway—react with carbon dioxide to form oxaloacetate, an intermediate metabolite of the Krebs cycle

(Figure 6.6). The oxaloacetate formed in this pathway can go into amino acid and nucleic acid biosynthesis, which will be discussed later in this chapter. Whereas all organisms fix carbon dioxide as part of their metabolism, heterotrophic organisms are unable to form a significant portion of their carbon skeleton from CO_2 fixation, and therefore, they still depend on organic compounds as substrates for forming macromolecules. The fixation of CO_2 is used by heterotrophic organisms to replenish intermediate compounds of the Krebs cycle consumed in biosynthetic reactions.

Figure 6.5

The conversion of glyceraldehyde 3-phosphate to glucose represents an anabolic pathway that effectively reverses the flow of carbon through glycolytic pathways. This pathway is important in the formation of glucose from noncarbohydrate substrates.

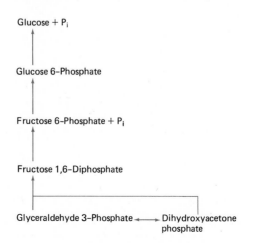

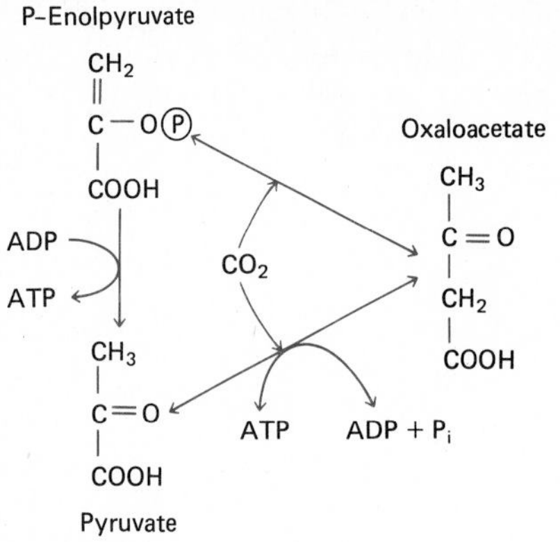

Figure 6.6

P–Enolpyruvate

$$CH_2$$
$$\|$$
$$C-O\,(P)$$
$$|$$
$$COOH$$

Oxaloacetate

$$CH_3$$
$$|$$
$$C=O$$
$$|$$
$$CH_2$$
$$|$$
$$COOH$$

ADP
ATP

CO_2

$$CH_3$$
$$|$$
$$C=O$$
$$|$$
$$COOH$$

ATP   ADP + P_i

Pyruvate

Figure 6.6

The reaction of pruvate or phosphoenolpyruvate with CO_2 forms oxaloacetate. This reaction is critical for replenishing the intermediates of the Krebs cycle that are used up in biosynthesis.

Discovery process

Winogradsky discovered chemoautotrophs in the 1880s; since then it has been known that autotrophic bacteria assimilate CO_2. It was not, however, until 1935 that it was accepted that heterotropic forms of life assimilate CO_2 or that CO_2 plays an important role as a chemical reactant in all heterotrophic organisms. The first experimental evidence of heterotrophic assimilation of CO_2 to form carbon-to-carbon linkages was presented by C. H. Werkman and H. G. Wood. In working with the propionic acid bacteria *Propionibacterium*, they found that the CO_2 present in the calcium carbonate, which had been added to neutralize acidity, was transformed into organic carbon. These results showed that nonphotosynthetic, typically heterotrophic organisms were able to assimilate CO_2. The initial findings were based on carbon balances determined by meticulous gravimetric analyses. Conclusive proof was provided in 1941 when the radioisotope technique showed the role CO_2 plays in heterotrophic metabolism. Werkman demonstrated that many heterotrophic bacteria, including anaerobes and aerobes, protozoa, fungi, yeasts, and animal tissues, assimilate CO_2. It is unlikely that life could be possible in the absence of available CO_2 because bacteria, as shown by Werkman and S. J.

Ajl in 1949, and protozoa, as shown by O. Rahn in 1941, die when deprived of CO_2 for several hours. It appears that all forms of life assimilate CO_2 and that this assimilation is an essential physiological function providing for the synthesis of indispensable metabolic intermediates. This principle has been used in the search for life on other planets; the *Viking*-Mars lander was equipped with a system for detecting the conversion of $^{14}CO_2$ to organic compounds.

Carbon dioxide conversion to methane

A few specialized types of bacteria, methanogens, are able to convert CO_2 to methane (CH_4). As already indicated, these methanogenic bacteria are distinct from other bacteria in several ways and as such are designated as members of the archaebacteria. Some methanogenic bacteria are able to use the flow of electrons from molecular hydrogen to reduce CO_2 and form CH_4, in a type of anaerobic respiration pathway (Figure 6.7). The conversion of carbon dioxide to methane, using molecular hydrogen as an electron donor, is an exergonic reaction with a ΔG^0 of -31 kcal/mole and an actual ΔG at cellular concentrations of reactants and products of about -15 kcal/mole. The synthesis of ATP in methanogens appears to involve an electron transport–oxidative phosphorylation mechanism rather than direct substrate-level phosphorylation. About one molecule of ATP can be synthesized for every molecule of CO_2 converted to CH_4. During the conversion of carbon dioxide to methane, it also appears that NADPH is generated. This NADPH probably is used for the incorporation of carbon dioxide into the

Figure 6.7

The conversion of CO_2 to CH_4 is carried out by a specialized group of archaebacteria. This is a strictly anaerobic pathway involving the flow of electrons from a hydrogen donor.

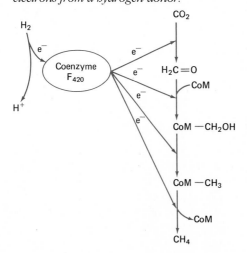

macromolecules of the cell. Approximately 90–95 percent of the carbon dioxide used by methanogens is converted to methane, presumably supporting ATP and NADPH synthesis, and the remainder is incorporated into cellular carbon. Methanogenic bacteria may have been among the earliest living organisms on earth because they can grow autotrophically on hydrogen and carbon dioxide under anaerobic conditions, and it is presumed that the early atmosphere of the earth would have permitted the growth of just such organisms.

Carbohydrate metabolism

After considering the ways in which autotrophic microorganisms can form organic compounds from carbon dioxide, we now turn to the ways in which organic compounds are converted into the macromolecules of the cell. Carbohydrates represent one of the four classes of compounds that compose the essential macromolecules of the cell. The formation of glucose through the Calvin cycle forms an important substrate for heterotrophic metabolism. We have already examined the use of glucose for the generation of ATP and indicated that glycolysis also provides essential precursors for biosynthetic pathways. Here we shall consider some additional aspects of carbohydrate metabolism from the viewpoint of the flow of carbon from substrate to macromolecular constituents of the cell.

Pentose phosphate pathway

As seen in the carbon dioxide fixation pathways, biosynthesis requires NADPH in addition to ATP. The glycolytic pathways of carbohydrate metabolism, which we discussed in Chapter 5, generate ATP but do not supply the cell with the source of reducing power, NADPH, needed for biosynthesis. In order to generate the required NADPH, heterotrophic microorganisms may carry out a metabolic pathway known as the **pentose phosphate pathway** (Figure 6.8). It should be noted, however, that there are alternative ways of generating NADPH, and *E. coli*, for example, can grow normally even without a pentose phosphate pathway. The intermediate metabolites of the pentose phosphate pathway are the same as the initial metabolites of the phosphoketolase pathway of glycolysis. In the pentose phosphate pathway, glucose is converted into ribulose 5-phosphate and carbon dioxide, a process that requires the hydrolysis of one ATP molecule and results in the generation of two NADPH molecules. This pathway provides a shunt to the glycolytic pathway that can be used to regulate the flow of carbon through the cell; it provides an alternate pathway for the flow of carbon from carbohydrates. The pentose phosphate shunt can supply the cell with NADPH and small carbon-containing molecules required for the synthesis of macromolecules, including a supply of pentose molecules needed to establish the backbone of the carbohydrate portion of nucleic acids, namely, the ribose and the deoxyribose of RNA and DNA.

The particular pathway employed for the flow of carbon depends on the need for NADPH, ATP, and carbon skeletons (small precursor molecules for incorporation into macromolecules). When a large amount of reducing power is required, the glucose molecule can be completely metabolized to carbon dioxide with the production of 12 molecules of reduced coenzyme NADPH (Figure 6.9). This series of reactions really involves a cyclic pathway in which glucose 6-phosphate is broken down and resynthesized and provides a large amount of reducing power needed by microorganisms during times of active growth. When the cell requires both NADPH and ATP, the phosphoglyceric acid molecule can enter the glycolytic pathway forming pyruvate, with NADPH being generated during the initial steps of the pentose phosphate pathway and ATP being generated as a result of the oxidation of the pyruvate (Figure 6.10). The combination of the pentose phosphate and glycolytic pathways supplies three-, four-, and five-carbon-containing molecules necessary for the carbon skeletons of a variety of essential macromolecules.

Disaccharides and polysaccharides as substrates for microbial growth

In addition to monosaccharides, microorganisms can use disaccharides and polysaccharides as substrates for growth. Common disaccharides that can be used as a source of carbon by microorganisms are maltose, which can be hydrolyzed to form glucose; sucrose, which can be hydrolyzed to form

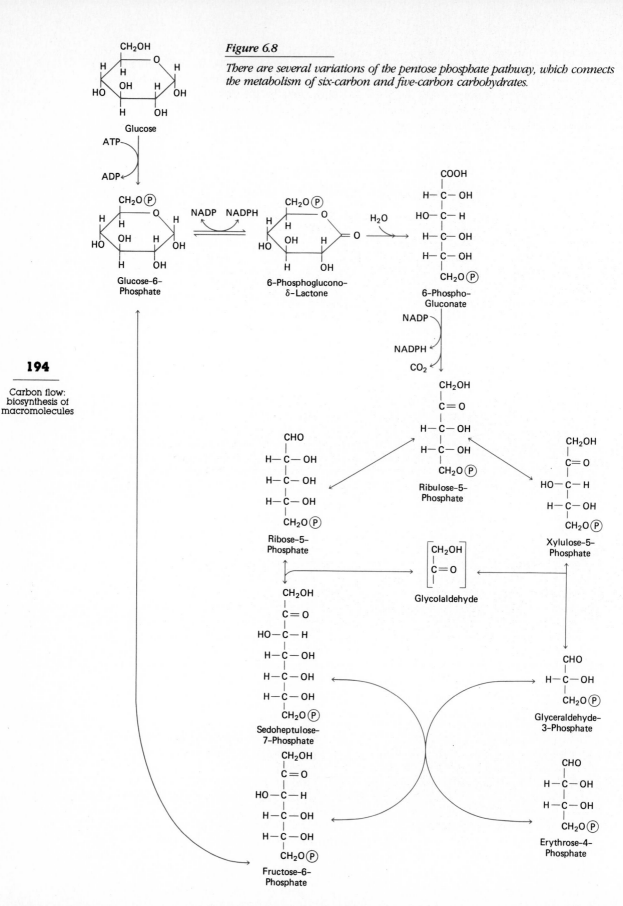

Figure 6.8

There are several variations of the pentose phosphate pathway, which connects the metabolism of six-carbon and five-carbon carbohydrates.

Figure 6.9

This version of the pentose phosphate pathway leads to the formation of 12 NADPH and is important when reducing power is needed for biosynthesis.

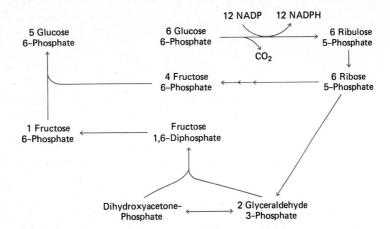

glucose and fructose by the action of the enzyme sucrase; and lactose, which can be hydrolyzed to form galactose and glucose by the action of the enzyme β-galactosidase. Not all microorganisms have the enzymes necessary to **hydrolyze** (break the bonds that hold these disaccharides together) and thus make use of these disaccharides, and the ability to use disaccharides is an important diagnostic characteristic used in the identification of various bacterial species.

The monosaccharides formed from these disaccharides can enter the glycolytic or pentose phosphate pathway (Figure 6.11). For example, the galactose derived from lactose can be converted to glucose 1-phosphate, which then can be transformed to glucose 6-phosphate, an intermediate in the glycolytic pathway. The glucose derived from lactose similarly can react to form glucose 6-phosphate. In these cases the same yields of ATP per six-carbon substrate molecule are obtained, as we discussed in Chapter 5 for glucose. In the case of sucrose, however, additional ATP is generated. When sucrose is hydrolyzed, the fructose is transformed to fructose 6-phosphate, but the glucose forms glucose 1-phosphate by a substrate-level phosphorylation reaction, and hence, ATP hydrolysis is not required for the initial step of glycolysis. The glucose 1-phosphate is then transformed, by the action of phosphoglucomutase, to glucose 6-phosphate that enters the normal glycolytic pathway. Because of the initiation of glycolysis without the need for ATP hydrolysis, there is an increase in the net production of ATP per six-carbon substrate molecule when sucrose is the substrate.

Like sucrose, the glucose derived from glycogen or starch undergoes a substrate-level phosphorylation reaction to form glucose 1-phosphate, which then enters a metabolic pathway of

glycolysis. Because the substrate-level phosphorylation reactions result in the formation of a glucose 6-phosphate molecule without the hydrolysis of ATP, there is a net gain of additional ATP molecules per six-carbon substrate molecule when these substrates are used. As indicated in Chapter 5, the use of many polysaccharides, such as starch and cellulose, generally requires exoenzymes. Extracellular enzymes produced by microorganisms include cellulases for the degradation of cellulose, amylases for the degradation of starches, chitinases for the degradation of chitin, and pectinases for the degradation of pectin. Only following the extracellular hydrolysis of the polysaccha-

Figure 6.10

This version of the pentose phosphate pathway produces limited amounts of ATP, pyruvate, and reduced coenzymes.

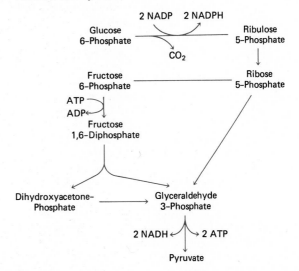

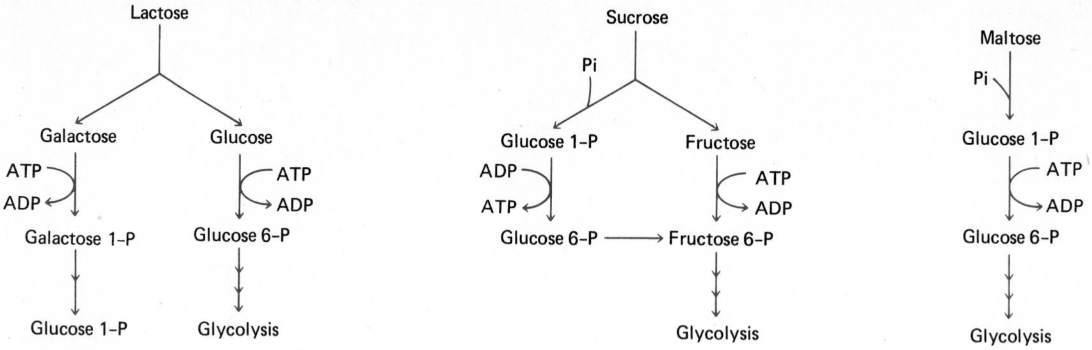

Figure 6.11

To enter the normal glycolytic pathways, disaccharides must be hydrolyzed and phosphorylated. In some cases, such as the metabolism of sucrose, the initial reaction is a substrate-level phosphorylation and thus additional ATP is generated compared to the metabolism of other disaccharides, such as lactose.

ride can smaller carbohydrate molecules enter the cell and serve as substrates for microbial metabolism. The ability of microorganisms to make use of polysaccharides for generating ATP and cellular carbon is very important because these are often the abundant available substrate materials in soil and aquatic habitats. The microbial degradation of these polymers, which make up a large part of the plant and animal residues that occur in these habitats, is essential for nutrient recycling and the continued productivity of many ecosystems.

Carbohydrate biosynthesis

In addition to using carbohydrates as substrates, microorganisms are able to synthesize carbohydrate molecules. This is essential because carbohydrates comprise portions of the macromolecules of the cell. We have already seen that carbohydrates can be synthesized from carbon dioxide by autotrophic microorganisms. Heterotrophs can also synthesize carbohydrates from various organic substrates. The biosynthesis of glucose from noncarbohydrate molecules, **gluconeogenesis**, involves the conversion of some substrate to pyruvate that is then converted to glucose (Figure 6.12). Gluconeogenesis is important because it allows microorganisms to convert noncarbohydrates to essential carbohydrate molecules. Proteins, for example, can be converted to pyruvates or phosphoenolpyruvates that are intermediary metabolites of the gluconeogenic pathway. Similarly, lipids in some organisms can be broken down into the three-carbon intermediates of the gluconeogenic pathway. The glucose synthesized in this pathway can be fed into the

pentose phosphate shunt to supply five-carbon carbohydrate molecules for the synthesis of molecules such as nucleic acid nucleotides.

The pathway of gluconeogenesis effectively reverses the flow of carbon occurring during glycolysis, and the intermediary metabolites of gluconeogenesis are identical to those of the glycolytic pathway. There are major differences between the two pathways, however, with several of the key steps involving different enzymes in order to achieve the effective reversal of carbon flow (Table 6.1). As indicated earlier, the enzymes mediating the key steps in this amphibolic pathway have different allosteric inhibitors that effectively regulate the direction of carbon flow. The amount of ATP in the cell acts to regulate the direction of carbon flow through the opposing pathways of glycolysis and gluconeogenesis. Gluconeogenesis is favored when the cell has an adequate supply of ATP because ATP is an allosteric activator for several of the key enzymes in gluconeogenesis and because ATP is an inhibitor for a key enzyme of glycolysis. Glycolysis is favored when ATP concentrations are relatively low. As already discussed, a high energy charge inhibits glycolysis, shutting off further production of ATP, and a low energy charge inhibits gluconeogenesis that requires ATP.

The glucose formed by gluconeogenesis may also be converted to larger polysaccharide molecules. Some of these polysaccharide molecules, such as starch or glycogen, may store carbon and energy within the cell. These polysaccharides can later be broken down to form glucose 1-phosphate or glucose 6-phosphate and thus enter the metabolic pathways of glycolysis and the pentose phosphate shunt. The glucose formed in gluco-

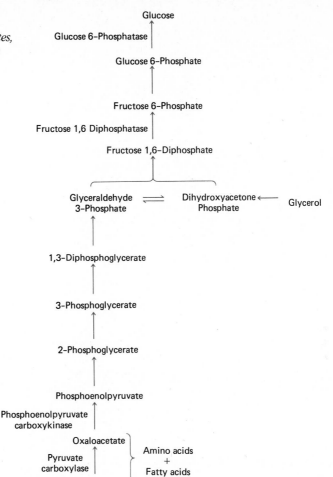

Figure 6.12

The conversion of noncarbohydrate substrates, such as amino acids, to carbohydrates, such as glucose, is accomplished via a gluconeogenic pathway.

neogenesis can also be used to synthesize the polysaccharides that make up the cell walls of microorganisms, including chitin and cellulose that make up the cell walls of many algae and fungi.

Glucose also is a precursor for *N*-acetylglucos-

amine and *N*-acetylmuramic acid, the repeating polysaccharide units of the glycan portion of the murein layer of the bacterial cell wall. The formation of *N*-acetylglucosamine involves the conversion of glucose to fructose 6-phosphate and

table 6.1

Some key enzymes in the glycolytic and gluconeogenic pathways

Enzyme	Reaction
Glycolysis	
Hexokinase	Glucose → glucose 6-phosphate
Phosphofructokinase	Fructose 6-phosphate → fructose 1,6-diphosphate
Pyruvate kinase	Phosphoenolpyruvate → pyruvate
Gluconeogenesis	
Glucose 6-phosphatase	Glucose 6-phosphate → glucose
Fructose 1,6-diphosphatase	Fructose 1,6-diphosphate → fructose 6-phosphate
Pyruvate carboxylase/phosphoenolpyruvate carboxykinase	Pyruvate → phosphoenolpyruvate

the subsequent reactions with glutamic acid and acetyl CoA. To form *N*-acetylmuramic acid, *N*-acetylglucosamine reacts with UTP (uridine triphosphate) to form *N*-acetylglucosamine-UDP, which then reacts with phosphoenolpyruvate to form *N*-acetylmuramic acid-UDP. These reactions form the necessary units of the glycan portion of the bacterial wall.

The actual synthesis of the peptidoglycan layer of the bacterial cell wall occurs in several stages (Figure 6.13). Initially, a peptide chain is added to *N*-acetylmuramic acid to form a monosaccharide peptide unit. A molecule of *N*-acetylglucosamine is then added to the *N*-acetylmuramic acid peptide unit to form a disaccharide peptide. An additional peptide unit, which forms the cross-linkage of the cell wall, is also added to the disaccharide unit. The disaccharides are then transferred to a growing peptidoglycan molecule. The details of the actual biosynthetic pathways involved in the synthesis of the peptidoglycan layer are quite complex and involve the activation of the carbohydrate molecules by UTP and the use of a membrane phospholipid (undecaprenol phosphate) as a carrier during biosynthesis of the wall. The synthesis of the bacterial cell wall can be viewed as occurring in four stages: synthesis of water-soluble carbohydrate precursors; attachment to a membrane lipid; formation of linear polymers outside the cytoplasmic membrane; and cross-linking of the polymers. Because the assembly steps occur outside the cell, it is important

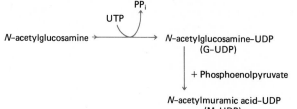

N-acetylglucosamine $\xrightarrow{\text{UTP} \quad \text{PP}_i}$ *N*-acetylglucosamine–UDP
(G-UDP)

$\downarrow$ + Phosphoenolpyruvate

N-acetylmuramic acid-UDP
(M-UDP)

ATP $\searrow$ L–ala

ATP $\searrow$ D–glu

ATP $\searrow$ L–lys

ATP $\searrow$ D–ala

M-UDP   D–ala
|
L–ala
|
D–glu
|
L–lys
|
D–ala
|
D–ala

M-carrier lipid
|
L–ala
|
D–glu $\xrightarrow{\text{UDP-G}}$ G — M — carrier lipid
|
L–lys
| Peptide $\longrightarrow$
D–ala
| Linkage
D–ala   of
              glycan
              +
        cross–linkage
           of
           peptide

Figure 6.13

The synthesis of peptidoglycan by bacteria is important to provide the backbone material of the cell wall. To form murein, amino acids must be added to N-*acetylmuramic acid, a repeating and alternating glycan layer of* N-*acetylmuramic acid and* N-*acetylglucosamine must be formed, and the peptide chains must be cross-linked.*

that the reactions be exergonic, as ATP is not available outside the cell to drive endergonic reactions. The activation of the cell wall subunits, using UTP, provides a mechanism for ensuring that the assembly reactions are exergonic and that ATP is not needed. In the case of the Gram negative cell wall, the synthesis and addition of lipopolysaccharide (LPS) also involves undecaprenol-P as a carrier. Synthesis of the LPS component of the cell wall occurs at the cytoplasmic membrane and involves the initial addition of fatty acids to a glucosamine disaccharide, followed by successive sugar additions. The assembled LPS component is then transferred to the cell wall structure.

Lipid metabolism

Like carbohydrates, lipid molecules can serve as substrates for supporting the growth of microorganisms and are also essential macromolecular constituents of cellular structures. Lipase enzymes can cleave the fatty acids from the glycerol portion of a triglyceride lipid molecule. The glycerol molecule can be metabolized to form dehydroxyacetone phosphate and then 3-phosphoglycerate and thereby enter the metabolic pathways that have already been discussed in the case of carbohydrate metabolism (Figure 6.14). In the case of a phospholipid, a phospholipase enzyme can cleave both the fatty acid and phosphate groups from the glycerol molecule, similarly converting the glycerol portion of the molecule to intermediate metabolites of the glycolytic/gluconeogenic pathways.

Fatty acid catabolism

The metabolism of the fatty acid portions of lipid molecules proceeds by a different pathway. **Fatty acids can be broken down into small two-carbon acetyl CoA units in the process of β-oxidation** (Figure 6.15). The fatty acid molecule initially reacts with coenzyme A to form a fatty acid-CoA molecule. The activation of the fatty acid with coenzyme A is coupled with the hydrolysis of an ATP molecule. Oxidation of this molecule releases acetyl CoA and forms a fatty acid-CoA complex that is two carbon atoms shorter than the parent fatty acid molecule. The formation of fatty acids that are successively two carbon atoms shorter than the parent molecule is characteristic of the β-oxidation process. The release of acetyl CoA

Figure 6.14

When triglycerides are metabolized, the glycerol can enter a glycolytic pathway via the ATP-driven formation of glycerol phosphate and the NADH-coupled reduction to form dihydroxyacetone phosphate.

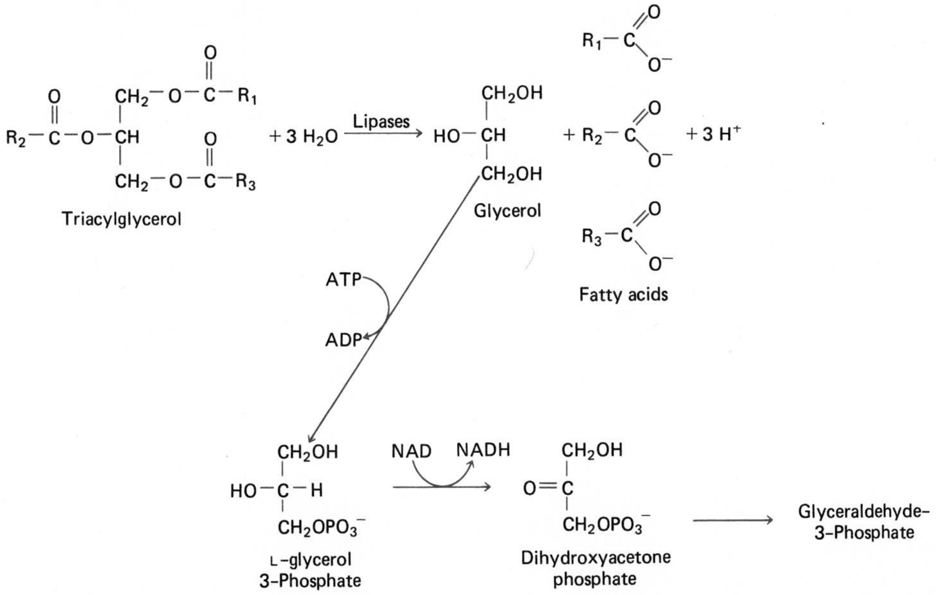

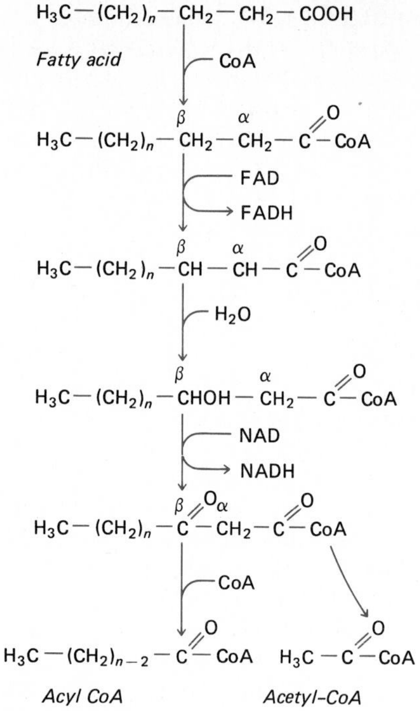

$H_3C - (CH_2)_n - CH_2 - CH_2 - COOH$

Fatty acid | CoA

$\underset{\beta}{H_3C} - (CH_2)_n - \underset{\beta}{CH_2} - \underset{\alpha}{CH_2} - C\overset{O}{\underset{}{\diagup}} CoA$

FAD → FADH

$H_3C - (CH_2)_n - \underset{\beta}{CH} - \underset{\alpha}{CH} - C\overset{O}{\underset{}{\diagup}} CoA$

H_2O

$H_3C - (CH_2)_n - \underset{\beta}{CHOH} - \underset{\alpha}{CH_2} - C\overset{O}{\underset{}{\diagup}} CoA$

NAD → NADH

$H_3C - (CH_2)_n - \underset{\beta}{C}\overset{O}{\diagup} - \underset{\alpha}{CH_2} - C\overset{O}{\underset{}{\diagup}} CoA$

CoA

$H_3C - (CH_2)_{n-2} - C\overset{O}{\underset{}{\diagup}} CoA \qquad H_3C - C\overset{O}{\underset{}{\diagup}} CoA$

Acyl CoA   *Acetyl-CoA*

Figure 6.15

The metabolism of fatty acids occurs via the β-oxidation pathway in which acetate and fatty acids that are progressively two carbon atoms shorter than the parent fatty acid are produced.

from the fatty acid is coupled with the formation of reduced coenzyme, as one molecule of $FADH_2$ and one molecule of NADH. The process is repeated continuously, forming fatty acid molecules which are successively two carbon atoms shorter, with the production each time of acetyl CoA, NADH, and $FADH_2$. The acetyl CoA can enter the Krebs cycle, through which the carbon of the acetyl group is converted to carbon dioxide. For each acetyl unit formed from the fatty acid, two molecules of carbon dioxide and additional reduced coenzyme molecules are formed during the Krebs cycle pathway. The reduced coenzyme formed in this process can be used to generate ATP via oxidative phosphorylation, thereby providing the energy required by the cell.

However, because fatty acid metabolism forms acetyl groups that contain only two carbon atoms, fatty acids cannot be converted into carbohydrates via the pathways that have already been discussed; gluconeogenesis requires the input of carbon skeletons with three carbon atoms, such as occur in a pyruvate molecule. In order to permit the flow of carbon from fatty acids to other classes of biochemicals, some microorganisms have an additional pathway, the **glyoxylate cycle** (Figure 6.16), that permits them to convert fatty acids into carbohydrates, proteins, and nucleic acids.

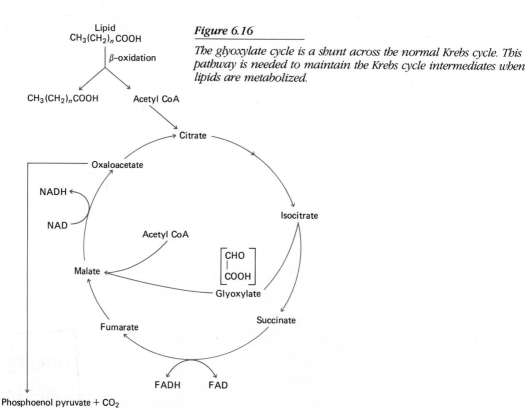

Figure 6.16

The glyoxylate cycle is a shunt across the normal Krebs cycle. This pathway is needed to maintain the Krebs cycle intermediates when lipids are metabolized.

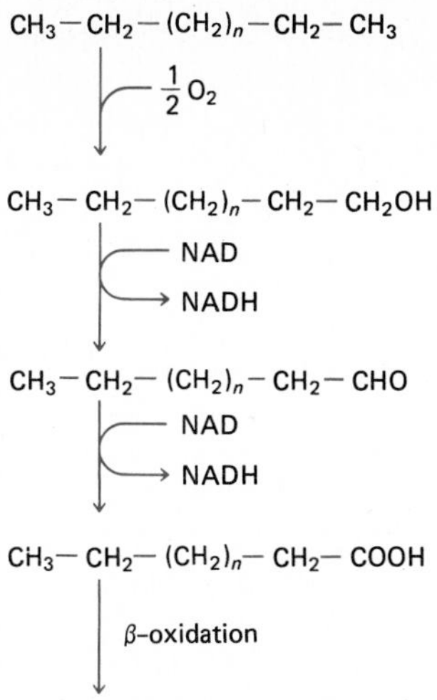

$$CH_3 - CH_2 - (CH_2)_n - CH_2 - CH_3$$

$$\downarrow \quad \tfrac{1}{2} O_2$$

$$CH_3 - CH_2 - (CH_2)_n - CH_2 - CH_2OH$$

$$\downarrow \quad \begin{array}{l} \text{NAD} \\ \text{NADH} \end{array}$$

$$CH_3 - CH_2 - (CH_2)_n - CH_2 - CHO$$

$$\downarrow \quad \begin{array}{l} \text{NAD} \\ \text{NADH} \end{array}$$

$$CH_3 - CH_2 - (CH_2)_n - CH_2 - COOH$$

$$\downarrow \quad \beta\text{-oxidation}$$

Figure 6.17

When hydrocarbons are metabolized, they first are oxidized to form fatty acids that are broken down via β-oxidation.

The glyoxylate cycle is really a shunt across the Krebs cycle. The conversion of oxaloacetate—formed in the glyoxylate cycle—to phosphoenolpyruvate—an intermediary metabolite of the gluconeogenic pathway—links the pathway of fatty acid metabolism with the pathway of carbohydrate metabolism. The formation of a six-carbon carbohydrate from a fatty acid requires the use of four acetyl CoA molecules, two carbons of which are released as carbon dioxide when oxyloacetate is transformed into phosphoenolpyruvate during the synthesis of the carbohydrate.

The metabolism of aliphatic hydrocarbons is closely related to fatty acid metabolism. Alkanes typically are successively oxidized to an alcohol, aldehyde, and then to a fatty acid (Figure 6.17). The fatty acids formed from these hydrocarbons are then further metabolized via β-oxidation. Some microorganisms, using this type of metabolism, are able to grow on the alkanes in crude oil and natural gas. Extensive branching of the alkane, however, can block β-oxidation, preventing growth on such hydrocarbons via this pathway. As a result many of the hydrocarbons found in petroleum are relatively resistant to microbial degradation.

Fatty acid biosynthesis

In **fatty acid biosynthesis** the flow of carbon is opposite that of β-oxidation and the catabolism of fatty acids. The synthesis of fatty acids proceeds by the sequential addition of two carbon units

derived from acetyl CoA (Figure 6.18). During fatty acid biosynthesis, the reactants are bound to a protein known as the acyl carrier protein. A key intermediate in the synthesis of fatty acids is malonyl CoA, which is formed from the ATP-driven reaction of acetyl CoA with carbon dioxide. It is the three-carbon malonyl CoA unit that successively contributes two carbon units for the elongation of the fatty acid. Carbon dioxide is released when the three-carbon malonyl-acyl carrier protein molecule contributes a two-carbon unit for the biosynthesis of fatty acids. The synthesis of a C_{16} saturated fatty acid, a common component of membrane phospholipids, can be described ac-

Figure 6.18

Fatty acid biosynthesis is not a simple reversal of β-oxidation. The synthesis of fatty acids involves two carbon additions from acetyl CoA and an acyl carrier protein.

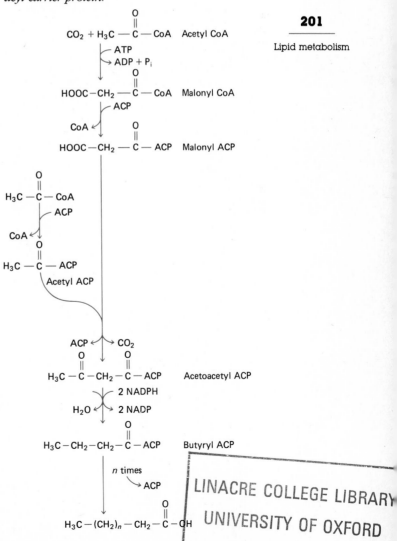

LINACRE COLLEGE LIBRARY
UNIVERSITY OF OXFORD

cording to the equation

8 Acetyl CoA + 7 ATP + 14 NADPH →
palmitate (C_{16}-fatty acid) + 14 NADP + 8 CoA
+ 6 H_2O + 7 ADP + 7 P_i

It is, thus, apparent that the synthesis of fatty acids requires both energy in the form of ATP and reducing power in the form of NADPH. Some of the required NADPH for this reaction comes from the pentose phosphate pathway. Additionally, NADPH can be formed in an ATP-driven series of reactions that can be summarized by the equation

NADP + NADH + ATP →
NADPH + NAD + ADP + P_i

Biosynthesis of poly-β-hydroxybutyric acid

The pathway for the **synthesis of poly-β-hydroxybutyric acid**, a common storage product of bacteria, is similar to the pathway for fatty acid biosynthesis (Figure 6.19). Acetyl coenzyme A reacts to form acetoacetyl CoA (a four-carbon derivative of CoA that can be reduced with NADH to form β-hydroxybutyryl CoA). Repetitive sequential addition of acetyl CoA results in chain-length elongation, and subsequent removal of the CoA portion of the molecule forms the poly-β-hydroxybutyric acid, which can accumulate in large

amounts in bacteria. Interestingly, unlike other biosynthetic reactions, the formation of poly-β-hydroxybutyrate uses the coenzyme NADH rather than NADPH.

Biosynthesis of phospholipids

The **synthesis of phospholipids** is required by both prokaryotic and eukaryotic organisms for incorporation into membranes. The formation of phospholipids involves the addition of fatty acids to glycerol phosphate. The glycerol 3-phosphate reacts with acyl CoA to form phosphatidate, a common intermediary metabolite in the synthesis of phospholipids and triglycerides (Figure 6.20). The fatty acid-acyl carrier protein molecules react with the phosphatidate to form elongated fatty acid derivatives.

Biosynthesis of sterols

In addition to phospholipids, the membranes of eukaryotic organisms contain sterols, such as cholesterol. Simple lipid molecules such as cholesterol are made up of repeating units of the unsaturated hydrocarbon isoprene (Figure 6.21). Isoprenoid hydrocarbons are synthesized from activated acetyl CoA molecules. The synthesis of isoprenoid hydrocarbons, which differs from fatty

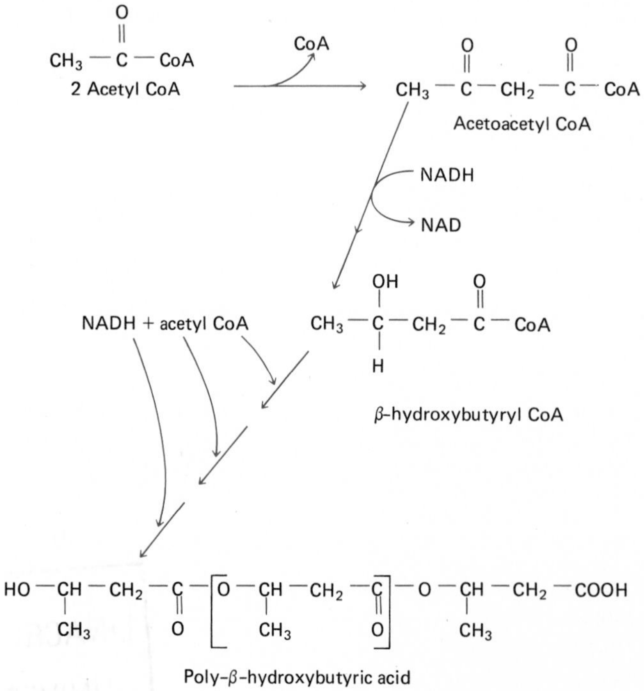

Figure 6.19

The synthesis of poly-β-hydroxybutyrate is used by bacteria to store carbon and energy reserves. This is an unusual biosynthetic pathway in that NADH, rather than NADPH, is used as a source of reducing power.

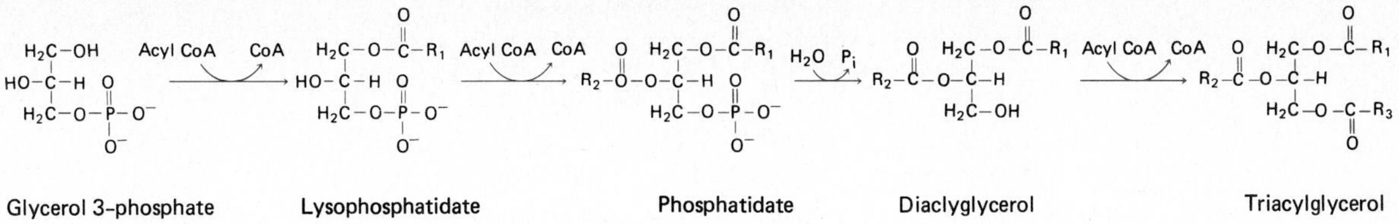

Glycerol 3-phosphate   Lysophosphatidate   Phosphatidate   Diaclyglycerol   Triacylglycerol

Figure 6.20

The formation of phosphatidate and lipids is necessary for the formation of cellular membranes.

acid biosynthesis in the mechanism of chain elongation, comes from reactions that initially form a six-carbon compound that is then decarboxylated. The synthesis of mevalonic acid from 3-hydroxy-3-methylglutaryl CoA, derived from the reaction of acetyl CoA with acetoacetyl CoA, is the key step in the formation of cholesterol. The activity of the enzyme 3-hydroxy-3-methylglutaryl CoA reductase regulates the rates of cholesterol biosynthesis. The biosynthesis of cholesterol is ex-

emplary of the fundamental mechanisms of long-chain carbon skeleton formation from five-carbon isoprenoid units. In addition to forming the backbone of sterols, isoprenoid hydrocarbons form the backbone of carotenoids, the brightly colored red, orange, and yellow pigment compounds of some microorganisms, and the phytyl portions of the chlorophyll molecules, which are essential for photosynthesis.

Figure 6.21

The synthesis of cholesterol involves the formation and subsequent condensation of isoentenyl pyrophosphate units.

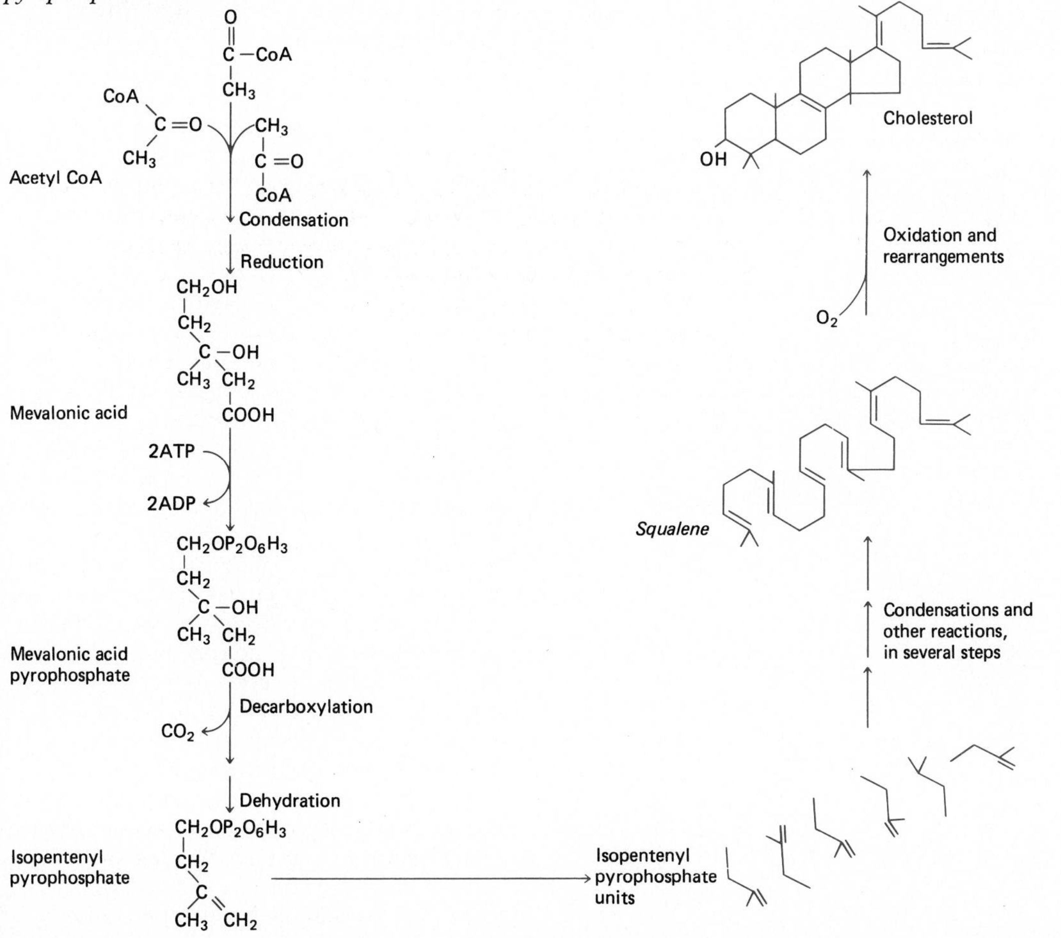

Protein metabolism

While carbohydrates and lipids can provide a cell with a source of carbon and energy, proteins additionally supply a source of nitrogen. Many microorganisms can use proteins as their sole source of carbon for generating ATP, reducing power, carbon skeletons of other required macromolecules, and nitrogen, an essential element in proteins and nucleic acids. Proteins, which are large macromolecules, are normally broken down extracellularly, by proteolytic enzymes (**proteases**), to form small peptides and individual amino acids that can enter the cell. The amino acids can be deaminated through the removal of the amino group to form a carboxylic acid. After deamination, the organic acids formed from the various families of amino acids can enter the Krebs cycle or glycolytic pathways, and thus the breakdown of proteins can establish a flow of carbon to form carbohydrate and lipid molecules. Additionally, several amino acids are precursors of nucleic acids, forming the necessary connection with the fourth class of macromolecules required by the cell.

The biosynthesis of proteins is a critical metabolic activity of all microorganisms. Much of the cell's activities involve synthesizing enzymes (proteins) that can catalyze specific metabolic activities. Indeed, it is through the synthesis of specific proteins that the genetic information of a microorganism is expressed. **Protein biosynthesis** can be viewed as occurring in two parts: the formation of the 20 essential amino acids and the linkage of the amino acids in the proper sequence to establish the primary structure of the protein molecule. The direction of the sequencing of amino acids to form the primary protein structure is under the direct control of the genetic informational macromolecules, and as such, the topic of protein biosynthesis will be covered in Chapter 7 as part of the discussion of genetic expression. In this section we shall consider the biosynthetic pathways for the amino acids that establish the carbon skeleton of proteins.

Biosynthesis of amino acids

Nitrogen fixation and the formation of ammonium ions

Unlike most lipid and carbohydrate molecules, amino acids contain nitrogen. The incorporation of inorganic nitrogen into organic molecules is essential for the synthesis of amino acids. Although molecular nitrogen is abundant in the atmosphere, most organisms are unable to use this as a source of nitrogen for incorporation into amino acids and the nitrogen-containing macromolecules of the cell. A limited number of microorganisms, however, have the ability to transform molecular nitrogen into ammonium nitrogen (Figure 6.22). These microorganisms possess the enzyme **nitrogenase**, which enables them to carry out nitrogen-fixing metabolic activities. The transformation of molecular nitrogen into ammonium nitrogen is endergonic and requires reducing power and ATP. The nitrogen-fixing microorganisms are quite important because they provide a supply of fixed nitrogen that can be assimilated by other organisms for incorporation into the formation of amino acids and other essential nitrogen-containing biochemicals. Biological **nitrogen fixation** is restricted to microorganisms and, except for one recent report of nitrogen fixation by a eukaryotic algal species, this process is strictly carried out by bacteria. Aside from the artificial production of nitrogen fertilizers, the only source of fixed nitrogen to support the growth of non-nitrogen-fixing organisms, including plants and animals, comes from these microorganisms.

Figure 6.22

The biological conversion of molecular nitrogen to ammonium, known as nitrogen fixation, is carried out by a restricted number of bacterial species. This conversion of molecular nitrogen to a fixed form that can be assimilated by other organisms often governs rates of productivity of plants, animals, and other microorganisms.

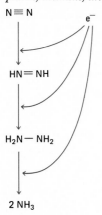

Figure 6.23

The reductive amination pathway can be used for the formation of L-glutamate when ammonium concentrations are high. In this pathway inorganic nitrogen is incorporated into an organic molecule.

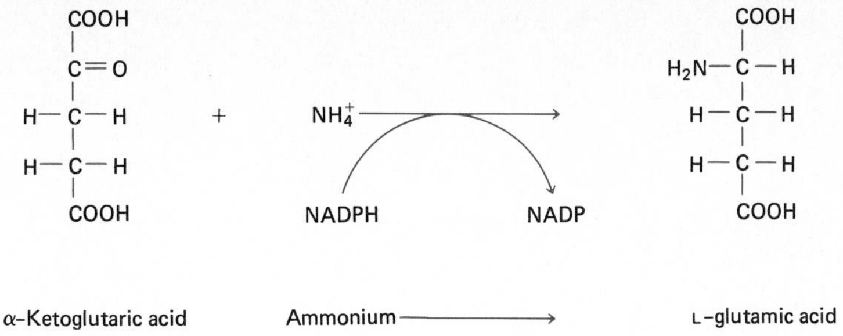

α-Ketoglutaric acid   Ammonium ⟶   L-glutamic acid

Incorporation of ammonium ions into amino acids

There are two pathways for the incorporation of inorganic ammonium nitrogen into organic amino acids. In both of these pathways ammonium ions are assimilated to form the amino acid L-glutamate that can then be transformed into the other amino acids. L-Glutamate can be formed from the reaction of ammonium ions with α-ketoglutarate, a Krebs cycle intermediate, in a pathway known as **reductive amination** (Figure 6.23). This pathway is catalyzed by the enzyme glutamate dehydrogenase and requires reducing power, which may be in the form of either NADPH or NADH. In the process of reductive amination, ammonium ions are combined directly with an α-ketocarboxylic acid to form an amino acid. Reductive amination occurs when ammonium ions are in relatively high concentrations and for most microorganisms does not appear to represent the main pathway for incorporating ammonium ions into amino acids.

When ammonium concentrations are low, as is most frequently the case, microorganisms resort to another pathway for the formation of L-glutamate, the **glutamine synthetase/glutamate synthase pathway** (Figure 6.24). In this pathway the amino acid L-glutamate reacts with ammonium ions to form the amino acid amide L-glutamine, a reac-

tion that requires energy in the form of ATP. The L-glutamine then reacts with α-ketoglutarate to form two molecules of L-glutamate in a reaction that requires reducing power as NADPH. Thus, the combined reactions catalyzed by the enzymes glutamine synthetase and glutamate synthase are described by the equation

$$\alpha\text{-Ketoglutarate} + NH_4^+ + NADPH + ATP \rightarrow$$
$$\text{L-glutamate} + NADP + ADP + P_i$$

The enzyme glutamine synthetase plays a key role in regulating the rates of intermediary metabolism because of the regulatory control it exerts on the flow of nitrogen into amino acids and consequently into proteins and nucleic acids. Glutamine synthetase is subject to cumulative feedback inhibition by each of the products of glutamine metabolism; that is, a series of different inhibitors can act additively to reduce the activity of this enzyme. Inhibitors of glutamine synthetase include tryptophan, histidine, alanine, glycine, carbamoyl phosphate, glucosamine 6-phosphate, CTP (cytidine triphosphate), and AMP. Each of these allosteric inhibitors appears to have its own binding site on the enzyme, and when all eight inhibitors are bound to the enzyme, the activity of glutamine synthetase is virtually shut off. AMP in

Figure 6.24

When the ammonium concentration is low, the conversion of inorganic ammonium ions to the amino acid L-glutamate occurs in a two-step process, the glutamine synthetase/ glutamate synthase pathway. This is the principal pathway used by bacteria for the formation of L-glutamate from the Krebs cycle intermediate α-ketoglutarate.

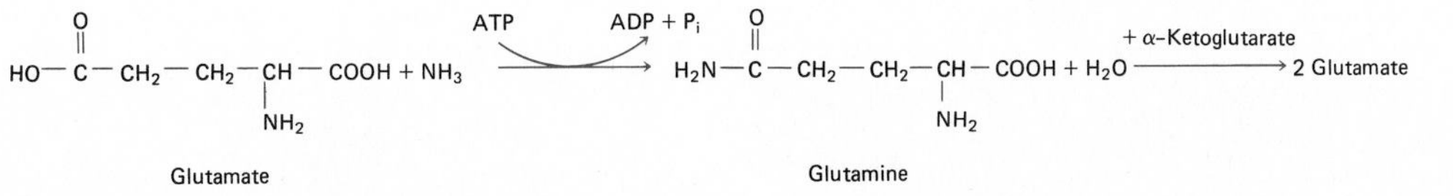

Glutamate   Glutamine

particular affects the sensitivity of this enzyme to feedback inhibition because the adenylated enzyme—that is, the enzyme with AMP attached—is more susceptible to cumulative feedback inhibition than the deadenylated form of the enzyme.

Biosynthesis of the major families of amino acids

The assimilation of nitrogen to form amino acid L-glutamate establishes the basis for the biosynthesis of the other essential amino acids found in protein macromolecules as well as for the biosynthesis of other essential nitrogen-containing compounds. The 20 amino acids originate from six different non-amino acid precursors, and as a result, there are only six **biosynthetic families of amino acids** (Figure 6.25). L-Glutamate is the parent molecule of one of the amino acid families and can be further metabolized to form the amino acids L-glutamine, L-proline, and L-arginine. L-Glutamate also serves as the nitrogen source for the other amino acids; that is, the amino group of all of the amino acids is derived from L-glutamate. The carbon skeletons for the various amino acids come from glycolytic, pentose phosphate, or Krebs cycle pathway intermediate metabolites that form the metabolic precursors for each amino acid family.

The ability to transfer the amino group of one amino acid to form another amino acid, a process known as **transamination**, is essential for the synthesis of all of the amino acids. This process involves specific transaminase enzymes and also requires the coenzyme pyridoxal phosphate (a derivative of vitamin B_6), which binds with the amino group being transferred during the transamination reaction. Glutamate transaminase is the most important of the transaminase enzymes be-

Figure 6.25

The biosynthesis of the 20 L-amino acids found in proteins represents the formation of six families of related amino acids.

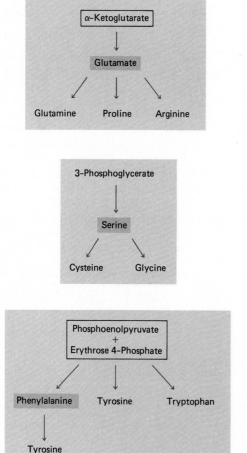
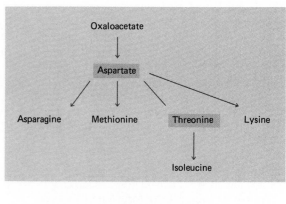
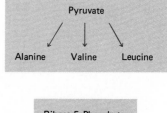
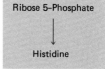

Figure 6.26

Once nitrogen is incorporated into an amino acid, the amino group can be transferred to an α-ketocarboxylic acid to form a new amino acid via transamination. This illustration demonstrates transamination to form L-aspartate.

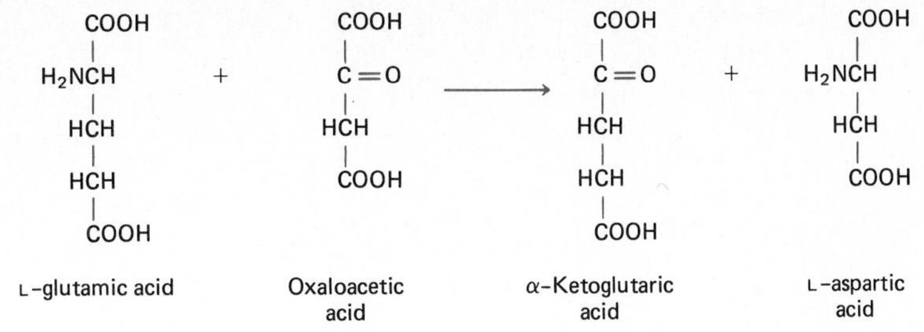

$$
\begin{array}{ccc}
\text{COOH} & & \text{COOH} \\
| & & | \\
\text{H}_2\text{NCH} & + & \text{C}=\text{O} \\
| & & | \\
\text{HCH} & & \text{HCH} \\
| & & | \\
\text{HCH} & & \text{COOH} \\
| & & \\
\text{COOH} & & \\
\text{L-glutamic acid} & & \text{Oxaloacetic acid}
\end{array}
\longrightarrow
\begin{array}{ccc}
\text{COOH} & & \text{COOH} \\
| & & | \\
\text{C}=\text{O} & + & \text{H}_2\text{NCH} \\
| & & | \\
\text{HCH} & & \text{HCH} \\
| & & | \\
\text{HCH} & & \text{COOH} \\
| & & \\
\text{COOH} & & \\
\alpha\text{-Ketoglutaric acid} & & \text{L-aspartic acid}
\end{array}
$$

cause it catalyzes the transfer of the amino group from L-glutamate to form the parental amino acids of the various amino acid families. L-Glutamate can transfer its amino group to an α-ketocarboxylic acid to form a new amino acid and α-ketoglutarate, the carboxylic acid precursor of L-glutamate. As an example of a transamination reaction, L-glutamate can react with oxaloacetate to form the amino acid L-aspartate (Figure 6.26). L-Aspartate is the parent molecule of another family of amino acids and can be further metabolized to form L-asparagine, L-methionine, L-threonine, L-lysine, or L-isoleucine.

Two intermediary metabolites in the conversion of L-aspartate to L-lysine, dihydrodipicolinic acid and meso-diaminopimelic acid (Figure 6.27), are used in prokaryotic organisms but do not form part of the structures of eukaryotic microorganisms. Diaminopimelic acid is one of the unusual amino acids that forms part of the peptide portion of the peptidolglycan molecule that makes up the bacterial cell wall. Dipicolinic acid occurs uniquely in bacterial endospores.

Transamination reactions can also be utilized to generate the amino acids L-alanine, L-valine, and L-leucine from reactions with pyruvate. The formation of L-alanine involves a single-step transamination reaction that converts pyruvate to L-alanine (Figure 6.28). Similarly, L-glutamate can react with 3-phosphoglycerate, an intermediate of the glycolytic pathway, to form the amino acid L-serine (Figure 6.29). Because 3-phosphoglycerate does not have a keto group that can react with the amino group of L-glutamate, the 3-phosphoglycerate must first be oxidized to 3-phosphohydroxypyruvate, a reaction that is coupled with the reduction of NAD to NADH. The initial product formed by the transamination reaction is 3-phosphoserine and the phosphate group is subsequently hydrolyzed to yield the final amino acid product of L-serine.

L-Serine is a precursor for the biosynthesis of the amino acids L-glycine and L-cystine. L-Cystine is one of the sulfur-containing amino acids, and the transformation of L-serine to L-cystine involves a reaction with hydrogen sulfide, which can be derived from the reduction of sulfate (Figure 6.30). The sulfur-containing amino acids are important in determining the secondary structure of a protein. The sulfur-containing amino acids also are often involved in establishing the active site of enzyme molecules.

The formation of L-histidine is far more complex than the previously described reactions (Figure 6.31). Histidine and nucleic acid purines arise from a common precursor molecule, ribose 5-phosphate, which is formed by the pentose phosphate cycle. The adenine unit of the molecule ATP provides one nitrogen and one carbon atom for the ring of L-histidine; the other nitrogen atom of the ring comes from the side chain of the amino acid L-glutamine. The amino group of L-histidine comes from a transamination reaction with L-glutamate.

The formation of the aromatic amino acids L-phenylalanine, L-tyrosine, and L-tryptophan originates with phosphoenolpyruvate (a glycolytic pathway intermediary metabolite) and erythrose 4-phosphate, which is formed via the pentose phosphate cycle. The details of the formation of the aromatic ring structure are relatively complex and will not be examined here. The amino group of the amino acids L-phenylalanine and L-tyrosine arise through transamination reactions with L-glutamate; L-tryptophan derives its amino group from a transamination reaction with L-serine.

Through these metabolic reactions the 20 essential amino acids of proteins can be synthesized. The fact that the precursors for the biosynthesis of amino acids are intermediary metabolites of the glycolytic, pentose phosphate, and Krebs cycle pathways establishes a unity of metabolic pathways that permits the flow of carbon between the major biochemical classes. The carbon from carbohydrates and lipids, for example, can be used to generate the carbon skeletons of the 20 essential amino acids and thus can form the backbone of protein molecules. The rates of amino acid biosynthesis are regulated in large part by feedback

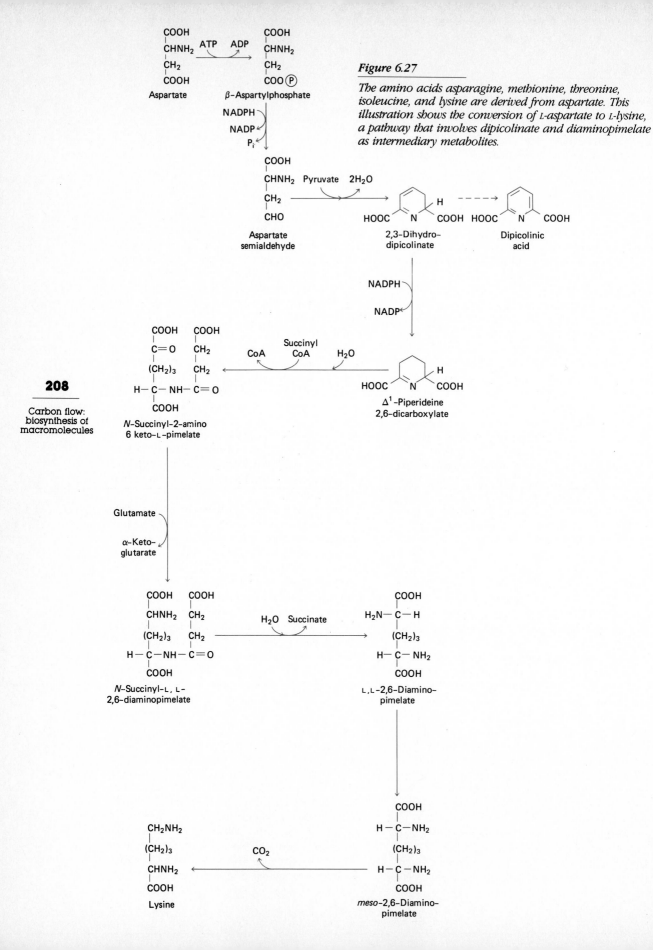

Figure 6.27

The amino acids asparagine, methionine, threonine, isoleucine, and lysine are derived from aspartate. This illustration shows the conversion of L-aspartate to L-lysine, a pathway that involves dipicolinate and diaminopimelate as intermediary metabolites.

Aspartate

β-Aspartylphosphate

Aspartate semialdehyde

2,3-Dihydro-dipicolinate

Dipicolinic acid

Δ^1-Piperideine 2,6-dicarboxylate

N-Succinyl-2-amino 6 keto-L-pimelate

N-Succinyl-L, L-2,6-diaminopimelate

L,L-2,6-Diamino-pimelate

meso-2,6-Diamino-pimelate

Lysine

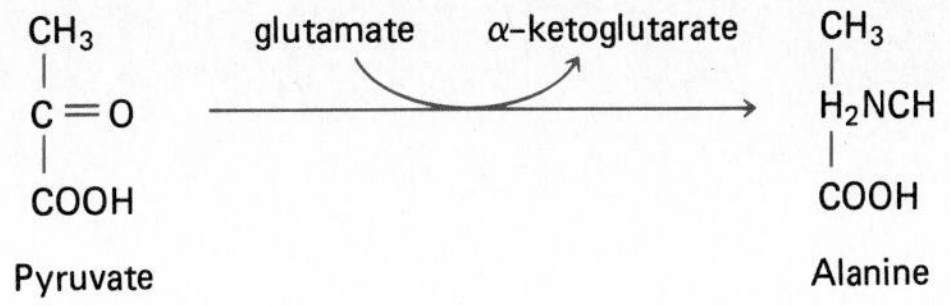

Figure 6.28

The transamination reaction of pyruvate with L-glutamate results in the formation of L-alanine.

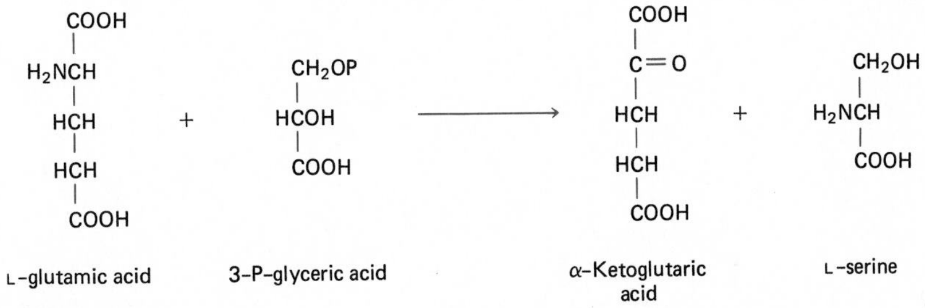

$$CH_3 - C{=}O - COOH \quad \xrightarrow[\text{glutamate} \quad \alpha\text{-ketoglutarate}]{} \quad CH_3 - H_2NCH - COOH$$

Pyruvate · Alanine

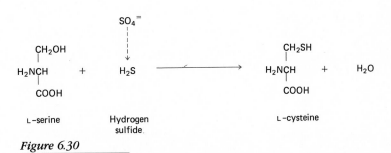

L-glutamic acid 　 3-P-glyceric acid 　 α-Ketoglutaric acid 　 L-serine

Figure 6.29

The transamination reaction of phosphoglycerate with L-glutamate results in the formation of L-serine.

$$CH_2OH - H_2NCH - COOH \;+\; H_2S \xrightarrow{\;SO_4^{=}\;} CH_2SH - H_2NCH - COOH \;+\; H_2O$$

L-serine 　 Hydrogen sulfide 　 L-cysteine

Figure 6.30

The formation of L-cystine, a sulfur-containing amino acid, involves the reaction of L-serine with hydrogen sulfide.

inhibition. Usually, there is a major regulatory step in each of the amino acid biosynthetic pathways. The rates of amino acid biosynthesis depend on the activities of the enzymes catalyzing these regulatory steps, and the final product of a pathway often acts as an allosteric inhibitor of the enzyme catalyzing the critical regulatory step. The control of amino acid biosynthesis is important in the overall regulation of metabolism because of the central role of enzymes that are composed of amino acids in catalyzing metabolic reactions.

Nucleic acid metabolism

As with proteins, the **biosynthesis of nucleic acid molecules** can be examined from the viewpoint of synthesizing the nucleotide bases that go into the biosynthesis of the nucleic acids or from the viewpoint of considering the sequencing of the nucleotides that make up the macromolecules DNA and RNA. In this chapter we will consider only briefly the biosynthesis of the nucleic acid bases, and in the next chapter we will examine how the nucleotides are aligned and linked together to form the DNA and RNA macromolecules. There are two classes of nucleic acid bases, purines and pyrimidines. The pathways for the biosynthesis of the purine and pyrimidine ring structures of nu-

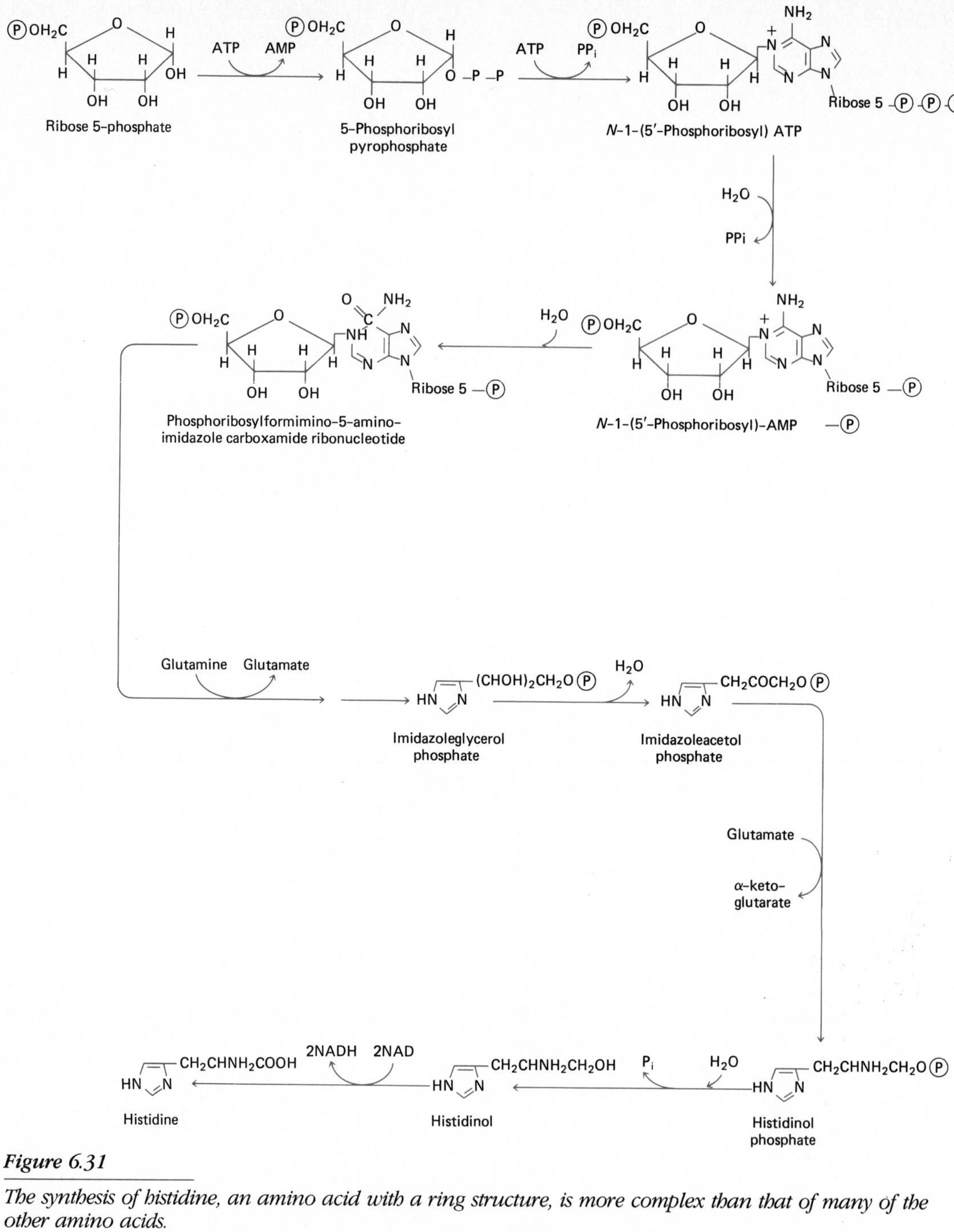

Figure 6.31

The synthesis of histidine, an amino acid with a ring structure, is more complex than that of many of the other amino acids.

cleotides are quite complex and only a cursory overview of the process will be given here. The biosynthesis of purine and pyrimidine nucleotides is important because these compounds are involved in many biochemical processes; they are the precursors of DNA and RNA. ATP is an adenine nucleotide, as are the major coenzymes, NAD, NADP, and CoA, and nucleotides are important activators and inhibitors that regulate the rates of metabolic reactions within the cell.

Biosynthesis of pyrimidines

The precursors of the atoms of the **pyrimidine ring** are ammonia, carbon dioxide, and L-aspartate (Figure 6.32). The synthesis of the pyrimidine ring begins with the formation of carbamoylaspartate from the reaction of carbamoyl phosphate and aspartate, catalyzed by the enzyme aspartate transcarbamoylase. This is the key regulatory step in the synthesis of the pyrimidine ring, and the enzyme aspartate transcarbamoylase is subject to allosteric feedback inhibition by the products of the reaction and to allosteric activation by ATP. The following steps in the formation of the pyri-

midine ring involve dehydration, cyclic ring formation, and oxidation to form orotate. After the pyrimidine ring is synthesized, ribose and phosphate are added to the molecule, using PRPP (5-phosphoribosyl 1-pyrophosphate). A carboxyl group is subsequently removed as carbon dioxide forming uridine monophosphate that is then phosphorylated to form uridine triphosphate, one of the nucleotides of RNA.

The uridine triphosphate can be modified to form cytidine triphosphate by the replacement of a keto group with an amino group in the pyrimidine ring (Figure 6.33). Cytidine triphosphate is a nucleotide in both RNA and DNA. The ribose

Figure 6.32

The formation of a pyrimidine ring is necessary for the biosynthesis of nucleic acids. This pathway results in the formation of uridine phosphate from aspartate and carbamoyl phosphate.

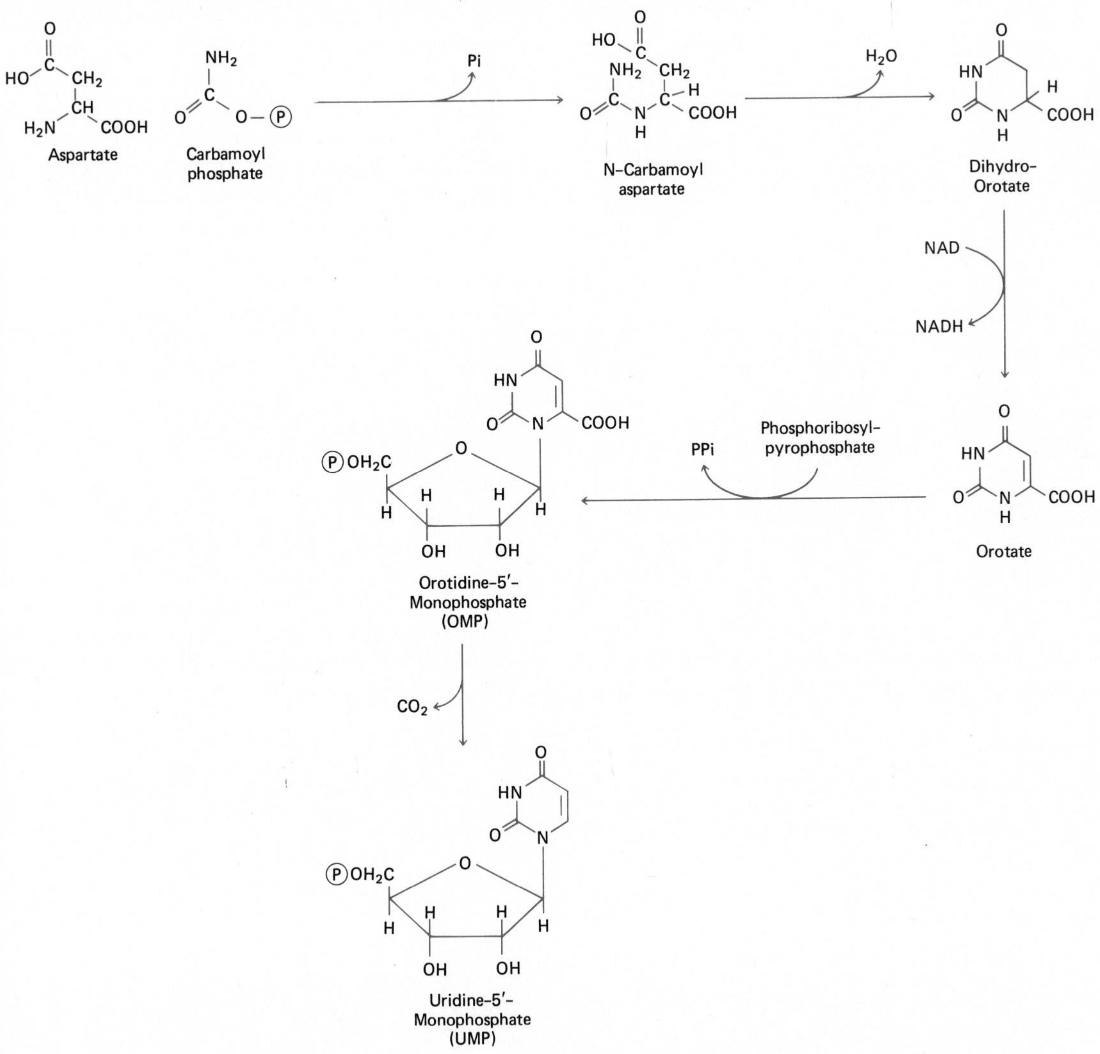

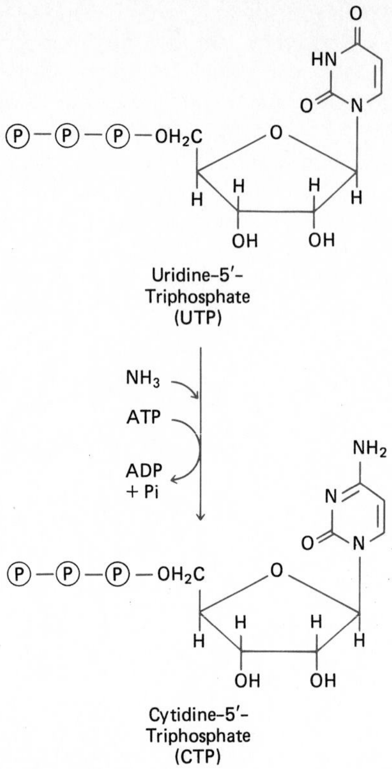

Uridine-5′-
Triphosphate
(UTP)

Cytidine-5′-
Triphosphate
(CTP)

Figure 6.33

The conversion of uridine triphosphate to cytidine triphosphate forms one of the nucleotides found in both DNA and RNA.

must be reduced, using NADPH to form the deoxyribose form of the nucleotide. Thus, the three pyrimidines of DNA and RNA, uracil, cytosine, and thymidine, are sequentially synthesized and the appropriate ribose and deoxyribose sugar moieties attached.

Biosynthesis of purines

The formation of the **purine nucleotides**, which have two rings, is somewhat more complex than that of the pyrimidines. Ten metabolic steps are involved in the formation of the basic purine ring structure. The purine ring is synthesized from a variety of amino acids, including L-aspartate, L-glycine, and L-glutamine (Figure 6.36). Additionally, carbon dioxide and a methyl group donated by folic acid are essential for the formation of the purine ring skeleton. The initial step in the synthesis of the purine ring involves the addition of an amino group to a phosphorylated ribose molecule, and the synthesis of the purine ring continues with the phosphorylated ribose already attached; this is in contrast to the synthesis of the pyrimidine ring, where the carbohydrate moiety is added after the formation of the ring structure. Once the basic purine ring is formed, it is modified to form the adenine and guanine nucleotides (Figure 6.37). The biosynthesis of the adenine ring involves the substitution of an amino group for a keto group. The biosynthesis of the guanine ring involves the addition of an amino group without the removal of a keto group. The adenosine phosphate molecule that is formed serves not only as a nucleotide base in DNA and RNA but also in the ATP, NAD, NADP, CoA, and FAD molecules. Thus, the synthesis of purine nucleotides is critical for the biosynthesis of many of the important molecules that act as energy carriers and coenzymes in the biochemical reactions of cellular metabolism.

sugar is reduced to the deoxyribose form of the nucleotide for DNA, using the reducing power of NADPH and the enzyme ribonucleoside reductase (Figure 6.34). Uridine triphosphate is also the precursor for thymidine triphosphate that occurs in DNA but not RNA (Figure 6.35). The formation of the nucleic acid base thymidine involves the addition of a methyl group, derived from folic acid, to the uridine ring structure. Because thymidine occurs exclusively in DNA, the ribose sugar

Figure 6.34

The conversion of ribose to deoxyribose produces the sugar backbone molecule of DNA.

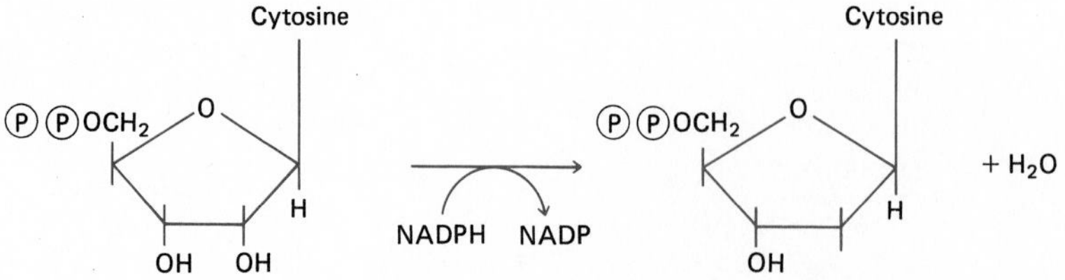

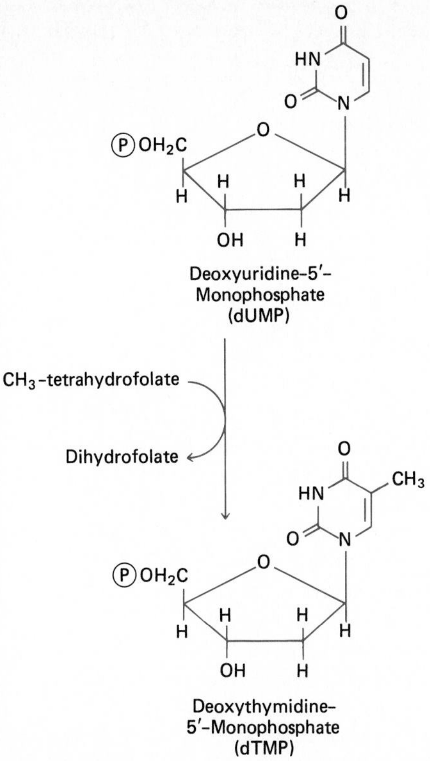

Deoxyuridine-5′-
Monophosphate
(dUMP)

CH₃-tetrahydrofolate

Dihydrofolate

Deoxythymidine-
5′-Monophosphate
(dTMP)

Glycine

CO_2

Aspartate

Methyl-Tetrahydrofolate

Formyl-
Tetrahydrofolate

Glutamine

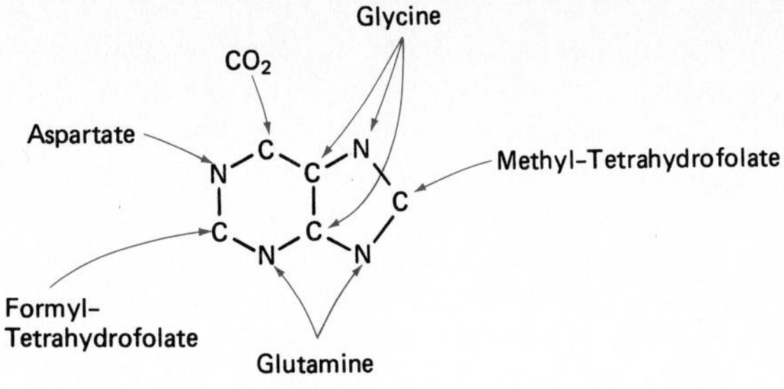

Purine ring

Figure 6.36

As shown in this illustration, the carbons involved in the formation of a purine ring are derived from several sources.

Figure 6.35

The conversion of uridine to thymidine produces a nucleotide found in DNA but not RNA.

Figure 6.37

The conversion of an inosine purine ring to adenine and guanine is needed to form the purine nucleotides found in both DNA and RNA. The adenosine monophosphate formed via this pathway is also used for the synthesis of ATP.

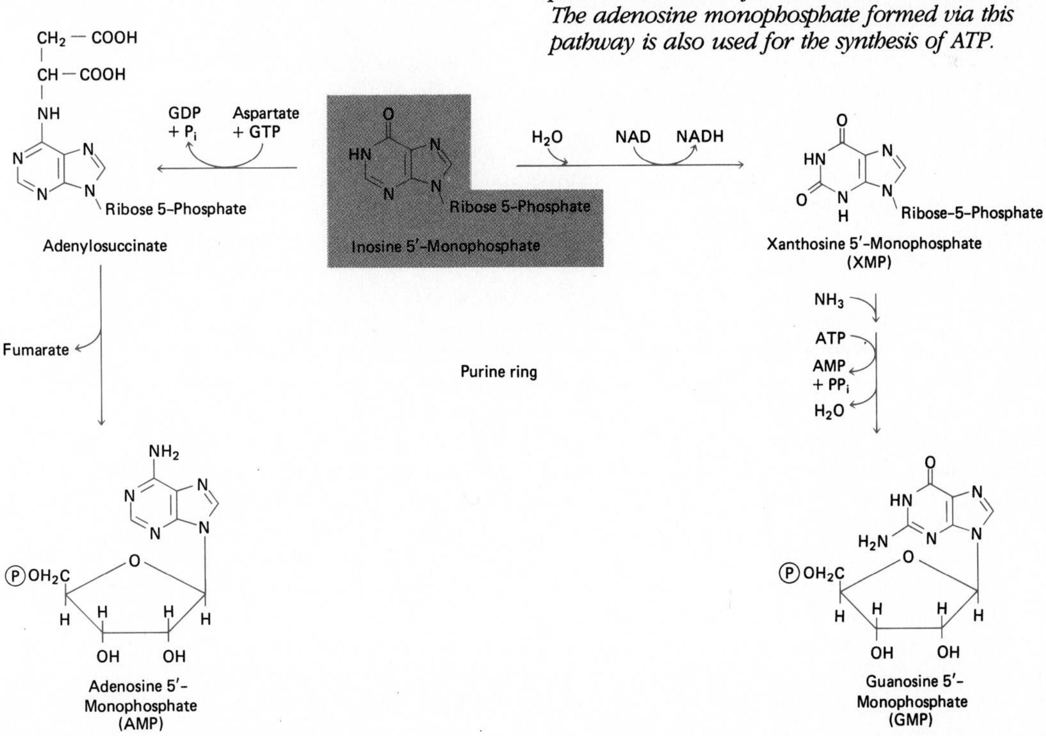

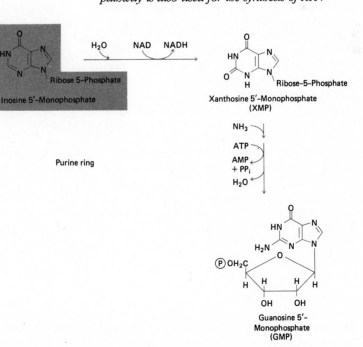

CH₂ — COOH

CH — COOH

NH

Ribose 5-Phosphate

Adenylosuccinate

GDP
+ Pᵢ

Aspartate
+ GTP

Ribose 5-Phosphate

Inosine 5′-Monophosphate

H_2O NAD   NADH

Ribose-5-Phosphate

Xanthosine 5′-Monophosphate
(XMP)

NH_3

ATP

AMP
+ PPᵢ

H_2O

Fumarate

Purine ring

Adenosine 5′-
Monophosphate
(AMP)

Guanosine 5′-
Monophosphate
(GMP)

The metabolic pathways of microorganisms form an interlocking network through which carbon flows. Some of the pathways involve the breakdown of organic molecules (catabolism), and other pathways involve the synthesis of the macromolecules that constitute microorganisms (anabolism). ATP and the reduced and oxidized forms of the coenzymes NAD and NADP play key roles in the flow of carbon through microorganisms because they supply energy and act as oxidizing and reducing agents, respectively, for microbial metabolism. The glycolytic and Krebs cycle pathways form a central focus not only for the generation of ATP but also for the flow of carbon through the microbial cell. There are several key shunts—for example the pentose phosphate pathway— that permit the cell to divert the carbon flow from one metabolic purpose, such as generating ATP, to another metabolic purpose, such as generating reducing power or generating small-carbon-skeleton molecules. The regulation of carbon flow through microbial metabolic pathways is subject to careful regulation, largely based on the ability to modify the configurations and activities of key regulatory enzymes. In many cases the energy charge of the cell determines the activities of these regulatory enzymes and thus the direction of metabolic activity.

The central metabolic pathways serve both for the catabolic breakdown of substrates—generating ATP, reduced coenzyme, and the precursors of anabolism—and for the biosynthesis of the cell's macromolecules. These amphibolic pathways use the same intermediary metabolites, and the regulation of key reactions, catalyzed by different enzymes, determines whether the pathway at a given time is catabolic or anabolic. The intermediary metabolites of the glycolytic, pentose phosphate, and Krebs cycle pathways form a central focus through which the metabolisms of carbohydrates, lipids, proteins, and nucleic acids are interconnected. Despite the complexity of the thousands of biochemicals that make up a microbial cell, there are relatively few junction points between the four major classes of biochemicals that occur in microorganisms; glucose 6-phosphate, pyruvate, and acetyl CoA form the key junction points that connect various metabolic pathways. The large macromolecules of the microbial cell are synthesized from relatively few, small organic compounds. A general theme of microbial metabolism is the conversion of a substrate molecule into a series of small organic compounds, which can then be utilized for the biosynthesis of many different large macromolecules. The metabolic pathways that generate the ATP and NADPH, required for biosynthesis, also generate the small organic molecules that establish the carbon skeletons of larger molecules.

Whereas the metabolites of the central metabolic pathways are involved in both anabolism and catabolism, there is a real distinction between carbon flow in biosynthetic and catabolic pathways. Part of the distinction is maintained by the use of different coenzymes in these opposing pathways. As a rule, NAD and NADH are involved in catabolic pathways aimed at the generation of ATP, and NADPH is generally involved in biosynthetic reactions. Biosynthetic pathways almost always require reducing power in the form of NADPH and energy in the form of ATP. Additionally, while the intermediary metabolites are identical, several different key enzymes are involved in biosynthetic and degradative pathways. The activities of these key enzymes regulate the rates and direction of carbon flow through the central amphibolic pathways. The key enzymes are usually subject to allosteric control so that the flow of carbon through a particular pathway can be accelerated or inhibited, depending on the metabolic needs of the cell. The allosteric activators and inhibitors act as feedback indicators that regulate the direction of carbon flow.

The ability of microorganisms to move carbon from one class of compounds to another enables some microorganisms to synthesize all of their required biochemicals from a single starting substrate molecule. In the case of autotrophic microorganisms, all of the biochemicals of the cell can be synthesized from carbon dioxide and water or other hydrogen donors. The Calvin cycle is the key metabolic pathway for the conversion of carbon dioxide to organic carbon by autotrophic microorganisms. The organic carbon formed by this pathway supports not only the autotrophic microorganisms that possess this pathway but also heterotrophic microorganisms that require preformed organic compounds for their existence. Regardless of their starting substrates, all microorganisms must synthesize or acquire the precursors for all four major classes of macromolecules. The biosynthesis of 20 essential amino acids forms the building blocks for proteins, and the biosynthesis of 8 nucleotides forms the basic building blocks of nucleic acids. In the case of proteins and nucleic acids, the ordering of these building blocks within the macromolecule (see discussion in Chapter 7) establishes its configuration and determines the macromolecule's functional properties. The informational nucleic acid macromolecules, and the functional enzymes they encode, determine the metabolic functioning of microorganisms.

Study Questions

1. What is the Calvin cycle, and what is its function? Why is this called a dark reaction cycle? What is required for the Calvin cycle to operate?

2. What is the pentose phosphate pathway, and why is it important?

3. What is an amphibolic pathway? What is a catabolic pathway? What is an anabolic pathway?

4. What is an exoenzyme, and why are they important for the metabolism of polysaccharides?

5. What is gluconegenesis? How would a cell make glucose starting with a protein?

6. What is β-oxidation? What is the problem with metabolizing compounds containing only two carbons, such as acetate ?

7. How is nitrogen incorporated into organic compounds to form amino acids?

8. How are allosteric effectors involved in regulating the flow of carbon through a cell? What enzyme is critical in regulating glycolysis?

9. Why are reduced coenzymes essential for biosynthesis?

10. What is the general strategy of a cell in terms of carbon flow?

Suggested supplementary readings

Suggested Supplementary Readings

Anderson, R. L., and W. A. Wood. 1969. Carbohydrate metabolism in microorganisms. *Annual Reviews in Microbiology* 33: 539–575.

Conn, E. E., and P. K. Stumpf. 1976. *Outlines of Biochemistry*. John Wiley & Sons, New York.

Dagley, S., and D. E. Nicholson. 1970. *An Introduction to Metabolic Pathways*. Blackwell Scientific Publications, Oxford, England.

Doelle, H. W. 1975. *Bacterial Metabolism*. Academic Press, New York.

Gottschalk, G. 1979. *Bacterial Metabolism*. Springer-Verlag, New York.

Lehninger, L. 1982. *Principles of Biochemistry*. Worth Publishers, Inc., New York.

Mandelstam, J., K. McQuillen, and I. W. Dawes. 1982. *Biochemistry of Bacterial Growth*. Blackwell Scientific Publications, Oxford, England.

Stryer, L. 1981. *Biochemistry*. W. H. Freeman and Co., San Francisco.

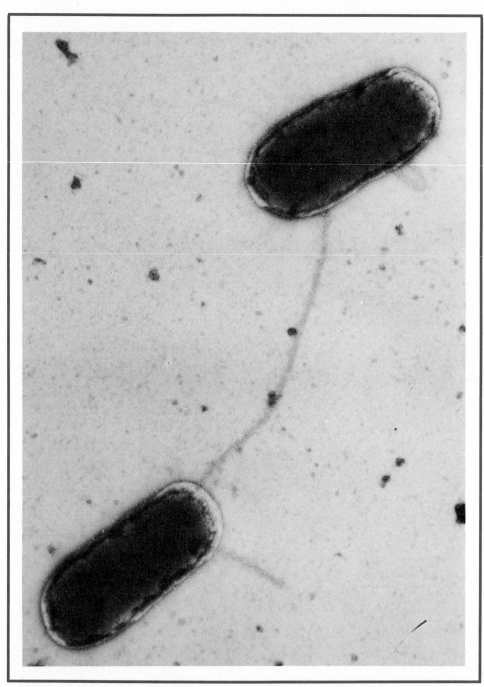

Electron micrograph of mating E. coli.
(Courtesy D. P. Allison, Oak Ridge National Laboratory.)

Microbial genetics

SECTION FOUR

DNA, RNA, and protein synthesis: the replication and expression of genetic information

7

The genetic information of prokaryotic and eukaryotic microorganisms, which determines the metabolic and structural nature of the organism, is encoded within the DNA molecule(s) of the cell. The transmission of hereditary information necessitates the faithful **replication of the DNA macromolecules** containing this genetic information. (In some viruses the genetic information is contained in RNA rather than DNA, and in these cases the transmission of RNA macromolecules from one generation to the next ensures the continuance of the hereditary information of the virus.) A single copy of the genetic information of a cell constitutes its **genome**. The genome of a microorganism is divided into segments that have specific functions; these segments of the DNA, consisting of sequences of nucleotides, are known as **genes**. Some genes code for the synthesis of RNA and proteins, respectively determining the sequences of the subunit nucleotide bases and amino acids in these macromolecules; such genes are known as **structural genes** or **cistrons**. Other genes have regulatory functions and act to control the activities of the cell.

The genetic information encoded in the genome of the microorganism is specified by the order or sequence of the nucleotides. The replication of the genome involves synthesizing new nucleic acid molecules that have the same nucleotide sequence as the genome of the parental organism, a process that requires great precision. Any change in the sequence of nucleotides within the informational macromolecules of the microorganism can greatly alter the characteristics

of that organism, and even a change in the position of a single nucleotide base can prove lethal for the progeny of a microorganism. The genome of the progeny must contain the appropriate information, which can be expressed through the metabolism of the cell, to permit the survival and growth of the organism.

The **expression of genetic information** involves using the information encoded within the genome (nucleic acids) to control the synthesis of proteins. Because enzymes are protein molecules, the genetic informational macromolecules, by controlling protein synthesis, exert control over the metabolic capabilities of microorganisms. The sequence of bases within the genome of the cell determines the sequence of amino acids within protein molecules and thus the functional properties of microbial enzymes. The expression of the genetic information of the organisms through the activities of the enzymes that are produced is reflected in the phenotypic features that distinguish one microorganism from another.

Transferring the information contained in DNA to form a functional enzyme occurs through protein synthesis, a process accomplished in two stages. The information in the DNA molecule is initially transcribed to form RNA molecules. Several different types of RNA molecules are involved in protein synthesis. One type of RNA molecule, **messenger RNA (mRNA)**, carries the information from the DNA molecule to the ribosomes, the anatomical sites of protein synthesis. The information encoded in the mRNA molecule is then translated into the sequence of amino acids that comprise the protein. The specific polypeptide sequences are specified by the sequences of nucleotide bases within the mRNA. In addition to messenger RNA, the process of translation requires **transfer RNA molecules (tRNA)** that help align the amino acids during the translational process. Further, the ribosomes themselves are largely comprised of **structural (ribosomal) RNA (rRNA)** molecules. Thus, in most microorganisms the nucleic acids of RNA act as informational mediators between the DNA where the genetic information is stored and the proteins that functionally express that information.

The expression of genetic information can be controlled at several levels. As we have already seen in Chapters 5 and 6, enzymes are subject to allosteric control and thus, once synthesized the activities of enzymes are subject to careful regulation. Microorganisms are also able to control their metabolic activities by regulating protein synthesis. In addition to encoding the information for the specific polypeptide sequences of proteins, the genome of the cell codes the information that regulates its own expression, and some sequences of the DNA are involved in regulatory functions rather than coding for specific polypeptide sequences. By controlling which of the genes of the organism are to be translated into functional enzymes, the cell is able to regulate its metabolic activities.

DNA replication

After establishing the importance of DNA for determining the properties of microorganisms, let us examine the replication of this macromolecule in more detail. (We have already considered the basic structure of the DNA macromolecule in Chapters 3 and 4.) **The replication of the double helical DNA molecule is a semiconservative process**, so-called because when a DNA molecule is replicated to form two double helical DNA molecules, each of the new daughter DNA molecules consists of one intact (conserved) strand from the parental double helical DNA and one newly synthesized complementary strand.

Semiconservative nature of DNA replication

The semiconservative nature of DNA replication was elegantly demonstrated in 1958 by **Matthew Meselson** and **Franklin Stahl** in a series of experiments that used the heavy isotope of nitrogen, ^{15}N (Figure 7.1). In these experiments, *Escherichia coli* was initially grown in a medium with a sole nitrogen source of ^{15}N ammonium ions. The bacteria incorporated the heavy nitrogen into their nucleic acids. The bacterial culture was then transferred to a medium with a nitrogen source of ^{14}N ammonium ions. During incubation the bacterial DNA was replicated and the bacterial cells reproduced. Cells were collected for analysis of the DNA after they had been allowed to grow for different generation times, and the DNA was then analyzed for the presence of ^{15}N and ^{14}N using density-gradient ultracentrifugation. In this ana-

lytical method heavy molecules move farther than lighter molecules, and thus, DNA containing ^{15}N moves a greater distance than DNA containing only ^{14}N. When Meselson and Stahl performed this experiment, all the initial DNA formed as a single band corresponding to heavy DNA. After one generation a single sedimentation band occurred corresponding to a hybrid DNA molecule of a mixture of ^{14}N- and ^{15}N-labeled DNA. After two generations, two bands occurred, one corresponding to light DNA (containing only ^{14}N) and the other corresponding to the ^{14}N-^{15}N hybrid DNA. These results are consistent with our understanding of a semiconservative mode of DNA replication. In the first generation the *E. coli* cells each contained one parental strand of DNA containing ^{15}N and one newly synthesized strand of DNA containing ^{14}N. In the second generation some of the cells contained the ^{15}N-labeled parental strand of DNA and a newly synthesized ^{14}N strand, and other cells contained the parental ^{14}N strand and a newly synthesized ^{14}N-containing strand.

The actual replication of the DNA molecule involves a complex series of coordinated enzymatic reactions. For semiconservative replication the two strands of the parental DNA must separate. The entire DNA molecule is not unwound prior to replication, but rather there is a localized unwinding of the DNA double helix, mediated by unwinding enzymes, that establishes a replication fork in the DNA molecule at the site of DNA synthesis (Figure 7.2). The parental DNA molecule acts as a template that codes for the synthesis of DNA. At the replication fork, free nucleotide bases are aligned opposite their base pairs in the pa-

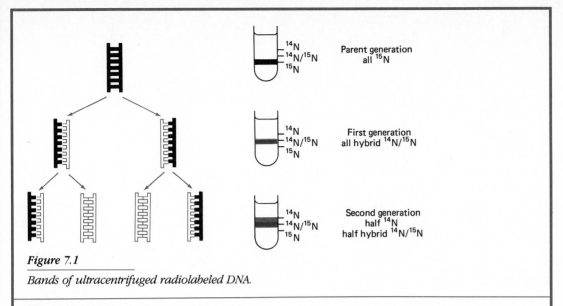

Figure 7.1
___

Bands of ultracentrifuged radiolabeled DNA.

Discovery process

The experiments of Meselson and Stahl relied on the ability specifically to label DNA in one generation by the incorporation of heavy nitrogen (^{15}N) into the DNA and to follow the fate of this tagged DNA from one generation to the next using density-gradient ultracentrifugation. The location of the bands was obtained by ultracentrifugation, in which the distance the DNA moves is a function of the molecular weight of the DNA, permitting the tracking of the fate of the heavy DNA when the cells were grown in the presence of normal, light nitrogen (^{14}N). Of several hypotheses, the only logical explanation for the banding pattern obtained in these experiments was that DNA replication occurs by a semiconservative method. The design of the experiments was eloquently simple and the results so clear-cut that an unambiguous answer to the fundamental question of how DNA is replicated was clearly achieved.

rental DNA molecules (A opposite T and G opposite C). At the replication forks there are four strands of DNA, two of which are conserved and two of which are newly synthesized.

DNA polymerase enzymes

The newly synthesized strands of DNA are established by linking the nucleotide bases together with phosphodiester bonds by the action of DNA polymerase enzymes (Figure 7.3). The action of the **DNA polymerase** enzyme, which results in the elongation of the nucleotide chain of the synthesized DNA molecule, can be likened to a zipper where the teeth of the zipper are initially aligned and progressively linked together in a continuous motion. The DNA polymerase enzymes have several interesting properties. It has been found that all DNA polymerase enzymes can add deoxynucleotides only to the free 3'-OH end of an existing nucleic acid polymer. These en-

zymes require an RNA primer molecule with a 3'-OH free end for DNA synthesis. The DNA polymerase enzymes are also DNA-dependent and require an existing DNA molecule to act as a tem-

Figure 7.2
___

Replication of the double-stranded DNA molecule requires localized unwinding and the establishment of a replication fork.

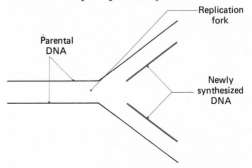

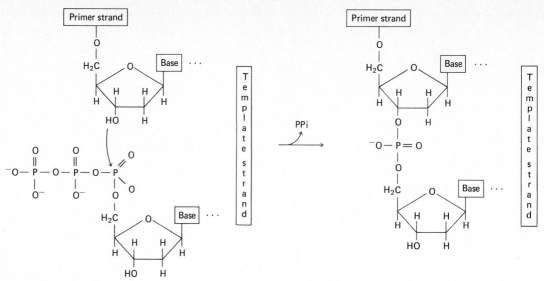

Figure 7.3

The action of DNA polymerase results in the formation of a diester linkage and the elongation of the DNA chain.

plate; such an enzymatic dependence on an existing template molecule is exceptional among biochemical reactions.

Several different DNA-dependent DNA polymerase enzymes have been isolated. The different polymerase enzymes serve somewhat different functions during DNA synthesis. The DNA polymerase enzymes of prokaryotes can remove the RNA primers from the DNA strand, which in the case of eukaryotes is accomplished by a ribonuclease enzyme. In prokaryotic microorganisms the DNA polymerase enzymes also exhibit exonuclease activity, that is, the ability to degrade or depolymerize a nucleic acid chain. Exonuclease activity is not associated with the DNA polymerase enzymes of eukaryotic organisms. The exonuclease activity of prokaryotic DNA polymerase enzymes allows them to correct errors in the DNA molecule, and if an inappropriate sequence of nucleotide bases has been inserted, the DNA polymerase enzyme can reverse direction, removing nucleotide bases from the free end of the DNA molecule and then renewing its polymerization activity. Some prokaryotic DNA polymerases can remove bases from the 3'-OH end; others have exonuclease activity from the 5'-OH end. Because the same enzyme involved in the synthesis of the DNA molecule also removes and corrects errors in the DNA molecule, prokaryotic organisms have an added safeguard for ensuring the accuracy of DNA replication.

Since the DNA polymerase enzymes can add nucleotides only to a 3'-OH free end, the direction of DNA synthesis is 5'-OH → 3'-OH. The fact that DNA polymerase enzymes can add deoxynucleotides only to the 3'-OH free end of a nucleic acid primer creates a paradox for our understanding of the replication of DNA. The two strands of the double helical DNA molecule are antiparallel; one strand runs from the 5'-OH to the 3'-OH free end, and the other complementary strand runs from the 3'-OH to the 5'-OH free end; therefore, synthesis of complementary strands requires that DNA synthesis proceed in opposite directions, while the double helix is progressively unwinding and replicating in one direction. One of the DNA strands can be continuously synthesized because it runs in the appropriate direction for the continuous addition of new free nucleotide bases to the free 3'-OH end of the primer molecule (Figure 7.4). Synthesis of this **continuous** or **leading strand of DNA** occurs simultaneously with the unwinding of the double helical molecule and progresses towards the **replication fork**. The other strand of the DNA, however, must be synthesized discontinuously (Figure 7.4). The initiation of the synthesis of the discontinuous strand of DNA begins only after some unwinding of the double helix has occurred and therefore lags behind the synthesis of the continuous strand; it is referred to as the **lagging strand**. Short segments of DNA, known as **Okazaki segments**, can be formed by the DNA polymerase enzyme running opposite to the direction of unwinding of the parental DNA molecule. The Okazaki fragments are joined together by the action of **ligase enzymes**, which establish phosphodiester bonds between the 3′-OH and 5′-OH ends of chains of nucleotides. The ligase enzymes are not involved in chain elongation, but rather they act as repair enzymes for sealing "nicks" within the DNA molecule. Thus, through the combined actions of DNA polymerase and DNA ligase enzymes both complementary strands of the DNA can be synthesized during DNA replication.

Differences in DNA replication in prokaryotes and eukaryotes

Although the basic mechanism of DNA polymerization is identical in all microorganisms, there are some differences between the actual replication of the DNA of prokaryotic and eukaryotic microorganisms. The bacterial chromosome occurs as a circular macromolecule, whereas the DNA macromolecules of eukaryotic chromosomes are linear. The replication of a circular DNA molecule must necessarily be different from the replication of a linear DNA molecule. Whereas DNA is the universal macromolecule for storing genetic information, and DNA replication is semiconservative in both eukaryotic and prokaryotic cells, it is now becoming apparent that there are major differences in how the information is stored within the DNA and how the information is processed and replicated in prokaryotic and eukaryotic organisms.

One major difference between prokaryotic and eukaryotic DNA replication concerns the number of sites at which DNA replication can begin. Those bacteria that have been examined all have a single point of origin of DNA replication. Polymerase enzymes move bidirectionally from the origin to the terminus of DNA replication, and this means that there are two replicating forks moving in opposite directions. The bidirectional movements of the replication forks produce a loop of DNA that extends out of the plane of the parental bacterial chromosome (Figure 7.5). In *E. coli* and presumably in other bacteria, the terminus for DNA replication is exactly opposite the origin in the circular bacterial chromosome, and the bidirectional replication of the DNA proceeds at identical speeds, meeting precisely at the termination site.

In contrast, the replication of eukaryotic DNA begins at multiple points of origin (Figure 7.6). The replication of eukaryotic DNA proceeds bi-directionally, utilizing polymerase and ligase enzymes that are analogous to those of prokaryotes. The rate of synthesis of DNA along a replicating fork may be slower in eukaryotic than in prokaryotic microorganisms, and although the replication of DNA in prokaryotes proceeds at a uniform rate, the rate of DNA synthesis can vary in a eukaryote. Despite these potential differences in the rates of DNA synthesis within a particular region of the DNA, the overall rate of DNA replication is higher in eukaryotes than in prokaryotes. This is because the DNA of eukaryotes has multiple replicons—a **replicon** being any DNA segment of a DNA macromolecule that has an origin and terminus—compared to the single replicon of the bacterial chromosome. Consequently, even though there is much more DNA in a eukaryotic chromosome than in a bacterial chromosome, the eukaryotic genome can be replicated much faster than the bacterial genome because of the multiple initiation points for DNA synthesis. This distinction between single and multiple origins of DNA synthesis appears to represent a fundamental difference between prokaryotic and eukaryotic microorganisms.

Figure 7.4

The unidirectional nature of DNA polymerase and the antiparallel nature of the double-stranded DNA molecule means that one strand of DNA is synthesized continuously and the other DNA strand discontinuously.

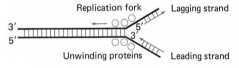

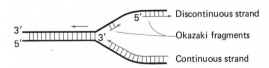

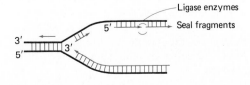

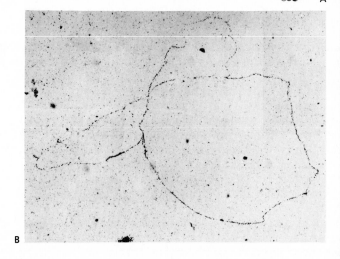

Figure 7.5

(A) The replication of a bacterial chromosome, showing the sequence of synthesizing a new circular loop of DNA. (B) Autoradiograph of the replicating chromosome of Escherichia coli. (Courtesy John Cairns, Harvard School of Public Health.)

Origin of replication marker

Rotation around axis

A

B

Discovery process

The circular nature of the bacterial chromosome and the fact that during synthesis the new circular loop of DNA grows out of the plane of the parental DNA was demonstrated by John Cairns using autoradiography. In this method, bacteria are grown in the presence of a radioactive compound, such as tritiated thymidine, which is incorporated into the DNA. A fine-grain photographic emulsion is placed over the bacterial cells and incubated; the areas of radioactivity are detected when the film is developed, as seen in this autoradiograph of the replicating chromosome of *Escherichia coli*. The visualization of the bacterial chromosome by autoradiography leaves no doubt as to its circular nature.

Replication of viral genomes

Unlike prokaryotic and eukaryotic cells where the genome is contained in a double helical DNA molecule, the genomes of viruses can be encoded in single- or double-stranded DNA or single- or double-stranded RNA. The replication of the viral genome that occurs within a prokaryotic or eukaryotic host cell can occur in a variety of ways, depending on the nature of the nucleic acid macromolecule in which the genetic information is stored.

DNA viruses

In some DNA viruses the DNA macromolecules are circular, resembling the bacterial chromosome, and in others the DNA macromolecules are linear. Some viruses can act like *E. coli*, exhibiting bidirectional DNA replication from a single point of origin, although in some cases, the terminus for DNA replication is offset from the origin. Some linear viruses exhibit multiple initiation points for

Figure 7.6

*(A) As shown in this
illustration, DNA replication
begins at multiple points in
eukaryotes. (B) The multiple
replication forks of a
eukaryotic chromosome are
visible in this electron
micrograph of a replicating
DNA molecule. (Courtesy D. R.
Wolstenholme, University of
Utah; reprinted by permission
of Springer-Verlag, from D. R.
Wolstenholme, 1973,
Chromosoma 43: 1–18.)*

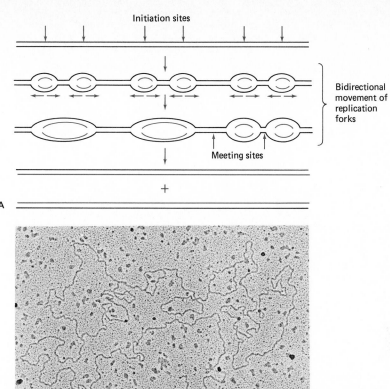

DNA synthesis and thus resemble eukaryotes. In other cases, the replication of viral DNA follows a rolling circle model in which a circular DNA molecule is used to spin off unidirectionally a linear DNA molecule (Figure 7.7). The rolling circle replication of a DNA molecule requires an endonuclease enzyme that can nick the circular DNA molecule, establishing a free end of the nucleotide chain that can "roll off" the circle. A variation of the **rolling circle model** of DNA replication appears in the synthesis of single-stranded DNA for those DNA viruses that lack the normal double helical DNA molecule.

RNA viruses

Several methods are employed by different RNA viruses for the replication of their genome (Figure 7.8). Some RNA viruses code for RNA-dependent RNA polymerase enzymes, and within a host cell these viruses are able to use their RNA genome as a template for RNA synthesis. The replication of genetic information in these RNA viruses is analogous to DNA replication in prokaryotic and eukaryotic organisms. In the case of some

Figure 7.7

*The rolling circle model provides an explanation
for how circular DNA replication occurs.
According to this mechanism, an endonuclease
nicks one strand of DNA, DNA polymerase adds
nucleotides, and a single strand of DNA falls off
the circle. The single-stranded DNA then serves as a
template for DNA synthesis to form double-
stranded DNA.*

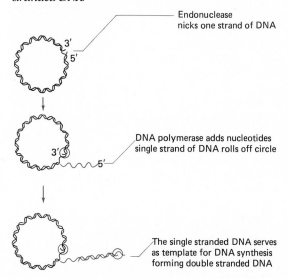

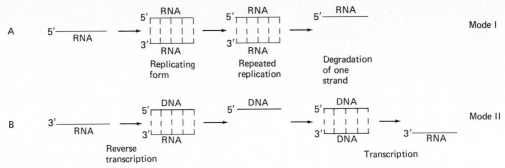

Figure 7.8

There are two quite different modes of replication used by different single-stranded RNA viruses for the replication of their genomes. (A) In some viruses a double-stranded RNA replicating form is produced which acts as a template for the synthesis of new RNA macromolecules. In single-stranded RNA viruses one of the strands then degenerates. (B) In other viruses the viral RNA genome acts as a template for the synthesis of a DNA molecule by reverse transcription. Normal transcription of the DNA then produces new copies of the RNA genome.

226

DNA, RNA, and
protein synthesis:
the replication
and expression
of genetic
information

single-stranded RNA viruses, such as the bacterio-phage Qβ, the genome of the virus (designated as a plus RNA strand) is used as a template for forming a replicative double-stranded form that has both plus and complementary minus RNA strands. When the RNA-dependent RNA polymerase enzyme uses the replicative form as a template the product includes mostly plus strands of RNA, which go into the genomes of the viral progeny. Another mechanism for RNA synthesis in some viruses employs reverse transcription. In **reverse transcription** the RNA viruses use their RNA genome as a template for an RNA-directed DNA po-

lymerase. These viruses produce replicates of DNA instead of RNA. The information in the DNA molecule is then used to direct the synthesis of RNA, which is accomplished by transcription. Some of the RNA is used for synthesizing proteins, and some of the RNA is put into the RNA genomes of the viral progeny. Thus, there are a variety of mechanisms used by different viruses for replicating their genetic information. Several of these mechanisms are distinct from those used in prokaryotic and eukaryotic cells for the replication of the genome.

RNA synthesis—transcription

Regardless of the mechanism used for replicating the genome, organisms depend on their abilities to use the genetic information in a functional manner. As indicated earlier, this expression of genetic information depends on a two-stage process in which the genome of the organism is used to direct the synthesis of RNA and thence proteins. **Transcription** is the process in which the information stored in the DNA molecule is used to code for the synthesis of RNA. In transcription the DNA serves as a template for the synthesis of RNA, accomplishing a critical transfer of information for the eventual expression of genetic information. The transcription of DNA results in the production of three classes of RNA: rRNA (ribosomal RNA), tRNA (transfer RNA), and

mRNA (messenger RNA). All three classes of RNA are required for the expression of genetic information.

Transcription is similar in several ways to DNA replication (Figure 7.9). The molecule of RNA that is synthesized is antiparallel to the strand of DNA that serves as a template. The process of transcription involves the unwinding of the double helical DNA molecule for a short sequence of nucleotide bases, the alignment of complementary ribonucleotides by base pairing opposite the nucleotides of the DNA strand that is being transcribed, and linkage of the nucleotides with phosphodiester bonds by a DNA-dependent RNA polymerase enzyme. (The binding of the RNA polymerase to the DNA may also be involved in the localized unwinding and the proper alignment of the complementary RNA nucleotides.) The RNA polymerase enzymes are able to link nucleotides only to the 3'-OH free end of the polymer and thus, the synthesis of RNA, like that of DNA, occurs in a 5'-OH → 3'-OH direction.

Although transcription is similar in several ways to DNA replication, there are some major differences between RNA synthesis and DNA synthesis. The RNA that is synthesized is single-stranded, and thus, for a given region only one strand of the DNA serves as a template; the strand of DNA coding for the synthesis of RNA is known as the **sense strand**. (Both strands of the DNA can serve as sense strands in different regions, and the term sense strand is applied only to the specific region of the DNA that is being transcribed.) Another difference is that the RNA polymerase enzyme is capable of linking two nucleotides as long as they are aligned opposite the complementary DNA template nucleotides, and therefore, unlike DNA replication, RNA synthesis does not require a primer. Prokaryotic microorganisms have one basic type of RNA polymerase enzyme that produces all three classes of RNA molecules. In *E. coli* there is only one form of this polymerase molecule, although other prokaryotes may possess several variants of the basic type of RNA polymerase. The evidence suggests that eukaryotic microorganisms have three distinct polymerase enzymes responsible for the synthesis of the three different classes of RNA.

Initiation and termination sites of transcription

The transfer of the information from DNA to RNA requires that transcription begin at precise locations. There are multiple initiation sites for transcription along the DNA molecule in both prokaryotes and eukaryotes (Figure 7.10). Different initiation sites are needed to begin the synthesis of different classes of RNA and the synthesis of messenger RNA for different polypeptide sequences. The DNA contains specific sequences of nucleotides, known as **promoter regions**, that serve as signals for the initiation for transcription. The promoter region of the DNA is the site where RNA polymerase binds for transcription. The presence of the promoter region specifies both the site of transcription initiation and which of the two DNA strands is to serve as the sense strand for transcription in that region.

The promoter regions in the DNA of all bacteria and viruses that have been examined consist

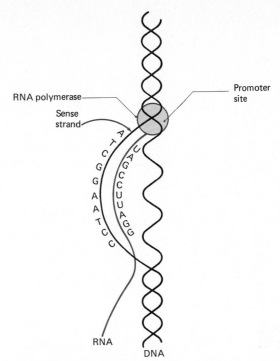

Figure 7.9

During transcription one strand of the DNA codes for the synthesis of RNA. The process of transcription begins at the promoter site, where RNA polymerase begins its job of linking the aligned nucleotides of the RNA molecule being synthesized.

of about 40 nucleotides and contain a seven-nucleotide sequence, known as the **Pribnow sequence**, which appears to be a key part of the recognition signal. A second recognition site, located about 35 bases from the start of the mRNA, has also been implicated in the initial binding of the RNA polymerase needed for the initiation of transcription. The binding of the **RNA polymerase enzyme** to the promoter region is dependent on the presence of a **sigma (σ) factor**, a subunit of the total RNA polymerase enzyme (Figure 7.11). (The entire RNA polymerase enzyme is designated $\alpha_2\beta\beta'\sigma$ to specify the five subunits that compose the complete enzyme.) Without the sigma subunit, the RNA polymerase enzyme fails to exhibit the necessary specificity for recognizing the initiation sites for transcription. The sigma factor ensures that RNA synthesis begins at the correct site. After the binding of the RNA polymerase molecule to the DNA, the sigma subunit dissociates from the RNA polymerase; the RNA polymerase without the sigma subunit is known as the **core enzyme**. The sigma subunit is then free to asso-

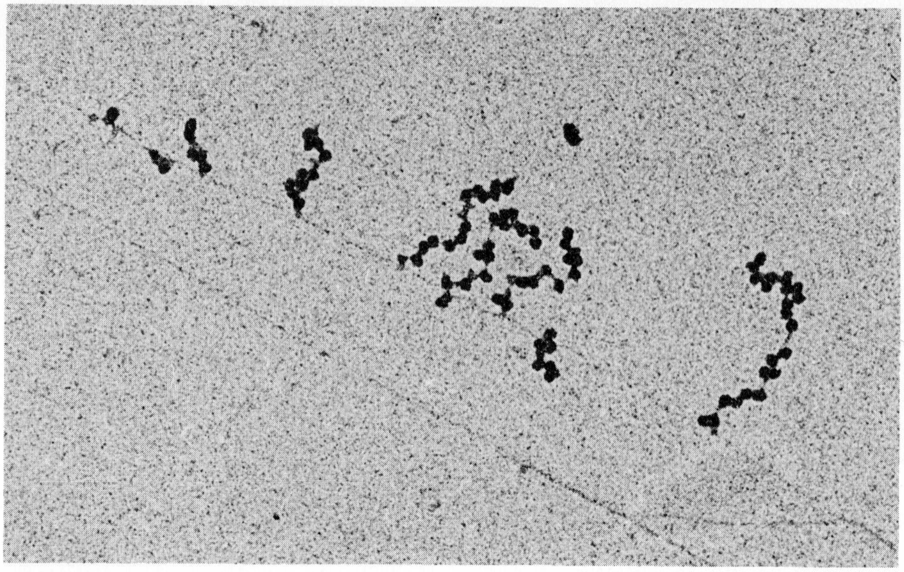

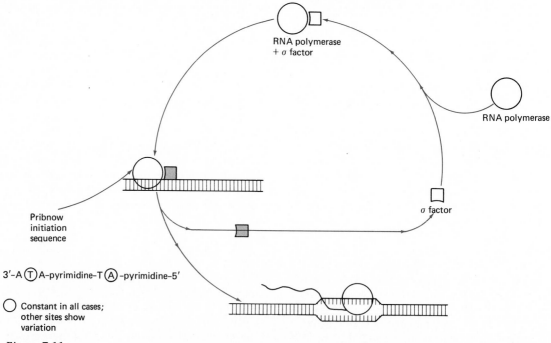

Figure 7.10

This autoradiograph shows transcription occurring at multiple sites along a region of the genome of E. coli *(165,000 ×). The double-stranded DNA is visible as thin parallel tracks and the RNA as dark chains growing away from the DNA. (Courtesy of O. L. Miller, Jr., University of Virginia; reprinted by permission of the American Association for the Advancement of Science, from O. L. Miller, Jr., B. A. Hankalo, and C. A. Thomas, Jr., 1970,* Science *169: 392–397.)*

RNA polymerase + σ factor

RNA polymerase

σ factor

Pribnow initiation sequence

3′-A Ⓣ A-pyrimidine-T Ⓐ -pyrimidine-5′

◯ Constant in all cases; other sites show variation

Figure 7.11

The site of binding of RNA polymerase to the promoter region is specified by the Pribnow sequence. Binding requires the addition of the sigma unit to the RNA polymerase.

ciate with another RNA polymerase molecule, completing that molecule and establishing the necessary specificity for the recognition of the transcriptional initiation site.

In addition to the specific initiation sites, there are specific **termination sites** for transcription. The termination sequence of nucleotides in the DNA contains a region with an abundance of GC bases followed by a region with an abundance of AT bases. The GC-rich region exhibits a symmetry that enables the synthesized RNA to fold back on itself, forming a stem and loop (Figure 7.12). The A bases in the AT-rich region code for a terminal sequence of several U nucleotides. These termination sequences cause the RNA polymerase to pause. In some cases, once the RNA polymerase pauses, transcription is terminated, but in some other cases a protein, referred to as the rho protein, is required to interrupt transcription.

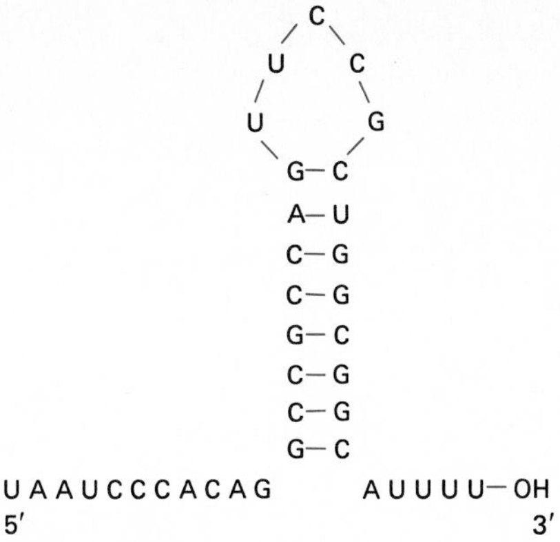

```
          C
     U        C
   U            G
      G— C
      A— U
      C— G
      C— G
      G— C
      C— G
      C— G
      G— C
U A A U C C C A C A G        A U U U U— OH
5'                                      3'
```

Figure 7.12

The stem-and-loop structure of mRNA results from the transcription termination sequence. The sequence of nucleotides at the 3'-OH end of the mRNA transcript allows a stable hairpin structure to form.

Differences in the synthesis of mRNA in prokaryotes and eukaryotes

It is the mRNA molecules formed by transcription that actually encode the information for protein synthesis. In prokaryotes, there is usually only one DNA sequence coding for a particular mRNA, whereas in eukaryotes there often are multiple copies of genes coding for the same mRNA molecules. In both prokaryotic and eukaryotic microorganisms, however, there can be multiple copies of a particular mRNA macromolecule, enabling establishment of multiple sites of synthesis for identical proteins using the multiple copies of the mRNA molecule. There is a drastic difference between prokaryotic and eukaryotic cells in the longevity of their mRNA molecules. In prokaryotic cells the mRNA molecules last for only a few minutes, whereas in eukaryotic cells a mRNA molecule can remain functional for hours or days. The long period of activity of a mRNA molecule in a eukaryotic cell imparts a relative stability in the protein complement, compared to the changing situation in a prokaryotic cell where the mRNA molecules are quickly degraded. The bacterial cell as a result can rapidly alter its metabolism in response to changing environmental conditions, whereas eukaryotic microorganisms are better adapted for continuous metabolism in a stable environment.

In addition to the differences in the half-lives of mRNA molecules in prokaryotic and eukaryotic cells, there is a fundamental difference in how the mRNA is formed and whether transcription

leads to the direct formation of mRNA or whether additional processing is necessary after transcription to form functional mRNA. The messenger RNA molecule in prokaryotic organisms is not modified between the time it is synthesized, through transcription of the DNA, and the time it is translated into the amino acid sequence of a polypeptide at the ribosome. In prokaryotic microorganisms the messenger RNA molecule is transcribed directly from the DNA and the DNA nucleotides coding for the mRNA occur as a contiguous sequence (Figure 7.13). Often the mRNA of prokaryotes is polycistronic; that is, it contains the information coding for several proteins; these proteins are coded for by a continuous region of the DNA and frequently have related functions. The mRNA molecule of bacteria contains a sequence of nucleotide bases at the beginning and the end of the molecule that do not code for specific amino acids in a polypeptide sequence. Rather, the beginning or leader sequence of nucleotides in the mRNA molecule is involved in the initiation of protein synthesis at the ribosomes. The nontran-

Figure 7.13

In prokaryotes mRNA is colinear with the region of the DNA that is transcribed.

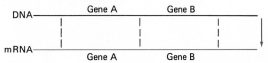

slated leader promotes binding of the mRNA to the ribosomes. The attachment of the mRNA to the ribosomes in prokaryotic cells often occurs before transcription of the mRNA is complete, indicating the close proximity of the transcriptional and translational processes.

In contrast to the formation of mRNA in prokaryotic cells, the RNA molecules of eukaryotic microorganisms are generally extensively modified after transcription from the DNA to form mRNA (Figure 7.14). The mRNA that is formed is **monocistronic** and contains only the information for one polypeptide sequence and the transcriptional and translational processes are spatially and temporarally separated. The precursor of messenger RNA in eukaryotes, known as **hnRNA (heterogeneous nuclear RNA)**, is several times larger than the mRNA molecule and is subjected to substantial **post-transcriptional modification** within the nucleus to form the mRNA. The processing of the hnRNA involves removing, adding, and rearranging sequences of nucleotides. The 5′-OH end of the hnRNA is capped with an inverted guanosine triphosphate (GTP) residue, some of the terminal adenine bases are methylated, and the 3′-OH end of the hnRNA molecule is modified by the addition of a sequence of adenosine nucleotides (poly-A tail). Perhaps the most surprising discovery concerning the transcription of DNA and the processing of hnRNA is the occurrence of intervening sequences or **introns**. (The regions that code for amino acid sequences are known as **exons**.) The introns do not code for amino acid sequences,

and the reason for their existence is unknown. (In one case within yeast mitochondria, an intron within the gene for one protein has been found to code for the removal of introns from genes that are subsequently transcribed.) Part of the processing of hnRNA involves the excision of introns to form the mature mRNA molecule. The rearrangement of the RNA molecules transcribed from the DNA to form the mature messenger RNA molecule involves cutting out and splicing together pieces of RNA. As a result, the mRNA molecules of eukaryotic microorganisms generally are not colinear with the DNA molecule; that is, the sequence of nucleotide bases in the mRNA is not complementary to the specific contiguous linear sequence of bases in the DNA molecule.

Synthesis of transfer and ribosomal RNA

Unlike messenger RNA, which is not modified after transcription in prokaryotic microorganisms, both ribosomal RNA and transfer RNA molecules are substantially modified in both prokaryotic and eukaryotic organisms. Although the RNA molecule is single-stranded, the molecule can fold back on itself, establishing double-stranded regions. Both tRNA and rRNA molecules have extensive "double-stranded" regions that arise from the folding of the primary RNA chain. Several different RNA molecules are found in ribosomes. In both prokaryotic and eukaryotic microorganisms, the RNA molecules transcribed from the DNA are larger than the rRNA molecules found in the ribosomes (Figure 7.15). The precursor rRNA molecules must therefore be processed in order to form the ribosomal RNA molecules. **The ribosomes of prokaryotes contain 5S, 16S, and 23S RNA.** The initial transcript from the DNA, however, is a large 30S molecule that can be cleaved by nuclease enzymes to form these different-size RNA molecules. **In eukaryotic microorganisms the ribosomes are composed of 5S, 5.8S, 18S, and 28S RNA molecules.** A large precursor RNA molecule is cleaved to form the 28S, 18S, and 5.8S rRNA molecules. In most cases a separate large precursor is used for the production of the 5S rRNA molecules.

The transfer RNA molecules are similarly synthesized as high-molecular-weight precursors that are then processed to produce the mature tRNA molecules (Figure 7.16). **Transfer RNA molecules have a multilobed structure** that is formed because of hydrogen binding between complementary regions of the molecule. All tRNA molecules

Figure 7.14

In eukaryotes extensive post-transcriptional modification of the primary RNA transcript, including the removal of introns, is needed to produce the mRNA. As a result of these modifications, the eukaryotic mRNA is not colinear with the DNA.

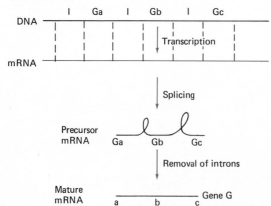

Figure 7.15

Post-transcriptional modification of RNA is needed to produce the rRNA molecules of the 70S and 80S ribosomes.

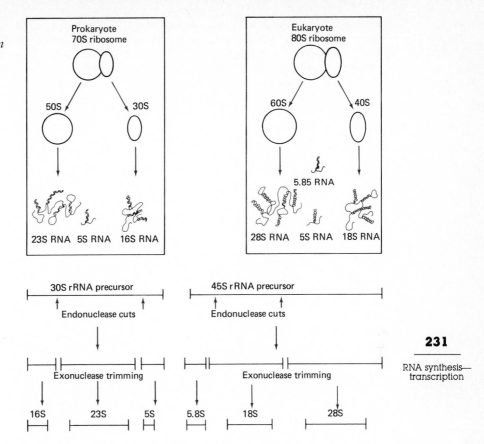

Figure 7.16

All tRNA molecules have a characteristic four-lobe structure that results from the internal base pairing of some of the nucleotides. Each lobe of the tRNA molecule has a distinct function. Several of the lobes are characterized by the inclusion of unusual nucleotides. One of the lobes, designated the DHU loop, contains dihydrouracil (DHU). The TψC loop contains the sequence ribothymine (T), pseudouracil (ψ), and cytosine (C). A third loop, which also contains modified purines, is designated the anticodon loop because it is complementary to the region of the mRNA, the codon, that specifies the amino acid to be incorporated during protein synthesis. The final loop always has the terminal sequence 3'-OH-ACC, which is where the amino acid binds.

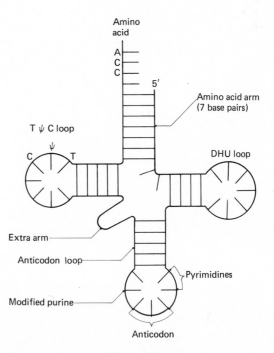

have a 3'-OH terminus with the nucleotide sequence CCA-3' that may be encoded in the primary nucleotide sequence or may be added enzymatically as a cap after transcription from the DNA template. Within the tRNA molecule several of the nucleotides are modified to form "unusual" nucleotide bases through the post-transcriptional modification of the normal RNA nucleotides. Some of the unusual nucleotide bases found in tRNA are pseudouridine, dihydrouridine, ribothymidine, and inosine. The specific functions of the unusual bases have not been established, but their hydrophobic nature may be important in the interactions of tRNA and ribosomes. The presence of the unusual nucleotides and the specific multilobe configuration of the tRNA molecule distinguishes it from other classes of RNA molecules.

Protein synthesis—the translation of the genetic code

All three types of RNA, rRNA, tRNA, and mRNA, are involved in transferring the information obtained from the sequence of nucleotide bases in the DNA to the sequence of amino acids in the polypeptide chain of a protein. The translation of the genetic code into protein molecules, which can functionally express the genetic information, occurs at the ribosomes. The ribosomes provide the spatial framework and structural support for aligning the translational process of protein synthesis. Distortion of the proper configuration of the ribosome can prevent the proper informational exchange and expression of the genetic information, and this forms the basis for the action of many antibiotics. The transfer RNA molecules carry the amino acids to the site of protein synthesis and properly align the amino acids for incorporation into the polypeptide chain. It is the messenger RNA molecule, however, that actually contains the coded information that specifies the sequence of the amino acids in the polypeptide chain.

The genetic code

Within the mRNA three sequential nucleotides are used to code for a given amino acid, and the **genetic code**, therefore, is termed a **triplet code** (Figure 7.17). Each of the triplet nucleotide sequences is known as a **codon**. Because there are four different nucleotides there are 64 possible codons; that is, there are 64 possible three-base combinations of the four different nucleotides. The genetic language, which is almost universal, can therefore be said to have four letters in the alphabet and 64 words in the dictionary, each word containing three letters. Although there are 64 possible codons, proteins in biological systems normally contain only 20 L-amino acids. (Some proteins do contain other amino acids, but in such cases the unusual amino acids are usually formed by post-translational modification.) Therefore, there are many more codons than are strictly needed for the translation of genetic information into functional proteins. More than one codon can code for the same amino acid, and therefore, the genetic code is said to be **degenerate**. Stated another way, the genetic code is redundant, with several codons coding for the insertion of the same amino acid into the polypeptide chain. Additionally, there are three codons that do not code for any amino acid. These codons have been referred to as **nonsense codons**. In actuality, the nonsense codons serve a very important function, acting as punctuators that signal the termination of the synthesis of a polypeptide chain.

Translating the information in mRNA

The translation of the information in the mRNA molecule, that is, reading the codons, is a directional process (Figure 7.18). **The messenger RNA is read in a 5'-OH → 3'-OH direction, and the polypeptide is synthesized from the amino terminal to the carboxyl terminal end.** The mRNA molecule is read one codon at a time, that is, three nucleotides, at a time, that specify a single amino acid. There are no spaces between the codons that are read three bases at a time, and therefore establishing a reading frame is critical for extracting the proper information. Within the mRNA molecule there are sequences of nucleotides that define the beginning and end of each encoded polypeptide chain. Here, we see the importance of having a mechanism for recognizing direction in the informational macromolecules. Just as we establish a convention for reading the

233

Protein
synthesis—
the translation
of the genetic
code

Figure 7.17

The codons of the genetic code. Each three-base codon calls for a specific amino acid or acts as a stop signal.

	Second letter				
First letter	U	C	A	G	Third letter
U	UUU ⎤ Phe UUC ⎦ UUA ⎤ Leu UUG ⎦	UCU ⎤ UCC ⎥ Ser UCA ⎥ UCG ⎦	UAU ⎤ Tyr UAC ⎦ UAA Stop UAG Stop	UGU ⎤ Cys UGC ⎦ UGA Stop UGG Trp	U C A G
C	CUU ⎤ CUC ⎥ Leu CUA ⎥ CUG ⎦	CCU ⎤ CCC ⎥ Pro CCA ⎥ CCG ⎦	CAU ⎤ His CAC ⎦ CAA ⎤ Gln CAG ⎦	CGU ⎤ CGC ⎥ Arg CGA ⎥ CGG ⎦	U C A G
A	AUU ⎤ AUC ⎥ Ile AUA ⎦ AUG Met	ACU ⎤ ACC ⎥ Thr ACA ⎥ ACG ⎦	AAU ⎤ Asn AAC ⎦ AAA ⎤ Lys AAG ⎦	AGU ⎤ Ser AGC ⎦ AGA ⎤ Arg AGG ⎦	U C A G
G	GUU ⎤ GUC ⎥ Val GUA ⎥ GUG ⎦	CGU ⎤ GCC ⎥ Ala GCA ⎥ GCG ⎦	CAU ⎤ Asp GAC ⎦ GAA ⎤ Glu GAG ⎦	GGU ⎤ GGC ⎥ Gly GGA ⎥ GGG ⎦	U C A G

Discovery process

Polyuridylic acid (...–U–U–U–U...) was the first synthetic mRNA created. It was made by reacting only uracil nucleotides with the RNA-synthesizing enzyme polynucleotide phosphorylase. In 1961 Marshall W. Nirenberg and J. Heinrich Matthei mixed this poly-U *in vitro* with the protein-synthesizing machinery of *E. coli* in a system that contained ribosomes, tRNAs, enzymes, amino acids, and ATP and GTP (as energy donors). They succeeded in synthesizing a protein, which was identified as polyphenylalanine, a string of phenylalanine molecules attached to form a polypeptide. They concluded that the triplet U–U–U must code for phenylalanine. By 1966 all 64 code words had been identified.

English language from left to right, the correct interpretation of the information stored in the mRNA molecule requires that it be read from the 5′-OH to the 3′-OH free end.

Initiating the translation of mRNA

The initiation of the reading of the mRNA molecule begins from a fixed starting point. The first codon normally read is AUG, which codes for the amino acid methionine. In eukaryotic microorganisms, methionine is always the first amino acid of the polypeptide sequence. In prokaryotic cells the codon AUG codes for *N*-formylmethionine to initiate the polypeptide chain, although the same codon (AUG) codes for methionine when it occurs elsewhere in the mRNA molecule. In some bacteria the codon GUG acts as the initiator codon that specifies *N*-formylmethionine to initiate the polypeptide chain, although when this codon occurs at an internal position of the mRNA molecule it codes for valine. Whereas either methionine or *N*-formylmethionine are the initial amino acids in all peptide chains, these terminal amino acids can later be enzymatically removed so that not all polypeptides found in microorganisms have methionine at their amino terminal ends.

The initiating codon establishes the **reading frame** for the rest of the mRNA molecule (Figure 7.19). Because the nucleotide bases are read three at a time along a continuous chain of nucleotides, changing the reading frame by a single nucleotide base can completely alter the informational con-

Figure 7.18

Protein synthesis (translation) is a directional process. The mRNA is read in the 5′-OH to 3′-OH direction, and the peptide chain is synthesized from the amino end to the carboxylic acid end.

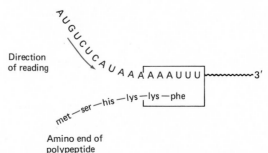

5′ end of mRNA

Direction of reading

Amino end of polypeptide

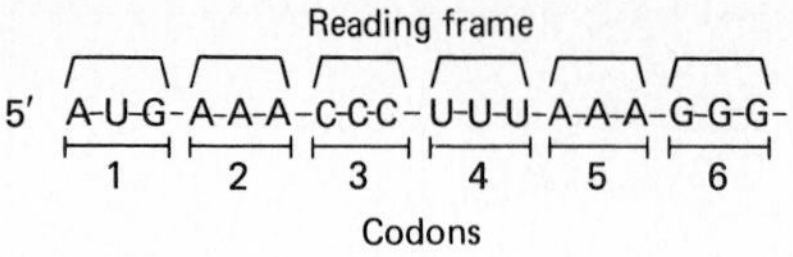

Reading frame

```
        ┌─┐ ┌─┐ ┌─┐ ┌─┐ ┌─┐ ┌─┐
5'  A-U-G-A-A-A-C-C-C-U-U-U-A-A-A-G-G-G-
        └─┘ └─┘ └─┘ └─┘ └─┘ └─┘
         1   2   3   4   5   6
```

Codons

Figure 7.19

The initiating codon of mRNA establishes the reading frame. The contiguous bases are read three at a time with no spacing between adjacent codons.

tent of the mRNA molecule. Several proteins are required as initiating factors to begin the translational process. The actual sequence for the initiation of translation may depend on the base pairing of a region of the leader sequence of the messenger RNA to the 16S rRNA in prokaryotes or the 18S rRNA in eukaryotes. The initiation of synthesizing the polypeptide begins with the association of the 30S or 40S ribosomal subunit with the mRNA, formylmethionyl tRNA or methionyl tRNA and GTP. The 50S or 60S ribosomal subunits then join to establish the 70S or 80S ribosomes in prokaryotes and eukaryotes, respectively. The association of the ribosomal subunits, which is essential for protein synthesis, requires magnesium ions.

The role of tRNA

Before going on we should consider the role of tRNA in the translation process. A tRNA molecule contains approximately 80 nucleotides, and about half of the nucleotides in the tRNA molecule exhibit base pairing. The tRNA molecule forms four lobes, some of which are characterized by the presence of unusual nucleotides. The transfer RNA brings the amino acids to the ribosomes and properly aligns them during translation. The

binding of the amino acid to the tRNA molecule occurs at the 3'-OH free hydroxyl end of the tRNA molecule. Attachment of an amino acid to its specific tRNA molecule is called **charging** (Figure 7.20); a tRNA molecule with its attached amino acid is said to be charged. Charging tRNA molecules requires ATP and an aminoacyl synthetase enzyme. There is at least one aminoacyl synthetase enzyme for each of the 20 amino acids. There are at least 20 different types of tRNA molecules, with each of the 20 amino acids that occur in proteins binding with a different tRNA molecule. The structure of the tRNA molecule plays a role in establishing the proper alignment of molecules during translation. One of the four lobes is attached to the amino acid; one contains a nucleotide sequence that interacts with the ribosomal RNA, establishing the proper orientation to the ribosome; one interacts with aminoacyl synthetase; and one lobe contains a region, the anticodon, that interacts with the messenger RNA.

Each of the tRNA molecules contains a different **anticodon** region, and the anticodon contains three nucleotides that are complementary to the three-based nucleotide sequence of the codon. It is the pairing of the codons of the mRNA molecules and the anticodons of the tRNA molecules that determine the order of amino acid sequence in the polypeptide chain. The third base of the anticodon does not always properly recognize the third base of the messenger RNA codon. As a result the tRNA molecules may **wobble**, resulting in the recognition by a tRNA anticodon of more than one codon.

Forming the polypeptide

Returning to the sequence of events during translation, there are two sites on the ribosome involved in protein synthesis, the **peptidyl site** and

Figure 7.20

Charging of tRNA with an amino acid is necessary so that the tRNA can properly align the amino acid during translation.

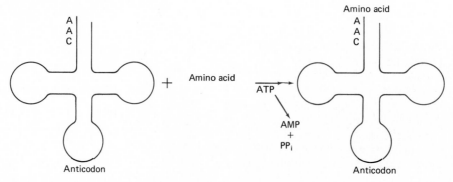

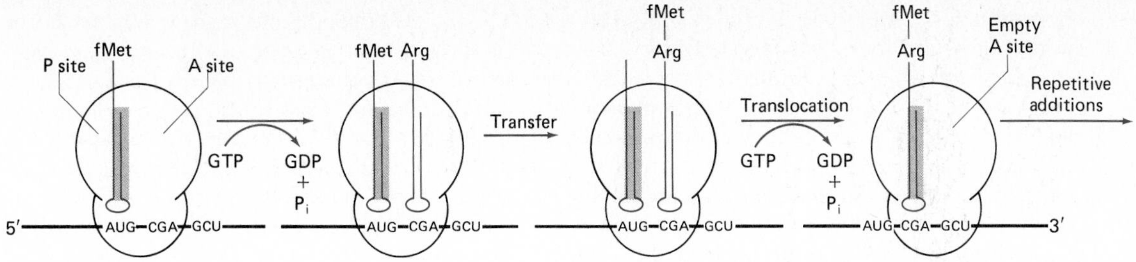

Figure 7.21

Translation is a multistep process in which the information in the mRNA is used to produce a polypeptide. This process occurs at the ribosomes, which provide the structural support for the proper alignment of the macromolecules involved in the translational process. During translation the anticodon of the charged tRNA pairs with the codon of the mRNA. Amino acids are transferred to the growing peptide chain as the mRNA is translocated along the ribosome, moving in three-base jumps.

the **aminoacyl site**. The aminoacyl site is the location where transfer RNA molecules bring individual amino acids to be sequentially inserted into the polypeptide chain. The peptidyl site is the location where the growing peptide chain is aligned. (At the initiation of protein synthesis, the formylmethionyl tRNA or methionyl tRNA occupies the peptidyl site on the ribosome rather than the aminoacyl site, which is normally where the charged tRNA molecules associate with the ribosome.) When tRNA molecules move to the aminoacyl site, the proper anticodon pairs with its matching codon (Figure 7.21). The polypeptide chain is then transferred from the tRNA occupying the peptidyl site to the aminoacyl site. A peptide bond is formed between the amino group of the amino acid attached at the aminoacyl site and the carboxylic acid group of the last amino acid that had been added to the peptide chain. At the initiation of protein synthesis, *N*-formylmethionine in prokaryotes and methionine in eukaryotes are transferred to the aminoacyl site, forming a peptide bond with the second amino acid coded for by the mRNA molecule. The messenger RNA then moves along the ribosome by three nucleotides. The movement of the messenger RNA, transfer RNA, and growing polypeptide chain along the ribosome, a process known as **translocation**, requires the input of energy from GTP. This is one of the rare biochemical reactions that requires a specific energy carrier other than ATP. Translocation moves the tRNA molecule with an attached peptide chain to the peptidyl site, leaving the aminoacyl site open for the anticodon of the next charged tRNA molecule to pair with the next codon of the mRNA. During translocation the tRNA molecule that has transferred its attached amino acids, and is thus no longer charged, returns to the cytoplasm where it can be recycled, that is, where it can once again be charged with an amino acid and returned to the aminoacyl site. The process is repeated over and over, resulting in the elongation of the polypeptide chain. Eventually, one of the nonsense codons appears at the aminoacyl site. Because no charged tRNA molecule pairs with the nonsense codon, the aminoacyl site is empty when the peptide chain tries to transfer to that site. Therefore, the translational process is terminated, and the polypeptide is released to the cytoplasm where it can play a functional role in mediating the metabolism and structure of the organism.

Regulation of gene expression

After considering the processes involved in converting the information stored in the DNA to functional proteins, we now can examine the regulatory mechanisms that control transcription and translation. The expression of the genetic information of microorganisms is subject to regulation at several levels. Some regions of the DNA are specifically involved in regulating transcription, and these regulatory genes can control the synthesis of specific enzymes. In some cases gene expression is not subject to specific genetic regulatory control, and the enzymes coded for by such regions of the DNA are continuously (constitutively) synthesized at a constant rate by the organism. In contrast to **constitutive enzymes**, some enzymes are synthesized only when the cell requires the enzymes, and such enzymes are either inducible or repressible. Often several enzymes that have related functions are controlled by the same regulatory genes.

The regulation of genetic expression in prokaryotes appears to be simpler than in eukaryotes. Eukaryotic organisms exhibit greater cell differentiation than occurs in prokaryotic microorganisms. The expression of genetic information in the eukaryotic cells of higher plants and animals is particularly complex; in such organisms the expression of genetic information must be different in the cells of tissues that perform differentiated functions. Genetic expression in eukaryotes, including eukaryotic microorganisms, appears to be controlled by a variety of mechanisms that do not exist in prokaryotes.

Regulating the metabolism of lactose—the lac operon

One of the best-studied regulatory systems concerns the enzymes produced by *E. coli* strain K-12 for the metabolism of lactose. Three enzymes are synthesized by *E. coli* for the metabolism of lactose, β-galactosidase, galactoside permease, and thiogalactoside transacetylase. The β-galactosidase enzyme cleaves the disaccharide lactose into the monosaccharides galactose and glucose. The permease is required for the transport of lactose across the bacterial cytoplasmic membrane; the role of the transacetylase is not yet established. The structural genes that code for the production of these three enzymes occur in a contiguous segment of the DNA. Although there is some basal level of gene expression in the absence of an inducer, these structural genes are appreciably transcribed only in the presence of an inducer.

The control of the expression of these structural genes for lactose metabolism is explained in part by the **operon model** (Figure 7.22). The regulation of these enzymes is controlled by the promoter region (*P*), a regulatory gene (*i*) that codes for the synthesis of a repressor protein, and an operator region (*O*) that occurs between the promoter and the structural genes involved in lactose metabolism. The regulatory gene codes for a **repressor protein**, which in the absence of lac-

tose binds to the **operator region** of the DNA. The binding of the repressor protein at the operator region blocks the transcription of the structural genes under the control of that operator region. In the case of the *lac* operon, this means that in the absence of lactose, the three structural *lac* genes, which code for the three enzymes involved in lactose metabolism, are not transcribed. The operator region is adjacent to or overlaps the promoter region. The binding of the repressor protein at the operator region interferes with the binding of RNA polymerase at the promoter region. An **inducer** substance such as allolactose, which is a derivative of lactose, binds to the repressor protein and alters the properties of the repressor protein so that it is unable to interact with and bind at the operator region. Thus, in the presence of an inducer that binds with the repressor protein, transcription of the *lac* operon is derepressed, and the synthesis of the three structural proteins needed for lactose metabolism proceeds. As the lactose is metabolized and its concentration diminishes, the concentration of the derivative allolactose also declines, making it unavailable for binding with the repressor protein; therefore, active repressor protein molecules are again available for binding at the operator region, and the transcription of the *lac* operon is repressed, ceasing further production of the enzymes involved in lactose metabolism that are controlled by this regulatory region of the DNA. The *lac* operon is typical of operons that control catabolic pathways; in the presence of an appropriate inducer the system is derepressed.

Catabolite repression

In addition to the specific operator-mediated regulation of transcription, there is a generalized type of repression known as **catabolite repression**. In the presence of an adequate concentration of glucose, a number of catabolic pathways are repressed, including those involved in the metabolism of lactose, arabinose, and galactose. When glucose is available for catabolism in the glycolytic pathway, disaccharides and polysaccharides need not be hydrolyzed to supply monosaccharides for the catabolic activities of the cell, and by blocking the metabolism of these more complex carbohydrates, the cell is able to conserve its metabolic resources.

Catabolite repression acts via the promoter region of DNA, and by doing so it complements the control exerted by the operator region. The efficient binding of RNA polymerase to promoter re-

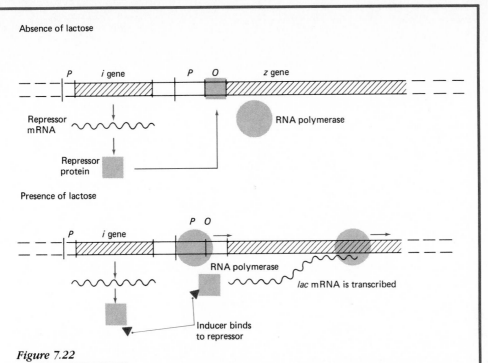

Absence of lactose

Presence of lactose

Figure 7.22

The operation of the lac *operon permits the turning on of the synthesis of the enzymes needed for lactose utilization when lactose is present.*

Discovery process

Biochemical studies on enzyme synthesis added much to our knowledge of the phenomenon of induction and repression, but it was the development of the techniques of bacterial genetics that permitted a more thorough understanding. Francois Jacob and Jacques Monod, working at the Pasteur Institute in the late 1950s, isolated and genetically analyzed many mutants in the lactose pathway. Their results led to the conclusion that induction and repression were under the control of specific proteins, which were coded for by regulatory genes. They proposed that the regulatory genes are closely associated with the structural genes coding for specific enzyme proteins, but that they are distinct from them. At Harvard University, Walter Gilbert and Benn Müller-Hill in 1966 isolated and purified the repressor molecule. *In vitro* studies showed that the repressor protein binds to DNA that contains the *lac* operon but does not bind when the DNA carries a deletion of the *lac* genes. Gilbert used the enzyme DNase to break apart DNA bound to repressor protein and recovered short DNA strands shielded from enzymatic activity by the repressor molecule; these were presumed to be the operator sequence. The sequence was determined, and each operator mutation was shown to involve a change in the sequence. Their results confirmed the identity of the operator sequence and showed the specificity of repressor–operator recognition, which can be disrupted by a single base substitution. Later, when the sequence of bases in the mRNA transcript of the *lac* operon was determined, the first 21 bases on the 5'-OH initiation end were shown to be complementary to the operator sequence Gilbert had determined, confirming the model of the system; the repressor protein binds to the operator, preventing initiation of transcription by the RNA polymerase bound at the promoter site.

gions subject to catabolite repression requires the presence of a catabolite activator protein (Figure 7.23). In the absence of the catabolite activator protein, the RNA polymerase enzyme has a greatly decreased affinity to bind to the promoter region. The catabolite activator protein in turn cannot bind to the promoter region unless it is bound to cyclic AMP. In the presence of glucose, cyclic AMP levels are greatly reduced, and thus, the catabolite activator protein is unable to bind at the promoter region. Consequently, the RNA polymerase enzymes are unable to bind to catabolite-repressible promoters, and transcription at a number of coordinated, regulated structural genes ceases. In the absence of glucose, cyclic AMP is synthesized from ATP, and there is an adequate supply of cyclic

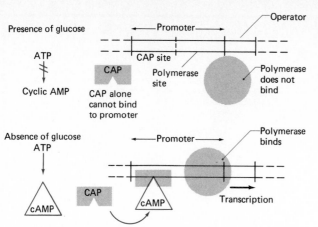

Figure 7.23

Catabolite repression explains why several catabolic pathways in the presence of glucose are shut off. As shown in this illustration, cyclic AMP plays a critical role in permitting RNA polymerase to bind to some promoters, and thus the concentration of cyclic AMP can regulate metabolism by controlling the synthesis of enzymes in operon regions adjacent to such promoters.

DNA, RNA, and protein synthesis: the replication and expression of genetic information

AMP to permit the binding of RNA polymerase to the promoter region. Thus, when glucose levels are low, cyclic AMP stimulates the initiation of many inducible enzymes.

Regulating the biosynthesis of tryptophan—the trp operon

In contrast to the inducible *lac* operon, which is involved in regulating the catabolism of lactose, some repressible operons have been found that control specific biosynthetic pathways. For example, the *trp* operon contains the genes that code for the enzymes required for the biosynthesis of the amino acid tryptophan (Figure 7.24). There are five structural genes in the *trp* operon, and as with other operons there is also an operator region, a promoter region, and a gene that codes for a regulator protein. The *trp* repressor protein is normally inactive and unable to bind at the operator region, but tryptophan can act as an allosteric effector or corepressor. In the presence of excess tryptophan, the *trp* repressor protein binds with tryptophan and subsequently is also able to bind to the *trp* operator region. When the *trp* repressor protein binds at the *trp* operator region, the transcription of the enzymes involved in the biosynthesis of tryptophan is repressed. In the case of the *trp* operon, tryptophan acts as the repressor substance that shuts off the biosynthetic pathway for its own synthesis when there is a sufficient supply of tryptophan. When the level of tryptophan in the cell declines, there is insufficient tryptophan to act a corepressor, and the transcription of the genes for the biosynthesis of tryptophan, therefore, resumes. The *trp* operon is typical of anabolic pathways; in the presence of

a sufficient supply of the biosynthetic product of the pathway, the system is repressed.

Regulation of transcription by translation

Another mechanism for regulating gene expression, which controls the synthesis of several enzymes involved in the biosynthesis of amino acids, functions through the regulation of the transcriptional process by the translational process. In this mode of regulating gene expression, the events that occur during translation affect the transcription of an operon region of the DNA. As indicated earlier, translation in prokaryotes normally begins before transcription of the messenger RNA is completed, with the leader portion of the messenger RNA attaching to the ribosome and translation beginning before the complete messenger RNA is transcribed. The leader sequence in prokaryotes occurs between the operator and the first structural genes. In the case of the *trp* operon, there is an **attenuator site** within the leader sequence of the operon where transcription can be interrupted (Figure 7.25). As described earlier, when tryptophan is plentiful it acts as corepressor and the operator region is blocked, but when the concentration of tryptophan diminishes, the repressor protein cannot bind at the operator region, and transcription of the *trp* operon begins.

Tryptophan is one of the amino acids coded for in the leader sequence of codons. When there is still some tryptophan available for incorporation into proteins, the peptide sequence coded for by this leader sequence can be successfully translated. When tryptophan reaches very low

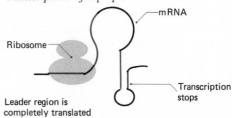

The figure at top shows the trp operon structure with labels: P, O, Attenuator, trp E, trp D, trp C, trp B, trp A (trp operon). Below: Leader, 3', trp mRNA, Structural enzymes. Arrows point to: Anthranilate synthetase, Anthranilate transferase, Indol glycerol phosphate transferase, Tryptophan synthetase B, Tryptophan synthetase A. Lower: P, R, P, O, trp E. Repressor + Co-repressor (tryptophan).

Figure 7.24

The operation of the trp *operon permits the cell to stop synthesizing the enzymes involved in the biosynthesis of tryptophan when there is a sufficient concentration of this amino acid.*

concentrations, however, translation is stopped at the leader codons that code for the insertion of tryptophan into the polypeptide; there simply is no tryptophan available for insertion into the polypeptide chain. When the translational process is halted, transcription proceeds through the entire sequence of the *trp* operon. If there is sufficient tryptophan present, however, to permit translation to proceed through the leader sequence, further transcription of the *trp* operon ceases at the attenuator site; that is, when the leader sequence of the *trp* mRNA is completely translated, a signal is transmitted to the RNA polymerase enzyme to cease further transcription.

In addition to the tryptophan operon, the histidine and the phenylalanine operons in *E. coli* also contain attenuator regions. As in the case of tryptophan, within the leader sequence of the mRNA molecule that codes for the synthesis of these other amino acids, there is a series of codons that code for the insertion of the amino acid to be synthesized. In the case of histidine, there is a sequence of seven contiguous codons for histidine in the leader sequence; in the case of phenylalanine, there are 15 codons for phenylalanine in the leader sequence. Only when the concentrations of these amino acids are very low does translation stall and transcription proceed through the attenuator site. The attenuator complements the regulation of gene expression by the operator gene; thus, there is a redundancy in the control

mechanisms for the biosynthesis of amino acids such as tryptophan. The leader sequence and associated attenuator site thus provide a mechanism for even finer control of transcription and the expression of genetic information than does the operator region.

Figure 7.25

Attenuation permits translational control of transcription of trp *operon.*

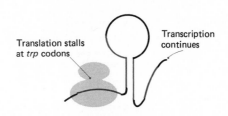

Ribosome
mRNA
Transcription stops
Leader region is completely translated

High tryptophan level

Translation stalls at *trp* codons
Transcription continues

Low tryptophan level

Gene expression in eukaryotic microorganisms

The regulation of genetic expression in eukaryotic microorganisms is more complex than in bacteria, and many of the mechanisms described for the regulation of gene expression may not operate in the same manner in prokaryotic and eukaryotic cells. Some enzymes in eukaryotic microorganisms are clearly inducible and others are repressible, but the control of the expression for these enzyme systems may not be under control mechanisms comparable to the *lac* operon in *E. coli*. Messenger RNA molecules in eukaryotes normally are not polycistronic as they are in prokaryotes, and thus, control over several different mRNA molecules may be required to achieve coordinated control in eukaryotes, whereas control of a single messenger RNA molecule, carrying the information for the expression of several genes in a prokaryotic cell, can regulate the synthesis of several enzymes with related functions. Additionally, in almost all cases the sequence of codons of a given gene in eukaryotic microorganisms is not colinear with the messenger RNA molecule nor with the polypeptide sequence of the protein. It is, therefore, unlikely that in eukaryotic microorganisms an operator region could regulate the transcription of a series of genes that were contiguous with the region of the regulator gene. This effectively precludes the existence of a unified operon region, such as occurs in prokaryotic microorganisms, although an analogous type of operon control over a gene cluster can still exist in eukaryotic cells.

There are a series of different control mechanisms that may be present in eukaryotic microorganisms to control genetic expression. These include the loss of genes, gene amplification, re-arrangement of genes, differential transcription of genes, post-transcriptional modification of RNA, and translational control (Figure 7.26). For example, in the protozoan *Oxytricha,* elimination of most of the DNA (gene loss) occurs in the vegetative cells. The elimination of some of the genetic information restricts the number of genes that can be expressed, and this mechanism allows for specialization in the vegetative cells, analogous to differentiation of somatic cells in higher organisms. (Most higher eukaryotic organisms do not seem to use the gene elimination model, although several insects do use gene elimination as a means of differentiation.) In contrast to gene loss, some eukaryotic microorganisms are capable of amplifying gene expression, thereby producing large amounts of the enzyme coded for by a given gene. Eukaryotes do so by increasing the amount of ribosomal RNA and thus the number of ribosomes that can be used to translate the information in a stable messenger RNA molecule. In the protozoan *Tetrahymena*, for example, there are hundreds of copies of the genes for rRNA in the vegetative cell that provide a mechanism for gene amplification. Within the eukaryotic genome, the position of some genes can be changed, thereby rearranging the location of genes within the eukaryotic chromosome. The rearrangement of genes can alter the expression of the information contained in those genes. For example, the rearrangement of genes has been found to result in altered mating types in yeasts and the production of altered surface biochemicals in protozoa. (This phenomenon has also been observed in prokaryotes, for example, in the case of *Salmonella*, flagellin protein.)

In addition to these mechanisms, promoter regions in eukaryotic organisms are involved in the binding of RNA polymerase enzymes and these could be sites for genetic regulation of the type exhibited in catabolite repression by prokaryotes. It also is likely that post-transcriptional modification of the hnRNA is involved in the control of genetic expression in eukaryotes. The leader sequences, introns, and trailer sequences of RNA molecules in eukaryotes probably affect the expression of genetic information in eukaryotic microorganisms. Different strains of the protozoan *Tetrahymena* have been found to have different introns in the genes that code for their rRNA. Further, the regulation of translation may be more important in controlling gene expres-

Figure 7.26

Several types of genetic control mechanisms appear to control metabolism in eukaryotes.

Mechanisms for controlling gene expression in eukaryotes

- Loss of genes
- Multiple copies of genes (gene amplification)
- Rearrangement of genes
- Control of transcription (differential transcription)
- Post-transcriptional modification of RNA
- Attenuation of transcription
- Attenuation of transcription
- Allosteric control of enzymes

sion in eukaryotes than in prokaryotes because of the relative longevity of messenger RNA in eukaryotic cells.

It is thus clear that there are a number of mechanisms for regulating gene expression in eukaryotic microorganisms that do not exist in prokaryotes. Unravelling the complexity of gene regulation in eukaryotes and developing a better understanding of genetic regulation in prokaryotes remain prime challenges for the microbial geneticist.

Postlude

In this chapter we have considered the mechanisms for the storage and expression of genetic information. In prokaryotic and eukaryotic microorganisms DNA is the molecule that stores the genetic information; in some viruses this function is performed by RNA. The replication of the genome of microorganisms is essential for the transmission of the hereditary information from one generation to the next, and the process of DNA replication is designed to ensure the fidelity of the DNA that is transmitted to the progeny. The replication of the double helical DNA molecule uses the parental DNA as a template for directing the sequencing of the nucleotides in the synthesis of new DNA molecules. DNA replication is semiconservative; the progeny receive one parental and one newly synthesized strand of DNA in their double helical DNA molecules. The DNA-dependent DNA polymerase enzymes, which catalyze the replication of DNA, add nucleotide bases only to the 3′-OH free end of a nucleotide chain. DNA polymerase enzymes form phosphodiester linkages only when the free nucleotides are correctly aligned opposite their base pairs. Because of the directional nature of the DNA polymerase enzymes, only one strand of the DNA can be continuously synthesized, and the synthesis of the opposing strand lags behind and involves the formation of discontinuous segments of DNA, which are then linked together by ligase enzymes. The directional nature of DNA is essential for its proper functioning, and the replication process reflects and ensures the continuance of directionality in this macromolecule.

The replication of DNA requires an unwinding of the double helical DNA molecule to form replication forks where complementary strands of DNA can be synthesized. Replication of the bacterial chromosome has a single initiation site, and DNA replication in the organisms so far examined proceeds bidirectionally from this origin, with the replication forks moving at identical rates in opposite directions around the circular bacterial chromosome until they meet at the termination point. In eukaryotic microorganisms DNA synthesis also occurs bidirectionally, but initiation occurs at multiple origin sites along the chromosome. Because there are multiple initiation sites for DNA replication, eukaryotes are able to replicate their genome more rapidly than prokaryotes. These differences represent fundamental distinctions between eukaryotic and prokaryotic cells.

The replication of the genome provides the mechanism for passing the genetic information from one generation to the next. For the genetic information to determine the structure and function of the organism, it must also be expressed and used to control the metabolism of the cell. The expression of genetic information involves transferring the information encoded in the genome to functional enzymes. This is accomplished by having the genome direct the process of protein synthesis, a two-stage process in which information encoded in the DNA is first transcribed to RNA molecules and then translated into the amino acid sequence of the polypeptide. Three types of RNA molecules are involved in protein synthesis, ribosomal RNA (rRNA), transfer RNA (tRNA), and messenger (mRNA). Transcription to form RNA uses the DNA as a template with only one strand of the DNA, the sense strand, serving as the template for the transcription of given RNA molecule. Specific initiation and termination sites for transcription are encoded in the DNA molecule. Transcription requires that RNA polymerase bind to the promoter region of the DNA to initiate RNA synthesis; the promoter region contains a sequence of nucleotides recognized by the RNA polymerase enzyme as the site of transcriptional initiation.

A critical function of transcription is the formation of mRNA because the information for the sequencing of the amino acids in a polypeptide is encoded within mRNA molecules. In prokaryotic microorganisms, messenger RNA is colinear with the DNA molecule. In contrast, eukaryotes have split genes, and a hnRNA molecule is initially synthesized and then extensively processed after transcription to produce mRNA molecules. Part of the processing of a hnRNA molecule involves the removal of intron regions, which occur between the nucleotides that actually code for the amino acids of a polypeptide.

The translation of the messenger RNA molecule occurs at the ribosomes. During translation the genetic code, consisting of 64 triplet nucleotide sequences known as codons, specifies the amino acids that are to be inserted into a polypeptide chain. Three of the codons do not specify any of the 20 amino acids that are found in proteins, but rather these nonsense codons act as termination signals for the translational process. Because more than one codon can specify the same amino acid, the genetic code is said to be degenerate. The genetic code appears to be almost universal in both prokaryotic and eukaryotic microorganisms, although some exceptions have been found in mitochondria where different nucleotide sequences are used to code for particular amino acids. The direction of reading of the messenger RNA molecule is from the 5'-OH to the 3'-OH free end, and the direction of synthesis of the polypeptide chain is from the amino to the carboxyl end. In the process of translation the ribosome supplies the structural framework for the proper alignment required for protein synthesis, the messenger RNA specifies the order of the amino acid sequence for the polypeptide chain, and the transfer RNA molecule brings the amino acid to the ribosome and aligns it in its proper sequence.

242

DNA, RNA, and
protein synthesis:
the replication
and expression
of genetic
information

Both prokaryotic and eukaryotic microorganisms can regulate the expression of genetic information in ways that permit them to carry out specialized activities during specific periods of time. The ability to direct the metabolic activities of the cell to the areas where they are needed is a critical conservation mechanism for microorganisms. The expression of genetic information can be controlled at several levels. In prokaryotic microorganisms some genes are clustered together, and their transcription is under the control of a regulatory operator region. The operator acts as a switch that can regulate transcription and thus can turn on and off protein synthesis, permitting the coordinated expression of genetic information. Some operons, such as the *trp* operon, are repressible and others, such as the *lac* operon, are inducible. The functioning of the operon can be likened to a building with multiple light fixtures and electrical switches. When a light switch is turned on, the lightbulbs function and emit light in a specified area of the building. In some rooms the light switch is turned on only when someone enters the room, a case that is analogous to the *lac* operon where the system is induced (derepressed) only in the presence of an inducer. In other areas of the building, such as the entrances, the lights are normally left on but may be shut off when they are not needed, a situation that is analogous to the *trp* operon, which is repressed only when the enzymes coded for by the *trp* structural genes are truly not needed.

In addition to the regulatory mechanisms within the DNA that control the process of transcription, the translation of the messenger RNA for several of the enzymes involved in the biosynthesis of amino acids in prokaryotic microorganisms can exert a regulatory control over the transcriptional process. The translational control of transcription involves an attenuator site in the leader sequence of the mRNA molecule. In eukaryotic microorganisms the control of gene expression is far more complex than in prokaryotes, and the post-transcriptional processing of mRNA in eukaryotes permits an additional level of regulation that does not occur in prokaryotic microorganisms.

Thus, the genome of the cell codes for its own replication, establishing a mechanism for hereditary transmission of genetic information; directs the synthesis of proteins, determining the metabolic capabilities and potential phenotype of the organism; and contains various control mechanisms for controlling genetic expression that allow microorganisms to finely regulate their metabolic activities. The elucidation of the molecular-level events involved in the transmission and expression of genetic information have revolutionized our understanding of microbial genetics. These advances in molecular genetics permit the understanding of many fundamental processes in microorganisms and the use of this understanding in many applied areas of microbiology.

Study Questions

1. How was the semiconservative nature of DNA replication demonstrated?

2. What is a DNA polymerase enzyme? An RNA polymerase enzyme? How are these enzymes different from the enzymes involved in microbial metabolism?

3. What are the similarities and differences between DNA replication in prokaryotes and eukaryotes?

4. We know it takes 60 minutes for *E. coli* to replicate its DNA; how can this bacterium divide every 20 minutes?

5. What are the steps in protein synthesis? What roles do different nucleic acids play in the expression of genetic information?

6. How is information encoded within nucleic acid molecules? What are the properties of the genetic code?

7. Why is it important that DNA and RNA have distinct 3'-OH and 5'-OH ends?

8. What signals the initiation and termination of transcription? Of translation?

9. How is the expression of DNA regulated at the level of transcription? How are operons and promoters involved in gene expression? Discuss the functioning of the lac and *trp* operons.

10. What is the glucose effect, and how does catabolite repression help explain its molecular basis?

11. What is a split gene, and why does it represent a fundamental difference between prokaryotes and eukaryotes?

Suggested Supplementary Readings

Ayala, F. J., and J. A. Kiger, Jr. 1980. *Modern Genetics*. Benjamin/Cummings Publishing Co., Menlo Park, California.

Birge, E. A. 1981. *Bacterial and Bacteriophage Genetics*. Springer-Verlag, New York.

Chambon, P. 1981. Split genes. *Scientific American* 244(5): 60–71.

Freifelder, D. 1978. *The DNA Molecule: Structure and Properties*. W. H. Freeman and Co., San Francisco.

Fristrom, J. W., and P. T. Spieth. 1980. *Principles of Genetics*. Chiron Press, Inc., Portland, Oregon.

Genetics: Readings from Scientific American. 1981. W. H. Freeman and Co., San Francisco.

Goldstein, L., and D. M. Prescott (eds.). 1980. *Gene Expression: The Production of RNA's*. Academic Press, New York.

Kaplan, A. S. (ed.). 1982. *Organization and Replication of Viral DNA*. CRC Press, Inc., Boca Raton, Florida.

Kornberg, A. 1980. *DNA Replication*. W. H. Freeman and Co., San Francisco.

Lewin, B. 1974. *Gene Expression—1: Bacterial Genomes*. John Wiley & Sons, New York.

Lewin, B. 1980. *Gene Expression—2: Eukaryotic Chromosomes*. John Wiley & Sons, New York.

Lewin, B. 1977. *Gene Expression—3: Plasmids and Phages*. John Wiley & Sons, New York.

Mays, L. L. 1981. *Genetics: A Molecular Approach*. Macmillan Publishing Co., Inc., New York.

Miller, J. H., and W. S. Reznikoff (eds.). 1980. *The Operon*. Cold Spring Harbor Laboratory, New York.

Rich, A., and S. H. Kim. 1978. The three-dimensional structure of transfer RNA. *Scientific American* 238(1): 52–62.

Stent, G. S., and R. Calendar. 1978. *Molecular Genetics: An Introductory Narrative*. W. H. Freeman and Co., San Francisco.

Stewart, P. R., and D. S. Letham (eds.). 1977. *The Ribonucleic Acids*. Springer-Verlag, New York.

Strickberger, M. W. 1976. *Genetics*. Macmillan Publishing Co. Inc., New York.

Stryer, L. 1981. *Biochemistry*. W. H. Freeman and Co., San Francisco.

Suzuki, D. T., A. J. F. Griffiths, and R. C. Lewontin. 1981. *An Introduction to Genetic Analysis*. W. H. Freeman and Co., San Francisco.

Watson, J. D. 1976. *Molecular Biology of the Gene*. Benjamin/Cummings Publishing Co., Menlo Park, California.

Yanofsky, C. 1981. Attenuation in the control of expression of bacterial operons. *Nature* 289: 751–758.

DNA, RNA, and
protein synthesis:
the replication
and expression
of genetic
information

Genetic variation: the introduction and maintenance of heterogeneity within the gene pool

8

Despite the biochemical mechanisms that are designed to ensure the fidelity of DNA replication, alterations in the genetic information can and do occur. The genomes of microorganisms are subject to change by a variety of mechanisms. Heritable changes in the sequence of nucleotide bases of an organism, known as **mutations**, alter the genome of the organism and introduce variability into the gene pool of microbial populations. For normally diploid eukaryotic microorganisms that contain pairs of chromosomes and thus a duplicity of genetic information, the variability introduced into the **gene pool** through mutation can result in differences in the information encoded in the corresponding genes (**alleles**) of the pairs of chromosomes. For prokaryotes that normally do not establish a diploid state, variability occurs within their single set of genes. Microorganisms may also contain small extrachromosomal elements (**plasmids**) that provide an additional site where specialized genetic information can be stored. Variability within the genomes of microorganisms provides a mechanism for selection and evolution. Variability also provides the heterogeneity needed within a population to adapt to environmental changes.

Once variability is introduced into the gene pool of a microbial population, there are several ways by which it is maintained. Genetic information can be transferred between microorganisms, and the exchange of genetic information ensures the passage of genetic variation from one generation to another. The exchange of genetic information between two individual microorganisms results in progeny that contain genetic information derived from two potentially different genomes. In the case of eukaryotic microorganisms, genetic exchange during sexual reproduction affords a mechanism for gene reassortment within the population and maintenance of genetic heterogeneity. Even in prokaryotic microorganisms, where reproduction is asexual or parasexual (not involving gamete formation or a long-lasting diploid state), there are genetic exchange processes that lead to the recombination of genetic information. The natural evolution of microorganisms is based on the selection of the genetic variants best suited for survival, and artificial manipulation of genetic exchange between organisms also can be used to "create" new microorganisms. The development of an understanding of the molecular basis of genetics has spawned the new and exciting field of genetic engineering, and applications of genetic engineering promise to revolutionize the industrial applications of microorganisms.

Mutation

We have previously emphasized that the mechanisms inherent in the DNA replication processes are designed to ensure that the process is error free; we will now see that despite all the safeguards some errors do occur, resulting in **mutations**. Any change in the sequence of nucleotide bases in the DNA can be reflected in the expression of the genetic information. Because the genetic code consists of triplet nucleotide sequences that determine the biochemical structure of proteins, changing the order of bases even slightly can alter the ability of the cell to produce enzymes that function properly. Similarly, because part of the microbial genome regulates the synthesis of proteins, mutations may alter the capacity of the cell to control properly its metabolic activity. Although mutations always change the genotype of the microorganism, often changing the functions of regulatory or structural genes, the fact that the genetic code is degenerate means that some mutations do not alter the phenotype of the organism.

Types of mutations

There are various classes of mutations. One type of mutation, **base substitution**, occurs when one pair of nucleotide bases in the DNA is replaced by another pair of nucleotides. There are two general types of base substitutions: transitions and transversions. **Transitions** involve the replacement of a purine on one strand by a purine and a pyrimidine on the other strand by a pyrimidine (Figure 8.1). The replacement of an adenine–thymine pair by a guanine–cytosine pair or the replacement of a guanine–cytosine pair by a adenine–thymine pair represent the possible transitional mutations. **Transversions** are base substitutions in which purines replace pyrimi-

dines and pyrimidines replace purines (Figure 8.2). The conversion of an adenine–thymine pair to a thymine–adenine or cytosine–guanine pair represents a transversion mutation. Similarly, the change of a guanine–cytosine pair to a cytosine–guanine or thymine–adenine pair establishes a transversion mutation. Because only one strand of the DNA acts as the sense strand, a transversion, such as from a GC to a CG pair, changes the sequence of nucleotides in the messenger RNA molecule that is transcribed.

Some base substitutions, known as **missense mutations**, result in a change in the amino acid inserted into the polypeptide chain. Missense mutations can result in the production of an inactive enzyme. Changes in a single amino acid within a polypeptide, though, often do not drastically reduce the activity of an enzyme and are rarely fatal to the microorganism. Because the genetic code is degenerate, the substitution of one nucleotide base for another may not even change the amino acid specified by the codon. Changes in the nucleotide sequence that alter the third base of a codon are most likely to produce such silent mutations because this is where most of the redundancy in the genetic code occurs. This process is expressed by the **wobble hypothesis**, which states that changes in the third position of the codon often do not alter the amino acid sequence of the polypeptide. (As discussed in Chapter 7, the tRNA molecules sometimes match only the first two nucleotides of the codon and thus are said to wob-

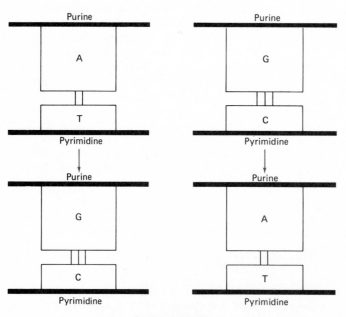

Transition mutation

Figure 8.1

In a transition mutation a purine is replaced by a different purine or a pyrimidine is replaced by a different pyrimidine.

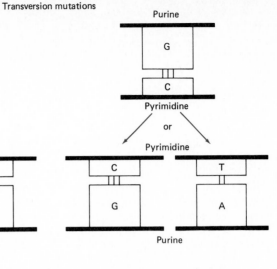

Transversion mutations

Figure 8.2

In transversion mutations a purine is replaced by a pyrimidine or a pyrimidine is replaced by a purine.

ble because of the variability in the third base position.) Silent mutations do not alter the phenotype of the organism and go undetected.

Whereas base substitutions sometimes have little effect on the phenotype of the organism, changing even a single nucleotide base within the large DNA macromolecule can radically change the genetic information of a cell. One special type of mutation that often has a major effect on the expression of the genetic information occurs when the alteration in the base sequence of the DNA results in the formation of a nonsense codon (Fig-

ure 8.3). Because the nonsense codons act as terminator signals during protein synthesis, the formation of a nonsense codon often signals the premature termination of a polypeptide chain, preventing the formation of a functional enzyme molecule. In bacteria, where the messenger RNA molecule often is polycistronic, a **nonsense mutation** can affect the synthesis of several polypeptides; such mutations that prevent the translation of subsequent polypeptides coded for in the same mRNA molecule are said to be **polar mutations**. Not all nonsense mutations within one cistron af-

Figure 8.3

A nonsense mutation can result in the premature termination of the synthesis of a polypeptide.

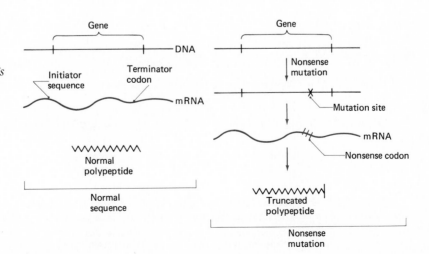

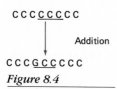

```
TTTATTTTT

        │
        ▼   Deletion

TTTTTTTT
```

```
CCCCCCCC

        │
        ▼   Addition

CCCGCCCCC
```

Figure 8.4

A frame shift mutation can be accomplished by deletion or addition.

fect the expression of other genes, and the degree of polarity depends on the relative locations of the termination and initiation sequences. Mutations may also alter the normal termination codons, and failure to terminate translation properly results in coding for an elongated polypeptide.

In addition to base substitutions, mutations can result from the insertion or deletion of nucleotide bases (Figure 8.4). A **deletion mutation** involves removal of one or more nucleotide base pairs from the DNA, and an **insertion mutation** involves the addition of one or more base pairs. Deletions of large numbers of base pairs, called **deficiencies**, can result in the loss of genetic information for one or more complete genes. Perhaps the greatest effect of deleting or adding a base occurs because of the nature of the translation process, which is dependent on establishing a proper reading frame for having the codons

specify the proper amino acids to be inserted into the polypeptide chain. Because adding or deleting a single base pair changes the reading frame of the transcribed messenger RNA, the deletion or addition of a single base pair can have as great an effect as a large deficiency. **Frame shift mutations** can result in the misreading of large numbers of codons.

In some cases a second mutation can occur that reestablishes the reading frame (Figure 8.5). Single base-pair additions and deletions can be reversed or reverted by such **suppressor mutations**. For example, the addition of a base pair after the deletion of a base pair can restore the reading frame, suppressing the expression of the first mutation. An intragenic suppressor mutation (mutation within one gene), such as one that reestablishes a reading frame, permits the successful synthesis of the polypeptide specified by the gene in which the mutation occurs. Suppressor mutations may also be intergenic, in which case a mutation within one gene affects another gene. Mutations that alter the anticodon region of tRNA molecules can be involved in such intergenic suppression of mutations (Figure 8.6). For example, a mutation in the anticodon of the tRNA molecule can suppress a nonsense mutation if the change in the anticodon results in the insertion of an amino acid where the mutant codon normally causes premature termination of the polypeptide chain.

In addition to altering structural genes, mutations may affect regulatory genes. Various types of mutations can alter the sequence of bases in the promoter or operator regions of the DNA, changing the ability of the organism to regulate protein synthesis at the level of transcription. Thus, for example, a mutation may change an inducible

Figure 8.5

A suppressor mutation can reestablish the proper reading frame.

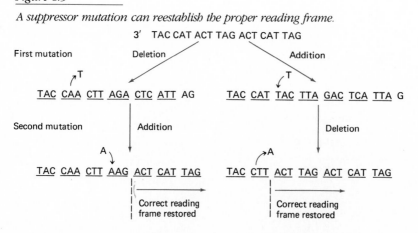

Figure 8.6

An intergenic suppression mutation changes the anticodon of tRNA and thereby negates a nonsense mutation.

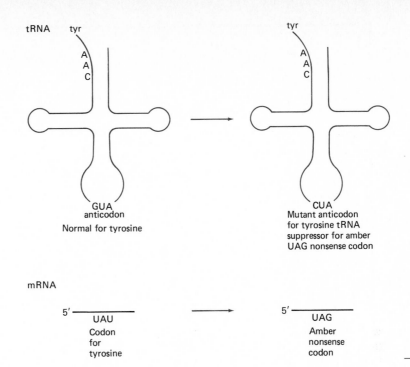

tRNA   tyr

A
A
C

GUA
anticodon

Normal for tyrosine

tyr

A
A
C

CUA
Mutant anticodon
for tyrosine tRNA
suppressor for amber
UAG nonsense codon

mRNA

5′ ————
UAU
Codon
for
tyrosine

5′ ————
UAG
Amber
nonsense
codon

or repressible enzyme system to a constitutive enzyme system. Changes in the leader sequence of a mRNA molecule also can modify the efficiency of translation.

Another way of viewing mutations is to consider their effect on the phenotype of the organism. When the mutation results in the death of the microorganism or its inability to reproduce, the mutation is said to be lethal. Such mutations may be conditional, causing the death of the organism only under certain environmental conditions, or may be unconditional, being lethal to the organism regardless of the environmental conditions. A **conditionally lethal mutation** causes a loss of viability only under some specified conditions where the organism would normally survive. **Temperature-sensitive mutations**, for example, alter the range of temperatures over which the microorganism may grow.

Nutritional mutations occur when a mutation alters the nutritional requirements for the progeny of a microorganism. Often, nutritional mutants will be unable to synthesize essential biochemicals, such as amino acids. Nutritional mutants (**auxotrophs**) require growth factors, such as specific amino acids, that are not needed by the parental (**prototroph**) strain. Nutritional mutants are often detected using the replica plating technique (Figure 8.7). In this method microorganisms can be repeatedly stamped onto media of differing composition. The distribution of microbial colo-

nies should be exactly replicated on each plate, and a colony that develops on a complete medium but not on a minimal medium (lacking a specific growth factor) indicates the occurrence of a nutritional mutation. The microorganisms that do not grow on the minimal medium represent auxotrophic strains. By determining which biochemicals permit the growth of the auxotroph, the site of the metabolic blockage can be determined. Many of the biochemical pathways described in Chapters 5 and 6 were elucidated in this way by using nutritional mutations; by examining the growth requirements of auxotrophic strains, it is possible to determine the sequential order of the intermediary metabolites in a metabolic pathway.

Factors that influence mutation rates

Mutations occur at relatively low natural rates. In *E. coli*, for example, the natural mutation rate is approximately one change per billion nucleotide pairs replicated. However, a variety of chemical and physical agents can increase the incidence of mutation.

Chemical mutagens

Various chemicals can modify the nucleotide bases and act as **chemical mutagens** (Figure 8.8). Hy-

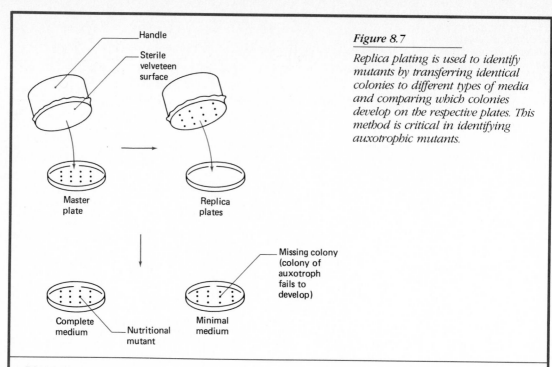

Figure 8.7

Replica plating is used to identify mutants by transferring identical colonies to different types of media and comparing which colonies develop on the respective plates. This method is critical in identifying auxotrophic mutants.

250

Genetic varia-
tion: the introduc-
tion and mainte-
nance of
heterogeneity
within the gene
pool

Discovery process

In 1952 Joshua and Esther Lederberg developed the replica plating technique in order to provide direct evidence for the occurrence of preexisting mutations prior to selection of those mutations by specific environments. They applied the replica plating method to the problem of the origin of bacterial variants of *Escherichia coli* to streptomycin or phage resistance. By using velveteen as an inoculator, they were able to copy the pattern of microbial growth from one agar plate (the master plate) to a series of other plates (replicas) without disturbing the spatial relationships of the bacteria. They replicated plates from plain agar to plates containing either streptomycin or phage and compared the locations of colonies on the master plate to any resistant colonies that appeared at the same position on the replica plates. This allowed them to identify the sites of resistant clones on the original plates and to demonstrate the preexistence of clones of adaptive mutants on the initial plates. By taking inocula from bacterial films at sites demonstrated by the replica plating method to contain mutants, they were able to isolate pure cultures of the mutants. The results of these experiments affirmed that microbial adaptation to deleterious agents results from preoccurring spontaneous mutations and populational selection in the heritable adaptation of bacteria to new environments.

droxylamine, for example, chemically modifies cytosine to uracil so that it pairs with adenine instead of its normal complementary base guanine. After one generation this change results in the replacement of a GC pair with an AT pair, that is, in a transition. Nitrosoguanidine and several other chemical mutagens can alkylate nucleotide bases, causing transitions that result in the substitution of GC for AT in the second generation. Nitrous acid oxidizes the amino group of cytosine or adenine, forming keto groups, converting cytosine to uracil and adenine to hypoxanthine, and resulting in a base substitution. There are several biochemicals that resemble the normal DNA nu-

cleotides and such chemicals are able to act as base analogs (Figure 8.9). For example, 5-bromo-uracil can replace thymine and pair with adenine or replace cytosine and pair with guanine, thus producing base substitutions in the DNA. Several chemical mutagens, such as acridine, result in base deletion or base addition mutations and cause frame shift mutations. Other mutagens, such as mitomycin C, form covalent cross-linkages between base pairs, preventing the replication of the DNA molecule. Thus, exposure to a variety of chemicals that act in different ways can greatly increase mutation rates.

The fact that microorganisms are susceptible to

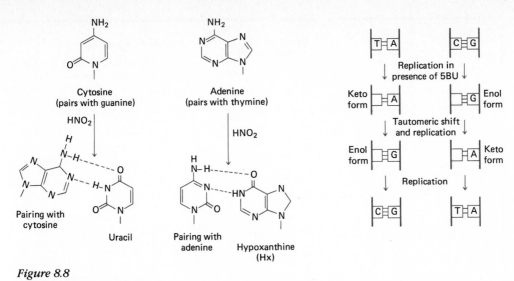

Figure 8.8

Exposure to mutagenic biochemicals, such as nitrous acid, causes changes within the DNA.

chemical mutagens can be used to determine the mutagenicity of various chemicals. In the **Ames test procedure** a strain of the bacterium *Salmonella typhimurium* is utilized as a test organism for determining chemical mutagenicity (Figure 8.10). The bacterial strain used in the Ames test procedure is an auxotroph that requires the amino acid histidine. The organisms are exposed to a gradient of the chemical being tested on a solid growth medium that lacks histidine. Normally, the bacteria cannot grow and in the absence of a chemical mutagen no colonies can develop. If the chemical is a mutagen, lethal mutations will occur in the areas of high chemical concentration, and no growth will occur in these areas. At lower chemical concentrations along the concentration gradient, however, fewer mutations will occur, and some of the cells may revert because of mutagenic action, producing histidine prototrophs that will be able to grow and produce visible bacterial

Figure 8.9

Exposure to a base analog, such as 5-bromouracil, results in mutation.

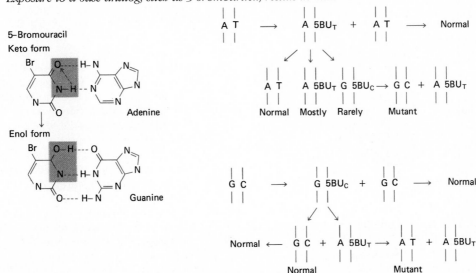

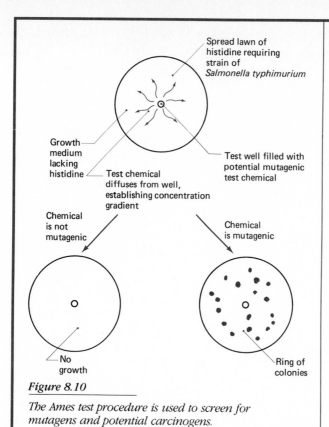

Spread lawn of
histidine requiring
strain of
Salmonella typhimurium

Growth
medium
lacking
histidine

Test chemical
diffuses from well,
establishing concentration
gradient

Test well filled with
potential mutagenic
test chemical

Chemical
is not
mutagenic

Chemical
is mutagenic

No
growth

Ring of
colonies

Figure 8.10

*The Ames test procedure is used to screen for
mutagens and potential carcinogens.*

252

Genetic varia-
tion: the introduc-
tion and mainte-
nance of
heterogeneity
within the gene
pool

Discovery process

The theoretical basis of the Ames test is
that nearly all proven carcinogens that
act directly, that is, by attacking the
DNA as opposed to indirect hormonal
action, are also mutagens. So rather
than screening chemicals directly for
carcinogenicity, Bruce Ames and his
coworkers thought it would be simpler
to test them first for mutagenicity, and
those chemicals not shown to be
mutagenic would be assumed to be
noncarcinogenic or indirect acting, and
those shown to be mutagenic would be
subjected to further testing. They
wanted a convenient bacterial genetic
system that involved a positive selection
for the desired event, so they chose a
reversion mutation, the reacquisition of
a genetically controlled trait—
specifically, the reversion of the
histidine mutations in *Salmonella
thypimurium*. They assembled a
collection of 13 mutant strains that
revert to the prototype very rarely and
that represent different types of
mutagenic events, including frame shift,
base substitution, and nonsense
mutations. Today, the use of this
bacterial assay greatly simplifies the
task of screening many potentially
dangerous chemicals, allowing us to
recognize potentially carcinogenic
compounds.

Figure 8.11

*Exposure to X rays greatly increases the rate of
mutation.*

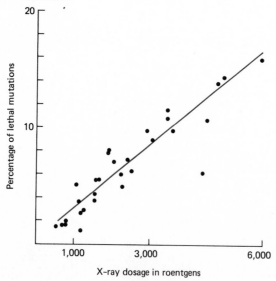

colonies on the medium. The appearance of bacterial colonies demonstrates that the chemical has
mutagenic properties. The Ames test is also useful
in determining if a chemical is a potential carcinogen (cancer-causing agent). There is a high
correlation between a chemical that is a mutagen
and whether that chemical also is a carcinogen.
Thus, determining whether a chemical has mutagenic activity is useful in screening large numbers of chemicals for potential mutagens, although the Ames test does not positively establish
whether a chemical causes cancer. In testing for
potential carcinogenicity, the chemical is incubated with a preparation of rat liver enzymes to
simulate what normally occurs in the liver, where
many chemicals are inadvertently transformed into
carcinogens in an apparent effort by the body to
detoxify the chemical. Following this activation
step, various concentrations of the transformed
chemical are incubated with the *Salmonella* auxotroph to determine whether it causes mutations
and is a potential carcinogen.

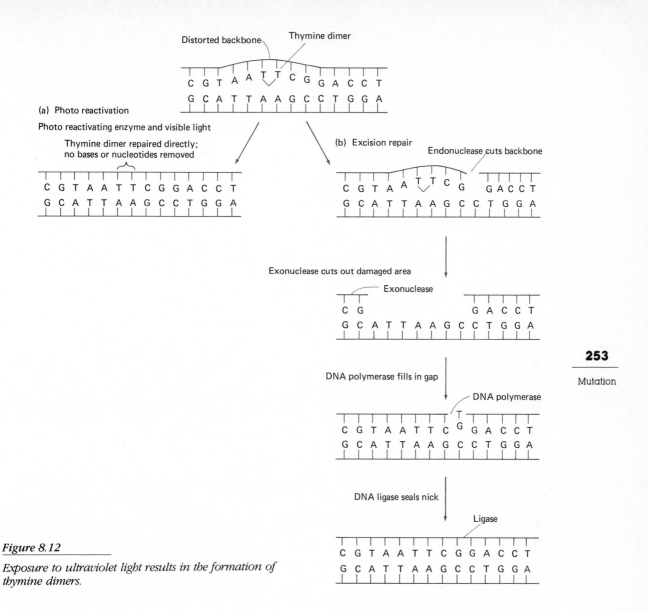

Distorted backbone — Thymine dimer

(a) Photo reactivation

Photo reactivating enzyme and visible light

Thymine dimer repaired directly;
no bases or nucleotides removed

C G T A A T T C G G A C C T
G C A T T A A G C C T G G A

(b) Excision repair

Endonuclease cuts backbone

C G T A A T T C G   G A C C T
G C A T T A A G C C T G G A

Exonuclease cuts out damaged area

Exonuclease

C G   G A C C T
G C A T T A A G C C T G G A

DNA polymerase fills in gap

DNA polymerase

C G T A A T T C G G A C C T
G C A T T A A G C C T G G A

DNA ligase seals nick

Ligase

C G T A A T T C G G A C C T
G C A T T A A G C C T G G A

Figure 8.12

Exposure to ultraviolet light results in the formation of thymine dimers.

Radiation

In addition to chemical mutagens, exposure to **radiation** can increase the rate of mutation. Exposure to high-energy radiation such as X rays can cause mutations because such high-energy ionizing radiation produces breaks in the DNA molecule. Exposure to gamma radiation, such as emitted by ^{60}Co, can be used for sterilizing objects, including plastic petri plates, because sufficient exposure to ionizing radiation results in lethal mutations and the death of all exposed microorganisms. The time and intensity of exposure determines the number of lethal mutations that occur (Figure 8.11), and thus establishes the required exposures when ionizing radiation is employed in sterilization processes.

Exposure to **ultraviolet light** also can result in base substitutions by creating covalent linkages between pyrimidine bases on the same strand of the DNA (Figure 8.12). A thymine dimer cannot act as a template for DNA polymerase and the occurrence of such dimers therefore prevents the proper functioning of polymerase enzymes. Exposure to ultraviolet light can cause lethal mutations and is sometimes used to kill microorganisms in sterilization procedures.

Microorganisms, though, have several mechanisms for repairing DNA containing **thymine di-**

mers. In one repair mechanism, exonuclease enzymes remove the dimers, polymerase enzymes fill in the gap, and ligase enzymes attach the newly inserted bases. The enzymes involved in this repair process are believed to be the same ones that carry out normal DNA replication. Thymine dimers can also be removed by a **photoreactivation** mechanism that breaks the covalent linkages between the thymine bases. The photoreactivation mechanism depends on an enzyme that functions only in the presence of light. Being able to remove thymine dimers is a particularly important adaptation in microorganisms that are normally exposed to high levels of solar radiation.

Another mechanism for repair of damaged DNA

is the **SOS system**. This mechanism of DNA repair is not restricted to damage induced by exposure to ultraviolet radiation. Rather, this system appears to be a radical repair system designed to save the cell when there is persistent DNA damage and is not restricted to the excision of thymine dimers. When the SOS system is activated cell division ceases, resulting in filamentous growth. The SOS system is activated by the proteolytic cleavage of repressors that normally keep the system shut off. The induction of this system occurs only after a delay during which incomplete replication of DNA has occurred. Once activated the SOS system fills in gaps in the DNA without copying the template so that errors are not promulgated.

Genome structure

Haploid and diploid cells

As indicated at the beginning of this chapter, prokaryotic and eukaryotic cells differ in their genome structure, and this has an important effect on how genetic variability introduced into the gene pool through mutation is maintained and transmitted. Bacterial cells have a single genome contained in the bacterial chromosome. Because there is a single set of genes, bacteria are **haploid**. In contrast, eukaryotic microorganisms generally have pairs of matching chromosomes and are **diploid**, having two copies of each gene during at least part of their life cycles. (Some eukaryotic microorganisms are haploid during much of their life cycle, whereas others are predominantly diploid.) Because diploid microorganisms contain two copies of each gene, a mutation in one of the copies may not be expressed immediately.

Different forms of the same gene that have arisen through mutational events are known as **alleles** (Figure 8.13). When both copies of the gene are identical, the microorganism is **homozygous**, and when the corresponding copies of the gene differ, the microorganism is **heterozygous**. Allelic forms of a gene often code for different amino acid sequences in polypeptides. The information encoded in one of the alleles may dominate over the information in the other allele; that is, one gene may be dominant and the other recessive. For example, a **recessive allele** may code for an inactive enzyme, whereas the **dominant allele** codes for a fully active enzyme. In such an organism, production of an enzyme could continue

even though a mutation had altered the genetic content encoded in one of the alleles. In some cases alleles may exhibit codominance, producing a hybrid state within an intermediate phenotype. For example, microorganisms with homozygous alleles may appear red or colorless, and those with heterozygous alleles may appear pink.

Even though a single diploid organism can have no more than two alleles, there may be more than two allelic forms of a given gene in the entire population. The number of alleles determines the number of potential **genotypes** for a given gene locus. If there are two alleles, there are three possible genotypes, two homozygotes and one heterozygote. If there are four alleles, however, there are ten possible genotypes, four homozygotes and six heterozygotes. The number of different genotypes that can arise from multiple alleles raises the potential for the degree of heterogeneity within the gene pool of a microbial population.

Plasmids

In addition to the genetic variability that can occur within the genome of the cell contained within the bacterial chromosomes of prokaryotes and the chromosomes of eukaryotes, some microorganisms contain plasmids (Figure 8.14). **Plasmids** are small extrachromosomal genetic elements that permit microorganisms to store additional genetic information. The acquisition or loss of plasmids alters the genomes of microorganisms because individuals of a microbial population that possess plasmids contain different genetic information than those individuals in the population which lack plasmids. Plasmids do not normally contain the genetic information for the essential metabolic activities of the microorganism but generally contain genetic information for specialized features. Plasmids are capable of self-replication and can be exchanged between bacteria, permitting the transfer of the information encoded within the plasmid DNA from one microbial population to another. The autonomous transmission of plasmids is a mechanism through which specialized information can be transferred from one bacterium to another. The fact that plasmids are self-reproducing, relatively small, and can be readily transmitted makes them useful tools in genetic engineering and the manipulation of plasmids in genetic engineering will be discussed later in this chapter.

Functions of plasmids

Although plasmids contain only a very small portion of the microbial genome, they are important. Plasmids can contain: the genetic information that determines the ability of bacteria to mate and whether a bacterial strain acts as a donor during conjugation; the information that codes for resistance to antibiotics and other chemicals, such as heavy metals, which are normally toxic to microorganisms; the genetic information for the degradation of various complex organic compounds, such as the aromatic hydrocarbons found in petroleum; and the genetic information for toxin production that renders some bacteria pathogenic for humans.

There are several different types of plasmids that serve different functions. The **F (fertility) plasmid** codes for mating behavior in *E. coli*. Strains of *E. coli* which have the F plasmid are donor strains, and those that lack the F plasmid are recipient strains. Bacteria that have the F plasmid are able to form pili of the F type that are involved in establishing mating pairs. The F plasmid may exist as an independent circular molecule of double helical DNA or may become incorporated into the bacterial chromosome. (The F plasmid will be discussed further in connection with the process of bacterial mating later in this chapter.)

Colicinogenic plasmids carry the genes for a protein toxic to enteric bacteria, such as *E. coli*; they also may carry the information necessary for bacterial conjugation. The acquisition of colicinogenic plasmids enables bacterial strains to enter into antagonistic relationships with other bacterial strains. For example, when strains of *E. coli*

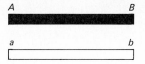

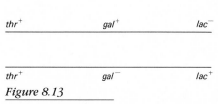

thr$^+$ *gal*$^+$ *lac*$^-$

thr$^+$ *gal*$^-$ *lac*$^+$

Figure 8.13

Alleles are different forms of the same genes.

Figure 8.14

This electron micrograph shows two different plasmid DNA molecules. The larger loop is plasmid pBF4 and was isolated from Bacteroides fragilis; *it encodes resistance to the antibiotic clindamycin. The smaller DNA loop is plasmid pSC101 from* Escherichia coli; *it encodes resistance to tetracycline (13,050×). (From BPS—Rod Welch, School of Medicine, Stanford University.)*

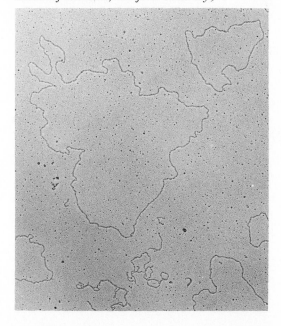

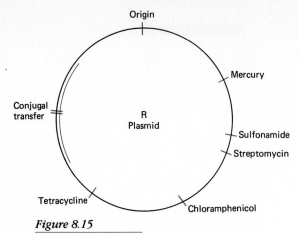

Figure 8.15

An R plasmid encodes the information for antibiotic resistance.

containing colicinogenic plasmids are mixed with other strains of *E. coli*, only one strain can survive.

The **R (resistance) plasmids** carry genes that code for antibiotic resistance (Figure 8.15). Some R factor plasmids also carry genes for conjugation. The enzymes coded for by the genes of R plasmids are able to degrade inactive antibiotics, thus conferring resistance on bacterial strains that possess them. R factors can be passed from one bacterial species to another, such as from *E. coli* to pathogenic strains of *Shigella* or *Salmonella*. Antibiotic-resistant strains of bacteria have become a serious health problem because R plasmids can occur in pathogenic bacteria, and the treatment of bacterial diseases of humans has been complicated by the occurrence of these pathogens that are resistant to multiple antibiotics.

Recombination

256

Genetic varia-
tion: the introduc-
tion and mainte-
nance of
heterogeneity
within the gene
pool

One of the ways that genetic variability is maintained within a population is through **recombination**, a process involving the exchange of genetic information between different DNA molecules that results in a reshuffling of genes. Recombination provides a mechanism for redistributing the informational changes that occur in the DNA as a result of mutation and can produce numerous new combinations of genetic information.

Types of recombinational processes

There are two different types of recombinational exchange processes. The classical type of exchange occurs between homologous regions of DNA molecules, as is seen in the crossing over of chromosomes where pairs of chromosomes containing the same gene loci pair and exchange allelic portions of the same chromosomes (Figure 8.16). (The term homologous indicates that the exchange is between alleles of the same gene and

does not imply that the DNA segments that are exchanged have the same nucleotide sequences. **Homologous recombination** can also be considered a general or reciprocal exchange of DNA.) Genetic recombination can occur as well between nonhomologous segments of DNA molecules. (**Nonhomologous recombination** can also be termed a nonreciprocal or site-specific exchange process. Nonhomologous recombination does not mean that there is no nucleotide homology in the segments of DNA that are exchanged but does imply that the extent of such homologous regions is limited. The difference between homologous and nonhomologous recombination is a matter a degree, and the distinction is not always clear.) Nonhomologous recombinations permit the joining together of DNA molecules from different sources. For example, viral DNA may become incorporated into a bacterial chromosome, plasmids may enter into bacterial chromosomes, and the locations of segments of DNA may be transposed in chromosomes. Both general and site-specific recombination alter

Figure 8.16

Chromosomal crossing over results in the recombination of genes.

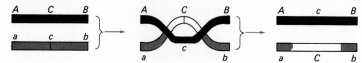

the arrangement of the genome and introduce variation into the gene pool of the population.

In **reciprocal recombination**, there is relatively good base pairing of corresponding regions of the DNA and the aligned chromosomes may establish duplexes between homologous regions of DNA. In eukaryotic microorganisms this often occurs during the process of meiosis, which results in the reduction of the number of chromosomes and the conversion of a diploid cell into a haploid cell. A similar homologous alignment of DNA molecules can occur when a bacterial chromosome or portion thereof is transferred from a donor to a recipient bacterium.

Events that occur during recombination

Several models have been developed to describe the events that occur during recombination. The experimental evidence indicates that single-stranded regions of DNA are involved in genetic recombination and that enzymes are intimately involved in the process. In one model, recombination is initiated by an endonuclease enzyme that nicks one of the strands of DNA; the free 3'-OH end of the DNA molecule that is formed by the action of the endonuclease enzyme acts as a primer for DNA synthesis involving a DNA polymerase enzyme (the synthesis of DNA produces a single strand of DNA as occurs during rolling circle replication); the newly synthesized region of the DNA pairs with the corresponding region of the homologous chromosome, establishing a heteroduplex; an endonuclease enzyme cleaves out the unpaired section of the DNA macromolecule; and finally, ligase enzymes join the free ends of the DNA strands (Figure 8.17). The formation of the heteroduplex is catalyzed by enzymes coded for by **rec (recombination) genes**; in organisms lacking rec genes, recombination does not occur. The result of this enzymatic action is the formation of a bridge between the two homologous DNA strands and the joining of chromosomes at a homologous region establishes a **chi form** (Figure 8.18). The chromosomes then rotate so that the two strands no longer cross each other but are still held together by covalent linkages. An **endonuclease** enzyme then cleaves the DNA strands, breaking the **heteroduplex** and establishing two independent chromosomes. In some cases the cleavage by the endonuclease enzymes only results in the exchange of DNA in the short region where the heteroduplex had formed, and this type of exchange is not considered to

establish recombinant DNA molecules. In other cases the action of the endonuclease enzyme results in the formation of chromosomes that have exchanged large portions of homologous DNA regions.

In *E. coli* it has been found that general recombination depends on the enzymes coded for by three recombination genes—designated *recA, recB,* and *recC.* The enzymes coded for by *recB* and *recC* are involved in unwinding double helical DNA and breaking the chains into small fragments. It appears that these enzymes generate single-stranded DNA that can then interact with a double-stranded DNA macromolecule. The *recA* protein catalyzes an ATP-driven assimilation reaction in which the single-stranded DNA invades a duplex (double-stranded) DNA macromolecule. In this process the single-stranded DNA locates a region of homology within the duplex, pairs with the complementary strand, and displaces the other strand.

Regardless of the mechanism, recombination results in an exchange of allelic forms of genes that can produce new combinations of alleles. Recombination, thus, provides a mechanism for generating diversity within the gene pool of a microbial population. The generation of heterogeneity, however, is restricted by the constraint that the exchanges must be between alleles of the same gene or between genes that have large sequences of corresponding nucleotides, and as a result reciprocal exchanges occur within species, but this process does not generally permit genetic exchange between different species.

In contrast to the general recombination process involving homologous segments of the DNA, **site-specific recombination** can occur between very different DNA segments and provides a mechanism for genetic exchange between species. **Nonreciprocal recombination** permits: the incorporation of viral DNA into a bacterial or eukaryotic chromosome; the insertion of plasmids, such as the F plasmid of *E. coli*, into the bacterial chromosome; and the translocation of antibiotic resistance genes from one plasmid location to another. There probably are several different mechanisms that can bring about nonreciprocal recombination. A number of enzymes appear to be involved in the site-specific transposition of genetic information that are different from those involved in reciprocal recombination. Microorganisms can be deficient in the enzymes coded for by the general reciprocal recombination genes and still carry out nonreciprocal transpositions. The ability to combine DNA from genetically dissimilar sources is really the basis for genetic en-

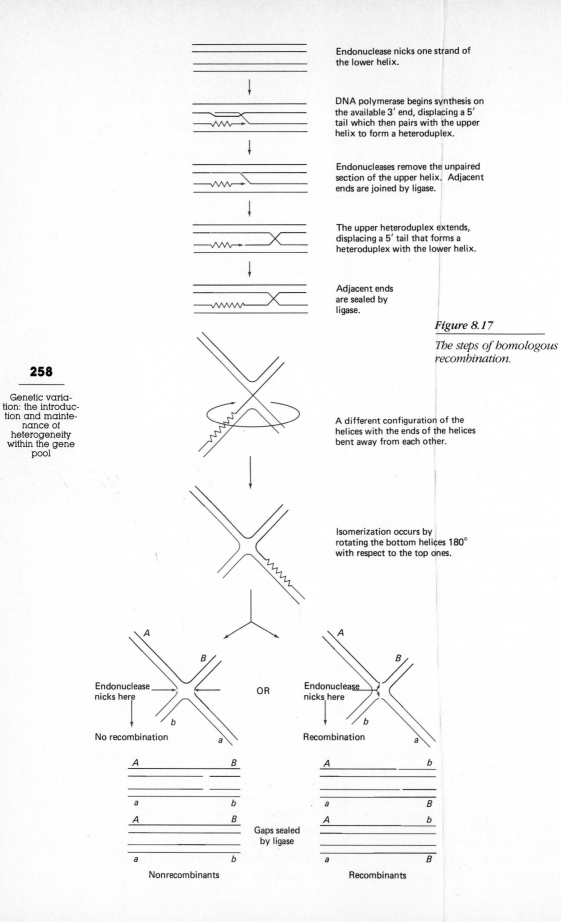

Endonuclease nicks one strand of the lower helix.

DNA polymerase begins synthesis on the available 3′ end, displacing a 5′ tail which then pairs with the upper helix to form a heteroduplex.

Endonucleases remove the unpaired section of the upper helix. Adjacent ends are joined by ligase.

The upper heteroduplex extends, displacing a 5′ tail that forms a heteroduplex with the lower helix.

Adjacent ends are sealed by ligase.

Figure 8.17

The steps of homologous recombination.

Genetic varia-
tion: the introduc-
tion and mainte-
nance of
heterogeneity
within the gene
pool

A different configuration of the helices with the ends of the helices bent away from each other.

Isomerization occurs by rotating the bottom helices 180° with respect to the top ones.

A

B

Endonuclease nicks here

OR

b

No recombination

a

A

B

Endonuclease nicks here

b

Recombination

a

A	*B*
a	*b*
A	*B*
a	*b*

Gaps sealed by ligase

A	*b*
a	*B*
A	*b*
a	*B*

Nonrecombinants

Recombinants

Figure 8.18

The chi form of DNA occurs during homologous recombination. (A) Diagrammatic representation. (B) Electron micrograph. (Courtesy of Huntington Potter and David Dressler, Harvard Medical School.)

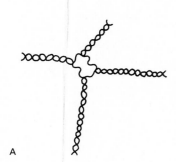

A

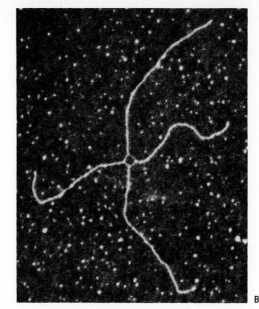

B

gineering. The development of recombinant DNA technology has gone hand in hand with the development of the understanding of the basic mechanisms involved in nonreciprocal DNA recombination.

Transposable genetic elements

The ability of specific segments of DNA to undergo nonreciprocal recombination, moving from one location to another, appears to depend in some cases on the nucleotide sequences at the ends of the transposable elements. The ends of the **transposable genetic elements** often contain inverted repetitive sequences of nucleotide bases that permit folding of the DNA stabilized by hydrogen bonding between complementary bases within the DNA macromolecule (Figure 8.19). The occurrence of these inverted sequences appears to be important in establishing the ability of genetic elements to exhibit nonreciprocal recombination. In nonreciprocal exchanges of DNA, the lack of large regions of DNA homology suggests that DNA–protein interactions, rather than base pairing, play a particularly important role in the insertion process.

There are several different types of transposable genetic elements. One type, an **insertion sequence (IS)**, can move around bacterial chromosomes, occurring at different locations on the chromosome. Insertion elements are small transposable genetic elements having about 1000 nucleotide bases. An IS is not homologous with the regions of the plasmids or the chromosomes into

which it is inserted. The IS has an identical nucleotide sequence that is repeated at each end of the DNA molecule. The nucleotide bases in the insertion sequence regions do not appear to code for structural proteins but may have a regulatory function and may be involved in specifying the sites at which site-specific recombination occurs. Insertion sequences, for example, can alter a promoter site and thereby affect the expression of nearby genes.

In contrast to insertion sequence elements, **transposons** are larger transposable genetic elements that contain genetic information for the production of structural proteins. Many of the transposons that have been studied code for antibiotic resistance. (The distinction between transposons and insertion elements is not always clear, and some insertion elements may code for pro-

Figure 8.19

Transposable genetic elements contain inverted repeated nucleotide sequences.

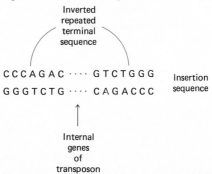

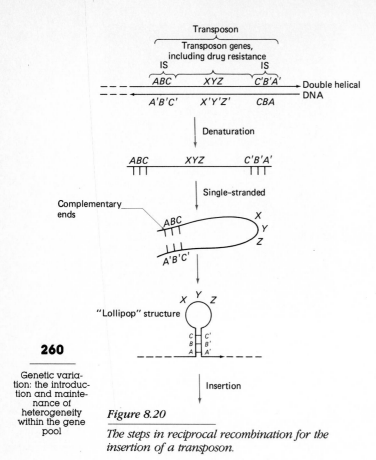

"Lollipop" structure

Insertion

Figure 8.20

The steps in reciprocal recombination for the insertion of a transposon.

Genetic varia-
tion: the introduc-
tion and mainte-
nance of
heterogeneity
within the gene
pool

teins whose functions have yet to be recognized.) Transposons, like insertion sequences, have identical nucleotide sequences that are repeated at each end of the DNA molecule, and these terminal nucleotide sequences appear to establish the basis for their enzymatic insertion (Figure 8.20). In some cases transposons may be constructed by the attachment of insertion sequences to structural genes. Recombination can occur between an insertion sequence on a bacterial chromosome and an insertion sequence on a transposon. For example, the F plasmid may be incorporated into the bacterial chromosome by recombination between the insertion sequence on the F plasmid and a homologous insertion sequence on the bacterial chromosome.

Like transposons and insertion sequences, a viral genome may be incorporated into a bacterial chromosome in the process of **lysogeny**. (The lytic and lysogenic life cycles of bacteriophage will be discussed in Chapter 9.) The genes of lysogenic phage can be expressed by the bacterial host, with the bacterium producing proteins that are coded for by the viral genes. In one bacterium, *Corynebacterium diphtheriae*, the presence of a lysogenic phage renders the bacterium pathogenic, and strains of *C. diphtheriae* that contain the viral genome produce proteins that are toxins and cause the disease symptoms of diphtheria, whereas strains of *C. diphtheriae* that lack the incorporated viral genome are harmless nonpathogens. Similarly, plasmids may be inserted into the bacterial chromosome. Plasmids or other transposable elements that become incorporated into a bacterial chromosome can later be excised. For example, bacteriophage that establish a lysogenic state may later reestablish an independent viral element that directs the normal lytic life cycle of the virus.

DNA transfer in prokaryotes

Genetic exchange and recombination in bacteria can occur by three different mechanisms: transformation, transduction, and conjugation (Figure 8.21). These mechanisms differ in the way the DNA is transferred between a donor and a recipient cell.

Transformation

In **transformation** a free DNA molecule is transferred from a donor to a recipient bacterium; that is, the donor bacterium leaks its DNA, generally

as a result of lysis of the bacterium, and the recipient bacterium is able to take up the DNA. In order to take up DNA, a recipient cell must be competent, and relatively few bacterial genera have been demonstrated to be capable of taking up naked DNA. When the DNA is taken up by a competent bacterium, one strand of the double helical DNA is enzymatically degraded (Figure 8.22). (In some species of *Bacillus* and *Streptococcus* it appears that only single-stranded DNA enters the recipient cell.) The intact strand of DNA forms a heteroduplex with the bacterial chromosome of the recipient bacterium. A nuclease degrades the

Figure 8.21

Genetic transfer in bacteria occurs by three distinct mechanisms: transformation, involving transfer of naked DNA; transduction, involving transfer of DNA by a phage; and conjugation, involving direct contact of bacteria during transfer of DNA.

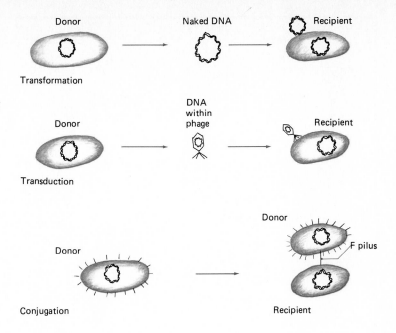

corresponding region of DNA in one of the strands of the recipient cell, and ligase enzymes join the donor DNA with the DNA of the recipient bacterial chromosome. This is an example of a reciprocal recombinational event, and if the allelic forms of the donor and recipient genes are not identical, the progeny of the recipient cell may have a composite (hybrid) genome different from either the donor or recipient strain.

A classic example of transformation involves the bacterium *Streptococcus pneumoniae* (formerly referred to as *Diplococcus pneumoniae*) (Figure 8.23). One strain of this bacterium produces a capsule and is a virulent pathogen, whereas another strain of the same bacterial species lacks the genetic information for capsule production and is avirulent. When dead cells of the virulent strain are mixed with avirulent live bacteria, transformation can occur, producing a mixture of avirulent and virulent bacteria; that is, the DNA containing the genes for capsule production can leak out of the dead bacteria and be taken up by the living bacteria that normally lack the genetic in-

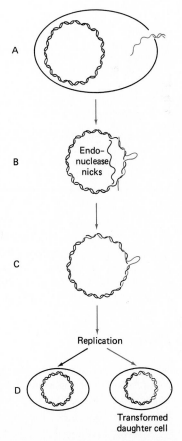

Figure 8.22

Diagram showing the steps of bacterial transformation. (A) A piece of double-stranded DNA from a wild type strain enters the bacterial cell. Exonuclease converts the transforming DNA to a single strand as it passes into the cell. (B) Heteroduplex forms as transforming DNA pairs with complementary host DNA. (C) Ligase completes the integration of transforming DNA. (D) One daughter cell is mutant like the recipient cell and one is transformed.

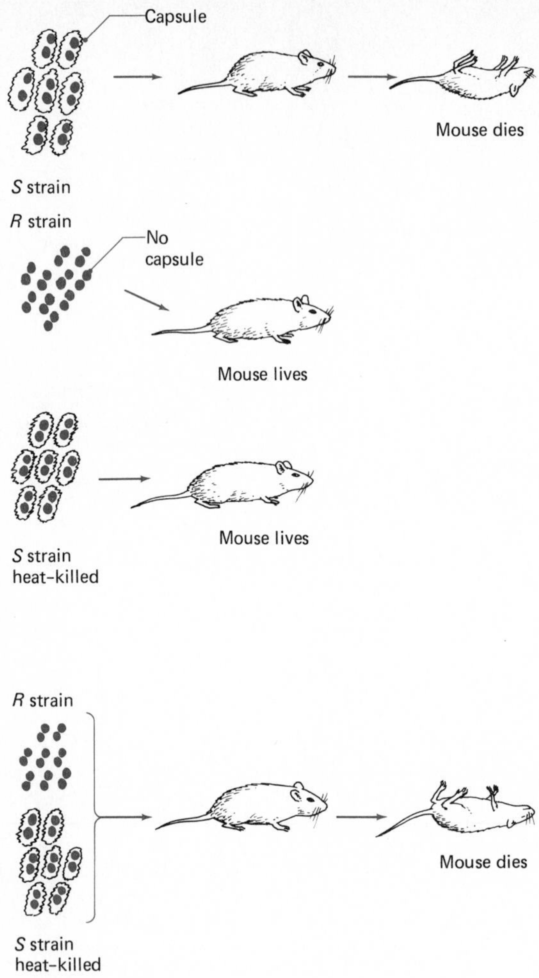

262

Genetic varia-
tion: the introduc-
tion and mainte-
nance of
heterogeneity
within the gene
pool

Figure 8.23

The transformation of Streptococcus pneumoniae *shows how the properties of a bacterial strain can be altered by this recombination process.*

Discovery process

The discovery of the transformation process provided the first direct proof that DNA is the genetic material. In the late 1920s a British scientist, Fred Griffith, was working with *Streptococcus pneumoniae*, a causative organism of pneumonia that produces a polysaccharide capsule. Griffith also isolated avirulent mutants, called *R* strains because they appeared as rough colonies on agar; the *R* strains lack a capsule, and Griffith demonstrated that they are unable to cause disease. Mice infected with even minimal doses of *S* (nonmutant, capsulated, smooth colony-producing) strains died a few days after exposure to the bacteria, whereas injection of even massive doses of *R* strains did not cause death. Griffith injected mixtures of heat-killed *S* strains and live *R* strains into mice; surprisingly, the mice died and Griffith was able to isolate live *S* strains from their corpses. The *R* strains had been transformed into a new, encapsulated, pathogenic strain by what appeared to be a genetic process, later called transformation. Oswald Avery, C. M. McCarty, and M. MacLeod provided the molecular explanation for this event in 1941 by separating the classes of molecules in the debris of the dead *S* cells and testing each one for its ability to cause transformation. First, they showed that it was not the polysaccharides in the capsules themselves that transform the *R* cells—this is merely the phenotypic expression of virulence. They found that only one class of molecules, deoxyribonucleic acid, induces transformation of DNA. They deduced that DNA is the agent that determines the polysaccharide character, and hence, the pathogenic characteristic, and that providing *R* cells with *S* DNA was the same as providing them with *S* genes. The unavoidable conclusion of this classic work is that the genetic information of the cell is contained within its DNA.

formation for capsule production. Recombination can occur and the progeny of the transformed bacteria become capable of producing capsules, and in this manner nonpathogenic strains can be transformed into deadly pathogens.

Transduction

Whereas in transformation free (naked) DNA is transferred, in **transduction** the DNA is transferred from a donor to a recipient cell by a viral carrier. For transduction a virus must acquire a portion of the genome of the host cell in which it reproduces. A bacteriophage, for example, can acquire bacterial DNA when it infects a bacterial host cell and can then transfer this acquired bacterial DNA to another bacterial cell. There are two different types of transduction, generalized and specialized, which as the names imply differ in whether they can bring about the general transfer of genes or only the transfer of specific genes.

Generalized transduction

Generalized transduction can result in the exchange of any of the homologous alleles. In generalized transduction, pieces of bacterial DNA are accidentally acquired by developing phage during their normal lytic growth cycle (Figure 8.24). (The normal lytic growth cycle of phage, which will be discussed in detail in Chapter 9, involves the invasion of a host cell, the reproduction of the virus within the host, and the lysis or bursting of the host cell to release the newly formed phage.) If a phage carries bacterial instead of viral DNA, it cannot reestablish a lytic growth cycle and cannot cause lysis in a recipient bacterium. Such a phage, however, can attach to and inject DNA into a recipient bacterium, permitting it to carry bacterial DNA from a donor cell and inject it into a recipient bacterial cell. Once inside the recipient cell, the DNA may be degraded by nuclease enzymes, in which case genetic exchange does not occur. The injected DNA, however, may undergo homologous recombination. If recombination occurs, the transduced recipients may possess new combinations of genes.

Specialized transduction

In contrast to generalized transduction, **specialized transduction** depends on the transmission of bacterial DNA from a donor to a recipient cell by a phage that does not kill the host cell. Such phage may incorporate their DNA into the bacterial chromosome rather than coding for the reproduction of the phage and the lysis of the host cell. The incorporation of the phage DNA into the bacterial chromosome is an example of nonhomologous recombination and is regulated by an operon type of control mechanism. Lambda (λ) and mu (μ) bacteriophage are capable of establishing a lysogenic growth cycle. Because they do not cause the lysis and death of the host cell, such phage are sometimes referred to as **temperate phage**. (These phage are capable of establishing both lytic and lysogenic growth cycles, and of the cells infected by a temperate phage only a minority become lysogens and the rest are lysed.) When the phage DNA is excised from the bacterial chromosome, reestablishing a lytic growth cycle, it occasionally carries with it adjacent bacterial genes and leaves behind some of the viral genes, making the phage defective in some viral genes.

The formation of such a defective phage also establishes a viral carrier of bacterial DNA. Because there is a specific site for incorporation of the phage DNA into the bacterial chromosome, only the genes that are adjacent to the site of insertion of the viral genome may be transferred by specialized transduction (Figure 8.25). As an

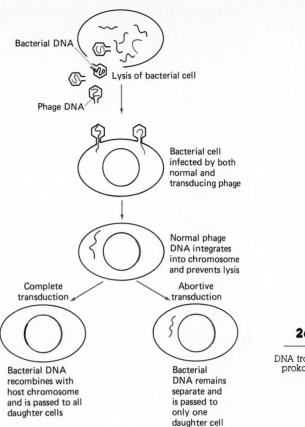

Bacterial DNA

Lysis of bacterial cell

Phage DNA

Bacterial cell infected by both normal and transducing phage

Normal phage DNA integrates into chromosome and prevents lysis

Complete transduction

Abortive transduction

Bacterial DNA recombines with host chromosome and is passed to all daughter cells

Bacterial DNA remains separate and is passed to only one daughter cell

Figure 8.24

Diagram showing how phage can acquire bacterial DNA for generalized transduction. The phage acts as a carrier to transport DNA from donor to recipient bacterium.

example, the λ phage may acquire from its host *E. coli* either the genes for galactose utilization or the genes for biotin synthesis because these are the two genes that flank the site where this phage inserts its DNA into the bacterial chromosome. Phage containing *E. coli* DNA may infect new host cells but are unable to insert DNA into the bacterial chromosome in the normal manner because they lack the necessary insertion sequence. However, successful insertion can occur in the presence of a second normal λ phage DNA molecule (Figure 8.26). This type of recombination produces a bacterial chromosome with a section containing both a normal and a defective phage DNA molecule and two loci for either the galactose or biotin gene. In this manner the recipient bacterium becomes diploid for either galactose or biotin. If these genes are of different allelic forms, the organism produced by specialized transduction is heterozygous.

A

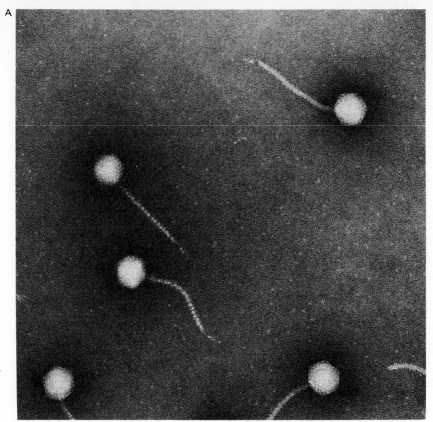

264

Genetic varia-
tion: the introduc-
tion and mainte-
nance of
heterogeneity
within the gene
pool

Integration

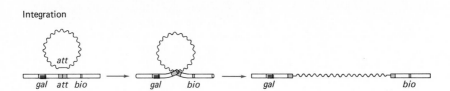

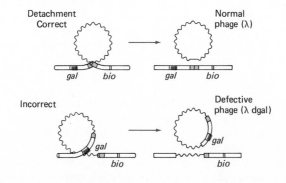

B

Figure 8.25

The bacteriophage lambda is incorporated into the bacterial genome at a specific site adjacent to the genes for galactose and biotin. (A) Electron micrograph of lambda (250,000×). (Courtesy R. C. Williams, University of California, Berkeley.) (B) Map showing site of insertion of lambda into the chromosome of E. coli.

Figure 8.26

Specialized transduction by lambda phage results in the transfer of a limited amount of genetic information.

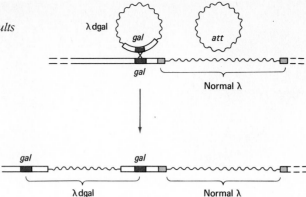

Conjugation

Unlike the already discussed genetic exchange mechanisms, **conjugation (mating)** requires the establishment of physical contact between the donor and recipient bacteria. To demonstrate that contact is needed for mating, two different auxotrophic strains can be placed in a U-shaped tube separated only by a membrane barrier (Figure 8.27). The barrier blocks physical contact, and thus conjugation; free DNA and phages, however, can pass through the membrane, and thus, the other recombinational processes of transduction and transformation can still occur. The physical contact between mating bacteria is established through the F pilus (Figure 8.28). As mentioned earlier, the F plasmid confers the ability to produce an F pilus needed to form mating pairs. Bacterial strains that produce F pili act as donors during conjugation, whereas those lacking the F plasmid are recipient strains. Recipient bacterial strains are designated F^- and donor strains are either designated F^+ if the F plasmid is independent or **Hfr (high-frequency recombination strain)** if the F plasmid DNA is incorporated into the bacterial chromosome. (The F plasmid may incorporate at different sites within the bacterial chromosome.) During bacterial mating, DNA from the donor strain is replicated by a rolling circle mechanism and is transferred into the recipient cell. Normally, only a portion of the donor bacterial chromosome is transferred during bacterial mating. The precise portion of the DNA that is transferred depends on the time of mating, that is, how long the F pilus maintains contact between the mating cells. When an F^+ cell is mated with an F^- cell, the F plasmid DNA is usually transferred from the donor to the recipient (Figure 8.29). The F plasmid does not normally recombine with the bacterial chromosome of the recipient bacterium, but rather

Figure 8.27

When direct contact between bacterial strains in a U-shaped tube is prevented by a membrane separating the two arms, genetic exchange by conjugation cannot occur.

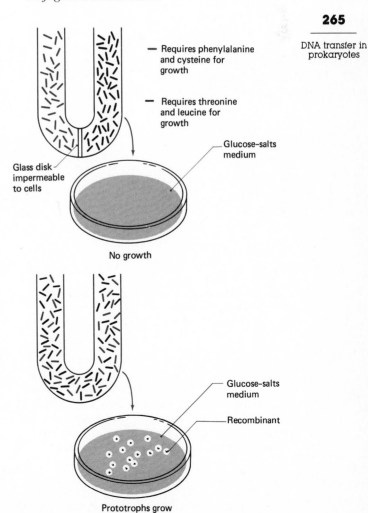

— Requires phenylalanine and cysteine for growth

— Requires threonine and leucine for growth

Glucose–salts medium

Glass disk impermeable to cells

No growth

Glucose–salts medium

Recombinant

Prototrophs grow

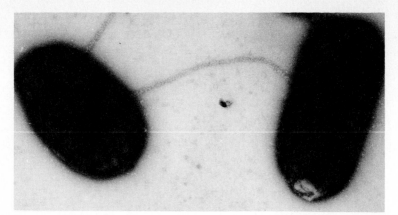

Figure 8.28

Electron micrograph showing the attachment of mating E. coli *cells by an F pilus (21,250×). (From BPS—D. P. Allison, Oak Ridge National Laboratory.)*

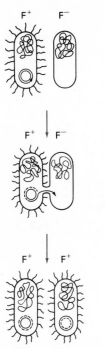

266

Genetic varia-
tion: the introduc-
tion and mainte-
nance of
heterogeneity
within the gene
pool

Figure 8.29

Diagram showing conjugation of F^+ and F^- strains.

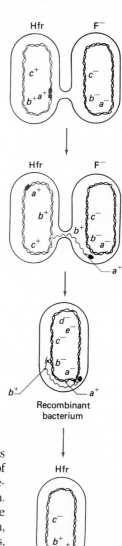

Figure 8.30

Conjugation of Hfr × F^- results in a high frequency of recombination.

Recombinant
bacterium

Hfr

Donor DNA
recombines with
recipient DNA.

the single-stranded linear DNA molecule that is transferred acts as a template for the synthesis of a complementary strand of DNA, and the double-stranded DNA then reestablishes a circular form. The independent circular F plasmid confers the genetic information for acting as a donor strain, and the offspring of the recipient of such a cross, therefore, are mostly donor strains. When an Hfr strain is mated with an F^- strain, the bacterial chromosome with the integrated F plasmid begins rolling circle DNA replication in response to attachment of the F pilus (Figure 8.30). A single strand of DNA is transferred from the donor Hfr strand to the F⁻ bacterium, and the DNA that is transferred may undergo homologous recombination with the recipient DNA. Because the F plasmid is often not near the beginning of the DNA that is replicated, only part of the F plasmid DNA normally is transferred in this type of mating cross,

and as a result the recipient cell normally remains F⁻. However, a relatively large portion of the bacterial chromosome is transferred from the donor to the recipient, and thus, there is a relatively high frequency of recombination of genes of the bacterial chromosome when Hfr strains are mated with F⁻ strains.

Genetic exchange in eukaryotes

In eukaryotic microorganisms genetic exchange normally occurs as a result of a sexual reproductive phase of the life cycle. As indicated earlier, the vegetative cells of many eukaryotic organisms are diploid, but in order to exchange genetic information these organisms normally form specialized reproduction gametes or haploid spores.

The conversion of a diploid to a haploid state occurs in the process of meiosis, which is also known as reductive division. **Meiosis** begins after DNA replication, so that the starting cell is actually tetraploid (Figure 8.31). During this tetraploid state the chromosomes are aligned side by side, and recombination can occur between homologous

Figure 8.31

Diagram showing the states of meiosis. Meiotic division results in a reduction of the chromosome number.

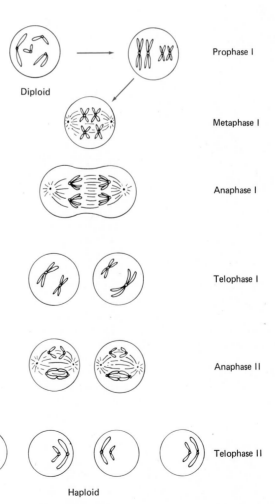

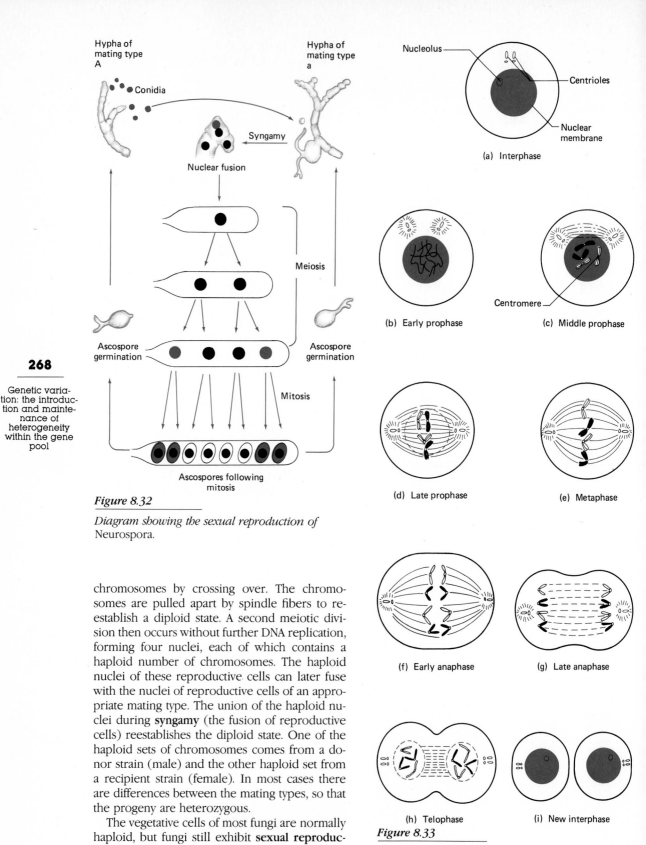

Hypha of mating type A

Conidia

Syngamy

Nuclear fusion

Hypha of mating type a

Meiosis

Ascospore germination

Ascospore germination

Mitosis

Ascospores following mitosis

Figure 8.32

Diagram showing the sexual reproduction of Neurospora.

(a) Interphase

Nucleolus

Centrioles

Nuclear membrane

(b) Early prophase

(c) Middle prophase

Centromere

(d) Late prophase

(e) Metaphase

(f) Early anaphase

(g) Late anaphase

(h) Telophase

(i) New interphase

Figure 8.33

Diagram showing the states of mitosis.

chromosomes by crossing over. The chromosomes are pulled apart by spindle fibers to reestablish a diploid state. A second meiotic division then occurs without further DNA replication, forming four nuclei, each of which contains a haploid number of chromosomes. The haploid nuclei of these reproductive cells can later fuse with the nuclei of reproductive cells of an appropriate mating type. The union of the haploid nuclei during **syngamy** (the fusion of reproductive cells) reestablishes the diploid state. One of the haploid sets of chromosomes comes from a donor strain (male) and the other haploid set from a recipient strain (female). In most cases there are differences between the mating types, so that the progeny are heterozygous.

The vegetative cells of most fungi are normally haploid, but fungi still exhibit **sexual reproduction** (Figure 8.32). In some fungi the vegetative cells can fuse to establish a **dikaryon**, a cell with two genetically different nuclei. In *Neurodispora*, for example, the heterokaryon formed from the fusion of cells arise from the fungal hyphae (filaments of cells constituting the vegetative form of the fungus). Shortly after the formation of the dikaryon, nuclear fusion occurs and establishes a diploid state. The diploid nucleus then undergoes meiosis, establishing four independent haploid nuclei. The haploid nuclei may undergo further division by mitosis to form eight haploid nuclei (Figure 8.33). Mitotic division does not alter the number of chromosomes, and thus the nuclei remain haploid. Maturation occurs to form eight ascospores (sexual spores that develop within a specialized structure, the ascus), and the ascospores can later germinate to form haploid vegetative cells. This type of reproduction is characteristic of many fungi. In some cases, though, meiosis is not followed by a mitotic division, and only four haploid spores are formed, and in others additional mitotic divisions occur, forming 16 or 32 haploid spores.

Sexual reproduction in eukaryotic microorganisms thus provides a mechanism both for the recombination of allelic genes on chromosomes and for the reassortment of chromosomes. This mechanism of genetic exchange ensures the maintenance of diversity in the gene pool of a population.

Genetic mapping

Mating of different strains of microorganisms that have different allelic forms of multiple genes can be used for **genetic mapping** of viruses, bacteria, and eukaryotic microorganisms. The frequency of recombination that results from mating is used to map the order and thus determine the relative locations (loci) of genes. Genes that are located near each other on a chromosome should segregate together with a high frequency, and the rates of recombination between such closely linked genes should be low. (If genes are close together, there is less chance that a break will occur between them than if they are far apart.) The circular nature of the bacterial chromosome and the linear nature of eukaryotic chromosomes were established by examining the frequencies of recombination. Similarly, the mobile nature of transposons has been established by careful genetic mapping.

In the case of *E. coli*, which has been well studied, mating between Hfr and F⁻ strains has been used to map large sections of the bacterial chromosome (Figure 8.34) . By vibrating a culture of synchronously mating bacteria, one can interrupt mating by breaking the F pilus, which stops further transfer of DNA. The order of genes on the bacterial chromosome can be determined by examining the times at which recombinants for given genes are found (Figure 8.35). In eukaryotic microorganisms mating is carried out between strains with differing allelic forms, and the frequency at which genes recombine together is used to determine whether the genes are closely linked (Figure 8.36). In mating experiments aimed at mapping the order of genes, the recovery of recombinants of marker genes are normally used as reference points for establishing the fine structure of the genome (Figure 8.37). If a gene of unknown location shows a high frequency of recombination along with the marker gene, it is likely that the marker and unknown genes are closely associated in the chromosome; if the genes are far apart, it is unlikely that recombinants of both the marker gene and the gene of unknown location will occur in the progeny.

Figure 8.34

Circular gene map of E. coli *showing only a few of the genes that have been mapped. (Based on B. J. Bachman and K. B. Low, 1980,* Microbiological Reviews *44:1–56.)*

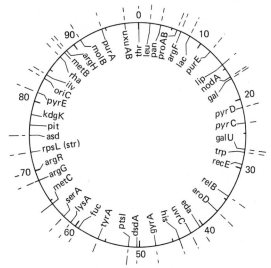

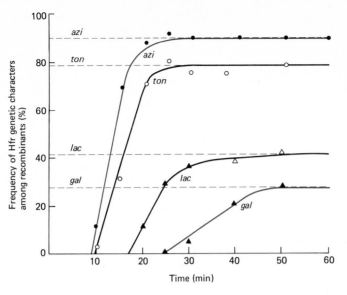

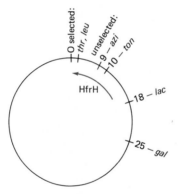

Genetic variation: the introduction and maintenance of heterogeneity within the gene pool

Transduction can similarly be used to establish the fine structure of the bacterial genome. In generalized transduction it is unlikely that genes will undergo cotransduction unless they are closely associated in the bacterial chromosome because the transducing phage carry a very small piece of the bacterial chromosome. Conversely, if two genes are closely linked, it is more likely that they will be cotransduced and recombine together than if they are not adjacently located on the bacterial chromosome. Cotransformation can similarly be used to map the microbial genome. Thus, using a variety of processes to achieve genetic exchange, the rates of recombination can be measured and the relative locations of the genes deduced, thus producing a detailed genetic map (Figure 8.38).

In addition to the detailed genetic maps of several genomes of prokaryotic and eukaryotic cells, the genetic maps of several viruses have been determined. Like the study of genetics of cellular organisms, the study of viral genetics is based on the ability to recognize mutants and to determine rates of recombination. Four primary types of mutations have been employed in viral genetic studies (Table 8.1). When two mutant viral strains are mixed in the same host cell, recombination can occur, and the frequency of recombinant types can be used to map the viral genome. Genetic recombination in viruses was discovered by Delbruck and Hershey in 1946, the same time that Lederberg demonstrated mating in bacteria. Delbruck and Hershey performed genetic crosses by infecting bacteria with a mixture of a host-range (*h*) and a rapid-lysis (*r*) mutant of a T-even phage (Figure 8.39). Four distinguishable viral types were found after incubation: the two parental types and two recombinant types with a marker from each parent (Figure 8.40). Determining viral genetic maps by such experiments is complicated because DNA molecules continue to recombine throughout the period of viral multiplication within a host cell. Because bacteriophage undergo several rounds of mating during an infection cycle,

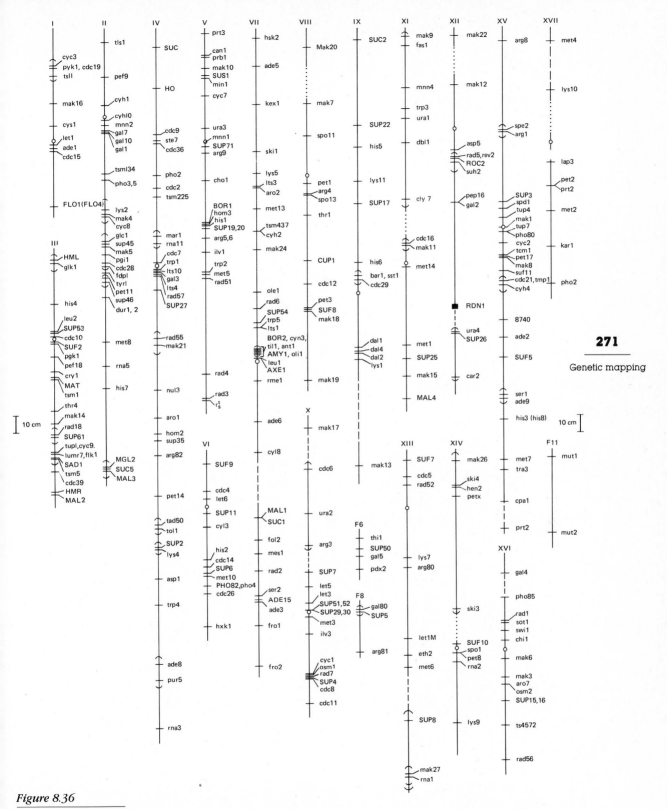

Figure 8.36

The genetic map for the 17 chromosomes of Saccharomyces cerevisiae. *(Based on R. T. Mortimer and D. Shild, 1980,* Microbiological Reviews *44:519–571.)*

Cross	Percentage of each genotype $r^- h^+$	$r^+ h^-$	$r^+ h^+$	$r^- h^-$
$r_a^- h^+ \times r_a^+ h^-$	34.0	42.0	12.0	12.0
$r_b^- h^+ \times r_b^+ h^-$	32.0	56.0	5.9	6.4
$r_c^- h^+ \times r_c^+ h^-$	39.0	59.0	0.7	0.9

Figure 8.37

The frequency of corecombination with marker genes permits determination of the relative positions of genes along the chromosome and the establishment of the genetic map.

Figure 8.38

Circular gene map of B. subtilis *showing complexity of genetic map. The inner ring shows the landmark loci of marker genes. (Based on D. Henner and J. A. Hoch, 1980,* Microbiological Reviews *44:57–82.)*

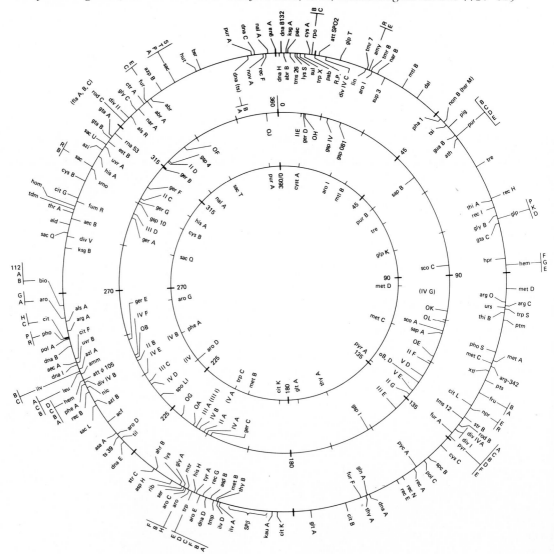

table 8.1

Types of mutations used in viral genetic studies

Class of mutation	Description
Host-range	A mutation that alters the host that a virus can infect, e.g., a wild-type bacteriophage (h^+) may be able to reproduce within *E. coli* strain B, while the mutant virus (h^-) may be able to reproduce only within *E. coli* strain B/2.
Plaque-type	When bacteriophage replicate within bacterial cells growing on a solid surface they produce zones of clearing, known as plaques, where bacterial cells are lysed when viruses are released from the host cell. The characteristics of the plaque, such as size and whether it is clear or turbid, are under the control of the viral genome. For example, wild-type r^+ bacteriophage exhibit rapid lysis and thus produce clear plaques, whereas r^- mutants produce turbid plaques.
Conditional-lethal Temperature-sensitive	Temperature-sensitive mutants are conditionally lethal because they permit a virus to replicate at one temperature but not at another within the normal growth range of the wild-type virus. For example, a temperature-sensitive (*ts*) mutant may replicate at 35°C but not at 43°C.
Nonsense	The formation of a nonsense codon within the viral genome can cause premature termination of transcription of polypeptide synthesis and thus the inability of the virus to replicate. These mutants can multiply, though, in bacterial strains carrying a supressor mutation that enables completion of the synthesis of the viral polypeptide.

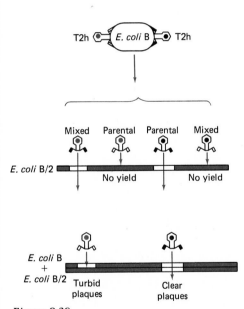

Figure 8.39

Mixing of wild-type and h *mutant bacteriophage in* E. coli *B produces four types of progeny. After infecting* E. coli *B/2, only phenotypically* h *phage produce progeny. Plating on a mixture of* E. coli *B and* E. coli *B/2 yields clear and turbid plaques. The turbid plaques are from genotypically* h^+ *phage, which do not infect strain B/2 cells, and clear plaques are from genotypically* h *phage, which infect both types of cells. When similar experiments are performed with phage carrying mutation at the* r *and* h *loci, four phenotypically distinguishable plaques are produced (see Figure 8.40).*

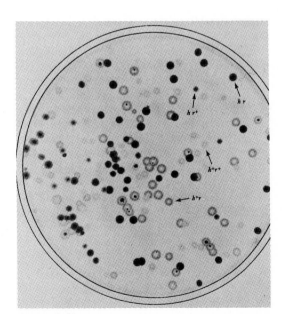

Figure 8.40

Plaques formed by a mixture of T2 phage carrying mutations at the h *and* r *loci. Phages with the* h^+ *allele produce turbid plaques (gray); phage with the* h *allele produce clear plaques (black); phage with the* r *allele are larger than those with the* r^+ *allele. (Reprinted by permission from Gunther Stent, 1963,* Molecular Biology of Bacterial Viruses, *Copyright © 1963, W. H. Freeman and Company.)*

i.e., before the phenotypes of the recombinants can be determined, the consequences of a mating between two phage DNA molecules cannot be inferred directly from the proportions of recombinants, but statistical analyses are required to account for the effect of repeated matings.

By appropriately interpreting recombination frequencies and by using genetic markers, it has been possible to determine detailed complete **genetic maps** for a variety of viruses. The genetic maps of several viruses show that genes are clustered according to their function. In some viruses, such as bacteriophage T7, the viral genome is lin-

ear (Figure 8.41); in other viruses, such as bacteriophage T4, the viral genome is circular (Figure 8.42). A particulaly interesting finding is that some viruses maximize the amount of information that is stored within the genome by using overlapping genes (Figure 8.43) and transcription of both strands of the DNA in opposite directions to code for different protein products (Figure 8.44). This is particularly important because the viral genome must be small and compact but must still code for a large number of gene products and regulatory functions.

Genetic modification and microbial evolution

The introduction of diversity into the gene pools of microbial populations establishes a basis for selection and evolution. There are numerous theories to explain evolutionary change. The basis of evolution lies with the ability to change the gene pool and to maintain favorable new combinations of genes. Mutations introduce variability into the genomes of microorganisms, resulting in changes in the enzymes that the organism synthesizes. The variations in the genome are passed from one generation to another and are disseminated through populations. Because of their relatively short generation times compared to higher organisms, changes in the genetic information of microorganisms can be widely and rapidly disseminated. Although in some cases the modification of the genome is harmful to the organism, some mutations being lethal or conditionally lethal, a mutation may change the genetic information in a favorable way. The occurrence of favorable mutations introduces information into the gene pool that can make an organism more fit for surviving in its environment and competing with other microorganisms for available resources. Over many generations natural selective pressures may result in the elimination of unfit variants and the continued survival of organisms possessing favorable genetic information. Change toward more favorable variants in a particular environment is the essence of evolution.

Although mutation is the basis for introducing variability into genetic information, it is genetic exchange and recombination that play critical roles in redistributing genetic information. Recombination creates new allelic combinations that may be adaptive. Altering the organization of genetic information within populations provides a basis

for directional evolutionary change. The exchange of genetic information can produce individuals with multiple attributes that favor the survival of a microbial population. The long-term stability of a population depends on its incorporating adaptive genetic information into the chromosome. Mutation and general recombination appear to provide a basis for the gradual selection of adaptive features. In particular, reciprocal recombination could be expected to produce an evolutionary link between closely related organisms, and nonreciprocal recombinational events appear to provide a mechanism for rapid stepwise evolutionary changes. The fact that unrelated genomes can recombine suggests that different lines of evolution can suddenly come together.

Plasmids and other transposable genetic elements may contribute to rapid changes in the genetic composition of a population, but the evolutionary stability of such changes is not clear. For example, the incidence of bacteria containing plasmids that code for antibiotic resistance has certainly increased since the widespread introduction of the use of antibiotics in medicine, particularly in hospital settings where antibiotic use is extensive (Figure 8.45). Possession of the genetic information that encodes for antibiotic resistance is adaptive for microorganisms trying to survive in the presence of a variety of antibiotics. The information contained in plasmids, however, can be readily lost, especially if selective pressure diminishes. It is too early to say whether possessing the information for antibiotic resistance will be of long-term evolutionary advantage to bacteria and if so, whether this information will be permanently incorporated into the bacterial chromosome.

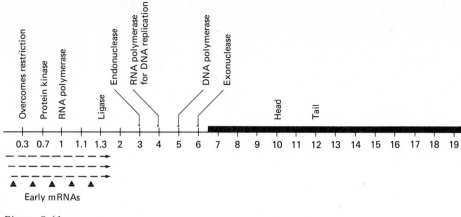

Figure 8.41

The linear genetic map of phage T7.

Figure 8.42

The circular genetic map of bacteriophage T4.

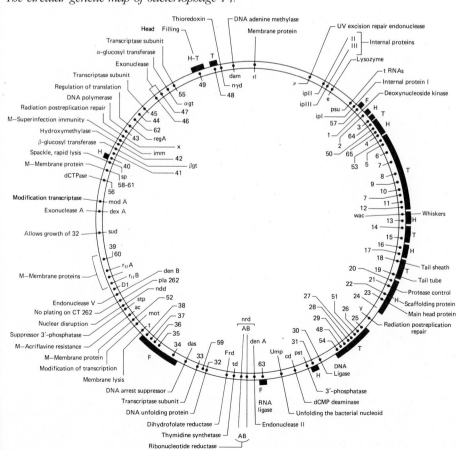

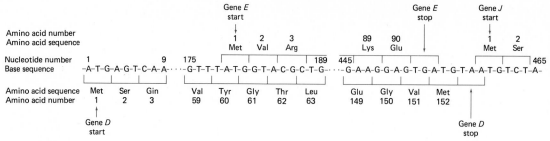

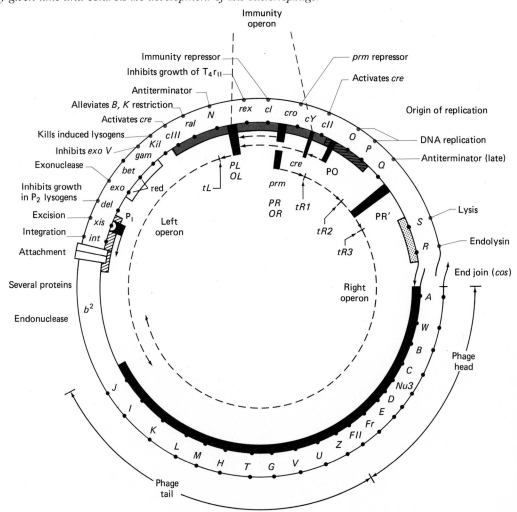

Figure 8.43

The genome of bacteriophage φX174, showing overlapping genes. The reading frame of gene E, which begins in the middle of the nucleotide sequence for gene D, is offset by one base from that of gene D. The stop codon for gene D also overlaps with the start codon for gene J. Similar overlapping genes occur in animal viruses such as SV40.

Figure 8.44

The genetic map of bacteriophage lambda shows that both left-hand and right-hand transcription are used for production of viral gene products. Regulation of gene expression determines which genes are expressed at any given time and controls the development of this bacteriophage.

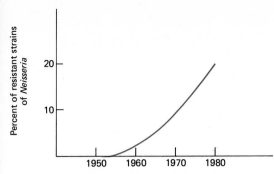

Figure 8.45

The incidence of antibiotic resistant strains has increased in recent years as a result of selective pressure. This graph shows the increase in the number of strains of Neisseria gonorrhoeae *that are resistant to 0.5 unit of penicillin/ml. New drugs must continually be developed to keep up with the rapid evolution of antibiotic-resistant strains.*

Incidence of strains of *Neisseria gonorrhoeae* which are resistant to 0.5 unit penicillin/ml

Genetic engineering

Besides the potential role of genetic recombination in natural evolutionary changes, it is possible to employ DNA recombination to direct or engineer the evolution of new microorganisms. Recombinant DNA technology promises to be a powerful tool for understanding basic genetic processes and has tremendous potential industrial applications.

Genetic engineering uses plasmids as carriers of unrelated DNA and the enzymes involved in the normal recombination and replication of DNA for splicing foreign DNA into the plasmid carriers (Figure 8.46). Plasmids can be isolated, and the plasmid DNA can be cut open by using a site-specific endonuclease, commonly called a **restriction enzyme**, to create a site where foreign donor DNA can be inserted. Restriction endonucleases normally function to prevent the incorporation of foreign DNA into a microbial genome by cutting both strands of a foreign DNA molecule. Type I restriction endonuclease enzymes cleave DNA at a random distance from a recognition site in the DNA nucleotide sequence. Type II restriction endonucleases cleave the DNA at the recognition site, and type III endonucleases cut the DNA at some precise distance from the recognition site. It is type II and III restriction endonucleases that are useful in genetic engineering. A type II restriction endonuclease cuts the DNA at a **palindromic** sequence of bases (a sequence of nucleotide bases that can be read identically in both the 3'-OH → 5'-OH and 5'-OH → 3'-OH direction), producing DNA with staggered single-stranded ends (Figure 8.47). These ends of the cut DNA can act as cohesive or sticky ends during recom-

bination, making them amenable for splicing with segments of DNA from a different source that has the been excised by using the same endonuclease. (Bacteria normally protect themselves against their own endonuclease enzymes by modifying DNA bases at the recognition sites where the endonucleases act, using specific modification enzymes for this purpose.)

If two different endonuclease enzymes are used, one to open the plasmid ring and another one to form a segment of donor DNA, it often is necessary to establish an artificial homology at the terminal ends of the donor and plasmid DNA molecules (Figure 8.48). This can be accomplished by adding poly-A (poly-adenine) tails to the plasmid and poly-T (poly-thymine) tails to the donor DNA. A transferase enzyme can then be employed to add poly-A tails to the 3'-OH ends of the DNA molecule. Pairing occurs between homologous regions of complementary bases, and ligase enzymes are used to seal the circular plasmid. The tails left by the action of the endonuclease enzyme are cleaved *in vitro*, using exonuclease enzymes. Virtually any source of DNA can be used as a donor, including human DNA. By adding a poly-T tail to the donor DNA after its excision with an endonuclease, the donor DNA can be made complementary to the poly-A tails of the plasmid DNA, permitting the formation of a circular plasmid molecule. If the same restriction endonuclease enzyme is used to cut both the donor and recipient plasmid DNA, the strands will have homologous ends and it will be unnecessary to add poly-A and poly-T tails (Figure 8.49). The sealing of the ends of the DNA molecules is accom-

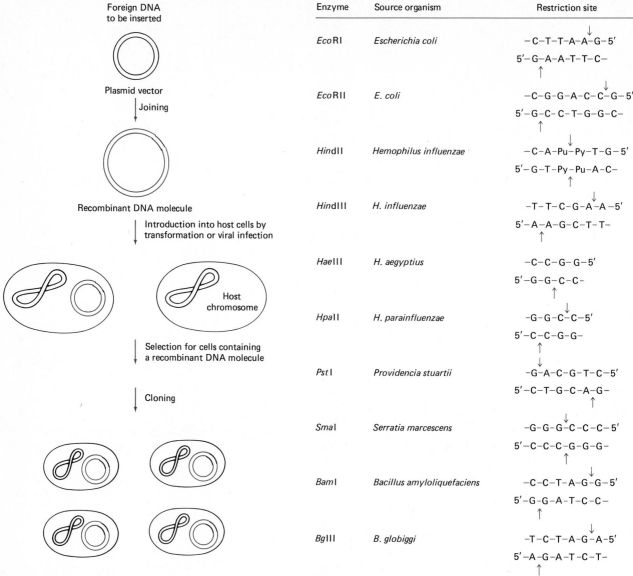

Foreign DNA
to be inserted

Plasmid vector

Joining

Recombinant DNA molecule

Introduction into host cells by
transformation or viral infection

Host
chromosome

Selection for cells containing
a recombinant DNA molecule

Cloning

Enzyme	Source organism	Restriction site
EcoRI	Escherichia coli	−C−T−T−A−A−G−5′ ↓ 5′−G−A−A−T−T−C− ↑
EcoRII	E. coli	−C−G−G−A−C−C−G−5′ ↓ 5′−G−C−C−T−G−G−C− ↑
HindII	Hemophilus influenzae	−C−A−Pu−Py−T−G−5′ ↓ 5′−G−T−Py−Pu−A−C− ↑
HindIII	H. influenzae	−T−T−C−G−A−A−5′ ↓ 5′−A−A−G−C−T−T− ↑
HaeIII	H. aegyptius	−C−C−G−G−5′ 5′−G−G−C−C− ↑
HpaII	H. parainfluenzae	−G−G−C−C−5′ ↓ 5′−C−C−G−G− ↑
PstI	Providencia stuartii	−G−A−C−G−T−C−5′ ↓ 5′−C−T−G−C−A−G− ↑
SmaI	Serratia marcescens	−G−G−G−C−C−C−5′ ↓ 5′−C−C−C−G−G−G− ↑
BamI	Bacillus amyloliquefaciens	−C−C−T−A−G−G−5′ ↓ 5′−G−G−A−T−C−C− ↑
BglII	B. globiggi	−T−C−T−A−G−A−5′ ↓ 5′−A−G−A−T−C−T− ↑

Figure 8.46

*Diagram showing a generalized outline for
genetically engineering a new organism. The ability
to create new organisms in this way holds great
promise for improving the quality of life by using
genetic engineering in medicine and industry to
produce useful products.*

Figure 8.47

*Illustration of the palindromic sequences at the sites of gene
cutting by several type II endonucleases.*

plished by using ligase enzymes, and in this manner a plasmid is created that contains a foreign segment of DNA.

Once the plasmid containing the desired additional DNA segments is formed, it can be added to a culture of a suitable recipient bacterium that will incorporate the plasmid. The plasmid is taken up by the bacterium, and then, regardless of its source, it comprises part of the bacterial genome. The plasmid DNA, including the foreign DNA segments, can be replicated and passed from one generation to another. It is possible to add several plasmids to a single bacterium so that the presence of the foreign DNA is amplified. When this approach is used, genes can be cloned (asexually reproduced), with bacteria acting as factories to

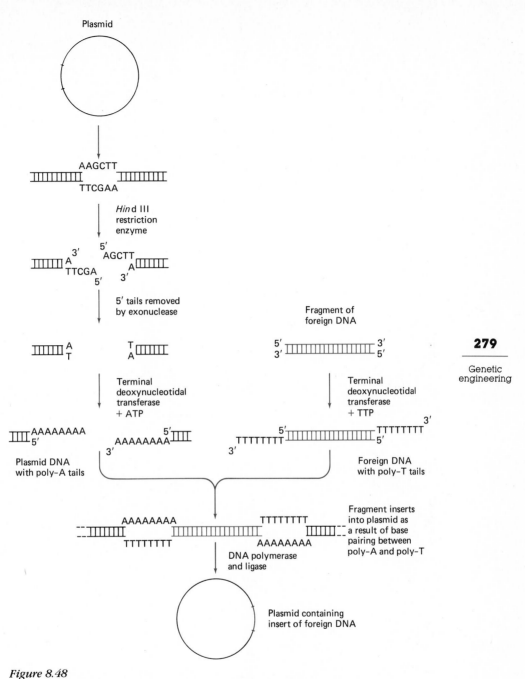

Figure 8.48

*Diagram showing the establishment of artificial homology when different
endonuclease enzymes are employed.*

produce multiple copies of identical genes. Because the genetic code is essentially universal, the information encoded in the DNA sequence can be expressed, and the polypeptide chain specified by the foreign DNA segment can be transcribed and translated to form an active protein molecule. The expression of the foreign genetic information, however, requires that the appropriate reading frame be established and that the transcriptional and translational control mechanisms be turned on to permit the expression of the DNA.

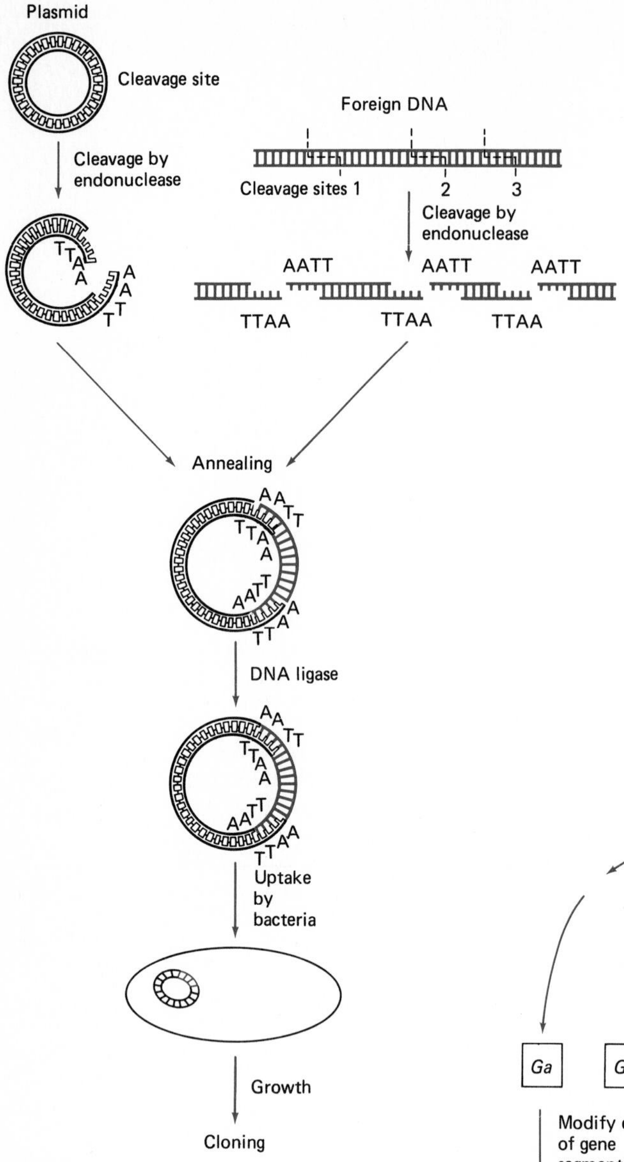

Figure 8.49

Diagram showing the formation of a recombinant DNA molecule when the same endonuclease enzyme is used for both the donor and recipient DNA.

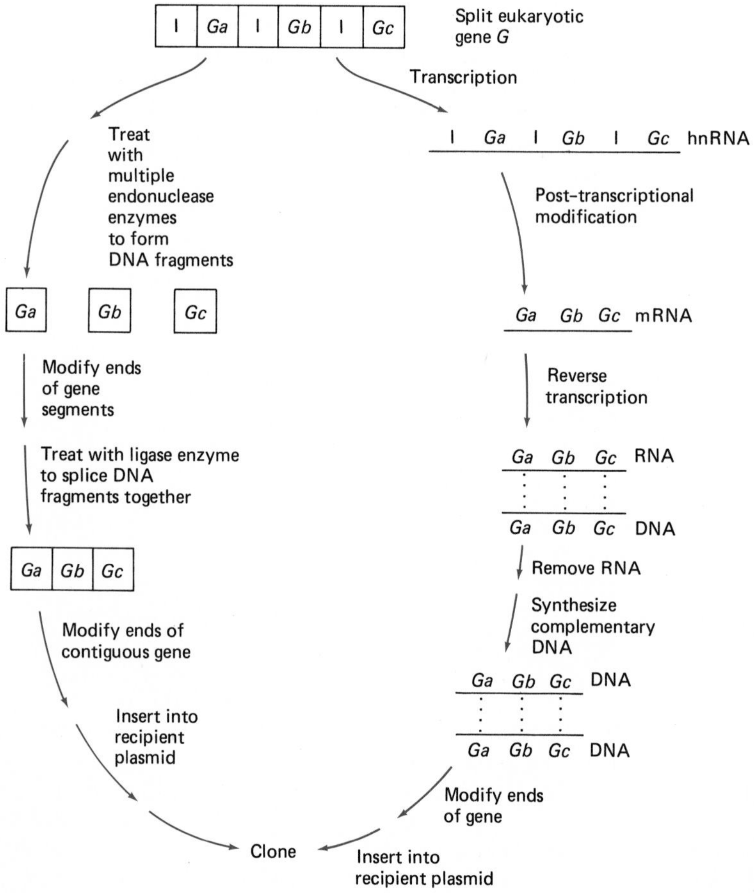

Figure 8.50

The establishment of a contiguous sequence of nucleotides is needed for the cloning of eukaryotic genes in bacterial cells. The need to remove intervening sequences is an unexpected complication in employing bacteria to produce eukaryotic gene products.

The construction of bacterial plasmids containing complete gene sequences, derived from eukaryotic organisms, is complicated by the fact that eukaryotic genes are generally split. In eukaryotic organisms post-transcriptional modification of the hnRNA is required in order to produce a mature messenger RNA molecule that can be properly translated, but bacteria do not possess the capacity to remove introns from eukaryotic DNA in order to form the mRNA needed for producing a functional protein molecule. Therefore, it is necessary to artificially cut and splice eukaryotic DNA or to use an alternative procedure to establish a contiguous sequence of nucleotide bases to define the protein that is to be expressed (Figure 8.50). The problem of the discontinuity of the eukaryotic gene can be overcome by using a messenger RNA molecule and a reverse transcriptase enzyme to produce a DNA molecule that has a contiguous sequence of nucleotide bases containing the complete gene. The single-stranded DNA molecule that is formed in this procedure is complementary to the complete messenger RNA molecule. The RNA can be removed by using a nuclease and a complementary strand of DNA synthesized by the reverse transcriptase enzyme. The double-stranded DNA molecule formed in this manner can then be inserted into a carrier plasmid. In this way many eukaryotic genes can be cloned in prokaryotic cells.

Despite the great precision of the DNA replication process, the genetic information of microorganisms is continuously changing. Various types of mutational events can introduce modifications into the DNA molecule, producing multiple allelic forms of the same gene, and recombinational processes permit further redistribution of genetic information. Heterogeneity within the gene pool of a microbial population is necessary for maintaining stability under fluctuating environmental conditions; genetic variability essentially provides protection against environmental uncertainty for the survival of the species. Whereas asexual reproduction could lead to rapid divergence toward a gene pool with temporarily advantageous restricted information, sexual reproduction ensures continued genetic heterogeneity that has long-term advantages when environmental conditions vary. The diversity introduced into the gene pool through mutation and recombination also establishes the basis for the selective evolution of microorganisms.

Recombination involves a restructuring of DNA molecules so that new genomes are formed containing information from different DNA sources. General recombination results in the exchange of corresponding genes, and nonreciprocal recombination permits the interchange of unrelated segments of DNA. A variety of mechanisms permit the transfer of genetic information from a donor to a recipient microorganism so that recombination can occur. In eukaryotes sexual reproduction results in the exchange of genetic information between closely related organisms. In bacteria, DNA may be transferred by: conjugation, which involves direct contact between the donor and recipient strain; transduction, which utilizes a virus as a vector for carrying the DNA from the donor to the recipient strain; and transformation, which involves the uptake of naked DNA by a competent recipient strain. DNA exchanges between microorganisms of differing allelic forms and the frequency of recombination can be used to map the microbial genome.

Recombination involving plasmids provides a mechanism for the particularly rapid dissemination of genes through a population. Plasmids are quite useful as carriers of foreign genetic information in genetic engineering. Recombinant DNA technology can be employed to create organisms that contain combinations of genetic information that do not occur naturally and, for example, bacteria containing genetically engineered plasmids can synthesize proteins that are normally produced only in eukaryotic organisms. In theory, recombinant DNA technology can be used to engineer organisms that can produce any desired combination of proteins. As a result genetic engineering holds great promise in industry and medicine because various proteins of economic importance or use in curing disease may be produced.

The potential of genetic engineering to short-circuit evolution has raised numerous ethical, legal, and safety questions. The Supreme Court of the United States has ruled, in a landmark decision, that genetic engineering can "create" novel living systems that can be patented as inventions. This ruling has established the precedent for future genetic engineering efforts. The safety question has been temporarily solved by using mutation and selection procedures to develop a fail-safe strain of *E. coli* that is unable to grow outside of a carefully defined culture medium. Nevertheless, the question remains: what if a genetically engineered plasmid were to enter another microorganism and spread? Could this represent a serious biological hazard? The debate also continues as to whether it is ethical to clone DNA and whether recombinant DNA technology will permit the cloning of higher eukaryotic organisms, including plants and animals. One must weigh these ethical questions against the benefits that can be derived through genetic engineering. There is little question that through genetic engineering the quality of human life can be improved. Scientists have a responsibility to see that the public is informed about genetic engineering, to see that research remains within acceptable guidelines, to use this powerful technique for examining basic questions about molecular-level genetics, and to develop genetically engineered organisms of agricultural, ecological, medical, and economic importance.

1. What is a mutation? How do mutations occur?

2. What effect does exposure to ionizing radiation have on rates of mutation? Exposure to a mutagen?

3. What is a plasmid, and what are some functions associated with plasmids?

4. What is recombination, and how is this process involved in maintaining heterogeneity within the gene pool?

5. Compare homologous and nonhomologous recombination.

6. What is a transposable genetic element?

7. How is DNA exchanged in prokaryotes?

8. Compare specialized and general transduction.

9. How is genetic variability related to the evolution of microorganisms?

10. What is recombinant DNA technology? How is genetic engineering able to create new organisms?

Ayala, F. J., and J. A. Kiger, Jr. 1980. *Modern Genetics.* Benjamin/Cummings Publishing Co., Menlo Park, California.

Birge, E. A. 1981. *Bacterial and Bacteriophage Genetics.* Springer-Verlag, New York.

Chakrabarty, A. M. 1978. *Genetic Engineering.* CRC Press, Inc., Boca Raton, Florida.

Cohen, S. N., and J. A. Shapiro. 1980. Transposable genetic elements. *Scientific American* 242(2): 40–49.

Fristrom, J. W., and P. T. Spieth. 1980. *Principles of Genetics.* Chiron Press, Inc., Portland, Oregon.

Genetics: Readings from Scientific American. 1981. W. H. Freeman and Co., San Francisco.

Gilbert, W., and L. Villa-Komanoff. 1980. Useful proteins from recombinant bacteria. *Scientific American* 242(4): 74–94.

Grobstein, C. 1977. The recombination debate. *Scientific American* 237(7): 22–33.

Kleckner, N. 1977. Translocatable elements in prokaryotes: review. *Cell* 11: 11–23.

Kolodny, G. M. (ed.). 1981. *Eukaryotic Gene Regulation*. CRC Press, Inc., Boca Raton, Florida.

Lewontin, R. C. 1974. *The Genetic Basis of Evolutionary Change*. Columbia University Press, New York.

Mays, L. L. 1981. *Genetics: A Molecular Approach*. Macmillan Publishing Co., Inc., New York.

Novick, R. 1980. Plasmids. *Scientific American* 243(6): 103–123.

Old, R. M., and S. B. Primrose. 1981. *Principles of Gene Manipulation: An Introduction to Genetic Engineering*. Blackwell Scientific Publications Ltd., Oxford, England.

Roughgarden, J. 1979. *Theory of Population Genetics and Evolutionary Ecology: An Introduction*. Macmillan Publishing Co., Inc., New York.

Shimke, R. T. 1980. Gene amplification and drug resistance. *Scientific American* 243(5): 60–69.

Stahl, F. W. 1979. *Genetic Recombination: Thinking About It in Phage and Fungi*. W. H. Freeman and Co., San Francisco.

Strickberger, M. W. 1976. *Genetics*. Macmillan Publishing Co., Inc., New York.

Suzuki, D. T., A. J. F. Griffiths, and R. C. Lewontin. 1981. *An Introduction to Genetic Analysis*. W. H. Freeman and Co., San Francisco.

Watson, J. D. 1976. *Molecular Biology of the Gene*. Benjamin/Cummings Publishing Co., Menlo Park, California.

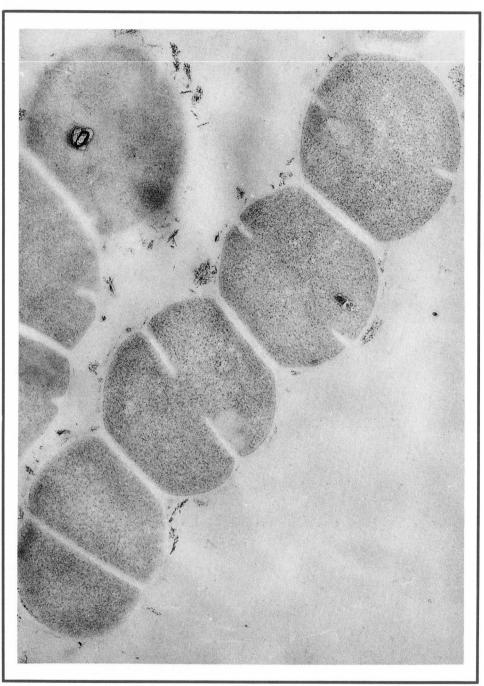

Dividing Streptococcus pyogenes *in chain formation. (From BPS—T. J. Beveridge, University of Guelph, Guelph, Ontario.)*

Microbial growth

SECTION FIVE

chapter 9
Growth and reproduction
of microorganisms

chapter 10
Control of microbial growth: influence
of environmental factors on growth
and death of microorganisms

Growth and reproduction of microorganisms

9

The growth and reproduction of microorganisms, essential for their continued existence, depends on the replication of their genetic information and the ability to convert available substrate resources into their required macromolecules. Much of the metabolic activity of a microorganism is aimed at: (1) replicating its genome so that it can be passed to the next generation; (2) using substrates to generate ATP, reducing power, and small precursors of macromolecules; and (3) synthesizing the macromolecules that constitute the essential structures of the cell, including proteins, nucleic acids, carbohydrates, and lipids. Biosynthetic activities for the synthesis of membranes and other boundary layers, which separate the organism from its surrounding environment, are indispensable and are central to the process of microbial reproduction. The ability of microorganisms to direct the replication of their own organizational structure is the critical activity permitting their reproduction as living systems.

At the microbial level, growth is essentially synonymous with reproduction. Although individual cells may increase in size, growth in biological systems normally occurs through cell multiplication during which genetic information is transmitted. In the case of unicellular microorganisms, growth thus is equivalent to reproduction because reproducing the cell also accomplishes the reproduction of the entire organism. Even in multicellular, eukaryotic microorganisms the growth of the whole organism can be viewed in most cases as an increase in the number of individual reproductive units within a biological assemblage.

Viral reproduction

Unlike all other organisms, the growth and **reproduction of viruses** occurs through an increase in the number of virion particles rather than through cellular multiplication. To reproduce, a virus must invade a suitable host cell, because only within such a cell do conditions permit viral replication. Viruses are essentially extensions of the host cells in which they reproduce, and there is a high degree of specificity between a particular virus and the specific host cell within which that virus can reproduce. Viruses that reproduce only within specific host bacterial cells are known as **bacteriophage** or simply phage; others that replicate within plant cells are known as **plant viruses**; and those that reproduce only within animal cells are known as **animal viruses**. For a specific virus to reproduce within the host cell: (1) the host cell must be permissive, and the virus must be compatible with the host cell; (2) the host cell must not degrade the virus; (3) the viral genome must possess the information for stopping the normal metabolism of the host cell; and (4) the virus must be capable of using the metabolic capabilities of the host cell for producing new virus particles containing replicated copies of the viral genome.

Although the specific details of viral replication vary from one virus to another, the general strategy for reproduction is the same for most viruses. The virus initially attaches to the outer surface of a suitable host cell. Generally, the **adsorption** of a virus to a host cell involves specific binding sites on the cell surface, which explains in part the high degree of specificity between the virus and the host cell. The nucleic acid of the virus then penetrates the cytoplasmic membrane. Within the host cell the viral genome achieves control of the cell's metabolic activities; in many cases the viral genome actually codes for the shutdown of those metabolic activities normally involved in host cell

reproduction. The virus then uses the needed biochemical components and anatomical structures of the host cell for the production of new viruses. In particular, the virus employs the host cell's ribosomes for producing viral proteins and the cell's ATP and reduced coenzymes for carrying out biosynthesis. The nucleic acid and protein capsid structures of the virus (the essential structural components of a virus) are synthesized separately and then assembled prior to release from the host cell. Generally, many virus particles are produced within a single host cell and are released together. The replication of a virus results in changes in the host cell, often causing the death of the cell within which the virus reproduces. The cycle of viral adsorption to a host cell, invasion, synthesis of viral nucleic acid and proteins, assembly of the virus structure, and release of viral progeny is repeated when a virus encounters another suitable host cell.

Figure 9.1

Drawing of the T-even phage, T4, showing its component parts. The phage tail fibers attach to a bacterial cell and the contraction of the sheath injects the viral DNA through the bacterial envelope. (A) Virion with extended tail fibers. (B) Virion with the tail sheath contracted and the spikes of the tail plate pointing against the bacterial cell wall. (After Simon and Anderson, 1967, Virology *32: 279.)*

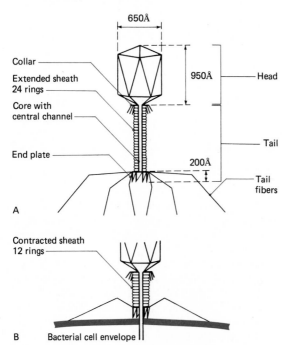

Lytic phage

Reproductive cycle

The normal reproductive cycle of bacteriophage results in the lysis of the host cell at the completion of the viral replication cycle, and so these bacteriophage are referred to as **lytic phage**. The lytic phage, of which there are numerous types, exhibit characteristic developmental sequences or stages in their reproductive cycles. The specific details of reproduction vary between the different types of lytic phage, but the general developmental sequence is similar to the reproduction of the T-even phage, such as T2, T4, and T6. The T-even phage are DNA viruses that have a complex binal symmetry with distinct head and tail structures (Figure 9.1). The general sequence of events in the lytic reproductive cycle of a T-even phage is depicted in Figure 9.2. The reproduction of T-even phages within cells of bacteria such as *E. coli* begins with the attachment of a T-even phage to the bacterial cell. There appear to be specific receptor sites on the bacterial cell surface where the phage may attach (Figure 9.3). The entire phage particle does not penetrate into the bacterial cell, but rather, the phage injects its DNA into the bacterium, using its tail structure like a syringe to inject the DNA into the bacterium. The contraction of the tail structure drives a tube through a pore in the cytoplasmic membrane of the bacterium, providing an entry tube for the phage DNA.

When the phage DNA enters the bacterial cell, it is not degraded by the exonucleases and en-

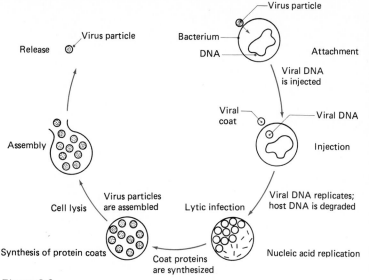

Figure 9.2

In the lytic phage reproduction cycle, the bacteriophage attaches to specific chemical sites on the surface of a bacterium and the viral nucleic acid penetrates the host cell. Once the viral nucleic acid is inside the bacterium, bacterial metabolism is halted by the destruction of the host genome. First viral DNA, then viral coat proteins appear; they then self-assemble, and the host cell lyses. The free virions may infect other susceptible host bacteria.

Discovery process

The use of bacteriophage as a model system has been important in the advance of the field of virology. A technically suitable virus–cell system has proven effective in spite of the large variety of viruses and their differences in structure and genetic information, because all viruses are similar in their mode of multiplication. Burnet in Australia, Schlesinger in Hungary, Lwoff in France, and Delbruck, Luria, and Cohen in the United States all used bacteriophage as model systems, even though it is now known that they are among the most complex of all viruses. It was this very complexity, however, that allowed the discovery of aspects of viral multiplication that might have been overlooked with simpler viruses. For example, the presence of 5-hydroxymethylcytosine, instead of cytosine, in the DNA of the bacteriophage made it possible to study the intracellular multiplication of viral DNA. Also, the presence in the infected cells of deoxycytidylate hydroxymethylase, which synthesizes that unusual base and its absence in uninfected cells, gave the first indication that virus-specified enzymes are made in the infected cells. Additionally, using these markers the replication of bacteriophage could be followed, permitting the elucidation of the lytic growth cycle.

donucleases of a compatible host cell. The DNA of T-even phage contains glucosylated hydroxymethyl cytosine instead of cytosine, a chemical modification of the DNA that prevents the nuclease enzymes of the bacterium from degrading the phage genome. For replication, the phage genome must code for DNA polymerase enzymes. The viral DNA is initially transcribed by using a bacterial RNA polymerase enzyme, and among the first proteins coded for are ones for the modification of the bacterial RNA polymerase. The phage-coded polypeptides replace or modify the sigma subunits of the bacterial RNA polymerase and by doing so alter the recognition sequence so that the RNA polymerase no longer binds at the normal Pribnow sequences of the bacterial DNA. Thus, transcription of bacterial genes ceases. The T-even phage also codes for a nuclease enzyme that degrades the host cell DNA, and the deoxynucleotides released by the degradation of the bacterial chromosome can be used as precursors for the synthesis of viral DNA.

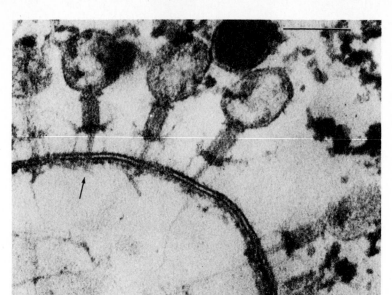

Figure 9.3

Electron micrograph showing the attachment of T-even phage to E. coli. *Different viruses attach at different receptor sites on the host cell surface. The tail structure penetrates the bacterial cell (arrow), and the phage DNA is injected into the bacterium as the sheath contracts. (Courtesy Lee Simon, Rutgers, The State University.)*

The entire sequence of penetration, shutting off host cell transcription and translation, and the degradation of the bacterial chromosome takes only a few minutes. The penetration of phage DNA into the bacterial cell and the stoppage of host cell macromolecular synthesis, involving several enzymes referred to as early proteins, represent the early developmental steps of viral replication. After these early events in the phage reproduction cycle, there is a further modification of the RNA polymerase enzyme, resulting in the cessation of further synthesis of the early phage proteins. This shift in the recognition site of the RNA polymerase coincides with the beginning of the synthesis of late proteins, that is, with the synthesis of proteins coded for late in the developmental sequence of the phage. The late genes coded for by the phage genome include some enzymes involved in the replication of the phage DNA. Special phage-coded enzymes are required for the production of the hydroxymethyl cytosine and the glucosylated nucleotides that occur in the phage genome. The late phage genes also code for the various proteins that make up the capsid structure of the phage. The tail, tail fiber, and head structures of the phage are made up of proteins coded for by different phage genes, with at least 32 genes involved in the formation of the tail structure and at least 55 genes involved in the formation of the head structure of the phage.

The transition from early to late gene transcription involves an interesting shift in which one of the two DNA strands of the phage genome serves as the sense strand (Figure 9.4). The **early genes** are transcribed in a counterclockwise direction, whereas the **late genes** are transcribed in a clockwise direction. Because transcription proceeds in the 5′-OH to 3′-OH direction, the change from counterclockwise to clockwise must mean that the opposing strands of the viral DNA code for the early and late proteins. The alteration of the recognitional subunit of the RNA polymerase enzyme accounts for the change in the base sequences of the DNA recognized as promoter sites for transcription of the viral genome, and permits changing the DNA strand that acts as the sense strand. By altering reading frames and by changing which DNA strand serves as the sense strand, the phage genome can encode the almost 150 genes involved in the replication of T4 phage.

After the production of the individual components of the virus, the virus is assembled with packaging of the nucleic acid genome within the protein capsid. The assembly of the T-even phage capsid is a complex process (Figure 9.5) but one which follows a sequential pathway. Assembly of the head and tail structures requires several enzymes that are coded for by the phage genome. The head, tail, and tail fiber units of the T-even phage capsid are assembled separately, and the tail fibers are added after the head and tail structures are combined. The small size of the viral particle means that the DNA must be tightly packed within the phage head assembly. Thus, packaging DNA into the head structure involves stuffing the head with DNA and cutting away the excess. When

Figure 9.4

Map of the known genes of the bacteriophage T4. The numbers on the inner dial are map distances in recombination units. The symbols in the rectangles indicate the functions altered by mutations in the corresponding genes. (neg = negative; del = delayed; arr = arrested; hd = head; lys = lysis; mat = maturation; def = defective.) (After Wood, 1974, Handbook of Genetics, *R. C. King, ed., vol. 1, 327.)*

the head structure is completely filled with DNA, any extra DNA is cleaved by a nuclease enzyme. The specific mechanism that is responsible for filling the phage head with DNA and folding the DNA within the assembled phage particle has not been totally elucidated.

Once the phages are assembled, they must be released from the host cell and encounter another host cell for further reproduction. One of the late proteins coded for by the phage genome is lysozyme, which catalyzes the breakdown of the peptidoglycan wall structure of bacteria. The action of the lysozyme results in sufficient damage to the cell wall that the wall is unable to protect the cell against osmotic shock, which results in the lysis of the bacterial cell and the release of the phage particles into the surrounding medium. The action of the lysozyme enzyme appears to be subject to phage-directed regulation, which ensures that the wall is not degraded prematurely before a sufficient number of phage particles has been completely assembled.

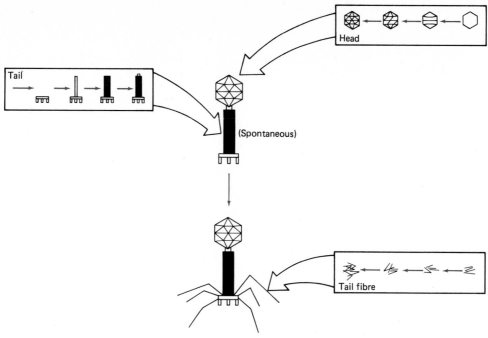

Figure 9.5

*The assembly of phage head and tail structures within the bacterial host cell. Three separate
subassembly lines are illustrated here, the head, sheath, hollow spike, and tail fiber components.
(Based on Wood et al., 1968,* Federation Proceedings *27: 1160.)*

Figure 9.6

*The one-step growth curve for the lytic
bacteriophage T1 on* E. coli *strain B grown in
nutrient broth at 37°C. Phage and bacteria were
mixed at a ratio of 1:10 at time 0. After 4 minutes,
45 percent of the phage had been adsorbed, at
which time the mixture was diluted. Assays were
made at intervals after dilution. Average phage
yield per infected cell = 138.*

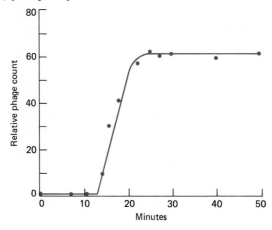

Growth curve of lytic phage

The lysis of the bacterial cell releases a large number of phage simultaneously, and consequently,
the lytic reproduction cycle exhibits a **one-step
growth curve** (Figure 9.6). The growth curve begins with an **eclipse period** during which there
are no complete infective phage particles, and the
naked DNA within the host cell is unable to infect
other cells to initiate a new reproductive cycle
(Figure 9.7). The end of the eclipse period is taken
as the time at which an average of one infectious
unit has been produced for each productive cell.
The eclipse period is part of a longer period, the
latent period. The **latent period**, which for a
T-even phage often is about 15 minutes, begins
when the phage injects DNA into a host cell and
ends when one viral infectious unit per productive cell, on average, has appeared extracellularly.
Because only complete phage particles are able
to initiate the lytic replication cycle, the latent period does not end until assembled phages begin
to accumulate within the bacterial cell. Completely assembled phage particles continue to accumulate within the bacterial cell until, at the time
of lysis, they reach a particular number known as
the burst size (Figure 9.8). The **burst size**, which

Figure 9.7

Detailed growth curve for bacteriophage T2, showing the eclipse, latent, and rise periods. To determine this viral growth curve, bacteria and phage were mixed for 2 minutes. The bacteria were recovered by centrifugation and plated to determine the number of productive bacteria. Other samples periodically were collected and used to assay for total phage after chloroform treatment to disrupt bacterial cells and extracellular phage in the supernatant.

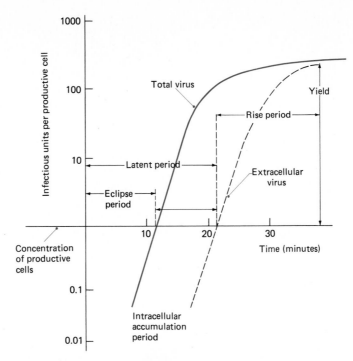

Figure 9.8

An assembled phage, T2, within the bacterial cell, E. coli *strain B. Note cell envelopes of the wall and cytoplasmic membrane; the clear area of the phage DNA pool containing many condensed phage DNA cores; and three phages attached to the cell surface, one empty and two still partially filled. The phage at the top shows the long tail fibers and the spikes of the tail plate in contact with the cell wall. The tail sheath is contracted, and the tail core has apparently reached the cell surface but has not penetrated it. (Courtesy Lee Simon, Rutgers, The State University.)*

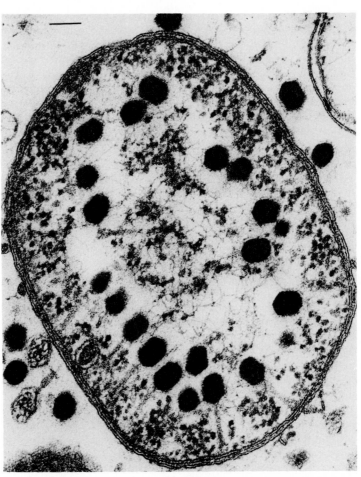

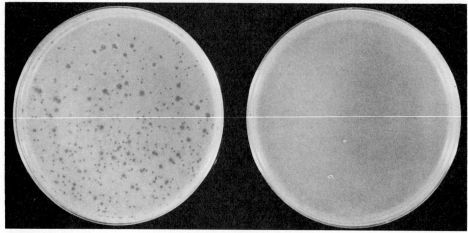

Figure 9.9

The lytic replication of phage on a lawn of bacteria growing on a solid medium produces zones of clearing known as plaques. Here, the plaques of bacteriophage λ *growing on* E. coli *are clearly visible. (From BPS—Richard Humbert, Stanford University.)*

varies from cell to cell, represents the average number of infectious viral units produced per cell; a typical burst size for a T-even phage is 200. As a result of the simultaneous release of a number of infective phages, the number of phages that can initiate a lytic reproduction cycle increases greatly in a single step. The entire lytic growth cycle for some T-even phage can occur in less than 20 minutes under optimal conditions.

In addition to the T-even DNA phages, there are also some RNA phages, such as Qβ and f2, that carry out a lytic reproductive cycle. RNA phages tend to be quite small. The RNA genome of these phage can serve as messenger RNA molecules for the production of viral proteins, but the replication of the viral genome requires the synthesis of a complementary replication strand of RNA that can then serve as a template for the production of the RNA genome. The complete assembly of the phage is followed by lysis of the host cell, catalyzed by a viral-coded lytic enzyme.

Assaying for lytic bacteriophage

The growth curve for a bacteriophage can be determined by inoculating a suspension of host bacterial cells with the phage and assaying for the number of infective phage particles at various time intervals. In one assay for infective bacteriophage, a lawn of bacteria is prepared on a suitable solid nutrient medium, and dilutions of the phage suspension are then spread over the same surface (Figure 9.9). In the absence of lytic bacteriophage, the bacteria form a confluent lawn of growth. Lysis by bacteriophage is indicated by the formation of a zone of clearing or **plaque** within the lawn of bacteria. Each plaque corresponds to the site where a single bacteriophage initiated its lytic reproductive cycle. The spread of infectious phage from the initially infected bacterial cell to the surrounding cells results in the lysis of the bacteria in the vicinity of the initial phage particle and hence the zone of clearing. The soft agar used in these assays permits diffusion of phage to nearby uninfected cells but does not permit the viruses that are produced to move to remote parts of the plate. Plaques do not continue to spread because when a cell is heavily reinoculated with phage before the time of normal lysis, the lysis of the cell is inhibited. The number of plaques that develop and the appropriate dilution factors can be used to calculate the number of bacteriophage in a sample.

Lysogeny—temperate phage replication

In addition to the normal lytic cycle, some temperate bacteriophage are capable of carrying out a lysogenic replication cycle (Figure 9.10). During the lysogenic replication of a virus, only the integrated genome of the virus is reproduced, and the virus does not code for the production of complete viral particles nor for the lysis of the host cell. The **lysogenic reproduction cycle** consists of three phases: (1) establishment of the **pro-**

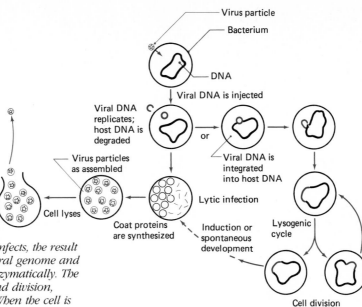

Figure 9.10

When a temperate bacteriophage infects, the result is usually a lysogenic cycle. The viral genome and the bacterial genome are fused enzymatically. The lysogenic cell continues growth and division, normally through binary fission. When the cell is induced, the viral DNA is excised, and a productive or lytic infection is carried out.

Labels in figure:
- Virus particle
- Bacterium
- DNA
- Viral DNA is injected
- Viral DNA replicates; host DNA is degraded
- or
- Viral DNA is integrated into host DNA
- Virus particles as assembled
- Lytic infection
- Cell lyses
- Coat proteins are synthesized
- Induction or spontaneous development
- Lysogenic cycle
- Cell division

phage (incorporation of the phage genome into the bacterial genome, generally into the bacterial chromosome but in some cases, e.g., the phage P1, into a bacterial plasmid); (2) maintenance of the prophage (replication of the phage DNA along with the normal replication of the DNA of the bacterium); and (3) release of the phage genome (excision of the phage DNA and initiation of the normal lytic reproduction cycle). The incorporation of the phage genome into a bacterial chromosome or plasmid is an example of nonreciprocal recombination. Once incorporated into the bacterial chromosome, the viral genome (referred to as a prophage) is replicated along with the bacterial DNA during normal host cell DNA replication. At a later time the prophage can be excised from the bacterial chromosome or plasmid DNA, reestablishing a normal lytic reproductive cycle. The release of the phage genome from the bacterial chromosome completes the lysogenic reproduction cycle.

The bacteriophage lambda (λ), for example, is a temperate phage that can alternate between the lytic reproductive cycle and the lysogenic reproductive cycle. The regulation of lambda reproduction, which determines whether the reproductive cycle is lytic or lysogenic, represents an interesting example of the ramifications of molecular-level control of gene expression. During the normal lytic reproduction cycle of lambda, transcription begins at two promoter sites during the early phase of reproduction (Figure 9.11). One of the promoter sites initiates clockwise transcription, and the other promoter site initiates counterclockwise transcription. The completion of the clockwise transcription of the phage genome requires the expression of a *Q* gene, which codes for a Q protein required for late gene expression. The complete counterclockwise transcription of the lambda genome requires an N protein. In the absence of N protein synthesis, transcription of the genes involved in the delayed early stage of phage reproduction cannot be made, and therefore, the replication of lambda phage DNA cannot proceed. Thus, both the *N* and *Q* genes must be expressed for the transcription of the complete gene.

The expression of both the N and Q proteins can be repressed. The lambda phage genome contains a *cI* gene that codes for a repressor protein that binds to the operator region of the phage genome that then controls the expression of the N protein, blocking the lytic reproductive cycle. The repressor protein also binds to another operator region, blocking the clockwise transcription of the lambda phage DNA and thus the production of the Q protein. In lysogenic reproduction only the repressor protein preventing the lytic reproduction of the phage is produced, and consequently, only the *cI* gene of the prophage is expressed. The expression of the *cI* gene is itself subject to regulation. Transcription of *cI* increases when the lambda repressor is present at low concentrations and high concentrations of repressor

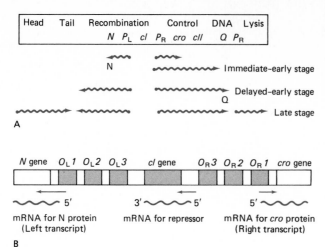

A

B

Figure 9.11

Diagram of lambda genome, showing promoter sites. (A) Three stages of transcription in the lytic growth of λ phage. The N protein formed in the immediate-early stage activates the delayed-early stage. In turn the Q protein is formed, which activates the late stage. Pr = promoter site. (After Echols, 1979, The Bacteria 8: 502.) (B) The O_L and O_R operator regions and the adjacent genes O_L1 and O_R1 have the highest affinity for the λ repressor. The repressor gene is cI. The left transcript starts with the N gene, whereas the right transcript starts with the cro gene.

protein inhibit further transcription. If the concentration of lambda repressor protein declines sufficiently to permit some further transcription of the phage genome, a protein, coded for by the *cro* gene, is produced. The cro protein represses further transcription of the *cI* gene, thus stopping synthesis of the repressor protein responsible for preventing complete expression of the phage genome. When this occurs the phage can carry out a normal lytic reproduction cycle. The reproductive cycle of bacteriophage is one of the few cases where details of the molecular-level events controlling development have been determined.

Plant viruses

Many plant viruses exhibit a reproductive cycle analogous to the lytic reproduction cycle of bacteriophage, involving adsorption of the virus onto a susceptible plant cell, penetration of the viral nucleic acid into the plant cell, assumption by the viral genome of control of the synthetic activities of the host cell, coding by the viral genome for the synthesis of viral nucleic acid and capsid components, assembly of the viral particles within the host cell, and finally, release of the complete viral particles from the host plant cell. In contrast to bacteriophage, both the capsid and nucleic acid core of viruses infecting eukaryotic cells may cross the cytoplasmic membrane by endocytosis with release of the nucleic acid from the capsid occurring within the host cell.

As an example of the reproductive cycle of a plant virus, let us examine tobacco mosaic virus (TMV). TMV has a single-stranded RNA genome that is contained within a helical array of protein subunits that comprise the viral capsid. Replica-

tion of TMV occurs within the cytoplasm of the infected cell. The RNA genome of TMV codes for an RNA-dependent RNA replicase enzyme that is used for the synthesis of a complementary RNA strand to serve as a template for the synthesis of the RNA genome of TMV. The RNA genome acts as a template for the synthesis of messenger RNA, which is subsequently translated at the plant cell ribosomes for the production of the protein coat subunits. Once the RNA and protein components of tobacco mosaic virus are synthesized, the assembly of the protein coat around the central RNA core can proceed spontaneously; that is, tobacco mosaic virus is self-assembled.

The initiation of assembly of TMV involves the attachment of the viral RNA to a protein disk subassembly of the core structure (Figure 9.12). The tobacco mosaic virus RNA is capable of binding with amino acids to initiate the assembly of the virus. The RNA molecule forms a loop, and the protein disk subunits are continuously added to the looped end of the RNA. The RNA overcomes the electrostatic forces that would prevent binding of the protein subunits, and without the RNA the protein subunits will not bind together at physiological pH and low ionic strength.

Within infected plant cells, the replicated tobacco mosaic viral particles form cytoplasmic inclusions (Figure 9.13). These viral inclusions are crystalline in nature. The chloroplast of a tobacco mosaic virus-infected leaf becomes chlorotic, leading to the death of the cell. The death of the plant cell releases completely assembled TMV particles and viral nucleic acid that has not been packaged with the protein subunits. Within plants both completely assembled viral particles and viral RNA can move from one cell to another, establishing new sites of infection. As a consequence

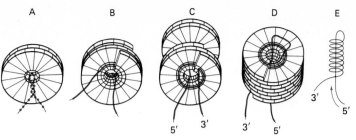

Figure 9.12

Diagram showing the self-assembly of TMV. (A) The initiation region of the RNA loops into the central hole of the protein disk and transforms it into (B) the helical lock-washer form; (C) additional disks add to the looped end of the RNA; (D) one of the RNA tails is continually pulled through the central hole to interact with incoming disks. (E) A schematic diagram of RNA in a partially assembled virus. (After P. J. G. Butler and A. Klug, 1978, Scientific American *239(5): 68–69.)*

Figure 9.13

Electron micrograph of a thin section of a tobacco leaf infected with TMV. (A) Note chloroplast at bottom and section of viral crystal showing alignment of multiple viral particles. (B) Note herringbone pattern of three-dimensional array of viral particles. (Courtesy Harry E. Warmke, University of Florida.)

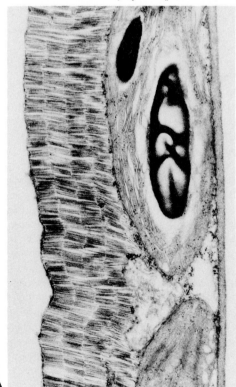

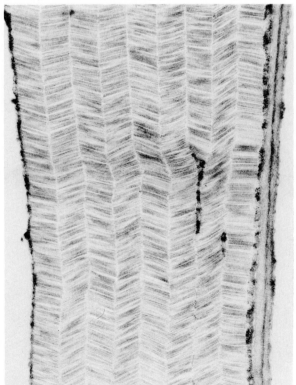

A

B

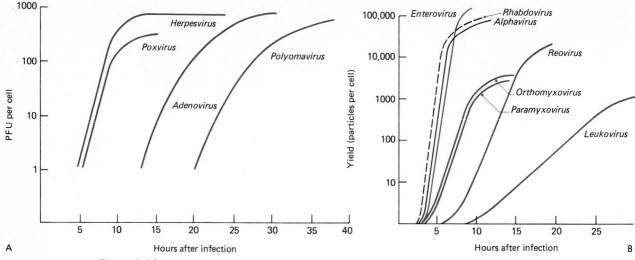

Figure 9.14

Growth curves for various DNA and RNA viruses. (A) Idealized multiplication curves (total infectious viruses per cell at intervals after infection at high multiplicity) of viruses representing the major genera of DNA viruses. (B) Idealized multiplication curves (total virus particles per cell at intervals after infection at high multiplicity) of viruses representing the major genera of RNA viruses (Enterovirus, poliovirus type 1; Alphavirus, Sindbis virus; Rhabdovirus, vesicular stomatitis virus; Orthomyxovirus, influenza type A; Paramyxovirus, Newcastle disease virus; Reovirus, reovirus type 3; Leukovirus, RAV-1). Latent periods, multiplication rates, and final yields are all affected by species and strain of virus and by cell strain and the kind of culture; latent period and multiplication rate, but not yield, are affected by multiplicity of infection.

of the reproduction of the viruses within the plant cells, the plant develops characteristic disease symptoms that include the appearance of a mosaic pattern of chlorotic spots on the leaves that gives both the disease and the virus their names.

Animal viruses

There are many types of animal viruses and many variations in the specific details of animal virus reproduction. In some cases the reproductive cycle of animal viruses closely resembles that of lytic bacteriophage, and in such instances there is a stepwise growth curve, with a burst of a large number of viruses released simultaneously. Unlike bacteriophage, however, the single-step growth curve for viruses occurs in hours rather than minutes (Figure 9.14). Though many viruses exhibit single-step growth curves characterized by the death of the host cell and the simultaneous release of a large number of viruses, some animal viruses characteristically do not kill the host cell and reproduce with a gradual slow release of intact viruses. Additionally, some animal viruses

transform the host cells, resulting in tumor formation, rather than death of the host cell.

The essential steps of the reproductive cycle for animal viruses are (1) **attachment** (adsorption of the virus to the surface of the animal cell); (2) **penetration** (entry of the intact virus or the viral genome into the host cell); (3) **uncoating** (release of the viral genome from the capsid); (4) **transcription** to form viral mRNA; (5) **translation** using viral-coded messenger RNA to form early proteins; (6) **replication of viral nucleic acid** to form new viral genomes; (7) **transcription** of messenger RNA to form late proteins needed for structural and other functions; (8) **assembly** of complete viral particles; and (9) **release** of new virions (Figure 9.15). Viruses appear to adsorb onto specific receptor sites on animal cell surfaces, and as a rule the entire viral particle enters the cell, often by endocytosis. Within the host cell, uncoating of animal viruses varies from one virus to another (Figure 9.16). The viral nucleic acids may be released at the cytoplasmic membrane, as occurs in enteroviruses, which are single-stranded RNA viruses; the virus may be uncoated in a series of complex steps within the host cell, as occurs in poxviruses, which are large double-stranded DNA viruses; or the virus may never be completely uncoated, as occurs in Reoviruses, which also are large, double-stranded RNA viruses. After uncoating, the genome of a DNA animal virus generally enters the nucleus, where it is replicated, whereas the genome of most RNA animal viruses need only enter the cytoplasm of the animal cell to be replicated. The diversity and complexity of animal virus reproduction make it impossible to go into further detail. Rather, a few representative examples of the reproduction of different types of animal viruses will be discussed.

Reproduction of DNA viruses

In the replication of Adenovirus, a representative

DNA virus, the host cell continues its normal metabolic activities for a short period of time after the entry of the virus into it. Uncoating the virus takes several hours, and during this period the viral nuclei acid is released from the capsid, entering the nucleus possibly through a nuclear pore. Within the nucleus, the viral genome codes for the inhibition of normal host cell synthesis of macromolecules. The viral genome also acts as a template for its own replication. Viral genes are transcribed, the resulting mRNA translated, and the proteins and enzymes needed for the assembly of the viral capsid are produced, with synthesis of viral proteins occurring at the ribosomes within the cytoplasm. The assembly of the Adenovirus particles occurs within the nucleus, and

Figure 9.15

Animal viral reproduction involves attachment, penetration, uncoating, transcription of viral nucleic acid, translation of early genes, replication of the viral genome, translation of late genes, assembly, and release.

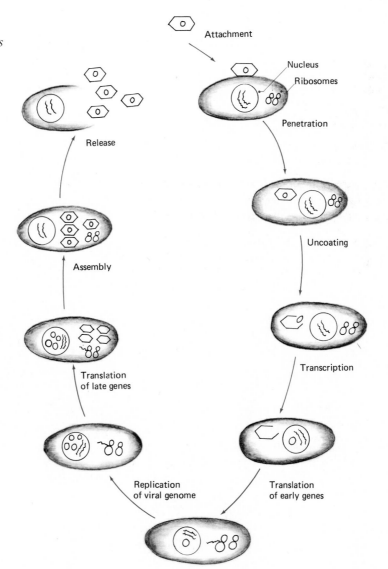

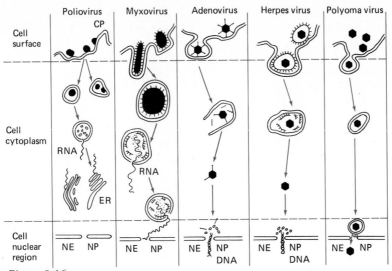

	Poliovirus	Myxovirus	Adenovirus	Herpes virus	Polyoma virus
Cell surface	CP				

Figure 9.16

Some of the mechanisms associated with the attachment, penetration, and uncoating of certain animal viruses as observed in the electron microscope from thin sections of infected cells sampled over known time courses. NE = nuclear envelope; NP = nuclear pore; CP = cytoplasmic membrane; ER = endoplasmic reticulum. (Based on S. Dales, 1973, Bacteriological Reviews *37:103–135.)*

therefore, the nucleus of an infected animal cell contains inclusion bodies consisting of crystalline arrays of densely packed Adenovirus particles (Figure 9.17). Upon lysis of the host cell, numerous Adenovirus particles are released.

Reproduction of RNA viruses

The RNA viruses exhibit many diverse strategies for reproduction. In the case of poliovirus, the RNA genome of the virus acts as a messenger RNA on entering the host cell, coding for the production of capsid proteins and an RNA-dependent RNA polymerase enzyme. Interestingly, the poliovirus RNA codes for a very large polypeptide chain, which is cleaved by protease enzymes, probably of the host cell, to form multiple different pro-

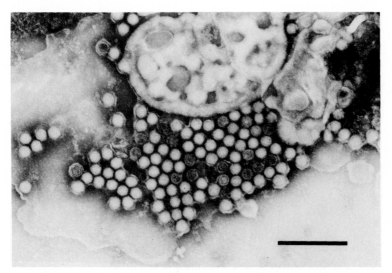

Figure 9.17

Electron micrograph of densely packed Adenovirus particles within the nucleus of an infected cell. (Courtesy F. Williams, EPA, Cincinnati.)

Figure 9.18

Poliovirus proteins are synthesized by multiple cleavages of a giant polypeptide precursor.

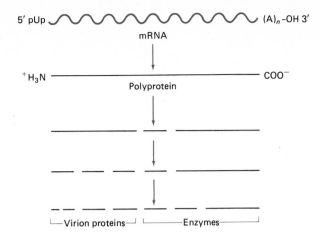

teins, including the RNA-dependent RNA polymerase enzyme and four proteins of the viral capsid (Figure 9.18). The RNA polymerase enzyme is used to produce a complementary **replicative RNA strand** that can act as a template for the synthesis of new viral genomes. The morphogenesis of the viral particle, that is, the assembly of the capsid and insertion of the RNA genome, is followed by the release of a large number of viral particles. The poliovirus is believed to code for enzymes that are involved in the cleavage of the cytoplasmic membrane of the host cell, and thus, the mechanism for release of poliovirus from the host cell resembles the release of lytic bacteriophage.

In the case of influenza viruses, the viral RNA genome serves as a template for the transcription of messenger RNA molecules within the nucleus of the host cell rather than acting as the messenger RNA, as occurs in poliovirus. The genome of the influenza viruses is **segmented**, consisting of eight different RNA molecules, which code for different monocistronic messenger RNA molecules, with each of the segments coding for separate mRNA molecules. The RNA-dependent RNA polymerase enzyme required for transcription of the viral genome is coded for by one of the RNA genome segments. The replication of the viral RNA genome involves the production of a complementary RNA strand that then serves as a template for the synthesis of new viral RNA genome molecules. The maturation of some viruses, such as influenza virus, occurs by **budding**, a process in which the viral particles are wrapped within a piece of the cytoplasmic membrane of the host cell (Figure 9.19). As a result, budding releases encapsulated (lipid-enveloped) influenza viruses slowly from infected host cells.

Reovirus, a double-stranded RNA virus, carries an RNA polymerase enzyme that is used for the synthesis of new viral genome molecules. The Reovirus genome is segmented, containing ten different double-stranded RNA molecules. Each of the ten RNA genome molecules codes for the production of a different protein. The proteins are then assembled into the viral capsid and the RNA genome is inserted prior to release of the completed Reoviruses.

Retroviruses are RNA viruses that use a reverse transcriptase to produce a DNA molecule within the host cell. The production of the DNA molecule requires an RNA-dependent DNA polymerase in order to carry out reverse transcription of the viral RNA. It is the DNA molecule "transcribed" from the viral RNA genome that actually codes for viral replication within the host cell. Retroviruses are released from host cells by budding, and viral replication is nondestructive to the host cell because it does not result in lysis and death of the infected cell. Thus, these viruses can be released slowly and continuously from infected host cells.

Transformation of animal cells

The DNA produced during the reproduction of retroviruses can also be incorporated into the host cell's chromosomes. Within the chromosomes of the host cell the viral genome can be transcribed, resulting in the production of viral-specific RNA and viral proteins. The DNA coded for by the virus, which is incorporated into the host cell genome, can be passed from one generation of animal cells to another. It is, therefore, possible for animals to inherit a viral genome. The presence of viral-derived DNA within the host cell can transform the animal cell. **Transformed cells** have altered surface properties and continue to grow even when they contact a neighboring cell, resulting in the formation of a tumor. Viruses that

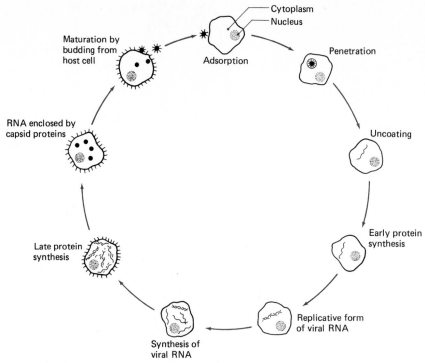

Cytoplasm
Nucleus

Maturation by
budding from
host cell

Adsorption

Penetration

RNA enclosed by
capsid proteins

Uncoating

Late protein
synthesis

Early protein
synthesis

Synthesis of
viral RNA

Replicative form
of viral RNA

Figure 9.19

*The life cycle of the influenza virus, showing budding release. The replication of
the influenza virus illustrates many of the features that characterize animal virus
interaction with host cells. The virion is taken into the cell intact; it is then
uncoated, which releases the single-stranded RNA genome in seven segments (only
one of which is shown here). A replicative form (Rf) is made for each segment. The
Rf in turn serves as the template for the synthesis of new genomes. As the late
proteins are synthesized, they come to lie just beneath the cell membrane. Virions
exist by budding through the membrane. A segment of the host cell's membrane
becomes their envelope. It is chemically modified by the inclusion of the viral
proteins hemagglutin and neuraminidase. In vitro virus production is eventually
brought to a halt by interferon. Virus replication does not lyse the cell.*

transform cells and cause cancerous growth are
called **oncogenic viruses**. Several different retroviruses produce malignancies within infected cells.
Rous sarcoma virus, for example, is an RNA tumor
virus that causes malignancies in chickens. Similarly, some DNA viruses, such as Simian virus
(SV40) and Polyoma virus, are capable of transforming host cells and producing malignant tumors. These oncogenic viruses reproduce in permissive hosts by using a normal lytic reproduction
cycle. Viral reproduction does not occur in non-
permissive host cells, but rather, part of the viral
genome is incorporated into the host cell genome, resulting in the transformation of the host
cell harboring the viral genome.

Quantitative assays for animal viruses

It is possible to assay quantitatively for animal
viruses in a method analogous to the plaque assay
for enumeration of bacteriophage. In a typical
procedure, a tissue culture monolayer of animal
cells growing on a plate surface is inoculated with
dilutions of a viral suspension and incubated for
various periods of time. Viral infection of the animal tissue culture cells may result in plaque formation, indicative of localized death of animal cells,
which can be observed microscopically (Figure
9.20). The number of plaques that form and the
dilution factors are employed to determine the
concentration of viruses in the sample. Additionally, virus-infected animal cells often develop ab-

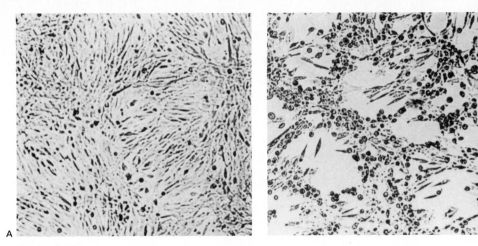

A B

Figure 9.20

Micrographs showing plaques in tissue culture. (A) Normal unstained bovine embryonic kidney cells (BEK, 40 ×), (B) Cytopathic effects (CPE) of infectious bovine rhinotracheitis virus (bovine herpes virus 1) in monolayers of BEK cells 48 hours after inoculation (unstained, 40 ×). (Reprinted by permission of Lea and Febiger, Philadelphia, from S. B. Mohanty and S. K. Dutta, 1981, Veterinary Virology.*)*

normally, which is visible as a change in their appearance, known as a **cytopathic effect (CPE)** (see Figure 16.5). It is possible to observe cytopathic effects in animal cell cultures and to count the number of cells exhibiting characteristic mor-

phological changes to determine the number of viruses. Hemagglutination, an immunological reaction that will be discussed in detail in Chapter 15, can also be used for determining the numbers of some viruses, such as influenza viruses. Hem-

Figure 9.21

Multiplication cycle for animal viruses that are released from the cell late and incompletely.

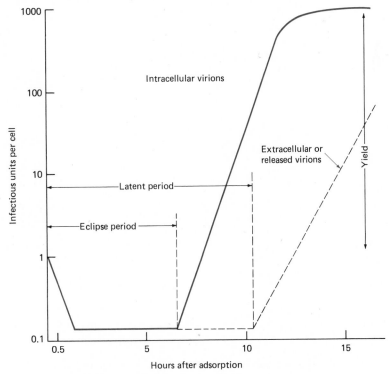

agglutination involves a reaction in which red blood cells clump together or agglutinate. By mixing red blood cells with dilutions of a viral suspension, it is possible to determine the greatest dilution that still brings about hemagglutination.

By using an appropriate method to quantitate the number of infectious animal viruses, a growth curve can be established (Figure 9.21). Many animal viruses exhibit a single-step growth curve for normal reproduction that includes an eclipse period during which infectious virions disappear and reproduction of viral particles occurs. At the end of the eclipse period, new viral progeny appear within the host cell, but often there is a further delay before the newly formed viruses are released from the host cell, except for viruses released by budding. Thus, the latent period, the time from the adsorption of the virus onto the host cell until the release of new viruses, generally exceeds the eclipse period.

Bacterial reproduction

Binary fission

Bacteria normally reproduce by **binary fission** with the formation of two equal-sized progeny cells (Figure 9.22). Reproduction requires the replication of the bacterial chromosome so that each daughter cell receives a complete genome. During cell division the bacterial chromosome appears to be attached to the cytoplasmic membrane and cell wall. The inward movement of the cytoplasmic membrane and cell wall, **septa formation**, pinches off and separates the two complete bacterial chromosomes, providing each of the progeny cells with a complete genome. The formation of septae physically cuts apart the bacterial chromosomes and distributes them to the two daughter cells. It may take twice as long, however, to duplicate the bacterial chromosome as

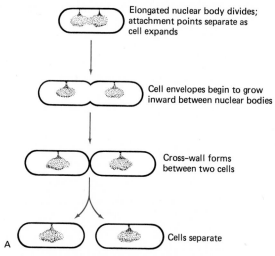

Elongated nuclear body divides; attachment points separate as cell expands

Cell envelopes begin to grow inward between nuclear bodies

Cross-wall forms between two cells

Cells separate

A

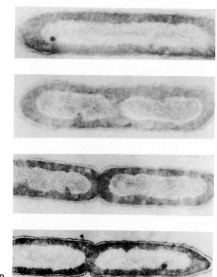

B

Figure 9.22

In binary fission ingrowth of the new wall material, septae formation, progresses to separate the two daughter cells, assuring the equal distribution of genetic information between the cells. (A) Schematic of the binary fission process. (B) Micrographs showing stages of cell division for synchronously grown Erwinia *showing invagination of cytoplasmic membrane and formation of cross-wall septum. (Reprinted by permission of the American Society for Microbiology, Washington, D.C., from E. A. Grula, and G. L. Smith, 1965,* Journal of Bacteriology *90:1054–1058.)*

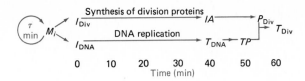

Figure 9.23

In the cell cycle of Escherichia coli, doubling of the initiation mass (M_i) takes place every mass doubling time (τ min). At each doubling two processes are initiated almost simultaneously. These are the initiation of DNA replication (I_{DNA}) and the initiation of the sequence of events leading to division (I_{Div}). Termination of chromosome replication (T_{DNA}) at 40 minutes induces the synthesis of termination protein (TP). The first 40 minutes of the division sequence involves protein synthesis, which is then followed by the initiation of assembly (IA). After 15–20 minutes more, the cell has reached a state (P_{Div}) where interaction between some septum "primordium" and termination protein leads to cell division (T_{Div}). Our picture of the timing of the main events in the cycle is therefore of periodic events that occur at intervals equal to the mass doubling time of the cell and at multiples of a constant cell mass (M_i). This event triggers two parallel but separate sequences of events that take constant periods of time to complete, largely independent of the rate of cell growth. One of these processes is chromosome replication and synthesis of the termination protein and requires 40–45 minutes to complete. The other is a sequence of protein synthesis, followed by another process, which may be assembly of some septum precursor. This process requires nearly 60 minutes, and at the end of it there is an interaction between the septum precursor(s) and the termination protein to give the final septum and cell division. This last event takes only a few minutes.

the divisional cell cycle in rapidly growing bacteria. In *E. coli*, for example, the replication of the complete genome takes 40 minutes, but the doubling time for *E. coli* under optimal conditions is only 20 minutes. In order to meet this need, the bacterium initiates a new round of DNA replication every 20 minutes, that is, every time the cell divides, even though the previously initiated replication forks have not yet reached the terminus. By initiating a new round of DNA replication every time the cells divide, the bacteria are able to produce completely duplicated genomes in time for cell division, with DNA replication occurring at the same frequency as cell reproduction.

To accomplish this, cell division must be synchronized with the replication of the bacterial chromosome. The regulatory mechanism is such that cell division takes place 20 minutes after the completion of the replication of the bacterial chromosome (Figure 9.23). Thus, the entire cell cycle for *E. coli* would be 60 minutes, 40 minutes for replication of the bacterial chromosome and a 20-minute interval after the completion of genome replication before cell division occurs. Completion of the replication of the bacterial chromosome is a prerequisite for cell division, and if the termination of DNA replication is blocked, the cell division that normally would occur 20 minutes later is prevented. The transcription of specific genes, which are required for cell division, occurs at or immediately after the termination of replication of the bacterial chromosome.

The control of the cell division cycle is complex, and there appear to be several different regulatory mechanisms that control cell division. Cell division appears to be under genetic control, and mutants have been isolated that fail to divide properly. Division also depends on many metabolic activities of the cell, and the actual step of cell division is related to the relative timing of the synthesis of various macromolecules. Cell division depends on DNA metabolism and protein synthesis. The bacterial genome must be replicated, and several proteins must be synthesized at or near the end of the DNA replication cycle for cell division to occur.

In addition to requiring duplicate bacterial chromosomes, cell division necessitates the synthesis of new cell wall and cytoplasmic membrane structures to surround and protect the daughter cells. During the reproductive cycle of bacteria, the parent cell elongates, and the cell wall grows inward from opposing sides of the bacterial cell, establishing a cross wall (Figure 9.24). Mesosomes associated with cross-wall formation may be involved in the transport of macromolecules, including peptidoglycan components, to the outside of the cell. The cross septum that divides the cells consists both of cell wall and cytoplasmic membrane. The synthesis of a new wall may occur only at the site of cross-wall formation or may occur at several sites within the cell wall of the parent cell (Figure 9.25). In several Gram positive bacteria, such as *Bacillus* species, cell wall synthesis appears to be restricted to the region near cross wall formation, but in several Gram nega-

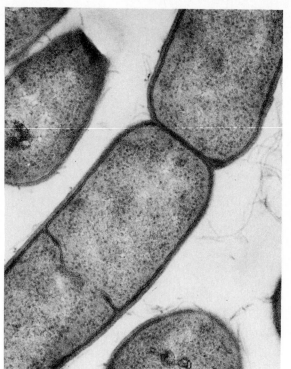

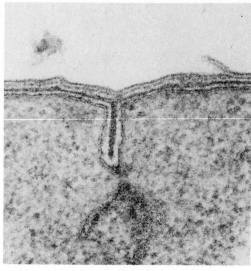

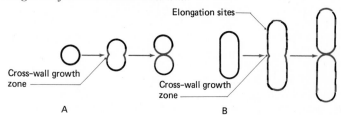

Figure 9.24

*Electron micrographs showing cross-wall
formation (A) in* Bacillus circulans *(48,000 ×);
(B) thin section of ingrowing septum of*
Sporosarcina ureae. *(From BPS—T. J. Beveridge,
University of Guelph.)*

tive bacteria cell wall synthesis appears to occur
at various sites throughout the cell wall of the
bacterium and is not limited to the site of cross-
wall formation. Upon completion of the cross-
wall formation, there are two equal-size cells that
can separate. Repeating the process results in the
multiplication of the bacterial population.

Other modes of reproduction

Although binary fission is the most common mode
of bacterial reproduction, some bacteria repro-

duce by other means. The predominant modes
of reproduction in all bacteria are asexual. The
various reproductive modes differ in how the cellular material is apportioned between the daughter cells and whether the cells separate or remain
together as part of a multicellular aggregation.
For example, *Hyphomicrobium* reproduces by
budding, a type of division characterized by an
unequal division of cellular material (Figure 9.26).
The daughter cell develops when a cross wall
forms, segregating a portion of the cytoplasm
containing a duplicate genome. Each daughter cell
is smaller in size than the parental "mother" cell.

Figure 9.25

*Sites of cell wall synthesis. (A) In a spherical Gram positive
coccus, new cell wall is added at the central zone where the
cross wall forms during cell division. Each daughter cell keeps
the coccus shape. (B) In a Gram negative rod-shaped bacterium,
new cell wall is added at multiple sites. A large area of new wall
originates from the added cross wall.*

Elongation sites

Cross-wall growth
zone

Cross-wall growth
zone

A

B

In the case of the actinomycetes, such as *Streptomyces*, reproduction involves the formation of **hyphae**. In this mode of reproduction the cell elongates, forming a relatively long and generally branched filament or hypha (Figure 9.27). Cross-wall formation results in the formation of individual cells containing complete genomes. Regardless of the mode of reproduction, bacterial multiplication requires replication of the genome and synthesis of new boundary layers, including cell-wall and cytoplasmic membrane structures.

Sporulation

In addition to binary fission and other modes of vegetative cell reproduction, some bacteria produce **spores** (specialized resistant resting cells). The production of spores represents an interesting variation of the normal binary fission mode of replication. Various types of spores, including **endospores** (heat-resistant spores formed within the cell), **myxospores** (resting cells of the myxobacteria formed within a fruiting body), **cysts** (resting or dormant cells sometimes enclosed in

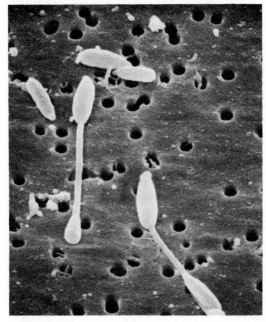

Figure 9.26

Micrograph of Hyphomicrobium, *showing budding form of replication. (Courtesy Richard L. Moore, University of Calgary.)*

307

Bacterial
reproduction

Figure 9.27

Micrograph of Streptomyces *showing hyphal filaments; some hyphae show branching and others are aerial (1584×). (From BPS—Centers for Disease Control, Atlanta.)*

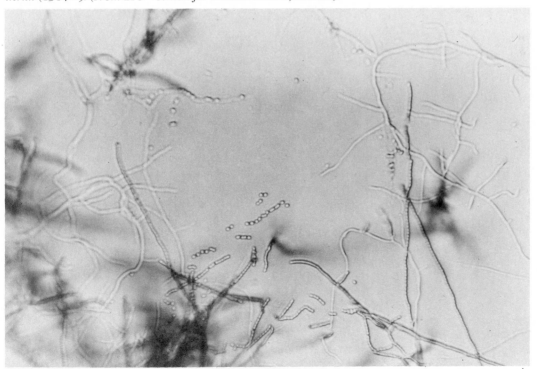

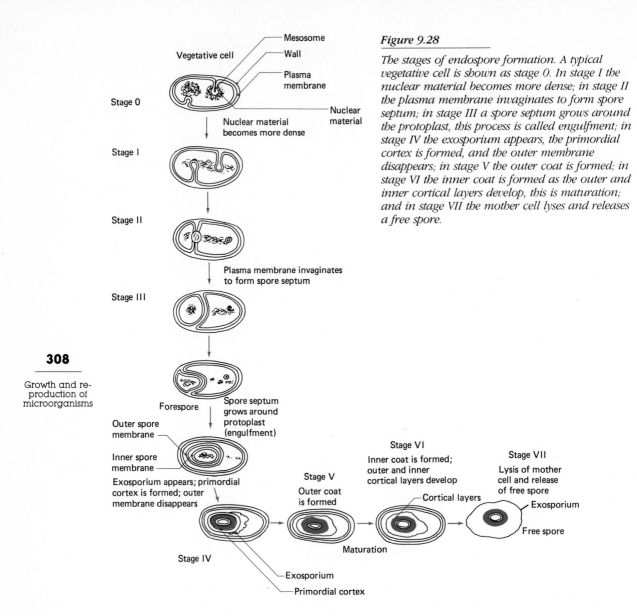

Figure 9.28

The stages of endospore formation. A typical vegetative cell is shown as stage 0. In stage I the nuclear material becomes more dense; in stage II the plasma membrane invaginates to form spore septum; in stage III a spore septum grows around the protoplast, this process is called engulfment; in stage IV the exosporium appears, the primordial cortex is formed, and the outer membrane disappears; in stage V the outer coat is formed; in stage VI the inner coat is formed as the outer and inner cortical layers develop, this is maturation; and in stage VII the mother cell lyses and releases a free spore.

Vegetative cell

Mesosome

Wall

Plasma membrane

Stage 0

Nuclear material

Nuclear material becomes more dense

Stage I

Stage II

Plasma membrane invaginates to form spore septum

Stage III

Forespore

Spore septum grows around protoplast (engulfment)

Outer spore membrane

Inner spore membrane

Exosporium appears; primordial cortex is formed; outer membrane disappears

Stage IV

Exosporium

Primordial cortex

Stage V

Outer coat is formed

Maturation

Stage VI

Inner coat is formed; outer and inner cortical layers develop

Cortical layers

Stage VII

Lysis of mother cell and release of free spore

Exosporium

Free spore

a sheath), and **arthrospores** (spores formed by the fragmentation of hyphae), are produced by different bacteria as part of their reproductive cycles.

For example, under certain conditions *Bacillus* and *Clostridium* do not divide by binary fission to form two equal cells but rather sporulate to form endospores. **Sporulation** and normal cell division are mutually exclusive processes. Normally sporulation is repressed, but it can be initiated during starvation conditions by derepression of transcription involving one or more nitrogen-containing phosphorylated repressor compounds. At some point the process of sporulation becomes irreversible and sporulating bacteria, committed to endospore formation, continue spore

formation even when starvation is relieved and conditions suitable for growth are restored. It is important that this occurs only when conditions are unfavorable for normal cell division so that the bacterium need only expend its metabolic energy on spore formation for survival when it cannot otherwise multiply. There are more than 30 sporulation gene loci controlling endospore formation. The triggering of these genes appears to involve a shift in RNA polymerase functions, mediated by the sigma subunit; the RNA polymerase of sporulating cells is inefficient at synthesizing mRNA for vegetative functions.

During the sporulation process there is an invagination of the cytoplasmic membrane within the cell to establish the site of endospore formation (Figure 9.28). A cross wall does not form during endospore production, but instead the endospore forms within the parent cell. A copy of the bacterial chromosome is incorporated into the endospore, and the various layers of the endospore are then synthesized around the bacterial DNA. The initial events in sporulation represent a slight modification of the normal vegetative cell division cycle, and no unique structural genes are required for the early events of the sporulation process. Many enzymatic activities increase, including alkaline phosphatase, exoprotease, and the enzymes of the Krebs cycle during this early period of sporulation. Late events in the formation of the endospore require the synthesis of biochemicals that are not made during the normal cell division cycle. These biochemicals are needed for the spore cortex and spore coat. They are specifically related to spore formation and include dipicolinic acid, involved in conferring heat resistance to the spore, and polypeptides composed almost exclusively of single amino acids, such as cystine. The formation of the completed spore involves the synthesis of two cell wall layers and the formation of a spore cortex. Once the endospore is formed within the parent cell, it can be released by lysis of the parent cell. The endospore is a highly resistant structure and can remain metabolically dormant for a prolonged period of time.

At a later time the spore can germinate (Figure 9.29). Upon suitable stimulation, such as heat shocking, the spore swells and breaks out of the spore coat, and the germ cell elongates. One of the striking features of spore **germination** is the speed at which metabolism shifts from a state of dormancy to the high activity levels that characterize an outgrowing cell. This shift in metabolic activity can occur within minutes. Spore germination is a degradative process in which enzymes become activated. During germination the endospore initially loses up to 30 percent of its dry weight. The endospore is metabolically self-sufficient. ATP generation, RNA synthesis, and protein synthesis can take place for at least 15 minutes using the endogenous energy and substrates, principally phosphoglycerate, contained in the spore. After spore germination, the organism renews normal vegetative growth.

Enumeration of bacteria

In order to assess rates of microbial reproduction, it is necessary to determine numbers of microorganisms. There are a variety of methods that can be employed for **enumerating bacteria**.

These include viable plate count, direct count, and most probable number determinations.

Viable count procedures

The **viable plate count** method is one of the most common procedures for the enumeration of bacteria (Figure 9.30). In this procedure serial dilutions of a suspension of bacteria are plated onto a suitable solid growth medium. Normally agar is used to solidify a medium containing the nutrients essential for the growth of the bacteria being enumerated. The suspension is spread over the surface of the agar plate (**surface spread technique**) or mixed with the agar prior to allowing it to solidify and pouring it into the plate (**pour plate technique**) (see Chapter 2). In some cases, when bacterial numbers in a sample are low, it is necessary to filter the suspension in order to concentrate the bacterial cells. The membrane filter can then be placed onto a suitable medium, and the colonies that develop on the filter can be counted. The agar plates are incubated under conditions suitable for bacterial growth. Multiplication of a bacterium on solid-media results in the formation of a macroscopic colony visible to the naked eye. It is assumed that each colony arises from an individual bacterial cell, and therefore, by counting the number of colonies that develop and by taking into account the dilution factors, the concentration of bacteria in the original sample can be determined. A major limitation of the viable plate count procedure is that it is selective. There is no set of incubation conditions and medium composition that permits the growth of all bacterial types. The nature of the growth medium and the incubation conditions determine which bacteria can grow.

Direct count procedures

Bacteria can also be enumerated by **direct counting procedures**. In this method dilutions of samples are observed under a microscope, and the numbers of bacterial cells in a given volume of sample are counted and used to calculate the concentration of bacteria in the original sample. Special counting chambers, such as a hemocytometer or Petroff-Hausser chamber, are frequently employed to determine the number of bacteria (Figure 9.31). In order to help visualize bacterial cells, it is often desirable to stain the cells. Fluorescent dyes are often used to stain bacteria in direct count procedures. Such dyes stain all cells, making it impossible to differentiate living from dead bacteria. The difficulty in establishing the metabolic status of the observed bacteria is a major limitation of this procedure.

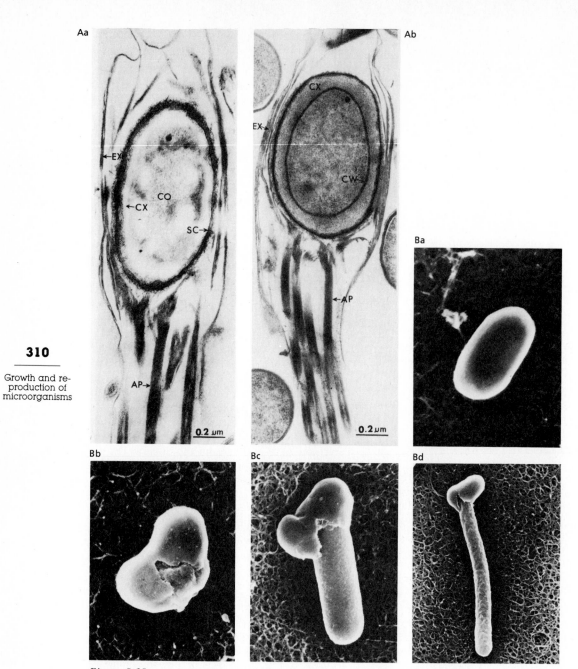

Figure 9.29

Micrographs showing spore germination. (A) (a) A thin section of a dormant spore of Clostridium
bifermentans; *the components include appendages (AP), core (CO), cortex (CX), spore coat layers (SC), and
exosporium (EX). After the stages of elongation and outgrowth, (b) the bacterial cell is constricted by the
spore coat as it emerges, and the cortex material (arrows) is squeezed from the remains of the spore.
(Reprinted by permission American Society of Microbiology from W. A. Samsonoff et al., 1970,* Journal of
Bacteriology *101: 1038–1045.) (B) (a) Scanning electron micrograph of a spore of* Bacillus subtilis *strain
(17,850×) 0 minutes following heat activation at 60°C for 1 hour; (b) after 90 minutes incubation on
nutrient agar at 37°C, a crack can be seen in an equatorial position (17,850×); (c) after 120 minutes on
an agar surface, the vegetative bacterium has started to grow (13,387×); (d) after 150 minutes incubation
on agar, the vegetative bacterium has grown extensively, the bacterium retains an empty spore shell on one
or both ends, and the cell is about 4–5 μm. (Courtesy of Akiko Umeda, Fukora University.)*

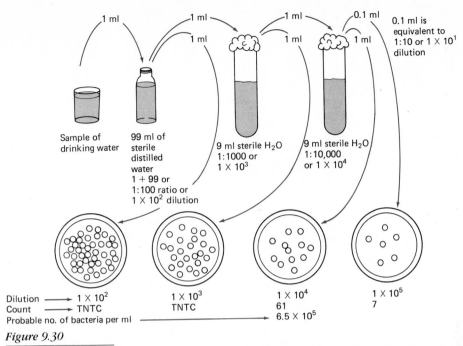

Dilution	1×10^2	1×10^3	1×10^4	1×10^5
Count	TNTC	TNTC	61	7
Probable no. of bacteria per ml			6.5×10^5	

Figure 9.30

The viable plate count procedure is used to determine the viable population in a bacterial culture. Here plating is accomplished using the pour plate technique.

Most probable number procedures

Another approach to bacterial enumeration, determination of the **most probable number (MPN)**, is a statistical method based on probability theory. In a most probable number enumeration procedure, multiple serial dilutions are performed, and a criterion, such as the development of cloudiness or turbidity in a liquid growth medium, is established for indicating whether a particular dilution replicate contains bacteria. The pattern of positive test results and statistical probability tables are then used to determine the concentration of bacteria in the original sample, that is, the most probable number of bacteria (Figure 9.32).

Procedures based on quantifying specific biochemical constituents

Various other procedures, based on the detection of specific microbial macromolecules or metabolic products, can be used to estimate numbers

Figure 9.31

The direct counting procedure, using the Petroff-Hausser counting procedure. The sample is added to a counting chamber of known volume. The slide is viewed and the number of cells in an area delimited by a grid determined. In the counting chamber shown, the whole grid has 25 large squares for a total area of 1 mm² and a total volume of 0.02 mm³.

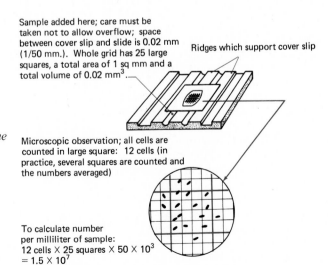

Sample added here; care must be taken not to allow overflow; space between cover slip and slide is 0.02 mm (1/50 mm.). Whole grid has 25 large squares, a total area of 1 sq mm and a total volume of 0.02 mm³.

Ridges which support cover slip

Microscopic observation; all cells are counted in large square: 12 cells (in practice, several squares are counted and the numbers averaged)

To calculate number per milliliter of sample:
12 cells × 25 squares × 50 × 10³
= 1.5 × 10⁷

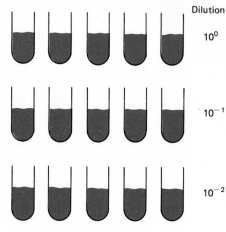

Dilution

10^0

10^{-1} } 5 replicate tubes at each dilution

10^{-2}

1 — Inoculate replicate tubes
2 — Incubate
3 — Record number of positive tubes at each dilution
4 — Consult MPN table
5 — Record most probable number of bacteria

Growth and re-
production of
microorganisms

Figure 9.32

In the five-tube most probable number (MPN) procedure, five replicate tubes are used at each dilution. (A) First inoculate replicate tubes and incubate an appropriate period of time; then record the number of positive tubes at each dilution; consult an MPN table (B); and finally, record the most probable number of bacteria.

Number of positive tubes at the stated dilution

10^0	10^{-1}	10^{-2}	MPN/100 ml	10^0	10^{-1}	10^{-2}	MPN
0	1	0	0.18	5	0	0	2.3
1	0	0	0.20	5	0	1	3.1
1	1	0	0.40	5	1	0	3.3
2	0	0	0.45	5	1	1	4.6
2	0	1	0.68	5	2	0	4.9
2	1	0	0.68	5	2	1	7.0
2	2	0	0.93	5	2	2	9.5
3	0	0	0.78	5	3	0	7.9
3	0	1	1.1	5	3	1	11.0
3	1	0	1.1	5	3	2	14.0
3	2	0	1.4	5	4	0	13.0
4	0	0	1.3	5	4	1	17.0
4	0	1	1.7	5	4	2	22.0
4	1	0	1.7	5	4	3	28.0
4	1	1	2.1	5	5	0	24.0
4	2	0	2.2	5	5	1	35.0
4	2	1	2.6	5	5	2	54.0
4	3	0	2.7	5	5	3	92.0
				5	5	4	160.0

B

of microorganisms. For example, murein can be quantitated, and because this biochemical occurs exclusively in the cell wall of bacteria, the concentration of murein can be used to estimate bacterial numbers. Similarly, lipopolysaccharide concentrations can be utilized to estimate numbers of Gram negative bacteria, as this biochemical occurs primarily in the outer cell wall structure of Gram negative bacteria. More general estimates of microbial numbers can be obtained indirectly by measuring protein, ATP, and DNA concentrations. All these biochemical approaches for determining bacterial numbers depend on the development of analytical chemical procedures for quantifying the particular biochemical and determining what proportion of a bacterial cell is composed of the specific biochemical constituent. Problems with these approaches involve devel-

oping appropriate conversion factors for equating the concentration of the particular biochemical with the actual number of bacterial cells and establishing that the particular biochemical is exclusively of bacterial origin.

Kinetics of bacterial growth

Because the optical density of the culture is proportional to the cell density (Figure 9.33), measuring the turbidity of the solution can be used for estimating numbers of bacterial cells. The increased turbidity of a liquid medium resulting from bacterial growth can be measured with a spectrophotometer (Figure 9.34) and used to establish a growth curve for a bacterium. Other methods that measure numbers of bacteria are

Figure 9.33

This graph shows relationship of optical density (OD) to cell number. Here, an optical density reading of 0.6 indicates a cell count of 2.0 × 10⁸.

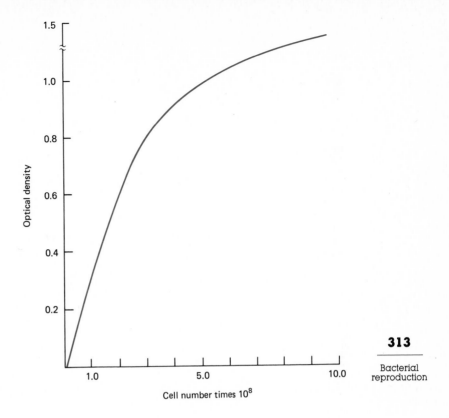

Figure 9.34

A spectrophotometer is used to measure turbidity by monitoring light transmission. The reading on the spectrophotometer is used to estimate numbers of cells in the microbial suspension.

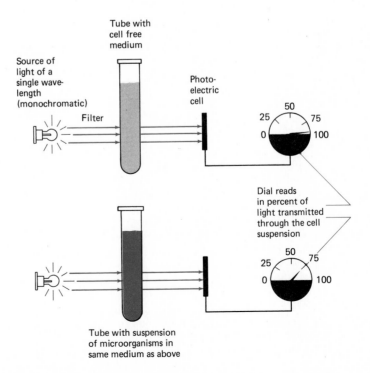

Geometric progression

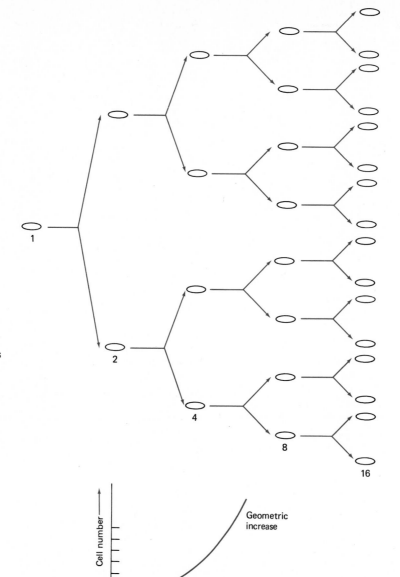

1

2

4

8

16

Cell number

Geometric increase

Time

also used for determining bacterial growth curves but are often more tedious than turbidometric procedures.

By its very nature, bacterial reproduction by binary fission results in doubling of the number of viable bacterial cells, and therefore, during active bacterial growth, the size of the microbial population is continuously doubling. Once cell division begins it proceeds exponentially, with one cell dividing to form two, each of these cells di-

viding so that four cells form, and so forth in a geometric progression (Figure 9.35). The time required to achieve a doubling of the population size, known as the generation time, is the unit of measure of microbial growth rate (Table 9.1). The generation time for bacteria can be expressed mathematically as

$$g = t/(\log_2 B_t - \log_2 B_0)$$

where g is the generation time, $\log_2 B_t$ is the log-

table 9.1

Growth rates for some representative bacteria under optimal conditions

Organism	Temperature °C	Generation time (min)
Bacillus stearothermophilus	60	11
Escherichia coli	37	20
Bacillus subtilis	37	27
Bacillus mycoides	37	28
Staphylococcus aureus	37	28
Streptococcus lactis	37	30
Pseudomonas putida	30	45
Lactobacillus acidophilus	37	75
Vibrio marinus	15	80
Mycobacterium tuberculosis	37	360
Rhizobium japonicum	25	400
Nostoc japonicum	25	570
Anabaena cylindrica	25	840
Treponema pallidum	37	1980

arithm to the base 2 of the number of bacteria at time t, $\log_2 B_0$ is the logarithm to the base 2 of the number of bacteria at the starting time, and t is the time period of growth. This equation can be converted to base 10 using the relationship $\log_2 B = 3.3 \log_{10} B$. The use of the natural logarithm is preferable conceptually because the unit of growth is then equal to one generation or doubling of cell number ($B_t = B_0 \times 2^n$, where n is the generation time, B_0 is the number of bacteria at the starting time, B_t is the number of bacteria at time t, and t is the time period of growth). By determining cell numbers during the period of active cell division, the generation time can be estimated (Figure 9.36). Comparing generation times, bacteria reproduce more rapidly than do higher organisms. A bacterium such as *E. coli* can have a generation time as short as 20 minutes under optimal conditions, although in nature many bacteria have generation times of several hours. Considering a bacterium with a 20-minute generation time, one cell would multiply to 1000 cells in 3.3 hours and 1,000,000 cells in 6.6 hours.

Normal growth curve

When a bacterium is inoculated into a new culture medium, it exhibits a characteristic growth curve (Figure 9.37). The **normal growth curve of bacteria** has four phases, the lag phase, the log or exponential growth phase, the stationary phase, and the death phase. During the **lag phase** there is no increase in cell numbers; rather, during this phase the bacteria are preparing for reproduction, synthesizing DNA and various inducible en-

zymes needed for cell division. During the **log phase** of growth, so named because the logarithm of the bacterial biomass increases linearly with time, bacterial reproduction occurs at a maximal rate for the specific set of growth conditions. During this phase of the growth curve, the **instantaneous growth rate** of a bacterium is proportional to the biomass of bacteria that is present (dB/dt

Figure 9.36

The estimation of the generation time of a bacterium based on observed cell numbers during a period of active division.

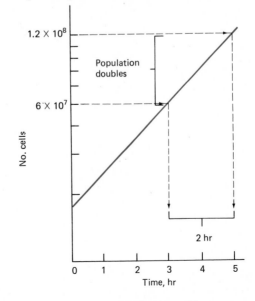

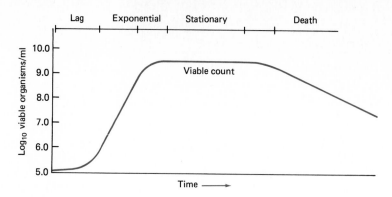

$= \alpha B$, where B is the bacterial biomass, t is time, and α is the instantaneous growth rate constant). During this period the generation time of the bacterium is determined. If a bacterial culture in the exponential growth phase is inoculated into an identical fresh medium, the lag phase is bypassed, and exponential growth continues. This occurs because bacteria are already actively carrying out the metabolism necessary for continued growth. If, however, the chemical composition of the new medium differs significantly from the original growth medium, the bacteria do go through a lag phase before entering the logarithmic growth phase when they synthesize the enzymes needed for growth in the new medium.

However, if the bacterium is not transferred to a new medium and no fresh nutrients are added, the **stationary growth phase** eventually is reached,

and there is no further net increase in bacterial cell numbers. During the stationary phase the growth rate is exactly equal to the death rate. A bacterial population may reach stationary growth when a required nutrient is exhausted, when inhibitory end products accumulate, or when physical conditions are appropriate. In all cases there is a feedback mechanism that regulates the bacterial enzymes involved in key metabolic steps. The duration of the stationary phase varies, with some bacteria exhibiting a very long stationary phase. Eventually, however, the number of viable bacterial cells begins to decline, signalling the onset of the **death phase**. The kinetics of bacterial death follow the same exponential kinetics as the logarithmic growth phase. This is true because the death phase really represents the result of the inability of the bacteria to carry out further re-

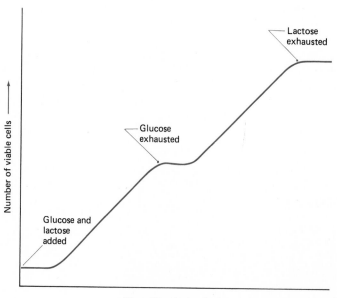

Figure 9.38

A biphasic growth curve reflects the preferential utilization of substrates and the phenomenon of diauxie.

production. The rate of the death phase need not, however, be equal to the rate of growth during the exponential phase. Modifying environmental conditions can alter both the rate of exponential growth and the death rate of a bacterium (see Chapter 10 for discussion of the effects of environmental parameters on microbial growth and death rates). The rate of death is proportional to the number of the survivors and can be expressed as

$$dt = \log_2 B_0 - \log_2 B_t$$

where d is the death rate, t is the time period, $\log_2 B_0$ is the natural logarithm of the number of bacteria at the starting time, and $\log_2 B_t$ is the natural logarithm of the survivors at time t.

Biphasic growth curve

Under some conditions microorganisms exhibit a **biphasic growth curve** rather than a normal growth curve (Figure 9.38). The biphasic growth curve reflects a sequential utilization of substrates. For example, cultures of E. coli exhibit biphasic growth when inoculated into a medium containing both glucose and lactose, indicative of the preferential metabolism of glucose before lactose as a substrate; the underlying regulatory mechanisms controlling the metabolism of lactose were discussed in Chapter 7. While growing on glucose, E. coli exhibits the normal lag, log, and stationary phases of growth. Rather than exhibiting a prolonged stationary phase, E. coli enters a secondary lag phase when the glucose has been completely exhausted. During this second lag phase allolactose acts as an inducer to derepress the lac operon system. The enzymes that are necessary for lactose metabolism are synthesized and the bacteria begin to grow exponentially by using the lactose substrate. When the lactose is completely used up, the bacteria enter a second stationary phase.

Batch and continuous culture

The normal bacterial growth curve is characteristic of bacteria in **batch culture**, that is, under conditions when a fresh medium is simply inoculated with a bacterium. A flask containing a liquid nutrient medium inoculated with a bacterium such as E. coli is an example of a batch culture. In batch culture growth nutrients are expended, and metabolic products accumulate in the closed environment. The batch culture models situations

such as occur when a canned food product is contaminated with a bacterium.

Bacteria may also be grown in **continuous culture** in a **chemostat** where nutrients are supplied and end products continuously removed so that the exponential growth phase is maintained. Because end products do not accumulate and nutrients are not completely expended, the bacteria never reach stationary phase. Continuous growth of bacteria can be accomplished by using a chemostat, a device in which a liquid medium is continuously fed into the bacterial culture (Figure 9.39). The liquid medium contains some nutrient in growth-limiting concentrations, and the concentration of the limiting nutrient in the growth medium determines the rate of bacterial growth. Even though bacteria are continuously reproducing, a number of bacterial cells are continuously being washed out and removed from the culture vessel.

Because the rate of cell washout is equal to the growth rate, the dilution rate is equal to the instantaneous growth rate of a bacterium growing in a chemostat. The instantaneous growth rate can be expressed as $\mu = 0.693k$, where μ is the instantaneous growth rate of the culture and k is the cell growth rate constant ($1/k$ = cell doubling or generation time). The relationship between the culture generation time and the concentration of the limiting substrate is

$$\mu = \mu_{max} \times s/(k_s + s)$$

Figure 9.39

A chemostat is a continuous culture device. In such a device the population density is controlled by the concentration of the limiting nutrient and the growth rate by the flow rate, which can be set arbitrarily.

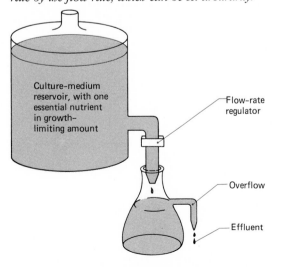

Culture-medium reservoir, with one essential nutrient in growth–limiting amount

Flow-rate regulator

Overflow

Effluent

where μ is the instantaneous culture generation time, μ_{max} is the maximal growth rate at saturating concentrations of substrate, s is the substrate concentration, and k_s is the saturation constant, defined as the substrate concentration at $\frac{1}{2} \mu_{max}$. This formula adequately describes what is found with many real chemostat cultures at equilibrium, that is, when steady state conditions prevail. A chemostat is probably a good model for bacterial growth in open systems, such as rivers and oceans, and by using chemostats and the appropriate mathematical calculations, the growth rates of bacteria in nature can be estimated.

Growth on solid media and bacterial colonies

The development of bacterial **colonies** on solid growth media follows the basic normal growth curve. However, the dividing cells do not disperse, and hence, the population is densely packed. Under these conditions nutrients rapidly become limiting at the center of the colony, and microorganisms in this area rapidly reach stationary phase. At the periphery of the colony, cells can continue to grow exponentially even while those at the center of the colony are in the death phase.

Figure 9.40

Typical edges of bacterial colonies, showing form (A), elevation (B), and margin (C).

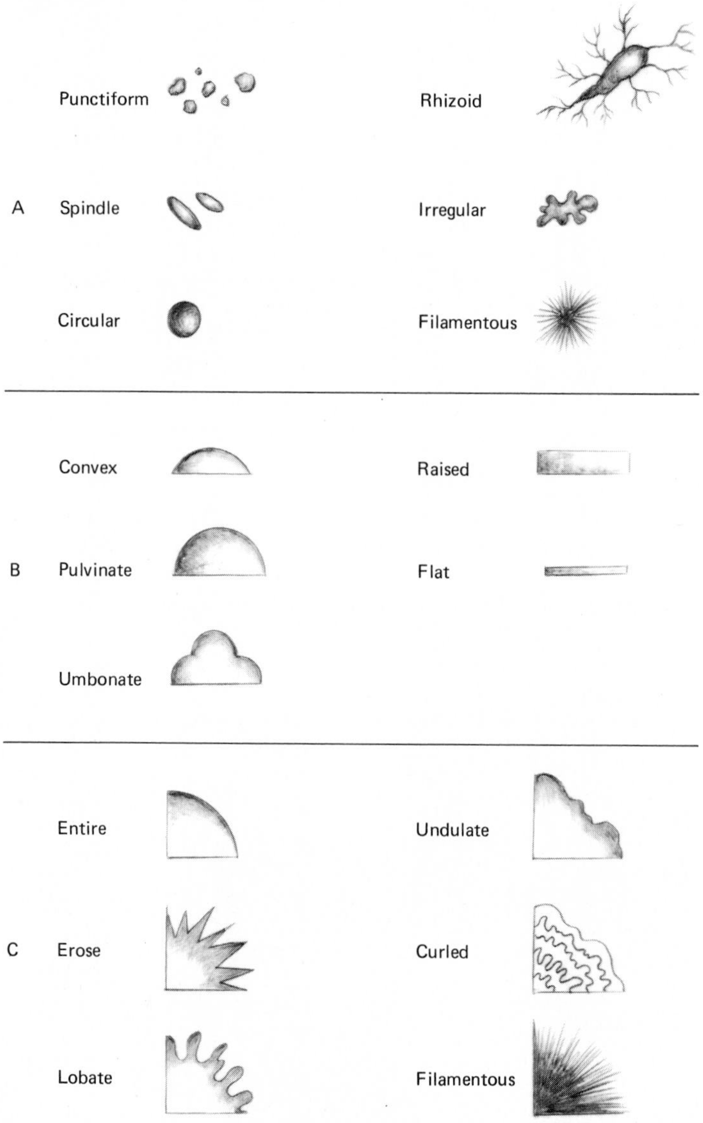

Bacterial colonies generally do not extend indefinitely across the surface of the media but have a well-defined edge, the shape of which is characteristic for a given bacterial species (Figure 9.40). Therefore, individual well-isolated colonies develop from the growth of individual bacterial cells. The fact that the bacteria have reproduced asexually by binary fission means that, barring muta-tion, all the bacteria in the colony should be genetically identical, i.e., each colony contains a clone of cells derived from a single parental cell. The isolation of bacteria from such a colony forms the basis of the streak plate technique, discussed in Chapter 2, for the isolation of pure bacterial cultures.

Reproduction of eukaryotic microorganisms

Eukaryotic microorganisms exhibit a wide variety of reproductive strategies. Some eukaryotic microorganisms have both sexual and asexual modes of reproduction, whereas others reproduce only asexually. For unicellular eukaryotic microorganisms, growth is equivalent to reproduction, but in multicellular eukaryotic microorganisms growth can occur without reproduction of the entire organism. Many eukaryotic microorganisms exhibit complex reproductive life cycles, often containing a sexual phase during which genetic exchange occurs (Figure 9.41). The sexual reproduction of eukaryotic microorganisms, important for maintaining heterogeneity within the gene pool, requires that they exhibit meiotic as well as mitotic division. During the life cycle of a eukaryotic microorganism, some of the stages are haploid, and others are diploid.

Fungi

Five basic types of life cycles occur in fungi (Figure 9.42). An asexual reproduction life cycle, indicating a complete lack of a sexual reproductive phase, characterizes and defines the fungi imperfecti. In a haploid-type life cycle, the diploid phase is of minimal duration with meiosis occurring immediately after fusion of nuclei. The haploid-dikaryotic life cycle is similar to the haploid life cycle, except that a dikaryon containing two nuclei exists for part of the life cycle within a given hyphal cell. In a haploid-diploid life cycle characteristic of relatively few fungi, the haploid and diploid phases alternate regularly. In the diploid life cycle, the haploid phase is restricted to the brief production of haploid gametes.

Yeasts
Because yeasts are fungi which exist predominantly as unicellular organisms, growth is equated with reproduction. The most common mode of reproduction for yeasts is by **budding** (Figure 9.43). The budding process involves the formation of a cross wall that separates the bud from the mother cell. The cross wall of *Saccharomyces*, for example, consists of chitin, which does not occur elsewhere in the cell wall. Budding follows mitotic division so that both the progeny and parent cell contain a complete genome. Budding can occur all around the mother cell (**multilateral budding**) or may be restricted to the end of the mother cell (**polar budding**) (Figure 9.44). The budding process leaves a **bud scar** on the mother cell, and consequently, only a limited number of progeny may be derived from an individual yeast cell. Although budding is the most common form of reproduction, various other reproductive strategies exist among the yeasts, including sexual reproduction and fission (Figure 9.45).

Filamentous fungi
In contrast to yeasts, the filamentous fungi normally develop multicellular structures known as **hyphae**. The growth of the fungus involves the elongation of the hyphae, generally with the formation of branches and cross walls separating individual cells of the fungus. Some fungi, however, form multinucleate **coenocytic mycelia** that lack cross walls, and therefore, these fungi are really one-celled multinucleate organisms (Figure 9.46). Dimorphism is characteristic of some fungi, which under some conditions are unicellular and appear yeast-like, and under other conditions grow as filamentous forms. Because filamentous fungi form elongated filaments, growth can occur in the absence of reproduction. Growth requires an increase in biomass, but unlike reproduction, cellular division does not always occur. In filamentous fungi, growth normally originates from the hyphal tip (Figure 9.47) The apical growth of a fungal hypha requires that the necessary polymers be transported to the area of new cell wall synthesis, and in eukaryotes the area of cell wall syn-

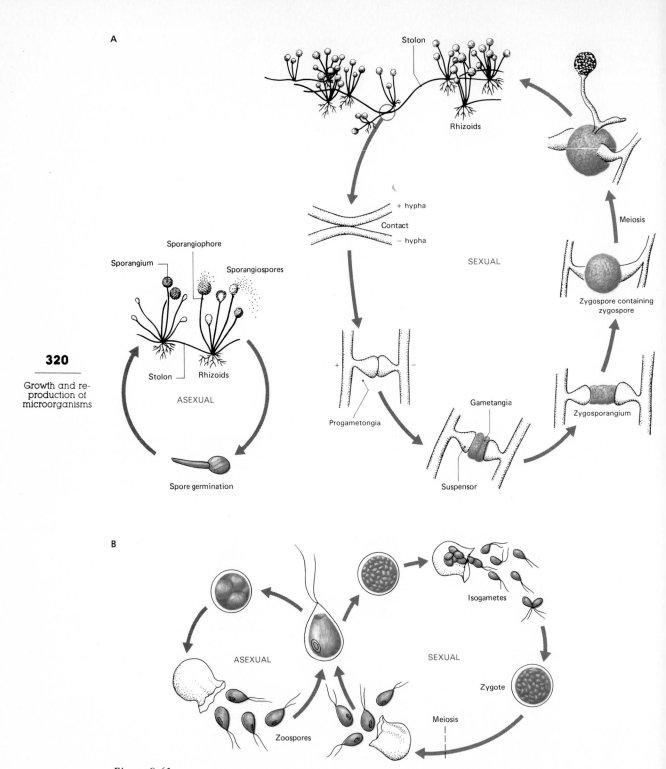

Figure 9.41

Some representative life cycles of eukaryotic microorganisms. (A) The life cycle of the bread-mold fungus Rhizopus stolonifer, *showing asexual and sexual reproduction. (B) Life cycle of the green alga* Chlamydomonas.

Figure 9.42

The five basic life cycles in fungi. Each circle represents a life cycle and should be followed clockwise. M = meiosis; ——— = a haploid phase; ▬▬▬ = a dikaryotic phase; and tinted line = a diploid phase. The life cycles shown are (1) asexual, (2) haploid, (3) haploid-dikaryotic, (4) haploid-diploid, and (5) diploid.

Figure 9.43

(A) Diagram illustrating reproduction by budding by a yeast. (B) Series of electron micrographs of Saccharomyces cerevesiae *at different stages of budding; as shown in these micrographs, vesicles (v) of uniform size accumulate preferentially at a cytoplasmic location where the bud grows. (Courtesy Bijan Ghosh, Rutgers, The State University.)*

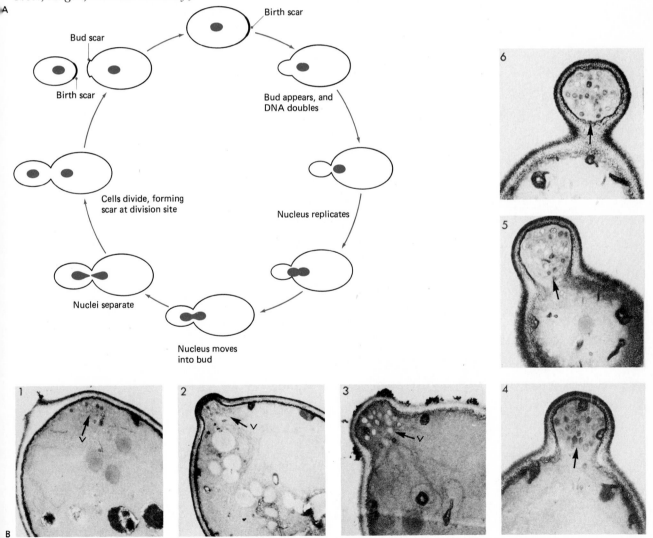

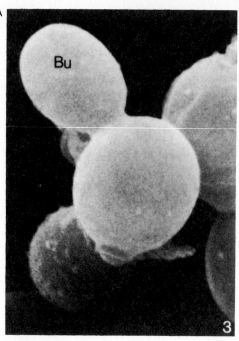

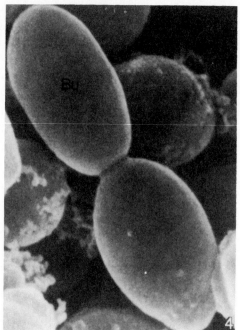

Figure 9.44

*Budding in yeasts can occur only at the pole (polar budding) or all around the cell
(multilateral budding). These electron micrographs illustrate the appearance of different yeasts
exhibiting these two forms of reproduction. Multilateral (A) and polar (B) budding in a yeast
(8,925× and 10,200×), Bu = bud. Reprinted by permission of Springer-Verlag, New York,
from R. G. Kessel and C. Y. Shih, 1976,* Scanning Electron Microscopy for Biology.*)*

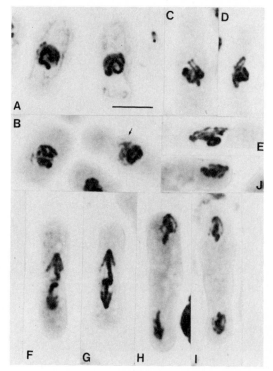

Figure 9.45

*Sequence of stages in the mitotic division of the
fission yeast* Schizosaccharomyces japonicus. *(A, B)
show resting nuclei each harboring a heavily
coiled nucleolar chromosome; (C, D, E) represent
the stages in the segregation of replicated
chromosomes; (F, G) anaphase; (G) a short cord
connects the thick, straight chromosome to the
nuclear pole; (H, I) telophase contraction of the
chromosomes; after the partitioning of the
chromosomes, a cross wall forms separating the
two daughter cells. (Courtesy J. Robinow,
University of Western Ontario.)*

Figure 9.46

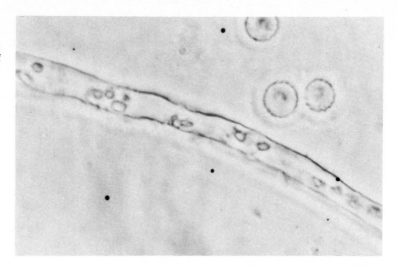

Micrograph of coenocytic hypha of Rhizopus *sp., showing multiple nuclei within one elongated cell and the lack of cross walls.*

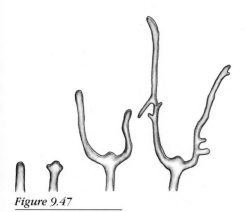

Figure 9.47

Filamentous fungi reproduce by extension of hyphae. Here the growth of a hyphal tip of Gelasinospora autosteira *is illustrated at 30-minute intervals. (Based on C. J. Alexopoulos and C. W. Mims, 1979,* Introductory Mycology, *John Wiley & Sons, New York.)*

thesis generally contains a large number of microfibrils that are involved in cell wall synthesis.

During the asexual phase of fungal reproduction a variety of spores may be produced (Figure 9.48). These include various **conidia** that are asexual spores borne externally on hyphae or specialized conidiophore structures. The conidia can be separated from the fungal hypha as single cells. One type of conidia, the **arthrospore**, represents fragmented hyphae. The fragmentation of multicellular eukaryotic microorganisms constitutes a form of reproduction because the individual fragments are each capable of reproducing the original organism. Other asexual fungal spores include **sporangiospores**, which are produced within a specialized structure known as the sporangium, and **chlamydospores**, which are thick-walled spores which occur within hyphal segments. These fungal spores can be dispersed from the fungal hyphae and later germinate to form new vegetative structures.

Fungi also can produce various types of sexual reproductive spores. Some sexual spores of fungi, **ascospores**, are formed within a specialized structure known as the **ascus** (Figure 9.49). The ability to produce ascospores distinguishes the ascomycete fungal group, which includes both yeasts and filamentous fungi, from other major fungal groups. In another major group of fungi, the basidiomycetes, the sexual spores are produced on a specialized structure known as the **basidium** (Figure 9.50). The sexual reproduction of basidiomycetes usually involves the fusion of hyphal cells. In several fungal groups, the Mastigomycota and Zygomycotina, sexual reproduction is generally by fusion of specialized gametes (Figure 9.51). These gametes are haploid, and their fusion reestablishes a diploid state. In many of the Gymnomycota the gametes are motile and are thus known as **zoospores**, but Zygomycotina species, such as bread molds, produce only nonmotile reproductive spores. The deuteromycetes, or fungi imperfecti, have no known sexual reproductive phase, and as far as we know they are restricted to asexual means of reproduction. If a sexual stage is discovered for a fungus that has been classified among the fungi imperfecti, it is reclassified into one of the other major fungal groups based on the type of sexual spores that are produced.

Enumeration of fungi

Defining numbers of fungi is often a difficult task because individual organisms frequently repre-

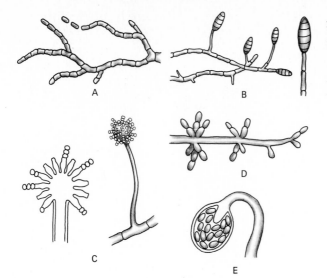

Figure 9.48

Fungi produce a variety of asexual spores. (A) Arthrospores. (B) Aleurospores. (C) Conidiospores and conidia. (D) Blastospores. (E) Sporangia and sporangiospores.

Growth and re-
production of
microorganisms

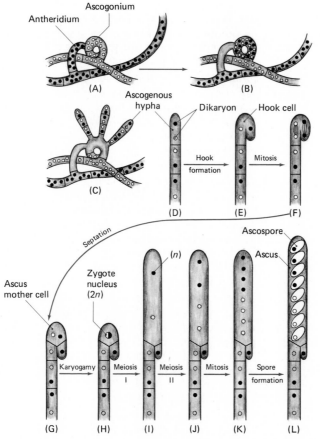

Figure 9.49

Sexual reproduction of one of the major groups of fungi is characterized by the formation of ascospores. This diagram shows the process of ascospore formation in a heterothallic ascomycete. Filled and unfilled nuclei represent the two compatible mating types.

Figure 9.50

Basidiospore formation, without crossing over, in a heterothallic basidiomycete. Filled and unfilled nuclei represent two compatible mating types.

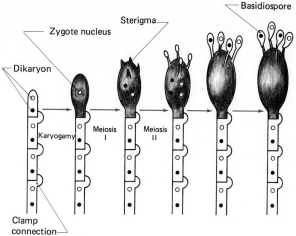

sent multicellular aggregations that can be considered as one or many individuals. The task is simplest when considering yeasts that can be enumerated by using viable count or direct count procedures analogous to the procedures used for enumeration bacteria. The task of enumeration is far more difficult when considering the filamentous fungi. Plate count enumeration procedures are biased toward the counting of fungal spores and underestimate the number of cells in a hyphal filament. The enumeration of filamentous fungi can be accomplished by determining the length of hyphae, which is considered a measure of fungal biomass, rather than the number of in-

Figure 9.51

Spore formation in Rhizopus. *(A) In asexual reproduction, a mycelium arises from the outgrowth of a fungal spore. Once the mycelium reaches a certain size, aerial hyphae arise, bearing sporangia filled with haploid spores. Upon dissemination, these spores give rise to new mycelial nets. (B) In sexual reproduction, when the subsurface rhizoids of a plus and minus type meet, they fuse and the result is zygospore formation.*

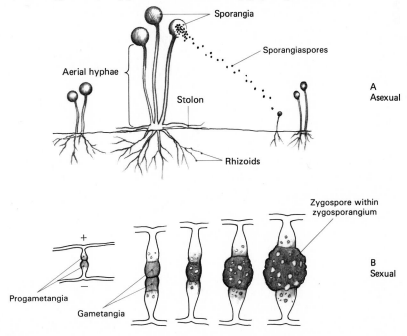

dividual cells. Direct microscopic observations with the aid of a micrometer can be used to measure the length of hyphae. However, this approach has some limitations. Fungi growing in an aqueous solution lacking growth nutrients can exhibit rapid growth of individual hyphae but minimal change in total biomass because the density of hyphae will be sparse. It is probably best, therefore, to determine the biomass of filamentous fungi by measuring the dry weight or a specific biochemical component, such as chitin, of the cell walls. In the case of filamentous fungi, changes in biomass is the appropriate measure of growth, rather than changes in cell numbers.

Algae

Algae, like fungi, exhibit complex life cycles in which spores play a key role (Figure 9.52). A variety of reproductive spores are produced by different algae (Table 9.2). Zoospores, aplanospores, and autospores are involved in asexual algal re-

production; these spores have the ability to give rise individually to new vegetative cells that resemble the parent organism. Algae also carry out several types of sexual reproduction. In **oogamy**, a small motile gamete unites with a larger nonmotile gamete. In **isogamy**, the donor and recipient gametes are morphologically indistinguishable. These are important characteristics of taxonomic groups of algae that will be discussed in Chapter 12, and much of algal taxonomy is based on the nature of the reproductive cycle and which spores are produced.

Algae can be enumerated by using viable count procedures, such as most probable number determination with the formation of visible "green" growth as the criterion for establishing positive test results. Algal biomass can be estimated by measuring the concentrations of chlorophyll *a*, the primary photosynthetic pigment in cyanobacteria and algae. Estimates of algal biomass, based on chlorophyll *a* determinations, can be made readily with a spectrophotometer and generally correlate well with estimates based on viable count procedures.

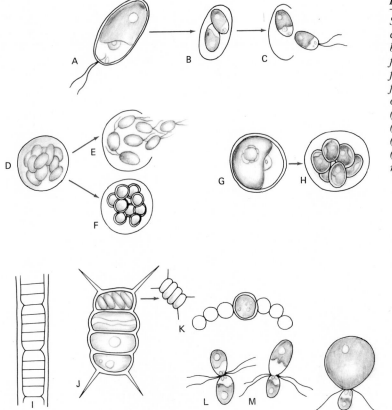

Figure 9.52

Some representative algal life cycles. (A–C) Bipartition or binary fission; (D,E) zoospore formation; (F) aplanospore formation; (G,H) autospore formation; (I) fragmentation or hormogonium formation; (J) autocolony formation; (K) akinete formation; (L) isogamy; (M) anisogamy; (N) oogamy. Note different methods of spore formation.

table 9.2

Descriptions of some algal reproductive spores

Spore type	Description
Androspore	A zoospore that grows into a dwarf male filament in some green algae
Aplanospore	A nonmotile spore that otherwise resembles a zoospore
Autospore	A nonmotile spore that is a miniature of the cell from which it is derived
Carpospore	A spore that arises by fertilization in the red algae
Microspore	The product of division of a diatom prior to spermatogenesis which undergoes meiosis with each microspore giving rise to four sperm cells
Monospore	A nonflagellate spore produced singly from a sporangium
Tetraspore	A spore produced within a tetrasporangium
Zoospore	A flagellated reproductive cell

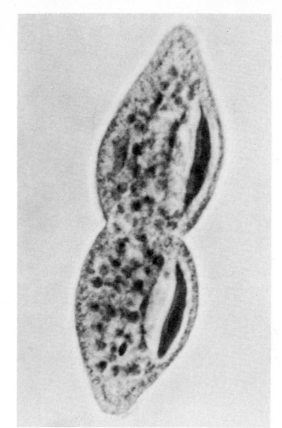

Figure 9.53

Micrograph showing binary fission of a Paramecium aurelia *(500×).*

Protozoa

Like other eukaryotic microorganisms, protozoa reproduce by a variety of asexual and sexual processes. Some protozoa, such as ameboid protozoans, reproduce by binary fission (Figure 9.53). In some flagellate protozoa, binary fission is longitudinal, occurring along the length of the cell, rather than by transverse binary fission, as occurs in bacteria and ameboid protozoa. Many ciliate protozoa reproduce by transverse fission. To avoid senescence, some protozoa exhibit autogamy, a nuclear reorganization process in which the macronucleus degenerates, the micronucleus is reorganized, and a new macronucleus is formed. In some protozoa the mother cell divides into many daughter cells, a process known as **multiple fission** (Figure 9.54). Multiple fission is important in the sporozoa, one of the four major taxonomic groups found in the protozoa.

Protozoa exhibit several modes of sexual reproduction. In some of the ciliate protozoa, conjugation occurs with the mating protozoa exchanging nuclei (Figure 9.55) and then carrying on further asexual reproduction. In some cases the conjugating protozoa fuse together and recombination occurs. Each of the conjugating protozoa is diploid and retains its identity throughout the conjugation process, exchanging micronuclei to accomplish genetic transfer and thereby to establish hybrids of the mating types. During sexual reproduction many protozoa produce haploid gametes. If the two gametes are morphologically similar, they are called **isogametes**, whereas if the donor and recipient gametes are morphologically dissimilar, they are known as **anisogametes**. The various reproductive strategies of the protozoa assure their successful reproduction and survival.

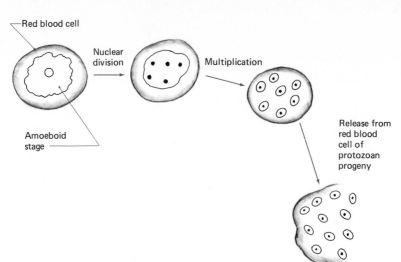

Figure 9.54

Drawing showing multiple fission in the protozoan sporozoan Plasmodium vivax, *the malarial parasite of humans. The reproductive cycle in human beings takes 48 hours and occurs within red blood cells.*

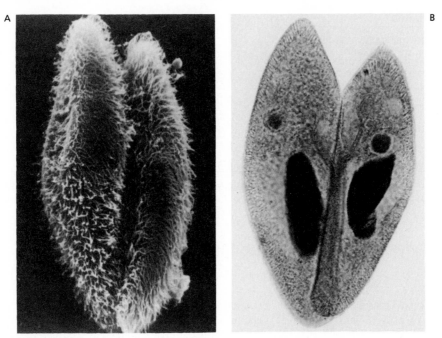

Figure 9.55

Conjugation, a form of parasexual reproduction, occurs in some protozoa. (A) Paramecium multimicronucleatum *(500×). (Reprinted with permission of Springer-Verlag, New York, from R. G. Kessel and C. Y. Shih, 1976,* Scanning Electron Biology for Biology.*) (B)* Paramecium caudatum. *(From Carolina Biological Supply.)*

Postlude

Reproduction assures that the genetic information of a microorganism is passed from one generation to the next. As a rule, reproduction in microorganisms is synonymous with microbial growth. Much of the metabolic activity of a microorganism is aimed at producing the macromolecules and structures needed to form progeny. Microorganisms exhibit a wide variety of reproductive strategies for accomplishing the transfer of the genome to new cells, including asexual reproductive modes that tend to limit genetic variability and sexual modes that foster heterogeneity within the population.

For viruses, reproduction can occur only intracellularly. These organisms lack the necessary metabolic capability and cellular structures to accomplish independent reproduction and therefore are dependent on their compatibility with host cells for continued existence. The key steps in the reproduction of viruses are (1) adsorption of the virus onto the host cell; (2) penetration of the viral nucleic acid across the host cell cytoplasmic membrane; (3) separation of the nucleic acid from the viral capsid; (4) shutdown of normal host cell synthesis of macromolecules; (5) production of proteins coded for by the viral genome; (6) replication of the viral genome; (7) assembly of the viral particles; and (8) release of the viral progeny. RNA and DNA viruses use different modes of reproduction for the expression and replication of the viral genome. In many viruses different reading frames and different directions of reading are used during transcription, enabling a small viral genome efficiently to code for the variety of proteins that are required for viral reproduction.

Often, the reproduction of a virus within a host cell results in the death of that cell. The lytic reproductive cycle of bacteriophage, for example, results in a one-step growth curve, with a large number of viruses released simultaneously from the lysed bacterial cell. Similarly, some plant and animal viruses reproduce with the simultaneous release of a large number of progeny from the host cell. Other viruses reproduce without killing the host cell, exhibiting a slow continuous release of progeny from the infected host cell, as occurs in the budding mode of reproduction. Some bacteriophage are temperate and are able to establish a prophage state that does not result in the death of the host cell. During lysogeny the viral DNA is incorporated into the bacterial chromosome and is replicated along with the bacterial DNA. At some later time the viral DNA can be excised from the bacterial chromosome, and a lytic reproduction cycle can be reestablished.

Whereas viruses can only reproduce intracellularly, the reproduction of other microorganisms involves cellular division. Most bacteria reproduce by binary fission, forming two equal-size cells with identical genomes. As a consequence of this reproductive mode, bacterial numbers double at regular time intervals. When inoculated into a fresh medium, bacteria exhibit a characteristic growth cycle, consisting of a lag phase, log or exponential phase, stationary phase, and death phase. The mathematics of the exponential growth and death phases reflects the binary fission mode of bacterial replication. The rates of bacterial growth and death are greatly influenced by environmental factors, the topic of the next chapter.

Eukaryotic microorganisms show a greater variety of reproductive strategies than bacteria, often exhibiting both asexual and sexual reproductive modes. This variety of reproductive strategies enhances the maintenance of a heterogeneity of genetic information. Meiosis and mitosis are essential processes for the redistribution of DNA during these life cycles. The fact that most eukaryotic microorganisms have several strategies for reproduction enables them to establish relatively complex life cycles, in many cases involving an alternation between haploid and diploid phases. As part of their reproductive strategies, eukaryotic microorganisms produce various sexual and asexual spores. Fungi, for example, produce many sexual and asexual spores, which are used to define the major fungal taxonomic groups.

The particular reproductive strategy of any organism must be adaptive or that organism will perish. Among microorganisms, reproduction is normally characterized by the formation of a large number of progeny. It is the ability of microorganisms to proliferate rapidly when conditions are appropriate that ensures survival of the species. The high reproductive capacity of microorganisms also allows for rapid evolution. Undoubtedly, many nonadaptive evolutionary lines rapidly become extinct. The many microbial species that continue to flourish represent the survivors whose reproductive capacities are high enough to overcome those factors that can lead to the extinction of a species. Various microbial species have developed specialized life stages, many times involving the production of spores, to persist during hostile periods and to facilitate the reproduction and dissemination of progeny to environments that can support continued growth of the species. Just as there are many species of microorganisms, there are numerous diverse reproductive strategies. The important thing for a microorganism is that the particular reproductive strategy ensure the passage of the genome of the organism to progeny that are able to continue reproducing, establishing the basis for the survival of the species.

1. Compare the lytic and lysogenic viral reproductive cycles. What are the similarities and differences in these two modes of replicating viral genomes?

2. How do we assay for lytic bacteriophage?

3. Discuss two ways that an RNA virus can replicate its genome.

4. How can viruses transform animal cells?

5. How do bacteria normally reproduce? Why are bacterial growth and reproduction considered synonymous?

6. How do we measure microbial growth? What units do we use to express microbial growth?

7. Discuss three approaches to the enumeration of bacteria. What are the advantages and disadvantages of each approach?

8. Describe the normal bacterial growth curve. What is occurring during each of the growth phases?

9. What is a biphasic growth curve? In the case of *E. coli* growing on a medium with both glucose and lactose, how is biphasic growth related to the regulation of gene expression?

10. Compare batch and continuous culture methods for growing microorganisms.

11. Compare the reproduction of a yeast with a filamentous fungus. How are yeasts and filamentous fungi enumerated?

12. What is the role of sexual reproduction in eukaryotic microorganisms?

Alexopoulos, C. J., and C. W. Mims. 1979. *Introductory Mycology*. John Wiley & Sons, New York.

Atlas, R. M. 1982. Enumeration and estimation of biomass of microbial components in the biosphere. In: *Experimental Microbial Ecology* (R. G. Burns and J. H. Slater, eds.), pp. 84–102. Blackwell Scientific Publications, Oxford, England.

Bazin, M. J. (ed.). 1982. *Microbial Population Dynamics*. CRC Press, Inc., Boca Raton, Florida.

Bold, H. C., and M. J. Wynne. 1978. *Introduction to the Algae*. Prentice-Hall, Englewood Cliffs, New Jersey.

Butler, P. J. G., and A. Klug. 1978. The assembly of a virus. *Scientific American* 239(5): 62–69.

Calendar, R. 1970. The regulation of phage development. *Annual Review of Microbiology* 24: 241–296.

Dawes, I. W., and J. N. Hansen. 1972. Morphogenesis in sporulating bacilli. *CRC Critical Reviews in Microbiology* 1: 479–520.

Dawson, P. S. S. (ed.). 1975. *Microbial Growth* (Benchmark Papers in Microbiology). Academic Press, New York.

Donachie, W. D., N. C. Jones, and R. Teather. 1973. The bacterial cell cycle. *Symposium of the Society for General Microbiology* 23: 9–44.

Fenner, F., B. R. McAnslan, C. A. Mims, J. Sambrook, and D. O. White. 1974. *The Biology of Animal Viruses*. Academic Press, New York.

Hanson, R. S., J. A. Peterson, and A. A. Youten. 1970. Unique biochemical events in bacterial sporulation. *Annual Review of Microbiology* 24: 53–90.

Hartwell, L. L. 1974. *Saccharomyces cerevisiae* cell cycle. *Bacteriological Reviews* 38: 164–198.

Higgins, M. L., and G. D. Shockman. 1971. Procaryotic cell division with respect to wall and membrane. *CRC Critical Reviews in Microbiology* 1: 29–72.

John, P. C. L. (ed.). 1981. *The Cell Cycle*. Cambridge University Press, New York.

Luria, S. E., J. E. Darnell, D. Baltimore, and A. Campbell. 1978. *General Virology*. John Wiley & Sons, New York.

Mandelstam, J., K. McQuillen, and I. W. Dawes. 1982. *Biochemistry of Bacterial Growth*. Blackwell Scientific Publications, Oxford, England.

Mazia, D. 1974. The cell cycle. *Scientific American* 230(1): 55–64.

Prescott, D. M. 1976. *Reproduction of Eukaryotic Cells*. Academic Press, New York.

Slater, M., and M. Schaechter. 1974. Control of cell division in bacteria. *Bacteriological Reviews* 38: 199–221.

Wood, W. B. 1978. Bacteriophage T4 assembly and the morphogenesis of subcellular structures. *Harvey Lectures* 73: 203–223.

Control of microbial growth: influence of environmental factors on growth and death of microorganisms

10

The rates of microbial growth and death are greatly influenced by a number of environmental parameters, some environmental conditions favoring rapid microbial reproduction and others not permitting any microbial growth. Conditions permitting the growth of one microorganism may preclude the growth of another. Not all microorganisms can grow under identical conditions. Each microorganism has a specific tolerance range for specific environmental parameters. Outside the range of environmental conditions under which a given microorganism can reproduce, it may either survive in a relatively dormant state or may lose viability; that is, it may lose the ability to reproduce and consequently die.

In both laboratory and natural situations some environmental parameter or interaction of environmental parameters controls the rate of growth or death of a given microbial species. In nature, where conditions cannot be controlled and many species coexist, fluctuating environmental conditions favor population shifts because of the varying growth rates of individual microbial populations within the community of a given location. In the laboratory it is possible to adjust conditions to achieve optimal growth rates for a given microorganism. Similarly, in industrial fermenters conditions can be adjusted to optimize microbial growth rates, thereby maximizing the accumulation of desired microbial metabolic products. Many laboratory and industrial applications use pure cultures of microorganisms, facilitating the adjustment of the growth conditions so that they favor optimal growth of the particular microbial species.

By adjusting environmental conditions one can also increase the death rate of microorganisms, an important consideration when trying to kill

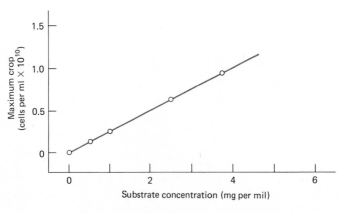

Figure 10.1

This graph shows the effect of substrate concentration on maximal growth of microorganisms. As substrate concentration increases, so does the maximal production of microorganisms.

microorganisms, as in sterilization procedures. Sterilizing products is very important in many areas, including food processing, medical procedures, and the pharmaceutical industry. It is also possible to alter environmental conditions so that microorganisms do not die, but do not reproduce, a method used for the preservation of microorganisms, such as in culture collections and food preservation. Many times the conditions needed to sterilize a product would ruin the desired product, and thus, for example, we preserve by freezing many foods whose taste and textural qualities would be destroyed if they were sterilized at high temperatures.

In this chapter we will consider the influence of a number of environmental factors on microbial growth and death. In later chapters we will see the practical importance of these factors as they are applied in environmental, industrial, food, and medical microbiology.

Microbial nutrition

As we discussed in Chapters 5 and 6, a variety of organic and inorganic nutrients are required for microbial metabolism. All organisms require carbon, nitrogen, oxygen, hydrogen, phosphorus, sulfur, and various other elements; these elements must be available in a usable chemical form to meet the nutritional requirements of a microorganism for that organism to grow. In nature the distribution of microorganisms is determined in part by the availability of specific nutrients in a given environment needed to support the growth of particular microbial species.

The specific **nutritional requirements** for different microorganisms vary greatly. For example, heterotrophic microorganisms require organic matter, but autotrophic microorganisms do not. Some microorganisms are **fastidious**, having complex specific growth factor requirements, and therefore, these microorganisms are able to reproduce only under greatly restricted conditions. Other microorganisms are capable of reproducing under less restricted conditions. Some microorganisms have simple nutritional requirements and are able to reproduce by using a single carbon source as their growth substrate. Such organisms have the metabolic capability of interconnecting the various biochemical pathways so that a particular carbon compound can be transformed into the carbon backbones of all the required macromolecular constituents of those organisms. The substrates that a microorganism can utilize and the specific growth factors it requires for reproduction are determined by the enzymatic capabilities of that microorganism as specified by its genome.

Defining culture media to support microbial growth

In order to grow a microorganism in the laboratory, it is necessary to establish conditions that will permit the reproduction of that organism. The extent of microbial growth, measured as the total microbial biomass produced, is often determined by the concentration of available substrate (Figure 10.1). Biomass production is also controlled by the qualitative composition of the medium and various environmental factors. Different types of microorganisms require culture media of differing composition to support their growth. For the growth of heterotrophic microorganisms, an organic substrate serves as the source of essential carbon and energy. Normal laboratory culture media contain proteins and/or carbohydrates as growth substrates (Table 10.1). For the growth of specific heterotrophic microorganisms, individual growth substrates may be included in the culture medium. For the growth of autotrophic microorganisms, the organic carbon source is omitted from the growth medium. For chemolithotrophs an inorganic chemical that can be oxidized to support the generation of ATP must be included. Autotrophic microorganisms also require a supply of carbon dioxide that they can convert to cellular biochemicals.

A **culture medium** must also include a variety of required inorganic chemicals to support the growth of either heterotrophic or autotrophic microorganisms (Table 10.2). Because molecular nitrogen can be used by only a few microorganisms, most microorganisms require fixed forms of nitrogen, such as organic amino nitrogen, ammonium ions, or nitrate ions. Hydrogen sulfide can be used by some microorganisms as a source of sulfur, but some microorganisms require sulfate as their source of sulfur. A nonselective growth medium normally contains a source of nitrogen, such as ammonium nitrate, phosphate in the form of phosphate buffer, sulfate, magnesium, sodium, potassium, and chloride ions. These inorganic chemicals are required for the biosynthesis of a variety of cellular biochemicals and for the maintenance of transport activities across the cyto-

334

Control of micro-
bial growth: in-
fluence of envi-
ronmental factors
on growth and
death of microor-
ganisms

table 10.1

The composition of some culture media for the growth of heterotrophic bacteria

Nutrient broth (general medium for cultivation of heterotrophic bacteria)	
Beef extract	3 g
Peptone	5 g
Water	1000 g

Glucose–inorganic salts medium (for growth of *E. coli*)	
Glucose	5 g
K_2HPO_4	7 g
KH_2PO_4	2 g
$MgSO_4$	0.1 g
$(NH_4)_2SO_4$	1 g
Water	1000 g

Glucose–yeast extract medium (for growth of bacteria with growth factor requirements)	
Glucose	5 g
K_2HPO_4	7 g
KH_2PO_4	2 g
$MgSO_4$	0.1 g
$(NH_4)_2SO_4$	1 g
$FeSO_4$	0.1 mg
Yeast extract (contains numerous vitamins, amino acids, and other growth factors)	5 g
Water	1000 g

table 10.2

Some common inorganic nutrients that are required for microbial growth

Ingredient	Amount
Water	1000 g
NH_4Cl	1 g
K_2HPO_4	1 g
$MgSO_4 \cdot 87H_2O$	200 mg
$FeSO_4 \cdot 87H_2O$	10 mg
$CaCl_2$	10 mg
Trace elements (Mn, Mo, Cu, Co, Zn) as inorganic salts	0.02–0.5 mg of each

plasmic membrane. Microorganisms generally have many other specific inorganic nutritional requirements. For example, various metals, including zinc, manganese, and copper among others, are generally required as trace elements. Some growth factors, such as vitamins and amino acids, may also be included in a medium designed to support the growth of microorganisms.

Not all microorganisms can be grown by using defined media. The nutritional requirements of many microorganisms are simply not known. These microorganisms are able to reproduce in nature where their nutritional needs are met, but we do not understand their growth requirements well enough to define the appropriate laboratory conditions needed for their growth. In a natural soil or water sample, we are typically able to culture less than 1 percent of the microorganisms that are present. In some cases the natural environ-

table 10.3

Some culture conditions that are employed for enrichment cultures

Carbon source	Oxygen	Nitrogen source	Inorganic sulfur source	Light	Organism enriched for
Nonfermentable organic carbon	+	N_2	—	—	*Azotobacter*
Nonfermentable organic carbon	—	NO_3^-	—	—	Denitrifying bacteria
Nonfermentable organic carbon	—	—	SO_4^{2-}	—	*Desulfovibrio*
Hydrocarbon	+	NH_4^+	SO_4^{2-}	—	Hydrocarbon oxidizers
—	+	NH_4^+	SO_4^{2-}	—	*Nitrosomonas*
—	+	NO_2^-	SO_4^{2-}	—	*Nitrobacter*
—	+	—	S^0	—	*Thiobacillus*
—	+	N_2	SO_4^{2-}	+	Cyanobacteria
—	—	NH_4^+	H_2S	+	Green sulfur bacteria

ment in which the microorganism reproduces can be established in the laboratory without actually defining the growth requirements. For example, soil extract is often added to growth media to support the reproduction of soil microorganisms. Microorganisms, such as viruses, that can reproduce only within cells can be grown in tissue culture or by using other methods that provide the necessary cells within which these microorganisms can reproduce. Often, the task of defining the proper medium for growing such organisms is tedious and taxes the creativity of the microbiologist.

In many cases, rather than defining the specific constituents, complex media are used. For example, many media contain beef extract (a complex mixture of proteins, carbohydrates, lipids, and other biochemical constituents obtained by extracting the water-soluble components from beef tissue), peptones (an enzymatic digest of protein-containing amino acids and other nitrogen-containing compounds, as well as vitamins and other compounds), and yeast extract (an aqueous extract of yeast cells containing vitamins and other growth factors). Such complex culture media will support the growth of many different types of heterotrophs but will exclude those fastidious organisms that have very specific growth requirements, and those organisms that cannot tolerate particular components in the undefined medium.

Enrichment culture technique

By taking into account the metabolic capabilities of specific microorganisms, it is possible to design growth media that will select for the growth of particular microorganisms based on their nutritional requirements. In the **enrichment culture technique** employed to isolate specific groups of microorganisms, the culture medium and incubation conditions are designed to favor the growth of a particular microorganism (Table 10.3). For example, in order to isolate microorganisms capable of metabolizing petroleum hydrocarbons, one can design a culture medium containing a hydrocarbon as the sole source of carbon and energy. By doing so, one establishes conditions whereby only microorganisms that are capable of metabolizing hydrocarbons can grow. Because other microorganisms cannot reproduce in this medium, one thereby selects for hydrocarbon-utilizing microorganisms. Similarly, a culture medium that favors the growth of chemolithotrophic ammonium oxidizers could be designed by providing ammonium ions and carbon dioxide in the medium, and omitting any organic carbon source. The enrichment culture technique models many natural situations in which the growth of a particular microbial population is favored by the chemical composition of the system and by environmental conditions.

Bioassay procedures based on nutritional requirements

The growth requirements of a given strain of microorganism can be used for **bioassay** procedures (Figure 10.2). For example, it is possible to employ vitamin-requiring microorganisms, such as auxotrophic bacterial mutants, to assay for vitamin concentrations. The growth of the test organism is limited by the availability of the required growth factor and is proportional to the concentration of the growth factor being assayed. By measuring the amount of growth, the concentration of the growth factor can be determined.

Figure 10.2

This graph illustrates the use of microorganisms for bioassay procedures. The bioassay procedure is based on the fact that the growth of an auxotrophic microorganism is proportional to the amount of a required growth factor; thus, by measuring the amount of bacterial growth, one can estimate the concentration of a growth factor available in the medium.

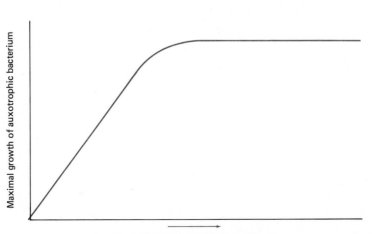

Concentration of required growth factor

Inhibitory organic and inorganic chemicals

Microbial reproduction is not only affected by the qualitative composition of the growth medium but also by the concentrations of the various components in the medium. Although a nutrient may be essential for growth in low concentrations, it may be toxic at higher concentrations. For example, some autotrophic microorganisms require trace amounts of growth factors but cannot grow in the presence of high amounts of organic matter. Similarly, many microorganisms may use hydrogen sulfide at low concentrations, although at higher concentrations hydrogen sulfide is very toxic to most microorganisms. Accordingly, in or-

der to achieve maximal growth rates, it is necessary to define the growth medium carefully.

Selective culture media

Inhibitory substances may intentionally be added to a culture medium to prevent the growth of particular microorganisms. For example, the stain methylene blue is more toxic to Gram positive bacteria than to Gram negative bacteria. By incorporating appropriate concentrations of methylene blue (about 0.5 percent) into a culture me-

table 10.4

Some common antimicrobial agents

Compound	Concentration	Application
Acids		
Acetic acid	1%	Food preservative
Benzoic acid	0.01%	Food preservative
Lactic acid	—	Food preservative (natural fermentation product)
Propionic acid	0.1%	Food preservative
Sorbic acid	0.01–0.1%	Food preservative
Sulfurous acid (added as SO_2)	0.05%	Food preservative
Alcohols		
Ethanol	75%	Topical antiseptic
Lysol	2%	Disinfectant
Dichlorophene	1%	Fungal inhibitor
Hexachlorophene (pHisoHex)	3%	Topical antiseptic
Aldehyde		
Formaldehyde	5%	Preservative
Halogens		
Chlorine	0.2–0.4 ppm	Water purification
Iodine	2%	Topical antiseptic
Heavy metals		
Copper sulfate	0.8 ppm	Algal inhibitor
Copper-8-hydroxy quinolate	1%	Fungal/bacterial inhibitor
Mercuric chloride	0.1%	Disinfectant
Phenylmercuric nitrate	0.002%	Fungal inhibitor
Mercuric naphthenate	0.1%	Fungal/bacterial inhibitor
Phenylmercuric acetate	10 ppm	Inhibitor of slime-producing microorganisms

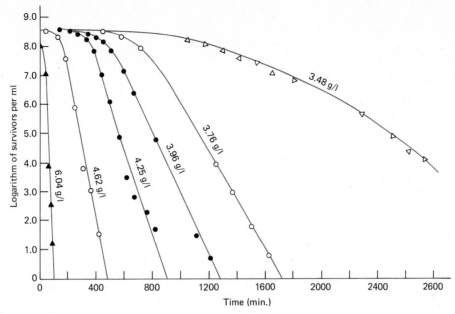

Figure 10.3

This graph charts the results of exposure of E.coli *to various concentrations of phenol at 35°C. The number of survivors, expressed logarithmically, is plotted against time. The concentrations of phenol used were (A) 5.04 g/l; (B) 4.62 g/l; (C) 4.26 g/l; (D) 3.96 g/l; (E) 3.76 g/l; and (F) 3.48 g/l. Increasing the concentration of disinfectant increases the speed with which bacterial cells are killed. (Based on R. C. Jordan and S. E. Jacobs, 1945,* Journal of Hygiene *44: 210, courtesy of Cambridge University Press.)*

dium, the growth of Gram positive bacteria may be inhibited, while not interfering with the growth of Gram negative bacteria. For example, the medium eosin methylene blue agar is frequently used for the selective culture of Gram negative bacteria, such as *E. coli*; the methylene blue in this medium inhibits the growth of Gram positive bacteria, and the other constituents of the medium favor the growth of the Gram negative microorganisms found in association with animal intestinal tracts.

Various antibiotics can be similarly incorporated into a medium to prevent the growth of specific types of microorganisms. Antibiotics are substances produced by microorganisms that inhibit or kill other microorganisms; these substances are used in medicine to inhibit the growth of pathogenic microorganisms, as will be discussed in Chapter 15 when we consider the use of antibiotics in the treatment of human diseases. Such culture media are frequently used in microbiological laboratories in order to select for the growth of specific microbial populations. For example, by incorporating an antibacterial antibiotic into a medium, fungi can be grown free of bacterial contamination. By incorporating the right inhibitory compound into a growth medium, the selectivity of that medium for a particular type of microorganism is established. **Selective media** are used in many areas of microbiology, including clinical microbiology where such media are employed for the isolation of pathogenic microorganisms; various examples of the specific design of such media and their uses in clinical microbiology will be discussed in Chapter 16.

Control of microbial growth by antimicrobial agents

In addition to their use in selective media, a number of inhibitory chemicals are employed for the control of microbial growth. Chemicals that kill microorganisms or prevent the growth of microorganisms are called **antimicrobial agents**. Some examples of common antimicrobial agents used to control microbial growth are listed in Table 10.4. Chemical inhibitors are widely used in various industries to prevent microbial growth, most notably as food preservatives. Concentration and contact time are critical factors that determine the effectiveness of an antimicrobial agent against a particular microorganism (Figure 10.3). Microorganisms vary in their sensitivity to particular

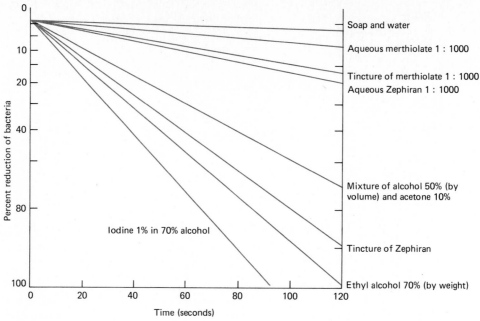

Control of microbial growth: influence of environmental factors on growth and death of microorganisms

Figure 10.4

This graph illustrates the comparative effectiveness of various antiseptic agents. In each test the bacterial microbiota before antiseptic application was considered as 100 percent. The residual biota immediately after use of the antiseptic is shown as a proportion of the original. The steeper the curve, the greater the effect. (Based on P. B. Price, 1957, Skin antisepsis, in Lectures on Sterilization, *J. H. Brewer, ed, Duke University Press, Durham, North Carolina.)*

antimicrobial agents. Generally, growing microorganisms are more sensitive than dormant stages, such as spores. Many antimicrobial agents are aimed at blocking active metabolism and preventing the organism from generating the macromolecular constituents needed for reproduction. Because resting stages are metabolically dormant and are not reproducing, they are not affected by such antimicrobial agents. Similarly, viruses are more resistant than other microorganisms to antimicrobial agents because they are metabolically dormant outside host cells.

Antimicrobial agents are used in a wide variety of applications. Antimicrobial agents are classified according to their application and spectrum of action. **Germicides** are chemical agents that kill microorganisms. Such chemicals may exhibit selective toxicity and, depending on their spectrum of action, may act as **viricides** (killing viruses), **bacteriocides** (killing bacteria), **algicides** (killing algae), or **fungicides** (killing fungi). Whereas germicides kill growing microorganisms, **microstatic agents** inhibit the growth of microorganisms but do not kill them. When a microstatic agent is removed, microorganisms resume their

growth. **Disinfectants** can be either germicides or microstatic agents that kill or prevent the growth of of pathogenic (disease-causing) microorganisms. Household cleaning agents often contain disinfectants to control the growth of microorganisms. Ammonia and bleach (hypochlorite) are widely used disinfectants. In general, agents that oxidize biological macromolecules, such as hypochlorite, are effective disinfectants. Disinfectants are not, however, considered safe for medical uses and are applied only to inanimate objects.

Antiseptics are similar to disinfectants but may be applied safely to biological tissues. These substances are used for topical (surface) applications and are not necessarily safe for consumption. Although not as effective as many of the other antiseptic agents, soap and water do reduce the number of microorganisms on the skin. Alcohol is far more effective than soap and water in reducing the numbers of microorganisms on the skin surface (Figure 10.4). Alcohol is probably the most widely used antiseptic and is used to reduce the numbers of microorganisms on the skin surface in the area of a wound as well as for the disinfection of various contaminated objects. Alcohol denatures proteins, extracts membrane lipids, and acts as a dehydrating agent, all of which contribute to its effectiveness as an antiseptic. Even viruses are inactivated by alcohol. Iodine is another effective antiseptic agent, killing all types of bacteria—including spores. Iodine is frequently applied to minor wounds to kill microorganisms contaminating the surface, preventing infection. Various dyes used in selective media, such as crystal violet, are similarly used as antiseptic agents. Such stains are normally effective bactericidal agents at concentrations of less than 1:10,000. Heavy metals are also used in antiseptic formulations. Mercuric chloride, copper sulfate, and silver ni-

trate are a few examples of heavy metal-containing compounds that are used to kill microorganisms. Silver nitrate, for example, is applied to the eyes of newborn human infants to kill possible microbial contaminants and is particularly important in precluding the transmission of gonococcal infections from an infected mother to the infant's eyes. Both cationic and anionic detergents are also used as antiseptics. Various other detergents are more effective, particularly those containing quaternary ammonium salts. A check of your pharmacy's shelves will illustrate the number of different agents and specific chemical formulations that are marketed as antiseptics.

Temperature

In addition to nutritional factors and chemical inhibitors, a number of other physical and chemical parameters greatly influence the rates of microbial growth. **Temperature** is one of the most important of these factors, affecting both the rates of microbial growth and death. As discussed in Chapter 3, temperature influences the rates of chemical reactions and the three-dimensional

configuration of proteins, thereby affecting the rates of enzymatic activities (Figure 10.5). At low temperatures, molecules move relatively slowly, and colliding molecules do not always have sufficient energy to bring about a reaction. Elevating the temperature increases the probability that molecules will collide and initiate a reaction. As long as the enzyme is not denatured, a rise of 10°C

Figure 10.5

The effect of temperature on rates of enzyme activities is represented in this graph. As temperature increases, the rate of enzyme activity increases.

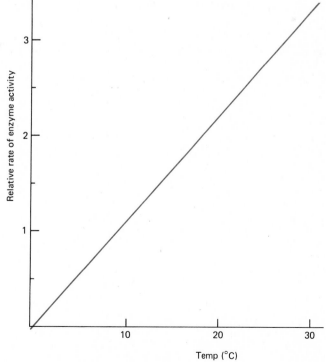

$$Q_{10} = \frac{\text{activity at temperature } T + 10°C}{\text{activity at temperature } T}$$

Enzyme	Q_{10} value (temp. range)
Catalase	2.2 (10–20°C)
Maltase	1.9 (10–20°C)
Urease	1.8 (20–30°C)

Figure 10.6

Formula for Q_{10} and some representative Q_{10} values. A rise of 10°C results in an approximate doubling of the rate of an enzymatic activity.

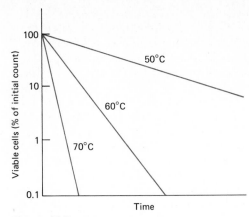

Figure 10.7

Effect of varying temperature on the death rate of a mesophilic bacterium. Above the maximal tolerance of the organism, increasing temperature greatly accelerates the death rate.

generally results in the approximate doubling of the rate of an enzymatic reaction. The Q_{10} of a reaction describes the actual change in reaction rate that occurs when the temperature is increased by 10°C (Figure 10.6). At elevated temperatures, however, proteins are denatured because of the disruption of their three-dimensional structure, and enzymatic activities decline above the specific temperature that is characteristic of the heat stability of the particular enzyme. As a result of protein denaturation at elevated temperatures, there is an upper temperature limit for microbial growth. At higher temperatures microorganisms are unable to survive because they cannot carry out their life-supporting metabolic activities.

Effect of high temperatures on microbial death rates

At temperatures above the maximal growth temperature, the death rate exceeds the growth rate; the higher the temperature above the maximal

growth temperature, the higher the death rate for that microorganism (Figure 10.7). Consequently, high temperatures can be used to kill microorganisms in order to control their proliferation. In many microbiological procedures high temperatures are used to sterilize materials. **Sterilization refers to the killing of all microorganisms in a sample.** Culture media in bacteriological laboratories are normally prepared by heat sterilization, killing all microorganisms, including even viruses and resting stages of other microorganisms (Figure 10.8). In the normal **heat sterilization** process, the medium is exposed to steam at 121°C, which corresponds to 15 pounds pressure, for 15 minutes in an autoclave (Figure 10.9). Autoclaving for this exposure period at this temperature kills all

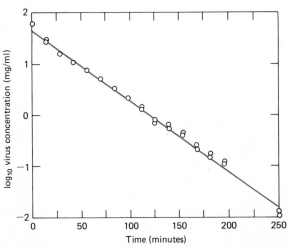

Figure 10.8

Heat inactivation of viruses is illustrated in this graph in which the heat denaturation of tobacco mosaic virus is plotted as remaining infective units with time during heating at 69.8°C in 0.2 M phosphate buffer, pH 7.0. (Based on M. A. Lauffer and W. C. Price, 1940, Journal of Biological Chemistry *133: 1–15, courtesy of American Society Biological Chemists.)*

Figure 10.9

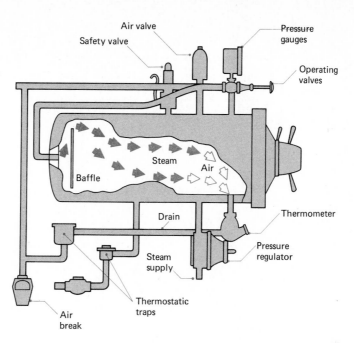

microorganisms, including endospores. Steam has great penetrating power and is able to kill microorganisms efficiently. Dry heat sterilization requires higher temperatures and much longer exposure periods, of the order of several hours, in order to kill all the microorganisms in a sample. Exposure in an oven for 2 hours at 170°C (328°F) is generally used for the dry heat sterilization of glassware and other items.

There are many practical applications of heat sterilization procedures. The use of elevated temperatures to sterilize or reduce the numbers of microorganisms in a sample is frequently employed in the food industry as a method of preserving food products. This topic will be discussed in greater detail in the section on food and industrial microbiology. Heat sterilization is also very important in medical microbiology, where sterile instruments are required for surgical procedures and where sterile culture media are required for diagnostic purposes. Not all products are amenable, however, to heat sterilization; in some cases the product is destroyed by high temperatures. Fortunately, there are other means of treating such materials to kill microorganisms. One such method is to remove bacteria and larger microorganisms by filtration, using a 0.2–0.45 μm pore size filter. Such filtration will remove most bacteria from a solution and is widely used for some industrial applications to sterilize liquids that cannot tolerate exposure to high temperatures.

The heat killing of microorganisms can be described by the **decimal reduction time** (*D*) (Figure 10.10). *D* is defined as the time required for a 10-fold reduction in the number of viable cells at a given temperature, that is, the time required for a log reduction in the number of microorganisms. As the temperature is increased above the

Figure 10.10

This is a hypothetical survivor curve for a decimal reduction time (D value) = 10 with an initial concentration of 10^4 cells/ml. After 10 minutes there is a log reduction in numbers of surviving cells. Increased process times are needed to sterilize highly contaminated materials. (Based on W. C. Frazier and D. C. Westhoff, Food Microbiology, *1978, McGraw-Hill Book Co., New York.)*

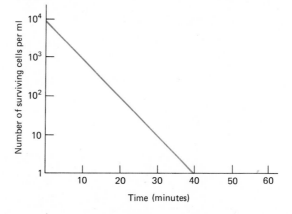

table 10.5

Effects of temperature on various microorganisms

Organism	Temperature growth range (°C)	Thermal death	
		°C	min
Acetobacter roseum	10–41	50	5
Bacillus subtilis	15–55	100	14
Clostridium botulinum	18–55	120	5
Clostridium tetani	14–50	105	10
Corynebacterium diphtheriae	15–40	54	10
Enterobacter aerogenes	12–44	60	30
Escherichia coli	10–45	60	10
Haemophilus influenzae	26–43	52	30
Klebsiella pneumoniae	12–43	55	30
Mycobacterium tuberculosis	30–42	65	15
Neisseria gonorrhoeae	25–40	55	<5
Neisseria meningitidis	25–42	50	<5
Proteus vulgaris	10–43	55	60
Pseudomonas aeruginosa	0–42	62	10
Salmonella typhosa	4–46	60	2
Shigella dysenteriae	10–40	60	10
Streptococcus pneumoniae	18–42	56	6
Streptococcus pyogenes	15–40	60	15
Vibrio cholerae	14–42	55	2
Yersinia pestis	0–45	55	5

Control of microbial growth: influence of environmental factors on growth and death of microorganisms

maximal growth temperature for a microorganism, the decimal reduction time is shortened. The decimal reduction time varies for different microorganisms (Table 10.5). In the food industry, where the decimal reduction time is important in establishing appropriate processing times for sterilizing food products, an 8D process is often used for the safe production of canned products, that is, exposing the sample at a given temperature for a period equal to eight times the decimal reduction time. Normally, for safety reasons the decimal reduction time for relatively heat-resistant microorganisms, such as a thermophilic *Bacillus* species, is used for designing processes in the canning industry.

Preservation of microorganisms at low temperatures

Besides having maximal growth temperatures, microorganisms also have minimal growth temperatures. At very low temperatures enzymatic activities are insufficient to permit microbial metabolism at the rates needed for reproduction, and consequently, microbial reproduction ceases at low temperatures. Unlike heat, low temperatures do not denature proteins. To the contrary, microorganisms are preserved at low temperatures in a dormant state. Therefore, freezing can be employed to preserve cultures of microorga-

nisms, and some strains of microorganisms in culture collections are maintained indefinitely by freezing them at very low temperatures, such as in liquid nitrogen. When freezing is used to preserve microorganisms, the rates of freezing and thawing must be carefully controlled to ensure the survival of the microorganisms because ice crystals formed during freezing can disrupt membranes. Glycerol is often employed as an "antifreeze" agent to prevent damage because of ice crystals and to ensure that microorganisms are not killed during the freeze and thaw procedures aimed at their preservation.

Effects of temperature on microbial growth rates

The **minimum and maximum temperatures** at which a microorganism can grow establish the **temperature growth range** for that microorganism (Figure 10.11). Within this growth range there will be an optimal growth temperature at which the highest rate of reproduction occurs. Because the generation time is the reciprocal of the instantaneous growth rate, the shortest generation time occurs at the optimal temperature. The **optimal growth temperature** is defined by the maximal growth rate, not the maximal cell yield. Sometimes, greater cell or product yields are achieved at lower or higher temperatures. Some

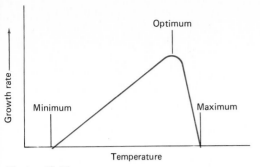

Figure 10.11

Effect of temperature on growth rate, showing growth ranges and optimal temperatures for microorganisms.

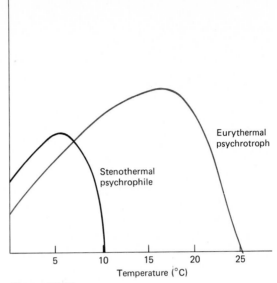

Figure 10.12

Tolerance growth curves of eury- and stenothermal organisms.

microorganisms, known as **stenothermal** organisms, grow only at temperatures near their optimal growth temperature, whereas **eurythermal** microorganisms grow over a wide range of temperatures (Figure 10.12). In the laboratory, **incubators** (controlled-temperature chambers) are normally used to establish conditions that permit the growth of a microbial culture at temperatures favoring optimal growth rates.

Different microorganisms have different optimal growth temperatures (Figure 10.13). Some microorganisms grow best at low temperatures. Such organisms, known as **psychrophiles**, have optimal growth temperatures of under 20°C. Psychrophilic microorganisms have enzymes that are

inactivated at even moderate temperatures, about 25°C. Some psychrophilic microorganisms are capable of growing below 0°C as long as liquid water is available. Psychrophilic microorganisms are commonly found in the world's oceans and are also capable of growing in a household refrigerator where they are important agents of food spoilage.

Figure 10.13

Temperature growth ranges and optima for psychrophilic, mesophilic, and thermophilic organisms.

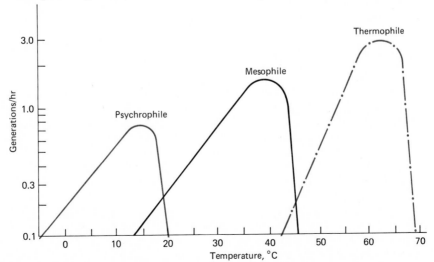

Other microorganisms, known as **mesophiles**, have optimal growth temperatures in the middle temperature range between 20 and 50°C. Most of the bacteria grown in introductory microbiology laboratory courses are mesophilic. Many mesophiles have an optimal temperature of about 37°C, which corresponds to human body temperature. Many of the normal resident microorganisms of the human body, such as *E. coli*, are mesophiles. Similarly, most human pathogenic microorganisms are mesophiles and thus are able to grow rapidly and establish an infection within the human body. Although some microorganisms are restricted to growth near the optimal growth temperature, others can grow over a wide range of temperatures. For example, psychrotrophic microorganisms are eurythermal mesophiles; their optimal growth temperature is between 20 and 50°C, but they are capable of growing at low temperatures (about 5°C), such as inside a refrigerator. Microorganisms can generally actively reproduce over a wider range of temperatures below the optimal growth temperature than above this temperature.

Whereas mesophiles grow in the middle temperature range, **thermophilic microorganisms**, such as *Bacillus staerothermophilus*, grow at higher temperatures, often growing only above 40°C. The upper growth temperature for thermophilic microorganisms, though, is about 99°C. Many thermophilic microorganisms have optimal growth temperatures of about 55–60°C. One finds thermophilic microorganisms in such exotic places as hot springs and effluents from laundromats. However, many thermophiles can survive very low temperatures, and viable thermophilic microorganisms are routinely found in frozen Antarctic soils. It should be made clear that the classification of an organism as a psychrophile, mesophile, or thermophile refers to the organism's optimal growth temperature and the temperature at which the organism can survive. Many *Bacillus* and *Clostridium* species, for example, are mesophiles (not thermophiles), even though their ability to produce endospores permits them to survive in a dormant state at very high temperatures.

An exciting new finding is that some thermophiles isolated from deep ocean thermal vents, the "black smoker bacteria," grow and actively reproduce at temperatures of at least 250°C. At 2500 meters, where these bacteria naturally occur, the pressure is 265 atmospheres and water boils at a temperature above 460°C. The existence of microbial life at these elevated temperatures suggests that living systems are limited by the availability of liquid water and not by high temperature. These bacteria, which are most likely archaebacteria, produce unusual membranes and thermostable proteins that permit them to survive and to reproduce under these extreme conditions.

As a rule, the maximal growth rates of thermophiles are greater than those of mesophiles, which in turn are greater than those of psychrophiles. The differences in optimal growth temperatures and temperature growth ranges between microorganisms result in a spatial separation of these different classes of organisms in nature. A microorganism can proliferate only when the environmental temperatures are restricted to the temperature growth range of that organism. The ability of a microorganism to compete for survival in a given system is favored when temperatures are near the optimal growth temperature of that organism.

Oxygen

Another factor that greatly influences microbial growth rates is the concentration of molecular oxygen. Microorganisms can be classified as aerobes, anaerobes, facultative anaerobes, or microaerophiles based on their oxygen requirements and tolerances (Figure 10.14). **Aerobic microorganisms** grow only when oxygen is available to support their respiratory metabolism, whereas anaerobic microorganisms, such as *Clostridium*, grow in the absence of molecular oxygen. **Anaerobic microorganisms** may carry out fermentation or anaerobic respiration to generate ATP. Some anaerobes have very high death rates in the presence of oxygen and such organisms are termed **oxylabile anaerobes**. Other anaerobic microorganisms, known as **oxyduric anaerobes**, although unable to grow, have low death rates in the presence of oxygen. In contrast to **obligate anaerobes**, which grow only in the absence of molecular oxygen, **facultative anaerobes**, such as *E. coli*, can grow with or without oxygen. As a rule, facultative anaerobes are capable of both fermentative and respiratory metabolism.

Figure 10.14

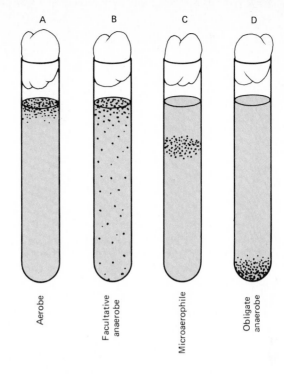

These test tubes show the relationship of the growth of various types of microorganisms to the presence of oxygen. (A) Aerobes grow in the presence of oxygen on the surface. (B) Facultative anaerobes grow throughout the tube. (C) Microaerophiles grow in a narrow band where the oxygen tension is reduced. (D) Obligate anaerobes grow at the bottom of the tube where there is no free oxygen.

Toxicity of oxygen to microorganisms

Although oxygen is required for the growth of many microorganisms, it can also be toxic. Some microorganisms, known as **microaerophiles**, grow only over a very narrow range of oxygen concentrations. Microaerophilic microorganisms require oxygen but exhibit maximal growth rates at reduced oxygen concentrations because higher oxygen concentrations are toxic to these organisms. As a consequence of the toxicity of oxygen to microorganisms, even aerobic microorganisms generally possess enzyme systems for detoxifying various forms of oxygen.

Oxygen can exist in a number of energetic states, some of which are more toxic than others (Figure 10.15). The common state of oxygen is the triplet form in which two of the electrons in the valance shell are unpaired and spin in parallel directions. In singlet oxygen, which has a higher energy level than the triplet form, these two electrons have antiparallel spins. Various enzymes are capable of catalyzing the conversion of triplet oxygen to singlet oxygen. The singlet oxygen form is chemically reactive and is extremely toxic to living organisms. Phospholipids can be oxidized by singlet oxygen, leading to a disruption of membrane

function and the death of microorganisms. Peroxidase enzymes in saliva and phagocyte cells (blood cells involved in the defense mechanism of the human body against invading microorganisms) generate singlet oxygen, accounting in part for the antibacterial activity of saliva and the ability of phagocytic blood cells to kill invading microorganisms.

Additionally, the conversion of oxygen to water, which occurs when oxygen serves as a terminal electron acceptor in respiration pathways, involves the formation of an intermediary form of oxygen known as the superoxide anion (O_2^-), in addition to forming singlet oxygen. Hydrogen peroxide and free hydroxyl radicals are also generated as intermediates in this process. These are extremely reactive oxidative chemicals that can irreversibly denature many biochemicals. Hydrogen peroxide is frequently utilized as an antiseptic because it is toxic and kills microorganisms. Microorganisms generally contain enzymes that protect them against the toxicity of these forms of oxygen. Both **catalase** and **peroxidase** enzymes are involved in the destruction of hydrogen peroxide. Catalase enzymes convert hydrogen peroxide to water and triplet, state oxygen. Microbial production of catalase enzymes can be demonstrated by adding a loopful of a microbial culture to a 3 percent solution of hydrogen peroxide

Form	Formula	Simplified electronic structure	Spin of outer electrons
Triplet oxygen (normal atmospheric form)	$^3O_2$	Ȯ – Ȯ	↑ ↑
	$^1O_2$	Ȯ – Ȯ	↑↓ ○ or ↑ ↓
Superoxide free radical	O_2^-	:Ö – Ȯ	↑↓ ↑
Peroxide	O_2^{2-}	:Ö – Ö:	↑↓ ↑↓

Figure 10.15

Oxygen occurs in a variety of electronic states. Some forms of oxygen are particularly toxic to microorganisms, and microorganisms have evolved various enzymes for removal of such toxic forms of oxygen.

(Figure 10.16). The evolution of gas bubbles, oxygen, is evidence for the action of the catalase enzymes. Peroxidase enzymes use the coenzyme NADH to convert hydrogen peroxide to water. In addition to these enzymes, the superoxide radical is converted to hydrogen peroxide and water by the action of the enzyme **superoxide dismutase**.

Both obligate aerobes and facultative anaerobes usually produce both catalase and superoxide dismutase enzymes (Figure 10.17). These enzymes permit such microorganisms to use oxygen and continue growing without accumulating toxic forms of oxygen that would kill the organism. In contrast, obligate oxylabile anaerobes generally lack both catalase and superoxide dismutase enzymes. The inability of these organisms to enzymatically remove toxic forms of oxygen probably accounts for the fact that they are obligately anaerobic and sensitive to oxygen. The oxyduric anaerobes generally produce superoxide dismutase but not catalase, evidence that they can survive in the presence of oxygen.

In addition to the various electron states of normal molecular oxygen (O_2), oxygen can also occur as ozone (O_3). Ozone is an extremely strong oxidizing agent and consequently is very toxic to microorganisms. Ozone is sometimes utilized as a disinfecting agent, for instance, in some water purification systems, to kill microorganisms. Ethylene oxide is another strong oxidizing agent that is used to sterilize many products that cannot tolerate high temperatures. Most hospitals have ethylene oxide sterilizers for sterilizing various medical items. Caution must be used when handling both ozone and ethylene oxide, as these compounds are quite toxic to humans as well as to microbes.

Achieving optimal oxygen concentrations for microbial growth

In nature the interactions between microbial populations often act to regulate oxygen concentrations. As a result, both aerobic and anaerobic populations are frequently found in close proximity. Under controlled laboratory conditions, it is possible to adjust the oxygen concentrations to maximize the growth rate of a particular microorganism. Because oxygen diffuses only slowly into liquid, the concentration of oxygen frequently limits the rate of microbial growth of aerobic and facultatively anaerobic microorganisms in liquid

Figure 10.16

A positive test for catalase shows bubbles arising from a Bacillus *bacterial colony upon addition of H_2O_2.*

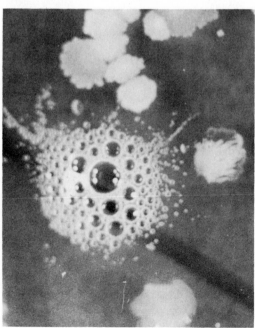

Figure 10.17

Enzymes produced by aerobes and anaerobes that remove toxic forms of oxygen.

	Catalase (2 $H_2O_2 \rightarrow$ 2 H_2O + O_2)	Superoxide dismutase (2 O_2^- + 2 $H^+ \rightarrow H_2O_2$ + O_2)
Aerobe	+	+
Facultative anaerobe	+	+
Oxyduric anaerobe	–	+
Oxylabile anaerobe	–	–

culture. (Facultative anaerobes are able to reproduce more rapidly under aerobic conditions, when they can generate ATP by respiratory metabolism, than under anaerobic conditions, when they utilize fermentation pathways to generate ATP.) In order to supply oxygen for the growth of aerobic microorganisms and overcome the growth-rate limitations caused by low oxygen concentrations, liquid cultures can be agitated at high speed on a shaker table or by an impeller within the culture vessel, or oxygen can be supplied to the culture vessel through forced aeration (Figure 10.18).

Forced aeration is frequently employed in industrial fermenters where large quantities of oxygen must be supplied to support rapid growth of very high numbers of microorganisms. Interrupting the supply of oxygen to an actively growing culture for even a brief period of time can lead to anaerobic conditions, causing in some cases a rapid die-off of the microorganisms; some microbial populations can lose viability if a rotary shaker is turned off for only a few minutes, such as may occur when changing flasks on the shaker table.

Whereas forced aeration is used to enhance the

Figure 10.18

(A) A three-tiered, variable-speed rotary shaker that can hold a variety of flasks for growth of aerobic microorganisms. (Courtesy New Brunswick Scientific.) (B) A benchtop fermenter in which aeration, temperature, agitation, and other factors can be carefully regulated to optimize conditions for microbial growth. (From BPS—David Cross.)

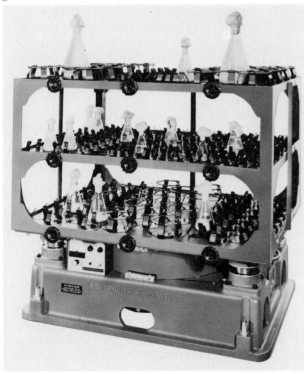

A

B

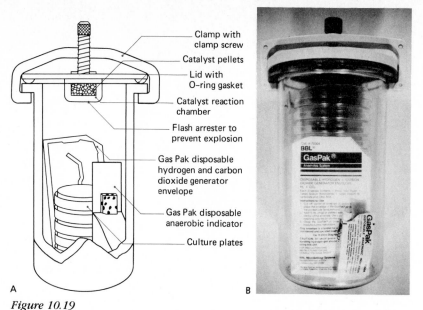

Figure 10.19

Anaerobic jars are used to maintain cultures under strict anaerobic conditions. (A) Schematic diagram and (B) photograph of the Gas Pak system.

Labels in figure:
- Clamp with clamp screw
- Catalyst pellets
- Lid with O-ring gasket
- Catalyst reaction chamber
- Flash arrester to prevent explosion
- Gas Pak disposable hydrogen and carbon dioxide generator envelope
- Gas Pak disposable anaerobic indicator
- Culture plates

A

B

rates of aerobic growth, oxygen must be excluded from the growth medium in order to permit the growth of obligate anaerobes. This can be accomplished by adding chemicals that react with and remove molecular oxygen from the growth medium. For example, sodium thioglycollate is frequently added to liquid culture media for the growth of anaerobes because it reacts with molecular oxygen, removing free oxygen from solution. Similarly, the amino acid cysteine and other compounds containing sulfhydryl groups can also be used to scavenge molecular oxygen from a growth medium. For liquid cultures nitrogen may be bubbled through the medium to remove air and traces of oxygen, and then the culture vessel is sealed tightly to prevent oxygen from reentering. Additionally, there are many types of **anaerobic culture chambers** that can be employed to exclude oxygen from the atmosphere (Figure 10.19). Common forms of anaerobic chambers, such as the Gas Pak system, generate hydrogen,

which reacts with the oxygen at a catalyst within the chamber to produce water. Carbon dioxide is also generated in this system to replace the volume of gas depleted by the conversion of oxygen to water. It is also possible to combine several approaches to ensure absolute anaerobic conditions. In the **roll tube method**, after sterilization of a prereduced medium (a medium from which oxygen is excluded by the incorporation of a chemical that scavenges the free oxygen) within a sealed test tube, the medium is rolled during cooling so that the medium covers the inside of a test tube; the medium is then inoculated with a microorganism under a stream of carbon dioxide or nitrogen and tightly sealed; the development of microbial colonies can be seen on the tube surface (see Figure 16.7), and individual cultures can be observed without disturbing other cultures, as must be done when a large anaerobic incubator is used.

Water activity

Although some microorganisms do not grow in the presence of air, all microorganisms require water for growth and reproduction. Water is an essential solvent and is needed for most biochemical reactions in living systems, and the availability of water has a marked influence on microbial growth rates (Figure 10.20). Pure distilled water has a **water activity** (A_w) of 1.0. Adsorption and solution factors, however, can reduce the availability of water and thus lower the water activity.

Water, for example, may be bound by a solute and hence be unavailable to the microorganisms. Because solute concentrations also affect osmotic pressures, the growth of microorganisms at low water activities can also be viewed in terms of osmotic pressure (see discussion later in this chapter). Adding high concentrations of sugars, such as sucrose, to a solution lowers the availability of water. For example, maple syrup has a water activity of 0.9. Similarly, adding salt (NaCl) to a solution can lower the availability of water. A saturated solution of sodium chloride has an A_w of 0.8. Seawater, though, which has a salt concentration of only about 3 percent, has a water activity of 0.98.

In the atmosphere availability of water is expressed as **relative humidity** (RH).

$$RH = 100 \times A_w$$

and thus a relative humidity of 90 percent corresponds to an A_w of 0.90. The relatively low availability of water in the atmosphere accounts for the inability of microorganisms to grow in the air. Microorganisms likewise are unable to grow on dry surfaces except when there is a relatively high humidity. Microbial growth on surfaces is a problem in tropical zones, where the available water in the atmosphere can support microbial growth, permitting microorganisms to grow on clothing, tents, and numerous other surfaces where microbial growth normally does not occur in temperate regions.

Water activity is an index of the water that is actually available for utilization by microorganisms. Most microorganisms require a water activity above 0.9 for active metabolism (Table 10.6).

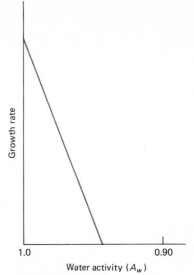

Figure 10.20

The effect of water activity (A_w) *on the growth rate of microorganisms. Normally, microorganisms grow best at high water activities and are severely inhibited by reduced activity.*

Some microorganisms, however, known as **xerotolerant** organisms, can grow at much lower water activities. Some yeasts grow on concentrated sugar solutions with an A_w of 0.60. As a rule, fungi are able to grow at lower water activities than other microorganisms, such as bacteria. Fungi, therefore, grow on many surfaces where the available water will not support bacterial growth. This is why fungal growth, but not bacterial growth, is commonly observed on the surface of bread.

table 10.6

Approximate limiting water activities for microbial growth

Water Activity (A_w)	Bacteria	Fungi	Algae
1.00	Caulobacter Spirillum		
0.90	Lactobacillus Bacillus	Fusarium Mucor	
0.85	Staphylococcus	Debaromyces	
0.80		Penicillium	
0.75	Halobacterium Halococcus	Aspergillus Chrysosporum	Dunaliella
0.60		Saccharomyces rouxii Xeromyces bisporus	

Figure 10.21

(A) This freeze dryer has a 6-liter capacity and is mobile. (Courtesy Virtis Co.) (B) This photograph shows the storage of freeze-dried cultures in a culture collection. (Courtesy American Type Culture Collection.)

Although lack of available water prevents microbial growth, it does not necessarily accelerate the death rate of microorganisms. Some microorganisms, therefore, can be preserved by drying. One can readily purchase active dried yeast for baking purposes, and after the addition of water, the yeasts begin to carry out active metabolism. Freeze-drying is a common means of removing water; this method can be used for preserving microbial cultures (Figure 10.21). During freeze-drying, water is removed by sublimation. This process generally eliminates damage to microbial cells from the expansion of ice crystals.

The fact that microorganisms are unable to grow at low water activities can be used for the preservation of many products. Salting was one of the early means of preserving foods and is still employed today. By adding high concentrations of salt, the A_w is lowered sufficiently to prevent the growth of most microorganisms. Many food products are also preserved by drying. This preservation method depends on maintaining the product

in a dry state, and exposure to high humidity can negate the factor limiting microbial growth, promoting microbial spoilage of food products preserved in this manner.

Whereas some microorganisms are relatively resistant to drying, other microorganisms are unable to survive desiccating conditions for even a short period of time. For example, *Treponema pallidum*, the bacterium that causes syphilis, is extremely sensitive to drying and dies almost instantly in the air or on a dry surface. Many microorganisms produce specialized spores that can withstand the desiccating conditions of the atmosphere. Such spores generally have thick walls that retain moisture within the cell. Many fungal spores can be transmitted over long distances through the atmosphere, with some spores even traveling from one continent to another through the air. The transmission of fungal spores through the air is a serious problem in agriculture because it permits the spread of fungal diseases of plants from one field to another.

Pressure

Osmotic pressure

Changing the solute concentration not only alters the availability of water but also alters the osmotic pressure. The cell wall structures of bacteria and other microorganisms make them relatively resistant to changes in osmotic pressure, but extreme osmotic pressures can result in the death of microorganisms. In hypertonic solutions microorganisms may shrink and become desiccated, and in hypotonic solutions the cell may burst. Organisms that can grow in solutions with high solute concentrations are called **osmotolerant** (Figure 10.22). These organisms are able to withstand high osmotic pressures and also grow at low water activities. Some microorganisms are actually **osmophilic**, requiring a high solute concentration for growth. For example, the fungus *Xeromyces* is an osmophile, and the optimum A_w for *Xeromyces* is approximately 0.9.

Salt tolerance

Tolerance to salt can also be viewed in terms of water activities and osmotic pressures. A special terminology is used to describe salt tolerance. Some microorganisms, known as **halophiles**, specifically require sodium chloride for growth (Figure 10.23). Moderate halophiles, which include

Figure 10.22

The growth of most microorganisms is restricted by high osmotic pressure, such as occurs when sugar concentrations are high. However, osmotolerant and osmophilic organisms can grow at relatively high sugar concentrations.

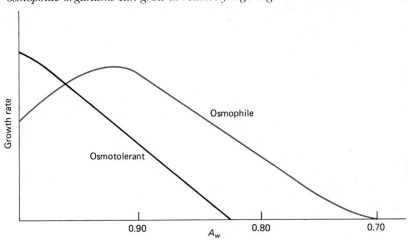

Figure 10.23

The effect of salt concentration (sodium ion concentration) on the growth of microorganisms of varying salt tolerances. Halophiles require salt for growth.

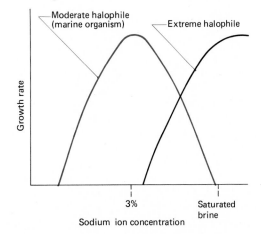

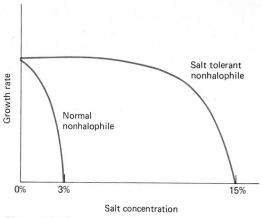

Figure 10.24

Diagram showing relative salt tolerance. Salt-tolerant microorganisms can grow on the skin and in brines where the growth of most microorganisms is restricted.

many unusual enzymes that require a high salt concentration for activity.

Most microorganisms, however, do not possess these physiological adaptations and are not tolerant to high salt concentrations. The degree of sensitivity to salt varies for different microbial species (Figure 10.24). Many bacteria will not grow at a salt concentration of 3 percent. Some strains of *Staphylococcus*, however, are salt-tolerant and grow at salt concentrations greater than 10 percent. This physiological adaptation in *Staphylococcus* is important, as some members of this genus grow on skin surfaces where salt concentrations can be relatively high.

many marine bacteria, grow best at salt concentrations of about 3 percent NaCl. Extreme halophiles exhibit maximal growth rates in saturated brine solutions. These organisms grow quite well in salt concentrations of greater than 15 percent NaCl and can grow in places such as salt lakes and pickle barrels. High salt concentrations normally disrupt membrane transport systems and denature proteins, and the extreme halophiles must possess physiological mechanisms for tolerating high salt concentrations. For example, the extreme halophilic bacterium, *Halobacterium*, possesses an unusual cytoplasmic membrane and

Hydrostatic pressure

In addition to osmotic pressure, **hydrostatic pressure** can influence microbial growth rates. Hydrostatic pressure refers to the pressure exerted by a water column as a result of the weight of the water column. Each 10 meters of water depth is equivalent to approximately 1 atmosphere pressure. Most microorganisms are relatively tolerant to the hydrostatic pressures in most natural systems but cannot tolerate the extremely high hydrostatic pressures that characterize deep ocean regions. High hydrostatic pressures of greater than 200 atmospheres generally inactivate enzymes and disrupt membrane transport processes. However, some microorganisms—referred to as **barotolerant**—can grow at high hydrostatic pressures, and there even appear to be some microorganisms—referred to as **barophiles**—that grow best at high hydrostatic pressures.

Acidity and pH

The **pH** of a solution describes the hydrogen ion concentration $[H^+]$. The pH is equal to $-\log[H^+]$ or $1/\log[H^+]$. A neutral solution has a pH of 7.0; acidic solutions have pH values less than 7; and alkaline or basic solutions have pH values greater than 7. Microbial growth rates are greatly influenced by pH values and are based largely on the nature of proteins. Because the charge interactions of the R groups of the amino acids in a polypeptide chain greatly influence the structure and function of proteins, enzymes are normally inactive at very high and very low pH values. The effects of pH and temperature on microbial growth

and death rates are interactive (Figure 10.25). In general, microorganisms are less tolerant of higher temperatures at low pH values than they are at neutral pH values.

As noted in Chapter 5, many fermentation pathways result in the production of acids. In fermenters and batch culture these acids can accumulate, drastically lowering the pH value and preventing continued microbial growth. In culture media and industrial fermenters, pH values must be controlled in order to achieve optimal growth rates. This is normally accomplished by buffering the solution. Buffers are used to maintain the pH value

Figure 10.25

The interactive effects of pH and temperature on microbial survival. The graph shows the influence of pH on the heat resistance of bacterial spores. The lower the pH, the lower the heat resistance of spore suspensions. (Based on N. W. Desrosier, 1970, The Technology of Food Preservation, *AVI Publishing Co., Westport, Connecticut.)*

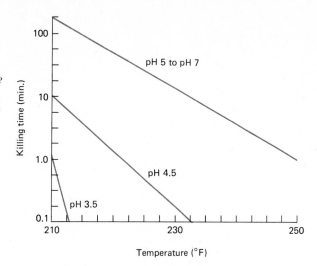

within a range, permitting continued microbial growth. Buffers are salts of weak acids or bases that keep the hydrogen ion concentration constant by maintaining an equilibrium with the hydrogen ions of the solution. Buffers thus dampen changes in pH. At neutral pH values a phosphate buffer may be used; at alkaline pH values borate buffers are often employed; and citrate buffers are frequently used for maintaining acidic conditions.

pH tolerance ranges

Microorganisms vary in their **pH tolerance ranges** (Table 10.7). Fungi generally exhibit a wider pH range, growing well over a pH range of 5–9, compared to most bacteria, which grow well over a pH range of 6–9 (Figure 10.26). Similarly, some fungi grow well at lower pH values, as low as 0. Some other eukaryotic microorganisms, including protozoa and algae, are able to grow at low pH values; the lower limit for growth of some protozoa is approximately 2 and for some algae approximately 1. Although most bacteria are unable to grow at low pH, there are some exceptional cases. Some bacteria tolerate pH values as low as 0.8. There are even some bacteria, called **acidophiles**, that are restricted to growth at low pH values. Some members of the genus *Thiobacillus* are acidophilic and grow only at pH values near 2. Acidophilic bacteria possess physiological adaptations that permit enzymatic and membrane transport activities. The cytoplasmic membrane of an acidophilic bacterium breaks down and cannot function at neutral pH values.

Differences in tolerance to acidic pH values can be used for designing selective growth media. A

table 10.7

Table of pH tolerances of various bacteria

Organism	Minimum	pH	
		Optimum	Maximum
Thiobacillus thiooxidans	1.0	2.0–2.8	4.0–6.0
Lactobacillus acidophilus	4.0–4.6	5.8–6.6	6.8
Escherichia coli	4.4	6.0–7.0	9.0
Proteus vulgaris	4.4	6.0–7.0	8.4
Enterobacter aerogenes	4.4	6.0–7.0	9.0
Clostridium sporogenes	5.0–5.8	6.0–7.6	8.5–9.0
Pseudomonas aeruginosa	5.6	6.6–7.0	8.0
Erwinia carotovora	5.6	7.1	9.3
Nitrobacter spp.	6.6	7.6–8.6	10.0
Nitrosomonas spp.	7.0–7.6	8.0–8.8	9.4

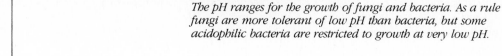

Figure 10.26

The pH ranges for the growth of fungi and bacteria. As a rule fungi are more tolerant of low pH than bacteria, but some acidophilic bacteria are restricted to growth at very low pH.

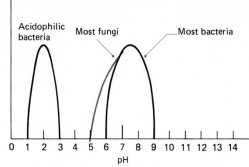

growth medium with a pH of 5.5 is favorable for the growth of most fungi but does not permit the growth of most bacteria. This factor is used in clinical isolation procedures; it also is employed in industry, where lowering the pH of a growth medium designed to support the growth of a fungus, such as *Saccharomyces*, can eliminate un-

wanted bacterial growth. Strong alkaline conditions can digest many cellular biochemicals. One way of killing microorganisms is to expose them to a strong base (alkaline digestion). The use of ammonia as a disinfectant in cleaning solutions is based on its ability to digest microorganisms.

Control of microbial growth: influence of environmental factors on growth and death of microorganisms

Radiation

Another way of killing microorganisms is by exposure to certain forms of radiation. There is a continuous spectrum of electromagnetic radiation (Figure 10.27). We divide the **electromagnetic spectrum** into certain categories of radiation, including: gamma rays (short wavelengths of 10^{-8} to 10^{-1} nm); X rays (wavelengths of 10^{-3} to 10^2 nm); ultraviolet light (wavelengths of 100–400 nm), visible light (wavelengths of 400–800 nm); infrared radiation (wavelengths of 10^3 to 10^5 nm); and microwaves (wavelengths of greater than 10^6 nm). High-energy, short-wavelength radiation disrupts DNA molecules, and exposure to short-wavelength radiations may cause mutations, many of which are lethal (see discussion in Chapter 8). Exposure to gamma, X-ray, and uv radiation increases the death rate of microorganisms and is used in various sterilization procedures to kill microorganisms.

Ionizing radiation

Gamma radiation and X radiation have high penetrating power and are able to kill microorganisms within a sample by inducing or by forming toxic free radicals. Free radicals are highly reactive

chemical species that can lead to polymerization and other chemical reactions disruptive to the biochemical organization of microorganisms. Viruses as well as other microorganisms are inactivated by exposure to **ionizing radiation** (Figure 10.28). Microorganisms, however, are generally more tolerant to ionizing radiation than are higher organisms.

Sensitivities to ionizing radiation vary (Table 10.8). Resistance to ionizing radiation is based on the biochemical constituents of a given microorganism. Nonreproducing (dormant) stages of microorganisms tend to be more resistant to radiation than are growing organisms. For example, endospores are more resistant than are the vegetative cells of many bacterial species. Exposure to 0.3–0.4 Mrads is necessary to cause a 10-fold reduction in the number of viable bacterial endospores. An exception is the bacterium *Micrococcus radiodurans*, which is particularly resistant to exposure to ionizing radiation. Vegetative cells of *M. radiodurans* tolerate as much as 1 Mrad of exposure to ionizing radiation with no reduction in viable count. It appears that efficient repair mechanisms are responsible for the high degree of resistance to radiation exhibited by this bacterium.

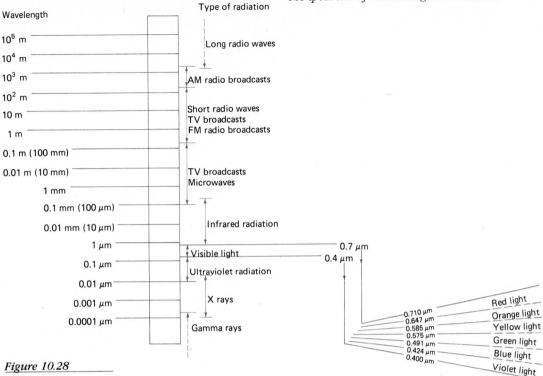

The Electromagnetic Spectrum

Wavelength | Type of radiation

- 10⁵ m
- 10⁴ m — Long radio waves
- 10³ m — AM radio broadcasts
- 10² m
- 10 m — Short radio waves / TV broadcasts
- 1 m — FM radio broadcasts
- 0.1 m (100 mm)
- 0.01 m (10 mm) — TV broadcasts / Microwaves
- 1 mm
- 0.1 mm (100 μm)
- 0.01 mm (10 μm) — Infrared radiation
- 1 μm — Visible light
- 0.1 μm — Ultraviolet radiation
- 0.01 μm
- 0.001 μm — X rays
- 0.0001 μm — Gamma rays

0.7 μm
0.4 μm

0.710 μm — Red light
0.647 μm — Orange light
0.585 μm — Yellow light
0.575 μm — Green light
0.491 μm — Blue light
0.424 μm — Violet light
0.400 μm

Figure 10.27

The spectrum of electromagnetic radiation.

Figure 10.28

Viral inactivation by exposure to ionizing radiation. The graph summarizes the results of three experiments exposing tobacco mosaic virus to X rays. (Based on W. Ginoza and A. Norman, 1957, Nature *179: 520–521.)*

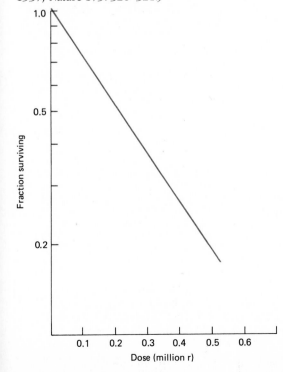

table 10.8

Radiation tolerances for various microorganisms

Organism	Dose (Mrad)
Bacteria	
Clostridium botulinum (type E)	1.5
Enterobacter aerogenes	0.16
Escherichia coli	0.18
Micrococcus radiodurans	6.0
Mycobacterium tuberculosis	0.14
Salmonella typhimurium	0.33
Staphylococcus aureus	0.35
Streptococcus faecalis	0.38
Viruses	
Polio virus	3.8
Vaccinia virus	2.5

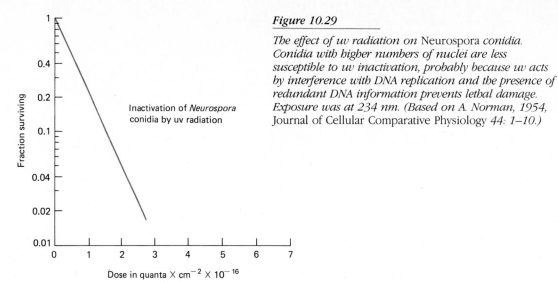

Figure 10.29

The effect of uv radiation on Neurospora *conidia. Conidia with higher numbers of nuclei are less susceptible to uv inactivation, probably because uv acts by interference with DNA replication and the presence of redundant DNA information prevents lethal damage. Exposure was at 234 nm. (Based on A. Norman, 1954,* Journal of Cellular Comparative Physiology *44: 1–10.)*

Ultraviolet light

Ultraviolet light does not have high penetrating power and is useful for killing microorganisms only on or near the surface of clear solutions (Figure 10.29). The strongest germicidal wavelength of 260 nm coincides with the absorption maxima of DNA, suggesting that the principal mechanism by which uv light exerts its lethal effect is through the formation of thymine dimers in the DNA. Microorganisms have enzymes that can repair the alterations in the DNA caused by exposure to uv light. The photoreactivation enzymes require exposure to light in the visible spectrum. In addition to initiating the formation of thymine dimers, exposure to ultraviolet light interferes with other

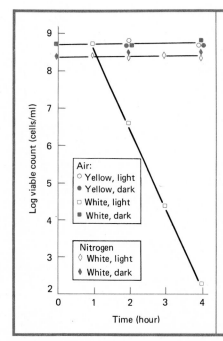

Figure 10.30

The graph illustrates the effect of sunlight on the viability of colorless and yellow-pigmented Micrococcus lutea. *Note that death occurs in the white strain but not in the pigmented strain. (Based on M. M. Matthews and W. R. Sistrom, 1959,* Nature *184: 1892.)*

Discovery process

Reasoning that carotenes might protect bacterial cells against the lethal effects of exposure to light, Matthews and Sistron exposed a colorless mutant of *Micrococcus lutea* to light in the air. Assuming that the presence of carotenes protects the bacteria against the lethal action of light, the carotene-less mutants should be killed, whereas the yellow-pigmented wild type should survive exposure to light. This was indeed their finding. After an exposure to direct sunlight of 2 hours, more than 99 percent of the colorless mutants are killed compared to less than 1 percent of the yellow wild-type cells. From this they inferred that the yellow pigment confers protection from sunlight. Further experiments showed that killing does not occur anaerobically but only in the presence of air. It was therefore properly deduced that light-induced killing is a photooxidation reaction requiring oxygen.

Figure 10.31

The rate of photosynthesis per unit cell is a function of light intensity for Chlorella vulgaris *adapted to high (30 Klux) light intensities. (Based on E. V. Steeman-Nielson, V. K. Hansen, and E. G. Jorgensen, 1962,* Scandinavian Society of Plant Physiologists *15: 505–517.)*

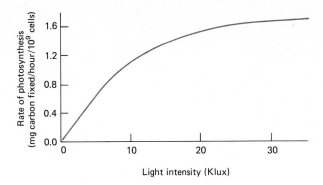

microbial biochemicals and functions. Exposure to ultraviolet light of a wavelength of 340 nm has a strong killing effect on microorganisms, although DNA does not strongly absorb light of this wavelength.

Visible light

Exposure to **visible light** can also cause the death of microorganisms. Visible light may be absorbed by various biochemicals of microbial cells, resulting in changes in the energetic states of those biochemicals. In some cases this can lead to interference with metabolic activities or damage to cellular structures. Additionally, exposure to visible light can lead to the formation of singlet oxygen. As indicated earlier in this chapter, singlet oxygen is chemically reactive and can result in the death of a microorganism. Some microorganisms produce pigments that protect them against the lethal effects of exposure to light (Figure 10.30). For example, yellow, orange, or red carotenoid pigments interfere with the formation and action of singlet oxygen, preventing its lethal action. Microorganisms possessing carotenoid pigments can tolerate much higher levels of exposure to sunlight than nonpigmented microorganisms. Pigmented microorganisms are often found on surfaces that are exposed to direct sunlight, such as on leaves of trees. Many viable microorganisms found in the air produce colored pigments. Likewise, some microbial spores that are principally used for aerial dispersal are pigmented.

Although exposure to high intensities of light can be lethal to some microorganisms, photosynthetic microorganisms require light in the visible spectrum to carry out oxidative photophosphorylation. The rate of photosynthesis is a function of light intensity (Figure 10.31). At some light intensities, rates of photosynthesis reach a maximum, and although light intensities above this level do not result in further increases in the rates of photosynthesis, light intensities below the optimal level result in lower rates of photosynthesis. One interesting effect of lowered light intensity levels is the changeover from oxygen-evolving to anoxygenic photosynthesis exhibited by some cyanobacteria. The wavelength of light also has a marked effect on the rates of photosynthesis. Different photosynthetic microorganisms are capable of using light of different wavelengths. For example, anaerobic photosynthetic bacteria use light of longer wavelengths than eukaryotic algae are capable of using. Many photosynthetic microorganisms have accessory pigments that enable them to use light of wavelengths other than the absorption wavelength for the primary photosynthetic pigments. The distribution of photosynthetic microorganisms in nature reflects the variations in the ability to use light of different wavelengths and the differential penetration of different colors of light into aquatic habitats.

Infrared and microwave radiations

Longer-wavelength infrared and microwave radiations have poor penetrating power and do not appear to kill microorganisms directly. Absorption of such long-wavelength radiation, however, results in increased temperature. Exposure to infrared or microwave radiations can thus indirectly kill microorganisms by exposing them to temperatures that are higher than their maximal growth temperatures. This is also true for microwaves. Because microwaves generally do not kill microorganisms directly, there is some concern in the food industry that cooking with microwave ovens may not adequately kill microorganisms contaminating food products.

There are many factors that influence microbial growth and death rates. Microorganisms exhibit ranges of tolerance to many abiotic factors. When a given parameter exceeds the tolerance range of a microorganism, the microorganism ceases to grow. In some cases exceeding a tolerance range increases the death rate, and the microorganism is unable to survive, whereas in other cases the inability to grow does not result in the death of the microorganism but rather in its preservation. Some microorganisms tolerate a wide range of values for a given parameter, whereas other microorganisms are stenotolerant and are restricted to growth near their optimal value for that parameter.

For each of the parameters affecting microbial growth and death rates, microorganisms exhibit optimal growth rates at specific values of the given parameter. In laboratory and industrial situations it is possible to adjust conditions to favor optimal growth rates or to preclude microbial growth. For example, cultures can be incubated at the optimal growth temperature if maximal growth rates are desired. If microbial growth is not desired, temperatures can be lowered to prevent microbial reproduction. Temperatures can also be raised to increase the death rate of microorganisms and sterilize various materials. The chemical composition of a growth medium can be adjusted to selectively favor the growth of a particular microorganism. This can be accomplished by including inhibitory substances in the medium to prevent the growth of undesirable microorganisms.

A culture medium must contain the nutrients that are required by the organism for growth. In some cases it is necessary to include a variety of growth factors, such as vitamins and amino acids. In other cases, microorganisms are capable of growing on simple media. In general, microorganisms require a source of carbon, nitrogen, phosphorus, iron, magnesium, sulfur, sodium, potassium, and chloride. Oxygen is also required by some microorganisms, but oxygen is toxic to others. Parameters that have a great influence on microbial growth and death rates include organic and inorganic nutrient concentrations, concentrations of organic and inorganic inhibitory substances, temperature, oxygen concentrations, water availability, radiation intensity, pressure, and pH. The combined effects of these factors generally determine the ability of a microorganism to survive and grow in a natural system. It is also these parameters that are manipulated in controlled situations to regulate the rates of microbial growth and death. A number of factors can be adjusted to increase microbial death rates and kill microorganisms. These include exposure to high heat and high irradiation levels. The modification of environmental parameters to control the rates of microbial growth and death is applied in many areas of microbiology and many of these applications will be discussed in later chapters. Understanding the environmental factors controlling microbial growth gives insight into the natural distribution of microorganisms in nature and provides the basis for developing methods to control microbial growth.

1. What nutrients are critical for the culture of both heterotrophic and autotrophic microorganisms? What additional nutrients are required for the growth of heterotrophs?

2. What is an enrichment culture? How would you use enrichment culture to isolate a hydrocarbon-utilizing microorganism?

3. What is the difference between a selective and differential culture medium? What would you add to a culture medium for the selective culture of *E. coli?*

4. What is an antimicrobial agent? Give some examples of how you routinely use such chemicals.

5. What is the effect of temperature on microbial growth and death rates? What is the difference between optimal and maximal growth temperature?

6. What is a mesophile? Psychrophile? Thermophile? Psychrotroph?

7. What effect does oxygen have on microorganisms? What enzymes are involved in the detoxification of reactive forms of oxygen?

8. Discuss the different methods for achieving optimal oxygen concentrations for growing different bacterial species. Consider an organism that is an obligate aerobe, one that is microaerophilic, and one that is a strict anaerobe.

9. What effect does high salt concentration have on microorganisms? What is the difference between a halophile and a salt-tolerant organism?

10. What mechanisms have microorganisms evolved for protection against radiation exposure?

Suggested Supplementary Readings

Brock, T. D. 1978. *Thermophilic Microorganisms and Life at High Temperatures*. Springer-Verlag, New York.

Brown, A. D. 1976. Microbial water stress. *Bacteriological Reviews* 40: 803–846.

Difco Manual of Dehydrated Culture Media and Reagents for Microbiological and Clinical Laboratory Procedures. 1971. Difco Laboratories, Detroit, Michigan.

Fridovich, I. 1977. Oxygen is toxic! *BioScience* 27: 462–466.

Friedman, E. I. 1982. Endolithic microorganisms in the Antarctic cold desert. *Science* 215: 1045–1053.

Gould, G. W., and J. E. Corry. 1980. *Microbial Growth and Survival in Extremes of Environment*. Academic Press, New York.

Gray, T. R. G., and J. R. Postgate (eds.). 1976. *The Survival of Vegetative Microbes*. Cambridge University Press, New York.

Hugo, W. B. (ed.). 1971. *Inhibition and Destruction of the Microbial Cell*. Academic Press, London.

Jannasch, H. W., and C. W. Wirsen. 1977. Microbial life in the deep sea. *Scientific American* 236(6): 42–52.

Kushner, D. J. 1968. Halophilic bacteria. *Advances in Applied Microbiology* 10: 73–97.

Kushner, D. J. (ed.). 1978. *Microbial Life in Extreme Environments*. Academic Press, London.

Larsen, H. 1967. Biochemical aspects of extreme halophilism. *Advances in Microbial Physiology* 1: 97–132.

Morita, R. Y. 1976. Psychrophilic bacteria. *Bacteriological Reviews* 39: 144–167.

Morris, J. G. 1975. The physiology of obligate anaerobiosis. *Advances in Microbial Physiology* 12: 169–246.

Russell, A. D., W. B. Hugo, and G. A. J. Ayliffe. 1982. *Principles and Practice of Disinfection, Preservation and Sterilisation*. Blackwell Scientific Publications Ltd., Oxford, England.

Skerman, V. D. B. 1969. *Abstracts of Microbiological Methods*. John Wiley & Sons, New York.

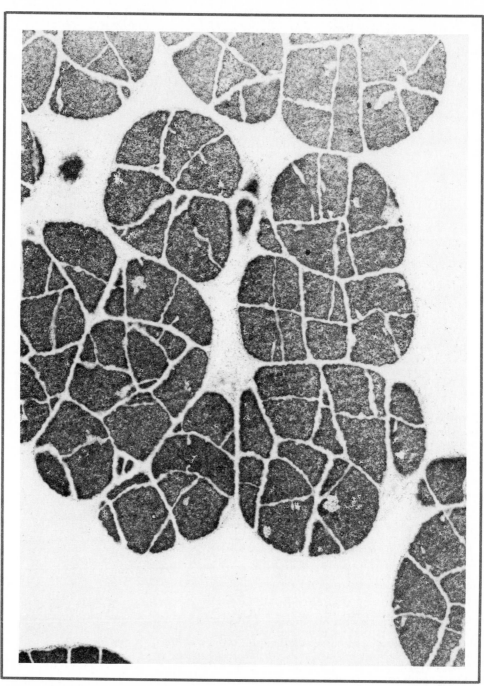

Electron micrograph of Methanococcus mazei.
(Courtesy Jack Pangborn and Robert Mah, University of California, Los Angeles.)

Microbial taxonomy

SECTION SIX

chapter 11
Classification and identification of microorganisms: systematics of bacteria

chapter 12
Classification and identification of microorganisms: systematics of fungi, algae, protozoa, and viruses

Classification and identification of microorganisms: systematics of bacteria

11

Microbial systematics is concerned with the ordered relationships that exist among the organisms comprising the microbial world. **Systematics** is the study of organisms, their diversity and interrelationships, with the aim of arranging them in an orderly manner. Accordingly, microbial systematics encompasses many areas of microbiology, including taxonomy, ecology, biochemistry, genetics, and morphology. Although systematics and taxonomy are sometimes used as synonymous terms, taxonomy is a subdivision of systematics. **Taxonomy** is the process, based on established procedures and rules, of describing the groups of organisms, their interrelationships, and the boundaries between microbial groups. Two functions of taxonomy are, first, to identify and describe the basic taxonomic units (species); and second, to devise an appropriate way of arranging and cataloging these units. The cardinal principles of taxonomy are that organisms exist as real separate groups and that there is a natural ordering of these groups. The rules, procedures, and boundaries of a taxonomic system for microorganisms must permit the unambiguous definition of a group of microorganisms and also show the relationships among groups of microorganisms. A microbial taxonomist is concerned with **classification** (ordering of organisms into groups based on their relationships), **nomenclature** (assigning names to the units described in a classification system), and **identification** (applying the system of classification and nomenclature to assign the proper name to an unknown organism and to place it in its proper position within the classification system).

Nomenclature

Because one function of a taxonomic system is to establish unambiguous names for organisms, a logical system of nomenclature is needed. Organisms are normally named according to a binomial system in which the organism is identified by its genus and species. Microorganisms are referred to by their unique **binomial name** consisting of the **genus** and **species** epithet of that organism. The names of microorganisms and all other organisms are given in Latin. Latin is used because it was the classical language of science when early classification systems were developed and formal names were first given to organisms on a systematic basis. When typed or handwritten, bacterial genus and species names are underlined to indicate that they are in Latin. In print the genus and species names are italicized, and the first letter of the genus name is capitalized; the species name is written in all lowercase letters. For example, we have already made frequent reference to the bacterial species *Escherichia coli*.

The rules of nomenclature for microorganisms are established by international committees. Different codes of nomenclature are used for different microbial groups: the code of nomenclature of bacteria applies to all bacteria; fungi and algae are covered by the botanical code; protozoa are named according to the zoological code; and viruses are named according to the viralogical code. In general, the codes of nomenclature are aimed at avoiding ambiguity and assuring that the name of a microorganism specifically and unambiguously designates that organism. In the field of bacteriology, a summary list of the approved names of bacteria was published in the January 1980 issue of the *International Journal of Systematic Bacteriology*. Only names published in that listing and those validated and published as supplements to that list in that journal are considered to be valid. Similar listings are published periodically in other journals establishing the validity of revised and new names of other microorganisms.

Classification of microorganisms

The second objective of taxonomy, **classification**, attempts to differentiate microbial taxa into structured groups so that members within a group are more closely related to each other than they are to members of any other group. The ordering of organisms into groups is based on an assessment of their similarities. Ideally, the classification of microorganisms should follow the natural ordering established by evolutionary processes, and therefore, taxonomic systems should be based on the genetic interrelationships among groups of microorganisms. Taxonomists frequently debate the validity of a taxonomic system, questioning whether it truly reflects the evolutionary relationships of the organisms and sometimes arguing vehemently whether the structure of the system contains the proper taxonomic units and reflects the appropriate ordering of those units. As a result, taxonomic systems are frequently revised. In Chapter 1 we considered the taxonomic position of microorganisms and examined three different classification systems that at one time or another have been considered to define properly the primary kingdoms and reflect the evolutionary relationships of living organisms. Obviously, the classification systems proposed by Haeckel, Whittaker, and Woese show the evolution of our understanding of **phylogenetic (evolutionary) relationships**.

Although microbial classification systems should reflect genetic similarities, the classification of microorganisms has been traditionally based on phenotypic characteristics. Phenotypic characteristics are readily determined by observing microorganisms growing in pure culture, whereas until recently methodology for directly analyzing the genome did not exist. Taxa based on observed phenotypic characteristics may not accurately reflect genetic similarities, and such a classification may not correspond to the evolutionary flow of events. It is possible for genetically dissimilar microorganisms (**homologously dissimilar**) to resemble each other phenotypically (**analogously similar**). For example, many genetically dissimilar bacteria produce yellow pigments, and a classification scheme based on such a phenotypic characteristic could produce a taxonomic group of genetically unrelated bacteria. In fact, classification systems are filled with errors made by using such phenotypic characteristics. Various groups of microorganisms that have been defined based on their apparent phenotypic relationship are now considered to be "groups of uncertain taxonomic affinity" because the taxonomic group may not be homologously similar and therefore may not accurately represent genetic similarities.

Hierarchical organization of classification systems

Although the **hierarchical organization** of a classification system should reflect the hierarchical branching of groups that is a natural consequence of evolution, evolutionary affinities among microorganisms are difficult to discern because there is no fossil record of most microorganisms. The lack of a fossil record makes examination of the remains of ancient microorganisms impossible. Additionally, the examination of microorganisms requires their culture in the laboratory, and many microorganisms that may prove to be critical evolutionary links have yet to be grown on defined media. As a result, microbial taxonomy generally requires many subjective decisions, resulting in an artificial systematic classification scheme. Some taxonomists are "lumpers," tending to lump together many similar organisms into large taxonomic units. In contrast, other taxonimists are "splitters," favoring small taxonomic groups that emphasize even minor differences between organisms. Arguments can be quite heated over the "proper taxonomic position of a microorganism," and these debates enhance the interest of microbiologists in the field of microbial taxonomy.

When classifying microorganisms, taxonomists use a hierarchy consisting of different organizational levels. The usual levels of a **taxonomic hierarchy**, from the highest to the lowest levels, are **kingdoms, phyla, classes, orders, families, genera,** and **species** (Table 11.1). Ideally, each level represents a different degree of homology, that is, differing degrees of genetic and evolutionary similarity. In reality, many levels represent varying degrees of analogous similarity (**phenotypic similarity**). The higher the taxonomic level, the greater the diversity of the organisms classified as belonging to that group. The species is the basic taxonomic unit of a classification system. Whereas species of higher organisms are readily recognized as a result of their reproductive isolation, a microbial species is difficult to define objectively. A microbial species can be considered as a group of isolate strains that have an overall similarity and are different from other fundamental groups. Defining a new microbial species requires that the organism be described and named and that it be different from previously described species. International committees have been established to rule on the validity of defining new species. Once an organism or group of organisms has been defined as representing a new species, a culture should be deposited in an appropriate culture collection. That type culture and the description of that culture become the reference for future identification.

Although species are the basic taxonomic units, the genetic variability of microorganisms permits a further division into subspecies or types. It is often important to identify the subspecies, or even a specific strain, of a given microorganism. For example, one strain of a bacterial species may produce a toxin and be a virulent pathogen, and other strains of the same species may be nonpathogenic. The ability to distinguish correctly between such strains and subspecies of a particular microbial species is of obvious importance in medical and industrial microbiology.

Approaches to the classification of microorganisms

A number of different approaches are used in microbiology for developing classification systems and establishing hierarchical relationships. The historical development of microbial classification systems was discussed in Chapter 1. Early approaches grouped microorganisms based on similarity of morphological appearance. Later, physiological properties dominated microbial classification. Modern microbial classification sys-

table 11.1

Hierarchy of taxonomic organization

Level	Description	Examples Fungus	Bacterium
Kingdom	A group of related phyla	Myceteae	Procaryotae
Phylum or division	A group of related classes	Amastigomycota	Eubacteria
Class	A group of related orders	Ascomycetes	—
Order	A group of related families	Xylariales	Spirochaetales
Family	A group of related tribes or genera	Sordariaceae	Spirochaetaceae
Tribe	A group of related genera	—	—
Genus	A group of related species	*Neurospora*	*Treponema*
Species	A group of organisms of the same kind	*N. crassa*	*T. pallidium*
Subspecies or type	Variants of a species	—	—

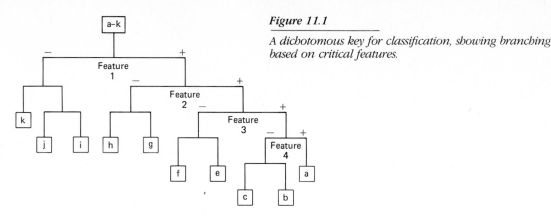

Figure 11.1

A dichotomous key for classification, showing branching based on critical features.

tems still include morphological and physiological characteristics, but molecular-level similarities (genetic relatedness) are given greater emphasis in modern taxonomic approaches. Modern classification systems may be **phenetic** (assessing similarity) or **phylogenetic** (assessing true evolutionary relationships). Organisms could have evolved quite differently and still have developed similar characteristics. Thus, microorganisms that are distantly related in evolutionary terms may still be similar and grouped together in a phenetic classification system but not in a phylogenetic classification system. Phenetic approaches require quantification and explicitness and do not permit *a priori* weighting of features. In contrast, phylogenetic approaches may consider evolution-

ary evidence and give added importance to those key features viewed as indicative of different evolutionary paths.

In a conventional or classical approach to classification, a number of features of a group of organisms are examined, and organisms are grouped based on their similarity with respect to selected features. In the development of such classification systems, certain features are generally considered more important than others for defining taxonomic units. These **key features** are those considered to be of primary importance in the separation of species. For example, the Gram stain reaction is considered a key test in many bacterial classification systems. A taxonomic system based on this approach employs a sequential series of hierarchical decisions to separate the taxonomic units called **taxa** or **taxons**. This type of classification system emphasizes the **branch points** between groups that are presumed to represent fundamental differences between taxa. Often the classification system is represented as a tree diagram showing the various key branch points and the individual taxa at the the ends of the branches (Figure 11.1).

An alternative approach, numerical taxonomy, does not emphasize points of branching but rather uses overall degrees of similarity between organisms to establish a taxon. In **numerical taxonomy** a single characteristic does not determine the taxonomic position of an organism. Instead, overall similarity and the definition of the taxa are based on several characteristics; a large number of characteristics are examined to ensure that no single characteristic achieves undue weighting. Various measures can be applied to assess similarity (Figure 11.2). The most common **indices of similarity** employed in microbiology are the simple matching (S_m) and the Jaccard (S_j) coefficients. The simple matching coefficient is based on all the measured characteristics. In con-

366

Classification and identification of microorganisms: systematics of bacteria

Figure 11.2

Formulas for simple matching and Jaccard coefficients for assessing similarity. Using these equations the similarities between microorganisms can be assessed based on phenotypic or genetic characteristics.

$$S_m = \frac{(++) + (--)}{(++) + (--) + (+-) + (-+)}$$

Where

S_m = simple matching coefficient

$++$ = positive matches

$--$ = negative

$(+-) + (-+)$ = mismatches

$$S_j = \frac{(++)}{(++) + (+-) + (-+)}$$

Where

S_j = Jaccard coefficient

$++$ = positive matches

$(+-) + (-+)$ = mismatches

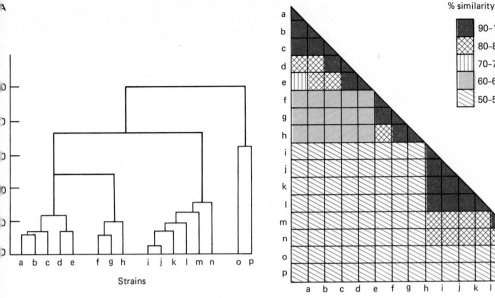

Figure 11.3

The similarities between microorganisms can be represented in several ways. In each, strains of highest similarity are grouped together. (A) Dendrogram showing the hierarchical relationships between organisms; the horizontal lines show the levels of similarity, and the vertical lines represent individual strains. (B) A similarity triangle does not show as much detail as a dendrogram, but groups of high similarities (taxa) are readily distinguished in this graphic representation.

trast, the calculation of the Jaccard coefficient does not use a characteristic when the organisms being compared are both negative for that feature; the assumption is made that such a feature may be an inappropriate description for the group under consideration. For example, a weight of over one ton may be an appropriate descriptive characteristic for elephants but is clearly inappropriate for microorganisms. When the simple matching coefficient is used, the inclusion of such an irrelevant feature in a classification system would artificially, even nonsensically, make organisms appear more similar than they really are; the Jaccard coefficient eliminates this problem.

The classification of organisms by using numerical taxonomy is normally represented graphically by a similarity matrix or a dendrogram (Figure 11.3). Organisms are grouped based on the overall similarity of all the tested characteristics by using a type of statistical analysis known as cluster analysis. In the graphic representations of these analyses organisms of high similarity occur in close geometric proximity, whereas organisms of low similarity are separated.

Criteria for classifying microorganisms

Phenotypic characteristics Regardless of the classification approach employed, it is necessary to characterize the organisms in order to assess their similarity. In order to characterize a microorganism, it is generally necessary to isolate and maintain pure cultures of the organism. Otherwise one might examine the combined features of several interactive organisms or an anomaly based on environmental variations. Determining the characteristics of a pure culture of a microorganism permits the description of that organism so that it can be classified. To exemplify the types of characteristics employed as microbial descriptors, some of the criteria frequently used in bacterial classification are listed in Table 11.2. The various criteria for classifying microorganisms represent a continuum from the analysis of the genome itself to the phenotypic expression of genetic information. Presumably, classification systems based on genetic analyses more accurately reflect a natural evolutionary hierarchy than other artificial classification systems based on phenotypic characteristics. For practical reasons, however, phenotypic characteristics are employed in most microbial classification and identification systems. The morphological, physiological, biochemical, and nutritional features normally examined include mode of reproduction, morphology, staining reactions, ability to form spores, motility, specific metabolic activities, and life cycle stages. Analysis of the biochemical constituents of the cell and particular structures, such as cell walls

table 11.2

Some criteria for classifying bacteria

Microscopic characteristics
Morphology
cell shape
cell size
arrangement of cells
arrangement of flagella
capsule
endospores
Staining reactions
Gram stain
acid fast stain

Growth characteristics
Appearance in liquid culture
Colonial morphology
Pigmentation

Biochemical characteristics
Cell wall constituents
Pigment biochemicals
Storage inclusions
Antigens
RNA molecules

Physiological characteristics
Temperature range and optimum
Oxygen relationships
pH tolerance range
Osmotic tolerance
Salt requirement and tolerance
Antibiotic sensitivity

Nutritional characteristics
Energy sources
Carbon sources
Nitrogen sources
Fermentation products
Modes of metabolism (autotrophic,
heterotrophic, fermentative, respiratory)

Genetic characteristics
DNA %G + C
DNA hybridization

368

Classification and
identification of
microorganisms:
systematics of
bacteria

and membranes among others, may also be valuable in revealing major differentiating features of groups of microorganisms.

Genetic characteristics In addition to phenotypic analyses, it is also possible to include genetic analyses in the development of classification schemes for microorganisms. One of the genetic analyses used for classifying microorganisms is to determine the relative proportion of guanine and cytosine compared to adenine and thymine base pairs in the DNA. Because there is pairing between complementary bases in DNA, specifying G + C also specifies A + T. Therefore, the composition of the DNA is normally described by specifying the **mole% G + C**. Measuring the pro-

portion of G + C in the DNA is a crude analysis of the microbial genome. Closely related organisms should have similar proportions of G + C in their DNA, and organisms with vastly differing proportions of G + C in their DNA can safely be said to be unrelated. Microorganisms that are otherwise similar, but differ by 2–3% G + C content in their genomes, represent different species. However, a similarity of G + C ratios does not necessarily establish the relatedness of organisms because the sequence of genes may be different even when the mole% G + C content is the same.

To determine the mole% G + C, the DNA can be slowly heated and the uv absorption measured spectrophotometrically at a wavelength of 260 nm. When double-stranded DNA melts, that is, when it is converted into two single strands, there is a change in its absorption characteristics. Ultraviolet absorption by the two single strands of DNA is about 40 percent greater than by the same strands in the double helical form. This change in uv absorption is known as a **hyperchromatic shift** (Figure 11.4). The midpoint temperature of the denaturation curve is known as the **melting temperature** (T_m) of the DNA. Because certain ions stabilize the hydrogen bonds of DNA, the melting temperature is influenced by the chemical composition of the solution in which the DNA is suspended. The empirical relationship between the melting temperature and the mole% G + C is given by the equation

$$T_m = 0.41\% \ (\text{mole}\% \ G + C) + 16.6 \log M + 81.5$$

where M is the concentration of monovalent cations within the range 0.0001–0.2 molar. Because guanine and cytosine base pairs form three hydrogen bonds, compared to only two such bonds for adenine and thymine pairs, the melting point of the DNA molecule is proportional to the mole% G + C of that molecule. (It takes more heat to break three hydrogen bonds than it does two such bonds.) The mole% G + C of the DNA may also be determined by measuring the buoyant density of the DNA. The density of the DNA can be accurately determined by using $CsCl_2$ density-gradient ultracentrifugation. The higher the mole% G + C, the greater the density of the DNA. (The molecular weight of guanine + cytosine is higher than adenine + thymine.) The **buoyant density** of the DNA is therefore a measure of the proportion of G + C in the DNA. The relationship between buoyant density (σ) and mole% G + C is given by the equation

$$\sigma = 1.66 + 0.00098 \ (\text{mole}\% \ G + C)$$

Thus, by two different approaches the mole%

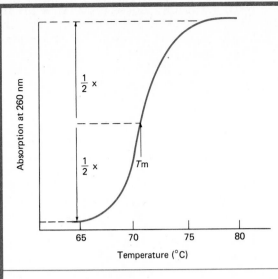

Figure 11.4

Melting curve for DNA, showing hyperchromatic shift and T_m. The hyperchromatic shift occurs because of separation of the two strands of DNA (melting). The temperature at which melting occurs is proportional to the mole% G + C. The mole% G + C is an important characteristic in assessing the genetic relatedness of two organisms.

Discovery process

The two strands of a DNA helix readily come apart when the hydrogen bonds that unite the base pairs are disrupted. Heating a solution of DNA or adding acid or alkali to ionize the bases will unwind the helix. The melting point of DNA (T_m) is defined as the temperature at which half of the helical structure is lost. The melting of DNA is easily followed by measuring its absorbance at 260 nm. The unstacking of the base pairs, that is, the melting of the double helix into single strands, results in increased absorbance. This effect is called hyperchromatism. The T_m of a DNA molecule depends heavily on its base composition. DNA molecules rich in GC base pairs have a higher melting temperature than those having an abundance of AT base pairs. GC base pairs are more stable than AT pairs because their bases are held together by three, rather than two, hydrogen bonds. Therefore, the AT-rich regions of the DNA are the first to melt, and the T_m of the DNA from many species varies linearly with GC content, rising from 77 to 100°C as the fraction of GC pairs increases from 20 to 78 percent. *In vivo* the double helix is melted by the action of specific proteins. Separated complementary strands of DNA spontaneously reassociate to form a double helix when the temperature is lowered, in a process called annealing. The facility with which double helices can be melted and reassociated is crucial for the biological functions of DNA.

G + C can be determined to assess the genetic relatedness of microorganisms.

Although the mole% G + C is a gross measure of the genome, a more precise measure of relatedness is the **DNA homology** between two organisms. Measuring the DNA homology accounts for the order of nucleotides as well as the overall composition of the genome. DNA homology measurements provide a powerful tool for classifying microorganisms. In one method for determining DNA homology, one of the organisms to be compared is grown in a medium containing tritiated thymine so that the radiolabelled thymine is incorporated into the DNA. The other organism is grown in a medium without the radiolabeled nucleotide so that its DNA is cold (nonradioactive). The DNA from the organisms being compared is then melted (converted to single-stranded DNA), mixed, and allowed to reanneal (reestablish double-stranded regions). Only homologous regions of the DNA (those regions having identical or nearly identical nucleotide sequences) will reanneal. Single-stranded DNA is removed chromatographically or by other procedures, and the degree of reannealing is then determined by measuring the amount of radioactivity in the double-stranded DNA form.

DNA homology may also be determined by density-gradient ultracentrifugation. One organism is grown in the presence of a heavy isotope of an element, such as ^{15}N, which is incorporated into the DNA, and another organism is grown in a medium containing the normal isotope, such as light ^{14}N. Thus, one of the organisms has heavy DNA, and the other organism has light DNA. The DNA of both organisms is then melted, mixed, and allowed to reanneal so that the hybrid DNA formed contains one strand of light and one strand of heavy DNA. The hybrid DNA is separated from the normal DNA by density-gradient ultracentrifugation and quantitatively measured. Data from such analyses of DNA homology permits the determination of the genetic relatedness and should facilitate the development of classification systems that truly reflect the natural phylogeny of microorganisms.

Epigenetic analyses Genetic analyses can also be extended to the analysis of gene products, namely, RNA molecules and proteins. The subunits of these macromolecules can be described explicitly. Proteins can be separated and analyzed by using electrophoresis and other analytical methods. The nucleotide sequences of RNA molecules can be determined and the similarity between RNA molecules assessed. In performing these analyses a particular type of RNA molecule, such as 16S ribosomal RNA, is broken into oligonucleotides by using a nuclease. The oligonucleotides are then separated and cataloged. By comparing the RNA oligonucleotide catalogs, the similarity between two organisms ($S_{\nabla\infty}$) can be expressed as $S_{\nabla\infty} = 2 \times$ the number of common oligonucleotides in organisms A and B/total number of oligonucleotides in organisms A and B. Because the order of nucleotides in RNA and amino acids in proteins is directly controlled by the genome, analyses of these gene products should permit combining phenetic and phylogenetic approaches to microbial taxonomy.

Such comparisons of gene products represent an intermediary stage between the analysis of genotype and phenotype. The examination of ribosomal RNA molecules is particularly useful for establishing the higher levels of taxonomic organization, for example, at the kingdom level. As has been discussed in previous chapters, analyses of rRNA molecules have been used to establish a new classification system composed of three primary kingdoms: Archaebacteria, Eubacteria, and Eukaryotes. Similar analyses of ribosomal RNA molecules are now used to assess the relatedness of microbial genera and species. As more is learned about the molecular-level composition of microorganisms and theories about the course of evolution are revised, phylogenetic classification systems will continue to change to reflect the prevailing evolutionary theories.

Identification of microorganisms

While classification attempts to establish an orderly arrangement of taxonomic groups, **identification** attempts to compare an unknown organism with previously established groups in order to determine whether that individual organism should be properly considered as a member of the group. The two processes, classification and identification, are not synonymous. Whereas classification defines taxonomic groups, identification is the process by which an organism of unknown taxonomic affinity is assigned to the correct group of organisms. One can view identification as a practical application of classification, permitting the correct identification of an unknown microorganism as a member of a preexisting taxonomic group. An identification system should permit the efficient and reliable identification of microorganisms.

Although classification and identification of microorganisms are distinct processes, both generally require that pure cultures be examined and characterized. The characteristics used in an identification system are limited to those significant features distinguishing one group from another. As a rule, identification systems try to use the minimal number of characteristics that will produce a reliable identification. Because the purpose of the identification system is to efficiently make a correct identification, different features often are included in the identification system than were used to establish the classification system.

Identification keys and diagnostic tables

The classical approach to microbial identification involves the development of keys or diagnostic tables. An **identification key** consists of a series of questions that lead through a classification system to the determination of the identity of the organism. In a **dichotomous key** a series of yes–no questions is asked that leads through the branches of a flow chart to the identification of a microorganism as a member of a specified microbial group (Table 11.3). The path to an identification in a true dichotomous key is unidirectional, and a single atypical feature or error in determining a feature will result in a misidentification. Most students in introductory microbiology laboratory courses use dichotomous keys for the identification of an unknown organism. Although dichotomous keys are frequently used in identification systems, other keys based on multiple-choice

table 11.3

___

Dichotomous key for the diagnosis of species of the genus Pseudomonas

___

1. Oxidase negative (Go to 2).
   Oxidase positive (Go to 4).
2. Lysine positive. . . . . *P. maltophilia*
   Lysine negative (Go to 3).
3. Motile. . . . . *P. paucimobilis*
   Nonmotile. . . . . *P. malleii*
4. Fluorescent (Go to 5).
   Nonfluorescent (Go to 8).
5. Pyocyanin and pyorubin positive. . . . . *P. aeruginosa*
   Pyocyanin and pyorubin negative (Go to 6).
6. Growth at 42°C . . . . *P. aeruginosa*
   No growth at 42°C (Go to 7).
7. Gelatinase positive . . . . *P. fluorescens*
   Gelatinase negative . . . . *P. putida*
8. Nonmotile. . . . . (probably not a pseudomonad)
   Motile (Go to 9).
9. Peritrichous . . . . (not a pseudomonad)
   Polar (Go to 10).
10. Glucose oxidation negative (Go to 11).
    Glucose oxidation positive (Go to 13).
11. 2 or more flagella. . . . . *P. diminuta*
    1 flagellum (Go to 12).
12. PHB positive. . . . . *P. testosteroni*
    PHB negative. . . . . *P. alcaligenes*
13. Mannose negative (Go to 14).
    Mannose positive (Go to 16).
14. Ornithine positive. . . . . *P. putrefaciens*
    Ornithine negative (Go to 15).
15. Mannitol positive . . . . *P. acidovorans*
    Mannitol negative. . . . . *P. pseudoalcaligenes*
16. Arginine positive (Go to 17).
    Arginine negative (Go to 19).
17. 6.5% NaCl positive. . . . . *P. mendocina*
    6.5% NaCl positive (Go to 18).
18. Gelatinase positive. . . . . *P. fluorescens*
    Gelatinase negative . . . . *P. putida*
19. Galactose negative (Go to 20).
    Galactose positive (Go to 21).
20. Mannose positive. . . . . *P. maltophilia*
    Mannose negative . . . . *P. vesicularis*
21. Lactose negative (Go to 22).
    Lactose positive (Go to 23).
22. 6.5% NaCl positive. . . . . *P. stutzeri*
    6.5% NaCl negative . . . . *P. pickettii*
23. Nitrogen production . . . . *P. pseudomallei*
    No nitrogen production (Go to 24).
24. Citrate positive . . . . *P. cepacia*
    Citrate negative. . . . . *P. paucimobilis*

___

questions also can be used. Regardless of whether the choices are dichotomous or not, the characteristics used in establishing an identification key must be constant for the particular group. For example, if the Gram stain is employed as a key feature in a dichotomous key, the groups separated by this characteristic must be either Gram positive or Gram negative; a group cannot contain both Gram positive and Gram negative members.

Besides identification keys, **diagnostic tables** can be developed to aid in microbial identification (Table 11.4). Such tables summarize the characteristics of the taxonomic groups but do not indicate a hierarchical separation of the taxa. Diagnostic tables generally give the appearance of being far more complicated than keys for the identification of microorganisms because they contain more information. However, in cases where some features are variable for different groups, diagnostic tables are better than keys for the successful identification of an unknown microorganism.

table 11.4

Diagnostic table for differentiating the species of Pseudomonas aeruginosa, P. fluorescens, *and* P. putida

Characteristic	P. aeruginosa	P. fluorescens	P. putida
Monotrichous polar flagella	+	−	−
Pyocyanin	+	−	−
Growth at 4°C	−	+	+
Growth at 42°C	+	−	−
Denitrification	d	−	−
Lecithinase	−	+	−
Gelatinase	d	+	−
Utilization			
Acetamide	+	−	
Creatinine	−		+
Benzylamine	−	−	+
Geranitol	+	−	−
Hippurate	−	−	+
Inositol	−	+	−
Phenylacetate	−	−	+
Trehalose	−	+	−

(+ over 90%, − less than 10%, d 10–90% strains tested are positive)

372

Classification and
identification of
microorganisms:
systematics of
bacteria

Computer-assisted identification

Computers greatly facilitate the identification of microorganisms. When computers are used, the data gathered on an unknown microorganism can

Figure 11.5

The probabilistic matrix approach to organism identification allows organisms of unknown affiliation to be identified as members of established taxa.

	Probability of positive results in test			
TAXA	a	b	c	Probability matrix
A	0.99	0.99	0.99	
B	0.99	0.01	0.10	
C	0.01	0.95	0.02	

	Test			
Organism X	a	b	c	Unknown test results
	+	+	+	

$P_A = (.99)(.99)(.99) = 0.9703$
$P_B = (.99)(.01)(.10) = 0.0001$
$P_C = (.01)(.95)(.02) = 0.0002$

Probability scores comparing X with taxa A, B, C

Sum = 0.9706

$$I_A = \frac{0.9703}{0.9706} = 0.9997$$

$$I_B = \frac{0.0001}{0.9706} = 0.0001$$

$$I_C = \frac{0.0002}{0.9706} = 0.0002$$

Normalized identification scores

Organism X identified as belonging to TAXON A

rapidly be compared to a data bank containing information on the characteristics of defined taxa. Keys and diagnostic tables can readily be programmed for computer-assisted identification of unknown isolates. Because computers rapidly perform large numbers of calculations and comparisons, computerized identification systems have also been developed to assess the statistical probability of correctly identifying a microorganism. In these methods the results of a series of phenotypic tests are scored and compared to the test results of organisms that have been classified as belonging to a particular taxonomic group. Unlike keys where individual tests are critical in reaching proper diagnostic identification, these identification systems assess the statistical likelihood of obtaining a particular pattern of test results.

Such identification systems often involve the development and use of **probabilistic identification matrices** (Figure 11.5). These probability matrices are developed by characterizing large numbers of strains belonging to each taxonomic group. In this way the variability of the group for a particular feature can be determined. Many of the commercial identification systems used in clinical laboratories are based on such probabilistic identification matrices. Some of these systems will be discussed in Chapter 16 when considering clinical identification systems. Several of the commercial systems simplify the process of identification by calculating a numerical profile to describe unambiguously the pattern of test results (Figure 11.6). The numerical profile of an unknown organism can be compared with the test pattern of a defined group to determine the probability that the test results could represent a member of that taxon. Because of the critical nature of making correct identifications in medical microbiology, a positive identification in a clinical identification system generally requires that the unknown organism be far more similar to one group to which it is identified as belonging than to any other group. For example, in some identification systems an unknown microorganism must be a thousand times more similar to one group than to all other groups in the system in order to establish a positive identification.

Bacterial systematics

After discussing the general aspects of microbial nomenclature, classification and identification, we now turn to the consideration of bacterial systematics. Extremely diverse groups of organisms exhibiting widely differing morphological, ecological, and physiological properties are included among the bacteria. The unifying feature of the bacteria is the fact that they all have prokaryotic cells. Bacterial systematics is always in a state of flux; descriptions of new bacterial genera and species, and revisions to older taxonomic classifications, frequently appear in the *International Journal of Systematic Bacteriology*. Periodically, the status of bacterial taxonomy has been summarized in a comprehensive volume, *Bergey's Manual of Determinative Bacteriology*. The eighth edition of *Bergey's Manual,* published in 1974, is the last comprehensive volume planned. Starting with the ninth edition, *Bergey's Manual* will appear in separate parts describing the taxonomic status of particular groups of microorganisms; volume 1 of the ninth edition deals only with Gram negative heterotrophs. Aside from its being the last unified volume, there are two very significant features of the eighth edition of *Bergey's Manual.* Unlike previous editions, it does not establish a hierarchical ordering of bacteria but rather simply describes 19 groups of bacteria (Table 11.5). These groups represent a classification of convenience. The groups do not have any official taxonomic status, and they are not integrated into any traditional hierarchy or classification system. To quote the editors: "The manual is meant to assist in the identification of bacteria. No attempt

Figure 11.6

A numerical profile can be created to identify an unknown organism. This approach is used in several widely used commercial systems for the identification of clinical isolates.

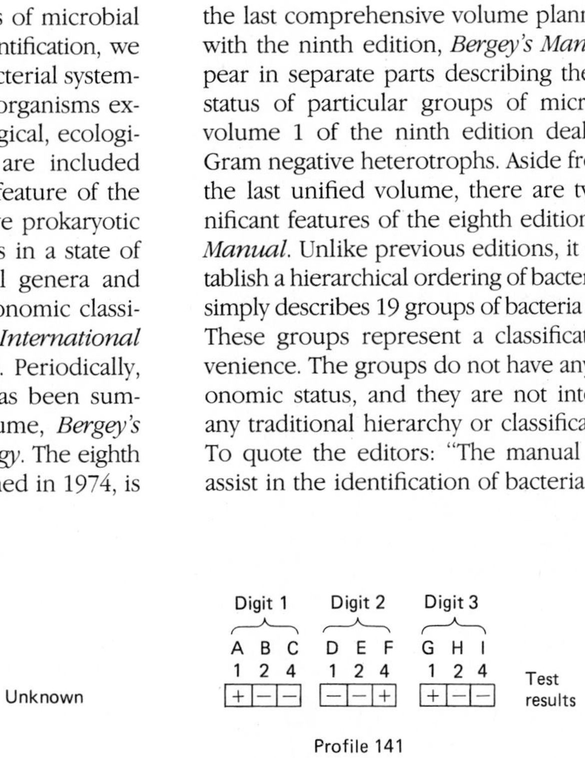

Profile 141

A — ornithine decarboxylase (ODC)
B — citrate utilization (CIT)
C — H_2S production (H_2S)
D — Voges Proskauer test (VP)
E — gelatin hydrolysis (GEL)
F — glucose utilization (GLU)
G — mannose utilization (MAN)
H — inositol utilization (INO)
I — sorbose utilization (SOR)

Probability matrix showing percent positive for all strains which have been classified

		ODC	CIT	H₂S	VP	GEL	GLU	MAN	INO	SOR	Profile
Escherichieae	*Escherichia coli*	75.7	0	1.0	0	0	99.9	99.0	0.5	94.4	145
	Shigella dysenteriae	0	0	0	0	0	94.1	2.0	0	18.6	040
	Sh. flexneri	0.9	0	0	0	0	99.0	91.8	0	24.0	041
	Sh. boydii	3.4	0	0	0	0	99.1	94.1	0	55.9	041
	Sh. sonnei	92.9	0	0	0	0	99.9	99.0	0	2.2	141

Unknown matches profile of *Shigella sonnei*

table 11.5

_____

The 19 groups of bacteria described in the eighth
edition of Bergey's Manual

Group

_____

The phototrophic bacteria
The gliding bacteria
The sheathed bacteria
Budding and/or appendaged bacteria
The spirochetes
Spiral and curved bacteria
Gram negative aerobic rods and cocci
Gram negative facultatively anaerobic rods
Gram negative anaerobic bacteria
Gram negative cocci and coccobacilli
Gram negative anaerobic cocci
Gram negative chemolithotrophic bacteria
Methane-producing bacteria
Gram positive cocci
Endospore-forming rods and cocci
Gram positive, asporogenous, rod-shaped
  bacteria
Actinomycetes and related organisms
The rickettsias
The mycoplasmas

374

_____

Classification and
identification of
microorganisms:
systematics of
bacteria

has been made to provide a complete hierarchy,
as in previous editions, because a complete and
meaningful hierarchy is impossible." Another major feature of the eighth edition of *Bergey's Manual* is the inclusion of the Cyanobacteria as a division of the kingdom Procaryotae, recognizing
that the "blue-greens" are properly considered
bacteria and not algae. The bacteria are placed
within a single kingdom (the Procaryotae) that is

separated into two divisions (Division I—the Cyanobacteria and Division II—the Bacteria). These
hallmarks of the eighth edition will be carried
forth in future editions of *Bergey's Manual.*

Following the lead of *Bergey's Manual,* this
section will discuss the major groups of bacteria
without attempting to place them into a hierarchical classification system. These groups of bacteria are defined primarily on physiological and
morphological criteria. In the remainder of this
chapter, we will consider the diversity of these
organisms.

Phototrophic bacteria

The **phototrophic bacteria** (Table 11.6) are distinguished from other bacterial groups by their
ability to use light energy to drive the synthesis
of ATP. Most of the organisms included in this
group are autotrophs capable of using carbon
dioxide as the source of cellular carbon. The various biochemical mechanisms of photosynthetic
metabolism were considered in Chapter 3. Some
of the phototrophic bacteria use water as an electron donor and liberate oxygen. These bacteria
can be considered as belonging to Oxyphotobacteria. The remainder of the photobacteria do not
evolve oxygen and with one exception can be
classified as belonging to Anoxyphotobacteria. The
exception, *Halobacterium,* belongs to the Archaebacteria and has a unique mode of phototrophic metabolism. *Halobacterium* will be considered together with the Archaebacteria later in
this chapter.

table 11.6

_____

Characteristics of the major groups of phototrophic bacteria

Metabolism	Taxonomic group	Photosynthetic pigments	Electron donors	Carbon source
Anoxygenic photosynthesis	Purple bacteria	Bacteriochlorophyll a or b, carotenoids	H_2, H_2S, S	Organic C or CO_2
Anoxygenic photosynthesis	Green bacteria	Bacteriochlorophyll a or b, carotenoids	H_2, H_2S, S	CO_2
Oxygenic photosynthesis*	Cyanobacteria	Chlorophyll a, phycobiliproteins	H_2O	CO_2
Oxygenic photosynthesis	Prochlorobacteria	Chlorophyll $a + b$, β-carotenes	H_2O	CO_2
Purple membrane-mediated	*Halobacterium*	Bacteriorhodopsin	—	Organic C

*Under some conditions photosynthesis is anoxygenic, and H_2S serves as the electron donor.

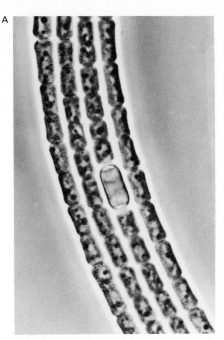
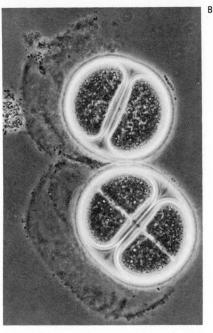

Figure 11.7

These micrographs of different cyanobacteria show the variety of morphologies they can exhibit. (A) An Anabaena *sp. (1420×). (B)* Chrococcus turgidus *(4860×). Both organisms were viewed under the phase contrast microscope. (From BPS—J. Robert Waaland, University of Washington.)*

Oxyphotobacteria

As just indicated, the **Oxyphotobacteria** are capable of splitting water to form oxygen as part of their photosynthetic metabolism. Two orders are contained in this subclass of the Photobacteria: **Cyanobacteriales** and **Prochlorales**. Both of these orders occupy intermediary positions between the phototrophic bacteria and the eukaryotic algae, indicating a probable evolutionary link to the higher photosynthetic organisms. The primary photosynthetic pigment in both cases is chlorophyll *a,* but the prochlorophytes also possess chlorophyll *b,* making them very similar to the green algae. Presumably, the prochlorobacteria are more closely related to the green algae than the cyanobacteria. Some cyanobacteria, on the other hand, are capable of anoxygenic photosynthesis (see discussion in Chapter 3), making them closely related to the Anoxyphotobacteria. Clearly, there is a phylogenetic relationship among the photosynthetic organisms, with the Oxyphotobacteria occupying an intermediate position between the Anoxyphotobacteria and the algae.

Cyanobacteriales
The **cyanobacteria**, or blue-green bacteria, are the most diverse and widely distributed group of photosynthetic bacteria. Over 1000 species of cyanobacteria have been reported, based largely on field observations. Field observations, however, leave many uncertainties about the variability of particular features and ambiguities concerning the separation of taxa. Examination of pure cultures has indicated that by eliminating ambiguous features, the 170 genera described based on field observations can be reduced to 22 genera. Among the cyanobacteria some genera characteristically are unicellular, whereas others are filamentous (Figure 11.7). The cell wall structures of cyanobacteria are of the Gram negative type. The cytoplasm of cyanobacteria is filled with photosynthetic membranes known as **thylakoids**. The primary photosynthetic pigment of the cyanobacteria is chlorophyll *a.* The outer surfaces of the photosynthetic membranes have associated granules known as **phycobilisomes**, which are composed of auxiliary photosynthetic pigments.

There are four major subgroups of cyanobacteria (Table 11.7). The **chroococcacean cyanobacteria** are unicellular rods or cocci. They reproduce either by binary fission (family Chroococcaceae) or by budding (family Chamesiphonaceae). Chroococcacean cyanobacteria are gener-

table 11.7

The subgroups of the cyanobacteria

Group	Description
Chroococcaean	Unicellular rods or cocci reproducing by binary fission or budding
Pleurocapsalean	Single cells enclosed in a fibrous layer; reproduce by multiple fission producing baeocytes
Oscillatorian	Cells form trichomes but do not form heterocysts
Heterocystous	Form trichomes with both vegetative cells and heterocysts

Figure 11.8

Heterocyst of the cyanobacterium Anabaena cylindrica *(14,000 ✕). (From BPS—Norma Lang, University of California, Davis.)*

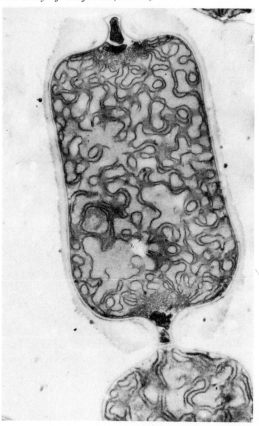

ally nonmotile. *Synechococcus, Synechocystis,* and *Chamaesiphon* are representative genera of chroococcacean cyanobacteria. One interesting genus in this group, *Gloeobacter,* lacks thylakoids and is purple in color; it could easily be confused with the anaerobic phototrophs, but in pure culture studies its biochemistry and metabolism have been shown to be typical of cyanobacteria.

The **pleurocapsalean cyanobacteria** are distinguished from the chroococcacean cyanobacteria by the fact that they exhibit multiple fission to produce small coccoid reproductive cells. In the phycological literature these reproductive cells are referred to as endospores, but to avoid confusion with endospore-forming bacteria, it has been proposed that the term **baeocyte** be used to describe the reproductive cells of the pleurocapsalean cyanobacteria. The pleurocapsalean cyanobacteria are unicellular, but the cells generally fail to separate completely following binary fission. Because binary fission does not result in complete separation of the cells, the pleurocapsalean cyanobacteria form multicellular aggregates. The **oscillatorian cyanobacteria** form filamentous structures, exclusively composed of vegetative cells, known as **trichomes**. In some cases the trichomes are straight, and in others they are helical. *Spirulina, Oscillatoria,* and *Pseudanabaena* are representative genera of oscillatorian cyanobacteria.

Like the oscillatorian cyanobacteria, the **heterocystous cyanobacteria** are filamentous. Unlike the oscillatorian cyanobacteria, however, the heterocystous cyanobacteria form differentiated cells known as **heterocysts** when growing in the absence of fixed forms of nitrogen. Heterocysts are nonreproductive cells that are distinguished from the adjoining vegetative cells by the presence of refractory polar granules and a thick outer wall (Figure 11.8). The ability to form heterocysts is associated with the physiological capability of fixing atmospheric nitrogen. The physiologically specialized heterocyst cells appear to be the anatomical site of nitrogen fixation in heterocystous cyanobacteria. *Nostoc* and *Anabaena* are probably the best-known genera of heterocystous cyanobacteria. Being able to carry out both oxygen-yielding photosynthesis and nitrogen fixation is a unique characteristic of cyanobacteria principally found among the heterocystous cyanobacteria. The heterocystous cyanobacteria are ecologically important because they can form both organic carbon and fixed forms of nitrogen that can support the nutritional requirements of other organisms.

Prochlorales The **prochlorales** are similar to the cyanobacteria except that they also synthesize chlorophyll *b*. Although they originally were considered to be cyanobacteria, their unique ability as prokaryotes to produce chlorophyll *b* is now considered significant enough to separate them into their own order. The only known genus, *Prochloron,* occurs as single-celled extracellular symbionts of marine invertebrates. These bacteria appear bright green on the surfaces of the animals with which they are associated. Various species of *Prochloron* have been recognized in field studies, but until the organisms are grown in pure culture, the validity of these species remains ambiguous.

Anoxyphotobacteria

Unlike the Cyanobacteria and Prochlorobacteria, the **Anoxyphotobacteria** do not evolve oxygen. The Anoxyphotobacteria require an electron donor other than water and carry out only one form of oxidative photophosphorylation. Physiologically, these bacteria carry out photosynthesis anaerobically. The anaerobic photosynthetic bacteria typically occur in aquatic habitats, often growing at the sediment water interface of shallow lakes where there is sufficient light penetration to permit photosynthetic activity; anaerobic conditions are sufficient to permit the existence of these organisms; and there is a source of reduced sulfur or organic compounds to act as elec-

tron donors for the generation of reduced coenzymes. The phototrophic bacteria include the **Rhodospirillaceae (purple nonsulfur bacteria), Chromatiaceae (purple sulfur bacteria), Chlorobiaceae (green sulfur bacteria), and Chloroflexaceae (green flexibacteria)** (Table 11.8). The green and purple sulfur bacteria utilize reduced sulfur compounds, such as hydrogen sulfide, as electron donors for generating reducing power. Most of the purple nonsulfur bacteria and green flexibacteria are unable to use reduced sulfur compounds, but rather these organisms utilize organic compounds to support photosynthetic growth.

Chromatiaceae Members of the family **Chromatiaceae** produce carotenoid pigments and may appear orange-brown, red-brown, purple-red, or purple-violet. The Chromatiaceae deposit elemental sulfur as a consequence of their utilization of reduced sulfur compounds as electron donors for generating reducing power (Figure 11.9). In all but one genus of organisms within this large family, the sulfur accumulates intracellularly. Members of the Chromatiaceae are potentially mixotrophic (capable of both photoautotrophic and heterotrophic growth), and all strains are capable of photoassimilating simple organic substrates such as acetate. The cells of *Chromatium,*

table 11.8

The phototrophic bacteria

Order Rhodospirillales	Cells capable of photosynthetic anaerobic metabolism
Suborder Rhodospirillineae (purple bacteria)	Cells contain bacteriochlorophyll *a* or *b*
Family Rhodospirillaceae (purple nonsulfur bacteria)	Cells photoassimilate simple organic substrates; most species unable to grow with sulfide as the sole electron donor
Genera: *Rhodospirillum, Rhodopseudomonas, Rhodomicrobium*	
Family Chromatiaceae (purple sulfur bacteria)	Cells able to grow with sulfide and sulfur as the sole electron donor; sulfur deposited inside or outside of cell
Genera: *Chromatium, Thiocystis, Thiosarcina, Thiospirillum, Thiocapsa, Lamprocystis, Thiodictyon, Thiopedia, Amoebobacter, Ectothiorhodospira*	
Suborder Chlorobiineae (green bacteria)	Cells contain bacteriochlorophyll *a, b,* or *c*
Family Chlorobiaceae (green sulfur bacteria)	Cells able to grow with sulfide and sulfur as the sole electron donor; sulfur deposited only outside of cell
Genera: *Chlorobium, Prosthecochloris, Chloropseudomonas, Pelodictyon, Clathrochloris*	
Family Chloroflexaceae (green flexibacteria)	Cells have flexible walls, gliding motility, form filaments, utilize organic carbon sources
Genera: *Chloroflexus, Chloronema, Oscillochloris*	

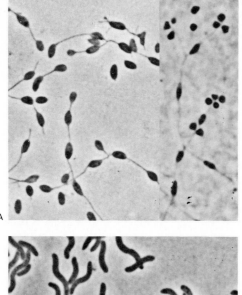

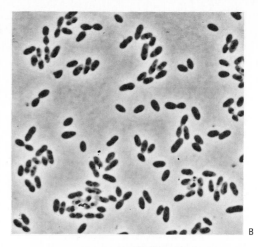

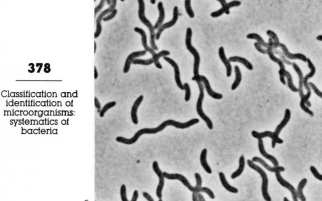

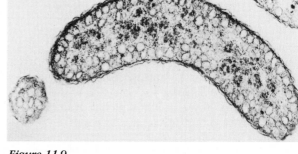

378

Classification and
identification of
microorganisms:
systematics of
bacteria

Figure 11.9

Morphologies of the purple sulfur bacteria.
(A) Rhodomicrobium vannielii *(1660×).*
(B) Rhodopseudomonas acidophila *(1660×).*
(C) Rhodospirillum molischianum *(1660×).*
(D) Rhodospirillum rubrum *(32,900×). (Reprinted by*
permission of the Bergey's Trust, John Holt, executor, from
Bergey's Manual of Determinative Bacteriology, Williams &
Wilkins Co., Baltimore.)

Thiocystis, Thiosarcina, Thiospirillum, and *Thiocapsa* do not contain gas vacuoles, but some genera of the family Chromatiaceae, such as *Thiodictyon* and *Thiopedia,* do contain gas vacuoles. The gas vacuoles permit an adjustment of cell buoyancy in a water column to a depth that is appropriate for light penetration and oxygen concentration, making anaerobic photosynthetic metabolism possible.

Chlorobiaceae The **Chlorobiaceae** produce green or green-brown carotenoid pigments and are obligately phototrophic. They assimilate carbon dioxide, utilizing sulfide or elemental sulfur as electron donors, and they deposit sulfur granules extracellularly. Some genera of Chlorobiaceae, such as *Pelodictyon,* produce gas vacuoles, but others, such as *Chlorobium* and *Chloropseu-*

domonas, do not contain gas vacuoles (Figure 11.10). All members of the Chlorobiaceae are nonmotile. These bacteria often occur in similar ecological situations as the Chromatiaceae.

Rhodospirillaceae Three genera are recognized within the **Rhodospirillaceae:** *Rhodospirillum* has spiral-shaped cells; *Rhodopseudomonas* has spherical or rod-shaped cells that do not form filaments; and *Rhodomicrobium* has oval cells that do form filaments and exhibits budding division (Figure 11.11). The Rhodospirillaceae generally produce red-purple carotenoid pigments. Members of the genera *Rhodospirillum* and *Rhodopseudomonas* are motile by means of polar flagella, whereas those of the genus *Rhodomicrobium* are peritrichously flagellated. Their photosynthetic development depends on the ability of the

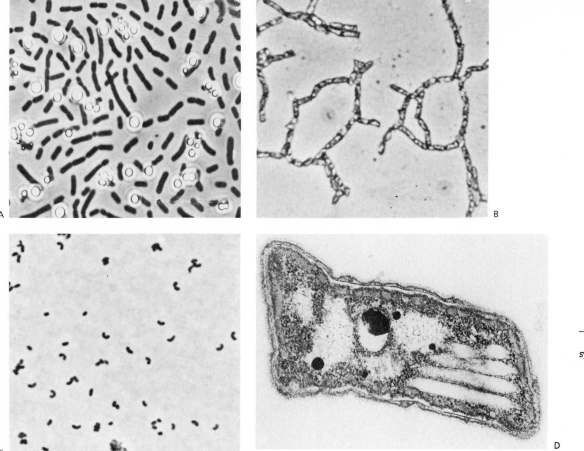

Figure 11.10

Morphologies of the green sulfur bacteria. (A) Chlorobium limicola *(1660×), note extracellular sulfur globules. (B)* Pelodictyon clathratiforme *(1660×), note gas vacuoles. (C)* Chlorobium phaeovibriodes *(2490×). (D)* Pelodictyon clathratiforme *(55,125×). (Reprinted by permission of the Bergey's Trust, John Holt, executor, from* Bergey's Manual of Determinative Bacteriology, Williams & Wilkins Co., Baltimore.)

cells to photoassimilate simple organic compounds. When sulfide or thiosulfate is utilized as electron donor, elemental sulfur is not deposited within the cell. The organic substrates utilized by the Rhodospirillaceae may serve as electron donors for generating reducing power or may be photoassimilated. Because they generally require preformed organic matter for growth and are able to utilize light energy for generating ATP, the type of metabolism carried out by these organisms is sometimes referred to as **photoheterotrophic metabolism.**

The Rhodospirillaceae convert carbon dioxide to organic matter by means of the Calvin cycle pathway. Some typical members of the Rhodo-spirillaceae use molecular hydrogen or sulfide as an electron donor and can grow without organic compounds. As such, it may be best to consider these organisms as photoautotrophs, generally requiring organic growth factor compounds. Indeed, most strains in the Rhodospirillaceae require one or more vitamins. Clearly, the Rhodospirillaceae occupy a boundary position between autotrophs and heterotrophs; the basic metabolic pathways are the same as those of other autotrophic microorganisms. Their ability to assimilate organic compounds and the requirement of many members of the Rhodospirillaceae for organic compounds establish a resemblance of these organisms to heterotrophs.

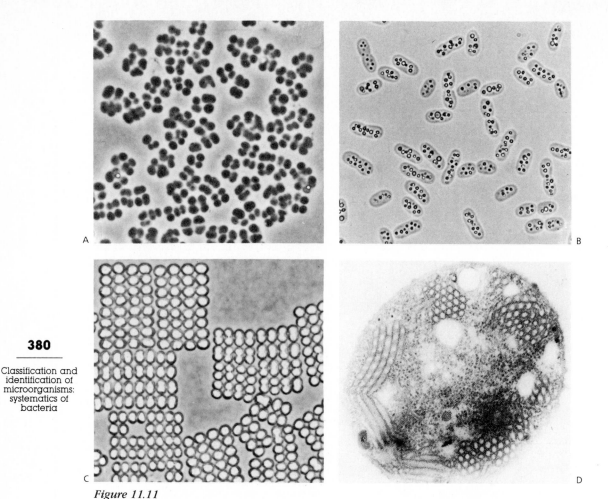

380

Classification and
identification of
microorganisms:
systematics of
bacteria

Figure 11.11

Morphologies of the purple nonsulfur bacteria. (A) Thiocapsa roseopersicina (1660×); note intracellular sulfur globules. (B) Chromatium vinosum (1600×). (C) Thiopedia rosea (1660×); note gas vacuoles. (D) Thiocapsa pfennigui (43,800×); note the bundled tubes of the intracytoplasmic membrane system. (Reprinted by permission of the Bergey's Trust, John Holt, executor, from Bergey's Manual of Determinative Bacteriology, *Williams & Wilkins Co., Baltimore.)*

Chloroflexaceae The **Chloroflexaceae** represent a relatively newly discovered family of anaerobic phototrophic bacteria. These bacteria have flexible walls, form filaments, and exhibit gliding motility. Gliding motility and formation of filaments were previously thought to be restricted among phototrophic bacteria to the cyanobacteria. Photosynthesis is anoxygenic, and some organic compounds are needed to achieve optimal growth. *Chloroflexus* is the only genus in this family that has been characterized in pure culture, although other genera have been assigned to the Chlorofexaceae based on field observations. *Chloroflexus aurantiacus* has been isolated from alkaline hot springs in various parts of the world. *Chloroflexus* resembles a green sulfur bacterium in cell ultrastructure and photosynthetic pigments but resembles a nonsulfur purple bacterium in its photosynthetic and catabolic metabolism. Both in terms of physiology and morphology, the Chloroflexaceae exhibit unique combinations of characteristic features of other phototrophic bacteria.

The gliding bacteria

We have already seen that some phototrophic bacteria are capable of gliding motility. In addi-

table 11.9

The gliding bacteria

Order Myxobacteriales	Produce fruiting bodies
Family Myxococcaceae	Vegetative cells tapered; microcysts spherical or oval
Genus: *Myxococcus*	
Family Archangiaceae	Vegetative cells tapered; microcysts rod-shaped, not in sporangia
Genus: *Archangium*	
Family Cystobacteraceae	Vegetative cells tapered; microcysts rod-shaped, in sporangia
Genera: *Cystobacter, Melittangium, Stigmatella*	
Family Polyangiaceae	Myxospores resemble vegetative cells
Genera: *Polyangium, Nannocystis, Chondromyces*	
Order Cytophagales	Fruiting bodies are not produced
Family Cytophagaceae	Pigmented; filaments not attached
Genera: *Cytophaga, Flexibacter, Herpetosiphon, Flexithrix, Saprospira, Sporocytophaga*	
Family Beggiatoaceae	Nonpigmented; filaments not attached; cells in flat filaments
Genera: *Beggiatoa, Vitreoscilla, Thioploca*	
Family Simonsiellaceae	Nonpigmented; filaments attached; cells in flat filaments
Genera: *Simonsiella, Alysiella*	
Family Leucotrichaceae	Filaments attached at one end
Genera: *Leucothrix, Thiothrix*	

tion to the cyanobacteria and Chloroflexaceae, the **Myxobacteriales** (fruiting myxobacteria) and the **Cytophagales** are grouped together based on their **gliding motility** on solid surfaces (Table 11.9). The mole% G + C of the gliding bacteria covers the entire range from 30 to 70 percent, and in all likelihood the gliding bacteria represent a phylogenetically heterogeneous group.

Myxobacteriales

The **myxobacteria** are small rods that are normally embedded in a slime layer. They lack flagella but are capable of gliding movement. A unique feature of the myxobacteria is that under appropriate conditions they aggregate to form fruiting bodies (Figure 11.12). The taxonomy of the myxobacteria is based largely on the fruiting body structures that are often brightly colored and visible without the aid of a microscope. Frequently, the **fruiting bodies** of myxobacteria occur on decaying plant material, on the bark of living trees, or on animal dung, appearing as highly colored slimy growths that may extend above the surface of the substrate. Within the fruiting body, the cells of the myxobacteria are dormant and are

termed **myxospores**. In some genera of myxobacteria, the myxospores cannot be distinguished from vegetative cells, but in other genera the myxospores are refractile and encapsulated, in which case they are known as **microcysts**. Most of the myxobacteria produce a variety of hydrolytic enzymes, such as cellulases, and many are capable of lysing other microorganisms.

Cytophagales

In contrast to the Myxobacteriales, the **Cytophagales** do not produce fruiting bodies. Members of both groups, however, do exhibit gliding motion. Genera in the order Cytophagales exhibit widely differing morphological forms (Figure 11.13) and modes of metabolism. They are unified only by the presence of gliding motion and lack of fruiting body formation. Some Cytophagales form filaments, and others do not. Some *Flexibacter* species, for example, may form very long filaments measuring as much as 100 µm. Some of the Cytophagales are chemolithotrophs. For example, *Beggiatoa* forms filaments, oxidizes hydrogen sulfide, and deposits sulfur intracellularly when growing on hydrogen sulfide. *Cyto-*

382

Classification and
identification of
microorganisms:
systematics of
bacteria

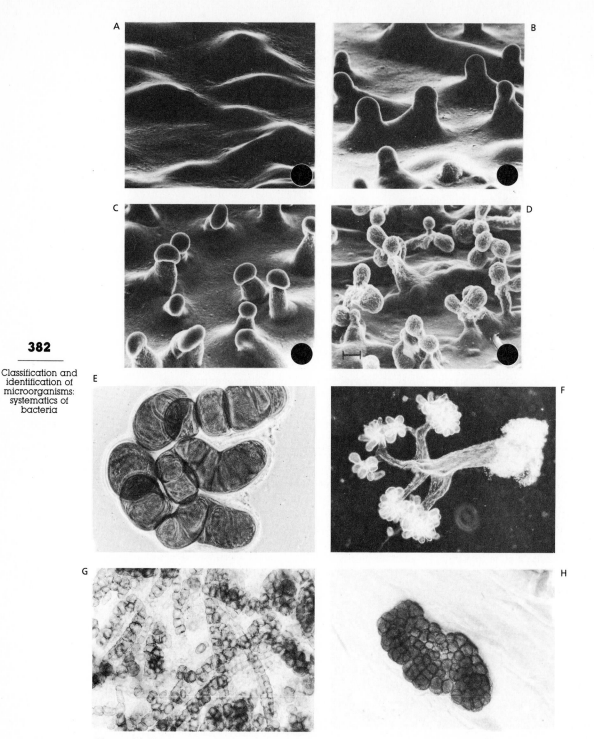

Figure 11.12

*Micrograph showing morphologies and fruiting of myxobacteria. The stages of fruiting body formation of
the myxobacterium* Stigmatella aurantiaca *are shown in (A) early aggregates; (B) early stalks; (C) late stalks;
and (D) mature fruiting body. (E)* Cystobacter fucus *fruiting bodies (120×). (F)* Chondromyces crocatus
(88×). (G) Sorangium cellulosum *fruiting bodies isolated from filter paper and aligned along dissolved
cellulose fibers. (H)* Sorangium cellulosum *fruiting body on agar surface (105×). (A–D from BPS—Karen
Stephens, Stanford Medical Center; E–H courtesy Hans Reichenbach, Gesellschaft für Biotechnologische
Forschung, Germany.)*

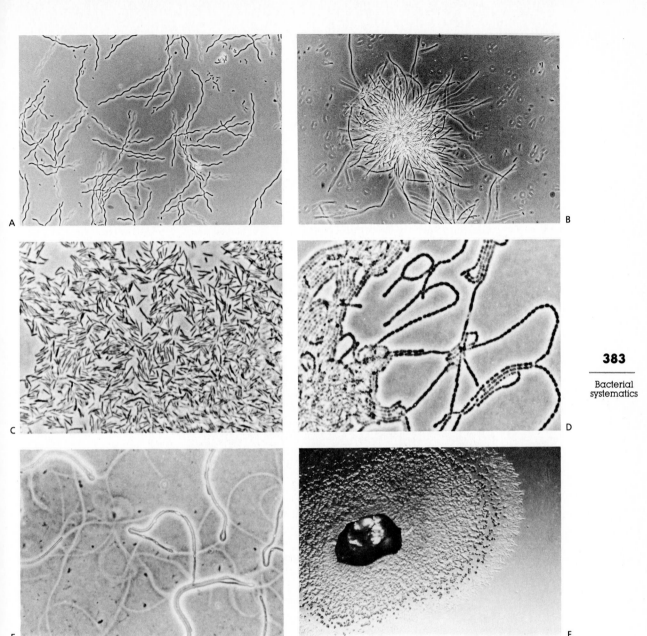

Figure 11.13

Micrographs showing morphologies of the Cytophagales. (A) Saprospira grandis *showing helical filaments (285×). (B)* Leucothrix mucor *showing a cluster of filaments from liquid culture and numerous cells released from their ends (380×). (C)* Cytophaga hutchinsonii *vegetative cells (1225×). (D)* Vitreoscilla stercoraria *showing filaments (980×). (E)* Herpetosiphon giganteus *showing filaments on agar surface with their slime trails (325×). (F) A spreading colony of* Herpetosiphon giganteus *on agar surface. A–E were viewed under the phase contrast microscope. (Courtesy Hans Reichenbach, Gesellschaft für Biotechnologische Forschung, Germany.)*

phaga species, on the other hand, do not form filaments. Cells of *Cytophaga* contain deep yellow-orange or red pigments and hydrolyze agar, cellulose, and chitin. As a consequence of their hydrolytic activities, these gliding bacteria play an important ecological role in the decomposition of organic matter.

The sheathed bacteria

The **sheathed bacteria** comprise those bacteria whose cells occur within a filamentous structure known as a sheath (Figure 11.14). The sheathed bacteria include the genera *Sphaerotilus, Lepto-*

thrix, Haliscomenobacter, Lieskeella, Phragmidio-thrix, Crenothrix, and *Clonothrix.* The formation of a sheath enables these bacteria to attach themselves to solid surfaces. This is important to the ecology of these bacteria because many sheathed bacteria live in low-nutrient aquatic habitats. By absorbing nutrients from the water that flows by the attached cells, these bacteria are able to conserve their limited energy resources. Additionally, the sheaths afford protection against predators and parasites. In some cases the sheaths may be covered with metal oxides. For example, in the genus *Leptothrix,* sheaths are encrusted with iron or manganese oxides. In other cases, such as in the

Classification and
identification of
microorganisms:
systematics of
bacteria

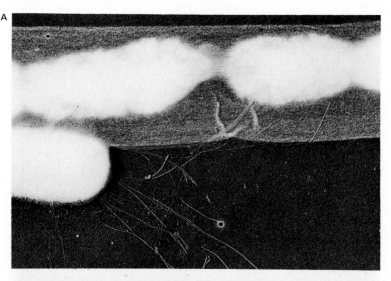
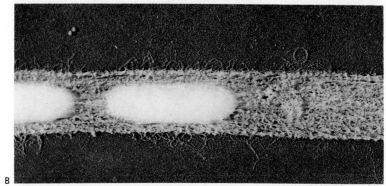
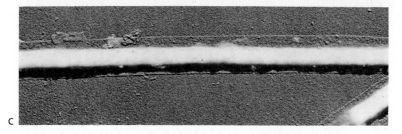

Figure 11.14

Micrographs showing morphologies of sheathed bacteria. (A) Sphaerotilus natans *(14,900 ×).
(B)* Leptothrix cholodnii *(14,900 ×).
(C)* Haliscomenobacter hyrossis *(14,900 ×). (Courtesy M. H. Deinema, Landbouwhogeschool Wageningen, Netherlands. A reprinted by permission of the American Society for Microbiology, Washington, D.C., from W. L. vanVeen, E. G. Mulder, and M. H. Deinema, 1978,* Bacteriological Reviews *42: 329–356; B reprinted by permission of Springer-Verlag, New York, from E. G. Mulder and M. H. Deinema, 1981,* The sheathed bacteria, *in* The Prokaryotes, *M. P. Starr, H. Stolp, H. G. Truper, A. Balows, and H. G. Schlegel, eds.)*

genus *Haliscomenobacter,* the sheath is not encrusted with metal oxides. In the genus *Sphaerotilus* the sheath is sometimes encrusted with iron oxides. *Sphaerotilus natans,* often referred to as the sewage fungus, is the only species in the genus *Sphaerotilus.* This organism normally occurs in polluted flowing waters, such as sewage effluents, where it may be present in high concentrations just below sewage outfalls.

Budding and/or appendaged bacteria

Like the sheathed bacteria, the **budding and/or appendaged bacteria** represent a heterogeneous group based on a particular morphological feature. The budding and/or appendaged bacteria have in common the formation of extensions or protrusions from the cell (Figure 11.15). In some

Figure 11.15

Micrographs of budding, stalked, and prothecate bacteria. (A) Planctomyces bekefil *rosette from pondwater, a negative-contrast TEM micrograph; b = bud, S = stalk, sp = spore appendages. (Courtesy Jean M. Schmidt, Arizona State University, reprinted by permission of Springer-Verlag, New York, from J. M. Schmidt, and M. P. Starr, 1980,* Current Microbiology *4:183–188.) (B) Interference light micrograph showing rosette formation by* Hyphomicrobium *at air–water interface. (C) Scanning electron micrograph showing typical morphology of* Hyphomicrobium. *(Courtesy Richard L. Moore, University of Calgary.) (D) Electron micrograph of the stalked bacterium* Caulobacter. *(Courtesy Jeanne Poindexter, City of New York Public Health Laboratories.) (E) Electron micrograph of* Prothecomicrobium pneumaticum *showing multiple prothecae. (Courtesy J. T. Staley, University of Washington.) (F) Micrograph of the flat six-pointed bacterium* Stella. *(Reprinted by permission of Springer-Verlag, New York, from P. Hirsch and H. Schlesner, 1981, The genus* Stella, *in* The Prokaryotes, *M. P. Starr, H. Stolp, H. G. Truper, A. Balows, and H. G. Schlegel, eds.)*

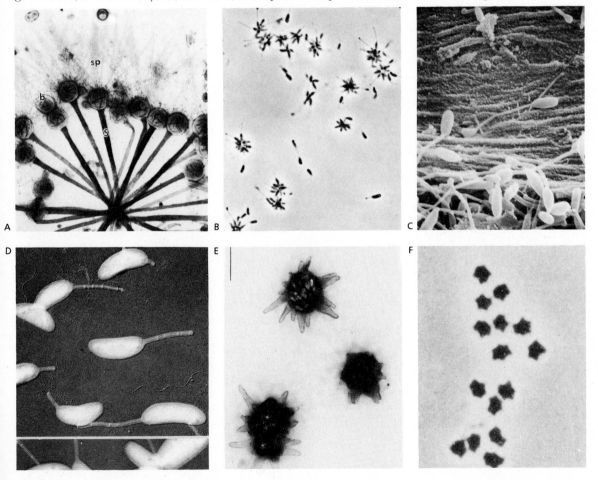

cases the cellular extensions have a reproductive function, but in other cases the protrusions serve a physiological rather than reproductive function. Some bacteria in this group reproduce by budding, but others reproduce by binary fission (Table 11.10). Several budding bacteria, such as *Rhodomicrobium,* are associated primarily with other groups based on their characteristic modes of metabolism. Budding and/or appendaged bacteria occur in all nutritional categories. Various studies from the time of Winogradsky have been aimed at determining if members of the genus *Gallionella* are capable of chemolithotrophic metabolism; they are probably facultatively chemolithotrophic because they are able to oxidize ferrous to ferric iron and fix carbon dioxide. *Gallionella* are sometimes considered to be sheathed bacteria because their "stalks" may be covered with iron hydroxide. Growth of *Gallionella* species often causes problems in iron pipes of water delivery systems.

The cell appendages of the bacteria in this group, known as **prosthecae**, afford the cell greater ef-

table 11.10

The budding and/or appendaged bacteria

Prosthecate bacteria
Prosthecae have a reproductive function; new cell formation by a budding process
Hyphomicrobium
Hyphomonas
Pedomicrobium
Thiodendron

Prosthecae have no reproductive function
Caulobacter
Asticcacaulis
Ancalomicrobium
Prosthecobacter
Prosthecomicrobium
Stella

Nonprosthecate bacteria; reproduction by a budding process
Pasteuria
Blastobacter
Seliberia

Bacteria with excreted appendages and holdfasts; reproduce by binary fission only
Gallionella
Nevskia

Reproduce by budding
Planctomyces

Genera of uncertain affiliation
Metallogenium
Caulococcus
Kusnezovia

ficiency in concentrating available nutrients. Many of the appendaged bacteria grow well at low nutrient concentrations. The appendages provide sufficient membrane surface to transport adequate nutrients into the cell to support the metabolic requirements of the organism. Many of the bacteria in this group primarily occur in aquatic habitats where concentrations of organic matter typically are low. *Caulobacter,* for example, is able to grow in very dilute concentrations of organic matter in lakes and even is able to grow in distilled water. The appendages of *Caulobacter* are referred to as **stalks**. In some cases the stalks of individual cells provide a **holdfast** by which the organisms can attach to a substrate. In other cases stalks do not function in attachment but may permit cells to adhere to each other, forming rosettes. Some of the appendaged bacteria form bizarre-looking cells. For example, members of the genus *Prosthecomicrobium* form prosthecae extending in all directions from the cell. *Seliberia* form radial clusters (star-like aggregates) of rod-shaped bacteria with a screw-like twisting of the rod surface and the formation of round reproductive cells by budding. At low nutrient concentrations, *Stella* forms flat cells resembling six-pronged stars. Many of these morphological forms are observed only at very low nutrient concentrations. The isolation of various new types of appendaged bacteria has greatly increased our views of morphological diversity among the bacteria and the relationship between morphology and nutritional status.

The spirochetes

The **spirochetes** are another group that has a distinct morphology. Spirochetes are helically coiled rods (Figure 11.16). The cell is wound around one or more central axial fibrils. The cell length varies in different genera from 3 to 500 μm. In addition to their characteristic morphology, the spirochetes exhibit a unique mode of motility. These bacteria move by a flexing motion of the cell, exhibiting greatest velocities in very viscous solutions where motility by bacteria with external flagella is slowest. Five genera are recognized in the family Spirochaetaceae, and several more have been proposed (Table 11.11). Members of the genus *Spirochaeta* are nonpathogens, occurring in aquatic environments, in hydrogen sulfide-containing mud, in sewage, and polluted waters. Many spirochetes, though, are human pathogens. Several members of the genus *Treponema,* for example, are human pathogens, with *Treponema*

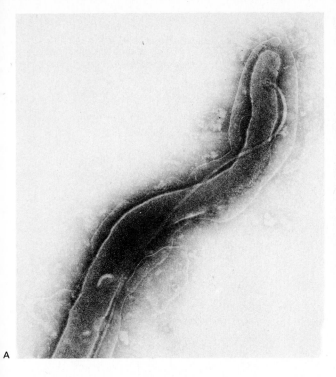

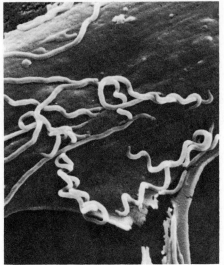

Figure 11.16

Micrographs of spirochetes. (A) Spirochete showing central axial filament (50,280×). (B) Treponema pallidum (4,760×). (From BPS—Stanley C. Holt, University of Massachusetts.)

pallidum causing syphilis and *Treponema pertenue* causing yaws. Likewise, some members of the genera *Cristispira, Borrelia,* and *Leptospira* are animal and human pathogens.

Spiral and curved bacteria

Members of the **spiral and curved bacteria** group are helically curved rods that may have less than one complete turn (comma-shaped) to many turns (helical), but unlike the spirochetes the cells are not wound around a central axial filament. The cells of members of this group are motile by means of polar flagella. Members of *Spirillum,* for example, have multiple polar flagella, usually at both ends (Figure 11.17). The genus *Microcyclus* was formed based on the observation of the unique closed ring-like morphology of these bacteria (Figure 11.18). Some members of *Campylobacter,* another genus affiliated with this group, are important pathogens of humans and other animals. *Bdellovibrio,* a genus of uncertain affiliation within this group, has the outstanding characteristic of being able to penetrate and reproduce within prokaryotic cells (Figure 11.19). All natu-

table 11.11

The spirochetes	
Order Spirochaetales	
Family Spirochaetaceae	
Spirochaeta	5–500 μm in length, 0.2–0.75 μm wide, free-living; anaerobic or facultative
Cristispira	30–150 μm in length, 0.5–3.0 μm wide, with 3 to 10 complete turns; not free-living
Treponema	5–15 μm in length, 0.09–0.5 μm wide; not free-living; anaerobic
Borrelia	3–15 μm in length, 0.2–0.5 μm wide; not free-living; anaerobic
Leptospira	6–20 μm in length, 0.1 μm wide; aerobic

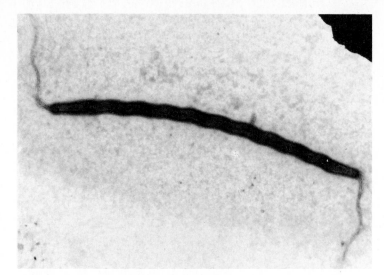

Figure 11.17

Negatively stained electron micrograph of Spirillum ostreae *isolated from an oyster (6,600 ×). (From BPS—P. Johnson, University of Rhode Island.)*

rally occurring strains of *Bdellovibrio* have been found to be bacterial parasites. Unlike viruses, *Bdellovibrio* do not lose their integrity when they reproduce within host cells. Within host cells, bacteria in the genus *Bdellovibrio* reproduce by binary fission. After reproduction of *Bdellovibrio* within a host cell, the host cell lyses, releasing the *Bdellovibrio* progeny.

Gram negative aerobic rods and cocci

The **Gram negative aerobic rods and cocci** encompass a large number of taxonomic units. Several major families are included in this group: **Pseudomonadaceae, Azotobacteraceae, Rhizobiaceae,** and **Methylomonadaceae** (Table 11.12).

Figure 11.18

Micrographs of Microcyclus, *showing ring formation. (A) In this phase photomicrograph of a gas vacuolate strain of* Microcyclus aquaticus, *the gas vacuoles appear as retractile areas within the cells. Note that before some cells separate from one another, they appear as a ring, hence the name* Microcyclus, *small circle. (Courtesy M. VanErt and J. T. Staley, University of Washington.) (B) This SEM micrograph shows the ring arrangements of* Microcyclus marinus *Raj (=* Cyanobacterium marinus *Raj). (Courtesy H. D. Raj, California State University, Long Beach.)*

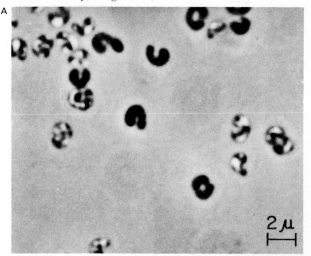

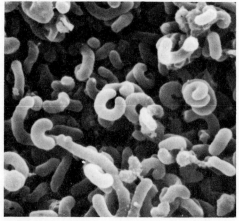

Members of the family Pseudomonadaceae are Gram negative straight or curved rods that are motile by means of polar flagella. Because their metabolism is respiratory, members of this family cannot carry out fermentative metabolism. Whereas the metabolism of the **Pseudomonadaceae** is obligately aerobic, some strains are able to carry out anaerobic respiration. The Pseudomonadaceae are unable to fix atmospheric nitrogen. Many *Pseudomonas* species are nutritionally versatile and are capable of degrading many natural and man-made organic compounds. Some *Pseudomonas* species produce characteristic fluorescent pigments, but others do not. For example, *Pseudomonas aeruginosa* produces yellow-green diffusible pigments that fluoresce at a wavelength of less than 260 nm. *Pseudomonas* species are widely distributed in soil and aquatic ecosystems, occurring as free-living bacteria or in association with plants and animals. Some *Pseudomonas* species are plant and animal pathogens. *Pseudomonas aeruginosa,* for example, can be a human pathogen and is commonly isolated from wound, burn, and urinary tract infections. All recognized species of *Xanthomonas,* a genus included in the Pseudomonadaceae, are plant pathogens. *Xanthomonas* species are Gram negative rods that are motile by means of polar flagella and in most cases produce yellow pigments.

Whereas nitrogen fixation does not occur in the Pseudomonadaceae, the family **Azotobacteraceae** is characterized by its capacity to **fix molecular nitrogen**. This family consists of Gram negative rods exhibiting pleomorphic morphology. The genera *Azotobacter* and *Beijerinckia* are particularly important free-living nitrogen-fixing bacteria. The practical importance of these bacteria will be considered later in the section on environmental microbiology. The **Rhizobiaceae** are also capable of fixing atmospheric nitrogen. *Rhizobium* species are able to infect plant roots, causing the formation of tumorous growths called nodules. Cells of *Rhizobium* stimulate nodule

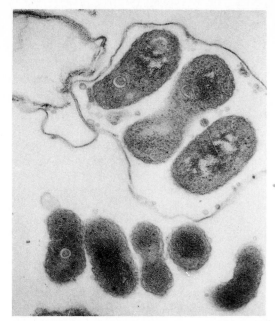

Figure 11.19

Bdellovibrio stolpii *is an elongated coiled* Bdellovibrio *and is shown here within the cell wall of its host* E. coli *and growing saprophytically adjacent. Note the curved shape of some of the cells. (Courtesy Jeffrey C. Burnham, Medical College of Ohio, and Sam F. Conti, University of Massachusetts.)*

production on the roots of leguminous plants. Free-living cells of *Rhizobium* are rod-shaped, but within the nodules they occur as pleomorphic (irregularly shaped) cells termed **bacteroids**. *Rhizobium* can fix atmospheric nitrogen only within root nodules and thus is considered an obligately symbiotic nitrogen fixer. Unlike *Rhizobium, Agrobacterium* species do not fix molecular nitrogen. *Agrobacterium* species, however, produce tumorous growths on infected plants, known as **galls**. *Agrobacterium tumefaciens* causes galls of many different plants and is an extremely important plant

table 11.12

Gram negative aerobic rods and cocci

Taxonomic group	Flagella	Carbon source	Denitrification	N_2 fixation
Pseudomonadaceae	Polar	Numerous	Some species	No
Azotobacteraceae	Peritrichous or polar	Numerous	No	Yes
Rhizobiaceae	Peritrichous or polar	Numerous	No	Yes
Methylomonadaceae	Polar or none	C_1 compounds only	No	No

pathogen causing large economic losses in agriculture.

The family **Methylomonadaceae** are bacteria that can utilize carbon monoxide, methane, or methanol as a sole source of carbon. The ability to use one-carbon-containing (C_1) organic compounds as the sole source of carbon and energy requires a special metabolic capability. Some of these bacteria are restricted to growth on C_1 compounds. The metabolism of these organisms is respiratory, using molecular oxygen as the terminal electron acceptor. Members of this group are of interest to industry as a potential source of protein for animal feed or as a human dietary supplement. The C_1 compounds, such as methane and methanol, are considered prime candidates as substrates for industrial processes aimed at growing microorganisms as a source of protein.

Several genera of uncertain affiliation are Gram negative aerobic rods. These include the genera *Brucella, Bordetella,* and *Francisella.* Some members of these three genera are important human pathogens. For example, *Bordetella pertussis* is the causative agent of whooping cough and *Francisella tularensis* the causative agent of tularemia. Other genera of uncertain affiliation in this group are *Alcaligenes, Acetobacter,* and *Thermus. Thermus* is an ecologically interesting genus growing well at temperatures over 70°C. Strains of this organism have been isolated from hot springs and the hot water tanks of laundromats.

Gram negative facultatively anaerobic rods

There are two major families of **Gram negative facultatively anaerobic rods**: the Enterobacteriaceae (motile by means of peritrichous flagella), and the **Vibrionaceae** (motile by means of polar flagella) (Table 11.13). The Enterobacteriaceae are divided further into five tribes: the Escherichieae, Klebsielleae, Proteeae, Yersinieae, and Erwinieae. These tribes are distinguished based on their metabolic pathways. Many of the bacteria studied in introductory microbiology laboratory courses belong to the Gram negative facultatively anaerobic rods.

The family **Enterobacteriaceae** includes the genera *Escherichia, Edwardsiella, Citrobacter, Salmonella, Shigella, Klebsiella, Enterobacter, Hafnia, Serratia, Proteus, Yersinia,* and *Erwinia.* Members of the genus *Escherichia* occur in the human intestinal tract. *E. coli* has achieved a special place in microbiology. It has been used as the test organism in many metabolic and genetic studies, and much of what we know about bacterial metabolism and bacterial genetics has been

table 11.13

The Gram negative facultatively anaerobic rods

Family Enterobacteriaceae	Peritrichous flagella; nonpigmented
Tribe Escherichieae	Mixed acid fermentation; optimal growth 37°C; G + C% 50–53
Genera: *Escherichia, Edwardsiella, Citrobacter, Salmonella, Shigella*	
Tribe Klebsielleae	Butanediol fermentation; optimal growth 37°C; G + C% 52–59
Genera: *Klebsiella, Enterobacter, Hafnia, Serratia*	
Tribe Proteeae	Optimal growth 37%C; G + C% 39–42
Genus: *Proteus*	
Tribe Yersinieae	Mixed acid fermentation; optimal growth 30–37°C; G + C% 45–47
Genus: *Yersinia*	
Tribe Erwineae	Mixed acid and butanediol fermentation; optimal growth 27–30°C; G + C% 50–58
Genus: *Erwinia*	
Family Vibrionaceae	Polar flagella; nonpigmented; G + C% 39–63
Genera: *Vibrio, Aeromonas, Pleisomonas, Photobacterium, Lucibacterium*	

Genera of uncertain affiliation *Chromobacterium, Flavobacterium, Zygomonas, Haemophilus, Pasteurella, Actinobacillus, Cardiobacterium, Streptobacillus, Calymmatobacterium*

table 11.14

The Gram negative anaerobic bacteria

Family Bacteriodaceae

Bacteroides	Produce mixtures of acids including succinic, acetic, formic, lactic, propionic
Fusobacterium	Produce butyric acid as major product
Leptotrichia	Produce lactic acid as the only major fermentation acid

Genera of uncertain affiliation

Desulfovibrio	Curved rods; anaerobic respiration using sulfate and producing hydrogen sulfide
Butyrivibrio	Curved rods; fermentative; produce butyrate
Succinovibrio	Curved rods; fermentative; produce succinate and acetate
Succinomonas	Straight rods; fermentative; produce large amounts of succinate and some acetate
Lachnospira	Curved rods; fermentative; produce mixture of acids
Selenomonas	Curved rods; fermentative; produce acetate, propionate, and lactate

elucidated in studies using *E. coli.* Additionally, *E. coli* is employed as an indicator of fecal contamination in environmental microbiology. The genera *Salmonella* and *Shigella* contain numerous different species, many of which are important human pathogens. In particular, typhoid fever and various gastrointestinal upsets are caused by *Salmonella* species, and bacterial dysentery is caused by *Shigella. Serratia marcesens,* once thought to be a nonpathogen, is now recognized as causing insect diseases and as an opportunistic human pathogen. *Serratia* strains can produce a red pigment known as prodiogiosin. All members of the genus *Erwinia* are plant pathogens. *Erwinia amylovora,* for example, causes fire blight of pears and apples.

The family **Vibrionaceae** includes the genera *Vibrio, Aeromonas, Plesiomonas, Photobacterium,* and *Lucibacterium.* Many of the *Vibrio* have curved rod-shaped cells. The habitat of *Vibrio* species is generally aquatic. *Vibrio cholerae* is an important human pathogen that causes cholera. *Photobacterium* and *Lucibacterium* are interesting because of their ability to luminesce. The mechanism of luminescence involves an ATP-driven reaction, an electron transport system, and a key reaction mediated by the enzyme luciferase. Some species of luminescent bacteria occur in association with fish; some of these fish are known as flashlight fish because of the light emitted by these bacteria. The association of luminescent bacteria with these fish is important in various behavioral aspects of these fish, including mating activities.

Several genera of uncertain affiliation are Gram negative facultatively anaerobic rods. These include the genera *Chromobacterium,* which produces violet pigments, and *Flavobacterium,* which produces a variety of insoluble yellow, orange, red, or brown pigments. Based on the mole% G + C, *Flavobacterium* clearly represents more than one group; in group I G + C% = 30–42 and in group II G + C% = 63–70. Also included in this group are the genera *Zymomonas, Haemophilus, Pasteurella, Actinobacillus, Cardiobacterium, Streptobacillus,* and *Calymmatobacterium.* Various species in these genera are important human pathogens.

Gram negative anaerobic bacteria

There is only one family, **Bacteroidaceae,** and relatively few genera in the **Gram negative anaerobic bacteria group** (Table 11.14). The genera included in the Bacteroidaceae are *Bacteroides, Fusobacterium,* and *Leptotrichia. Bacteroides* are important members of the normal microbiota of humans. They may be the dominant microorganisms occurring in the intestinal tract of humans. *Bacteroides* species characteristically form pleomorphic rods (Figure 11.20). Some members of the family Bacteroidaceae are human pathogens. Some species of *Leptotrichia,* for example, produce abscesses in the oral cavity. Various *Bacteroides* and *Fusobacterium* species also cause human infections.

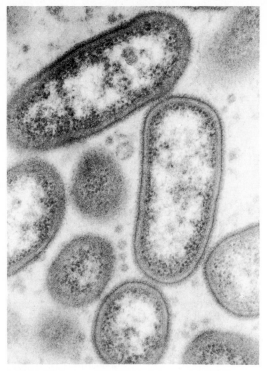

392

Classification and
identification of
microorganisms:
systematics of
bacteria

Figure 11.20

Bacteriodes asaccharolyticus *is a Gram negative rod, isolated from the oral cavity, where it causes gingivitis. (From BPS—Stanley C. Holt, University of Massachusetts.)*

Figure 11.21

Electron micrograph of Neisseria gonorrhoeae, *showing diplococci (61,500 ×). (From BPS— Centers for Disease Control, Atlanta.)*

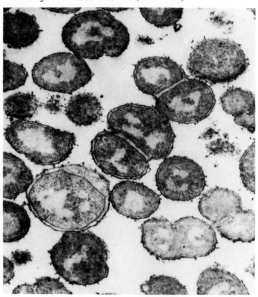

The genus *Desulfovibrio* is considered as a genus of uncertain affiliation within the group of Gram negative anaerobic rods. Members of the genus *Desulfovibrio* are curved rods capable of reducing sulfates, or other reducible sulfur compounds, to hydrogen sulfide. *Desulfovibrio desulfuricans,* the type species for this genus, is normally found in bogs and anaerobic sediments, where it plays an important role in the biogeochemical cycling of sulfur. Other genera affiliated with the Gram negative anaerobic rods include *Butyrovibrio, Lachnospira, Succinovibrio, Succinomonas,* and *Selenomonas.* Species of these genera occur in the rumen (a compartment of the stomach of cows and related animals), where they play critical metabolic roles in the digestion of cellulosic materials and the nutrition of the animal, producing low-molecular-weight fermentation carbohydrate substrates.

Gram negative cocci and coccobacilli

Only one family of Gram negative cocci and coccobacilli, Neisseriaceae, is recognized (Figure 11.21). The family **Neisseriaceae** includes the genera *Neisseria, Branhamella, Moraxella,* and *Acinetobacter.* The cells of *Neisseria* and *Branhamella* are cocci, whereas the cells of *Moraxella* and *Acinetobacter* are coccobacilli (oval-shaped) (Table 11.15). Members of the genera *Neisseria, Branhamella,* and *Moraxella* are parasitic, and some are important human pathogens. *Acinetobacter* species are saprophytic, although some are opportunistic pathogens. Members of the genus *Acinetobacter* are nutritionally versatile and can utilize a variety of organic compounds as sole source of carbon and energy. Several species of the genus *Neisseria* are important human pathogens. For example, *Neisseria gonorrhoeae* causes gonorrhea, and *Neisseria meningitidis* causes meningitis.

Gram negative anaerobic cocci

The **Gram negative anaerobic cocci** include only four genera, *Veillonella, Acidaminococcus, Megasphaera,* and *Gemmiger.* Each of these genera contains very few species. The cells of species in all of these genera typically occur as diplococci (pairs of cocci). *Veillonella* species have complex nutritional requirements and are unable to grow on individual organic substrates. They also require carbon dioxide for growth. Although these

table 11.15

The Gram negative cocci and coccobacilli

Family Neisseriaceae

Neisseria	Cocci; divide in two planes; sensitive to penicillin; oxidase positive; G + C% 47–52
Branhamella	Cocci; divide in two places; oxidase positive; reduce nitrates; G + C% 40–45
Moraxella	Coccobacilli; divide in one plane; oxidase positive; sensitive to penicillin; G + C% 40–46
Acinetobacter	Coccobacilli; divide in one plane; oxidase negative; penicillin resistant; G + C% 39–47

Genera of uncertain affiliation

Paracoccus	Cocci; divide in one plane; aerobic; G + C% 64–67
Lampropedia	Cells rounded or cubical; aerobic; divide in two planes; G + C% 61

organisms are fastidious in their nutritional requirements, they comprise part of the normal human microbiota, representing, for example, 5 to 16 percent of the bacteria found in the oral cavity. Some *Veillonella* species are human pathogens causing infections in the oral cavity and intestinal and respiratory tracts.

Gram negative chemolithotrophic bacteria

The metabolic activities of the Gram negative chemolithotrophic bacteria are extremely important in biogeochemical cycling reactions (Table 11.16). These bacteria oxidize inorganic compounds in order to generate ATP. Because their ATP-generating metabolism is inefficient, they metabolize large amounts of substrate to meet their energy requirements. The metabolic transformations of inorganic compounds mediated by these organisms cause global-scale cycling of various elements between the air, water, and soil.

The family **Nitrobacteraceae** oxidizes ammonia or nitrite in order to generate ATP. Organisms in this family, commonly referred to as nitrifying bacteria, are commonly found in soil, freshwater, and seawater. Many of the nitrifying bacteria have extensive internal membrane systems (Figure 11.22). There are two physiological groups in the

table 11.16

The Gram negative chemolithotrophic bacteria

Oxidize ammonia or nitrite	Family Nitrobacteraceae
Oxidize nitrite to nitrate	*Nitrobacter* *Nitrospina* *Nitrococcus*
Oxidize ammonia to nitrite	*Nitrosomonas* *Nitrosospira* *Nitrosococcus* *Nitrosolobus*
Oxidize sulfur and sulfur compounds	*Thiobacillus* *Sulfolobus* *Thiobacterium* *Macromonas* *Thiovulum* *Thiospira*
Oxidize iron or manganese	Family Siderocapsaceae
Iron or manganese oxides deposited	*Siderocapsa* *Naumanniella* *Ochrobium*
Iron but not manganese deposited	*Siderococcus*

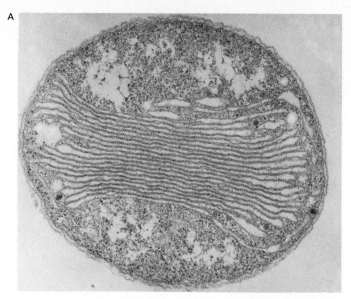

A

394

Classification and
identification of
microorganisms:
systematics of
bacteria

Figure 11.22

Micrographs showing morphologies and extensive internal membrane systems of nitrifying bacteria. (A) Nitrosococcus oceanus (44,530×). (B) Nitrobacter multiformis (33,950×). (C) Nitrobacter winogradsky (69,619×). (Courtesy Stan Watson, Woods Hole Oceanographic Institute, Woods Hole, Massachusetts. A reprinted by permission of Academic Press, New York, from S. W. Watson, and C. C. Remsen, 1970, Journal of Ultrastructure Research *33: 148–160; B reprinted by permission of Springer-Verlag, New York, from S. W. Watson, L. B. Graham, C. C. Remsen, and F. W. Valois, 1971,* Archiv für Mikrobiologie *76: 183–203; and C reprinted by permission of the American Society for Microbiology, Washington, D.C., from S. W. Watson, 1971,* International Journal of Systematic Bacteriology *21: 254–270.)*

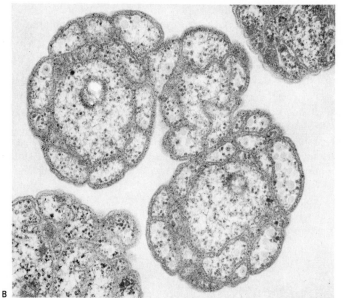

B

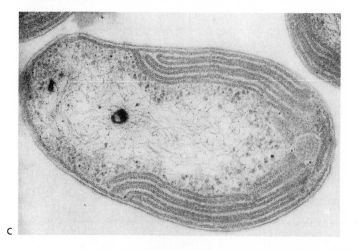

C

family Nitrobacteraceae; the first group oxidizes nitrite to nitrate, and the second group oxidizes ammonia to nitrite. Most members of the family Nitrobacteraceae are obligate chemolithotrophs. There are seven genera in the family Nitrobacteraceae: *Nitrobacter, Nitrospina, Nitrococcus, Nitrosomonas, Nitrosospira, Nitrosococcus,* and *Nitrosolobus.* The first three genera, whose names begin with the prefix *Nitro-,* oxidize ammonia; the remaining four genera, whose names begin with the prefix *Nitroso-,* oxidize nitrite. *Nitrobacter* species are extremely important nitrifiers in soil, oxidizing nitrite to nitrate. *Nitrosomonas* species, likewise, are important nitrifiers in soil, oxidizing ammonia to nitrite. The combined actions of the members of the genera *Nitrosomonas* and *Nitrobacter* permit the conversion of ammonia to nitrate. The change in electronic charge between NH_4^+ and NO_3^- alters the mobility of these nitrogenous ions in soil and has a major influence on soil fertility.

Several different genera of **chemolithotrophic bacteria** metabolize sulfur and sulfur-containing inorganic compounds. The genus *Thiobacillus* derives energy from the oxidation of reduced sulfur compounds; they are Gram negative rods, motile by means of polar flagella. Some members of the genus *Thiobacillus* oxidize only sulfur compounds, whereas others, such as *Thiobacillus ferrooxidans,* can also oxidize ferrous iron to ferric iron in order to generate ATP. *Thiobacillus* species can be used in the recovery of minerals, including uranium, and its oxidation of reduced iron and sulfur compounds mobilizes various metals so that they can be extracted from even low-grade ores. *Thiobacillus thiooxidans,* frequently used in biological metal recovery, is an acidophile. Optimum growth for this species occurs in the pH range 1–3.5. *Thiobacillus thiooxidans* is often found in association with waste coal heaps. The metabolic activities of this organism produce acid mine drainage, a serious ecological problem associated with some coal mining operations.

Members of the family **Siderocapsaceae** are able to oxidize iron or manganese, depositing iron and/or manganese oxides in capsules or in extracellular material. Members of the genus *Siderocapsa,* for example, have spherical cells embedded in a common capsule partially encrusted with iron and/or manganese oxides. The taxonomic status of the entire family and of the genus *Siderocapsa* in particular has been questioned frequently. The description of these bacteria as unicellular, non-thread-forming or non-stalk-forming iron and/or manganese bacteria that under natural conditions are able to deposit metal oxides on or in extracellular mucoid material, is taxonomically rather imperfect and undoubtedly the source of the controversy. Although their proper taxonomic position is in doubt, these bacteria do exist and are ecologically important. They are widely distributed in nature, and their metabolic activities are of geological importance. Members of this family are found in iron-bearing waters, forming high concentrations in the lower portions of some lakes.

Gram positive cocci

The **Gram positive cocci** include three families: the Micrococcaceae, the Streptococcaceae, and the Peptococcaceae (Table 11.17). The coccoid cells of the **Micrococcaceae** may occur singly or as irregular clusters (Figure 11.23). For example, the genus *Staphylococcus* typically forms grape-like clusters. Most strains of *Staphylococcus* can grow in the presence of 15 percent NaCl. Species of *Staphylococcus* commonly occur on skin surfaces. *Staphylococcus aureus* is a potential human pathogen, infecting wounds and also causing food poisoning.

In the family **Streptococcaceae**, Gram positive cocci occur as pairs or chains (Figure 11.24). The metabolism of the Streptococcaceae is fermentative. Even though their metabolism is anaerobic, the Streptococcaceae are listed as being facultatively anaerobic because they grow in the presence of air. Indeed, many species of *Streptococcus* occur in the oral cavity, where they are continuously exposed to air. Some members of *Streptococcus* are human pathogens. For example, rheumatic fever is caused by *Streptococcus pyogenes.* Several *Streptococcus* species are also responsible for the formation of dental caries.

The **Peptococcaceae** have complex nutritional requirements. Cells of organisms of this family may occur singly, in pairs, or in regular or irregular masses. They are obligately anaerobic and produce low-molecular-weight volatile fatty acids, carbon dioxide, hydrogen, and ammonia as the main products of amino acid metabolism.

Endospore-forming rods and cocci

The **endospore-forming rods and cocci** are extremely important because of the heat resistance of the endospore structure. The genera *Bacillus, Sporolactobacillus, Clostridium, Desulfotomaculum, Sporosarcina,* and *Thermoactinomyces* are all characterized by the formation of endospores

Figure 11.23

Scanning electron micrograph of Staphylococcus aureus, *showing grape-like clusters (4560 ×). (Courtesy Robert Apkarian, University of Louisville.)*

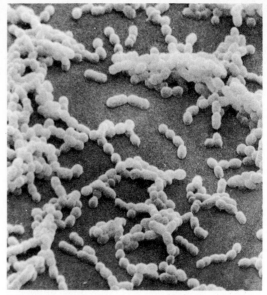

Figure 11.24

Micrograph of Streptococcus. *Note chains of coccal-shaped bacteria. This particular strain,* S. mutans, *occurs on tooth surfaces, where it produces acid that erodes the enamel, leading to dental caries. (From BPS—Z. Skobe, Forsythe Dental Center.)*

table 11.17

The Gram positive cocci

Family Micrococcaceae

Micrococcus	Irregular clusters; G + C% 60–75
Staphylococcus	Irregular clusters; G + C% 30–40
Planococcus	Tetrads; G + C% 39–52

Family Streptococcaceae

Streptococcus	Pairs or chains; homofermentative; G + C% 33–42
Leuconostoc	Pairs or chains; heterofermentative; G + C% 38–44
Pediococcus	Pairs or tetrads; homofermentative; G + C% 34–44
Aerococcus	Pairs or tetrads; homofermentative; G + C% 36–40
Gemella	Single or pairs; G + C% 31–35

Family Peptococcaceae

Peptococcus	Irregular clusters; no growth at pH 2.5; cellulose not degraded; G + C% 36–37
Peptostreptococcus	Chains; no growth at pH 2.5; cellulose not degraded; G + C% 33–34
Ruminococcus	Irregular clusters or chains; no growth at pH 2.5; cellulose degraded; G + C% 40–46
Sarcina	Tetrads or octads; growth at pH 2.5; cellulose not degraded; G + C% 28–31

table 11.18

Endospore producers

Family Bacillaceae	
Bacillus	Rods; aerobic or facultative; catalase usually produced
Sporolactobacillus	Rods; microaerophilic; catalase not produced
Clostridium	Rods; anaerobic; sulfate not reduced to sulfide
Desulfotomaculum	Rods; anaerobic; sulfate reduced to sulfide
Sporosarcina	Cocci; tetrads or octads
Family Micromonosporaceae	
Thermoactinomyces	Filamentous thermophiles

(Table 11.18). These genera occur in the family Bacillaceae. Most endospore formers are Gram positive rods. Only members of the genus *Desulfotomaculum* are Gram negative endospore formers. With the exception of members of the genus *Sporosarcina*, the other endospore formers are rod-shaped.

The two most important genera of endospore-forming bacteria are the genera *Bacillus* and *Clostridium*. *Bacillus* species are strict aerobes or facultative anaerobes. *Clostridium* species are obligately anaerobic. The endospore-forming bacteria are extremely important in food, industrial, and medical microbiology. Food spoilage by *Bacillus* and *Clostridium* species is of great economic importance. Several *Clostridium* species are important human pathogens. For example, *Clostridium botulinum* is the causative agent of botulism, *Clostridium tetani* causes tetanus, and *Clostridium perfringens* causes gas gangrene.

Gram positive, asporogenous, rod-shaped bacteria

The **Gram positive, asporogenous (nonsporulating), rod-shaped bacteria** include the family Lactobacillaceae. These are Gram positive rods that produce lactic acid as the major fermentation product; they occur in fermenting plant and animal products that have available carbohydrate substrates; they are also found as part of the normal human microbiota, in the oral cavity, vaginal, and intestinal tracts. *Lactobacillus* is the only genus within the family Lactobacillaceae. *Lactobacillus* is extremely important in the dairy industry; cheese, yogurt, and many other fermented products are made by the metabolic activities of *Lactobacillus* species.

There are several genera of uncertain affiliation that are Gram positive non-spore-forming rod-shaped bacteria. These include the genera *Listeria, Erysipelothrix,* and *Caryophanon.* The *Listeria* are Gram positive rods tending to produce chains. Several species of *Listeria* are animal pathogens. *Caryophanon latum,* the type species for the genus *Caryophanon,* produces large rods or filaments up to 3 μm in diameter. This organism is normally found on animal fecal matter. The filaments of *Caryophanon latum* are divided by closely spaced cross walls into numerous disk-shaped cells less than 1 μm long (Figure 11.25). The unusual morphology of *Caryophanon latum* is quite striking.

Actinomycetes and related organisms

Coryneform bacteria

The **coryneform group** of bacteria is a heterogeneous group defined by the characteristic irregular morphology of the cells and the tendency of the cells to show incomplete separation following cell division. The coryneform bacteria exhibit pleomorphic morphology. They do not form true filaments. The irregular morphology and the association of the cells after division, however, indicates a relationship to the filament-forming actinomycetes. This group includes the genera *Corynebacterium, Arthrobacter, Brevibacterium, Cellulomonas,* and *Kurthia* (Table 11.19). Many species of *Corynebacterium* are plant or animal pathogens. For example, *Corynebacterium diphtheriae* is the causative agent of diphtheria. As noted in an earlier chapter, *Corynebacterium diphtheriae* causes diphtheria only when it is infected with a lysogenic phage. Cells of *Coryne-*

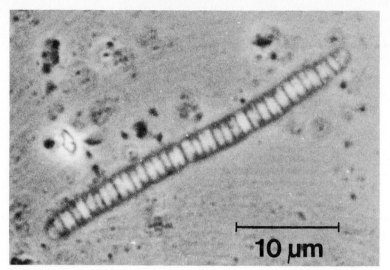

Figure 11.25

This phase contrast micrograph of Caryophanon latum, *as found in cow dung enrichment cultures, shows the multicellularity of the trichome. (Courtesy William Trentini, Mount Allison University.)*

10 μm

398

Classification and
identification of
microorganisms:
systematics of
bacteria

table 11.19

The coryneform bacteria

Corynebacterium	Gram positive rods frequently showing club-shaped swellings; snapping division produces angular arrangement of cells; G + C% 57–60
Arthrobacter	Gram positive rods showing a marked change in form; exhibiting rudimentary life cycle; G + C% 60–72
Cellulomonas	Gram positive rods which attack cellulose; G + C% 71–73
Kurthia	Gram positive rods in young culture, cocci in old culture

Figure 11.26

This micrograph of Corynebacterium diphtheriae *shows the snapping division typical of the species. (From BPS—Centers for Disease Control, Atlanta.)*

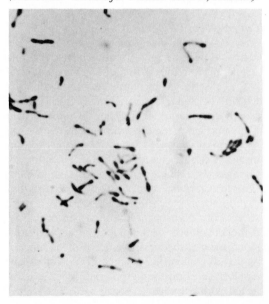

bacterium exhibit snapping division; that is, after binary fission the cells do not completely separate (Figure 11.26) and appear to form groups resembling "Chinese letters" when viewed under the microscope. The genus *Arthrobacter,* which is widely distributed in soils, is interesting because it exhibits a simple life cycle (Figure 11.27), in which there is a change from rod-shaped cells to coccoid cells. The sequence of morphological changes in the growth cycle distinguishes *Arthrobacter* from other genera. The coccoid cells present during the stationary growth phase are sometimes referred to as arthrospores and cystites. The formation of arthrospores represents the beginning of a regular life cycle that is characteristic of eukaryotic microorganisms but is rare among the prokaryotes.

The family **Propionibacteriaceae** is also considered to be associated with the coryneform bacteria. The cells are Gram positive rods that produce propionic acid, acetic acid, or mixtures of organic acids by fermentation; lactic acid is not a major fermentation product but is itself used as a fermentation substrate. There are two genera in the family Propionibacteriaceae: *Propionibacte-*

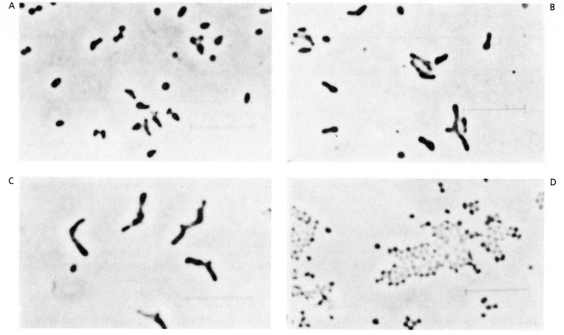

Figure 11.27

Micrographs showing life cycle of Arthrobacter. *A medium was inoculated with coccoid cells and incubated at 25°C. (A) After 6 hours, showing outgrowth of rods from coccoid cells. (B) After 12 hours, the rods are longer. (C) After 24 hours, rods predominate. (D) After 3 days, the rods have given rise to coccoid cells. (Reprinted by permission of Springer-Verlag, from R. M. Keddie and D. Jones, 1981, Saprophytic, aerobic coryneform bacteria, in* The Prokaryotes, *M. P. Starr, H. Stolp, H. G. Truper, A. Ballows, and H. G. Schlegel, eds.)*

rium and *Eubacterium*. Species of *Propionibacterium* are important in the dairy industry; they normally carry out a propionic acid fermentation. Species of *Eubacterium* usually produce mixtures of organic acids, including large amounts of butyric, acetic, and formic acid; they do not produce propionic acid, lactic acid, succinic acid, or acetic acid as major fermentation products.

Actinomycetes

The order **Actinomycetales** contains bacteria characterized by the formation of branching filaments. Many of the more evolved actinomycetes resemble the fungi in appearance, but their cells are prokaryotic, and they are clearly bacteria. There are eight families in the order Actinomycetales: Actinomycetaceae, Mycobacteriaceae, Frankiaceae, Actinoplanaceae, Dermatophilaceae, Nocardiaceae, Streptomycetaceae, and Micromonosporaceae (Table 11.20). The various families of the order Actinomycetales are distinguished from one another by the nature of their mycelia and spores (Figure 11.28). Various types of spores are produced by actinomycete species. Many of these spores are involved in the dispersal of actino-

mycetes. Only *Thermoactinomyces* produces endospores. Spore production is an important diagnostic characteristic for the identification of actinomycetes.

The actinomycetes are widely distributed in nature. The oxidative forms are numerous and occur primarily in soils. The main ecological role of actinomycetes is in the decomposition of organic matter in soil. Fermentative types are primarily found in association with humans and other animals. Some actinomycetes are human pathogens. For example, *Actinomyces israelii* is the causative agent of actinomycosis. *Mycobacterium tuberculosis* is the causative agent for tuberculosis. The genus *Mycobacterium* is considered an actinomycete, although the formation of mycelia is rudimentary. *Mycobacterium* is acid fast; that is, stained cells resist decolorization with acid alcohol. The separation between mycobacteria, corynebacteria, and various pleomorphic bacteria is not easy. Different observers may classify the same strain as belonging to the genus *Corynebacterium, Arthrobacter, Nocardia,* or *Mycobacterium*. Many of the Actinomycetaceae produce antibiotics. The production of antibiotics by actinomy-

table 11.20

The actinomycetes

Order Actinomycetales

Family Actinomycetaceae	Mycelium not formed; no spores formed; not acid fast
Genera: *Actinomyces, Arachnia, Bifidobacterium, Bacterionemia, Rothia*	
Family Mycobacteriaceae	Mycelium not formed; no spores formed; acid fast
Genus: *Mycobacterium*	
Family Frankiaceae	Mycelium formed; symbionts in plant nodules with free stage in soil
Genus: *Frankia*	
Family Actinoplanaceae	Mycelium formed; saprophytes or facultative parasites; spores borne inside sporangia
Genera: *Actinoplanes, Spirillospora, Streptosporangium, Amorphosporangium, Ampullariella, Pilimelia, Planomonospora, Planobispora, Dactylosporangium, Kitasatoa*	
Family Dermatophilaceae	Mycelium divides transversely to form motile cocci; saprophytes or facultative parasites; spores not borne in sporangia
Genera: *Dermatophilus, Geodermatophilus*	
Family Nocardiaceae	Mycelium fragments to form nonmotile cells; saprophytes or facultative parasites; spores not borne in sporangia; aerial spores usually absent
Genera: *Nocardia, Pseudonocardia*	
Family Streptomycetaceae	Mycelium tends to remain intact; saprophytes or facultative parasites; spores not borne in sporangia; usually abundant aerial mycelia and long spore chains
Genera: *Streptomyces, Streptoverticillium, Sporichthya, Microellobosporia*	
Family Micromonosporaceae	Mycelium remains intact; saprophytes or facultative parasites; spores not borne in sporangia; spores formed singly or in short chains
Genera: *Micromonospora, Thermoactinomyces, Actinobifida, Thermomonospora, Microbispora, Micropolyspora*	

cetes, such as *Streptomyces griseus,* is extremely important in the pharmaceutical industry. The availability of such antibiotics has revolutionized medical practice. Many previously fatal diseases are now easily controlled by using antibiotics produced by *Streptomycetes* and other actinomycetes. This topic will be discussed further in the medical and industrial microbiology sections.

The rickettsias

The **rickettsias** include two orders of bacteria: **Rickettsiales** and **Chlamydiales.** The rickettsias are intracellular parasites. The majority of members of the Rickettsiales are Gram negative and multiply only within host cells. Within host cells the rickettsias reproduce by binary fission. The rickettsias lack the enzymatic capability of producing

sufficient amounts of ATP to support their reproduction; they are able to obtain the ATP from the host cells in which they grow.

Many species of Rickettsiales cause disease in humans and other animals. Some members of the family Rickettsiaceae are adapted to existence in arthropods but are capable of infecting vertebrate hosts, including humans. Many members of the genus *Rickettsia* are carried by insect vectors and cause diseases in humans. For example, *Rickettsia rickettsii* is transmitted by ticks and causes Rocky Mountain spotted fever. Selected diseases transmitted by arthropod vectors will be discussed in the medical microbiology section.

The **Chlamydiales** are obligate intracellular parasites whose reproduction is characterized by the change of the small, rigid-walled infectious form of the organism (elementary body) into a larger, thin-walled noninfectious form (initial body)

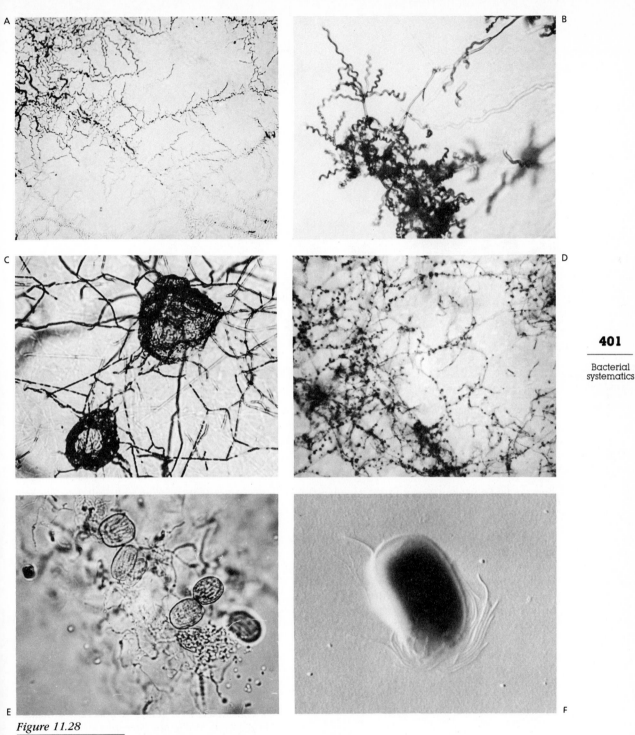

Figure 11.28

Micrographs of actinomycetes, showing hyphae and spores. (A) Hyphae of Nocardia *occurring in a zigzag formation. They fall apart easily into rod-shaped elements (370×). (B) Aerial hyphae of a* Streptomyces *sp., showing formation of coils of chains of conidia (495×). (C) The aerial hyphae of this* Streptomyces *sp. form sclerotia (495×). (D) This* Thermomonospora *sp. forms single conidia (495×). (E) Sporangium of an* Actinoplanes *sp. in water; note that the sporangium is releasing spores (495×). (F) Transmission electron micrograph of a geranium-shadowed preparation of* Actinoplanes *sp. (30,000×). (Courtesy Hubert Lechevalier, Rutgers, The State University.)*

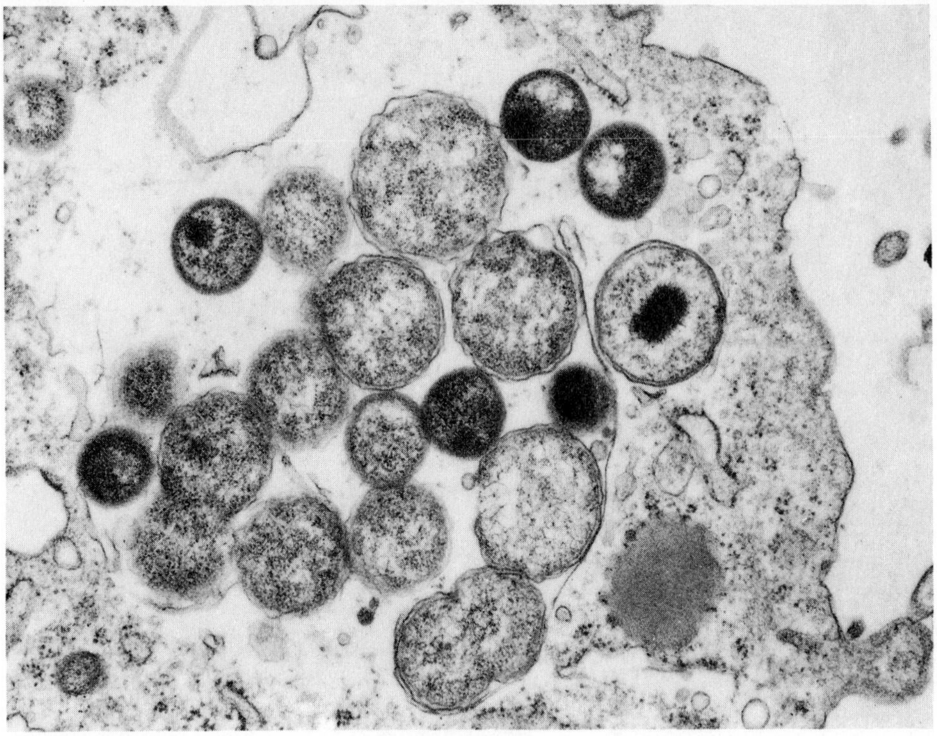

402

Classification and
identification of
microorganisms:
systematics of
bacteria

Figure 11.29

Micrograph of Chlamydia psittaci *in the cytoplasm of a cell (24,940×). (From BPS—R. C. Cutlip, National Animal Disease Center, Ames, Iowa.)*

that divides by fission (Figure 11.29). Members of the order Chlamydiaceae are metabolically limited. These organisms are unable to generate sufficient ATP to support their reproduction. The chlamydias have sometimes been referred to as large viruses, but they are truly bacteria. Members of the genus *Chlamydia* are Gram negative bacteria. The reproductive cycle for these organisms takes about forty hours. *Chlamydia* cause human respiratory and urogenital tract diseases, and in birds they cause respiratory diseases and generalized infections. For example, the disease psittacosis, parrot fever, is caused by *Chlamydia psittaci.*

The mycoplasmas

The **mycoplasmas** are classified as **Mollicutes** (soft skin) because they **lack a cell wall**. The mycoplasmas are bacteria that are bounded by a single triple-layered membrane. Although it is difficult to define the mycoplasmas in a way that clearly

distinguishes them from cell wall-deficient mutants of other bacterial groups, the mycoplasmas appear to be fundamentally different from other bacterial groups. They are the smallest organisms capable of self-reproduction. When growing on artificial media, mycoplasmas form small colonies that have a characteristic "fried egg" appearance (Figure 11.30). Members of the genus *Mycoplasma* require sterols for growth. Several members of this genus cause diseases in humans. For example, some forms of pneumonia are caused by *Mycoplasma* species.

Like the colonies of *Mycoplasma,* the colonies of *Spiroplasma* species exhibit a typical biphasic "fried egg" appearance. The genus *Spiroplasma* is considered as a genus of uncertain affiliation in the class Mollicutes. *Spiroplasma* species lack a cell wall, have a triple-layered cytoplasmic membrane, and require sterols for growth. Members of the genus *Spiroplasma* cause diseases in plants and animals. For example, *Spiroplasma citri* causes "Stubborn" diseases of citrus plants. Suckling mouse cataract disease is also caused by a *Spiroplasma* species.

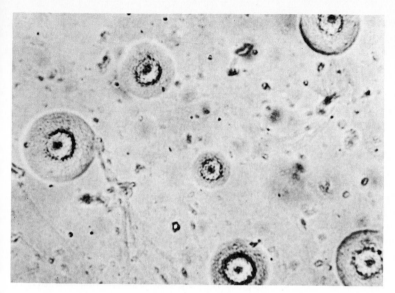

Figure 11.30

Micrograph of colonies of Mycoplasma, *showing "fried egg" appearance. (Courtesy Leonard Hayflick, University of Florida.)*

Endosymbionts

Several bacterial genera have been recognized that are **obligate endosymbionts of invertebrates**; that is, they live within the cells of invertebrate animals without adversely affecting the animal. For example, the protozoan *Paramecium aurelia* can harbor a variety of endosymbiotic bacteria (Table 11.21). Additionally, new genera of endosymbionts have recently been described for other protozoa, insects, and various other invertebrates. Although these bacteria can be readily seen within host cells, difficulties in culturing them outside of the host have hampered efforts to determine their proper taxonomic status. Developing an understanding of the nutritional requirements of

table 11.21

Symbionts of Paramecium aurelia

Genus	Common names	Description
Caedobacter	Kappa	Varying in size and distinguished by the presence of a 0.5-μm-diameter inclusion within the host cell; exhibits killing of sensitive strains
Pseudocaedobacter	Pi	Slender rod until recently considered a mutant of kappa; nonkilling symbiont
	Nu	A nonkilling symbiont similar in appearance to pi and mu
	Mu	Slender rod, often elongated; distinguished because its killing action is wholly dependent on cell–cell contact between mating paramecia
	Gamma	A diminutive bacterium, frequently appearing as doublets, strong killing of other strains is shown by gamma bearers
Tectobacter	Delta	Rod distinguished by an electron-dense material surrounding the outer of its two membranes
Lyticum	Lambda	Appears as a typical motile bacterium with peritrichous flagella although its movement within the cytoplasm has not been observed
	Sigma	Largest of all endosymbionts of *Paramecium aurelia*; curved flagellated rod resembling lambda

these bacteria has permitted the creation of complex media for their culture and identification.

The archaebacteria

As indicated earlier, the members of this group have been shown to be phylogenetically related based on analysis of their 16S ribosomal RNA molecules. They also share in common several morphological and physiological features that make them distinct from other bacteria, including the lack of murein in their cell walls and the unusual ether linkage that occurs in their phospholipid molecules. They appear to be "primitive" bacteria and, as members of the kingdom **Archaebacteria,** are considered to be distantly related to other prokaryotes.

Methane-producing bacteria

The **methane-producing,** or **methanogenic, bacteria** represent a highly specialized physiological group. Members of the family **Methanobacteriaceae** form methane by the reduction of carbon dioxide. The methanogenic bacteria are very strict obligate anaerobes. In order to produce methane,

these organisms utilize electrons generated in the oxidation of hydrogen or simple organic compounds, such as acetate and methanol. Methanogenic bacteria are unable to use carbohydrates, proteins, or other complex organic substrates. Improvements in anaerobic isolation techniques have permitted the isolation of many new species of methanogens. The methanogens often form consortia in association with other microorganisms. The microorganisms associated with the methanogens maintain the low oxygen tensions and provide the carbon dioxide and fatty acids required by the methanogenic bacteria. Such associations are extremely important in the rumen of animals such as cows. A major source of atmospheric methane comes from the rumen of such animals. The U.S. Environmental Protection Agency once issued an indelicate report stating that the burping cow was the major source of atmospheric hydrocarbon pollutants. Clearly, though, in urban areas hydrocarbon pollutants originate primarily from automotive exhausts and not cows.

The family Methanobacteriaceae has been extensively revised recently to accommodate several new taxa. These genera of methane-gener-

table 11.22

The methanogenic bacteria

Order Methanobacteriales	Cells short, lancet-shaped cocci to long filaments; cells appear Gram positive but lack murein; strict anaerobes, oxidize hydrogen; reduce CO_2 to methane
Family Methanobacteriaceae	
Genera *Methanobacterium*	Slender rods, often forming filaments
Methanobrevibacter	Short rods or lancet-shaped cocci, often in pairs or chains
Order Methanococcales	Cells cocci; oxidize hydrogen or formate; reduce CO_2 to methane; Gram negative with protein units external to cytoplasmic membrane
Family Methanococcaceae	
Genus *Methanococcus*	Cocci single, in pairs, or clumps
Order Methanomicrobiales	Cells cocci to rods; Gram negative or Gram positive; motile or nonmotile; strict anaerobes; oxidize hydrogen or formate; reduce CO_2 to methane or form methane via fermentation of methanol and related compounds
Family Methanomicrobiaceae	Gram negative cocci to rods
Genera *Methanomicrobium*	Short motile rods
Methanogenium	Irregular coccoid cells
Methanospirillum	Curved slender motile rods, often forming filaments
Family Methanosarcinaceae Genus *Methanosarcina*	Gram positive coccoid cells occurring in packets

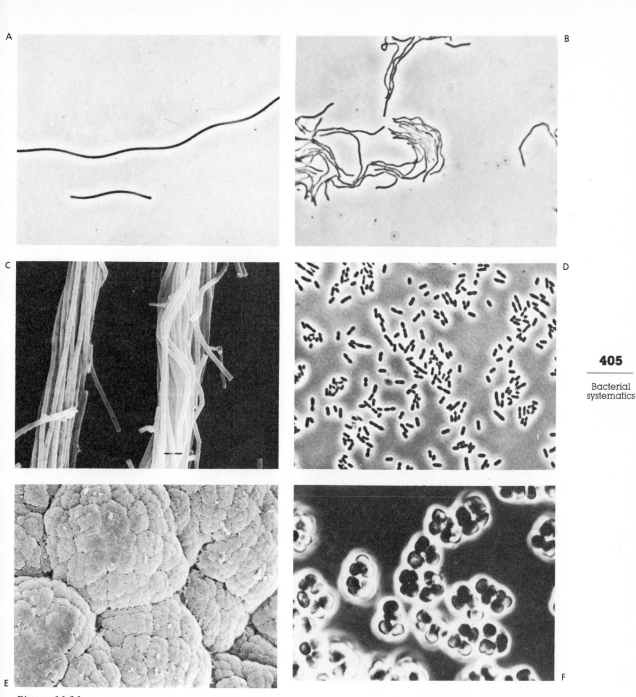

Figure 11.31

Micrographs of methanogens. (A) Phase contrast micrograph of Methanospirillum hungatei. *(B) Phase contrast micrograph of* Methanobacterium thermoautotrophicum. *(C) Scanning electron micrograph of* Methanobacterium soehngenii. *(D) Phase contrast micrograph of* Methanobrevibacter ruminatium. *(E) Scanning electron micrograph of* Methanococcus mazei. *(F) Phase contrast micrograph of* Methanosarcina *sp. (A–F Courtesy Robert Mah, C with Robert Sleat, E with Jack Pangborn, University of California Los Angeles.)*

ating bacteria can be differentiated based on relatively few morphological and physiological features (Table 11.22). Among these methanogens, *Methanobacterium* are rods or lancet-shaped cocci, and *Methanosarcina* are large cocci occurring in packets; *Methanococcus* occur as cocci arranged singly or in irregular clusters (Figure 11.31).

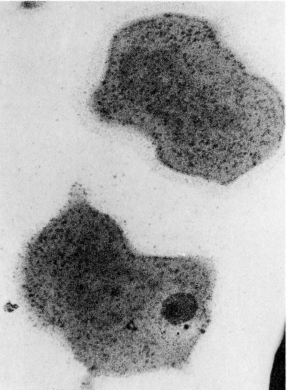

Figure 11.32

Micrographs of Sulfolobus. *(A) Scanning electron micrograph showing* Sulfolobus *on sulfur particles (7,550×); (B) Transmission electron micrograph of S.* brierleyi *(69,825×). (Courtesy C. L. and J. A. Brierley, New Mexico Technical University.)*

406

Classification and
identification of
microorganisms:
systematics of
bacteria

Halobacteriaceae

The genera *Halobacterium* and *Halococcus,* included in the family **Halobacteriaceae,** have the characteristic properties of the archaebacteria. The unusual bacteriorhodopsin-mediated phototrophic metabolism of *Halobacterium* was discussed earlier. In addition to this unique feature, all members of the family Halobacteriaceae are obligate halophiles growing only in media containing at least 15 percent NaCl. Members of this family are found in ecosystems that have extremely high sodium chloride concentrations, such as some salt lakes, the Dead Sea, and foods preserved by salting.

Sulfolobus

Based on their staining reactions, the bacteria in the sulfur-oxidizing genus *Sulfolobus* have been characterized as Gram negative spherical cells (Figure 11.32). The cell walls of *Sulfolobus,* however, lack peptidoglycan. They also share other common properties with the Archaebacteria, including rRNA homology and membrane structure. Members of this genus are thermophiles, with an optimum growth temperature of 70–75°C. *Sulfolobus* species occur in hot, acidic environments. Distribution in such unusual habitats appears to be another common characteristic of many Archaebacteria.

Postlude

Microbial systematics is a very broad area of microbiology. The more we know about an organism the better we are able to understand its relationships to other organisms, so that the organism can be properly classified. The classification and nomenclature of microorganisms is important in all fields of microbiology. One purpose of a taxonomic system is to provide a basis for nomenclature. Microorganisms are named by using a binomial system in which they are specified by their genus and species names. The naming of microorganisms permits one to refer to a particular organism unambiguously, which is essential for communication regarding the properties of a particular microorganism. In this chapter and throughout the rest of this book you will find the names and descriptions of microbial taxa, including the practical importance of particular species. Associating the proper name with an organism is critical for its classification.

Microbial classification systems attempt to describe the natural organization of the microbial world, showing the similarities among organisms in one taxonomic group and the diversity of microbial taxa. Ideally, classification systems should reflect the evolutionary separation of organisms. At the microbial level, however, this has been nearly impossible because of the lack of a fossil record. Arbitrary decisions traditionally have had to be made in classifying microorganisms, with some taxonomists emphasizing differences between microbial groups and others emphasizing similarities. All microbial classification systems developed so far are artificial, and the classification of microorganisms is subject to constant review and revision. Although some microbiologists consider taxonomy dull, the field is marked by constant strife and differing opinions as to whether particular systems of classification assign microorganisms to their proper taxonomic places. Taxonomy is the only area of microbiology where it has been necessary to establish a formal judicial system to arbitrate and rule on the validity of differing opinions.

A variety of approaches are used to classify microorganisms, including phenetic approaches aimed at assessing group similarities and phylogenetic approaches aimed at reflecting natural evolutionary patterns. Most classification systems for microorganisms are based on phenotypic characteristics, sometimes leading to the formation of genetically heterogeneous groups of uncertain affiliation. Many different phenotypic features are used in classifying microorganisms. To determine the characteristics of a microorganism so that it can be unambiguously classified, the organism must be grown in pure culture. Modern culture techniques, including the development of methods for isolating strict anaerobes in pure culture, have permitted the proper classification of many microorganisms that were previously unknown.

The classical approach to microbial classification emphasizes the points of separation between microbial groups. In contrast, the numerical taxonomic approach examines overall similarity. The numerical taxonomic approach is less sensitive to individual features than approaches that establish key features for distinguishing microbial taxa. Each system has its own merits. Modern approaches to microbial taxonomy include genetic analyses, and such analyses promise to revolutionize microbial classification systems. Already we have seen the recognition of the Archaebacteria, and as genetic and epigenetic approaches are used to examine other groups, a clearer picture of the phylogenetic relationships among microorganisms will certainly develop. It is now routine in describing microbial taxa to determine the mole% $G + C$ of the DNA. Studies on DNA and RNA homologies to determine the true relatedness of member species of particular taxa are clarifying the taxonomic status of many groups of microorganisms. Although genetic analyses are emphasized in modern classification systems, physiological and morphological features are still extremely important in microbial taxonomy. The inclusion of the cyanobacteria together with other groups of bacteria represents an example of a major recent taxonomic revision based on the recognition of the fundamental difference between organisms having prokaryotic and eukaryotic cells.

In addition to establishing the relationships among microorganisms, a taxonomic system provides a basis for identifying microorganisms. Classification and identification are distinct processes. Identification involves the comparison of an unknown organism with previously described taxa. Microbial identification systems should permit the unambiguous identification of an unknown microorganism by using the minimal number of tests; the system should be accurate and efficient. Computers have facilitated the identification of microorganisms greatly by allowing the rapid comparison of numerous characteristics and the assessment of the statistical probability of making a correct identification. Computer-assisted identification systems have become an integral part of the clinical microbiology laboratory, where rapid and accurate identification of microorganisms is essential for the proper diagnosis and treatment of infectious diseases.

In addition to considering the general aspects of microbial systematics, the features of the major groups of bacteria are reviewed in this section. The major groups, organized based on unifying morphological and/or physiological properties, represent the great diversity of microbial morphological forms and physiological functions. Only the highlights of each group could be considered in this section. The discussion of various applied topics in many of the following chapters will refer to and add information about these bacteria. The suggested supplementary readings should be consulted for more detailed consideration of a microbial group of particular interest.

1. How are bacteria named?

2. What is the difference between classification and identification?

3. Compare phenetic and phylogenetic approaches to microbial classification.

4. Why has the use of DNA homology caused the reclassification of many bacterial species? Why is DNA homology a better measure of relatedness than phenotypic characteristics for developing classification systems?

5. Why does comparing the mole% G + C permit the assessment of genetic relatedness? Why does the DNA homology better describe genetic relatedness than the proportion of G + G in the DNA?

6. What is an identification key? How is a dichotomous key used for identifying an unknown? Create a key for separating the following bacterial genera: *Bacillus, Clostridium, Escherichia, Enterobacter, Streptococcus,* and *Staphylococcus.*

7. How do computers aid in the identification of bacteria?

8. Describe the similarities and differences between cyanobacteria and other phototrophic bacteria.

9. How are the archaebacteria different from other bacterial groups? How has the use of ribosomal RNA analysis helped define the relationships among these organisms?

10. Which bacterial genera are characterized by endospore formation?

11. What phenotypic characteristics are used to distinguish major groups of bacteria?

12. How would you go about finding a description of a bacterial genus that had been described ten years ago? A description of one that had been described for the first time this year?

Balch, W. E., G. E. Fox, L. J. Magrum, C. R. Woese, and R. S. Wolfe. 1979. Methanogens: reevaluation of a unique biological group. *Microbiological Reviews* 43: 260–296.

Barksdale, L., and K. S. Kim. 1977. *Mycobacterium. Bacteriological Reviews* 41: 217–372.

Becker, Y. 1978. The chlamydia: molecular biology of procaryotic obligate parasites of eucaryocytes. *Microbiological Reviews* 42: 274–306.

Buchanan, R. E., and N. E. Gibbons (eds.). 1974. *Bergey's Manual of Determinative Bacteriology,* 8th ed. Williams & Wilkins Co., Baltimore.

Colwell, R. R. 1973. Genetic and phenetic classification of bacteria. *Advances in Applied Microbiology* 16: 137–176.

Cowan, S. T., and L. R. Hill (eds.). 1978. *A Dictionary of Microbial Taxonomy.* Cambridge University Press, New York.

Cross T., and M. Goodfellow. 1973. Taxonomy and classification of the actinomycetes. In: *Actinomycetales: Characteristics and Practical Importance* (G. Sykes and F. A. Skinner, eds.), pp. 11–112. Academic Press, London.

Gibbons, N. E., and R. G. E. Murray. 1978. Proposals concerning the higher taxa of bacteria. *International Journal of Systematic Bacteriology* 28: 1–6.

Goodfellow, M., and D. E. Minnikin. 1977. Nocardioform bacteria. *Annual Review of Microbiology* 31: 159–180.

Hastings, J. W., and K. H. Nealson. 1977. Bacterial bioluminescence. *Annual Review of Microbiology* 31: 549–595.

Henriksen, S. D. 1976. *Moraxella, Branhamella,* and *Acinetobacter. Annual Review of Microbiology* 30: 63–83.

Holt, S. C. 1978. Anatomy and chemistry of spirochetes. *Microbiological Reviews* 42: 114–160.

Johnson, R. C. 1976. The spirochetes. *Annual Review of Microbiology* 31: 39–61.

Krieg, N. R. 1976. Biology of the chemoheterotrophic spirilla. *Bacteriological Reviews* 40: 55–115.

Lapage, S. P., P. H. A. Sneath, E. F. Lessel, V. B. D. Skerman, H. P. R. Seeliger, and W. A. Clark (eds.). 1975. *International Code of Nomenclature of Bacteria.* American Society for Microbiology, Washington, D.C.

Laskin, A. I., and H. A. Lechevalier. 1977. *Handbook of Microbiology: Bacteria.* CRC Press, Inc., Boca Raton, Florida.

Macy, J. M., and L. Probst. 1979. The biology of gastrointestinal bacteriodes. *Annual Review of Microbiology* 33: 561–594.

Pfennig, N. 1977. Phototrophic green and purple bacteria: a comparative systematic survey. *Annual Review of Microbiology* 31: 275–290.

Poindexter, J. S. 1981. The caulobacters: ubiquitous unusual bacteria. *Microbiological Reviews* 45: 123–179.

Razin, S. 1978. The mycoplasmas. *Microbiological Reviews* 42: 414–470.

Sanderson, K. E. 1976. Genetic relatedness in the family Enterobacteriaceae. *Annual Review of Microbiology* 30: 327–349.

Shapiro, L. 1976. Differentiation in the *Caulobacter* cell cycle. *Annual Review of Microbiology* 30: 377–408.

Skerman, V. B. D. 1967. *A Guide to the Identification of the Genera of Bacteria.* Williams & Wilkins Co., Baltimore.

Skerman, V. B. D., V. McGowan, and P. H. A. Sneath (eds.). 1980. *Approved Lists of Bacterial Names.* American Society for Microbiology, Washington, D.C.

Skinner, F. A., and D. W. Lovelock. 1980. *Identification Methods for Microbiologists.* Academic Press, New York.

Sneath, P. H. A. 1978. Classification of microorganisms. In: *Essays in Microbiology* (J. R. Norris, ed.), chap. 9. John Wiley & Sons, Chichester, England.

Sneath, P. H. A. 1978. Identification of microorganisms. In: *Essays in Microbiology* (J. R. Norris, ed.), chap. 10. John Wiley & Sons, Chichester, England.

Sneath, P. H. A., and R. R. Sokal. 1973. *Numerical Taxonomy: The Principles and Practice of Numerical Classification.* W. H. Freeman and Co., San Francisco.

Stanier, R. Y., and G. Cohen-Bazire. 1977. Phototrophic prokaryotes: the cyanobacteria. *Annual Review of Microbiology* 31: 225–274.

Starr, M. P., H. Stolp, H. G. Truper, A. Ballows, and H. G. Schlegel (eds.). 1981. *The Prokaryotes: A Handbook on Habits, Isolation, and Identification of Bacteria.* Springer-Verlag, Berlin.

vanVeen, W. L., E. G. Mulder, and M. H. Deinema. 1978. The *Sphaerotilus-Leptothrix* group of bacteria. *Microbiological Reviews* 42: 329–356.

Whitcomb, R. F. 1980. The genus *Spiroplasma. Annual Review of Microbiology* 34: 677–709.

Woese, C. R. 1981. Archaebacteria. *Scientific American* 244(6): 98–122.

Zeikus, J. G. 1977. The biology of methanogenic bacteria. *Bacteriological Reviews* 41: 514–541.

410

Classification and
identification of
microorganisms:
systematics of
bacteria

Classification and identification of microorganisms: systematics of fungi, algae, protozoa, and viruses

12

Classification of fungi

As with other microorganisms, the classification of the fungi is difficult, sometimes ambiguous, vehemently debated, and subject to constant revision. The diversity of the fungi and the interrelationships between the fungi and other groups of eukaryotic microorganisms make fungal taxonomy particularly difficult. The fungi are eukaryotic heterotrophic microorganisms. They are nonphotosynthetic and typically form reproductive spores. Many fungi exhibit both sexual and asexual forms of reproduction. Some fungi are unicellular, but many fungi form filaments of vegetative cells known as mycelia. Mycelia are integrated masses of individual tube-like filaments of hyphae. The mycelia usually exhibit branching and are typically surrounded by cell walls containing chitin and/or cellulose. By a concise definition: **the fungi are achlorophyllous, saprophytic or parasitic, with unicellular or more typically filamentous vegetative structures usually surrounded by cell walls composed of chitin or other polysaccharides, propagating with spores and normally exhibiting both asexual and sexual reproduction.** If this definition seems complex and vague, it is; of necessity a broad definition is needed to accommodate all the morphological and physiological anomalies that occur among the fungi.

The **classification of fungi** is based largely on the means of reproduction, including the nature of the life cycle, reproductive structures and reproductive spores. The primary taxonomic groupings are based on the sexual reproductive spores. To a lesser extent, fungal systematics relies on the morphological characteristics of the vegetative cells. Whereas most classical approaches to fungal systematics are based largely on observing the morphology of the reproductive forms, physiological, biochemical, and genetic characteristics are included in some modern classification systems. Physiological features are particularly important in the classification of yeasts, which are primarily unicellular fungi.

In a formal systematic sense, **yeasts** are not recognized as being separate from the rest of the fungi and are classified along with their filamentous counterparts. In practice, however, the yeasts are typically treated separately from the filamentous fungi in both classification and identification systems. Separate classification and identification systems, for example, have been developed that include only the yeasts. Revisions concerning yeast systematics are published in the *International Journal of Systematic Bacteriology* rather than in the mycological literature. Commercial clinical systems are also available for identifying pathogenic yeasts. Similarly, separate classification and identification systems have been developed for other groups of fungi, such as the mushrooms. Such systems are very important because of their functional utility for identification purposes, even if they do not follow a formal taxonomic scheme based on phylogenetic relationships.

The fungi have traditionally been split in formal classification systems into two large groups: the **slime molds** and the **true fungi**. The slime molds represent a borderline case between the fungi and the protozoa. They can just as well be classified with the protozoa, but traditionally they have been studied by mycologists. The vegetative cells of the slime molds are ameboid and lack a cell wall, making them similar to protozoa. The slime molds and the true fungi are similar because they both produce spores that are surrounded by wall structures.

412

Classification and identification of microorganisms: systematics of fungi, algae, protozoa, and viruses

table 12.1

The major taxonomic divisions of the kingdom Myceteae (Fungi)

Division I.   Gymnomycota
  Subdivision 1. Acrasiogymnomycotina
    Class 1. Acrasiomycetes
  Subdivision 2. Plasmodiogymnomycotina
    Class 1. Protosteliomycetes
    Class 2. Myxomycetes
      Subclass 1. Ceratiomyxomycetidae
      Subclass 2. Myxogastromycetidae
      Subclass 3. Stemonitomycetidae

Division II. Mastigomycota
  Subdivision 1. Haplomastigomycotina
    Class 1. Chytridiomycetes
    Class 2. Hyphochytridiomycetes
    Class 3. Plasmodiophoromycetes
  Subdivision 2. Diplomastigomycotina
    Class 1. Oomycetes

Division III. Amastigomycota
  Subdivision 1. Zygomycotina
    Class 1. Zygomycetes
    Class 2. Trichomycetes
  Subdivision 2. Ascomycotina
    Class 1. Ascomycetes
      Subclass 1. Hemiascomycetidae
      Subclass 2. Plectomycetidae
      Subclass 3. Hymenoascomycetidae
      Subclass 4. Laboulbeniomycetidae
      Subclass 5. Loculoascomycetidae
  Subdivision 3. Basidiomycotina
    Class 1. Basidiomycetes
      Subclass 1. Holobasidiomycetidae
      Subclass 2. Phragmobasidiomycetidae
      Subclass 3. Teliomycetidae
  Subdivision 4. Deuteromycotina
    Class 1. Deuteromycetes
      Subclass 1. Blastomycetidae
      Subclass 2. Coclomycetidae
      Subclass 3. Hyphomycetidae

Gymnomycota

In the classification system described by Alexopoulos and Mims, the slime molds are placed in the division **Gymnomycota** of the Kingdom Myceteae (fungi) (Table 12.1). The vegetative cells of organisms in the division Gymnomycota lack cell walls and their nutrition is phagotrophic; that is, they engulf and ingest nutrients. All **slime molds** exhibit characteristic life cycles, a feature used to subdivide this division (Figure 12.1). Organisms in the subdivision **Acrasiogymnomycotina (Acrasiales)**, which often feed largely on bacteria, are known as cellular slime molds. The Acrasiomycetes, the single class within the subdivision Acrasiogymnomycotina, form a fruiting (spore-bearing) body known as a **sporocarp**. The sporocarps of Acrasiomycetes are generally stalked structures. The stalks normally consist of walled cells, and this characteristic forms the basis for designating these organisms as the cellular slime molds. The sporocarp releases spores that germinate, forming **myxamoebae** (ameboid cells that form pseudopodia). The myxamoebae swarm together or aggregate to form a pseudoplasmodium. Within the pseudoplasmodium, the cells of Acrasiomycetes do not lose their integrity. The pseudoplasmodium undergoes a developmental sequence (differentiation), culminating in the formation of a **sporocarp**, which is a special type of fruiting body that bears a mucoid droplet at the tip of each branch, containing spores with cell walls.

The pseudoplasmodium formation of the Acrasiomycetes is of special interest because of the biochemical communication involved in initiating swarming activity. The swarming behavior of *Dictyostelium discoideum* has been extensively studied. Under appropriate conditions, when food sources become limiting, the myxamoebae cease their feeding activity and swarm to an aggregation center. The swarming activity is initiated when one or more cells at the aggregation center release cyclic AMP (**acrasin**). At the biochemical level cyclic AMP is responsible for communication between the myxamoebae. The myxamoebae move along the concentration gradient of cyclic AMP until they reach the center of aggregation. When the myxamoebae reach the center of aggregation, they mass together to form a pseudoplasmodium. Swarming occurs as a pulsating wave motion in which the chemical stimulus, cyclic AMP, is transmitted from cells that are proximal to the aggregation center to distant cells. Different species of Acrasiomycetes exhibit different waveforms in their

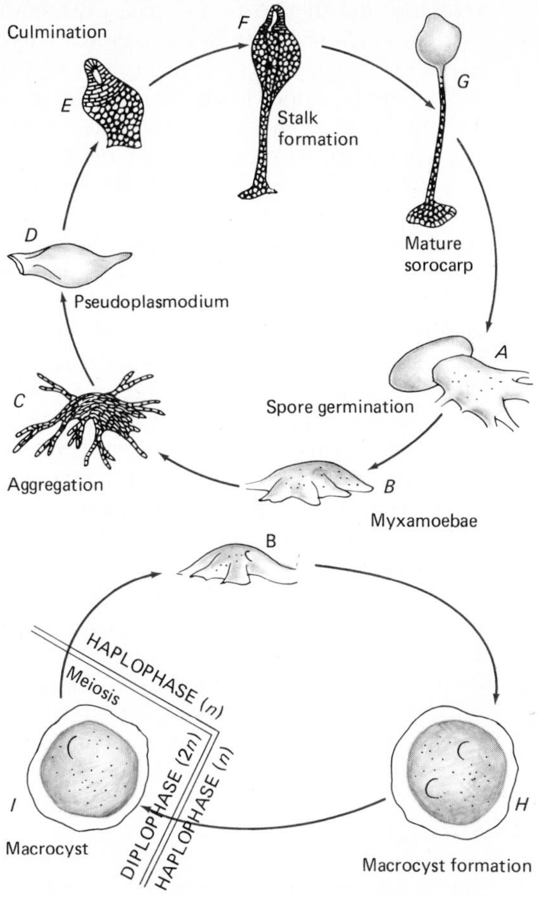

Figure 12-1

The life cycle of a slime mold, Dictyostelium discoideum. *(A) Germination of a spore with a single myxamoeba issuing from it. (B) Myxamoebae. (C) Streams of aggregating myxamoebae. (D) Pseudoplasmodium, grex. (E) Beginning of culmination, myxamoebae at the front end of the grex pushing down the middle to the stalk cylinder. (F) Later stage, showing stalk formation. (G) Mature sorocarp with cellular stalk. (H) Young macrocyst in which karyogamy occurs. (I) Mature macrocyst where meiosis takes place. Myxamoebae form upon germination of the mature macrocyst.*

Discovery process

The observation that slime mold aggregates under certain conditions led to a long series of fascinating experiments to elucidate the mechanism of developmental behavior. In using quite different methods, such as the attraction of amebae by aggregation centers located across semipermeable barriers and the interruption of aggregation patterns by currents of water, it was argued that the gathering of amebae to a central collection point occurred by chemotaxis. The chemotactic behavior was postulated by J. T. Bonner in the late 1940s to be the response to a gradient of a substance, which was called "acrasin," that somehow orients the cells toward areas of higher concentration. In 1953 B. M. Schaffer devised an ingenious test for acrasin in which sensitive *Dictyostelium discoideum* were sandwiched under a small agar block and arcrasin water applied to the outside meniscus. When this was done at a rate of about every 10 seconds, the amebae would be attracted to the outside edge; lengthening the intervals produced no effect. The interpretation of this was that the acrasin disappeared rapidly, thus requiring frequent repeat applications. Schaffer also showed that the acrasin was degraded enzymatically. A quantitative assay was needed, and two were developed. T. M. Konijn put sensitive *D. discoideum* in small drops in washed agar, drops of test solution were placed close by, and if acrasin was present the amebae burst out of the original drop. This test was made quantitative by dilution of the test solutions and by varying the distance between the test solution drops and the amebae. The cellophane square test placed amebae on a small square of dialysis membrane, which was then placed on the surface of unwashed agar containing the test substance. The rate at which the amebae moved away from the square was measured; the faster the rate, the greater the power of the attractant, for the amebae do not actually move faster, but their paths are more perfectly directed away from the square. This test has shown both vegetative and aggregating amebae to be subject to chemotaxis. Now that it was clear that bacterial extracts attract slime mold amebae, Konijn and T. J. Barkley tested many biochemicals for their ability to mediate aggregation, including cyclic 3′,5′-cyclic AMP, which they found to be extraordinarily active with *D. discoideum*. This was found to be the component secreted by *E. coli* that affects aggregating amebae. The cells of *D. discoideum* also secrete large quantities of a specific phosphodiesterase that converts cyclic AMP to 3′,5′-cyclic AMP. By using a variety of different techniques to prevent this phosphodiesterase from immediately converting the slime mold's own cyclic AMP, it was possible to show that *D. discoideum* does synthesize its own cyclic AMP. From this it is deduced that "acrasin" is cyclic AMP; the production of 3′,5′-cyclic AMP by an ameba that finds a food source acts as a chemotactic signal, communicating to other amebae to migrate to that aggregation center.

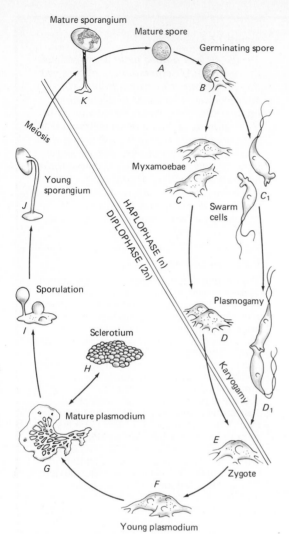

Mature sporangium

Mature spore

Germinating spore

A

B

K

Meiosis

Young
sporangium

J

HAPLOPHASE (n)

DIPLOPHASE (2n)

Myxamoebae

C

Swarm
cells

C₁

Sporulation

I

Plasmogamy

D

Sclerotium

H

Karyogamy

D₁

Mature plasmodium

G

E

Zygote

F

Young plasmodium

Figure 12.2

The life cycle of a Myxomycete. (A) Mature haploid spore. (B) Germinating spore. (C) Myxamoeba. (C₁) Swarm cells. (D) Fusing myxamoebae. (D₁) Fusing swarm cells. (E) Young zygote. (F) Young plasmodium. (G) Mature plasmodium. (H) Sclerotium. (I) Sporulation, sporangial initials. (J) Young premeiotic spongium with spores. (K) Mature postmeiotic sporangium.

414

Classification and
identification of
microorganisms:
systematics of
fungi, algae,
protozoa, and
viruses

swarming behavior, some moving in linear wave-like motion and others exhibiting spiral wave motion.

The second subdivision of the Gymnomycota, the **Plasmodiogymnomycotina**, includes two classes: the **Protosteliomycetes** and the **Myxomycetes**. The Myxomycetes are known as the **true slime molds**. Some species of Myxomycetes form myxamoebae, and others form flagellated cells,

known as **swarm cells**. The myxamoebae or swarm cells fuse together to form a true plasmodium. The **plasmodium** of the Myxomycetes is a multi-nucleate protoplasmic mass that is devoid of cell walls and is enveloped in a gelatinous slime sheath. The plasmodium gives rise to brilliantly colored fruiting bodies. The classification of the Myxomycetes is based largely on the structure of the fruiting body. In many species of Myxomycetes spores are formed inside the fruiting body. These spores are sometimes referred to as endospores, but they do not bear any resemblance to the endospores of bacteria. The spores of myxomycetes generally have a definite thick wall. In the life cycle of Myxomycete the spores are released from the sporangia and disseminated. At a later time they germinate, producing myxamoebae and/or swarm cells (Figure 12.2). The myxamoebae and/or swarm cells later unite in a sexual fusion process to initiate formation of the plasmodium. The brightly colored fruiting bodies of Myxomycetes are often seen on decaying logs or other moist areas of decaying organic matter (Figure 12.3). Myxomycetes are often conspicuous on grass lawns, appearing as large blue-green colonies. To remove these unsightly blemishes from an otherwise luxuriant lawn, simply mow the grass.

Mastigomycota

The **Mastigomycota** comprise the second major division of the kingdom Myceteae. Some Mastigomycota are unicellular, but most form extensive filamentous coenocytic mycelia. Mastigomycota typically produce motile cells with flagella during part of their life cycle. Asexual reproduction by species in this division normally involves motile spores called **zoospores**. As opposed to the phagotrophic mode of nutrition exhibited by members of the Gymnomycota, nutrition is accomplished in the Mastigomycota by absorption of nutrients. The division Mastigomycota includes four classes: Chytridiomycetes, Hyphochytridiomycetes, Plasmodiophoromycetes, and Oomycetes (Table 12.2).

The **Chytridiomycetes** are differentiated from all other fungi by the production of **zoospores**, which are motile with a single posterior flagellum of the whiplash type. Members of the order Chytridiale are referred to as **chytrids**. Many chytrids are parasitic on other fungi, algae, and plants. A few, such as *Olpidium brassicae,* which infects cabbage roots, are responsible for plant diseases. The **Hyphochytridiomycetes**, of which only about 15 species are known, produce uniflagellate zoospores of the tinsel type. The tinsel-type flagellum has many short projections extending down its length. The **Plasmodiophoromycetes** are known as the **endoparasitic slime molds** and are obligate parasites of other fungi, algae, and plants. They cause an abnormal enlargement of the host cells, called hypertrophy, and often an abnormal multiplication of the host cells called hyperplasia. The Plasmodiophoromycetes form a plasmodium that develops within the host cells. Various species of Plasmodiophoromycetes cause diseases of plants that are of economic significance. For example, *Plasmodiophora brassicae* causes clubfoot of cabbage, and *Spongospora subterranea* causes powdery scab of potatoes.

The **Oomycetes**, known as the **water molds**, reproduce by using flagellated zoospores. The zoospores typically have two flagella, one of the tinsel type and the other of the whiplash type. Sexual reproduction in the Oomycetes typically involves the formation of **oospores**, which are thick-walled spores that develop by contact with specialized **gametangia** (structures containing differentiated cells involved in sexual reproduction). The gametangia of Oomycetes normally occur at the terminal ends of the mycelia. During gametangial contact, male gametes pass through a fertilization tube into the female gametangia (Figure 12.4). The male gametangium is referred to as an antheridium and the female gametangium as an oogonium. Members of the Saprolegniales, an order of the Oomycetes, are abundant in aquatic ecosystems. Several species in the order are Perenosporales plant pathogens. For example *Phytophthora infestans* causes potato blight and was responsible for the great Irish potato famine of 1845 and 1846, which resulted in the great wave of immigration from Ireland to the United States.

Amastigomycota

The vegetative cells of the **Amastigomycota**, the third division of the fungi, vary from single cells to mycelia that may be coenocytic (multinucleate) or may have extensive septation (cross walls separating cells within the mycelia). Unlike the Gymnomycota and Mastigomycota, the Amastigomycota do not produce motile cells. There are four subdivisions in the Amastigomycota: Zygomycotina, Ascomycotina, Basidiomycotina, and Deuteromycotina (Table 12.3).

Zygomycotina

The **Zygomycotina** typically have coenocytic my-

A

B

Figure 12.3

(A) The slime mold Lycogala epidendrum *growing on a log. (Courtesy Varley Wiedeman, University of Louisville.) (B)* Lycogala epidendrum *(aethalium). (Courtesy Orson K. Miller, Jr., Virginia Polytechnical Institute and State University.)*

415

Classification of fungi

table 12.2

The four classes in the division Mastigomycota

Class 1. *Chytridiomycetes* Vegetative form varied, produces posteriorly uniflagellate motile cells with whiplash flagella

Class 2. *Hyphochytridiomycetes* A small group of fungi, produces motile anteriorly uniflagellate cells with tinsel flagella

Class 3. *Plasmodiophoromycetes* Parasitic fungi with multinucleate thalli (plasmodia) within the cells of their hosts. Resting cells (cysts) produced in masses but not in distinct sporophores; motile cells with two anterior whiplash flagella

Class 4. *Oomycetes* Vegetative form varied, usually filamentous, with a coenocytic, walled mycelium; hyphal wall, produces zoospores, each with one whiplash and one tinsel flagellum; sexual reproduction ooagamous resulting in the formation of oospores.

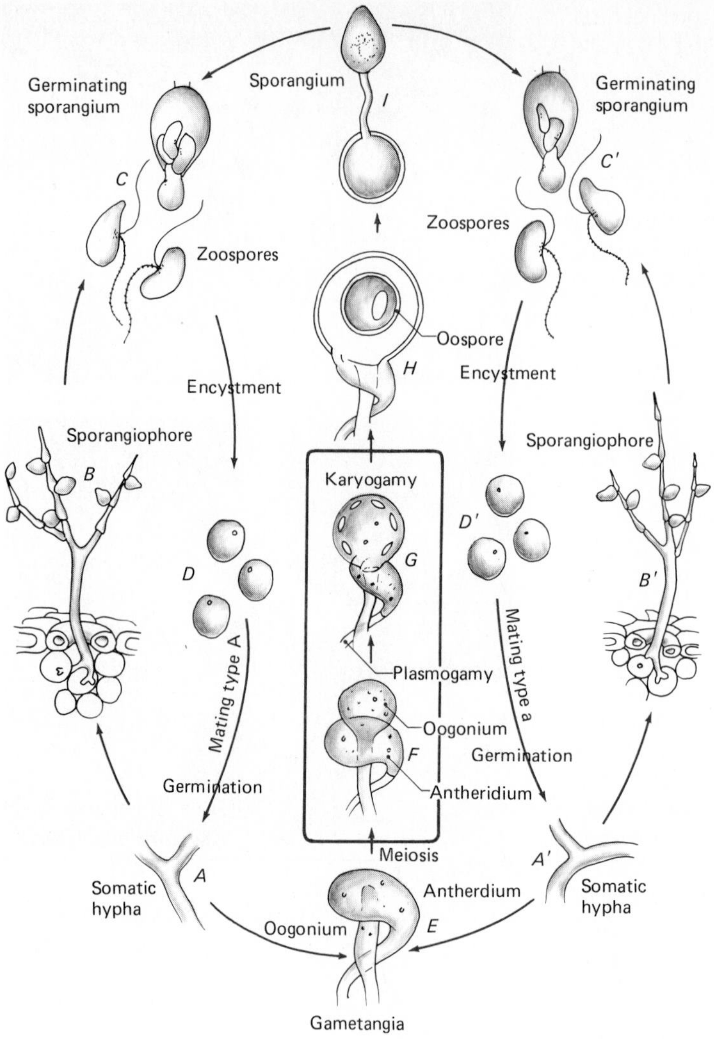

Figure 12.4

The life cycle of an Oomycete, Phytophthora infestans.

416

Classification and
identification of
microorganisms:
systematics of
fungi, algae,
protozoa, and
viruses

table 12.3

The four subdivisions of the Amastigomycota

Subdivision I. *Zygomycotina* Saprophytic, parasitic or predatory fungi, coenocytic mycelium; asexual reproduction usually by sporangiospores; sexual reproduction, where known, by fusion of equal or unequal gametangia resulting in the formation of zygosporangia containing zygospores	Subdivision III. *Basidiomycotina* Saprophytic, symbiotic or parasitic fungi; unicellular or, more typically, with a septate mycelium, producing basidiospores on the surface of various types of basidia
Subdivision II. *Ascomycotina* Saprophytic, symbiotic or parasitic fungi; unicellular or with a septate mycelium, producing ascospores in sac-like cells (asci)	Subdivision IV. *Deuteromycotina* Saprophytic, symbiotic, parasitic or predatory fungi; unicellular or, more typically, with a septate mycelium, usually producing conidia from various types of conidiogenous cells; sexual reproduction unknown

celia and are characterized by the formation of a **zygospore**, a sexual spore that forms within a zygosperangium. For sexual reproduction some species are **heterothallic**, requiring gametangia of two different mating types, whereas others are **homothallic**, requiring only one type of gametangia for zygote formation. In addition to sexual reproduction, the Zygomycotina characteristically produce asexual sporangiospores within a sporangium. Many members of the Zygomycotina are

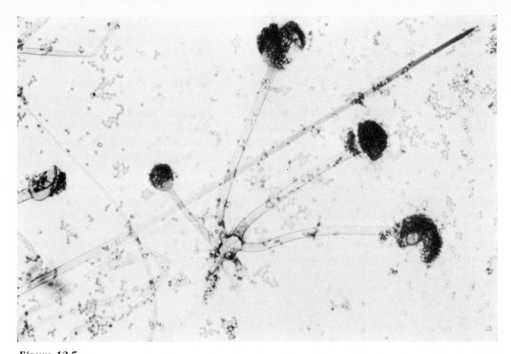

Figure 12.5

Micrograph of the bread mold Rhizopus. *(From BPS—C. Emmons, Phoenix, Arizona.)*

plant or animal pathogens. There are two classes within the subdivision Zygomycotina: Zygomycetes and Trichomycetes. Species of the Trichomycetes are obligately associated with arthropods and normally grow within the guts of these animals, where they attach to the chitinous lining of the gut by means of a specialized structure known as a holdfast. The order Mucorales (the bread molds) occurs within the class Zygomycetes. *Rhizopus stolonifer* is the common bread mold (Figure 12.5). Some *Rhizopus* species are important in the food industry. For example, *Rhizopus oryzae* is used for the production of fermented Oriental foods, such as tempeh.

Several species of Mucorales exhibit an interesting morphological change known as **dimorphism**, occurring under some conditions as filamentous mycelia and under other conditions as yeast-like unicellular forms. For example, *Mucor rouxii* grows in a yeast-like form in atmospheres with a high percentage of carbon dioxide but produces filamentous mycelia at normal atmospheric concentrations of CO_2. *Mucor* species are important opportunistic human pathogens and can cause serious infections in burn wounds. The genus *Pilobolus* also occurs in the order Mucorales. *Pilobolus* species are interesting because of their forceful mode of spore discharge (Figure 12.6). *Pilobolus* species can shoot their spores several

meters into the air with the entire spore cluster ejected in the direction of highest light intensity. In this way the spores of *Pilobolus* are released to an area of open air where air currents are likely to further disperse the spores.

Figure 12.6

In the spore discharge by Pilobolus, *the entire spore mass is shot toward an area of high light intensity.*

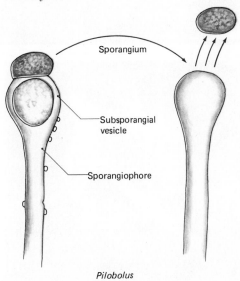

Sporangium

Subsporangial
vesicle

Sporangiophore

Pilobolus

Ascomycotina

Members of the subdivision **Ascomycotina** produce sexual spores within a specialized sac-like structure known as the **ascus**. These ascomycetes are sometimes called the sac fungi. The ascomycetes normally produce a definite number of ascospores within the ascus. The mycelia of ascomycetes are composed of septate hyphae and the cell walls of the hyphae of most ascomycetes contain chitin. Asexual reproduction in the ascomycetes may be carried out by fission, fragmentation of the hyphae, formation of chlamydospores (thick-walled spores within the hyphal filaments), and production of conidia (nonmotile spores produced on a specialized spore-bearing cell). During sexual reproduction the ascomycetes normally exhibit a short-lived dikaryotic stage (having cells containing two nuclei) between the time of fusion of gametes (**plasmogamy**) and the time of fusion of the two nucleii (**karyogamy**).

Ascosporogenous yeasts The members of the subclass Hemiascomycetidae, in which the ascomycetous yeasts occur, are morphologically simple ascomycetes that generally lack hyphae. Many yeasts are ascomycetes (Figure 12.7). The morphology of the ascospore is a critical taxonomic feature for classifying yeasts to the genus level (Table 12.4). Classification of the yeasts to the species level normally employs numerous biochemical and physiological characteristics as well as morphological features. The metabolic activities of the ascosporogenous yeasts have many industrial applications. *Saccharomyces cerevisiae* is used

418

Classification and
identification of
microorganisms:
systematics of
fungi, algae,
protozoa, and
viruses

Figure 12.7

Micrographs of ascospores of various yeasts showing characteristic morphologies. (A) Smooth spheroidal ascospores of Pichia carsonii. *(B) Hat-shaped ascospores of* Pichia spartinae. *(C) Saturn-shaped ascospore of* Schwanniomyces occidentalis. *(D) Ascospore of* Saccharomyces kloeckerianus. *(Courtesy C. P. Kurtzman, U.S. Department of Agriculture, Northern Regional Center, Peoria, Illinois.)*

table 12.4

Descriptions of ascospores found in different genera of yeasts

Genus	Number of ascospores	Shape of ascospores
Citeromyces	1	Spheroidal
Coccidiascus	8	Fusiform
Debaryomyces	1–4	Spheroidal, ovoidal, warty
Dekkera	1–4	Hat-shaped
Endomycopsis	1–4	Spheroidal, hat-shaped, saturn-shaped, sickle-shaped
Hanseniaspora	1–4	Hat-shaped, helmet-shaped, walnut-shaped
Hansenula	1–4	Hat-shaped, spheroidal, hemispheroidal, saturn-shaped
Kluyveromyces	1–many	Crescentiform, reniform, spheroidal, ellipsoidal
Lipomyces	1–16	Ellipsoidal, lenticular
Lodderomyces	1–2	Oblong, ellipsoidal
Metschnikowia	1–2	Needle-shaped
Nadsonia	1–2	Spheroidal, warty
Nematospora	8	Spindle-shaped
Pachysolen	4	Hemispheroidal
Pichia	1–4	Spheroidal, hat-shaped, saturn-shaped, warty
Saccharomyces	1–4	Spheroidal, ellipsoidal
Saccharomycodes	4	Spheroidal
Saccharomycopsis	1–4	Ovoidal, double-walled
Schizosaccharomyces	4–8	Spheroidal, ovoidal
Schwanniomyces	1–2	Walnut-shaped, warty
Wickerhamia	1–16	Cap-shaped
Wingea	1–4	Lens-shaped

as baker's yeast. Many fermented beverages are also produced by using members of the genus *Saccharomyces*. Most commonly, *Saccharomyces carlsbergensis* and *Saccharomyces cerevisiae* are used for the production of beer, wine, and spirits.

Taphrinales In addition to the ascosporogenous yeasts, the subclass Hemiascomycetidae includes the order Taphrinales. The Taphrinales resemble yeasts in that they reproduce asexually by budding and sexually by the production of ascospores; they differ from the yeasts in that they produce a definite true mycelium. Members of the Taphrinales are parasitic on plants. *Taphrina deformans*, for example, causes peach leaf curl, *Taphrina cerasi* causes witches' broom of cher-

ries, *Taphrina pruni* causes prune pockets, and *Taphrina coerulescens* causes puckering of oak leaves.

Euascomycetidae The **Euascomycetidae** (**true ascomycetes**) produce asci that normally develop from dikaryotic hyphae. The asci are produced in or on a specific structure known as the ascocarp. The euascomycetes are divided according to the structure of the ascocarp into the Plectomycetes, in which the ascocarp has no special opening; the Pyrenomycetes, in which the ascocarp is shaped like a flask; and the Discomycetes, in which the ascocarp is cup-shaped (Figure 12.8). Species of Pyrenomyctes are important for their roles in basic scientific investigations. For example, studies on

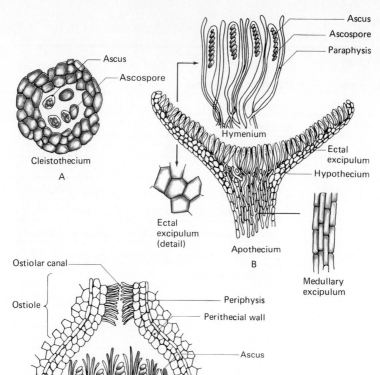

Hymenium

Ascus
Ascospore

Cleistothecium
A

Ectal
excipulum

Hypothecium

Ectal
excipulum
(detail)

Apothecium
B

Medullary
excipulum

Ostiolar canal
Ostiole

Periphysis
Perithecial wall

Ascus

Ascospore
Stroma

Perithecium
C

Figure 12.8

Asci and ascocarps occur in various characteristic shapes. This figure shows three types of ascospores: (A) Clentothecium, (B) Apothecium, and (C) Perithecium.

420

Classification and
identification of
microorganisms:
systematics of
fungi, algae,
protozoa, and
viruses

the genetics of *Neurospora*, a pyrenomete, have greatly added to our understanding of recombinational processes. *Neurospora* is useful in genetic studies because the spores can be isolated from the ascus and the genotypes readily determined. Besides their usefulness in basic microbiological studies, Pyrenomycete species are of practical importance. Some Pyrenomycetes are important plant and animal pathogens. The powdery and black mildews occur in this taxonomic group. *Claviceps purpurea* causes the ergot of rye. Cattle and other such animals are poisoned when grazing on grasses contaminated with the resting bodies (sclerotia) of the fungus. Sclerotia are hard resting bodies that are resistant to unfavorable conditions and may remain dormant for prolonged periods. Various alkaloid biochemicals are produced by *Claviceps purpurea*. Some of these have hallucinogenic properties, whereas others, such as those used to induce labor for childbirth,

are useful medicinals. *Endothia parasitica*, another Pyrenomycete, is the causative agent of chestnut blight. This organism was introduced into North America from eastern Asia in the early twentieth century and quickly devastated the chestnut trees of the United States and Canada.

The Plectomycetes also include some important plant and animal pathogens. This group includes the black molds, blue molds, and ringworms. Several species of *Ceratocystis* are responsible for blue stain that reduces the commercial value of lumber. *Ceratocystis ulmi* is the causative agent of Dutch elm disease, a great threat to elm trees in North America. The fungus that causes the human disease histoplasmosis, *Emmonsiella capsulata*, also belongs to the subclass Plectomycetes. *Emmonsiella capsulata* was formerly known as *Histoplasma capsulatum* before the sexual reproductive stage of the organism was known, and some medical mycologists still retain the name of the imperfect (nonsexual) form when referring to this organism. This fungus commonly occurs in soils that are contaminated with fecal droppings from birds. Other fungi known by the names of their imperfect forms associated with the Plectomycetes include the well-known genera *Penicillium* and *Aspergillus*.

The Discomycetes include the cup fungi, morels, and truffles. Species of the genus *Morchella* (morels) are gastronomical delights (Figure 12.9). All morels are edible and delicious. Truffles occur in the order Tuberales and, like morels, are considered edible delicacies.

Basidiomycotina

The subdivision **Basidiomycotina** contains the single class Basidiomycetes. This class contains the most complex fungi, including the smuts, rusts, jelly fungi, shelf fungi, stinkhorns, bird's nest fungi, puffballs, and mushrooms. The Basidiomycetes are distinguished from other classes of fungi by the fact that they produce sexual spores, known as **basidiospores**, on the surfaces of specialized spore-producing structures, known as **basidia**. (Figure 12.10). The Basidiomycetes are also known as the

Figure 12.9

Photograph of a morel, Morchella esculenta. *These fungi are considered gastronomic delights. (Courtesy Orson K. Miller, Jr., Virginia Polytechnical Institute and State University.)*

Figure 12.10

There are various types of basidia. (A) Typical holobasidium. (B) Tuning fork basidium of Dacrymyces. *(C) Basidium of* Tulasnella. *(D) Basidium of* Tremella. *(E) Basidium of* Auricularia. *(F) Basidium of* Puccinia. *(G) Micrograph of mature basidia of* Gomphidius glutinosus *in all stages (300×). (Courtesy Orson K. Miller, Jr., Virginia Polytechnical Institute and State University.)*

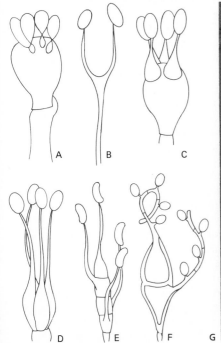

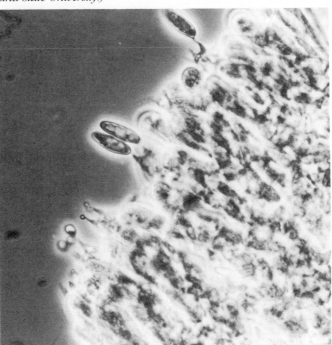

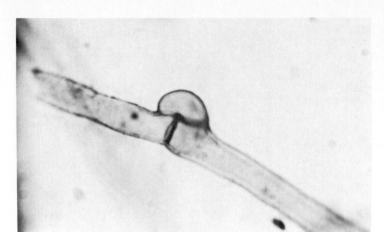

Figure 12.11

Basidiomycetes characteristically exhibit clamp cell connections. This micrograph of Chroogomphus loculatus *shows clamped vegetative hyphae. (Courtesy Orson K. Miller, Jr., Virginia Polytechnical Institute and State University.)*

club fungi based on the typical shape of the basidia. The mycelia of Basidiomycetes typically form **clamp connections** between cells (Figure 12.11). The clamp cell connections are generally indicative of a dikaryotic mycelium. Additionally, the mycelia of many Basidiomycetes are character-

ized by specialized cross walls between connecting cells, known as **dolipore septa** (Figure 12.12). The dolipore septum has a central pore surrounded by a barrel-shaped swelling of the cross wall. Effectively, the clamp cell connections and the dolipore septa permit enhanced communication of chemicals through the mycelia of the organism.

Aphyllophorales The Basidiomycete order **Aphyllophorales** contains the **shelf** or **bracket fungi**. These are some of the most conspicuous fungi, often seen growing on trees (Figure 12.13). The fruiting bodies of these fungi are tough and leathery. In addition to the bracket fungi, the Aphyllaphorales includes the cantharelles, the coral fungi, the tooth fungi, and the pore fungi. The majority of these fungi are saprophytic, growing on dead and living plant materials. The growth of fungi in this group on wood results in two characteristic types of decay, called brown rots and white rots because of the color of the rotted wood. In brown rot only the cellulose component of wood is decomposed, leaving the brown lignins. In white rot both cellulose and lignin are degraded, producing white-colored wood.

Agaricales (mushrooms) The Basidiomycete order **Agaricales** includes the **mushrooms**. The mushrooms that we see are the fruiting bodies (**basidiocarps**) of basidiomycetes. The spore-bearing structures (**basidia**) of mushrooms are borne on the surface of the gills of the basidiocarp (Figure 12.14). In the boletes, which also are included in the Agariales, the basidia are not borne on gills but rather within tubes. Some mushrooms are edible, but others are extremely poisonous. The proper identification of mushrooms is critical lest one become the victim of

Figure 12.12

Dolipore septum. Septa of this form have been reported from basidiocarps of Auricularia, Calvatia, Dacrymyces, Exidia, Merilius, *and* Polyporus, *and are suggested in the mature* Calvatia gigantea *mycelium.*

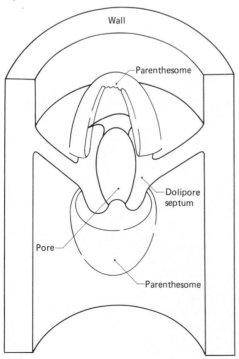

mushroom poisoning. It is sometimes easy to confuse an edible species with one that is deadly poisonous.

The order Agaricales is divided into several families (Table 12.5). The family Russulaceae includes two genera, *Russula* and *Lactarius*. Many species of *Russula* produce brilliantly colored beautiful caps. Some of these attractive *Russula* species are poisonous. *Russula* species produce white spores and brittle fruiting bodies. The family Amanitaceae includes the genus *Amanita*. The genus *Amanita* is characterized by free gills and the presence of an annulus and a volva (Figure 12.15). Many *Amanita* species are quite beautiful, but most are deadly poisonous. *Amanita phalloides* is known as the death cap because most deaths due to mushroom poisoning have been attributed to the ingestion of this species. *Agaricus* species, which occur in the family Agaricaceae, produce white to brown mushrooms, with free gills and an annulus but no volva. Several mushrooms of this genus, such as *Agaricus bisporus*, are grown commercially. The family Coprinaceae contains the genus *Coprinus*, which is known as the ink cap mushroom. Autodigestion (self-decomposition) causes mushrooms of the genus *Coprinus* to dissolve into a black ink-like liquid.

Figure 12.13

The bracket fungi often are conspicuous, growing on trees. (From BPS—J. Robert Waaland, University of Washington.)

Figure 12.14

Mushrooms are visible macroscopic fruiting bodies of basidiomycetes. Some mushrooms are edible, but others are quite poisonous. (A) This Pleurotus ostreatus *is edible and good. (B) This* Amanita virosa *is deadly poisonous. (Courtesy Orson K. Miller, Jr., Virginia Polytechnical Institute and State University.) (C) This* Amanita pantherina *is also a poisonous mushroom. (From BPS— J. Robert Waaland, University of Washington.)*

A

B

C

table 12.5

The families in the order Agaricales

Family Agariaceae	Possess free gills, an annulus, but no volva; produce brown spores
Family Amanitaceae	Characterized by free gills and the presence of an annulus and volva
Family Boletaceae	Possess vertically arranged tubes instead of gills; the basidia occur within the tubes
Family Coprinaceae	Possess gills that dissolve into a black inky liquid at maturity
Family Cortinariaceae	Possess attached gills, may or may not have an annulus; produce light brown spores
Family Lepiotaceae	Some members resemble the amanitas but lack a volva
Family Rodophyllaceae	Possess attached gills, produce annular spores
Family Russulaceae	Possess attached gills, produce white spores, and are distinguished by the presence of specialized spherical cells known as spherocysts
Family Strophariaceae	Possess attached gills, produce dark spores
Family Tricholomataceae	Produce white spores, possess attached gills
Family Volvariaceae	Produce salmon to pink-colored basidiospores

424

Classification and
identification of
microorganisms:
systematics of
fungi, algae,
protozoa, and
viruses

Gasteromycetes Several orders comprise the **gasteromycetes**, another basidiomycete group, including the puffballs, earthstars, stinkhorns, and bird's nest fungi. Unlike other Basidiomycetes, the spores of the gasteromycetes are not forcefully discharged. Members of this taxonomic group are commonly seen in a number of habitats. The order Phallales (stinkhorns) produces a green gelatinous ooze and a foul smell when the basdiocarp undergoes autodigestion, releasing the basidiospores. Although humans find the odor quite offensive, flies are attracted by the smell. Some of the ooze containing the basidiospores adheres to the flies, providing a mechanism for the dissemination of the basidiospores of these fungi.

Teliomycetidae (rusts, smuts, and basidiomycetous yeasts) There are a few yeasts that are Basidiomycetes. These are placed into the subclass Teliomycetidae along with the rusts and smuts. *Rhodosporidium* and *Leucosporidium*, which have been placed in this subclass, are basidiomycetous yeasts found in marine environments. The numerous species of rust and smut fungi are the most serious fungal plant pathogens. The members of the subclass Teliomycetidae do not produce a typical basidium. There are over 20,000 species of rust fungi and over 1,000 species of smut fungi. Rusts and smuts are characterized by the production of a resting spore known as a **teliospore**, which is thick-walled and binucleate. The rusts, all of which are plant pathogens, occur in the order *Uredinales*. The rust fungi require two unrelated hosts for the completion of their normal life cycle. For example, white pine blister rust uses gooseberry bushes as its alternate host. Important plant diseases are caused by rust fungi, and these fungal plant pathogens cause great economic losses in agriculture. Members of the genus *Puccinia* cause rust of cereals. Rust of beans is caused by *Uromyces* species; pine blister rust

Figure 12.15

Diagram of an Amanita *mushroom, showing annulus and volva. Recognizing members of this genus is important because they are among the most deadly and can be mistaken for edible forms.*

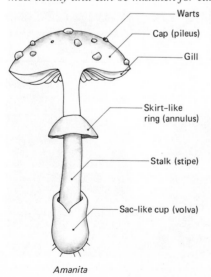

Amanita

Figure 12.16

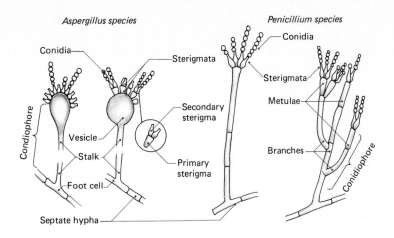

is caused by a *Cronartium* species. The smuts occur within the order Ustilaginales. The smut fungi also cause serious economic losses in agriculture. Members of the genus *Ustilago* cause smut of corn, wheat, and other plants; *Tilletia* species cause stinking smut of wheat; *Sphacelotheca* species cause loose smut of sorghum; and *Urocystis* species cause smut of onion. These are but a few examples of the common plant diseases caused by smut fungi. Many of the diseases caused by rusts and smuts will be considered in Chapter 22 when we discuss agricultural microbiology and diseases of crops.

Deuteromycotina

The final subdivision of the fungi is the **Deuteromycotina**. Members of this subdivision are known as the **Deuteromycetes** or **Fungi Imperfecti**. There are about 15,000 species in the Fungi Imperfecti.

The vegetative structures of most of the fungi in this class resemble Ascomycetes, although the vegetative structures of a few members of the Fungi Imperfecti resemble Basidiomycetes. Sexual forms of reproduction in the Deuteromycetes, however, do not occur or have not been detected, and it is the lack of observed sexual spores that places fungal species into this group.

Without observation of the sexual reproductive stage, it is impossible to place members of the Deuteromycetes into either the Ascomycetes or Basidiomycetes. Many Fungi Imperfecti, though, are obviously closely related to the ascomycetes or basidiomycetes. Whereas the sexual stages of several fungi have now been detected, a wholesale reclassification of fungi that have traditionally been placed in the subdivision Deuteromycetes has fortunately been avoided. For example, the sexual or perfect stages for several members of

table 12.6

Some representative genera of Deuteromycetes

Alternaria   Soil saprophytes and plant pathogens, muriform spores fit together like bricks of a wall

Arthrobotrys   Soil saprophytes, some form organelles for capture of nematodes

Aspergillus   Common molds, radially arranged, colored, often black, conidiospores

Aureobasidium (Pullularia)   Short mycelial filaments, lateral blastospores; often damage painted surfaces

Candida   Common yeast, some cause mycoses; some species able to grow in concentrated sugar solutions; some species able to grow on hydrocarbons

Coccidioides   *C. imminitis* causes mycotic infections in humans and animals

Cryptococcus   Yeasts, saprophytic in soil, but some may cause mycoses in animals and humans

Geotrichum   Common soil fungus; older mycelial filaments break up into arthrospores

Helminthosporium   Cylindrical, multiseptate spores; many are economically significant plant pathogens

Penicillium   Common mold with colored, often green, conidiospores arranged in brush shape

Trichoderma   Common soil saprophyte with highly branched conidiophores

the genera *Penicillium* and *Aspergillus* have been found to involve the production of ascospores, but the members of these genera are still classified among the Fungi Imperfecti. It is likely, though, that many genera of Deuteromycetes will eventually be reclassified into other subdivisions.

The Fungi Imperfecti are classified largely on the basis of the morphological structure of the vegetative phase and on the types of asexual spores produced (Figure 12.16). The Deuteromycetes include many important genera of filamentous fungi,

such as *Penicillium*, and yeasts, such as *Candida*. Representative genera of the Deuteromycetes are listed in Table 12.6. Some Deuteromycetes cause diseases in plants and animals. For example, *Candida albicans* can cause serious human infections. Some species of Fungi Imperfecti are important in food and industrial microbiology. The antibiotic penicillin, for example, is produced by *Penicillium* species, and both *Aspergillus* and *Penicillium* species are used in the production of various foods, such as blue cheese and soy sauce.

Classification of algae

Whereas the fungi are nonphotosynthetic, the **algae are eukaryotic photosynthetic organisms**. As such, they are differentiated from all other microorganisms. The algae are separated from the plants by their lack of tissue differentiation. In Whittaker's five-kingdom classification system some of the algae are placed in the kingdom Protista along with the protozoa, and other algae exhibiting more extensive organizational development are placed in the kingdom Plantae. Indeed, some organisms that are classified as algae are borderline cases with higher plants, and others are borderline cases with protozoa. There are some algae that lose their ability to carry out photosynthetic metabolism, rendering them indistinguishable from the

protozoa. Some motile unicellular algae have traditionally been studied by both protozoologists and phycologists. This has led to an inevitable confusion in the literature because zoologists and botanists typically use different features and criteria for establishing classification systems. Most traditional algal classification systems include the blue-green algae, but as was discussed in Chapter 11, these organisms are properly considered as cyanobacteria because of their prokaryotic cells. The reclassification of the "blue-greens" as cyanobacteria is still considered controversial and is opposed by many phycologists.

The algae are classified into seven major divisions based largely on the types of photosynthetic

table 12.7

The major divisions of algae

Chlorophycophyta (green algae). Photosynthetic pigments: chlorophylls *a* and *b*, carotenes, several xanthophylls. Storage product: starch. Cell wall: cellulose, xylans, mannans, absent in some, calcified in some. Flagella: 1, 2–8, many, equal, apical.

Chrysophycophyta (golden and yellow-green algae, including diatoms). Photosynthetic pigments: chlorophylls *a* and *c*, carotenes, fucoxanthin and several other xanthophylls. Storage product: chrysolaminaran. Cell wall: cellulose, silica, calcium carbonate. Flagella: 1–2, unequal or equal, apical.

Cryptophycophyta (cryptomonads). Photosynthetic pigments: chlorophylls *a* and *c*, carotenes, xanthophylls (alloxanthin, crocoxanthin, monadoxanthin), phycobilins. Storage product: starch. Cell wall: absent. Flagella: 2, unequal, subapical.

Euglenophycophyta (euglenoids). Photosynthetic pigments: chlorophylls *a* and *b*, carotenes, several xanthophylls. Storage product: paramylon. Cell wall: absent. Flagella: 1–3, apical, subapical.

Phaeophycophyta (brown algae). Photosynthetic pigments: chlorophylls *a* and *c*, carotenes, fucoxanthin and several other xanthophylls. Storage product: laminaran. Cell wall: cellulose, alginic acid, sulfated mucopolysaccharides. Flagella: 2, unequal, lateral.

Pyrrophycophyta (dinoflagellates). Photosynthetic pigments: chlorophylls *a* and *c*, carotenes, several xanthophylls. Storage product: starch. Cell wall: cellulose or absent. Flagella: 2, one trailing, one girdling.

Rhodophycophyta (red algae). Photosynthetic pigments: chlorophyll *a* (also *d* in some), phycocyanin, phycoerythrin, carotenes, several xanthophylls. Storage product: Floridean starch. Cell wall: cellulose, xylans, galactans. Flagella: absent.

pigments that are produced, the types of reserve materials that are stored intracellularly, and the morphological characteristics of the cell (Table 12.7). The relative concentrations of photosynthetic pigments give the algae characteristic colors. Many of the major algal divisions have common names based on these characteristic colors, such as the green algae, red algae, and brown algae.

Chlorophycophyta

The **Chlorophycophyta**, or **green algae**, are widely distributed in aquatic ecosystems. The cellular organization in different species of the Chlorophycophyta may be unicellular, colonial, or filamentous. Most cells are uninucleate, but several orders of green algae are characterized by the formation of coenocytic filaments. The chloroplasts of many unicellular green algae contain a pigmented region known as the stigma or red eyespot. Some green algae contain contractile vacuoles that serve an osmoregulatory function, protecting the cell against osmotic shock. The Chlorophycophyta normally store starch as a reserve material. The cell walls of different species of Chlorophycophyta are composed of cellulose, mannans, or xylans, but a high proportion of the cell wall also may be composed of protein.

Unicellular and colonial green algae

The green algae can be divided into a number of orders. The order Volvocales includes the unicellular green algae, which are normally motile by means of flagella. Several well-known genera of algae occur in this order, including *Chlamydomonas* and *Volvox*. The genus *Chlamydomonas* contains several hundred species. Members of this genus are unicellular and biflagellate.

Members of the genus *Volvox* form sphaeroidal colonies (Figure 12.17). The colonies contain many small vegetative cells and relatively few re-

Figure 12.17

Colony of Volvox aureus. *(Courtesy Carolina Biological Supply Co.)*

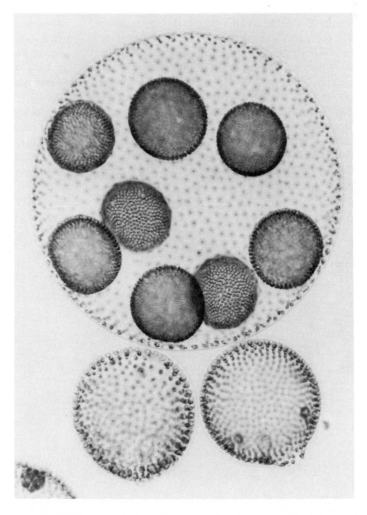

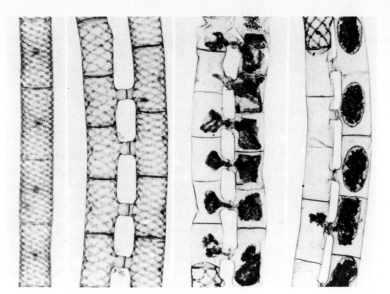

Figure 12.18

Micrograph of Spirogyra *in conjugation. (Courtesy Carolina Biological Supply Co.)*

productive cells. The reproductive cells lack flagella and are called gonidia. The cells within a colony of *Volvox* act in a cooperative fashion so that the entire colony behaves as a "superorganism." The flagella of the vegetative cells face outward and are able to move the entire colony in a unified manner. The colonies act as reproductive individuals, some colonies producing male gametes and others producing female gametes. The colonies of some species thus exhibit sexual differentiation. Reproduction is dependent on the intact colony. The colonies of *Volvox* approach the level of tissue differentiation. It would appear that algae having such complex levels of organization represent an evolutionary link between

microorganisms and higher plants. Based on ultrastructural analyses of the microtubules involved in mitotic division, however, *Volvox* does not appear to represent the missing evolutionary link between the green algae and green plants.

Filamentous green algae

There are several filamentous types of green algae, including members of the genera *Ulothrix* and *Spirogyra*. Probably the best-known genus of green algae, *Spirogyra*, occurs in the order Zygnemales. *Spirogyra* is an example of a filamentous green alga. The walls of the filament are continuous. The chloroplasts of *Spirogyra* form a spiral within the filaments (Figure 12.18). Another well-

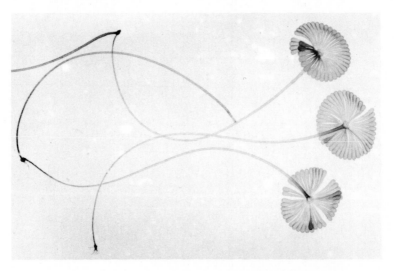

Figure 12.19

Photograph of Acetabularia, *showing its delicate shape. (Courtesy Carolina Biological Supply Co.)*

Figure 12.20

Scanning electron micrograph of a Trachelomonas *sp.*, a freshwater euglenoid flagellate. Based on morphological appearance, these euglenoid organisms have been classified by both phycologists and protozoologists (2000×). (From BPS—G. A. Antipa and E. B. Small.)

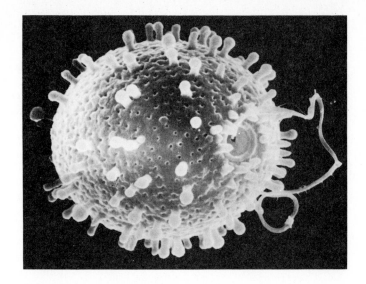

known algal genus, *Ulva*, occurs in the order Ulvales. Species of *Ulva*, commonly known as sea lettuce, are restricted to marine habitats. *Ulva* grows in marine habitats attached to rocks and other surfaces. Another marine form, the genus *Acetabularia,* is a tubular green alga (Figure 12.19). This organism is known as the mermaid's wine goblet. There are several nonmotile coccoid types of green algae. These include members of the genera *Chlorococcum* and *Chlorella*, which are unicellular types, and members of the genus *Scenedesmus*, which form colonies of 4–8 laterally united cells.

Euglenophycophyta

The **Euglenophycophyta** are similar to the Chlorophycophyta in that they contain chlorophylls *a* and *b* and typically appear green in color. The Euglenophycophyta differ from the Chlorophycophyta with respect to their cellular organization, and their intracellular reserve storage products. The Euglenophycophyta are unicellular. They lack a cell wall but normally are surrounded by an outer layer, known as a pellicle, composed of lipid and protein. The Euglenophycophyta do not store starch but rather store paramylon, a β-1,3-glucose polymer. The Euglenophycophyta appear to be closely related to the protozoa (Figure 12.20). Members of this division that lose their photosynthetic apparatus are indistinguishable from protozoa. Reproduction in the Euglenophycophyta is normally by longitudinal division. These algae are widely distributed in aquatic and soil habitats. *Euglena* is the best-known genus of Euglenophycophyta (Figure 12.21). *Euglena* species have two flagella for locomotion and normally contain a contractile vacuole to protect them against osmotic shock.

Figure 12.21

Micrograph of Euglena, *Fl = flagellum (3600×). (Reprinted by permission of Springer-Verlag, New York, from R. G. Kessel and C. Y. Shih, 1976,* Scanning Electron Microscopy for Biology.*)*

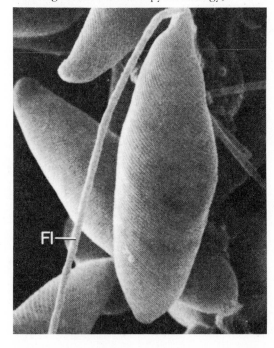

Chrysophycophyta

The division **Chrysophycophyta** includes the classes **Xanthophyceae** (yellow-green algae), **Chrysophyceae** (golden algae), and **Bacillariophyceae** (the diatoms). Chrysophycophyta species produce a diversity of pigment biochemicals, cell wall biochemicals, and cell types. The Chrysophycophyta are unified by the production of the same reserve storage material, chrysolaminarin. Chrysolaminarin is a β-linked polymer of glucose. Chrysophycophyta species produce carotenoid and xanthophyll pigments that tend to dominate over the chlorophyll pigments; this confers the golden-brown colored hues on members of this division. Most members of the Chrysophycophyta are unicellular, although some are colonial.

Xanthophyceae and Chrysophyceae

Genera in the family Xanthophyceae include *Botrydiopsis*, a unicellular form; *Tribonema*, a filamentous form; and *Vaucheria*, a coenocytic tubular form. *Vaucheria* is known as the water felt and is widely distributed in moist soils and aquatic habitats. The Xanthophyceae generally reproduce asexually by zoospores (motile spores) or aplanospores (nonmotile spores). Members of the class Chrysophyceae are typically motile by means of flagella. Asexual reproduction in the Chrysophyceae is by cell division, zoospores, or statospores (silicified cysts). Many members of the Chrysophyceae form siliceous or carbonaceous walls.

Bacillariophyceae (diatoms)

The Bacillariophyceae or diatoms produce distinctive cell walls known as **frustules**. There are approximately 200 genera in the Bacillariophyceae. The diatoms typically are golden brown in color. The frustules of diatoms, which are also known as **valves**, have two overlapping halves; the larger portion is referred to as the **epitheca** and the smaller as the **hypotheca** (Figure 12.22). The halves of the frustule fit together like a petri dish. The geometric appearance of diatoms renders them aesthetically attractive (Figure 12.23). Pennate diatoms have bilateral symmetry, whereas centric diatoms have radial symmetry. Some diatoms are benthic, living at the bottom of aquatic ecosystems at the sediment layer, and other diatoms are planktonic, living suspended in open water bodies. The growth of diatoms is dependent on the concentrations of available silica because the cell walls of diatoms are impregnated with silica. Holes in the silica walls, called puntae, allow exchange of nutrients and metabolic wastes between the cell and its surroundings. The frustules of diatoms are resistant to natural degradation and accumulate over geologic periods. As a result diatoms are preserved in fossil records dating back to the Cretaceous period 65 million years ago. There are significant deposits of diatom frus-

430

Classification and
identification of
microorganisms:
systematics of
fungi, algae,
protozoa, and
viruses

Figure 12.22

Diagrams of diatom frustules. Cellular organization of Pinnularia*: (A) surface view of frustule—note raphe and central and polar nodules; (B) living cell of* Pinnularia *in girdle view—note massive chloroplasts, central nucleus, and overlapping valves; (C) girdle view of recently divided cell. (D) During the division process of a diatom, the progeny synthesize the inner frustula, leading to the progressive size decrease of one cell line.*

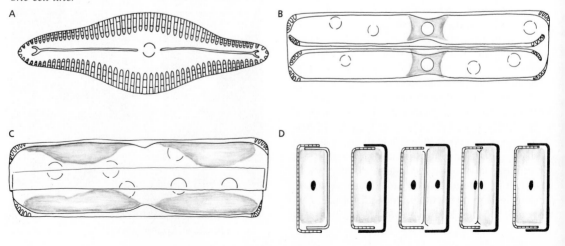

Figure 12.23

Micrographs of a few diatoms. The distinctive morphologies of the diatoms make their identification possible based only on microscopic observation. (A) Fragilaria, *a colonial diatom (480×). (B) Front and side views of* Cymbella, *a stalked diatom (570×). (From BPS—J. Robert Waaland, University of Washington.)*

A

B

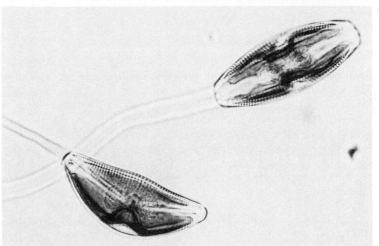

tules in the world. Such deposits are known as diatomaceous earth and are mined for a number of commercial uses. Diatomaceous earth is sometimes used as an abrasive in toothpaste and metal polish. The most extensive use of diatomaceous earth is in the filtration of liquids, especially those from sugar refineries.

Reproduction in diatoms is normally by the formation of uneven-size cells. Asexual reproduction involves the synthesis of a new cell wall structure, where each daughter cell reconstructs the smaller half of the frustule, regardless of which half of the parent frustule it receives. Therefore, continued asexual reproduction tends to result in diatoms of progressively smaller size. Occasionally, environmental conditions or the severe reduction in cell size leads to sexual reproduction with the production of **auxospores**, which are larger cells that act to reestablish larger-size diatoms. The valves of some diatoms have an opening along the apical axis known as the **raphe** (Figure 12.24).

Figure 12.24

Micrograph showing raphe structure of the diatom Navicula *(2300×). (Reprinted by permission of Springer-Verlag, New York, from R. G. Kessel and C. Y. Shih, 1976,* Scanning Electron Microscopy for Biology.*)*

Figure 12.25

432

Classification and identification of microorganisms: systematics of fungi, algae, protozoa, and viruses

Diatoms that have a raphe exhibit gliding motility, with the direction of movement depending on the shape of the raphe. For example, owing to differences in the shapes of their raphes, *Navicula* species exhibit straight movement, and those of *Nitzschia* exhibit curved movements. The gliding motility of diatoms permits these organisms to exhibit phototaxis, allowing them to move toward or away from light.

Pyrrophycophyta

The **Pyrrophycophyta** or **fire algae** are generally brown or red in color because of the presence of xanthophyll pigments. The Pyrrophycophyta are unicellular, biflagellate, and store starch or oils as their reserve material. The **dinoflagellates**, which are members of the Pyrrophycophyta, are characterized by the presence of a transverse groove that divides the cell into two semicells (Figure 12.25). The two flagella of the dinoflagellates emerge from an opening in the groove. Reproduction in the Pyrrophycophyta is primarily by cell division. The cell walls of Pyrrophycophyta contain cellulose and sometimes form structured plates, called **theca**. Because of their cell wall structures, some dinoflagellates that produce thecal plates are referred to as armored dinoflagellates (Figure 12.26).

Several dinoflagellates exhibit **bioluminescence**, the characteristic on which the designation fire algae is based. Some species also exhibit regular 24-hour behavioral patterns, known as circadian rhythms. For example, *Gonyaulax polyedra* exhibits a cyclic expression of luminescence with peak luminescence occurring in the middle of the dark period. The luminescent capacity of *Gonyaulax polyedra* allows this organism to glow at night. The glow rhythm is associated with a nightly increase in the level of luciferin and luciferase, the same enzyme substrate system that is operative in fire flies. *Gonyaulax polyedra* also exhibits circadian rhythms with respect to its photosynthetic activities and cell division, with maximal cell divisions occurring at dawn and maximal photosynthetic activities occurring at midday.

Species of *Gonyaulax* and other dinoflagellates are extremely important because they produce the toxic blooms known as **red tides** that

Figure 12.26

Diagrams of armored dinoflagellates, so named because of their thecal plates.

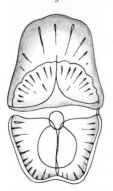

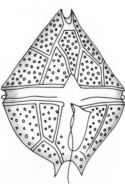

Gymnodinium   *Peridinium*

tend to color water red or red-brown in the vicinity of the bloom. The toxins of dinoflagellates during such blooms may kill invertebrate organisms. Although the blooms kill relatively few marine organisms, their toxins are concentrated in the tissues of filter-feeding molluscs, such as clams and oysters. Ingestion of shellfish containing dinoflagellates results in paralytic shellfish poisoning. This is a serious form of food poisoning. In order to prevent outbreaks of paralytic shellfish poisoning in humans, the collection of shellfish is banned during outbreaks of red tide.

Some dinoflagellates tend to enter into mutually beneficial (symbiotic) relationships with a variety of marine invertebrates. Such associations are termed **zooxanthellae**. Within such associations the animal cell provides protection and carbon dioxide for photosynthesis for the dinoflagellates, and the algae provide the animal with oxygen and organic carbon for its nutritional needs. Often dinoflagellates grow on ingested bacteria and other algal species. As such, dinoflagellates are mixotrophic, capable of both heterogeneous and photoautotrophic metabolism. Some microbiologists hypothesize that such symbiotic relationships are responsible for the evolution of higher organisms.

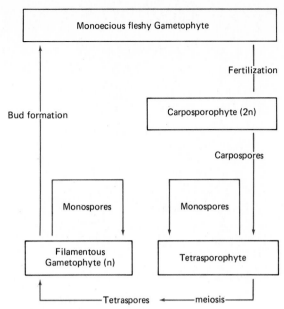

Figure 12.27

The reproductive cycle in the red algae, Pseudogloiophloea confusa. *Meiosis occurs during formation of the tetraspores.*

Cryptophycophyta

The **Cryptophycophyta** are a small group of unicellular algae that normally reproduce by longitudinal cell division, typically producing two flagella of equal length. These algae normally appear brown in color. The cryptomonads have asymmetric cells that are flattened and bounded by an outer covering called the **periplast**. Representative genera of this group are *Cryptomonas* and *Chroomonas*.

Rhodophycophyta

The **Rhodophycophyta** or **red algae** contain **phycocyanin** and **phycoerythrin** in addition to chlorophyll pigments. The red color of the Rhodophycophyta is due to the phycoerythrin. The primary reserve material in the Rhodophycophyta is **Floridean starch**, a polysaccharide similar to amylopectin found in higher plants. Rhodophycophyta exhibit a specialized type of oogamous sexual reproduction involving specialized female cells called **carpogonia** and specialized male cells

called **spermatia** (Figure 12.27). Tetraspores are formed during the life cycle of some red algae, and the tetraspores eventually differentiate into the male and female gametes. *Nemalion, Callithamnion, Delesserica, Anthithamnion, Callophyllis,* and *Porphyridium* are representative genera of red algae (Figure 12.28). Most red algae occur in marine habitats. Red algae typically have a bilayered cell wall with an inner microfibrillar rigid layer and an outer mucilaginous layer. Various biochemicals, including agar and carrageenin, occur in the cell walls of red algae. The agar and carrageenin of red algae are widely used as thickening agents and binders in various food products. Agar is also used as a solidifying agent in culture media, upon which the cultivation of bacteria largely depends. The carrageenin of *Chondrus crispus* is utilized in puddings. The red alga *Porphyra* is cultivated and harvested by the Japanese as a source of food.

Phaeophycophyta

The division **Phaeophycophyta** or **brown algae** includes over 200 genera and 1,500 species. The Phaeophycophyta produce **xanthophylls** that dominate over the carotenoid and chlorophyll

A

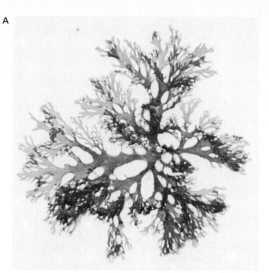

B

Figure 12.28

Micrographs of some red algae. (A) Callophyllis flabellulata. *(B)* Delesseria decipiens. *(From BPS—J. Robert Waaland, University of Washington.)*

434

Classification and
identification of
microorganisms:
systematics of
fungi, algae,
protozoa, and
viruses

Figure 12.29

Photograph of the kelp, Lamanaria saccharina.
(From BPS—J. Robert Waaland, University of Washington.)

pigments and impart a brown color to these organisms. The main reserve materials for the brown algae are laminarin and mannitol. The cell walls of the Phaeophycophyta are generally composed of two layers, an inner cellulosic layer and an outer mucilaginous layer. Alginic acid is normally found as a biochemical constituent of the cell wall. The Phaeophycophyta are almost exclusively marine organisms and are found primarily in coastal zones. The **kelps**, which are brown algae, can form macroscopic structures up to 50 meters in length (Figure 12.29). It is difficult to consider organisms of that size as members of the microbial world. Most kelps have vegetative structures consisting of a holdfast, stem, and blade (Figure 12.30). These are histologically complex organisms that exhibit a degree of cellular differentiation. The genera *Fucus* and *Sargassum* are important representatives of the Phaeophycophyta. Large populations of *Sargassum natans* occur in the Atlantic Ocean in the region known as the Sargasso Sea. Species of *Fucus* commonly occur along rocky shores, attached to the rocks by disk-like holdfasts. These brown algae clearly are the most complex organisms classified as algae or for that matter as microorganisms, representing a borderline case between algae and plants. Green plants, however, do not appear to have evolved directly from brown algae. Rather, parallel evolution appears to have occurred in which organisms in different evolutionary lines developed similar adaptive organized structures.

Figure 12.30

Drawings showing the complex structure of brown algae. (A) Postelsia palmaeformis, *the sea palm. (B)* Laminaria agardhii.

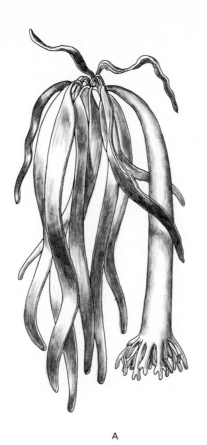

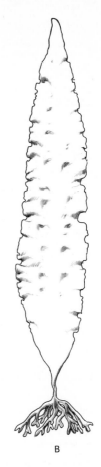

A

B

Classification of protozoa

Protozoa occur in the kingdom Protista. This subkingdom includes over 65,000 species. According to the newly revised 1980 classification of the protozoa reported by the Committee on Systematics and Evolution of the Society of Protozoologists, the protozoa, which are essentially unicellular organisms, are divided into seven phyla (Table 12.8). This new classification system departs from earlier classification systems that recognized four phyla primarily based on their means of motility: the **Mastigophora (flagellates)**; **Sarcodina (pseudopodia formers)**; **Ciliophora (ciliates)**; and **Sporozoa (spore formers)**. The new system encompasses the classification of several groups claimed by other disciplines, including most of the algae and the lower fungi, but for purposes of overall classification of microorganisms within this text, the slime molds are still considered as fungi and photosynthetic microorganisms as algae. There are several noteworthy aspects to the new protozoa classification system. First, the Sarcodina and the Mastigophora are included in one phylum, the Sarcomastigophora. Within this phylum the Mastigophora and Sarcodina are treated as separate subphyla. The other major change in this classification system is the division of the Sporozoa into four separate phyla: the Apicomplexa, Microspora, Acetospora, and Myxospora. The last three have spores, but many members of the Apicomplexa do not; the Apicomplexa have an apical complex visible by electron microscopy.

Sarcomastigophora

Sarcodina

The **Sarcodina** are motile by means of pseudopodia. **Pseudopodia** are cytoplasmic extensions, sometimes referred to as false feet or rhizopods.

table 12.8

The classification of the protozoa

Old system:

Ciliiphora. Locomotion: cilia. Reproduction: asexual, transverse fission; sexual, conjugation. Nutrition: ingestive.

Mastigophora. Locomotion: usually paired flagella. Reproduction: asexual, longitudinal fission. Nutrition: heterotrophic, absorptive.

Sarcodina. Locomotion: pseudopodia (false feet). Reproduction: asexual, binary fission. Nutrition: phagocytic.

Sporozoa. Locomotion: usually none, some stages with flagella. Reproduction: asexual, multiple fission; sexual, within host; spores formed. Nutrition: absorptive.

New system:

Sarcomastigophora. Locomotion: flagella, pseudopodia, or both. Reproduction: Sexually, when present, essentially syngamy. Representative genera: *Monosiga, Bodo, Leishmania, Trypanosoma, Giardia, Opalina, Amoeba, Entamoeba, Difflugia.*

Labyrinthomorpha. Synonymous with the net slime molds. Produce ectoplasmic network with spindle-shaped or spherical nonameboid cells; in some genera ameboid cells move within network by gliding.

Apicomplexa. Produce apical complex visible with electron microscope; all species parasitic. Representative genera: *Eimeria, Toxoplasma, Babesia, Theileria.*

Microspora. Unicellular spores, each with imperforate wall; obligate intracellular parasites. Representative genus: *Metchnikovella.*

Ascetospora. Spore multicellular; no polar capsules or polar filaments; all species parasitic. Representative genus: *Paramyxa.*

Myxospora. Spores of multicellular origin with one or more polar capsules; all species parasitic. Representative genera: *Myxidium, Kudoa.*

Ciliophora. Cilia produced at some stages in life cycle. Reproduce by binary transverse fission, budding and multiple fission also occur. Sexuality involving conjugation, autogamy, and cytogamy; most free-living heterotrophs. Representative genera: *Didinium, Tetrahymena, Paramecium, Stentor.*

436

Classification and identification of microorganisms: systematics of fungi, algae, protozoa, and viruses

The false feet may occur in a variety of forms, including extensions of the ectoplasm that include flow of endoplasm (**lobopodia**); filamentous projections composed entirely of ectoplasm (**philopodia**); filamentous projections with branching (**rhizopodia**); and axial rods within a cytoplasmic envelope (**axopodia**) (Figure 12.31). The pseudopodia are used for engulfing and ingesting food as well as for locomotion. Members of the Sarcodina may move at rates of 2–3 cm per hour under optimal conditions. Some members of the Sarcodina are important because they cause human diseases. For example, *Entamoeba histolytica* causes amebic dysentery, a serious debilitating human disease.

The **Amoebida** form lobopodia and lack a skeletal structure. Members of the genus *Amoeba* have no distinct shape because the flow of cytoplasm continuously changes the shape of true ameba. The giant ameba, *Amoeba proteus*, is normally about 250 μm in length (Figure 12.32). This or-

ganism is readily found and visualized microscopically in pondwater samples. *Amoeba* feed on a number of smaller organisms, including bacteria and other protozoa. For example, *Amoeba proteus* can ingest the protozoa *Tetrahymena* and *Paramecium.*

The **Heliozoida** typically produce numerous radiating axopodia. The heliozoans are freshwater forms. The **Radiolaria**, like the heliozoans, are members of the Actinopodea. Radiolarians typically have axopodia, with a skeleton of silicon or strontium sulfate (Figure 12.33). The radiolarians occur in marine ecosystems. The silica-containing exoskeletons of the radiolarians are quite attractive when viewed microscopically.

The **Foraminiferida** are also marine members of the Sarcodina. Foraminiferans form one or many chambers composed of siliceous or calcareous tests (Figure 12.34). A **test** is a skeletal or shell-like structure. The tests of the Foraminferida accumulate in marine sediments and are preserved in the geological record. The white cliffs of Dover are composed largely of the test structures of foraminiferans. Many of the foraminiferans are recognized in fossil records, whereas there is no fossil record for many microorganisms.

Mastigophora

The Mastigophora are the flagellate protozoa. Because some members of the Mastigophora are able to produce pseudopodia in addition to possessing flagella, these organisms are now classified together with the Sarcodina. It is in this subphylum that protozoologists place the dinoflagellates, euglenoids, and other algae. The members of the Mastigophora here recognized as protozoa are heterotrophic and occur within the class zoomastigophorea. Many members of the Mastigophora are plant and animal parasites. The genera *Trypanosoma* and *Leishmania* contain species that produce serious human diseases (Figure 12.35). *Trypanosoma gambiense*, for example, causes African sleeping sickness, and *T. cruzi* is the causative agent of Chagas' disease. Infections with the flagellate protozoan *Giardia* can cause severe diarrhea. *Leishmania donovani* is the causative agent for Kala-azar disease, also known as dum dum fever. Human diseases caused by flagellate protozoa are normally transmitted by arthropods, and control of many of these diseases rests with controlling the carrier rather than eliminating the disease-causing protozoa.

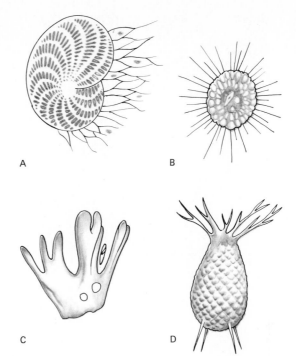

A   B

C   D

Figure 12.31

Illustrations of members of the Sarcodina *with different types of false feet. (A)* Elphidium crispa *with rhizopodia. (B)* Actinosbaerium eichhorni *with axopodia. (C)* Chaos carolinensis *with lobopodia. (D)* Euglypha alveolata *with filopodia.*

Figure 12.32

Micrographs of Amoeba, *showing extensions of pseudopodia. (A)* Amoeba proteus. *(Courtesy Robert Apkarian, University of Louisville.) (B)* Pelomyxa carolinensis, Lo = lobopodia (80×). *(Reprinted by permission of Springer-Verlag, New York, from R. G. Kessel, and C. Y. Shih, 1976,* Scanning Electron Microscopy for Biology.*)*

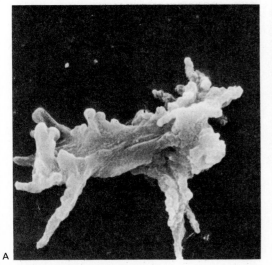

A

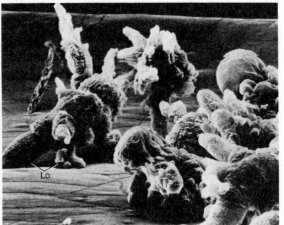

B

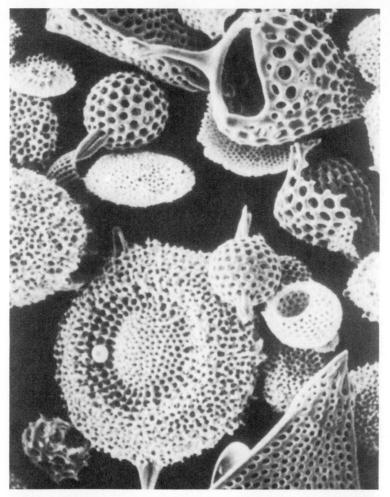

Figure 12.33

Micrograph of radiolarian shells (400×). (Reprinted by permission of Springer-Verlag, New York, from R. G. Kessel and C. Y. Shih, 1976, Scanning Electron Microscopy for Biology.*)*

Figure 12.34

A larger foraminiferan, Heterostegina depressa, *showing the chambers of the test (test size = 2 mm). The calcareous test consists of many chambers coiled in a plane. Each chamber is subdivided into chamberlets. As in all larger foraminferans, it harbors unicellular symbiotic algae in its protoplasm, which give a yellow coloration to the living specimen. These larger foraminiferans occur in tropical and subtropical shallow seas. (Courtesy Rudolf Rottger, Universität Kiel.)*

Ciliophora

Ciliophora are motile by means of cilia. Some members of the Ciliophora have a mouth-like region known as a **cytostome** (Figure 12.36). In addition to their role in locomotion, the cilia move food particles into the cytostome. The ciliate protozoa reproduce by various asexual and sexual means. Asexual reproduction is often by binary fission, and sexual reproduction is usually by conjugation. The Ciliophora normally contain two nuclei, a macronucleus and a micronucleus, both of which are diploid. The macronucleus is involved in asexual reproduction and the micronucleus in sexual reproduction. *Paramecium* is perhaps the best-known genus of ciliate protozoa. Other genera of Ciliophora include *Stentor, Vorticella, Tetrahymena,* and *Didinium.* Ciliate protozoa consume other microorganisms, including other protozoa as their food source. For example, members of the genus *Didinium* can consume *Paramecium* species, providing a dramatic picture of the microbial world when viewed by scanning electron microscopy (Figure 12.37).

Sporozoa

The **Sporozoa**—a group comprising four phyla—are parasites, exhibiting complex life cycles. The adult forms are nonmotile, but immature forms

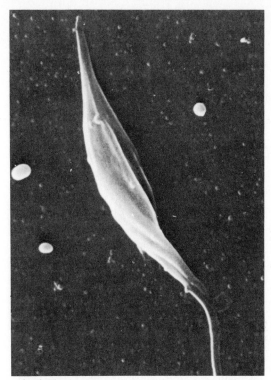

Figure 12.35

Micrograph of Trypanosoma cruzi, *the causative agent of Chaga's disease. (From BPS—Stephen Baum, Albert Einstein College of Medicine, New York.)*

Figure 12.36

Scanning electron micrograph of the ciliate protozoan Paramecium *showing the cytostome region. Note ciliary rows along the dorsal side in top photo. The bottom picture shows the ventral side and the vestibular opening into the oral area. (Courtesy Eugene W. McArdle, Northeastern Illinois University.)*

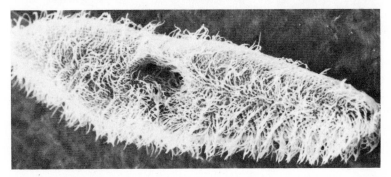

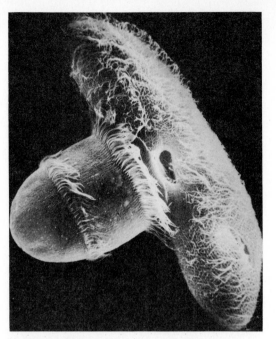

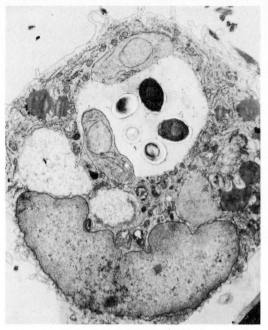

440

Classification and
identification of
microorganisms:
systematics of
fungi, algae,
protozoa, and
viruses

Figure 12.37

Micrograph of one ciliate protozoan, Didinium
nasutum, *consuming another ciliate,* Paramecium
multimicronucleatum, *at an early stage of the
ingestion (500 ×). (From BPS—H. S. Wessenberg
and G. A. Antipa, 1970,* Journal of Protozoology
17:250–270.)

Figure 12.38

The protozoan Encephalitozoon cuniculi
*(Microsporida) developing within the vacuole of
macrophage. The vacuole contains spores and
earlier stages, possibly sporonts, the latter adhering
to the host cell cytoplasm. (Courtesy V. Sprague
and S. H. Vernick, reprinted by permission of
Society of Protozoologists, from V. Sprague and S.
H. Vernick, 1971,* Journal of Protozoology *18:562.)*

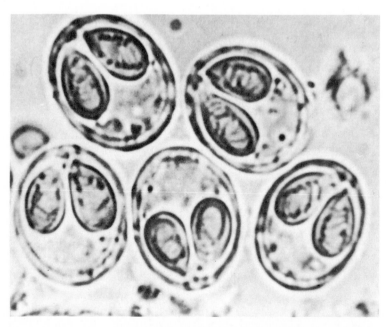

Figure 12.39

Micrograph of the spores of
Myxosoma *growing in the
tissues of the bluegill (3200 ×).
(Reprinted by permission of the
editor from J. Lom and G. L.
Hoffman, 1971,* Journal of
Parasitology *57:1307.)*

Figure 12.40

Diagram of life cycle of the sporozoan, Monocystis. Monocystis *lives in the common earthworm. (A) Spores are eaten by the earthworm and sporozoites are released. (B) Young trophozoite in sperm morula. (C) Trophozoite grows into worm, earthworm sperm develop from morula cells and attach to trophozoite. (D) Trophozoites associate in pairs and form a cyst wall (gametocytes). (E) Nuclei of gametocytes divide. (F) Gametes are produced. (G) Gametes fuse, i.e., fertilization occurs, to form zygotes. (H) Zygotes secrete spore walls, sporocysts. (I) Nuclei divide to form eight sporozoites in each spore. (J) Spores liberated when earthworm dies. Spores may also be transferred to another worm during intercourse and infect testes of other worms. They may also be transferred to worm cocoon, entering the ground when the cocoon disintegrates.*

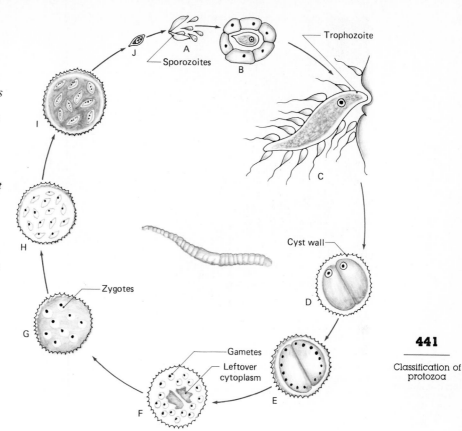

and gametes may be motile. The sporozoans derive nutrition by absorption of nutrients from the host cells they inhabit (Figure 12.38). The immature stages are referred to as **sporozoites** (Figure 12.39). Reproduction of the **trophozoite**, the adult stage of a Sporozoan, occurs asexually by multiple fission (Figure 12.40). The cells produced by multiple fission eventually mature into gametes that can be involved in sexual reproduction. The multiple fission process can result in the production of thousands of spores. *Plasmodium vivax* and various other species of *Plasmodium* cause malaria. During the course of a malarial infection, life cycles of *Plasmodium* involve reproduction within human red blood cells, with the periodic release of large numbers of protozoa. As discussed earlier, members of the Apicomplexa, which include *Plasmodium* species, are characterized by the formation of an apical complex (Figure 12.41); it is interesting that such morphological observations, achievable only by using electron microscopy, are now employed for defining major taxonomic groupings.

The ultrastructural organelles of the sporozoans in the phylum Apicomplexa, as revealed by electron microscopy, include a **polar ring** within which there is a **conoid**. The conoid is an electron-dense structure composed of coiled filaments; this structure is characteristic of all members of this phylum except *Plasmodium, Babesia,* and *Thieleria.* Extending posteriorly from this area are paired tubular organelles called **rhoptries**. Elongate, electron-dense organelles known as **micronemes** are found in the anterior end of the protozoan and further characterize the morphology of the sporozoans. The combination of anterior organelles—the polar ring (usually with a conoid), rhoptries, and micronemes—is known as the **apical complex.** The apical complex appears to play a role in cell penetration, a most important function in the sporozoa because they are all parasitic, reproducing intra- or intercellularly.

The typical reproductive cycle of the sporozoa involves three phases: (1) The host is invaded by an infective stage known as a sporozoite; within

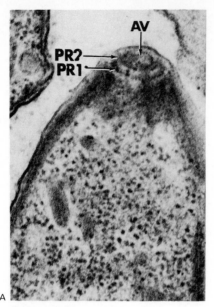

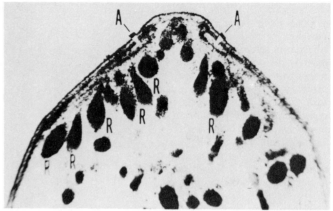

Figure 12.41

The apical structure of some members of the Apicomplexa. (A) Micrograph of Eimeria labbeana, *showing apical vesicles (AV) and preconoidal rings (PR1, PR2) (33,500×). (Courtesy T. Varghese, reprinted by permission of the Society of Protozoologists, from T. Varghese, 1975,* Journal of Protozoology *22:68.) (B) Transmission electron micrograph of the apical complex of* Babesia bigemina, *showing rhoptries, (R), arranged in rows alongside the "ribs" of the apical umbrella, A (45,000×). (Reprinted by permission of the Society of Protozoologists, from G. Weber, 1980,* Journal of Protozoology *27:65.)*

442

Classification and
identification of
microorganisms:
systematics of
fungi, algae,
protozoa, and
viruses

the host cell the sporozoite undergoes multiple fission to produce uninucleate individuals known as merozoites—the process of merozoite formation is called **merogony**. (2) The merozoites can continue to multiply by multiple fission; eventually the merozoites cease reproducing in this manner and become differentiated into sexual stages known as gametocytes—the process of gametocyte formation is known as **gametogeny**. (3) The gametocytes, or the gametes they produce,

fuse by syngamy to form a zygote; the zygotes form oocytes within which multiple fission occurs to produce sporozoites—this final process in the life history of a sporozoan is known as **sporogony**. This complex reproductive strategy of the sporozoa insures their survival; in particular, the involvement of multiple fission throughout the life history insures numerous infective individuals and contributes to the abundance of disease-causing sporozoans.

Classification of viruses

Viruses have generally been classified nonsystematically, the name of the virus based generally on the disease caused by that virus. Formal systems for the classification and nomenclature of viruses are relatively new, and most viruses are still referred to by their common names, as opposed to using a name assigned in a formal classification system. For example, we commonly refer to the measles virus, the poliovirus, and others, even

though these viruses have been classified and given other formal names. An International Committee on Nomenclature of Viruses is still working on establishing an acceptable formal classification system that can gain wide acceptance.

Viruses are often classified into large groups according to the type of host cell they infect. Three groups of viruses have usually been recognized: **animal viruses**, **plant viruses**, and **bacteriophage**.

Animal viruses traditionally refer to those viruses that infect only vertebrate animals. Viruses that infect invertebrate animals and microorganisms other than bacteria are not covered in classification systems that rely on these three lines of division. As a result separate taxonomic consideration has been given to the insect viruses and to the viruses that infect fungi, protozoa, and algae. Viral taxonomists working with viruses that infect different types of host cells have developed separate taxonomic schemes for classifying the viruses. Various subcommittees of the International Committee on Nomenclature of Viruses have been working to develop classification systems for each of these viral groups, with separate subcommittees developing criteria for classifying animal viruses, plant viruses, and viruses of microorganisms. The result is that a unified system has yet to be developed.

Formal classification systems are largely based on the nature of the nucleic acid molecule and the arrangement of the capsid (Table 12.9). Critical questions include whether the virus is a DNA or RNA virus, whether the nucleic acid is single-stranded or double-stranded, and whether the virus is naked or enveloped. Important features used in classifying viruses include the molecular weight of the nucleic acid, the symmetry of the capsid, the number of capsomeres, and the site of capsid assembly. Plant viruses are still grouped largely on the basis of the host cells infected by the virus. Nomenclature of animal viruses is based largely on molecular considerations. Similarly, bacteriophage are characterized by their molecular structure.

Viruses can be separated into groups on the basis of the type and form of the nucleic acid genome, and the size, shape, structure, and mode

table 12.9

The characteristics of various families of viruses

Viridae	Nucleic acid	Symmetry	Nucleocapsid	Virions with cubic symmetry			Virions with helical symmetry		Mol wt (× 10⁶) of nucleic acid	Number of nucleic acid strands
				Number of capsomers	Diameter of the nucleocapsid (nm)	Diameter of the envelope (nm)	Diameter and length of the nucleocapsid (nm)	Dimensions of the enveloped virions (nm)		
Ino-	D	H	N				0.5 × 85		1.7–3	1
Pox-	D	H?	E				?	250 × 160 300 × 230	160–240	2
Micro-	D	C	N	12	25				1.7	1
Parvo-	D	C	N	32	22				1.8	1
Denso-	D	C	N	42	20				160–240	2
Papilloma-(papova)	D	C	N	72	45–55				3–5	2
Adeno-	D	C	N	252	70				20–25	2
Irido-	D	C	N	812	130				126	2
Herpes-	D	C	E	162	77	150–200			54–92	2
Uro-	D	BC	N							2
Rhabdo-	R	H	N				2 × 13 1 × 125			1
Myxo-	R	H	E				9 × ?	100	2–3	1
Paramyxo-	R	H	E				18 × ?	120	7.5	1
Stomato-(rhabdo)	R	H	E				18 × ?	175 × 68	6	1
Thylaxo-	R	H	E				?	1000	10	1
Napo-	R	C	N	32	22–27				1.1–2	1
Reo-	R	C	N	92	70				10	2
Cyano-	R	C	N	32 or 42	54					2
Encephalo-	R	C	E	?	?		60–80		2–3	1

Code: D = DNA; R = RNA; H = helical; C = cubic; B = binal; N = naked; E = enveloped.
Based on A. Lwoff and P. Tournier, 1971, Remarks on the classification of viruses, in *Comparative Virology*, K. Maramorosch and E. Kurstak (eds.), Academic Press, New York.

of replication of the virus particle. Within each group, subgroupings can be based on immunological properties, the specific biochemicals (antigens) associated with the virus that evoke an immune response. Groups of viruses can be assigned at the genus and species level. In fact, the genus is probably the most common taxonomic level for grouping viruses. The fact that viruses exhibit great host-cell specificity probably means that at the genus and species level, viruses infect only one type of cell. Viruses are really genetic extensions of the cells within which they reproduce. As such, it is probably legitimate to consider the host cell in which the virus is capable of reproducing as a legitimate taxonomic criterion for classifying viruses.

The real problems in the classification of viruses come at the higher levels of organization. Viruses are not included in the recognized kingdoms of living organisms. In most cases the validity of genus-level designations has been recognized, but relatively few families of viruses have been recognized as being legitimate taxonomic groupings. Phylogenetic relationships among the viruses are unclear, and establishing a hierarchy of taxonomic relationships beyond the genus level

Figure 12.42

The families and genera of the DNA viruses of vertebrates. (Reprinted by permission of CRC Press, Boca Raton, Florida, from A. I. Laskin and H. A. Lechevalier, eds., 1978, Handbook of Microbiology.*)*

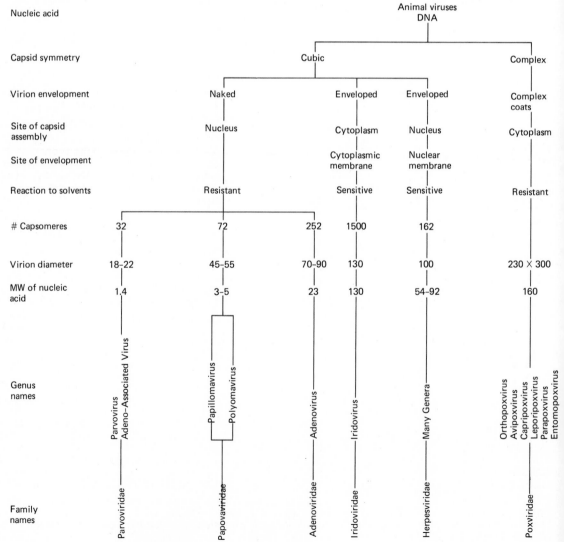

appears to add little conceptually to viral systematics. In the section below, brief descriptions of the major groups of viruses are given.

Animal viruses

DNA viruses

The DNA animal viruses are divided into six major families (Figure 12.42) following the criteria established by the International Committee on Taxonomy of Viruses.

Parvoviridae The members of this group are small viruses that contain single-stranded DNA. They have cubical symmetry with 32 capsomeres and have no envelope. Replication and capsid assembly takes place in the nucleus of the host cell. Members of the genus *Parvovirus* are resistant to heat inactivation and can withstand exposure to 60°C for 30 minutes. This group includes some interesting "defective" viruses that cannot multiply alone but can reproduce with the aid of another virus. For example, *Adeno-associated* satellite viruses can multiply only in the presence of replicating adenoviruses.

Papovaviridae The papovaviruses are small viruses containing double-stranded circular DNA. They have cubical symmetry with 72 capsomeres. Papovaviruses are assembled within the nucleus of the host cell. This group has been recognized in the viral family Papovaviridae and causes tumors in vertebrate animals. One genus included in this group is *Papillomavirus* (wart virus), which causes human warts. The other genus in this group, *Polyomavirus,* contains some viruses that cause tumor development.

Adenoviridae Adenoviruses are medium-size viruses containing double-stranded DNA. They exhibit cubic symmetry and have 252 capsomeres. Adenoviruses normally have spikes projecting from the capsid that give these viruses a characteristic shape (Figure 12.43). The spikes are involved in the adsorption of the virus to the host

Figure 12.43

(A) Micrograph of adenovirus type 18 within host cell (132,000×). (From BPS—C. Garon and J. Rose, National Institute of Health.) (B) Micrograph of an adenovirus, showing projecting spikes (Courtesy H. C. Pereira, reprinted by permission of Academic Press, from R. C. Valentine and H. C. Pereira, 1965, Journal of Molecular Biology *13:13–20.)*

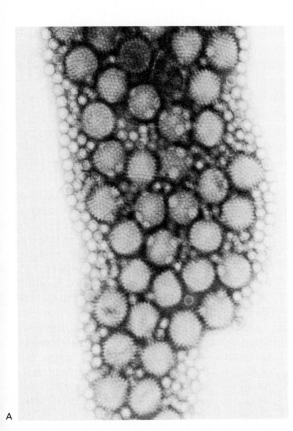

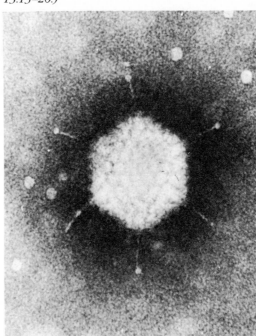

A

B

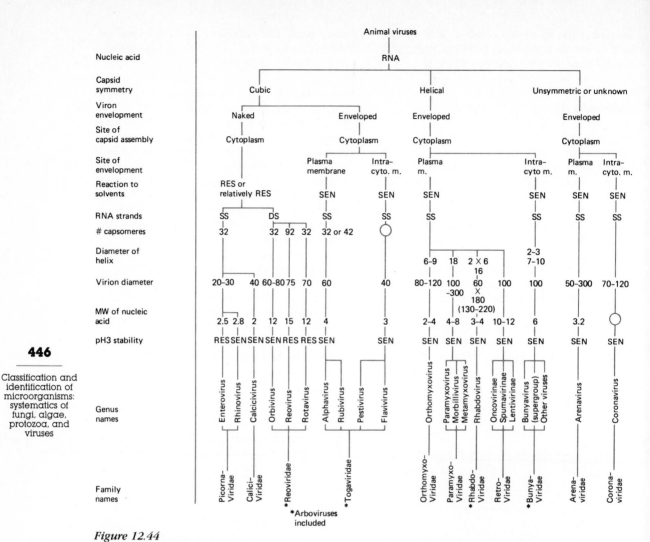

Figure 12.44

The families and genera of the RNA viruses of vertebrates. (Reprinted by permission of CRC Press, Boca Raton, Florida, from A. I. Laskin and H. A. Lechevalier, eds., 1978, Handbook of Microbiology.*)*

446

Classification and
identification of
microorganisms:
systematics of
fungi, algae,
protozoa, and
viruses

cell. The adenoviruses multiply within the nuclei of infected cells, where they produce an array of crystalline particles. Several viruses in this group are able to reproduce within human cells, causing disease. Members of the genus *Adenovirus*, for example, are associated with acute respiratory tract infections.

Iridoviridae   The iridoviruses may be divided into mammalian, fish-amphibian, and insect genera. Members of the genus *Iridovirus* contain double-stranded DNA. The virions are isometric, and some are enveloped. These are among the largest viruses.

Herpesviridae   The herpesviruses are medium-size viruses containing linear double-stranded DNA. The nucleocapsid is about 190 nm in diameter. The capsid has cubical symmetry with 162 capsomeres. Herpesviruses are assembled in the nucleus and are surrounded by a lipid envelope. They acquire membrane lipids by budding through the nuclear membrane. Herpesviruses establish latent infections within host animals that can last for the entire life of the host.

Herpesviruses, including *Herpes simplex* types 1 and 2, cause a number of diseases in humans. One form of venereal disease is caused by a strain of *Herpes simplex* virus, and there is some indication that sexually transmitted herpesviruses may also be associated with cervical cancer. The *Herpes simplex* viruses can cause inflammations of the mouth and gums, known as gingivostomatitis. The Epstein-Barr virus, the causative agent for infectious mononucleosis, and the Varicella-Zoster virus, the causative agent for chicken pox and shingles, are also included among the herpesvirus group.

Poxviridae Poxviruses are large viruses containing double-stranded DNA. Poxviruses may also include enzymes, such as RNA polymerase, within the viral particle. The inclusion of enzymes within the virus particle appears to be quite unusual. The poxviruses reproduce within the cytoplasm of host cells. Many poxviruses are human pathogens primarily reproducing within skin tissues. The formation of vesicular lesions in superficial body tissues is symptomatic of many diseases caused by poxviruses. Smallpox, cowpox, monkeypox, and fowlpox are examples of diseases caused by members of the poxvirus group.

RNA viruses
The RNA animal viruses are divided into 11 major families (Figure 12.44).

Picornaviridae The picornaviruses are small, single-stranded RNA viruses. The nucleocapsid has cubical symmetry and is nonencapsulated. The maturation of members of the picornavirus group occurs in the cytoplasm of the host cell. Two genera within the Picornaviridae, *Enterovirus* and *Rhinovirus*, have members that infect humans. The Enteroviruses include *Poliovirus, Echovirus*, and *Coxsackievirus*. Diseases caused by members of this group include poliomyelitis and foot-and-mouth disease. Picornaviruses also cause mild infections of the gastrointestinal and respiratory tracts.

Caliciviridae The members of the genus *Calicivirus* are isometric single-stranded RNA viruses. This group is closely related to the Picornaviridae and is sometimes included as part of that family. Diseases caused by members of this group include respiratory disease of cats and vesicular exanthema of swine and sea lions.

Reoviridae Member viruses contain segmented double-stranded RNA genomes. Three genera are included in this group: *Reovirus, Orbivirus*, and *Rotavirus*. Rotaviruses have been identified as an important causative agent of human infantile diarrhea. Several orbiviruses also cause important diseases that include bluetongue of sheep and cattle, epizootic hemmorrhagic disease of deer, African horse sickness, and human Colorado tick fever.

Togaviridae Many viruses in the family Togaviridae were formerly known as arboviruses because these viruses are transmitted by arthropod vectors. The members of the Togavirus group are single-stranded RNA viruses that have an enveloped viral particle. Some of the diseases caused by members of the Togavirus group include equine encephalitis, St. Louis encephalitis, dengue fever, and yellow fever.

Orthomyxoviridae The orthomyxoviruses are single-stranded RNA viruses that exhibit helical symmetry. The viral particles are contained within an envelope (Figure 12.45). Most orthomyxovirus

Figure 12.45

Micrograph of influenza A virus, strain USSR. (Courtesy E. Palmer, Centers for Disease Control, Atlanta.)

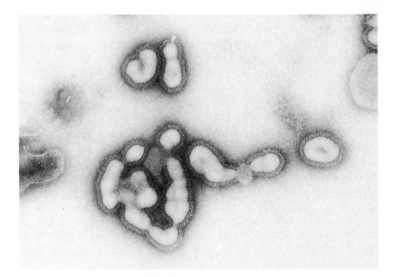

Figure 12.46

Micrograph of a rabies virus particle (330,000×). (From BPS—Alyne K. Harrison, Centers for Disease Control, Atlanta.)

particles have surface projections or spikes as part of their outer wall. The viruses that cause influenza are members of this group. Many influenza viruses are referred to by common names that indicate their geographic origins, such as Hong Kong flu virus.

Paramyxoviridae The paramyxoviruses are similar to the orthomyxoviruses but generally are somewhat larger. The paramyxovirus group has several viruses that cause diseases in humans, including mumps virus and measles virus.

Rhabdoviridae Rhabdoviruses are single-stranded RNA viruses that have a characteristic rod shape, resembling a bullet (Figure 12.46). Members of the rhabdovirus group are known to infect vertebrate animals, insects, and plants. Rabies is probably the best-known example of the diseases caused by a member of the rhabdovirus group.

Bunyaviridae Viruses in this family contain single-stranded RNA in three linear segments. The virions are formed by budding from organelles within the host cells, primarily by budding from intracytoplasmic membranes of the Golgi apparatus. Rift Valley fever and Nairobi sheep disease are among the diseases caused by bunyaviruses.

Arenaviridae The arenaviruses are single-stranded RNA viruses with the genome arranged in segments. Lassa fever is caused by members of the genus *Arenavirus*.

Coronaviridae Members of the genus *Coronavirus* are pleomorphic with bulbous projections from the membrane envelope that surrounds the virions. Acute respiratory disease and diarrheal diseases of humans and other animals are caused by viruses in this family.

Insect viruses

The insect viruses can be classified into a number of groups or genera. Some of these genera appear to be coincident with those assigned for viruses that infect vertebrate animal cells. For example, the entomopoxviruses are poxviruses of invertebrates. The insect viruses include the family Iridoviridae (Figure 12.47). These are very large double-stranded DNA viruses and include *Aedes iridescent* virus and *Chironomus iridescent* virus. It appears likely that some members of the iridovirus group are also able to infect vertebrate animal cells.

Plant viruses

The formal systematics of plant viruses has attempted to retain information concerning the infected host cells within the classification system. As a result plant viruses are still primarily grouped according to the type of disease they cause. The classification of plant viruses has tended to maintain smaller groups of viruses.

The recognized groups or genera of plant viruses include: *Cauliflower mosaic* virus (DNA virus of higher plants), *Tobravirus* (tobacco rattle virus), *Nepovirus* (tobacco ringspot virus), *Tobamovirus* (tobacco mosaic virus), *Tombusvirus* (tomato bushy stunt virus), *Tobacco necrosis* virus, *Bromovirus* (brome mosaic virus), *Cucumovirus* (cucumber mosaic virus), *Potyvirus* (potato virus Y), *Watermelon mosaic* virus, *Potexvirus* (potato virus X), *Tymovirus* (turnip yellow mosaic virus),

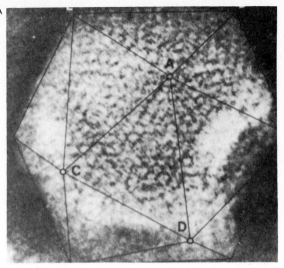

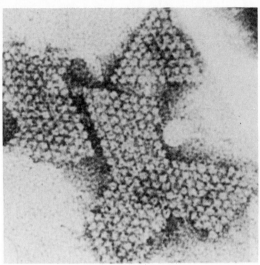

Figure 12.47

(A) The icosahedral structure of an iridovirus virion. (Reprinted by permission of the Society for General Microbiology, from N. G. Wrigley, 1969, Journal of General Virology 5: 123.) (B) The triangular faces of an iridovirus virion. (Reprinted by permission of the Society for General Microbiology, from N. G. Wrigley, 1970, Journal of General Virology 6: 169.)

Luteovirus (barley yellow dwarf virus), and *Ilarvirus* (necrotic ringspot). Some major properties of each of the plant virus groups are listed in Table 12.10. No attempt will be made here to further describe the groups of plant viruses. The names of the viral groups themselves are a key to the types of diseases caused by the viruses in that group, including the nature of the host plant and the symptoms of the diseases. It is obvious from this listing that the taxonomy of plant viruses has not yet established a unified systematic scheme comparable to the families of animal viruses.

table 12.10

The groups of plant viruses

Bromovirus (brome mosaic virus) Small, icosahedral RNA viruses	*Tobacco necrosis* virus Isometric RNA viruses
Cauliflower mosaic virus (DNA virus of higher plants) Double-stranded DNA; reproduce in cytoplasm	*Tobamovirus* (tobacco mosaic virus) Rod-shaped, helical symmetry, single-stranded RNA
Cucumovirus (cucumber mosaic virus) Naked, icosahedral, RNA viruses	*Tobravirus* (tobacco rattle virus) Rod-shaped, nematode transmitted, positive-stranded RNA viruses, segmented genome
Luteovirus (barley yellow dwarf virus) Small isometric virus, RNA genome	*Tombusvirus* (tomato bushy stunt virus) Small RNA viruses, cubic symmetry, resistant to elevated temperatures and organic solvents
Nepovirus (tobacco ringspot virus) Polyhedral, nematode transmitted, RNA viruses	*Tymovirus* (turnip yellow mosaic virus) Icosahedral virus, RNA genome, transmitted by flea beetles
Potexvirus (potato virus X) Flexous rods, 480–580 nm, RNA genome	*Watermelon mosaic* virus Flexous rods, 700–950 nm, RNA genome
Potyvirus (potato virus Y) Flexous, rod-shaped, helical symmetry, single-stranded RNA	

table 12.11

The characteristics of some of the groups of viruses that infect microorganisms

Tailed bacteriophage.	Genome: DNA, double-stranded. Virion: complex shape, binary symmetry, variable number of capsomeres. The tails of the phage are long and contractile in Group A (Myoviridae), long and noncontractile in Group B (Styloviridae), and very short in Group C (Pedoviridae). Example: T-even coliphages.
Cubic bacteriophage.	Group 1 (Microviridae)—Genome: DNA, single-stranded. Virion: icosahedral, cubic symmetry, 12 capsomeres. Example: φx174 Group 2 (Corticoviridae)—Genome: DNA, double-stranded. Virion: cubic symmetry, enveloped, Example: PM-2 Group 3 (Leviviridae)—Genome: RNA, single-stranded. Virion: icosahedral, cubic symmetry, 32 capsomeres. Example: f$_2$ Group 4 (Cystoviridae)—Genome: RNA, double-stranded. Virion: cubic symmetry, enveloped. Example: φ6
Filamentous bacteriophage (Inoviridae).	Genome: DNA, single-stranded. Virion: rod-shaped, helical symmetry. Example: fd

450

Classification and
identification of
microorganisms:
systematics of
fungi, algae,
protozoa, and
viruses

Viruses of microorganisms

The viruses that multiply within the host cells of microorganisms are divided initially according to the group of microorganisms within whose cells they can reproduce, into **bacteriophage** (infect bacteria), **mycoviruses** (infect fungi), **phycoviruses** (infect algae), and viruses of protozoa. The viral groups are largely defined and separated based on molecular considerations. Of the microbial viruses the bacteriophage have been most extensively studied. The characteristics of some of the groups of bacteriophages are listed in Table 12.11. These viruses exhibit great morphological diversity (Figure 12.48; see pages 452–453).

The viruses of microorganisms exhibit great host cell specificity that may in part be due to the nature of the receptor sites that are required for viral adsorption. Bacteriophage, therefore, are often named according to the species of bacteria that they infect, for example, as coliphage for viruses that infect *E. coli*. An even finer division of the coliphage can be made because some viruses infect only donor strains. Such viruses adsorb onto the F pili of bacteria, which explains the host cell's specificity. These viruses can be classified separately, and the International Committee on Nomenclature of Viruses at one point recommended the establishment of the viral group Masculovirus for viruses that infect male or donor bacteria.

Postlude

From the discussion of the systematics of the various groups of microorganisms, it is clear that each major group is classified on a totally different basis. In some cases organisms have been classified in more than one group, such as the dinoflagellates that are treated by phycologists and protozoologists alike. Basing classification systems on phenotypic characteristics of necessity leads to some ambiguities. Without question the phylogenetic relationships among eukaryotic microorganisms will be clarified in future years, and the prevailing classification systems will be revised to include genetic analyses. For the most part, however, the taxonomy of eukaryotic microorganisms is still dominated by the classical approach based on field observations of the different morphological forms that occur during the life cycle of the organism.

The fungi, for example, are classified largely on the basis of their modes of reproduction. The sexual spores of fungi are the most important features that are used in their classification and identification. Additionally, the asexual spores and vegetative structures of the fungi are used for the finer definition of taxa. Although fungi typically form filamentous mycelia, one group, the yeasts, are characteristically unicellular. The fungi include the slime molds, a taxonomic group having interesting life cycles, especially those slime molds that exhibit communication and coordinated behavior.

The protozoa are classified into four major groups based largely on their means of locomotion. Some protozoa form extensions of the cytoplasm known as pseudopodia or false feet. This group is known as the Sarcodina and includes the genus *Amoeba*. The pseudopodia are involved in both locomotion and the ingestion of food. The Ciliophora are protozoa that are motile by means of cilia. The genus *Paramecium* is an example of a ciliate protozoan. The Mastigophora are protozoa that are motile by means of flagella. The Sporozoa, all of which are parasites, are generally nonmotile. Members of this group produce spores during their life cycles.

The algae are classified into seven groups based largely on pigment production and the biochemical nature of the storage reserve materials. The seven groups of algae are the green algae, euglenoids, brown algae, golden and yellow-green algae, dinoflagellates, cryptomonads, and red algae. Some of these groups, such as the euglenoids, are closely related to protozoa. Other algae, such as the brown algae, are closely related to plants. Many algae produce complex macroscopic structures; the kelps, for example, are often 50 meters in length. The algae include the diatoms, organisms that produce frustules containing silica that are highly symmetric and quite beautiful.

The viruses are classified both systematically, based largely on their molecular properties (modern approach), and nonsystematically, based largely on the host cells they reproduce within and the diseases they cause (classical approach). Different classification systems are used for vertebrate animal viruses, insect viruses, plant viruses, bacteriophages, and viruses that infect other microorganisms. The nature of the genome and capsid structure are important characteristics used in the formal classification of viruses. In the case of the viruses, it seems appropriate to consider the host cell as a primary basis for classification because viruses may have evolved as genetic extensions of prokaryotic and eukaryotic cells rather than as a direct evolutionary line from one viral species to another.

Finally, a word of advice for those interested in the taxonomy of eukaryotic microorganisms. We have discussed only the basic characteristics of the more common organisms. These organisms exhibit a diversity of form and mycologists, phycologists, and protozoologists have developed separate and elaborate languages of specialized terminology (jargon) to describe the characteristics used in the classification of eukaryotic microorganisms. It is first necessary to learn the proper terminology before taxonomic keys and diagnostic tables become useful aids for identifying these microorganisms. Additionally, because many of the key features used for identifying eukaryotic microorganisms involve the observation of morphologic forms, gaining practical laboratory and field experience with an expert taxonomist is critical for developing the skills needed to identify these organisms. Introductory microbiology laboratory courses traditionally are heavily biased toward the bacteria and often omit the examination of eukaryotic microorganisms. With some practical experience one can readily learn to recognize and properly identify many of these organisms in their natural habitats. In the ensuing chapters we will consider the practical aspects of the microorganisms whose systematics (including physiology, morphology, and taxonomy) we have examined in this and previous chapters.

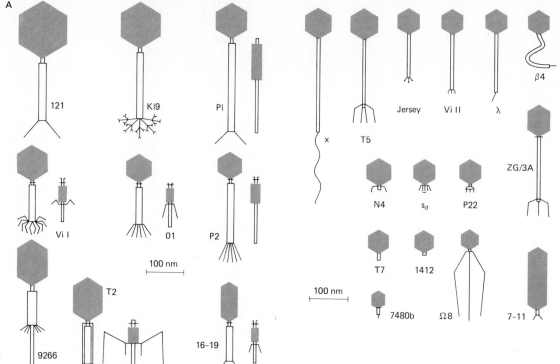

A

Classification and
identification of
microorganisms:
systematics of
fungi, algae,
protozoa, and
viruses

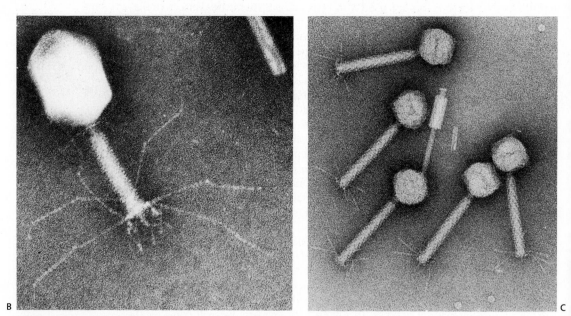

B   C

Figure 12.48

Varied morphologies of some bacteriophage. (A) Schematic representation of the morphologies of phages of enterobacteria. (B–I) Electron micrographs of some bacteriophage with differing morphologies: (B) T4 phage (217,000×); (C) P2 phage (164,000×); (D) λ phage (99,000×); (E) Fd phage (34,000×); (F) T5 phage (66,000×); (G) MS2 phage (173,000×); (H) T7 phage (117,000×); (I) φX174 phage (150,000×). (Courtesy Robley Williams, University of California, Berkeley.)

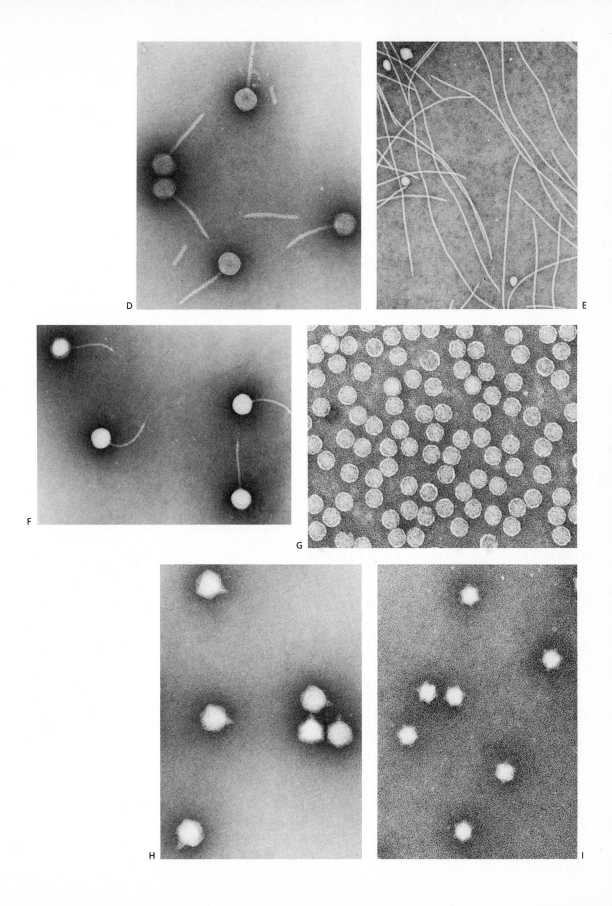

1. On what basis are the major groups of fungi defined? What is the importance of spore formation to fungal classification?

2. On what basis are the major groups of algae defined? What is the importance of photosynthetic pigments and reserve materials to algal classification?

3 On what basis are the major groups of protozoa defined? What is the importance of mode of locomotion to protozoan taxonomy?

4. Why are dinoflagellates treated by both protozoan and algal taxonomists?

5. What are the differences between the ascomycetes and the basidiomycetes?

6. On what formal basis are viruses classified? Why is it particularly difficult to establish a unified system for viral taxonomy?

7. Is the nature of the host cell a valid criterion for classifying a virus?

8. What is a diatom? What is unique about the structure of a diatom?

9. Should the brown algae be considered as plants or microbes?

10. Does our concept of a species differ between viruses, bacteria, and eukaryotic microorganisms? If so how?

Classification and identification of microorganisms: systematics of fungi, algae, protozoa, and viruses

Ainsworth, G. C., and A. S. Sussman (eds.). 1965–1973. *The Fungi: An Advanced Treatise* (5 volumes). Academic Press, New York.

Alexopoulos, C. J., and C. W. Mims. 1979. *Introductory Mycology*. John Wiley & Sons, New York.

Barnett, J. A., R. W. Payne, and D. Yarrow. 1979. *A Guide to Identifying and Classifying Yeasts*. Cambridge University Press, New York.

Bold, H. C., and M. J. Wynne. 1978. *Introduction to the Algae: Structure and Reproduction*. Prentice-Hall, Englewood Cliffs, New Jersey.

Chapman, V. J., and D. J. Chapman. 1975. *The Algae*. St. Martin's Press, New York.

Cole, G. T., and B. Kendrick (eds.). 1981. *Biology of Conidial Fungi*. Academic Press, New York.

Corliss, J. O. 1979. *The Ciliated Protozoa: Characterization, Classification and Guide to the Literature*. Pergamon Press, New York.

Dalton, A. J., and F. Haguenau (eds.). 1973. *Ultrastructure of Animal Viruses and Bacteriophages: An Atlas*. Academic Press, New York.

Farmer, J. N. 1980. *The Protozoa: Introduction to Protozoology*. C. V. Mosby Co., St. Louis.

Fenner, F. 1976. The classification and nomenclature of viruses. *Journal of General Virology* 31: 463–470.

Fenner, F., B. R. McAuslan, C. A. Mims, J. Sambrook, and D. O. White. 1974. *The Biology of Animal Viruses*. Academic Press, New York.

Fraenkel-Conrat, H., and R. R. Wagner (eds.). 1977. *Comprehensive Virology: Regulation and Genetics—Plant Viruses*. Plenum Press, New York.

Fraenkel-Conrat, H., and R. R. Wagner (eds.). 1978. *Comprehensive Virology: Newly Characterized Protist and Invertebrate Viruses*. Plenum Press, New York.

Fraenkel-Conrat, H., and P. C. Kimball. 1982. *Virology*. Prentice-Hall, Englewood Cliffs, New Jersey.

Gray, W. D., and C. J. Alexopoulos. 1968. *Biology of the Myxomycetes*. The Ronald Press Co., New York.

Hollings, M. 1978. Mycoviruses: viruses that infect fungi. *Advances in Virus Research* 22: 2–54.

Jahn, T. L., E. C. Bovee, and F. F. Jahn. 1979. *How to Know the Protozoa*. Wm. C. Brown Co., Dubuque, Iowa.

Kudo, R. R. 1977. *Protozoology*. Charles C. Thomas, Publisher, Springfield, Illinois.

Laskin, A., and H. A. Lechevalier. 1979. *Handbook of Microbiology: Fungi, Algae, Protozoa, and Viruses*. CRC Press, Inc., Boca Raton, Florida.

Lee, R. E. 1980. *Phycology*. Cambridge University Press, Cambridge, England.

Lemke, P. A., and C. H. Nash. 1974. Fungal viruses. *Bacteriological Reviews* 38: 29–56.

Levine, N. D., J. O. Corliss, F. E. G. Cox, G. Deroux, J. Grain, B. M. Honigberg, G. F. Leedale, A. R. Loeblich, J. Lom, D. Lynn, E. G. Meringeld, F. C. Page, G. Poljansky, V. Sprague, J. Vavra, and F. G. Wallace. 1980. A newly revised classification of the Protozoa. *Journal of Protozoology* 27: 37–58.

Lodder, J., and N. Kreger-van Rij. 1970. *The Yeasts: A Taxonomic Study*. North Holland Publications, Amsterdam.

Luria, S. E., J. E. Darnell Jr., D. Baltimore, and A. Campbell. 1978. *General Virology*. John Wiley & Sons, New York.

Maramorosch, K. (ed.). 1977. *Insect and Plant Viruses: An Atlas*. Academic Press, New York.

Miller, O. K. 1979. *Mushrooms of North America*. E. P. Dutton Co., New York.

Moore-Landecker, E. 1982. *Fundamentals of the Fungi*. Prentice-Hall, Englewood Cliffs, New Jersey.

Palmer, E. L., and M. L. Martin. 1982. *An Atlas of Mammalian Viruses*. CRC Press, Inc., Boca Raton, Florida.

Phaff, H. J., M. W. Miller, and E. M. Mrak. 1978. *The Life of Yeasts*. Harvard University Press, Cambridge, Massachusetts.

Pickett-Heaps, J. D. 1975. *Green Algae: Structure, Reproduction and Evolution in Selected Genera*. Sinauer Associates, Publishers, Sunderland, Massachusetts.

Rose, A. H., and J. S. Harrison (eds.). 1969. *The Yeasts*. Academic Press, New York.

Trainor, F. R. 1978. *Introductory Phycology*. John Wiley & Sons, New York.

Westphal, A. 1976. *Protozoa*. Blackie & Son Ltd., Glasgow.

Wildy, P. 1971. Classification and nomenclature of viruses. *Monographs in Virology* 5: 1–81.

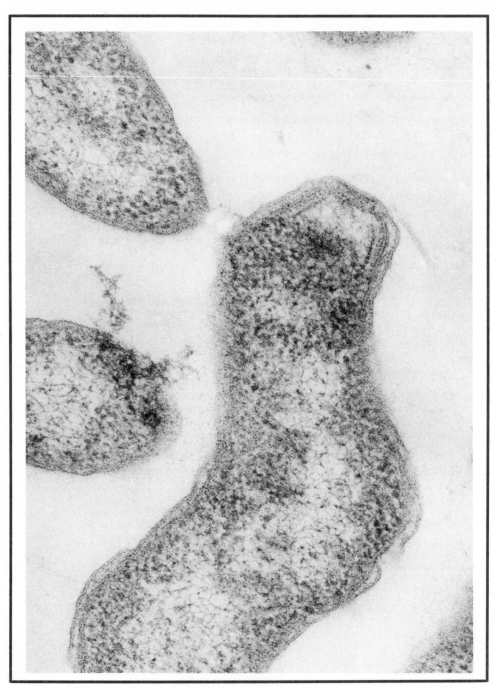

Electron micrograph of Campylobacter fetus.
(Courtesy R. G. E. Murray, University of Western Ontario.)

Medical microbiology

SECTION SEVEN

Interactions between microorganisms and humans

13

From the moment of birth, human beings are continuously exposed to microorganisms. Some of the interactions between human beings and microbial populations are essential for the well-being of the individual, but others result in diseases. The nature of the interaction between a specific microorganism and a human host depends on the physiological properties of both the microorganism and the host. Mutually beneficial relationships between microbial populations and human hosts are termed **synergistic**. A relationship is one of **commensalism** when the microbial population benefits without adversely nor beneficially affecting the human host. **Parasitism** occurs when the microbial population benefits, but the human host is adversely affected. Pathogenic microorganisms can establish a parasitic relationship with a host organism that results in host disease. The interactions of microorganisms and humans thus represent a continuum from those that are needed to maintain good health to those that cause human disease and even death.

The normal human microbiota

Although you cannot see them, you are literally covered with microorganisms. In fact, the average adult human has 10^{13} eukaryotic animal cells and 10^{14} associated prokaryotic and eukaryotic cells of microorganisms, or put another way, the normal human being is composed of just over 10^{14} cells, only 10 percent of which are human cells and the remaining 90 percent of which are microbial. The body surfaces of most animals, including humans, are populated by microorganisms, with distinct microbial populations inhabiting the surface tissues of the skin, oral cavity, respiratory tract, gastrointestinal tract, and genitourinary tract. The microbial populations most frequently found in association with particular tissues are referred to as the indigenous microbial populations, **normal microflora**, or **normal microbiota**. Although the term microflora is used extensively, the term microbiota is preferable as it avoids any inference that microorganisms are little plants. The concept of a normal microbiota does not imply a constant, identical association between particular microbial populations and human beings. Rather, the normal microbiota qualitatively describes the species that are generally found within the stable mixture of microbial populations (microbial community) associated with particular body tissues, and within this microbial community the relative concentrations of individual populations can fluctuate.

Usually, the normal microbiota of the human body are nonpathogenic (do not cause disease). However, the relationships between microorganisms, humans, and disease are dynamic, and microorganisms that are normally nonpathogenic can cause disease under appropriate conditions. Virtually any microorganism is a potential opportunistic pathogen capable of causing disease when the right set of conditions occur. For example, a person's normal defense mechanisms are subject to failure and a host with compromised defense mechanisms is subject to infection. The common adage "when you are tired and run down you are more prone to infection," though an obvious oversimplification, has much validity because infectious diseases are as much the result of the failure of the human physiological and immunological defense systems as they are the result of the special properties (**virulence factors**) of pathogenic microorganisms.

In most cases the relationships between animals and their normal microbiota are mutually beneficial. Germ-free animals develop abnormalities of the gastrointestinal tract and are more susceptible to disease than animals with normal associated microbiota. The normal associated microbiota of animals contribute in part to the normal defense mechanisms that protect animals against infection by pathogens. To determine the role of the indigenous microbiota, it is possible to deliver an animal by caesarean section and raise that animal in the absence of microorganisms (Figure 13.1). Such **germ-free animals** provide suitable experimental models for investigating the interactions of animals and microorganisms. Comparing animals possessing normal associated microbiota with germ-free animals permits the elucidation of the complex relationships between microorganisms and host animals.

Animals reared in germ-free environments are especially susceptible to disease. Apparently, such

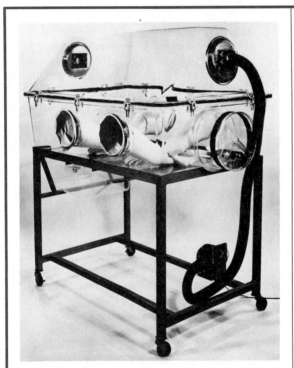

Figure 13.1

Photograph of an isolette used for the rearing of germ-free animals. Sterile food and water can be fed to the animals so that they are never exposed to microorganisms. (Courtesy Germ Free Laboratory, Miami, Florida.)

Discovery process

Germ-free experimentation extends the microbiologist's pure culture concept to *in vivo* studies. It has been learned from using germ-free animals that a lack of exposure of an animal to microbes results in a complex of deleterious effects. Germ-free animals differ from other members of their species. Their metabolic rates and cardiac outputs are reduced. Structures that are designed to defend against bacterial invasion—such as the lymphatic system, the antibody-forming system, and the mononuclear phagocyte system—are poorly developed in these animals. Some of the animal's organs that would normally have natural populations of bacteria are often reduced in size or capacity. Their nutritional requirements also differ; they require vitamin K that other animals do not because the *E. coli* that colonize the normal animal's intestinal tract synthesize this vitamin, helping to meet the animal's nutritional requirements. As would be expected, germ-free animals are more susceptible to bacterial infection. Organisms—such as *Bacillus subtilis* and *Micrococcus luteus*—that are harmless to other animals cause disease in germ-free animals; more exotic pathogenic microorganisms, such as *Vibrio cholerae* and *Shigella dysenteriae*, are far more readily able to establish infections where there are no normal microbiota that have a competitive advantage within the intestinal tract. At the same time, though, germ-free animals are resistant to *Entamoeba histolytica*, the causative organism of amebic dysentary, because the protozoan requires the normal intestinal bacteria as a food source. Likewise, tooth decay is no problem to germ-free animals, even those on high-sugar diets, because of the lack of lactic acid bacteria in their oral cavities.

animals fail to develop adequate defenses against pathogenic microorganisms; the lymphatic tissues, which play an important role in protecting the body against infections, typically are poorly developed in germ-free animals. Additionally, some members of the normal indigenous microbiota exhibit antagonism toward potential pathogens. For example, acid production by the indigenous microbiota of the vaginal tract lessens the probability of infection with *Neisseria gonorrhoeae*. Other mechanisms of antagonism by the indigenous microbiota that enhance host resistance to disease include alteration of the oxygen tension, production of antibiotics, and competition for available nutrients. Besides their role in preventing certain diseases, the indigenous microbiota contribute to the nutrition of the animal by synthesizing nutrients essential to the welfare of the host. For example, germ-free animals require vitamin K, which normally is synthesized by the resident microbiota of the gastrointestinal tract. The microbiota of the gastrointestinal tract also synthesize biotin, riboflavin, pantothenate, and pyridoxine, supplying these vitamins to the animal host. Thus, the maintenance of a "healthy" indigenous microbiota is essential to the maintenance of a healthy individual.

The acquisition of a resident microbiota by human beings occurs in stages and therefore is termed a successional process. Whereas some parasites can migrate through the placenta, the human fetus is normally sterile, with colonization of body tissues beginning during the birth process. The different tissues of the body provide distinct habitats with varying environmental conditions for the growth of differing microbial populations. The growth of microorganisms on body tissue surfaces alters the localized environmental conditions, leading to the successional changes in the populations of microorganisms associated with the tissues until a relatively stable normal microbiota is established.

Not all body tissues, though, provide suitable habitats for the growth of microorganisms. For example, most of the urinary tract lacks a resident microbiota. Only the distal end of the urinary tract has a resident microbiota and urine that does not contact these extremities is considered a sterile body fluid. Similarly, blood is considered a sterile body fluid because the circulatory system does not possess a resident microbiota. In reality, various microorganisms frequently enter the bloodstream but normally do not establish growing populations within the circulatory system. For example, a segment of the circulatory system associated with the liver, known as the hepatic portal

system, normally contains low numbers of bacteria that pass through the intestinal wall as a result of abrasions to the lining of the intestinal tract caused by food particles. These bacteria are routinely eliminated from the circulatory system by specialized cells in the liver, known as **Kupffer cells**. Such defense cells and other antimicrobial factors in blood prevent the establishment of a resident microbiota within the circulatory system. Transient **bacteremia** (bacteria in the bloodstream), however, can occur throughout the circulatory system, even in healthy individuals. For example, following some dental procedures, such as oral prophalaxis (cleaning of teeth) and tooth extractions, bleeding of the gums results in a transient systemic bacteremia lasting about 24 hours.

Microorganisms indigenous to the skin

Factors influencing growth of microorganisms on the skin

Human **skin surfaces** are not especially favorable habitats for the growth of microorganisms (Figure 13.2). Even though the skin is continuously exposed to microorganisms, most of these microorganisms are unable to reproduce there. A major factor that determines the distribution of microbial populations on the skin surface is the microenvironment. Environmental factors, such as temperature, water activity, pH, and salinity, represent a severe environmental stress that can prevent many microbial populations from colonizing the skin surface. The lack of available water is a major limiting factor controlling the extent of such

Figure 13.2

Drawing of the surface of the skin, indicating some of the indigenous microbiota found on the human skin surface.

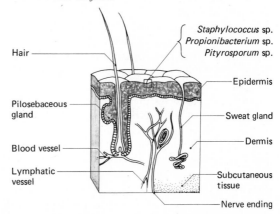

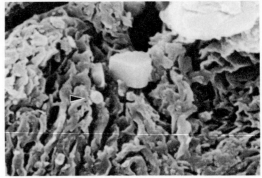

Figure 13.3

Scanning electron micrographs of bacteria on the surface of human skin. (A) Skin showing coccal-shaped bacteria (3900 ×). (B) Skin showing coccal-shaped bacteria and crystalline deposits from sweat (3900 ×). (Courtesy Robert Apkarian, University of Louisville.)

microbial growth. Although sweat glands secrete fluids containing water onto the skin surface, sweat contains high concentrations of salt and organic wastes that reduce the water activity (a_w) sufficiently to inhibit most microbial growth. Even if able to grow at low water activities, the microorganisms that become established within the resident microbiota of the skin must tolerate the osmotic stress caused by the high salt concentrations there.

Additionally, the presence of antimicrobial agents, including free fatty acids produced by the host animal and antimicrobial compounds produced by those microorganisms that do successfully establish themselves as the normal microbiota of the skin, act to prevent foreign microorganisms from growing on the skin surface. Lipids on the skin surface are derived mainly from the secretion of sebum from the sebaceous glands. Sebum is secreted into hair follicles and whereas some bacteria are able to colonize these depressions in the skin surface, many of the unsaturated free fatty acids in sebum have antimicrobial activity. The microorganisms inhabiting the skin surface must tolerate such natural antimicrobial biochemical secretions and also metabolize these compounds as their source of nutrients. Furthermore, many microorganisms that successfully inhabit the skin surface produce antimicrobial substances; these compounds, some of which are low-molecular-weight fatty acids, act to prevent the invasion of the skin surface by other microbial populations.

Resident microbiota of the skin

The dominant microbial populations on the skin surface are Gram positive bacteria, which are normally relatively resistant to desiccation as compared to Gram negative bacteria. Members of the genera *Staphylococcus* and *Micrococcus* are frequently the most abundant microorganisms on the skin surface (Figure 13.3) because they are generally salt-tolerant and can utilize the lipids present on the skin surface. Other Gram positive bacteria usually found as part of the normal microbiota of the skin surface include *Corynebacterium, Brevibacterium*, and *Propionibacterium* species. Gram negative bacteria generally occur primarily in the moister regions of the skin surface, such as in the armpits and between the toes. Relatively few fungi are included in the normal microbiota of the skin surface. Two yeasts, *Pityrosporum ovale* and *P. orbiculare*, though, are able to metabolize the lipids found on the skin surface and normally occur on scalp tissues. Additionally, some dermatophytic fungi occasionally grow on the skin surface, producing diseases such as athlete's foot and ringworm.

Effects of washing on the indigenous microbiota of the skin

Proper cleanliness habits and good hygienic practices tend to prevent the establishment of non-indigenous microorganisms among the natural skin microbiota by preventing the buildup of excessive concentrations of organic matter. The presence of high concentrations of organic matter can

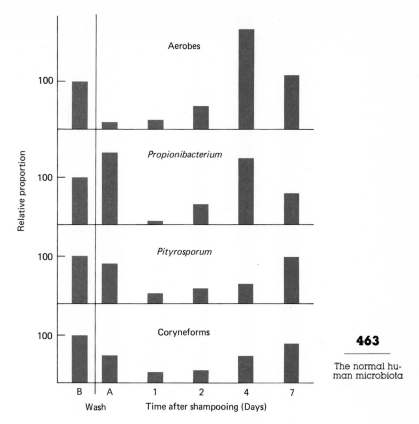

Figure 13.4

The changes in microbial populations on the human scalp after shampooing. (Reprinted by permission of the Society for Applied Bacteriology, from R. P. Marples, 1974, Effects of germicides on skin flora, in The Normal Microbial Flora of Man, *F. A. Skinner and J. G. Carr, eds.)*

make the skin surface a more favorable environment for microorganisms, permitting the growth of many microorganisms that otherwise could not grow there. Washing the skin surface, however, removes excess organic matter and temporarily reduces the numbers of both resident and transient microorganisms (Figure 13.4). Shortly after the skin surface is washed, the populations of indigenous microbiota on the skin surface begin to return to their original numbers.

Oral cavity

In contrast to the relatively few microbial populations that flourish on the skin surface, an abundance of microorganisms develops within the oral cavity. The oral cavity contains a great variety of surfaces, with differing environmental conditions providing the varied habitats for colonization by diverse microbial populations. Within the oral cavity, microorganisms can grow on various surfaces, including the gums (gingiva) and teeth (Figure 13.5). The availability of water and nutrients, including growth factors, provides an environ-

ment favorable for the proliferation of microorganisms.

Resident microbiota of the oral cavity

Streptococcus species normally constitute a high proportion of the **normal microbiota of the oral cavity**. Various *Streptococcus* species, including *S. mutans*, produce slime layers and adherence factors that allow them to stick onto tooth surfaces. Other *Streptococcus* species, such as *S. sanguis*, colonize saliva. Some of these lactic acid bacteria form dental plaque on the surfaces of teeth and are implicated in the formation of dental caries (Figure 13.6). Although *Lactobacillus* species are normally present in low numbers, they are frequently found in association with dental caries. Besides *Streptococcus* species, obligate anaerobic bacteria are found in high numbers in the oral cavity, including Gram negative coccoid members of the genus *Veillonella* and Gram positive species of *Bacteroides* and *Fusobacterium*. Additional members of the normal microbiota of the oral cavity include species of the bacteria *Actinomyces, Neisseria, Staphylococcus, Micrococcus, Vibrio, Leptotrichia*, and *Rothia*; fungi, such as the

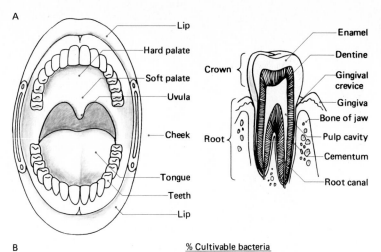

A

- Lip
- Hard palate
- Soft palate
- Uvula
- Cheek
- Tongue
- Teeth
- Lip

Crown {
Root {

- Enamel
- Dentine
- Gingival crevice
- Gingiva
- Bone of jaw
- Pulp cavity
- Cementum
- Root canal

Figure 13.5

(A) The human oral cavity and detail of a tooth. (B) The cultivatable bacteria found in its various parts.

B

	% Cultivable bacteria		
	Tooth	Tongue	Gingival crevice
Streptococcus	25	45	25
Bacteroides	7	5	17
Fusobacterium	5	1	4
Corynebacterium	—	4	17
Veillonella	13	15	11
Actinomyces	33	2	9
Peptostreptococcus	11	5	9

Figure 13.6

This scanning electron micrograph shows the formation of dental plaque (2200×) in the human mouth just 3 days after the subject stopped brushing the teeth. (From BPS—Z. Skobe, Forsythe Dental Center.)

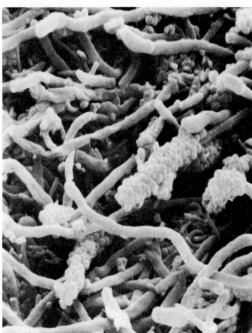

yeast *Candida albicans*; and protozoa, such as species of *Entamoeba* and *Trichomonas*. The somewhat surprisingly high incidence of obligate anaerobes within the resident microbiota of the oral cavity results from the high rates of metabolism by the facultatively anaerobic members of this microbial community that scavenge the free oxygen, producing conditions that favor the growth of obligate anaerobes.

Effects of brushing

Whereas we routinely attempt to cleanse tooth surfaces of food particles and dental plaque, daily brushing represents only a temporary disturbance to the normal microbiota of the oral cavity. Within minutes of even the complete removal of microorganisms from the tooth surface, populations of *Streptococcus sanguis, S. salivarius,* and *Actinomyces viscosus* that colonize the tongue and saliva once again cover the tooth surface. Within a week or two after a thorough dental cleaning, the microbial community colonizing the tooth surface includes species of *Bacteroides, Veillonella, Neisseria, Fusobacterium,* and *Rothia,* in addition to *Streptococcus* and *Actinomyces* species. Although cleaning teeth does not eliminate this resident microbiota, brushing does remove substrates that permit the overgrowth of microorga-

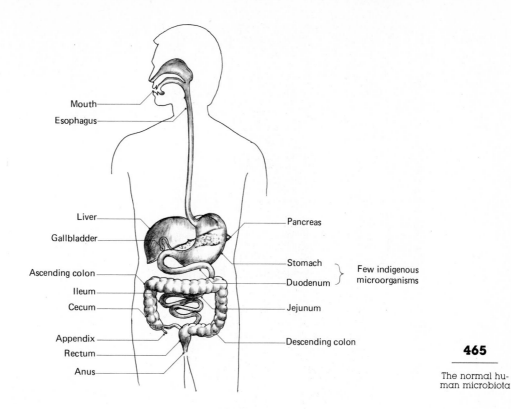

Resident
microbiota
species of
Bacteroides
Fusobacterium
Clostridium
Lactobacillus
Escherichia
Proteus
Klebsiella
Enterobacter
Streptococcus

Mouth
Esophagus
Liver
Gallbladder
Ascending colon
Ileum
Cecum
Appendix
Rectum
Anus

Pancreas
Stomach
Duodenum
Jejunum
Descending colon

Few indigenous
microorganisms

Figure 13.7

The human gastrointestinal tract and its resident microbiota.

nisms, the acidic products of microbial metabolism, and dental plaque that are responsible for dental caries and periodontal disease.

Gastrointestinal tract

Many microorganisms are washed from the oral cavity into the remainder of the **gastrointestinal tract**. Additionally, we regularly ingest viable microorganisms with our food. Although some microorganisms are able to reproduce within food particles in the stomach, this is a transitory phenomenon. Exposure to the gastric juices secreted into the stomach kills most microorganisms. The low pH of the stomach also precludes the existence of an indigenous microbiota associated with the lining of that organ. In contrast to the stomach, there is an abundant indigenous microbiota associated with the lower regions of the small intestine and the colon or large intestine (Figure 13.7). The constant temperature, 37°C, and availability of water and nutrients renders the large intestine a favorable habitat for the growth of a variety of microbial populations. Food, partially digested by the body's enzymes, is continuously supplied to the microorganisms inhabiting the intestinal tract, supporting the growth of a large resident microbiota. Consequently, the highest numbers of resident microbiota are associated with the large intestine (Figure 13.8).

Resident microbiota of the gastrointestinal tract

The initial residents of the intestinal tract in breast-fed infants are members of the genus *Bifidobacterium*, whereas in bottle-fed infants *Lactobacillus* species initially colonize the intestinal tract. These initial colonizers of the intestinal tract are later displaced by other bacterial species, which include both obligate and facultative anaerobes and members of the genera *Lactobacillus, Streptococcus, Clostridium, Veillonella, Bacteroides, Fusobacterium*, and coliform bacteria. The enteric bacteria, including members of the genera *Escherichia, Proteus, Klebsiella*, and *Enterobacter*, are facultative anaerobes. Although these facultative anaerobes were once thought to comprise

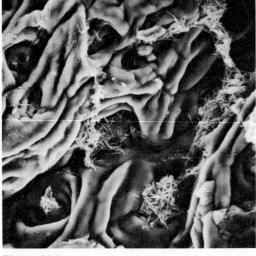

Figure 13.8

The microbiota of the intestine are shown in the mucous membrane of the large intestine. The bacteria (366×) are in crypts. (From BPS—D. C. Savage, University of Illinois, Champaign-Urbana.)

Interactions between microorganisms and humans

the majority of the microorganisms inhabiting the intestinal tract, methodological improvements for the isolation and culture of obligate anaerobes have revealed that up to 99 percent of the intestinal microbiota may be obligate anaerobes of the genera *Bacteroides* and *Fusobacterium*. The actual proportions of the individual bacterial populations within the indigenous microbiota of the intestinal tract vary, depending in part on the diet of the host.

Respiratory tract

The **normal microbiota of the respiratory tract** are quite different from the indigenous microorganisms of the gastrointestinal tract. Even though the respiratory tract has various defense mechanisms for preventing microbial infections, the upper respiratory tract, including the nasal cavity and nasopharynx, is normally inhabited by various species of the genera *Streptococcus, Staphylococcus, Branhamella, Neisseria, Haemophilus, Bacteroides,* and *Fusobacterium* as well as members of the spirochete and coryneform groups. The lower respiratory tract, though, lacks a normal resident microbiota. An abundance of blood cells capable of engulfing and digesting microbial contaminants are present in the lower respiratory tract, preventing the establishment of an indigenous microbiota. The lower respiratory tract may be colonized temporarily by exogenous pathogens inhaled on dust particles or water droplets and the same microorganisms that normally colonize the upper respiratory tract.

Genitourinary tract

Like the lower respiratory tract, most of the genitourinary tract, including the kidneys and urinary bladder, are normally free of microorganisms. The external genital regions, though, of both sexes contain various indigenous microbial populations. For example, the terminal areas of the urethra in both males and females are colonized with bacteria. Because of its relatively large surface area

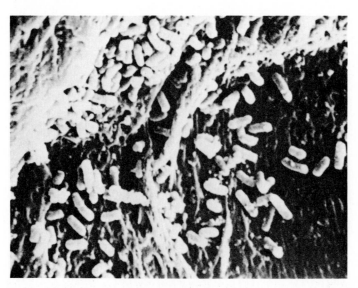

Figure 13.9

The normal microbiota of the vaginal tract is represented by this bacterial microcolony (9,500×) at the intercellular borders on a rat vaginal epithelium. (Reprinted by permission of the American Society for Microbiology, Washington D.C., from B. Larsen, A. J. Markovetz, and R. P. Galask, 1977, Applied and Environmental Microbiology *34: 85.)*

and the nature of mucoidal secretions, the vaginal tract harbors an especially large variety of resident microbiota. The normal microbiota of the vaginal tract include bacteria such as *Streptococcus, Lactobacillus, Bacteroides,* and *Clostridium*; coliforms; spirochetes; yeasts, including members of the genus *Candida*; and flagellated protozoans of the genus *Trichomonas* (Figure 13.9). The low pH of the vaginal tract, 4.4–4.6, limits the species of microorganisms that can reproduce there, and the metabolic activities of the resident microbiota contribute to maintaining the low pH. Thus, the microorganisms that multiply within the vaginal tract are normally acid-tolerant. The resident microbiota of the vaginal tract is a carefully balanced complex microbial community in which large fluctuations occur during the normal menstrual cycle. Occasionally, population imbalances occur, for instance, the proliferation of the yeast *Candida albicans* or the protozoan *Trichomonas vaginalis*, leading to inflammation or vaginitis.

Virulence factors of pathogenic microorganisms

After seeing that microorganisms naturally occur in association with the surface tissues of healthy individuals, we now turn to the **host-parasite interactions** responsible for **human diseases** caused by microorganisms. What is it about some microorganisms that allow them to establish the infections that cause human disease? To answer this question we need to examine the host defense mechanisms that normally preclude infectious diseases and the intrinsic properties of pathogenic microorganisms that allow them to overcome these defenses and initiate the physiological changes inherent in the disease process. An **infection** of the human body by a pathogenic microorganism results in disease when the potential of the microorganism to disrupt normal bodily functions is fully expressed. In many cases, though, infections with potentially pathogenic microorganisms are attenuated so that the overt clinical manifestation of the disease does not occur. Many healthy individuals are carriers of potentially pathogenic microorganisms. Although we understand today that pathogenicity is not a property of the microorganism alone, it is still useful to examine the intrinsic properties of pathogenic microorganisms that contribute to their potential for causing human disease.

Some disease-causing microorganisms possess properties, referred to as **virulence factors**, that enhance their pathogenicity and allow them to invade human tissues and disrupt normal bodily functions. The virulence of pathogenic microorganisms, that is, their ability to induce human disease, depends in large part on two properties of the microorganisms, invasiveness and toxigenicity. **Invasiveness** refers to the ability of the microorganisms to invade human tissues, attach to cells, and multiply within the cells or tissues of the human body. Most of the microorganisms of the normal microbiota of humans are noninvasive and do not invade the tissues on whose surfaces they grow. Microorganisms that possess invasive properties are able to establish infections within host cells and tissues.

Toxigenicity refers to the ability of a microorganism to produce biochemicals, known as toxins, that disrupt the normal functions of cells or are generally destructive to human cells and tissues (Table 13.1). **Toxins** cause discernible damage to human systems and in some cases cause death. Some toxin-producing microorganisms can grow outside of the host and still cause disease symptoms if the toxins enter human tissues. These toxin-producing strains need not establish an infection within the human body to cause disease. In the past we have classified toxins produced by microorganisms as endotoxins if they comprise part of the microbial cell, and as exotoxins if they are secreted by the cell, but we now know that some exotoxins are not released until the cell is disrupted, and substances classified as endotoxins are sometimes released from the cell without lysis. Therefore, a better classification system for toxins is based on the biochemical nature of the toxin.

Lipopolysaccharide toxins

The term **bacterial endotoxin** refers to the **lipopolysaccharide (LPS)** portion of the cell wall of Gram negative bacteria. When Gram negative bacteria die, their cell walls disintegrate, releasing the lipopolysaccharide toxin. Some growing Gram negative bacteria also release lipopolysaccharide toxin. The phospholipid portion of the LPS molecule, termed lipid A, does not have the same structure of membrane phospholipids but rather is composed of a sequence of phosphorylated

table 13.1

Some toxins produced by microorganisms causing disease in humans

Microorganism	Toxin	Disease	Action or activity
Clostridium botulinum	Several neurotoxins	Botulism	Paralysis—blocks neural transmisssion
Clostridium perfringens	α-Toxin κ-Toxin θ-Toxin	Gas gangrene	Lecithinase Collagenase Hemolysin
Clostridium tetani	Neurotoxin (tetanospasm)	Tetanus	Spastic paralysis interferes with motor neurons
	Tetanolysin		Hemolytic cardiotoxin
Corynebacterium diphtheriae	Diphtheria toxin	Diphtheria	Blocks protein synthesis at level of translation
Streptococcus pyogenes	Streptolysin O Streptolysin S Erythrogenic	Scarlet fever	Hemolysin Hemolysin Causes rash of scarlet fever
Shigella dysenteriae	Neurotoxin	Bacterial dysentery	Hemorrhagic, paralytic
Staphylococcus aureus	Enterotoxin	Food poisoning	Intestinal inflammation
Aspergillus flavus	Aflatoxin B_1	Aflatoxicosis	Blocks protein synthesis at level of transcription
Amanita phalloides	α-Amanitin	Mushroom food poisoning	Blocks protein synthesis at level of transcription

glucosamine residues. It is the lipid A portion of the LPS molecule that is primarily responsible for the physiological effects of the endotoxin. The physiological effects of lipopolysaccharide toxins include fever, circulatory changes, and other general symptoms, such as weakness and nonlocalized aches. The injury to the circulatory system by the lipopolysaccharide of the Gram negative cell is basic to the action of this toxin, but the mechanism of its action is not yet understood. The effects of lipopolysaccharide toxins are not specific to the particular species of Gram negative bacteria, and thus, there is no specific characteristic disease symptomology associated with the endotoxin of a particular bacterial species. **Lipopolysaccharide toxins** of *Salmonella* and *Shigella* species are responsible in part for diseases, such as gastroenteritis, caused by these pathogens, but these pathogens also produce protein toxins that are largely responsible for the pathogenicity of these organisms.

Protein toxins

In contrast to lipopolysaccharide toxins, the effects of **protein toxins (exotoxins)** are specific to the microorganism producing the toxin, and these toxins cause distinctive clinical symptoms. Whereas lipopolysaccharide toxins are produced exclusively by Gram negative bacteria, protein toxins are produced by both Gram negative and Gram positive bacteria. Protein toxins are more readily inactivated by heat than are lipopolysaccharide toxins. A protein toxin can normally be inactivated by exposure to boiling water for 30 minutes, whereas lipopolysaccharide toxins can withstand autoclaving (Table 13.2). Typically, protein toxins are excreted into the surrounding medium. For example, *Clostridium botulinum*, the causative organism for botulism, can secrete a potent exotoxin into canned food products, the ingestion of even minute amounts of which is lethal. Protein toxins are generally more potent than lipopolysaccharide toxins, with far less protein toxin needed to produce serious disease symptoms than is required for disease symptoms due to lipopolysaccharides. As examples of the potency of protein toxins, about 30 grams of diphtheria toxin could kill 10 million people and 1 gram of botulinum toxin could kill everyone in the United States (over 225 million people).

Often, protein toxins are referred to by the disease they cause, such as diphtheria toxin or botulinum toxin. They may also be categorized according to the symptoms that they cause. As examples, neurotoxins affect the nervous system, enterotoxins cause an inflammation of the tissues of the gastrointestinal tract, and cytotoxins interfere with cellular functions.

Neurotoxins

Some of the toxins produced by *Clostridium botulinum*, the bacterium that causes botulism, are **neurotoxins**. The neurotoxins responsible for the symptomology of botulism bind to nerve synapses, blocking the release of acetylcholine from nerve cells of the central nervous system and causing the loss of motor function. The binding of the toxin to the myoneural junction appears to be a two-step process with toxin first binding to one receptor, followed by binding to another receptor. The binding of the toxin blocks conduction of the nerve impulse at or near the end of the point of final branching between the nerve filaments and the presynaptic part of the end plate complex of the muscle cell, preventing the proper transmission of the nerve impulse to the muscle. The inability to transmit impulses through motor neurons can cause respiratory or cardiac failure, resulting in death.

The neurotoxin **tetanospasmin** produced by *Clostridium tetani*, the causative agent for the disease tetanus, interferes with the peripheral nerves of the spinal cord. Tetanospasmin blocks the ability of these nerve cells to properly transmit signals to the muscle cells, causing the symptomatic spasmatic paralysis of tetanus. Like the neurotoxin produced by *C. botulinum*, the neurotoxin of *C. tetani* paralyzes motor neurons, but unlike botulinum toxin, tetanospasmin acts only on the nerves of the cerebrospinal axis. It is postulated that tetanus toxin inhibits the release of glycine from the inhibitory neurons (interneurons) in the anterior horn of the spinal cord. Because glycine is the inhibitory neurotransmitter in these interneurons, the result is convulsions similar to those produced by strychnine, which is known to compete with glycine for receptor sites.

The neurotoxin produced by *Shigella dysenteriae*, the so-called shiga toxin, differs from the neurotoxins produced by *C. botulinum* and *C. tetani* in that it interferes with the circulatory vessels that supply blood to the central nervous system rather than affecting the nerve cells directly. The neurological effects of the shiga toxin are thus secondary to the primary action of the toxin on the vascular circulatory system.

Enterotoxins

The **enterotoxin cholaragen** produced by *Vibrio cholerae*, the causative agent for the disease cholera, blocks the conversion of cyclic AMP to ATP by increasing the activity of adenylcylase. The resulting elevated concentrations of cyclic AMP cause the release of inorganic ions, including chloride and bicarbonate ions, from the mucosal cells that line the intestine into the intestinal lumen. Although the exact mechanism of toxin action on adenylcyclase is not understood, the change in the ionic balance resulting from the action of this toxin causes the movement of large amounts of water into the lumen in an attempt to balance the osmotic pressure, leading to severe dehydration that sometimes results in the death of infected individuals.

Cytotoxins

Cytotoxins include those proteins that cause lysis of blood cells and those that interfere with protein synthesis. For example, diphtheria toxin, produced by *Corynebacterium diphtheriae*, inhibits protein synthesis in mammalian cells. This toxin blocks transferase reactions during the translation

table 13.2

Comparison of selected characteristics of bacterial lipopolysaccharide toxins (endoxins) and protein toxins (exotoxins)

Characteristic	Endotoxin	Exotoxin
Chemical composition	Lipopolysaccharide-protein complex	Protein
Source	Cell walls of Gram negative bacteria; released upon death and autolysis of the bacteria	Gram negative and Gram positive bacteria; excretion products of growing cells, or in some cases, substances released upon autolysis and death of the bacteria
Effects on host	Nonspecific	Generally affects specific tissues
Thermostability	Relatively heat-stable (may resist 120°C for 1 hour)	Heat-labile; most are inactivated at 60–80°C

of mRNA, preventing the addition of amino acids and thus the elongation of the peptide chain. The production of diphtheria toxin is particularly interesting because only lysogenized cells of *C. diphtheriae* produce diphtheria toxin proteins. The protein toxin is coded for by the phage genome. Thus, only when *C. diphtheriae* is infected with a virus does a human infection with *C. diphtheriae* result in disease.

In addition to bacterial protein toxins, some eukaryotic microorganisms produce potent cytotoxins. For example, dinoflagellates produce highly potent toxins that cause paralytic shellfish poisoning. Many mushrooms are highly poisonous because of the potency of the **mycotoxins (fungal toxins)** they produce. The term mycotoxin actually encompasses a wide variety of biochemicals, with varying modes of action. Some cause ultrastructural changes in the host, whereas others interfere with the metabolic activities of host cells. Although there is no generalized mechanism that applies to all mycotoxins, the mode of action of most mycotoxins appears to be based primarily on their ability to interact with macromolecules, subcellular organelles, and organs of animals. In the most infamous of the poisonous mushrooms, *Amanita*, the toxin alpha amanitin blocks transcription of DNA by interfering with RNA polymerase enzyme. Similarly, *Aspergillus* species produce aflatoxins that bind to DNA and prevent transcription of genetic information, resulting in a variety of adverse effects on human beings and animals.

Some proteins produced by pathogenic microorganisms contribute to their virulence by causing the lysis of red blood cells (Table 13.3). Cytotoxins that cause the lysis of human erythrocytes are termed **hemolysins** because their action results in the release of hemoglobin from these red blood cells. For example, *Streptococcus* species produce various hemolysins, including streptolysin O, an oxygen-labile and heat-stable protein, and streptolysin S, an acid-sensitive and heat-labile protein. The hemolytic action of these cytotoxins produces zones of clearing when these bacteria are grown on blood agar plates. A complete zone of clearing around a bacterial colony growing on a blood agar plate is referred to as **beta hemolysis** and the partial clearing of the blood agar plate around the bacterial colony is referred to as **alpha hemolysis** (Figure 13.10). Alpha hemolysis involves the conversion of hemoglobin to methemoglobin, with the production of a zone of greening (partial clearing) around the colony. Hemolytic activity is associated not only with *Streptococcus* species but also with various other bacterial genera, including *Staphylococcus* and *Clostridium* as well as other bacterial species. In addition to red blood cells, white blood cells are killed by some microbial cytotoxins. For example, leukocidin produced by *Staphylococcus aureus* causes lysis of leukocytes, contributing to the pathogenicity of this organism.

Enzymes as virulence factors

Some enzymes produced by microorganisms also interfere with normal mammalian functions, and some of these enzymes contribute to the virulence of microbial pathogens. For example, various **phospholipase** enzymes produced by microorganisms can destroy animal cell membranes. Phospholipases can act as hemolysins, causing the lysis of red blood cells. Indeed, some substances that have been classified as toxins are now known to be toxic enzymes. For example, the alpha toxin of *C. perfringens* is a **lecithinase** (also known as phospholipase C or phosphatidylcholine phosphohydrolase). This enzyme hydrolyzes lecithin

table 13.3

Some extracellular enzymes involved in microbial virulence

Enzyme	Action	Examples of bacteria producing enzyme
Hyaluronidase	Breaks down hyaluronic acid (spreading factor)	*Streptococcus pyogenes*
Coagulase	Blood clots, coagulation of plasma	*Staphylococcus aureus*
Phospholipase	Lyses red blood cells	*Staphylococcus aureus*
Lecithinase	Destroys red blood cells and other tissue cells	*Clostridium perfringens*
Collagenase	Breaks down collagen (connective tissue fiber)	*Clostridium perfringens*
Fibrinolysin (kinase)	Dissolves blood clots	*Streptococcus pyogenes*

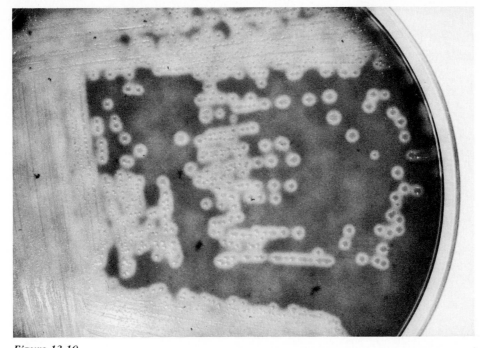

Figure 13.10

Beta-hemolytic growth of Streptococcus pyogenes *on blood agar plates, showing zones of clearing where hemolysis has occurred around the colonies. (From BPS—L. M. Pope and D. R. Grote, University of Texas, Austin.)*

which is a lipid component of eukaryotic membranes. Lecithinase activity, associated with *C. perfringens*, is in part responsible for the ability of this organism to cause gas gangrene. The hydrolysis of lecithin by the alpha toxin of *C. perfringens* destroys many cells and is the primary cause of the extensive tissue damage in this disease. Lecithinase also acts as a hemolysin, causing lysis of red blood cells in addition to destroying cells of various other tissues.

Some *Staphylococcus* and *Streptococcus* species produce **fibrinolysin (kinase)**. The fibrinolytic enzymes staphylokinase and streptokinase catalyze the lysis of fibrin clots. The action of these two fibrinolytic enzymes may enhance the invasiveness of pathogenic strains of *Staphylococcus* and *Streptococcus* by preventing fibrin in the host from walling off the area of bacterial infection. Without the action of fibrin, the pathogens are free to spread to surrounding areas. In a somewhat different way, the production of **coagulase** enhances the virulence of some *Staphylococcus* species. The enzyme coagulase converts fibrinogen to fibrin. Some *Staphylococcus* species, such as *Staphylococcus aureus*, produce this enzyme, and the deposition of fibrin around the staphy-

lococcal cells presumably protects the cells against the circulatory defense mechanisms of the host. Coagulase negative strains of *Staphylococcus aureus*, however, still have been found to be virulent pathogens. It is thus difficult to associate virulence with the activity of a single enzyme, even though these enzymes appear to play a role in the virulence of a variety of pathogenic microorganisms.

Several other enzymes produced by microorganisms can destroy body tissues. For example, **hyaluronidase** breaks down hyaluronic acid, the substance that holds together the cells of connective tissues. Pathogens that produce hyaluronidases spread through body tissues, and therefore, hyaluronidase is referred to as the "spreading factor." Various species of *Staphylococcus, Streptococcus,* and *Clostridium* produce hyaluronidase enzymes. Some *Clostridium* species also produce collagenase, an enzyme that breaks down the proteins of collagen tissues. The k toxin of *C. perfringens*, for example, is a collagenase that contributes to the spread of this organism through the human body. The breakdown of fibrous tissues enhances the invasiveness of pathogenic microorganisms. Thus, the actions of some micro-

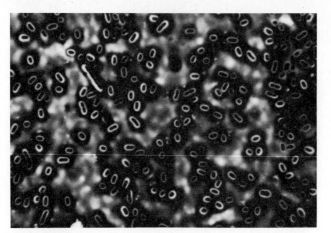

Figure 13.11

Micrograph showing encapsulated bacteria. The presence of a capsule increases the potential virulence of microbial pathogens because they are more resistant to host defense mechanisms. This is Klebsiella pneumoniae *(700 ×). (From BPS—Leon LeBeau, University of Illinois Medical Center.)*

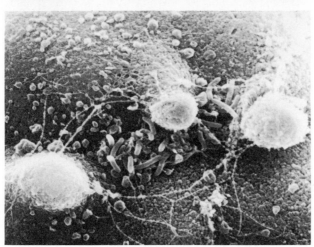

Figure 13.12

Micrograph showing Vibrio cholerae *attached to the lining of the gut of this infant mouse at the tip of the villus (3600 ×). The ability to attach to the intestine is an important characteristic of many pathogens that enter the body through the gastrointestinal tract. (From BPS—Garry T. Cole, University of Texas, Austin.)*

bial enzymes contribute to the virulence of pathogens by enhancing the ability of the microorganisms to proliferate within body tissues and by interfering with the normal defense mechanisms of the host organism.

Factors that interfere with phagocytosis

Several other factors contribute to the virulence of microorganisms, including the production of surface layers that interfere with the ability of **phagocytic blood cells** to engulf and destroy bacteria that invade the human body. As discussed in earlier chapters, **capsules** protect some bacteria against the host defense mechanism of phagocytosis. Capsules surrounding the cells of strains of *Streptococcus pneumoniae*, for example, permit these bacteria to evade the normal defense mechanisms of the host, allowing them to reproduce

and cause the disease symptomology of pneumonia. The virulence of other bacteria, including *Haemophilus influenzae* and *Klebsiella pneumoniae*, is also enhanced by capsule production (Figure 13.11).

Adhesion factors

In addition to their role in avoiding phagocytosis, capsules and slime layers contribute to the ability of bacteria to attach or adhere to particular host cells or tissues. Many pathogenic bacteria must adhere to mucous membranes in order to establish an infection. Specific factors that enhance the ability of a microorganism to attach to the surfaces of mammalian cells are termed **adhesins**, and the production of such substances is another important factor that determines the virulence of particular pathogens. The pili of several pathogenic bacteria and their associated adhesins appear to play a key role in permitting the bacteria to adhere to host cells and establish infections. For example, enteropathogenic strains of *E. coli* have particular adhesins associated with their pili that permit them to bind to the mucosal lining of the intestine. In a similar manner, *Vibrio cholerae* is able to adhere to the mucosal cells lining the intestine, allowing the establishment of an infection (Figure 13.12).

Likewise, the adsorption of certain viruses onto specific receptor sites of human cells establishes the necessary prerequisite for the uptake of the viruses by those cells, leading to the reproduction of the viruses, disruption of normal host cell function, and the production of disease symptoms by the invading viral pathogens. Some viruses, such as adenoviruses, have external spikes that aid in their attachment to host cells. Similarly, the spikes of orthomyxoviruses and paramyxoviruses attach to receptors of *N*-acetylneuraminic acid on the surfaces of human red blood cells. The ability of pathogenic microorganisms, including viruses, to attach to and invade particular cells and tissues establishes specific tissue affinities for pathogenic microorganisms.

Host defense mechanisms

Outer barriers

Intact body surfaces represent the first line of defense against microorganisms (Figure 13.13). Most microorganisms, including the normal microbiota of humans, are generally noninvasive and so do not penetrate the skin and mucous membranes. The outer surface of the skin layer is composed of **keratin**, a protein not readily enzymatically degraded by microorganisms. Keratin resists the penetration of water and thus proves a formidable external barrier for microorganisms to penetrate. We have already discussed the contributions of saline sweat and sebum to the prevention of microbial growth on skin surfaces. Those microorganisms that are able to tolerate the environmental conditions of the skin and proliferate there, themselves produce antimicrobial substances (also termed allelopathic substances) that act to prevent the establishment of infection on or through the skin by pathogenic microorganisms.

The respiratory tract (Figure 13.14) is protected in part against the invasion of pathogenic microorganisms by secretions of **mucus**. Mucus is secreted from goblet cells and subepithilial glands. Microorganisms tend to become entrapped in mucus and are swept out of the body by **ciliated epithelial cells** composing the lining of much of the respiratory tract. The ciliated epi-

Figure 13.13

The first lines of defense against infection: protection by external body surfaces acting as barriers.

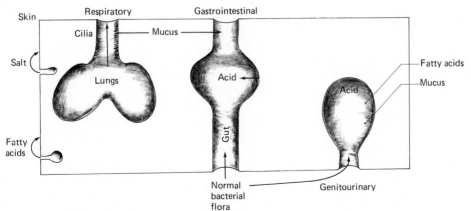

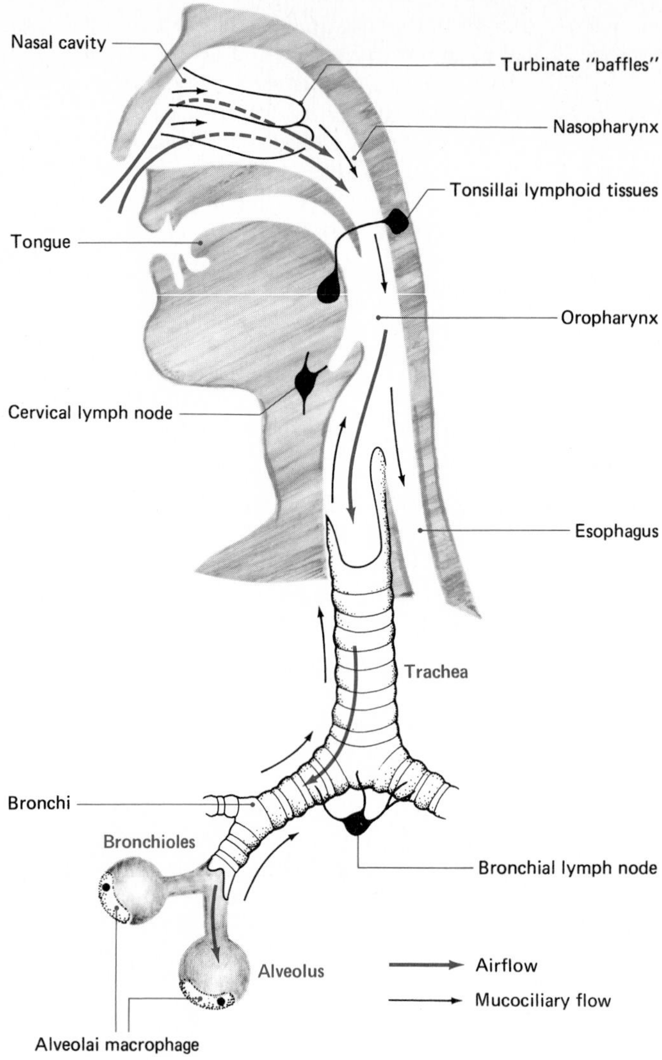

Figure 13.14

The respiratory tract has several mechanisms of nonspecific resistance that protect against invasion by pathogenic microorganisms.

Nasal cavity

Turbinate "baffles"

Nasopharynx

Tongue

Tonsillai lymphoid tissues

Oropharynx

Cervical lymph node

Esophagus

Trachea

Bronchi

Bronchioles

Bronchial lymph node

Alveolus

Airflow

Mucociliary flow

Alveolai macrophage

thelial cells effectively act as filters to prevent potential pathogens from penetrating the surface tissues of the respiratory tract (Figure 13.15). Some of the mucus and microorganisms are swept out of the body through the oral and nasal cavities, and other entrapped microorganisms are swept to the back of the throat where they are swallowed and enter the digestive tract. Sneezing and coughing tend to remove many of these microorganisms from the respiratory tract (Figure 13.16). Additionally, the swallowing reflex moves most particulates, including microorganisms that become attached to mucus, into the digestive, rather than into the respiratory tract.

The very low pH of the stomach creates a barrier for many microorganisms entering the digestive tract, as most microorganisms are unable to tolerate the acidity. Thus, the numbers of viable microorganisms are greatly reduced during passage through the stomach. Bile and digestive enzymes in the intestines further reduce the numbers of surviving microorganisms. The mucous membranes of the intestinal tract make it difficult for pathogenic microorganisms to attach to and penetrate the tract lining. Additionally, the large populations of indigenous microbiota in the lower intestinal tract protect the host against invasion by pathogens; the natural microbiota of the gastrointestinal tract enter into antagonistic relationships with nonindigenous microorganisms. As a result, most nonindigenous microorganisms entering the intestinal tract are degraded during passage through the gastrointestinal system or are removed along with large numbers of indigenous microorganisms in the passage of fecal material from the body.

Figure 13.15

Electron micrograph showing filtering of influenza viruses by the cilia lining the respiratory tract. The influenza viruses appear as small black particles. Many microorganisms are trapped on the surface mucus of the projecting cilia and fail to penetrate the respiratory tract, but in the case of influenza viruses, some virions may penetrate, leading to the onset of influenza. (Courtesy R. Dourmashkin, New York University.)

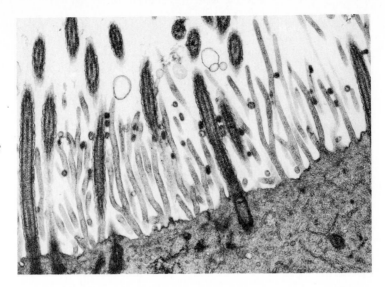

Other body tissues are also protected against invasive microorganisms. The genitourinary tract is protected from pathogenic microorganisms by a variety of mechanisms that include the outward flow of mucus, low pH, high salt concentrations, the presence of antimicrobial enzymes, and antagonistic relationships with the natural microbiota of the lower genitourinary tract. The tissues of the eye are protected against the undesirable growth of microorganisms by the presence of various enzymes in tears, including lysozyme. Lysozyme, which also is found in other body fluids including saliva and mucus, degrades the cell walls of bacteria, conferring antimicrobial activity on these body fluids. The continuous washing of the eye with tears, which contain this antimicrobial substance, generally prevents the growth of microorganisms on the tissues of the eye. In a similar way swallowing, coughing, and sneezing expose bacteria to body fluids with antimicrobial activity, thus reducing the number of potential pathogens.

Figure 13.16

A violent cough expels microorganisms from the respiratory tract. Unfortunately, this defense mechanism propels droplets into the air, and thus potentially toward the respiratory tract of another individual. (A) Note the distant spread of droplets as this young boy coughs. (B) Note how the spread of droplets is inhibited by this little girl's placing her hand over her mouth.

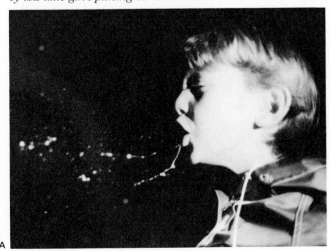

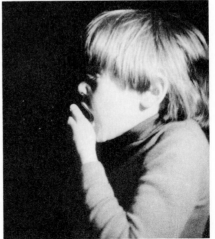

A

B

Phagocytosis

Should microorganisms penetrate these outer barriers, they are subject to **phagocytosis** by a variety of cells of the blood, the **mononuclear phagocyte system** (formerly known as the reticuloendothelial system) and the lymphatic system. Phagocytosis is a highly efficient host defense mechanism against the invasion of microorganisms. Such defense mechanisms are innate properties of the host organism that are active without prior exposure to microbial pathogens. As a rule, such innate defense mechanisms exhibit relatively low specificity toward particular species of microorganisms. Whereas phagocytic activities are part of the nonspecific defense

Figure 13.17

Blood contains many different types of cells, many of which are involved in nonspecific protection against invasion by microorganisms. (A) The different types of blood cells. (B) The normal cellular composition of adult human blood.

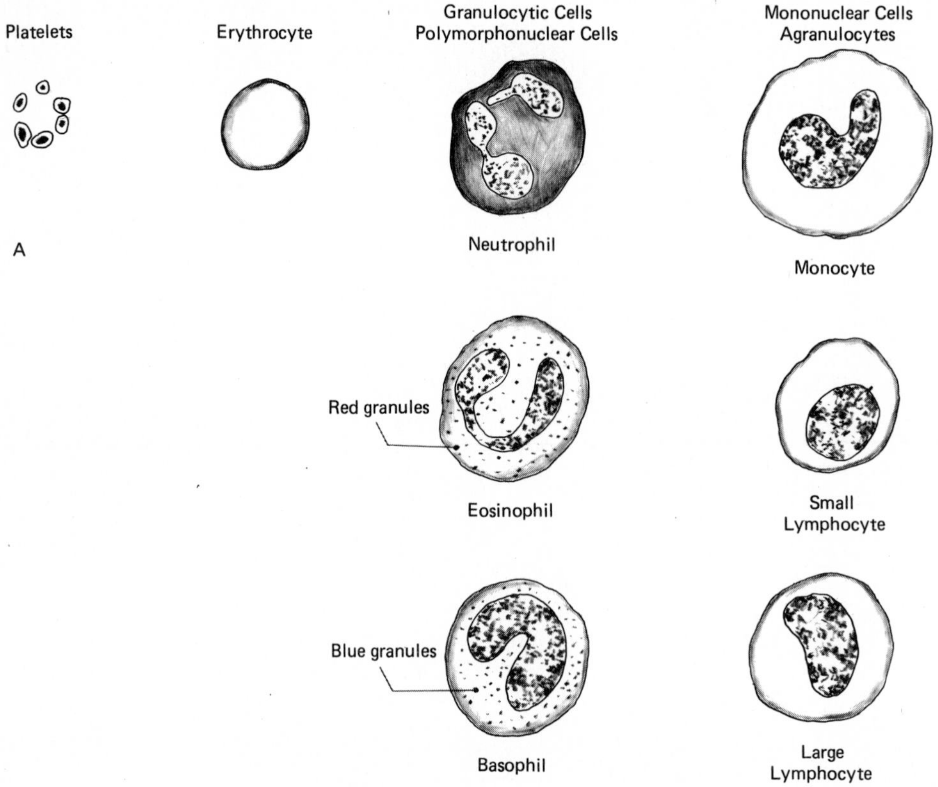

B   Normal Cellular Composition of Adult Human Blood

Cell Type	No. of cells per µl	Approximate % (leukocytes)
Leukocytes (white cells)	4,500–9,000	
Granulocytes		
Neutrophils	3,000–6,750	59 ± 15
Basophils	25–90	< 1
Eosinophils	100–360	2 to 4
Mononuclear cells		
Lymphocytes	1,000–2,700	34 ± 10
Monocytes	150–170	3 to 8
Platelets	145,000–375,000	
Erythrocytes (red cells)	Men = 4.2–5.4 × 10^6 Women = 3.6–5.0 × 10^6	

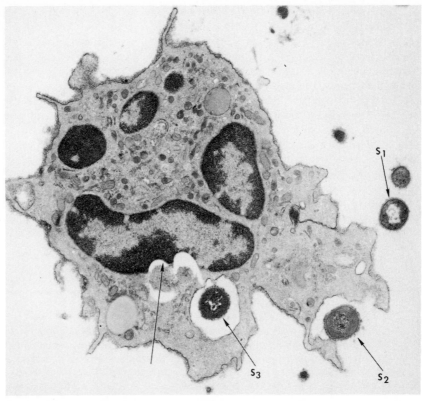

Figure 13.18

Micrograph of a polymorphonuclear leukocyte in the presence of Streptococcus
pyogenes. S_1 *is free,* S_2 *is being phagocytized, and* S_3 *has been phagocytized. The
arrow shows where the nucleus is being digested. (From BPS—John G. Hadley,
Battelle Pacific Northwest Laboratories, Richland, Washington.)*

mechanism against infection, these same activities
play a role in the specific immune defense system
that will be discussed in Chapter 14. In fact, phagocytosis is most efficient at removing bacterial cells
when it is activated by components of the specific
immune response system.

In blood, several types of white blood cells
(**leukocytes**) are involved in nonspecific resistance against pathogenic microorganisms (Figure
13.17). Some of the leukocytes, called granulo-
cytes, contain cytoplasmic granules. These include: **basophils**, leukocytes stained with basic dyes;
eosinophils, leukocytes that react with acidic dyes,
becoming red when stained with the dye eosin;
and **neutrophils** (also called **polymorphonuclear
neutrophils**, **polymorphs**, or **PMNs**) that contain
granules that exhibit no preferential staining—
that is, they are stained by neutral, acid, and basic
dyes. The leukocytes that do not contain granular
inclusions (agranulocytes) include the monocytes
and the lymphocytes. The **monocytes** are partic-

ularly important in the nonspecific immune response, and **lymphocytes** are especially important in the specific immune response.

Neutrophils, the most abundant phagocytic cells
in blood, are produced in the bone marrow and
are continuously present in circulating blood, affording protection against the entry of foreign
materials. These white blood cells exhibit chemotaxis and are attracted to foreign substances, including invading microorganisms, which they engulf and digest along with particulate matter (Figure
13.18). Neutrophils are short-lived in the body,
living for only a few days, but are replenished to
the blood from the bone marrow in high numbers.

Phagocytic blood cells can have numerous ly-
sozomes that contain hydrolytic enzymes capable
of digesting microorganisms. During phagocytosis the microorganism is engulfed by the pseudopods of the phagocytic cell and is transported
by endocytosis across the membrane of the

phagocytic cell where it is contained within a vacuole called a **phagosome**. The phagosome migrates to and fuses with a lysosome, thus exposing the microorganisms to the enzymes within the lysosome, including degradative enzymes and enzymes that catalyze the production of biochemicals such as hydrogen peroxide and superoxide radicals (Figure 13.19). During phagocytosis there is an increase in oxygen consumption by the phagocytic cells associated with elevated rates of metabolic activities of the hexose monophosphate shunt. As a consequence, oxygen is converted to the superoxide anion, hydrogen peroxide, singlet oxygen, and hydroxyl radicals, all of which have antimicrobial activity. Additionally, phagocytic cells contain digestive enzymes, including some that degrade D-amino acids, that are involved in killing ingested bacteria. The degraded microorganisms are transported to the cytoplasmic membrane within a vacuole and are removed from the phagocytic cells by exocytosis or are consumed within the phagocytic cell.

In addition to neutrophils, **monocytes** are formed by the stem cells in the bone marrow. These mononuclear cells, which are larger than neutrophils, are the precursors of macrophages and are able to move out of the blood to tissues that are infected with invading microorganisms. Outside the blood monocytes become enlarged, forming **phagocytic macrophages**. In contrast to neutrophils, macrophages are long-lived, persisting in tissues for weeks or months. Once these differentiated macrophages are formed, they are capable of reproducing to form additional macro-

Figure 13.19

The phagocytic killing of pathogenic microorganisms involves the engulfment and digestion of the microorganisms by phagocytic blood cells. First, the prey become attached to the surface of the phagocyte membrane, the attachment is assisted by complement and/or antibodies (A). Numerous pseudopodia surround the prey, and the prey enters the cell by phagocytosis (B). The membranes of the pseudopodia fuse to enclose the prey within a phagosome (C). Enzymes from many granules are dumped into the phagosome, and the prey is often digested (D). The fluids and solutes are absorbed by the cell or removed by exocytosis (E).

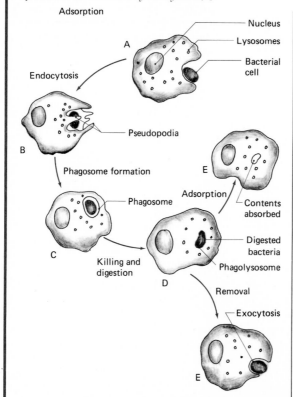

Discovery process

Eli Metchnikoff in 1884 described phagocytosis in his studies on the infection of the common waterflea, *Daphnia*, by a fungus (*Monospora bicuspidata*). This was an ideal system for him to use in his effort to determine how the bacterial cells that he saw inside white blood cells in the body got there, because the *Daphnia* is simple and transparent and the fungus is large and easily seen without staining. Metchnikoff was able to observe all stages of the *Monospora* in the abdominal cavity of the sick *Daphnia* where he saw spores penetrating the intestinal wall as a result of peristalsis, whereupon blood corpuscles immediately began to surround and attach themselves to the spores. These blood corpusles are circulating, colorless, phagocytic cells adapted to the uptake of solid particles. Metchnikoff was able to observe and describe the changes the spore underwent until it was destroyed and separated into irregular grains. Today, we recognize the phagocytic activity of human white blood cells as a primary line of defense against invasion of the body by pathogenic microorganisms.

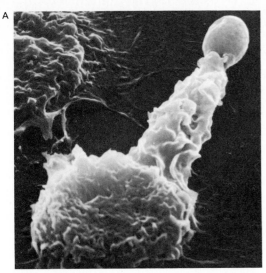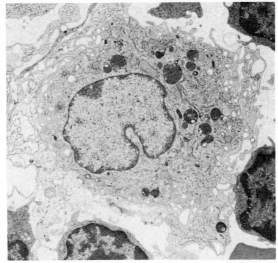

Figure 13.20

(A) This aveolar macrophage is utilizing an extended pseudopod for ingesting a yeast cell (3000 ×). (From John G. Hadley, Battelle Pacific Northwest Laboratories, Richland, Washington.) (B) This macrophage is from the medulla of a rat lymph node (4,600 ×). (From BPS—R. Rodewald, University of Virginia.)

phages, whereas neutrophils are terminal cells that are nonreproductive and must be replenished from the bone marrow. Macrophages, like neutrophils, are able to engulf and digest microorganisms, but some of the lysosomal enzymes of macrophages are different from those of polymorphs. Some microorganisms are resistant to the enzymatic activities of neutrophils and/or macrophages. For example, *Mycobacterium tuberculosis* survives and even multiplies within macrophages. As a result some microorganisms survive, continuing to grow and cause infection because of the failure of these phagocytic cells to kill invading pathogens. This is one of the reasons that tuberculosis is a persistent disease and is difficult to treat.

Although not as numerous as neutrophils, macrophages are distributed throughout the body, including at fixed sites within the **mononuclear phagocyte** system (Figure 13.20). The mononuclear phagocyte refers to a systemic network of phagocytic cells distributed through a network of loose connective tissue and the endothelial lining of the capillaries and sinuses of the human body. The phagocytic cells associated with the linings of the blood vessels in bone marrow, liver, spleen, lymph nodes, and sinuses constitute this host defense system. Some of the macrophages in the mononuclear phagocyte system occur at fixed sites and are designated with particular names. For example, **microglia** refer to macrophages of the

central nervous system; **Kupffer cells** are phagocytic cells which line the blood vessels of the liver; **dust cells** are macrophages fixed in the alveolar lining of the lungs; and fixed macrophages in connective tissues are referred to as **histiocytes** (Figure 13.21). Other macrophages of the mononuclear phagocytic system are called **wandering cells** because they move freely into tissues where foreign substances have entered, eliciting a chemotactic response to chemical stimuli formed by these phagocytic cells. Wandering macrophages occur in the peritoneal lining of the abdomen and the alveolar lining of the lung as well as in other tissues. The presence of relatively high numbers of macrophage in the respiratory tract is important in preventing the establishment of both pathogens and a normal indigenous microbiota within the tissues of the lower respiratory tract.

The lymphatic system also plays a major role in host resistance to pathogens through the lymphocytes, cells involved in the specific immune response, the role of which will be discussed in Chapter 14.

Interferons

In addition to phagocytic white blood cells, some blood factors, including interferons, are involved in the nonspecific host resistance to invading

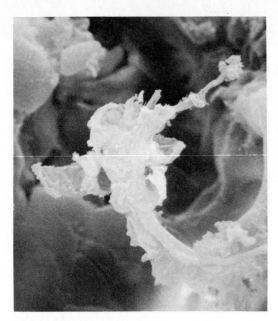

Figure 13.21

Kupffer cells are some of the fixed macrophage cells of the reticuloendothelial system. In this scanning electron micrograph of a Kupffer cell (3000×), the pseudopodia of this fixed macrophage of the rat liver are visible. The topmost pseudopod appears to be attached to a foreign body. (Courtesy Robert Apkarian, University of Louisville.)

480

Interactions be-
tween microor-
ganisms and
humans

Figure 13.22

The time course for interferon and serum antibody production after viral infection in the lungs of mice is illustrated in this graph.

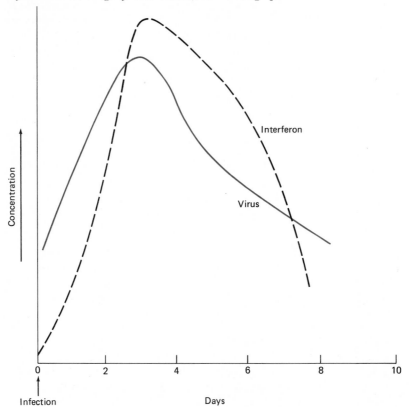

microorganisms. **Interferon** is a family of inducible glycoproteins produced by eukaryotic cells in response to viral infections and other stimuli. Interferon production is also elicited by other microbial pathogens that reproduce within host cells, including bacteria, such as rickettsias and chlamydias, and protozoa, such as those causing malaria. The production of interferon occurs shortly after such infections (Figure 13.22). Interferon is produced both by infected tissue cells and by certain lymphocytes that are part of the lymphatic defense system. Interferon glycoproteins are of relatively low molecular weight and are normally produced only in low concentrations. These glycoproteins are released from infected cells and migrate to uninfected cells, protecting the cells from viral infections (Figure 13.23). Interferons do not block the entry of the virus into a cell but rather prevent the replication of viral pathogens within protected cells.

The synthesis of interferons is regulated at the level of transcription. It appears that an infection with a virus induces synthesis of interferon glycoproteins with double-stranded RNA viruses, such as myxoviruses, the most potent inducers of interferon synthesis. When interferons move to uninfected cells, the synthesis of specific proteins that inhibit translation is derepressed. It is hypothesized that these translational inhibitory proteins block the translation of viral mRNA without preventing the translation of host cell mRNA molecules. The interferon system thus involves a complex series of molecular events that induce an antiviral state. These events include: recognition of an interferon-inducing molecule; derepression and synthesis of interferon proteins; modification and secretion of the interferon molecules; interaction of interferon with susceptible cells; activation and synthesis of previously repressed genetic information; and the alteration of the cell's metabolism as expressed by some identifiable interferon interaction, such as resistance to viral infection. It should be emphasized, though, that interferons themselves are not antiviral substances. They have no direct effect on viruses, and their antiviral action is mediated by cells in which they induce an antiviral state.

Interferon production is considered a nonspecific resistance factor because interferon proteins do not exhibit specificity toward the particular pathogen, which means that interferon produced in response to one virus is also effective in preventing the replication of other types of viruses. Although interferon glycoproteins are not specific for a species of invading viruses or other intracellular microbial parasites, they are specific for the host organism that produced them; that is, interferons produced by human cells are effective only in human cells and do not exert a protective effect against intracellular parasites in other animal species. Interferons appear to be the main defense mechanism against viral infections, with the production of interferons playing a key role in preventing viral infections such as the common cold. Additionally, interferons hold some promise as antitumor agents preventing proliferation of cancer cells. This topic is an area of intensive investigation, but it has yet to be demonstrated to have clinical significance.

Complement

In addition to interferons, blood contains other glycoproteins called **complements** that play a role in the removal of invading pathogens. Complement glycoproteins include at least 11 different proteins in the blood serum. The complement system normally acts in association with the specific immune response that will be discussed in Chapter 14. The complement system, however, can act in an alternate pathway that does not require antibody molecules of the specific immune response system (Figure 13.24). Some biochemicals, such as lipopolysaccharide toxins, can trigger the nonspecific response of the complement system. The nonspecific initiation of the complement system involves the activation of the C_3 protein of the complement system. The C_3 complement molecule is split into C_{3a}, which acts as a chemotactic stimulus for neutrophils, and C_{3b}, which adheres to the bacterial cell. The conversion of C_3 also permits the addition of other complement proteins in a cascade fashion. The action of the complement system increases the susceptibility of microorganisms to phagocytosis and also promotes the inflammatory response.

Inflammatory response

The **inflammatory response** represents a generalized response to an infection and is designed to localize the invading microorganisms and arrest the spread of the infection. **The inflammatory response is characterized by four symptoms: reddening of the localized area; swelling; pain; and elevated temperature. The redness** results from capillary dilation that allows more blood to flow. The term dilation is really a misnomer because the capillaries are no more dilated than they normally are; during "dilation" there are simply fewer

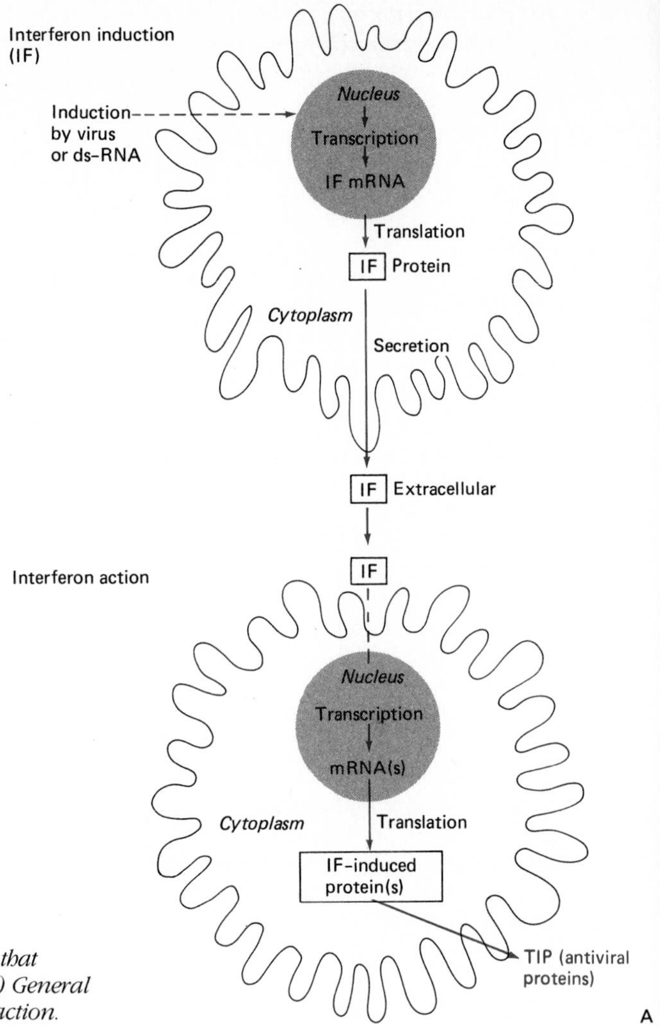

Interferon induction
(IF)

Induction——by virus
or ds-RNA

Nucleus

Transcription

IF mRNA

↓ Translation

IF Protein

Cytoplasm

Secretion

IF Extracellular

IF

Interferon action

Nucleus

Transcription

mRNA(s)

↓ Translation

IF-induced
protein(s)

TIP (antiviral
proteins)

A

Figure 13.23

*The production interferon activates a system that
prevents viral replication within host cells. (A) General
sequence of events. (B) Details of interferon action.*

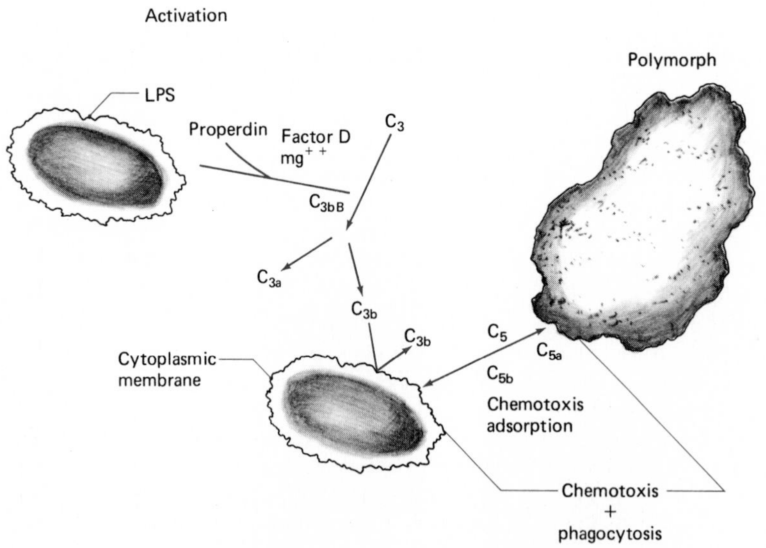

Activation

Polymorph

LPS

Properdin   Factor D
mg^{++} C_3

C_{3bB}

C_{3a}

C_{3b}

Cytoplasmic
membrane

C_{3b} C_5

C_{5a}

C_{5b}

Chemotoxis
adsorption

Chemotoxis
+
phagocytosis

Figure 13.24

*The alternate complement
system involves a cascade of
complement molecules to
the surface of an invading
microorganism, leading to
enhanced phagocytosis and
death of the pathogen by
lysis.*

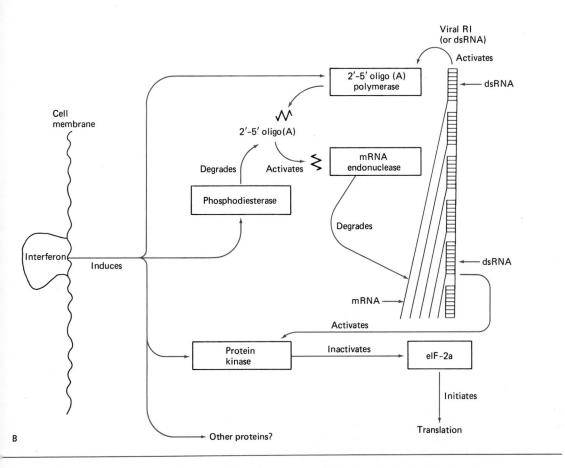

constricted capillaries, permitting increased blood circulation through more open capillaries. The **elevated temperature** also results because the capillary dilation permits increased blood flow through these vessels. The dilation of blood vessels is accompanied by increased capillary permeability, causing **swelling** as fluids accumulate in bases surrounding the tissue cells. Actually, the swelling is due to increased permeability of the venules, but the term increased capillary permeability is entrenched in the clinical terminology to describe this phenomenon. **Pain** is due to lysis of blood cells, triggering the production of bradykinin and prostaglandins. Bradykinin decreases the firing threshhold for pain fibers and the prostaglandins, PGE_1 and PGE_2, intensify this effect. Aspirin, which we often use to decrease pain, antagonizes prostaglandin formation but has little

or no effect on bradykinin formation. Thus, aspirin can decrease but not eliminate the pain associated with the inflammatory response.

After considering the characteristic symptoms of the inflammatory response, let us discuss how it helps defend against infections by pathogenic microorganisms. As indicated, the dilation of blood vessels in the area of the inflammation increases blood circulation, allowing increased numbers of phagocytic blood cells to reach the affected area. Neutrophils are initially most abundant, but in the later stages of inflammation, monocytes and macrophages of the mononuclear phagocyte system predominate (Figure 13.25). The phagocytic cells are able to kill many of the ingested microorganisms. Phagocytic blood cells stick to the lining of the blood vessels and migrate to the affected tissues, passing between the endothelial

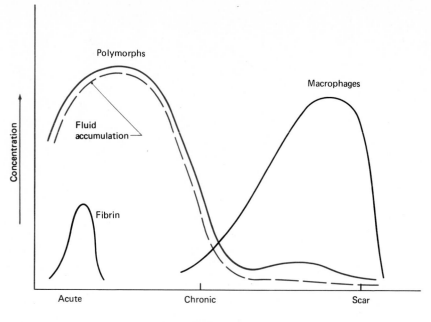

Figure 13.25

During the early stages of the inflammatory response, polymorphs are particularly important, and in the later stages of an inflammatory response the concentration of polymorphs declines and the number of macrophage increases.

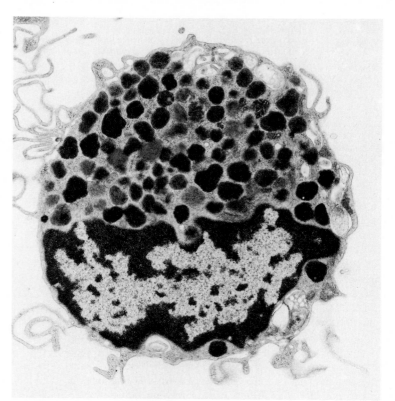

Figure 13.26

A human mast cell from the lung, showing granule inclusions of histamines. The release of histamine from these cells occurs as part of the immune response, leading to altered vascular permeability. (Courtesy Ann Dvorak, Beth Israel Hospital, Boston.)

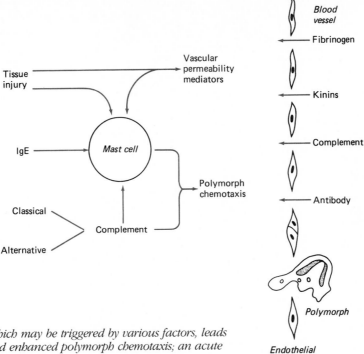

Figure 13.27

The degranulation of mast cells, which may be triggered by various factors, leads to altered vascular permeability and enhanced polymorph chemotaxis; an acute protective inflammatory reaction results.

cells of the blood vessel by a process known as diapedesis. The death of phagocytic blood cells involved in combating the infection results in the release of histamine, prostaglandins, and bradykinins which, in addition to their other effects, are vasodilators. Additionally, mast cells react with complement, leading to the release of large amounts of histamine contained within these blood cells (Figure 13.26). These substances are biochemical mediators of altered circulation during the inflammatory response (Figure 13.27). Thus, the death of some phagocytic cells and the release of partially degraded microorganisms enhances the inflammatory response.

The area of the inflammation also becomes walled off as result of fibrinous clots that develop. The deposition of fibrin isolates the inflamed area, cutting off normal circulation. The fluid that forms in the inflamed area is known as the inflammatory exudate, commonly called pus. This exudate con-

tains dead cells and debris in addition to body fluids. After the removal of the exudate, the inflammation can terminate, and the tissues may return to their normal functioning. The elevation of body temperature (fever) aids in the ability of the inflammatory response to protect against invading microorganisms. Fever is a response to pyrogens, including bacterial endotoxins, that enter the blood as a result of the death of microorganisms or that are released by phagocytic blood cells. Elevated body temperatures favor enhanced rates of phagocytic activity and other aspects of the immune response. The elevated temperatures also often reduce the rates of reproduction of pathogenic microorganisms, many of which have optimal temperatures coincident with the normal body temperature. Thus, the complex reactions of the inflammatory response work in an integrated fashion to contain and eliminate infecting pathogenic microorganisms.

Postlude

The interactions between microorganisms and human beings represent a delicate balance between biological populations. The normal microbiota of humans play a role in maintaining a "healthy" condition, but various microorganisms can act as pathogens causing disease. The normal microbiota of human beings are usually indigenous to the surfaces of body tissue, most notably the oral cavity, gastrointestinal tract, vaginal tract, and skin surface. These resident microorganisms are generally noninvasive. The environmental conditions of particular body tissues and the physiological properties of particular microbial populations determine the species composition of the resident microbial community of a given body tissue. In contrast to the normal human microbiota, pathogenic microorganisms that cause human disease generally are either invasive, toxigenic, or both. There are a variety of factors that confer virulence on microbial populations, including the ability to produce toxins, which disrupt the normal functioning of mammalian cells; invasiveness factors, such as enzymes that permit the microorganisms to spread through body tissues; and adaptations, such as capsules, that permit the pathogenic microorganisms to evade the normal host defense mechanisms.

The ability of a microorganism to reproduce within the human body, producing an infection, requires that such a pathogenic microorganism overcome or evade a variety of host defense mechanisms. The surfaces of body tissues are generally resistant to penetration by microorganisms. In addition to acting as a physical barrier, human surface tissues often contain biochemicals with antimicrobial activity, such as the fatty acids of skin tissues and the enzyme lysozyme in tears. Microorganisms that penetrate the surface barriers are confronted by several nonspecific resistance mechanisms. Several substances in the circulatory system, such as interferon and complement glycoproteins, have antimicrobial activities. Interferons in particular prevent the reproduction of viruses within human cells.

Additionally, various cells that circulate through the body, such as monocytes, neutrophils, and macrophages have phagocytic activity. The ability to engulf and ingest invading microorganisms eliminates most would-be pathogens. The inflammatory response brings together several different mechanisms for preventing the proliferation and spread of infecting pathogenic microorganisms. The inflammatory response tends to limit the spread of an infection, to direct phagocytic cells to the site of infection, and to enhance the effectiveness of phagocytic removal of infecting pathogens. These host defense mechanisms are physiological in nature and as such are subject to individual variation. Numerous factors can act to compromise the host defense mechanisms, rendering an individual susceptible to disease-causing pathogenic microorganisms. For example, a wound breaks the normal surface barrier, permitting invasion by microorganisms; emotional and physiological stress (e.g., improper nutrition, exposure to extreme temperatures, and tension) lower the physiological potential for successful phagocytic removal of infecting microorganisms; and age may influence the physiological ability of an individual to resist infection, with young children and the elderly being particularly prone to various infectious diseases.

In addition to the mechanisms already described for maintaining a balance between microbial populations and human beings without the occurrence of disease, there is a specific or acquired immune defense mechanism that protects individuals against diseases caused by microorganisms. The topic of specific acquired immunity will be discussed in the next chapter. It should be noted, though, that the specific immune defense mechanisms are interactive with several of the nonspecific host resistant mechanisms. The complement proteins and phagocytic cells, for example, play a role in both nonspecific and specific immune defense mechanisms. Clearly, the preclusion of microbial diseases of humans involves an elaborate and complex integrated system of host resistance mechanisms.

1. What is meant by the normal human microbiota?

2. What body surfaces are normally colonized by microorganisms?

3. What body fluids normally are free of bacteria? Why?

4. How does the resident microbiota of the skin differ from the resident microbiota of the gastrointestinal tract? What are some of the major bacterial genera occurring in these two habitats?

5. What factors influence which bacterial genera can establish themselves within the indigenous microbial community of the skin?

6. How does the presence of an indigenous microbiota benefit humans?

7. What attributes contribute to the virulence of pathogens? What is the difference between toxigenicity and invasiveness?

8. What is a toxin? What is the difference between an endotoxin and an exotoxin?

9. What is phagocytosis? How does it contribute to our resistance to pathogenic microorganisms?

10. What is an inflammatory response? How does inflammation act to prevent the spread of pathogens throughout the body?

11. How is the respiratory tract protected against invasion by pathogenic microorganisms?

12. How is interferon involved in protection against viral infections?

Suggested Supplementary Readings

Bernheimer, A. W. 1976. *Mechanisms in Bacterial Toxicology*. John Wiley & Sons, New York.

Bowden, G. H. W., D. C. Elwood, and I. R. Hamilton. 1979. Microbial ecology of the oral cavity. *Advances in Microbial Ecology* 3: 135–218.

Burke, D. C. 1977. The status of interferon. *Scientific American* 236(4): 42–62.

Christensen, C. M. 1975. *Molds, Mushrooms and Mycotoxins*. University of Minnesota Press, Minneapolis.

Chu, F. S. 1978. Mode of action of mycotoxins and related compounds. *Advances in Applied Microbiology* 22: 83–143.

Clarke, R. T. J., and T. Bauchop (eds.). 1978. *Microbial Ecology of the Gut*. Academic Press, London.

Collee, J. G. 1981. *Applied Medical Microbiology*. Blackwell Scientific Publications, Oxford, England.

Cuatrecasas, P. (ed.). 1977. *The Specificity and Action of Animal, Bacterial and Plant Toxins*. Chapman & Hall, London.

Friedman, R. M. 1981. *Interferons: A Primer*. Academic Press, New York.

Gilman, A. G., L. S. Goodman, and A. Gilman (eds.). 1980. *Goodman and Gilman's The Pharmacological Basis of Therapeutics*. Macmillan Publishing Co., New York.

Goren, M. 1977. Phagocyte lysosomes: interactions with infectious agents, phagosomes, and experimental perturbations in function. *Annual Review of Microbiology* 31: 507–533.

Gotze, O., and H. J. Muller-Eberhard. 1976. The alternative pathway of complement activation. *Advances in Immunology* 24: 1–35.

Immunology: Readings from Scientific American. 1976. W. H. Freeman and Co., San Francisco.

Isenberg, H. D., and A. Balows. 1981. Bacterial pathogenicity in man and animals. In: *The Prokaryotes* (M. P. Starr, H. Stolp, H. G. Truper, A. Balows, and H.G. Schlegel, eds.), Volume 2, pp. 83–122. Springer-Verlag, Berlin.

Kass, E. H., and S. M. Woolf. 1973. *Bacterial Lipopolysaccharides: The Chemistry, Biology and Clinical Significance of Endotoxins*. The University of Chicago Press, Chicago.

Marples, M. J. 1969. Life on the human skin. *Scientific American* 220(1): 108–129.

Mims, C. A. (ed.). 1976. *The Pathogenesis of Infectious Disease*. Academic Press, London.

Robbins, S. L., and R. S. Cotran. 1974. *The Pathological Basis of Disease*. W. B. Saunders, Philadelphia.

Rosebury, T. 1962. *Microorganisms Indigenous to Man*. McGraw-Hill Book Co., New York.

Savage, D. C. 1977. Microbial ecology of the gastrointestinal tract. *Annual Review of Microbiology* 31: 107–133.

Schlesinger, R. B. 1982. Defense mechanisms of the respiratory system. *BioScience* 32(1): 45–50.

Shilo, M. 1967. Formation and mode of action of algal toxins. *Bacteriological Reviews* 31: 18–193.

Skinner, F. A., and J. G. Carr (eds.). 1974. *The Normal Microflora of Man*. Academic Press, London.

Smith, H. 1968. Biochemical challenge of microbial pathogenicity. *Bacteriological Reviews* 32: 164-184.

Smith, H., and J. H. Pearce (eds.). 1972. *Microbial Pathogenicity in Man and Animals*. Society for General Microbiology, Cambridge, England.

Stewart, W. E. (ed.). *Interferons and Their Actions*. 1977. CRC Press, Inc., Boca Raton, Florida.

Wilkinson, P. C. 1974. *Chemotaxis and Inflammation*. Churchill Livingstone, Edinburgh.

Wogan, G. N. 1975. Mycotoxins. *Annual Review of Pharmacology* 15: 436–451.

Zweifach, B. W., L. Grant, and R. T. McCluskey. 1974. *The Inflammatory Process*. Academic Press, New York.

The immune response

14

The **immune response system** of human beings and other higher animals provides a mechanism for a specific response to the invasion of particular pathogenic microorganisms and other foreign substances. It is largely this specific physiological response that protects us against disease. In contrast to nonspecific defense mechanisms, **the immune response is characterized by specificity, memory, and the acquired ability to detect foreign substances.** The human immune response is able to recognize macromolecules that are different in some way from the normal macromolecules of the body. The ability to differentiate "self" from "nonself" at the molecular level is an underlying necessity for the development of the specific immune response. The specificity of the immune response permits the recognition of even very slight biochemical differences between molecules, and consequently, the macromolecules of one microbial strain can elicit a different response than the macromolecules of even a very closely related strain of the same species. The specific immune response is adaptive or acquired, in that having responded once to a particular macromolecule, called an **antigen**, a memory system is established that permits a rapid and specific secondary response upon reexposure to that same substance (Figure 14.1). By being able to rapidly recognize and respond to pathogenic microorganisms, a state of immunity results that precludes infection with those specific pathogens. Because we remember which microorganisms have previously elicited an immune response, we can acquire or develop immunity to

Figure 14.1

The primary and secondary immune responses. A rabbit was injected on two separate occasions with staphylococcal toxoid. The antibody response to the second contact with antigen is more rapid and intense.

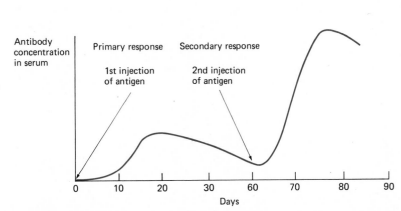

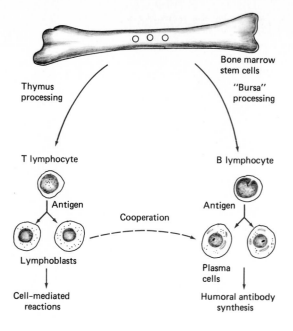

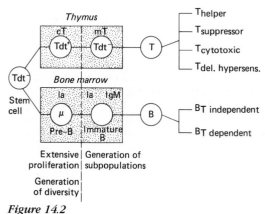

Figure 14.2

The differentiation of T and B cells forms the basis for the two arms of the immune defense network.

Examination of the basis of immunity reveals that there are two different forms of the immune response. In one response mode, called **antibody-mediated immunity**, specific proteins (**antibodies**) are made when foreign **antigens** are detected. In antibody-mediated immunity, plasma cells derived from certain white blood cells (**B lymphocytes**) synthesize antibodies in response to the detection of a foreign macromolecule with antigenic properties. By definition, an antigen can be any macromolecule that elicits the formation of antibody and that can subsequently react with antibody. In old medical terminology blood and other vital body fluids were considered as "humors," and thus, antibody-mediated immunity is often referred to as humoral immunity because the antibody molecules flow extracellularly through the body fluids. In the second form of the immune response, called **cell-mediated immunity**, certain cells of the body acquire the ability to destroy other cells that are recognized as foreign or abnormal. In contrast to antibody-mediated immunity, cell-mediated immunity depends on another class of lymphocyte cells, known as **T lymphocytes**. Sensitized T lymphocytes interact with "foreign" cells to bring about the destruction of those cells.

Immunity thus depends on B and T lymphocytes, which are cells differentiated in the lymphatic system (Figure 14.2). Both B and T lymphocytes originate from bone stem cells and become differentiated during maturation. The precursors for T cells pass through the liver and spleen before reaching the thymus gland, where they are processed. Although differentiated in the thymus gland, T cells are inactive until later maturation in other lymphoid tissues. The thymus-dependent differentiation of T cells or thymocytes occurs during childhood, and by adolescence the secondary lymphoid organs of the body generally contain their full complement of T cells. In human beings the B lymphocytes appear to develop in the bone marrow. The term B lymphocyte actually refers to bursa-dependent lymphocytes, so-named because these lymphocytes are differentiated in chickens and other birds in the lymphoid organ known as the bursa of Fabricius. Even though we do not possess a bursa of Fabricius, the designation B lymphocyte is applied to lymphocytes that can differentiate into antibody-synthesizing cells in humans. Within lymphatic tissues, B lymphocytes give rise to antibody-secreting plasma cells in response to antigenic stimulation.

specific diseases. As a consequence of **acquired immunity**, we usually suffer from many diseases, such as chicken pox, only once. We can also intentionally expose ourselves to specific foreign macromolecules, through the use of vaccines, to artificially establish a state of immunity. As discussed in Chapter 1, the use of vaccination to prevent disease was an ancient practice in the Far East that gained scientific credibility during the eighteenth century as a result of the pioneering work of Edward Jenner.

Antibody-mediated immunity

Antigens

A wide variety of macromolecules can act as **antigens**, including all proteins, most polysaccharides, nucleoproteins, lipoproteins, and various small biochemicals if they are attached to proteins or polypeptides. The two essential properties of an antigen are its immunogenicity (ability to stimulate antibody formation) and its specific reactivity with antibody molecules. The antigen molecule consists of a reactive portion, the **hapten**, that chemically reacts with the antibody molecule and a carrier portion that is necessary for the stimulation of antibody production. A hapten can react with antibody molecules but is unable to elicit the formation of antibody, whereas a complete antigen molecule is both reactive and elicits the production of specific antibodies. Antigenic molecules may be multivalent, having multiple reactive sites, or monovalent, having only one reactive site. Generally, multivalent antigens elicit a stronger immune response than monovalent antigens. In some cases a multivalent antigen, termed a **heterophile**, **heterologous antigen**, or **Forssman antigen**, can react with antibodies produced in response to a different antigen.

In many cases, antigens are associated with cell surfaces and are therefore called **surface antigens**. Human cells have specific surface antigens, and, for example, human blood types are determined by the presence or absence of antigens designated A and B on the surfaces of red blood cells. Microorganisms, including viruses and bacteria, also have many surface antigens, some of which may be associated with particular anatomical structures. For example, strains of *Salmonella* have specific antigens associated with their flagella proteins, called **flagella** or **H antigens**, and other specific antigens associated with the surface lipopolysaccharides of the cell wall, called **somatic** or **O antigens**.

One of the reasons that the immune response is effective in preventing disease is that the toxins contributing to the virulence of pathogenic microorganisms usually have antigenic properties. Antibodies produced against toxins are referred to as **antitoxins**. Most protein toxins are highly antigenic, eliciting the synthesis of high titers (concentrations) of antibody. Similarly, bacterial lipopolysaccharide toxins are antigenic and are responsible for the initiation of antibody pro-

duction against many Gram negative bacteria. Denatured proteins often retain their antigenic properties. Of particular importance are denatured proteins called **toxoids**, which retain their antigenic properties but do not cause the onset of disease symptoms; they are antigenically active macromolecules but are no longer toxins. Toxoids are useful for eliciting antibody-mediated immune responses without producing the diseases caused by toxins and toxin-producing pathogenic microorganisms. Toxoids are used as vaccines for protecting individuals against diphtheria, tetanus, and various other diseases caused by pathogenic microorganisms that produce protein toxins.

Antibody molecules

Antibody molecules, often referred to as **immunoglobulins**, are proteins found in the serum fraction of blood, that is, in the liquid portion after blood plasma is clotted. The synthesis of antibody molecules occurs in response to antigenic stimulation. **There are five classes of immunoglobulins: IgG, IgA, IgM, IgD, and IgE.** The characteristics of the five major classes of immunoglobulin molecules, all of which are globular proteins, are summarized in Table 14.1. The various classes of immunoglobulins serve different functions in the immune response; they are difficult to separate but can be differentiated by using electrophoresis, a method that separates proteins based on the charge, size, and shape of the macromolecule (Figure 14.3). The classes of immunologlobulins form relatively broad electrophoretic bands, indicating the heterogeneity of molecules within each class and making it impossible to use routine electrophoresis for the quantitation of individual classes of immunoglobulins.

Structure of immunoglobulin macromolecules

Immunoglobulin molecules all have the same basic structure consisting of four peptide chains, two identical heavy chains and two identical light chains, joined by disulfide bridges linking the chains (Figure 14.4). Differences in the heavy chains are responsible for the five major classes of immunoglobulins. There are five types of heavy chains referred to as alpha (α), mu (μ), gamma

table 14.1

Properties of five classes of immunoglobulins

Property	IgG	IgA	IgM	IgD	IgE
Molecular weight	150,000	160,000 and dimer	900,000	185,000	200,000
Number of basic 4-peptide units	1	1, 2	5	1	1
Heavy chains	γ	α	μ	δ	ϵ
Light chains	$\kappa + \lambda$	$\kappa + \lambda$	$\kappa + \lambda$	$\kappa + \lambda$	$\kappa + \lambda$
Valency for antigen binding	2	2 (4)	5 (10)	2	2
Concentration range in normal serum	8–16 mg/ml	1.4–4 mg/ml	0.5–2 mg/ml	0–0.4 mg/ml	17–450 mg/ml
% total immunoglobulin	80	13	6	0–1	0.002
Complement fixation: Classical	+ +	–	+ + +	–	
Alternative	–	±	–	–	

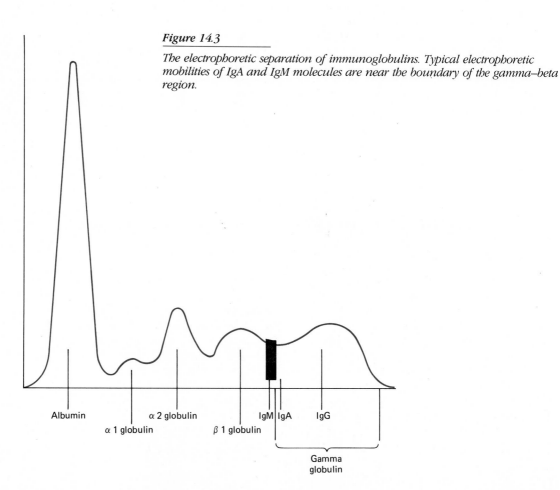

Figure 14.3

The electrophoretic separation of immunoglobulins. Typical electrophoretic mobilities of IgA and IgM molecules are near the boundary of the gamma–beta region.

Albumin

α 1 globulin

α 2 globulin

β 1 globulin

IgM IgA IgG

Gamma globulin

Property	IgG	IgA	IgM	IgD	IgE
Cross placenta	+	−	−	−	−
Fix to homologous mast cells and basophils	−	−	−	−	+
Binding to macrophages and neutrophils	+	±	−	−	−
Major characteristics	Most abundant Ig of body fluids; combats infecting bacteria and toxins	Major Ig in sero-mucous secretions; protects external body surfaces	Effective agglutinator produced early in immune response	Mostly present on lymphocyte surface	Protects external body surfaces; responsible for atopic allergies

Figure 14.4

(A) An immunoglobulin molecule, showing light and heavy chains. (B) A representation of the folded structure of the immunoglobulin molecule.

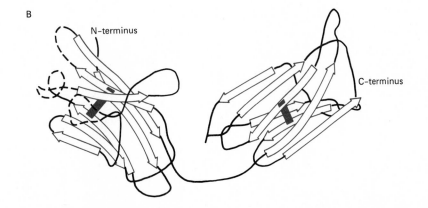

A

N-terminal   C-terminal

Light

S
|
S

Heavy

S
|
S

Heavy

S
|
S

Light

B

N-terminus

C-terminus

(γ), delta (δ), and epsilon (ϵ). The light chains of the immunoglobulin molecules may be either kappa (κ) or lambda (λ). The immunoglobulin molecule can be split by the enzyme papain to form two identical Fab (antigen-binding) fragments and an additional Fc (crystalizable) fragment (Figure 14.5). The enzyme pepsin cleaves the immunoglobulin molecule in another location, forming a divalent antibody binding fragment but not forming Fc fragments. The two portions of the immunoglobulin macromolecule serve different functions. It is the Fab fragment that actually binds to antigen molecules, whereas the Fc fragment augments the action of the immunoglobulin molecule by binding to complement molecules and/or phagocytic cells.

The flexibility and variability of the immunoglobulin molecules permits the specificity of their reactions with antigenic biochemicals. Allosteric changes in the immunoglobulin macromolecule result in a conformation having a three-dimensional structure that allows the antigen and antibody molecules to fit properly together. The ends of the Fab fragments are said to be hypervariable, containing different amino acid sequences that give rise to varying three-dimensional structures

for different immunoglobulin molecules (Figure 14.6). The variable portions at the ends of the peptide chains of the Fab fragments contain about 110 amino acids coded for by genes in the gamma (γ) region. About 10 percent of the total human genome could code for all the types of immunoglobulin molecules. The constant regions of the heavy chains, coded for by genes in the C region, are involved in secondary biological functions after the binding of antigen to the antibody molecule. The genetic variability coding for a diversity of antibody molecules provides the basis for the immune response to many different antigens.

Types of immunoglobulins

IgG is generally the largest fraction of the body's immunoglobulin molecules, generally comprising approximately 80 percent. IgG, the predominant circulating antibody, readily passes through the walls of small vessels (venules) into extracellular body spaces where it can react with antigen and stimulate the attraction of phagocytic cells to invading microorganisms. Reactions of IgG with surface antigens on bacteria activate the complement system and attract additional neutrophils to the site of the infection. IgG also crosses the placenta and confers immunity upon the fetus through the first months after birth. IgG plays a major role in preventing the systemic spread of infection through the body and in the recovery from many infectious diseases.

IgA occurs in mucus and in secretions such as saliva, tears, and sweat. IgA is important in the respiratory, gastrointestinal, and genitourinary tracts, where it plays a major role in protecting surface tissues against invasion by pathogenic microorganisms. IgA also is secreted into human breast milk and plays a role in protecting nursing newborns against infectious disease. The IgA molecules bind with surface antigens of microorganisms, preventing the adherence of such antibody-coated microorganisms to the mucosal cells lining the respiratory, gastrointestinal, and genitourinary tracts. The IgA molecules do not initiate the classical complement pathway but do play a role in activating the alternate complement pathway discussed in Chapter 13. Plasma contains relatively high concentrations of monomeric IgA molecules, but it is the dimers of IgA that bind to receptors on the surface of secretory cells, leading to their secretion into body fluids (Figure 14.7). In the mucus secretions, IgA is the major immunoglobulin molecule involved in the immune response that protects external body surfaces. In these mucus secretions IgA is present as a dimer linked to a secretory component.

Figure 14.5

Splitting of the immunoglobulin macromolecule by papain and pepsin produces characteristic fractions with distinct functions.

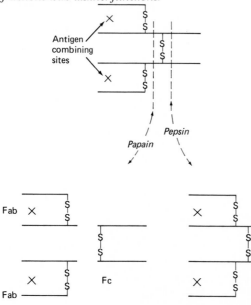

Figure 14.6

Illustration of immunoglobulin, showing constant (C_L and C_H) and variable (V_L and V_H) amino acid composition regions (L = light chain, H = heavy chain). The amino acid residues are numbered, starting from the N-terminal end. C_L starts at residue 108 for κ types and 109 for λ types. The high degree of variation in amino acid residues can be seen in the hypervariable region.

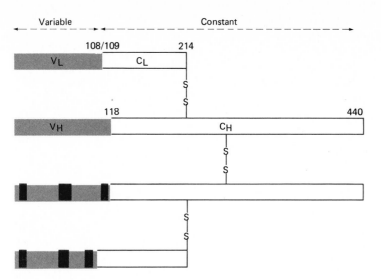

IgM is a high-molecular-weight immunoglobulin, occurring as a pentamer; that is, IgM contains five monomeric units of the basic four-peptide-chain immunoglobulin molecule. IgM molecules are formed prior to IgG molecules in response to exposure to an antigen. Because of its high number of antigen-binding sites, the IgM molecule is effective in attaching to multiple cells that have the same surface antigens. As such, it is important in the initial response to a bacterial infection. During the later stages of infection, IgG molecules are more important. IgM molecules occur primarily in the blood serum and, together with IgG molecules, are important in preventing the circulation of infectious microorganisms through the circulatory system.

IgD antibody molecules are present on the surface of some lymphocyte cells together with IgM. Although its precise role remains to be defined, it appears that IgD may play a role in lymphocyte activation and suppression. Within blood plasma, IgD molecules are short-lived, being particularly susceptible to proteolytic degradation.

IgE molecules are normally present in the blood serum as a very low proportion of the immunoglobulins. The ratio of IgG:IgE is normally 50,000:1. IgE serum levels, though, are elevated in individuals with allergic reactions, such as hay fever, and in some persons with chronic parasitic infections. The main role of IgE appears to be the protection of external mucosal surfaces by mediating the attraction of phagocytic cells and the initiation of the inflammatory response. IgE molecules are important because they bind to mast cells and assist in mediating hypersensitivity reactions, including allergic responses such as hay fever. The tight binding of IgE to cell membranes in circulating mast cells and leukocytes, based on the charac-

Figure 14.7

The mechanism of IgA secretion is exemplified by the transfer of circulating IgA into the bile. Dimeric IgA in the sinusoid bonds to surface secretory piece, thereby activating the uptake and secretion of the immunoglobulin. There is perhaps an analogy with the uptake of IgG into macrophage through stimulation of the surface IgG receptors.

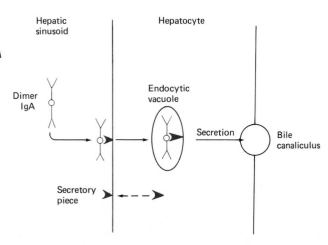

teristics of the Fc portion of the IgE molecule, is an unusual property of IgE.

Within each of the major classes of immunoglobulins, there are variants. For example, IgG can be grouped into four subclasses, IgG1, IgG2, IgG3, and IgG4; each subclass has differences in the heavy chains of the immunoglobulin molecule. The variations of the immunoglobulin molecule can be classified as: **isotypes**, referring to variations in the heavy chain constant regions associated with different classes and subclasses of immunoglobulins; **allotypes**, referring to genetically controlled allelic forms of the immunoglobulin molecule; and **idiotypes**, referring to the individual specific immunoglobulin molecules differing in the hypervariable regions of the Fab fragments. These numerous variations of immunoglobulins establish the needed diversity of macromolecules for the immune response.

Synthesis of antibody

The ability to synthesize the great diversity of antibody molecules that are needed to exhibit the specificity of the immune response can be explained in large part by the **clonal selection theory** (Figure 14.8). According to the clonal selection theory, B lymphocyte cells are programmed to synthesize a single specific immunoglobulin molecule. The particular immunoglobulin molecule for which the individual lymphocyte is genetically competent is integrated into the cytoplasmic membrane of that B lymphocyte cell and acts as a specific surface receptor for an antigen molecule. Antibodies formed by B lymphocyte cells are secreted into their membranes, rather than being released extracellularly. The reaction of the antigen molecule with the surface antibody receptor initiates the differentiation and multipli-

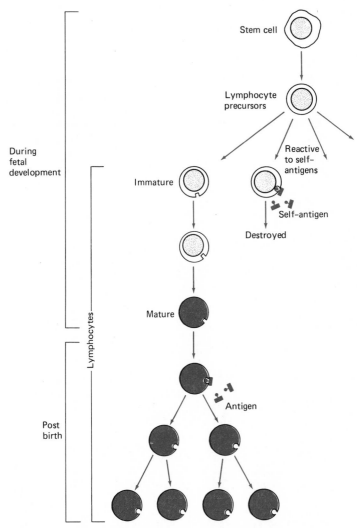

Figure 14.8

Illustration of the clonal selection theory. As shown, there is a critical shift in how lymphocytes react with antigen before and after birth. During fetal development the immune response network loses the ability to react with self-antigens. After birth the detection of foreign antigens leads to the cloning of lymphocytes and differentiation to destroy the foreign substance.

Figure 14.9

The response of B cells to antigen results in the production of a memory system and development of antibody-synthesizing plasma cells.

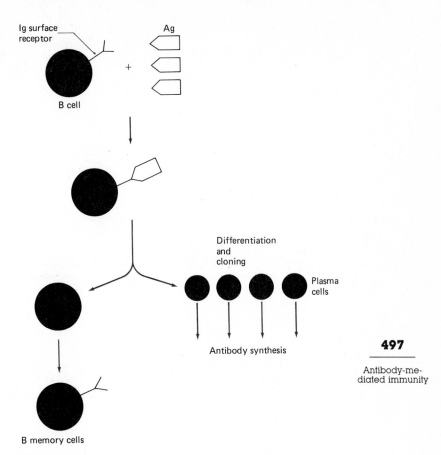

497

Antibody-mediated immunity

cation of that cell to form two different cell populations: plasma cells that are able to synthesize a specific antibody and secondary B cells (Figure 14.9). The secondary B lymphocytes constitute memory cells capable of initiating the antibody-mediated immune response upon detection of the specific foreign antigen molecule for which they are genetically programmed. Compared to the primary B cells, these secondary B cells circulate more actively from blood to lymph and live longer.

Thus, in addition to leading to the formation of antibody-secreting cells, the reaction of B cells with antigen leads to the establishment of an increased population of memory cells. Although the B lymphocyte memory cells do not secrete antibody themselves, they are the precursors for **plasma cells**. The plasma cells derived from the cloning and differentiation of B lymphocytes are responsible for the secretion of extracellular antibodies into the blood serum. Each clone of a plasma cell line secretes a single specific antibody molecule. Thus, according to this mechanism, antibody production is under genetic control coded within the B lymphocyte cells prior to exposure to antigen molecules. The antigen molecules sim-

ply trigger the response but do not act as templates for the synthesis of the antibody molecules, nor are the antigen molecules responsible for modifying and determining the functional properties of the antibody molecules.

Events during fetal development

How, though, can specific immunoglobulins be synthesized in response to specific foreign antigens, and how can we avoid synthesizing immunoglobulins that react with our own antigens? To explain this, we hypothesize that interactions between B lymphocytes and antigens during fetal development are quite different from those that occur after birth. The clonal selection theory depends on the development and differentiation of a large population of B lymphocytes during fetal development. It assumes that during the development of the fetus, a complete set of lymphocytes are developed with each individual lymphocyte cell containing the genetic information for initiating an immune response to a single specific antigen, with receptors for each antigen located within the cytoplasmic membranes of the differentiated B lymphocytes. According to the clonal selection theory, an individual lymphocyte responds to only one antigen because it has only one type of receptor, and the correspondence between lymphocyte receptor and the antigenic determinant accounts for the specificity of the antibody-mediated immune response.

To explain the development of **tolerance to "self" antigens**, the clonal selection theory says that during fetal development surface receptors of B cells react with antigen but that lymphocyte cells are unable to divide to form a clone of cells. Rather, the reaction of B lymphocytes with antigens during fetal development leads to the destruction of those lymphocytes. As such, there is a negative selection for lymphocytes capable of reacting with the antigens of the human body, and only lymphocytes that fail to react with antigens during the fetal development period survive. Theoretically, any antigen that is detected and reacts with the appropriate fetal lymphocyte is recognized as "self." A state of tolerance to "self" antigens develops because all the B lymphocytes that could produce antibody to "self" antigens are destroyed during development of the fetus. Thus, at birth the remaining genetically differentiated B lymphocytes should be competent to react only with foreign antigens, that is, with "nonself" antigens. Accordingly, one should not exhibit autoimmunity; that is, one should not show an immune response against one's own antigens. **Autoimmunity** can occur, though, if B cells are not exposed to some human antigens, such as sperm antigens, during fetal development. Additionally, autoimmunity will occur if mutations reestablish B cell lines that were eliminated, and/or if some B cells programmed for reacting with "self" antigens survive fetal development. Indeed, it appears that some B cell lines that are genetically programmed for reacting with "self" antigens are present in the body, but that these B cells are normally held in check by the action of T lymphocyte suppressor cells. Thus, the development of self-tolerance does not require the complete elimination of self-reactive precursor lymphocytes, but it is necessary that suppression be dominant.

Postnatal events

After completion of fetal development, the reaction of an antigen with the surface-bound receptor of a lymphocyte cell initiates a totally different response than occurs during the fetal development period (Figure 14.9). The postnatal interaction of an antigen and lymphocyte initiates rapid cell division leading to the cloning of B cells, that is, the positive selection of lymphocyte cells. This

hypothesis assumes that a full complement of B cells for binding all possible antigens with which the individual is capable of reacting are formed prior to actual exposure to the antigen.

The clonal theory requires a very large number of genetically different lymphocyte cells. More than 1 million genetically different lymphocyte types may be present at the time of birth. Whereas there is initially a large diversity of B cells, there are only a few that can react with each given antigen. Because relatively few B cells with the appropriate receptors are present upon the first exposure to a given antigen, the primary response is characteristically slow, producing relatively low yields of antibody. There is a long lag period in the primary response, during which selection, differentiation, and cloning of appropriate B cell lines must occur.

After an antigen has elicited an antibody-mediated immune response, there is an increase in the number of B lymphocytes capable of reacting with that antigen. Because antigen binds to and selects for cells having receptors of the highest affinity, this process results in an increase in the number of lymphocyte cells with receptors of high affinity for the particular antigenic molecule. The cloning of these cells establishes a bank of memory cells, so that upon subsequent exposure to the same antigen there is a larger population of B cells that can rapidly and efficiently initiate the secondary immune response.

In addition to the activation of B lymphocytes by an antigen, a cooperative interaction between B and T cells is necessary to effectively stimulate antibody production (Figure 14.10). **Helper T cells** are required before an activated B cell can initiate cloning and extensive antibody synthesis. Usually a B cell cannot be triggered to initiate antibody synthesis without interacting with helper T cells. These helper T cells provide a necessary second signal for the initiation of cloning. Macrophage are also involved in the cooperative antibody response. Mononuclear monocyte and macrophage cells are involved in presenting the antigen to the T lymphocyte cells in the proper orientation so that the T cells can recognize the antigen.

Culture of hybridomas

The fact that each specific antibody is synthesized by a different cell line of B lymphocytes and their derived plasma cells makes it possible to culture cells producing monoclonal antibodies. This can be accomplished by fusing myeloma cells—tumor cells of the immune system that produce large

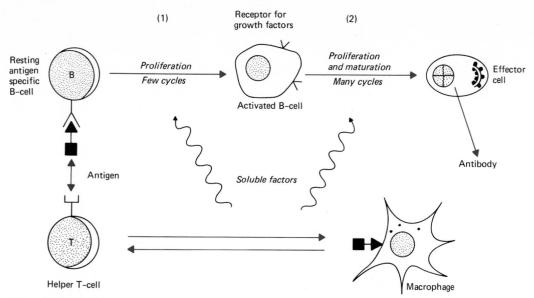

Figure 14.10

Once thought to be totally independent systems, we now recognize that there is extensive cooperative interactions between B and T cells in response to an antigen. Stage 1 is antigen-specific; helper T cells are activated by macrophage-processed antigen to stimulate the resting B cell, which has bound antigen through its hapten-specific Ig receptors, to transform after a small number of divisions to a blast cell with receptors for soluble growth factors. Stage 2 is nonantigen-specific; these soluble growth factors are produced by interactions between the T cell and macrophage, possibly different from the original type initially representing the antigen, and stimulate the blast cells to repeated division and maturation to become antibody-producing effector cells.

amounts of unusual immunoglobulins and that can be readily cultured—with lymphocyte cells programmed for the synthesis of a single antibody. The resulting hybrid cell, known as a **hybridoma**, can be cultured and can synthesize specific antibodies (Figure 14.11). The highly specific monoclonal antibodies produced this way are useful in clinical procedures and may prove useful in the treatment of some diseases. If, for example, monoclonal antibodies could be made for antigens that were unique to particular pathogens, these antibodies could be used to treat specific diseases.

Antigen–antibody reactions

The fundamental basis of antibody-mediated immunity depends on the reactions of antigen molecules with antibody molecules. Antibody-mediated immunity plays an important role in preventing and eliminating bacterial infections. The reactions of IgA molecules with bacteria and vi-

ruses in the fluids surrounding surface tissues prevent the adsorption of many potential pathogens onto these surface barriers. In this way IgA antibody molecules prevent the establishment of infections. IgG antibody molecules, acting as antitoxins, are able to neutralize toxin molecules by combining with the toxin molecules, blocking their reactions, and preventing the onset of disease symptomology. Even poisonous cobra venom can be neutralized by reaction with appropriate specific antibody molecules. The interactions of antibody with surface antigens of bacterial cells render many pathogenic bacteria susceptible to phagocytosis. In fact, the ingestive phagocytic attack on most bacteria requires an initial antigen–antibody reaction before phagocytic blood cells can engulf the invading bacteria. Antigens that activate phagocytosis are called **opsonizing antigens** and the phagocytic destruction of pathogenic bacteria, **opsonization**. Antigen–antibody reactions are required to overcome infections by bacteria, such as *Haemophilus influenzae*, that are inherently resistant to phagocytosis. These *in vivo* reactions between antigen and antibody mole-

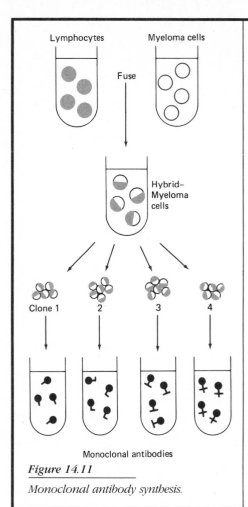

Lymphocytes   Myeloma cells

Fuse

Hybrid–
Myeloma
cells

Clone 1   2   3   4

Monoclonal antibodies

Figure 14.11

Monoclonal antibody synthesis.

Discovery process

Hybridomas are produced by fusing B cells, which are genetically programmed for producing a particular (single) antibody, with a cancer cell, which is capable of proliferating rapidly under laboratory conditions. The resulting hybridoma (hybrid cell) and the technique for making them was described in 1975 by Cesar Milstein and Georges Kohler in England. They injected antigen into a mouse whose B cells began to produce antibodies against that antigen. They then removed the mouse's spleen, which contained many B cells. These cells were incubated with cancerous tumor cells derived from myeloma tumors of bone marrow; some of the B cells fused with the tumor cells to produce the hybrid cells. The hybrids are screened to select ones that produce a desired antibody. Grown in culture, the descendants of "clones" of a single hybridoma cell continue to produce only the type of antibody characteristic of the parent B cell, hence the name "monoclonal" antibody. The hybridoma formed is a permanent union; hybridoma cells are uniform, highly specific, and can readily be produced in large quantities. Monoclonal antibodies are already used for diagnosis of allergies and infectious diseases, such as hepatitis, rabies, and some venereal diseases. Early stages of cancer may also be detectable with monoclonal antibodies, because cancer cells have surface antigens that differ from those of normal cells. It is possible that in the future, monoclonal antibodies could be used to treat cancer and infectious disease. Monoclonal antibodies used alone or chemically attached to drugs could specifically locate and destroy a cancer cell or pathogenic microorganism without damaging surrounding healthy tissue. The production of hybridomas capable of monoclonal antibody synthesis holds great promise for improved treatment of many diseases.

cules constitute a major line of defense against invading bacteria and other microorganisms.

In addition to their importance *in vivo*, the reactions between antigen and antibody molecules are important in many laboratory procedures, including many used for the diagnosis of disease. Some specific uses of antigen–antibody reactions in clinical diagnosis will be discussed in Chapter 16. Here we will consider some general types of antigen–antibody reactions.

Precipitin reactions

When antigen and antibody molecules react, they sometimes form large polymeric macromolecules whose solubility properties result in their **precipitation**. This precipitation of the antigen–antibody complex can be readily visualized by allowing the antigen and antibody molecules to mix in an agar gel. The formation of insoluble complexes that aggregate and precipitate is de-

pendent on the relative concentrations of the antigen and antibody molecules. In general, precipitation occurs when there is an excess of antibody molecules, whereas complexes of antigen–antibody tend to be soluble when formed in the presence of excess antigen molecules (Figure 14.12). Thus, to achieve maximal precipitation of the antigen–antibody complex, the reacting antigen and antibody molecules must be present in the proper relative concentrations.

Diffusion methods

In order to establish the appropriate relative concentrations for precipitation reactions, the antigen and antibody molecules are generally allowed to diffuse together, establishing a concentration gradient. A characteristic precipitation band occurs at the appropriate relative proportions of antigen and antibody. Some precipitin tests employ a single diffusion gradient. In **single diffusion methods**, an antigen is allowed

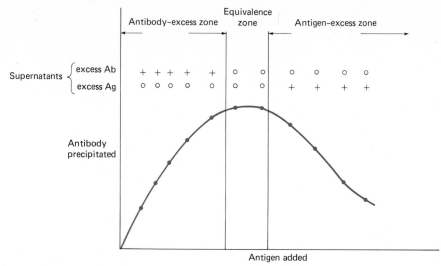

Figure 14.12

This graph illustrates the effects of antigen and antibody concentrations on precipitation. Maximal precipitation occurs when there is an optimal ratio between the concentrations of antigen and antibody.

to diffuse unidirectionally into a tube containing a uniform concentration of soluble antibody so that the antigen establishes a concentration gradient through the tube (Figure 14.13). If the antigen and antibody react to form an insoluble complex, a band of precipitation will occur at a particular level in the tube. The formation of a precipitant requires that the antigen and antibody match; that is, the antibody must be reactive specifically with the antigenic determinants of the particular biochemical applied to the tube.

In **double diffusion procedures**, such as the Ouchterlony method, both antigen and antibody are allowed to diffuse toward each other (Figure 14.14). In this method antigen and antibody molecules are placed in separate wells cut into an agar gel. The antigen and antibody molecules dif-

fuse toward each other and, if they react, they form a precipitation band in the region of optimal relative proportions of antigen and antibody molecules. This method can be used both to identify antigens and to compare the immunological relationships between antigens. In order to assess the relationship between antigens, multiple adjacent wells are filled with the antigens and allowed to diffuse toward a single well filled with antibody molecules. The formation of confluent precipitation bands indicates identity; the formation of a spur indicates partial identity; and a lack of confluent precipitation indicates a lack of identity between the antigen molecules. The arced shape of the precipitation band formed is characteristic of the antigen molecule and is generally concave to the well containing the reactant of

Figure 14.13

In the single diffusion method antibody is allowed to diffuse into a tube containing dispersed antigen, leading to a zone of precipitation in the region where the antigen and antibody achieve the proper relative concentrations.

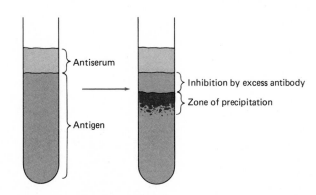

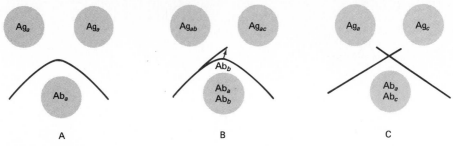

A   B   C

Figure 14.14

*In the double diffusion method the antibody and antigen diffuse toward each other.
(A) The line of confluence obtained with two antigens that cannot be distinguished by
the antiserum used. (B) Spur formation by partially related antigens having a common
determinant* a *but individual determinants* b *and* c *reacting with a mixture of
antibodies directed against* a *and* b. *The antigen with determinants* a *and* c *can only
precipitate antibodies directed to* a. *The remaining antibodies (Ab_b) cross the precipitin
line to react with the antigen from the adjacent well that has determinant* b, *giving rise
to a "spur" over the precipitin line. (C) Crossing over of lines formed with unrelated
antigens.*

higher molecular weight because the large-size
molecules tend to diffuse more slowly through
an agar gel.

Immunoelectrophoresis Antigen–antibody reactions can also be used in conjunction with electrophoresis. Electrophoresis permits the separation of complex mixtures of macromolecules. The
precipitation of antigen–antibody complexes is an
extremely specific reaction. **Immunoelectropho-**

resis combines the efficiency of electrophoretic
separation with the selectivity and sensitivity of
detection of the immunological reaction between
antigen and antibody molecules. To improve the
resolution of separation of the components of a
complex mixture, two separate electrophoretic
runs can be made in differing directions. In two-
dimensional immunoelectrophoresis the component antigenic compounds are initially separated on the basis of their electrophoretic mobil-

Figure 14.15

*Two-dimensional immunoelectrophoresis allows a finer separation of antigens than can be
achieved in one direction. Antigens are separated on the basis of electrophoretic mobility. The
second run, at right angles to the first, drives the antigens into the antiserum-containing gel to
form precipitin peaks; the area under the peak is related to the concentration of antigen.*

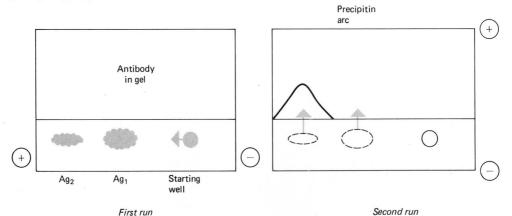

First run   *Second run*

ity, a function of the size, shape, and charge of the molecules (Figure 14.15). A second electrophoretic run, perpendicular to the original run, drives these antigenic compounds into a gel containing antibody. Precipitation bands occur at the site of antigen–antibody reaction.

Radioactive binding techniques

Radioimmunoassay techniques also permit the sensitive quantitation of antigen molecules (Figure 14.16). The radioimmunoassay method was initially developed by using ^{125}I radioactive iodine for the detection of very low levels of hormones, such as insulin. The steps of the radioimmunoassay are (1) reaction of specific antibody with a sample containing an unknown quantity of antigen; (2) the addition of radiolabeled antigen, which combines with any unreacted antibody; (3) the separation of radiolabeled antigen–antibody complex from uncombined radiolabeled antigen; (4) determination of the amount of radiolabeled antigen–antibody complex formed; and (5) the calculation of the concentration of the unknown antigen. Quantitation of the concentration of the antigenic compound is based on the fact that the unknown antigen competes with the radiolabeled antigen for antibody binding sites, and thus the quantity of radiolabeled antigen–antibody complex is diminished in direct proportion to the amount of unlabeled antigen–antibody complex formed.

The ability to detect and measure low levels of radioactivity coupled with the specificity of the antigen–antibody reaction makes the radioimmunoassay a very sensitive procedure for the quantitation of a variety of antigenic compounds. Because of the problems inherent in working with radioactive chemicals, several other related procedures employing nonradioactive-labeled antigens have been developed. These include the **enzyme-linked immunosorbant assay (ELISA)**, in which the antigen is bound to an enzyme, such as peroxidase or phosphatase, which can produce colored reaction products from an appropriate substrate. The enzyme-linked immunosorbant assay is similar to the radioimmunoassay in that it is based on the competition between a tagged antigen molecule of known quantity and an unlabeled antigen of unknown quantity. Similar assay procedures may also employ dyes specifically to label the antigen molecules. These various procedures differ in their sensitivity and applications. Today, many different labeled immunoassay procedures are run routinely in clinical laboratories for the diagnosis of the underlying causes of various human disease conditions.

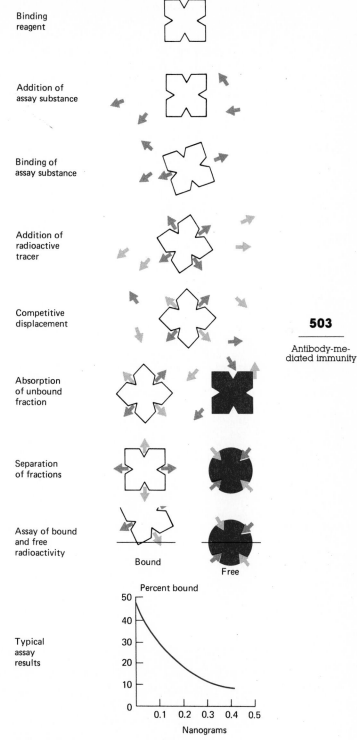

Binding reagent

Addition of assay substance

Binding of assay substance

Addition of radioactive tracer

Competitive displacement

Absorption of unbound fraction

Separation of fractions

Assay of bound and free radioactivity

Bound

Free

Typical assay results

Percent bound

Nanograms

Figure 14.16

Radioimmunoassay (RIA) has become an important technique for quantitating polypeptides, steroids, thyroid hormones, and vitamin B_{12}.

Immunofluorescence

Among the dyes that can be used to tag antibodies are fluorescent dyes, such as fluorescein and rhodamine, which can be coupled readily to antibody molecules to form **conjugated fluorescent dyes**. Such dyes can be used specifically to stain the antigenic reactive sites of microorganisms (Figure 14.17). The stained preparations are visualized by using a fluorescence microscope with an excitation wavelength appropriate for the dye moiety of the conjugated fluorescent antibody stain. The conjugated fluorescent antibody dye molecules retain the specificity of the antigen binding site of the antibody molecule. Thus, these dyes can be used to react specifically with microorganisms possessing particular antigens, providing a powerful tool for the visualization and identification of specific microorganisms.

This technique is particularly useful in identifying specific strains of microorganisms within a mixed microbial community; it is also invaluable in identifying pathogenic microorganisms that are difficult or impossible to culture in the laboratory. Immunofluorescent microscopy permits the visualization and specific identification of *Treponema pallidum*, the causative organism of syphilis, in tissues of infected individuals. This type of microscopy also permits the detection of pathogens such as *Legionella pneumophila*, the causative organism of Legionnaire's disease, in both infected individuals and environmental samples.

Agglutination reactions

ABO blood groups   When the antigen is located on a cell surface, the reaction of antigen and antibody can produce clumping or **agglutination** of

Figure 14.17

Fluorescent antibody staining allows the detection of specific microorganisms. This method is useful in clinical identification of pathogenic microorganisms. (A) Diagram of conjugated antibody fluorescent staining. (B) Photograph showing conjugated antibody fluorescent staining of Samonella typhimurium *in a human fecal sample. (C) The same human fecal sample stained with acridine orange to show the total microbial content. Comparing B and C, the specificity of immunofluorescence is obvious. (Courtesy B. Bohlool, reprinted by permission of Plenum Press, from B. B. Bohlool and E. L. Schmidt, 1980. The* immunofluorescence approach in microbial ecology, Advances in Microbial Ecology *4: 203–241.)*

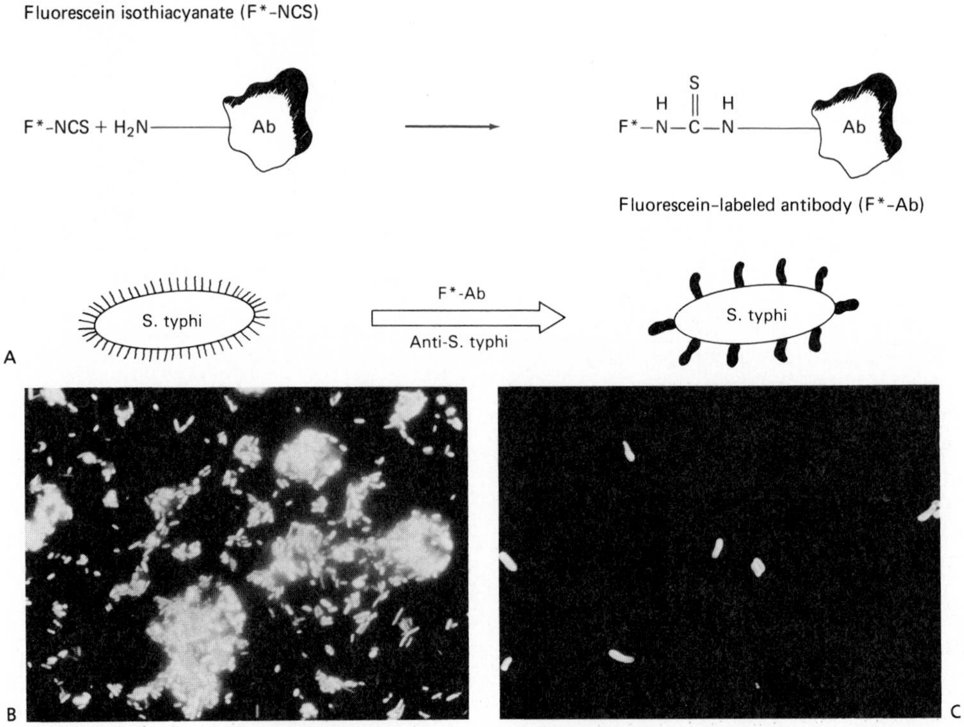

Fluorescein isothiacyanate (F*-NCS)

F*-NCS + H$_2$N— Ab → F*-N-C-N— Ab

Fluorescein-labeled antibody (F*-Ab)

S. typhi → S. typhi

F*-Ab
Anti-S. typhi

Figure 14.18

Agglutination occurs when antigens are
located on the surface of a cell; the reaction of
antibody with the surface antigens causes the
cells to clump or agglutinate.

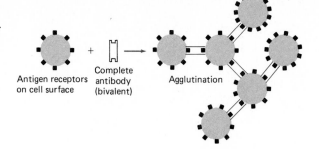

Antigen receptors on cell surface

Complete antibody (bivalent)

Agglutination

the cells. The agglutination reaction is probably best known for its use in blood typing (Figure 14.18). Human red blood cells contain antigens on their surfaces. The membranes of human red blood cells may possess either type A or type B polysaccharide antigens, both type A and type B antigens, or neither of these two antigens. Blood types are determined by mixing known antisera (anti-A and anti-B antibodies) with a blood sample (Figure 14.19). An agglutination reaction indicates the presence of the corresponding antigen. Type A blood has type A but not type B surface antigens on the red blood cells. Type B blood has type B but not type A antigens. Type AB blood has both type A and type B red blood cell surface antigens. A person with type O blood has neither A nor B antigens on their red blood cell surfaces.

An individual's blood serum also contains antibody to any antigens that do not occur in the cytoplasmic membranes of the red blood cells of that individual. For example, type O blood contains both anti-A and anti-B antibodies. The antigen–antibody reactions involving these red blood cell surface antigens establish the primary compatibility of blood types; a transfusion with an incompatible blood type can lead to the death of the individual receiving the incompatible blood type. An individual with type O blood is a universal donor but can receive only transfusions of type O blood. If type O blood is given to an individual with type A, B, or AB blood, there may be some side reactions because of the reaction of antibody in the donor serum, but these reactions are normally fairly minor because of the relatively low concentrations of these antibodies when diluted through the circulatory system of the transfusion recipient. Under ideal conditions, assuming the availability of all blood types, type O blood is used only for transfusions of recipients who have type O blood, but in emergencies, type O blood may be used regardless of the blood type of the recipient.

Other blood cell antigens In addition to the A and B antigens of the red blood cell, various other antigens, including M, N, and Rh, can be identified on the red blood cell surface. Some of these antigens can create incompatibilities between blood types, and thus, before transfusions the blood of donor and recipient is screened for about 8–10 of the antigens most commonly involved in transfusion reactions. The detailed serotyping of red blood cell antigens can also be used for other purposes, such as determining paternity. Because the antigens of the red blood cells are genetically determined, the genetic information is heritable, and the pattern of minor antigens on the red blood cell surface—while not unique—is sufficiently characteristic to reflect genetic linkage.

The Rh (rhesus) antigen of the red blood cell

Figure 14.19

Illustration of blood typing to determine the major
blood groups (A, B, AB, and O) based on
agglutination reactions.

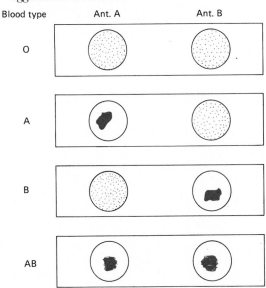

Blood type · Ant. A · Ant. B

O

A

B

AB

membrane is especially important for determining the compatibility of fetal and maternal antigen–antibody interactions. Rh positive blood indicates the presence of the Rh antigen, whereas Rh negative blood cells lack this antigen. If the red blood cells of the fetus are Rh positive, having the Rh antigen, an immune reaction may occur within an Rh negative mother, resulting in the production of anti-Rh antibodies. Generally, such interactions do not occur until the birth of a first child, when bleeding for the first time exposes the mother to the child's antigens and initiates a primary immune response. During subsequent pregnancies, however, when maternal anti-Rh IgG molecules cross the placenta, they may react with the Rh antigens of the fetal red blood cells, resulting in a condition known as hemolytic disease of the newborn (HDN), once called erythroblastosis fetalis. The development of the immune response in the mother, and thus this condition, can be prevented by injecting the mother with anti-Rh antibodies after each delivery to clear the Rh positive fetal cells rapidly from the maternal bloodstream, minimizing priming of the mother's immune response system. This procedure is carried out in high risk cases where the father is Rh positive and the mother is Rh negative.

Serotyping of microorganisms In addition to blood typing, agglutination reactions can be used for the rapid identification of microorganisms. **Serotyping** provides essential information for the identification of many pathogenic bacterial species. For example, species of *Salmonella*—including *S. typhi*, the causative organism of typhoid fever, and *Neisseria gonorrheae*, the causative organism of gonorrhea, as well as many other pathogenic bacteria —can be identified rapidly by using agglutination reactions.

Antigen-coated latex beads Because the agglutination test requires that the antigens be on particle surfaces, soluble antigens are not suitable for direct agglutination tests. It is possible, however, to attach soluble antigens or antibodies to latex beads, forming antigen-coated particles that can then be employed for agglutination testing. Some rapid screening tests for detecting pregnancy use antigen-coated latex beads. These tests detect the antibodies produced in response to the condition of pregnancy.

Hemagglutination inhibition It is also possible to assess the inhibition of agglutination in order to determine the quantities of certain antigens. For example, some viruses, such as influenza virus, will cause agglutination of red blood cells. The binding of serum antibodies to viral-coated red blood cells prevents agglutination. The degree of inhibition, hemagglutination inhibition, can be used to assay for antiviral antibodies.

Complement

In addition to their role in the alternate complement pathway discussed in Chapter 13, complement molecules can react with antigen–antibody

Figure 14.20

The sequence of classical complement activation by membrane-bound antibody showing formation of C3a and C5a fragments chemotactic for polymorphs and with anaphylatoxin activity causing histamine release. Immune adherence through C3 to macrophage, platelets, or red cells facilitates phagocytosis. Fixation of C8 and C9 generates cytolytic activity. Fragments released during the activation of C4 and C2 are thought to have chemotactic and kinin-like activity, respectively.

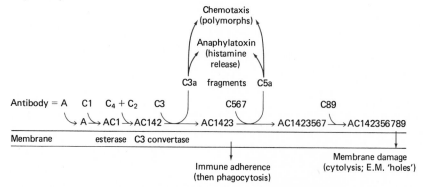

complexes to initiate the **classical complement pathway** (Figure 14.20). The complex series of complement reactions eventually leads to the lysis of microbial cells. Complement represents a family of glycoprotein molecules designated C1, C2, C3, and so forth, with the numbers assigned based on the order of their discovery. In the alternate complement pathway, the sequence of complement reactions leading to cell lysis begins with C3 (see Figure 13.23). Additional complement molecules are involved in the initiation of the classical complement pathway. As a result of the interaction of complement with antigen–antibody complex, reactions can occur that otherwise would

not, and thus, the initiation of the complement sequence by antigen–antibody complex augments the specific immune response.

When IgM or IgG antibodies combine with an antigen on a cell surface, complement C1 can bind to the Fc region of the antibody molecule, forming an antigen–antibody–complement complex. This complex initiates the cascade of complement molecules with several consequences. At each stage of the complement cascade, different complement molecules are activated so that they consequently activate the next complement molecule in the pathway. The initiation of the classical complement pathway involves C1, C2, and C4. Three

Figure 14.21

Illustration of complement fixation test procedure. (A) Typical positive serum, containing antibody. (B) Typical negative serum, antibody absent.

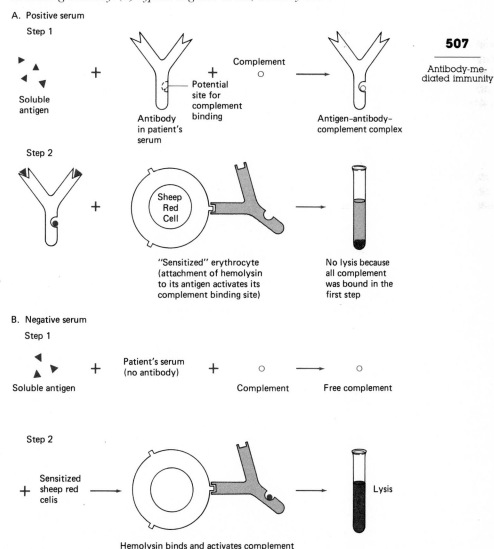

proteins (C1q, C1r, C1s), comprising the C1 complex, are triggered by the antigen–antibody complex to become active proteases. The proteolytic action of the activated C1 complex in turn activates C2 and C4. The complement cascade has several consequences, including the chemotactic attraction of neutrophils and the initiation of the inflammatory response. The adherence of complement molecules to the surfaces of microorganisms also facilitates the adherence of macrophages and neutrophils to the microbial cells, enhancing phagocytosis. Additionally, the complement sequence results in damage to the cytoplasmic membrane, resulting in the lysis and death of the cell on which the complement cascade occurs. The full complement system leads to membrane damage and can cause lysis in Gram negative bacteria by permitting the enzyme lysozyme to reach the cytoplasmic membrane. The ability to bring about cell lysis is important for the destruction of bacterial pathogens.

Complement fixation

Complement molecules are fixed as a consequence of their reaction in both the classical and alternate complement pathways, and it is possible to measure the degree of **complement fixation** for diagnostic procedures (Figure 14.21). In a typical complement fixation test an individual's serum, containing an unknown amount of antibody and free complement, is mixed with a known antigen, a specific virus, for example. The reactant mixture is then added to a test indicator system, such as sensitized sheep red blood cells (RBC), which are sheep RBC that have been agglutinated by reaction with anti-sheep RBC antibody. If, after reaction with antigenic compounds, there is free complement available from the individual's serum, the sheep red blood cells will undergo hemolysis. Thus, in this system free human complement will augment the antigen–antibody reaction that involves sheep red blood cells, resulting in lysis of the sheep red blood cells. This can be easily visualized and indicates that the individual is probably not suffering from a disease caused by that virus. Had a virus been eliciting an immune response in the individual, the antibody and complement molecules available in the individual's serum would have reacted with the viral antigens, resulting in the fixation of the complement molecules. In the absence of free complement, the sheep red blood cells in the indicator system would have remained agglutinated, and no lysis would have occurred. A positive reaction for complement fixation (the failure to observe lysis of the sensitized sheep red blood cells) is indicative of an infection in the individual being examined.

Cell-mediated immune response

Whereas the antibody-mediated immune response system recognizes substances that are outside of host cells, the **cell-mediated immune response** is effective in eliminating infections by pathogenic microorganisms that develop within host cells. Cell-mediated immunity is important in controlling infections where the pathogens are able to reproduce within human cells, including infections caused by: viruses; some bacteria, such as rickettsias and chlamydias; and some protozoa, such as trypanosomes. **The cell-mediated immune response depends on the actions of T lymphocytes**. When stimulated by an appropriate antigen, T lymphocytes respond by dividing and differentiating into cytotoxic T cells, and various other T cells that release biologically active soluble factors that mediate the responses of other cells involved in the immune response (Figure 14.22). The soluble factors, collectively termed lymphokines, are effective in mediating the responses of monocytes and macrophages. Like the B lymphocytes, the T lymphocytes have surface receptors that can react with antigen, triggering the cell-mediated response. These surface receptors, however, are not immunoglobulin molecules as they are in B lymphocytes.

Lymphokines

As part of the cell-mediated immune response, **lymphokines** are released by T lymphocytes after antigenic stimulation. These soluble factors exhibit different activities (Table 14.2). The lymphokines produced by T cells include: (1) macrophage chemotaxis factor, a lymphokine that causes the attraction and accumulation of macrophage to the site of lymphokine release; (2) migration inhibition factor, which inhibits macrophage cells from migrating farther once they have reached the site of lymphokine attraction; (3) macrophage activating factor, which results in an alteration of

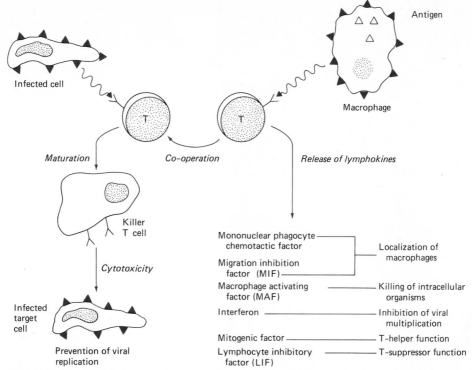

Infected cell

Antigen

Macrophage

Maturation

Co-operation

Release of lymphokines

Killer
T cell

Cytotoxicity

Infected
target
cell

Prevention of viral
replication

Mononuclear phagocyte
chemotactic factor

Migration inhibition
factor (MIF)

Localization of
macrophages

Macrophage activating
factor (MAF)

Killing of intracellular
organisms

Interferon

Inhibition of viral
multiplication

Mitogenic factor

T-helper function

Lymphocyte inhibitory
factor (LIF)

T-suppressor function

Figure 14.22

The cell-mediated immune response, the response of T cells to antigen, showing lymphokines and cytotoxic cell formation. This operates through the generation of cytotoxic T cells and the release of lymphokines through the stimulation of two distinct T subpopulations. Different lymphokines may be produced by different lymphocyte subsets. The intense proliferation induced by antigenic stimulation has not been shown but is an essential factor in the amplification of the response.

macrophage cells that increases their lysosomal activities and thus their ability to kill and ingest organisms; (4) skin reactive factor, which enhances capillary permeability and thus movement of monocytes across the vascular spaces; and (5) immune interferon, which activates antiviral proteins, preventing further intracellular multiplication of viruses. In addition to these lymphokines, T helper cells are involved in the stimulation of B cells as part of the antibody-mediated immune response, and T suppressor cells are involved in shutting off the antibody-mediated response by suppressing synthesis of antibody. It is thus apparent that lymphokines and T lymphocytes play a major role in regulating the activities of other cells involved in the immune response.

One of the specific lymphokines, **immune or gamma (γ) interferon**, is secreted by lymphocytes in response to a specific antigen to which they have been sensitized or stimulated to divide. This immune interferon is different in some of its

table 14.2

Effects of lymphokines

Cells affected	Lymphokine
Macrophage	Migration inhibitory factor Macrophage activating factor Chemotactic factor
Neutrophils	Leukocyte inhibitory factor Chemotactic factor
Lymphocytes	Mitogenic factor T cell replacing factor Suppressor factors Chemotactic factor
Eosinophils	Chemotactic factor Migration stimulatory factor
Basophils	Chemotactic factor
Other cells	Cytotoxic and cytostatic factors (lymphotoxin) Interferons Osteoclast activating factor Colony stimulating factor

physiological effects from other interferon molecules and may kill tumor cells. Like other interferons, immune interferon molecules have antiviral activities, stimulating the synthesis of antiviral proteins, including 2,5-adenylate polymerase. 2,5-Adenylate polymerase, when bound to double-stranded RNA, a viral replicative intermediate, activates an endonuclease that can cleave viral RNA. The primary function of immune interferon, though, appears to be different from that of other interferons. Immune interferon may regulate the proliferation of the lymphoid cells that are stimulated to divide in response to interactions with antigenic biochemicals. Immune interferon may also enhance phagocytosis by macrophage and enhance the cytotoxicity of lymphocytes and the activities of killer T cells.

Cytotoxic T cells

In addition to the antigen-induced release of lymphokines from T cells, intracellular infections, such as viral infections, can result in the generation of killer T cells that are specifically cytotoxic for intracellularly infected host cells (Figure 14.23). The **killer T cells**, also known as **cytotoxic T cells** and **T aggressor cells**, appear to direct their activity against the major histocompatibility antigens of the cell surface. These histocompatibility antigens are genetically controlled by the **major histocompatibility complex (MHC)** region of the genome. The MHC antigens are highly polymorphic because of the fact that allelic genes each code for different antigens. These MHC antigens play an important role in the recognition of "self"

and "nonself" tissues and are especially important in the immune response to tumors and tissue transplants.

The ability of cytotoxic T cells to recognize and to react with antigens on cell surfaces depends on the gene products of the major histocompatibility complex. Small variations in the MHC antigens, which occur in tumor cells and cells infected with viruses, can be detected by T cells, leading to the appropriate cytotoxic response by these killer T cells. The ability to recognize antigens on the surfaces of cytoplasmic membranes permits T cells to monitor and regulate the proliferation of other types of cells. This appears to be extremely important not only for preventing the unchecked development of cells harboring intracellular pathogens but also for inhibiting tumor development. As a result of the ability of cytotoxic T cells to recognize transformed human cells, many would-be tumor cells do not lead to malignant growths.

In a number of intracellular viral infections, antibodies are ineffective. In such cases, cell-mediated immunity augments the antibody-mediated immune response by controlling infections. Although antibody molecules can neutralize free viruses, antibodies are unable to penetrate host cells and attack viruses multiplying within host cells. It is the cell-mediated immune response that has the capability of eliminating cells infected with viruses. The sensitized T aggressor cells are able to recognize virally modified histocompatibility antigens or a complex of histocompatibility antigen with virally associated antigen in order to direct properly the cytotoxic attack of these lymphocytes against only those cells that are infected

Figure 14.23

The interaction of cytotoxic T cells with cells having foreign MHC antigens. The major histocompatibility complex (MHC) and surface antigens in this region of the genome play an important role in the recognition of incompatible cells.

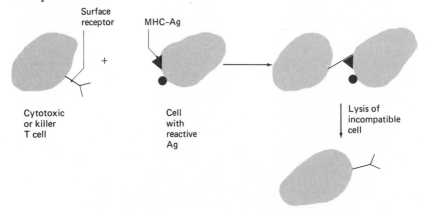

with a virus. Additionally, the T helper and T suppressor cells, which interact with the B cells to regulate antibody-mediated immunity, depend on their ability to recognize antigens coded for by the *IA* and *IR* genes of the MHC region. The MHC gene products, in association with T lymphocytes, thus seem to provide an integrated mechanism for regulation of the immune response. Indeed, as the complexity of the specific immune response is unraveled, it becomes increasingly clear that humans have evolved an interactive network of various cells and macromolecules formed by these cells to regulate the immune response in a functional manner. It is also clear that antigens and receptors on cell surfaces permit molecular-level communication between cells so that cells can act in an interactive and cooperative fashion.

Dysfunctional immunity

Whereas the specific immune response plays a critical role in protecting humans and other higher vertebrate animals from numerous infectious diseases, failures of the immune response can compromise the ability of humans to resist infection, leaving an individual susceptible to a variety of diseases. Indeed, one can consider that many infectious diseases result from failures of the immune response to protect the individual adequately against the invasion of and/or toxicity associated with pathogenic microorganisms. **Failures of the immune response** can also result in autoimmune diseases in which the inability to differentiate properly between "self" and "nonself" antigens results in reactions with "self" antigens and the killing of some of one's own cells. The development of tumor cells can also be viewed as a failure of the immune response, but in this case the failure to recognize and to respond properly to inappropriate cells within the body allows malignant cells to proliferate in an uncontrolled manner. The normal active immune response can also be undesirable in some cases, such as in transplants where the immunological recognition and response to the foreign antigens of a donor results in tissue rejection. Additionally, allergies are the result of physiological changes resulting from certain types of immune responses, and individuals suffering from allergies know too well that immune responses may be occasionally dysfunctional.

Immunodeficiencies

There are several types of deficiencies that can occur within the immune system, resulting in the failure of the system to recognize and respond properly to the antigens of pathogenic microorganisms. Individuals with **immunodeficiencies** are more prone to infection than those who are capable of a complete and active immune response. Immunodeficiencies can affect the variety of cells that are interactive in the immune response system (Table 14.3).

The most severe type of immunodeficiency, se-

table 14.3

Types of immunodeficiencies

	Example	Immune response		Infection	Treatment
		Humoral	Cellular		
Complement	C_3 deficiency	Normal	Normal	Pyogenic bacteria	Antibiotics
Myeloid cell	Chronic granulomatous disease	Normal	Normal	Catalase positive bacteria	Antibiotics
B cell	Infantile sex-linked a-γ-globulinaemia (Bruton)	Absent	Normal	Pyogenic bacteria; *Pneumocystis carinii*	γ-Globulin
T cell	Thymic hypoplasia (DiGeorge)	Lower	Absent	Certain viruses; *Candida*	Thymus graft
Stem cell	Severe combined deficiency (Swiss type)	Absent	Absent	All the above	Bone marrow graft

vere combined deficiency, results from a failure of stem cells to differentiate properly. Individuals with severe combined deficiency, having neither B nor T lymphocytes, are incapable of any immunological response. Any exposure of such individuals to microorganisms can result in the unchecked growth of the microorganisms within the body, resulting in certain death. Individuals suffering from severe combined deficiency can be kept alive in sterile environments where they are protected from any exposure to microorganisms (Figure 14.24). Bone marrow grafts may be employed to establish normal immune functions, but the grafts must come from siblings with histocompatible bone marrow.

Less severe immunodeficiencies occur when only B cell or only T cell functions are lacking. The **DiGeorge syndrome** results from a failure of the thymus to develop properly, so that T lymphocytes do not become properly differentiated.

Figure 14.24

David, born in 1971, is the oldest surviving person with untreated severe combined immune deficiency. He is shown here just before his ninth birthday playing with a fish tank from inside his isolation bubble, where he has lived since birth. He also has a sterile isolator "space suit" that gives him some very limited mobility outside of this bubble and protects him from infections his body could not fight off. (Courtesy United Press International Photo.)

Individuals suffering from this condition do not exhibit cell-mediated immunity and thus are prone to viral and other intracellular infections. Additionally, because T helper cells are involved in enhancing antibody production by B cells, the antigen-mediated or humoral response is depressed in individuals suffering from DiGeorge syndrome. The complete absence of the thymus is rare, and partial DiGeorge syndrome—in which some T cells are produced although in lower numbers than are formed in individuals with fully functional thymus glands—is more common.

Bruton's congenital a-γ-globulinemia results in the failure of B cells to differentiate and produce antibodies. This immunodeficiency disease is sex-limited and affects only males. The cell-mediated response in individuals suffering from this disease is normal. Children with Bruton's globulinemia are particularly subject to bacterial infections, including those by pyogenic (pus-producing) bacteria, such as *Staphylococcus aureus, Streptococcus pyogenes, S. pneumoniae, Neisseria meningitidis*, and *Haemophilus influenzae*. The treatment of this disease involves the repeated administration of IgG to maintain adequate levels of antibody in the circulatory system.

The most common form of immunodeficiency also involves a failure of B cells. This immunodeficiency is known as **late-onset hypogammaglobulinemia**. In this condition there is a deficiency of circulating B cells and/or B cells with IgG surface receptors. Such individuals are unable to respond adequately to antigen through the normal differentiation of B cells into antibody-secreting plasma cells. Other immunodeficiencies may affect the synthesis of specific classes of antibodies. For example, some individuals exhibit IgA deficiencies, producing depressed levels of IgA antibodies. Such individuals are prone to infections of the respiratory tract and body surfaces normally protected by mucosal cells that secrete IgA.

Immunodeficiencies may also affect the complement system. Individuals who fail to produce sufficient C3 complement are unable to respond properly to bacterial infections. The lack of an active complement system limits the inflammatory response and lytic killing of pathogenic bacterial cells. Immunodeficiencies may also affect the functioning of monocytes, neutrophils, and macrophage. Phagocytic cells lacking enzymes that produce hydrogen peroxide and other antimicrobial forms of oxygen do not have proper lysosomal functions that kill bacteria, and pathogenic bacteria are able to multiply within such metabolically deficient phagocytic cells. Antibiotics can be used to protect individuals with deficiencies of both the complement system and active phagocytic cells from invading pathogenic bacteria.

Acquired immune deficiency syndrome

In 1979, a new disease—acquired immune deficiency syndrome or AIDS— started to appear among homosexual men in major cities of the United States. By 1983 the incidence of AIDS in the gay community of San Francisco was 1 in 350, an alarmingly high rate. AIDS also showed an initially high incidence among Haitians. This disease has spread from its initial focal groups to the general populus. AIDS is transmitted by direct contact, through the placenta from mother to fetus, and via blood transfusions. Great concern must be expressed about the safety of the blood donor supply for transfusions.

Individuals with AIDS exhibit severe immunosuppression. Specifically, individuals with AIDS have more T suppressor cells than T helper cells, effectively shutting off the immune response network. The most likely cause is the infection of T helper cells by a virus, leading to the death and depletion of these cells that form a necessary part of the healthy immune defense network. As a result of the immunodeficiency, individuals with AIDS are subject to infection by a wide variety of disease-causing microorganisms and to development of a form of cancer (Kaposi's sarcoma). Generally, one infection follows another in victims of this disease. The onset of the disease is sudden; early symptoms include low-grade fever, swollen glands, and general malaise. Early treatment with antibiotics can offer protection against infection, and interferon is being tested for control of Kaposi's sarcoma. Meanwhile, microbiologists are trying to discover the cause and cure of AIDS.

Hypersensitivity reactions

An excessive immunological response to an antigen can result in tissue damage and a physiological state known as **hypersensitivity**. Hypersensitivity reactions occur when an individual is sensitized to an antigen so that further contact with that antigen results in an elevated immune response. The hypersensitivity reaction may be immediate, occurring shortly after exposure to the antigen, or delayed, occurring a day or more after contact with the antigen. There are several types of hypersensitivity reactions, each mediated by different aspects of the immune response.

Type 1 or **anaphylactic hypersensitivity** occurs

Figure 14.25

(A) Electron micrograph of a typical mast cell, showing extensive granular inclusions of the potent vasoactive agent histamine (10,500×); (B) Higher magnification of granular inclusions of a mast cell, showing their characteristic "scrolls," which consist of coils of a crystalline or fibrillar structure (104,000×). (Courtesy of D. Zucker-Franklin, New York University Medical Center, reprinted by permission of Grune & Stratton, Inc., from D. Zucker-Franklin, 1980, Ultrastructural evidence for the common origin of human mast cells and blastophils, Blood 56: 536.)

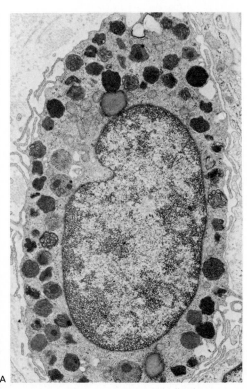

A

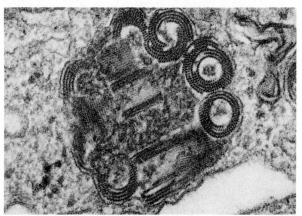

B

when an antigen reacts with antibody bound to mast or basophil blood cells (Figure 14.25), leading to disruption of these cells with the release of vasoactive mediators, such as histamine (Figure 14.26). In anaphylactic hypersensitivity the mast cells are coated with antibody whose Fc region binds to specific sites on the mast cell surface. IgE antibodies are generally involved in such binding to mast cells. The release of the contents of mast cells, after the reaction of antigen with antibody that is bound to the mast cell surface, establishes the basis for a severe physiological response. The sudden release of a large amount of histamine, a potent vasodilator, into the bloodstream can produce anaphylactic shock, causing respiratory or cardiac failure. Bee stings and the administration of antibiotics, such as penicillin, can trigger such anaphylactic hypersensitivity reactions.

Figure 14.26

Illustration of type 1 hypersensitivity reaction: an immune system response to an allergen. Allergens come into contact with specific IgE antibody molecules on the surface of a basophil or a mast cell. A subsequent membrane response causes the release of chemical mediators from granules in the basophil or mast cell.

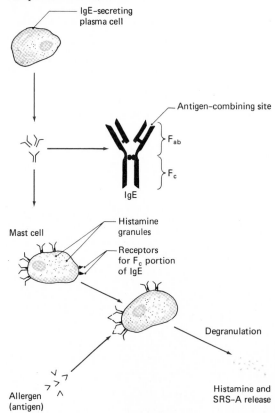

Atopic allergies result from a localized expression of type 1 hypersensitivity reactions. The interaction of antigens (allergens) with cell-bound IgE on the mucosal membranes of the upper respiratory tract and conjunctival tissues initiates a localized type 1 hypersensitivity reaction. The symptoms of such reactions include coughing, sneezing, and respiratory difficulties, in severe cases producing asthma. Hay fever and allergies to certain foods are examples of such atopic allergies. Atopic allergies can be diagnosed by using skin testing (Figure 14.27). The subcutaneous or intradermal injection of antigens results in a localized inflammation reaction if the individual is allergic to that antigen, that is, if that individual exhibits a type 1 hypersensitivity reaction. In this way allergens for a particular individual are identified. The symptoms of atopic allergies can be controlled, at least in part, by the use of antihistamines.

Type 2 hypersensitivity, or **antibody-dependent cytotoxic hypersensitivity** reactions, occur by a different mechanism. In type 2 hypersensitivity reactions an antigen present on the surface of the cell combines with an antibody, resulting in the death of that cell by stimulating phagocytic attack or by initiating the sequence of the complement pathway. This type of antibody-dependent cytotoxic response occurs after transfusions with incompatible blood types. Rh incompatibility between mother and fetus is another example of type 2 hypersensitivity.

Type 3 hypersensitivity, or **complex-mediated hypersensitivity reactions**, involves antigens, antibodies, and complement, which initate an inflammatory response. Such an inflammatory response is part of the normal immune response, but if there are large excesses of antigen, the antigen–antibody–complement complexes may circulate and become deposited in various tissues. Inflammatory reactions from such deposition of immune complexes can cause physiological damage to kidneys, joints, and skin. Various examples of type 3 hypersensitivity or Arthus immune-complex reactions are listed in Table 14.4.

Type 4 hypersensitivity reactions are known as **cell-mediated**, or **delayed hypersensitivity**. Delayed hypersensitivity reactions involve T lymphocytes and, as the name implies, occur only after a prolonged delay after exposure to the antigen, such reactions often reaching maximal intensity 24–72 hours after initial exposure. Delayed hypersensitivity reactions occur as allergies to various microorganisms and chemicals. Contact dermatitis, resulting from exposure of the skin to various chemicals, is a typical delayed hypersen-

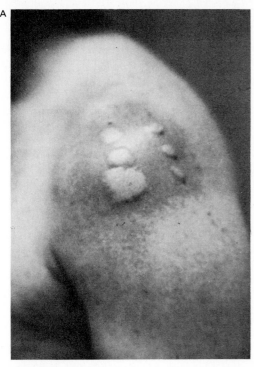

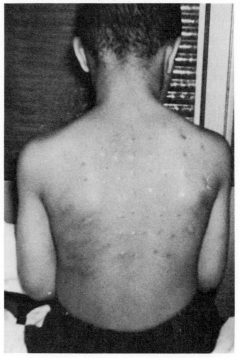

Figure 14.27

Photographs showing skin testing for allergies. The positive reactions show inflammation and development of pustules; results are rated on a scale of 1–4+. (A) Results of intradermal inoculation into the arm of several allergens. Each test shows erythema and a wheal; the pseudopods on the bottom right indicate a clear 4+ reaction to intradermal antigens. (B) The scratch test is performed on the back or forearm. The scratches should be 1 cm long and 2.5 cm apart and the allergens then applied to the torn skin surface. As shown in this photograph, a large number of tests against different allergens, ca. 50, can be performed simultaneously on the back of the patient; in the series of tests shown, positive scratch tests—with erythema, wheal, and pseudopods—are evident in row 9, and negative tests are exemplified in the first row and column. (Courtesy B. Issacs, Norton's-Kosair-Children's Hospital, Louisville.)

sitivity reaction. Poison ivy is one of the best-known examples of such contact dermatitis. Exposure of sensitized individuals to the oils of the poison ivy plant result in a characteristic rash that occurs after an initial delay. The delayed hypersensitivity reaction involves the release of lymphokines from activated T cells. These lymphokines initiate a typical series of cell-mediated immune responses, including the attraction of phagocytic cells and the initiation of the inflammatory response. The delayed type of hypersensitivity reaction is a normal reaction to the intracellular invasion of host

table 14.4

Examples of type 3 hypersensitivy reactions

Disease condition	Caused by exposure to
Farmer's lung	Actinomycete spores
Cheese washer's disease	*Penicillium casei* spores
Furrier's lung	Fox fur protein
Maple bark stripper's disease	*Cryptostroma* spores
Pigeon fancier's disease	Pigeon antigens
Serum sickness	Foreign blood serum

table 14.5

Some types of autoimmunity

Disease	Antigen
Hashimoto's thyroiditis; primary myxedema	Thyroglobulin
Thyrotoxicosis	Cell surface TSH receptors
Pernicious anemia	Intrinsic factor; parietal cell microsomes
Addison's disease	Cytoplasm adrenal cells
Premature onset of menopause	Cytoplasm steroid-producing cells
Male infertility (some)	Spermatozoa
Juvenile diabetes	Cytoplasm of islet cells; cell surface
Goodpasture's syndrome	Glomerular and lung basement membrane
Myasthenia gravis	Skeletal and heart muscle; acetyl choline receptor
Autoimmune hemolytic anemia	Erythrocytes
Idiopathic thrombocytopenic purpura	Platelets
Ulcerative colitis	Colon "lipopolysaccharide"
Sjögren's syndrome	Ducts, mitochondria, nuclei, thyroid; IgG
Rheumatoid arthritis	IgG; collagen
Systemic lupus erythematosus	DNA; nucleoprotein; cytoplasmic soluble Ag; array of other Ag including elements of blood-clotting factors

cells by a variety of pathogens. This reaction can also be used for diagnostic testing. For example, in the tuberculin test, the delayed onset of an inflammatory response after the subcutaneous injection of an appropriate antigen is indicative of a positive delayed hypersensitivity reaction.

Transplantation

As indicated earlier, tissues contain surface antigens that are coded for by the major histocompatibility complex. Reactions to these antigens generally preclude successful transplantation of tissues unless the transplant tissues come from a compatible donor or unless the normal immune response is suppressed. There are a number of drugs that may be used to suppress the normal immune response; in particular, cyclosporin A—a microbial product—obviates rejection problems with organ transplants by specifically suppressing cell-mediated immunity. Suppression of the immune response, however, renders the individual susceptible to a variety of infectious pathogens that are normally excluded by the host immune defense mechanisms. In a sense, the normal immune defense mechanism is dysfunctional in transplantation and grafting because in these cases it does not serve the desired or useful function.

The ability to transplant major organs, such as kidneys and hearts, is dependent on being able to control the normal immune response and prevent rejection of the transplanted tissues.

Autoimmune diseases

There are a number of autoimmune diseases that result from the failure of the immune response to recognize properly "self" antigens. Such **autoimmunity diseases** often result in the progressive degeneration of tissues. A number of such diseases are summarized in Table 14.5. In systemic lupus erythematosus, numerous autoantibodies are produced that react with "self" antigens, including some directed at DNA molecules. In this disease antigen–antibody complexes often circulate and settle in the glomeruli of the kidney, causing kidney failure. In cases of myasthenia gravis, antibodies react with nerve–muscle junctions. In autoimmune hemolytic anemia antibodies react with red blood cells, causing anemia. Various other disease conditions may reflect the failure of the immune system to recognize "self" antigens. These diseases can be treated by using immunosupressive substances to prevent the "self-destruction" of body tissues by the body's own immune response.

The defense mechanisms of a host against pathogenic microorganisms involve a complex network of interactions between various host cells (Figure 14.28). The specific immune response represents a major line of protection against invading microorganisms and other foreign biochemicals. The immune response is activated by exposure to certain biochemical determinants known as antigens. Immunity depends on the ability to detect and respond rapidly to such antigens.

Two types of lymphocyte cells, B lymphocytes and T lymphocytes, are both involved in conferring immunity but differ in their modes of action. B cells are primarily involved in the antibody-mediated immune response. Upon exposure to an appropriate antigen, B cells undergo differentiation, forming plasma cells that are able to secrete antibodies. There is a diversity of B lymphocytes, each containing

Figure 14.28

Illustration of the complex network of interactions of host defense mechanisms. Reactions influenced by T cells are indicated by broken lines.

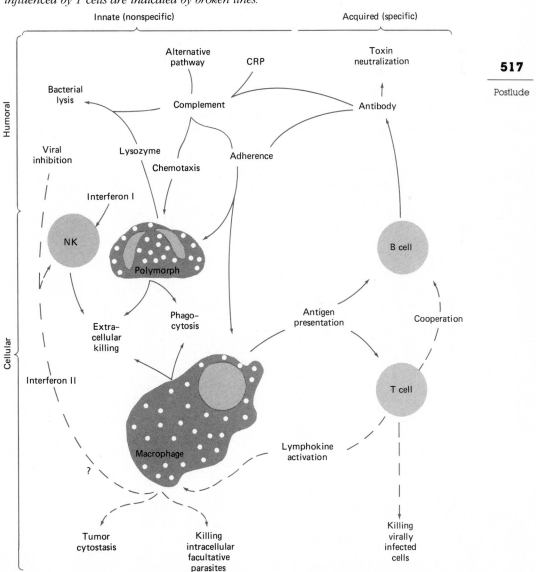

different surface receptors, and each B cell can establish clones of plasma cells capable of responding to a particular antigenic determinant. The lymphocytes contain the genetic information for producing specific antibodies, and lymphocyte cell lines maintain a memory system for recognizing and responding to foreign antigens. The clonal selection theory explains how they come to possess the lymphocytes that are able to distinguish "self" from "nonself" antigens. Failures of this recognition process can render one susceptible to infectious diseases or may result in autoimmunity.

Five major classes of antibodies are secreted, each serving a somewhat different function in the immune response. IgG antibodies are the major circulating antibodies, reacting with antigens, neutralizing toxins, and preventing the spread of pathogenic microorganisms through the circulatory system. IgA antibodies involved in the protection of surface cells from microbial attachment are secreted from mucosal cells. The reaction of antigen and antibody can form a complex that can react further with complement molecules to initiate the classical complement pathway, resulting in the lysis or death of invading microorganisms. The classical complement pathway also enhances phagocytosis and initiates the inflammatory response, both of which contribute to the elimination of invading microorganisms.

There are various possible reactions between antigen and antibody molecules, many of which are used *in vitro* for diagnostic purposes. Precipitation reactions, tagged antibody or tagged antigen assays, radioimmunoassays, and agglutination reactions are employed frequently in the clinical laboratory for diagnostic purposes. Many pathogenic microorganisms can be identified by their specific reactions with defined antibodies. Agglutination reactions are used most commonly in blood typing and in other procedures where antigens are bound to particle surfaces.

Within the human body, antibody-mediated immunity constitutes a major line of defense against toxins and the development of bacterial infections outside host cells. The immunological defense against intracellular pathogens, such as viruses, depends, in large part, on cell-mediated immunity. The cell-mediated immune response involves T cells that may release lymphokines that initiate a variety of defense mechanisms, including increased antibody synthesis, enhanced phagocytosis, and inflammation. Cell-mediated immunity protects the body against a variety of intracellular parasites, most notably viruses. In addition to releasing other lymphokines, activated T cells release interferon, preventing the multiplication of viruses in uninfected cells. T cells may also be differentiated into cell lines with cytotoxic properties that are able to recognize and kill cells harboring viruses. T killer cells eliminate infected cells, and this appears to be the main mechanism by which viral infections are eliminated. The ability of T cells to recognize altered host cells depends on their interaction with surface antigens coded for by the major histocompatibility complex. The MHC antigens are also important in the ability of cytotoxic T cells to recognize tumor cells, preventing the formation of malignancies, and the ability of T cells to recognize foreign tissues, determining the compatibility of transplanting graft tissues.

Failures of the immune response to produce the necessary differentiated tissue lines, such as macrophages, B lymphocytes, and T lymphocytes, result in immunodeficiency diseases in which an individual is unable to respond properly and thus defend against invasion by microorganisms. Many human infections occur when the host defense mechanisms are compromised and the immune response fails to respond optimally to potential pathogens. The regulation and expression of the immune response is an extremely complex system that is developed only in higher vertebrate animals, such as human beings. The primary function of this highly developed and integrated system seems to be the protection of "self" from invasion by disease-causing foreign agents.

1. What is an antigen?

2. What is an antibody?

3. How does the clonal selection theory explain the ability of the immune response network to distinguish between "self" and "nonself" antigens?

4. Discuss the differences between a primary and a secondary (memory) immune response.

5. Discuss the differences between the antibody-mediated and cell-mediated immune response systems. Why do we need two such elaborate defense systems?

6. What is an agglutination reaction? How is this reaction used in blood typing? What is meant by compatible blood for transfusion purposes?

7. How can agglutination reactions be carried out with soluble antigens?

8. What is complement? How is it involved in the immune response to a bacterial infection?

9. What are autoimmune diseases? How can such diseases occur?

10. What is an allergy? How are allergies related to the immune response?

11. What is a lymphokine? What functions do lymphokines play in the immune response?

12. How is the MHC region of the genome related to tissue compatibility?

13. What are the five major classes of immunoglobulins? Compare the role of each immunoglobulin class in the immune response.

**Suggested
Supplementary
Readings**

Allen, J. C. 1980. *Infection and the Compromised Host.* Williams & Wilkins Co., Baltimore.

Baglioni, C., and T. W. Nilsen. 1981. The action of interferon at the molecular level. *American Scientist* 69: 392–399.

Bellanti, J. (ed.). 1978. *Immunology II.* W. B. Saunders Co., Philadelphia.

Benacerraf, B. 1981. Role of MHC gene products in immune regulation. *Science* 212: 1229–1238.

Benacerraf, B., and E. R. Unanue. 1979. *Textbook of Immunology.* Williams & Wilkins Co., Baltimore.

Bigley, N. J. 1980. *Immunologic Fundamentals.* Year Book Medical Publishers, Chicago.

Doria, G., and A. Eshkol. 1980. *The Immune System.* Academic Press, New York.

Eisen, H. N. 1980. *Immunology.* Harper & Row, Hagerstown, Maryland.

Fudenberg, H., D. Sites, J. Caldwell, and J. Wells (eds.). 1980. *Basic and Clinical Immunology.* Lange Medical Publications, Los Altos, California.

Immunology: Readings from Scientific American. 1976. W. H. Freeman and Sons, San Francisco.

Kimball, J. W. 1983. *Introduction to Immunology.* Macmillan Publishing Co., New York.

Milstein, C. 1980. Monoclonal antibodies. *Scientific American* 243(4): 66–74.

Nahmias, A. J., and R. J. O'Reilly (eds.). 1981. *Immunology of Human Infection*. Plenum Publishing Co., New York.

Roitt, I. M. 1980. *Essential Immunology*. Blackwell Scientific Publications, Oxford, England.

Rose, N. R. 1981. Autoimmune diseases. *Scientific American* 244(2): 80–103.

Stansfield, W. D. 1981. *Serology and Immunology*. Macmillan Publishing Co., New York.

Prevention and treatment of human diseases caused by microorganisms

15

Despite the elaborate defense mechanisms aimed at preventing the invasion of the human body and the establishment of diseases by pathogenic microorganisms, human beings are subject to a variety of infectious diseases. The development of a basic understanding of the interrelationships between human beings and microorganisms, particularly the immune defense system and the virulence factors of microbial pathogens, has led to appropriate practices for preventing and diminishing the incidence of microbial diseases of human beings. Many infectious diseases, when they do occur, are relatively easily treated with antibiotics. We need only think about the number of times a physician has prescribed an antibiotic for us or our children to appreciate the wide use of antibiotics in medical practice. The control of microorganisms pathogenic to human beings is fundamental to the practice of modern medicine. Many once widespread deadly diseases, such as plague, are rare today because of the effective use of antibiotics. The use of antimicrobial agents for treating infectious diseases has led to greatly increased life expectancies.

Epidemiology

In order to control the spread of infectious diseases, it is necessary to understand the nature of the pathogens and their routes of transmission. The examination of disease transmission is part of the field of **epidemiology**. The epidemiologist considers the **etiology of the disease** and the factors involved in **disease transmission**, especially as these factors relate to populations. With this information the epidemiologist evaluates how a disease outbreak in a population can be effectively controlled.

Epidemiology is based on the statistical probability that a susceptible individual will be exposed to a particular pathogen that would result in disease transmission. The likelihood of disease transmission depends on the concentration and virulence of the pathogen, the distribution of susceptible individuals, and the potential sources of exposure to the pathogenic microorganisms. In many cases a disease outbreak can be traced to a single source of exposure. The epidemiologist often acts as a detective to locate the origin of a disease outbreak, in some cases, searching for a source of tainted food, in others, direct contact with infected individuals and so forth. The epidemiologist is aided by the knowledge that **disease transmission normally occurs via the air, the ingestion of contaminated food or water, and by direct contact with infected individuals or contaminated inanimated objects (fomites)**. By determining where and what people have eaten, where they have been, and with whom they have

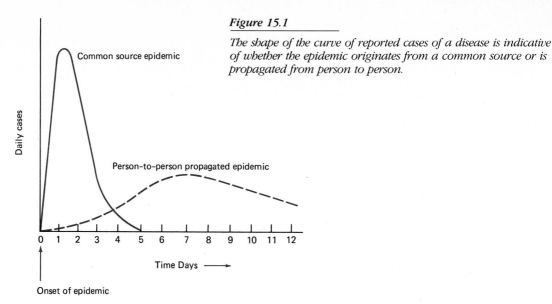

Figure 15.1

The shape of the curve of reported cases of a disease is indicative of whether the epidemic originates from a common source or is propagated from person to person.

been in contact, the epidemiologist can establish a pattern of disease transmission.

The number of cases reported each day and the locations of disease occurrences enables the epidemiologist to distinguish between a **common source outbreak**, which is characterized by a sharp rise and rapid decline, and a **person-to-person epidemic**, which is characterized by a relatively slow and prolonged rise and decline (Figure 15.1). In the United States the Centers for Disease Control (CDC) in Atlanta, Georgia, compiles the statistics necessary for such determinations. When common sources (**reservoirs**) or carriers (**vectors**) of pathogens are identified, action can be taken to break the chain of disease transmission, for instance by recalling potentially contaminated foods from the marketplace. When person-to-person (propagated) transmission is responsible for disease outbreaks in a population, steps can be taken to reduce the numbers of susceptible individuals, for example, by immunization, thereby breaking the chain of transmission. Indeed, it is the aim of the epidemiologist to identify the sources of disease outbreaks and to advise public health officials on the steps that should be taken to prevent such disease outbreaks.

Prevention of diseases by avoiding exposure to pathogenic microorganisms

Perhaps the most effective way of preventing diseases caused by microorganisms is to avoid exposure to pathogenic microorganisms. A total avoidance of microorganisms, though, is not practical because we are continuously exposed to microorganisms carried through the air, in water and foods, and on the surfaces of virtually all objects that we contact. Only in the rarest of cases, when the immune system is totally nonfunctional, is absolute avoidance of contact with microorganisms practiced (see Figure 14.24). It is possible, though, to control microbial populations and our interactions with them in ways that reduce the probability of encountering pathogenic microorganisms, thus reducing the incidence and spread of infectious diseases. A greatly diminished incidence of many diseases caused by microorganisms is the consequence of an adequate understanding of the modes of transmission of pathogenic microorganisms and preventive measures to reduce exposure to disease-causing microorganisms. Many modern sanitary practices are aimed at reducing the incidence of diseases by preventing the spread of pathogenic microorganisms or by reducing their numbers to concentrations that are insufficient to cause disease.

The methods employed for **preventing exposure to specific disease-causing microorganisms** vary depending on the particular route of transmission (Figure 15.2). Proper sewage treatment and drinking water disinfection programs reduce the likelihood of contracting a disease through contaminated water. Recognition of the fact that many serious diseases, such as typhoid, are transmitted through water contaminated with fecal material is the basis for strict water quality control. The topic of assessing and maintaining water quality will be discussed in Chapter 23. Quality control measures are also applied throughout the food industry to prevent the transmission of disease-causing microorganisms through food products. A number of methods are used to preserve food products, preventing the growth of microorganisms that both spoil food and may cause human disease (see discussion in Chapter 19). Pasteurization of milk is a good example of a process designed to reduce exposure to pathogenic microorganisms that occur and proliferate in untreated milk. Chlorination of municipal water supplies is widely used to prevent exposure to the pathogenic microorganisms that occur in water supplies and thus to ensure the safety of drinking water. The aim of washing one's hands before eating is to avoid the accidental contamination of one's food with soil or other substances that may

harbor populations of disease-causing microorganisms.

Failure to maintain quality control of water and food supplies often results in outbreaks of disease; for example, cholera outbreaks often occur when sewage is allowed to mix with drinking water supplies, such as frequently occurs in the Far East when monsoon rains cause flooding, resulting in contamination of drinking water supplies. Outbreaks of botulism are associated with improperly canned food products, when food products have not been heated long enough to kill contaminating spores of *Clostridium botulinum*, resulting in the exposure of the individuals who ingest them to the lethal toxins produced by this bacterium. Extensive quality control testing is required in most countries to prevent the outbreaks of disease associated with contaminated water and food supplies.

It is relatively more difficult to control the transmission of disease-causing microorganisms through the air than it is to ensure water and food quality control. Several steps, though, can be taken to reduce the likelihood of exposure to airborne pathogenic microorganisms. Covering one's nose and mouth with a handkerchief when sneezing and coughing reduces the number of potential

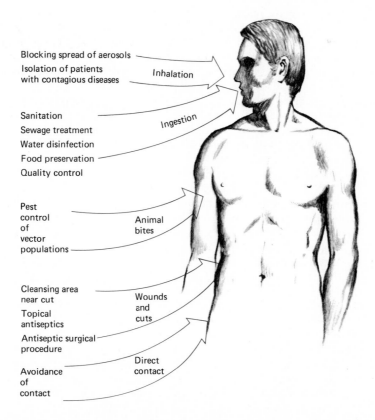

Figure 15.2

Routes of disease transmission and steps that may be taken to prevent exposure to pathogenic microorganisms.

Blocking spread of aerosols
Isolation of patients with contagious diseases
Inhalation

Sanitation
Sewage treatment
Water disinfection
Food preservation
Quality control
Ingestion

Pest control of vector populations
Animal bites

Cleansing area near cut
Topical antiseptics
Antiseptic surgical procedure
Wounds and cuts

Avoidance of contact
Direct contact

A
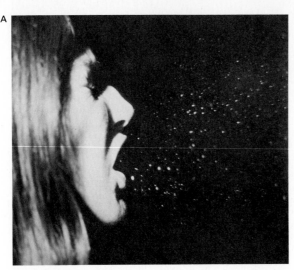

B

Figure 15.3

These photographs show droplet spread by sneezing and the effect of covering one's nose and mouth with a handkerchief when sneezing. (A) An unstifled sneeze. (B) The use of a handkerchief blocks the spread of droplets.

pathogens that may become airborne in aerosols and that therefore may be transmitted from an infected individual to a susceptible host (Figure 15.3). The use of surgical masks when visiting individuals who are particularly susceptible to airborne pathogenic microorganisms, such as newborn infants, is an important precaution for preventing the spread of infectious diseases. Likewise, preventing the exposure of individuals whose immunological defense mechanisms are compromised as a result of a variety of conditions—such as treatment for cancer or a recent organ transplant—to airborne pathogenic microorganisms is an important aspect of patient management practice. Similarly, masks should be worn in the presence of patients with tuberculosis or other pulmonary diseases or else other precautions, such as maintaining a safe distance from the infected individual, should be taken to minimize exposure to the pathogenic microorganisms. Isolation of individuals with contagious microbial diseases, in which the infectious agent is airborne, is often practiced. For example, children with measles, chicken pox, or mumps are often kept away (**isolated**) from other children who are not immune to these diseases. Such practices decrease the probability of exposure to pathogenic organisms and prevent, or at least reduce, the transmission of disease.

Avoiding direct contact with infected individuals is also important for preventing the spread of diseases when the pathogen is transmitted by direct contact. Historically, the isolation of leprosy patients in remote colonies is an example of the extreme steps that have been taken to prevent contact of such individuals with the general population. Today, this extreme practice is not needed to prevent the spread of leprosy because of the use of antimicrobial agents. Avoiding sexual contact, though, with individuals suffering from venereal diseases, such as syphilis and gonorrhea, interrupts the transmission of the pathogens that cause these diseases; such avoidance of contact with infected individuals is essential for and will undoubtedly remain the main method for controlling the spread of sexually transmitted diseases.

In the case of pathogens that enter the body through breaks in the skin surface, a variety of procedures are employed to reduce the probability of exposure to such pathogenic microorganisms. Great care, for example, is taken during surgical procedures to prevent contamination by accidental introduction of microorganisms into the exposed tissues. Sterile operating rooms, instruments, garments, gloves, and masks are all used by a hospital surgical staff (Figure 15.4). Wounds are cleansed to prevent the introduction of foreign material that may harbor potential pathogens, and antiseptics are applied to skin surfaces to minimize the entry of pathogenic microorganisms into tissues normally protected by an intact skin covering.

Practices are normally employed to control in-

Figure 15.4

Compare this modern hospital operating theater to that pictured from the 1860s in Figure 1.15. Carbolic acid is no longer used to prevent contamination, but the operating room surgical staff take many precautions to preclude microbial contamination of tissue exposed during the surgical procedure. (A) Everyone in the room is masked, gloved, capped, and gowned. (B) Note how much of the face is covered by the mask. (C) Note that the parts of the patient not involved in the operation are also covered. (Courtesy of Michelle Ising, University of Louisville.)

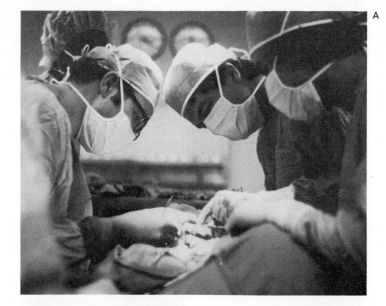
A

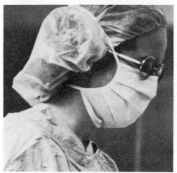
B

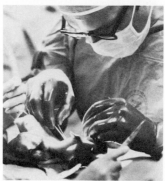
C

sect and other animal populations that act as vectors for the transmission of diseases caused by pathogenic microorganisms. **Vectors** are carriers of pathogenic microorganisms involved in the transmission of disease. The most notable vectors of pathogenic microorganisms are mosquitos, lice, ticks, and fleas. Some public health measures, such as mosquito control programs, are aimed at reducing the sizes of these vector populations and thus lowering the probability of exposure to the pathogenic microorganisms capable of causing diseases such as plague, typhus fever, yellow fever, malaria, and a variety of other diseases transmitted by animal vectors.

Preventing disease by immunization

Before considering how immunization can be used to prevent disease, we need consider how an epidemic occurs and the kinetics of disease transmission among susceptible individuals (Figure 15.5). **Epidemics** are outbreaks of disease in which unusually high numbers of individuals in a population contract the disease. As individuals recover from a disease, they are immune, that is, not susceptible, and thus no longer participate in the chain of disease transmission. The number of individuals who must be immune to prevent epidemic outbreak of disease is a function of the infectivity of the disease (I), the duration of the disease (D), and the proportion of susceptible individuals in the population (s). When the triple product sID is low because of a high proportion

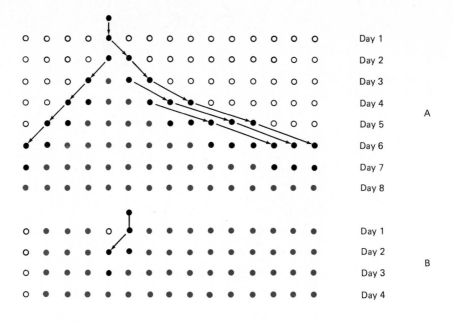

○ Susceptible   ● Sick

● Immune   ⟶ Routine of transmission

Figure 15.5

The kinetics of the spread of an infectious disease and the effect of increasing the number of immune individuals in the population in limiting epidemic outbreaks of disease. (A) The individuals in this population are all susceptible. The introduction of a sick individual initiates an epidemic outbreak of the diseases. At the height of the epidemic, 50 percent of the population are ill. Eventually, all individuals develop the disease and are subsequently immune. (B) Eighty percent of the population are immune to the disease when a sick individual enters the population. No epidemic occurs, and even a susceptible individual fails to contact and contract the disease.

of immune individuals (about 70 percent of the population is immune), the whole population is protected, a concept known as **herd immunity**. The proportion of a population that must be immune to prevent an epidemic varies depending on the effectiveness with which the pathogen is transmitted from one individual to another and the virulence of the pathogen.

Herd immunity can be established by artificially stimulating the immune response system through the use of vaccines, rendering an individual unsusceptible to a particular disease and protecting the entire population. **Vaccines** are preparations of antigens designed to simulate the normal primary immune response, resulting in a proliferation of the memory cells and the ability to exhibit a secondary memory or anamnestic response upon subsequent exposure to the same antigens. **Immunization** has become a worldwide practice for preventing various infectious dis-

eases. The antigens within the vaccine need not be associated with active virulent pathogens. The antigens in the vaccine need only elicit an immune response, with the production of antibodies possessing the ability to cross-react with the critical antigens associated with the pathogens against which the vaccine is designed to confer protection. The usefulness of vaccines rests with the fact that they can confer immunity; that is, they render an individual unsusceptible to a disease without actually producing the disease or at least not a serious form of the disease.

Some of the vaccines that are useful in preventing diseases caused by a variety of microorganisms are listed in Table 15.1. Vaccines may contain antigens prepared by killing or inactivating pathogenic microorganisms, or vaccines may use attenuated or weakened strains that are unable to cause the onset of severe disease symptoms. Some vaccines are prepared by denaturing microbial exotoxins to produce toxoids. Protein exotoxins, such as those involved in the diseases tetanus and diphtheria, are suitable for toxoid preparation, and the vaccines for preventing these diseases employ toxins inactivated by treatment with formaldehyde. These toxoids retain the antigenicity of the protein molecules; that is, they elicit the formation of antibody and are reactive with antibody molecules, but because the proteins are denatured, they are unable to initiate the biochemical reactions associated with the active toxins that cause disease conditions. In some cases whole microorganisms rather than individual protein toxins are used for preparing vaccines. When a variety of other microorganisms are killed by treatment with chemicals, radiation, or heat, the antigenic properties of the pathogen are retained without the risk that exposure to the vaccine could cause the onset of the disease associated with the virulent live pathogens. The vaccines used for the prevention of whooping cough (pertussis) and influenza are representative preparations containing antigens prepared by inactivating, that is, killing, pathogenic microorganisms.

In contrast to the above vaccines, some vaccine preparations contain living but attenuated strains of microorganisms. Pathogens are attenuated by a variety of procedures, including moderate use of heat, chemicals, desiccation, and growth in tissues other than the normal host. The Sabin vaccine for poliomyelitis, for example, uses viable polioviruses attenuated by growth in tissue culture. These viruses are capable of multiplication within the digestive tract and the salivary glands but are unable to invade the nerve tissues and thus do not produce the symptoms of the disease polio. The vaccines for measles, mumps, rubella, and yellow fever similarly utilize viable but attenuated viral strains. Attenuated strains of rabies virus can be prepared by desiccating the virus after growth in the central nervous system tissues of a rabbit or following growth in a chick or duck embryo. Vaccines containing viable attenuated strains require relatively low amounts of the antigens because the microorganism is able to replicate after administration of the vaccine, resulting in a large increase in the amount of antigen available within the host to trigger the immune response mechanism. Quality control is extremely important in preparing vaccines, particularly those using attenuated strains, so as to prevent disease outbreaks resulting from inadequately attenuated pathogens or from microorganisms that accidentally enter the vaccine preparation.

The effectiveness of a vaccine depends on a

table 15.1

Some vaccines that are useful in preventing diseases caused by microorganisms

Disease	Vaccine
Smallpox	Attenuated live virus
Yellow fever	Attenuated live virus
Hepatitis B	Attenuated live virus
Measles	Attenuated live virus
Mumps	Attenuated live virus
Rubella	Attenuated live virus
Polio	Attenuated live virus (Sabin)
Polio	Inactivated virus (Salk)
Influenza	Inactivated virus
Rabies	Inactivated virus
Tuberculosis	Attenuated live bacteria
Pertussis	Inactivated bacteria
Cholera	Inactivated bacteria
Diphtheria	Toxoid
Tetanus	Toxoid

number of factors including the antigens within the vaccine, the other chemicals in the vaccine preparation, and the route of administration of the vaccine. Some vaccines must contain multiple antigens to prevent diseases caused by a variety of virulent strains of a given pathogen. For example, the Sabin polio vaccine contains types 1, 2, and 3 antigens derived from the three predominant strains of poliovirus. Other vaccines, such as those used to prevent influenza, are directed at particular strains of the pathogens, and different vaccines are needed to protect against different forms of the same disease.

The greatest success in preventing disease through the use of vaccines can be seen in the case of smallpox (Figure 15.6). The vaccine used for preventing smallpox contains a live strain of pox virus. The vaccine most commonly used for preventing smallpox is prepared from scrapings of lesions from cows or sheep. The scrapings are treated with 1 percent formaldehyde to kill bacterial contaminants and 40 percent glycerol to stabilize the viral antigens. The viral antigens are quite labile, which is why live viral preparations are required for successful vaccination to achieve a state of immunity. Various viral strains have been used for the production of commercial vaccines. Although the commercial strains used for vaccine production were presumed to have been derived from cowpox virus, it now appears, based on its antigenic properties, that an attenuated strain of smallpox virus may have been inadvertently substituted for the cowpox virus. Because of the length of time that this virus has been cultivated, it is difficult to positively identify its original source,

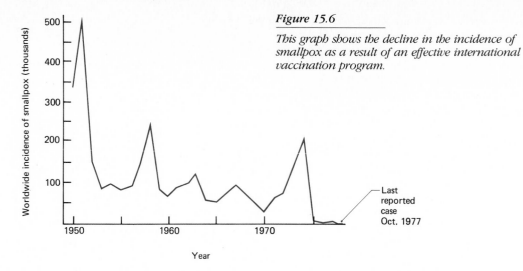

Figure 15.6

This graph shows the decline in the incidence of smallpox as a result of an effective international vaccination program.

but certainly the pox virus used for vaccine preparation differs from the cowpox viruses found in nature.

Regardless of the origins of the viral strain used in its vaccines, smallpox, a once dreaded disease, has been completely eliminated through an extensive worldwide immunization program conducted under the auspices of the World Health Organization (WHO). The success of the WHO program depended on the use of lyophilized vaccines to overcome the problem of inactivation of the viral antigens in hot climates. The program

Figure 15.7

Adjuvants enhance antigenicity, as is illustrated by this graph.

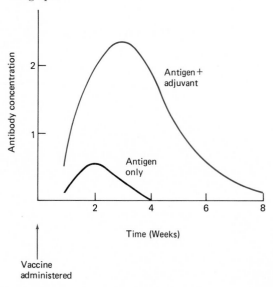

was not without risks because the virus used for vaccination was virulent enough to cause a fatality rate of 1 in one million vaccinations. By immunizing a sufficient portion of the world's population against smallpox, though, it was possible to interrupt the normal transmission of smallpox virus from infected individuals to susceptible hosts. A consequence of the success of this immunization program is that it is no longer necessary to vaccinate against smallpox. The successful elimination of smallpox through a vaccination program was dependent on the fact that human beings are the only known host for the smallpox virus and that the virus has a relatively short survival time outside human host tissues. Smallpox presumably has been permanently eliminated and as such is the only infectious human disease known to have been eliminated.

Some chemicals, known as **adjuvants**, greatly enhance the antigenicity of other biochemicals (Figure 15.7). The inclusion of adjuvants in vaccines therefore can greatly increase the effectiveness of the vaccine. When protein antigens are mixed with aluminum compounds, for example, a precipitate is formed that is more useful for establishing immunity than the proteins alone. Alum-precipitated antigens are released slowly, enhancing the stimulation of the immune response. The use of adjuvants can eliminate the need for repeated booster doses of the antigen—which increase the intracellular exposure to antigens to establish immunity—and also permits the use of smaller doses of the antigen in the vaccine.

The antigens in the vaccine may be introduced into the body by a number of routes: **intradermally** (into the skin), **subcutaneously** (under the skin), **intramuscularly** (into the muscle), **intravenously** (into the bloodstream), and into the mucosal cells lining the respiratory and gastrointestinal tracts. The effectiveness of a given vaccine depends in part on the normal route of entry for the particular pathogen. For example, polioviruses normally enter via the mucosal cells of the respiratory or gastrointestinal tracts and therefore, the Sabin polio vaccine is administered orally, enabling the attenuated viruses to enter the mucosal cells of the gastrointestinal tract. It is likely that vaccines administered in this way stimulate secretory antibodies of the IgA class in addition to other immunoglobulins. Intramuscular administration of vaccines, such as the Salk polio vaccine, a vaccine containing inactivated polioviruses, is more likely to stimulate IgG production, which is particularly effective in precluding the spread of pathogenic microorganisms and toxins produced by such organisms through the circulatory system.

Multiple exposures to antigens are sometimes needed to ensure the establishment and continuance of a memory response. Periodic **booster vaccinations** are necessary, for example, to maintain immunity against tetanus. Several administrations of the Sabin vaccine are needed during childhood to establish immunity against poliomyelitis. No booster vaccinations, though, are needed to establish immunity against measles, mumps, and rubella. For each of these diseases, a single administration of an attenuated virus strain is sufficient to establish lifelong immunity. The recommended schedule for the administration of

some vaccines is shown in Table 15.2. Not all diseases can be prevented by using vaccines. Some antigens confer immunity that lasts only weeks or months. Such short-lived immunity may be effective in preventing disease if there is a known likelihood of exposure to a given pathogen, but it is not feasible to attempt to use vaccines that confer only short-term immunity on a wide scale for preventing disease.

There are also several other immunological procedures that may be used to prevent disease. For example, a variety of **antitoxins** can be used to neutralize toxins of microbial or other origin that are responsible for disease symptomology. Antitoxins are used to neutralize the toxins in snake venom, saving the victims of snake bites. The toxins in poisonous mushrooms can also be neutralized by administration of appropriate antitoxins. The administration of antitoxins and immunoglobulins to prevent disease occurs after exposure to a toxin and/or an infectious microorganism.

It is also possible to establish **passive immunity** by the administration of IgG. Passive immunity lasts for a limited period of time and does not involve the establishment of a memory immune response capability. Such passive immunity is conferred naturally upon an infant by the passage of IgG molecules across the placenta during fetal development. IgG and other immunoglobulins are found in the colostrum and milk of nursing mothers, protecting newborns against infectious diseases during the early period of life. The administration of IgG is also particularly useful

table 15.2

Recommended schedule for vaccine administration

Disease	Primary immunization	Booster doses
Diphtheria, pertussis, tetanus	Intramuscular DPT injections at 2, 4, 6, and 18 months	One intramuscular booster at 3–6 yrs (tetanus every 10 yrs)
Influenza	Seasonally for high-risk elderly and chronically ill	
Measles	One subcutaneous injection at 15 months	None
Mumps	One subcutaneous injection at 1 yr	None
Polio	Sabin vaccine: oral at 2, 4, and 15 months	One oral dose at 5 yrs
	Salk vaccine: intramuscular injections at 2, 3, 4, and 15 months	Intramuscular injections every few years
Rubella	One subcutaneous injection at 1 yr	None

therapeutically in preventing disease in persons with immunodeficiencies and other high-risk individuals.

Although vaccines are normally administered prior to exposure to antigens associated with pathogenic microorganisms, some vaccines are administered after suspected exposure to a given infectious microorganism. In these cases the purpose of vaccination is to elicit an immune response before the onset of disease symptomology. For example, tetanus vaccine is administered after puncture wounds that may have introduced

Clostridium tetani into deep tissues, and rabies vaccine is administered after animal bites that may have introduced rabies virus. The effectiveness of vaccines administered after the introduction of the pathogenic microorganisms depends on the relatively slow development of the infecting pathogen prior to the onset of disease symptoms and the ability of the vaccine to initiate antibody production before the active toxins are produced and released to the site where they can cause serious disease symptoms.

Treating diseases with antimicrobial agents

When normal host defense mechanisms and preventive measures fail to protect an individual against the establishment of a particular pathogenic microorganism, there are a large number of antimicrobial agents available for treating diseases caused by microorganisms. Such drugs have become an essential part of modern medical practice. The **antimicrobial agents** used in medical practice are aimed at eliminating the infecting microorganisms or at preventing the establishment of an infection. These chemicals should not be confused with the large number of drugs used in medical practice for alleviating the symptoms of disease or for treating diseases not caused by

microorganisms. To be of therapeutic use, an antimicrobial agent must exhibit **selective toxicity**. A therapeutically useful antimicrobial agent must inhibit infecting microorganisms and exhibit greater toxicity to the infecting pathogens than to the host organism. A drug that kills the patient is of no use in treating infectious diseases, whether or not it also kills the pathogens! Even selective, therapeutically useful antimicrobial agents, though, can produce side effects (Table 15.3). As a rule, antimicrobial agents are of most use in medicine when their mode of action involves biochemical features of the invading pathogens not possessed by normal host cells.

table 15.3

Side effects of some antibiotics of therapeutic value

Antimicrobial agent	Main side effects
Penicillin G	Hypersensitivity reactions
Cephalothin	Hypersensitivity reactions
Erythromycin	Gastrointestinal upset
Streptomycin	Hypersensitivity reactions; vestibular toxicity; auditory impairment
Gentamicin	Renal toxicity; gastrointestinal upset
Tetracycline	Gastrointestinal irritation; hepatic toxicity
Sulfonamides	Hypersensitivity reactions; renal injury; anemia and other blood disturbances; hepatic toxicity
Amphotericin B	Hypersensitivity reactions; anemia; renal function impairment; pain; convulsions; fever
Chloramphenicol	Hypersensitivity reactions; gastrointestinal upset
Chloroquine	Headache; visual disturbances; gastrointestinal upset

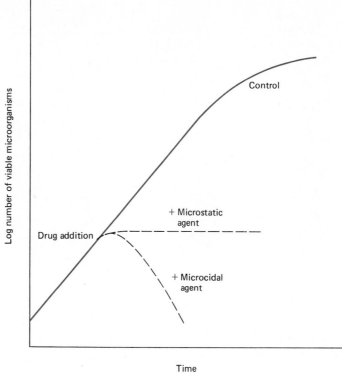

Classification and selection of antimicrobial agents

Some antimicrobial agents are **microcidal** (they kill microorganisms), and others are **microstatic** (they inhibit the growth of microorganisms but do not actually kill them) (Figure 15.8). Microstatic agents can prevent the proliferation of infecting microorganisms, holding populations of pathogens in check until the normal immune defense mechanisms can eliminate the invading pathogens. **Antibiotics** represent a major class of antimicrobial agents. By definition, antibiotics are biochemicals produced by microorganisms that inhibit the growth of, or kill, other microorganisms (Figure 15.9). By their very nature, antibiotics must exhibit selective toxicity because they are produced by one microorganism and exert varying degrees of toxicity against others. The formal definition of an antibiotic distinguishes biochemicals that are produced by microorganisms from organic chemicals that are synthesized in the laboratory. This distinction is no longer meaningful because organic chemists can synthesize biochemical structures of many naturally occurring antibiotics. Additionally, many antibiotics in current medical use are chemically modified forms

of microbial biosynthetic products. Therefore, in this chapter the term antibiotic is used in the broad sense to refer to both natural and semisynthetic antimicrobial agents that are available to the physician for treating infectious diseases. The discovery and use of antibiotics have revolutionized medical practice in the twentieth century.

The **selection of a particular antimicrobial agent** for treating a given disease depends on several factors, including: (1) the sensitivity of the infecting microorganism to the particular antimicrobial agent; (2) the side effects of the antimicrobial agent, relative to direct toxicity to mammalian cells and to the microbiota normally associated with human tissues; (3) the biotransformations of the particular antimicrobial agent that occur *in vivo*, relative to whether the antimicrobial agent will remain in its active form for a sufficient period of time to be selectively toxic to the infecting pathogens; and (4) the chemical properties of the antimicrobial agent that determine its distribution within the body, relative to whether or not adequate concentrations of the active antimicrobial chemical will be able to reach the site of infection in order to inhibit or kill the pathogenic microorganisms causing the infection. For example, although many antibiotics possess antimicrobial ac-

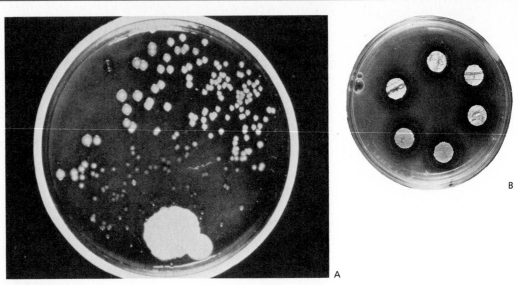

Figure 15.9

These photographs show the growth of antibiotic-producing and sensitive strains demonstrating an inhibitory effect. (A) The original culture plate of Alexander Fleming showing the growth of a Penicillium *colony and the lack of colonies of* Staphylococcus *near the fungal colony. Colonies developing nearest the* Penicillium *show evidence of bacteriolysis. (From BPS—E. Chain and H. W. Florey, 1944,* Endeavor *3:9.) (B) Photograph showing zones of inhibited bacterial growth of a normal antibiotic-sensitive bacterium surrounding agar plugs selected from six different* Streptomyces *strains. (Courtesy Boyd Woodruff, Merck, Sharpe and Dohme, Rahway, New Jersey, reprinted by permission of the Society of General Microbiology, from H. B. Woodruff and L. E. McDaniel, 1958, The antibiotic approach, in* Symposia of the Society of General Microbiology, Number VIII, The Strategy of Chemotherapy, *p. 49.)*

Discovery process

Fleming's chance discovery of the effect of mold contamination on a bacterial culture plate was possible only because Fleming's background and knowledge had been advanced by the growth of general scientific awareness such that what he saw made sense to him and fitted into a historical pattern of scientific investigation. Similar observations of the inhibition of one microorganism by another had been reported for more than 50 years. In 1876 Tyndall described the antagonistic action of a species of *Penicillium* on bacterial growth. Gosio recorded the first isolation of an antibiotic in crystalline form in 1896. Crude antibiotic preparations were attempted in the early twentieth century but produced erratic results. So what was it about Fleming's discovery of penicillin that made modern antibiotics a reality? Fleming himself pointed out that the pioneering work on antibiotics was thwarted because of the failure of those microbiologists to pursue the chemical investigations necessary to separate and purify the active agents in their extracts. By the 1930s, however, microbiology had become far more chemistry-oriented, and this approach culminated in the preparation of solid penicillin by Chain, Florry, and others at Oxford University in 1940.

tivities effective against the pathogenic bacteria that cause urinary tract infections, only a limited number of antibiotics are effective in treating these infections because relatively few antibiotics can reach and be concentrated in the tissues of the urinary tract in their active form. Additionally, one antimicrobial agent can influence the effects of another antimicrobial agent. In some cases treatment with two drugs enhances the effectiveness of treatment, and in other cases one drug can interfere with the inhibitory effects of a second antimicrobial agent (Figure 15.10).

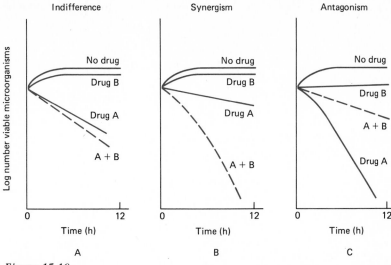

Figure 15.10

This diagram illustrates the possible interactions between antimicrobial agents. (A) indifference; (B) synergism; and (C) antagonism. Interactions are reflected in the log numbers of microorganisms that are still viable after application of these agents.

Antibiotics effective against bacterial infections

The biochemical differences in the cell structures of bacterial cells and eukaryotic cells (discussed in Chapter 4) form the basis for the effective use of antibiotics against bacterial infections. The bacterial cell wall with its unique peptidoglycan and the 70S ribosome represent two major sites against which antibacterial agents may be directed. Most of the common antibiotics used in medicine for treating bacterial infections are inhibitors of cell wall or protein synthesis. Some antibiotics are more selective than others with respect to the bacterial species that they can inhibit. A **narrow-spectrum antibiotic** may be targeted at a particular pathogen, for instance, at Gram positive cocci, or at a particular bacterial species (Table 15.4). In contrast, some antibiotics have a **broad spectrum of action**, inhibiting a relatively wide range of bacterial species, including both Gram positive and Gram negative types. The choice of a particular antibiotic depends in part on the biochemical properties of the specific infecting bacterial strain. In most cases; physicians make an educated guess as to which antibiotic is appropriate for treating a particular infection, and the selection of the antibiotic is based on the most likely pathogen causing the given disease symptomology and the antibiotics generally known to be effective against such pathogens. Many times a physician will select a broad-spectrum antibiotic in order in ensure effective treatment. Only in special cases, such as when a patient fails to respond to a particular antibiotic and an infection persists, is an attempt normally made to isolate the pathogenic bacterium and to determine the range of specific antibiotic sensitivity of that organism.

However, concern is mounting in the medical field about the overuse of antibiotics because an undesired side effect of such "drug abuse" is the

table 15.4

Some common antibiotics and whether they have a narrow or broad spectrum of action

Antibiotic	Spectrum of action
Penicillin G	Narrow
Methicillin	Narrow
Gentamicin	Narrow
Erythromycin	Narrow
Isoniazid	Narrow
Amphicillin	Broad
Cephalosporins	Broad
Streptomycin	Broad
Kanamycin	Broad
Tetracyclines	Broad
Chloramphenicol	Broad
Sulfonamides	Broad

selection for disease-causing antibiotic-resistant strains. It is now considered proper medical practice to perform culture and sensitivity studies to determine the proper antibiotic for treating a patient. Only in cases of life-threatening infections should antibiotics be used without such testing to avoid the selective pressure for the development of antibiotic-resistant pathogens. The importance of the problem of excessive use of antibiotics was recently underscored when the American Medical Association called upon physicians to avoid unnecessary use of antibiotics.

The reason for concern about how we use antibiotics is that numerous bacterial strains have acquired the ability to resist the effects of some antibiotics, with some bacterial strains, generally containing R plasmids, having **multiple antibiotic resistance**. The basis of resistance in some cases rests with the ability of the particular strain to produce enzymes that degrade the antibiotics, preventing the active form of the antibiotic from reaching the bacterial cells where they could be inhibitory. For example, some bacterial strains produce penicillinase enzymes (β-lactamases) that are able to degrade the antibiotic penicillin, making such strains resistant to penicillin. Resistance may also be due to decreased drug uptake, metabolic transformation of the drug to its nontoxic forms, decreased transformation of the drug to its active form, and/or decreased sensitivity of the microbial structure against which the drug is directed.

Modes of action of antibacterial agents

In this section we will consider the properties and modes of action of many of the common antibiotics currently in use. It should be noted that only the generic names of the antibiotics will be used because the same compound may be produced by several companies under several different trade names (Table 15.5). Many municipalities now require that pharmacies fill a prescription by its generic name to encourage competitive pricing, unless a specific brand name is specified. Such laws are controversial because the pharmaceutical manufacturers lose the incentive to develop new antibiotics and to maintain high and costly quality control when a less stringently regulated and less expensive product will be used to fill a prescription. The public wants cost-effective but quality antibiotics.

Cell wall inhibitors   The **penicillins** and **cephalosporins** are two widely used classes of antibiotics that **inhibit the formation of bacterial cell wall structures**. Penicillins are synthesized by strains of the fungus *Penicillium*. The cephalosporins are produced by members of the fungal genus *Cephalosporium*. The penicillins and cephalosporins both contain a β-lactam ring and thus have related biochemical structures (Figure 15.11). There are various specific penicillin and cephalosporin antibiotics containing differing biochemical substituent groups. The different antibiotics included in these broad classes of compounds

table 15.5

Generic and trade names for some common antibiotics

Generic name	Trade name
Tetracycline	Acromycin; Panmycin; Tetracyn; Tetrachel; Rexamycin
Oxytetracycline	Teramycin
Chlorotetracycline	Aureomycin
Demeclocycline	Declomycin
Methacycline	Rondomycin
Doxycycline	Vibramycin
Minocycline	Minocin; Vectrin
Penicillin G	Crysticillin; Duracillin
Ampicillin	Amcill; Omnipin; Penbritin; Polycillin
Cephalothin	Keflin
Cephalexin	Keflex
Chloramphenicol	Chloromycetin; Mychel
Gentamicin	Garamycin
Kanamycin	Kantrex
Erythromycin	Ilotycin
Nystatin	Mycostatin; Nilstat
Trimethoprim-sulfamethoxazole	Bactrin; Septra
Chloroquine	Aralen; Avloclor; Resochin

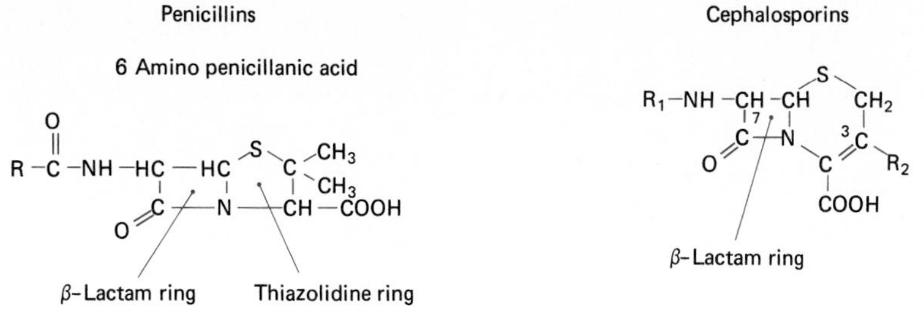

Penicillins

6 Amino penicillanic acid

Cephalosporins

β–Lactam ring   Thiazolidine ring

β–Lactam ring

Penicillins	R-Side chain	Cephalosporins	R_1	R_2
Penicillin G	$-CH_2-$ (phenyl)	7–Aminocephalosporanic acid	$H-$	$-CH_2-O-C(=O)-CH_3$
Phenoxymethyl penicillin (Pen V)	$-OCH_2-$ (phenyl)	Cephalothin	(thiophene)$-CH_2-C(=O)-$	$-CH_2-O-C(=O)-CH_3$
Methicillin	(phenyl with two OCH_3)	Cefazolin	(tetrazole)$N-CH_2-C(=O)-$	$-CH_2-S-$(thiadiazole)$-CH_3$
Oxacillin	(phenyl, isoxazole with CH_3)	Cephapirin	(pyridine)$-S-CH_2-C(=O)-$	$-CH_2-O-C(=O)-CH_3$
Nafcillin	(naphthyl with OC_2H_5)	Cephalexin	(phenyl)$-CH(NH_2)-C(=O)-$	$-CH_3$
Ampicillin	(phenyl)$-CH(NH_2)-$	Cephradine	(cyclohexadienyl)$-CH(NH_2)-C(=O)-$	$-CH_3$
Amoxicillin	$HO-$(phenyl)$-CH(NH_2)-$	Cefoxitin	(thiophene)$-CH_2-C(=O)-$	$-CH_2-O-C(=O)-NH_2$
Carbenicillin	(phenyl)$-CH(CO_2Na)-$	Cefamandole	(phenyl)$-CH(OH)-C(=O)-$	$-CH_2-S-$(tetrazole)$-CH_3$

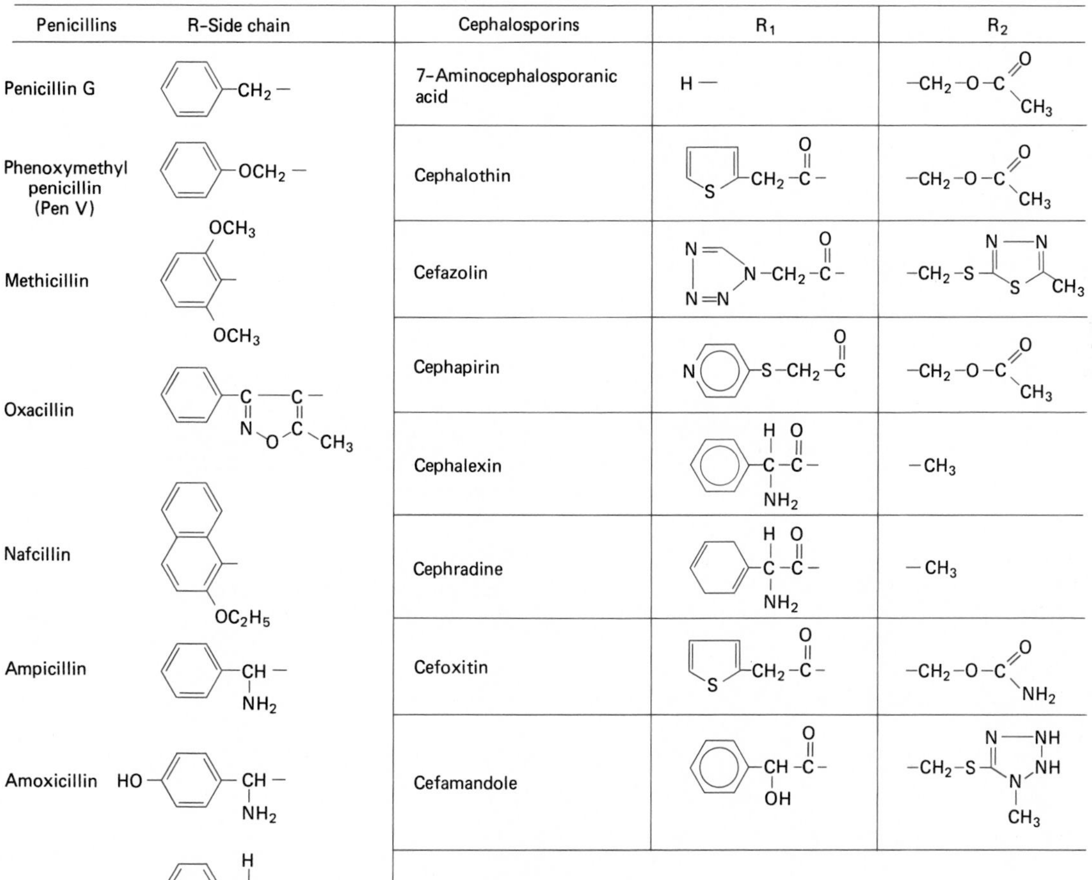

Figure 15.11

The biochemical structures of penicillins and cephalosporins, all of which contain a β-lactam ring.

exhibit differing spectrums of antibacterial activity (Table 15.6). Both the penicillins and cephalosporins inhibit the formation of peptide cross-linkages within the peptidoglycan backbone of the cell wall. These antibiotics specifically inhibit the enzymes involved in the cross-linkage for transpeptidase reactions (Figure 15.12). It appears that the β-lactam portion of cephalosporin and penicillin antibiotics binds to the transpeptidase enzyme, preventing the binding of the enzyme to the normal substrate, D-alanine. Bacterial cell walls lacking the normal cross-linking peptide chains are subject to attack by **autolysins** (autolytic enzymes produced by the organism that degrade the cell's own cell wall structures). The result is that, in the presence of cephalosporins or penicillins, growing bacterial cells are subject to lysis because without functional cell wall structures,

table 15.6

Some diseases and their causative organisms for which penicillins and cephalosporins are recommended

Causative organism	Disease	Drug of choice
Gram positive cocci		
Staphylococcus aureus	Abscesses Bacteremia Endocarditis Pneumonia Meningitis Osteomyelitis Cellulitis	Pencillin G A penicillinase-resistant penicillin
Streptococcus pyogenes	Pharyngitis Scarlet fever Otitis media, sinusitis Cellulitis Erysipelas Pneumonia Bacteremia Other systemic infections	Penicillin G Penicillin V
Streptococcus (*viridans* group)	Endocarditis Bacteremia	Penicillin G (±) streptomycin
Streptococcus agalactiae (group B)	Septicemia Meningitis	Ampicillin or penicillin G (±) an aminoglycoside
Streptococcus faecalis (enterococcus)	Endocarditis Urinary tract infection Bacteremia	Penicillin G + an aminoglycoside Ampicillin Penicillin G + an aminoglycoside
Streptococcus bovis	Endocarditis Urinary tract infection Bacteremia	Penicillin G
Streptococcus (anaerobic species)	Bacteremia Endocarditis Brain and other abscesses Sinusitis	Penicillin G
Streptococcus pneumoniae (pneumococcus)	Pneumonia Meningitis Endocarditis Arthritis Sinusitis Otitis	Penicillin G
Gram negative cocci		
Neisseria gonorrhoeae (gonococcus)	Genital infections Arthritis-dermatitis syndrome	Ampicillin or amoxicillin Penicillin G Ampicillin or amoxicillin Penicillin G
Neisseria meningitidis (meningococcus)	Meningitis Bacteremia	Penicillin G
Gram positive rods		
Bacillus anthracis	"Malignant pustule" Pneumonia	Penicillin G
Corynebacterium diphtheriae	Pharyngitis Laryngotracheitis Pneumonia Other local lesions	Penicillin G

Causative organism	Disease	Drug of choice
Corynebacterium species, aerobic and anaerobic diphtheroids	Endocarditis	Penicillin G (±) an aminoglycoside
Listeria monocytogenes	Meningitis Bacteremia Endocarditis	Ampicillin (±) an aminoglycoside
Erysipelothrix rhusiopathiae	Erysipeloid	Penicillin G
Clostridium perfringens	Gas gangrene	Penicillin G
Clostridium tetani	Tetanus	Penicillin G
Gram negative rods		
Haemophilus influenzae	Otitis media Sinusitis Bronchitis Epiglottitis	Amoxicillin Ampicillin
Enterobacter aerogenes	Urinary tract infection	Cephamandole
Klebesiella pneumoniae	Urinary tract infection Pneumonia	Cephalosporin
Pasteurella multocida	Wound infection Abscesses Bacteremia Meningitis	Penicillin G
Bacteroides sp.	Oral disease Sinusitis Brain abscess Lung abscess	Penicillin G
Fuscobacterium nucleatum	Ulcerative pharyngitis Lung abcess Empyema Genital infections Gingivitis	Penicillin G
Streptobacillus moniliformis	Bacteremia Arthritis Endocarditis Abscesses	Penicillin G
Spirochetes		
Treponema pallidium	Syphilis	Penicillin G
Treponema pertenue	Yaws	Penicillin G
Leptospira	Weil's disease Meningitis	Penicillin G
Actinomycetes		
Actinomyces israelli	Cervical, facial, abdominal, thoracic, and other lesions	Penicillin G

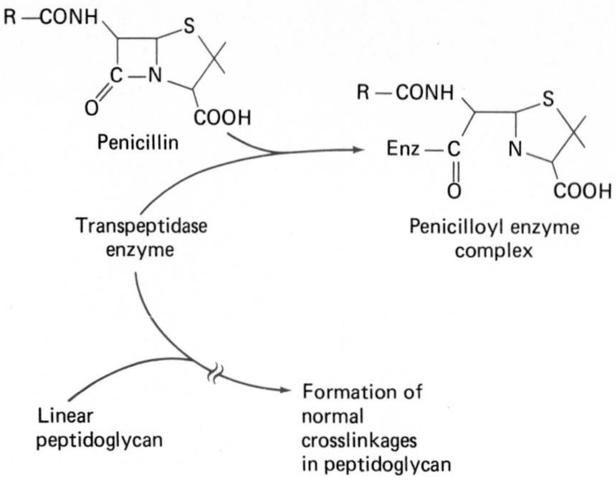

Figure 15.12

The mode of action of penicillin and cephalosporin antibiotics involves blockage of the normal cross-linkages in the peptidoglycan layer of the bacterial cell wall.

Prevention and treatment of human diseases caused by microorganisms

the bacterial cell is not protected against osmotic shock. It should be noted that the penicillin and cephalosporin antibiotics do not themselves remove intact cell walls and thus are ineffective against resting or dormant cells.

Many of the penicillins, such as penicillin G, have a relatively narrow spectrum of activity, being most effective against Gram positive cocci, including *Staphylococcus* species. Other penicillin antibiotics, such as ampicillin, have a broader spectrum of activity, inhibiting some Gram negative as well as Gram positive bacteria. Ampicillin, an amino-substituted penicillin, is active against many Gram negative rods, including *Escherichia coli, Haemophilus influenzae, Shigella* sp., and *Proteus* sp. To inhibit peptidoglycan synthesis effectively in Gram negative bacteria, the antibiotic must pass through the outer lipopolysaccharide layers to reach the peptidoglycan located at the inner portion of the cell wall. The broad-spectrum activity of ampicillin appears to be based on its ability to penetrate to the site of action of the transpeptidase enzyme, whereas narrow-spectrum penicillins, such as penicillin G, are relatively inefficient at reaching this site.

Penicillin G and various other β-lactam antibiotics are subject to inactivation by penicillinase enzymes (β-lactamases). In fact, none of the broad-spectrum penicillins is penicillinase-resistant. These antibiotics are ineffective against penicillinase-producing bacterial strains. For example, penicillin G is normally effective against *Neisseria gonorrhoeae*, a Gram negative coccus that causes gonorrhea, but some penicillinase-producing strains of *N. gonorrhoeae* have now been found, requiring the use of antibiotics other than penicillin G in the treatment of cases of gonorrhea caused by these penicillin-resistant strains. There may also be other causes for the penicillin re-

sistance of *N. gonorrhoeae*. About one in 10^9 cells of *N. gonorrhoeae* is resistant to penicillin and thus high enough antibiotic concentrations must be given for a long enough time to allow the natural body defense mechanisms to eliminate all the infecting bacteria. Structural modifications of penicillin G, such as occur in methicillin, can render the molecule resistant to penicillinases but may also narrow the spectrum of action, limiting the primary use of such antibiotics to the treatment of infections caused by penicillinase-producing *Staphylococcus* species.

In contrast to the penicillins, the cephalosporins generally have a broad spectrum of action, and many of the cephalosporins, such as cefoxitin and cephalothin, are relatively resistant to penicillinase. As such, the cephalosporins are useful in treating a variety of infections caused by Gram positive and Gram negative bacteria. Many physicians are now using broad-spectrum cephalosporins where the use of narrow-range and more specifically directed penicillins would be adequate. Cephalosporins are most prudently used as alternatives to penicillins in cases where the patient is allergic to penicillin and in cases where the pathogen is not penicillin sensitive. Cephalothin is often the antibiotic of choice for treating severe staphylococcal infections, such as endocarditis, to avoid complications in cases where the infecting *Staphylococcus* species produces β-lactamases. Cefamandole, another one of the cephalosporins, is widely used in treating pneumonia, as it is active against *Haemophilus influenzae, Staphylococcus aureus*, and *Klebsiella pneumoniae*, which are frequently the causative agents of respiratory tract infections resulting in pneumonia. The cephalosporins may also be used in place of penicillins for the prophylaxis of infection by Gram positive cocci after surgery.

In addition to the penicillins and cephalosporins, several other antibiotics inhibit cell wall synthesis, including vancomycin, bacitracin, and cycloserine (Figure 15.13). These antibiotics do not block the enzymes involved in the formation of peptide cross-linkages in the murein component of the wall but rather block other reactions involved in the synthesis of the bacterial cell wall. Cycloserine is a structural analog of D-alanine and can prevent the incorporation of D-alanine into the peptide units of the cell wall (Figure 15.14). In the presence of D-cycloserine, the subunits that

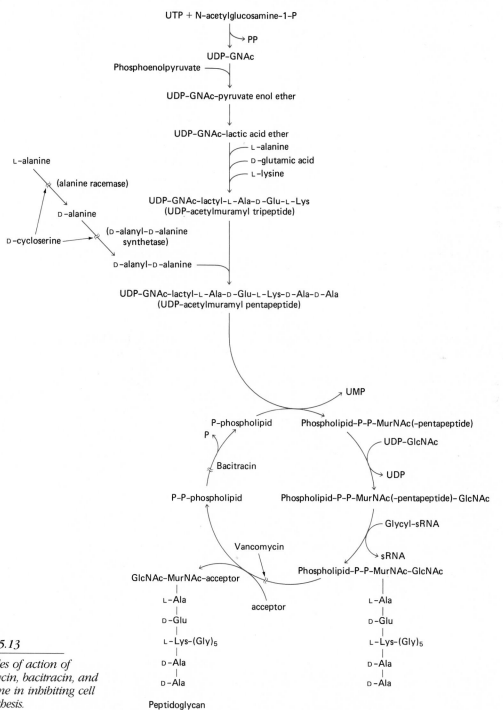

Figure 15.13

The modes of action of vancomycin, bacitracin, and cycloserine in inhibiting cell wall synthesis.

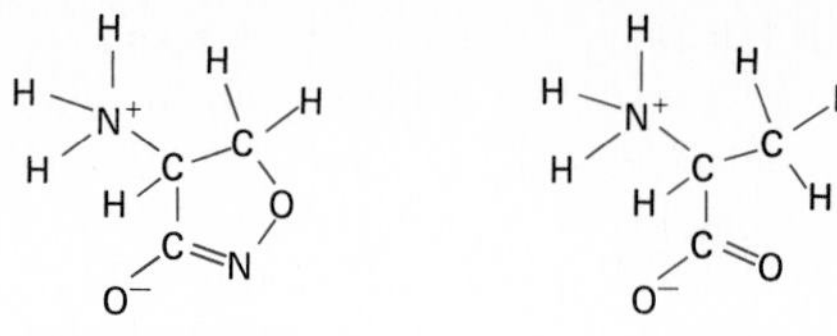

D-Cycloserine   D-Alanine

Figure 15.14

The structures of D-cycloserine and D-alanine. One can easily see how cycloserine can act as an analog of the normal amino acid needed for proper cell wall formation.

are necessary for cell wall synthesis cannot be adequately synthesized. Cycloserine is a broad-spectrum antibiotic produced by *Streptomyces orchidaceus*. Its therapeutic use is limited by its toxic reactions involving the central nervous system. Cycloserine is inhibitory for *Mycobacterium tuberculosis* and has been used in conjunction with other antibiotics in the treatment of tuberculosis.

Vancomycin and bacitracin prevent the linkage of the *N*-acetylglucosamine and *N*-acetylmuramic acid moities that compose the peptidolglycan molecule. Bacitracin is produced by strains of *Bacillus subtilis*. The use of bacitracin is restricted to topical application because this antibiotic causes severe toxicity reactions. Vancomycin is produced by *Streptomyces orientalis* and is especially effective against strains of *Staphyloccocus aureus*. Vancomycin is used to treat only serious infections caused by penicillin-resistant strains of *Staphyloccocus* and/or when the patient exhibits allergic reactions to penicillins and cephalosporins.

Inhibitors of protein synthesis The antibiotics streptomycin, gentamycin, neomycin, kanamycin, tobramycin, and amikacin are **inhibitors of bacterial protein synthesis** (Figure 15.15). These **aminoglycoside** antibiotics contain amino sugars linked by glycosidic bonds. The aminoglycosides are used almost exclusively in the treatment of infections caused by Gram negative bacteria. These antibiotics are relatively ineffective against anaerobic bacteria and facultative anaerobes growing under anaerobic conditions, and their action against Gram positive bacteria is also limited. The aminoglycoside antibiotics are produced by actinomycetes. For example, streptomycin is produced by *Streptomyces griseus*, neomycin by *Streptomyces fradiae*, kanamycin by *Streptomyces kanamyceticus*, and gentamicin by *Micromonospora purpurea*; amikacin is a semisynthetic derivative of kanamycin.

The aminoglycoside antibiotics bind to the 30S ribosomal subunit of the 70S prokaryotic ribosome, blocking protein synthesis and decreasing the fidelity of translation of the genetic code. Aminoglycosides disrupt the normal functioning of the ribosomes by interfering with the formation of initiation complexes, the initial step of protein synthesis that occurs during translation. Additionally, aminoglycosides induce misreading of the messenger RNA molecules, leading to the formation of nonfunctional enzymes. The interference of protein synthesis results in the death of the bacterium. Various mutations, though, can occur that reduce the effect of misreading some mRNA molecules, in some cases even leading to a dependence on streptomycin-induced misreading of the genetic information.

Figure 15.15

Structures of representative aminoglycoside antibiotics, streptomycin and neomycin.

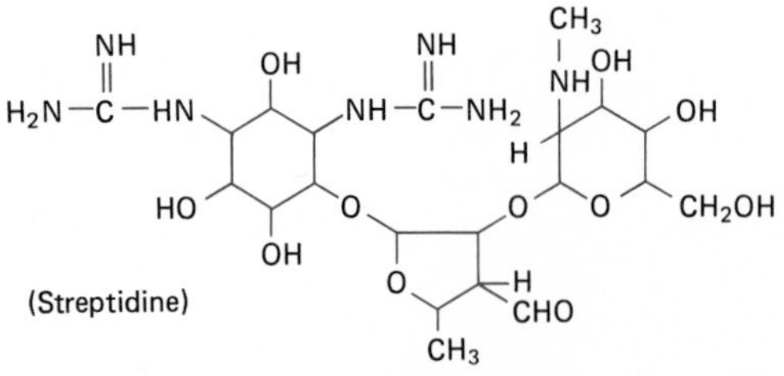

Streptomycin

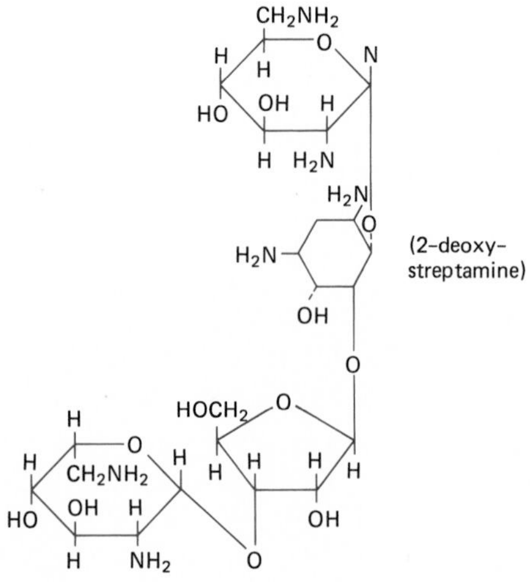

Neomycin

To be effective, the aminoglycoside antibiotics must be transported across the cytoplasmic membrane. Although sensitive bacteria transport the aminoglycosides across the membrane, accumulating these antibiotics intracellularly, resistant strains may lack a mechanism for aminoglycoside transport across the membrane. Resistant strains may also produce enzymes that degrade or transform the aminoglycoside molecules. For example, various enzymes associated with the cytoplasmic membranes of some bacterial strains are capable of adenylating, acetylating, or phosphorylating aminoglycoside antibiotics. Additionally, mutations can occur that alter the site at which the aminoglycosides normally bind to the bacterial ribosomes. Some *Pseudomonas aeruginosa* strains, for example, possess ribosomes to which streptomycin is unable to bind.

The aminoglycoside antibiotics are useful in treating a variety of diseases (Table 15.7). Streptomycin is used in the treatment of a limited number of bacterial infections. It is, for example, sometimes used in the treatment of brucellosis, tularemia, endocarditis, plague, and tuberculosis. Gentamicin is effective in treating urinary tract infections, pneumonia, and meningitis. Gentamicin is, however, extremely toxic and thus is used only in cases of severe infections that may prove lethal if unchecked, particularly when the infecting bacteria are not sufficiently sensitive to other, less toxic antibiotics. Tobramycin has properties similar to gentamicin, but *Pseudomonas aeruginosa* is particularly sensitive to tobramycin and thus this antibiotic is sometimes used for the treatment of pneumonia and other infections when caused by *Pseudomonas* species. Neomycin, which

is active against a broad spectrum of Gram negative bacteria, is primarily used in topical application for various infections of the skin and mucous membranes. Kanamycin, a narrow-spectrum antibiotic, is frequently used by pediatricians for infections due to *Klebsiella, Enterobacter, Proteus,* and *E. coli.* Amikacin, which has the broadest spectrum of activity of the aminoglycosides, is the antibiotic of choice for treating serious infections by Gram negative rods acquired in hospitals because such nosocomial infections are often caused by bacterial strains that are resistant to multiple antibiotics, including other aminoglycosides such as gentamicin.

In addition to the aminoglycoside antibiotics, a number of other antibiotics inhibit bacterial protein synthesis. These antibiotics include the tetracyclines, chloramphenicol, erythromycin, lincomycin, clindamycin, and spectinomycin. Some recommended therapeutic uses of these antibiotics are shown in Table 15.8. Unlike the aminoglycoside antibiotics that are bacteriocidal, these inhibitors of bacterial protein synthesis are generally bacteriostatic.

The **tetracyclines**, like the aminoglycosides, bind specifically to the 30S ribosomal subunit, apparently blocking the receptor site for the attachment of aminoacyl transfer RNA to the messenger RNA ribosome complex and thus preventing the addition of amino acids to a growing peptide chain. The sensitivity to such tetracyclines depends on the transport of the tetracycline molecules across the cytoplasmic membrane. Some tetracyclines, such as doxycycline, appear to pass directly across the membrane, whereas other tetracyclines can enter the cell only by active transport. Resistance

table 15.7

Some diseases and their causative organisms for which aminoglycoside antibiotics are recommended

Causative organism	Disease	Drug of choice
Gram negative rods		
Enterobacter aerogenes	Urinary tract; other infections	Gentamicin; tobramycin
Proteus sp.	Urinary tract; other infections	Gentamicin; tobramycin
Pseudomonas aeruginosa	Bacteremia	Gentamicin; tobramycin
Acinetobacter	Various nosocomial infections; bacteremia	Gentamicin
Yersinia pestis	Plague	Streptomycin (±) tetracycline
Serratia	Variety of nosocomial and opportunistic infections	Gentamicin
Mycobacterium tuberculosis	Tuberculosis	Streptomycin + other antibiotics

table 15.8

Some therapeutic uses of tetracyclines, chloramphenicol, erythromycin, and clindamycin

Causative organism	Disease	Drug of choice
Gram negative rods		
Salmonella	Typhoid fever Paratyphoid fever Bacteremia	Chloramphenicol
Haemophilus influenzae	Pneumonia Meningitis	Chloramphenicol
Haemophilus ducreyi	Chacroid	A tetracycline
Brucella	Brucellosis	A tetracycline ($\pm$) streptomycin
Vibrio cholerae	Cholera	A tetracycline
Flavobacterium meningosepticium	Meningitis	Erythromycin
Pseudomonas mallei	Glanders	Streptomycin + a tetracycline
Pseudomonas pseudomallei	Melioidosis	A tetracycline ($\pm$) chloramphenicol
Campylobacter fetus	Enteritis Bacteremia	No treatment or erythromycin Chloramphenicol
Bacteroides fragilis	Brain abscess Lung abscess Intra-abdominal abscess Empyema Bacteremia Endocarditis	Chloramphenicol Clindamycin
Legionella pneumophila	Legionnaire's disease	Erythromycin
Spirochetes		
Borrelia recurrentis	Relapsing fever	A tetracycline
Miscellaneous agents		
Mycoplasma pneumoniae	"Atypical pneumonia"	Erythromycin A tetracycline
Rickettsia	Typhus fever Murine typhus Brill's disease Rocky Mountain spotted fever	Chloramphenicol A tetracycline
Chlamydia trachomatis	Trachoma Inclusion conjunctivitis Nonspecific urethritis	A sulfonamide + a tetracycline A tetracycline A tetracycline

Prevention and treatment of human diseases caused by microorganisms

to the tetracyclines is often plasmid-mediated and involves an alteration of the mechanisms of membrane transport of the tetracycline molecules.

There are a variety of tetracycline antibiotics that have the same basic four-ring structure with some variations in the substituents (Figure 15.16). The tetracyclines are produced by various *Streptomyces* species. For example, chlortetracycline (aureomycin) is produced by *S. aureofaciens*, oxytetracycline by *S. rimosus*, and demeclocycline by *S. aereofaciens*; methacycline, doxyclycline, minocycline, and tetracycline are all semisynthetic derivatives. The tetracyclines are effective against a variety of pathogenic bacteria, including rickettsia and chlamydia species. Tetracylines, for example, are used therapeutically in treating the rickettsial infections of Rocky Mountain spotted fever, typhus fever, and Q fever and the chlamydial diseases of lymphogranuloma venereum, psittacosis, inclusion conjunctivitis, and trachoma.

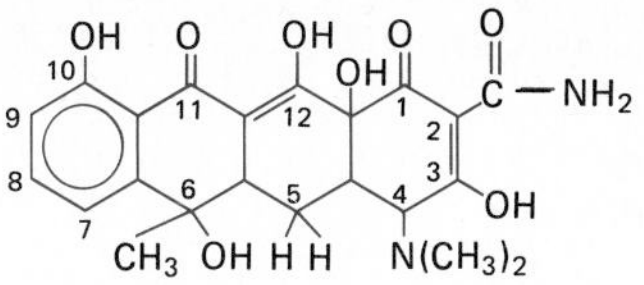

Tetracylines are useful in treating a variety of other bacterial infections, including pneumonia caused by *Mycoplasma pneumoniae*, brucellosis, tularemia, and cholera.

Unlike the antibiotics discussed so far that inhibit bacterial protein synthesis, **chloramphenicol** acts primarily by binding to the 50S ribosomal subunit, preventing the binding of tRNA molecules to both the aminoacyl and peptidyl binding sites of the ribosome. Consequently, peptide bonds are not formed when chloramphenicol is present in association with the bacterial ribosome. Chloramphenicol, which is produced by *Streptomyces venezuelae*, is a fairly broad-spectrum antibiotic, active against many species of Gram negative bacteria. Resistance to chloramphenicol is generally associated with the presence of an R plasmid that codes for enzymes able to transform the chloramphenicol molecule. The production of an acetyl transferase enzyme can inactivate the chloramphenicol molecule because acetylated derivatives of chloramphenicol do not bind to bacterial ribosomes. This appears to be the main mechanism by which resistance to chloramphenicol occurs. Chloramphenicol has a number of toxic effects, limiting its therapeutic uses to those where the benefits of chloramphenicol use outweigh the dangers associated with toxic reactions. Chloramphenicol is the antibiotic of choice for treating typhoid fever as well as various other infections caused by *Salmonella* species; it is also effective against anaerobic pathogens and can be used effectively in treating diseases, such as brain abscesses, normally caused by anaerobic bacteria.

Like chloramphenicol, **erythromycin** acts by binding to 50S ribosomal subunits, blocking protein synthesis. Erythromycin, produced by *Streptomyces erythreus*, is a macrolide antibiotic, so named because it contains a multimembered lactone ring attached to deoxy sugar moieties (Figure 15.17). This antibiotic is most effective against Gram positive cocci, such as *Streptococcus pyogenes*. Erythromycin is not active against most aerobic Gram negative rods but does exhibit antibacterial activity against some Gram negative organisms, such as *Pasteurella multocida, Bordetella pertussis,* and *Legionella pneumophila.* Therapeutically, erythromycin is recommended for the treatment of Legionnaire's disease and is also effective in treating pneumonia caused by *Mycoplasma pneumoniae,* diphtheria, and whooping cough. Erythromycin may also be used as an alternative to penicillin in treating staphylococcal infections, streptococcal infections, tetanus, syphilis, and gonorrhea.

Like erythromycin, the antibiotics lincomycin

Figure 15.16

The structures of tetracycline antibiotics.

Compound	Substituent(s)	Position(s)
Chlortetracycline	—Cl	(7)
Oxytetracycline	—OH, —H	(5)
Demeclocycline	—OH, —H; —Cl	(6; 7)
Methacycline	—OH, —H; =CH$_2$	(5; 6)
Doxycycline	—OH, —H; —CH$_3$, —H	(5; 6)
Minocycline	—H, —H; —N(CH$_3$)$_2$	(6; 7)

and clindamycin bind to the 50S ribosomal subunit, blocking protein synthesis. Lincomycin is produced by *Streptomyces lincolnensis*, and clindamycin is a semisynthetic derivative of lincomycin. The use of these antibiotics is restricted by their side effects, such as severe diarrhea. Clindamycin is particularly effective against Gram positive bacteria, including anaerobes, and infections due to *Bacteroides* and *Fusobacterium* species can be treated effectively with clindamycin.

Several other antibiotics that inhibit protein synthesis are not useful in treating bacterial infections because they inhibit protein synthesis in mammalian cells to the same extent that they do in bacterial cells. If the mode of action of these antibiotics is not specific toward bacteria, they are not therapeutic antibacterial agents. For example, puromycin is an analog of the transfer RNA mol-

Figure 15.17

The biochemical structure of erythromycin.

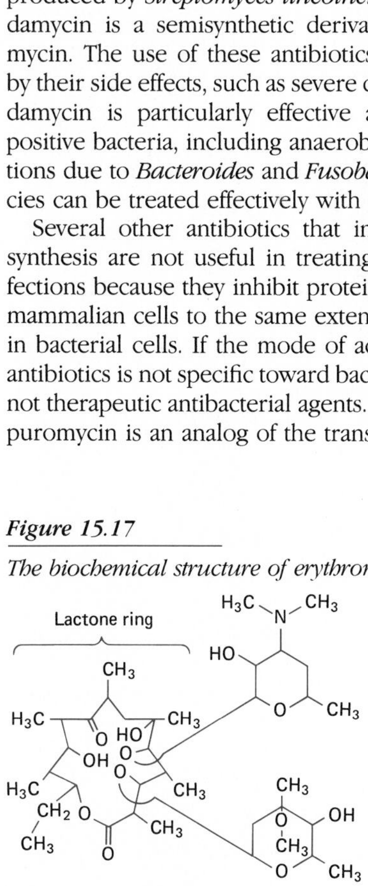

Erythromycin

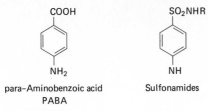

para-Aminobenzoic acid
PABA

Sulfonamides

Figure 15.18

Sulfonamide, showing its similarity to para-aminobenzoic acid.

ecule and can compete with tRNA molecules for binding to ribosomes. The mode of action of this antibiotic does not distinguish between inhibiting eukaryotic and prokaryotic protein synthesis. Similarly, dactinomycin (actinomycin D) blocks protein synthesis in both bacterial and eukaryotic cells; this antibiotic binds to double-stranded DNA, blocking transcription of the genetic information to form a messenger RNA molecule. Although not useful in treating bacterial infections, dactinomycin does have a therapeutic role in treating some malignancies where it is desirable to block the rapid division of cancer cells. Rifampin, a semisynthetic derivative of rifamycin B, also blocks protein synthesis at the level of transcription. Rifampin inhibits DNA-dependent RNA polymerase enzymes and thus can block transcription; this antibiotic is more effective against bacterial RNA polymerase enzymes than mammalian RNA polymerases and therefore can be used therapeutically in treating some bacterial diseases. Rifampin is used in combination with other antibiotics in

the treatment of mycobacterial diseases, such as tuberculosis.

Inhibitors of membrane transport

The cytoplasmic membrane is the site of action of some antimicrobial agents. The polymyxins, such as polymyxin B, interact with the cytoplasmic membrane, causing changes in the structure of the bacterial cell membrane and leakage from the cell. Polymyxin B is bactericidal, and its effectiveness is restricted to Gram negative bacteria. The action of polymyxin B is related to the phospholipid content of the cell wall and membrane complex. Sensitive bacteria take up more polymyxin B than resistant strains. The principal use of polymyxin B and colistin (polymyxin E) is in the treatment of infections caused by *Pseudomonas* species and other Gram negative bacteria that are resistant to penicillins and the aminoglycoside antibiotics. Both polymyxin B and colistin are useful in treating severe urinary tract infections caused by *Pseudomonas aeruginosa* and other Gram negative bacteria, particularly when the infecting bacteria are resistant to other antibiotics.

Inhibitors with other modes of action

There are several other antibiotics that act in other ways, some acting as antimetabolites, and others whose modes of action are unknown. Some of these antibiotics are useful in treating specific infections, such as tuberculosis, and others are particularly useful in treating infections of particular tissues, such as urinary tract infections. For example, the sulfonamides, sulfones, and para-aminosalicylic acid are **structural analogs** of para-

Figure 15.19

Folic acid metabolism, showing inhibition by analogs of para-aminobenzoic acid.

p-Aminobenzoic acid (PABA)   (Pteridine)   (PABA)   (Glutamic acid)

Sulfanilamide

Folic acid

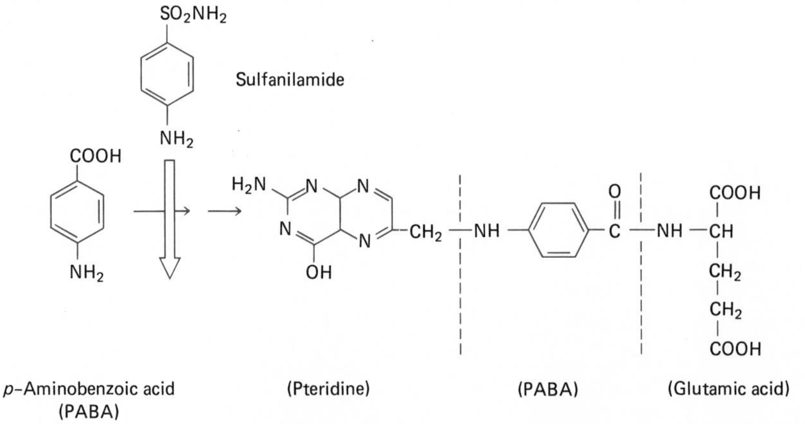

aminobenzoic acid, which makes them useful antibacterial agents (Figure 15.18). Mistakenly adding an analog, such as sulfonamide, in place of the normal substance, para-aminobenzoic acid in this case, results in the formation of molecules that are unable to perform their essential metabolic functions; in this case there is a failure of critical coenzyme functions. Folic acid is an essential coenzyme composed in part of para-aminobenzoic acid. Mammalian cells are unable to synthesize folic acid and require an intake of folic acid as part of their diet and cellular uptake via an active transport system. Bacterial cells, in contrast, normally synthesize their required folic acid and are unable to transport folic acid across their cytoplasmic membranes (Figure 15.19). The analogues of para-aminobenzoic acid are effective competitors with the natural substrate for the enzymes involved in the synthesis of folic acid and as such are able to inhibit the formation of this required coenzyme, causing a bacteriostatic effect. The sulfones are useful in treating leprosy. The use of sulfonamides and para-aminosalicylic acid has declined as the result of the occurrence of resistant strains and the development of more effective antibiotics with fewer toxic side effects.

Trimethoprim is an inhibitor of dihydrofolate reductase, especially in prokaryotes. Dihydrofolic acid is a coenzyme required for 1-carbon transfers, such as occur in the synthesis of thymidine and purines (Figure 15.20). Trimethoprim is effective in blocking bacterial growth by preventing the formation of the active form of the required coenzyme. The effectiveness of trimethoprim is enhanced when antibiotic therapy is coupled with sulfamethoxazole. The combined formulation of trimethoprim and sulfamethoxazole greatly enhances the antibacterial activities of these antimetabolites. Trimethoprim is a broad-spectrum antibacterial agent and is effective in the treatment of many urinary and intestinal tract infections. It is used primarily for the treatment of urinary infections due to *E. coli*, *Proteus*, *Klebsiella*, and *Enterobacter*. The trimethoprim-sulfamethoxazole mixture is also effective in treating typhoid fever.

In addition to the use of trimethoprim-sulfamethoxazole, several other compounds are used as antiseptics in treating urinary tract infections. These compounds, which inhibit the growth of many bacterial species, include methenamine, nalidixic acid, oxolinic acid, and nitrofurantoin. The usefulness of these drugs, though, depends on the fact that they are concentrated in the urinary tract tissues and thus can act as antiseptics specifically within that tract. Nalidixic acid acts by

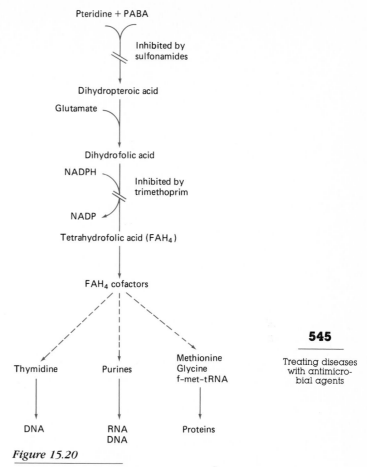

Figure 15.20

The effects of trimethoprim and sulfomethoxazole on thymidine and purine synthesis.

blocking DNA synthesis, but transcription and translation (protein synthesis) can still occur. This antimicrobial agent is effective against most Gram negative bacteria that cause urinary tract infections. Nitrofurantoin inhibits several bacterial enzymes, but the specific mode of action of this antimicrobial agent is unknown. Nitrofurantoin is used only in a limited number of cases because it is generally not as effective as other antibiotics, including sulfanomides.

Isoniazid is sometimes used in the treatment of tuberculosis, but it is particularly effective against *Mycobacterium tuberculosis* when used alone and is generally used in association with other antibiotics in treating tuberculosis. The specific mode of action of isoniazid is not known, but its primary action appears to involve the inhibition of mycolic acid biosynthesis. Mycolic acids are unique components of the cell walls of mycobacteria, and blockage of the biosynthesis of these compounds could specifically inhibit mycobacteria. Thus, a large number of antibacterial agents with varying modes of action and differing activity spectrums are available for treating infectious diseases caused by bacterial pathogens.

Antimicrobial agents effective against infections caused by eukaryotic microorganisms

In contrast to the relative ease with which bacterial infections can be treated, the fact that fungi, protozoa, and humans all have eukaryotic cells limits the sites against which antimicrobial agents can be selectively directed to control eukaryotic pathogens of human beings. Consequently, there are relatively few antimicrobial agents of therapeutic value effective against infections caused by eukaryotic microorganisms, with particularly few effective for treating systemic infections.

Antifungal agents

There are sufficient differences between a fungal cell and human cell so that some therapeutically useful compounds with antifungal activity have been discovered. Some of the therapeutic uses of **antifungal agents** are listed in Table 15.9. The **polyene antibiotics**, amphotericin B, and nystatin are used in treating a variety of fungal diseases. The polyene antibiotics act by altering the permeability properties of the cytoplasmic membrane, leading to the death of the affected cells. Interactions of polyenes with the sterols in the cytoplasmic membranes of eukaryotic cells appear to form channels or pores in the membrane, allowing leakage of small molecules through the membrane (Figure 15.21). Differences in the sensitivity of various organisms is determined by the concentrations of sterols in the membrane. Because mammalian cells, like fungi, contain sterols in their cytoplasmic membranes, it is not surprising that polyene antibiotics also cause alterations in the membrane permeability of mammalian cells and toxicity to mammalian tissue as well as the death of fungal pathogens.

Nystatin, which is produced by *Streptomyces noursei*, is primarily used in the treatment of topical infections by members of the fungal genus *Candida*. Vaginitis, thrush, and *Candida* infections of the gastrointestinal tract are effectively treated by using nystatin. Amphotericin B, which is produced by *Streptomyces nodosus*, has a relatively broad spectrum of activity and is used in the treatment of systemic fungal infections. Amphotericin B is the most effective therapeutic agent for treating systemic infections due to yeast and fungi. The potential toxic side effects of amphotericin B usage, however, such as kidney damage, require careful supervision of its administration. Patients requiring administration of amphotericin B must be hospitalized so that the initial reaction to the therapy can be carefully supervised. Patients who have received amphotericin B almost invariably exhibit some toxic side effects, but without the administration of this drug, systemic fungal infections are almost invariably fatal. Amphotericin B is used in the treatment of cryptococcosis, histoplasmosis, coccidioidomycosis, blastomycosis, sporotrichosis, and candidiasis.

In addition to the polyene antibiotics, such imidazole derivatives as miconazole and clotrimazole have a broad spectrum of antifungal activities and are used in the topical treatment of superficial mycotic infections. These two antimicrobial

table 15.9

Some therapeutic uses of antifungal agents

Causative organism	Disease	Drug of choice
Candida albicans	Skin and superficial mucus membrane lesions	Amphotericin B Nystatin
Cryptococcus neoformans	Meningitis	Amphotericin B + flucytosine
Candida albicans *Aspergillus* *Mucor*	Pneumonia Meningitis Skin lesions	Amphotericin B
Histoplasma capsulatum	Lung lesions Histoplasmosis	Amphotericin B
Coccidioides immitis	Coccidiomycosis (desert fever)	Amphotericin B
Blastomyces dermatitidis	Blastomycosis	Amphotericin B

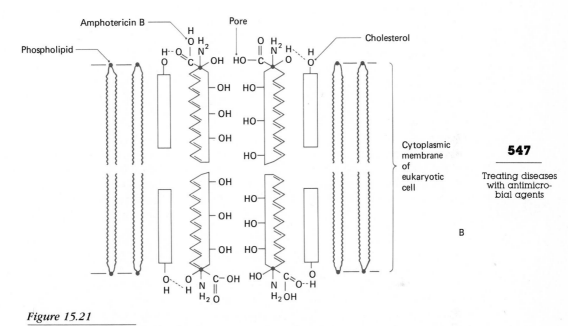

Figure 15.21

(A) The structure of the polyene antibiotic, amphotericin B, and (B) its effect on the cytoplasmic membrane of a eukaryotic cell.

agents appear to alter membrane permeability, leading to the inhibition and/or death of selected fungal species. Flucytosine, a fluorinated pyrimidine (Figure 15.22), is also effective in treating systemic fungal infections. Flucytosine is less effective, but also less toxic, than amphotericin B and is primarily used in combination with amphotericin B treatment. Within fungal cells flucytosine is converted to fluorouracil and further metabolized to form an inhibitor of thymidylate synthetase, causing an inhibition of normal nucleic acid synthesis. Mammalian cells do not convert as much flucytosine to fluorouracil as do fungal cells, accounting for the selective toxicity of this antifungal agent.

Griseofulvum is another antibiotic that is effective against some fungal infections. This antibiotic is produced by *Pencillium griseofulvum* and causes a disruption of mitotic spindles, inhibiting fungal mitosis. Griseofulvum is used in the treatment of fungal diseases of the skin, hair, and nails caused by various species of dermatophytic fungi, such as *Microsporum, Epidermophyton*, and *Trichophyton*. These dermatophytes concentrate griseofulvum by an active uptake process, and their sensitivity is correlated directly with their ability to concentrate the antibiotic.

Antiprotozoan agents

Treatment of human protozoan diseases with antimicrobial agents presents a special problem because many of the pathogenic protozoa exhibit a complex life cycle, often including stages that develop intracellularly within mammalian cells. Different antimicrobial agents are generally needed for use against different forms of the same path-

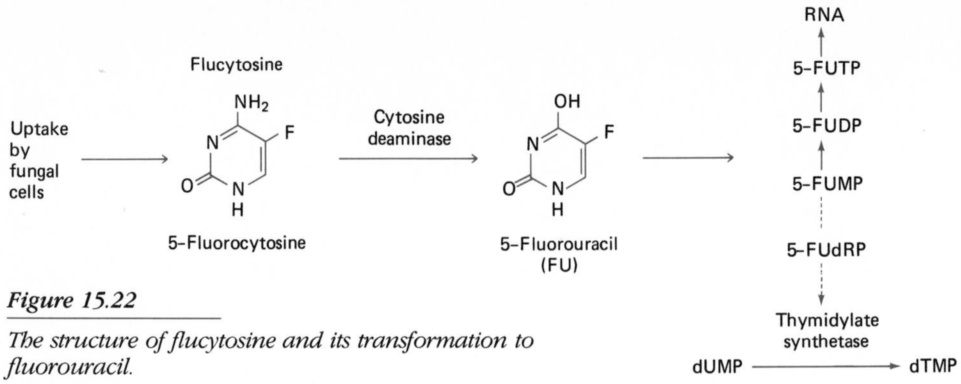

Figure 15.22

The structure of flucytosine and its transformation to fluorouracil.

ogenic protozoan, depending on the stage of the life cycle and the involved tissues (Table 15.10). For example, the protozoan species of the genus *Plasmodium* that cause malaria exhibit complex life cycles, parts of which are carried out in the liver and blood of human beings (Figure 15.23). The erythrocytic stage of the *Plasmodium* life cycle that occurs within human blood cells is the most sensitive to **antimalarial drugs**. The life stages that occur within the liver are difficult to treat and the sporozoites injected into the bloodstream by mosquitos are not affected by antimalarial drugs. The antimalarials effective against the erythrocytic forms of the protozoan include chloroquine and amodiaquine. These drugs are most widely used for suppressing the malarial infection, but neither is effective against the hepatic stages of the *Plasmodium* that occur in the liver (Figure 15.24). These antimalarial agents appear to interfere with DNA replication. The effect of these drugs is a rapid schizontocidal action, that is, the rapid interruption of schizogony or multiple division that occurs within red blood cells. The sensitivity of malarial protozoa to these drugs depends on the active transport of these compounds into the protozoa and the selective accumulation of the drugs intracellularly.

Chloroguanide is also used in the suppression of malaria. This drug is transformed within the body to a triazine derivative that inhibits the enzyme dihydrofolate reductase and thus interferes with the essential metabolic reactions involving

this coenzyme, which are required for the proliferation of the malaria protozoa. Chloroguanide is sometimes used concurrently with sulfonamide compounds that also interfere with folate metabolism. Chloroguanide binds more strongly to the plasmodial enzyme than to the comparable mammalian dihydrofolate reductase, accounting for its selective inhibition. In addition to affecting the schizont stage, chloroguanide effects the sterilizing action of gametocytes. Because resistance to the synthetic antimalarial drugs is increasing, quinine, one of the early drugs used for the effective treatment of malaria, is once again being used to treat this disease.

For the radical cure of malaria, that is, the eradication of both the erythrocytic and liver stages of the protozoan, primaquine is normally used. This drug is used in conjunction with chloroquine and chloroguanide. The precise mode of action of primaquine has not been elucidated. Because of the toxic side effects of primaquine, it is primarily used in the treatment of relapsing malarial infections. Pyromethamine, which also inhibits folic acid metabolism, has also been used in the treatment of malaria. Many *Plasmodium* strains, however, have developed resistance to pyromethamine, limiting its usefulness in treating malaria.

Several other drugs are used in the treatment of various other protozoan infections (Table 15.11). As in the case of malaria, the life cycle of the particular protozoan determines which agents will

table 15.10

Antimalarial agents used for treating the different stages of malaria

Drug	Stage inhibited	Use
Chloroquine; amodiaquine	Erythrocytic	Prophylaxis and treatment of acute attack
Primaquine	Hepatic	Radical cure

Figure 15.23

The complex life cycle of the malarial protozoan makes treatment of the disease with antibiotics extremely difficult.

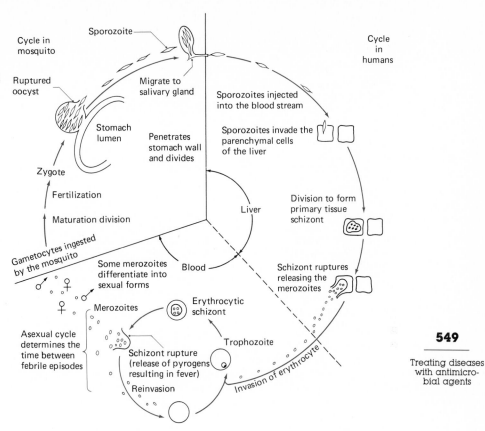

Cycle in mosquito

Sporozoite

Cycle in humans

Ruptured oocyst

Migrate to salivary gland

Sporozoites injected into the blood stream

Stomach lumen

Penetrates stomach wall and divides

Sporozoites invade the parenchymal cells of the liver

Zygote

Fertilization

Liver

Division to form primary tissue schizont

Maturation division

Gametocytes ingested by the mosquito

Some merozoites differentiate into sexual forms

Blood

Schizont ruptures releasing the merozoites

Merozoites

Erythrocytic schizont

Asexual cycle determines the time between febrile episodes

Schizont rupture (release of pyrogens resulting in fever)

Trophozoite

Reinvasion

Invasion of erythrocyte

be effective in controlling the infection. Only a few of these antiprotozoan agents will be discussed further here. Metronidazole is used in the treatment of dysentery caused by the protozoan *Entamoeba histolytica*. Metronidazole interferes with hydrogen transfer reactions, specifically inhibiting the growth of anaerobic microorganisms, including anaerobic protozoa. Pentamidine and related diamidine compounds are useful in treating infections by members of the protozoan genus *Trypanosoma*. Compounds of this type interfere with DNA metabolism. Melarsoprol is an arsenical, that is, an arsenic-containing compound useful in treating some stages of human trypanosomiasis, particularly because of its ability to penetrate into cerebrospinal fluid (Figure 15.25). Arsenicals react with the sulfhydryl groups of proteins, inactivating a large number of enzymes. It appears that mammalian cells can metabolize these compounds to nontoxic forms more rapidly than protozoan cells, accounting for the selective toxicity of melarsoprol to trypanosome protozoans. The drug sodium stibogluconate, an antimony-containing compound, is useful in treating diseases caused by members of the protozoan genus *Leishmania*. Antimony compounds of this type in-

hibit the enzyme phosphofructokinase in some life history stages of the leishmanias, accounting for its inhibitory effects. Other antiprotozoan agents useful in the chemotherapy of protozoan diseases include suramin, a nonmetallic compound that inhibits a wide variety of enzymes, and nifurtimox, which is effective against *Trypanosoma cruzi*, the causative organism of Chagas' disease.

Figure 15.24

The biochemical structures of chloroquine and amodiaquine, which are used to treat malaria.

Chloroquine

Amodiaquine

table 15.11

Some drugs used in the treatment of diseases caused by protozoan pathogens

Infecting organism-disease	Drug of choice
Entamoeba histolytica	Diiodohydroxyquin
Asymptomatic cyst passer	
Mild intestinal disease	Metronidazole
Severe intestinal disease	Metronidazole
Hepatic abscess	Metronidazole
Giardia lamblia	Quinacrine hydrochloride
Balantidium coli	Oxytetracycline
Dientamoeba fragilis	Diiodohydroxyquin
Trichomonas vaginalis	Metronidazole
Pneumocystis carinii	Trimethoprim-sulfamethoxazole
Toxoplasma gondii	Pyrimethamine plus trisulfapyrimidines
Leishmaniasis	
Leishmania donovani (kala-azar, visceral leishmaniasis)	Sodium stibogluconate
Leishmania tropica (oriental sore, cutaneous leishmaniasis)	Sodium stibogluconate
Leishmania braziliensis (American mucocutaneous leishmaniasis)	Sodium stibogluconate
African trypanosomiasis	
Trypanosoma gambiense	Pentamidine
Trypanosoma rhodesiense	Suramin
Either *T. gambiense* or *T. rhodesiense* in late disease with central nervous system involvement	Melarsoprol
South American trypanosomiasis (Chagas' disease)	
Trypanosoma cruzi	Nifurtimox

550

Prevention and
treatment of hu-
man diseases
caused by micro-
organisms

Antiviral agents

The search for **antiviral drugs** comparable to the antibiotics used to control bacterial infections has been fruitless for the most part. There are no broad-spectrum antiviral agents currently in clinical use, and most viral infections cannot be treated effectively by using antiviral chemicals. The integral role of the mammalian host cell in the process of viral replication complicates the difficulty of finding compounds that specifically inhibit viral replication. Most compounds that prevent the reproduction of viruses also interfere with mammalian cell metabolism, resulting in adverse effects on human cells so as to preclude the therapeutic use of such agents. Very few antiviral

agents have been found with clinical applicability, and these generally have a narrow spectrum of antiviral activity.

The best "treatment" for many viral diseases is prevention through the appropriate use of vaccines and by the controlling the vectors that act as carriers for viruses. Generally, recovery from a viral infection is dependent on the natural immune defense response of the body. For most viral infections treatment is aimed at maintaining a physiological state in which an effective immune response can be ensured, by following the sage advice, "rest and drink plenty of fluids." The discovery of the role of interferons in the natural immune response to viruses holds promise for the future if interferons can be produced commercially and if they can be administered with therapeutic value. Genetic engineering would seem to provide the greatest hope for the commercial production of interferons, making these antiviral substances available for medical treatment of viral infections. Because, as we will see in Chapters 17 and 18, viruses cause many human diseases, the development of antiviral agents of therapeutic value is very important for improving our ability to treat numerous diseases.

Only a few chemical agents have been developed so far for therapeutic use in treating specific

Figure 15.25

The structure of melarsoprol, an effective antiprotozoan agent.

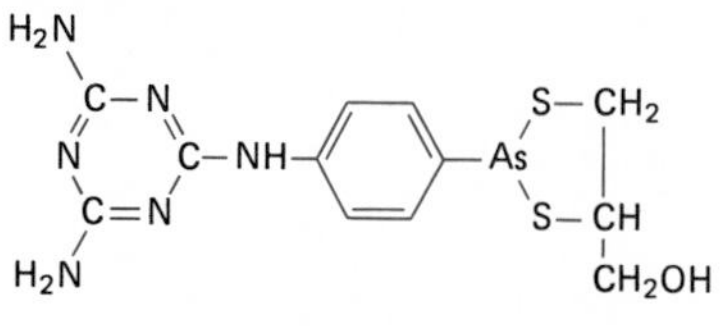

Melarsoprol

table 15.12

Some antiviral agents and their therapeutic uses

Causative organism	Disease	Drug of choice
Herpes simplex virus	Keratoconjunctivitis	Vidarabine
		Trifluridine
		Acyclovir
	Encephalitis	Vidarabine
	Cold sores	Acyclovir
	Genital herpes	Acyclovir
Influenza virus A	Influenza	Amantadine

viral infections (Table 15.12). Among these anti-viral compounds, idoxuridine, an analog of thymidine that interferes with DNA metabolism in both viral and mammalian cells, has clinical uses in the treatment of herpes simplex keratitis. Amantadine is recommended for the prophylaxis of high-risk patients in cases of documented influenza A virus epidemics. Amantadine appears to interfere with the absorption and uptake of influenza A viruses. There are several compounds that have antiviral activity against herpes viruses. Trifluridine is a nucleic acid base analog that inhibits DNA synthesis. This fluorinated pyrimidine nucleoside is effective against epithelial keratitis of the eye caused by *Herpes simplex* virus. Vidarabine was originally developed for the treatment of leukemia but has proven to be more effective in treating herpes simplex encephalitis and keratoconjuntivitis. Vidarabine (vira-A) is an adenine arabinoside. Within mammalian cells vidarabine is phosphorylated to the corresponding nucleotide that acts as an inhibitor of viral DNA polymerase. The selectivity of vidarabine occurs because this drug inhibits mammalian DNA synthesis

to a lesser extent than viral DNA replication.

Acyclovir (9-[2-hydroxyethoxymethyl]guanine) has also been found to be an effective inhibitor of herpes viruses. Acyclovir has been reported to be more effective than idoxuridine and vidarabine for the treatment of herpetic ocular disease. This drug appears to be effective in treating cold sores and genital herpes caused by *Herpes simplex*. It also may be of value in treating chicken pox and shingles. Acyclovir, a nucleoside analog, is converted *in vivo* to an acylguanosine triphosphate that inhibits herpes simplex viral DNA polymerase, thus blocking viral DNA replication. The activation of acyclovir is initiated by a viral-directed thymidine kinase enzyme that converts this compound to an acycloguanosine monophosphate, which is subsequently converted to the acycloguanosine di- and triphosphates. In an uninfected cell there is only very limited conversion of acyclovir to the phosphorylated acylguanosines. Because an enzyme coded for by the herpes virus is required to activate acyclovir, this compound exhibits selective antiviral activity, making it therapeutically valuable.

Postlude

The prevention of infectious diseases is clearly desirable, and various measures can be employed for the prophylaxis of diseases caused by microorganisms. The probability of contracting an infectious disease can be greatly reduced by using measures that control the levels of the populations of pathogenic microorganisms and the probability of exposure to such organisms. These measures often rely on proper sanitation, food handling and preservation, prevention of wound contamination, and minimizing the spread of aerosols that contain microorganisms. Public health measures aimed at controlling infectious diseases are also frequently aimed at controlling vector populations, such as mosquitos, fleas, ticks, and lice, that can act as carriers of pathogenic microorganisms.

The use of vaccines to render the host unsusceptible to specific pathogenic microorganisms is of great use in controlling some microbial diseases. Smallpox, for example, has been eliminated as a human disease as a consequence of the extensive use of vaccination. Polio, measles, mumps, rubella, yellow fever, tetanus, and rabies are all diseases that can be prevented by the appropriate use of vaccines. The effectiveness of vaccines depends on the normal memory response of the human immune system and on establishing active immunity through the exposure to antigens that elicit the desired immune response without causing disease. Both inactivated and attenuated microorganisms are employed in different vaccines used for immunization purposes.

When host defense mechanisms fail, and exposure to a pathogenic microorganism results in the development of an infection, there are a variety of antimicrobial agents that can be employed for the treatment of disease. The bacterial diseases are perhaps the easiest of the infectious diseases to treat because the differences between prokaryotic and eukaryotic cells provide sites against which chemicals can exert selective inhibition against bacteria without killing or severely inhibiting host mammalian cells. The discovery and use of antibiotics, which are antimicrobial agents produced by microorganisms, have revolutionized medical practice. The choice of a particular antibiotic for treating a disease rests on the nature of the infecting microorganism and the tissues in which the infection occurs. In general, the murein component of the bacterial cell wall and the 70S ribosomes of bacteria provide targets against which many of the therapeutically useful antibiotics are aimed. Prudent medical practice requires that antibiotics be specifically aimed at the infecting microorganisms, that is, wherever possible specific or narrow-spectrum antibiotics be employed, so as to avoid the selective pressure for the development of antibiotic-resistant strains of pathogenic microorganisms. The occurrence of antibiotic-resistant strains is becoming a very serious problem in treating infectious diseases.

The treatment of infections by viruses and eukaryotic microorganisms presents a greater problem than bacterial infections for finding specific inhibitors that are not also directed at mammalian cells. Systemic mycotic and protozoan infections are particularly difficult to treat, often persisting for long periods of time. Many of the drugs used in the treatment of such infections also cause side effects toxic to human tissues.

Microbiologists associated with the pharmaceutical industry are continuously screening for new antibiotics or derivatives of known antimicrobial agents that might be of therapeutic value in treating infectious diseases. The activities of the pharmaceutical industry in discovering and producing antimicrobial agents will be considered in Chapter 20. Advances continue to be made in our ability to treat diseases caused by microorganisms as new antimicrobial agents are discovered and developed. We have made great advances in our ability to prevent and treat infectious diseases and look forward to achieving a higher standard of human health in the future.

1. How should you avoid exposure to airborne pathogens?

2. What is a vaccine? How is vaccination used to prevent disease?

3. How does a physician select an antibiotic for treating an infectious disease?

4. Is penicillin useful in treating the common cold? Explain.

5. Why is it easier to find antibacterial agents than it is to discover useful antifungal agents?

6. Why is it so difficult to find antimicrobial agents for treating viral diseases?

7. What antibiotics should you prescribe for each of the following conditions?
   a. urinary tract infection
   b. upper respiratory tract bacterial infection
   c. fungal infection of the vaginal tract
   d. herpes encephalitis
   e. malaria

8. Discuss the mode of action of penicillin.

9. Why is penicillin ineffective against bacteria that produce β-lactamases?

10. Why is an inhibitor of transcription not useful in treating bacterial infections of humans?

11. Discuss the mode of action of streptomycin.

12. Discuss the mode of action of vira-A.

Suggested Supplementary Readings

Benenson, A. S. (ed.). 1975. *Control of Communicable Diseases in Man.* American Public Health Association, Washington, D.C.

Berdy, J. (ed.). 1980–1982. *Handbook of Antibiotic Compounds.* CRC Press, Inc., Boca Raton, Florida.

Bryan, L. E. 1982. *Bacterial Resistance and Susceptibility to Chemotherapeutic Agents.* Cambridge University Press, New York.

Gilman, A. G., L. S. Goodman, and A. Gilman (eds.). 1980. *Goodman and Gilman's The Pharmacological Basis of Therapeutics.* Macmillan Publishing Co., New York.

Kagan, B. M. 1974. *Antimicrobial Therapy.* W. B. Saunders, Philadephia.

Lynn, M., and M. Solotorovsky (eds.). 1981. *Chemotherapeutic Agents for Bacterial Infections* (Benchmark Papers in Microbiology). Academic Press, New York.

Physician's Desk Reference. Published annually. Medical Economics Co., Oradell, New Jersey.

Pratt, W. B. 1977. *Chemotherapy of Infection.* Oxford University Press, New York.

Roitt, I. M. 1980. *Essential Immunology.* Blackwell Scientific Publications, Oxford, England.

Smith, H. (ed.). 1977. *Antibiotics in Clinical Practice.* Pitman Medical Publishing Co., Kent, England.

Stansfield, W. D. 1981. *Serology and Immunology.* Macmillan Publishing Co., New York.

553

Suggested supplementary readings

Clinical microbiology: identification of disease-producing microorganisms and determination of antibiotic susceptibility

16

554

Clinical microbiology: identification of disease-producing microorganisms and determination of antibiotic susceptibility

Clinical microbiology and immunology laboratories are involved in determining the causes of disease and assisting the physician in selecting the appropriate treatment methods. A major task of a clinical laboratory is to determine whether a given disease condition is caused by a microbial infection, with the objective of documenting the presence or absence of infectious agents in samples collected from an ailing individual. If a disease is of microbial origin, the clinical laboratory has the responsibility of **identifying the causative microorganism (etiologic agent)**. Further, the clinical laboratory is responsible for **assessing the effectiveness of antimicrobial agents** that may be selected for treatment of the infection. The clin-

ical microbiology laboratory applies the full scope of our basic understanding of microorganisms, the immune response, and the relationships between host and pathogen with regard to the disease process. Diagnostic methods include various types of microscopic observation, isolation, and culture methods for the identification of pathogenic microorganisms. Clinical identification schemes employ a variety of morphological, physiological, and biochemical features to provide accurate and timely identification of a variety of human pathogens. Immunological (*in vivo*) and serological (*in vitro*) antigen–antibody reactions are also employed for the detection and identification of various etiologic agents.

Indicators that diseases are of microbial etiology

The characteristics of the immune response system provide the basis for determining whether a disease is of probable microbial etiology. In most cases a microbial infection elicits an inflammatory response, characterized by fever, pain, swelling, and redness. Although an inflammatory response does not necessarily reflect an infectious disease, an elevated body temperature (fever) is often used as presumptive evidence of a microbial infection. The physician observing a patient with a red sore throat, and who is also running a fever, assumes that the symptoms are the result of a microbial infection. In many such cases, when the pre-

sumptive evidence strongly indicates pharyngitis (infection of the pharynx), treatment is administered without rigorous clinical diagnosis and confirmation of the cause.

Differential blood counts

In other cases, where the identification of a microbial infection is not clear-cut, additional presumptive evidence of an infectious process can be obtained by performing a differential blood count in which the relative concentrations of dif-

Plate 5

Various differential and selective media are used for the isolation of pathogenic microorganisms. The characteristic appearance of a pathogen on such media aids the clinical microbiologist in determining the causative organism. (A) Salmonella typhimurium *on xyline-lysine-desoxycholate agar (XLD); the black color of the colonies indicates hydrogen sulfide production. (B)* Vibrio cholerae *on TCBS agar; the yellow color indicates acid production. (C)* Salmonella enteritidis *on Hektoen agar; the black colonies indicate hydrogen sulfide production, the lack of yellow color around the colonies indicates that acid is not produced. (D)* Klebsiella pneumoniae *on MacConkey agar showing characteristic pink colonies.*

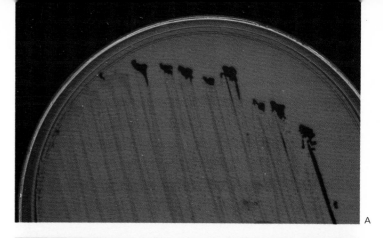

A

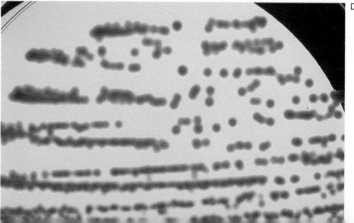

B

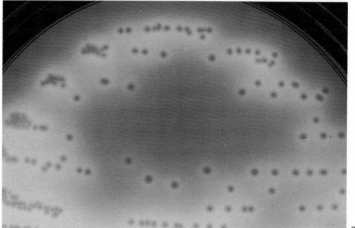

C

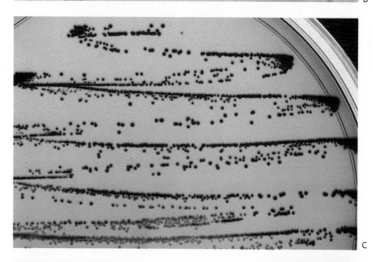

D

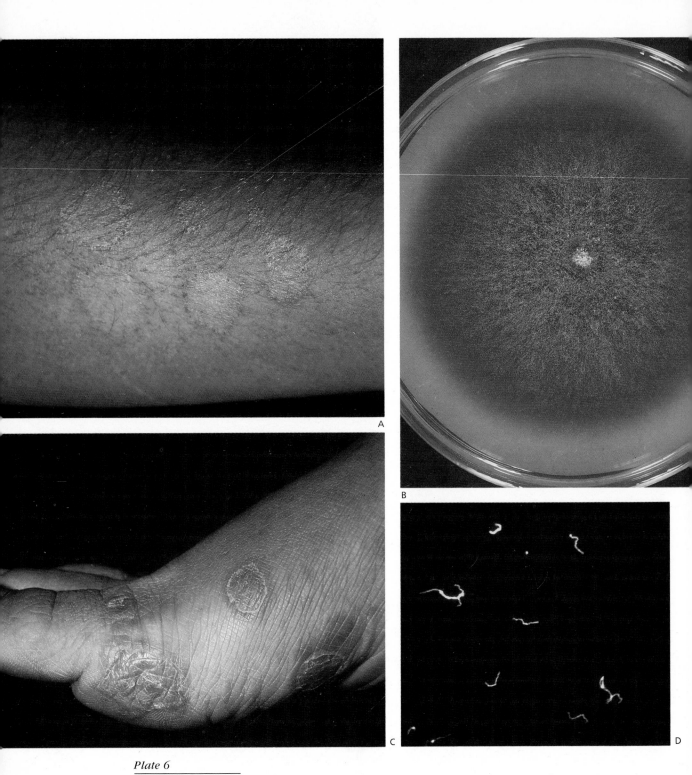

Plate 6

In the diagnosis of a disease, the physician sees the manifestation of the disease, and the clinical microbiologist sees the causative organism. Here we see both views. (A) The forearm of a patient with tinea corporis (ringworm of the body). (B) Trichosporon rubrum, *a fungus that causes various forms of tinea, grown on cornmeal glucose agar. (C) Secondary syphilis lesions of a foot. (D)* Treponema pallidum, *the bacterium that causes syphilis, viewed after fluorescent antibody staining. (A,B,C from Camera M. D. Studios—Carroll Weiss; D from BPS—Centers for Disease Control.)*

Plate 7

The differing views of infectious diseases as seen by the physician and the clinical microbiologist—both views contribute to the diagnosis of disease and the selection of the appropriate treatment. (A) Characteristic appearance of impetigo of the cheek. (B) Staphylococcus aureus *causes several human diseases including impetigo. When grown on blood agar, colonies of* S. aureus *are surrounded by a zone of clearing indicating β-hemolysis. (C) Pharyngitis, inflammation of the pharynx, is shown in this view of a "red throat." (D)* Streptococcus pyogenes *causes pharyngitis or "strep throat." On blood agar* S. pyogenes *form small pinpoint colonies; such group A streptococci also exhibit β-hemolysis. (E) This Gram stain of a sputum sample from a patient with pneumonia shows the presence of* Streptococcus pneumoniae. *(F) On blood agar* S. pneumoniae *exhibits partial clearing (greening) indicating α-hemolysis. (A,C from Camera M. D. Studios—Carroll Weiss; E from BPS—Leon Lebeau.)*

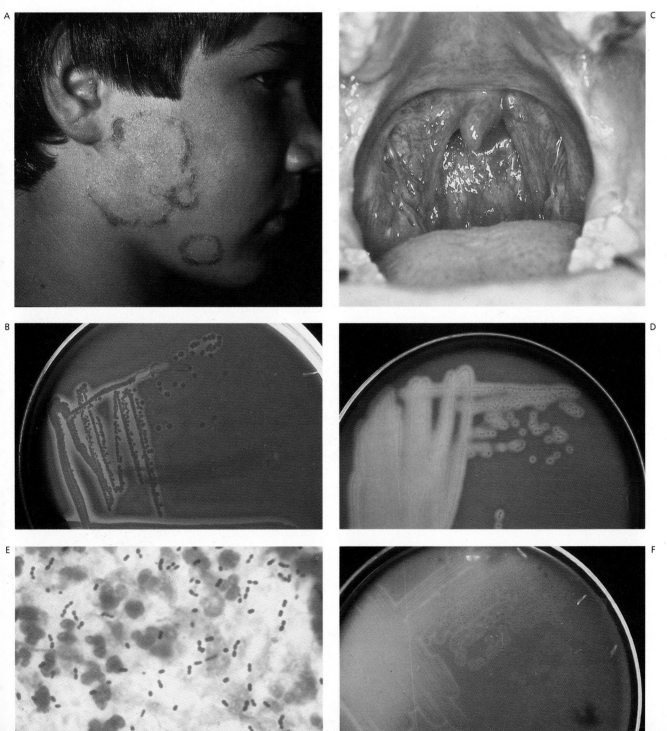

Plate 8

The clinical microbiologist employs various systems for identifying pathogens and for determining the effectiveness of different antibiotics for treating infectious diseases. (A) The enterotube is one of several widely used commercial systems available for identifying fermentative bacteria, e.g., enteric pathogens. (B) The Flow Laboratory system is used in some clinical laboratories for identifying nonfermentative pathogens, e.g., Pseudomonas *species. (C) The streptex system uses passive agglutination for identifying* Streptococcus *groups. Many other pathogens are rapidly identified using similar immunological systems. (D) The effectiveness of various antibiotics against a* Staphylococcus *species is revealed by examining the zones of clearing around discs impregnated with the antibiotics. (A,B from BPS—Leon Lebeau; C courtesy Wellcome Diagnostic.)*

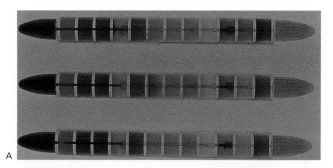

A

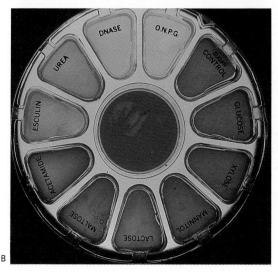

B

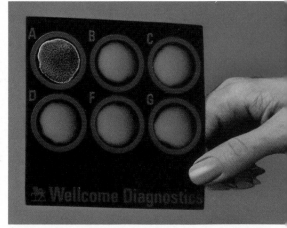

C

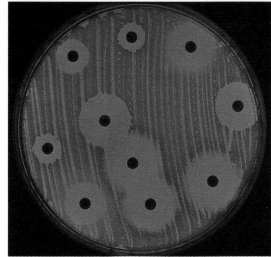

D

Total leukocytes 7500	Baso-philes 0–1	Eosino-philes 2–4	Normal Neutrophils				Lympho-cytes 21–30	Mono-cytes 4–8
			Myelo-cytes 0	Juveniles 0–1	Stabs 3–5	Segments 58–66		
Scarlet fever								
16,680	2	0	84	1	15	58	18	7
Appendicitis								
13,800	0	0	0	10	59	20	10	0
Staphylococcus septicemia								
34,950	0	0	0	12	31	46	8	3
Tularemia								
19,550	1	0	0	6	53	23	12	5

Figure 16.1

Some representative differential white blood cell counts for various infections.

ferent types of blood cells are determined. This clinical procedure can provide a general indication of the nature of the infecting agent, that is, if the disease is mediated by a virus, bacterium, fungus, or protozoan. Changes in the composition of the blood usually occur as a consequence of a microbial infection. Such changes generally result from the immune response and are reflected in shifts in the relative quantities and types of white blood cells. An **elevated white blood cell count** (**leukocytosis**) is characteristic of many systemic infections. A systemic bacterial infection, for example, is normally characterized by a progressive **neutrophilia** (**neutrophilic leukocytosis—increase in neutrophil cells**), particularly by an increase of young neutrophil cells known as **stab** or **band** cells (Figure 16.1). As compared to mature neutrophils, stab cells have a U-shaped nucleus that is slightly indented but not segmented. The increase in stab cells, indicative of neutrophilia, is known as a shift to the left, which refers to a blood cell classification system in which immature blood cells are positioned on the left side of a standard reference chart, and mature blood cells are placed on the right. The recovery phase of an infection is characterized by a reduction in fever, a decrease in the total number of leukocytes, and an increase in the number of monocytes. Gradually, the relative numbers of the various white blood cells return to their respective normal ranges.

In addition to systemic infections, some localized infections, such as abdominal abscesses, result in neutrophilia. Not all bacterial infections, though, show this characteristic leukocytosis. Some bacterial infections, such as typhoid fever, paratyphoid fever, and brucellosis, actually result in a persistant depression in the numbers of neutrophil cells (**neutropenia**). Many viral infections similarly result in lowered numbers of white blood cells (**leukopenia**). A general indication of whether a disease is of bacterial or viral origin, therefore, may be obtained by performing a white blood cell count and determining whether there is a significant shift in the quantity of neutrophils.

Changes in quantities of **eosinophils** may also indicate the nature of the infection. The number of eosinophil cells generally declines during systemic bacterial infections. **Eosinophilia** (increased numbers of eosinophils) is symptomatic of allergic diseases and parasitic infections, including those mediated by protozoans, and thus, the observation of elevated numbers of eosinophils is useful in the preliminary diagnosis of such diseases.

In some diseases, such as infectious mononucleosis, there are characteristic changes in the white blood cells (Figure 16.2). There is a transient leukocytosis because of an increase in lymphocytes, which characteristically are enlarged and show obvious changes in the nucleus, including the shape, size, and density of the nuclear region. These changes are useful in the diagnosis of this disease. A few microbial infections, malaria for example, result in decreased numbers of red blood cells (**anemia**). Thus, a simple examination of the blood often gives a preliminary indication of the etiology of a disease condition, establishing the

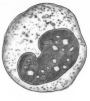

Normal

Abnormal

Figure 16.2

There are various characteristic changes in white blood cells that occur as a result of infectious mononucleosis. (A) A normal monocyte. (B) An atypical monocyte, indicative of infectious mononucleosis; the monocyte is enlarged, pleomorphic, and vacuolated.

556

Clinical microbiology: identification of disease-producing microorganisms and determination of antibiotic susceptibility

direction of additional test procedures for positively identifying the causative agent of a disease.

Skin testing

Skin testing, based upon delayed hypersensitivity reactions, is another useful procedure in the presumptive diagnosis of several diseases. Skin tests, however, are merely screening methods used as aids in the diagnosis of infectious diseases with regard to prior exposure and cannot be used for positive diagnosis of a disease. In skin testing, antigens derived from a test organism are injected intradermally. The development of redness in 24–72 hours is evidence for a delayed hypersensitivity reaction, indicating that the patient had previously been exposed and became sensitized to that specific antigen. A positive skin test may indicate an active infection caused by the organism from which the antigens are derived but usually reflects an earlier exposure to that organism.

The classic skin test for a microbial infection is the tuberculin reaction for detecting probable cases of tuberculosis (Figure 16.3). A purified protein derivative extract (PPD) from *Mycobacterium tuberculosis* is injected subcutaneously, and the area near the injection is observed for evidence of a delayed hypersensitivity reaction. A positive test results in **erythema** (reddening) and **induration** (hardening) of the skin with the peak reaction occurring between 48 and 72 hours. The reliability of the tuberculin test, though, depends on how the antigen is administered. In the Montoux test, commonly used in the United States, an appropriate dilution of PPD is injected intradermally into the superficial layers of the skin of the forearm. Other test procedures, including the once widely used Tine test, employ various mechanical devices and multiple punctures to expose the individual to the antigen and are not as reliable as the Montoux PPD procedure. Similar skin tests are available for the diagnosis of coccidioidomycosis, using coccidiodin, an antigen derived from *Coccidioides immitis*; histoplasmosis, using histoplasmin, a crude filtrate from *Histoplasma capsulatum*; leprosy, using lepromin derived from *Mycobacterium leprae*; brucellosis, using brucellergen obtained from a *Brucella* species; and the venereal disease lymphogranuloma venereum, using lygranum from *Chlamydia* species. In many cases skin tests are used to screen a population for individuals who are infected with a pathogen but who have not developed clinical symptoms, thus identifying cases where additional rigorous test procedures should be performed. For example, tuberculosis testing is routinely carried out on schoolchildren to identify possible carriers of *M. tuberculosis*.

Screening and isolation procedures

Whereas disease symptomology, changes in blood composition, and skin testing can serve as indicators of microbial infection and possibly the nature of the microorganisms causing the disease, many other nonmicrobiological factors may produce similar symptoms and clinical findings. The positive diagnosis of an infectious disease, therefore, requires the isolation and identification of the pathogenic microorganism or the identification of antigens specifically associated with a given microbial pathogen.

A variety of procedures are employed for the **isolation of pathogenic microorganisms** from different tissues. Very different procedures are required for the isolation of different types of microorganisms. For example, the procedures used for the isolation of pathogenic bacteria are not applicable for the isolation of viruses, and the

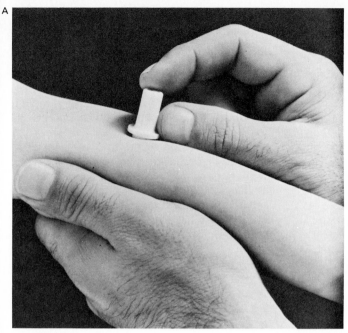

Figure 16.3

The results of a skin test, showing a positive tuberculin reaction. Development of inflammation after a delay is indicative of a positive reaction. (A) Application of the antigen to the skin in the Tine test. (B) Record card for Tine test showing development of positive reaction. (Courtesy Lederle Laboratories, Wayne, New Jersey.)

procedures designed for the recovery of aerobic microorganisms preclude the isolation of obligate anaerobes. Most clinical procedures have been designed to screen and facilitate the recovery of those etiologic agents of disease that predominate within specific tissues. When the symptomology suggests that the disease may be caused by a rare pathogen and/or routine screening fails to detect a probable causative microorganism, additional specialized isolation procedures may be required. In the pages following, we will examine some of the routine procedures employed for the isolation of pathogens from various sites of the human body so that the etiologic agents of infectious diseases can be identified.

Upper respiratory tract cultures

For the isolation of pathogens from the upper respiratory tract, **throat and nasopharyngeal cultures** are collected using sterile cotton swabs. The cotton swabs are placed in sterile transport media to prevent desiccation during transit to the laboratory. These cultures are streaked onto blood agar plates—which were prepared by using defibrinated sheep red blood cells— and incubated in an atmosphere of 5–10 percent CO_2 for isolation of microorganisms. Human red blood cells are not used in the preparation of blood agar because the natural antibodies inhibit the recovery of bacteria, particularly *Streptococcus* species. Blood agar plates permit the detection of alpha (greening of the blood around the colony), beta (zones of clearing around the colony), and gamma (no clearing) hemolysis. *Streptococcus pyogenes*, which forms relatively small colonies and demonstrates beta hemolysis on blood agar, is the predominant pathogen detected by using throat swabs and blood agar plates. The detection of β-hemolytic streptococci is important because these organisms can cause rheumatic fever. β-Hemolytic *Staphylococcus aureus* cultures as well as any other hemolytic bacteria may also be detected by using this procedure.

Haemophilus influenzae, Neisseria meningitidis, and *N. gonorrhoeae* can also be detected by using throat swabs and plating on various media. These organisms grow better on chocolate agar (a medium prepared by heating blood agar until it turns a characteristic brown color) than on plain blood agar. Thayer-Martin medium is preferred for the isolation of *N. gonorrhoeae* because the growth of normal throat microbiota is inhibited on this medium. Infection of the upper respiratory tract with these pathogens can lead to serious diseases—such as meningitis—if the infection spreads, making early diagnosis important. Also, *Haemophilus influenzae* may cause acute epiglottitis, the rapid diagnosis of which is important, because death can result within 24 hours. Nasopharyngeal swabs may also be used for determining the presence of *Bordetella pertussis*, the causative organism of whooping cough. The isolation of *B. pertussis* requires plating on a special medium, such as Bordet-Gengou potato medium. Cultures of *B. pertussis* also may be obtained by having the patient cough directly onto a plate of Bordet-Gengou medium.

When diphtheria is suspected, additional special procedures must be carried out. The presence of bacteria demonstrating typical snapping division (Chinese letter formation) indicates the possible presence of *Corynebacterium diphtheriae* (Figure 16.4). However, this is not a positive diagnosis because other bacteria with similar morphologies may be present. Usually, several media are employed in the culture of *C. diphtheriae*. Colonies of *C. diphtheriae* on tellurite serum agar—a medium used for the culture of *Corynebacterium* species—appear smooth, glistening, and gray-black. Gram stains prepared on colonies that develop on tellurite agar can confirm the presence of morphologically typical *C.*

558

Clinical microbiology: identification of disease-producing microorganisms and determination of antibiotic susceptibility

diphtheriae, giving an early indication of the presence of this bacterium before more rigorous identification procedures can be performed.

In addition to culturing for bacterial pathogens, throat swabs can be used to collect viral pathogens. The laboratory growth of viruses employs tissue cultures rather than bacteriological media. Primary rhesus monkey kidney cells are used most frequently for viral tissue culture. Antibiotics are added to the tissue culture to prevent bacterial and fungal growth. Viruses are washed from throat swabs by using appropriate synthetic medium such as Hanks' or Earls' basal salt solutions, and the solution is then added to tissue culture tubes or plates. The viral infection of the tissue culture cells normally produces morphological changes, known as **cytopathic effects** (CPE), which can be observed readily by microscopic observation. Some viruses exhibit a characteristic CPE that can be used in the identification of the virus (Figure 16.5), but in other cases, serological procedures and/or electron microscopy are necessary to identify viral isolates.

Lower respiratory tract cultures

Isolating microbial pathogens from the lower respiratory tract is a more formidable task than culturing organisms from the upper respiratory tract. Sputum, an exudate containing material from the lower respiratory tract, is frequently used for the culture of lower respiratory tract pathogens. Unfortunately, sputum samples vary greatly in quality and should therefore be examined microscopically before screening is carried out in order to determine whether they are suitable for culturing lower respiratory tract organisms. Acceptable sputum samples should have a high number of neutrophils, should show the presence of mucus, and should have a low number of squamous epithelial cells (Table 16.1). A large number of epithelial cells generally indicates contamination with oropharyngeal secretions, and such samples are not suitable. In order to ensure the quality of lower respiratory tract specimens, transtracheal aspiration may be employed. In this technique a needle is passed through the neck, a catheter is extended into the trachea, and samples are then collected through the catheter tube. Collection of sputum by using the transtracheal procedure avoids contact with oropharyngeal microorganisms so that the clinician is certain of the source of any isolates that are obtained. It is important to know where the isolated strains originate because some microorganisms are not likely to be associated

Figure 16.4

The typical Chinese letter formation of Corynebacterium diphtheriae *is shown in this micrograph. (From BPS—Centers for Disease Control, Atlanta.)*

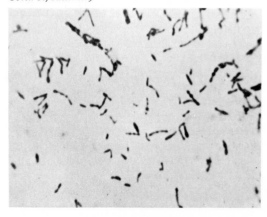

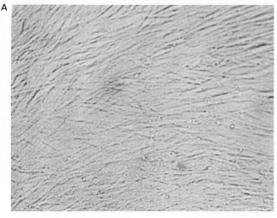
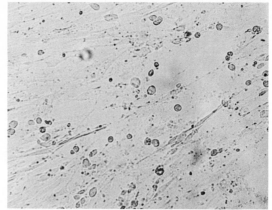

A

B

Figure 16.5

The characteristic cytopathic effect (CPE) of Rhinovirus *in tissue culture is indicated by cell shrinkage, nuclear pyknosis, and loss of adsorption to glass. (A) Micrograph of normal uninfected cell line (60×). (B) Micrograph showing CPE in HDL cells after infection with* Rhinovirus. *(Courtesy Jack Schieble, California Department of Health Services.)*

with disease when they occur among the normal microbiota of the upper respiratory tract, but if these same organisms are found in the lower respiratory tract they are prime candidates for the etiologic agents of disease.

The routine examination of sputum involves plating on appropriate selective, differential, and enriched media, for example, on blood agar, chocolate agar, and MacConkey's agar (a medium used for the isolation of Gram negative enteric bacteria). The bacterial pathogens normally detected by using this technique include *Streptococcus pneumoniae, Staphylococcus aureus, Streptococcus pyogenes, Klebsiella pneumoniae,* and *Haemophilus influenzae,* among the other pathogens that cause lower respiratory tract infections. In cases of suspected tuberculosis and diseases caused by fungi, additional procedures are necessary. Examination of stained smears of the sputum are useful in detecting such infections. Yeast-like cells of *Histoplasma capsulatum, Coccidioides immitis,* and *Candida albicans* may be observable in such smears. In cases where these organisms appear to be present, fungal culture media, such as Sabouraud's agar supplemented with antibacterial antibiotics, should be employed for culturing the suspected fungal pathogens. For the diagnosis of tuberculosis, the acid-fast stain procedure can reveal the presence of *M. tuberculosis* in sputum samples. Members of the genus *Mycobacterium* are acid-fast and appear red when stained by this procedure. Sputum showing presumptive evidence of the presence of *M. tuberculosis* should be cultured by using Lowenstein-Jensen medium—or other suitable medium that supports the growth of *M. tuberculosis*—for the positive diagnosis of this organism.

Cerebrospinal fluid cultures

In cases of suspected infection of the central nervous system, **cerebrospinal fluid (CSF)** can be obtained by performing a lumbar puncture. It is important to determine rapidly whether there is a microbial infection of the cerebrospinal fluid because such infections (meningitis) can be fatal if not rapidly and properly treated. Several chemical tests can be performed immediately to determine whether a CSF infection is of probable bacterial or viral origin. Most bacterial infections of

table 16.1

Rating system for determining quality of sputum samples

Number of cells per low-power field		Score
Epithelial	Leukocytes	
>25	<10	−3
>25	10–25	−2
>25	>25	−1
10–25	>25	+1
<10	>25	+2*

*High-quality specimen that is acceptable for complete identification of microbiota.

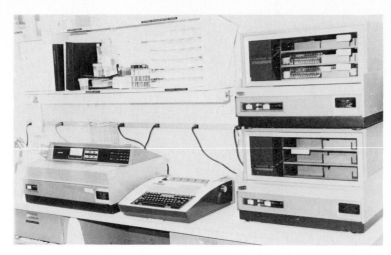

Figure 16.6

An autobac system for rapid screening of body fluids for bacterial contamination. The use of such automated systems has decreased the time needed by the clinical laboratory to make a diagnosis and recommend to the physician an appropriate course of treatment.

Clinical microbiology: identification of disease-producing microorganisms and determination of antibiotic susceptibility

the cerebrospinal fluid greatly reduce the level of glucose, whereas viral infections do not alter the glucose level, and this difference can be determined rapidly by measuring glucose and lactic acid concentrations in the cerebrospinal fluid.

The cerebrospinal fluid can be screened further for possible bacterial infection by observing Gram stained slides and by culture techniques. The growth of bacteria in a liquid enrichment medium can easily and rapidly be detected as an increase in turbidity (Figure 16.6). Cultures can also be obtained from CSF by plating on blood and chocolate agar. In order to provide a sufficient innoculum, the CSF is routinely centrifuged to concentrate the bacteria, and the sediment is used for inoculation. Bacteria commonly associated with cases of bacterial meningitis include *Streptococcus pneumoniae, Neisseria meningitidis, Haemophilus influenzae, Streptococcus pyogenes, Staphylococcus aureus, Escherichia coli, Klebsiella pneumoniae*, and *Pseudomonas aeruginosa. Streptococcus pneumoniae* and *Neisseria meningitidis* probably are the most frequently found etiological agents of bacterial meningitis in adults, whereas *Haemophilus influenzae* most frequently is found in children but rarely occurs in cases of adult bacterial meningitis. It should be noted that various other microorganisms can cause meningitis, including anaerobic bacteria, fungi, and protozoa. Anaerobic bacteria can be cultured in thioglycollate broth or other media in the absence of free oxygen. Fungi and protozoa can be observed microscopically and identified by using both cultural and serological test procedures. In cases of tubercular CSF infections, additional culture techniques are required for the isolation of *Mycobacterium tuberculosis*. Detection of neurosyphilis requires serological diagnosis because

Treponema pallidum cannot be cultured. Definitely, in cases of suspected infections of the cerebrospinal fluid, rapid and accurate diagnosis of the infecting agent is crucial for determining the appropriate modes of treatment.

Blood cultures

The detection of bacteria in blood, likewise, is important in diagnosing a number of diseases, including **septicemia** (systemic bacterial infection of the bloodstream—blood poisoning). For culturing bacteria, blood should be collected by venal puncture, using aseptic technique. The numbers of infecting bacteria in the blood are often low and may vary with time, and therefore, it is necessary to collect and examine blood samples at various time intervals. For example, in respiratory infections bacteria may follow a 45-min to 1-h cycle of entry into the blood, removal, and reentry, whereas in endocarditis the numbers of bacteria in the blood generally remain relatively constant. Both aerobic and anaerobic culture techniques are needed to ensure the growth of any bacteria present in the blood. There are several effective methods employed in clinical laboratories for isolating anaerobes, ensuring that exposure to air is avoided. The roll tube-streak method, using a VPI inoculating system and pre-reduced media, or equivalent procedures, may be required for the culture of very strict anaerobic bacteria (Figure 16.7).

Another rapid screening procedure for the presence of bacteria in blood employs a medium containing radiolabeled glucose. The conversion of glucose to radiolabeled carbon dioxide is rap-

A

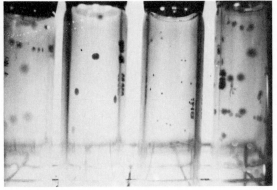 B

Figure 16.7

The culture of obligate anaerobes requires special inoculation and culture procedures. (A) The VPI anaerobic culture transfer system and (B) roll tube setup, showing colony development of anaerobes. (Courtesy Donald E. Hash, Anaerobe Laboratory, Virginia Polytechnical Institute and State University.)

idly determined with great sensitivity by using automated instrumentation (Figure 16.8). Initial screening of the blood for bacterial contaminants can be accomplished in liquid media, both aerobically and anaerobically, through using an assay such as increased turbidity as an index. Liquid media can also serve as an enrichment culture prior to plating on such solid media as blood agar, chocolate agar, and MacConkey agar. The blood sample should be diluted (blood:broth ratio of 1:10 or greater) and an inhibitor of coagulation and phagocytosis added to remove residual bactericidal factors in the blood. Additionally, if there was previous antibiotic treatment, the blood specimens may require further dilution, or an appropriate antibiotic inactivator may be added to permit growth of bacterial pathogens present in the patient's blood.

Figure 16.8

The Bactec system for detecting microbial growth uses a radiolabeled substrate. Growth of microorganisms on the labeled substrate releases radiolabeled gaseous products that are detected in this instrument. The Bactec system is advantageous because of its sensitivity and speed. (Courtesy Johnston Laboratories, Cockeysville, Maryland.)

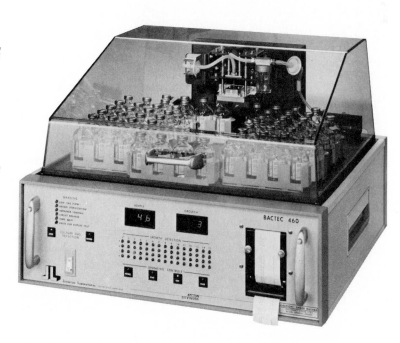

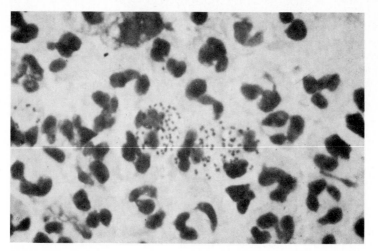

Figure 16.9

Micrograph of Neisseria gonorrhoeae *in pus; the presence of bean-shaped diplococci is diagnostic of gonorrhea. (From Centers for Disease Control, Atlanta.)*

Urine cultures

In order to detect urinary tract infections, a midstream catch of voided urine is usually employed to minimize contamination with the normal mi-

562

Clinical microbiology: identification of disease-producing microorganisms and determination of antibiotic susceptibility

Figure 16.10

Micrograph of Treponema pallidum, *using dark-field microscopy, showing its characteristic corkscrew shape. (From BPS—Centers for Disease Control, Atlanta.)*

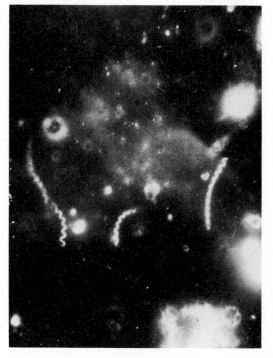

crobiota of the genitourinary tract. Precautions are normally taken to avoid contamination with exogenous bacteria during voiding of the urine sample, for instance, by washing the area around the opening of the urinary tract. Urine, though, normally becomes contaminated with bacteria during discharge through the urethra, particularly in females. Therefore, culture of the urine should be performed both qualitatively and quantitatively. High numbers of a given microorganism are indicative of infection rather than contamination of the urine during discharge. In general, greater than 10^5 bacteria/ml indicates a urinary tract infection. Plating should be performed on a general medium, such as blood agar, and on selective media, such as cysteine lactose electrolyte-deficient agar (CLED), MacConkey agar, or eosine-methyline blue (EMB) agar. Bacteria of clinical significance that may be found in the urine include *Escherichia coli, Klebsiella, Proteus, Pseudomonas, Salmonella, Serratia, Streptococcus*, and *Staphylococcus* species. Gram negative enteric bacteria are found most frequently to be the etiologic agents of urinary tract infections.

Urethral and vaginal exudate cultures

The examination of urethral and vaginal exudates centers on the detection of microorganisms that cause sexually transmitted diseases, most notably *Neisseria gonorrhoeae, Chlamydia* sp., and *Treponema pallidum*. In males, one symptom of gonorrhea is the painful release of a urethral exudate. Gram stained slides of the exudate are made, and if Gram negative, kidney bean-shaped diplococci are present, this suggests a diagnosis of gonorrhea (Figure 16.9). In females, gonorrhea is more difficult to detect because of the high numbers and variety of normal microbiota associated with the vaginal tract. Culture techniques using inoculation with swabs collected from the cervix, vagina, and anal canal can be employed for detecting *Neisseria gonorrhoeae* in females. Thayer-Martin medium and chocolate agar, incubated under an atmosphere of 5–10 percent carbon dioxide, are employed for the culture of *N. gonorrhoeae*. Screening for pathogens infecting the genital tract such as *Haemophilus ducreyi*, *Streptococcus pyogenes*, *Staphylococcus aereus*, and *Candida albicans* is accomplished by plating on blood agar, chocolate agar, MacConkey agar, and Sabouraud's agar. *Treponema pallidum*, the causative organism of syphilis, cannot be cultured on laboratory media, but exudates from primary or secondary lesions can be examined by using dark-field microscopy and fluorescent antibody staining for the detection of *Treponema pallidum* (Figure 16.10). The characteristic morphology and movement of spirochetes can be readily detected by dark-field microscopic observation, but caution should be employed in the diagnosis because nonpathogenic spirochetes may also be present.

Fecal cultures

Stool specimens are normally used for the isolation of microorganisms that cause intestinal tract infections. The common enteric bacterial pathogens are *Salmonella* spp., *Shigella* spp., enteropathogenic *E. coli*, *Vibrio cholerae*, *Campylobacter* spp., *Yersinia enterocolytica*, and *Staphylococcus aureus*. Because fecal matter contains a large number of nonpathogenic microorganisms, it is necessary to employ selective and differential media for the isolation of intestinal tract pathogens. Usually, selective media contain some toxic factor that selectively inhibits some microorganisms, often preventing the overgrowth of pathogenic microorganisms that are present in low numbers by other more numerous microbial populations. Selective media contain components that select for the growth of particular microorganisms. Differential media contain indicators that permit the recognition of microorganisms with particular metabolic activities. For example, pH indicators are often incorporated into media for the detection of acidic metabolic products. Common selective and differential media that

are employed for the isolation of intestinal tract pathogens include *Salmonella-Shigella* (SS), Hektoen enteric (HE), xylose-lysine-desoxycholate (XLD), brilliant green, eosin methylene blue (EMB), Endo, and MacConkey agars (Table 16.2). A combination of a differential medium, such as MacConkey agar, and a selective medium, such as Hektoen enteric agar, is often used for the isolation of intestinal tract pathogens.

It is often necessary to carry out an **enrichment culture** before *Salmonella* and *Shigella* species can be isolated by using differential or selective solid media. For example, in cases of suspected typhoid fever it may be necessary to carry out an enrichment in appropriate selective medium, such as GN (Gram negative) broth or selenite F broth, before isolation of *Salmonella typhi* can be achieved (Table 16.3). In cases of suspected viral infections, tissue cultures can be inoculated with fecal matter for the culture of enteric viral pathogens. Characteristic cytopathic effects and serological procedures can then be employed in the identification of viral pathogens. Additionally, electron microscopy can be used for the direct detection of viruses, such as *Rotavirus*, that cause viral gastroenteritis (Figure 16.11).

Eye and ear cultures

Fluids collected from eye and ear tissues can be inoculated onto blood agar, chocolate agar, and MacConkey's agar or other defined media to culture bacterial pathogens commonly found in these tissues. Additionally, a Gram stain slide can be prepared and observed to identify the presence of bacteria. The microscopic observation of stained slides can give an indication as to whether the infection is due to bacterial pathogens. In the case of eye infections, it is particularly important to differentiate between bacterial and viral infections in order to determine the appropriate treatment.

Skin lesion cultures

Material from skin lesions, including wounds and boils, can be collected for culture purposes with swabs, aspirates or washings. Such material, though, often is contaminated with endogenous bacteria. A variety of bacteria can infect wounds and cause localized skin infections. Both aerobic and anerobic culture techniques are required for the screening of wounds for potential pathogens. Par-

table 16.2

Media for isolation of Enterobacteriaceae

MacConkey agar

MacConkey agar is a differential plating medium for the selection and recovery of Enterobacteriaceae and related enteric Gram negative rods. Bile salts and crystal violet are included to inhibit the growth of Gram positive bacteria and some fastidious Gram negative bacteria. Lactose is the sole carbohydrate. Lactose-fermenting bacteria produce colonies that are varying shades of red because of the conversion of the neutral red indicator dye (red below pH 6.8) from the production of mixed acids. Colonies of non-lactose-fermenting bacteria appear colorless or transparent.

Eosin methylene blue (EMB) agar

Eosin methylene blue (EMB) agar is a differential plating medium that can be used in place of MacConkey agar for the isolation and detection of the Enterobacteriaceae and related coliform rods from specimens with mixed bacteria. The aniline dyes (eosin and methylene blue) in this medium inhibit Gram positive and fastidious Gram negative bacteria. They also combine to form a precipitate at acid pH, thus also serving as indicators of acid production.

Desoxycholate-citrate agar (DCA)

Desoxycholate-citrate agar is a differential plating medium used for the isolation of members of the Enterobacteriaceae from mixed cultures. The medium contains about three times the concentration of bile salts (sodium desoxycholate) as MacConkey agar, making it most useful in selecting species of *Salmonella* from specimens overgrown or heavily contaminated with coliform bacteria or Gram positive organisms. Sodium and ferric citrate salts in the medium retard the growth of *Escherichia coli*. Lactose is the sole carbohydrate, and neutral red is the pH indicator and detector of acid production.

564

Clinical microbiology: identification of disease-producing microorganisms and determination of antibiotic susceptibility

Endo agar

Endo agar is a solid plating medium used to recover coliform and other enteric organisms from clinical specimens. The medium contains sodium sulfite and basic fuchsin, which serve to inhibit the growth of Gram positive bacteria. Acid production from lactose is not detected by a pH change, rather from the reaction of the intermediate product, acetaldehyde, which is fixed by the sodium sulfite.

Salmonella-Shigella agar (SS)

SS agar is a highly selective medium formulated to inhibit the growth of most coliform organisms and permit the growth of species of *Salmonella* and *Shigella* from clinical specimens. The medium contains high bile salts concentration and sodium citrate, which inhibit all Gram positive bacteria and many Gram negative organisms, including coliforms. Lactose is the sole carbohydrate and neutral red the indicator for acid detection. Sodium thiosulfate is a source of sulfur, and any bacteria that produce H_2S gas are detected by the black precipitate formed with ferric citrate.

Hektoen enteric agar (HE)

HE agar is devised as a direct plating medium for fecal specimens to increase the yield of species of *Salmonella* and *Shigella* from the heavy numbers of normal microbiota. The high bile salt concentration of this medium inhibits the growth of all Gram positive bacteria and retards the growth of many strains of coliforms. Acids may be produced from three carbohydrates, and acid fuchsin reacting with thymol blue produces a yellow color when the pH is lowered. Sodium thiosulfate is a sulfur source, and H_2S gas is detected by ferric ammonium citrate, producing a black precipitate.

Xylose lysine desoxycholate agar (XLD)

XLD agar is less inhibitory to growth of coliform bacteria than HE and was designed to detect *Shigella* species in feces after enrichment in GN broth. Bile salts in relatively low concentration make this medium less selective than the other media included in this table. Three carbohydrates are available for acid production and phenol red is the pH indicator. Lysine-positive organisms, such as most *Salmonella enteriditis* strains, produce initial yellow colonies from xylose utilization and delayed red colonies from lysine decarboxylation. The H_2S detection system is similar to HE agar.

ticular concern must be given to the possible presence of *Clostridium tetani* and *Clostridium perfringens* because these anaerobes cause serious diseases. Various fungi may also be involved in skin infections, and appropriate fungal culture media are required for the isolation of these organisms. Dermatophytic fungi and actinomycetes that cause skin infections can also be detected by direct microscopic examination of skin tissues because the characteristic morphological appearance of filamentous fungi and bacteria often permits rapid presumptive diagnosis of the disease.

table 16.3

Enrichment broths for recovery of Enterobacteriaceae

Selenite broth

Peptone, 5 g; lactose, 4 g; sodium selenite, 4 g; sodium phosphate, 10 g; distilled water, 1 l, pH, 7.0.
Selenite broth is recommended for the isolation of *Salmonella* species from specimens such as feces and urine. Sodium selenite is inhibitory to *E. coli* and many other coliform bacilli, including many strains of *Shigella*.

Gram negative (GN) broth

Polypeptone, 20 g; glucose, 1 g; D-mannitol, 2 g; sodium citrate, 5 g; sodium desoxycholate, 0.5 g; dipotassium phosphate, 4 g; monopotassium phosphate, 1.5 g; sodium chloride, 5 g; distilled water, 1 l; pH, 7.0.
GN broth is designed for the recovery of *Salmonella* and *Shigella* species when they are in small numbers in fecal specimens. Because of the relatively low concentration of desoxycholate, GN broth is less inhibitory to *Escherichia coli* and other coliforms. Most strains of *Shigella* grow well. The desoxycholate and citrate are inhibitory to Gram positive bacteria. The increased concentration of mannitol over glucose limits the growth of *Proteus* species while encouraging growth of *Salmonella* and *Shigella* species, both of which are capable of fermenting mannitol.

Figure 16.11

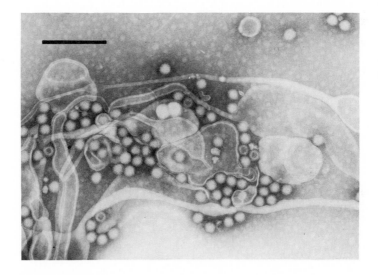

Virions of Rotavirus *in the stool of an infant with acute gastroenteritis are shown in this negatively stained preparation. Bar = 400 nm. (Courtesy F. P. Williams, Jr., U.S. Environmental Protection Agency, Cincinnati.)*

Identification of pathogenic microorganisms

Determining that a disease is of microbial etiology (is caused by microorganisms) and isolating microorganisms are only part of the job of diagnosing a specific disease. It is also necessary to identify the specific causative organism. A wide range of biochemical and serological procedures are available for the definitive **identification of microbial isolates of clinical significance.** Accuracy, reliability, and speed are important factors governing the selection of clinical identification protocols. The selection of the specific proce-

dures to be employed for the identification of pathogenic isolates is guided by the presumptive identification of the organism at the genus or family level, based on the observation of colonial morphology and other growth characteristics on the primary isolation medium, and on the microscopic observation of stained specimens. Some examples of the protocols and criteria used for the identification of various clinical isolates will be discussed in the following pages.

Both classical dichotomous key and probabilistic matrix approaches are used for the identification of pathogenic microorganisms. Pathogenic filamentous fungi and protozoa are generally identified based on morphological characteristics of the organism, growth appearance, and a limited number of biochemical tests. Various other morphological and biochemical characteristics are used for identifying other pathogens. Conventional identification schemes for bacteria and yeasts rely on the determination of a variety of biochemical features exhibited by growing isolates (Figure 16.12). In general, less than 20 tests are required to identify clinical bacterial isolates at the species level. The tests are aimed at specifically distinguishing the isolates present in the specimen, using the minimal number of tests to define distinct taxa accurately. We will now discuss some approaches that are used in the clinical identifica-

tion of representative groups of disease-causing microorganisms.

Miniaturized commercial identification systems

Several commercial systems have been developed for the identification of members of the family Enterobacteriaceae. These systems are widely used in clinical microbiology laboratories because of the high frequency of isolation of Gram negative rods indistinguishable except for characteristics determined by detailed biochemical and/or serological testing. Systems commonly used in clinical laboratories include the Enterotube, API 20-E, Minitek, Micro-ID, Enteric Tek, and r/b enteric systems. The pattern of test results obtained in

Clinical microbiology: identification of disease-producing microorganisms and determination of antibiotic susceptibility

Figure 16.12

Dichotomous key approach for the identification of the Gram negative rods, Enterobacteriaceae. Such keys have been developed for identifying virtually all known bacterial and mycotic human pathogens.

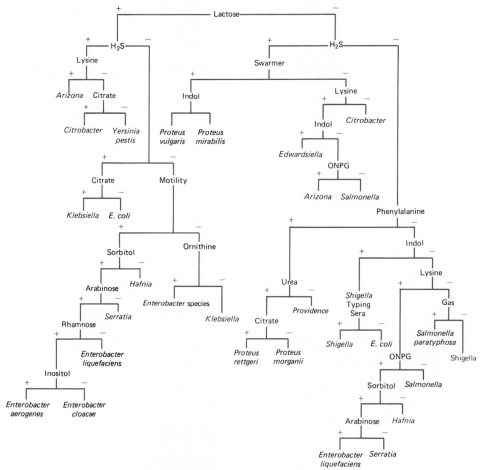

table 16.4

Features examined in the Enterotube I system

Characteristics	Visual reactions	
	Positive	Negative
Dextrose	Yellow	Red
Gas in dextrose chamber	Bubbles	No bubbles
Lysine decarboxylase	Purple-blue	Yellow
Ornithine decarboxylase	Purple-blue	Yellow
Hydrogen sulfide	Black media	No blackening
Indole (add Kovac's reagent to H_2S chamber)	Red ring	No red ring
Lactose	Yellow	Red
Phenylalanine (add 10% $FeCl_3$)	Brown	Light green
Dulcitol (read in phenylalanine chamber)	Yellow	Light green
Urease	Red	Light yellow
Citrate	Deep blue	Light green

these systems is converted to a **numerical code** that can be used to calculate the identity of the isolate. The numerical code describing the test results obtained for a clinical isolate is compared with results in a data bank describing test reactions of known organisms. Some of the commercial systems list a series of possible identifications indicating the statistical probability that a given organism (**biotype**) could yield the observed test results. All of the commercial systems employ miniaturized reaction vessels, and some are designed for automated reading and computerized processing of test results. The systems differ in how many and which specific biochemical tests are included. They also differ in whether they are restricted to identifying members of the family Enterobacteriaceae or whether they can be utilized for identifying other Gram negative rods. The test results obtained with all these systems show excellent correlation with conventional test procedures, and these package systems yield reliable identifications as long as the isolate is one of the organisms the system is designed to identify.

The Enterotube I system contains eight solid media from which 11 different features can be determined (Table 16.4). This system has a self-contained inoculating needle that is touched to a colony on the isolation plate and drawn through the tube. The characteristics determined in the Enterotube are used to generate a four-digit biotype number from which bacterial identifications

can be made (Figure 16.13). A newer Enterotube II system employs additional tests and generates a five-digit number.

The **API 20-E system**, as the name implies, utilizes 20 miniature capsule reaction chambers. Twenty-one tests, though, are run in this system (Table 16.5). A suspension of bacteria is used to inoculate each of the reaction chambers. The results of the API 20-E test system yield a seven-digit biotype number from which a computer-assisted identification can be made (Figure 16.14). The results of the API system can also be used in the identification of some nonfermentative Gram negative rods and for the identification of anaerobic bacteria. For the identification of nonfermentative Gram negative rods, six additional tests are run to generate a nine-digit biotype identification number (Figure 16.14). Over 100 taxa of Gram negative rods can be identified by using the API 20-E system.

The **Minitek system** consists of reagent-impregnated paper disks to which a broth suspension is added for the determination of characteristic reactions. The number of disks and tests employed is variable; generally, 17 tests at a time are used to differentiate clinical isolates. The profile of test results obtained with the Minitek system is useful in identifying obligate anaerobes as well as facultative enteric bacteria.

The **Micro-ID system**, in contrast, which is based upon constituitive enzymes, is designed primarily for identifying members of the Enterobacteri-

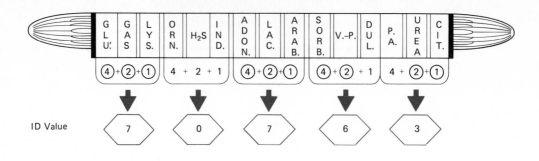

ID Value   Organism
70763   *Klebsiella pneumoniae*

Figure 16.13

Conversion of Enterotube test results to a numerical code. The speed and ease of identification makes these systems very popular in clinical laboratories.

table 16.5

Features examined in the API 20-E system

568

Clinical microbiology: identification of disease-producing microorganisms and determination of antibiotic susceptibility

Characteristics	Visual reactions	
	Positive	Negative
ONPG (β-galactosidase)	Yellow	Colorless
Arginine dihydrolase	Red-orange	Yellow
Lysine decarboxylase	Red-orange	Yellow
Ornithine decarboxylase	Red-orange	Yellow
Citrate	Dark blue	Light green
Hydrogen sulfide	Blackening	Colorless
Urease	Cherry red	Yellow
Tryptophan deaminase (add 10% $FeCl_3$)	Red-brown	Yellow
Indole	Red ring	Yellow
Voges-Proskauer (add KOH plus α-naphthol)	Red	Colorless
Gelatin	Pigment diffusion	No pigment diffusion
Glucose	Yellow	Blue-green
Mannitol	Yellow	Blue-green
Inositol	Yellow	Blue-green
Sorbitol	Yellow	Blue-green
Rhamnose	Yellow	Blue-green
Sucrose	Yellow	Blue-green
Melibiose	Yellow	Blue-green
Amygdalin	Yellow	Blue-green
Arabinose	Yellow	Blue-green

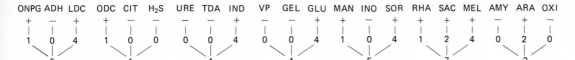

ONPG	ADH	LDC	ODC	CIT	H₂S	URE	TDA	IND	VP	GEL	GLU	MAN	INO	SOR	RHA	SAC	MEL	AMY	ARA	OXI
+	−	+	+	−	−	−	−	+	−	−	+	+	−	+	+	+	+	−	+	−
1	0	4	1	0	0	0	0	4	0	0	4	1	0	4	1	2	4	0	2	0

5   1   4   4   5   7   2

Normal 7 digit code 5144572 = *E. coli.*

Construction of a 9 Digit Profile

To the seven–digit profile described above, two digits are added corresponding to the following characteristics:

NO₂: Reduction of nitrate to nitrite only
N₂ GAS: Complete reduction of nitrate to N₂ gas or amines
MOT: Observation of motility
MAC: Growth on MacConkey medium
OF/O: Oxidative utilization of glucose (OF–open)
OF/F: Fermentative utilization of glucose (OF–closed)

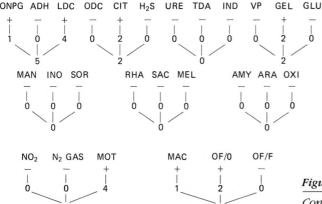

ONPG	ADH	LDC	ODC	CIT	H₂S	URE	TDA	IND	VP	GEL	GLU
+	−	+	−	+	−	−	−	−	−	+	−
1	0	4	0	2	0	0	0	0	0	2	0

5   2   0   2

MAN	INO	SOR		RHA	SAC	MEL		AMY	ARA	OXI
−	−	−		−	−	−		−	−	−
0	0	0		0	0	0		0	0	0

0   0   0

NO₂	N₂ GAS	MOT		MAC	OF/O	OF/F
−	−	+		+	+	−
0	0	4		1	2	0

4   3

9 digit code 520200043 = *Pseudomonas maltophilia*

Figure 16.14

Conversion of API 20-E test results to a numerical code for both fermentative and nonfermentative bacteria.

aceae, although it can also be used for biotyping *Haemophilus influenzae* and *H. parainfluenzae*. Bacteria must be screened for oxidase activity, and only oxidase negative strains are tested with this system. All commercial ID systems actually recommend performing the oxidase test before attempting an identification. This system employs 15 reaction chambers, and the test results are used to generate a five-digit identification code number. The Micro-ID system lists possible identifications and probabilities based on the results of the 15 biochemical test reactions (Table 16.6). Identifications with the Micro-ID system can be accomplished in as little as 4 hours.

The API 20-E and Minitek systems have been expanded to identify obligate anaerobes. In order to use these systems to identify anaerobes, the isolates are grown in a liquid culture medium suitable for the growth of anaerobes, and the tests are carried out under anaerobic conditions. There is also a 20-test API 20-C system, Automicrobic system, and a Uni-Yeast-Tek system with 11 test chambers that can be used in clinical laboratories for yeast identification. Several other systems are also available for the identification of nonfermentative Gram negative rods, such as the Automicrobic (AMS), Oxi-Ferm, and Oxitech systems. The Oxi-Ferm system is similar in design to the Enterotube, but eight different media are used in the tube for the identification of nonfermentative bacteria. Additional other systems are becoming available for identifying chemically important pathogens. API also has a system (API SerImm-Sure) for the specific identification of *Streptococcus, Salmonella*, and *Shigella* strains based on serological reactions.

table 16.6

Features examined in the Micro-ID system

Characteristics	Visual reactions	
	Positive	Negative
Voges-Proskauer	Pink-red	Light yellow
Nitrate reduction	Red	Colorless to light pink
Phenylalanine	Green	Light yellow
H_2S	Brown-black	White
Indole	Pink-red	Light yellow to orange
Ornithine decarboxylase	Gray-purple	Yellow
Lysine decarboxylase	Gray-purple	Yellow
Malonate	Green-blue	Yellow
Urease	Orange-red	Yellow
Esculin hydrolysis	Brown-black	No color change
ONPG (β-galactosidase)	Light or dark yellow	No color change
Arabinose	Yellow-amber	Purple
Adonitol	Yellow-amber	Purple
Inositol	Yellow-amber	Purple
Sorbitol	Yellow-amber	Purple

Clinical microbiology: identification of disease-producing microorganisms and determination of antibiotic susceptibility

Figure 16.15

The tube test for coagulase production is used for the detection of pathogenic Staphylococcus aureus. *Both control and positive reactions are shown in this photograph. (From BPS—Leon J. LeBeau, University of Illinois Medical Center.)*

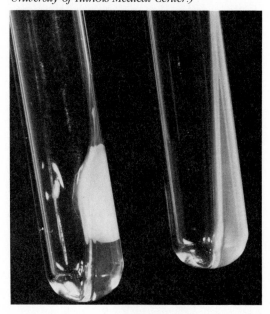

Conventional tests for diagnosing pathogenic microorganisms

In contrast to these widely used miniaturized commercial systems for the identification of Gram negative rods, tests for the identification of Gram positive bacteria normally employ conventional media and do not rely on Adonsonian taxonomy for identifying the isolates. The identification of *Staphylococcus* species, for example, is based on the observation of typical grape-like clusters, catalase activity, coagulase production, and mannitol fermentation. Members of the genus *Staphylococcus* are strongly catalase positive, whereas *Streptococcus* species are catalase negative. Catalase activity can be readily observed by adding a drop of hydrogen peroxide to a colony and observing the production of gas bubbles. Pathogenic strains of *Staphylococcus aureus* are generally coagulase positive and produce acid from mannitol, in contrast to strains of *Staphylococcus epidermidis*, which are negative for both coagulase production and mannitol fermentation. Coagulase activity, the ability to clump or clot blood plasma (Figure 16.15), is associated with virulent strains of *S. aureus* and other staphylococci.

Streptococcus species are separated into serological groups based on antigenic differences. The pathogenically important *Streptococcus* groups are A, B, D, F, and G. Biochemical test procedures are also routinely employed to distinguish species and groups of streptococci. Presumptive identification of the particular streptococcal group can be made based on the type of blood hemolysis, bacitracin susceptibility, optochin susceptibility, hippurate hydrolysis, esculin hydrolysis, and tolerance of 6.5 percent NaCl (Table 16.7).

A novel approach to the identification of obligate anaerobes used in clinical laboratories involves the gas-liquid chromatographic (GLC) detection of metabolic products. The anaerobes are grown in a suitable medium, and the short-chain volatile fatty acids produced are extracted in ether. Fatty acids detected in this procedure include acetic, propionic, isobutyric, butyric, isovaleric, valeric, isocaproic, and caproic acids (Figure 16.16). The pattern of fatty acid production can be used to differentiate and identify various anaerobes. When coupled with observations of colony and cell morphology and a limited number of biochemical tests, the common anaerobes isolated from clinical specimens can be identified (Table 16.8).

Serological testing in the identification of pathogenic microorganisms

In addition to using growth and biochemical characteristics for the identification of pathogenic microorganisms, a variety of **serological test procedures** are employed for identifying disease-causing microorganisms. Serology refers to immunological reactions *in vitro*. The immunological reactions used in these tests have been discussed in Chapter 14. Some specific examples will be discussed here to illustrate the application of these serological procedures for clinical diagnostic purposes. Serological tests are particularly useful in identifying pathogens that are difficult or impossible to isolate on conventional media and for identifying pathogenic strains of microorganisms of which there are many varieties not easily distinguished by biochemical testing. For example, over 2000 serotypes in the genus *Salmonella* are defined by the O (somatic cell) and H (flagella) antigens, with each serotype defined by a constellation of O and H antigens. The identification of pathogenic viruses and nonculturable bacteria, such as *Treponema pallidum*, generally depend on serological testing.

Agglutination

Heterophile antibodies A number of diseases can be rapidly diagnosed by employing agglutination tests. For example, infectious mononucleosis, caused by the Epstein-Barr virus, is routinely diagnosed through using serological agglutination tests. The blood serum of individuals infected with this virus contains antibodies that will agglutinate sheep red blood cells. Such antibodies are called **heterophile antibodies** because they cross-react with antigens other than the ones that elicited their formation. Heterophile antibodies may be present in normal sera, but in individuals with infectious mononucleosis, the concentration of such antibodies is greatly elevated. By testing the ability of serial dilutions of blood serum to cause agglutination of sheep red blood cells, the titer of heterophile antibodies can be determined. In performing these tests the pa-

table 16.7

Some diagnostic reactions for identification of Streptococcus *species*

Hemolysis	Beta	Beta	Beta	Alpha Beta None	Alpha None	Alpha None	Alpha
Bacitracin susceptibility	+	−	−	−	−	−	±
Hippurate hydrolysis	−	+	−	−	−	−	−
Bile esculin hydrolysis	−	−	−	+	+	−	−
Tolerance to 6.5% NaCl	−	±	−	+	−	−	−
Optochin susceptibility or bile solubility	−	−	−	−	−	−	+
Presumptive identification	Group A	Group B	Beta hemolytic not groups A, B, or D	Group D enterococci	Group D not enterococci	*Viridans*	*Pneumoniae*

572

Clinical microbiology: identification of disease-producing microorganisms and determination of antibiotic susceptibility

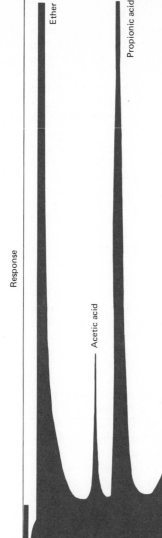

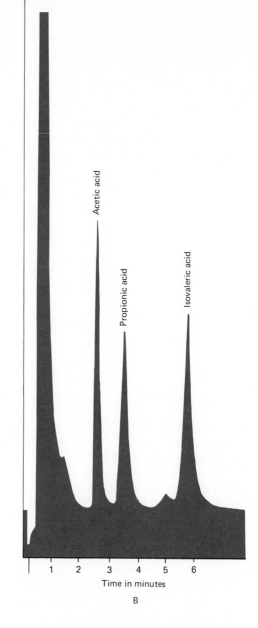

Figure 16.16

Gas chromatographic tracing, showing the detection of the fatty acids produced by anaerobes: (A) Proprionibacterium *sp.; (B)* Bacteriodies fragilis.

tient's serum is heated to 56°C for 30 minutes to destroy complement that could result in lysis rather than agglutination. A titer of 1:224 or greater is considered as presumptive evidence of infectious mononucleosis. Additional differential tests for infectious mononucleosis can be carried out to confirm the diagnosis. In these tests the patient's serum is mixed with guinea pig kidney tissue in one tube and beef erythrocyte antigen in another.

The guinea pig tissue does not absorb heterophile antibodies produced in response to the Epstein-Barr virus and does not reduce the ability of the serum to agglutinate sheep cells, whereas erythrocyte antigens do absorb these heterophile antibodies, lowering the ability of the serum to agglutinate red blood cells (Figure 16.17).

The detection of heterophile antibodies, that is, cross-reactive antibodies, is also useful in di-

table 16.8

Diagnostic features of some anaerobic bacteria

Organism	Indole produced	Esculin hydrolyzed	Catalase produced	Glucose fermented	Fatty acids produced
Actinomyces israelii	−	+	−	+	A,L,S
Propionibacterium acnes	+	−	+	+	A,P
Peptostreptococcus anaerobius	−	−	−	−	A,B,IB,IV,IC
Streptococcus intermedius	−	+	−	−	(A),L
Veillonella alcalescens	−	−	+	−	A,P
Clostridium botulinum	−	±	−	+	A,(P),(IB),B,IV, (V),(IC)
C. perfringens	−	±	−	+	A,B,(P)
C. tetani	±	−	−	-	A,B,(P)

A = acetic acid, B = butyric acid, P = propionic acid, IB = isobutyric acid, IC = isocaproic acid, IV = isovaleric acid, L = lactic acid, S = succinic acid, ( ) = variable

Figure 16.17

An agglutination slide test for the diagnosis of infectious mononucleosis provides a simple, rapid, and accurate method for detecting the presence of antibodies formed in response to infection with the Epstein-Barr virus. In this photograph, I is a positive reaction and II is a negative reaction.

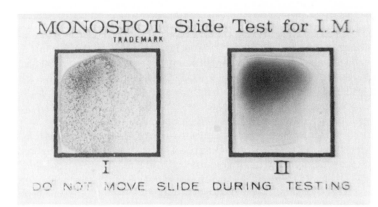

MONOSPOT Slide Test for I.M.
TRADEMARK

I   II

DO NOT MOVE SLIDE DURING TESTING

573

Identification of
pathogenic
microorganisms

agnosing several other diseases. For example, the Weil-Felix test is used to diagnose some diseases caused by *Rickettsiae* species. In these tests antibodies produced in response to a particular rickettsial infection agglutinate bacterial strains of *Proteus*, designated OX-19, OX-2, and OX-K. The results of the Weil-Felix test are considered to give presumptive diagnoses of rickettsial diseases (Table 16.9). Febrile agglutinins, which are anti-

bodies produced in response to various fever-producing bacteria, can similarly be detected by agglutination tests. The Widal test uses antigens from *Salmonella* species for performing these tests, permitting the diagnosis of typhoid and paratyphoid fevers.

Atypical cases of pneumonia caused by *Mycoplasma* can also be detected by several different types of agglutination tests. In one test procedure

table 16.9

Some typical Weil-Felix reactions

Causative organism	Agglutination with protein strain			Diseases
	OX-19	OX-2	OX-K	
Rickettsia prowazekii	+ + + +	+	−	Epidemic typhus
R. mooseri	+ + + +	+	−	Murine typhus
R. rickettsii	+/+ + + +	+/+ + + +	−	Rocky Mountain spotted fever
R. tsutsugamushi	−	−	+ +	Scrub typhus
R. akari	−	−	−	Rickettsialpox
Rochalimaea quintana	−	−	−	Trench fever
Coxiella burnetii	−	−	−	Q fever

the patient's own red blood cells are used to test for agglutination. Antibodies produced in response to pneumonia caused by *Mycoplasma pneumoniae* cause hemagglutination at 4°C but not at the normal body temperature of 37°C. The detection of cold agglutinins, antibodies that cause agglutination at refrigerator temperatures but not at body temperatures, can indicate several disease conditions other than atypical pneumonia, but the disease symptoms normally permit definition of the disease etiology. Atypical pneumonia can also be detected in the *Streptococcus* MG agglutination test. In this procedure the heterophile antibodies in the patient's serum, produced in response to the *Mycoplasma* infection, are cross-reactive with a nonhemolytic strain of *Streptococcus* and agglutination of the *Streptococcus* cells can be assayed.

Coagglutination *Streptococcus* species can also be serotyped by using **coagglutination** procedures. This approach is used in Streptex and Phadabac systems. In coagglutination, dead *Staphylococcus* cells are coated with IgG and mixed with the streptococci and anti-streptococcal antibodies. The reaction forms a lattice matrix of agglutinated cells. The larger clumps of cells that are produced by coagglutination provide greater test sensitivity than simple agglutination tests.

Hemagglutination inhibition Some viruses cause **hemagglutination** of red blood cells, a fact used in **hemagglutination inhibition (HI)** tests for determining whether a patient has been exposed to a specific virus. In the HI test the patient's serum is incubated with viral antigens, and the mixture is then added to red blood cells. If the patient's serum contains antibody for the specific virus, the viral antigens are neutralized or inactivated, and

agglutination of the red blood cells is reduced or absent. Agglutination occurs if the patient does not have antibodies to the antigens of the particular virus. The degree of hemagglutination can thus be employed to diagnose viral diseases. Dilutions of the antigen-treated sera are used to determine the end point of hemagglutination inhibition (Figure 16.18), a useful procedure for diagnosing diseases such as rubella and influenza.

Passive agglutination The agglutination test procedures discussed so far depend on the occurrence of antigens on cell surfaces. In cases where the antigen to be detected is soluble and not associated with a cell or other particle, **passive agglutination** tests can be employed. In such tests the antigen is attached to a particle surface before running an agglutination test. The particle may be a cell or, more commonly, a latex bead. Very small quantities of antigen-coated beads are required for these tests.

Passive agglutination using latex beads is rapidly replacing various older procedures and is now employed for detecting *Haemophilus influenzae, Streptococcus pneumoniae, Neisseria meningitidis, Staphylococcus aureus,* and rubella virus (Figure 16.19). Latex beads may also be coated with soluble antibodies and used in agglutination tests. Pneumonia caused by *Streptococcus pneumoniae* can be detected in this manner. Blood serum from patients with pneumococcal pneumonia possesses a protein called C-reactive protein (CRP). The CRP reacts with the C-polysaccharide of the capsular material of *S. pneumoniae.* Anti-CRP attached to latex beads will react with C-reactive protein, causing readily observed agglutination. A positive agglutination test of the patient's serum with anti-CRP-coated latex beads gives presumptive evidence of pneumococcal pneu-

Clinical microbiology: identification of disease-producing microorganisms and determination of antibiotic susceptibility

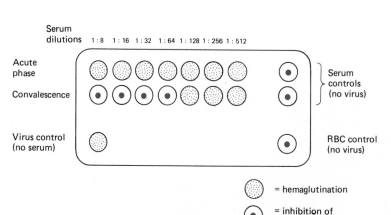

Figure 16.18

This is a diagram of a hemagglutination test for a virus in a Microtiter plate. Titer of virus hemagglutination is 128. This quantitative test can detect acute and convalescent cases of viral diseases such as hepatitis.

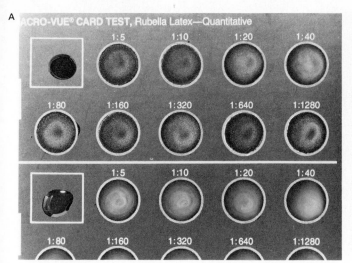

Figure 16.19

Passive latex agglutination tests for the diagnosis of infection. (A) This is a Rubascan card test for the qualitative detection of serum antibodies to rubella virus. Controls and serum specimen are placed onto appropriate circles on the test card, using a micropipet. A drop of the antigen reagent is added to each circle. The card is then rotated on a mechanical rotator at 100 rpm for 8 min under a humidifying cover. The results are then observed under a high-intensity incandescent light. Sera showing any agglutination with the latex antigen are reported as positive. (Courtesy Hynson, Westcott and Dunning, Baltimore, Maryland.) (B) This rapid latex slide agglutination test is for the simultaneous detection of clumping factor and protein A to identify Staphylococcus aureus. *Staphylococci can be differentiated in this procedure, using latex particles coated with plasma. Plasma contains both fibrinogen, required for detecting clumping factor activity, and immunoglobulin G, necessary for detection of protein A. When a specimen culturally resembling* S. aureus *is mixed with the reagent, an agglutination pattern of large visible clumps appears within 45 sec, if it is positive. Uninoculated saline with added reagent is the negative control. (Courtesy Scott Laboratories, Fiskeville, Rhode Island.)*

monia. The inflammatory response is monitored by measuring the CRP and is considered more reliable than other methods.

Passive agglutination can also be used to detect *Treponema pallidum* in the diagnosis of syphilis. The antigens from *T. pallidum* can be conjugated with red blood cells in passive agglutination tests run to detect the presence of antibodies against *T. pallidum* in a patient's serum. Another test used for the diagnosis of syphilis is designed to detect the presence of an IgM antibody produced in individuals infected with *T. pallidum*. This IgM antibody, known as reagin, reacts with phospholipids and can be detected by using cardiolipin, an alcoholic extract of beef hearts. The reaction of reagin with cardiolipin results in an agglutination or flocculation, which can be visualized by using low-power microscopy (Figure 16.20). The VDRL (Venereal Disease Research Laboratory) and

Figure 16.20

Degrees of flocculation reaction for syphilis diagnosis as microscopically observed in the cardiolipin microflocculation test.

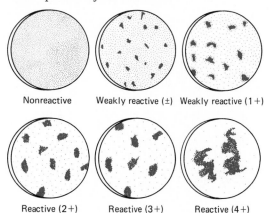

Nonreactive   Weakly reactive (±)   Weakly reactive (1+)

Reactive (2+)   Reactive (3+)   Reactive (4+)

Figure 16.21

Preciptin test in capillary tubes for identifying Lancefield Streptococcus *groups. This classic procedure has been largely replaced by other diagnostic procedures, such as passive agglutination. (Courtesy Leon J. LeBeau, University of Illinois Medical Center.)*

576

Clinical microbiology: identification of disease-producing microorganisms and determination of antibiotic susceptibility

Figure 16.22

Micrograph showing a positive quellung reaction for Streptococcus pneumoniae *(500 ×). (Courtesy Centers for Disease Control, Atlanta.)*

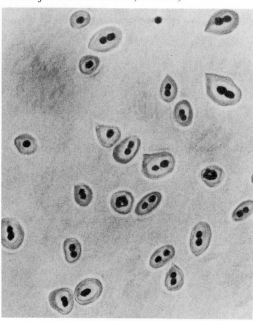

the RPR (rapid plasma reagin) tests, which are conveniently performed through using a prepared kit, are widely used for the detection of syphilitic reagin.

Precipitin test

Precipitin tests are used in serotyping various bacteria and in diagnosing the causative agents of some diseases. The clinically significant *Streptococcus* species, for example, can be separated into Lancefield antigenic groups, using precipitin tests. In this procedure, cells of an unknown β-hemolytic *Streptococcus* species are collected by centrifugation and autoclaved to release the antigens. The cellular debris is removed by centrifugation and the supernatant fluid is added to a capillary tube containing antiserum for a known antigenic *Streptococcus* group. A precipitation reaction indicates that the organism tested is homologous to the indicated *Streptococcus* antiserum group (Figure 16.21).

Neufield quellung test

The **Neufield quellung test**, although no longer routinely used in clinical laboratories, has been employed for identifying *Streptococcus pneumoniae* in body fluids. In this procedure, a specimen, such as cerebrospinal fluid, is mixed with defined pneumococcal antisera, and the mixture is examined microscopically. The quellung reaction results in an increased prominence (swelling) of the capsule, permitting easy recognition of the quellung positive *S. pneumoniae* serotypes (Figure 16.22).

Counter immunoelectrophoresis

Counter immunoelectrophoresis (CIE) is also used in a number of diagnostic procedures. CIE can be useful in detecting the presence of various microorganisms in body fluids such as cerebrospinal fluid, urine, and blood. This procedure depends on a precipitin reaction to identify homology between antigen and antibody, relying on immunodiffusion with electrophoresis driving the antigen and antibody toward each other. CIE is used to detect both antigens and antibodies in body fluids and antigens from microbial cultures. *Streptococcus pneumoniae, Haemophilus influenzae, Neissera meningitidis, Pseudomonas aeurginosa, Klebsiella pneumoniae, Escherichia coli, Staphylococcus aureus, Mycoplasma,* and *Legionella pneumophila* all can be detected by clinically using counter immunoelectrophoresis (Figure 16.23).

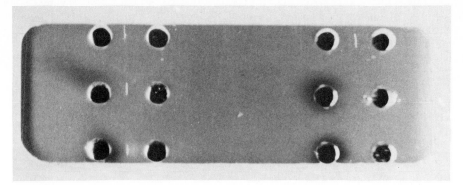

Figure 16.23

Counter immunoelectrophoresis testing is used for the rapid diagnosis of various infections. For example, CIE is used to screen spinal fluid for bacteria as shown in this photograph of actual test results. At left are positive controls, showing precipitation bands between wells filled with antibodies and antigens: top, *Haemophilus* influenzae; *middle,* Neisseria meningitidis; *bottom,* Streptococcus pneumoniae. *At right are reactions with patient's serum, showing a positive test for the presence of* H. influenzae *and negative results for the other two antigens.*

Complement

Measurement of **complement fixation** is another serological approach to diagnosing the etiologic agents of disease. Although it has been replaced with more reliable methods for diagnosing syphilis, the Wassermann test still represents a classic example of a diagnostic complement fixation (CF) test. The complement fixation test employs an indicator system, normally consisting of sheep red blood cells, and homologous antibodies for the red blood cell antigens. The patient's serum is heated to 56°C for 30 minutes to inactivate any available free complement. The serum is then mixed with a known antigen and free complement and the mixture added to the indicator system. Hemolysis of the sheep red blood cells indicates that there is free complement, that an antigen–antibody complement complex did not form when the patient's serum was added to the known antigen and complement. This constitutes a negative test for the diagnosis of the disease caused by the organism from which the known antigen was derived. If, on the other hand, lysis of the sheep red blood cells does not occur, it indicates that complement was fixed in the reaction, and thus, there is presumptive evidence that the patient had the given infection.

In the case of the Wassermann test, it was found that the IgM antibody, reagin, was responsible for fixing complement. Reagin is released from damaged liver cells during syphilis infection, and the test was therefore diagnosing a symptom of the disease rather than specific antibodies that are produced for *Treponema pallidum*. Although no longer used for the diagnosis of syphilis, complement fixation tests are used clinically for the diagnosis of various diseases, including a number of fungal diseases such as histoplasmosis and coccidioidomycosis. Complement fixation tests can also be used for the detection of viral and rickettsial antigens and antibodies.

Enzyme-linked immunosorbent assay

Still another approach to serological diagnosis of disease-causing microorganisms is the **enzyme-linked immunosorbent assay (ELISA)**. This method permits the detection of antibody by combining antigen–antibody with an enzyme conjugant that gives a color-producing reaction that can be read spectrophotometrically (Figure 16.24). The ELISA assays can be used for the detection of antibodies, such as those produced in response to infections with *Salmonella, Yersinia, Brucella, Rickettsia, Treponema, Legionella, Mycobacterium,* and *Streptococcus* species. The ELISA assay can also be used to detect the toxins (antigens) produced by *Vibrio cholerae, Escherichia coli,* and *Staphylococcus aureus.*

Radioimmunoassays

Radioimmunoassays (RIA) are also applicable to the detection of pathogenic bacteria. In this approach a radioisotopic label, rather than a color reaction, is used to assay the extent of anti-

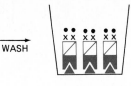

| Antigen ▲ is adsorbed to solid phase. | Serum containing specific antibody ⋈ is added. | Anti-species antibody-enzyme conjugate ⊠ˣˣ is added. | Substrate ·· is added to the sandwich and following incubation a color reaction occurs. |

A Illustration of antibody detection by enzyme-linked immunosorbent assay.

| Specific antibody ■ is bound to the solid phase.. | Antigen-containing specimen is added ▲ . | Enzyme conjugated specific antibody ⊠ˣˣ is added. | Enzyme substrate ∞ is added and following incubation a color reaction occurs. |

B Double antibody sandwich technique for antigen detection by enzyme-linked immunosorbent assay.

578

Clinical microbiology: identification of disease-producing microorganisms and determination of antibiotic susceptibility

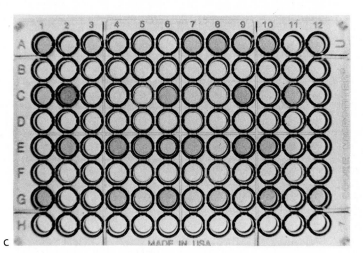

C

Figure 16.24

ELISA (enzyme-linked immunoassay) procedures are used for the detection of various pathogens. (A) Antibody detection by enzyme-linked immunosorbent assay. (B) Double antibody sandwich technique for antigen detection by enzyme-linked immunosorbent assay. (C) Results of ELISA procedure for the detection of Rubella *virus.* Rubella *antigen is attached to the surfaces of the wells of a plastic microtiter plate. The patient's serum is added to the well. Specific antibody, if present in the patient's serum, binds to the attached antigen. Enzyme-conjugated antihuman IgG is added next. It binds to the antibody–antigen complex. Then the enzyme substrate is added and is hydrolyzed by the bound enzyme conjugate. After a specified time the enzyme substrate reaction is stopped with sodium hydroxide. The absorbance of the hydrolyzed substrate is measured at 405 nm with a spectrophotometer and is directly proportional to the amount of antibody in patient's serum. In this photograph the darker the well, the greater the amount of antibody in the patient's serum. (Courtesy Karen Cost, Norton's Hospital, Louisville.)*

gen–antibody reaction. The RIA assays are extremely sensitive. RIA assays can be carried out for the detection of *Staphylococcus* enterotoxins, *Clostridium botulinum* toxin, *Neiserria meningitidis, Haemophilus influenzae, Pseudomonas aeuruginosa*, and *Escherichia coli.*

Immunofluorescence

Direct fluorescent antibody staining (FAB) uses a defined antibody conjugated with a fluorescent dye to stain the unknown microorganisms. In this procedure microorganisms are stained only if the antibody reacts with the surface antigens of the microorganisms on the slide. Viewing with a fluorescence microscope permits the specific detection of microorganisms that react with the fluorescent antibody stain. In this procedure specific pathogenic microorganisms can be identified even in the presence of other microorganisms. The fluorescence antibody staining method is useful for

Figure 16.25

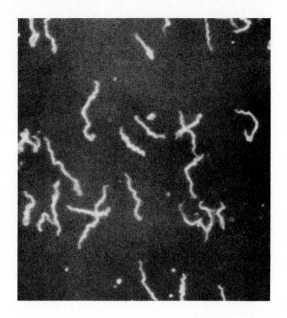

both the identification of clinical isolates and for detecting pathogens, such as *Treponema pallidum* (Figure 16.25) or *Legionella pneumophila*, within exudates from infected tissues (see color plate 6D).

Indirect immunofluorescence tests are also employed in identifying bacteria, such as *T. pallidum*. In the fluorescent test for treponemal antibodies (FTA), dead *T. pallidum* cells are fixed to a slide, and a patient's serum is added. The slide is washed and a fluorescent anti-immunoglobulin is added. If the antibodies in the patient's serum react with the *T. pallidum* on the slide, the bacteria will be stained (positive test), whereas if no reaction occurs, the fluorescent antibody is washed off the slide, and no fluorescing bacteria are visible (negative test).

Antimicrobial susceptibility testing

In addition to identifying the etiologic agents of infectious diseases, the clinical microbiology laboratory is expected to determine the susceptibility of disease-causing microorganisms to antimicrobial agents. The determination of the **antimicrobial susceptibility** of a pathogen is important in aiding the clinician to select the most appropriate agent for treating that disease. It does no good to prescribe an antibiotic that is ineffective against the microorganism causing the disease. Physicians want to avoid indiscriminate administration of antibiotics because the selective pressures of excessive antibiotic usage can induce problematical antibiotic-resistant strains of pathogenic microorganisms. The clinical microbiology laboratory provides information, through standardized *in vitro* testing, with regard to the activities of antimicrobial agents against microorganisms that have been isolated and identified as the probable etiologic agents of disease. Antibiotic susceptibility testing, relying on the observation of antibiotics inhibiting the growth and/or killing cultures of microorganisms *in vitro*, provides the physician with the information needed to prescribe the proper antibiotics for treating infectious diseases.

Agar diffusion methods

The qualitative susceptibility of microorganisms to antimicrobial agents can be determined on agar plates by using filter paper disks impregnated with antimicrobial agents. The **Bauer-Kirby test** system is a standardized antimicrobial susceptibility procedure in which a culture is inoculated onto the surface of Mueller-Hinton agar, followed by the addition of antibiotic-impregnated disks to the agar surface. In **agar diffusion methods** for determining antibiotic susceptibility, the antibiotics diffuse into the agar, establishing a concentration gradient. Inhibition of microbial growth is indicated by a clear zone (zone of inhibition) around the antibiotic disk (Figure 16.26). The diameter of the zone of inhibition reflects the solubility properties of the particular antibiotic—that is, the concentration gradient established by diffusion of the antibiotic into the agar—and the sensitivity of the given microorganism to the specific antibiotic. Standardized zones for each antibiotic disk have been established to determine whether the microorganism is sensitive (S), intermediately sensitive (I), or resistant (R) to the particular antibiotic (Table 16.10). The results of Bauer-Kirby

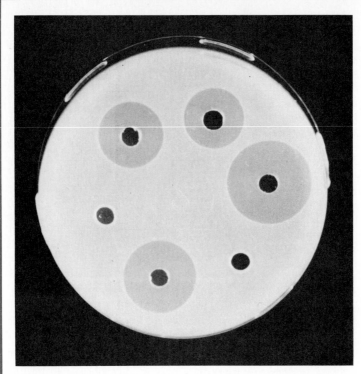

Figure 16.26

Zone of inhibition around an antibiotic disk on an agar plate. This is an antibody sensitivity testing plate showing sensitivity of a bacterial isolate to various antibiotics. (Courtesy Bernard Abbott, Eli Lilly Research Laboratories, Indianapolis.)

580

Clinical microbiology: identification of disease-producing micro-organisms and determination of antibiotic susceptibility

Discovery process

Before World War II penicillin production was limited and extremely expensive, but during and after the war it became necessary to develop some means for predicting when the use of penicillin was the appropriate treatment for an individual with an infectious disease. As soon as antibiotics became commercially available to medical practitioners, the need for antimicrobial susceptibility testing immediately became apparent. The first method for performing antibiotic susceptibility testing was developed by Alexander Fleming. In Fleming's "ditch plate" technique a strip of agar in the form of a ditch is removed from a petri plate and replaced with medium containing the mold extract penicillin (see Figure 15.9). Multiple streak inocula of the organisms to be tested are made at right angles to the ditch. Strains that grow up to the ditch are resistant to penicillin; species that do not grow in the region adjacent to the ditch are sensitive. As numerous new antibiotics were discovered and became available for use by physicians, more rapid and reliable methods were needed for determining which antibiotic to use in the treatment of a particular patient. In 1943 J. W. Foster and H. B. Woodruff first reported the use of antibiotic-impregnated filter paper strips on agar plates inoculated with the test organism (see Figure 15.9). Zones of inhibition occur around strips impregnated with antibiotics to which the bacterial strain is sensitive. J. G. Vincent and H. W. Vincent improved the technique that same year by introducing antibiotic-impregnated paper disks, thus increasing the number of antibiotics that could be tested simultaneously against a bacterial isolate in one petri dish. In 1944 D. C. Morely devised a method for drying the paper disks after soaking them in an antibiotic solution. These early methods for testing antibiotic susceptibility were not designed to produce quantitative results. At the end of the 1950s, the status of antimicrobial susceptibility testing was marked by a lack of an acceptable standardized procedure that led to variable results depending on which laboratory performed the tests. In response to this situation, a World Health Organization committee was formed to evaluate antibiotic susceptibility testing; the decisions of this WHO committee formed the groundwork for the development of the Kirby-Bauer standardized technique. In 1961 T. G. Anderson developed the steps that Bauer and Kirby incorporated into a standardized method. In the Kirby-Bauer procedure the following items are standardized: the amount of each antibiotic impregnated into the disks used in the test procedure; the composition of the test medium; the nature of the inoculum; the incubation time; and the measurement of the size of the zone of inhibition needed to establish antibiotic susceptibility. The Kirby-Bauer antibiotic susceptibility test procedure is the widely accepted standard test employed in most clinical laboratories.

table 16.10

Interpretation of zones of inhibition for Bauer-Kirby antibiotic susceptibility testing

Antibiotic	Disk conc.	Inhibition zone diameter (mm)		
		Resistant	Intermediate	Susceptible
Amikacin	0.01 mg	13 or less	12–13	14 or more
Ampicillin	0.01 mg	11 or less	12–13	14 or more
Bacitracin	10 units	8 or less	9–11	13 or more
Cephalothin	0.03 mg	14 or less	15–17	18 or more
Chloramphenicol	0.03 mg	12 or less	13–17	18 or more
Erythromycin	0.015 mg	13 or less	14–17	18 or more
Gentamicin	0.01 mg			13 or more
Kanamycin	0.03 mg	13 or less	14–17	18 or more
Lincomycin	0.002 mg	9 or less	10–14	15 or more
Methicillin	0.005 mg	9 or less	10–13	14 or more
Nalidixic acid	0.03 mg	13 or less	14–18	19 or more
Neomycin	0.03 mg	12 or less	13–16	17 or more
Nitrofurantain	0.3 mg	14 or less	15–16	17 or more
Penicillin G Staphylococci	10 units	20 or less	21–28	29 or more
Penicillin Other organisms	10 units	11 or less	12–21	22 or more
Polymyxin	300 units	8 or less	9–11	12 or more
Streptomycin	0.01 mg	11 or less	12–14	15 or more
Sulfonamides	0.3 mg	12 or less	13–16	17 or more
Tetracycline	0.03 mg	14 or less	15–18	19 or more
Vancomycin	0.03 mg	9 or less	10–11	12 or more

testing indicate whether a particular antibiotic has the potential for effective control of an infection caused by a particular pathogen.

The Bauer-Kirby agar diffusion test procedure is designed for use with rapidly growing bacteria. This method is not directly applicable for use with filamentous fungi, anaerobes, or slow-growing bacteria, although modifications of the media composition and incubation conditions can be made for testing the antibiotic susceptibility of such microorganisms. Different standardized systems are used for performing antibiotic sensitivity testing in these cases. For example, prereduced Wilkins-Chalgren agar, anaerobic transfer techniques, and anaerobic incubation can be used for determining the antibiotic sensitivities of anaerobic bacteria.

Liquid diffusion methods

Besides Bauer-Kirby and other agar diffusion testing, many clinical laboratories use the Autobac system for measuring microbial growth, based on light-scattering or equivalent automated liquid diffusion methods for antibiotic sensitivity testing. This system uses Eugonic broth and/or low-thymidine broth culture and standardized antibiotic disks, but the antibiotics in the disks elute into the broth rather than into agar, as in the Bauer-Kirby procedure. The concentrations of the antibiotics and the density and growth phase of the cultures are adjusted so that uniform interpretive guidelines can be used for assessing antibiotic sensitivities. A normalized light-scattering index is generated to determine S, I, and R; the light-scattering index for resistance is 0.00–0.50, intermediate sensitivity 0.51–0.60, and susceptible 0.60–1.00, except for penicillin G, where the index for resistance is 0.00–0.90. Other automated systems are also available for performing this procedure, including the Microscan, BBL Sceptre, Vitek AMS, and Abbott MSII. These automated systems simplify and enhance the reliability of antimicrobial susceptibility testing.

Minimum inhibitory concentration procedures

Another approach to antimicrobial susceptibility testing is to determine the **minimum inhibitory concentration (MIC)** using tube dilution procedures. This procedure determines the concentration of an antibiotic that is effective in preventing the growth of the pathogen and gives an indication of the dosage of that antibiotic that should be effective in controlling the infection in the pa-

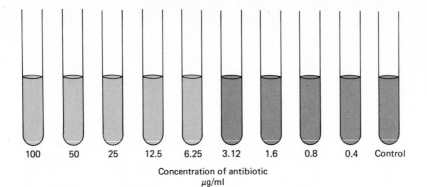

100 50 25 12.5 6.25 3.12 1.6 0.8 0.4 Control

Concentration of antibiotic
μg/ml

Figure 16.27

Tube dilution procedure for determining break point in MIC (minimum inhibitory concentration) determination. The minimal inhibitory concentration for the test illustrated here is 6.25 μg/ml. As compared with the Bauer-Kirby procedure, the MIC provides quantitative information that can be used to determine the dose rate and administration schedule for an antibiotic.

582

Clinical microbiology: identification of disease-producing microorganisms and determination of antibiotic susceptibility

tient. A standardized microbial inoculum is added to tubes containing serial dilutions of an antibiotic, and the growth of the microorganism is monitored as a change in turbidity. In this way, the break point or minimum inhibitory concentration of the antibiotic that prevents growth of the microorganism *in vitro* can be determined (Figure 16.27). The MIC may also be determined by using a single concentration of an antibiotic by comparing the rate of growth of the microorganisms in control and antibiotic-treated tubes. The MIC indicates the minimal concentration of the antibiotic that must be achieved at the site of infection to inhibit the growth of the microorganism being tested. By knowing the MIC and the theoretical levels of the antibiotic that may be achieved in body fluids, such as blood and urine, the physician can select the appropriate anti-

table 16.11

Achievable levels of some common antibiotics in various body fluids

Antibiotic	Achievable peak blood levels (μg/ml)	Achievable urine levels (μg/ml)	Dose/6–12 hours
Clindamycin	1–4	>20	Oral 150–300 mg
	6–10	>60	IV 300–600 mg
Erythromycin	1–2		Oral 250–500 mg
	10–20		IV 300 mg
Penicillin	2–3	>300	Oral 500 mg
	6–8	>300	IM 1×10^6 units
	4–7	>300	IV 500 mg
Ampicillin	1–3	>50	Oral 250–500 mg
	2–6	>20	IM 250–500 mg
	10–25	>100	IV 1–1.5 g
Cephalothin	3–18	>300	Oral 250–500 mg
	9–24	>1000	IM 0.5–1 g
	30–85	>1000	IV 1–2 g
Gentamicin	2–10	>20	IM/IV 1–2 mg
Tetracycline	1–2	>200	Oral 250–500 mg
	10–20	>200	IV 500 mg
Chloramphenicol	10–12	>100	Oral 1 g
	20–30	>200	IV 1 g
Nitrofurantoin		>100	Oral 50–100 mg

IV = intravenous; IM = intramuscular.

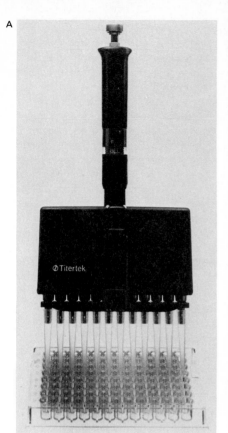 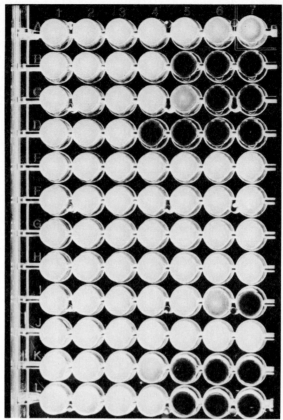

Figure 16.28

Test plates used for automated MIC determination. (A) This Titertek automatically fills the microtiter plates. (Courtesy Flow Laboratories, McLean, Virginia.) (B) In this photograph the results of MIC testing are shown. Lack of growth, i.e., inhibition, appears as clear wells. (From BPS—Leon J. LeBeau, University of Illinois Medical Center, Chicago.)

biotic, the dosage schedule, and the route of administration (Table 16.11). Generally, a margin of safety of ten times the MIC is desirable to ensure successful treatment of the disease.

MIC determinations can be used for determining the antibiotic sensitivity of both aerobic and anaerobic microorganisms. The use of microtiter plates and automated inoculation and reading systems makes the determination of MIC's feasible for use in the clinical laboratory (Figure 16.28). MIC's can even be performed on normally sterile body fluids without isolating and identifying the pathogenic microorganisms. For example, blood or cerebrospinal fluid containing an infecting microorganism can be added to tubes containing various dilutions of an antibiotic and a suitable growth medium. An increase in turbidity would indicate the growth of microorganisms and the fact that the antibiotic at that concentration was ineffective in inhibiting microbial growth, whereas

a lack of growth would indicate that the pathogenic microorganisms were susceptible to the antibiotic at the given concentration. By determining the minimum inhibitory concentration, the appropriate dosage as well as the right antibiotic can be selected for treating an infectious disease.

Minimal bactericidal concentration

MIC determinations are aimed at determining the concentration of an antibiotic that will inhibit growth but are not designed to determine whether the antibiotic is microcidal. It is, however, also possible to determine the **minimal bactericidal concentration (MBC)**. The minimal bactericidal concentration, also known as the minimal lethal concentration (MLC), refers to the lowest concentration of an antibiotic that will kill a defined proportion of viable organisms in a bacterial suspension during a specified exposure period. Generally, a 99.9 percent kill of bacteria at an initial concentration of 10^5–10^6/ml during an 18–24-hour exposure period is used to define the MBC. In order to determine the minimal bactericidal concentration, it is necessary to plate the tube suspensions showing no growth in tube dilution (MIC) tests to determine whether the bacteria are indeed killed or whether they survive exposure to the antibiotic at the concentration being tested. Although determination of the MIC is adequate for establishing the appropriate concentration of an antibiotic that should be administered for controlling the infection in patients with normal immune response levels, the MBC is essential in cases of endocarditis and is particularly useful in determining the appropriate concentration of an antibiotic for use in treating patients with lowered immune defense responses, such as in patients receiving chemotherapy treatment for cancer.

Serum killing power

Another approach to determining the effectiveness of an antibiotic is to measure **serum killing power (Schlicter test)**. Instead of adding dilutions of an antibiotic to suspensions of bacteria in a growth medium, a bacterial suspension is added to dilutions of the patient's serum. The ability of the bacteria to grow in the patient's blood is assessed by measuring changes in turbidity. Assuming that the patient is being treated with an antibiotic, no bacterial growth should occur. The break point in the dilutions where bacterial growth occurs reflects the concentration of the antibiotic in the patient's blood and the *in vivo* effectiveness of the antibiotic in controlling the infection. Inhibition at dilutions of the patient's serum of greater than or equal to 1:8 is considered as an acceptable level.

584

Clinical microbiology: identification of disease-producing microorganisms and determination of antibiotic susceptibility

Postlude

It has not been possible in this chapter to consider all the body tissues and fluids that may be examined in the clinical laboratory for evidence of microbial infection or to consider the variety of techniques that may be employed for the isolation and identification of many specific microbial pathogens. The general procedures for diagnosing microbial infectious diseases have been discussed, including various methods for the isolation and identification of microbial pathogens. The clinical identification of the etiologic agents of disease relies on both biochemical and serological testing. The specific culture methods and identification systems that are employed depend on the preliminary diagnosis of the disease and the microorganisms that are presumed to be the causative agents of the disease based on preliminary screening procedures.

The procedures used in clinical microbiology and immunology laboratories are subject to change as new and better screening procedures aimed at the isolation of pathogenic microorganisms are developed. Speed and accuracy in determining the nature of the disease, the disease etiology, and the appropriate antimicrobial agents that may be effective in controlling the infection are of the essence because an error or delay in diagnosing a disease can prove fatal. Automation has greatly improved speed and reliability in the clinical laboratory, and pathogenic microorganisms can now be isolated and identified within hours. The development of various serological procedures has greatly improved the accuracy and speed of many diagnoses. Similarly, antibiotic susceptibility testing can be accomplished rapidly. At present, hospital clinical laboratories attempt to provide a complete diagnostic report, identifying the pathogen and its antibiotic sensitivities, within 24 hours. In the future, serological testing may permit diagnosis of many diseases at bedside, and improved antibiotic susceptibility testing may permit *in vivo* monitoring of the effectiveness of antibiotic treatment. Indeed, it has been predicted that by the year 2000 virtually all diagnoses will be made at bedside by using serological methods, providing the physician with immediate information needed to initiate proper therapy.

Study Questions

1. How are differential blood counts used to determine if a disease is caused by a microorganism?

2. How are microorganisms isolated from the lower respiratory tract to identify pathogens? What precautions must be taken to ensure the origin of the culture?

3. How are miniaturized identification systems used in the clinical microbiology laboratory? Describe the tests used in three different systems.

4. Why is it important to quickly and accurately identify cultures sent to the clinical microbiology laboratory?

5. What are serological tests? How are they used in the identification of pathogens?

6. How is immunofluorescence used to identify *Treponema pallidum*? Why are serological methods critical for identifying this pathogenic bacterium?

7. Describe the hemagglutination test.

8. What is a cytopathic effect? How is it used in the diagnosis of infectious mononucleosis?

9. Why is it essential to perform antimicrobial susceptibility testing on pathogenic isolates?

10. How do we go about determining the proper antibiotic for treating a disease? Describe several approaches to determine the sensitivity of a pathogen to antibiotics. Is antimicrobial sensitivity the sole criterion for selecting an antibiotic? Discuss.

11. What is an MIC? Why is this an increasingly common test in clinical microbiology laboratories?

Suggested Supplementary Readings

Bailey, W. R., and E. G. Scott. 1975. *Diagnostic Microbiology*. C. V. Mosby Co., St. Louis.

Davidsohn, I., and J. B. Henry (eds.). 1974. *Clinical Diagnosis by Laboratory Methods*. W. B. Saunders, Philadelphia.

Fudenberg, H. H., D. P. Sites, J. L. Caldwell, and J. V. Wells (eds.). 1980. *Basic and Clinical Immunology*. Lange Medical Publications, Los Altos, California.

Koneman, E. W., S. D. Allen, V. R. Dowell, Jr., and H. M. Sommers (eds.). 1979. *Color Atlas and Textbook of Diagnostic Microbiology*. J. B. Lippincott Company, Philadelphia.

Lennette, E. H., A. Balows, W. J. Hausler, Jr., and J. P. Truant (eds.). 1980. *Manual of Clinical Microbiology*. American Society for Microbiology, Washington, D.C.

Milgrom, F., C. J. Abeyounis, and K. Kano (eds.). 1981. *Principles of Immunological Diagnosis in Medicine*. Lea & Febiger, Philadelphia.

Ravel, R. 1978. *Clinical Laboratory Medicine: Clinical Application of Laboratory Data*. Year Book Medical Publishers, Chicago.

Rose, N. R., and H. Friedman (eds.). 1980. *Manual of Clinical Immunology*. American Society for Microbiology, Washington, D.C.

Rytel, M. J. 1979. *Rapid Diagnosis in Infectious Diseases*. CRC Press, Boca Raton, Florida.

Schneierson, S. S. 1965. *Atlas of Diagnostic Microbiology*. Abbott Laboratories, North Chicago, Illinois.

Stansfield, W. D. 1981. *Serology and Immunology*. Macmillan Publishing Co., New York.

Human diseases caused by microorganisms: the respiratory, gastrointestinal, and genitourinary tracts as portals of entry

17

586

Human diseases caused by microorganisms: the respiratory, gastrointestinal, and genitourinary tracts as portals of entry

Despite the extensive defense mechanisms of the human body and the preventive measures that can be taken to avoid infections with microbial pathogens, humans remain susceptible to a variety of infectious diseases. The dynamics of the interactions between microorganisms and human beings precludes the total elimination of all, or even most, **infectious diseases**. Our increased understanding of the interrelationships between pathogenic microorganisms and their human hosts, however, has resulted in the reduced incidence of many infectious diseases, better management of sick individuals, and decreased mortality from many once deadly diseases. In this and the following chapter we will consider some of the diseases of humans caused by microorganisms, with emphasis on the epidemiology of the disease, considering the characteristics of various diseases including the interrelationships between pathogenic microorganisms, human host populations, and disease symptomology.

The discussion of human diseases caused by microorganisms is organized according to the **portal of entry** of the pathogen because this approach focuses on the route of transmission and the factors governing the spread of disease. The portal of entry indicates the tissue through which the pathogen is able to penetrate the surface protective barriers of the body. In many cases pathogenic microorganisms establish localized infections in the region of the portal of entry, but in other cases pathogens are able to spread systemically and establish infections involving other body tissues. Many pathogens possess adaptive features, known as **invasive properties**, that permit them to penetrate the body's defense mechanisms through a particular portal of entry. Most pathogens can establish an infection through only a single portal of entry, although some can cause disease through several different routes of entry. The normal body openings that serve as portals of entry for pathogenic microorganisms are the respiratory, gastrointestinal, and genitourinary tracts, and the diseases caused by pathogens that enter via these routes will be discussed in this chapter. Additionally, in Chapter 18, we will consider those infective processes where pathogens establish infections in surface areas of the body, such as the skin and eye, or enter the body when the normal surface barriers are physically disrupted as the result of surgery, wounds, or animal bites.

Diseases caused by pathogens that enter via the respiratory tract

It should not be surprising, as we inhale 10,000–20,000 liters of air per day containing microorganisms, that **the respiratory tract provides a portal of entry for many human pathogens**. Potential pathogens freely enter the respiratory tract through the normal inhalation of air and when substances are placed in the mouth. Various viruses, bacteria, and fungi are able to multiply within the tissues of the respiratory tract, sometimes causing localized infections and sometimes entering the circulatory system through the numerous blood vessels associated with the respiratory tract and spreading through the bloodstream to other sites in the body. In order to establish an infection via the respiratory tract, a pathogen must overcome the natural immunological defense mechanisms that are particularly extensive in the lower respiratory tract where there are numerous phagocytic cells. The microorganisms generally establishing infections via the upper respiratory tract are different from those pathogens that are able to move past the cilia and mucus secretions designed to restrict the movement of particles, including microorganisms, to the regions of the lower respiratory tract. Several factors contribute to the fact that the respiratory tract is a major portal of entry for pathogenic microorganisms. The upper respiratory tract is in continuous contact with air that contains many microorganisms. The respiratory system has a very large surface area to facilitate gas exchange and there is a high degree of interfacing between the respiratory tract and circulatory system to permit reoxygenation of blood in the lower respiratory tract. Thus, the potential for respiratory infection is great.

Transmission through the air is undoubtedly the main route of transmission of pathogens that enter via the respiratory tract. **Airborne transmission** often occurs when droplets containing pathogenic microorganisms are transferred from an infected to a susceptible individual (Figure 17.1). Droplets regularly become airborne during normal breathing, but the coughing and sneezing associated with respiratory tract infections are responsible for the spread of pathogens in aerosols and thus the airborne transmission of disease. As discussed in Chapter 15, the incidence of these diseases can be reduced by covering one's nose and mouth while coughing and sneezing, and avoiding contact with contagious individuals, a practice we are taught to follow at an early age.

Diseases caused by viral pathogens

The common cold

The **common cold** is the most frequent infectious human disease, and it is safe to assume that we all have had a cold at some time. More than 200 million work and school days are lost each year in the United States alone because of the common cold. The common cold actually refers to a cluster of diseases primarily caused by viruses, characterized by a localized inflammation of the upper respiratory tract and the release of mucus secretions and generally accompanied by sneezing and sometimes coughing. The etiologic agents re-

Figure 17.1

Routes of transmission of pathogens to the respiratory tract. Close contact with an individual emitting pathogens from the respiratory tract increases the probability of successful transmission.

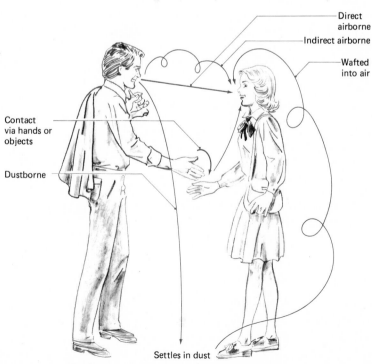

Contact via hands or objects

Dustborne

Direct airborne

Indirect airborne

Wafted into air

Settles in dust

sponsible for the majority of cases of the common cold have yet to be identified, but we do know that in adults approximately 25 percent of all colds are caused by **rhinoviruses**, while only about 10 percent of colds in children are caused by rhinoviruses. Rhinoviruses are members of the Picornaviridae and are distinguished from other members of this family, which includes poliovirus, by their inactivation at low pH and ability to maintain infectivity at 50°C. There are at least 90 immunologically distinct types of rhinoviruses capable of causing the common cold, and hence, it is not surprising that immunity does not offer continuous protection against all of the antigenically distinct viruses capable of causing the common cold.

Viruses causing the common cold infect the cells lining the nasal passages and pharynx, producing an inflammatory response with associated tissue damage in the infected region. The initial viral infection can be followed by a secondary bacterial infection as the normal microbiota of the upper respiratory tract invade the damaged tissues. Symptoms during the course of the common cold include nasal stuffiness, sneezing, coughing, headache, malaise (a vague feeling of discomfort), sore throat, and sometimes a slight fever. There is no specific treatment for the common cold, which is a self-limiting clinical syndrome. Recovery usually occurs within one week without complications as a result of the natural immune defense response. As with many other respiratory diseases, colds occur primarily during the winter months, in part because of the physiological stress posed by exposure to cold temperatures, and in part because of increased contact of individuals during indoor winter activities.

Influenza

Compared to the common cold, **influenza** often is a more serious and debilitating disease. Three antigenic types of *Myxovirus influenzae*, designated A, B, and C, are the causative agents of influenza. Influenza is transmitted by inhalation of droplets containing *M. influenzae*, released into the air as droplets originating from the respiratory tracts of infected individuals. Outbreaks of influenza are cyclical, with major outbreaks caused by type A virus occurring every two to four years, those caused by type B virus every four to six years, and outbreaks caused by type C virus occurring only rarely. Within each of the major types of influenza viruses there are various antigenic subtypes that are responsible for different outbreaks of influenza (Figure 17.2). Minor antigenic changes occur because of mutations, but the ma-

jor antigenic shifts probably result from recombination, that is, from gene reassortment between an animal and a human strain. Such antigenic shifts have been experimentally demonstrated. The antigenic shift, resulting from the addition of new genes, produces new strains of influenza virus. Specific strains of influenza virus are often designated according to the location where outbreaks of the disease associated with new varieties of the influenza virus showing major antigenic changes are first detected. For example, the Hong Kong strain of influenza A virus was first seen in the Orient in 1968 and the London strain of influenza A virus appeared in England in 1972.

The disease influenza is characterized by the sudden onset of a fever, with temperatures abruptly reaching 102–104°F approximately one to three days after actual exposure and onset of infection. The development of the disease is further characterized by malaise, headache, and muscle ache. In uncomplicated cases of influenza, the viral infection is self-limiting and recovery occurs within a week. However, infection with influenza virus can lead to complications, such as a secondary bacterial infection, causing pneumonia. Complications associated with influenza infections are prevalent among the elderly and individuals with compromised host defense responses. Such individuals should be immunized against the prevalent strain of influenza virus prior to the outbreak of influenza epidemics because complications can result in death.

One serious complication associated with outbreaks of influenza is the development of **Reye's syndrome**, an acute pathological condition affecting the central nervous system. Reye's syndrome also occurs after infections with other viruses, and the specific relationship to influenza virus is not clear. Occurrences of Reye's syndrome are highly, but inexplicably, correlated with outbreaks of influenza B virus. Reye's syndrome principally is associated with children. For reasons that have yet to be elucidated, there is a high incidence of Reye's syndrome when aspirin is used for treating the symptoms of a viral infection. Consequently, some pediatricians have warned against the use of aspirin for children with influenza and other viral infections of the respiratory tract. Because a cause-and-effect relationship to the use of aspirin during viral infections has not been established, the issuance of this warning, which could be required on bottles of aspirin, is very controversial.

Another condition involving the central nervous system associated with influenza infections is **Guillain-Barré syndrome**. In 1976 there was an increased incidence of this syndrome associated

588

Human diseases caused by microorganisms: the respiratory, gastrointestinal, and genitourinary tracts as portals of entry

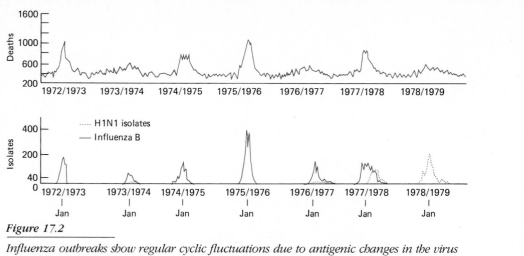

Figure 17.2

Influenza outbreaks show regular cyclic fluctuations due to antigenic changes in the virus and associated variations in the susceptibility of the population. The highest incidence of influenza occurs during the winter, as shown in this graph of reported deaths in 121 U.S. cities compared with influenza isolates reported to WHO (World Health Organization) Collaborating Centers in the United States from September 1972 to August 1979.

Discovery process

Outbreaks of influenza occur each year. In some years the severity and incidence of influenza are greater than others. It has been discovered that major outbreaks of influenza, which easily can become epidemic, occur when new strains of influenza viruses arise through genetic modification and when a large proportion of the human population lacks immunity to that particular strain. Epidemics depend on the emergence of these new strains. Influenza viruses display a marked tendency to mutate, and so new variants of the virus are constantly arising against which the majority of people do not possess the required immune mechanisms. Therefore, epidemics are always a threat. Epidemics of influenza are usually short and subside quickly, followed several weeks later by a secondary wave, a free period, and then a tertiary wave. In secondary and tertiary outbreaks the number of people attacked is less than in the primary outbreak, but the disease tends to be more severe, complications are more common, and the mortality rate is higher. By detecting the primary outbreak and identifying the particular strain of influenza virus responsible for that outbreak, steps can be taken to minimize the extent of the influenza outbreak. For example, the Asian flu epidemic of 1957 was the most severe outbreak of influenza since the great pandemic of 1918–1919, but far fewer people died in 1957 because the 1957 outbreak did not come as a surprise. A new and potent strain of type A influenza virus appeared in southern China early in 1957. It reached Amsterdam, Sydney, and San Francisco within weeks, but even before it had spread from the Far East a worldwide network of scientists had been mobilized. Researchers worked in about 30 laboratories to identify the new strain and to trace the pattern of influenza outbreaks. Epidemiologists at World Health Organization centers in the United States and Great Britain organized the information and mapped the worldwide course of the disease. The early detection of this outbreak of a new form of influenza, Asian flu, allowed time for the preparation of an effective vaccine. Within 10 months of detecting the first cases of Asian flu, pharmaceutical manufacturers had produced 10 million doses of vaccine. The use of the vaccine to immunize large numbers of individuals who otherwise would have been susceptible to influenza and the availability of antibiotics to combat secondary bacterial infections minimized the mortality due to this serious health threat. The Surgeon General of the United States announced that "for the first time in history the medical community was ahead of an impending epidemic." The ability to predict impending influenza outbreaks, to identify the causative viral strain, and rapidly to produce effective vaccines allows us to minimize the effects of outbreaks of influenza.

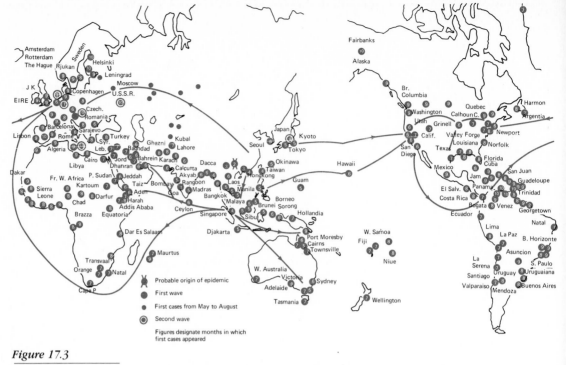

Amsterdam
Rotterdam
The Hague
Rjukan
Helsinki
Leningrad
Moscow
J K
Copenhagen
U.S.S.R.
EIRE
Czech.
Romania
Barcelona
Sarajevo
Lisbon
Rome
Turkey
Algeria
Leb.
Jord
Ghazni
Kubal
Cairo
Bahrein
Karach
Dhanran
Libya
P. Sudan
Jeddah
Dakar
Fr. W. Africa
Kartoum
Taiz
Bombay
Sierra
Leone
Darfur
Aden
Chad
Harah
Addis Ababa
Brazza
Equatoria
Ceylon
Dar Es Salaam
Transvaal
Orange
Maurtus
Natal
Cape P

Fairbanks
Alaska
Br.
Columbia
Washington
Quebec
Calhoun C.
Harmon
Utah
Grinell
Argentia
Valley Forge
Newport
Calif.
Louisiana
Norfolk
San
Diego
Texas
Florida
Cuba
Mexico
San Juan
Jam
Guadeloupe
El Salv.
Panama
Trinidad
Costa Rica
Bogata
Venez
Georgetown
Ecuador
Natal
Lima
La Paz
B. Horizonte
La
Serena
Asuncion
Santiago
Uruguay
S. Paulo
Uruguaiana
Valparaiso
Mendoza
Buenos Aires

Japan
Kyoto
Seoul
Tokyo
Okinawa
Hawaii
Taiwan
Hong Kong
Guam
Borneo
Sorong
Hollandia
Djakarta
Port Moresby
W. Samoa
Cairns
Fiji
Townsville
Niue
W. Australia
Victoria
Adelaide
Sydney
Tasmania
Wellington

Bahdad
Dacca
Calcutta
Akyab
Madras
Rangoon
Bangkok
Laos
Manila
Malaya
Singapore
Sibu
Brunei
Lahore
Goa

Probable origin of epidemic
First wave
First cases from May to August
Second wave
Figures designate months in which
first cases appeared

Figure 17.3

Map showing worldwide spread of influenza during an epidemic outbreak. This epidemic originated in southeast Asia and spread in waves to all other population centers of the world.

Human diseases caused by micro-organisms: the respiratory, gastrointestinal, and genitourinary tracts as portals of entry

with an active immunization program against a predicted outbreak of swine flu, which did not occur. Swine flu is caused by an influenza virus that normally occurs in pigs but which can be transmitted to humans with fatal consequences. The outbreak of Guillain-Barré syndrome appears to have resulted from contamination of some batches of vaccine with viable influenza viruses.

There have been several major or **pandemic** outbreaks of influenza during the nineteenth and twentieth centuries. During 1918–1919, outbreaks of influenza resulted in the deaths of over 20 million people. Many of these deaths may have been the result of secondary infections with *Streptococcus pneumoniae*, rather than from the primary influenza infection. The continued genetic drift with major antigenic shifts in the virus resulting from genetic recombination permits these periodic epidemic outbreaks. The introduction of a virus into a population most of whose members are susceptible to infection establishes the conditions needed for the establishment of an epidemic. Influenza outbreaks spread worldwide via airborne transmission from the site of an initial outbreak with a new strain, and it is possible to watch the disease spread from one area to another (Figure 17.3). Each year epidemiologists

make prognostications about the severity of influenza outbreaks, and public health officials take the necessary steps of immunizing high-risk individuals and warning the public about the dangers of this disease. Even in a nonepidemic year, influenza causes a significant number of deaths; for example, the death rate due to influenza in 1980—a nonepidemic year—in the United States was 0.3/100,000.

The clinical diagnosis of influenza depends on the isolation of an influenza virus by using tissue culture with a cell line such as monkey kidney cells and the serological detection of increased titers of antibody in the patient's serum that are reactive with *Myxovirus* antigens. As with most other viral diseases, treatment of uncomplicated cases of influenza centers on treating the symptoms, with recovery from the disease dependent upon the immune response of the infected individual. The antiviral drug amantadine is effective against influenza A viruses, but only during the incubation period. Once the symptoms of the disease have appeared, amantadine is ineffective against influenza viruses, and thus, this drug is of limited clinical use. Primary control of influenza is achieved by vaccinating individuals who are prone to the complications resulting from this

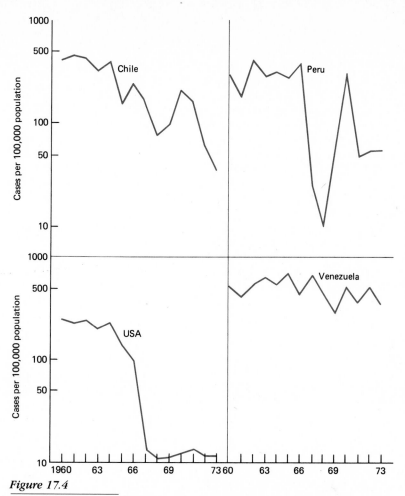

Figure 17.4

Reported cases of measles per 100,000 population in four countries from 1960 to 1973. Although the incidence of this disease has been greatly reduced in the United States through immunization practices, it remains a major problem in other countries.

disease, leaving others unprotected to periodically suffer from influenza.

Measles

Unlike influenza, which occurs at any age, **measles** is a highly contagious disease occurring almost exclusively in children. The measles virus is readily transmitted from an infected child to a susceptible host, as illustrated by the fact that there is greater than a 90 percent incidence of an acute infection after exposure to measles virus by susceptible children. Measles can be prevented by childhood immunization, and the rate of measles infections in the United States, at least, has been declining regularly in recent years, although in other countries measles has not declined in the same way (Figure 17.4). In 1983 there was a major

outbreak of measles at Indiana University because a large proportion of the student body had not been immunized and lacked immunity to measles.

After initial viral multiplication in the mucosal lining of the upper respiratory tract, measle viruses appear to be disseminated to lymphoid tissues where further multiplication occurs. Before the onset of symptoms, large numbers of measles viruses are shed in secretions of the respiratory tract and eye and in urine, promoting the rapid epidemic spread of this disease. Infection with measles viruses can involve a number of organs, and there is a high rate of mortality associated with measles in regions of the world where malnutrition and limited medical treatment facilities predominate. When measles is fatal, the virus generally invades the central nervous system.

Measles is characterized by the eruption of a skin rash approximately 14 days after exposure to the measles virus. The rash generally appears initially behind the ears, spreading rapidly to other areas of the body during the next three days. Disease symptomology often begins a few days before the onset of the characteristic measles rash. These initial symptoms include high fever, coughing, sensitivity to light, and the appearance of **Koplik's spots** (red spots with a white dot in the center that occur in the oral cavity, generally first appearing on the inner lip). Treatment is normally supportive, including rest and the intake of sufficient fluids. In uncomplicated cases, the fever disappears within two days, and the individual returns to normal activities a few days later. If the fever persists for more than two days after the eruption of the rash, it is likely that a complication, such as bronchitis or pneumonia, has developed. In these cases additional treatment is needed to cure secondary infections.

German measles

Like measles, transmission of **rubella virus**, the causative agent of **German measles (rubella)**, appears to be via droplet spread, with the initial infection occurring in the upper respiratory tract. In contrast to the measles virus, however, the rubella virus exhibits a relatively low rate of infectivity, and thus, prolonged exposure appears to be needed for establishment of infection. After multiplication in the mucosal cells of the upper respiratory tract, rubella viruses appear to be disseminated systemically through the blood. Approximately 18 days after initiation of the infection by the rubella virus, a characteristic rash, appearing as flat pink spots, occurs on the face and subsequently spreads to other parts of the

592

Human diseases caused by microorganisms: the respiratory, gastrointestinal, and genitourinary tracts as portals of entry

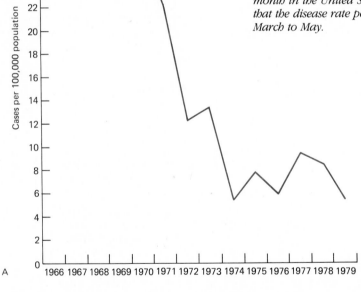

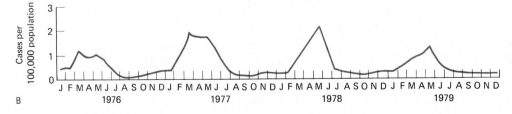

Figure 17.5

Incidence of rubella (German measles) in the United States. (A) Reported case rates by year in the United States from 1966 to 1979, showing the great decline in the incidence of this disease after the introduction of a vaccine for its prevention. (B) Reported case rates by month in the United States from 1976 to 1979, showing that the disease rate peaks in the spring of each year, March to May.

body. Enlarged and tender lymph nodes and a low-grade fever characteristically precede the occurrence of the German measles rash. In children and adolescents, rubella is usually a mild disease. If it is acquired during pregnancy, however, the fetus can become infected with the rubella virus, resulting in congenital rubella syndrome, characterized by the development of multiple abnormalities in the infant. There is a very high rate of mortality, exceeding 25 percent, in cases of congenital rubella syndrome. Vaccination has greatly reduced the incidence of rubella (Figure 17.5) in children and is also used to confer immunity on women of childbearing age who had not contracted the disease at an earlier age.

Mumps

Mumps, another disease primarily occurring in childhood, is caused by a *Paramyxovirus*, an enveloped RNA virus with numerous spikes projecting from the capsid. The ribonucleic acid-protein core comprises a complement-fixing antigen, and the envelope has neuraminidase hemagglutinating and hemolytic activities. The mumps virus is transmitted via contaminated droplets of saliva. The initial infection appears to occur in the upper respiratory tract, with subsequent dissemination to the salivary glands and other organs. Mumps is characterized by the enlargement of one or more of the salivary glands, resulting from viral replication within those glands. Swelling on both sides (**bilateral parotitis**) occurs in about 75 percent of patients, with swelling in one gland usually preceding swelling on the other side by about five days. The average incubation period for mumps is 18 days, and the swollen salivary glands generally persist for less than two weeks. The mumps virus may spread to various body sites, and although the effects of the disease are normally not long-lasting, there can be several complications; for example, mumps is a major cause of deafness in childhood. In males past puberty, the mumps virus can cause **orchitis** (inflammation of the testes), but old wives' tales to the contrary, mumps rarely results in male sterility.

Chicken pox and shingles

Chicken pox (**varicella**) is caused by the **varicella-zoster virus**, a member of the herpesvirus group, which also causes shingles (Herpes zoster). Ninety percent of all cases of chicken pox occur in children under 9 years of age. In children, chicken pox is generally a relatively mild disease, but when this disease occurs in adults the symptoms are characteristically severe. Chicken pox is a highly contagious disease and probably is transmitted via contaminated droplets and direct contact with vesicle fluid containing varicella-zoster viruses. The initial site of viral replication has not been positively established but appears to be in the upper respiratory tract. Local lesions occur in the skin after dissemination of the virus through the body. These skin lesions become encrusted, and the crusts fall off in about one week. Vesicles also occur on mucous membranes, especially in the mouth. In some cases, the varicella-zoster virus spreads to the lower respiratory tract, resulting in pneumonia, and in this way several other tissues, including the central nervous system, can also be involved in complicated cases of chicken pox. Unlike the other childhood viral diseases, vaccination practices have not yet been introduced, and outbreaks of chicken pox continue to show regular seasonal cycles of the same magnitude (Figure 17.6).

In adults, **shingles** is the principal disease resulting from infections with the varicella-zoster virus. Shingles affects individuals who have developed circulating antibodies in response to infection with this virus. It is the varicella-zoster virus acquired in childhood that can remain as a **provirus** within the body; that is, the viral DNA can be incorporated into human chromosomes in a state of lysogeny. The presence of such proviruses can later become manifest as a new disease. In cases of shingles the virus reaches the sensory ganglia of the spinal or cranial nerves, producing an inflammation. There is usually an acute onset of pain and tenderness along the affected sensory nerves. A rash develops along the affected nerves usually lasting for two to four weeks, but the pain may last for weeks or months.

Infectious mononucleosis

Infectious mononucleosis is caused by the **Epstein-Barr (EB) virus**, a member of the herpesvirus group. The EB virus occurs in oropharyngeal secretions of infected individuals, and the virus appears to be transmitted by exchange of oropharyngeal secretions containing the EB virus; although transmission can also occur by droplet spread. Infectious mononucleosis most commonly occurs in young adults 15–25 years of age, a fact that is explained by the exchange of saliva during kissing, a prevalent activity often involving more partners for this age group than others. In the course of the infection, mononuclear white blood cells are affected, leading to the characteristic changes in the white blood cells used in diagnosing the disease. The symptoms of this disease include a sore throat, low-grade fever that generally peaks in the early evening, enlarged and

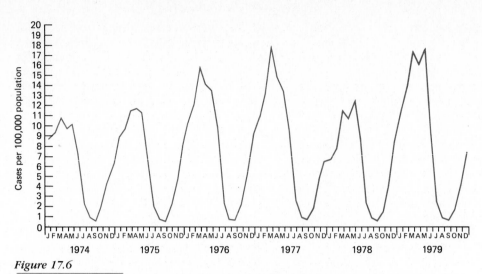

Figure 17.6

Reported case rates of chicken pox by month in the United States from 1974 to 1979. Chicken pox continues to be the second most frequently reported infectious disease in the United States. The seasonal pattern has remained fairly constant, with the peak incidence occurring between March and May. At present there is no licensed vaccine for preventing this disease.

594

Human diseases caused by micro-organisms: the respiratory, gastrointestinal, and genitourinary tracts as portals of entry

tender lymph nodes, general tiredness, and weakness. In most cases of infectious mononucleosis the symptoms are relatively mild and the acute stage of the illness lasts less than three weeks.

Diseases caused by bacterial pathogens

Pneumonia

Pneumonia, an inflammation of the lungs involving the alveoli, can be caused by a number of viral and bacterial agents. Pneumonia is often a complication that occurs when the host defense mechanisms are compromised as a result of other diseases. Frequently, pneumonia is a **nosocomial infection** (an infection acquired while a hospital patient) after surgery or during the course of treatment for another disease, when patients are "run down" and their physiologically impaired state depresses the effectiveness of the immune response system. The lack of movement and deep breathing in post-surgical patients reduce the efficiency of the normal defense mechanisms to clear the lungs of mucus and bacteria, and the accumulation of fluids favors the establishment of a microbial infection. There is a high rate of mortality in cases of pneumonia (Figure 17.7); more than half of all pneumonia cases are caused by bacteria, and this disease ranks among the top causes of death from infectious diseases. Bacteria

that cause pneumonia most frequently enter the lungs in air, although transport of pathogens to the lungs through the bloodstream can also result in pneumonia.

The most frequent etiologic agent of bacterial pneumonia is *Streptococcus pneumoniae* (pneumococcus), a Gram positive, capsule-forming coccus (Table 17.1). Often, pneumococcal pneumonia is an endogenous disease, with the infection originating from the individual's normal throat microbiota. Several other bacteria, including *Staphylococcus aureus, Haemophilus influenzae,* and *Klebsiella pneumoniae,* are also responsible for a significant number of cases of pneumonia. The symptoms of pneumococcal pneumonia, which is most prevalent in adult males, include the sudden onset of a high fever, production of colored, purulent sputum, and congestion. In most patients an upper respiratory tract infection with the characteristic symptom of a sore throat precedes the development of pneumococcal pneumonia.

During the development of pneumonia, bacteria reproduce in the lung tissue, forming a lesion. The phagocytic portion of this inflammatory response results in decreased numbers of bacteria within the lesion. Bacteria spread through the alveoli and into the pulmonary system. The exudate that develops during pneumonia interferes with gas exchange in the lungs. Without treatment the death rate from pneumococcal pneumonia is

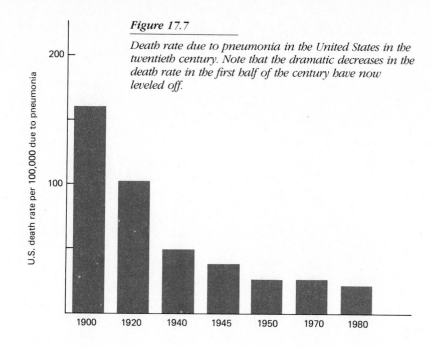

Figure 17.7

Death rate due to pneumonia in the United States in the twentieth century. Note that the dramatic decreases in the death rate in the first half of the century have now leveled off.

about 30 percent. Antibiotic treatment is effective for curing bacterial pnuemonia, with penicillin the antibiotic of choice for treating pneumonia caused by *Streptococcus pneumoniae*. The specific antibiotic treatment for pneumonias caused by bacteria other than *S. pneumoniae* varies in accord with their specific antibiotic sensitivities.

Atypical pneumonia

Primary atypical pneumonia Several bacteria cause **atypical pneumonias** requiring special treatments. Identification of the pathogen and determination of its antibiotic susceptibility are essential for selecting the best antimicrobial agents in cases of atypical pneumonia. *Mycoplasma pneumoniae*, for example, causes an atypical self-limiting pneumonia (primary atypical pneumonia) that has a low death rate. *M. pneumoniae*

lacks a cell wall structure and is not sensitive to penicillin. This organism, however, is sensitive to tetracycline and erythromycin, which can be used effectively in the treatment of this type of atypical pneumonia. Unlike other mycoplasmas, *M. pneumoniae* can attach to the epithelial surface of the respiratory tract. This bacterium does not penetrate the epithelial cells nor does it produce a protein toxin, but the hydrogen peroxide released by the bacterium causes cell damage, including loss of the cilia lining the respiratory tract and death of surface epithelial cells.

Legionnaire's disease Another atypical form of pneumonia is caused by *Legionella pneumophila* and related species in this genus. In addition to the typical symptoms of pneumonia, **Legionnaire's disease** is often characterized by kidney

table 17.1

Frequency of major types of bacterial pneumonia

Clinical entity	Etiologic agent	Percent of cases	Indicated antimicrobial drug
Pneumococcal lobar pneumonia	*Streptococcus pneumoniae*	Over 90	Penicillin
Klebsiella (Friedlander's) pneumonia	*Klebsiella pneumoniae*	1–5	Gentamicin
"Flu" pneumonia	*Haemophilus influenzae* type b	1–5	Ampicillin

and liver involvement and by an unusually high incidence of associated gastrointestinal symptoms. The fever associated with this disease starts low but then typically reaches 104–105°F. If untreated the fatality rate is about one in six. *L. pneumophila* produces β-lactamase enzymes and is not sensitive to most penicillins and cephalosporins, but it is sensitive to other antibiotics, such as erythromycin and tetracycline. Erythromycin is the antibiotic of choice when Legionnaire's disease is diagnosed. *L. pneumophila* is a Gram negative, fastidious, rod-shaped organism, whose nutritional requirements for growth complicated early isolation attempts, initially confounding epidemiologists who were trying to discover the etiologic agent of Legionnaire's disease.

The disease syndrome caused by these organisms is referred to as Legionnaire's disease because the first detected outbreak of this disease occurred during a convention of the American Legion in Philadelphia during July 1976. In the investigation of this outbreak of the disease, the first 90,000 man-hours of investigation, costing over $2 million and employing virtually all conventional isolation procedures, failed to reveal the causative agent of this disease. Finally the breakthrough revealing that this disease is of bacterial etiology came by using indirect immunofluorescent staining with antibodies from the sera of affected individuals. Later, it was discovered that the bacterium, subsequently named *L. pneumophila* (lung-loving), could be grown on a chocolate agar medium if iron and cysteine were included as growth factors. It was also later found by examining stored blood sera that a 1968 outbreak of a disease in Pontiac, Michigan, the etiology of which had not been identified, had been caused by a different strain of *L. pneumophila*. Various other outbreaks of this disease have since been identified.

Species of *Legionella* appear to be natural inhabitants of bodies of water. During periods of rapid evaporation, such as occur during summer, the bacteria can become airborne in aerosols, and inhalation of contaminated aerosols can lead to the onset of the illness. In some cases, outbreaks of Legionnaire's disease have been traced to air-conditioning cooling systems. These bacteria multiply in the cooling system waters, which are rapidly evaporated to provide cooling, and inadvertently are permitted to become airborne and circulate through the air-conditioning system.

Psittacosis Psittacosis, which is also known as ornithosis or parrot fever, is another type of atypical pneumonia. This disease is caused by an ob-

596

Human diseases caused by micro-organisms: the respiratory, gastrointestinal, and genitourinary tracts as portals of entry

ligate intracellular bacterium, *Chlamydia psittaci.* Birds act as a reservoir for *C. psittaci*, from which the name of the disease is derived. This infection is contracted through inhalation of *C. psittaci*, and accordingly, the bacteria enter the body through the respiratory tract. The primary route of transmission of psittacosis is from birds to human beings via aerosol dispersal of droplets and contaminated dust particles. Parakeets, canaries, other pet birds, and domestic fowl are frequently the sources of human infection. *C. psittaci* multiplies in the cells of the mononuclear phagocyte system prior to systemic dissemination through the bloodstream. The symptoms of psittacosis include fever, headache, malaise, and coughing. Psittacosis is generally a mild disease and in uncomplicated cases recovery normally occurs within one week, aided by the use of tetracyclines in the treatment of this disease.

Bronchitis

Bronchitis, an inflammatory disease involving the bronchial tree that does not extend into the pulmonary alveoli, can be caused by several microorganisms, including *Streptococcus* sp., *Staphylococcus* sp., *Haemophilus influenzae*, *Mycoplasma pneumoniae*, and various types of viruses. *M. pneumoniae* and various viruses appear to be the most frequent causative organisms of bronchitis, but it is difficult to define the specific etiologic agent for this disease because bronchitis almost always occurs as a complication of another disease condition, such as **pharyngitis** (**sore throat**). The symptoms of bronchitis are normally preceded by those associated with a normal upper respiratory tract infection, such as malaise, headache, and sore throat. The onset of bronchitis is marked by the development of a cough that eventually yields mucopurulent sputum, reflective of the development of bronchial congestion. Acute bronchitis can be effectively treated with antibiotics, such as penicillin and tetracycline. The development of chronic bronchitis is not because of microbial infection alone, but rather the etiology of this disease appears to depend on irritation of the bronchii by repeated microbial infections and/or the inhalation of irritants, such as cigarette smoke. These irritations compromise the normal secretory and ciliary function of the bronchial mucosa, and the resulting excessive mucus secretion in the bronchii favors bacterial growth and the establishment of infection.

Rheumatic fever

Streptococcus species, which cause a variety of diseases, are normally transmitted through the air in contaminated droplets and establish the primary infection in the tissues of the upper respiratory tract. In some cases the infection is limited to these tissues, causing conditions such as pharyngitis and tonsilitis. In other cases the streptococci or protein exotoxins produced by streptococci enter the circulatory system and spread systemically. In the case of **scarlet fever**, for example, the systemic spread of hemolysins produced by *S. pyogenes* is manifest as a rash of pinhead red spots, and in rheumatic fever the systemic spread of *S. pyogenes* involves multiple body sites.

Rheumatic fever is generally the most serious consequence of *Streptococcus pyogenes* infections. The symptoms of this disease vary, but characteristically there is a high fever, painful swelling of various body joints, and cardiac involvement, including subsequent development of heart murmurs from childhood occurrences of rheumatic fever. The symptoms of rheumatic fever normally begin to occur a little over two weeks after a characteristic sore throat, associated with an upper respiratory tract infection with *S. pyogenes*. Because of the serious manifestations of rheumatic fever, it is important to diagnose the etiologic agents of sore throats in children. Throat swabs plated on blood agar can readily be screened for the presence of β-hemolytic streptococci, and when they are detected, serological and/or biochemical tests are carried out to determine if group A streptococci, the group that includes *S. pyogenes*, are present. Penicillin is effective in treating group A streptococcal infections, and its use in treating streptococcal pharyngitis can prevent the occurrence of rheumatic fever.

The specific causal relationship between *S. pyogenes* and the symptoms of rheumatic fever has not been established. It is possible that antibodies produced in response to a group A streptococcal infection are cross-reactive with cardiac antigens and that it is an "autoimmune" response that actually results in cardiac damage. The treatment of rheumatic fever, therefore, includes the use of anti-inflammatory drugs to reduce tissue damage as well as the use of antibiotics to remove the infecting streptococci.

Otitis media

Otitis media is an inflammatory disease of the mucosal lining of the middle ear. Bacterial infections of the middle ear normally originate from an upper respiratory infection, with the bacteria entering the ear through the auditory (eustachian) tube, the principal portal of entry to the ear. *Streptococcus pneumoniae* is the etiologic agent in over 50 percent of the cases of otitis media, and ampicillin is effective in treating such infections. *Streptococcus pyogenes* and *Haemophilus influenzae* are also frequently the causative agents of middle ear infections. Manifestations of middle ear infections normally include severe pain and fever and in cases caused by *Streptococcus* sp., the tympanic membrane is usually fiery red.

Carditis

In addition to cardiac involvement as one manifestation of rheumatic fever, streptococcal species are among the most prevalent bacteria responsible for inflammation of the heart muscle, **carditis**. For example, over 50 percent of the reported cases of endocarditis in the United States during the 1960s were attributed to streptococcal infections, although rickettsia and fungi also cause carditis. **Endocarditis** (inflammation of the endocardium, a specialized membrane of epithelial and connective tissue that lines the cardiac chambers and forms much of the heart valve structure) most frequently occurs when normally nonpathogenic members of the microbiota associated with body surfaces enter the circulatory system and attach to cardiac tissues. *Streptococcus viridans*, present in high numbers in the oral cavity, is the most frequently identified etiologic agent of endocarditis, and dental treatments are often implicated in the mobilization of *S. viridans*, initiating entry of this bacterium into the bloodstream through the tissues of the oropharyngeal cavity. **Myocarditis** (inflammation of the myocardium, which is principally cardiac muscle) can also result from microbial infections, most commonly *Streptococcus* infections. Microbial growth on the heart tissues causes serious abnormalities, and if untreated is often fatal (Figure 17.8). Treatment of infections of the heart tissues usually employs broad-spectrum antibiotics that are not inactivated by β-lactamases and often requires prolonged administration of high doses of antibiotics to ensure removal of the infecting bacteria.

Diphtheria

Diphtheria, caused by strains of *Corynebacterium diphtheriae*, results from a protein toxin produced by strains of *C. diphtheriae* harboring a lysogenic phage. Diphtheria toxin is a potent protein toxin exhibiting toxicity against almost all mammalian cells. The bacteria generally do not invade the tissues of the respiratory tract, but it is rather the toxin disseminating through the body that causes the severe symptoms of this disease. *C. diphtheriae* is normally transmitted via droplets from an infected individual to a susceptible host,

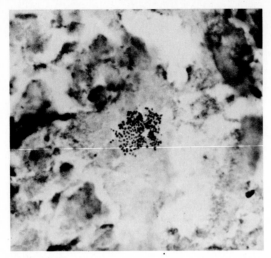

Figure 17.8

This micrograph shows a section of heart tissue infected with Streptococcus viridans. *(Courtesy John J. Bochino, Norton's Kosair Children's Hospital, Louisville, Kentucky.)*

598

Human diseases caused by microorganisms: the respiratory, gastrointestinal, and genitourinary tracts as portals of entry

establishing a localized infection on the surface of the mucosal lining of the upper respiratory tract.

There is generally a localized inflammatory response, pharyngitis, in the vicinity of bacterial multiplication in the upper respiratory tract. In severe infections with *C. diphtheriae*, symptoms include low-grade fever, cough, sore throat, difficulty in swallowing, and swelling of the lymph glands. Complications from diphtheria can block respiratory gas exchange and result in death due to suffocation. The extensive use of vaccines to prevent diphtheria has greatly reduced the incidence of this disease but has not altered the case fatality ratio (Figure 17.9). In immunized individuals, infection with toxicogenic strains of *C. diphtheriae* are generally restricted to a localized pharyngitis with no serious complications. Diphtheria, however, remains a serious problem in socioeconomically depressed regions of the world where extensive immunization is not practiced. Treatment of diphtheria involves the use of antitoxin to block the cytopathic effects of diphtheria toxin, and the use of antibodies in the treat-

Diphtheria—reported annual incidence and mortality rates, and case fatality ratio, United States — 1920-1980

Figure 17.9

Incidence, mortality rates, and case fatality ratio of diphtheria in the United States from 1920 to 1980. The rates of this disease have declined dramatically since effective vaccination programs were begun. However, the case fatality ratio has remained relatively constant, underscoring the importance of prevention of this disease and the difficulty in its treatment.

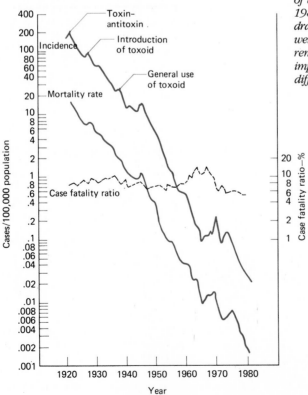

ment of diphtheria prevents the occurrence of serious symptoms associated with diphtheria toxin. This immunological treatment is augmented by the use of antibiotics, such as erythromycin, to eliminate the bacterial infection.

Whooping cough

Whooping cough or **pertussis** derives its name from the distinctive symptomatic cough associated with this disease. Other symptoms resemble those of the common cold, although vomiting often occurs after severe coughing episodes. Whooping cough is caused by *Bordetella pertussis*, a Gram negative cocco-bacillus, which exhibits fastidious nutritional requirements. *B. pertussis* is capable of reproducing within the respiratory tract, and high numbers of *B. pertussis* are found on the surface tissues of the bronchi and trachea. *B. pertussis* produces several toxins that establish the pathogenicity of this organism. Erythromycin and tetracyclines are effective in eliminating the infecting bacteria, although treatment of whooping cough primarily involves maintenance of an adequate oxygen supply. The administration of pertussis vaccine has greatly reduced the occurrence of whooping cough, and prevention of the disease is accomplished by routine immunization of infants.

Tuberculosis

Tuberculosis, caused by *Mycobacterium tuberculosis* and related mycobacterial species, is primarily transmitted via droplets from an infected to a susceptible individual, although tuberculosis can also be transmitted through the ingestion of contaminated food. Before the extensive use of pasteurization, milk contaminated with *M. tuberculosis* was associated with outbreaks of this disease. The principal portal of entry, however, for *M. tuberculosis* is through the respiratory tract because much lower numbers of bacteria are required to establish an infection via the respiratory tract than are needed to initiate tuberculosis through the gastrointestinal system.

The common form of tuberculosis involves an infection of the pulmonary system, with multiplication of *M. tuberculosis* occurring in the lower respiratory tract despite the phagocytic activity of macrophage that protects the lower respiratory tract from infection by most potential bacterial pathogens. The pulmonary form of tuberculosis involves inflammation and lesions of lung tissue, which can be detected by chest X rays (Figure 17.10). The bacteria spread from the primary lesions to the draining lymph and then through lymph and blood to other parts of the body. In-

fection with *M. tuberculosis* elicits a cellular immune response because bacteria are able to reproduce within phagocytic cells, and a delayed hypersensitivity reaction is typical of infection with *M. tuberculosis*. Dormant mycobacteria can remain within the body, and the infectious process can be reactivated at a later time, various physiological factors probably contributing to reactivation of the disease.

The course of tuberculosis varies greatly between infected individuals; in some cases the infection is restricted to the area of primary lesions, and in others it spreads into various other tissues. Disease symptoms, including fatigue, weight loss, and fever, generally do not appear until there are extensive lesions in the lung tissues. As a result of the slow growth rate of *M. tuberculosis* and the ineffectiveness of phagocytic cells in killing this bacterial species, tuberculosis is generally a persistent and progressive infection. Without treatment, tuberculosis is often fatal. Effective treatment of tuberculosis is generally prolonged and is achieved by using multiple antibiotics such as

Figure 17.10

X ray showing development of tuberculosis in lungs of an adult. The left lung is normal, but the right lung shows a large calcified tubercle and smaller tubercles and infiltration in the apex of the right lumg. (From BPS—R. B. Morrison, M.D., Austin, Texas.)

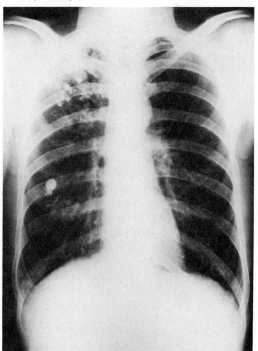

streptomycin, rifampin, and isoniazid. Malnutrition and stress are important factors relating to the resistance to tuberculosis and the course of the disease. Extrapulmonary tuberculosis commonly develops as a result of reactivation of dormant lesions established during a primary infection. The common locations of extrapulmonary involvement are the genitourinary system, bones and joints, lymph nodes, pleura, and peritoneum. In some cases the central nervous system is also involved, as appeared in the case of novelist Thomas Wolfe who, after a severe case of influenza or pneumonia, developed a reactivated case of tuberculosis that spread through his bloodstream, leading to his death from meningeal tuberculosis. (It should be noted that the disease was not clinically confirmed, and some mycologists claim that coccidioidomycosis rather than tuberculosis was responsible for Wolfe's death.)

Meningitis

Meningitis, an inflammation of the meninges (membrane surrounding the brain and spinal cord), can be caused by various viruses and bacteria. The typical transmission of bacteria causing meningitis is via droplet spread, with the initial infection occurring in the respiratory tract followed by transmission via the bloodstream to the meninges. Injuries that expose the central nervous system to bacterial contaminants provide an alternate portal of entry. *Neisseria meningitidis, Haemophilus influenzae, Streptococcus pneumoniae,* and *Escherichia coli* are the most common etiologic agents of bacterial meningitis. There is a high degree of correlation between the age of the patient and the specific etiologic agent (Table 17.2). For example, *Neisseria meningitidis,* also known as the meningococcus, is often the causative agent of bacterial meningitis in patients between 5 and 40 years of age but is rarely found in cases of meningitis in younger children. Similarly, *E. coli* often causes meningitis in infants but not in adults.

Meningitis is characterized by sudden fever, severe headache, painful rigidity of the neck, nausea, vomiting, and frequently by convulsions, delirium, and coma. If untreated, bacterial meningitis is usually fatal. Because death may occur within hours of recognition of the infection, accurate and swift diagnosis and speedy initiation of treatment is essential. The diagnosis of the particular etiologic agent of meningitis depends on the isolation and identification of the pathogens from the cerebrospinal fluid. A number of antibiotics are used in the treatment of meningitis, and the specific antibiotic of choice is determined by the anti-

biotic susceptibility of the causative agent of the disease; in many cases it is a penicillin.

Q fever

Q fever, caused by the rickettsia *Coxiella burnetii,* is a unique rickettsial disease because it does not manifest itself as a rash and is the only one normally transmitted via the respiratory tract. The disease can also be transmitted through direct contact of the bacteria with the eyes and via the gastrointestinal tract. *C. burnetii* is normally maintained within nonhuman animal populations, such as cattle and sheep, where it is transmitted via tick vectors. The bacteria can become airborne on fomites (inanimate objects), such as hair and dust particles, and establish human infections by invading the lower respiratory tract, leading to a systemic infection. The symptoms of Q fever often include fever, headache, chest pain, nausea, and vomiting. Tetracyclines and cloramphenicol are normally used in treating this disease.

Diseases caused by fungal pathogens

Histoplasmosis

Histoplasmosis, one of several fungal diseases of humans, is caused by the fungus *Histoplasma capsulatum* (*Emmonsiella capsulata*), entering the respiratory tract through the inhalation of spores that are then deposited within the lungs. *H. capsulatum* is a dimorphic fungus exhibiting yeastlike growth at 37°C and filamentous growth at 25°C. Normally, histoplasmosis is a self-limiting disease in which the symptoms may be absent or resemble a mild cold. In some cases, however, the systemic distribution of the fungus to different organs of the body may prove fatal. The fungus is able to persist within the lung and the tissues and cells of the mononuclear phagocyte system. Amphotericin B is used in the treatment of cases of progressive disseminated histoplasmosis.

Histoplasmosis is endemic to certain regions of the world, such as the Ohio and Mississippi River Valleys of the United States (Figure 17.11). The fungus is found in soils contaminated with bird droppings, and dust particles released from abandoned bird roosts appear to be involved in some outbreaks of this disease. The apparent association of bird roosts with histoplasmosis has been used as the justification for large-scale kills of blackbirds, but the usefulness of this procedure has not been conclusively demonstrated.

Coccidioidomycosis

Coccidioidomycosis is caused by *Coccidioides im-*

600

Human diseases caused by microorganisms: the respiratory, gastrointestinal, and genitourinary tracts as portals of entry

table 17.2

Correlation of age with the etiologic agents of meningitis

Etiologic agent	Percent of isolates			
	Under 2 months	2–60 months	5–40 years	Over 40 years
Neisseria meningiditis	—	20	45	10
Haemophilus influenzae	—	60	5	2
Escherichia coli and other Enterobacteriaceae	55	—	—	10
Pseudomonas aeruginosa	2	—	—	—
Streptococcus pneumoniae and other *Streptococcus* sp.	28	12	25	55
Staphylococcus sp.	5	—	10	13
Other	10	8	10	10

mitis, which—like *Histoplasma capsulatum*—exhibits dimorphism. Coccidioidomycosis is also referred to as Valley or San Joaquin Fever because of the geographic distribution of *C. immitis* and the associated areas of occurrence of this disease. Because of the association of the spores of *C. immitis* with the arid soils of the southwest United States, these soils must be disinfected before ship-

ping to other regions. Many individuals living in this region show a positive skin test, indicating that they have been infected at some time with this fungus. Visitors to the southwestern states of Nevada, California, Utah, Arizona, and New Mexico often develop symptoms of a mild cold because of infection with *C. immitis*. Normally, *C. immitis* occurs in soil and transmission of coccid-

601

Diseases caused by pathogens that enter via the respiratory tract

Figure 17.11

This map, based on the incidence of skin reactivity to histoplasmin among Naval recruits between 1958 and 1965, shows that histoplasmosis in the United States occurs primarily in the Mississippi and Ohio Valleys. In southern Kentucky, middle Tennessee, and southern Missouri, the incidence of skin reactivity is as high as 90 to 95 percent. (From L. B. Edwards et al., 1969, An atlas of sensitivity of tuberculin, PPD-B, and histoplasmosis in the United States, American Review of Respiratory Diseases *99: 1.)*

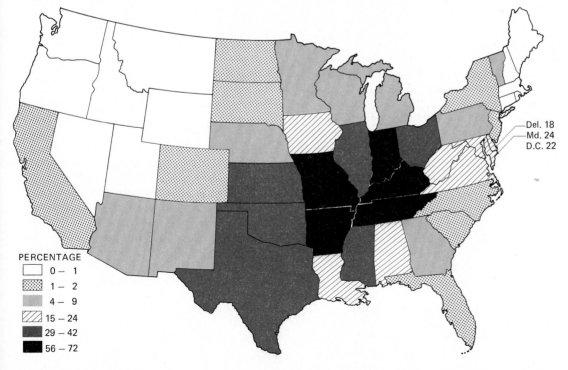

PERCENTAGE
- 0 – 1
- 1 – 2
- 4 – 9
- 15 – 24
- 29 – 42
- 56 – 72

Del. 18
Md. 24
D.C. 22

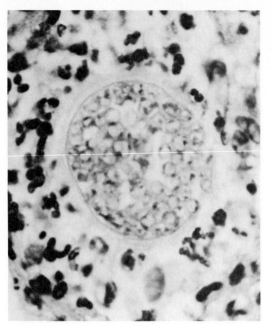

ioidomycosis involves inhalation of dust particles containing conidia of this fungus. When deposited in the bronchi or alveoli, the conidia of *C. immitis* elicit an inflammatory response (Figure 17.12). Within host tissues, *C. immitis* appears as spherules containing multiple spores. In some cases, *C. immitis* remains localized in the area of the primary lesion, but the organism can be distributed to other parts of the body. Symptoms of coccidioidomycosis include chest pain, fever, malaise, and a dry cough. In most cases, no special treatment is required for the cure of localized coccidioidomycosis, and upon recovery the individual is immune to this disease. However, when there is evidence of systemic dissemination of *C. immitis*, treatment with amphotericin B is normally used in effectively treating the disease.

Blastomycosis

North American blastomycosis is a systemic disease caused by *Blastomyces dermatidis*. The primary site of infection is the lungs, from which the fungus can be disseminated to many other body tissues. *B. dermatidis* is a dimorphic fungus that normally inhabits soils. Blastomycosis occurs most frequently in the southeastern United States. Skin lesions are common in cases of blastomycosis. As with other systemic mycosis, blastomycosis can be treated with amphotericin B.

602

Human diseases caused by microorganisms: the respiratory, gastrointestinal, and genitourinary tracts as portals of entry

Figure 17.12

Micrograph showing a section of lung infected with Coccidioides immitis. *The fungus occurs as endospores within a body known as a spherule. Such fungal infections initiate a major inflammatory response. (Courtesy American Society of Clinical Pathologists.)*

Diseases caused by pathogens that enter via the gastrointestinal tract

Microorganisms routinely enter the gastrointestinal tract in association with ingested food and water. A large resident microbiota develops in the human intestinal tract after birth and is important for the maintenance of good health and not involved in disease processes. The resident microbiota of the gastrointestinal tract are normally noninvasive and are associated with the surface tissues and ingested food material. Some pathogenic microorganisms, however, possess toxigenic or invasive properties that permit them to cause disease when they enter the gastrointestinal tract.

There are two distinct processes that can initiate the onset of disease through the gastrointestinal tract. In one case microorganisms growing in food or water can produce toxins, and the ingestion of these microbial toxins initiates a disease process. Such diseases are classified as **food poisoning** or **intoxication** because the etiologic agents of the disease need not grow within the body; that is, there is no true infectious process. Toxins absorbed through the gastrointestinal tract can cause neural damage and death in some cases as well as localized inflammation and gastrointestinal upset in others. In the second type of disease-causing process, invasive pathogens establish an initial infection through the gastrointestinal tract and cause localized gastrointestinal upset or systemic disease symptomology. Generally, the establishment of infection through the gastrointestinal tract requires a relatively large **infectious dose**; that is, a relatively large number of pathogenic microorganisms are required to successfully overcome the inherent defense mechanisms of the gastrointestinal tract. Quite different measures are required to prevent and treat infectious gastrointestinal diseases and those specific microorganisms responsible for food poisoning.

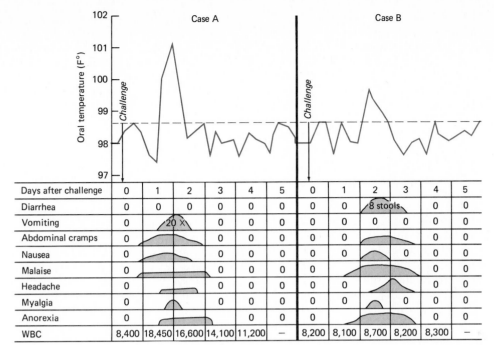

	Case A						Case B					
Days after challenge	0	1	2	3	4	5	0	1	2	3	4	5
Diarrhea	0	0	0	0	0	0	0	0	8 stools	0	0	
Vomiting	0	20 X	0	0	0	0	0	0	0	0	0	
Abdominal cramps	0			0	0	0	0	0			0	0
Nausea	0			0	0	0	0	0		0	0	0
Malaise	0			0	0	0	0				0	0
Headache	0			0	0	0	0	0			0	0
Myalgia	0			0	0	0	0	0		0	0	0
Anorexia	0			0	0	0	0			0	0	
WBC	8,400	18,450	16,600	14,100	11,200	—	8,200	8,100	8,700	8,200	8,300	—

Figure 17.13

The time course of symptoms in viral gastroenteritis are shown in this illustration of the response of two volunteers to oral administration of stool filtrate derived from a volunteer who received original Norwalk rectal swab specimen. The height of the shaded curve is roughly proportional to the severity of the sign or symptom. Essentially, this sort of experiment both confirms the cause of the disease by Koch's postulates and demonstrates the variability of symptomology; in viral diseases such as this, human subjects rather than experimental animals are used because of the specificity of the virus–host relationship. (Reprinted by permission of the University of Chicago Press, from R. Dolin et al., 1971, Transmission of acute infectious nonbacterial gastroenteritis to volunteers by oral administration of stool filtrates, Journal of Infectious Diseases *123: 307.)*

Diseases caused by viral pathogens

Gastroenteritis

Gastroenteritis involves an inflammation of the lining of the gastrointestinal tract. In most cases viral gastroenteritis is a self-limiting disease, often referred to as the 24-hour or intestinal flu. Viral gastroenteritis is not caused by an influenza virus and is not related to true cases of flu, but rather this disease can be caused by several different viruses including adenoviruses, coxsackieviruses, polioviruses, and members of the **ECHO virus group**. The **Norwalk agent**, a small DNA virus identified as being responsible for an outbreak of "winter vomiting disease" that occurred in Norwalk, Ohio, in 1968, appears to be an important etiologic agent of various viral gastroenteritis outbreaks. **Rotavirus**, a large RNA virus, also appears to be a very common etiologic agent for diarrhea in infants, particularly in socioeconomically depressed regions of the world.

Viruses causing gastroenteritis normally replicate within cells lining the gastrointestinal tract, and large numbers of viruses are released in fecal matter. Contamination of food with fecal matter is an important route of transmission of viral gastroenteritis as well as many other diseases caused by microorganisms that enter via the gastrointestinal tract. The characteristic symptoms of a viral gastroenteritis include sudden onset of gastrointestinal pain, vomiting, and/or diarrhea (Figure 17.13). Recovery normally occurs within 12 to 24 hours of the onset of disease symptomology. As a result of the vomiting and diarrhea, there can be a severe loss of body fluids and dehydration. The loss of water and resultant imbalances in electrolytes can have serious consequences, particularly in infants, where viral gastroenteritis is sometimes fatal.

Hepatitis

Hepatitis is a systemic viral infection that primarily affects the liver. There are several types of hepatitis viruses designated types A, B, and C. Type A hepatitis virus normally enters the body via the gastrointestinal tract and causes infectious hepatitis. Although there are some documented cases of foodborne hepatitis B, type B hepatitis virus, which will be discussed in Chapter 18, normally enters the body through skin punctures and causes serum hepatitis. Hepatitis type A virus is usually transmitted by the fecal-oral route and is prevalent in areas with inadequate sewage treatment. Several outbreaks of viral hepatitis have been associated with contaminated shellfish that have concentrated viruses from sewage effluents.

The initial symptoms of infectious hepatitis include fever, abdominal pain, and nausea, followed by jaundice, the yellowing of the skin indicative of liver impairment caused by the virus (Figure 17.14). Damage to liver cells also results in increased serum levels of enzymes, such as transaminases, normally active in liver cells. The detection of increased serum levels of these enzymes is used in diagnosing this disease. Infectious hepatitis caused by the type A virus tends to be less serious than serum hepatitis. In most cases of infectious hepatitis, the infection is self-limiting, and recovery occurs within four months.

Poliomyelitis

Polioviruses may enter the body either through the gastrointestinal or respiratory tracts. Transmission through the ingestion of food and water containing polioviruses is considered very important. Polioviruses are able to multiply within the tissues of the oropharynx and intestines. Viruses entering the bloodstream are disseminated, and further viral replication occurs within lymphatic tissues. Polioviruses have the ability to cross the blood–brain barrier, where they continue to multiply within neural tissues and cause varying degrees of damage to the nervous system.

604

Human diseases caused by microorganisms: the respiratory, gastrointestinal, and genitourinary tracts as portals of entry

Figure 17.14

The time course of symptoms in infectious hepatitis shows that jaundice occurs beginning one week after the initial fever. (After W. P. Havens, Jr., and Paul in Viral and Rickettsial Infections of Man, *F. L. Horsfall, Jr., and I. Tamm, eds., J. B. Lippincott Co., Philadelphia.)*

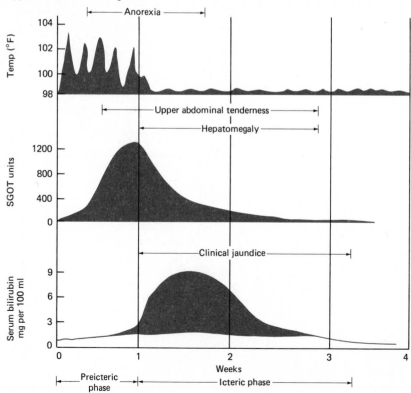

The initial symptoms of poliomyelitis (commonly referred to as **polio**) include headache, vomiting, constipation, and sore throat. In many cases, these early symptoms are followed by obvious neural involvement, including paralysis due to the injury of motor neurons. Although the paralysis can affect any motor function, in over half of the cases of paralytic poliomyelitis, the arms and/or legs are involved. Fortunately, the paralytic symptoms are 1000 times less frequent than nonparalytic infections, and many cases of poliovirus infection fail to show any evidence of clinical symptomology.

Poliomyelitis is prevalent in children and as such is also called **infantile paralysis**. The disease also strikes adults and in fact the fatality rate in adults is much higher than in children (Table 17.3). The use of the **Salk** and **Sabin polio vaccines** has dramatically reduced the incidence of this disease (Figure 17.15). It is important that preschool children be immunized because major outbreaks of poliomyelitis have traditionally been associated with transmission among children in close contact together in a schoolroom. Despite the ability to prevent this serious disease, many children are not immunized voluntarily, even in economically

affluent countries such as the United States. Many school systems now require evidence of polio vaccination before a child may be enrolled. This is essential because the lowered rate of occurrence of this disease means that there are fewer paralyzed individuals to serve as visible reminders of the seriousness of this disease. Constant efforts to reinforce parental awareness of the importance and success of vaccination against poliomyelitis are worthwhile.

table 17.3

Age-specific case fatality rates of paralytic poliomyelitis in the United States, 1960–1968

Age group	Cases	Deaths	Case fatality rate (%)
0–4	1899	92	4.8
5–9	959	65	6.8
10–14	379	33	8.7
15–19	205	18	8.8
20–29	462	71	15.4
30–39	275	65	23.6
40 and over	136	41	30.1

Figure 17.15

Incidence of poliomyelitis per 100,000 population in the Americas, 1955 to 1973. Note that whereas the rates of incidence were similar in 1955, the occurrence of this disease in North America dramatically declined after the introduction of vaccines; no similar decline has occurred in Central and South America, where extensive vaccination has not been carried out.

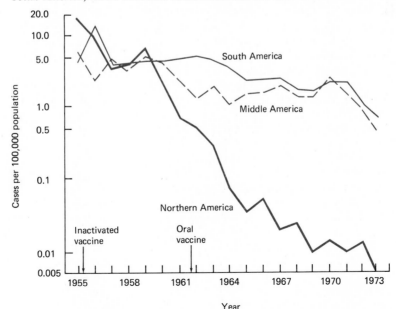

Diseases caused by bacterial pathogens

Botulism

Botulism, the most serious form of bacterial food poisoning, is caused by neurotoxins produced by *Clostridium botulinum*. The toxins are absorbed from the intestinal tract and transported via the circulatory system to motor nerve synapses where their action blocks normal neural transmissions. Various strains of *C. botulinum* elaborate different toxins. Types A, B, and E toxins cause food poisoning of humans. Type E toxins are associated with the growth of *C. botulinum* in fish or fish products, and most outbreaks of botulism in Japan are caused by type E toxins because large amounts of fish are consumed there. Type A is the predominant toxin in cases of botulism in the United States, and type B toxin is most prevalent in Europe.

Over 90 percent of the cases of botulism involve improperly home canned food. Of 236 outbreaks of botulism in the United States between 1899 and 1974, 57 percent were caused by contaminated vegetables, 15 percent by contaminated fish, and 12 percent by fruit containing *C. botulinum*. The endospores of *C. botulinum* are heat-resistant and can survive prolonged exposure at 100°C. Contaminated canned foods provide an optimal anaerobic environment for the growth of *C. botulinum*. This bacterium, though, cannot grow and produce toxin at low pH and thus is not a problem in acidic food products. Symptoms of botulism can appear 8–48 hours after ingestion of the toxin, and an early onset of disease symptomology normally indicates that the disease will be severe. Type A toxin botulism is generally more severe than the disease caused by other types of toxin. In severe cases of botulism, there is paralysis of the respiratory muscles, and despite improved medical treatment the mortality rate is still about 25 percent. The use of trivalent ABE antibodies is useful in treating this disease, but it is of paramount importance in the treatment to ensure continued respiratory functioning.

Clostridium botulinum is normally not capable of establishing an infection in adults because of the low pH of the stomach and upper end of the small intestine. However, in infants, prior to the colonization of the intestinal tract by *Lactobacillus* species, *C. botulinum* can reproduce and elaborate neurotoxin from the gastrointestinal tract tissues. There is evidence that some cases of sudden infant death syndrome, or crib death, can be attributed to *C. botulinum*. Accordingly, additional concern is being given to food products that infants consume with respect to the possible ingestion of *C. botulinum* endospores.

Perfringens food poisoning

Another clostridial species, *Clostridium perfringens*, is a major cause of a less severe form of food poisoning. *Clostridium perfringens* generally accounts for over 10 percent of the outbreaks of food-borne disease in the United States. The ingestion of food containing toxin produced by *C. perfringens* and the adsorption of the toxin into the cells lining the gastrointestinal tract initiate this disease. Toxin type A of *C. perfringens* is associated with most cases of clostridial food poisoning, particularly with cooked meats if a gravy is prepared with the meat. The spores of *C. perfringens* type A can survive the temperatures used in cooking many meats, and if incubated in a warm gravy, there is sufficient time for the spores to germinate and the growing bacteria to elaborate sufficient toxin to cause this disease.

The symptoms of food poisoning associated with *C. perfringens* generally appear within 10 to 24 hours after ingestion of food containing toxin. Unlike botulism, in cases of food poisoning caused by *C. perfringens*, recovery generally occurs within 24 hours. The symptoms of food poisoning by *C. perfringens* include abdominal pain and diarrhea, but vomiting, headache, and fever normally do not occur.

Staphylococcal food poisoning

Strains of *Staphylococcus aureus* which cause food poisoning produce an exotoxin that is also an enterotoxin. The release of toxin from the bacterial cells causes an inflammation of the lining of the gastrointestinal tract. The staphylococcal enterotoxins do not exert a direct local effect but must be absorbed through the bloodstream and reach the digestive tract in order to initiate a staphylococcal food poisoning syndrome. *S. aureus* is able to reproduce within many different types of food products. Enterotoxin-producing strains of *S. aureus* often enter foods from the skin surfaces of people handling food. Custard-filled bakery goods, dairy products, processed meats, potato salad, and various canned foods are frequently found to be the source of the toxin in cases of staphylococcal food poisoning. Salads prepared for a summer picnic can easily be contaminated (inoculated) with *S. aureus*, and when left in the sun in a wicker basket (incubated), the bacteria can multiply, producing a sufficient amount of enterotoxin to provide an unexpected nighttime encore to the day's fun.

The symptoms of staphylococcal food poison-

606

Human diseases caused by microorganisms: the respiratory, gastrointestinal, and genitourinary tracts as portals of entry

ing occur relatively rapidly after ingestion of toxin containing food, usually within 1 to 6 hours. The symptoms of staphylococcal food poisoning generally include nausea, vomiting, abdominal pain, and diarrhea. Symptoms generally subside within 8 hours of their onset, and complete recovery usually occurs within a day or two. The prevention of staphylococcal food poisoning depends on proper handling and preservation of food products to prevent contamination and subsequent growth of enterotoxin-producing strains of *Staphylococcus.*

Gastroenterocolitis

Gastroenterocolitis, an inflammation of the intestinal lining, can result from various bacterial infections. *Clostridium perfringens* and *Staphylococcus aureus*, for example, frequently involved in cases of food poisoning, can also establish infections in the gastrointestinal tract and cause gastroenterocolitis.

Enterotoxin-producing strains of *Escherichia coli* are also capable of causing both mild and severe forms of enterocolitis. In most cases, enterotoxin-producing strains of *E. coli* do not invade the body through the gastrointestinal tract, but rather toxin released by cells growing on the surface lining of the gastrointestinal tract causes diarrhea. Aside from diarrhea, abdominal cramps is normally the only other clinical symptom of this disease. Travelers from the United States to Mexico often suffer severe diarrhea as a result of ingestion of strains of *E. coli* foreign to their own microbiota and therefore generally avoid drinking the water.

In some cases, enteropathogenic strains of *E. coli* invade the body through the mucosa of the large intestine to cause a serious form of dysentery. Many cases of severe diarrhea in children are caused by noninvasive, enterotoxin-producing strains of *E. coli*. Invasive strains of *E. coli* are primarily associated with contaminated food and water in Southeast Asia and South America. The ability to invade the mucosa of the large intestine depends on the presence of a specific K antigen in enteropathogenic serotypes of *E. coli*. The enterotoxins produced by *E. coli* cause a loss of fluids from intestinal tissues. With proper replacement of body fluids and maintenance of the essential electrolyte balance, infections with enterotoxic *E. coli* normally are not fatal.

Salmonellosis **Salmonellosis**, caused by members of the genus *Salmonella*, which are Gram negative short rods, is commonly manifested as gastroenterocolitis. Various *Salmonella* species, especially the numerous serotypes of *Salmonella enteritidis*, are commonly the etiologic agents of salmonellosis. Like many enteropathogenic bacteria, *Salmonella* species have pili that enable them to adhere to the lining of the gastrointestinal tract (Figure 17.16). Whereas *Salmonella* species are able to reproduce within the intestines, causing inflammation, they do not normally penetrate the mucosal lining and enter the bloodstream, but in some cases *Salmonella* species can gain access to the circulatory system, causing bacteremia. For example, paratyphoid fever, which is caused by strains of *S. paratyphi* and *S. typhimurium*, is characterized by gastroenteritis and a relatively high rate of bacteremia (Figure 17.17). *Salmonella* species causing gastroenteritis are normally trans-

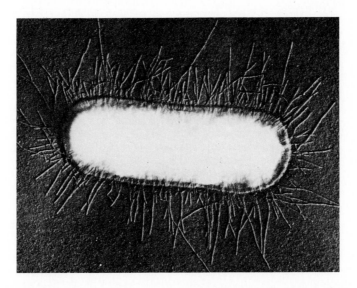

Figure 17.16

Micrograph of Salmonella anatum *surrounded by pili. The pili, or fimbriae as they are often called, permit the bacteria to adhere to the lining of the gastrointestinal tract. (Courtesy J. P. Duguid, reprinted by permission of the editor,* Journal of Medical Microbiology, *from J. P. Duguid et al., 1966,* Journal of Pathology and Bacteriology *92: 107.)*

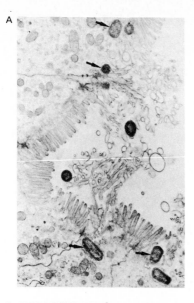

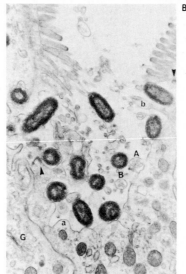

Human diseases
caused by micro-
organisms: the
respiratory, gas-
trointestinal, and
genitourinary
tracts as portals
of entry

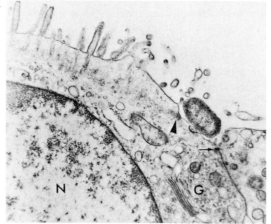

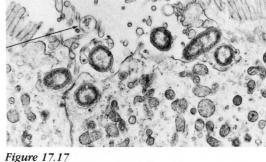

Figure 17.17

Micrographs showing invasion of the intestine by Salmonella. *(A) Note extensive degeneration of the microvilli terminal cell and apical cytoplasm that is localized at site of bacterial penetration (arrows). Other cytoplasmic components and adjacent cells are unaltered (5,000×). (B) Note the projections (referred to as blebs) with (A) or without (a) vesicles that arise from the host cell cytoplasm and are pinched off (B, C and b, c) into a cavity that also contains degenerating microvilli. Blebs with vesicles (A, B, and C) are similar to the multivesicular bodies in Figure 17.17D. There are increasing numbers of small vesicles around the Golgi (G). An intercellular junctional complex is laterally displaced (12,000×). (C) A Salmonella cell lies close to a junctional complex (thin arrow). The terminal web (thick arrow) of the cell at the left ends abruptly at the lumen surface. The corresponding terminal web of the cell at the right is absent. The degeneration of the microvilli and apical cytoplasm is severe near the bacterium (9,000×). G = Golgi; N = nucleus. (D) The microvilli, terminal web, and apical cytoplasm are replaced by a shallow cavity in which degenerated microvilli, blebs, and vesicles are present. The adjacent intercellular tight junctions are laterally displaced. The arrow indicates a striking similarity between a multivesicular body (MVB) and a bleb containing small vesicles. The remaining cytoplasmic organelles are intact (6,000×). (A reprinted by permission of the American Society for Microbiology, Washington, D. C., from A. Takeuchi, 1975, in* Microbiology—1975, *p. 176; B–D reprinted by permission of Hoeber Medical Division of Harper & Row, Hagerstown, Maryland, from A. Takeuchi, 1967,* American Journal of Pathology *50: 125, 127, 123.)*

mitted by ingestion of contaminated food. Birds and domestic fowl, especially ducks, turkeys, and chickens, including their eggs, are commonly identified as the sources of *Salmonella* infections. Inadequate cooking of large turkeys and the ingestion of raw eggs contribute to a significant number of cases of salmonellosis.

Enterocolitis from *Salmonella* infections is normally characterized by abdominal pain, fever, and diarrhea that lasts 3–5 days. The onset of disease symptoms normally occurs 8–24 hours after ingestion of contaminated food. Nausea and vomiting may be initial symptoms but usually do not persist once pain and diarrhea begin. The feces may contain mucus and blood. Generally, the disease is self-limiting, with recovery occurring within one week. During acute salmonellosis the feces may contain 1 billion *Salmonella* cells per gram. Fecal contamination of water and food supplies can contribute to the transmission of this disease. The treatment of salmonellosis normally does not require the use of antibiotics because the prognosis is normally for relatively rapid recovery.

Shigellosis Shigellosis, or bacterial dysentery, is an acute inflammation of the intestinal tract caused by species of the Gram negative genus *Shigella*, including *S. flexneri, S. sonnei,* and *S. dysenteriae*. The transmission of *Shigella* species normally occurs by the direct anal-oral route, although water and food supplies are involved in some outbreaks of bacterial dysentery. *Shigella* species are able to penetrate the mucosal cells of the large intestine and multiply in the submucosa. Areas of intense inflammation develop around the multiplying bacteria and microabscesses form and spread, leading to bleeding ulceration (Figure 17.18).

The symptoms of *Shigella* infections include abdominal pain, fever, and diarrhea, with mucus and blood in the excretion. Bacterial dysentery normally is a self-limiting disease, with recovery occurring 2–7 days after onset. The severe dehydration associated with this disease can cause shock and lead to death in children, for whom the incidence of bacterial dysentery is highest. In cases of diagnosed shigellosis in children, antibiotics are used in treating the disease; however, many *Shigella* strains now contain plasmids coding for antibiotic resistance, and the spectrum of antibiotics needed to effectively combat shigellosis has increased as a result and tetracycline, ampicillin, and nalidixic acid are now employed for treating bacterial dysentery caused by different *Shigella* species.

Other *Campylobacter fetus* var. *jejuni* has been found to be the causative agent of many cases of gastroenteritis in infants. In fact, *C. fetus* may be more important in juvenile gastroenteritis than *Salmonella* species. The transmission of *C. fetus* appears to be via contaminated food or water. *C. fetus* is a Gram negative, motile, spiral-shaped bacterium, formerly known as *Vibrio fetus*, which also causes fetal abortion in cattle and sheep.

Vibrio parahaemolyticus is responsible for a large number of cases of gastroenteritis in Japan and perhaps in the United States. *V. parahaemo-*

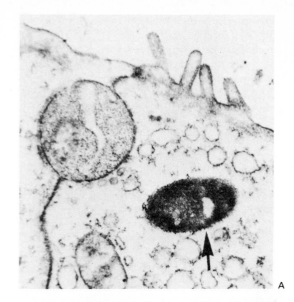

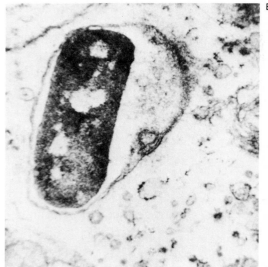

Figure 17.18

Micrographs showing intestine infected with Shigella. *(A) Apical portion of intestinal epithelium 24 hours after* Shigella *infection. A* Shigella *organism (arrow) lies free in the cytoplasm near the lumen surface. The microvilli and terminal web have undergone regressive changes. The endoplasmic reticulum is swollen. Membrane-bound intracytoplasmic inclusions contain osmiophilic granular material resembling ribosomes and a mitochondrion. A similar structure is present in the intracellular space (6000 ×). (B) A composite of intra-epithelial membrane-enclosed* Shigella *organisms (9,000 ×). The membranes frequently enclose cytoplasmic components. G = Golgi apparatus. (Reprinted by permission of Hoeber Medical Division of Harper & Row, Hagerstown, Maryland, from A. Takeuchi et al., 1965,* American Journal of Pathology 47:1037, 1941.)

lyticus occurs in marine environments, and the ingestion of contaminated seafood, particularly the eating of raw fish, is the main route of transmission. Gastroenteritis caused by *V. parahaemolyticus* requires establishment of an infection within the gastrointestinal tract, rather than simple ingestion of an enterotoxin. The symptoms of gastoenteritis generally appear 12 hours after ingestion of contaminated food and include abdominal pain, diarrhea, nausea, and vomiting. Recovery from this form of gastroenteritis normally occurs in 2 to 5 days, and the mortality rate is very low.

Bacillus cereus species are responsible for a low proportion of gastoenterocolitis cases. Strains of this bacterial species cause a relatively mild form of gastroenteritis and recovery normally occurs in less than a day. The occurrence of gastoenteritis due to *B. cereus* requires the ingestion of a large number of spores. The symptoms of *B. cereus* gastrointestinal tract infections include abdominal pain, profuse diarrhea, and nausea.

Yersinia enterocolitica and related species produce a severe form of enterocolitis. The symptoms of an infection with *Y. enterocolitica* resemble appendicitis and include abdominal pain, fever, diarrhea, vomiting, and leucocytosis. Often an appendectomy is performed before this disease is properly diagnosed as **yersiniosis**. Outbreaks of yersiniosis are most common in Western Europe but have also been confirmed in the United States. In an outbreak of yersiniosis in New York involving over 200 school children, the source of infection was traced to a common source of chocolate milk. Ten children underwent unnecessary appendectomies before the true etiology of the disease was established. *Y. enterocolytica* is widely distributed and has been found in water, milk, fruits, vegetables, and seafoods. This organism is psychrotrophic and thus is able to reproduce within refrigerated foods where it can multiply and reach an infectious dose. In fact, *Y. enterocolytica* grows better at 25°C than at 37°C.

Typhoid fever

Outbreaks of **typhoid fever**, a systemic infection caused by *Salmonella typhi*, are associated with contaminated water supplies and the handling of food products by individuals infected with *Salmonella typhi*. Although the portal of entry for *S. typhi* normally is the gastrointestinal tract, infections with *S. typhi* do not initially cause gastroenteritis, but rather the bacteria simply enter the body via this route and cause infections at other sites. In the course of the disease, however, the intestines become involved along with various other organs. A relatively low infectious dose is required for *S. typhi* to establish an infection. The infecting bacteria rapidly enter the lymphatic system and are disseminated through the circulatory system. Phagocytosis by neutrophil cells does not kill *S. typhi*, and the bacteria continue to multiply within phagocytic blood cells. The surface Vi antigen of *S. typhi* apparently interferes with phagocytosis, and elimination of infecting cells depends on the antibody-mediated immune response.

After invasion of the mononuclear phagocyte system, the infection with *S. typhi* becomes localized in lymphatic tissues, particularly in Peyer's patches of the intestine, where ulcers can develop. Localized infections always develop and cause damage to the liver and gall bladder and sometimes also to the kidneys, spleen, and lungs.

The symptoms of typhoid fever include fever (104°F), headache, apathy, weakness, abdominal pain, and a rash with rose-colored spots. The symptoms develop in a stepwise fashion over a three-week period. If no complications occur, the fever begins to decline at the end of the third week. However, if untreated the mortality rate averages 10 percent. Chloramphenicol is effective in the treatment of typhoid fever, and its use and other antibiotics have reduced the death rate to approximately 1 percent.

Cholera

Cholera is caused by the Gram negative, curved rod, *Vibrio cholerae*, serotypes *cholerae* and *eltor*. Although we typically associate cholera with Asia, sometimes referring to the disease as Asiatic cholera, this disease also occurs in the United States, primarily in the Gulf Coast region where cases have been traced to contaminated shellfish. Cholera is a particular problem in socioeconomically depressed countries where there is poor sanitation and inadequate sewage treatment, and where medical facilities have a limited capacity to deal with outbreaks of this disease (Figure 17.19). This disease is endemic in the Ganges Delta, and there are annual epidemic outbreaks of cholera in India and Bangladesh. In endemic areas of Asia the death rate is normally 5–15 percent. Seasonal outbreaks of cholera often occur in Southeast Asia when monsoon rains wash sewage material into drinking water supplies. During sudden epidemic outbreaks of cholera, the mortality rate may reach 75 percent.

V. cholerae is able to multiply within the small intestine and produces the enterotoxin responsible for the symptoms of cholera, which occur suddenly and include nausea, vomiting, abdominal pain, diarrhea with "rice water stools," and

610

Human diseases caused by microorganisms: the respiratory, gastrointestinal, and genitourinary tracts as portals of entry

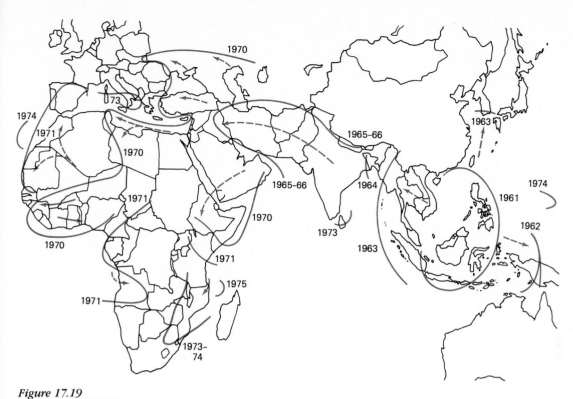

Figure 17.19

Map showing worldwide incidence of cholera from 1961 to 1975.

611

Diseases caused
by pathogens
that enter via the
gastrointestinal
tract

severe dehydration, followed by collapse, shock, and in many cases death. *V. cholerae* itself does not invade the body and is not disseminated to other tissues. It is the enterotoxin produced by *V. cholerae* that binds irreversibly to the epithelial cells of the small intestine, stimulating the formation of cyclic AMP within the mucosal cells of the intestine. The accumulation of cyclic AMP initiates secretion of water and electrolytes into the lumen of the small intestine because of changes in membrane permeability of the mucosal cells. The rapid loss of fluid from the cells of the gastrointestinal tract associated with this disease often produces shock, and if untreated there is a high mortality rate. The initial diarrhea that results from infection with *V. cholerae* can cause the loss of several liters of fluid within a few hours. The treatment of cholera centers on replacing fluids and maintaining the electrolyte balance, that is, on combating shock. Treatment with tetracycline generally reduces the duration of the disease.

Appendicitis

There are more than 200,000 cases per year in the United States of **appendicitis**, an inflammation of the appendix occurring when there is an obstruction of the lumen of the appendix. Appendicitis is caused by a mixture of bacterial populations that constitute the normal microbiota of the intestinal tract, and virtually all members of the normal microbiota of the gastrointestinal tract can contribute to appendicitis. The symptoms of appendicitis normally include abdominal pain, localized tenderness, fever, nausea, vomiting, and leukocytosis. Treatment of appendicitis usually involves the surgical removal of the appendix. Serious complications from appendicitis arise if the infection is permitted to progress and the appendix ruptures. In such cases, systemic bacteremia ensues, and there is a high rate of mortality. Antibiotics are not effective in treating uncomplicated acute appendicitis and are unable to stop the development of infection before rupture of the appendix can occur. In cases where appendicitis is complicated by rupture of the appendix, antibiotic treatment is, however, essential in controlling the spread of infection to other tissues.

Fungal food poisoning

Although fungi normally do not cause infections via the gastrointestinal system, **mycotoxins** produced by some fungi are responsible for some

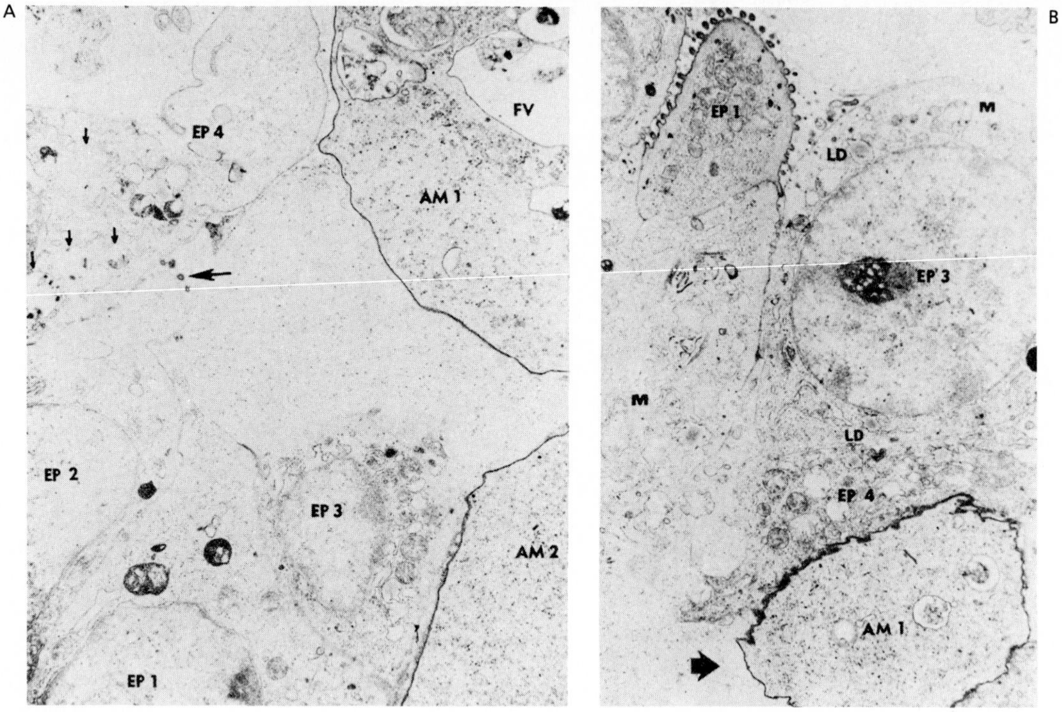

612

Human diseases
caused by micro-
organisms: the
respiratory, gas-
trointestinal, and
genitourinary
tracts as portals
of entry

Figure 17.20

Micrograph of intestinal epithelium of a guinea pig infected with Entamoeba histolytica. *(A) Amebae (AM) in contact with the luminal surface of interglandular epithelial cells (EP) (7,000×). The varying degrees of cytoplasmic change range from nearly normal in one cell (EP1) to severe alterations (EP4). The microvilli and terminal web have disappeared for the most part, with a remnant remaining (large arrow) in a host cell (EP2). The cell membrane of amebae is thicker and denser than the host plasmalemma. AM1 has adhered to the projected apical cytoplasm of host cell EP3. Note spectrum of mitochondrial changes (small arrows) within a single cell (EP4). (B) A portion of ameba cytoplasm (AM1), apparently a pseudopodium, extending from ameba AM2 toward the basal lamena (dotted line). The basal cytoplasm of a host cell lacks cellular organelles and shows cytoplasmic projection (arrow) toward the pseudopodium (AM1). Note close contact between membrane of host cell (EP4) and ameba (AM1). Increasing degeneration process ranging in degree from EP1 to EP4 is evident. A third ameba (AM3) is in the lumen (6,000×). M = altered mitochondria; LD = liquid droplet. (Reprinted by permission of the American Society for Tropical Medicine and Hygiene, from A. Takeuchi and B. P. Phillips, 1975,* American Journal of Tropical Medicine and Hygiene *24: 39, 41.)*

serious cases of food poisoning. Many mycotoxins are potent neurotoxins. Various species of mushrooms contain toxins that can be absorbed through the gastrointestinal tract and the ingestion of **poisonous mushrooms**, such as *Amanita phalloides*, is normally fatal. The amatoxins and phallotoxins produced by *A. phalloides*, and other species of *Amanita*, produce symptoms of food poisoning 8–24 hours after their ingestion. Initial symptoms include vomiting and diarrhea; later, degenerative changes occur in liver and kidney cells, and death may occur within a few days of ingesting as little as 5–10 mg of toxin.

Some filamentous fungi, other than mushrooms, also produce toxins that can cause human disease. *Aspergillus* species growing on peanuts and grains produce **aflatoxins**, which are potent carcinogens as well as toxic. They are known to cause death in sheep and cattle and may be involved in some human disease conditions. Aflatoxins are the only known carcinogens for which the United States government has set permissible levels; all other products with carcinogenic activity are banned outright. Ergotism, another disease caused by fungi, results from ingesting grain containing **ergot alkaloids** produced by *Claviceps* *purpurea*. The toxins of *C. purpurea* cause degeneration of the capillary blood vessels, and this type of fungal food poisoning has a relatively high mortality rate. Symptoms of ergotism may include vomiting, diarrhea, thirst, hallucinations, convulsions, and lesions at the body extremities. Various outbreaks of mass hallucinations have been traced to contamination of food with ergot alkaloids, and there even are theories that the Salem "witch hunts" were related to grain contamination and widespread ergotism.

Algal food poisoning

Algae are rarely considered as the etiologic agents of disease, but **paralytic shellfish poisoning** is caused by toxins produced by dinoflagellate *Gonyaulax*. Blooms of *Gonyaulax* cause red tides in coastal marine environments. During such algal blooms the algae and the toxins they produce can be concentrated in bivalve shellfish, such as clams and oysters. The ingestion of shellfish containing algal toxins can lead to symptoms that resemble botulism. Shellfishing is banned in areas of *Gonyaulax* blooms to prevent this form of food poisoning.

Diseases caused by protozoan pathogens

Amebic dysentery

Amebic dysentery or **amebiasis** is caused by the protozoan *Entamoeba histolytica*. Infections with *E. histolytica* may be asymptomatic or may involve mild or severe diarrhea and abdominal pain. Amebic dysentery occurs as a result of inadequate sewage treatment and contamination of water with *E. histolytica*, whose cysts are not killed by normal chlorination methods used to treat municipal drinking water. Infection is acquired by ingesting contaminated food or water containing cysts of *E. histolytica*. Infestation occurs in the small intestine without causing any disease symptoms. The **trophozoites** (motile flagellate forms) multiply within the colon where they appear to ingest red blood cells, yeasts, and bacteria (Figure 17.20). When immune surface defense mechanisms are lowered, the trophozoites are able to invade the epithelial cells of the colon. The invasion of the colon lining results in the formation of ulcers. The trophozoites spread through the submucosa, cutting off the blood supply through the mucosal lining, which leads to sloughing off of mucosal

cells and enlargement of ulcers. Mucus and blood characteristically appear in the feces from these lesions. Cysts are shed with the fecal matter and can enter and contaminate water supplies. In some cases *E. histolytica* can spread to the liver, lung, or skin, causing ulcer or abscess formation in these tissues. Several antiprotozoan drugs, such as metronidazale, are effective in treating amebiasis, and the choice of which one depends on whether the infection is restricted to the intestinal tract or whether other organs, such as the liver, are involved.

Giardiasis

Giardia lamblia, a flagellated protozoan, is responsible for most cases of infectious diarrhea caused by protozoa. *G. lamblia* forms both motile trophozoites and nonmotile cysts, and the cysts are the infective form. The cysts of *G. lamblia* can enter the gastointestinal tract through contaminated water. A high incidence of giardiasis occurred among groups touring Leningrad during the 1970s as a result of contaminated water supplies. In 1973 a major outbreak of giardiasis occurred in upstate New York, with an estimated 4,800 individuals developing symptoms of the disease. The following year the disease was the most common waterborne disease in the United States. *G. lamblia* can live saprophytically within the small intestine without causing any symptoms of giardiasis, and in the United States almost 4 percent of the population appear to be infected by this organism. Excessive growth of the organism, however, can cause disease symptoms that include diarrhea, dehydration, mucus secretion, and flatulence. Metronidazole is generally used in the treatment of this disease.

Toxoplasmosis

Toxoplasmosis is caused by the protozoan *Toxoplasma gondii*, a member of the sporozoa. Meat containing cysts of *T. gondii* is frequently involved in the transmission of this disease, although *T. gondii* can also be transmitted congenitally (Figure 17.21). The protozoa normally multiply at the site of entry and then are disseminated via the bloodstream to other organs and tissues. Growth of the trophozoites of *T. gondii* causes the death of infected cells. Cell-mediated immunity is involved in containing infections of *T. gondii*. The symptoms and prognosis of toxoplasmosis depend on the virulence of the infecting strain of *T. gondii* and which organs are involved. In most cases toxoplasmosis is asymptomatic, but when multiple organs are involved,

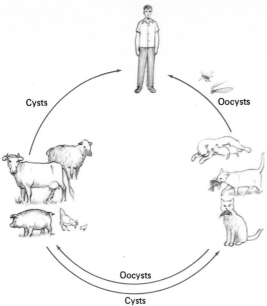

Figure 17.21

Life cycle of Toxoplasma gondii. *Cats play an important role in transmission of this disease, appearing to be the definitive host.*

Cysts

Oocysts

Oocysts

Cysts

the consequences are serious and mortality can occur. Some infections with *T. gondii* have been successfully controlled with antimalarial drugs, such as pyrimethamine. Serious consequences of toxoplasmosis occur when there is central nervous system involvement, which is particularly prevalent in the congenital transmission of this disease.

614

Human diseases caused by microorganisms: the respiratory, gastrointestinal, and genitourinary tracts as portals of entry

Diseases caused by pathogens that enter via the genitourinary tract

The extremities of the genitourinary tract in both males and females contain an indigenous microbiota, but the inner tissues of the genitourinary tract, including the bladder and kidneys, are normally sterile. These tissues are subject to infection with **opportunistic pathogens** if they become contaminated by microorganisms from the indigenous microbiota of the extremities of the genitourinary tract or the gastrointestinal tract. Additionally, the genitourinary tract provides the portal of entry for pathogens that are directly transmitted during sexual intercourse. Infections with such pathogens are known as venereal, or sexually transmitted, diseases.

Urinary tract infections

Urinary tract infections can be caused by a wide variety of microorganisms. The urethra and urinary bladder are most frequently the sites of infection within the urinary tract, with the resulting infections referred to as **urethritis** and **cystitis**, respectively. The kidney is also subject to microbial infections leading to the disease **pyelonephritis**. Although many microorganisms can cause urinary tract infections, the most common etio-

logic agents are Gram negative bacteria, particularly those normally occurring in the gastrointestinal tract. Accidental contamination of the urinary tract with fecal matter appears to be one of the most important means of transmission of such infections. Although *Escherichia coli* probably is the most common cause of urinary tract infections, members of the genera *Klebsiella, Enterobacter, Serratia, Proteus,* and *Pseudomonas* are also isolated relatively frequently (Table 17.4). Infections with *Serratia* species and *Pseudomonas aeruginosa* most often occur after catheterization procedures because the insertion of a catheter can carry microorganisms from the extremities of the genitourinary tract, contaminating the inner tissues and resulting in urethritis and cystitis.

The symptoms of urethritis normally include pain and a burning sensation during urination. Cystitis is characterized by suprapubic pain and the urge to urinate frequently. Growth of bacteria in the urinary tract can cause an obstruction. Various antibiotics, such as nitrofurantoin and nalidixic acid, are used in treating infections of the lower urinary tract. Depending on whether there is an obstruction or another underlying cause involved in the establishment of an infection, the prognosis is normally good.

table 17.4

Representative relative proportions of bacterial isolates from hospital-acquired and community-acquired urinary tract infections

Community-acquired organisms	%	Hospital-acquired organisms	%
Escherichia coli	57	Escherichia coli	39
Klebsiella pneumoniae	13	Proteus mirabilis	18
Proteus mirabilis	8	Pseudomonas aeruginosa	17
Streptococcus sp.	8	Streptococcus sp.	11
Pseudomonas aeruginosa	7	Klebsiella pneumoniae	10
Proteus rettgeri	5	Serratia marcescens	8
Others	2		

Infections of the urethra may spread to the kidney, causing serious life-threatening disease. **Pyelonephritis** can cause kidney failure and death, but as with other urinary tract infections, the use of antibiotics is generally effective in curing this disease. *Escherichia coli* is the most frequent etiologic agent of pyelonephritis in patients who develop infections outside of hospital settings. *Proteus mirabilis, Enterobacter aerogenes, Klebsiella pneumoniae, Pseudomonas aeruginosa, Streptococcus* sp., and *Staphylococcus* sp. are additional causative agents of pyelonephritis. Infections with *Proteus* sp. are especially significant because they have the enzymatic capacity to decompose urea in the urine to ammonia and carbon dioxide. Urgent and frequent urination is normally symptomatic of pyelonephritis. Pyelonephritis and other urinary tract infections are more frequent in adult women than in males or young females, in part because of anatomical differences in the structure of the urinary tract of females, and also because of the physiological changes that occur during menstruation and pregnancy that favor the establishment of microbial infections.

Female genital tract infections

Vaginal tract infections
The female genital tract is subject to infection by a variety of microorganisms. The trauma associated with sexual intercourse, menstrual bleeding, normal changes during pregnancy, and trauma due to childbirth contribute to occurrences of infections of the female genital tract. Infections of the vulva and vagina are common and generally not serious. Vulvovaginitis is an inflammation of both the vulva and vagina and can be caused by viruses, bacteria, fungi, and protozoa. In about 20 percent of the cases of vulvovaginitis, the flagellate protozoan *Trichomonas vaginalis* is the etiologic agent. The fungus *Candida albicans* is also frequently implicated. Bacterial and fungal overgrowth of the vaginal tract is common during pregnancy when the pH of the vaginal tract increases. The use of birth control pills also tends to result in higher pH values in the vagina, and the administration of antibacterial antibiotics, which can adversely affect the indigenous bacterial populations, also favors the development of *Candida* infections of the vaginal tract.

The excessive growth of *Trichomonas vaginalis, Candida albicans*, or various bacteria cause changes in the mucosal cells lining the vaginal tract. The symptoms of vulvovaginitis include increased vaginal discharge and burning. In cases where *T. vaginalis* is the causative agent, there is normally a profuse, greenish, odorous discharge, and when vulvovaginitis is caused by *Candida* infections, the discharge is normally thicker and cheesier and is released in lesser quantity. Antibiotic treatment is effective in controlling vulvovaginitis, with the specific drug choice depending on the identification of the specific etiologic agent. Metronidazole can be used in cases of protozoan infections, and nystatin generally is used for *Candida* infections.

Infections of the lower parts of the female genital tract may also spread to higher structures, including the cervix and fallopian tubes. Infections of the upper regions of the genital tract are particularly serious during pregnancy and can lead to septic abortion. Some infections of the gestating female genital tract, such as with the Gram positive, acid-fast bacterium *Listeria monocytogenes*, can also be transferred to the fetus, resulting in congenital abnormalities. Microbial infection with *L. monocytogenes* can occur either through the placenta or during childbirth. In some cases of female genital tract infections, it is necessary to perform a Caesarean section delivery in order to protect the infant from contamination.

Toxic shock syndrome

Toxic shock syndrome is caused by strains of *Staphylococcus aureus* that contain a lysogenic phage. This bacterium can enter the body via the genital tract, and the elaboration of its toxins causes high fever, nausea, vomiting, and in many cases death. This disease is not restricted to women and can occur after introduction of *S. aureus* via other portals of entry, including surgical wounds. The occurrence of toxic shock syndrome, though, is especially correlated with the use of tampons during menstruation, particularly if these devices are left in place for a long period of time. The association of this disease with the use of tampons received a great deal of publicity in the early 1980s, forcing one major manufacturer to remove its product from the market. It is likely that the trauma associated with the use of tampons produces lesions in the vaginal wall that provide a portal of entry for *S. aureus*. The hormonal changes that occur during menstruation also favor the proliferation of bacteria in the vaginal tract.

616

Human diseases caused by microorganisms: the respiratory, gastrointestinal, and genitourinary tracts as portals of entry

Puerperal fever

Puerperal or **childbed fever** is a systemic bacterial infection that may be acquired via the genital tract during childbirth or abortion. The most frequent etiologic agents of postpartum sepsis are β-hemolytic group A and B *Streptococcus* species. *Staphylococcus, Pseudomonas, Bacteroides, Peptococcus, Peptostreptococcus*, and *Clostridium* species as well as other bacteria can also cause this disease. The source of infection normally comes from the obstetrician, obstetrical instruments, or bedding. The bacteria causing puerperal fever are not normally part of the resident microbiota of the vaginal tract. Prior to the introduction of aseptic procedures, puerperal fever was often a fatal complication after childbirth and remains an important complication following childbirth and abortion procedures. It was the leading cause of maternal death in Massachusetts in the mid-1960s. Penicillin is usually effective in treating postpartum sepsis. The use of proper obstetric procedures generally prevents this disease.

Sexually transmitted diseases

Sexually transmitted diseases are contracted by direct sexual contact with an infected individual, generally during sexual intercourse. The physiological properties of the pathogens causing these diseases restrict their transmission, for the most part, to direct physical contact because the etiologic agents of the sexually transmitted diseases have very limited natural survival times outside infected tissues. The rate of incidence of sexually transmitted diseases reflects contemporary sexual behavioral patterns but may also in part reflect changes in reporting and recording cases of these diseases. At present, outbreaks of some sexually transmitted diseases are considered to be reaching epidemic proportions. The social implications inherent in the transmission of these diseases often overshadow the fact that they are infectious diseases and must be treated as medical problems with emphasis on curing the patient and reducing the incidence of disease by preventing spread of the infectious agents. The overall control of sexually transmitted diseases rests with breaking the network of transmission, which necessitates public health practices that seek to identify and treat all sexual partners of anyone diagnosed as having one of the sexually transmitted diseases.

Gonorrhea

Gonorrhea is a sexually transmitted disease caused by the Gram negative diplococcus, *Neisseria gonorrhoeae*, often referred to as gonococcus. *N. gonorrhoeae* is a fastidious organism readily killed by drying and exposure to metals. The sensitivity of *N. gonorrhoeae* to desiccation makes the chances of transmission of gonorrhea through inanimate objects, such as toilet seats in public restrooms, negligible. The adherence of *N. gonorrhoeae* to the lining of the genitourinary tract during sexual transmission and the ensuing spread of the infection depends on the pili of this bacterial species. *N. gonorrhoeae* is able to penetrate to the subepithelial connective tissue during spread of the infection (Figure 17.22). *N. gonorrhoeae* infects the mucosal cells lining the epithelium. This bacterium is able to infect the urethra, cervix, anal canal, pharynx, and conjunctivae.

There has been an alarming increase in the number of cases of gonorrhea in the United States since 1960 (Figure 17.23). This increase coincides with the widespread introduction of birth control pills, the hormones in which cause pH values in the vaginal tract to increase, removing the normal protection afforded against infections by acid-sensitive bacteria, including *N. gonorrhoeae*. Gonorrhea is normally contracted from someone who is asymptomatic or who has symptoms but does not seek treatment. The rate of gonorrhea acquisition for males is about 35 percent after a single exposure to an infected female and rises to 75 percent after multiple sexual contacts with the same individual.

In most cases gonorrhea is a self-limiting disease, but in both sexes the infection may spread

617

Diseases caused
by pathogens
that enter via the
genitourinary
tract

Figure 17.22

Electron micrographs of gonococcus in the genital tracts of patients with gonorrhea. (A) Gonococcus (solid arrow) and commensal bacteria lying on the cervical epithelium of a female (25,650×). Colloidal thorium treatment reveals well-defined acidic capsules on the commensal bacteria (hollow arrows). Glycoproteins of the epithelial cell coat and in cervical mucus have also been stained. By contrast, there is no evidence of a capsule in the gonococcal surface. Light staining with colloidal thorium can be attributed to contaminating host cell material. (B) Gonococcus lying between epithelial cells in a human fallopian tube organ culture 8 h after challenge (21,900×). The gonococcus is tightly enclosed by the surrounding host cell surfaces (arrows) that are linked in places by desmosomes. (C) Gonococci approaching the surfaces of a urethral epithelial cell from a male with early symptomatic gonorrhea (29,200×). Microvilli from the host cell have made contact with the bacteria. The vesicles (V) are characteristically associated with host-grown gonococci and appear at some points (arrows) to be budded off from the bacterial surface. (Reprinted by permission of the American Society for Microbiology, Washington, D.C., from M. E. Ward et al., 1975, in Microbiology—1975, *D. Schlessinger, ed., pp. 188–199.)*

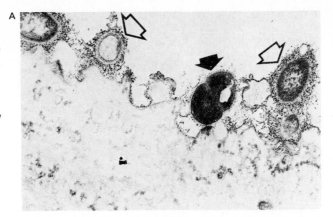

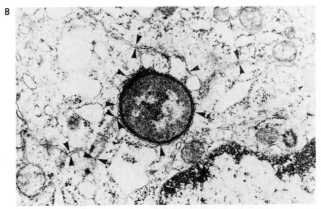

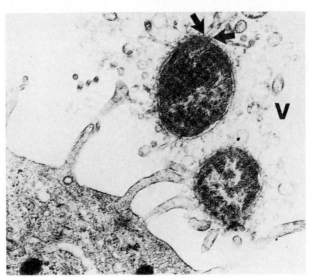

to contiguous parts of the genitourinary tract, and *N. gonorrhoeae* may be disseminated to other parts of the body. Infections of the prostate in males and the fallopian tubes in females produces sterility in some cases of gonorrhea. In men gonorrhea results in a characteristic painful, purulent urethral discharge. In women the most common site of infection is the cervix. Various other tissues may be involved if the infection spreads. Female infections with *N. gonorrhoeae* exhibit a wide variety of symptoms. Most commonly, symptomatic cases exhibit inflammation with some pain and

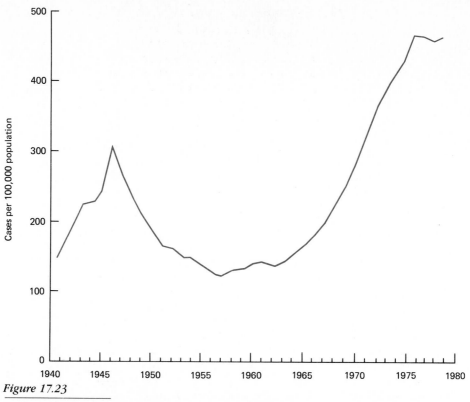

Figure 17.23

The incidence of gonorrhea shows an alarming, epidemic increase in recent years, as shown by these reported civilian cases in the United States from 1941 to 1978. Note the sharp decline from 1946 to 1956, followed by the dramatic increase since then. The precise cause for this rise is unknown; some point to increased sexual activity in a permissive society; it seems likely that the use of oral contraceptives that lead to altered vaginal pH has played a role in the increased ability of Neisseria gonorrhoeae *to survive, resulting in increased transmission rates of this disease.*

618

Human diseases caused by micro-organisms: the respiratory, gas-trointestinal, and genitourinary tracts as portals of entry

swelling, abnormal vaginal discharge, and abnormal menstrual bleeding. Because many women with gonorrhea are asymptomatic, the eyes of all infants are routinely washed immediately after birth with a 1 percent solution of silver nitrate to prevent infections of the eye, which could result from the transmission of *N. gonorrhoeae* from mother to infant during passage through the vaginal tract.

Gonorrhea is readily treated with antibiotics, with penicillin being the antibiotic of choice. Other antibiotics, such as tetracycline, are also effective. In recent years there has been an increase in the tolerance of *N. gonorrhoeae* to antibiotics, creating a major concern in the treatment of gonorrhea (Figure 17.24). The recent identification of β-lactamase-producing strains of *N. gonorrhoeae* in some cases may lead to a change in the preference of penicillin for treating gonorrhea because such strains are resistant to most penicillins. In cases of penicillin-resistant strains of *N.*

gonorrhoeae, streptomycin is the antibiotic of choice.

Syphilis

Syphilis is a sexually transmitted disease caused by *Treponema pallidum*. This organism is a bacterial spirochete, fastidious in its growth requirements and readily killed by drying, heat, and disinfectants such as soap, arsenicals, and mercurial compounds. The inability of *T. pallidum* to survive long outside the body makes transmission through inanimate objects (fomites) virtually nonexistent. Transmission depends on direct contact with the infective syphilitic lesions containing *T. pallidum*. Historically hot bath spas and arsenic- and mercury-containing compounds have been used in the treatment of syphilis. *Treponema pallidum* enters the body via abrasions of the epithelium and by penetrating into mucous mem-

Figure 17.24

Histogram showing increased resistance of Neisseria gonorrhoeae *to penicillin from 1950 to 1969. The increasing occurrence of resistant strains coincides with increased use of antibiotics and reflects selective pressure.*

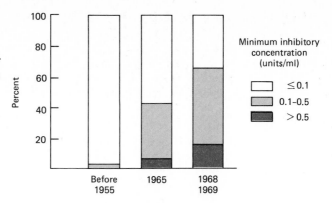

branes. The bacteria migrate to the lymphatic system shortly after penetrating the dermal layers.

Syphilis manifests itself in three distinct stages. During the **primary stage** of syphilis, a chancre develops at the site of *Treponema* inoculation. Primary lesions generally occur on the genitalia. The average incubation time for the manifestation of primary syphilis is 21 days after infection. The primary lesions typically heal within 3 to 6 weeks. The **secondary stage** of syphilis normally begins 6–8 weeks after the appearance of the primary chancre. During this stage there are cutaneous lesions and lesions of the mucus membranes that contain infective *Treponema pallidum*. Lesions may appear on the lips, tongue, throat, penis, vagina, and numerous other body surfaces (Figure 17.25). There may be additional symptoms of systemic disease during this stage, such as headache, low-grade fever, enlargement of the lymph nodes, etc. After the secondary stage, syphilis enters a char-

acteristic **latent period** during which there is an absence of any clinical symptoms of the disease. The latent phase can be detected by serological means because the Wassermann and similar tests are still positive even in the latent period. The latent phase marks the end of the infectious period of syphilis.

The **tertiary phase** of syphilis, also known as late syphilis, usually does not occur until years after the initial infection. During tertiary syphilis damage can occur to any organ of the body. In about 10 percent of the cases of untreated syphilis, the tertiary phase involves the aorta, and damage to this major blood vessel can result in death. In approximately 8 percent of the cases of untreated syphilis, there is central nervous system involvement with a variety of neurological manifestations, including personality changes and paralysis.

If untreated, approximately 25 percent of the

619

Diseases caused
by pathogens
that enter via the
genitourinary
tract

Figure 17.25

Secondary syphilis lesions on the palms of hands resulting from syphilis infection. (From BPS—Leonard Winograd, Stanford University.)

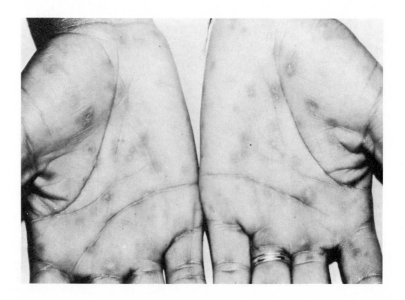

individuals who contract syphilis will suffer one or more relapses of the secondary stage during the first four years of illness, 15 percent will develop tertiary benign lesions, 10 percent cardiovascular lesions, and 8 percent central nervous system lesions. The risks of debilitating symptoms and death make this a very serious form of venereal disease. Fortunately, syphilis can be treated with penicillin and other antibiotics, particularly during the early stages. Unlike gonorrhea, the number of cases of syphilis in the United States has been relatively constant (Figure 17.26). There has been a large increase in reported cases of syphilis among homosexual men and, for example, over 60 percent of the cases of syphilis reported in the state of Washington during 1971 involved homosexual transmission. As with other sexually transmitted diseases, the overall control of syphilis, however, rests with finding and treating all sexual contacts who may have contracted this disease and may be involved in the further transmission of syphilis. No long-term immunity develops after infections with *Treponema palli-*

dum, and individuals who are cured by treatment with antibiotics remain susceptible to contracting syphilis again.

In addition to sexually transmitted syphilis, *Treponema pallidum* can be transmitted across the placenta of pregnant women with syphillis, infecting the fetus and causing stillbirth or congenital syphilis in the newborn. Stillbirth is likely if pregnancy occurs during the primary or secondary stages of syphilis. Congenital syphilis is most likely during the latent period of the disease. Congenital syphilis has very serious consequences, usually resulting in mental retardation and neurological abnormalities in the infant; the probability of such infants surviving depends on the specific nature of the neurological impairment.

Lymphogranuloma venereum

Lymphogranuloma venereum is normally transmitted by sexual contact and is caused by species of *Chlamydia trachomatis*, which can also cause less serious urethritis. *Chlamydia* species are small, obligate intracellular parasitic bacteria. The

Human diseases caused by microorganisms: the respiratory, gastrointestinal, and genitourinary tracts as portals of entry

Figure 17.26

Incidence of primary and secondary syphilis among the civilian population in the United States from 1941 to 1979. The incidence of this sexually transmitted disease has not shown the same dramatic increase as has occurred for gonorrhea.

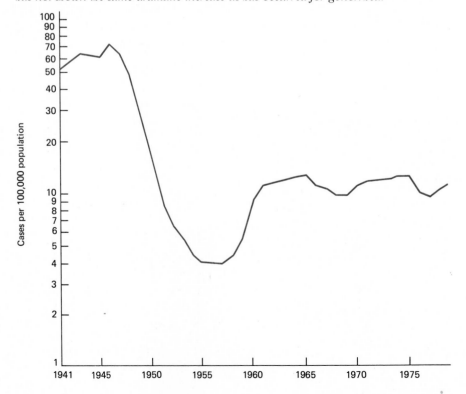

mechanism by which the bacteria enter the body is not known but appears to be through small abrasions in the genitourinary tract. The *Chlamydia* cells are phagocytized and carried to lymph nodes. The lymph nodes are the primary sites affected in this disease, with swelling and tenderness usually occurring 5–21 days after healing of the primary lesions in the area of initial infection. In about 25 percent of the cases, there is a genitoanorectal syndrome, characterized by a bloody, mucopurulent rectal discharge. Lymphogranuloma venereum can cause serious consequences, but this rarely occurs. Although treatment with tetracycline and other antibiotics is usually effective in preventing the onset of late and serious symptoms, it does not seem to shorten the usual time of 4-6 weeks required for enlarged lymph nodes to heal.

Chancroid

Chancroid is a relatively rare sexually transmitted disease caused by *Haemophilus ducreyi*. Chancroid occurs most frequently in the underdeveloped nations of Africa, the Caribbean, and Southeast Asia, but the incidence of this disease has been increasing in the United States. Soft chancres develop 3–5 days after sexual exposure, and untreated lesions may persist for months. Chancroidal ulcers heal quickly but often leave deep scars. Sulfanilamide and tetracycline are used in the treatment of this disease.

Granuloma inguinale

Granuloma inguinale, also known as granuloma venereum, is caused by a Gram negative rod-shaped bacterium, *Calymmatobacterium granulomatis*. Granuloma inguinale most frequently occurs in India, the west coast of Africa, islands of the South Pacific, and some South American countries. The disease is relatively rare in the United States. It appears to be sexually transmitted, as evidenced by the fact that initial lesions occur on the genitalia and the first lesions appear 9–50 days after sexual intercourse with an infected individual. There appears to be a very low rate of infection for this sexually transmitted disease, and in many cases sexual partners do not contract this disease. The genitalia develop characteristic ulcers, older portions of which exhibit loss of pigmentation. Chloramphenicol, erythromycin, and tetracycline, as well as other antibiotics, are used in the treatment of this disease.

Genital herpes

Genital herpes is caused by a herpes simplex type 2 virus. Diseases caused by herpes simplex type 1 virus will be discussed in the next chapter because herpes simplex type 1 does not appear to be sexually transmitted, and the genitourinary tract need not be the primary portal of entry. Herpes simplex 2 is most frequently transmitted by sexual contact and causes infection of the genitalia. It is estimated that 20 million Americans now have genital herpes and that there will be at least a half a million new cases per year unless effective means are found for controlling this disease. In women the primary site of herpes infection is the cervix but may also involve the vulva and vagina. In men the herpes simplex virus frequently infects the penis. The primary infection exhibits symptoms of genital soreness and ulcers in the infected areas. The virus and manifestations of infection may be transmitted to other areas of the body, most notably the mouth and anus. Genital herpes may have particularly serious repercussions in pregnant women because the virus can be transmitted to the infant during vaginal delivery, causing damage to the infant's central nervous system and/or eyes. Herpes is lethal to up to 60 percent of infected newborns, and for surviving babies there is a 50 percent risk of blindness or neurological damage.

The ulcers produced from herpes simplex type 2 infection generally heal spontaneously in 10 to 14 days. However, because of the budding mode of reproduction of herpesviruses, the infection is not eliminated when the ulcers heal, but rather a reservoir of infected cells remains in the body within nerve cells. At subsequent times multiplication of the viruses can produce new ulcers, even in the absence of additional sexual activity. It is not known exactly what initiates subsequent attacks of herpes, but such recurrences may be triggered by sunlight, sexual activity, menstruation, and stress. Although a direct relationship between genital herpes viruses and cancer has not been established, adolescent females who have had extensive sexual contacts and have developed genital herpes infections have an elevated rate of development of cervical cancer. The disease remains transmissible, which interferes with establishing stable sexual relationships; there are many adverse psychological effects associated with genital herpes. Genital herpes disrupts marital relationships, and the epidemic outbreak of genital herpes may reverse the sexual revolution. Several of the newly developed antiviral drugs should be useful in the treatment of herpes viral infections. In particular, acyclovir and interferon show promise of reducing the severity of the symptoms; neither of these drugs, though, promises a cure for the disease.

In this chapter we have considered a number of the diseases of humans caused by microbial pathogens that enter the body via respiratory, gastrointestinal, and genitourinary tracts. Figure 17.27 compares the relative incidence of 14 diseases discussed in this chapter in various regions of the world. The pathogens involved in causing human disease are generally restricted to a particular portal of entry through which they can enter the body and establish an infection.

Airborne transmission, particularly in the form of droplet spread of aerosols containing pathogenic microorganisms from infected to susceptible individuals, is important in outbreaks of diseases caused by pathogenic microorganisms that enter through the respiratory tract. Crowded conditions favor airborne transmission of pathogens that enter the body this way. As a result, outbreaks of diseases caused by such pathogens often affect individuals who work or live together, such as children attending the same school.

Contaminated food and water supplies play an important role in disease transmission in cases where the pathogens or toxins produced by microorganisms enter the body through the gastrointestinal tract. Food poisoning occurrences often affect a number of individuals who have eaten together. Many times all members of a family, or patrons of a restaurant, will be stricken after ingestion of contaminated food. There have been occasional outbreaks of food-borne infection or poisoning among the passengers on a ship or plane who have been served the same meal. Major outbreaks of food-borne disease represent a frequent problem in institutions that serve food prepared in large quantities, such as prisons and nursing homes.

In the case of pathogens that enter the body through the genitourinary tract, contamination with microorganisms comprising the normal microbiota of the gastrointestinal tract and sexually transmitted pathogens are principally involved in such infections. The treatment of sexually transmitted diseases requires public health measures to identify the source of infection in order to interrupt the chain of transmission. Recognition that sexually transmitted diseases represent a medical, rather than a social, problem is important for effectively treating and reducing the incidence of these diseases.

Generally, infections that are localized near the point of entry are less serious than those that spread systemically. In many cases, microbial infections are self-limiting, and recovery usually occurs without serious complications. Some diseases, however, are associated with a high mortality rate if left untreated, but fortunately, antibiotics are effective in treating many such diseases. The key to control of microbial diseases rests with lowering the incidence of infection wherever possible by taking preventative measures. By knowing the etiologic agents and the routes of transmission of specific diseases, public health measures can be taken to control many diseases. These measures include vaccination, water treatment, sanitation improvement, and educational programs, as well as treatment of infected individuals to reduce reservoirs of pathogenic microorganisms. The effective elimination of smallpox as a disease of human beings can be repeated for several other diseases, and in almost all cases more effective treatment and preventative measures can reduce the rate of occurrence of infectious diseases. We should remember that we have only known how to treat infectious disease on a rational basis for a very few decades. We have made tremendous advances in the post-World War II era in reducing the incidence of infectious diseases and the fatalities due to pathogenic microorganisms. We have the medical means to reduce the destructiveness of many diseases in many areas of the world but lack the socioeconomic determination to mobilize the resources necessary to do so. We need to apply our efforts to preventing and/or developing effective treatment methods for the various remaining infectious diseases.

622

Human diseases caused by microorganisms: the respiratory, gastrointestinal, and genitourinary tracts as portals of entry

A = Africa
C = South-East Asia
E = Eastern Mediterranean

B = Americas
D = Europe
F = Western Pacific

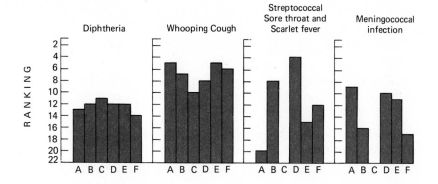

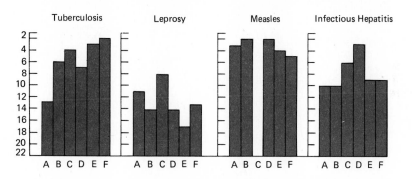

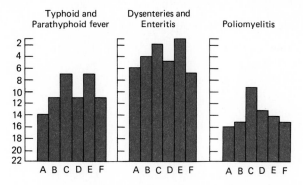

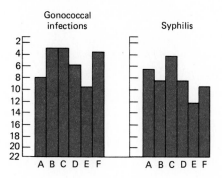

623

Postlude

623

Figure 17.27

*Histograms showing relative rankings of 14
infectious diseases in six regions of the
world.*

1. What is a portal of entry?

2. How are pathogens mainly transmitted that enter the body through the gastrointestinal tract? How do we control the transmission of these pathogens?

3. How are pathogens mainly transmitted that enter the body through the respiratory tract? How can we control the transmission of these pathogens?

4. What is the difference between a food-borne infection and food poisoning (intoxication)? How is this difference reflected in how we control different types of food-borne disease?

5. Name three viral, three bacterial, and three protozoan diseases that enter through the respiratory tract.

6. For each of the following diseases, what are the causative organisms and characteristic symptoms?
   a. influenza
   b. tuberculosis
   c. rheumatic fever
   d. histoplasmosis
   e. poliomyelitis
   f. Legionnaire's disease
   g. whooping cough
   h. typhoid fever
   i. botulism
   j. cholera

7. What organisms often cause meningitis?

8. What organisms often cause urinary tract infections?

9. What organisms normally cause vaginitis?

10. What is a sexually transmitted disease? What are some examples of these diseases and what are the causative organisms? Why is it difficult to control the spread of these diseases, considering that many are readily treated with antibiotics?

11. Name five diseases that are routinely prevented today through the use of prophylactic immunization.

624

Human diseases caused by micro-organisms: the respiratory, gastrointestinal, and genitourinary tracts as portals of entry

Boyd, R. F., and B. G. Hoerl. 1981. *Basic Medical Microbiology*. Little, Brown and Co., Boston.

Boyd, R. F., and J. J. Marr. 1980. *Medical Microbiology*. Little, Brown and Co., Boston.

Braude, A. I. (ed.). 1980. *Medical Microbiology and Infectious Disease*. W. B. Saunders Co., Philadelphia.

Dubos, R. J., and R. G. Hirsch (eds.). 1965. *Bacterial and Mycotic Infections of Man*. J. B. Lippincott Co., Philadelphia.

Davis, B. D., R. Dulbecco, H. N. Eisen, H. S. Ginsberg, and W. B. Wood. 1981. *Microbiology*. Harper & Row, Publishers, Hagerstown, Maryland.

Duguid, J. P. (ed.). 1979. *Medical Microbiology*. Churchill Livingstone, New York.

Emmons, C. W., C. H. Binford, J. P. Ulz, and K. J. Kwon-Chung. 1977. *Medical Mycology*. Lea & Febiger, Philadelphia.

Gilbert, D. N., and J. P. Sanford (eds.). 1979. *Infectious Diseases*. Grune & Stratton, New York.

Hoeprich, P. D. (ed.). 1977. *Infectious Diseases: A Modern Treatise of Infection Processes*. Harper & Row, Publishers, Hagerstown, Maryland.

Jawetz, E., J. L. Melnick, and E. A. Adelberg. 1980. *A Review of Medical Microbiology*. Lange Medical Publications, Los Altos, California.

Joklik, W. K., and H. P. Willett (eds.). 1980. *Zinsser Microbiology*. Appleton Century Crofts, New York.

Riemann, H., and F. L. Bryan (eds.). 1979. *Food-Borne Infections and Intoxications*. Academic Press, New York.

Stuart-Harris, C. 1981. The epidemiology and prevention of influenza. *American Scientist* 69: 166–172.

Top, F. H., and P. Wehrle. 1976. *Communicable and Infectious Diseases*. C. V. Mosby, St. Louis.

Volk, W. A. 1978. *Essentials of Medical Microbiology*. J. B. Lippincott Co., Philadelphia.

Wilson, G. S., and A. A. Miles (eds.). 1975. *Topley and Wilson's Principles of Bacteriology and Immunity*. Williams & Wilkins, Baltimore.

Youmans, G. P., P. Y. Paterson, and H. M. Sommers. 1980. *The Biological and Clinical Basis of Infectious Disease*. W. B. Saunders Co., Philadelphia.

625

Suggested
supplementary
readings

Human diseases caused by microorganisms: the skin as a portal of entry and diseases of superficial body tissues

18

Relatively few microorganisms are able to invade the body through the skin, and therefore, **the skin normally acts as an effective barrier to microbial infection**. If, however, the skin tissues are mechanically interrupted as a result of wounds, burns, animal bites, or surgical procedures, a portal of entry is opened whereby pathogenic microorganisms may enter the body. The process of breaking the skin surface not only provides a portal of entry for potentially pathogenic microorganisms but also often inoculates microorganisms directly into the circulatory system and inner body tissues. For example, when a child scrapes a hand or knee on the ground, the wound is frequently contaminated with dirt and associated microorganisms. Even a puncture wound with a sterile hypodermic syringe can pick up microorganisms from the vicinity of the puncture and carry them through the skin surface. The routine cleansing of wounds and use of topical antiseptics after minor skin punctures and abrasions are accepted prophylactic measures taken to prevent the establishment of infections. Many animals are carriers of microorganisms, and their bites simultaneously disrupt the skin barrier and inoculate the wound with microorganisms whose pathogenic potential may be life threatening. Arthropods, in particular, commonly act as vectors of some very dangerous human pathogens, and their bites can establish some serious infections.

Paradoxically, although the superficial tissues of the body serve as barriers to penetration by pathogenic microorganisms, sometimes they themselves are also subject to infection. The eyes, for example, are normally protected from infections by the antimicrobial activities of tears, but some microorganisms can establish infections of the eye. The skin is extremely resistant to microbial infection, but some microorganisms nevertheless manage to establish cutaneous infections. The oral cavity, including the tooth surfaces and gingiva, is colonized by a large resident microbiota that under particular conditions cause diseases within the oral cavity, such as dental caries and periodontal diseases. Additionally, a limited number of microorganisms are able to penetrate the skin, and direct contact between these microorganisms and the skin surface can initiate a disease syndrome.

Diseases transmitted through animal bites

Many human infections that are transmitted via animal bites have a reservoir of infections in lower animals. These diseases, termed **zoonoses**, are defined as infectious diseases of lower animals transmittable to humans. In these instances, the lower animal populations and humans act as alternate hosts for the proliferation of the pathogens. It is important in considering the transmission of such diseases to examine the **reservoirs** and **vectors** of the infectious agents, as well as considering the nature of the specific etiologic agent of the disease (Table 18.1). The prevention of infectious diseases that enter the body through animal bites often involves maintaining control of infected reservoir and carrier animal populations. For example, mosquito control programs are employed to prevent the spread of many arthropod-borne diseases such as yellow fever and malaria. It should be remembered, however, that even though animals play a critical role in the transmission of such diseases, it is the viruses, bacteria, or protozoa that are the actual etiologic agents.

Diseases caused by viral pathogens

Rabies

Rabies is principally a disease of carnivorous animals other than humans. The rabies virus, a bullet-shaped, single-stranded RNA virus, can be transmitted to people through the bite of an infected animal (Figure 18.1). In urban settings, dogs are most frequently involved in the transmission of rabies. Other wild animals involved in the transmission of rabies include foxes, skunks, jackals, mongooses, squirrels, raccoons, coyotes, badgers, and vampire bats (Figure 18.2). A 1983 urban outbreak of rabies in the Washington, D.C.,

table 18.1

Representative diseases of humans transmitted by arthropod bites

Disease	Etiologic agent	Biological vector	Reservoir
Yellow fever	Yellow fever virus	Mosquito (*Aedes aegypti*, *Haemagogus* spp.)	Humans, monkeys
Dengue fever	Dengue fever virus	Mosquitos (*Aedes* spp., *Armigeres obturbans*)	Humans
Eastern equine encephalitis	Encephalitis viruses	Mosquito (*Aedes* spp., *Culex* spp., *Mansonia titillans*)	Humans, horses, birds
Colorado tick fever	Colorado tick fever virus	Wood ticks (*Dermacentor andersoni*)	Golden-mantled ground squirrel
Plague	*Yersinia pestis*	Rodent fleas (*Xenopsylla cheopis*), human fleas (*Pulex irritans*)	Rodents (rats)
Tularemia	*Francisella tularensis*	Ticks (*Dermacentor* spp., *Amblyomma* spp.), deerflies (*Chrysops discalis*)	Rodents, ticks
Rocky Mountain spotted fever	*Rickettsia rickettsii*	Ticks (*Dermacentor* spp., *Amblyomma* spp., *Ornithodoros* spp., etc.)	Rodents
Endemic typhus fever	*Rickettsia typhi*	Fleas (*Xenopsylla cheopis* and others)	Humans
Relapsing fever	*Borrelia recurrentis* and other species	Body louse (*Pediculus humanus*)	Humans, ticks
Chagas' disease	*Trypanosoma cruzi*	Cone-nosed bugs (*Triatoma* spp., *Panstrongylus* spp., *Rhodnius* spp.)	Dogs, cats, opossums, rats, armadillos
African trypanosomiasis (sleeping sickness)	*Trypanosoma gambiense*; *T. rhodesiense*	Tsetse flies (*Glossina* spp.)	Humans, wild mammals
Malaria	*Plasmodium vivax*; *P. malariae*; *P. falciparum*; *P. ovale*	Mosquitos	Humans
Leishmaniasis	*Leishmania donovani*; *L. tropica*; *L. braziliensis*	Sandflies (*Phlebotomus* spp.)	Dogs, foxes, rats, mice, two-toed sloth, gerbils, humans

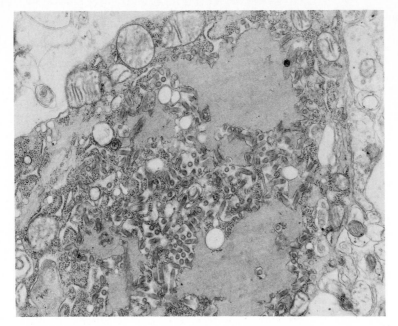

Figure 18.1

Micrograph of rabies virus from the neuron of an inoculated mouse (15,000x). Note how the cell is crowded with virus particles. (From BPS—A. Harrison, Centers for Disease Control, Atlanta.)

Human diseases caused by microorganisms: the skin as a portal of entry and diseases of superficial body tissues

area was associated with raccoons rather than dogs. Rabies viruses multiply within the salivary glands of infected animals, and normally enter humans in the animal's saliva through the portal of entry established by the animal's bite. The rabies virus is not able to penetrate the skin by itself, and deposition of infected saliva on intact skin does not necessarily result in transmission of the disease. Transmission from bats can also occur via the respiratory tract involving transmission via aerosols formed in the atmosphere around dense populations of infected bats.

When rabies viruses enter through an animal bite, they are normally deposited within muscle tissues where they subsequently multiply. The rabies viruses reach peripheral nerve endings and migrate to the central nervous system. Cytoplasmic inclusion bodies, known as **Negri bodies**, develop within the neurons of the brain. Multiplication of rabies viruses within the nervous system causes a number of abnormalities manifested as the symptoms of this disease. The initial symptoms of rabies include anxiety, irritability, depression, and sensitivity to light and sound. These symptoms are followed by development of hydrophobia (fear of water) and difficulty in swallowing. As the infection progresses, there is paralysis, coma, and death.

Once the clinical symptoms of rabies begin, the disease is considered to be invariably fatal, and therefore, treatment of rabies requires vaccination before the symptoms become manifest. The

Figure 18.2

Incidence of rabies in wild and domestic animals in the United States from 1953 to 1978.

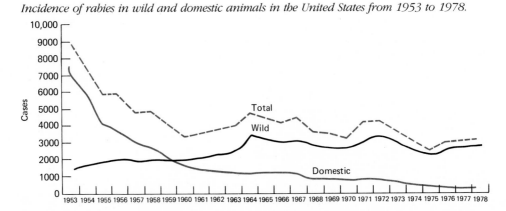

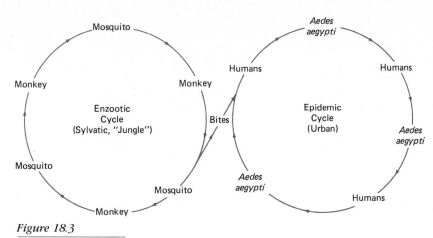

Figure 18.3

Transmission cycle patterns for yellow fever, showing the relationship between enzootic and epidemic transmission cycles of the disease.

rabies vaccine stimulates antibody synthesis that prevents proliferation of the virus before it can cause irreversible damage to the central nervous system. It is critical to examine the animal that has bitten a human for the presence of rabies virus, to diagnose the disease in the animal to determine if vaccination is necessary. Frequently, the public is asked to help identify and locate a dog that has bitten a child in order to determine if it is necessary to carry out the vaccination procedure. Only if the animal is conclusively shown not to be infected with rabies can the immunization procedure be safely omitted for the bitten individual. In a bizarre publicized case, a rock singer bit off the head of a bat during a 1982 concert and had to receive rabies vaccinations to prevent contraction of this disease. The vaccination procedure employed for many years used a vaccine prepared from rabies virus propagated in embryonated duck eggs and inactivated by β-propiolactone. The treatment involved 21 daily injections followed by booster inoculations 10 and 20 days later. Recent improvements in the vaccine have reduced the number of administrations that are required to establish immunity. The new rabies vaccine, prepared by growing the virus in human diploid fibroblast tissue culture, has a higher concentration of the necessary antigens for eliciting an immune response and only three injections over a seven-day period are required to establish immunity.

Yellow fever

Yellow fever, caused by a small RNA *Flavivirus* of the family Togaviridae, is transmitted by mosquito vectors, predominantly by *Aedes aegypti*. There are two epidemiologic patterns of transmission.

Urban transmission involves vector transfer by *Aedes aegypti* from an infected to a susceptible individual, and jungle yellow fever normally involves transmission by mosquito vectors among monkeys, with transfer via **mosquito vectors** to human beings representing an occasional deviation from the normal transmission cycle (Figure 18.3). Outbreaks of yellow fever were a major problem in the construction of the Panama Canal, leading to Walter Reed's instrumental work in establishing the relationship between yellow fever and mosquito vectors in 1901. Today, yellow fever occurs primarily in remote tropical regions, and current outbreaks of yellow fever occur primarily in Central America, South America, the Carribean, and Africa (Figure 18.4).

The onset of yellow fever is marked by anorexia (loss of appetite), nausea, vomiting, and fever. The multiplication of the virus results in liver damage, causing the jaundice from which the disease derives its name. The symptoms generally last for one week after which either recovery begins or death occurs. The fatality rate for yellow fever is about 5 percent. There is no effective antiviral drug at present for treating yellow fever; however, the disease can be prevented by vaccination, using the 17D strain of yellow fever virus. The urban form of transmission has been largely controlled by effective mosquito eradication programs, but the jungle form of transmission cannot easily be interrupted because of the large natural reservoir of yellow fever viruses maintained within monkey populations.

Dengue fever

Dengue fever also is caused by a *Flavivirus* transmitted by the mosquito *Aedes aegypti*. Like other

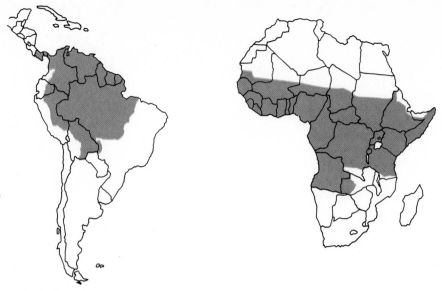

Figure 18.4

630

Human diseases
caused by micro-
organisms: the
skin as a portal
of entry and dis-
eases of superfi-
cial body tissues

The geographic distribution of yellow fever, showing regions where it is endemic in South America and Africa.

Figure 18.5

Incidence of dengue fever in Puerto Rico in 1977 and 1978.

members of the family Togaviridae, the dengue virus multiplies within the cytoplasm. Outbreaks of dengue fever occur in the Caribbean, South Pacific, and Southeast Asia (Figure 18.5). There are several different serotypes of dengue virus and three distinct disease syndromes: classic, hemorrhagic, and shock. A characteristic rash and fever develop during all forms of this disease. The classic form of dengue is a self-limiting disease, but in the hemorrhagic and shock syndromes the fatality rate can reach 8 percent in children.

The dengue virus replicates within the circulatory system, causing viremia (viral infection of the blood stream), that persists for 1–3 days during the febrile period. Multiplication of the dengue virus within the circulatory system causes vascular damage. It has been postulated that immune reactions contribute to the formation of complexes that initiate intravascular coagulation or hemorrhagic lesions. Previous exposure to dengue virus and the presence of cross-reacting antibody seem to be important in determining the severity of the disease symptomology. Damage to the circulatory vessels appears to occur when antigen–antibody complexes activate the complement system with the release of vasoactive compounds.

Encephalitis

Encephalitis, a disease defined by an inflammation of the brain, can be caused by various viruses, many of which are arthropod-borne and belong to the family Togaviridae. Viruses capable of causing encephalitis in humans are maintained in populations of various vertebrates, particularly birds and rodents, as well as within populations of arthropods. Transmission to human beings, via an arthropod vector in which the virus has multiplied, represents a dead end in the transmission cycle (Figure 18.6). Viral multiplication within the arthropod initially occurs in the gut and is followed by dissemination through the hemolymph and multiplication within the salivary glands. Viruses accumulate in the saliva of the arthropod, and this facilitates transfer to a person bitten by the arthropod. Outbreaks of viral encephalitis exhibit seasonal cycles, with increased numbers of cases occurring during summer when the vector populations are at their peak (Figure 18.7).

The different forms of viral encephalitis include: **eastern equine, western equine, Venezuelan equine, St. Louis, Japanese B, Murray valley, California, and tick-borne encephalitis** (Table 18.2). The specific viral etiologic agent, arthropod vector, and geographic distribution are different for each of these forms of encephalitis. Infections with encephalitis-causing viruses begin with viremia, followed by localization of the viral infection within the central nervous system, where lesions develop. The locations of the lesions within the brain are characteristic for each type. With the exception of St. Louis encephalitis, where kidney damage also occurs, the pathologic changes in cases of encephalitis are normally restricted to the central nervous system.

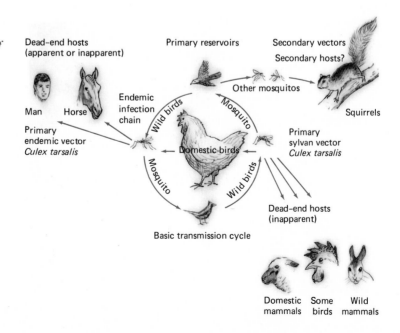

Figure 18.6

Epidemiological transmission of equine encephalitis. The chains for rural St. Louis encephalitis are similar, except that horses are inapparent, rather than apparent, hosts. EEE infections also have a similar summer infection chain, but a few significant differences exist: the identity of the vector infecting humans is unknown; domestic birds do not appear to be a significant link in the chain; and it has a bird-to-bird secondary cycle in pheasants whose role is unclear.

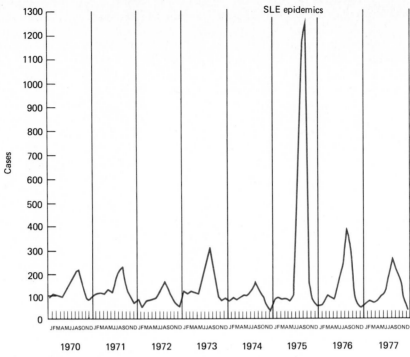

Figure 18.7

Incidence of encephalitis in the United States from 1970 to 1977.

Human diseases
caused by micro-
organisms: the
skin as a portal
of entry and dis-
eases of superfi-
cial body tissues

table 18.2

Togaviridae that cause encephalitis in humans

Genus	Subgroup	Viral species	Vector	Geographic distribution
Alphavirus	I	Eastern equine encephalitis (EEE)	Mosquito	Eastern USA, Canada, Brazil, Cuba, Panama, Dominican Republic, Trinidad, Philippines
		Venezuelan equine encephalitis (VEE)	Mosquito	Brazil, Colombia, Ecuador, Trinidad, Venezuela, Mexico, Florida, Texas
		Western equine encephalitis (WEE)	Mosquito	Western USA, Argentina, Canada, Mexico, Guyana, Brazil
Flavivirus	I	St. Louis encephalitis	Mosquito	USA, Trinidad, Panama
		Japanese B encephalitis	Mosquito	Japan, Guam, Eastern Asian mainland, India, Malaya
		Murray Valley encephalitis	Mosquito	Australia, New Guinea
		Ilheus	Mosquito	Brazil, Guatemala, Honduras, Trinidad
	IV	Tick-borne group (Russian spring-summer encephalitis group),	Tick	USSR, Canada, Malaya, USA, Central Europe, Finland, Japan, India, Great Britain
		Rio Bravo (bat salivary gland)	Unknown	California, Texas

Encephalitis symptoms are often subclinical. When the disease is symptomatic, the manifestation of encephalitis begins with fever, headache, and vomiting, followed by stiffness and then paralysis, convulsions, psychoses, and coma. Different forms of viral encephalitis have different outcomes. For example, in symptomatic cases of eastern equine encephalitis the mortality rate is approximately 80 percent, whereas in western equine encephalitis, the fatality rate is under 15 percent. Individuals who recover from symptomatic encephalitis may have permanent neurological damage.

No antiviral drug has been developed yet for the treatment of encephalitis caused by arthropod-borne viruses. Vaccines have been developed, though, that are effective in establishing immunity against the specific viruses that cause encephalitis, and there is an effective vaccine against Japanese B virus. As a rule, control of arthropod-borne viral encephalitis depends on control of vector populations. Insecticides have been widely used in public health programs to reduce the population levels of mosquito and tick vectors of viral encephalitis. Mosquito eradication programs are often intensified when positive diagnoses of encephalitis raise the possibility of widespread outbreaks.

Not all forms of encephalitis are arthropod-borne. For example, *Herpes simplex* virus, which is not normally transmitted through vectors, can cause both meningitis and encephalitis. In recent clinical trials vidarabine, an adenine arabinoside, has been demonstrated to be effective in treating *Herpes simplex* virus encephalitis, but this drug has not been proven effective against other encephalitis-causing viruses.

Another form of encephalitis, Kuru, has a very different mode of transmission. The specific etiologic agent of Kuru has not been identified, but transmission has been found to occur through the ingestion of human brain tissue during ritual cannibalism in New Guinea. Kuru is caused by a slow-growing virus that may take years to establish disease symptomology. The slow-growing viruses have several unusual properties, including their very long generation times and their ability to survive autoclaving. In Kuru there is a progressive vacuolation of the neurons that produces a progressive inability to control voluntary muscular movement (ataxia). The symptoms of Kuru are similar to several other degenerative nervous diseases, including scrapie and Creutzfeldt-Jakob diseases, suggesting that slow-growing viruses may be the causative organisms for these diseases that at present are of unknown etiology. Indeed, it is possible that several of these diseases may be caused by the same etiologic agent.

Colorado tick fever

Colorado tick fever is caused by a double-stranded RNA flavi-togavirus. The virus is maintained largely within populations of the golden-mantled ground squirrel, which acts as a reservoir, and is transmitted to human beings through a tick vector (*Dermacentor andersoni*). Colorado tick fever occurs mainly in the western United States. The disease symptoms characteristically show two episodes of fever, separated by a few days. Malaise, muscle aches, and vomiting are also symptomatic manifestations. The disease is self-limiting and recovery normally occurs within 1–2 weeks without treatment. Prevention of Colorado tick fever involves avoiding tick bites, for example, by wearing protective clothing when in tick-infested areas and by using arthropod repellents.

Diseases caused by bacterial pathogens

Plague

Plague is caused by *Yersinia pestis*, a Gram negative, nonmotile, pleomorphic rod. *Y. pestis* is normally maintained within populations of wild rodents and is transferred from infected to susceptible rodents by fleas. *Y. pestis* is able to multiply within the gut of the flea, which blocks normal digestion, causing the flea to increase the frequency of feeding attempts and so to bite more animals, increasing the probability of disease transmission. Plague is endemic to many rodent populations and, for example, *Y. pestis* is permanently established in rodent populations from the Rocky Mountains to the west coast of the United States.

The transmission of plague was extremely widespread during the Middle Ages because of poor sanitary conditions and the abundance of infected rat populations in areas of dense human habitation (see Figure 1.7). The development of rat control programs and improved sanitation methods in urban areas has greatly reduced the incidence of this disease. It is not possible, however, to completely eliminate plague in human beings because of the large number of alternate hosts in which *Y. pestis* is maintained. In rural environments, for example, *Y. pestis* is found in ground squirrels, prairie dogs, chipmunks, rabbits, mice, rats, and other animals.

The introduction of *Y. pestis* into humans

through flea bites initiates a progressive infection that can involve any organ or tissue of the body. Phagocytosis is effective in killing many of the invading bacteria, but some cells of *Y. pestis* are resistant to phagocytosis and continue to multiply and spread through the circulatory system. In **bubonic plague**, *Y. pestis* becomes localized and causes inflammation in the regional lymph nodes. The enlarged lymph nodes are called **buboes**, from whence the name of the disease is derived. The symptoms of bubonic plague include malaise, fever, and pain in the areas of the infected regional lymph nodes. Severe tissue necrosis can occur in various areas of the body, and it appears as blackened skin. It was this symptom that gave the name "black death" to the disease in the Middle Ages. As the infection progresses, the symptoms become quite severe, and without treatment the fatality rate is 60 to 100 percent. The disease can also progress in humans to involve the pulmonary system, leading to pneumonic plague. Transfer of *Y. pestis* then can occur through droplet spread, establishing outbreaks of primary pneumonic plague. The pneumonic form of plague has been very important in plague epidemics. If the bacteria invade the lungs, as in primary pneumonic plague, the disease often progresses rapidly and is manifest by severe prostration, respiratory difficulties, and death within a few hours of onset. Plague can be effectively treated with antibiotics, and streptomycin generally is the drug of choice against *Y. pestis*, although other antibiotics, such as chloramphenicol and tetracycline, can also be used.

Relapsing fever

Relapsing fever is caused by various species of *Borrelia*, a Gram negative spirochete (Figure 18.8). These pathogenic bacteria are normally transmitted to people through a bite by the body louse, *Pediculus humanus*. Epidemics of louse-borne relapsing fever occur under conditions that favor the proliferation of body lice, such as crowded conditions with relatively poor sanitation. The lice act strictly as vectors and do not develop the disease symptomology when they acquire *Borrelia* from an infected human. Crushing an infected louse can release *Borrelia* into the break in the skin caused by the louse bite. *Borrelia* species can also be transmitted through tick vectors, accounting for periodic outbreaks of relapsing fever in the western United States.

The clinical manifestations of relapsing fever occur intermittently. There is normally a sudden onset of fever approaching 105°F that lasts for 3–6 days. The fever then falls rapidly and remains normal for 5–10 days, followed by a second onset of fever that generally lasts 2–3 days. Additional relapses frequently occur in tick-borne disease but do not normally occur when the disease is louse-borne. During the course of relapsing fever, *Borrelia* sp. are able to multiply in the blood and various other body tissues. The natural removal of *Borrelia* from the body depends on an antibody-mediated immune response rather than on phagocytosis. Anatomical abnormalities develop in the spleen, and lesions may also occur in various other body organs. Tetracycline and chloramphenicol are effective against *Borrelia* and are used in treating this disease. With proper treatment over 95 percent of patients with relapsing fever recover. Control of this disease depends on avoiding contact with vectors carrying *Borrelia*, particularly body lice and ticks. Pesticides are useful in many areas of the world in maintaining low populations of rodents and other animals that act as nonhuman vertebrate hosts of vector ticks.

Bartonellosis

Bartonellosis is an infection caused by *Bartonella bacilliformis*, a small Gram negative rod, motile by multiple polar flagella. This bacterium is a rickettsia and as such is an obligate intracellular parasite. Bartonellosis can be manifest in two stages:

Figure 18.8

Micrograph of Borrelia hermsi *from a smear of mouse blood stained by the Giemsa technique, followed by counterstaining with crystal violet (2200x). The characteristic morphology of the bacteria is shown in this preparation. (Courtesy Richard T. Kelly, Baptist Memorial Hospital, Memphis.)*

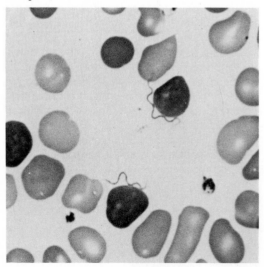

Oroya fever, which is an acute febrile stage, and a later cutaneous stage, known as Verruga peruana. Bartonellosis is transmitted to people through the bites of infected sand flies. Sand fly vectors occur only in the valleys of the Andes Mountains, and thus, this disease occurs only in geographically restricted regions of Peru, Ecuador, and Columbia. *B. bacilliformis* is unusual in its restricted geographic distribution and its affinity for red blood cells.

B. bacilliformis is able to proliferate within the endothelial cells of blood vessels after inoculation by the bite of the sand fly vector. The bacteria reenter the circulatory system and infect erythrocytes, generally causing severe anemia accompanied by fever, headache, and delirium. The mortality rate from Bartonellosis can be as high as 40 percent during the Oroya fever stage. Chloramphenicol is effective in treating Oroya fever and greatly reduces the fatality rate. If the patient survives the Oroya fever stage, there is a subclinical period that may be followed by the development of reddish lesions at various sites on the body, particularly the extremities. Prevention of Bartonellosis depends on avoiding contact with infected sand fly vectors. The sand flies carrying *Bartonella* are nocturnal, and the use of insect repellents and protective clothing after dark are useful for people in the Andes.

Rocky Mountain spotted fever

Rocky Mountain spotted fever is caused by a rickettsia and is the first of several diseases to be discussed in this section that are caused by rickettsias transmitted to humans via biting arthropod vectors (Table 18.3). In the case of Rocky Mountain fever, *R. rickettsii* is transmitted to human beings through the bite of a tick. *R. rickettsii* is normally maintained within various tick populations, such as the wood tick and dog tick (Figure 18.9). The bacteria multiply within the mid-gut of the tick and are passed congenitally from one generation of ticks to the next. Humans are accidental hosts of *R. rickettsii* as a result of occasional bites of infected ticks that allow transfer of *R. rickettsii* to them.

Rocky Mountain spotted fever occurs in areas of North and South America, most commonly in spring and summer when ticks and people are most likely to come in contact and normally occur in well-defined localized regions. During the mid-twentieth century many cases of Rocky Mountain spotted fever occurred in the United States in the region of the Rocky Mountains, but relatively few cases have been reported there in recent years. On the other hand, outbreaks of Rocky Mountain spotted fever have risen dramatically in the eastern United States, where most cases now occur (Figure 18.10).

table 18.3

Summary of certain important epidemiologic and clinical characteristics of rickettsial diseases

Disease	Epidemiologic features		
	Usual mode of transmission to humans	Reservoir	Geographic occurrence
Spotted fever group:			
Rocky Mountain spotted fever	Tick bite	Ticks; rodents	Western Hemisphere
Tick typhus	Tick bite	Ticks; rodents	Mediterranean, littoral Africa, Asia
Rickettsialpox	House mouse mite bite	Mites; mice	USA, Russia, Korea
Typhus group:			
Primary louse-borne typhus	Infected louse feces rubbed into broken skin or as aerosol to mucous membranes	Humans	Worldwide
Brill-Zinsser disease	Recrudescence months or years after primary attack of louse-borne typhus	—	Worldwide
Murine typhus	Infected flea feces rubbed into broken skin or as aerosols to mucous membranes	Rodents	Worldwide (scattered pockets)
Scrub typhus	Mite bite	Mites; rodents	SE Asia, W and SW Pacific, Japan

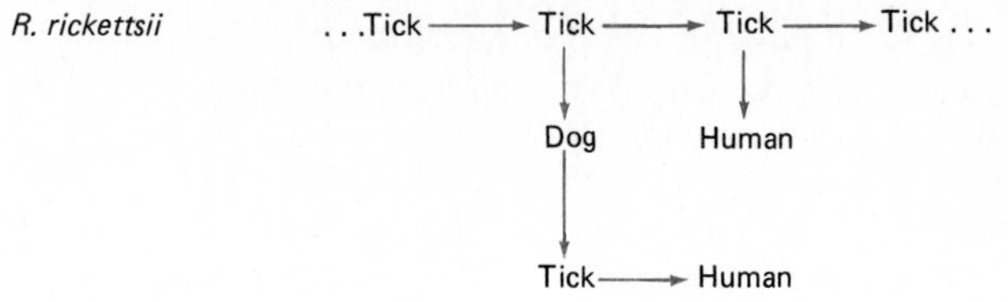

R. rickettsii

...Tick ⟶ Tick ⟶ Tick ⟶ Tick...

Dog → Human

Tick ⟶ Human

Figure 18.9

Transmission pattern of Rocky Mountain spotted fever.

When injected into human beings, *R. rickettsii* multiply within the endothelial cells lining the blood vessels. Vascular lesions occur and account for the production of the characteristic skin rash associated with this disease. The rash is most prevalent at the extremities, particularly on the palms of the hands and the soles of the feet. Lesions probably also occur in the meninges, causing severe headaches and a state of mental confusion. If treated with antibiotics, such as chloramphenicol and tetracycline, the disease is rarely fatal, but if untreated the overall mortality rate is probably greater than 20 percent. Prevention of Rocky Mountain spotted fever primarily involves control of population levels of infected ticks and the avoidance of tick bites. However, control is difficult to achieve, and in 1975 there were almost 900 cases of Rocky Mountain spotted fever in the United States.

Typhus fever

There are several types of **typhus fever**, all of which are caused by rickettsias transmitted to humans via biting arthropod vectors (Figure 18.11). **Infectious or classical typhus fever** is caused by *Rickettsia prowazekii* and is transmitted to humans via the body louse. The chain of transmission is restricted to humans and lice. The lice contract the disease from infected humans and in turn pass the rickettsia on to susceptible human hosts. *R. prowazekii* multiplies within the epithelium of the mid-gut of the louse. When an infected body louse bites another human it defecates at the same time, depositing feces containing *R. prowazekii*, which enter through the wound created by the bite. The name **epidemic typhus** is derived from the fact that this form of typhus is transmitted from person to person only by the body louse. Under crowded conditions with lice

Figure 18.10

Map showing incidence of Rocky Mountain spotted fever in the continental United States in 1977. Note that the highest rate of incidence is in the eastern half of the country.

Legend
- • 1–5 Cases
- ▲ 6–10 Cases
- ■ > 10 Cases

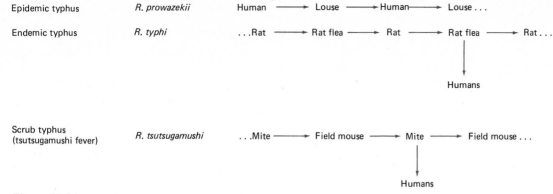

Epidemic typhus   *R. prowazekii*   Human → Louse → Human → Louse . . .

Endemic typhus   *R. typhi*   . . .Rat → Rat flea → Rat → Rat flea → Rat . . .

     ↓   Humans

Scrub typhus (tsutsugamushi fever)   *R. tsutsugamushi*   . . .Mite → Field mouse → Mite → Field mouse . . .

     ↓   Humans

Figure 18.11
_____

Transmission patterns for different types of typhus fever.

infestation, the disease can spread easily and cause large numbers of cases. For example, millions of cases occurred in World War I and in the concentration camps during World War II. The onset of epidemic typhus involves fever, headache, and rash. The heart and kidneys are frequently the site of vascular lesions. If untreated, the fatality rate in persons 10–30 years old is approximately 50 percent. The Weil-Felix test is useful in diagnosing classic typhus fever (see Table 16.9). Chloramphenicol, tetracycline, and doxycycline are effective in treating epidemic typhus.

Relapses of louse-borne typhus can occur years after the primary attack. Such recurrences are referred to as the **Brill-Zinsser disease**. *R. prowazekii* apparently can survive in some cells of the body for a prolonged period of time and give rise to this later infective phase. The factors that initiate recurrence of typhus are unknown, and there is no known way of preventing these recurrences. Prevention of louse-borne typhus fever involves controlling louse infestation, and delousing infected patients effectively interrupts the chain of transmission.

Murine or **endemic typhus fever** is caused by *Rickettsia typhi* and is transmitted to humans by rat fleas. Murine typhus is normally maintained in rat populations endemically through transmission by rat fleas. Occasionally, rat fleas will attack people, and if they are infected with *R. typhi* the disease can be transmitted. As with louse-borne typhus, the flea deposits pathogenic bacteria in the fecal matter, which is rubbed into the flea bite by the host because of the local irritation caused by the bite. The symptoms of murine typhus are similar to those of classical typhus fever but are generally milder. Chloramphenicol and tetracycline are effective in treating this disease, and there

is a relatively low mortality rate. Prevention of murine typhus depends on limiting rat populations, which also limits the size of the vector rat flea population.

Scrub typhus, caused by *Rickettsia tsutsugamushi*, is almost exclusively restricted to Japan, Southeast Asia, and the western Pacific Islands. The disease is transmitted to people through the bite of mite vectors. *R. tsutsugamushi* is normally transmitted congenitally in mite populations and only accidently is introduced into the human population. In humans multiplication of *R. tsutsugamushi* occurs at the site of inoculation and is later distributed through the circulatory system, and eventually the infection becomes localized in the lymph nodes. In scrub typhus there is a characteristic lesion of the skin, known as the **eschar**, that occurs at the site of initial infection. The eschar may be difficult to find because mite bites may occur anywhere on the body. The symptoms of scrub typhus include fever, severe headache, and rash. Chloramphenicol and tetracycline are effective in treating this form of typhus. Controlling vector populations and minimizing the chance of being bitten by an infected mite are means by which this disease can be prevented.

Rickettsialpox

Rickettsialpox, caused by *Rickettsia akari*, is transmitted to humans by the mouse mite (Figure 18.12). *R. akari* is congenitally maintained in populations of the mouse mite, and in addition to serving as a reservoir for *R. akari*, the mouse mite acts as a vector. Normally, mouse mites do not attack people, but under certain conditions humans can be accidental recipients of *R. akari* from an infected mouse mite, leading to the onset of the disease rickettsialpox. Inoculation occurs

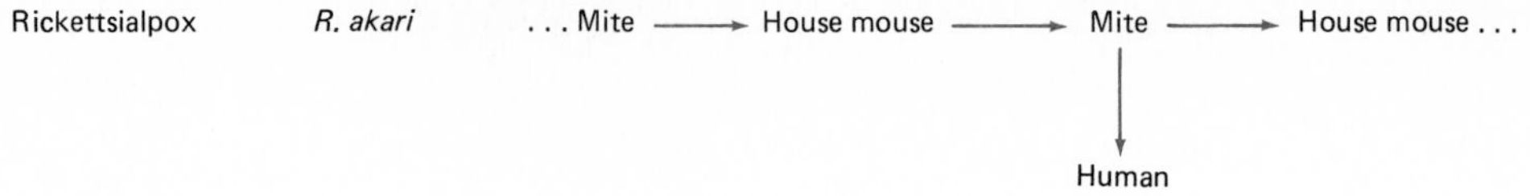

Rickettsialpox   *R. akari*   . . . Mite ⟶ House mouse ⟶ Mite ⟶ House mouse . . .

Human

Figure 18.12

Transmission pattern for rickettsialpox.

through the oral secretions of the mite. Rickettsialpox occurs almost exclusively in urban environments of the United States and the Soviet Union, where there are insufficient mice for the mites to feed on, and it is under these conditions that infected mites will bite people.

The infection of *R. akari* leads to the formation of a primary lesion at the site of the bite. As with other rickettsial diseases, the bacterial infection can spread along the vessels of the circulatory system, causing vascular lesions. Systemic spread of the infection is followed by the onset of the illness, which is manifest by fever, headache, and secondary lesions. The rash associated with this disease may cover any part of the body, but unlike Rocky Mountain spotted fever, the rash rarely develops on the palms or soles. Rickettsialpox is a benign disease, and symptoms disappear without treatment beginning approximately ten days after the onset of disease. The use of tetracyclines and other antibiotics are effective in ending the manifestation of disease symptomology within one day of beginning antimicrobial therapy. Prevention of rickettsialpox depends on effective rodent control programs; the elimination of mice from urban settings can effectively end this disease.

Diseases caused by protozoan pathogens

Malaria

On a worldwide basis malaria is one of the most common human infectious diseases (Figure 18.13). The annual incidence of malaria is about 150,000,000 cases. Malaria has been largely eliminated from North America and Europe but remains the most serious problem of infectious dis-

Figure 18.13

Incidence of malaria in the Americas based on reported cases in 15 countries.

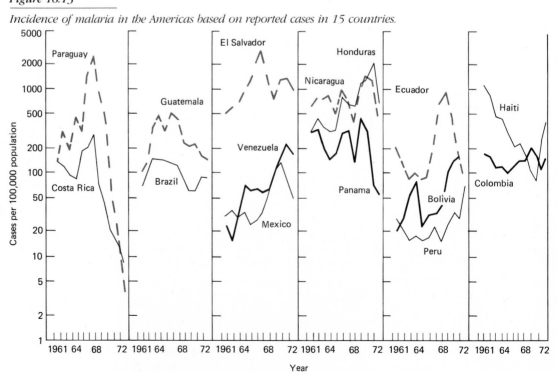

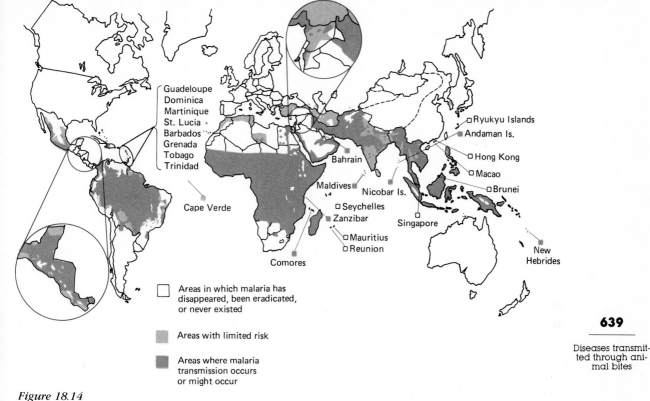

Figure 18.14

Map of worldwide incidence of malaria, illustrating the epidemiological assessment of the status of the disease in December 1976. Malaria remains one of the most serious infectious diseases in the world today.

ease in tropical and subtropical regions of the world (Figure 18.14). **This disease is caused by four species of *Plasmodium*** (Table 18.4). *P. vivax* and *P. falciparum* are most frequently involved in human infections. The *Anopheles* mosquito is the vector responsible for transmitting malaria to human beings. After inoculation into the body, the sporozoites of *Plasmodium* begin to reproduce within liver cells (Figure 18.15). Multiplication of the *Plasmodium* sporozoites occurs by schizogony by which a single sporozoite can produce as many as 40,000 merozoites. The

table 18.4

Summary of important characteristics of human malarias

Etiologic agent	*P. falciparum*	*P. vivax*	*P. ovale*	*P. malariae*
Incidence	Common	Common	Uncommon	Uncommon
Primary hepatic schizogony	1–40,000 in 5.5–7 days	1–10,000 in 6–8 days	1–15,000 in 9 days	1–2000 in 13–16 days
Secondary hepatic schizogony	1–8 to 24 (av. 16) in 48 hr.	1–12 to 24 (av. 16) in 48 hr.	1–6 to 16 in 48 hr.	1–6 to 12 (av. 8) in 72 hr.
Incubation period	8–27 days (av. 12)	8–27 days (av. 14) (rarely months)	9–17 days (av. 15)	15–30 days
Mortality	High in nonimmunes	Uncommon	Rarely fatal	Rarely fatal

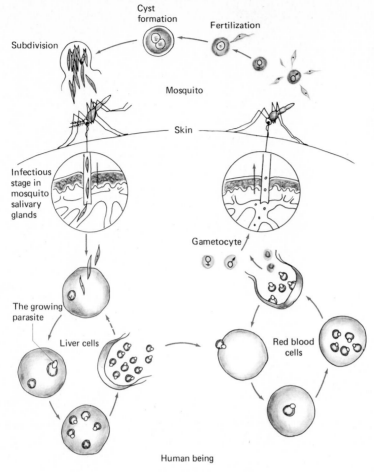

Figure 18.15

Reproduction of a malaria parasite within the mosquito vector and the human body. The protozoan exhibits two distinct reproductive phases.

Cyst formation

Fertilization

Subdivision

Mosquito

Skin

Infectious stage in mosquito salivary glands

Gametocyte

♀ ♂

The growing parasite

Liver cells

Red blood cells

Human being

merozoites are capable of invading blood erythrocyte cells. Invasion of erythrocytes by hepatic merozoites begins the erythrocytic phase of malaria. The *Plasmodium* merozoites are able to reproduce within the red blood cells, causing anemia.

The symptoms of malaria begin approximately two weeks after the infection established by the mosquito bite. The symptoms include chills, fever, headache, and muscle ache. The symptoms of malaria are periodic with symptomatic periods, generally lasting less than six hours. Schizogony occurs every 48 hours for *P. vivax* and *P. ovale*, and every 72 hours for *P. malariae*, resulting in a synchronous rupture of infected erythrocytes. The symptoms of the disease are caused by the asexual erythrocytic cycle, but the periodic onset of disease symptoms coincide with the rupture of infected erythrocytes.

Malarial infections persist for long periods of time and are rarely fatal, except when the disease is caused by *P. falciparum*. There is no vaccine for malaria, but the disease can be prevented by drug prophylaxis. Individuals traveling to areas with high rates of malaria, such as Southeast Asia and Africa, often use antimalarial drugs, such as chloroquine, to avoid contracting this disease. The use of insect netting and other measures to prevent being bit by an infected mosquito are extremely important in avoiding contracting malaria. In the United States control measures have been effective, but periodic morbidity increases have occurred after overseas military ventures (Figure 18.16).

Leishmaniasis

Leishmaniasis, caused by infections with members of the protozoan genus *Leishmania*, is transmitted to human beings by sand fly vectors. *Leishmania* species reproduce in people and other animals intracellularly as a nonmotile form, the amastigote. In sand flies the protozoa exists in a flagellated form, the promastigote. Four species of *Leishmania* cause leishmaniasis in humans (Table 18.5). *L. mexicana* and *L. tropica*, sometimes referred to as Old World cutaneous leishmaniasis or oriental sore, cause infections limited to the skin. *L. braziliensis* causes infections of skin and

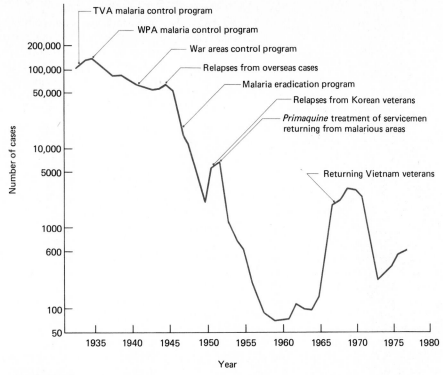

Figure 18.16

Incidence of malaria in the United States from 1933 to 1977. The total number of cases has fluctuated with the application of control measures and the return of military personnel from areas of the world where malaria is common.

table 18.5

Epidemiology of Leishmaniasis

Etiologic agent	Disease	Geographic distribution
Leishmania donovani	Kala-azar, visceral leishmaniasis	Mediterranean Southern Europe Central Asia China India Sudan
Leishmania tropica	Oriental sore, Old World cutaneous leishmaniasis	Mediterranean Basin, Central Asia, India
Leishmania mexicana	New World cutaneous leishmaniasis	Mexico Guatemala Central America Amazonia
Leishmania braziliensis	New World cutaneous leishmaniasis	Central America South America

mucocutaneous junctions. The most serious form of leishmaniasis is caused by *L. donovani*, which multiplies throughout the mononuclear phagocyte system and causes **kala-azar disease**.

Leishmaniasis is geographically restricted to regions where sand flies can reproduce and acquire *Leishmania* species from infected canines and rodents. The protozoa are able to reproduce within the sand flies and are present in the saliva. The lesions caused by *Leishmania* infections may be minor or extensive. The formation of extensive ulcers can produce permanent scars. The cutaneous leishmaniasis syndromes are self-limiting. The kala-azar syndrome, however, can be fatal. Untreated cases of kala-azar disease normally last several years and in the terminal stages can involve liver and heart damage and hemorrhages. The cutaneous forms of leishmaniasis can be treated with amphotericin B. The kala-azar syndrome can be treated with sodium stibogluconate, an antimony-containing drug, or with amphotericin B. The prevention of leishmaniasis involves controlling the vector and reservoir populations. The use of DDT has been effective in some regions in eliminating the sand fly vector. In other cases infected rodent populations have been controlled, greatly reducing the incidence of leishmaniasis.

Trypanosomiasis

Trypanosomiasis is caused by infections with species of the protozoan genus *Trypanosoma*. American trypanosomiasis, or **Chagas' disease**, occurs in Latin America and is caused by *T. cruzi*, usually transmitted to humans by infected tryatomine (cone-nosed) bugs. *T. cruzi* is a flagellate protozoan, but in vertebrate hosts it forms a nonflagellate form, the amastigote (Figure 18.17). Dogs and cats are reservoirs of *T. cruzi*. The vectors of Chagas' disease normally live in the mud and wood houses of South America, and construction of better housing eliminates the habitat for vector populations that brings them into close contact with humans.

When *T. cruzi* infects human hosts, the protozoa initially multiply within the mononuclear phagocyte system. Later, the myocardium and nervous systems are invaded. Damage to the heart tissue occurs as a result of this infection. In 90 percent of the cases, there is spontaneous remission of the disease, but 10 percent of the hospitalized patients die during the acute phase because of myocardial failure. Death due to heart disease as a result of Chagas' disease may also occur well after recovery from the acute phase. Chagas' disease is the leading cause of cardiovascular death in South America, and the incidence

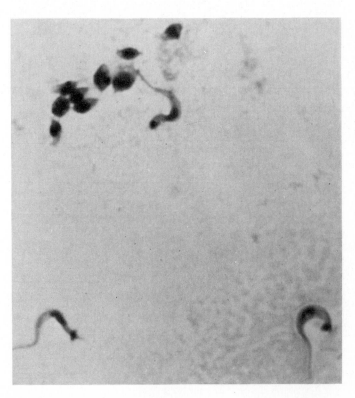

Figure 18.17

Micrograph showing flagellate (trypomastigote) and nonflagellate (spheromastigote) forms of Trypanosoma cruzi. *The trypomastigote, formerly known as the trypanosomal stage, occurs within the blood and within the insect vector. The spheromastigote stage, formerly known as the amastigote and prior to that as the leishmanial stage, occurs within tissues such as heart muscle and brain. Each spheromastigote has one kinetoplast and one nucleus that appear as the two dark bodies within each cell. (Courtesy Centers for Disease Control, Atlanta.)*

of this disease in Brazil is extraordinarily high. Several antiprotozoan drugs, such as aminoquinoline, are effective in treating Chagas' disease if the symptoms are recognized early, but once the progressive stages have begun, treatment is supportive rather than aimed at eliminating the infecting agent.

African trypanosomiasis, also known as **African sleeping sickness,** is caused by infections with *T. gambiense* and *T. rhodesiense*. There is some debate over the taxonomic position of these two species, and it is possible that both really are strains of *T. brucei*. The etiologic agents of African trypanosomiasis are transmitted to people through the tsetse fly vector. The tsetse flies acquire *Trypanosoma* species from various vertebrate animals, such as cows, which act as reservoirs of the pathogenic protozoa. In humans, infections with *T. gambiense* or *T. rhodesiense* are disseminated through the mononuclear phagocyte system, and there is evidence of localization within regional lymph nodes. Multiplication of the protozoa can produce damage to heart and nerve tissues. Progression through the central nervous system takes from months to years. If untreated the initially mild symptoms, which include headaches, increase in severity and lead to fatal meningoencephalitis. If the disease is diagnosed before there is central nervous system involvement, it can be successfully treated with antiprotozoan agents, such as suramin. If there is central nervous system involvement, melarsoprol, an arsenical, is used for treating the disease. Prevention of African trypanosomiasis involves controlling population levels of the tsetse fly, accomplished by clearing vegetation to destroy the natural habitats of the tsetse fly.

Diseases transmitted through direct contact

In some cases the deposition of pathogenic microorganisms on the skin surface can lead to an infectious disease. Some diseases transmitted in this manner are restricted to superficial skin infections, but in other cases the pathogens are able to enter the body and spread systemically. Although relatively few microorganisms possess the enzymatic capability to establish infections through the skin surface, some microorganisms are able to enter the subcutaneous layers through the channels provided by hair follicles. The transmission of some **contact diseases** may follow minor abrasions that allow the pathogens to circumvent the normal skin barrier. Many of the human contact diseases can be transmitted in other ways, such as through the respiratory or gastrointestinal tracts or via animal bites. They are grouped together here based on the relative frequency and/or importance of this mode of transmission.

Diseases caused by viral pathogens

Warts
Warts are benign tumors of the skin caused by papillomaviruses, which are small icosahedral DNA viruses. Transmission of wart viruses appears to occur primarily by direct contact of the skin with wart viruses from an infected individual, although indirect transfer also may occur through fomites. The human papillomaviruses appear to infect only humans and no other animals, with children developing warts more frequently than adults. The development of warts can occur on any of the body surfaces and the appearance of the warts varies, depending on its location. At present there is no effective antiviral treatment for human warts, and therapy often involves destruction of infected tissues by applying acid or freezing. In general, human warts is a self-limiting disease and recovery can be expected without treatment within two years.

Herpes simplex infections
As indicated in Chapter 17, there are two types of *herpes simplex* viruses. *Herpes simplex* **1 virus** is most frequently involved in nongenital herpes infections. Infections with *herpes simplex* 1 virus most commonly involve the skin above the waist, lips, and mouth, and the virus appears to be transmitted through direct contact of the surface epithelial tissues with the virus. A focal infection develops around the site of inoculation, and dissemination of the virus occurs from the primary focal lesion. In most cases the lesions are limited to the epidermis and surface mucous membranes; however, in some cases the herpes simplex virus can spread systemically, causing central nervous system infections.

The development of herpes viruses in the mouth and on the lips is seen as the development of lesions known as cold sores or fever blisters. Sim-

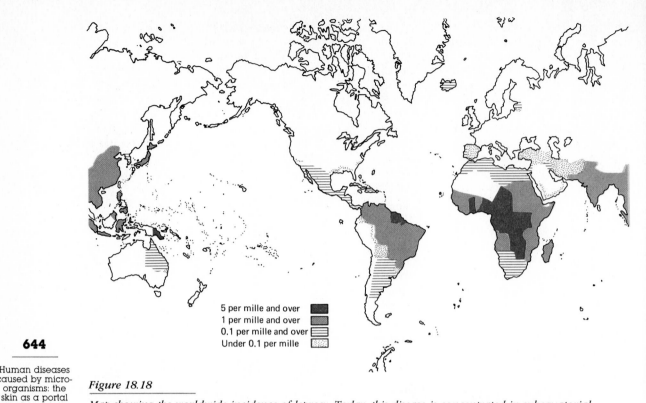

644

Human diseases
caused by micro-
organisms: the
skin as a portal
of entry and dis-
eases of superfi-
cial body tissues

Figure 18.18

Map showing the worldwide incidence of leprosy. Today, this disease is concentrated in subequatorial zones.

ilar lesions also develop in infected regions of the skin. The lesions generally heal within several days of their appearance, but new lesions develop periodically, indicative of the persistence of *herpes simplex* infections. Several of the newly developed antiviral drugs, such as adenine arabinoside (vira-A), are effective against herpes viruses and may prove useful in treating cutaneous herpes infections.

Diseases caused by bacterial pathogens

Leprosy

In 1980 there were about 3 million cases of **leprosy**, caused by *Mycobacterium leprae*, in the world (Figure 18.18). There are two forms of leprosy. The lepromatous or cutaneous form is characterized by granulomatous nodules in the skin, whereas the tubercular or neural form involves lesions around the peripheral nerves. Bacteria present in the lesions are infective and can be transmitted through the skin or mucous membranes of susceptible individuals. There may be one billion viable cells of *M. leprae* per gram of

skin in advanced cases of lepromatous leprosy, and direct skin contact appears to be very important in the transmission of this disease.

Leprosy may also be spread through droplets, which can have as many as 10^7 cells/ml of *M. leprae,* from infected individuals to susceptible hosts where the primary infection occurs within the lung tissues. There is an extremely long incubation period for leprosy, usually 3–5 years, before the onset of disease symptomology. The symptoms of leprosy vary, but the earliest detectable symptoms generally involve skin lesions. Unlike other mycobacterial species, *M. leprae* is able to reproduce within nerve tissues, with damage to the nervous system occurring as a result of the ability of *M. leprae* to reproduce within certain nerve cells (Schwann cells).

During the course of this disease, many organs and tissues of the body may be infected in addition to the infection of nerve cells characteristic of all forms of leprosy. In the tubercular form there are relatively few nerves and skin areas involved, but in lepromatous leprosy multiplication of *M. leprae* is not contained by the immune defense mechanisms, and the bacteria are disseminated through many tissues. Treatment of leprosy can be achieved by using dapsone, which is bacteriostatic, or rifampin, which is bactericidal. Prolonged treatment with antimicrobial agents is needed to maintain control of leprosy infections. Leprosy is rarely fatal, and complete recovery occurs after treatment in many cases of leprosy (Figure 18.19).

Anthrax

Anthrax is primarily a disease of animals other than humans, but this disease can occasionally be transmitted to people. The disease is caused by *Bacillus anthracis*, a Gram positive, endospore-forming rod. Transmission to humans can occur by direct contact of the skin with the endospores of *B. anthracis*, via the respiratory tract through inhalation of spores and via the gastrointestinal tract through the ingestion of spores. The cutaneous route of transmission accounts for 95 percent of the cases of anthrax in the United States. Contact with animal hair, wool, and hides containing spores of *B. anthracis* is often implicated in transmission of anthrax, and therefore, it is also known as woolsorter's disease. Deposition of spores of *B. anthracis* under the epidermis per-

mits germination with subsequent production of toxin by the growing bacteria.

The localized accumulation of toxin causes necrosis of the tissue in the formation of a blackened lesion. The development of cutaneous anthrax can initiate a systemic infection, and untreated cutaneous anthrax has a fatality rate of 10–20 percent. Cutaneous anthrax can be treated with penicillin and other antibiotics, reducing the death rate to under 1 percent. Avoiding contact with infected animals and preventing the development of anthrax in farm animals through the use of anthrax vaccine have effectively reduced the incidence of this disease.

Tularemia

Tularemia is caused by *Francisella tularensis*, a Gram negative, fastidious coccobacillus. *F. tularensis* causes infections in many mammals and arthropod vectors. Transmission to humans can occur through ingestion of contaminated material and handling of infected animals. *F. tularensis* may gain entry to the body directly through the skin, particularly through minute openings, such as at hair follicles and near the fingernails. Tularemia

Figure 18.19

Throughout history leprosy has been treated as a dread disease, and individuals with this disease have been outcast from societies. Although recovery is slow, this disease can be effectively treated, as shown in these photographs. (A) A young boy with borderline leptomatous leprosy. (B) The same child two years later after treatment with appropriate medication. (Courtesy American Leprosy Missions, Bloomfield, New Jersey.)

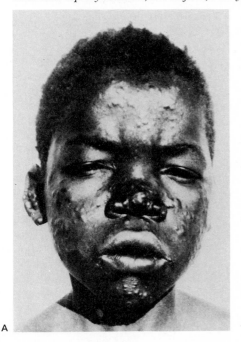

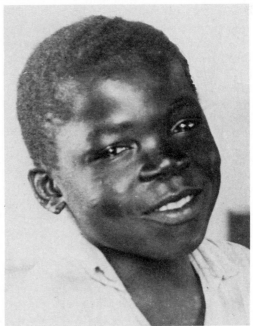

A

B

is a particular occupational problem for individuals who handle potentially contaminated animals, such as hunters, butchers, cooks, agricultural workers, etc.

Often, ulcers of the fingers are symptomatic of tularemia that is contracted through the handling of infected animals. The localized ulcers result from the multiplication of *F. tularensis*, as does a systemic infection with localization within the regional lymph nodes. In most cases of tularemia, there is an elevated fever, development of skin ulceration, and enlargement of the lymph nodes. The death rate in untreated cases of tularemia acquired through the skin is approximately 5 percent, but higher mortality rates approaching 30 percent occur when the disease is contracted through inhalation. Streptomycin, tetracycline, and chloramphenicol are effective in controlling this disease. Prevention of tularemia can be achieved by avoiding contaminated animals, and because rabbits have an especially high rate of infection with tularemia, special precautions should be taken in their handling.

Brucellosis

Brucellosis is caused by several *Brucella* species, all of which are Gram negative, small, nonmotile, aerobic rods. Brucellosis is an infectious disease of nonhuman animals that can be transmitted to humans through ingestion of contaminated milk and the handling of infected animals. The use of pasteurization has greatly reduced the transmission of brucellosis via the gastointestinal tract, and most human infections today result from direct contact with infected animals. Brucellosis is an occupational hazard in farming, veterinary practice, and meat packing plants, where *Brucella* species can enter through the skin, particularly in areas of minor abrasions.

Once *Brucella* species enter the body, the bacteria are able to spread rapidly through the mononuclear phagocyte system and can multiply within phagocytic cells. Infecting *Brucella* species normally become localized in the regional lymph nodes. There may be enlargement of the spleen and liver, and other symptoms include weakness, chills, malaise, headache, backache, and fever. The fever may rise and fall in some cases, and thus, this disease is often referred to as **undulant fever**. The death rate in untreated cases of brucellosis is about 3 percent. The use of tetracyclines, however, is effective in treating this disease and reduces the death rate to near zero percent. The prevention of brucellosis involves eliminating the disease in animals such as cattle, sheep, and goats, the reservoirs of *Brucella*. It is important to seg-

regate and treat infected animals to prevent the spread of this disease through animal herds. Vaccines are effective in limiting the spread of brucellosis through some animal populations, such as cattle, but there are no effective vaccines for other animals, such as hogs.

Leptospirosis

Leptospirosis, or **Weil's disease**, is principally a disease of nonhuman animals caused by *Leptospira interrogans* (Figure 18.20). Leptospires are long spirochetes that can enter the human body through the skin, especially if there are small abrasions. The disease can also be transmitted through the respiratory tract. Infected animals excrete leptospires in their urine and people generally contract the disease through contact with contaminated material harboring viable *Leptospira* species.

During the acute phase of this illness, the symptoms normally include a high spiking fever, chills, headache, muscle ache, malaise, abdominal pain, nausea, and vomiting. Various body organs can be involved, and most fatal cases result from infection of the kidney and subsequent renal failure. In most cases, leptospirosis is subclinical, and the prognosis for complete recovery is excellent. When severe symptoms develop, however, the death rate from Weil's disease is about 10–40 percent. Penicillin may provide a useful treatment and may shorten the duration and severity of the illness. Avoidance of potentially contaminated environments, such as sewers and contaminated streams, is effective in preventing outbreaks of this disease.

Pyoderma and impetigo

Pyoderma is an infection of the skin caused primarily by *Staphylococcus* and *Streptococcus* species. **Impetigo**, which occurs almost exclusively in children during warm weather, is a type of pyoderma caused by infections with *Staphylococcus aureus* and/or *Streptococcus pyogenes*. Direct contact with infected material appears to be important in the transmission of impetigo. Strains of *S. pyogenes* are deposited and remain viable on normal skin surfaces, and minor trauma to the skin permits invasion by the streptococci. *S. pyogenes* produces hyaluronidase, which permits spread of the bacteria. The infection results in the formation of a lesion that can be invaded by *S. aureus* from the skin surface, establishing a secondary infection. Typical manifestations of impetigo are localized lesions that progress to form pustules. Pyoderma caused by *Streptococcus* and *Staphylococcus* species tends to be benign, al-

Figure 18.20

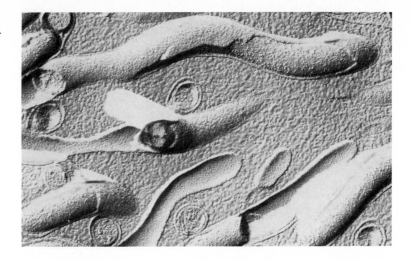

A freeze-etched micrograph of the spirochete Leptospira interrogans. *(From Stanley C. Holt, University of Massachusetts.)*

though the pustules may persist and spread, forming satellite infections. The disease normally begins on the extremities and intense itching is associated with the lesions that develop. Scratching facilitates spread to other parts of the body.

Glomerulonephritis

In some cases systemic infections can result from bacterial pyoderma. The spread of streptococcal skin infections are especially problematical if there are wounds that open a portal of entry for systemic dissemination. The spread of *S. pyogenes* to the kidneys is a particular problem that can result in **glomerulonephritis**. The best evidence now suggests that glomerulonephritis is an autoimmune disease that follows a streptococcal infection and is not due to an active kidney infection. Streptococcal and staphylococcal skin infections can be treated by the application of topical antimicrobial drugs, such as bacitracin, if there are relatively few lesions. In cases where the pyoderma is widespread, systemic administration of antimicrobial agents, such as penicillin or erythromycin, are effective in treating the disease. Pyoderma caused by streptococcal and staphylococcal skin infections can be prevented for the most part by washing the skin to prevent accumulation of large numbers of nonresident microbiota and by cleansing and applying antiseptics to minor skin abrasions.

Acne

Acne is a common problem during adolescence. The disease occurs when microbial invasion at the base of hair follicles initiates excessive secretions by the sebaceous glands. Various bacteria, including *Propionibacterium acnes* (*Corynebac-*

terium acnes) and *Staphylococcus epidermidis*, have been implicated as the etiologic agents of acne. Outbreaks of acne most commonly occur on the face, upper back, and chest. The disease is characterized by inflammatory papules, pustules, and cysts. Acne most frequently occurs during puberty because of the hormonal changes that occur during that period. Once adaptation to mature levels of sex hormones occurs, adults are normally not susceptible to this inflammatory disease.

Mastitis

Mastitis is an infection of the mammary glands. This disease is normally caused by infections with *Staphylococcus aureus*, although other bacteria can also cause this disease. The bacteria are able to penetrate the skin surface through the nipples of nursing mothers and multiply in the mammary glands. Treatment of mastitis involves administration of penicillin or other antibiotics that are effective against *Staphylococcus* and other Gram positive bacteria. The advisability of continuing nursing during mastitis is currently being debated among physicians, some advising cessation of nursing and others considering that the worst thing to do.

Mycetoma

Mycetoma, an infectious disease of the skin, can progress to involve subcutaneous tissues and bones. This disease is caused by various species of actinomycytes, including *Nocardia madurae*, and by various fungal species. One form of mycetoma, known as madura foot, is important in the Madura Province of India. The feet are the most common sites of involvement in cases of mycetoma because the actinomycetes and fungi causing mycetoma appear most frequently to enter the skin through abrasions acquired while walking barefoot on soil, although the infectious agents may penetrate the skin in other ways. The multiplication of the bacteria or fungi within the infected area results in swelling, which is frequently grotesque. It is necessary to identify the specific etiologic agent of mycetoma in order to determine the appropriate treatment. When the causative organism is a fungus, amphotericin B can be used. If a *Nocardia* species or other actinomycyte is responsible for the disease, sulfonamides are normally effective in the treatment. The remaining worldwide incidence of this disease is largely a function of socioeconomic conditions because the occurrence of mycetoma can be greatly reduced by wearing shoes.

Diseases caused by fungal pathogens

As just discussed, fungi as well as bacteria can cause mycetoma. Additionally, various fungal species are responsible for a number of superficial infections of the skin. Many of the fungi that cause superficial skin infections are dermatophytes; that is, they only infect the skin and its appendages, such as hair and nails. Even though the growth of dermatophytic fungi is restricted to the skin, the host–parasite interaction can cause serious manifestations. Contact of the skin with a spore of a dermatophyte can initiate infection, with the dermatophytic fungi growing filamentously within the dead kerratin-containing layers of the skin. The colonization of the skin by dermatophytic fungi initiates a cell-mediated immune response that generally occurs 10–35 days after infection. The

development of the cell-mediated immune response causes inflammatory damage to the skin tissue; however, it also prevents further lateral spread of the fungus.

Diseases caused by **dermatophytic fungi** are manifest as **tinea** or **ringworm**. These diseases are normally well localized and never fatal. The specific diseases are distinguished based on which regions of the body are infected (Table 18.6). Most dermatophytic fungi are members of the genera *Microsporum* and *Trichophytum*. Transmission of dermatophytic fungi is enhanced by conditions of high moisture and sweating, and retention of moisture increases the probability of contracting superficial infections of the skin. The transmission of athlete's foot, for example, is often associated with the high moisture levels and bare feet of athletes in a locker room, although it is now known that this disease is not acquired unless the individual has skin abrasions that the fungus can infect. Drying feet and using antifungal agents, however, can reduce the spread of this disease. It is virtually impossible to protect all body areas against potential infection with superficial dermatophytic fungi, resulting in a high incidence of dermatomycoses.

Sporotrichosis

In contrast to the superficial fungal diseases just discussed, **sporotrichosis** is a subcutaneous mycosis caused by the fungus *Sporothrix schenckii*. Inoculation of *S. schenckii* into the skin as a result of a minor injury initiates this infection. Sporotrichosis occurs worldwide and is especially widely distributed in South Africa, France, and Mexico. The infection begins with the formation of a subcutaneous nodule, with secondary nodules de-

table 18.6

Epidemiology of dermatomycoses

Disease	Causative agent	Transmission	Examples of sources
Tinea capitis (ringworm of the scalp)	*Microsporum* sp. *Trichophyton* sp.	Direct or indirect contact	Lesions, combs, toilet articles, headrests
Tinea corporis (ringworm of the body)	*Epidermophyton* *Microsporum* sp. *Trichophyton* sp.	Direct or indirect contact	Lesions, floors, shower stalls, clothing
Tinea pedis (ringworm of the feet—athlete's foot)	*Epidermophyton* *Trichophyton* sp.	Direct or indirect contact	Lesions, floors, shoes and socks, shower stalls
Tinea unguium (ringworm of the nails)	*Trichophyton* sp.	Direct contact	Lesions
Tinea cruris (ringworm of the groin—jock itch)	*Trichophyton* sp. *Epidermophyton*	Direct or indirect contact	Lesions, athletic supports

veloping later as the infection spreads through the lymphatic system. Lesions due to sporotrichosis normally occur on the extremities and spread to other parts of the body. The spread of the disease occurs slowly, permitting time for diagnosis and therapy. As a result, the death rate due to sporotrichosis is very low. The cutaneous form of sporotrichosis can be treated with a solution of potassium iodine, but if the fungus is widely disseminated, administration of amphotericin B is necessary.

table 18.7

Bacterial causes of wound infections recorded at an urban hospital

Staphylococcus aureus	48%
Enteric and other bacteria associated with the gastrointestinal tract	49%
Escherichia coli	
Proteus spp.	
Klebsiella spp.	
Enterobacter spp.	
Bacteroides spp.	
Streptococcus faecalis	
Streptococcus pyogenes, group A	3%

Infections associated with wounds and burns

Infections after wounds

Wounds disrupt the protective barrier of the skin and provide a portal of entry through which microorganisms can enter the circulatory system and deep body tissues. Microorganisms on the skin surface can readily pass through the opening of a wound, and as a result, many infections associated with wounds are caused by opportunistic pathogens derived from the normal microbiota of the skin. To avoid entry of bacteria into wounds, the area is usually covered with gauze to protect against contamination.

Staphylococcus aureus is the most frequent etiologic agent of wound infections, and *Streptococcus pyogenes* is also often associated with infections of wounds. In cases of severe wounds, where the integrity of the gastrointestinal tract is disrupted, enteric bacteria are frequently the causative agents of wound infections (Table 18.7). In many cases wound infections are localized at the site of the wound, but infections established in this manner can spread systemically and may involve many body tissues and organs. For example, infections with *Staphylococcus aureus* established through skin wounds can spread and form abscesses in bone marrow (osteomyelitis) and other body tissues, including the spine and brain. Superficial wounds can generally be treated with topical antiseptics or antibiotics to prevent the establishment of infections. Serious deep wounds, however, may require the prophylactic use of systemic antibiotics to prevent the onset of serious infections. Infections of deep wounds may involve anaerobic bacteria of the genera *Clostri-*

dium, Bacteroides, and *Fusobacterium*, as well as *Staphylococcus* and *Streptococcus* species.

Gas gangrene

Deep wounds not only provide a portal of entry for microorganisms, but the tissue damage often interrupts circulation to the area, creating conditions that permit the growth of obligately anaerobic bacteria. **Gas gangrene** is a serious infection that may result from the growth of *Clostridium perfringens* and other *Clostridium* species. The development of gas gangrene is dependent on the deposition of endospores of *Clostridium* in the wound tissue and the development of anaerobic conditions that permit the germination and multiplication of these obligately anaerobic bacteria.

The *Clostridium* species that cause gas gangrene produce toxins, the diffusion of which extends the area of dead and anaerobic tissues. The exotoxins produced by these *Clostridium* species are tissue necrosins and haemolysins that account in part for the rapid spread of infection. The growing *Clostridium* species produce carbon dioxide and hydrogen gases and the formation of odoriferous low-molecular-weight metabolic products. The gas that accumulates is primarily hydrogen because it is less soluble than CO_2. In most cases, the onset of gas gangrene occurs within 72 hours of the occurrence of the wound, and if untreated the disease is fatal. Even with antimicrobial treatment, there is a high rate of mortality, and therefore, radical surgery—amputation—is often employed to prevent the spread of infection. If treated rapidly enough, localized areas of

Figure 18.21

This painting of a soldier dying of tetanus, by Charles Bell, shows the characteristic position of the jaw and neck associated with this disease. (By the kind permission of the President and Council of the Royal College of Surgeons of Edinburgh.)

650

Human diseases caused by micro-organisms: the skin as a portal of entry and diseases of superficial body tissues

necrotic tissue can be excised and high doses of penicillin administered to block the spread of the infection. The prevention of gas gangrene rests with ensuring that wounds are not suitable for the growth of the anaerobic *Clostridium* species. This requires that wounds have adequate drainage to prevent establishment of anaerobic conditions and that foreign material and dead tissue be removed.

Tetanus

Tetanus is caused by *Clostridium tetani*. This organism is able to produce a neurotoxin that can produce severe muscle spasms. Tetanus is sometimes referred to as lockjaw because the muscles of the jaw and neck contract convulsively so that the mouth remains locked closed, making swallowing difficult (Figure 18.21). *C. tetani* is widely distributed in soil. Transmission to humans nor-

table 18.8

Portals of entry of Clostridium tetani *in cases of tetanus*

Portal of entry	Frequency (%)
Wounds	31
Punctures	27
Lacerations	8
Abrasions	3
Crush	3
Surgical and obstetrical	3
Injections	4
Ulcers	3
Other wounds	14
Unknown	7

mally occurs as a result of a puncture wound that inoculates the body with spores of *C. tetani*. If anaerobic conditions develop at the site of the wound, the endospores of *C. tetani* germinate, and the multiplying bacteria can produce neurotoxin. *C. tetani* is noninvasive and multiplies only at the site of inoculation. The neurotoxin produced by *C. tetani*, however, spreads systemically, causing the symptoms of this disease.

Virtually any type of wound into which foreign material carrying spores of *C. tetani* is introduced may lead to the development of tetanus (Table 18.8). Tales of the association of rusty nails with this disease probably originated because farmers often developed tetanus after stepping on such nails that were contaminated with soil and endospores of *C. tetani*, but clearly rusty nails are not the cause of this disease. If untreated, tetanus is frequently fatal, but if recovery does occur, there are no lasting effects. Tetanus can be treated by the administration of tetanus antitoxin to block the action of the neurotoxin. The disease can be prevented by immunization with tetanus toxoid to preclude the development of infection with *C. tetani*, and tetanus booster vaccinations are frequently given after wound injuries to ensure immunity against this disease.

Infections after burns

Burns remove the protective skin layer, exposing the body to numerous potential pathogens. Microbial infection after extensive burns, where

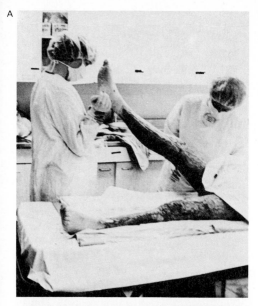

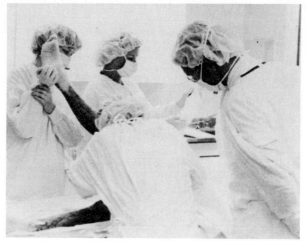

Figure 18.22

Burn victims are especially vulnerable to infection because of the destruction of the surfce skin barrier. The open air procedure is currently the method of choice used at many burn units; in this procedure the burn wound is left open to the air, except when dressings are applied. As a result extreme precautions are taken to minimize exposure to microorganisms, including keeping burn victims in isolation rooms and ensuring that individuals in contact with such patients are properly attired to prevent the transmission of potentially pathogenic microorganisms. (A) The vulnerability of burn tissues to infection is apparent in this photograph showing the leg of a man with severe burns over most of the body. (B) When applying dressings all staff must be masked, gloved, and gowned. (A–B courtesy Norton, Kosairs, Childrens Hospital, Louisville, Kentucky.) (C) As shown here even more extensive protective measures are taken in some burn facilities to protect patients from infection. (Courtesy I. Feller, National Burn Center, Ann Arbor, Michigan.)

a large portion of the skin is damaged, is a very serious complication that often results in the death of the patient. *Staphylococcus aureus, Streptococcus pyogenes, Pseudomonas aeruginosa, Clostridium tetani,* and various fungi often cause infections in burn victims. It is important to avoid contamination of the burn area that can introduce opportunistic pathogens into exposed tissue (Figure 18.22). To ensure that infections are detected in time to permit treatment with antimicrobial agents, microbiological tests are frequently performed; typically two times a week a moist swab

is run over a 1 cm^2 area and cultured for quantitation of bacteria and fungi; a count of greater than 10^5 is indicative of colonization and pending invasion. For prophylaxis, silver-containing antimicrobials, such as silvidene, are frequently used. When infections are detected, specific topical agents are selected based on the sensitivity of the infecting agent. If infections do develop, the prognosis depends on the size of the burn, the extent of infection, and the physiological state of the patient.

652

Human diseases
caused by micro-
organisms: the
skin as a portal
of entry and dis-
eases of superfi-
cial body tissues

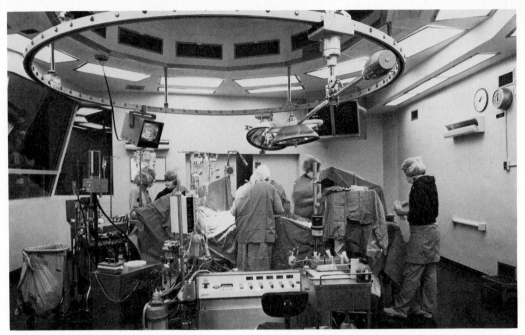

Figure 18.23

In operating theaters, walls are disinfected, instruments sterilized, and staff appropriately garbed to prevent infection of exposed tissues. Note that the surgical staff are all attired so as to minimize bringing any microorganisms in from the outside. Surgical staff wear hats, masks, gloves, gowns, and even shoe covers, and extensively scrub (wash) their skin before entering the operating room. (Courtesy National Institutes of Health, Washington, D.C.)

Discovery process

In the 1850s, when Joseph Lister was beginning his career as a surgeon, the infamous hospital diseases of erysipelas, pyanemia, septicemia, and gas gangrene made surgical wards nightmares pervaded by death rather than healing. Although it was clear that dirt and overcrowding increased the incidence of disease, there was no adequate understanding of the role of microorganisms in the spread of disease or the particular susceptibility of surgical wounds to infection. Anesthetics, which were introduced to surgical practice at the time Lister began his career, made the pain of surgery more tolerable, and hence, more people were willing to submit to the surgeon's blade. However, anesthetics did not lower the mortality rate attributable to postoperative infections. Even in a successful year a third of Lister's patients who had undergone major surgery, such as amputation, would die from infections. When the work of Pasteur—which demonstrated that microorganisms carried in the air are responsible for putrefaction of animal flesh—became known to Lister, it made sense to him that airborne microbes were responsible for postoperative infections. Lister knew that Pasteur killed germs by heating, but this did not help with the problem of how to prevent airborne contamination of open wounds. Lister reasoned that if he covered a wound with a dressing that did not exclude air but that killed the microorganisms floating in the air, he might be able to reduce the death rate. He remembered a newspaper article describing the use of carbolic acid to treat sewage and that the acid also killed certain parasites on cattle. Lister in 1865 decided to try out his theory on a patient with a compound fracture; complications from such cases frequently resulted in infection and death. Although his first trial was a failure, Lister continued his experimental trials on the use of carbolic acid to protect wounds against infection; these experiments soon met with success. "Though hardly expecting success, I tried the application of carbolic acid to the wound, to prevent decomposition of the blood, and so avoid the fearful mischief of suppuration throughout the limb. Well, it is now 8 days since the accident, and the patient has been going as if there was no external wound." Lister extended his work to the use of aerosol sprays of carbolic acid to prevent infection during surgery. He published his work, ensuring that others knew of his discoveries. Lister's work marked the beginning of the era of antiseptic surgery. His success depended on having the right background, steadfast determination, and scientific reasoning.

Infections associated with medical procedures

Medical procedures are designed to cure diseases, but **some procedures used in the treatment of disease can inadvertently introduce pathogenic microorganisms into the body and initiate an infectious process**. Several types of nosocomial infections—including pneumonia acquired in hospitals, urinary tract infections that develop as a result of the insertion of a catheter, and puerperal fever that develops from gynecological procedures—have been discussed in Chapter 17. In this section some additional problems will be considered concerning the development of infections that develop as a result of medical procedures that penetrate the skin, as for example, in association with surgical incisions.

Infections after surgical procedures

Surgical procedures often expose deep body tissues to potentially pathogenic microorganisms. A surgical incision circumvents the normal body defense mechanisms. Great care is therefore taken in modern surgical practices to minimize microbial contamination of exposed tissues (Figure 18.23). These practices include the use of clean operating rooms with minimal numbers of airborne microorganisms, sterile instruments, masks, and gowns, all of which prevent the spread of microorganisms from the surgical staff to the patient, and the application of topical antiseptics prior to making the incision in order to prevent accidental contamination of the wound with the indigenous skin microbiota of the patient. After many surgical procedures antibiotics are given for several days as a prophylactic measure.

Despite all of these precautions, infections still sometimes occur after surgery (Table 18.9). Infections after surgery can be quite serious, as the patient is already in a debilitated state. The onset of such infections is generally marked by a rise in fever. A purulent lesion may develop around the wound. Particularly, serious complications may follow open heart surgery if the patient develops endocarditis, caused by *Staphylococcus* or *Streptococcus* species. In surgical procedures involving cutting the intestines, the normal gut microbiota may contaminate other body tissues, causing peritonitis and infections of other tissues caused by enteric bacteria, unless great care is taken to minimize such contamination, and antibiotics are used to prevent microbial growth (Table 18.10). The specific microorganisms causing infections of surgical wounds and the specific tissues that may be involved depend on the nature of the surgery and the tissues that are exposed to potential contamination with opportunistic pathogens.

Serum hepatitis

Serum hepatitis is caused by hepatitis B virus. An estimated 80,000–100,000 new cases of hepatitis B viral infection occur in the United States each year. The incidence is much higher in Africa and Asia. Although the principal means of transmission of serum hepatitis involves transfusions with contaminated blood, this virus may also be transmitted by various other routes. There is a high rate of transmission of serum hepatitis among drug addicts who frequently use contaminated syringes. Perhaps 10 percent of those infected with hepatitis B become chronic carriers. The carrier rate of hepatitis B virus in blood donors in the United States appears to be between 0.5 and 1 percent, but in other countries the carrier rate may be as high as 5 percent. It is estimated that there are 2 million carriers of hepatitis B virus in the world. The surface envelope antigen of hep-

table 18.9

Some factors in the development of surgical wound infections

Situation	Staphylococcus aureus (%)	Enteric bacteria (%)	Streptococcus pyogenes (%)
Emergency operation	10	21	1
In wound of second operation	11	12	0
First operation of day	17	16	0
Other	10	0	2
Total:	48	49	3

table 18.10

Bacteriology of intra-abdominal infections

Surgical operations preceding intra-abdominal abscess

Surgical operation	Percentage of cases resulting in infections
Stomach and duodenum: plication of perforated ulcer; gastrotomy; gastric resection; vagotomy; duodenal diverticulectomy	35%
Biliary tract: cholecystectomy; cholecystostomy; cholecystectomy with common-duct exploration; common-duct exploration	22
Appendectomy	12
Colon: sigmoid resection	3
Small intestine: segmental resection	6
Splenectomy	3
Splenorenal shunt	2
Nephrolithotomy (with drainage)	2
No preceding operation	15
Total	100%

General bacteriological results

Mixed flora (aerobes and anaerobes)	78%
Aerobic or facultative bacteria only	17
Anaerobic bacteria only	5
Total	100%

Anaerobic bacteria recovered

Bacteroides fragilis; Bacteroides spp.; *Fusobacterium* spp.; *Clostridium perfringens; Clostridium* spp.; *Peptococcus* spp.; *Peptostreptococcus* spp.; *Eubacterium* spp.; other bacteria

Aerobic or facultative organisms recovered

Escherichia coli; Proteus spp.; *Klebsiella* spp.; *Pseudomonas* spp.; other Gram negative bacteria; group D streptococci; other streptococci; *Staphylococcus aureus; Candida* spp.

Human diseases caused by microorganisms: the skin as a portal of entry and diseases of superficial body tissues

atitis B virus, the Australian antigen, can be detected in the blood serum of an infected individual. The blood of an infected individual may remain infective for months or years.

The clinical manifestations of serum hepatitis may include the development of jaundice and generally are very similar to those described for hepatitis A infections in Chapter 17. When the viral infection is transmitted through blood transfusions, the fatality rate associated with serum hepatitis is about 10 percent, reflecting the high dose of viruses normally transmitted via this route and the fact that the patient receiving the transfusion is in a debilitated state. One of the most important ways of preventing the transmission of this disease is to screen blood donors and to eliminate contaminated blood from blood banks. Avoiding reuse of syringe needles among drug addicts also reduces the spread of this disease.

The United States Food and Drug Administration has approved a new vaccine for hepatitis B produced from viral particles isolated from the blood of human carriers of the disease. This vaccine, made available in 1982, was the first completely new viral vaccine in a decade and the first ever licensed in the United States made directly from human blood. The production of this vaccine involves a 65-week cycle of purification and safety testing to preclude inclusion of viruses or other undesired factors from the blood of hepatitis B carriers. The cost is about $100 for the three doses needed to establish immunity. Several recent advances in genetic engineering, however, indicate that a second-generation hepatitis B vaccine may be available soon at a considerably lower price. At present the approved hepatitis B vaccine is intended for high-risk individuals, including health care workers and drug addicts.

Infections of the eye

Conjunctivitis

Conjunctivitis is an inflammation of the conjunctiva (mucous membrane) covering the cornea and under the eylids and results from bacterial or viral infections of the eye. Bacterial conjunctivitis can be caused by a variety of bacteria, including *Haemophilus* sp., *Moraxella* sp., *Staphylococcus aureus, Streptococcus pneumoniae, Pseudomonas aeruginosa, Corynebacterium* sp., *Neiserria gonorrhoeae*, and *Chlamydia* sp. *Haemophilus aegyptius* is the common etiologic agent of epidemic conjunctivitis or pink eye among school children. The establishment of an infection of the eye requires not only inoculation with opportunistic pathogens but also that the bacteria overcome the natural defenses of the eye. Thus, bacteria causing conjunctivitis must resist washing out by tears and inactivation by lysozyme and antibodies. Contamination of the eye with foreign material and injuries of the eye tissue favor the establishment of infections.

The manifestations of conjunctivitis and the seriousness of this disease depend on the particular pathogenic microorganisms. The symptoms of conjunctivitis generally include swelling and reddening of the eyelids and the formation of purulent discharges. The eye normally becomes red and itchy because of extensive dilation of the capillaries, and consequently, the disease is referred to as pink eye. There may also be some photophobia and blurring of vision. Many cases of bacterial conjunctivitis are self-limiting, and the disease symptoms disappear within one week after onset. In some cases, however, such as when the infection is caused by *Pseudomonas aeruginosa*, the disease may progress to involve the cornea. When the cornea is infected with *P. aeruginosa*, perforation may result within 24 hours after onset of infection. Other bacterial and viral species may also cause ulceration and damage of the cornea.

Some forms of conjunctivitis can be transmitted from infected mother to child during birth. Both *Neisseria gonorrhoeae* and *Chlamydia trachomatis*, the causative agents of trachoma and inclusion conjunctivitis, can be transferred in this manner. The use of silver nitrate to treat the eyes of newborns can prevent the transmission of *N. gonorrhoeae* but not of *C. trachomatis*. **Inclusion conjunctivitis** caused by *C. trachomatis* can also be acquired by adults and children by direct contact with infected individuals and from swimming in nonchlorinated waters contaminated with this bacterial species from genital sources. In cases of bacterial conjunctivitis, the use of antimicrobial agents are useful in limiting the severity of the disease. Sulfacetamide, neomycin, and gentamicin are frequently used in treating infections of the eye.

In addition to bacterial conjunctivitis, several viruses cause acute conjunctivitis. Adenoviruses are the etiologic agents of epidemic keratoconjunctivitis or shipyard eye. The disease derives its name from the fact that outbreaks have occurred in shipyards and industrial plants where trauma to the eye may occur as an occupational hazard. In this form of conjunctivitis, the conjunctivae become inflamed and the cornea may become keratinized, covered with dead keratin protein.

Epidemic hemorrhagic conjunctivitis is caused by an enterovirus. This disease is associated with epidemic outbreaks in Africa, Asia, and Europe. The disease is transmitted by direct or indirect contact with discharges from infected eyes and by droplets containing the infective enteroviruses. This form of conjunctivitis is characterized by subconjunctival hemorrages. The disease is self-limiting, and antimicrobial agents are not useful in treating this form of conjunctivitis. Viral conjunctivitis can also be caused by herpes simplex virus type 1. Ulcerations caused by infection with herpes simplex virus can cause blindness in some cases. The use of antiviral drugs is effective in controlling herpes simplex virus but not other viral infections of the eye.

Trachoma

Trachoma is a form of keratoconjunctivitis caused by *Chlamydia trachomatis*. Trachoma occurs worldwide but is most prevalent in dry regions of the world, including the southwest United States, where this disease is a particular problem on Indian reservations. Transmission of the disease is through direct contact of the eye with infectious material. The development of the chlamydial infection in cases of trachoma initially involves a localized infection of the conjunctivae, but this is followed by spread to other areas of the eye, including the cornea. In *C. trachomatis* infections vascular papillae and lymphoid follicles are formed,

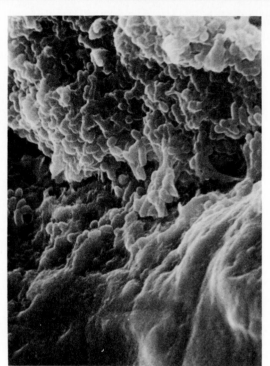

656

Human diseases
caused by micro-
organisms: the
skin as a portal
of entry and dis-
eases of superfi-
cial body tissues

Figure 18.24

Streptococcus mutans *in dental plaque, the
bacterial accumulation on the tooth, is seen in
this carious lesion of tooth enamel (3100x).
These bacteria produce polymers and an
enzyme, dextran sucrase, that creates a sticky
film on the tooth surface, providing a favorable
environment for bacteria to grow and attack
tooth enamel; it also restricts the movement of
their metabolic product, lactic acid, away from
the tooth surface, resulting in tooth decay. Of
particular importance is sucrose in the diet,
which is necessary for formation of the plaque
polymer. Reducing intake of sucrose and
cleaning the tooth surface to remove plaque
buildup reduces dental caries. (From BPS—
Z. Skobe, Forsythe Dental Center.)*

Discovery process

The ecological principles that have evolved from studies of the adherence of bacteria to surfaces of
the mouth, particularly the teeth, appear to have broad applicability to host–parasite relationships in
other areas of the body and may provide a basis for understanding the natural resistance or
susceptibility of hosts and tissues to pathogenic agents. Although a variety of bacterial types are
indigenous to the mouth, probably because of the diversity of surfaces available for colonization and
its relatively unselective conditions, streptococci and Gram positive filamentous organisms are most
prevalent. *Streptococcus mutans* is suspected to be of etiologic importance in dental caries. R. J.
Gibbons and J. van Houte found that highly pathogenic strains of *S. mutans* characteristically form
large bacterial masses on the surfaces of teeth, whereas most of the other nonpathogenic species
they tested did not. The formation of dental plaques by *S. mutans* was significantly enhanced by
adding sucrose to the diets of test animals. The addition of other carbohydrates, such as glucose,
maltose, and fructose, did not have this effect. *S. mutans* can synthesize extracellular glucans and
fructans from sucrose, which in turn allow it to attach and accumulate on the walls of culture vessels,
immersed teeth, and other solid objects. The mouth is subjected to fluid flow, and therefore,
colonization requires either that bacteria attach themselves to the surface or else that bacterial
multiplication occur at a rate faster than the dilution rate caused by the flowing secretions. The
available evidence does not support the latter theory. Until recently, though, little has been known
about the mechanisms responsible for bacterial attachment to surfaces. Studies by Gibbons,
van Houte, and coworkers suggest that reversible sorption occurs and that only a very small
percentage of the cells that do sorb to surfaces in the mouth become sufficiently firmly bound to
initiate colonization. In the case of *S. mutans*, this firm attachment to a tooth is promoted by the *de
novo* synthesis of extracellular polysaccharides from sucrose while the organisms are weakly
associated with the surface. This appears to be the mechanism by which dietary sucrose fosters the
implantation of *S. mutans* into the mouths of rodents and humans, because *in vitro* experiments and
experimentation with animals have shown that the presence of preformed glucans on the bacterial
surface prior to sorption does not promote firm attachment. Electron microscopic studies have shown
that oral *Streptococcus* species possess a fibrillar coating; in *S. salvarius, S. mitis,* and *S. pyogenes*
these fibrils mediate the attachment to epithelial surfaces. It also appears that substances in saliva
mediate the establishment of the primary film that leads to the formation of dental plaque, with
subsequent bacterial growth resulting in formation of dental caries. It is hoped that once the
mechanism of attachment is fully understood, ways will be found to prevent plaque formation with
concomitant elimation of dental caries.

eye tissues can become deformed and scarred, and a chronic infection can lead to partial or total loss of vision. Several antibiotics, including sulfonamides, tetracyclines, and erythromycin, are useful in treating trachoma.

Infections of the oral cavity

Dental caries

The surfaces of the oral cavity are heavily colonized by microorganisms. Excessive growth of microorganisms in the mouth can cause diseases of the tissues of the oral cavity. One of the most common human diseases caused by microorganisms is **dental caries**. Caries is initiated at the tooth surface as a result of the growth of *Streptococcus* species. These streptococci can initiate caries because they have the following essential properties: (1) they can adhere to the tooth surface; (2) they produce lactic acid as a result of their fermentative metabolism, thereby dissolving the dental enamel surface of the tooth; and (3) they produce a polymeric substance, which causes the acid to remain in contact with the tooth surface. *S. mutans, S. sanguis*, and *S. salivarius* are implicated as the causative agents of dental caries. These bacteria produce a dextran sucrose enzyme that catalyzes the formation of extracellular glucans from sucrose, responsible for the accumulation of dental plaque (Figure 18.24). Dental plaque is an accumulation of a mixed bacterial community in a dextran matrix. The accumulation of bacteria in plaque may be 500 cells thick. There is a high degree of structure within plaque, indicative of sequential colonization by different bacterial populations and the different positions of each population within this complex bacterial community.

Dental caries can be prevented by limiting dietary sugar substrates from which the bacteria produce acid and plaque and by removing accumulated food particles and dental plaque by periodic brushing and flossing of the teeth. Such hygienic practices are particularly effective if performed after eating meals and snacks. Additionally, tooth surfaces can be rendered more resistant to microbial attack by including calcium in the diet, as by drinking milk, and by fluoride treatments of the tooth surface. The administration of fluorides in the diet, such as by consumption of fluoridated water, during the period of tooth formation can reduce dental caries by as much as 50 percent.

Periodontal disease

In addition to causing dental caries, microorganisms growing in the oral cavity can cause diseases involving the supporting tissues of teeth (periodontal disease). **Periodontal disease** can be manifest as **gingivitis** (inflamation of the gingiva), **necrotizing ulcerative gingivitis** (trench mouth),

Figure 18.25

(A) The anatomy of a tooth, showing sites of microbial attack. (B) Diagram showing development of periodontal disease involving the tissues surrounding the tooth.

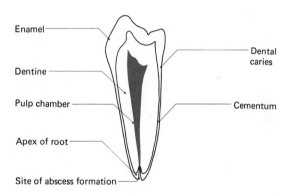

periodontitis (inflammation of the periodontium, also known as pyorrhea), and **periodontosis** (noninflammatory degeneration of the periodontium leading to bone regression) (Figure 18.25). Periodontal disease develops when food particles and dental plaque are allowed to accumulate between the tooth and surrounding tissues.

Different bacterial populations are involved in different stages of development of periodontal disease. Members of the genus *Actinomyces*, such as *A. vicosus*, are involved in the formation of early gingivitis, the most common form of periodontal disease. As the inflammation of the gingiva progresses, anaerobic Gram negative bacteria, such as species of *Veillonella, Campylobacter,* and *Fusobacterium*, become prevalent, but it is still unknown which bacterial species are important in causing this disease condition. The further progression of the gingival inflammation into the periodontal membrane involves the development of a mixed community of microorganisms. The specific bacterial populations vary in different syndromes of periodontitis. Many spirochetes and unknown bacteria are involved in these infections. Reports indicate that the bacterial species involved in periodontidis may include *Bacteroides asaccharolyticus, Treponema denticola, Eikenella corrodens, Fusobacterium nucleatum, Actinomyces viscosus,* and *Actinomyces naeslun-*

dii. The development of trench mouth (Vincent's infection) is accompanied by the proliferation of populations of *Borrelia vincentii, Fusobacterium fusiforme,* and unidentified spirochetes. Clinical manifestations of trench mouth include fever, swelling, ulceration, and bleeding of the gums with tissue necrosis, causing ulcers between the teeth. Trench mouth can be treated with penicillin and metronidazole. Periodontosis involves a limited number of species of Gram negative anaerobic rods, including *Capnocytophaga (Bacteroides ochraceus), Eikenella corrodens, Fusobacterium nucleatum, Bacteroides melaninogenicus,* and *Selenomonas sputigena.*

The development of advanced forms of **periodontal disease**, involving tissues beyond the gingiva, is serious and can lead to the loss of teeth. For example, in chronic destructive periodontitis, the microbial infection of the subgingiva can establish pockets between the tooth and supporting tissues in which microorganisms can proliferate (periodontal pocket). Microorganisms growing in such protected pockets erode the alveolar bone and destroy periodontal membranes, causing loosening of the teeth that can lead to tooth loss. The control of periodontal disease can be achieved by frequent removal of accumulations of dental plaque, by daily flossing, and periodic professional dental cleanings.

Postlude

In this chapter we have examined a number of human diseases caused by microbial pathogens that enter the body through the skin and that cause diseases of superficial body tissues. Trauma, especially if associated with the introduction of foreign material, is important in establishing infections of surface body tissues. Direct contact and fomites contaminated with pathogenic microorganisms are often involved in the transmission of infections of the eye and skin, and proper hygienic practices, which lower the likelihood of contact with pathogenic microorganisms, greatly reduce the incidence of these diseases.

In the case of diseases of the oral cavity caused by microorganisms indigenous to the mouth, it is not possible to avoid contact with the pathogens. Consequently, the prevention of dental caries and periodontal disease depends on the effective removal of dental plaque with its associated microbial populations. Essentially, the prevention of these diseases depends on the elimination of microbial metabolic products before they can accumulate in concentrations sufficient to cause diseases involving the tissues of the oral cavity. Inclusion of elements in the diet, such as fluoride, that render the host tissues more resistant to microbial attack, and the reduced intake of foods containing high carbohydrate levels, which microorganisms can readily utilize as substrates to proliferate and form dental plaque, can greatly reduce the incidence of dental caries and periodontal disease.

Wounds and animals bites that disrupt the protective skin layers provide the main portals of entry by which pathogenic microorganisms penetrate the skin and gain entry to the body. Minimizing contamination of wounds with foreign matter and associated microorganisms is important in controlling infections that may occur when the skin barrier is broken. Surgical practices utilize elaborate aseptic procedures to minimize potential infection, but nevertheless, infections sometimes occur after surgery, attesting to the vulnerability of the body to microbial infection when the skin barrier is disrupted and host defense mechanisms are impaired.

Animal bites are extremely important in the transmission of various infectious diseases. Many arthropod populations act as vectors, providing both the microbial inoculum and the portal of entry for initiating an infectious process. Reducing population levels of vectors, such as mosquitos, ticks, fleas, and lice are important in controlling the morbidity rates of diseases transmitted in this manner. Additionally, because many diseases transmitted via arthropod vectors are maintained in nonhuman animal populations, it is important to recognize and control the levels of infection in reservoir populations.

In the preceding six chapters the principles of interaction between microorganisms and humans that produce human disease have been considered. The establishment of infectious disease processes in human beings involves dynamic interactions between microbial pathogens and human host cells and tissues. The human body possesses an elaborate defense network that acts to prevent microbial infections, including the antibody-mediated and cellular immune defense mechanisms, which generally preclude the growth of microorganisms in body tissues. However, as we have seen, pathogenic microorganisms possess metabolic capabilities that allow them to overcome or escape these defense mechanisms and produce disease. The manifestations of disease depend on the dynamics of the interactions between the host and parasite.

In this and the preceding chapter the epidemiology of infectious diseases has been emphasized. Microbial pathogens are generally restricted to specific portals of entry that determine their potential routes of transmission. Understanding the transmission processes permits the development of effective means of preventing or at least reducing the incidence of many diseases. When prevention fails and an infectious disease occurs, the clinical microbiology laboratory employs various methods for diagnosing the etiology of the disease and determining the applicability of the various antimicrobial agents available for treating infectious diseases.

At least in developed countries, such as the United States, the prevention and control of infectious diseases have greatly reduced morbidity and mortality rates. Similar advances in controlling infectious diseases have yet to be fully realized in socioeconomically depressed regions of the world where poor nutrition, poor sanitation, and lack of sufficient medical facilities allow the continuation of high rates of morbidity and mortality from infectious diseases. Advances in our understanding of host–parasite relationships and improvements in the delivery of health services based on our current understanding of disease processes can alleviate much of the suffering and death associated with infectious diseases.

Study Questions

1. What is a vector? What are some common vectors of infectious diseases?

2. For each of the following diseases, what is the causative organism, the reservoir for that etiologic agent, and the vector for disease transmission?
  a. plaque
  b. Rocky Mountain spotted fever
  c. malaria
  d. smallpox
  e. endemic typhus

3. How is rabies treated?

4. What is the etiologic agent of malaria, and why is this disease difficult to treat?

5. Why are serious *Clostridium* infections associated with deep wounds?

6. What are the differences between the toxins produced by *Clostridium botulinum* and *C. tetani*?

7. Why do urinary tract infections frequently occur after catheterization procedures?

8. What is the cause of serum hepatitis? Why is this disease associated with drug addicts?

9. Why are burn victims prone to fatal microbial infections?

10. Discuss the etiology of dental caries. Why should removing sucrose from the diet lower the rate of this disease?

Suggested Supplementary Readings

Boyd, R. F., and B. G. Hoerl. 1981. *Basic Medical Microbiology*. Little, Brown and Co., Boston.

Boyd, R. F., and J. J. Marr. 1980. *Medical Microbiology*. Little, Brown and Co., Boston.

Braude, A. I. (ed.). 1980. *Medical Microbiology and Infectious Disease*. W. B. Saunders Co., Philadelphia.

Burgdorfer, W., and R. Anacker. 1982. *Rickettsiae and Rickettsial Diseases*. Academic Press, New York.

Burnett, G. W., H. W. Scherp, and G. S. Schuster. 1976. *Oral Microbiology and Infectious Disease*. Williams & Wilkins Co., Baltimore.

Dubos, R. J., and R. G. Hirsch (eds.). 1965. *Bacterial and Mycotic Infections of Man*. J. B. Lippincott Co., Philadelphia.

Davis, B. D., R. Dulbecco, H. N. Eisen, H. S. Ginsberg, and W. B. Wood. 1981. *Microbiology*. Harper & Row, Publishers, Hagerstown, Maryland.

Duguid, J. P. (ed.). 1979. *Medical Microbiology*. Churchill Livingstone, New York.

Emmons, C. W., C. H. Binford, J. P. Ulz, and K. J. Kwon-Chung. 1977. *Medical Mycology*. Lea & Febiger, Philadelphia.

Faust, E. C., P. C. Beaver, and R. C. Jung. 1975. *Animal Agents and Vectors of Human Disease*. Lea & Febiger, Philadelphia.

Hoeprich, P. D. (ed.). 1977. *Infectious Diseases: A Modern Treatise of Infection Processes*. Harper & Row, Publishers, Hagerstown, Maryland.

Jawetz, E., J. L. Melnick, and E. A. Adelberg. 1980. *A Review of Medical Microbiology*. Lange Medical Publications, Los Altos, California.

Joklik, W. K., and H. P. Willett (eds.). 1980. *Zinsser Microbiology*. Appleton Century Crofts, New York.

Nolte, W. A. (ed.). 1980. *Oral Microbiology*. C. V. Mosby Co., St. Louis.

Top, F. H., and P. Wehrle. 1976. *Communicable and Infectious Diseases*. C. V. Mosby, St. Louis.

Volk, W. A. 1978. *Essentials of Medical Microbiology*. J. B. Lippincott Co., Philadelphia.

Wilson, G. S., and A. A. Miles (eds.). 1975. *Topley and Wilson's Principles of Bacteriology and Immunity*. Williams & Wilkins, Baltimore.

Yamaguchi, T. (ed.). 1981. *Color Atlas of Clinical Parasitology*. Lea & Febiger, Philadelphia.

Youmans, G. P., P. Y. Paterson, and H. M. Sommers. 1980. *The Biological and Clinical Basis of Infectious Disease*. W. B. Saunders Co., Philadelphia.

661

Suggested
supplementary
readings

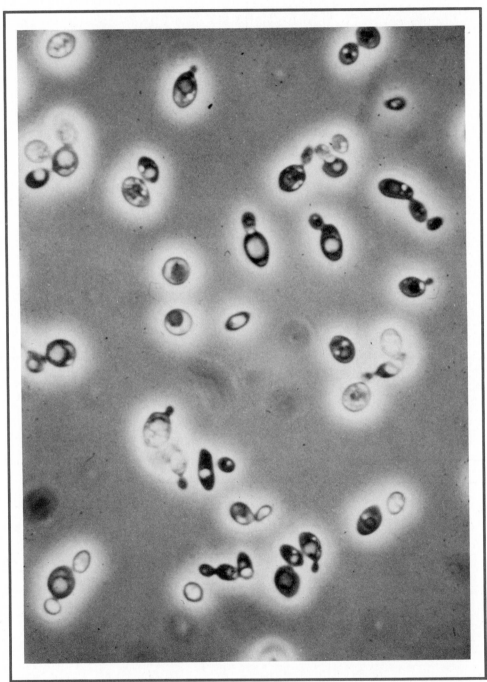

Phase contrast micrograph of Saccharomyces cerevisiae.
(Courtesy Cletus Kurztman, USDA Northern Regional Research Center.)

Food and industrial microbiology

SECTION EIGHT

Food microbiology

19

Food products serve not only as sources of nutrition for humans and other animals but also as substrates for the growth of microorganisms. The uncontrolled growth of microorganisms in food causes spoilage, a serious problem accounting for sizable losses of food products that are critically needed to meet global nutritional requirements. A major concern in food microbiology is the development of an understanding of the processes involved in food spoilage and the appropriate measures for the preservation of food products in order to control the undesirable activities of microorganisms that result in food spoilage. In particular, it is necessary to prevent contamination of food products with human pathogens and to control the potential proliferation of toxin-producing microorganisms, which can result in food poisoning and the transmission of food-borne pathogens discussed in Chapter 17. Careful quality control in the food industry is essential for preventing outbreaks of food-borne disease. Additionally, growth of microorganisms in foods can alter the quality of the product, causing sizable economic losses to the food industry. As we have already considered in some detail food-borne human diseases, it is this latter aspect of food spoilage that we will examine in this chapter. Also, in this chapter we will examine the uses of microorganisms for producing foods. Although the uncontrolled growth of microorganisms in food products is deleterious, the controlled growth of microorganisms on or in foods is used to produce numerous foods and beverages that are considered gastronomic delights. For example, the growth of microorganisms on milk produces cheese; on fruits, wine; and on grains, beer. In these cases the growth of microorganisms and the accumulation of their metabolic products is desirable.

Food spoilage

Food spoilage is defined as any change in a food that renders it undesirable or unsafe for human consumption, and the growth of microorganisms in a food represents only one process that may cause food to spoil. The growth of pathogenic microorganisms in a food is certainly undesirable, as it can make that food unsafe to eat, but other microbially induced changes in food, such as decreased nutritional content and altered taste, odor, color, and texture, can also make a food undesirable for human consumption. Such food spoilage is difficult to define because it depends on the culture of the consumer; for example, Eskimos bury fish to produce a stinky mess and in Britain meat should have the strong aroma of aging, but these same foods would likely be rejected as spoiled in most parts of the United States. Foods are classified according to their susceptibility to microbial spoilage as nonperishable, semiperishable, or perishable. Perishable foods, such as meats, fish, poultry, most fruits and vegetables, eggs, and milk, readily spoil because of microbial activities and generally have short shelf lives unless steps are taken to remove, kill, or prevent the growth of associated microorganisms. Semiperishable foods, such as potatoes and apples, generally remain unspoiled for prolonged periods of time unless improperly handled. Stable or nonperishable foods, such as sugar, flour, and numerous dry products, do not spoil under normal conditions, but even nonperishable foods can spoil under appropriate conditions. For example, if cereals, grains, and flours are stored under conditions of high moisture, various fungal and bacterial populations are able to grow and spoil the product. Undoubtedly, we have all observed the growth of fungi on bread that has been stored too long.

The changes that occur in a food during microbial spoilage depend on the particular microbial populations involved, their enzymatic activities, environmental conditions, and the nature of the food. The microorganisms involved in the spoilage process generally do not originate within the food, the inner tissues of most plants and animals being sterile, but rather come from the surface tissues or are a result of contamination during processing. Several factors controlling food spoilage are intrinsic to the food, but others are extrinsic environmental parameters. The most important **extrinsic factors influencing food spoilage** are (1) temperature of storage; (2) relative humidity; and (3) oxygen concentration. By controlling these parameters it is possible to alter the rates of food spoilage. It is somewhat more difficult to control the inherent properties of animal and plant tissues that influence microbial spoilage of foods. The intrinsic moisture content, pH, and physical and chemical nature of the food, in part, control the numbers and types of microorganisms involved in the spoilage process. The bio-

chemical composition of a food, in particular, has a marked influence on which microbial populations are involved in the spoilage process and the microbial decomposition products associated with the spoilage of that food (Table 19.1).

Representative spoilage processes

Fruits, vegetables, meats, poultry, seafoods, milk and dairy products, and various other food products differ in their biochemical composition (Table 19.2), and therefore are subject to spoilage by differing microbial populations (Table 19.3).

Spoilage of fruits and vegetables

Fruits and vegetables are subject to rot because of the microbial degradation of pectin, the biochemical responsible for maintaining the firmness and texture of fruits and vegetables (Figure 19.1). Microbially produced pectinesterases and polygalacturonases hydrolyze pectins, resulting in the formation of **soft spots** in fruits and vegetables. About 20 percent of the harvested crops of fruits and vegetables are lost to spoilage primarily because of the activities of bacteria and fungi. Carbohydrates, which are present in high concentrations in fruits, vegetables, and other foods, are readily degraded by numerous microorganisms, resulting in the production of various degradation products, such as low-molecular-weight acids and alcohols. The accumulation of such products of microbial metabolism can alter the taste of a food, generally causing the food to sour. The taste of sour milk, for example, is associated with the accumulation of lactic acid from the microbial transformation of carbohydrates. The accumulation of low-molecular-weight acids may also alter the smell of the product; sour milk often can be recognized just by smelling it. In canned foods, production of acid and no gas is referred to as **flat-sour spoilage** because the food becomes sour, but the can shows no evidence of food spoilage because no gas is produced; that is, the can re-

table 19.1

Some food spoilage processes

Process	Substrate	Products
Putrefaction	Proteins	Amino acids, amines, ammonia, and hydrogen sulfide
Souring	Carbohydrates	Acids, alcohols, and carbon dioxide
Rancidity	Lipids	Fatty acids and glycerol
Soft rot	Pectin	Methanol, galacturonic acid, and polygalacturonic acid

table 19.2

Composition of some representative foods by percent weight

Food	Carbohydrate	Protein	Fat	Ash	Water
Cereals					
Wheat flour, white	74	11	2	2	12
Rice, milled, white	79	7	1	1	13
Maize (corn), whole grain	73	10	4	1	12
Earth vegetables					
Potatoes, white	19	2	+	1	78
Potatoes, sweet	27	1	+	1	70
Vegetables					
Carrots	9	1	+	1	89
Radishes	4	1	+	1	94
Asparagus	4	1	+	1	93
Beans, snap, green	8	2	+	1	89
Peas, fresh	17	7	+	1	75
Lettuce	3	1	+	1	95
Fruits					
Banana	24	1	+	1	74
Orange	11	1	+	1	87
Apple	15	+	+	+	84
Strawberries	8	1	1	1	90
Melon	6	1	+	+	93
Meat					
Beef, medium fat	—	18	22	1	60
Veal, medium fat	—	19	14	1	66
Pork, medium fat	—	12	45	1	42
Lamb, medium fat	—	16	28	1	56
Zebra, medium fat	1	20	4	1	74
Poultry					
Chicken	—	20	13	1	66
Duck	—	16	30	1	53
Turkey	—	20	20	1	58
Fish					
Nonfatty filet	—	16	1	1	82
Fatty fish filet	—	20	10	1	69
Crustaceans	3	15	2	2	79
Dried fish	—	60	21	15	4
Milk					
Cow, whole	5	4	4	1	87
Goat, whole	5	4	5	1	86
Cheese					
Hard, whole milk	2	25	31	5	37
Soft, low-fat milk	5	15	7	3	70

+ Less than 0.5

mains flat. When gas is produced during spoilage of canned foods, the can swells, as can be seen by examining the lids of the can which are designed to push outward if pressure builds within the can because of microbial action.

Spoilage of meats

Meats and other proteinaceous products can be decomposed by anaerobic bacteria, resulting in **putrefaction**. Putrefaction of meat is the result of the breakdown of proteins by proteinases. The

table 19.3

Some spoilage organisms of important foods

Class of food products	Genera dominating when spoilage occurs during standard conditions of storage
Cereal grains	*Aspergillus, Fusarium, Monilia, Penicillium, Rhizopus*
Bread	*Bacillus, Aspergillus, Endomyces, Neurospora, Rhizopus*
Vegetables	*Achromobacter, Pseudomonas, Flavobacterium, Lactobacillus, Bacillus*
Fruits and juices	*Acetobacter, Lactobacillus, Saccharomyces, Torulopsis*
Fresh meat	*Micrococcus, Cladosporium, Thamnidium, Achromobacter, Pseudomonas, Flavobacterium*
Sausage, bacon, ham, etc.	*Micrococcus, Lactobacillus, Streptococcus, Debaromyces, Penicillium*
Whale meat	*Streptococcus, Clostridium, Achromobacter, Pseudomonas, Flavobacterium*
Poultry	*Achromobacter, Pseudomonas, Flavobacterium, Micrococcus, Salmonella*
Fish	*Achromobacter, Pseudomonas, Flavobacterium, Micrococcus, Vibrio*
Shellfish	*Achromobacter, Pseudomonas, Flavobacterium, Micrococcus, Vibrio*
Milk and milk products	*Streptococcus, Lactobacillus, Microbacterium, Achromobacter, Pseudomonas, Flavobacterium, Bacillus*
Eggs	*Pseudomonas, Cladosporium, Penicillium, Sporotrichum*

Figure 19.1

This rotting fruit shows soft spot formation. Such spoilage of fruits and vegetables causes great economic losses and limits the useful shipping times and shelf lives of many products. (From BPS—W. Rosenberg, Iona College.)

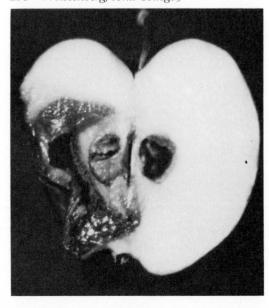

subsequent degradation of amino acids produces foul-smelling, low-molecular-weight sulfur- and nitrogen-containing compounds, such as mercaptans, hydrogen sulfide, ammonia, and amines (Table 19.4). The evolution of noxious odoriferous compounds from putrefying proteins renders food unacceptable for human consumption. The characteristic odor of hydrogen sulfide, for example, renders rotting eggs inedible, and the development of off odors and slime on poultry and beef reflect increased microbial populations and spoilage of the meat (Figure 19.2). In canned foods such spoilage is referred to as sulfide-stinkers. Under aerobic conditions the decomposition of proteins generally does not result in the production of compounds with obnoxious odors. The spoilage of meat under aerobic conditions, though, can result in the accumulation of surface slime, generally because of the growth of *Pseudomonas, Achromobacter, Streptococcus, Leuconostoc, Bacillus, Micrococcus*, and *Lactobacillus* species. The physical state of the meat influences which microbial species may be involved in spoilage. The spoilage of fresh whole meats is normally associated with lactic acid bacteria, particularly *Lactobacillus, Leuconostoc*, and *Streptococcus* species, whereas the spoilage of ground beef is primarily due to the growth of *Pseudomonas, Acromobacter*, and *Micrococcus*.

table 19.4

Products from the microbial decomposition of amino acids

Chemical process	Products
Oxidative deamination	Keto acid + NH_3
Hydrolytic deamination	Hydroxy acid + NH_3
Reductive deamination	Saturated fatty acid + NH_3
Dehydrogenation deamination	Unsaturated fatty acid + NH_3
Decarboxylation	Amine + CO_2
Hydrolytic deamination + decarboxylation	Primary alcohol + NH_3 + CO_2
Reductive deamination + decarboxylation	Hydrocarbon + NH_3 + CO_2
Oxidative deamination + decarboxylation	Fatty acid + NH_3 + CO_2

Spoilage of other foods

Spoilage by microbial degradation of fats and oils produces rancidity, caused by oxidation or hydrolysis of lipids. Spoiled butter, for example, becomes **rancid** because of the hydrolysis of butterfat with the production of free fatty acids and glycerol and the accompanying development of undesirable flavors. Fish with high lipid levels, such as salmon and mackerel, may also become rancid because of the hydrolysis of fats and oils.

The growth of some microorganisms can alter the color of the food. For example: the growth of *Serratia marcescens* can produce bright red bread; the combined growth of *Pseudomonas* sp. and *Streptococcus lactis* produces blue milk; the growth of fungi on meat surfaces can produce off colors, such as black spot caused by *Cladosporium herbarum*, white spot caused by *Sporotrichum carnis*, and green patches produced by *Penicillium* species; and the growth of heterolactic fermenters on cured meats, such as frankfurters, causes greening because of the action of peroxidases on

Figure 19.2

The increases in the numbers of microorganisms during meat spoilage, including the time of onset of off odor and slime formation, are shown in this graph. Storage was at 5°C. Note that although chicken initially has a considerably higher microbial content than beef, the proliferation of microorganisms in the beef begins after only one day, and the sharp increases in the chicken do not begin until day 3.

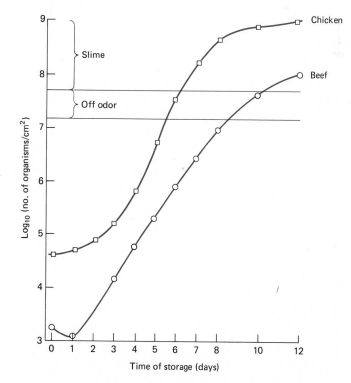

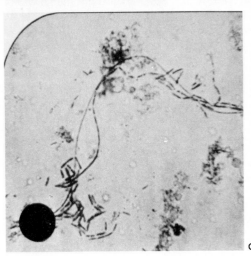

Figure 19.3

The development of rope in bread is shown in these photographs. (A) Infected bread breaks down to a sticky mass that forms strings when pulled apart; it smells like a ripe canteloupe and its color is dark. (B) Slices of ropy bread show the improper consistency of the loaf. (C) The problem is due to the growth of microorganisms, some of which are shown in this micrograph. Rope development is effectively controlled by using chemical inhibitors, by shortening the processing time for bread production, and by adjusting environmental conditions in the bakery, especially lowering temperature and humidity, so that undesirable microbial growth does not occur in the dough before the bread is placed in the oven. (Courtesy American Institute of Baking, Manhattan, Kansas.)

the meat pigments. The development of "unnatural" colors in a food reduces its acceptability, regardless of whether there is any associated development of off taste or smell. Microbial growth can also alter the texture of a food. For example, the growth of *Bacillus* species in products such as milk and dough can produce **ropiness**, a textural change resulting from the growth of encapsulated *Bacillus* species, together with the hydrolysis of starch and proteins (Figure 19.3).

Indicators of human pathogens associated with foods

Perhaps the most serious food spoilage process involves the growth of pathogenic and toxin-pro-

ducing microorganisms. As discussed in Chapter 17, the ingestion of foods containing toxins and human pathogens can cause a variety of gastrointestinal diseases. In some cases the growth of pathogenic microorganisms in a food is accompanied by obvious signs of spoilage, such as the production of gas or foul-smelling compounds, which give a clear indication that the product should not be eaten. In other cases there are no obvious signs of spoilage to indicate the presence of pathogens or toxins. It is therefore necessary to carry out inspection programs aimed at detecting the possible contamination of food products with pathogenic microorganisms. As a consequence of the high quality control procedures employed in the food industry, few occurrences of food-borne disease are traced back to the food processor. Rather, most outbreaks of food poisoning are caused by improper post-process handling of the food, often in a food service setting.

Just as the hospital clinical laboratory performs tests to detect pathogenic microorganisms in a patient, the quality control laboratory in the food industry performs tests to detect pathogens in food products and to ensure that the numbers and types of microorganisms associated with a food are not likely to cause serious food spoilage and/or health problems. In some cases quality control procedures are aimed at detecting the presence of specific microorganisms, but in most cases the test procedures rely on examining the food for indicator organisms. For example, coliform counts are routinely performed on representative samples of many food products as an indication of possible fecal contamination, since food contaminated with fecal material has a relatively high probability of containing human pathogens. Foods, such as hamburger, often have 10^5 total bacteria per gram, but as long as they do not contain any *Salmonella* or other pathogens, they are considered safe for human consumption. Most municipalities have mandated testing programs and standards of acceptable microbiological quality for establishing the safety of food products. In the United States, agencies involved in the surveillance and regulation of foods are the Federal Food and Drug Administration (FDA), the United States Department of Agriculture (USDA), and state boards of health. The established standards, however, do not always correlate well with the safety of the food. For example, many food microbiologists now feel that the coliforms are not a good indicator of food-borne pathogens, yet this still is used in many situations of an index of food safety. Standards based on the presence or absence of pathogens, such as those set by Congress and regulated by the USDA and FDA, do provide the public with the safeguards needed for quality assurance of the safety of food products.

Food preservation methods

Food preservation methods are aimed at preventing the microbial spoilage of food products and the growth of food-borne pathogens. Thus, **the two principal goals of food preservation methods are (1) increasing the shelf life of the food, and (2) ensuring the safety of food consumption.** There are a variety of food preservation methods (Table 19.5), all of which attempt

table 19.5

Principles and principal methods of food preservation

Principles of food preservation
In accomplishing the preservation of foods by the various methods, the following principles are involved:

1. Prevention or delay of microbial decomposition
   a. by keeping out microorganisms (asepsis)
   b. by removal of microorganisms, e.g., filtration
   c. by reducing rate of microbial growth, e.g., by low temperature, drying, anaerobic conditions, and chemical inhibitors
   d. by killing microorganisms, e.g., by heat or radiation

2. Prevention or delay of self-decomposition of the food
   a. by inactivation of food enzymes, e.g., blanching
   b. by prevention of chemical reactions, e.g., by using antioxidants

Principal methods of food preservation

1. Asepsis, or keeping out of microorganisms
2. Removal of microorganisms
3. Maintenance of anaerobic conditions
4. Use of high temperatures
5. Use of low temperatures
6. Drying
7. Irradiation
8. Use of chemical preservatives

to eliminate microorganisms from foods and/or limit their growth rates. Although some of these preservation methods can increase shelf life indefinitely, others only extend the usefulness of the food for a limited period of time.

The choice of a food preservation method depends on the nature of the food. It is critical that the procedures used for preserving foods not render the product unacceptable for human consumption. For example, although boiling milk for a long period of time could greatly extend its shelf life, the product would no longer be palatable. Food preservation methods must include processing necessary to establish conditions that will prevent microbial growth during storage. Understanding the principles employed in the preservation method is important so that subsequent storage and handling of the food do not negate the effectiveness of the preservation method. We will now consider some of the methods commonly used for food preservation.

Asepsis

Preventing the contamination of food products with pathogenic and spoilage microorganisms is an important part of food preservation methods. The use of **sanitary methods** minimizes microbial contamination of foods and is critical whenever food products are handled. This includes both the commercial handling of foods during processing and the handling of foods in the home during preparation. Washing of utensils, such as cutting instruments that may contaminate foods minimizes the possibility of microbial contamination. It should be remembered that anything that contacts the food, including hands and work surfaces, may be a source of microbial contamination and that once the food is inoculated, the microorganisms may reproduce rapidly under favorable conditions.

The use of appropriate methods for processing food products can reduce the levels of microbial contamination and prevent the spread of microorganisms through the food, but aseptic methods do not eliminate microorganisms from the food. For example, the careful processing of animal products in the abatoir (slaughterhouse) is critical in order to minimize microbial contamination of meats by microorganisms associated with the animal surfaces and gut contents, but even with aseptic handling the meat surface retains microbial contaminants. Trimming away spoiled portions of a food is an obvious way of reducing microbial contamination, but the spread from the

spoiled area to the remaining food may still occur unnoticed. Washing fruits and vegetables can be important for removing soil particles, animal fertilizers, and associated microorganisms that may cause food spoilage, but caution must be taken when washing products to make sure that the washing process itself does not add or spread spoilage microorganisms through the food. Chlorine is normally added to water to ensure that the wash water does not act as a source of microbial contaminants.

The packaging of food products is extremely important in preventing microbial contamination during transport and storage. Many foods have a natural protective covering that prevents or delays microbial decomposition of the sterile inner tissues, such as the outer skins of fruits and vegetables and the shells of eggs. Processing of foods, however, often removes or disrupts these protective layers, exposing the product to microbial spoilage. Most processed food products are wrapped or placed in sealed containers to preclude microbial contamination and spoilage of the food. Quality control in packaging is an important aspect of the food processing operation.

Filtration and centrifugation

Filtration and **centrifugation** can be used to remove microorganisms from some food products. Filtration is used to eliminate completely (sterilize) or reduce the numbers of microorganisms in drinking water, beer, wine, soft drinks, and fruit juices. Industrial filtration methods employ various types of bacteriological filters, including sintered glass, cellulose, and nitrocellulose with pore sizes small enough to trap bacteria. Because of problems with clogging of the filters, only clear liquids can effectively be processed by filtration. Centrifugation can be used to clarify a liquid prior to filtration. Generally, centrifugation will not completely remove all microorganisms from a food, but the reduction in microbial numbers that can be achieved by centrifugaion can extend the shelf life of a food.

High temperatures

Killing food-borne microorganisms by exposure to high temperatures is a very effective means of preventing food spoilage. The ability to use high-temperature preservation methods for a given food is dependent on the temperature sensitivity of that food. If a food is destroyed by exposure to the

Figure 19.4

(A) Illustration showing the steps in a commercial canning procedure. (B) Photograph of a commercial canning operation; the cans are carried along a conveyor belt through the filling and sealing operations and into the retort; in modern automated systems the speed of the conveyor determines the exposure time of the cans within the retort. (Courtesy American Can Company, Barrington, Illinois.)

Discovery process

The method of canning to preserve foods had its origin in 1795 when the French government offered a prize of 12,000 francs for the development of a practical method of food preservation. In 1809 Francois (Nicolas) Appert succeeded in preserving meats in glass bottles that had been kept in boiling water for varying periods of time. Appert was issued a patent for his process in 1810. As early as 1820 the commercial production of canned foods was begun in the United States by W. Underwood and T. Kensett. Appert's discovery that foods could be preserved for prolonged periods of time when they were heated and stored in the absence of oxygen came at the same time that the questions of spontaneous generation and the role of microorganisms in fermentation and putrefaction were being debated by the premier scientists of the day. Spallanzani in 1765 had shown that beef broth that had been boiled for an hour and sealed did not spoil. Appert applied to home economics the results of Spallanzani's experiments. Not being a scientist, Appert probably did not understand either why his method worked or its long-range significance. It was not until almost a half-century later that Pasteur, in disproving the theory of spontaneous generation, provided the scientific basis for understanding why Appert's canning method worked. Pasteur pointed out that Appert's method, even when modified by using temperatures below 100°C and relatively short incubation times, was a practical method for preventing undesirable ferments. Today, the method of canning, begun over a century ago in the absence of an understanding of why it worked, that is, without the knowledge of its scientific rationale, is one of the most widely used methods for preserving foods.

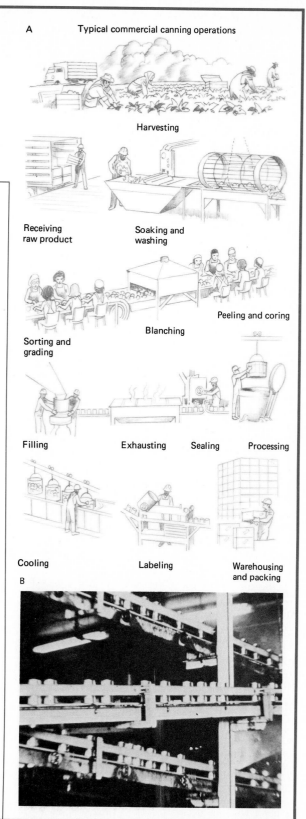

A Typical commercial canning operations

Harvesting

Receiving raw product

Soaking and washing

Peeling and coring

Blanching

Sorting and grading

Filling   Exhausting   Sealing   Processing

Cooling   Labeling   Warehousing and packing

B

high temperatures necessary to kill its associated microorganisms, high-temperature methods obviously cannot be used. High-temperature preservation methods may involve sterilization, exposing the food at a given elevated temperature for a period of time sufficient to kill all microorganisms associated with that food, followed by asepsis to prevent recontamination of the food. Canning, for example, uses heat for sterilizing the food and hermetic sealing under anaerobic conditions to prevent food spoilage. In this preservation method the high-temperature exposure kills all the microorganisms in the product, the can or jar acts as a physical barrier to prevent recontamination of the product, and the anaerobic conditions prevent oxidation of the biochemicals in the food (Figure 19.4).

Pasteurization

Alternatively, exposure to high temperatures may be used just to reduce the numbers of microorganisms associated with the food, as long as human pathogens are eliminated. **Pasteurization** uses relatively brief exposures to moderately high temperatures to reduce the numbers of viable microorganisms and to eliminate human pathogens, but a pasteurized food retains viable microorganisms, which means that additional preservation methods are needed to extend the shelf life of the product. Such processing of foods is normally followed by other preservation methods, often refrigeration, to reduce the growth rates of the surviving microorganisms. The temperatures and exposure times required for the effective preservation of foods depend on the nature of the food and the heat resistance properties of the microorganisms associated with that food. Pasteurization of milk, for example, is largely aimed at eliminating a limited number of nonspore-forming pathogenic bacteria, namely *Brucella* sp., *Coxiella burnetii*, and *Mycobacterium tuberculosis*, which are associated with the transmission of disease via contaminated milk. These microorganisms are relatively sensitive to elevated temperatures, and the pasteurization of milk is therefore normally achieved by exposure to 62.8°C for 30 minutes (**low-temperature, long-time hold or LTH process**) or 71.7°C for 15 seconds (**high-temperature, short-time or HTST process**). Although milk in the United States is normally preserved by pasteurization and requires refrigeration to extend its shelf life, exposure to 141°C for two seconds can be used to sterilize the milk. Sterilized milk is currently marketed in several European countries and has a indefinite shelf life provided that the container remains sealed.

Canning

Unlike pasteurization, **canning** of foods normally involves exposure for longer periods of time to higher temperatures in order to kill endospore-forming microorganisms (Table 19.6). Heat-canned foods do not have to be sterile because other factors such as low intrinsic pH of the food or the removal of oxygen (lowering of the oxidation/reduction, O/R, potential) can prevent the growth of microorganisms. They should be "commercially sterile," which means that no viable microorganisms can be detected by conventional cultural methods or that numbers of survivors are too low to be significant under conditions of storage. Exposure to 115°C for 15 minutes is generally considered necessary in home canning to ensure killing of endospore formers in medium-acid (pH 4.5–5.3) to low-acid (pH >5.3) foods. Particular concern must be given to ensuring sterility of such foods because of the possible growth of *Clostridium botulinum* and the seriousness of the disease botulism. Somewhat lower temperatures, for example, exposure to 100°C for 10 minutes, are often employed in the home canning of acidic (pH <4.5) foods, in part because of the lowered thermal resistance of microorganisms under acidic conditions, and because of the fact that *C. botulinum* is unable to grow at low pH values (Table 19.7). Several commercial canning operations that have not adhered to the necessary standards and whose operations resulted in outbreaks of botulism have been put out of business. Most problems occur with home canning where a lack of care can result in insufficient heating. A particular problem is canning without heating, the cold-pack method, which should not be used for meats, vegetables, or other foods that might contain dangerous microorganisms.

table 19.6

Recommended process times for foods in number 2 cans

Food	Temperature (°F)		Time (min.)
	Initial	Final	
Asparagus spears	120	240	25
Green beans	120	240	20
Lima beans	140	240	35
Baked beans in sauce	140	240	95
Beets in brine	140	240	30
Carrots in brine	140	240	30
Corn, cream style	140	240	105
Corn, whole kernel	140	240	50
Peas in brine	140	240	35
Spinach, whole leaf	140	252	50

table 19.7

Classification of canned foods and their processing requirements

Acidity class	pH	Representative foods	Spoilage agents	Processing
Low acid	7	Ripe olives, eggs, milk, poultry, beef, oysters	Mesophilic *Clostridium* species	High temperature 240–250°F
	6	Beans, peas, carrots, beets, asparagus, potatoes	Thermophiles, plant enzymes	High temperature 240–250°F
Medium acid	5	Figs, tomato soup, ravioli	Limit of growth of *C. botulinum*	High temperature 240–250°F
Acid	4	Potato salad, pears, peaches, oranges, apricots, tomatoes	Aciduric bacteria	Boiling water 212°F
		Sauerkraut, apples, pineapples, grapefruits, strawberries	Plant enzymes	Boiling water 212°F
Highly acid	3	Pickles, relish, lemon juice, lime juice	Yeasts, fungi	Boiling water 212°F

Determining the appropriate temperature and exposure time for a food preservation process involves establishing thermal death time curves for the most heat-resistant microorganisms that may be present in the food. The **D value (decimal reduction time)** is the exposure time at a given temperature needed to reduce the number of viable microorganisms by 90 percent (Figure 19.5). In the canning industry the relatively heat-resistant *Bacillus stearothermophilus* or *Clostridium* PA 3679 is used for determining an acceptable D value.

The D values at different temperatures can be used to construct a thermal death time curve from which the z value can be calculated (Figure 19.6). The **thermal death time (TDT)** is the time required at a specific temperature to kill a specific number of microorganisms. The **z value** describes the temperature change, in °F, needed to reduce the D value (exposure time) by a factor of 10 while still achieving the same effective kill of microorganisms. The D values can also be used to calculate an **F value**, the exposure time in min-

Figure 19.5

A survivor curve for endospores heated at 240°F in canned pea brine at pH 6.2. The D value as determined from this curve is 8 minutes. This curve establishes the rate of destruction of microorganisms at a given temperature. Obviously, the higher the initial microbial concentration in the food, the longer the heating time needed to reduce the numbers of survivors to safe levels.

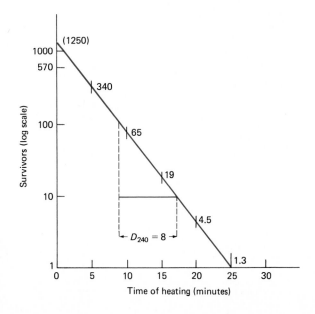

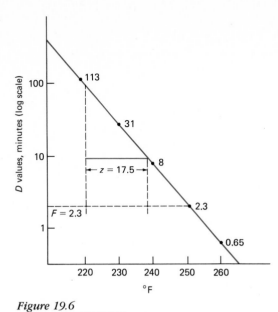

Figure 19.6

This diagram illustrates the z and F values for endospores heated at 240°F in canned pea brine at pH 6.2.

utes at 250°F (121°C), needed to kill the spores or vegetative cells of a particular organism (Table 19.8). By knowing the *D, F,* and *z* values, the process time required for preservation of a food can be determined. For example, the endospores of *Clostridium botulinum* have an *F* value of 0.21 minutes. The canning industry must employ a 12*D* process for nonacid foods. Heating the food for 2.52 minutes, 12 times the *F* value, reduces the probability of the survival of *Clostridium botulinum* endospores to 10^{-12}. Thus, if there were one spore in every can, the probability of contamination remaining after processing should be reduced to one in every million million cans. Heat-

table 19.8

Thermal death times of bacterial cells and spores

Organism	Time (min.)	Temperature (°C)
Vegetative cells		
Salmonella typhi	4	60
Staphylococcus aureus	19	60
Escherichia coli	25	60
Lactobacillus bulgaricus	30	71
Endospores		
Bacillus subtilis	18	100
Clostridium botulinum	200	100
Flat-sour bacteria	Over 1000	100

ing at 121°C for 2.52 minutes therefore should ensure the safety of canned foods with respect to possible contamination with *Clostridium botulinum*. Some endospore-forming microorganisms, such as flat-sour bacteria (endospore-forming bacteria that produce acid but little or no gas in canned foods), have *F* values of 4 to 5 minutes, and therefore, heating for longer periods of time is essential for killing flat-sour and other resistant bacteria.

The exposure times needed to kill microorganisms varies for different food products. Differences in the required exposure times and temperatures depend on the intrinsic properties of the food, such as pH, and on the numbers and types of microorganisms that are normally associated with that food. For example, whereas milk is pasteurized at 71.7°C for 15 seconds, pasteurization of ice cream mix normally requires 82.2°C for 20 seconds. It is necessary to determine the appropriate process times for each individual food. A general formula for calculating the process time needed to destroy a certain number of microorganisms in a food is given as: $\log t/F = (250 - T)/z$, where t = the time in minutes needed to kill a specified number of test organisms, T = the temperature in °F, and z and F represent the thermal death relationship values for the given test microorganism. This formula can be used for determining the appropriate exposure times at a given temperature that are needed to preserve food products. It is necessary to take into account the heating and cooling times in order to determine the appropriate process time (Figure 19.7).

Radiation

Ionizing radiation is used to pasteurize or sterilize some food products. Exposure to ultraviolet light, for example, is sometimes used to sterilize the surfaces of fruitcakes prior to wrapping. This method, though, cannot be widely employed because of the poor penetrating capacity of ultraviolet light and the possibility of oxidative changes in the food that may result. Exposure to gamma radiation can also be used for food preservation because exposure to such radiation can kill cells directly and form free radicals from water that indirectly kill cells and inactivate viruses. Most sterilization procedures involving exposure to radiation employ gamma radiation from ^{60}Co (cobalt-60) or ^{137}Ce (cesium-137). Bacon, for example, can be sterilized by radappertization, a process of sterilizing foods by exposure to radiation, using radiation doses of 4.5–5.6 Mrads. Raduriza-

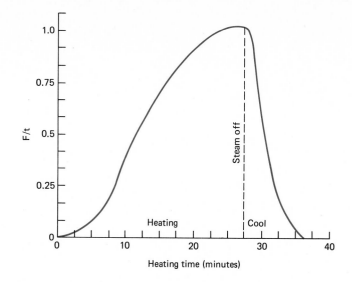

Figure 19.7

A lethality curve for a canned food heat process. The lethal value of the process is described by the area under the curve. The entire heating and cooling time must be considered when establishing the proper process time.

tion, functionally equivalent to pasteurization, is used to kill nonspore-forming human pathogens that may be present in the food. Radurization can be used to increase the shelf life of seafoods, vegetables, and fruits (Table 19.9).

As with thermal processing, **D values can be calculated for radiation exposures** in order to determine the appropriate radiation dosage level for preserving food products (Figure 19.8). But in contrast to the thermal process, only the total dosage is important, and the exposure time required to reach that dosage is irrelevant. In designing a preservation process that uses radiation, it is necessary to consider the types of microorganisms that may be present in the food and their radiation resistance. As a rule, bacterial spores are more resistant to radiation exposure than vegetative cells, with the exception of *Micrococcus radiodurans,* which is exceptionally resistant to radiation exposure. Developing a radiation preservation process by using *M. radiodurans* as a test organism ensures a margin of safety against spores, including those of *Clostridium botulinum.*

Low temperatures

Refrigeration and **freezing** are widely used for the preservation of foods (Table 19.10). Low temperatures restrict the rates of growth and enzymatic activities of microorganisms. Most mesophilic microorganisms are unable to grow at refrigerator temperatures (5°C) and thus, though most pathogenic microorganisms are unable to grow in refrigerated foods, *C. botulinum* type E will

grow and produce toxin (Figure 19.9). Psychrophilic and psychrotrophic microorganisms, however, are able to grow slowly at 5°C, and thus, although refrigeration extends the shelf life of the product, it does not do so indefinitely. Also, delicate foods, such as fish and chicken, spoil rapidly at refrigerator temperatures. Freezing at temperatures of −20°C or lower precludes microbial growth entirely. Not all food products can be preserved by freezing because of the damage that may occur to the food as a result of ice formation. Desiccation of frozen foods (freezer burn), although not a microbial spoilage process, causes serious quality defects.

table 19.9

Potential useful radiation dosages for extending the shelf life of food products

Product	Dose (krads)	Shelf life (days)
Fishery products		
Atlantic haddock fillets	100–250	30–37
Fresh shrimp	100–200	21–28
Pacific cod fillets	100	16–18
Pacific oysters	100	31
King crab meat	100	21
Meats and poultry		
Fresh meat and poultry	50–100	21
Fruits		
Cherries	250	14–20
Oranges	200	90
Peaches, nectarines	200	14
Pineapples	300	14
Strawberries	200	14–18

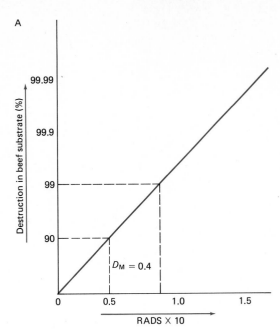

A

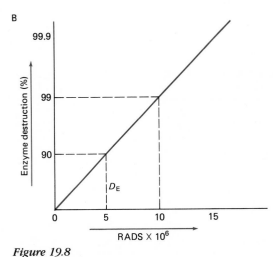

B

Figure 19.8

D *values for radiation treatments of*
(A) Clostridium botulinum *and (B) enzymes.*

table 19.10

Recommended storage temperatures and
approximate storage life of various fresh,
dried, and processed foods

Food	Storage temp. (°F)	Storage life (approx.)
Vegetables		
Asparagus	32	3–4 weeks
Beans (green)	45	8–10 days
Brussel sprouts	32	3–4 weeks
Carrots	32	10–14 days
Cauliflower	32	2–3 weeks
Celery	32	2–4 months
Corn, sweet	32	4–8 days
Cucumbers	47	10–14 days
Lettuce	32	3–4 weeks
Onions	32	6–8 months
Peas, green	32	1–2 weeks
Potatoes, sweet	58	4–6 months
Spinach	32	10–14 days
Squash	35	10–14 days
Tomatoes	32	7 days
Fruits		
Apples	32	2–7 months
Figs	30	5–7 days
Grapefruit	40	4–8 weeks
Lemons	50	1–4 months
Oranges	33	8–12 weeks
Peaches	32	2–4 weeks
Blueberries	32	3–6 weeks
Strawberries	32	7–10 days
Dairy products		
Butter	34	2 months
Eggs, fresh	30	8–9 months
Eggs, dried	35	6–12 months
Meat, poultry, fish		
Beef	32	1–6 weeks
Ham	32	7–12 weeks
Poultry	32	1 week
Fish	35	5–20 days

Commercially frozen foods are often prepared by quick freezing, a process that decreases damage to some food products. **Blanching** of fruits and vegetables, by scalding with hot water or steam prior to deep freezing, inactivates plant enzymes that may produce toughness, change in color, and loss of flavor and nutritive value. A brief heat treatment prior to freezing also reduces the numbers of microorganisms on the food surface, up to 99 percent, enhances the color of green vegetables, and displaces air trapped in food tissues.

Quick freezing generally denotes a freezing time of 30 minutes or less, using small units of food. It can be accomplished (1) by direct immersion of food in a refrigerant, as in freezing fish in brine; (2) by indirect contact with the refrigerant, where the food or package is in contact with the passage through which the refrigerant at -17.8 to $-45.6°C$ flows; or (3) by air-blast freezing, where frigid air at -17.8 to $-34.4°C$ is blown across the materials being frozen. For **slow or sharp freezing**, where only natural air circulation is used, the temperature usually ranges between -15 and $-29°C$, and freezing takes from 30 to 72 hours. Quick freez-

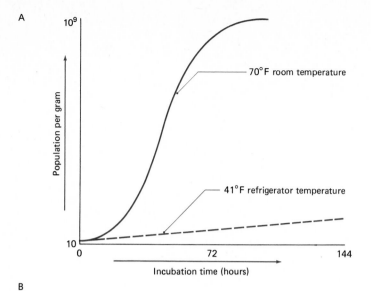

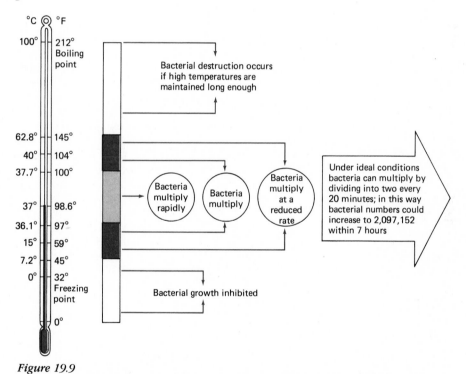

Figure 19.9

Temperature growth relationships of microorganisms. (A) Diagram shows that the growth rate of microorganisms is much lower at refrigerator than at room temperatures. (B) Diagram shows that microorganisms grow over a wide range of temperatures. Knowing the temperature growth ranges for organisms, such as those capable of causing spoilage or disease, is important to selecting the proper temperature for the storage of different foods. The safe and growth-supportive temperatures for bacteria are indicated.

ing is usually the superior method because smaller ice crystals are formed and so there is less destruction of food cells. The shorter solidification period means less diffusion of soluble materials and separation of ice. The prevention of microbial growth also is more prompt, and there is a more rapid slowing of enzyme action.

Freezing does not kill most microorganisms, although some microbial death may occur during freezing and during thawing as a result of ice crystal damage to microbial membranes. In fact, freezing at extremely low temperatures is routinely used for preserving microorganisms in type culture collections. Therefore, when food is thawed, the microorganisms associated with that food can grow, leading to food spoilage and potential accumulation of microbial pathogens and toxins if the food is not promptly prepared or consumed. Once frozen, it is generally not advisable to thaw and refreeze food products. Refreezing of food products also alters the texture, flavor, and nutritional qualities of the food, and even though these changes are not related to microbial spoilage, the changes lower the acceptability of the food. Freezing, thawing, and refreezing generally disrupts the texture of the food and in addition permits invasion of the food by microorganisms that are normally restricted to the food surfaces. When thawed a second time, refrozen food products are more prone to microbial spoilage than are foods that are only allowed to thaw once. This problem can be avoided if proper care is taken during handling to prevent excessive microbial contamination and growth.

Desiccation

The **drying of foods** is a very effective means of preservation, provided that the food does not become moist during storage. The establishment of bacteriostatic environmental conditions is the general principle involved in using dehydration to preserve food products. Desiccation need not kill the microorganisms in food products, but microbial growth requires water. Bacteria generally will not grow below an a_w (water activity) value of 0.9, and fungi generally will not grow below an a_w value of 0.65 (Table 19.11). If food can be maintained at an a_w value of 0.65 or less, spoilage is unlikely for several years. Products preserved by drying include fruits, vegetables, eggs, cereals, grains, meat, and milk. Drying can be accomplished by many different methods, for example, the ancient natural practice of sun drying, where food is placed on trays and placed outdoors with no control of temperature or humidity. This method is limited to warm dry climates and to fruits such as raisins, prunes, figs, apricots, peaches, and pears. Artificial drying can be accomplished by the passage of heated air with controlled humidity over the food.

Foods can also be dehydrated by the evaporation of its water by using a heated drum dryer, in which the product is passed over a heated drum, with or without a vacuum, or by using a spray dryer (Figure 19.10), in which the product is sprayed into dry heated air. Spray drying is normally used for the preparation of dried eggs. Milk can be dehydrated by using either the spray drying or heated drum processes. Evaporated milk is prepared by removing 60 percent of the water from whole milk. Powdered milk generally has over 85 percent of the water in whole milk removed. Freeze drying is less destructive and yields higher-quality foods than drying at elevated temperatures but is much more expensive (Figure 19.11). Therefore, this process, which sublimes water directly from frozen foods under high vacuum, is used only for high-value products, such as meats, camping rations, and coffee.

Anaerobiosis

Packaging of food products under anaerobic conditions is effective in preventing aerobic spoilage processes. Vacuum packing is used to eliminate air, and ground coffee, for example, is preserved by using this method of packaging. The absence of oxygen not only prevents the growth of aerobic microorganisms but also prevents autoxidation of the food as a result of intrinsic enzymatic activities, greatly increasing the shelf life of the product. Establishing anaerobic conditions, however, also creates ideal conditions for the growth of

table 19.11

Minimum a_w *for the growth of various food spoilage fungi*

Organisms	Minimum a_w
Candida utilis	0.94
Botrytis cinerea	0.93
Rhizopus nigricans	0.93
Mucor spinosus	0.93
Candida scotti	0.92
Trichosporon pullulans	0.91
Candida zeylanoides	0.90
Endomyces vernalis	0.89
Alternaria citri	0.84
Aspergillus glaucus	0.70
Aspergillus echinulatus	0.64

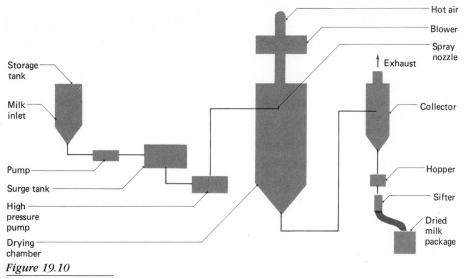

Figure 19.10

Diagram of spray dryer system used in milk industry to produce powdered milk by dehydration. Desiccated milk has a greatly extended shelf life compared to normal milk.

obligate anaerobes, such as *Clostridium botulinum*, and therefore is normally used in conjunction with additional preservation methods. For example, in canning, heat sterilization precedes packaging under anaerobic conditions.

Controlled atmospheres

Controlled atmospheres containing 10 percent CO_2 sometimes are used to preserve stored food products. This preservation method is commonly employed in several countries for the preservation of fruits, such as apples and pears. Carbon dioxide at this concentration retards fungal rotting of fruits. In addition to CO_2, ozone can be added to storage chambers to preserve foods. Because of the high oxidizing activity of ozone, however, this procedure cannot be used with high-lipid foods because it would cause rancidity. These controlled atmosphere preservation techniques are receiving increased use today in the food industry.

Figure 19.11

This commercial freeze dryer has a 500-lb condensate capacity. Such commercial units are used in the production of freeze-dried coffee and other freeze dried foods. (Courtesy Virtis Co., Gardner, New York.)

Chemical preservatives

The use of **chemical additives** is widespread and represents an important means of preserving foods (Table 19.12). The chemical preservative used depends on the nature of the food and the likely spoilage microorganisms. Although there is great concern today over the addition of any chemicals to foods because of the finding that some chemicals that have been used as food additives, such as red dye number 2, are potential carcinogens, it must remembered that the effective preservation of food prevents spoilage and the transmission of food-borne diseases. In the United States the federal Food and Drug Administration is responsible for determining and certifying the safety of food additives and must approve any chemicals that are added to foods as preservatives.

Salt and sugar
The addition of salt or sugar to a food reduces the amount of available water and alters the osmotic pressure. High salt concentrations, such as occur in saturated brine solutions, are bacteriostatic, and the shrinkage of microorganisms in brine solutions can cause loss of viability. Salting is effectively used for the preservation of fish, meat, and other foods. However, because of the association of high levels of salt in the diet with high blood pressure and heart disease, there is currently great interest in lowering the salt contents

of foods. Sugars, such as sucrose, also act as preservatives and are effective in preserving fruits, candies, condensed milk, and other foods. Some foods, including maple syrup and honey, are preserved naturally by their high sugar content.

Organic acids
Various low-molecular-weight carboxylic acids are inhibitors of microbial growth. Lactic, acetic, propionic, citric, benzoic, and sorbic acids or their salts are effective food preservatives. An examination of the lists of food additives in the various foods in your pantry will rapidly convince you of the wide use of organic acids as preservatives. The effectiveness of a particular **acidulant** depends on the pH of the food that determines the degree of dissociation of the acid. For example, at the same pH the order of effectiveness as a preservative is HCl < citric acid < lactic acid < acetic acid.

Propionates are primarily effective against filamentous fungi. The calcium and sodium salts of propionic acid are used as preservatives in bread, cake, and various cheeses, and as propionates are effective inhibitors of rope formation in bread dough and milk. Besides their intentional addition to various foods, propionates form naturally during the production of Swiss cheese and act as a natural preservative.

Lactic and acetic acids also are effective preservatives that form naturally in some food prod-

table 19.12

Some representative chemical food preservatives

Preservatives	Concentration	Target organisms	Foods
Propionic acid and propionates	0.32%	Fungi	Bread, cakes, some cheese
Sorbic acid and sorbates	0.2%	Fungi	Cheeses, syrups, jellies, cakes
Benzoic acid and benzoates	0.1%	Fungi	Margarine, cider, relishes, soft drinks, catsup
Sulfur dioxide, sulfites, bisulfites, metabisulfites	200–300 ppm	Microorganisms	Dried fruits, grapes, molasses
Ethylene and propylene oxides	700 ppm	Fungi	Spices
Sodium diacetate	0.32%	Fungi	Bread
Sodium nitrite	200 ppm	Bacteria	Cured meats, fish
Sodium chloride	None	Microorganisms	Meats, fish
Sugar	None	Microorganisms	Preserves, jellies
Wood smoke	None	Microorganisms	Meats, fish

ucts. Cheeses, pickles, and sauerkraut contain concentrations of lactic acid that normally protect the food against spoilage. Vinegar contains acetic acid, an effective inhibitor of bacterial and fungal growth. Acetic acid is used to pickle meat products and is added as a preservative to various other products, including mayonnaise and catsup. Both of these preservatives, however, will prevent surface fungal growth on a food only if molecular oxygen is excluded.

Benzoates, including sodium benzoate, methyl-p-hydroxybenzoate (methylparaben), and propyl-p-hydroxybenzoate (propylparaben) are extensively used as food preservatives. Benzoates are primarily effective in the undissociated form, and thus, their use generally is restricted to highly acid foods. Benzoates are used as preservatives in such products as fruit juices, jams, jellies, soft drinks, salad dressings, fruit salads, relishes, tomato catsup, and margarine.

Sorbic acid, used primarily as calcium, sodium, or potassium salts (e.g., sodium sorbate), is most effective in acid foods. It is more effective as a preservative at pH 4–6 than the benzoates. Sorbates inhibit fungi and bacteria, such as *Salmonella, Staphylococcus*, and *Streptococcus* species. Sorbates are frequently added as preservatives to cheeses, baked goods, soft drinks, fruit juices, syrups, jellies, jams, dried fruits, margarine, and various other products. Citric acid is used as a preservative in some soft drinks.

Nitrates and nitrites

Nitrates and nitrites are added to cured meats to preserve the red meat color and protect against the growth of food spoilage and poisoning microorganisms. Nitrates also effectively inhibit *Clostridium botulinum* in meat products such as bacon and ham. Recently, however, there has been great concern over the addition of nitrates and nitrites to meats because these salts can react with secondary and tertiary amines to form nitrosamines, which are highly carcinogenic.

Sulfur dioxide, ethylene oxide, and propylene oxide

Sulfur dioxide and various sulfites have antimicrobial activities. Such fruit products as lemon juice, wine, and dried fruit can be preserved by fumigating with sulfur dioxide and by adding liquid sulfites to these products. Ethylene and propylene oxides are microcidal and can be used to sterilize food products. These compounds are used as fumigating agents in the food industry and are primarily applied to dried fruits, nuts, and spices as antifungal agents.

Wood smoke

Smoking of foods has been traditionally used both to impart a desirable flavor to the food and to deposit preservative chemicals from the smoke on the food. This method of food preservation was used by the American Indians, who hung meat on the tops of the teepee poles while fires were going inside to smoke foods, and smoking is still used for preserving various meats, such as smoked hams. Wood smoke contains a large number of volatile compounds possessing bacteriostatic and bactericidal activities, including formaldehyde, phenol, cresols, and low-molecular-weight fatty acids.

Antibiotics

Antibiotics are among various other chemicals used as preservatives for some food products. Tetracyclines, for example, have been used for preserving poultry and fish products. Concern over the increased occurrence of antibiotic-resistant strains has caused regulatory agencies in many countries to ban the use of antibiotics as food preservatives.

Microbiological production of food

Although microbial growth is a problem when it results in food spoilage, microorganisms are used beneficially in the food industry for food production. Many of the foods and beverages we commonly enjoy, such as wine and cheese, are examples of the products of microbial enzymatic activity. For the most part, it is the fermentative metabolism of microorganisms that is exploited in the production of food products. The accumulation of fermentation products, such as ethanol and lactic acid, is desirable because of their characteristic flavors and other properties. Only a few processes, such as the production of vinegar, make use of microbial oxidative metabolism. The microbial production of foods can be viewed as an exercise in harnessing microbial biochemistry to produce desired, rather than adverse, changes in food products.

The production of **fermented foods** requires the proper substrates, microbial populations, and environmental conditions to obtain the desired end product. Quality control is essential in food fermentations to ensure that the product is of high quality. A fermented food may require additional preservation to prevent the spoilage of that product, and additional uncontrolled microbial growth could further transform the food into an inedible product. For example, once wine is produced it must be maintained under anaerobic conditions in order to prevent its further oxidation to acetic acid.

The microbial processes in the production of food traditionally employ microbial enzymatic activities to transform one food into another, with the microbially produced food product having vastly differing properties than the starting material. In addition to using microorganisms to produce fermented food products, microbial biomass is now considered as a potential source of protein for meeting the food needs of an expanding world population. Some microorganisms, such as mushrooms, have been used as food products for centuries. The growth of bacteria, algae, and fungi as proteinatious food, however, is a relatively new concept. Microbial biomass can be used as an animal feed supplement or may be developed as a direct source of protein for human consumption.

Fermented dairy products

Numerous products are made by the **microbial fermentation of milk**, including buttermilk, yogurt, and many cheeses. The fermentation of milk is primarily carried out by lactic acid bacteria. The lactic acid fermentation pathway and the accumulation of lactic acid from the metabolism of the milk sugar lactose is common to the production of fermented dairy products. The accumulation of lactic acid in these products acts as a natural preservative. The differences in flavor and aroma of the various fermented dairy products are due to additional fermentation products that may be present in only relatively low concentrations.

Buttermilk, sour cream, kefir, and koumis

Different fermented dairy products are produced by using different strains of lactic acid bacteria as starter cultures and different fractions of whole milk as the starting substrate (Table 19.13). Sour cream, for example, uses *Streptococcus cremoris* or *S. lactis* for the production of lactic acid, and *Leuconostoc cremoris* or *S. lactis diacetilactis* for production of the characteristic flavor compounds. Cream is used as the starting substrate for this product. If skim milk is used as the starting material, cultured buttermilk is produced.

table 19.13

Some foods produced from fermented milks

Fermented product	Microorganisms responsible for fermentation	Description
Sour cream	*Streptococcus* sp., *Leuconostoc* sp.	Cream is inoculated and incubated until the desired acidity develops.
Cultured buttermilk	*Streptococcus* sp., *Leuconostoc* sp.	Made with skimmed or partly skimmed pasteurized milk.
Bulgarian buttermilk	*Lactobacillus bulgaricus*	Product differs from commercial buttermilk in having higher acidity and lacking aroma.
Acidophilus milk	*Lactobacillus acidophilus*	Milk for propagation of *L. acidolphilus* and the milk to be fermented is sterilized and then inoculated with *L. acidophilus*. This milk product is used for its medicinal therapeutic value.
Yogurt	*Streptococcus thermophilus*, *Lactobacillus bulgaricus*	Made from milk in which solids are concentrated by evaporation of some water and addition of skim milk solids. Product has consistency resembling custard.
Kefir	*Streptococcus lactis*, *Lactobacillus bulgaricus*, yeasts	A mixed lactic acid and alcoholic fermentation; bacteria produce acid, and yeasts produce alcohol.

Bulgarian buttermilk is produced by using *Lactobacillus bulgaricus* for the production of both lactic acid and flavor compounds. Butter is normally made by churning cream that has been soured by lactic acid bacteria. *S. cremoris* or *S. lactis* are used to produce lactic acid rapidly, and *Leuconostoc citrovorum* produce the necessary flavor compounds. The *Leuconostoc* enzymes attack citrate in milk, producing diacetyl, which gives butter its characteristic flavor and aroma. Kefir and koumis, which are popular in some European countries, are fermentation products of *S. lactis, S. cremoris*, other *Lactobacillus* spp., and yeasts. Lactic acid, ethanol, and carbon dioxide are formed during the fermentation and give these products their characteristic flavors.

Yogurt

Over 550,000 pounds of **yogurt** are produced annually in the United States. Yogurt is produced by fermenting milk with a mixture of *L. bulgaricus* and *S. thermophilus*. Yogurt fermentation is carried out at 40°C. The characteristic flavor of yogurt is due to the accumulation of lactic acid and acetaldehyde produced by *L. bulgaricus*. Because of the tart taste of acetaldehyde, most yogurt produced in the United States is flavored by adding fruit.

Cheese

A wide variety of **cheeses** are produced by microbial fermentations (Figure 19.12). Cheeses consist of milk curds that have been separated from the liquid portion of the milk (whey). The curdling of milk is accomplished by using the enzyme rennin (casein coagulase or chymosin) and lactic acid bacterial starter cultures. Rennet is obtained from calf stomachs or by microbial production. Cheeses are classified as: (1) soft, if they have a high water content (50–80 percent); (2) semihard, if the water content is about 45 percent; and (3) hard, if they have a low water content (less than 40 percent). Cheeses are also classified as unripened if they are produced by a single-step fermentation, or as ripened if additional microbial growth is required during maturation of the cheese to achieve the desired taste, texture, and aroma (Table 19.14). Processed cheeses are made by blending various cheeses to achieve a desired product. If the water content is elevated during processing, thereby di-

Figure 19.12

A wide variety of cheeses are produced in many areas of the world. This photograph shows some of the cheeses produced in Wisconsin. (Courtesy American Dairy Association of Wisconsin, Madison.)

table 19.14

The classification of some cheeses

Cheese		Microorganisms used in production	
Soft, unripened			
Cottage	*Streptococcus lactis*	*Leuconostoc citrovorum*	
Cream	*Streptococcus cremoris*		
Neufchatel	*Streptococcus diacetilactis*		
Soft, ripened, 1–5 months			
Brie	*Streptococcus lactis*	*Penicillium camemberti*	*Brevibacterium linens*
	Streptococcus cremoris	*Penicillium candidum*	
Camembert	*Streptococcus lactis*	*Penicillium camemberti*	
	Streptococcus cremoris	*Penicillium candidum*	
Limburger	*Streptococcus lactis*	*Brevibacterium linens*	
	Streptococcus cremoris		
Semisoft, ripened, 1–12 months			
Blue	*Streptococcus lactis*	*Penicillium roqueforti* or	
	Streptococcus cremoris	*Penicillium glaucum*	
Brick	*Streptococcus lactis*	*Brevibacterium linens*	
	Streptococcus cremoris		
Gorgonzola	*Streptococcus lactis*	*Penicillium roqueforti* or	
	Streptococcus cremoris	*Penicillium glaucum*	
Monterey	*Streptococcus lactis*		
	Streptococcus cremoris		
Muenster	*Streptococcus lactis*	*Brevibacterium linens*	
	Streptococcus cremoris		
Roquefort	*Streptococcus lactis*	*Penicillium roqueforti* or	
	Streptococcus cremoris	*Penicillium glaucum*	
Hard, ripened, 3–12 months			
Cheddar	*Streptococcus lactis*	*Lactobacillus casei*	
	Streptococcus cremoris		
	Streptococcus durans		
Colby	*Streptococcus lactis*	*Lactobacillus casei*	
	Streptococcus cremoris		
	Streptococcus durans		
Edam	*Streptococcus lactis*		
	Streptococcus cremoris		
Gouda	*Streptococcus lactis*		
	Streptococcus cremoris		
Gruyere	*Streptococcus lactis*	*Lactobacillus helveticus*	*Propionibacterium shermanii* or
	Streptococcus thermophilus		*Lactobacillus bulgaricus* and
			Propionibacterium freudenreichi
Swiss	*Streptococcus lactis*	*Lactobacillus helveticus*	*Propionibacterium shermanii* or
	Streptococcus thermophilus		*Lactobacillus bulgaricus* and
			Propionibacterium freudenreichi
Very hard, ripened, 12–16 months			
Parmesan	*Streptococcus lactis*	*Lactobacillus bulgaricus*	
	Streptococcus cremoris		
	Streptococcus thermophilus		
Romano	*Lactobacillus bulgaricus*	*Streptococcus thermophilus*	

luting the nutritive content of the product, the product is called a "processed food" rather than a cheese.

The natural production of cheeses involves a lactic acid fermentation with various mixtures of *Streptococcus* and *Lactobacillus* species used as starter cultures to initiate the fermentation. The flavors of different cheeses result from the use of different microbial starter cultures, varying incubation times and conditions, and the inclusion or omission of secondary microbial species late in the fermentation process.

Ripening of cheeses involves additional enzymatic transformations after the formation of the cheese curd, using enzymes produced by lactic acid bacteria or enzymes from other sources. Unripened cheeses do not require the additional enzymatic transformations. Cottage cheese and cream cheese are produced by using a starter culture similar to the one used for the production of cultured buttermilk and are soft cheeses that do not require ripening. Sometimes a cheese is soaked in brine to encourage the development of selected bacterial and fungal populations during ripening. Limburger is a soft cheese produced in this manner. During ripening the curds are softened by proteolytic and lipolytic enzymes, and the cheese acquires its characteristic aroma. The production of Parmesan cheese also involves brine curing.

Swiss cheese formation involves a late propionic acid fermentation, with ripening accomplished by *Propionibacterium shermanii* and *P. freudenreichii*. The propionic acid yields the characteristic aroma and flavor, and the carbon dioxide produced during this late fermentation forms the holes in Swiss cheese. Various fungi are also used in the ripening of different cheeses. The unripened cheese is normally inoculated with fungal spores and incubated in a warm moist room to favor the growth of filamentous fungi (Figure 19.13). For example, blue cheeses are produced by using *Penicillium* species. Roquefort cheese is produced by using *P. roqueforti* and camembert and brie are produced by using *P. camemberti* and *P. candidum*.

Fermented meats

Several types of **sausage**, such as Lebanon bologna, the salamis, and the dry and semidry summer sausages, are produced by allowing the meat to undergo a heterolactic acid fermentation during curing. The fermentation has both a preservative effect and also adds a tangy flavor to the meat. Various lactic acid bacteria are normally involved in the fermentation, but *Pediococcus cerevisiae* can be used for controlled production of these types of meats.

Leavening of bread

Yeasts are added to bread dough to ferment the sugar, producing the carbon dioxide that leavens the dough and causes it to rise. The principal yeast used in bread baking is *Saccharomyces cere-*

Figure 19.13

Photograph showing cheese production room with fungi growing on the surface of the ripening Camembert cheese. (Courtesy of Foods and Wines of France, New York.)

visiae, **Bakers' yeast**. Bakers' yeast is produced in large quantities for the baking industry (Figure 19.14). The yeast is normally grown in a molasses-mineral salts medium at a pH of 4.3–4.5 and temperature of 30°C, with the molasses substrate added gradually to maintain a sugar concentration of 0.5–1.5 percent. The concentration of sugar in the fermentor is critical, as too high a sugar concentration represses respiratory enzymes and alcohol production even under highly aerobic conditions. The yeasts are generally collected by centrifugation and pressed through a filter to remove excess liquid. For the baking industry, the yeasts are normally formed into cakes or they are dried further to form active dry yeast containing less than 8 percent water and are frequently used for home baking purposes.

In the baking process, the yeast is used strictly as a source of enzymes to carry out an alcoholic fermentation. No growth of the yeast occurs during the first two hours after addition to the dough, by which time the baking process is normally completed. Amylases in the dough convert starch to sugars, and the yeasts metabolize the sugars that are formed, producing carbon dioxide and ethanol. Besides *Saccharomyces cerevisiae*, various other microorganisms, including coiform bacteria and *Clostridium* species, can be employed for leavening bread. The microorganisms used for bread leavening must produce carbon dioxide from the fermentation of sugars to be useful.

In modern home and commercial baking processes, excess amounts of yeast are normally added so that the fermentation time is very short. Older, more traditional bread-making processes use less yeast and longer fermentation times. When

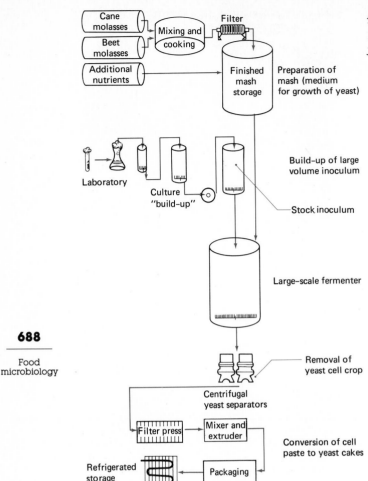

Figure 19.14

The steps in the commercial production of bakers' yeast.

the process time exceeds two hours, however, there can be an undesirable growth of fungi and bacteria. During fermentation the dough becomes conditioned as a result of the action of proteases on the flour protein, gluten. The enzymes are produced by the yeasts or may be added from other sources. As a result of this conditioning, the gluten matures, becoming elastic and capable of retaining the carbon dioxide gas produced by the yeasts during the fermentation. Sugar, or amylase to convert starches to sugar, is normally added to the flour in order to increase the rate of gas production by the yeast. Addition of increased amounts of yeast and various salts to support yeast metabolism also increases the rate of gas production. The leavening process is normally carried out at 27°C, which is optimal for fermentation. Too high or too low a temperature can result in reduced rates of gas production. After leavening, the bread is baked. The carbon dioxide bubbles are trapped in the dough and give rise

to the honeycomb texture and increased volume of the baked bread. Although the interior of the bread does not reach 100°C, the heating is sufficient to kill the yeasts, inactivate their enzymes, expand the gas, evaporate the ethanol produced during the fermentation, and establish the structure of the bread loaf. During the baking there is also a gelatinization of the starch, which results in setting of the bread. In the dough the gluten gives structural support, but in the baked bread the structural support comes from the gelatinized starch.

In addition to leavening bread, microbes produce the characteristic flavors of some breads. For example, the production of San Francisco sour dough bread utilizes the yeast *Torulopsis holmii* and a heterofermentative *Lactobacillus* species to sour the dough and give this bread its characteristic sour flavor. Rye bread is also produced by initially souring the dough; cultures of *Lactobacillus plantarum*, *L. brevis*, *L. bulgaricus*, *Leucon-*

oscoc mesenteriodes, and *Streptococcus thermophilus* are employed as starter cultures in making different rye breads. The action of heterofermentative lactic acid bacteria produces the bread's characteristic flavor.

Alcoholic beverages

Microorganisms, principally yeasts in the genus *Saccharomyces*, are used to produce various types of **alcoholic beverages**. The production of alcoholic beverages relies on the alcoholic fermentation, that is, the conversion of sugar to alcohol by microbial enzymes. The flavor and other characteristic differences between various types of alcoholic beverages reflect differences in the starting substrates and the production process, rather than differences in the microbial culture or the primary fermentation pathways employed in the production of alcoholic beverages.

Beer and ale

Beer is a very popular beverage with a high per capita consumption rate. The worldwide production of beer is over 18 billion gallons per year. Beer and ale are malt beverages, so-named because the initial preparation of the substrate for microbial fermentation involves barley malt and the production of beer begins with the **malting** of the barley (Figure 19.15). Malt is a mixture of amylases and proteinases prepared by germinating barley grains for about a week and crushing the grains to release the plant enzymes. Some beers, particularly those produced in Europe, are prepared entirely from malted barley. In the production of most beers, however, the malt is added to adjuncts in a process known as mashing. The

Figure 19.15

The steps in the brewing of beer. The production of beer begins with the malting of barley, in which the grain is induced to sprout to produce enzymes that will catalyze the breakdown of starch. The malt is ground and mixed with warm water, and often other cereals such as corn, before going to the mash tun, where over a period of a few hours enzymes break down the long chains of starch into smaller molecules of carbohydrate. The aqueous extract called wort is separated from the mix and boiled with hops in a brew kettle. The boiling extracts flavors from the hops and stops the enzyme action in the wort. The hops are removed and the wort is put in a fermenting vessel, where it is pitched, or seeded, with yeast. After fermentation the beer may go to a lagering tank to mature, and then it is pasteurized and bottled.

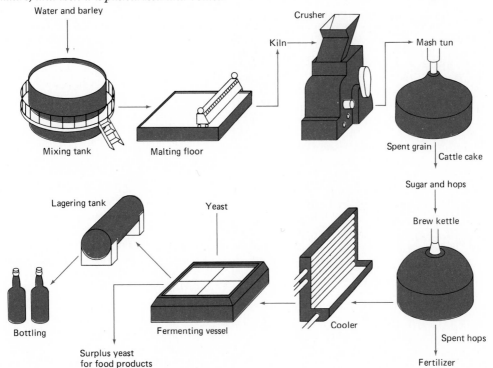

malt adjuncts, such as corn, rice, and wheat, provide carbohydrate substrates for ethanol production. During the mashing process the amylases from the barley malt hydrolyze the starches, other polysaccharides, and the proteins in the malt adjunct. The mash is heated to reach temperatures of about 70°C, which favors the rapid enzymatic conversion of starch to sugars. The insoluble materials are allowed to settle from the mash and serve as a filter. The clear liquid that is produced in this process is called **wort**.

The wort is then cooked with hops, the dried flowers of the hop plant. This cooking concentrates the mixture, inactivates the enzymes, extracts soluble compounds from the hops, and greatly reduces the numbers of microorganisms prior to the fermentation process. Additionally, compounds in the hops extract, principally resins such as humulone, have antibacterial properties and protect the wort from the undesirable growth of Gram positive bacteria that could sour the beer.

The fermentation of the wort to produce beer in most countries is carried out by the yeast *Saccharomyces carlsbergensis*, a bottom fermenter. This means that at a late stage in fermentation, the yeasts flocculate or aggregate and settle, partially clarifying the beer. *Saccharomyces cerevisiae* is also sometimes used in beer production, particularly in Great Britain and parts of the United States, but it is a top yeast and rises to the surface during fermentation. The inoculation of the yeast into the cooled wort, known as **pitching**, uses a heavy inoculum of about one pound of yeast per barrel of beer. The wort is initially aerated to favor the reproduction of the yeast but is then allowed to become anaerobic, favoring the fermentative production of alcohol and carbon dioxide. Usually the fermentation process is carried out at low temperatures and may take one to two weeks to reach completion. During fermentation the yeasts convert the sugars in the wort to alcohol and carbon dioxide and also produce small amounts of glycerol and acetic acid from the fermenation of the carbohydrates. Proteins and lipids are converted to small amounts of higher alcohols, acids, and esters, which contribute to the flavor of beer. The active fermentation process is accompanied by extensive foaming of the mixture because of the production of carbon dioxide. Foaming decreases as the fermentation nears completion (Figure 19.16). The product is then known as a green beer and requires aging to achieve the characteristic flavor and aroma of the finished product.

The commercial production of beer is usually a **batch process**, in which the substrates and inoculum are added to a brewing kettle. When the fermentation is completed, the products are collected as a single batch. In some countries the production of beer is carried out in a continuous flow-through process, in which fresh substrate is continuously or periodically added to the fermentation and product is continuously collected. This production process is analogous to the operation of a chemostat.

During the aging process precipitation of proteins, yeasts, and resins occurs, resulting in a mellowing of the flavor. The mature beer is removed and filtered. The finished product is carbonated to achieve a carbon dioxide content of 0.45–0.52 percent. In the commercial production of beer, the carbon dioxide is normally collected during the fermentation phase and reinjected during the finishing process. In home production of beer, a small additional amount of sugar is usually added

Figure 19.16

Photograph showing beer being produced. The foaming is due to the rapid action of the yeast, which produces carbon dioxide as well as alcohol. In the traditional European manner the beer is carefully inspected by the brewmaster at all stages of the fermentation to ensure production of a quality product. (Reprinted by permission of Running Press from M. Jackson, 1977, The World Guide to Beer.)

to each bottle to permit limited additional fermentative production of CO_2, achieving carbonation within the bottle. Most bottled or canned beers are pasteurized at 60–61°C or filtered to remove viable yeast. Commercially produced beer has an alcohol content of about 3.8 percent.

In addition to normal beer, there are a number of other malt beverages. **Light or low-carbohydrate beers** are produced by using a wort prehydrolyzed with fungal glucoamylases and amylases. The prehydrolysis of dextrin in the wort to maltose and glucose permits the yeasts to ferment the carbohydrates completely to alcohol and carbon dioxide, greatly reducing the concentration of residual carbohydrates in the beer. These low-calorie beers are particularly popular today for those who consume large amounts of beer and don't wish to develop a "beer belly."

Ale is produced by using *Saccharomyces cerevisiae*. The fermentation is carried out at temperatures of 12–25°C, permitting the fermentation to reach completion in only five to seven days. The yeast cells are carried upward with the carbon dioxide, and excess cells are skimmed off the top during the fermentation period. A higher concentration of hops is used in the production of ale than in beer, contributing to the particularly tart taste of ale; additionally, some ales have higher alcohol concentrations than most beers.

Saki, a Japanese beverage, is a yellow rice beer.

The alcohol concentration of saki normally is 14–17 percent. In the production of saki, a starter culture—normally *Aspergillus oryzae*—is used as a source of fungal enzymes to hydrolyze the rice starch to sugars that can then be converted to alcohol by *Saccharomyces* species during fermentation. The *Aspergillus* spores are mixed with steamed rice and incubated at approximately 35°C for five to six days before inoculation with yeast. The yeast fermentation of the rice mash to produce saki takes several weeks. Sonti, a similar product produced in India, uses the fungus *Rhizopus sonti* to convert the rice starch to sugars, which are subsequently fermented by yeasts.

Distilled liquors

Distilled liquors or spirits are produced in a manner similar to beer, except that after the fermentation process the alcohol is collected by distillation, permitting the production of beverages with much higher alcohol concentrations than could accumulate during the fermentation process (Figure 19.17). The initial steps in the production of distilled spirits are analogous to those in beer production, beginning with a mashing process in which the polysaccharides and proteins in a starting plant material are converted to sugars and other simple organic compounds that can be readily fermented by yeasts to form alcohol.

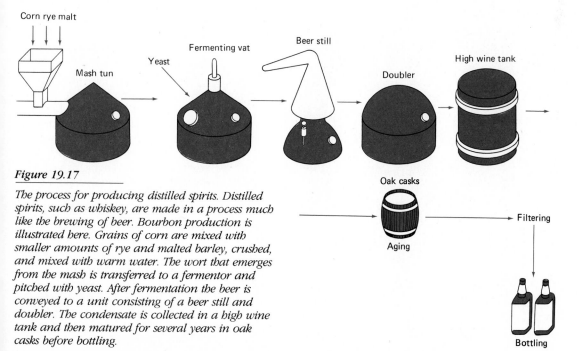

Figure 19.17

The process for producing distilled spirits. Distilled spirits, such as whiskey, are made in a process much like the brewing of beer. Bourbon production is illustrated here. Grains of corn are mixed with smaller amounts of rye and malted barley, crushed, and mixed with warm water. The wort that emerges from the mash is transferred to a fermentor and pitched with yeast. After fermentation the beer is conveyed to a unit consisting of a beer still and doubler. The condensate is collected in a high wine tank and then matured for several years in oak casks before bottling.

Various starting plant materials are used for the production of different distilled liquor products. Rum is produced by using sugar cane syrup or molasses as the initial substrate, rye whiskey is produced from the fermentation of a rye mash, bourbon or corn whiskey uses corn mash, and brandy comes from the fermentation of grapes. The yeasts used in the production of distilled liquors typically are special distiller strains of *Saccharomyces cerevisiae*, which yield relatively high concentrations of alcohol. The yeasts produced during fermentation are collected, dried, and used as animal feed. The mash is sometimes soured prior to the yeast fermentation process by allowing a lactic acid fermentation to occur initially in order to prevent the growth of undesired microorganisms that might interfere with the fermentative action of the yeast.

The alcoholic product formed from the fermentation of the wort, known as a beer or wine, is heated in a still and the alcohol is collected. In addition to the alcohol, various volatile organic compounds, fusel oils, are collected with the distillate and contribute to the characteristic flavors of the different distilled liquor products. Distilled products also differ from one another in the nature of the distillation process; Scotch whiskey, for example, is distilled in batches by using small pot stills, whereas many other distilled whiskeys are produced by using continuous distillation processes. The distilled alcohol product is normally aged to yield a mellow-tasting alcoholic beverage.

Wines

Wine is fermented primarily from grapes, although other fruits are sometimes used (Figure 19.18). Red wines are produced by using red grapes, whereas white wines are made from white grapes or from grapes that have had their skins removed. The initiation of wine begins when the grapes are crushed to form a juice or **must** (Figure 19.19). In the classical European method of wine production, wild yeasts from the surface skins of the grapes are the only inoculum for the fermentation. In modern wine production, however, the natural microbiota associated with the grapes are removed by sulfur dioxide fumigation or by the addition of metabisulfite so that the "wild microorganisms" do not compete with the defined yeast strains used to ferment the grapes in this process. The grape must is then inoculated with a specific strain of yeast, normally a variety of *Saccharomyces cerevisiae*. By using specific yeasts strains and controlled fermentation conditions, a consistent-quality product can be produced. Initially, the grape must and yeasts are stirred to increase aeration and permit the proliferation of the yeasts. The mixing is later discontinued, permitting anaerobic conditions to occur that favor the production of alcohol. The sugar content of the grapes determines the final ethanol concentration. Because 1 mole of glucose can yield only 2 moles of ethanol via glycolysis, the final ethanol concentration is half the initial sugar concentration. It usually ranges from 15 to 25 percent and limits the alcohol concentration of the wine to 7.5–12.5 percent. The sugar content of the grapes depends on the grape variety and its ripeness and varies from season to season, accounting in part for the reason that some years and vineyards are better than others for the production of quality wines.

The fermentation of red wines typically is carried out at 24–27°C for three to five days, and white wines take seven to fourteen days at 10–21°C. During fermentation wine is periodically racked, that is, filtered through the bottom sediments and added back to the top of the fermentation vat. Carbon dioxide produced during the fermenta-

Figure 19.18

As shown in this photograph, there are many varieties of wine produced around the world. The shape of the bottle is characteristic of the production region. The labels give valuable information for predicting the quality of the wine inside, but cannot replace the true test of taste. (Stock, Boston.)

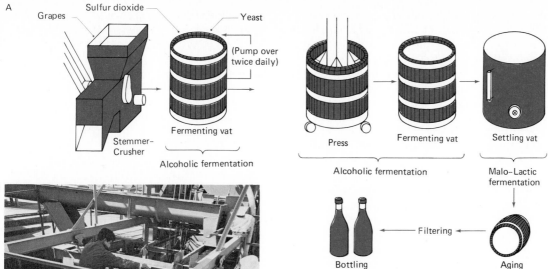

Figure 19.19

The wine production process. (A) Diagram of the batch process, the commonest way of making wine, although some of the cheaper wines are manufactured by a process of continuous fermentation. Grape juice is fed steadily into a fermenting stage and wine is steadily removed. The processes that produce red and white wines differ somewhat. (B) The commercial presses gently extract as much juice as possible without breaking the seeds. The corkscrew conveyor brings the grape must from the crusher. (C) The wine is aged in casks before bottling. (D) Bottles are mechanically filled with careful quality control. (B–D Courtesy The Wine Institute, San Francisco.)

tion process forces the skins and other debris to the surface. The color of red wine originates from the extraction of the pigments from the grape skin by the alcohol produced during the fermentation. At the end of fermentation, wines typically have an alcohol content of 11–16 percent by volume. Wines are aged after the fermentation process to achieve their final bouquet and essence of flavor. During aging some fermentation of the malic acid of grape juice is carried out by lactobacilli (malolactic fermentation), reducing the acidity of the wine.

By using similar processes, a variety of wines can be produced. Dry wines contain little or no sugar, whereas sweet wines contain some residual unfermented sugar. Fortified or desert wines are distilled to achieve an alcohol content of 19–21 percent. Normally, the carbon dioxide produced during the alcoholic fermentation is allowed to escape and the wine is therefore still. In the case of champagne and other sparkling wines, however, the carbonation is essential. In some commercially produced champagne, carbon dioxide is reinjected into the wine after fermentation. In the classic French method of producing champagne, the wine is fermented in the bottle. After fermentation is complete, the bottles are inverted, and the yeast sediments into the neck of the specially shaped champagne bottles. The yeasts are frozen and removed as a plug without excessive loss of carbon dioxide. Wines stoppered with a cork must be stored on their side to prevent the cork from drying out, which would permit air to enter and allow the alcohol to be oxidized by bacteria to form acetic acid. The spoilage of wines, with the formation of vinegar, sour wine, is a serious problem. In the United States, most wine bottles are sealed with a plastic stopper and therefore need not be stored on their side to preclude the souring of the wine.

Vinegar

The production of **vinegar** involves an initial anaerobic fermentation to convert carbohydrates by *Saccharomyces cerevisiae* to alcohol, followed by a secondary oxidative transformation of the alcohol to form acetic acid by *Acetobacter* and *Gluconobacter*. The starting materials for the production of vinegar may be fruits, such as grapes, oranges, apples, pears; vegetables, such as potatoes; malted cereals, such as barley, rye, wheat, and corn; and sugary syrups such as molasses, honey, and maple syrup. The type of vinegar is determined by the starting material. For example, wine vinegar comes from grapes and cider vinegar from other fruits. The history of the commercial production of vinegar shows an interesting progression in fermenter design to accomplish the necessary transfer of oxygen to the bacteria, and we will examine some of the types of vinegar generators that have been developed.

In slow methods for the production of vinegar—still used in some small European operations—an initial natural alcoholic fermentation achieves an alcohol concentration of 11–13 percent. After production of the alcoholic liquid, acetic bacteria are seeded into the solution and allowed to convert the alcohol slowly to acetic acid. In the **Orleans process** for producing vinegar, a barrel is filled about one-fourth full with raw vinegar from a previous run to provide the active inoculum. A wine, hard cider, or malt liquor is then added as a substrate. Sufficient air is left in the barrel to permit oxidative metabolism, acetic acid bacteria grow as a film on the top of the liquid, and the conversion of alcohol to acetic acid takes from several weeks to several months to complete at 21–29°C. The rate of vinegar production is limited primarily by the transfer of oxygen.

To increase the rate of acetic acid production, a vinegar generator can be used in which the alcohol-containing liquid is trickled over a surface film of acetic acid bacteria (Figure 19.20). In a typical vinegar generator the acetic acid bacteria are maintained as a film on wood chips. The alcohol liquid is sprinkled over the wood chips, and during the slow trickling of the liquid down through the generator, the alcohol is converted to acetic acid. Air enters the generator from the bottom, favoring the oxidative process. In order to control any excessive heat that may be generated during this process, cooling coils are normally required. One or two runs of the alcoholic liquid through the generator are sufficient to produce high-quality vinegar.

Today, though, industrial producers of vinegar use **submerged culture reactors** (Figure 19.21). Forced aeration is used to maximize the rate of acetic acid production, and the bacteria grow in the fine suspension created by the air bubbles and fermenting liquid. Through using the submerged method for vinegar production, an 8–12 percent alcoholic liquid is inoculated with an *Acetobacter* species at 24–29°C with carefully controlled aeration. Using a 10 percent alcohol solution as substrate, the acetic acid yield can be 13 percent. Once the vinegar is formed, it is clarified by passage through a filter and allowed to age to

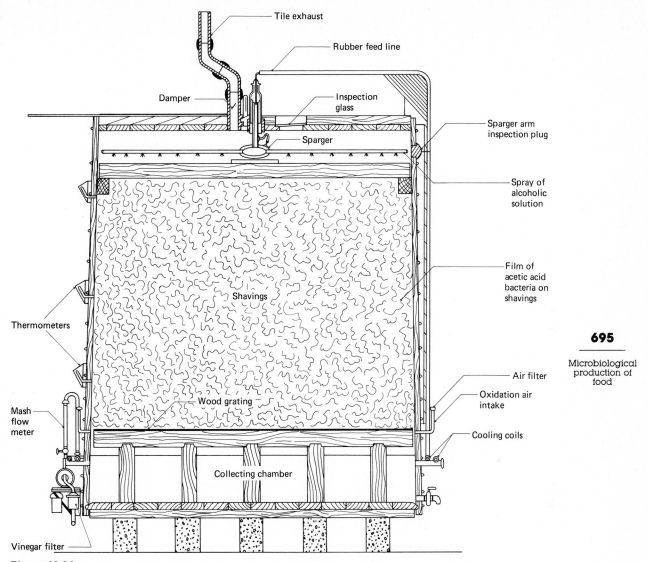

Tile exhaust

Rubber feed line

Damper

Inspection glass

Sparger arm inspection plug

Sparger

Spray of alcoholic solution

Shavings

Film of acetic acid bacteria on shavings

Thermometers

Air filter

Oxidation air intake

Mash flow meter

Wood grating

Cooling coils

Collecting chamber

Vinegar filter

Figure 19.20

The problem with the production of vinegar is supplying the needed oxygen and substrate to the acetic acid bacteria. In the classic vinegar generator the bacteria develop as a film on wood chips, and the aerated liquid containing the ethanol substrate is dripped over the bacteria.

achieve its final body, taste, and bouquet. The vinegar may be pasteurized at 60–66°C for a few seconds to remove any remaining viable bacteria.

Fermented vegetables

Vegetables, such as cabbage, carrots, cucumber pickles, green tomatoes, leafy vegetables, greens, and olives, are fermented by using lactic acid bac-

teria as a means of preservation. Other fermentations, particularly of soybeans, are carried out to produce specially desired flavors, aromas, and textures in food products.

Sauerkraut

Sauerkraut is produced from a lactic acid fermentation of wilted, shredded cabbage (Figure 19.22). Salt, 2.25–2.5 percent, is added to shredded cabbage to help extract the plant juices, control the microbiota during the fermentation, and maintain

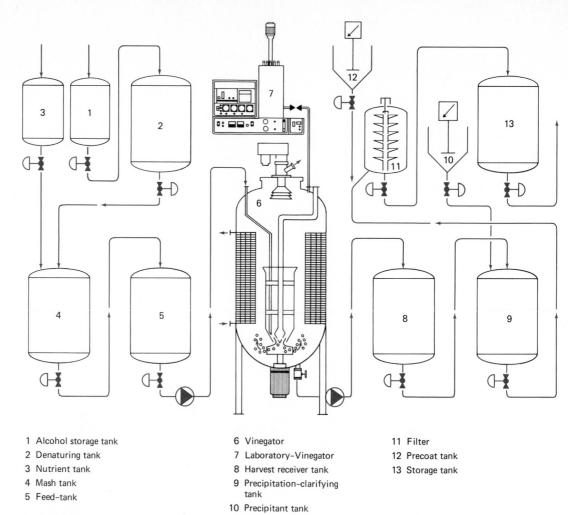

1 Alcohol storage tank
2 Denaturing tank
3 Nutrient tank
4 Mash tank
5 Feed-tank

6 Vinegator
7 Laboratory-Vinegator
8 Harvest receiver tank
9 Precipitation-clarifying
   tank
10 Precipitant tank

11 Filter
12 Precoat tank
13 Storage tank

Figure 19.21

In a modern submerged culture vinegar generator a controlled level of oxygen is achieved. The alcohol in storage tank 1 is denatured with a little acetic acid and is diluted in denaturing tank 2. In mash tank 4 the substrate is prepared and readied from the denatured alcohol through the addition of water and nutrients. This readied mash is stored in feed-tank 5. In the Vinegator (6) the mash is oxidized into vinegar in batches by Acetobacter *under controlled aeration, constant temperature, and controlled level of residual O_2. When the concentration of residual alcohol reaches near zero, a part of the vinegar is transferred to harvest receiver tank 8. Fresh substrate is then pumped into the Vinegator at a preset level. In clarifying tank 9 the new vinegar is mixed with the precipitant that brings down the present clouding. The precipitated cloud is removed by filtration and the clarified vinegar is stored in tank 13.*

an even dispersal of the bacteria. Anaerobic conditions develop in the salted shredded cabbage and surrounding juice, primarily as a result of continued respiration of plant cells, but also because of some bacterial metabolism.

The production of sauerkraut involves a succession of bacterial populations (Figure 19.23). Coliform bacteria, such as *Enterobacter cloacae*, are prominent in the initial mixed community and

produce gas and volatile acids as well as some lactic acid. The accumulating lactic acid exerts a selective pressure on the microbial community, causing population shifts and continued succession. As a result, after the initial fermentation there is a shift in the microbial community, and *Leuconostoc mesenteroides*, which grows well at 21°C and is not inhibited by 2.5 percent salt, becomes the dominant microbial population. Up to 1 per-

Figure 19.22

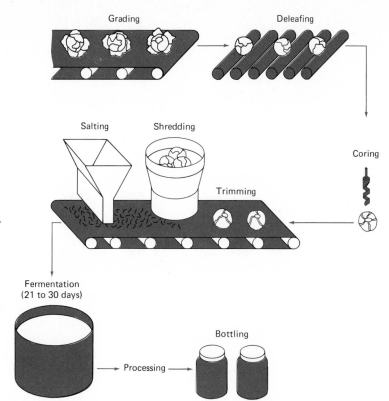

Grading

Deleafing

Salting Shredding

Coring

Trimming

Fermentation
(21 to 30 days)

Bottling

Processing

Preservation of vegetables, illustrated here by the preservation of cabbage by the dry salt method (sauerkraut production), is an application of fermentation that long predates canning and freezing and is still widely practiced on a commercial scale. In this method a brine is generated by osmotic gradients arising from the interaction of the salt and the natural fluids of the cabbage. In the brine the lactic acid bacteria originating on the fresh cabbage become the dominant species in the extended process of fermentation.

cent lactic acid may accumulate—and yeasts and various bacteria may grow as a surface film—during this phase of the fermentation.

The continuing succession of bacterial populations next favors the development of *Lactobacillus plantarum*, which produces acid but no gas. During this phase of the fermentation the concentration of lactic acid reaches 1.5–2 percent. Growth of *L. plantarum* also removes mannitol that is produced by *Leuconostoc* and that has an undesirable bitter flavor. The fermentation can be stopped at this stage by canning or refrigerating

the sauerkraut. If there is any residual sugar and mannitol after the action of *L. plantarum*, the successional process can continue with the development of *Lactobacillus brevis*, a gas-producing species. The growth of *L. brevis* can increase the lactic acid concentration to 2.4 percent and also imparts a bitter acid flavor to the sauerkraut. High-quality sauerkraut has a lactic acid concentration of about 1.7 percent and a clean acid flavor, with low concentrations of diacetyl contributing to the aroma and flavor of the final product.

Figure 19.23

During sauerkraut production there are successional changes in the bacterial populations in response to changing environmental conditions. The production of high-quality sauerkraut depends on the contributions of several different microbial populations.

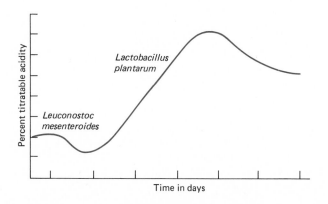

Percent titratable acidity

Lactobacillus plantarum

Leuconostoc mesenteroides

Time in days

Pickles

The traditional method for producing **pickles** by fermentation uses the natural microbiota associated with the cucumber and controlled temperature and salt concentrations to regulate the fermentation process (Figure 19.24). Controlled fermentation of pickles can also be achieved by inoculation with *Lactobacillus plantarum* and *Pediococcus cerevisiae*. The traditional process takes six to nine weeks to reach completion. During this period the salt concentration is gradually increased to reach a final concentration of about 15.9 percent NaCl. At the beginning of the fermentation, when the salt concentration is low, many bacterial genera are able to grow, including *Pseudomonas, Flavobacterium,* and *Bacillus.* As the salt concentration is increased, the populations that become favored include the lactic acid bacteria *Leusonostoc mesenteroides, Streptococcus faecalis,* and *Pediococcus cerevisiae.* As the lactic acid and salt concentrations increase, *Lactobacillus plantarum* becomes the dominant bacterium, beginning several days after the fermentation and continuing until the salt concentration surpasses 10 percent. The completion of the fermentation process involves yeasts that grow at high salt concentrations. During the final yeast fermentation stage some carbohydrates are converted to alcohol. The growth of film-forming yeasts, such as *Debaryomyces, Pichia, Endomycopsis,* and *Candida,* lowers the lactic acid concentration.

Because of the complexity of changes in the microbial community during this natural fermentation, the process often goes awry and yields unmarketable pickles, such as: (1) floaters and bloaters that float because of excessive gas accumulation within the cucumber; (2) hollow pickles, in which the cucumber contents have shriveled because of excessive salt or the formation of high concentrations of acetic acid; (3) stinkers, due to the accumulation of H$_2$S; (4) black pickles, due to bacterial pigment production; (5) soft pickles, due to fungal proteases; and (6) slippery pickles, due to the surface growth of encapsulated bacteria. Controlled fermentation conditions—and a pure inoculum of *Pediococcus cerevisiae* and *Lactobacillus plantarum* after the removal of the natural microbiota by fumigation or chlorination—can be used to increase the likelihood of producing a quality pickle.

The sourness of the pickle reflects the amount of lactic acid that accumulates during the fermentation. Several varieties of pickles are produced by using modification of the basic fermentation process. In the production of dill pickles a brine of 7.5–8.5 percent NaCl is used. The dill herb is added for flavoring, and vinegar also is normally added to prevent undesirable fermentation reactions. Because of the low concentration of salt, various indigenous soil bacteria on cucumber surfaces are able to grow during the initial stages of fermentation. As lactic acid accumulates, the bacterial community becomes dominated by *Leuconostoc messenteroides, Streptococcus faecalis, Pediococcus cerevisiae,* and *Lactobacillus plantarum.* The final concentration of lactic acid in dill pickles is in the range 1–1.5 percent.

Olives

The production of **green olives** involves a lactic acid fermentation. The harvested olives are washed with a solution of sodium hydroxide that removes most of the oleuropein, a very bitter phenolic glucoside, which gives unfermented olives a very undesirable flavor. The olives are next placed in a brine solution and a lactic acid fermentation of six to ten months is permitted to occur. During the first two weeks of the fermentation, the brine becomes stabilized as compounds are leached from the olives and microbial populations begin to multiply. At the intermediate stage, which occurs during the following two to three weeks, *Leuconostoc* is the dominant bacterial species and lactic acid accumulates. The final stage of fermentation is dominated by *Lactobacillus plantarum* and *L. brevis*; yeasts and various bacteria also occur during this stage. The final acidity of the olives is approximately 7.1 percent lactic acid.

Soy sauce

Several Oriental foods are prepared by fermenting soybeans or rice. **Soy sauce,** a brown, salty tangy sauce, which in Japanese is called shoyu, is produced from a mash consisting of soybeans, wheat, and wheat bran. Soy sauce is used as a condiment or as an ingredient in other sauces (Figure 19.25). The starter culture for the production of soy sauce is produced by a **koji fermentation,** a dry fermentation, in which a mixture of soybeans and wheat is inoculated with spores of *Aspergillus oryzae* (Figure 19.26). The mixture is moistened but is not submerged in liquid. The fungi grow on the surface of the soybeans and wheat, accumulating various enzymes including proteinases and amylases. Various bacterial populations, normally dominated by lactic acid bacteria, also develop during this koji fermentation. After the starter culture develops, it is dried and extracted.

The extract is mixed with a mash consisting of autoclaved soybeans, autoclaved and crushed wheat, and steamed wheat bran. The mash with

Figure 19.24

Photograph showing pickle production. Problems occasionally arise during the fermentative production of pickles, leading to the formation of bloaters, stinkers, and other undesired products. (Courtesy Paramount Pickles, Louisville.)

the koji is incubated in flat trays for several days at approximately 30°C and is then soaked with concentrated brine. The resulting mixture is called maromi. The mash is then incubated from 10 weeks to over a year, depending on the incubation temperature. During this incubation period the proteinases, amylases, and other enzymes of the koji are active, and there is a succession of microbial populations. The maturation begins with lactic acid bacteria, including lactic acid production by *Pediococcus soyae*, and later involves alcoholic fermentations by yeasts, such as *Saccharomyces rouxii*, *Zygosaccharomyces soyae*, and *Torulopsis* species. The most important organisms during the fermentation process are *Aspergillus oryzae*, which produce proteinases and amylases; *Lactobacillus*

Figure 19.25

Soy sauce production is a two-stage fermentation process, involving a dry fermentation followed by a submerged fermentation.

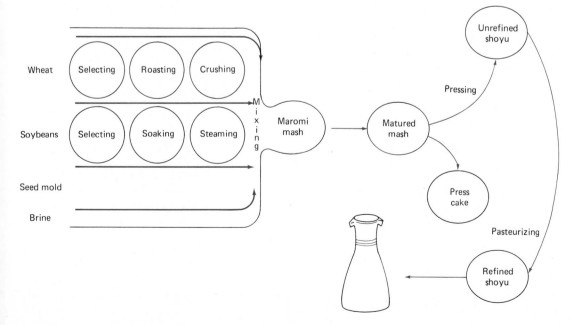

Figure 19.26

Photograph of koji fermentation tray, showing growth of Aspergillus sojae *on a mixture of soybeans and wheat. (Courtesy Kikkoman Foods, Walsworth, Wisconsin.)*

species, which produce sufficient amounts of lactic acid to prevent spoilage by other microorganisms; and yeasts, which produce sufficient alcohol to increase the flavor content.

An interesting problem was encountered when soy sauce production was begun in the United States. In Japan the maturation process is carried out in concrete tanks, and the necessary microbial populations are maintained in the porous concrete surface. In the United States, where sterilized stainless steel tanks are used for the secondary fermentation process, it was difficult to define, maintain, and add at the proper times the necessary cultures that are needed for the successional process involved in producing quality soy sauce. Eventually, the process was perfected, and only a soy sauce connoisseur can tell the difference between the U.S. and Japanese products.

Miso

Miso is also produced by using a koji fermentation with *Aspergillus oryzae*. Steamed polished rice, placed in shallow trays, is used in the production of the starter culture. The koji is mixed with a mash of steamed soybeans, and after the addition of salt, the fermentation is allowed to proceed at 28°C for one week, at 35°C for two months, and at room temperature for several additional weeks. The miso is normally ground into a paste to be combined with other food before eating.

Tempeh

Tempeh is an Indionesian food produced from soybeans. The soybeans are soaked at 25°C, dried, and inoculated with spores of various species of *Rhizopus*. The mash is incubated at 32°C for 20 hours during which there is mycelial growth. The product is then salted and fried before eating.

Tofu and sofu

Tofu (Japanese) or **sofu** (Chinese) is a cheese-like product produced by fermenting soybeans with *Mucor* spp. The soybeans are soaked, ground to a paste, and curdled by adding calcium or magnesium salts. The pressed curd blocks are placed in trays at 14°C and incubated for one month during which time the fungal populations develop.

Natto

Natto also is produced from boiled soybeans and involves the incubation of *Bacillus subtilis* with soybeans for one to two days, during which proteinase enzymes soften and add flavor compounds to the soybeans. Various other Oriental foods are also produced by similar fermentations.

Poi

Poi is a fermented food product from the Hawaiian Islands. In the production of poi, the stems of the taro plant are steamed, ground, and subjected to fermentation for one to six days. During the first few hours coliforms, *Pseudomonas*, and various other microorganisms predominate. Then a successional process occurs with *Lactobacillus, Streptococcus*, and *Leuconostoc* becoming the dominant populations, and finally, yeasts and the fungus *Geotrichum candidum* flourish. The fermentation products, principally lactic acid, acetic acid, formic acid, ethanol, and carbon dioxide, contribute to the characteristic texture, flavor, and aroma of poi.

Single cell protein

In addition to using microorganisms to enzymatically transform substrates into desired food products, microorganisms can be grown as a source of **single-cell protein (SCP)**, so-named because the microorganisms are single-celled organisms rich in protein. Microorganisms grow rapidly and produce a high-yield, high-protein crop. The proteins of selected microorganisms contain all the essential amino acids. Various bacteria, fungi, and algae are potential sources of large amounts of single-cell protein. The algae *Scenedesmus* and *Spirulina*, for example, have been cultured in various warm ponds as a food source. The production of single-cell protein from algae is advantageous because these organisms are able to utilize solar energy, greatly reducing the amount of fuel resources required to produce single-cell protein. Some algae currently are harvested as a source of food.

Research on the concept of single-cell protein production was begun during the 1960s by oil companies when petroleum was inexpensive and appeared to be an economically attractive substrate for growing SCP. The Imperial Chemical Works in Britain produces Pruteen, the SCP product of *Methylophilus methylotrophus*, a bacterium that grows on C_1 compounds. Indeed, most processes designed for producing SCP rely on growing microorganisms on C_1 compounds. *M. methylotrophus* is grown on methanol, derived from methane, and the cell crop is harvested, centrifuged, dried, and sold in pellet or granular form (Figure 19.27). Because of dramatic increases in the price of oil, petroleum hydrocarbons are no longer considered as the primary substrates for producing SCP. The product simply could not be economically competitive with soybean and fish meal. Future less expensive sources of methanol, perhaps derived from cellulose, will likely revive the prospects for large-scale production of microbial SCP.

Yeasts are excellent prospects for development as commercial sources of single-cell protein. Yeast-based SCP has a high vitamin content. Various species of yeast, including members of the genera *Saccharomyces, Candida*, and *Torulopsis*, can be grown on waste materials, recycling these substances into useful sources of food. The growth of yeasts on waste materials serves a dual function: the removal of the unwanted substances and

Figure 19.27

The production of bacterial protein from methanol. If a cheap source of methanol can be found, perhaps by microbial fermentation of waste products, the production of single-cell protein (SCP) could play a significant role in meeting world food needs.

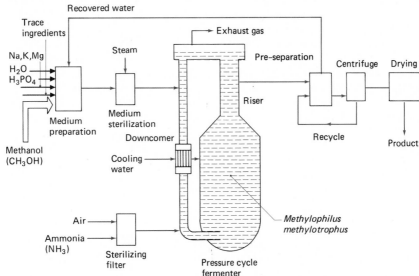

Production of bacterial protein from methanol

the production of much needed protein rich foods. In the U.S.S.R. there is huge commercial production of *Candida* yeast protein from hydrolyzed peat. Approximately 1.1 million tons of yeast protein per year are being produced in a rapidly expanding Soviet industry that aims to reduce Soviet dependence on imported grain.

Single-cell protein is primarily produced as an animal feed. There are problems with using SCP for direct human consumption because of high concentrations, 6–11 percent, of nucleic acids. This may result in increased serum levels of uric acid, causing kidney stone formation or gout, possible allergic reactions, and possible gastrointestinal reactions including diarrhea and vomiting. Chickens and other animals, however can be grown on single-cell protein rather than on plant materials, helping to meet world food needs. Researchers are still trying to find the proper microorganism and set of production conditions to produce SCP that can be fed directly to humans.

Postlude

The field of food microbiology can be viewed as an application of the principles of controlling microbial growth to prevent food spoilage and the food-borne transmission of diseases of humans, together with the harnessing of microbial activities to produce food products for human consumption. On the one hand uncontrolled microbial food spoilage processes reduce the value of food products, and on the other hand controlled microbial spoilage processes are used to produce valuable food products. Various food preservation methods are employed to increase the shelf life of numerous food products. Several modern preservation methods, including canning and freezing, can extend the shelf life of the product indefinitely. Increasing the shelf life of products is extremely important because much of the world's population now lives in urban centers, requiring increased storage and transport, and because many foods are grown seasonally but desired year-round. The shipping and distribution of food to urban population centers, which are distant from agricultural areas of food production, necessitates increased storage times for foods, compared to agricultural societies where the food can be consumed quickly after harvesting. Proper preservation of foods is important for public health reasons as well as for saving needed resources for meeting human nutritional needs.

The use of microorganisms to produce food products is also important to meet the food demands of the world's expanding population. Microbial fermentation products, such as cheeses, are part of the normal diet of most individuals. Various beverages, such as beer and champagne, are also frequently consumed. Many of these microbially produced foods are considered delicacies, adding enjoyment to eating. The use of microorganisms for producing single-cell protein represents an important potential inexpensive source of protein-rich food for the future. Developments in the field of food microbiology represent important ways in which our basic understanding of microbial processes can be applied for helping overcome food shortages and aid in the economic development of various parts of the world. Today's challenge to the food industry is to increase food production, to increase the shelf life of foods, and to respond to public desire for foods that are nutritious, safe, and tasty.

Study Questions

1. What is food spoilage?

2. What is the difference between an intrinsic and an extrinsic factor with respect to food spoilage processes?

3. What is the difference between canning and pasteurization?

4. Discuss three processes for drying foods to prevent spoilage.

5. Why is adding salt to a food useful for preventing spoilage? Why are we trying to limit the use of this preservation method?

6. What are the differences in the methods used to produce beer, wine, and distilled liquors?

7. How is soy sauce made?

8. How is cheese made? What is ripening? When we consider the great variety of cheeses, how can they all be made from essentially the same starting material?

9. How is sauerkraut produced? Discuss the role of microbial succession in the production of sauerkraut.

10. What can go wrong in the production of pickles?

11. What is a chemical food preservative? Should these be added to food products?

12. What is single-cell protein? What are the useful candidate substrates for SCP production? Can recombinant DNA technology help create a microbial strain that will solve world hunger? Discuss.

Suggested Supplementary Readings

Ayres, J. C., J. O. Mundt, and W. E. Sandine. 1980. *Microbiology of Foods*. W.H. Freeman and Co., San Francisco.

Desrosier, N. W. 1977. *The Technology of Food Preservation*. AVI Publishing Co., Westport, Connecticut.

Fennema, O. 1976–. *Principles of Food Science* (4 volumes). Marcel Dekker, New York.

Frazier, W. C., and D. C. Westhoff. 1978. *Food Microbiology*. McGraw-Hill Book Co., New York.

Gaman, P. M., and K. B. Sherrington. 1981. *The Science of Food: An Introduction to Food Science, Nutrition, and Microbiology*. Pergamon Press, New York.

Hobbs, B. C., and J. H. Christian. 1974. *The Microbiological Safety of Foods*. Academic Press, New York.

International Commission on Microbial Specifications for Foods (ICMSF). 1974. *Microorganisms in Foods: Sampling for Microbiological Analysis—Principles and Specific Applications*. University of Toronto Press, Toronto.

International Commission on Microbial Specifications for Foods (ICMSF). 1978. *Microorganisms in Foods: Their Significance and Methods of Enumeration*. University of Toronto Press, Toronto.

International Commission on Microbial Specifications for Foods (ICMSF). 1980. *Microbial Ecology of Foods* (2 volumes). Academic Press, New York.

Jay, J. M. 1970. *Modern Food Microbiology*. Van Nostrand Reinhold Co., New York.

Kharatyan, S. G. 1978. Microbes as food for humans. *Annual Reviews of Microbiology* 32: 301–327.

Pederson, C. S. 1979. *Microbiology of Food Fermentations*. AVI Publishing Co., Westport, Connecticut.

Rose, A. H. 1981. The microbiological production of food and drink. *Scientific American* 245(3): 126–139.

Sharpe, A. N. 1980. *Food Microbiology*. Charles C. Thomas Publisher, Springfield, Illinois.

Weiser, H. H., G. J. Mountney, and W. A. Gould. 1971. *Practical Food Microbiology and Technology*. AVI Publishing Co., Westport, Connecticut.

Industrial microbiology

20

table 20.1

Types of microbial products

Acidulants
Alkaloids
Amino acids
Animal growth promoters
Antibiotics
Antihelminthic agents
Antimetabolites
Antioxidants
Antitumor agents
Coenzymes
Converted sterols and steroids
Emulsifying agents
Enzymes
Enzyme inhibitors
Fatty acids
Flavor enhancers
Herbicides
Insecticides
Ionophores
Iron transport factors
Lipids
Nucleic acids
Nucleosides
Nucleotides
Organic acids
Pesticides
Pharmacological agents
Pigments
Plant growth promoters
Polysaccharides
Proteins
Solvents
Starter cultures
Sugars
Surfactants
Vitamins

Industrial microbiology, in its broadest sense, is concerned with all aspects of the economy that relate to microbiology. In a more restricted sense, industrial microbiology is concerned with employing microorganisms to produce a desired product and also preventing microorganisms from diminishing the economic value of various products. This duality of purpose is clearly seen in the food industry, a major area of industrial microbiology discussed in Chapter 19. **Quality control** to prevent microbial contamination of various industrial products and limiting microbial corrosion and biodeterioration are important concerns of industrial microbiology, as is producing and profiting from the sale of quality microbial products. Various commercial products of important economic value made by microorganisms are (1) **pharmaceuticals**, including antibiotics, steroids, vaccines, and vitamins; (2) **organic acids**; (3) **amino acids**; (4) **enzymes**; (5) **organic solvents**; and (6) **synthetic fuels** (Table 20.1). Many of these products can be produced both microbially and by chemical synthesis. The choice of which process to employ generally depends on economics, and it is not surprising that some products that historically have been produced by microorganisms are now produced chemically and vice versa. The decision is dictated by variable costs, and so changes in the costs of raw materials and the market value of a particular product influence the feasibility of microbial production of that product.

The fermentation industry

In industrial microbiology the term **fermentation** is not used in its restricted scientific sense, referring to metabolic pathways that proceed by fermentation rather than respiration, but rather is used in a wider sense to include any chemical transformation of organic compounds carried out by using microorganisms and their enzymes. Industrial processes using microorganisms exploit the enzymatic activities of the microbe to produce substances of commercial value. Production methods in industrial microbiology bring together the raw materials (substrates), microorganisms (specific strains or microbial enzymes), and a controlled favorable environment (created in a fermentor) to produce the desired substance. The essence of an industrial process is to combine the right organism, inexpensive substrate, and proper environment to produce high yields of a desired product (Figure 20.1).

Critical activities of industrial microbiologists

Figure 20.1

This diagram of a generalized fermentation process shows the steps involved in transforming a raw substrate into a final product.

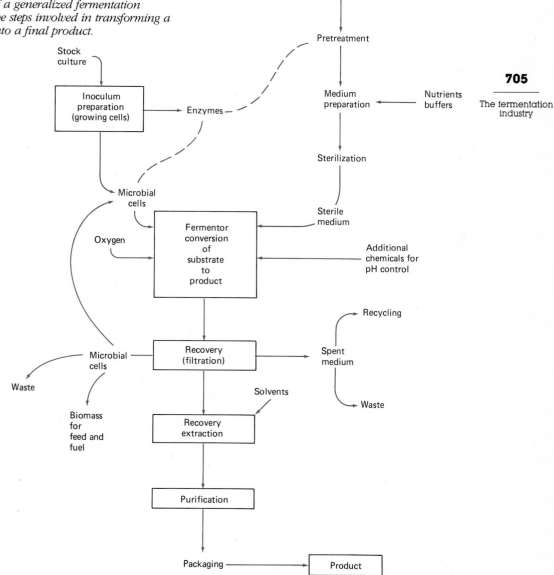

include (1) the search for microorganisms that carry out biotransformations of commercial importance, with the emphasis on finding or creating specific strains of microorganisms that will yield sufficient quantities of the desired product to permit commercial production on an economically favorable basis; (2) defining the substrate mixture—containing the least expensive components—that will produce the highest yield of the desired product, recognizing that many times the presence or absense of even trace amounts of a component will vastly alter the yield of the desired product; (3) designing fermentors to optimize the environmental conditions in order to achieve maximal product yields; and (4) developing recovery methods that achieve separation of the desired product from microbial cells, residual substrate, and other metabolic products in

table 20.2

Some microbial species used for producing commercial products

Industrial chemicals	
Saccharomyces cerevisiae	Ethanol (from glucose)
Kluyveromuces fragilis	Ethanol (from lactose)
Clostridium acetobutylicum	Acetone and butanol
Aspergillus niger	Citric acid
Xanthomonas campestris	Polysaccharides
Amino acids and flavor-enhancing nucleotides	
Corynebacterium glutamicum	L-Lysine
Corynebacterium glutamicum	5'-Inosinic acid and 5'-guanylic acid
Corynebacterium glutamicum	MSG
Vitamins	
Ashbya gossypii	Riboflavin
Eremothecium ashbyi	Riboflavin
Pseudomonas denitrificans	Vitamin B_{12}
Propionibacterium shermanii	Vitamin B_{12}
Enzymes	
Aspergillus oryzae	Amylases
Aspergillus niger	Glucamylase
Trichoderma reesii	Cellulase
Saccharomyces cerevisiae	Invertase
Kluyveromyces fragilis	Lactase
Saccharomycopsis lipolytica	Lipase
Aspergillus	Pectinases and proteases
Bacillus	Proteases
Mucor pussilus	Microbial rennet
Mucor meihei	Microbial rennet
Polysaccharides	
Leuconostoc mesenteroides	Dextran
Xanthomonas campestris	Xanthan gum
Pharmaceuticals	
Penicillum chrysogenum	Penicillins
Cephalosporium acremonium	Cephalosporins
Streptomyces	Amphotericin B, kanamycins, neomycins, streptomycin, tetracyclines, and others
Bacillus brevis	Gramicidin S
Bacillus subtilis	Bacitracin
Bacillus polymyxa	Polymyxin B
Rhizopus nigricans	Steroid transformation
Arthrobacter simplex	Steroid transformation
Mycobacterium	Steroid transformation
Escherichia coli (via recombinant-DNA technology)	Insulin, human growth hormone, somatostatin, interferon

the most economical manner. The complexity of achieving even a simple single-step transformation of a molecule is great when considered on the large industrial scale.

Selection of industrial microorganisms

The **selection of microorganisms for use in the fermentation industry** begins with screening to find the right microorganism. Of the many species of microorganisms, relatively few possess the genetic information needed for producing economically useful products (Table 20.2). The screening procedures employed in industry are designed to separate potentially useful microorganisms, possessing the potential for producing a commercially useful product, from the rest. Large-scale industrial screening procedures incorporate assays that permit identification of these microorganisms.

Screening of antibiotic producers

The search for antibiotics in the pharmaceutical industry presents a good example of how **screening procedures** are employed to select microorganisms for industrial applications. A useful antibiotic-producing strain must produce metabolites that inhibit the growth or reproduction of pathogens. This essential property can be assayed for using test strains and examining whether the isolate being screened produces substances that inhibit the growth of that test organism. If a sus-

pension of the test organism is applied to the surface of an agar plate, the zone of inhibition around a colony may indicate that the organisms in that colony are producing an antibiotic (Figure 20.2). Alternatively, the crude filtrate of a broth-grown microbial culture can be added to a culture of a test organism to determine whether substances with antimicrobial activity are produced by the organism being screened. A positive result in such a primary screening procedure does not in any sense ensure the discovery of an industrially useful antibiotic-producing strain, but it simply identifies those strains of microorganisms having the potential for further development. **Secondary screening procedures** are then carried out to determine whether the organism is indeed producing a substance of industrial interest that merits further investigation and development. The secondary screening may include both qualitative assays, aimed at identifying the nature of the substance being produced and determining whether it is a new compound not previously considered for industrial production, and quantitative assays, aimed at determining how much of the substance is being produced.

In the case of screening for antibiotic producers, the crude filtrate from a broth culture may be separated chromatographically and the antimicrobial activities of the separated components determined (Figure 20.3). The individual active components can then be isolated and used for further screening against additional test organisms to determine the microbial inhibition spectrum. This additional screening is useful in

Figure 20.2

This petri plate shows zones of inhibition around bacterial colonies as part of testing during primary antibiotic screening procedures. (Courtesy Eli Lilly and Co., Indianapolis.)

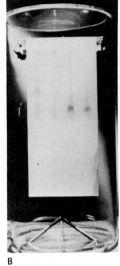
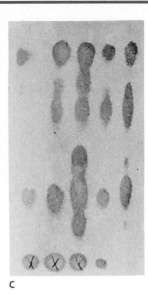

A　　B　　　　　　　　C

Figure 20.3

Paper chromatography is used during secondary screening procedures to identify active antibiotic compounds. (A) This room has many large tanks for paper chromatography. (B) An individual tank showing separation of a potentially active compound. (C) The resultant paper chromatogram is then placed onto a large agar plate so that the separated products diffuse into the agar. A lawn of bacteria is applied to the agar, and after incubation zones of inhibition around isolated spots indicate the locations of potentially useful products. (Courtesy Marvin Hoehn and Bernard Abbott, Eli Lilly and Co., Indianapolis.)

Discovery process

The discovery of new antibiotics results from laborious hours of searching. Samples from many sources, including soils from around the world, have been examined as potential sources of antibiotic-producing microorganisms; countless strains of microbial isolates have been tested by pharmaceutical laboratories. Of the numerous investigations, few studies yield evidence of promising new compounds of potential clinical importance. Identification of compounds with antimicrobial activity is an essential step in the screening process. Paper chromatography permits the relatively easy separation of compounds from a complex mixture. Compounds are separated based on their relative polarities in this form of affinity chromatography. By placing the paper chromatogram directly onto an agar plate, the resolved compounds are absorbed into the agar. Zones of inhibition indicate the presence and locations of antimicrobial compounds. Many compounds can be screened in this way to identify those few that may prove useful. Compounds showing antimicrobial activity can be isolated and tested to determine that they are truly new compounds that merit further study. Tests include determining the effectiveness of a new compound against a wide variety of pathogenic microorganisms and elucidating potential toxicity and other untoward side effects that may occur in mammals. Only a limited number of compounds identified in the original screening procedure as being of potential clinical use ever reach clinical trials, and still fewer are ever marketed. The development and successful marketing of antibiotics traditionally is the result of serendipity and hours of laborious screening rather than any great theoretical insight.

determining whether the substance has a broad or narrow range of activity and if it is particularly effective against specific pathogens. Assuming that an organism is indeed identified as possessing the potential for producing a useful new antibiotic, many additional tests are required to determine whether sufficient quantities of the substance can be produced to permit industrial production.

The screening program should identify the optimal incubation conditions for maximal economic yield of the product. Usually, toxicity testing must be performed to determine whether the product really has the potential for selectively inhibiting pathogens without causing severe side effects that would preclude its therapeutic use. The secondary screening procedure thus yields a

Figure 20.4

Induced mutations in a penicillin-producing strain were used to produce a new commercially useful strain. Repeated mutations were necessary to create a strain of the mold Penicillium chrysogenum *that synthesized enough penicillin to form the basis of a commercial process. Radiation and chemical agents were employed to induce mutations in the mold. They were S = spontaneous mutation; X = X radiation; UV = ultraviolet radiation; and NM = nitrogen mustard. Selection of the superior strains ultimately gave rise to strain E-15.1, which yielded 55 times as much penicillin as laboratory strains. Simultaneous improvements in fermentation techniques increased yields still further; yield figures in this chart reflect both kinds of increase. Classical genetic techniques such as these are still important in the antibiotics industry, as the complexity of antibiotic synthesis makes it impractical to develop new strains by directly altering simple genes. Current fermentation methods yield more than 20 g/l.*

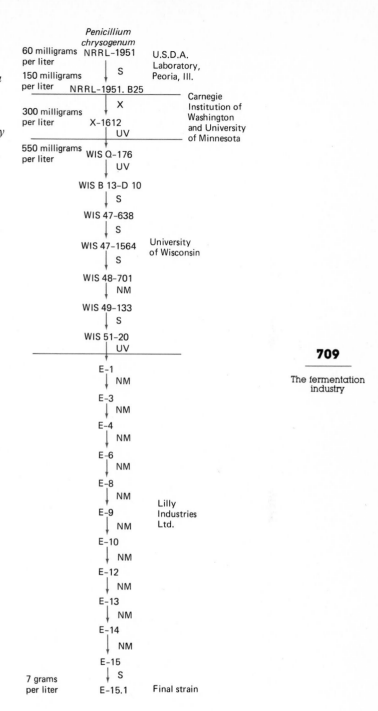

great deal of information about potentially useful microorganisms, allowing emphasis to be placed on the development of processes employing microorganisms likely to produce economically valuable substances.

Screening for microorganisms that possess the potential for producing industrially important substances examines both naturally occurring microorganisms and genetic variants. The classical approach used for finding new antibiotic-producing strains has been to screen large numbers of isolates from soil samples for microorganisms that naturally produce antimicrobial substances. Additionally, mutations can be induced by exposure to radiation or mutagenic chemicals to increase the genetic variability within the populations being screened, with the hope of isolating a unique microbial strain capable of producing a novel metabolite with the desired properties or a strain that produces large quantities of a valuable substance. Often, once a microorganism is identified as possessing the genetic information for producing a potentially useful substance, it is then necessary to carry out successive stages of mutation before isolating a strain of that organism that can be employed for commercial production. For example, the *Penicillium* species observed by Alexander Fleming to inhibit the growth of *Staphylococcus* had obvious potential for commercial development but did not produce sufficient quantities of penicillin to permit industrial production. Extensive screening of soil samples from around the world led to the isolation of a potentially useful strain from soil collected at Peoria, Illinois. Multiple successive mutations, though, were nec-

essary to develop a strain of *Penicillium chrysogenum* capable of producing nearly 100 times the concentration of penicillin produced by the original strain, making production of penicillin commercially feasible (Figure 20.4). The mutation and screening approach has been important in the successful development of various strains of microorganisms currently used in the fermentation industry.

Genetic engineering has opened up many new possibilities for employing microorganisms to produce economically important substances. Whereas the mutation and selection approach is hit-and-miss, the use of recombinant DNA technology permits the purposeful manipulation of genetic information to engineer a microorganism possessing the capability of producing high yields

of a great variety of products. Until the recent breakthroughs in the techniques of genetic engineering, a bacterium could produce only substances coded for in its bacterial genome, but it is now possible to engineer bacterial strains that produce plant and animal gene products (Figure 20.5). Thus, bacteria now exist that produce human interferon, insulin, and other hormones. The development of microorganisms producing high yields of such substances promises to revolution-

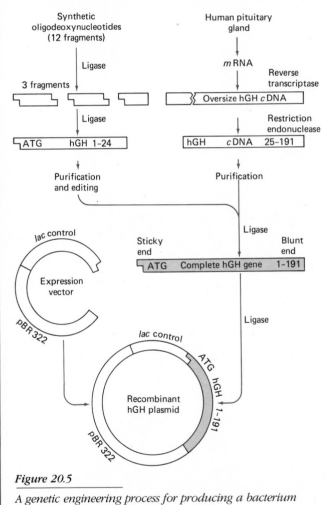

Discovery process

The recognition that genes are split in eukaryotic cells has forced genetic engineers to develop special techniques for incorporating human genes into bacteria in order to produce useful gene products. Through a series of novel procedures involving a combination of chemical synthesis and isolation of the natural molecules, a gene for a human growth hormone has been created and placed into *E. coli*. Human growth hormone is a polypeptide 191 amino acid units long elaborated by tissues of the pituitary gland, and its absence leads to a form of dwarfism that can be cured by administration of the hormone. The segment of the gene that codes for the first 34 amino acids of the peptide was constructed chemically from blocks of nucleotides. To obtain the rest of the gene, a series of enzymes were used. Reverse transcriptase was employed to copy the gene for the hormone from messenger RNA obtained from human pituitary tissues. The use of reverse transcriptase simplifies the job of cutting and splicing, because the DNA produced in this way is colinear with the sequence of nucleotides in the mRNA. Restriction endonucleases cut out the needed fragment. DNA ligase was then used to join the natural and synthetic fragments. The complete gene produced in this manner has been inserted into a modified version of plasmid pBR322 incorporating the *lac*

Figure 20.5

A genetic engineering process for producing a bacterium capable of synthesizing human growth hormone.

operon. The synthetic part of the growth hormone gene had been constructed earlier with its own initiation codon (ATG). The hormone could therefore be produced independently in bacterial cells without the need for attachment to a bacterial protein. Clinical trials of human growth hormone produced by genetically engineered bacteria are quite promising; children receiving injections of bacterially produced human growth hormone for a few months are approaching normal heights for their age group with no significant side effects.

ize the economics of the pharmaceutical industry. The seemingly unlimited potential for engineering microorganisms capable of producing economically lucrative products has spawned a major new industry, genetic engineering.

It is critical in both the mutation-screening and genetic engineering approaches for the development of microbial strains of industrial importance to consider genetic regulatory mechanisms. The development of a commercially useful microbial strain often involves overcoming natural regulatory mechanisms that limit the amount of the gene product produced. In nature it is advantageous for microorganisms to produce the minimal amount of a substance they require because doing so gives that organism a competitive edge for survival. As a result various genetic regulatory mechanisms have evolved to conserve available carbon and energy resources and to avoid production of excessive amounts of any product. In the fermentation industry, however, the valuable microbial strains are those that produce excessively high amounts of the desired product. Many mutant or genetically engineered strains used for industrial production no longer possess the genetic regulatory mechanisms for conserving their resources and producing limited amounts of a substance. Whereas such organisms would not do well in natural environments where competition for available resources dictates which organisms survive, they do quite well in fermentors where competition is eliminated and optimal conditions are created to favor the growth of that microbial strain in pure culture.

Because industry relies on specific microbial strains, it is important to maintain those specific genetic variants and protect against further spontaneous mutations that could alter the economics of the fermentation process. Industrial strains of microorganisms are therefore maintained in culture collections, generally in a dormant state where they are protected against mutational processes. The maintenance of these stock cultures is an essential, although sometimes boring, part of industrial microbiology. Checks are periodically run on production strains to ensure that they have retained their essential genetic capabilities. If undesirable alterations in the production strain are detected, new cultures are initiated from those maintained in the stock culture collection.

Production methods

After discovering or engineering a microorganism that produces a commercially valuable prod-

uct, it is necessary to develop a fermentation process that optimizes conditions for the desired microbial activity and that yields maximal amounts of product with the highest economic profit. A balance must be achieved between production costs and the price of the product because excessive costs may preclude the economic feasibility of commercial production. The development of a commercial process occurs in a stepwise fashion, initially using small flasks, then small fermentors (under 10 gallons), intermediate-size fermentors (up to several hundred gallons), and finally, large-scale fermentors (thousands of gallons) (Figure 20.6). At each stage of production, development conditions are adjusted to produce maximal yield at minimal cost. The organic and inorganic composition of the medium, pH, temperature, and oxygen concentration are the main factors that are varied to maximize the efficiency of the production process. Even in a batch process, conditions are often varied during the fermentation to achieve the maximal product yield, and conditions are monitored during the fermentation process to ensure that critical parameters remain within allowable limits. It is necessary that the reaction chambers and substrate solutions be sterilized prior to the addition of the microbial strain being used in the production process. This is particularly important because the strains of microorganisms used in industrial fermentations are selected for their ability to produce the desired product in high yield, rather than for their ability to compete with other microorganisms. Infections of fermentation reactions with microbial contaminants can easily lead to a competitive displacement of the strain being employed to produce the product, with obviously deleterious results.

Fermentation medium

The composition of the **fermentation medium** must contain the nutrients essential to support growth of the microbial strain and the formation of the desired product. Essential nutrients for microbial growth include a source of carbon, nitrogen, and phosphorus (Table 20.3). The choice of a particular nutritive source is made on economic as well as biological grounds. Depending on the nature of the fermentation process, all of the raw materials may be added at the beginning of the fermentation, or nutrients may be fed to the microorganisms gradually during the course of the fermentation. Often, plant materials, such as molasses, are used as a carbon source. Some pretreatment of the raw material is frequently necessary to convert complex carbohydrate materials

A

Figure 20.6

(A) A small-scale fermentor used in the design of an antibiotic fermentation process. In this dual fermentor unit, rates of aeration, pH, temperature, and other factors can be varied while growth and product formation are monitored until optimal conditions are determined. (Courtesy New Brunswick Scientific Company.) (B) Intermediate-size fermentors are used in a pilot plant to scale up the fermentation. (C) Large-scale fermentors are used for commercial antibiotic production. (B–C Courtesy Marvin Hoehn and Bernard Abbott, Eli Lilly Research Laboratories.)

B

C

table 20.3

Nutrient sources for industrial fermentations

Nutrient	Raw material
Carbon source:	
Glucose	Corn sugar
Sucrose	Molasses
	Starch
	Cellulose
Fats	Vegetable oils
Hydrocarbons	Petroleum fractions
Nitrogen source:	
Protein	Soybean meal
	Cornsteep liquor (from corn milling)
	Distillers' solubles (from alcoholic-beverage manufacture)
Ammonia	Pure ammonia or ammonium salts
Nitrate	Nitrate salts
Molecular nitrogen	Air (for nitrogen-fixing organisms)
Phosphorus source:	Phosphate salts

into relatively simple sugars that can be readily metabolized by microorganisms. Either organic nitrogen, sometimes in the form of cornsteep liquor, or inorganic nitrogen, such as ammonia, may be used to meet the nutritional needs of the microbial strain. Phosphorus is usually added as an inorganic salt. Since crude raw materials are normally employed in the medium, many of the minor nutritional requirements of microorganisms are met because they naturally occur in appropriate concentrations in the raw material. In some fermentation processes, however, trace elements, such as heavy metals, must be present in specific concentrations to achieve acceptable yields of the desired product. The quality of the water used in the fermentation and the nature of the pipes used to supply solutions to the fermentation reaction can be especially important. In some cases, metals leaching from pipes can inhibit microbial production of fermentation products, and in other cases such leached metals may be essential for achieving optimal yields of the desired product.

Aeration

Many industrial fermentations are aerobic processes, and therefore, it is important to achieve the **optimal oxygen concentration** to permit microbial growth with maximal product yield. The transfer of oxygen to microorganisms in **large-scale fermentors** is particularly difficult because the microorganisms must be well mixed and the oxygen dispersed to form relatively uniform concentrations in order to support maximal production rates. The development of fermentors for the growth of obligately aerobic microorganisms in a broth (submerged aerobic culture) requires careful design in order to achieve optimal oxygen concentrations throughout the solutions contained in high-volume fermentors. Many fermentor designs have mechanical stirrers to mix the solution, baffles to increase turbulence and ensure adequate mixing, and forced aeration to provide needed oxygen (Figure 20.7). It should be noted that a high concentration of microbial cells, as is achieved in a fermentor, can rapidly deplete the soluble oxygen in an aqueous solution, creating anaerobic conditions that may not be favorable to microbial production of the desired product. Forced aeration and mechanical mixing, though, are relatively expensive because of high energy costs and must be economically justified for use in industrial fermentation processes.

pH

The **pH** of the reaction is also critical. The enzymes involved in forming the desired product all have optimal pH ranges for maximal activity and limited pH ranges in which activity is maintained. The rapid growth of microorganisms in a fermentor can quickly alter the pH of the reaction

Figure 20.7

A batch reactor is employed for most current applications of industrial microbiology. In essence the reactor is a vessel in which a medium and a biological catalyst are mixed and then given an optimum environment in which to react. The temperature and the pH are regulated. Filtered air, sometimes enriched with oxygen, is bubbled through the mixture. Samples are removed for chemical and biological assay. To prevent contamination, steam is directed through the various inlets to keep them sterilized, and the pressure inside the vessel is maintained at a value greater than atmospheric pressure. At the end of a specified period, ranging from hours to days, the batch is drained from the vessel and the product is isolated and purified.

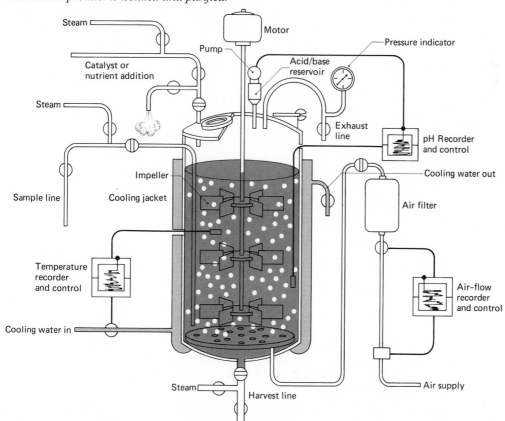

medium. For example, if the microorganisms accumulate acid, which in fact may be the desired product, the pH of a nonbuffered medium can decline precipitously, halting microbial production of the desired fermentation product. To prevent such changes, fermentation media are often buffered to dampen changes in the pH. Additionally, the pH of the reaction normally is continuously monitored and acid or base added as needed to maintain the pH of the reaction solution within acceptable tolerance limits.

Temperature

The **temperature** of the reaction must also be carefully regulated to achieve optimal yields of product. Rapidly growing microorganisms can generate a large amount of heat that must be dissipated to prevent inactivation of enzymes. Cooling coils are often employed in fermentors to regulate temperature in order to maximize rates of product accumulation. Heating coils are used in some fermentors where elevated temperatures are required for achieving optimal rates of product formation. The heating coils are also used for periodic sterilization of the fermentor chamber.

Batch vs continuous processes

In addition to considering nutritional and envi-

ronmental parameters, a fermentation process may be designed as a **batch or continuous flow process**. The choice of a particular process design depends on the economics of both production and recovery of the desired product. As compared to batch processes, flow-through fermentors are more prone to contamination with undesired microorganisms, and thus, quality control is difficult to maintain. The flow-through design, however, has the advantage of producing a continuous supply of product that can be recovered at a constant rate for commercial distribution (Figure 20.8). By their very nature, batch processes require significant startup times to initiate the fermentation process, incubation times to allow fermentation products to accumulate, and recovery times during which the product is separated from the spent medium and microbial cells.

Immobilized enzymes

The use of **immobilized enzymes** is an interesting alternative method for producing a desired product. In this process microbial enzymes and/or microbial cells are adsorbed onto a solid surface support, such as cellulose (Figure 20.9). The bonded microbial enzymes act as a solid surface catalyst. A solution containing the biochemicals to be transformed by the enzymes is then passed across the solid surface. Temperature, pH, and oxygen concentration are set at optimal levels to achieve maximal rates of conversion. This type of process is very useful when the desired transformation involves a single metabolic step, but it is more complex when many different enzymatic activities are required to convert an initial substrate into a desired end product. The use of immobilized enzymes makes an industrial process far more economical, avoiding the wasteful expense of having to continuously grow microorganisms and discard the unwanted biomass. It is essential in such immobilized enzyme systems to maintain enzymatic activity, that the enzymes not be washed off the surface nor become inactivated during the process. When whole cells, rather than cell-free enzymes, are employed in such immobilized systems, it is necessary to maintain viability of the microorganisms during the process, and this generally involves adding necessary growth substrates.

Recovery methods

There are various methods employed in both continuous and batch culture processes for the **recovery of fermentation products**. The commonly used recovery methods include distillation, centrifugation, filtration, and chromato-

Figure 20.8

Flow-through fermentors are becoming increasingly popular in industry. The economics of using such fermentors is favorable, although it is more difficult to maintain quality control of the product. In this diagram a simple flow-through fermentor is shown. More elaborate designs permit control of the same environmental factors as are controlled in batch processes. The main difference is that the substrates are continuously added and the products continuously harvested.

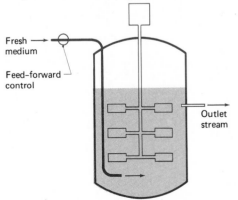

Fresh medium

Feed-forward control

Outlet stream

graphic separation. The use of distillation for the recovery of ethanol, for example, was discussed in Chapter 19 when considering the production of distilled liquors. Examples of other recovery methods will be seen when considering the production of specific products later in this chapter. After the successful recovery of a microbial fermentation product, the substance is packaged and marketed. Packaging is an another important aspect of the production process because it must protect the product from contamination. Thus, every step in a fermentation process, from the isolation of the strains to the delivery of the product, must be carefully controlled to ensure the delivery of a high-quality product.

Production of pharmaceuticals

The **pharmaceutical manufacturing industry**—a major source of employment for industrial microbiologists—is primarily concerned with dis-

Figure 20.9

As shown in this diagram, immobilized enzymes are prepared by binding an enzyme to a carrier such as carboxymethylcellulose.

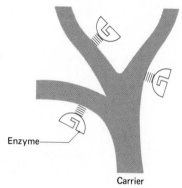

ease processes (some of which are caused by microorganisms), and with making drugs (many of which are produced by microorganisms) to control disease processes. The world's supply of pharmaceuticals, including many antibiotics, steroids, vitamins, and vaccines, is produced in substantial part by microorganisms. The microbial production of pharmaceuticals is a major industry; antibiotic sales alone accounted for approximately $5 billion in worldwide sales in 1980. The role of microorganisms in producing these pharmaceuticals is economically important for industry and is essential for making these compounds available at a cost low enough to permit their wide use in preventing and treating numerous diseases. In this section some representative examples will be discussed to illustrate the processes involved in the production of various pharmaceuticals.

Antibiotics

Of the thousands of different **antibiotic compounds** made in nature by various microorganisms, relatively few are produced commercially. The major antibiotics used in medicine, their structures and modes of action, have been discussed in Chapter 15, and the microorganisms used for producing these antibiotics are shown in Table 20.4.

Penicillin In a typical process for manufacturing **penicillin**, an inoculum of *Penicillium chrysogenum* is produced by inoculating a dense suspension of spores of the fungus onto a wheat bran-nutrient solution. The cultures are allowed to incubate for approximately one week at 24°C and are then transferred to an inoculum tank. In some cases these spores are germinated to produce mycelia for inoculation into these tanks. The

table 20.4

Some antibiotics produced by microorganisms

Antibiotic	Produced by
Amphotericin-B	*Streptomyces nodosus*
Bacitracin	*Bacillus licheniformis*
Carbomycin	*Streptomyces halstedii*
Chlorotetracycline	*Streptomyces aureofaciens*
Chloramphenicol	*Streptomyces venezuelae* or total chemical synthesis
Erythromycin	*Streptomyces erythreus*
Fumagillin	*Aspergillus fumigatus*
Griseofulvin	*Penicillium grisofulvin* *Penicillium nigricans* *Penicillium urticae*
Kanamycin	*Streptomyces kanamyceticus*
Neomycin	*Streptomyces fradiae*
Novobiocin	*Streptomyces niveus* *Streptomyces spheroides*
Nystatin	*Streptomyces noursei*
Oleandomycin	*Streptomyces antibioticus*
Oxytetracycline	*Streptomyces rimosus*
Penicillin	*Penicillium chrysogenum*
Polymyxin-B	*Bacillus polymyxa*
Streptomycin	*Streptomyces griseus*
Tetracycline	Dechlorination and hydrogenation of chlorotetracycline; direct fermentation in dechlorinated medium
Vira-A (adenine arabinoside)	*Streptomyces antibioticus*

Figure 20.10

Time course changes in carbohydrate, nitrogen, penicillin, and biomass concentrations during penicillin fermentation. During the period of biomass accumulation lactose and ammonia concentrations decline and there is little penicillin production. After one day of incubation penicillin begins to accumulate. During this period glucose and nitrogen are fed to the cells at rates that optimize penicillin production.

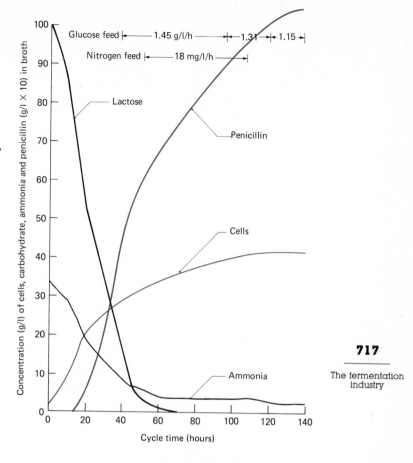

inoculum tanks are agitated with forced aeration for one to two days to provide a heavy mycelial growth for inoculation into a production tank. In some cases additional step-up procedures are employed in which sequentially larger tanks are used to achieve larger amounts of mycelial inoculum for the production tanks. The typical medium used for the production of penicillin has changed radically in the last few decades. Whereas in 1945 the typical medium contained 3.5 percent cornsteep liquor solids, 3.5 percent lactose, 1 percent glucose, 1 percent calcium carbonate, 0.4 percent potassium phosphate, 0.25 percent vegetable oil, and a penicillin precursor, such as phenylacetic acid, the medium used today typically uses 10 percent total glucose or molasses by continuous feed, 4–5 percent cornsteep liquor solids, 0.5–0.8 percent total phenylacetic acid by continuous feed, and 0.5 percent total vegetable oil by continuous feed. The major change is the elimination of lactose from the medium and the use of continuous-feed substrate addition to increase the efficiency of penicillin production. The phenylacetic acid is the precursor used to form the benzyl side chain of penicillin G. The addition of this precursor steers the fungal metabolic reactions to form increased amounts of penicillin. The pH of the medium after sterilization is approximately 6, which is critical because penicillin is inactivated at both low and high pH values. The pH is maintained near neutrality during the course of the fermentation by the addition of alkali to the medium as needed. Incubation temperature for the fermentations is maintained at approximately 25–26°C, and aeration is provided during the production process.

The typical course of a penicillin fermentation takes seven days, although longer times may be required when using very large fermentors (Figure 20.10). During the first day of the fermentation, there is a large increase in the biomass of *Penicillium* mycelia. Glucose is rapidly used during this early phase, providing the necessary carbon and energy for the production of fungal mycelia. At a later stage reducing the glucose concentration provides the necessary nutritional starvation conditions that favor penicillin production. The nitrogen required to support fungal growth comes from the cornsteep liquor. The production of penicillin, a secondary metabolite

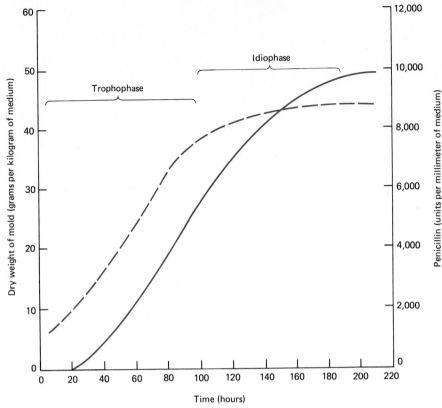

Figure 20.11

This graph shows how penicillin production lags behind production of fungal biomass. A secondary metabolite is not formed as a direct result of the metabolism that keeps the cells alive. Hence, the accumulation of a secondary metabolite in a reactor vessel lags behind the growth of the cells that produce it. The graph shows the accumulation of mold cells (dashed line) and the subsequent accumulation of penicillin (solid line). The values of temperature and pH that are best for the growth of cells are seldom best for the synthesis of a secondary metabolite. In a batch process one seeks a compromise between the two sets of optimum conditions.

(**idiolite**) not required for the growth of the fungus, lags behind the accumulation of fungal biomass (**trophophase**). Penicillin begins to accumulate in the medium on the second day (**idiophase**) and reaches its maximal concentration a few days later (Figure 20.11).

When the fermentation is completed, the concentration of penicillin having reached maximal achievable levels, the liquid medium containing the penicillin is separated from the fungal cells by using a rotating vacuum filter (Figure 20.12). The fungal biomass is scraped from the surface of the filter drum, dried, and marketed as an animal feed supplement. The penicillin is recovered from the filtrate, using various extraction procedures (Figure 20.13). The penicillin is extracted from the solution by using inorganic solvent and then extracted back into aqueous solution. The

exchange of penicillin back and forth between organic and aqueous solvents is accomplished by altering the pH and results in the partial purification of the antibiotic. Spent solvents used in the extraction of the penicillin are recycled. Potassium ions are then added to the aqueous solution, resulting in the formation of the crystalline potassium salt of penicillin G, which can be recovered by filtration or centrifugation. The filtered and dried penicillin salt is over 99.5 percent pure.

The penicillin G produced in this process can be further modified to form various penicillin derivatives (Figure 20.14). The modification of penicillin may be accomplished chemically or by using microbial enzymes. For example, 6-aminopenicillanic acid can be formed by fermentation, using bacterial acylase enzymes in an aqueous solution

Figure 20.12

A rotating vacuum drum separates mycelial biomass from the broth containing antibiotic. Machines like this one are used in the commercial manufacture of such antibiotics as penicillin and streptomycin. The mycelia collect on the outside of the drum and are scraped off, and the liquid is sucked through the filter. (Courtesy Marvin Hoehn and Bernard Abbott, Eli Lilly Research Laboratories, Indianapolis.)

Figure 20.13

The procedure for the commercial recovery and purification of antibiotics, such as penicillin, is presented in this flow chart.

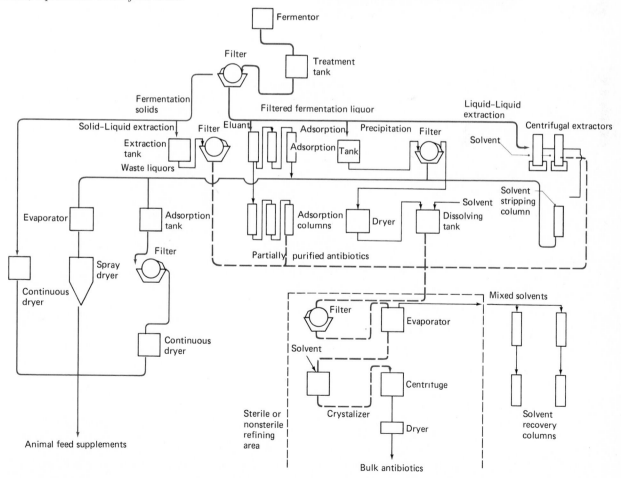

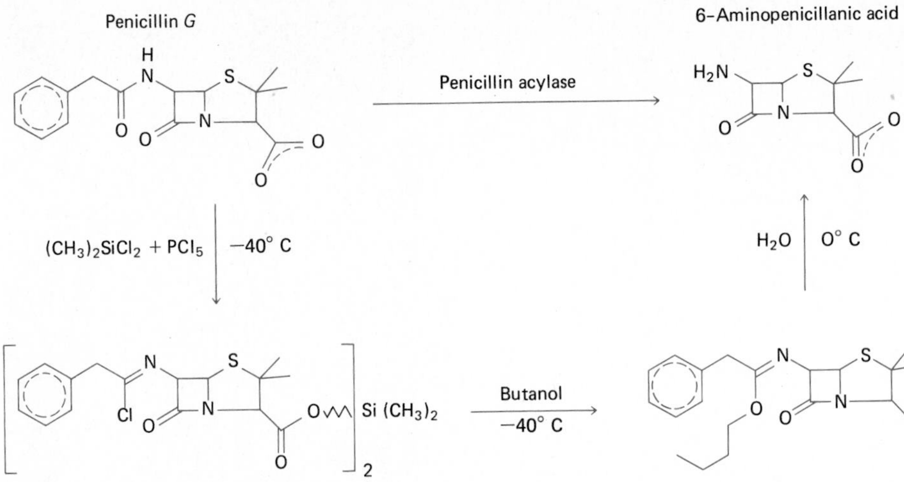

Figure 20.14

The modification of penicillin G to 6-aminopenicillanic acid (6-APA) can be accomplished both chemically and microbiologically. The microbiological conversion has fewer steps and is cheaper. In the production of semisynthetic penicillins, 6-APA forms a chemical nucleus to which side chains can be attached, yielding new antibiotics. In the chemical conversion three steps must be carried out at low temperature and under strictly anhydrous conditions with a number of chemical solvents. The biological process relies on bacteria that make acylases, which are enzymes that cleave away the benzyl group, leaving 6-APA. The fermentation can be carried out in water at 37°C.

at 37°C. The same transformation of penicillin G to form 6-aminopenicillanic acid can also be accomplished chemically in three steps by using various chemical solvents, anhydrous conditions,

Figure 20.15

The conversion of cephalosporin C molecules to 7-aminocephalosporanic acid.

7-Aminocephalosporanic acid

and low temperatures. Similar transformations of the basic penicillin structure can also yield other penicillin derivatives. For example, in 1981 piperacillin was approved as a broad-spectrum antibiotic, and in 1982 azlocillin was also introduced for use against strains that are resistant to earlier-generation penicillins.

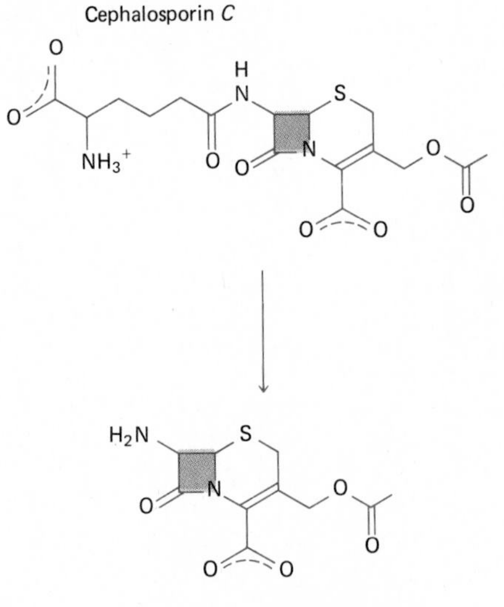

Cephalosporins Similar semisynthetic approaches can be used for manufacturing other antibiotics. For example, cephalosporin C is made as the fermentation product of *Cephalosporium acremonium*, but this form of the antibiotic is not potent enough for clinical use. The cephalosporin C molecule, however, can be transformed by removal of an α-aminoadipic acid side chain to form 7-α-aminocephalosporanic acid, which can be further modified by adding side chains to form clinically useful products with relatively broad spectra of antibacterial action (Figure 20.15). Various side chains can be added as well as removed from both 6-aminopenicillanic and 7-amino-cephalosporanic acids to produce antibiotics with varying spectra of activities and varying degrees of resistance to inactivation by enzymes produced by pathogenic microorganisms. We are now into the so-called third-generation cephalosporins, such as moxalctam, which have been developed to combat bacteria that produce β-lactases.

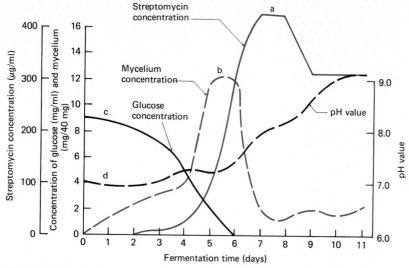

Figure 20.16

The time course of a streptomycin production process: (a) Streptomycin concentration; (b) mycelium concentration; (c) glucose concentration; and (d) pH value.

Streptomycin   Other antibiotics are produced by using different microbial strains, media, incubation conditions, and recovery methods. Streptomycin and various other antibiotics are produced using strains of *Streptomyces griseus*. As in the case of the penicillin fermentation, spores of *S. griseus* are inoculated into a medium to establish a culture with a high mycelial biomass for introduction into an inoculum tank with subsequent use of the mycelial inoculum to initiate the fermentation process in a production tank. The basic medium for the production of streptomycin contains soybean meal as the nitrogen source, glucose as the carbon source, and sodium chloride. The optimum temperature for this fermentation is approximately 28°C, and the maximal rate of streptomycin production is achieved in the pH range 7.6–8.0. High rates of aeration and agitation are required to achieve maximal production of streptomycin. The course of the fermentation lasts approximately 10 days and yields of streptomycin exceed 1 gram per liter.

The classic fermentation process for the production of streptomycin involves three phases (Figure 20.16). During the first phase there is rapid growth of *S. griseus*, with production of mycelial biomass. The proteolytic enzymatic activity of *S. griseus* releases ammonia to the medium from the soybean meal, causing a rise in pH. During this initial fermentation phase there is little production of streptomycin. During the second phase

there is little additional production of mycelia, but the secondary metabolite streptomycin accumulates in the medium. The glucose added in the medium and the ammonia released from the soybean meal are consumed during this phase of the fermentation. The pH remains fairly constant, between 7.6 and 8.0. In the third and final phase of the fermentation, after depletion of carbohydrates from the medium, streptomycin production ceases and the bacterial cells begin to lyse. There is a rapid increase in pH because of the release of ammonia from the lysed cells, and the fermentation process normally is ended by the time the cells begin to lyse.

After completion of the fermentation, the mycelium is separated from the broth by filtration and the streptomycin recovered. Streptomycin is a water-soluble basic substance and is insoluble in most organic solvents. One method of recovery and purification is to adsorb the streptomycin onto activated charcoal and elute with acid alcohol. The antibiotic is then precipitated with acetone and further purified by using column chromatography. Several other chemical procedures can be employed for recovering and purifying streptomycin.

Steroids

The use of microorganisms to carry out biotransformations of **steroids** is very important in the pharmaceutical industry. Steroid hormones regulate various aspects of metabolism in animals, including humans. One such hormone, cortisone, has found to relieve the pain associated with rheumatoid arthritis. Various cortisone derivatives are also useful in alleviating the symptoms associated with allergic and other undesired inflammatory responses of the human body. Additionally, various steroid hormones regulate human sexuality, and some of these steroids are manufactured as oral contraceptives. Thus, there is a great demand for cortisone and related steroid compounds because of these therapeutic values. The physiological properties of a steroid depend on the nature and the exact position of the chemical constituents on the basic steroid ring structure. The chemical synthesis of steroids is very complex because of the requirement to achieve the necessary precision of substituent location.

For example, cortisone can be synthesized chemically from deoxycholic acid (Figure 20.17), but the process requires 37 steps, many of which must be carried out under extreme conditions, with the resulting product costing over $200 per gram. The major difficulty in chemically synthesizing cortisone is the need to introduce an oxygen atom at the number 11 position of the steroid ring, but this can be accomplished by microorganisms. The fungus *Rhizopus arrhizus*, for example, hydroxylates progesterone, forming another steroid with the introduction of oxygen at the number 11 position (Figure 20.18). The fungus *Cunninghamella blakesleeana* similarly can hydroxylate the steroid cortexolone to form hydrocortisone with the introduction of oxygen at the number 11 position. Other transformations of the steroid nucleus carried out by microorga-

nisms include hydrogenations, dehydrogenations, epoxidations, and removal and addition of side chains. The use of such microbial transformations in the formation of cortisone has lowered the original cost over 400-fold, so that in 1980 the price of cortisone in the United States was less than 50 cents per gram, compared to the original $200.

In a typical steroid transformation process the microorganism, such as *Rhizopus nigricans*, is grown in a fermentation tank, using an appropriate growth medium and incubation conditions to achieve a high biomass. In most cases aeration and agitation are employed to achieve rapid growth. After the growth of the microorganisms, the steroid to be transformed is added, and so, for example, progesterone is added to a fermentor containing *R. nigricans* that has been growing for approximately one day, and the steroid is hydroxylated at the number 11 position to form 11-α-hydroxyprogesterone. The product is then recovered by extraction with methylene chloride or various other solvents, purified chromatographically, and recovered by crystallization. A large number of similar transformations are carried out to produce a great variety of steroid derivatives for different medicinal uses (Figure 20.19).

Vaccines

As discussed in the medical microbiology section, the use of **vaccines** is extremely important for preventing various serious diseases. The development and production of these vaccines constitute an important activity of the pharmaceutical industry. The production of vaccines involves growing microorganisms possessing the antigenic properties needed to elicit a primary im-

Figure 20.17

By chemical synthesis the conversion of deoxycholic acid to cortisone requires 37 separate steps.

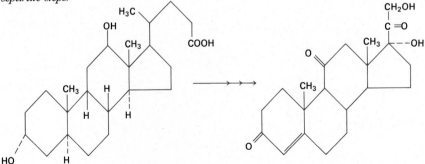

Deoxycholic acid   Cortisone

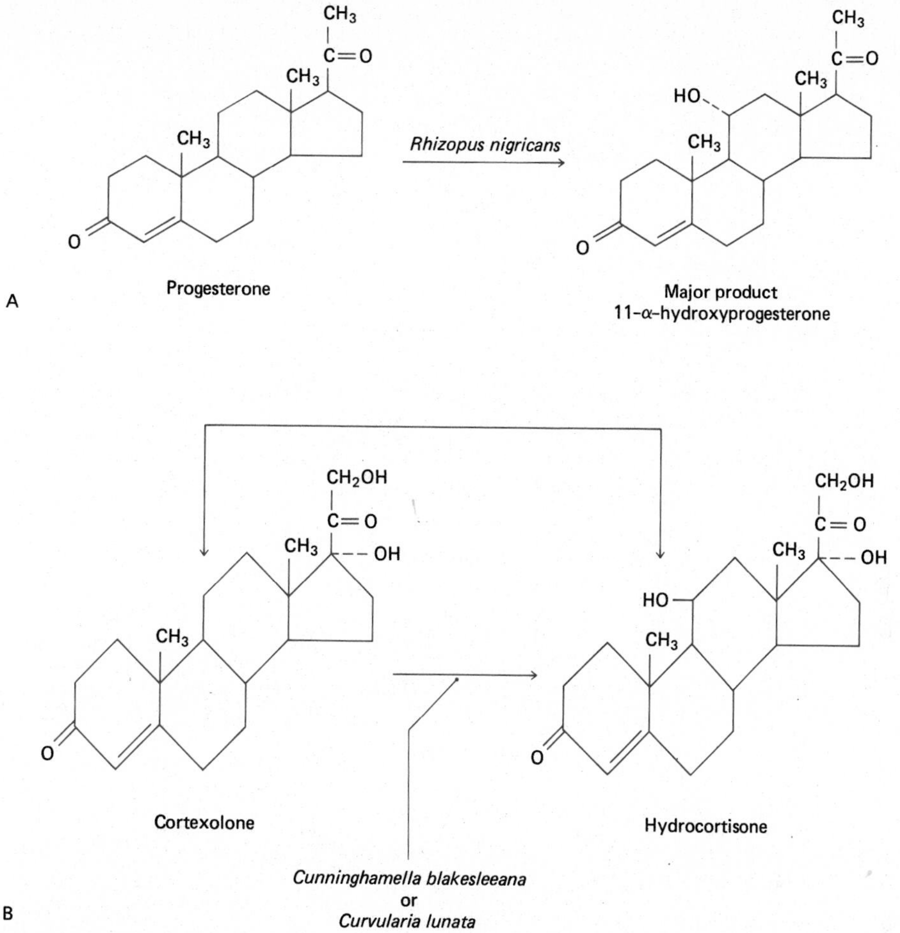

Figure 20.18

The specificity of microbial enzymes allows them to be used for transformations of various cortisone derivatives. The hydroxylation of (A) progesterone by Rhizopus arrhizus *and (B) cortexolone by* Cunninghamella blakesleeana *are reactions that illustrate the use of microorganisms to hydroxylate the cortisone ring to form new cortisone products.*

mune response. Vaccines are produced either by using mutant strains of pathogens or by attenuating or inactivating virulent pathogens without removing the necessary antigens for eliciting the immune response.

For the production of vaccines against viral diseases, strains of the virus often are grown by using embryonated eggs (Figure 20.20). Individuals who are allergic to eggs cannot be given such vaccine preparations. Viral vaccines may also be produced by using tissue culture. The production of vaccines that are effective for preventing diseases caused by bacteria, fungi, and protozoa generally involves growing the microbial strain on an artificial medium, and problems with allergic responses are therefore minimized. In the case of

the commercial production of toxoids, relatively mild treatment of an exotoxin with formalin or heat can be used to destroy the toxic activities without removing the antigenic properties of the molecule. Commercially produced vaccines must be tested and standardized before use. It is critical that the vaccine not contain active forms of a virulent pathogen, lest the vaccine act as a source of transmission of the disease it aims to prevent. Unfortunately, there have been several outbreaks of disease associated with improperly prepared vaccines. High standards of quality control and appropriate safety test procedures can prevent such incidents.

In addition to vaccines, immunoglobulins can be produced by injecting antigens into suitable

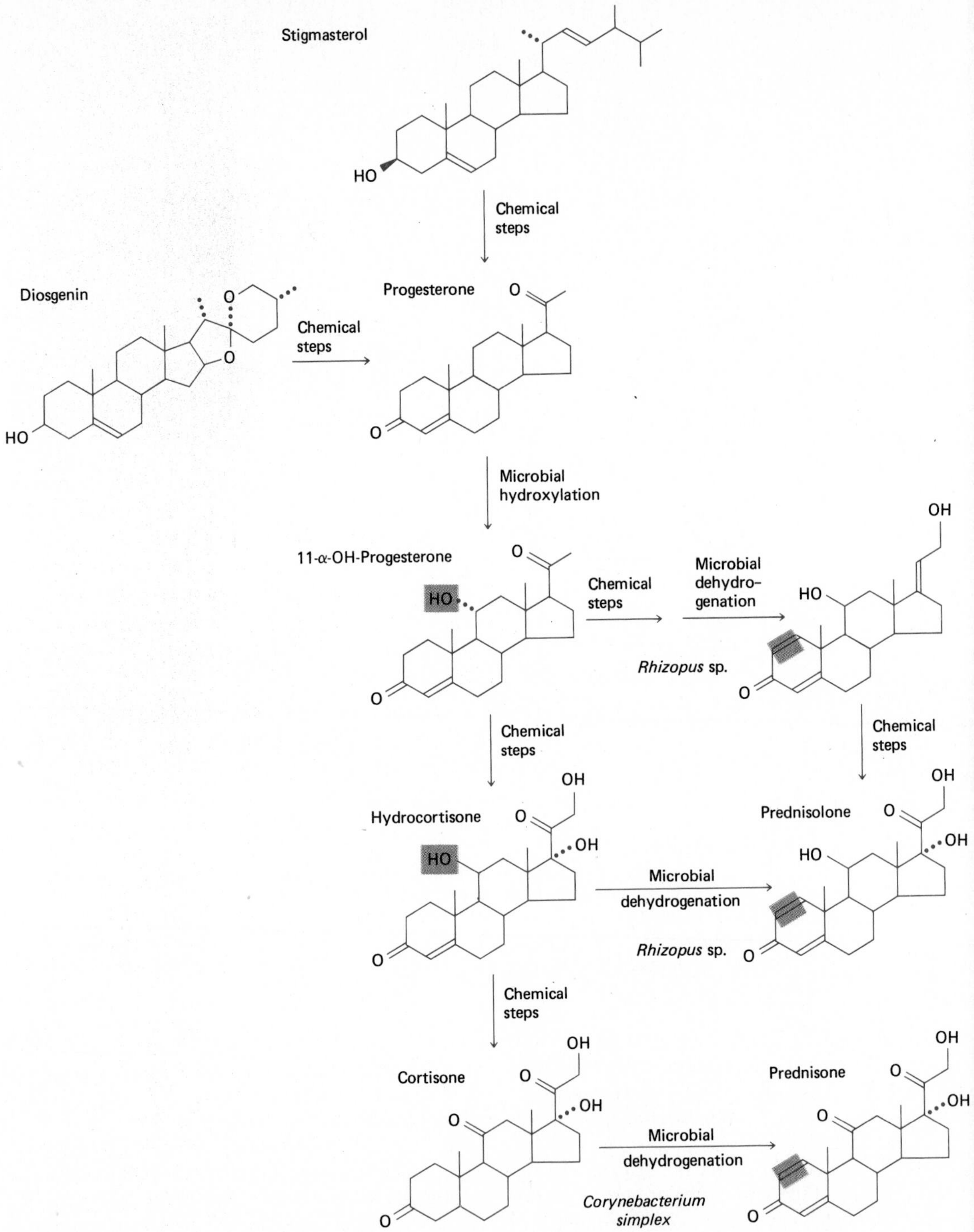

Figure 20.19

Examples of steroid transformations. Note that some steps in these reactions are chemically induced and that others are microbial transformations.

Figure 20.20

This photograph illustrates the large-scale production of vaccines using inoculated embryonated eggs. (Courtesy Wyeth Laboratories, Philadelphia.)

animals, such as horses, and collecting their blood serum after a period of time sufficient to permit antibody formation. Gammaglobulin can be prepared in this manner, and injections of this immunoglobulin are sometimes useful to augment the natural immune defense, particularly if the normal response is impaired.

Vitamins

Various **vitamins**, which are essential animal nu-tritional factors, can be produced by microbial fermentations (Table 20.5). Vitamin B_{12}, for example, can be produced as a by-product of *Streptomyces* antibiotic fermentations (Figure 20.21). A soluble cobalt salt is added to the fermentation reaction as a precursor to vitamin B_{12}. Relatively high amounts of this vitamin accumulate in the medium at concentrations that are not toxic to the *Streptomyces*. Vitamin B_{12} can also be produced commercially by direct fermentation, using

table 20.5

Production of some vitamins by using microorganisms

Vitamin	Culture	Medium	Fermentation conditions	Yield
Riboflavin	*Ashbya gossypii*	Glucose, collagen, soya oil, glucine	6 days, 36°C, aerobic	4.25 g/l
L-Sorbose (in vitamin C synthesis)	*Gluconobacter oxidans* subsp. *suboxidans*	D-Sorbitol, 30% cornsteep	45 h, 30°C, aerobic	70% based on substrate used
5-Ketogluconic acid (in vitamin C synthesis)	*Gluconobacter oxidans* subsp. *suboxidans*	Glucose $CaCO_3$, cornsteep	33 h, 30°C, aerobic	100% based on substrate used
Vitamin B_{12}	*Propionibacterium shermanii*	Glucose, corn-steep, ammonia, cobalt, pH 7.0	30°C, anaerobic (3 days) + aerobic (4 days)	23 mg/l

Figure 20.21

(A) The structure of vitamin B$_{12}$, cyanocobalamin. (B) Photograph showing column chromatography used for the purification of vitamin B$_{12}$. (Courtesy Merck, Sharp and Dohme, Rahway, New Jersey.)

Cobinamide

Nucleotide
(5,6 dimethyl
benziminazole)

A

B

Propionibacterium shermanii or *Pseudomonas denitrificans*, and these are the organisms used today for the production of this antibiotic. *P. shermanii* can be grown in anaerobic culture for three days and aerobic culture for four days to produce streptomycin. The growth medium for streptomycin production by this organism contains glucose, cornsteep liquor, and cobalt chloride. The medium is maintained at pH 7 by using ammonium hydroxide. *Pseudomonas denitrificans* is grown for two days in aerated culture for streptomycin production, using a medium containing sucrose, betaine, glutamic acid, cobalt chloride, 5,6-dimethylbenzimidazole, and salts. Cobalt chloride is added to the medium as a precursor for vitamin B$_{12}$ formation, generally in concentrations of 2–10 ppm.

Riboflavin can also be produced as a fermentation product by using various microorganisms. Riboflavin is a by-product of the acetone butanol fermentation and is produced by various *Clostridium* species. Commercial production of riboflavin by direct fermentation often uses the fungal species *Eremothecium ashbyii* or *Ashbya gossypii*. Riboflavin production using such fungi employs a medium containing glucose and/or corn oil. Corn oil may be added even when glucose is used as the primary growth substrate to increase yields of riboflavin. The fermentation using *Ashbya gossypii* to produce riboflavin is normally carried out

at 26–28°C, pH 6–7.5, for approximately 4–5 days. After growth of the yeast, the cells are recovered and used as a feed supplement for animals to supply needed riboflavin. Various other vitamins can also be produced by fermentation, but relatively low yields often limit their economic potential.

Production of organic acids

Several **organic acids**, including acetic, citric, lactic, itaconic, gluconic, and gibberellic acids, can be produced by microbial fermentation (Table 20.6). The production of acetic acid or vinegar has been discussed in Chapter 19. Acetic acid can also be used for other commercial purposes, for example, as a stop bath in photographic processing, though, for economic reasons most production for such purposes today is accomplished by using chemical syntheses.

Gluconic acid
Gluconic acid also has a variety of commercial uses. Calcium gluconate, for example, is used as a pharmaceutical to supply calcium to the body; ferrous gluconate similarly is used to supply iron in the treatment of anemia; and gluconic acid in dishwasher detergents prevents spotting of glass surfaces due to the precipitation of calcium and

table 20.6

Some organic acids produced by fermentation

Product	Culture	Substrate (yield-%)	Process
Acetic acid	*Acetobacter* sp.	Ethanol (98–99)	Continuous aerated process using an alcoholic solution containing (%): glucose 0.9; ammonium phosphate 0.4; magnesium sulfate 0.1; potassium citrate 0.1; pantothenic acid 0.00005. Extraction by filtration.
Lactic acid	*Lactobacillus delbrueckii*	Milk whey, molasses, pure sugars (90)	10–15% glucose, 5–6 days, 50°C in corrosion-resistant fermentor pH 5.5–6.0 buffered with $CaCO_3$; no aeration, growth factors provided by malt. Extraction by precipitation after heating to 80°C and the addition of chalk (calcium lactate is formed); extraction with solvents; esterification with methanol followed by distillation.
Fumaric acid	*Rhizopus* sp.	Glucose (60)	3 days at 30°C with aeration; pH 5–6 maintained by the addition of NaOH. Extraction by acidification of media and crystallization.
Gluconic acid	*Aspergillus niger*	Glucose and corn-steep liquor (90)	36 h at 30°C with aeration; pH 6.5. Extraction by filtration and purification using cation exchange column.

magnesium salts. Gluconic acid is produced by various bacteria, including *Acetobacter* species, and by several fungi, including *Penicillium* and *Aspergillus* species. *Aspergillus niger*, for example, converts glucose to gluconic acid in a single enzymatic reaction (Figure 20.22). The commercial production of gluconic acid, using *A. niger*, employs a submerged culture process. *A. niger* is initially grown to form a sufficient amount of mycelia, after which the conversion of glucose to gluconic acid, mediated by the fungal enzyme glucose oxidase, is purely an enzymatic reaction. A typical growth medium for the production of gluconic acid contains approximately 25 percent glucose, various salts, calcium carbonate, and a compound containing the element boron. The boron in the medium stabilizes calcium gluconate, maintaining this compound in solution and preventing its precipitation, permitting the use of excess calcium carbonate to neutralize most of the gluconic acid produced, and keeping the pH within acceptable limits. The fermentation is conducted at 30°C with aeration and agitation. Cooling coils are used to control the heat evolved in this oxidative process. The growth of fungal mycelium is limited by the concentration of nitrogen in the medium. The gluconic acid is recovered from the fermentation by addition of calcium hy-

Figure 20.22

The conversion of glucose to gluconic acid by oxidation by the glucose oxidase of Aspergillus niger.

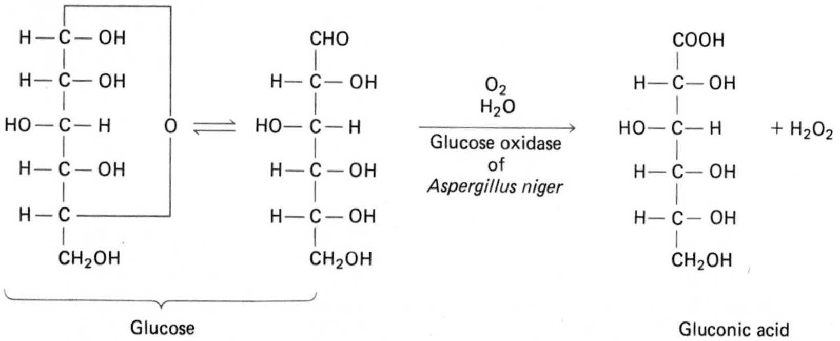

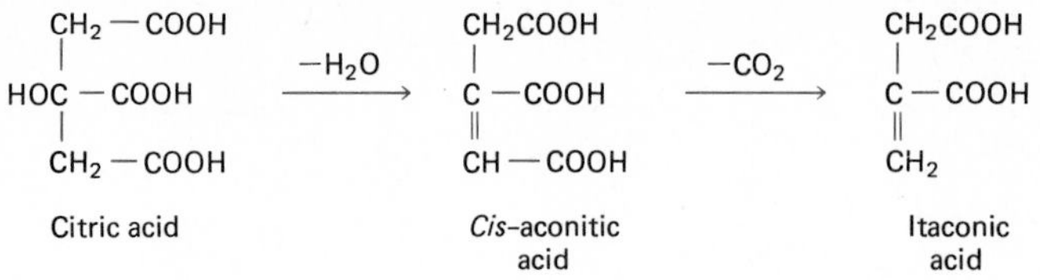

$$\underset{\text{Citric acid}}{\begin{array}{c} CH_2-COOH \\ | \\ HOC-COOH \\ | \\ CH_2-COOH \end{array}} \xrightarrow{-H_2O} \underset{\substack{\textit{Cis}\text{-aconitic} \\ \text{acid}}}{\begin{array}{c} CH_2COOH \\ | \\ C-COOH \\ \| \\ CH-COOH \end{array}} \xrightarrow{-CO_2} \underset{\substack{\text{Itaconic} \\ \text{acid}}}{\begin{array}{c} CH_2COOH \\ | \\ C-COOH \\ \| \\ CH_2 \end{array}}$$

Figure 20.23

The conversion of citric acid to itaconic acid.

droxide to form crystaline calcium gluconate. Free gluconic acid can then be recovered by addition of acid.

Citric acid

Citric acid is also produced by cultures of *Aspergillus niger*. Commercially produced citric acid is used in various ways, including as a food additive, especially in the production of soft drinks, as a metal chelating and sequestering agent, and as a plasticizer. The composition of the fermentation medium is extremely critical for obtaining high yields of citric acid. It is essential to limit the growth of the fungus for high levels of citric acid to accumulate, which can be accomplished by having a deficiency of trace metals or phosphate in the medium. A typical medium for the production of citric acid contains molasses, ammonium nitrate, magnesium sulfate, and potassium phosphate. Acid

is added to achieve a low pH, and some of the metals in the medium are complexed with ferricyanide, removing them from solution, or alternatively, metals are removed using cation exchange resins.

Citric acid is produced commercially by using both aerated submerged culture and stationary tray cultures. The fermentation is conducted at approximately 30°C. In the tray fermentation method spores are blown onto the surface of a shallow medium, the spores float on the surface, and air is blown across the mycelial mat that forms. Providing proper aeration levels is important but is difficult in this tray fermentation technique because it also is necessary to prevent contamination with other fungal spores. The trays are harvested after 7–10 days incubation and the citric acid is recovered. Recovery of citric acid is accomplished by precipitation of calcium citrate from hot neutral aqueous solution and acidification to remove the calcium.

Itaconic acid

Itaconic acid is used as a resin in detergents. The transformation of citric acid by *Aspergillus terreus* can be used for the fermentative production of itaconic acid, although chemical procedures for producing this compound are also available (Figure 20.23). The fermentation process uses a well-aerated molasses-mineral salts medium at very low pH, below pH 2.2. At higher pH values *A. terreus* degrades itaconic acid, and the desired product obviously would not accumulate. As in the case of the citric acid fermentation, low levels of trace metals must be removed to achieve acceptable product yields (Table 20.7). The development of fungal mycelia in this fermentation is intentionally limited, often by using a low inoculum size, in order to produce high accumulations of itaconic acid. Recovery is accomplished by evaporation of the fermentation medium to crystallize the itaconic acid.

Gibberellic acid

Gibberellic acid and related gibberellins are plant hormones and are extensively used as growth-

table 20.7

Effects of the concentrations of some metals in the fermentation medium on itaconic acid production by a mutant of Aspergillus terreus

Element (mg/l)	Yield (% conversion)
Zinc	
0	16
0.5	43
6	50
Copper	
0.5	55
1	52
3	53
6	55
Calcium	
0	9
337	43
2700	59
Iron	
0	57
1	25
2	17
4	17

promoting substances to stimulate plant growth, flowering, and seed germination and to induce the formation of seedless fruit. The commercial production of gibberellins can be used to enhance agricultural productivity. Gibberellic acid is formed by the fungus *Gibberella fujikuroi* (= *Fusarium moniliforme*) and can be produced commercially, using aerated submerged culture. A glucose-mineral salts medium, incubation temperature of approximately 25°C, and slightly acidic pH conditions are employed for the production of gibberellic acid. Production normally takes two to three days, with accumulation of gibberellic acid lagging behind the growth of the fungus.

Lactic acid

Lactic acid has various commercial uses. It is used in foods as a preservative, in leather production for deliming hides, and in the textile industry for fabric treatment. Various forms of lactic acid are also used for other purposes, in resins as poly-lactic acid, in plastics as various derivatives, in electroplating as copper lactate, and in baking powder and animal feed supplements as calcium lactate. The formation of lactic acid in making various fermented dairy products has been discussed in Chapter 19. *Lactobacillus delbrueckii* is widely used in the commercial production of lactic acid, but various other *Lactobacillus, Streptococcus,* and *Leuconostoc* species also are of industrial importance for the production of this compound.

A typical medium for the production of lactic acid contains 10–15 percent dextrose or other fermentable sugar, 10 percent calcium carbonate to neutralize the lactic acid formed, and ammonium phosphate and trace amounts of other nitrogen sources. Corn sugar, beet molasses, potato starch, and whey are often used as sources of carbohydrates for this fermentation. A typical production process for lactic acid uses an incubation temperature of 45–50°C and pH of 5.5–6.5. The fermentor is agitated to suspend the calcium carbonate but is not aerated because this is an anaerobic process. The fermentation is normally completed within five to seven days with approximately 90 percent of the sugar converted to lactic acid. After the fermentation, calcium carbonate is added to raise the pH to 10, and the medium is heated and filtered. This procedure kills the bacteria, coagulates proteins, removes excess calcium carbonate, and decomposes residual carbohydrates.

Various additional procedures are then used to recover and purify the lactic acid. In one proce-dure the lactic acid is recovered by evaporating the medium to concentrate and crystallize calcium lactate. Then the lactic acid is purified by acid treatment to remove the calcium and by passage through an activated charcoal filter to remove organic impurities. In another procedure the lactic acid is extracted with an organic solvent using continuous counter-current flow. In still another procedure the lactic acid is methylated and the methyl ester recovered by distillation, followed by hydrolysis of the ester in boiling water to yield lactic acid. There are still other alternative approaches used commercially for the extraction and recovery of lactic acid. The choice of recovery methods depends on the cost and the purity of lactic acid required for the particular application. The recovery of lactic acid of high enough purity for some applications is difficult to achieve, and the cost of recovery has forced the replacement of lactic acid with alternative chemicals for some commercial uses.

Production of amino acids

Microbial production of the **amino acids** lysine and glutamic acid presently accounts for over $1 billion in annual worldwide sales (Figure 20.24). Animals require various amino acids in their diets. Lysine and methionine are essential amino acids but are not present in sufficient concentrations in grains to meet animal nutritional needs. Lysine produced by microbial fermentation and methionine produced synthetically are used as animal feed supplements and as additives in cereals. Glutamic acid is principally made for use as monosodium glutamate (MSG), a widely used ingredient in soup production. The flavoring industry in the United States consumed more than 30,000 tons of MSG in 1980, some of which was imported from Japan, a major producer of amino acids by fermentation, as well as some from Taiwan and South Korea.

In the microbial production of amino acids, only the desired L-isomer is formed, and chemical syntheses produce a racemic mixture that requires costly separation procedures to remove the biologically inactive D-isomer half of the mixture. The major problem in using microbial fermentation for commercial production of amino acids is overcoming the natural microbial regulatory control mechanisms that limit the amount of amino acid produced and released from the cells. Commercial amino acid production processes have successfully overcome these restrictions, and in

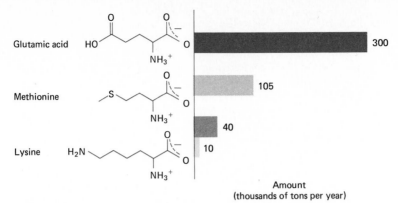

Figure 20.24

Amino acids are commercially important products. Worldwide sales of amino acids were $1.7 billion in 1980. Glutamic acid, methionine, and lysine are the ones now made in the largest quantities. Methionine and lysine are nutritionally essential amino acids used as animal feed additives. Glutamic acid and 80 percent of the lysine are made by fermentation. Methionine is manufactured by chemical synthesis. All amino acids have two isomers, only one of which can participate in biological reactions. Fermentation yields only the biologically active isomer. In chemical synthesis half of the yield is the inactive one. This specificity makes the biological means more efficient, but it has not always been possible to exploit it. With increased understanding of cellular metabolism, all industrially valuable amino acids may soon be made by fermentation.

the future genetically engineered strains with defective control mechanisms and membranes will undoubtedly permit the economic production of a variety of amino acids by microbial fermentation.

Lysine

The direct production of L-lysine from carbohydrates uses a homoserine-requiring auxotroph of *Corynebacterium glutamicum*. The blocking of homoserine synthesis at homoserine dehydrogenase results in feedback inhibition leading to the accumulation of lysine (Figure 20.25A). Cane molasses is generally used as the substrate, and the pH is maintained near neutrality by adding ammonia or urea. As the sugar is metabolized, lysine accumulates in the growth medium. Through using the homoserine-requiring auxotroph about 50 g/l of lysine can be produced in two to three days (Fig 20.25B).

Glutamic acid

L-Glutamic acid and MSG can be produced by direct fermentation, using strains of *Brevibacterium, Arthrobacter*, and *Corynebacterium*. Cultures of *Corynebacterium glutamicum* and *Brevibacterium flavum* are widely used for the large-scale production of MSG. The fermentation process employs a glucose-mineral salts medium and periodic additions of urea as a nitrogen source during the course of the fermentation; the pH is maintained at 6–8, the temperature is about 30°C, and the medium is well aerated. The difficulty in the production of glutamic acid, as well as other amino acids, by direct fermentation is getting the cells to secrete sufficient quantities of the amino acid to permit commercial production. There are several methods for inducing leaky membranes that permit excretion of the amino acid product from the cell. One approach is to grow *C. glutamicum* in a medium with suboptimal concentrations of biotin (Figure 20.26). Without an adequate supply of biotin, the cells form membranes that are deficient in phospholipids, and the glutamic acid is secreted through these leaky membranes. Another approach is to add fatty acids or surface active agents (detergents) to disrupt the membranes and release the glutamic acid from the cells. Still another way of causing the cell to excrete amino acids is to add penicillin to the medium during the log phase of growth, causing the bacteria to become leaky and release glutamic

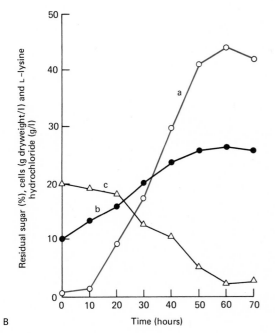

Aspartate → Aspartyl phosphate → Aspartic semialdehyde →∦ L–Homoserine → L–Threonine

No. 901

Aspartic semialdehyde → Dihydrodipicolinate → Diaminopimelate → L–Lysine

L–Homoserine → L–Methionine

L–Threonine → α–Oxobutyrate → L–Isoleucine

A

B

Figure 20.25

The production of L-lysine by Corynebacterium
glutamicum. *(A) The regulation of lysine biosynthesis
by* C. glutamicum. *Dotted lines = operation of
feedback inhibition; dashed lines = repression
of enzyme synthesis. (B) A time-course graph
for this fermentation: line a = changes in the
concentration of L-lysine; line b = changes in the
concentration of cells; and line c = changes in the
concentration of residual sugar.*

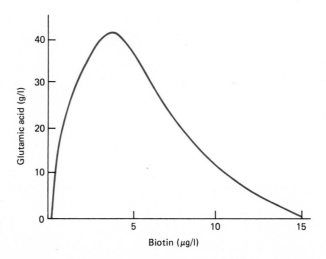

Figure 20.26

*The effect of biotin concentration on the yield
of glutamic acid.*

table 20.8

Important uses for enzymes produced by microorganisms

Industry	Application	Enzyme	Source
Analytical	Sugar determination	Glucose oxidase	Fungi
	Glycogen determination	Galactose oxidase	Fungi
	Uric acid determination	Urate oxidase	Fungi
Baking	Bread baking	Amylase	Fungi
		Protease	Fungi
Brewing	Mashing—making beer	Amylase	Bacteria
		Glucamylase	Fungi
Carbonated beverages	Oxygen removal	Glucose oxidase	Fungi
Cereals	Breakfast foods	Amylase	Fungi
Chocolate, cocoa	Syrups	Amylase	Fungi, bacteria
Coffee	Coffee bean fermentation	Pectinase	Fungi
Confectionery	Soft-center candies	Invertase	Bacteria, fungi
Dairy	Cheese production	Rennin	Fungi
Dry cleaning	Spot removal	Protease, amylase	Bacteria, fungi
Eggs, dried	Glucose removal	Glucose oxidase	Fungi
Fruit juices	Clarification	Pectinases	Fungi
	Oxygen removal	Glucose oxidase	Fungi
	Debittering of citrus	Narigninase	Fungi
Laundry	Spot removal	Protease, amylase	Bacteria
	Cold-soluble laundry starch	Amylase	Bacteria
Leather	Bating	Protease	Bacteria, fungi
Meat	Meat tenderizing	Protease	Fungi, bacteria
Mayonnaise and salad dressings	Oxygen removal	Glucose oxidase	Fungi
Paper	Starch modification for paper coating	Amylase	Bacteria
Pharmaceutical and clinical	Digestive aids	Amylase	Fungi, bacteria
		Protease	Fungi, bacteria
		Lipase	Fungi
		Cellulase	Fungi
	Wound debridement (tissue removal)	Streptokinase-streptodornase Protease	Bacteria
Photographic	Recovery of silver from spent film	Protease	Bacteria
Starch and syrup	Corn syrups	Amylase, dextrinase	Fungi
		Glucose isomerase	Fungi, bacteria
	Production of glucose	Glucamylase, amylase	Fungi, bacteria
Textile	Desizing of fabrics	Amylase	Bacteria
		Protease	
Wine	Clarification	Pectinases	Fungi

acid to the surrounding medium. Adjusting the pH and adding sodium chloride can be used to convert glutamic acid to the desired MSG.

Production of enzymes

Enzymes have a variety of commercial applications, some of which are shown on Table 20.8.

Enzymes for industrial applications can be obtained from plants, animals, and microorganisms. Microbial production of useful industrial enzymes is advantageous because of the large number of different enzymes and the virtually unlimited supply of enzymes that can be produced by microorganisms. A generalized scheme for the microbial production of commercial enzymes is shown in Figure 20.27. Enzymes produced for in-

Figure 20.27

Generalized schemes for the production of (A) crude (extracellular) enzymes and (B) purified enzymes (endocellular).

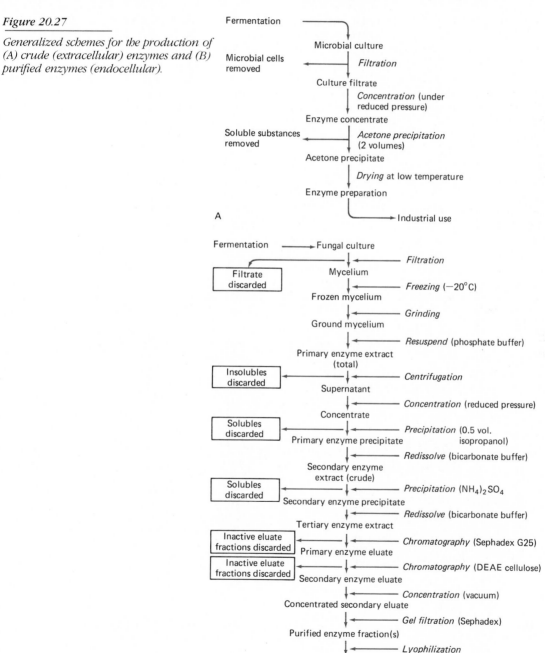

dustrial processes include proteases, amylases, glucose isomerase, glucose oxidase, rennin, pectinases, and lipases. The four extensively produced microbial enzymes are protease, glucamylase, alpha-amylase, and glucose isomerase (Figure 20.28).

Proteases

Proteases represent a class of enzymes that attack the peptide bonds of protein molecules, forming small peptides. Proteases produced by different bacterial species are used for different industrial purposes. The largest commercial application of bacterial alkaline proteases is in the laundry industry, principally in modern detergent formulations. In the cleaning industry proteases are used as spot removers in dry cleaning, as presoak treatments in laundering, and in laundry detergents. The action of the enzyme degrades various proteinaceous materials, such as milk and eggs, forming small peptide fragments that can be washed out readily. In dry cleaning, proteases are effective spot removers and are even effective in removing blood spots. These protease enzymes are relatively heat stable and are able to remain active in warm-hot water long enough for them to degrade the proteinaceous materials contaminating the fabric being washed. When used as a presoak, protease enzymes have sufficient time to act to degrade insoluble proteinaceous materials staining the fabric. Proteases for detergents are largely produced by *Bacillus licheniformis.*

Another major use of microbial proteases is in the baking industry. Protease enzymes are used to alter the properties of the gluten proteins of flour. Fungal protease is added in the manufacture of most commercial bread in the United States to reduce the mixing time and improve the quality of the loaf. Either fungal or bacterial protease is used in the manufacture of crackers, biscuits, and cookies. Fungal proteases are principally obtained from *Aspergillus* species, and bacterial proteases are primarily produced, using *Bacillus* species. Fungal proteases have a wider range of pH tolerance than bacterial proteases. Both dry surface and submerged aerated culture methods are used for growing protease-producing bacterial and fungal strains.

Proteases are also used for various other purposes, including as digestive aids. Adding protease enzymes to beef can soften or tenderize the meat, making the meat more edible. A typical meat tenderizer contains 5 percent fungal protease as well as monosodium glutamate and other ingredients. In the leather industry microbial proteases are used for bating of hides, which improves the quality by softening the leather. In the textile industry protease enzymes are used for removing proteinaceous sizing and freeing silk fibers from the proteinaceous material in which they are embedded.

Amylases

Amylases are used for the preparation of sizing agents and removal of starch sizing from woven cloth, preparation of starch sizing pastes for use in paper coatings, liquefaction of heavy starch pastes that form during heating steps in the manufacture of corn and chocolate syrups, production of bread, and removal of food spots in the dry cleaning industry, where amylase functions in conjunction with protease enzymes. Amylases are

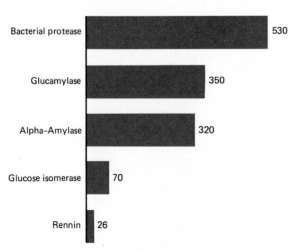

Figure 20.28

Commercial sales of enzymes were $300 million in 1980. The five enzymes illustrated here are produced by microbiological methods. Bacterial protease, which degrades proteins by cleaving peptide bonds, is used as a cleaning aid. Glucamylase, alpha-amylase, and glucose isomerase convert starch into high-fructose corn syrup sweetener, which is replacing sucrose in soft drinks. Amylases break down starch to yield glucose; glucose isomerase converts glucose into fructose. Renin, employed in making cheese, can be extracted from the fourth stomach of a cow or can be made biologically.

also sometimes used to replace or augment malt for starch hyrolysis in the brewing industry, as in the production of low-calorie beers.

There are various types of amylases including: (1) alpha-amylase, which converts starch to oligosaccharides and maltose; (2) beta-amylase, which converts starch to maltose and dextrins; and (3) glucamylase, which converts starch to glucose. All three enzymes are used in the production of syrup and dextrose manufacture from starch. Fungal production of amylases uses *Aspergillus* species. For example, *A. oryzae* is used to produce amylases from wheat bran in stationary culture, and *A. niger* is used for producing amylases in aerated submerged culture, using a starch-mineral salts medium. *Bacillus subtilis* and *B. diastaticus* are used for the commercial production of bacterial amylases. Both aerated submerged and surface culture methods may be employed for the production of bacterial amylase enzymes. There are several ways of recovering amylase enzymes from bacterial and fungal cultures. In the case of dry culture production, the amylases can be extracted with water from the fungal mycelium and wheat bran medium. When liquid culture is employed, evaporation can be used to concentrate the amylase enzymes. The amylases can then be precipitated from solution by the addition of ethanol, acetone, or ammonium sulfate.

The conversion of starch to a high-fructose corn syrup sweetener, using microbial enzymes, represents an economically significant and relatively new industrial process, producing over two million tons of high-grade sweetener per year. This sweetener, produced in a three-step process using the enzymes alpha-amylase, glucamylase, and glucose isomerase, is rapidly replacing sucrose as the primary sweetener in soft drinks. In the final step glucose (approximately 50 percent) is converted into fructose by the enzyme glucose isomerase. The use of mutation-screening methods combined with genetic recombination techniques has permitted the development of strains of *Bacillus subtilis* with greatly enhanced abilities to produce high yields of alpha-amylase. The development of such strains markedly increases the economic feasibility of producing sweeteners and similar products using microbial enzymes.

Other enzymes

Various other microbial enzymes are produced for industrial applications. As indicated in Chapter 19, **rennin** is used in the production of cheese, and *Mucor pussilus* or *M. meihei* can be used for the commercial production of rennin for curdling milk in cheese production. Fungal pectinase enzymes are used in the clarification of fruit juices. Glucose oxidase, produced by fungi, is used for removing glucose from eggs prior to drying, since powdered dried eggs brown because of the chemical reaction of proteins with glucose, and removing the glucose stabilizes and prevents deterioration of the dried egg product. Glucose oxidase is also used to remove oxygen from various products, such as soft drinks, mayonnaise, and salad dressings, preventing oxidative color and flavor changes.

Microbial enzymes may also be used for the production of synthetic polymers. For example, the plastics industry now uses chemical methods for producing alkene oxides used in the production of plastics. It is possible to synthesize alkene oxides by using microbial enzymes (Figure 20.29), and the use of genetically engineered microbial strains can make such synthesis commercially feasible. The synthesis of alkene oxides from alkenes can be accomplished by the sequential action of three enzymes: pyranose-2-oxidase from the fungus *Oudmansiella mucida*, a haloperoxidase from the fungus *Caldariomyces*, and an epoxidase from a *Flavobacterium* species. The production of propylene oxide, using microbial enzymes, could revolutionize the plastics industry, altering the cost of producing this widely used material.

Production of solvents

Several organic solvents can be produced by using microbial fermentation, but **organic solvents** are produced today by chemical synthesis. For example, although **ethanol** is produced by fermentation for beverages, industrial alcohol is produced by chemical synthesis. Fermentation processes have been important in the past in the industrial production of organic solvents and, as economic conditions change, will likely be used again in the future. The process for producing **acetone** and **butanol** was discovered by Chaim Weizmann in England. During World War I the microbial production of acetone was very important for the production of the explosive cordite, and microbially produced butanol was converted to butadiene and used for making synthetic rubber. After the war the demand for acetone declined, but the need for *n*-butanol increased. Butanol is used in brake fluids, urea-formaldehyde resins, and in producing protective coatings, such as lacquers used on automobiles.

The microbial production of acetone and butanol uses anaerobic *Clostridium* species. The fermentation discovered by Weizmann was based on

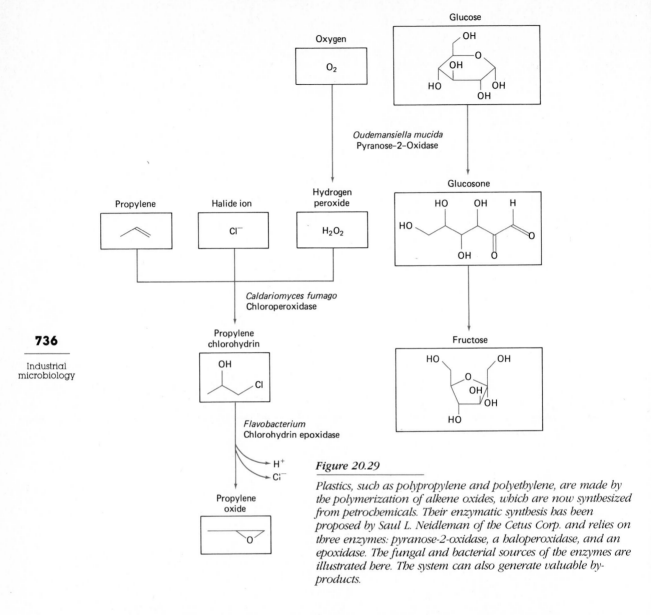

Figure 20.29

Plastics, such as polypropylene and polyethylene, are made by the polymerization of alkene oxides, which are now synthesized from petrochemicals. Their enzymatic synthesis has been proposed by Saul L. Neidleman of the Cetus Corp. and relies on three enzymes: pyranose-2-oxidase, a haloperoxidase, and an epoxidase. The fungal and bacterial sources of the enzymes are illustrated here. The system can also generate valuable by-products.

the conversion of starch to acetone by *C. aceto-butylicum*. Other species, such as *C. saccharoa-cetobutylicum*, are able to convert the carbohydrates in molasses to acetone and butanol. These *Clostridium* species synthesize butyric and acetic acids that are then converted to butanol and acetone. The yields of these neutral solvents are typically low, approximately 2 percent by weight of the fermentation broth, representing a 30 percent conversion of carbohydrates to neutral solvents. The accumulation of higher concentrations of these solvents is limited by the toxicities of the compounds. The solvents produced by fermentation are recovered by distillation. In South Africa, because of the scarcity of petroleum and the abun-

dance of plant residues as substrates for fermentation, butanol and acetone are produced by microbial fermentation employing this process. Elsewhere, these solvents are produced from petroleum; however, as the costs of petrochemicals increase, the feasibility of more extensive use of fermentation for producing *n*-butanol is enhanced.

Today, **glycerol** production is also primarily by chemical synthesis, based on the saponification of fats and the chemical oxidation of propane and propylene. Glycerol is used as: (1) a solvent in flavorings and food coloring agents; (2) a lubricant in the manufacture of pet food, candy, cake icings, toothpaste, glue, cellophane, and other products; and (3) an emollient and demulcent in pharmaceuticals and cosmetics. Glycerol is also used in the production of explosives and propellants. The production of glycerol by fermentation by Germany was an important factor during World War I because the glycerol was used for the production of munitions. The microbial production of glycerol is accomplished by adding sodium sulfite to a yeast-ethanol fermentation process. The sodium sulfite reacts with the carbon dioxide to produce sodium bisulfite, which prevents the reduction of acetaldehyde to ethanol. This blockage results in a divergence of the metabolic pathway with the accumulation of glycerol. Glycerol can be produced by using yeasts, such as *Saccharomyces cerevisiae*, and bacteria, such as *Bacillus subtilis*. The microbial production of glycerol may be renewed as a result of the finding that some yeasts can synthesize glycerol without the need for adding sodium sulfite, thus making the process economically competitive with chemical methods of glycerol production.

Production of fuels

Limited petroleum resources are forcing many industrialized nations to seek alternative fuel resources. **Microbial production of synthetic fuels** has the potential for helping meet world energy demands. Useful fuels produced by microorganisms include ethanol, methane, hydrogen, and hydrocarbons. The use of microorganisms to produce commercially valuable fuels depends on finding the right strains of microorganisms that are able to produce the desired fuel efficiently and having an inexpensive supply of substrates available for the fermentation process. It is obviously imperative that the production of synthetic fuels not consume more natural fuel resources than are produced. Microbial production of fuels can be a particularly attractive process when waste materials, such as sewage and municipal garbage, are used as the fermentation substrate.

Ethanol
The microbial production of **ethanol** has become an important source of a valuable fuel, particularly in regions of the world having abundant resource supplies of plant residues. Brazil produces and uses large amounts of ethanol as an automotive fuel and plans to replace gasoline with ethanol by the 1990s. **Gasohol**, a 9:1 blend of gasoline and ethanol, has become a popular fuel in the Midwestern United States. At present about 100 million gallons of ethanol per year are used as a fuel in the United States, but 12 billion gallons of ethanol per year would be required to completely replace gasoline use in the United States. There are several major limitations to the successful production of sufficient quantities of ethanol to serve as a major fuel source, including: (1) ethanol is relatively toxic to microorganisms, and therefore, only limited concentrations of ethanol can accumulate in a fermentation process; (2) carbohydrate substrates normally used for the production of ethanol in the food industry are relatively expensive, making the cost of fuel produced by fermentation high; and (3) distillation to recover ethanol requires a substantial energy input, reducing the net gain of fuel as an energy resource produced in this process.

Despite the problems with the economics of ethanol production, several processes can be employed for the commercial production of ethanol as a fuel (Figure 20.30). The finding that the bacterium *Zymomonas mobilis* ferments carbohydrates, forming alcohol twice as fast as yeasts, appears to represent a significant advance in the search for a microbial strain for producing ethanol as a fuel. *Thermoanaerobacter ethanolicus*, a thermophilic bacterium, may be even more efficient than the organisms currently used for the fermentative production of ethanol. Corn sugar and plant starches are currently used as substrates for the production of ethanol, but the prices of these substrates vary greatly, depending on plant harvests, and they also are needed as food resources. Biomass produced by growing photosynthetic microorganisms is a potential source of an inexpensive substrate for ethanol production, but cellulose from wood and other plant materials is probably the most promising substrate. A two-step fermentation process can be used for the conversion of cellulose to ethanol in which cellulose is first converted to sugars, generally by *Clostridium* species, and the carbohydrates are then converted to ethanol by yeasts, *Zymomonas* or *Thermoanaerobacter* species. It is very likely that genetic engineering can create a microbial species that can efficiently convert cellulose directly to ethanol, which will also tolerate high concentrations of ethanol. Such an organism should permit the commercial production of ethanol as a fuel.

Methane
Methane produced by methanogenic bacteria is another important potential energy source. Methane can be used for the generation of mechanical, electrical, and heat energy. Large amounts of

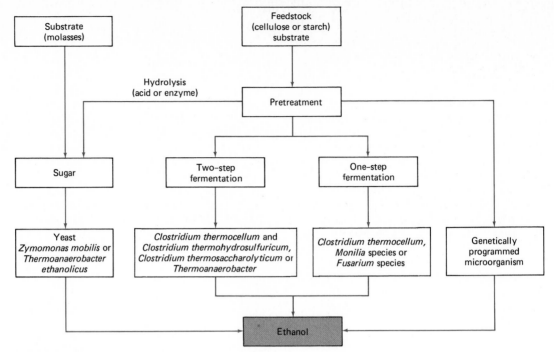

Figure 20.30

The biological production of ethanol as a fuel is accomplished by much the same method as the production of alcoholic beverages. The substrate is either crude sugar (from sugar cane or beet molasses) or starch (from corn, wheat, rye, or cassava) that has been converted into sugar. Yeasts are usually the fermenting organism, but bacteria, such as Zymomonas mobilis *and* Thermoanaerobacter ethanolicus, *may prove to be more efficient. Because the prices of crude sugar and starch vary so much and they are needed as foodstuffs, cellulose and related polymers of wood, which are abundant, renewable, and cheaper, are also used as substrate. The cellulose can be fermented in either one or two steps. In the two-step process one microorganism breaks down the cellulose into its component sugar units, which are subsequently fermented into ethanol by another microorganism. In the one-step process a single microorganism both breaks down the cellulose and ferments the resulting sugar into ethanol. Microorganisms that have this capability are rare, and so another strategy is to genetically program a yeast or a bacterium so that it converts cellulose into ethanol in one operation.*

methane can be produced by anaerobic decomposition of waste materials. Many sewage treatment plants are able to meet their own energy needs from the production of methane in their anaerobic sludge digesters. Excess methane produced in such facilities is sufficient to supply power for some municipalities. Efficient generation of methane can be achieved by using algal biomass grown in pond cultures, sewage sludge, municipal refuse, plant residue, and animal waste. Methanogenic bacteria are members of the archaebacteria; they are obligate anaerobes and produce methane from the reduction of acetate and/or carbon dioxide. The production of methane generally requires a mixed microbial community, with some bacterial populations converting the available organic carbon into low-molecular-weight fatty acids that are substrates for methanogens.

Other fuels

Hydrogen is another potential fuel source that can be produced by microorganisms. Hydrogen is not currently employed as a major fuel but could develop into one if an efficient production process were found. Photosynthetic microorganisms are capable of evolving hydrogen and growing by using solar energy. Such organisms are able to convert solar energy into a fuel that can be stored and used to power electrical generators and provide a conventional source of energy. Such photoproduction of hydrogen is an intriguing, but as yet far from practical, idea.

Various microorganisms, including algae, also are capable of producing higher-molecular-weight hydrocarbons, but the potential use of such hydrocarbons as a fuel source has received relatively little attention. Although physicochemical processes are crucial, microorganisms are believed to play a role in the formation of petroleum deposits. A more thorough understanding of the basic mechanisms of microbial hydrocarbon formation and the formation of petroleum deposits should permit the development of genetically engineered microorganisms and fermentation processes to artificially produce synthetic sources of petroleum hydrocarbons.

Microbially enhanced recovery of mineral resources

As the technological level of an ever-increasing world population rises, so does the need for industrially important minerals. As the easily accessible high-grade deposits of ores are depleted, it becomes increasingly important to find innovative and economical methods to recover such metals from lower-grade deposits, which for technical or economic reasons have not been used. Microorganisms can play an important role in recovering valuable minerals from low-grade ores, meeting some of the needs of industrialized society. The use of microorganisms in such recovery processes employs microbial metabolic activities to gain access to, rather than to actually produce, desired products.

Bioleaching of metals

Microbial mining recovers metals from ores with low metal content that are not suitable for direct smelting by the process of **bioleaching** (Figure 20.31). It is possible to extract the metal from low-grade sulfide or sulfide-containing ore economically by using the metabolic activities of thiobacilli, particularly *Thiobacillus ferrooxidans*. Under optimal conditions in the laboratory, as much as 97 percent of the copper in low-grade ores has been recovered by bioleaching, but such high yields are seldom achieved in actual mining operations. Even a 50–70 percent recovery of copper by bioleaching, from an ore that would otherwise be completely unproductive, would be an important achievement. The process is currently applied on a commercial scale to low-grade copper and uranium ores and laboratory-scale experiments indicate that the process also has promise for the recovery of nickel, zinc, cobalt, tin, cadmium, molybdenum, lead, antimony, arsenic, and selenium from their low-grade sulfide-containing ores. The leaching process can also be used for separation of the insoluble $PbSO_4$ from other metals that occur in the same ore.

$MS + 2O_2 \rightarrow MSO_4$ represents the equation for the general process carried out by *Thiobacillus ferrooxidans* and related species, where M represents a divalent metal. Because the metal sulfide is insoluble and the metal sulfate is usually water soluble, this transformation produces a readily leachable form of the metal. *Thiobacillus ferrooxidans* is a chemolithotrophic bacterium that derives energy through oxidation of either a reduced sulfur compound or ferrous iron. It exerts its bioleaching action by oxidizing the metal sulfide being recovered directly, converting S^{2-} to SO_4^{2-}, and/or indirectly by oxidizing the ferrous iron content of the ore to ferric iron. The ferric iron, in turn, chemically oxidizes the metal to be recovered to a soluble form that can be leached from the ore.

If the ore formation is sufficiently porous and overlays a water-impermeable strata, it is possible to leach the ore *in situ* without first mining the ore. An appropriate pattern of boreholes is established, with some of the holes used for the injection of the leaching liquor and others used for the recovery of the leachate (Figure 20.32). More frequently, though, this bioleaching process is accomplished after the ore is mined, broken up, and piled in heaps on a water-impermeable formation or on a specially constructed apron (Figure 20.33). Water is then pumped to the top of the ore heap and trickles down through the ore to the apron. A continuous reactor-type leaching operation for recovery of copper from its low-grade sulfide ore is shown in Figure 20.34. The leaching water and ore usually supply enough dissolved mineral nutrients to satisfy the needs of *Thiobacillus ferroxidans*, but in some cases mineral nutrients, such as ammonia and phosphate, must be added. In most of these bioleaching operations, the leached metal is then extracted with an organic solvent and subsequently removed from the solvent by stripping. Both the leaching liquor and the solvent are recycled.

Copper, which is in high demand for the electrical industry, is generally in short supply. A typical low-grade copper ore contains 0.1–0.4 percent copper. The "pregnant" leaching solution may contain 1–3 g of copper per liter. In copper leach-

A

C

B

Figure 20.31

Bioleaching process for mineral recovery in a leaching dump of a large open-pit copper mine in Bingham Canyon, Utah. Very high numbers of bacteria are involved, more than a million per gram of ore. They are mainly members of the genus Thiobacillus *and assist in the leaching operation by converting the iron in various compounds from the ferrous to the ferric form, which is an effective oxidizing agent. The bacteria oxidize pyrite to form sulfuric acid, to maintain the high acidity of the leaching solution, and to oxidize the insoluble copper-containing sulfide minerals in order to produce soluble copper sulfate, which migrates to the bottom of the dump where it collects in catch basins. The solution is periodically pumped out to facilities where the copper is recovered. These photographs show large-scale bacterial leaching operations. (A) Aerial overview of mining operation at Bingham Canyon, Utah. (B) The leaching solution is sprayed onto the field and recycled. (C) The leachate drains into a catch basin; the copper is recovered from the basin. (Courtesy K. C. Hochstetler, Salt Lake City.)*

Discovery process

As early as 1000 B.C., copper was recovered from the drainage water of mines in the Mediterranean Basin, but it was not until just 25 years ago that the role of bacteria in this traditional method of mineral extraction was realized. Ten percent of the U.S. copper production today is the result of the deliberate exploitation of bacteria. In 1957 the relationship between the presence of *Thiobacillus ferrooxidans* and the dissolution of metals in copper-leaching operations was recognized. But even that short a time ago, the need for this biomining was not apparent, so the technology for exploiting bacterial leaching for the recovery of metals was not developed. The development of such technology is as much a function of economic need as it is a matter of scientific capability. Now that energy is so costly and we have realized that the supply of accessible, high-grade metal ores is limited, there is a need to extend the study of bacterial leaching beyond those few investigators who have been examining these organisms and the reactions they carry out. Increasing the rate of bacterial oxidation reactions, through techniques such as genetic engineering, would be an important improvement; the slowness of the biological process is one of its main drawbacks compared to chemical extraction methods. The greatest potential application of bacterial leaching for mineral recovery lies in the controlled treatment of vast quantities of solid materials, such as discarded waste rock, overburden, and tailings, which contain small quantities of widely dispersed valuable metals. For example, in the United States there is some 7 billion tons of sulfide deposit material, with an average nickel content of 0.2 percent, worth approximately $60 billion. This resource is not currently being exploited because of the inefficiency of current mining and extractive technologies and also because of possible detrimental effects on the environment. Advances in biomining technology would make it possible to recover this nickel and along with it some 400 million pounds of cobalt found in the same sulfide-rich material. Such a process must emphasize low capital investment, greater recovery of metal, and minimal environmental damage.

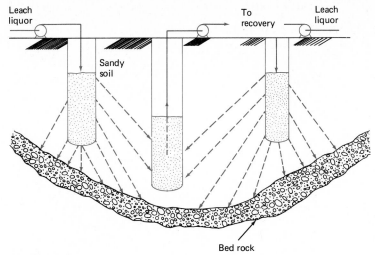

Figure 20.32

The hole-to-hole bioleaching process is used for low-grade ores. The process employs Thiobacillus *species and is practiced where the ore is relatively porous and overlies an impermeable bedrock. Leaching is from multiple injection wells to a central collection well. (After J. Zajic, 1969,* Microbial Biogeochemistry, *Academic Press, New York.)*

Figure 20.33

The heap bioleaching process is also used for the recovery of low-grade ore. The ore is mined, crushed, and heaped in the shape of a cone with its point cut off on an asphalt pad. The leaching liquor is pumped to the top of the heap and percolates through the ore. The leachate is collected, processed, and recycled.

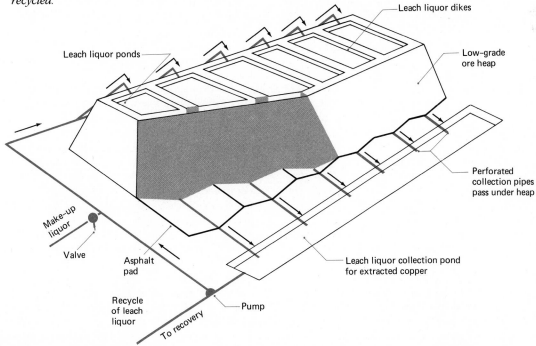

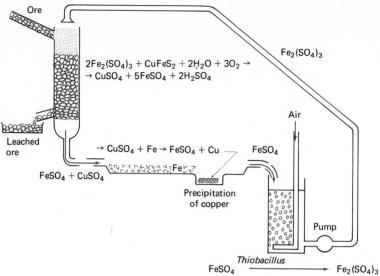

Figure 20.34

A continuous reactor leaching operation extracts copper from low-grade ore. The oxidation of sulfide and ferrous iron is carried out by Thiobacillus ferrooxidans *generating the acid for leaching. Copper is precipitated by exchange with iron, using scrap iron. The Kennecott Copper Co. holds the patent on the process.*

ing operations, the action of *Thiobacillus* involves both direct oxidation of CuS and indirect oxidation of CuS via generation of ferric ions from ferrous sulfide. Ferrous sulfide is present in most of the important copper ores, such as chalcopyrite (CuFeS$_2$). The copper is recovered either by solvent extraction or by using scrap iron. In the latter case, copper replaces iron according to the equation

$$CuSO_4 + Fe^0 \rightarrow Cu^0 + FeSO_4$$

and is more advantageous for bioleaching because the organic solvent residues in the leaching liquor may inhibit the continued activity of *T. ferrooxidans*.

The **recovery of uranium**, a fuel required by the nuclear power generation industry, can also be enhanced by using microbial activities. The microbial recovery of uranium from otherwise useless low-grade ores is helpful in overcoming the so-called energy crisis. Nuclear safety and waste disposal problems as well as the limited supply of uranium render current nuclear fission generators controversial, and so for all these reasons they may very well be only a stopgap solution to the international energy shortage. Although bioleaching cannot influence safety considerations,

this process can have an immediate and direct bearing on the economics of nuclear power production by providing a mechanism for commercial utilization of low-grade uranium deposits and for the recovery of uranium from low-grade nuclear wastes. Recovery of uranium from radioactive wastes is extremely important because it overcomes the problem of waste disposal, a major shortcoming of using nuclear power generators.

Insoluble tetravalent uranium oxide (UO$_2$) occurs in low-grade ores. Although there is no evidence for the direct oxidation of UO$_2$ by *T. ferrooxidans*, UO$_2$ can be converted to the leachable hexavalent form (UO$_2$SO$_4$) indirectly by the action of *T. ferrooxidans*. *T. ferrooxidans* oxidizes the ferrous iron in pyrite (FeS), which often accompanies uranium ores, to ferric ions. The oxidized iron acts as an oxidant, converting UO$_2$ chemically to UO$_2$SO$_4$, which can be recovered by leaching. The technical and economic feasibility of employing *Thiobacillus* for the recovery of uranium and copper minerals depends on a variety of factors. The particular form of the naturally occurring mineral is very important; for example, pyritic uranium oxide ores are suitable for bioleaching, whereas those ores not in close association with iron-containing minerals are not. Bacterial leaching of uranium is most feasible in geological strata where the ore is in the tetravalent state and is closely associated with reduced sulfur and iron minerals. The geological formation in which the minerals occur is also important in determining the suitability of the bioleaching process. *In situ* bioleaching is ideal when there is a natural drainage system, as through a fault with an impermeable basin, that will permit an economical recovery of the minerals. If these conditions do not exist, mining must precede the heap leaching process described previously.

The characteristics of the ore have an important effect on its susceptibility to bioleaching. When the chemical form of the element within the ore is resistant to microbial attack, or when toxic products are produced as a result of microbial activity, the metal is difficult to recover through the use of bioleaching. Fortunately, many ore minerals exist in chemical forms that are readily subject to bioleaching. The rate of leaching is determined in large part by the size of the mineral particles. Increasing the surface area, accomplished by crushing and/or grinding, generally speeds production efficiency. Environmental factors must also be conducive for efficient bioleaching to occur. Optimal conditions for bioleaching using *T. ferrooxidans* are (1) temperature 30–50°C, (2) pH 2.3–2.5, and (3) iron concentration 2–4 g/l of leach liquor; available oxygen and nutrients such as ammonium nitrogen, phosphorus, sulfate, and magnesium, are essential for the growth of *T. ferrooxidans*.

The oxidative activities of *Thiobacillus* can produce high temperatures in some mineral deposits, which may exceed the tolerance limits of the species being used. Obviously, this would lead to decreased bioleaching activity and mineral production. Because of these high temperatures, thermophilic sulfur-oxidizing microorganisms may be useful for some bioleaching processes. Members of the genus *Sulfolobus* are obligate thermophiles that can oxidize ferrous iron and sulfur in a manner similar to that of the members of the genus *Thiobacillus*. These acid-tolerant thermophilic bacteria can oxidize inorganic substrates and are used in the bioleaching of metallic sulfides. *Sulfolobus* has been used for the bioleaching of molybdenite (molybdenum sulfide), whereas *Thiobacillus* is intolerant of high concentrations of this metal as well as mercury and silver.

In addition to bioleaching to recover metals, some microorganisms are able to concentrate metals in their cells at concentrations several orders of magnitude higher than occur in the surrounding solution. Such bioconcentration has the potential for extracting rare metal ores from dilute solutions and for recovering metals from industrial effluents. Theoretically, microorganisms can be used to recover gold from the sea. It is not clear, though, whether such bioconcentration procedures will permit the development of economically feasible industrial processes for the recovery of valuable metals. In one such process, *Rhizopus* binds uranium from low-grade ores and nuclear wastes, contributing not only to the production of nuclear fuels but also to solving the nuclear waste disposal problem.

Oil recovery

Besides their role in the recovery of metals, microorganisms can be used to increase the recovery of petroleum hydrocarbons. The **tertiary recovery of petroleum** and the enhanced recovery of hydrocarbons from oil shales have become increasingly important as readily recoverable oil supplies have diminished and the price of oil has continued to increase dramatically. Tertiary recovery of oil employs solvents, surfactants, and polymers to dislodge oil from geological formations. The use of tertiary recovery methods has the potential for recovering 60–120 billion barrels of oil in United States reserves alone that otherwise could not be recovered. Xanthan gums produced by bacteria, such as *Xanthomonas campestris,* are promising compounds for the tertiary recovery of oil. These polymers have high viscosity and flow characteristics that allow them to pass through small pores in the rock layers containing oil deposits. When added during water flooding operations, xanthan gums help push the oil toward the production wells. These polymers are produced by conventional fermentation processes in which *X. campestris* is grown and the xanthan gums are recovered.

Bioleaching of oil shales also has the potential of greatly enhancing recovery of hydrocarbons. Many oil shales contain large amounts of carbonates and pyrites, and the removal of these minerals increases the porosity of the shale, enhancing recovery of the oil. Acid dissolves the carbonates and can be produced by *Thiobacillus* species growing on the sulfur and iron in the pyrite. Such microbial leaching appears to have the potential for making recovery of hydrocarbons from oil shales economically feasible.

Quality control of industrial products

The term **quality control** in industrial microbiology has several meanings. In the case of a production process, the concern is with the quality of the product. In microbial fermentation processes quality control procedures must include checks on the fermentation medium and conditions, microbial production strain, recovery efficiency, and packaging methods. Quality is determined by the yield and purity of the product. A low yield of product is economically unfavorable and may reflect a lack of proper control during the fermentation process, and a low-quality product may preclude the use of the product for its intended purpose. Environmental parameters must be maintained as close to optimal as possible to maximize the yield and quality of the product. The producing microbial strain, which in many processes is genetically unstable, must retain its essential genetic and physiological features for producing high yields of the desired product. Checks must be run and when necessary the strain must be replaced with a new stock culture. The extraction and packaging procedures must en-

sure a lack of excess contamination with substances other than the desired product.

Additionally, concern often must be given to preventing microbial contamination of the product. It is essential that many products be free of microbial contaminants even if they are not produced using microbial fermentation processes. Plastic and glass petri plates, pipettes, and many other items used in the microbiology laboratory must be sterile. Likewise, many pharmaceuticals, medical supplies, and surgical instruments must be free of microbial contaminants for their safe use. A variety of sterilization techniques are employed in different industrial processes to remove viable microorganisms from various products. The choice of a particular sterilization method depends on the nature of the product. Plastic products, for example, which cannot withstand exposure to high temperatures, are often sterilized by exposure to radiation or chemical oxidizing agents, such as ethylene oxide. Liquid products, such as antibiotic solutions, may be filtered to remove microorganisms, and the bottling of many drugs

Figure 20.35

Photographs of sterile drug filling operation for producing medicinals in ampules. (A) Workers at Giza, Egypt, manufacturing plant preparing pharmaceuticals in an area completely isolated and supplied with a change of filtered, sterile air each minute. (B) Modern high-speed equipment at pharmaceutical manufacturing plant in Regensburg, West Germany, washes, sterilizes, and fills ampules for hypodermic solutions. (Courtesy E. R. Squibb & Sons, Inc., Princeton, New Jersey.)

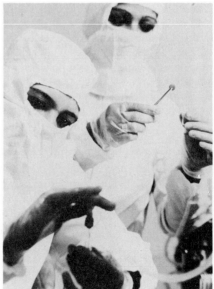

A

B

is accomplished using aseptic techniques, often with the aid of sterile air laminar flow hoods (Figure 20.35). Industrial microbiologists in many industries are involved with designing procedures to sterilize or reduce the numbers of microorganisms associated with the final product to acceptable levels.

It is especially critical that some products, such as surgical suturing material, be free of microbial contaminants. Checking representative samples of products is used to monitor product quality. The discovery of a contaminated sample can lead to the recall and destruction of an entire production lot. Sometimes, notification of a product recall, such as for contaminated food items, is given through the mass media. The removal of contaminated products from use is costly but often essential. Quality control procedures must ensure that the numbers of microorganisms associated with a product are within the prescribed allowable limits for their safe use.

Biodeterioration

Several billion dollars in losses occur every year as a result of the microbial **biodeterioration** of materials, excluding food products. Microbial metabolic activities are involved in the corrosion of metals and the deterioration of various materials, including wood, paper, paint, and textile products. Various control measures are employed in different industries for reducing losses due to microbial biodeterioration. It is estimated that without control measures microbial biodeterioration would cause losses exceeding $50 billion per year. The magnitude of these figures attests to the significance of microbial biodeterioration in reducing the economic value of various products and the importance of controlling microbial activities to reduce these losses.

Paper

The production of paper is a major industry. In the United States paper products are consumed at a rate in excess of 500 pounds per person per year. Losses to the pulp and paper industry through the activities of microorganisms are substantial. Microorganisms can affect all stages of the paper-making process; microorganisms are able to degrade wood, the raw material for making paper. Growth of microorganisms during paper production reduces the quality of the final product, and microbial growth on paper products can preclude their intended use by consumers. In the paper-making process, pulpwood—broken into chips—is cooked under high pressure in a digester for a sufficient period of time to solubilize the lignins, pectins, and other adhesive biochemicals in the wood. Either acids, in the acid sulfite process, or bases, in the kraft sulfate process, are added during cooking to aid in the digestion of the wood. When the cooking process is completed, the digester is suddenly opened, releasing the pressure and causing the wood chips literally to explode and separate into loosely bonded cellulose fibers. The cooking solution is recovered and used for various purposes. For example, sulfite liquor is processed and used as a nutrient for growing yeasts and as an animal feed supplement. The fibers recovered from the cooking process are often bleached to remove pigments. A slurry of cellulose fibers, 5–15 percent fiber in water, is milled to form sheets of paper.

Microbial biodeterioration can occur at various stages during this paper-making process. It is estimated that 10 percent of all pulpwood that is cut is lost to biodeterioration. Cellulolytic fungi are particularly important in the degradation of wood. Many different fungi cause wood rots, reducing the value of wood for manufacture of paper (Figure 20.36). Microbial growth in pulpwood results in (1) loss of usable cellulose fibers; (2) loss of strength in the paper; (3) and discoloration and imperfections, including holes, spots, and various other blemishes to the finished product. The severity of these problems depends on the particular manufacturing process and the specific enzymatic activities of the contaminating microorganisms. Pulp stored for long periods of time for processing can be subject to very large losses because of microbial biodeterioration. Addition of copper sulfate or other antimicrobial agents to the pulp is used to minimize such losses.

During the paper-making process growth of microorganisms on the surfaces of the equipment used in the paper mill can reduce the quality of the paper produced. **Slimes** can develop from the growth of masses of microorganisms on paper mill equipment and in the solutions used in making paper, causing significant losses because of the production of unacceptable quality paper. It is estimated that microbial growth in the pulp and

Figure 20.36

Photographs of wood showing fungal wood rot. The growth of the fungus causes biodeterioration of the wood and loss of integrity and usefulness of the wood for structural support. (A) Typical brown arbical decay with white fungal mycelia on the surface. (B) Brown rot caused by Gloeophyllum trabeum *in a railroad cross-tie. (C) Cross section of douglas fir with brown rot. (Courtesy W. Eslyn, USDA Forest Products Laboratory, Madison, Wisconsin.)*

paper mill cause losses of $1–5 per ton of paper produced. The losses due to slime formation include production time lost to clean equipment, decreased performance of equipment, and the loss of finished products not acceptable for consumer use (Figure 20.37). Several environmental param-

eters can be modified to control slime deposits, including periodic washing of equipment with hot sodium hydroxide to kill contaminating microorganisms.

Various chemicals can be used as **microbiocides** in the paper industry. The use of chlorine is probably of greatest value in the paper industry for controlling slime. Chlorine is added to the water coming into the paper mill to control bacterial growth. This is the same chlorination process used to maintain acceptable levels of microorganisms in drinking water supplies. Various antifungal agents, including phenolic and organosulfite compounds, are also used in the paper mill to control undesired microbial growth. Mercurials have been used extensively in the paper-making industry but are now banned by the United States Food and Drug Administration for use in paper for food packaging.

Wood

In addition to paper, other wood products are subject to microbial biodeterioration. As indicated above, many fungi produce cellulase enzymes that are able to decompose wood cellulose. To prevent biodeterioration, many wood products are preserved from microbial attack by coating the surface of the wood with paint or lacquer. Paint protects wood surfaces by preventing microorganisms and the water needed for their growth from reaching the surface of the wood. Various wood products are protected from biodeterioration by impregnation with antimicrobial agents. For example, railroad ties and telephone poles are normally impregnated with creosote, an antimicrobial phenolic compound, and marine wood pilings are treated with copper-naphthalene to prevent biodeterioration.

Paint

Although paints are used to protect wood surfaces against biodeterioration, paints themselves are organic compounds subject to microbial degradation. Paint components, including primary pigments, as well as finished paint products are subject to microbial attack. The growth of microorganisms in a paint product can alter the color and binding properties of the paint (Figure 20.38). The growth of microorganisms in the paint can lead to peeling as a result of a loss of adhesion between the paint film and the surface being coated. A large number of fungal and bacterial species have been isolated from paints. Therefore, various microcidal agents are routinely added to paint products to prevent microbial growth and biodeterioration of the paint.

The only effective means of preventing microbial attack on liquid paints and paint films is through the use of chemical inhibitors. Relatively few microcidal and microstatic agents have been found to be effective in paint products. The choice of the paint preservative is dictated by the effec-

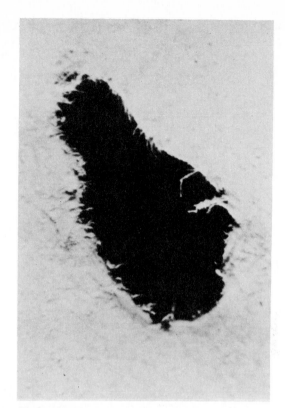

Figure 20.37

Photograph of paper, showing imperfection (hole) due to slime formation resulting from microbial growth. (Courtesy Nalco Chemical Company, Oak Brook, Illinois.)

tiveness of the particular chemical in controlling the growth of contaminating microorganisms and also on the particular application. Mercury and lead compounds have been widely used in paints to control microbial biodeterioration, but the human toxicity of many of the compounds used as preservatives in paints has led to a severe curtailment in their use. Other antimicrobial agents used in paints include chlorinated phenols, inorganic pigments, such as barium metaborate, quaternary ammonium compounds, and various other microbiocides.

Marine paints present a special problem for the prevention of biodeterioration. Such paint products must prevent microbial degradation of the wood and metal surfaces of ships and other marine structures. Marine paints should also resist marine fouling. The growth of microorganisms on a ship's hull can greatly reduce the performance characteristics of that vessel, and loss of speed can be a serious problem. Various biocides are routinely incorporated into marine paints to

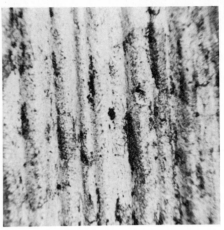

Figure 20.38

These photographs show the detrimental effects of microbial growth on paint surfaces. (A) Fungal growth on the exterior of this building has caused discoloration of the painted surface. (B) A close-up view of the discolored area of the painted building. (C) Filaments of Aureobasidium pullulans *that grow on paint film (14×). (Courtesy Amy Z. Leathers, PPG Industries, Coatings and Resins Division, Pittsburgh, Pennsylvania.)*

reduce microbial development of surface films that can support attachment and growth of higher organisms, such as barnacles. Various biocidal agents may be employed to control microbial fouling of surfaces. Copper oxides most frequently have been used as antibacterial, antialgal, and antifungal agents to prevent the development of slime on marine surfaces. Antifouling paints often contain 55 percent cuprous oxide, but such high levels of antifouling agents can cause corrosion problems. High concentrations of copper oxides, which slowly leach from the paint, are needed to control microbial growth over a long period of time.

Textiles

Textiles, particularly those composed of natural organic fibers, such as cotton and flax, are readily attacked by microorganisms. Most fabrics are not subject to extensive biodeterioration because they are dehydrated and there is insufficient water available to support microbial growth. In the tropics, however, there is adequate moisture in the air (humidity) to support microbial growth on various textiles. Fungi are the most important microorganisms in such textile biodeterioration processes. **Mildew**, due to the growth of various fungi, is a particular problem in the deterioration of products such as canvas tents (Figure 20.39). The growth of fungi on a textile surface causes discoloration and weakening of fiber strength. Adding waterproofing agents containing small quantities of phenolic compounds to textiles can be used to reduce losses due to mildew formation. Storing camping equipment and other textile products in dry places is important for the prevention of the biodeterioration of the fabric. Despite preventative measures, annual losses due to textile biodeterioration are in the millions of dollars.

Metal corrosion

Corrosion of iron and steel pipes can occur as a result of a variety of chemical reactions that establish an electrochemical gradient, leading to loss of metal from the pipe due to electrolysis (Figure

20.40). The growth of several microbial species is also important in corrosion processes (Table 20.9). The growth of the sulfate-reducing bacterium *Desulfovibrio desulfuricans* is especially important in corrosion of metals. Sulfate-reducing bacteria do not attack the metal directly but accelerate the electrolytic corrosion process by promoting depolarization of the anodic and cathodic surfaces, thereby driving the corrosive reactions and causing anaerobic corrosion. The reduction of sulfate to hydrogen sulfide results in cathodic depolarization. The hydrogen sulfite formed from the reduction of sulfate reacts with ferrous iron to form sulfide; the effect of this reaction is anodic depolarization. Additionally, the metabolic activities of **anaerobic sulfate-reducing bacteria** result in the formation of iron hydroxides, which are corrosion products. Sulfate-reducing bacteria are active within the pH range 6–8. Serious corrosion of iron and steel structures occurs in anaerobic waterlogged soils, near neutral pH. Anaerobic corrosion of water mains, buried in land reclaimed from the sea north of Amsterdam, forced their replacement every two to three years. Anaerobic microbial corrosion of cast iron causes **graphitization**, a process in which a pipe loses much of its iron, resulting in the formation of a

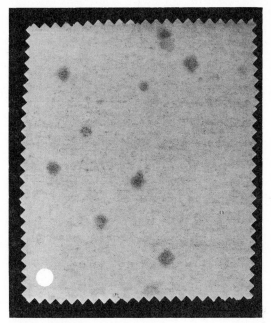

Figure 20.39

This cotton duck tent fabric has mildew spores (black spots) that are clearly visible after being weathered for 12 months. (Courtesy David A. Yeadon and Robert J. Harper, Jr., USDA Southern Regional Research Center, New Orleans.)

Figure 20.40

Corrosion of metal pipes. (A) This cast iron pipe shows graphitization resulting from the activities of sulfate-reducing bacteria. The pipes that appear to be in good condition are actually the same pipes as those with the holes, except that the corrosion products have not been removed. (Courtesy Melvin Romanoff and Warren Iverson, National Bureau of Standards.) (B) SEM micrograph of the bacteria associated with an area of corroding pipe. (Courtesy Edwin E. Geldreich, USEPA, Cincinnati.)

A

B

table 20.9

Some bacteria involved in corrosion processes

Organism	Oxygen requirement	Inorganic components	Metabolic end products	Habitat	Optimal range Temp.	pH
Sulfate reducing *Desulfovibrio desulfuricans*	Anaerobic	Sulfate, thiosulfate	Hydrogen sulfide	Water, soil, mud, oil-reservoir	25–30	6–7.5
Sulfur oxidizing *Thiobacillus thiooxidans*	Aerobic	Sulfur, thiosulfate	Sulfuric acid	Soil, water	28–30	2–4
Thiosulfate oxidizing *Thiobacillus thioparus*	Aerobic	Thiosulfate, sulfur	Sulfur, sulfuric acid	Soil, water, mud, sewage	28–30	7
Iron bacteria *Crenothrix Leptothrix Gallionella*	Aerobic	Iron, manganese	Ferric or manganese oxides	Water	25	8
Nitrate reducing *Thiobacillus dentrificans*	Facultative	Thiosulfate, sulfur, sulfide	Sulfate	Soil, mud, peat, water	30	7–9
Hydrogen utilizing *Hydrogenomonas*	Microaerophilic	Hydrogen	Water	Soil, water	28–30	7

soft and brittle pipe that is easily broken. This is what happened to the buried pipes in the Netherlands. Steel and aluminum pipes are also subject to anaerobic corrosion. Anaerobic microbial corrosion of steel results in more localized pitting, sometimes causing perforation of the pipe.

Aerobic microbial corrosion is not as serious an economic problem as anaerobic corrosion. Aerobic corrosion processes are carried out by various bacteria, such as *Gallionella, Crenothrix,* and *Leptothrix* species, that oxidize metals and form metalic oxides as corrosion products. The

Figure 20.41

Diagram illustrating the corrosion process as it occurs on the internal surface of a well-aerated water delivery pipe when a tubercle is formed by iron bacteria.

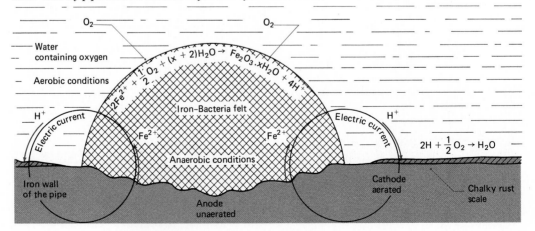

gelatinous sheaths of these bacteria form a barrier to oxygen diffusion, causing the development of an electrical potential difference between the surface covered by the bacteria (anodic region) and the exposed surface (cathodic region). This electrical differential can be extensive, leading to accelerated loss of metal and corrosion (Figure 20.41). Sulfur-oxidizing *Thiobacillus* species, which produce high concentrations of sulfuric acid, are also important in aerobic corrosion processes.

The microbially mediated corrosion of metal pipes is a serious problem in oil and gas delivery systems. One method of protecting pipelines against the action of sulfate-reducing microorganisms is to increase the pH above 9.5. Buried pipes can be coated to prevent contact between the metal surface, water, and soil microorganisms. Additionally, electric currents can be applied to the pipe to preclude corrosion processes. Various bacterial inhibitors are employed to control microbial corrosion. The most commonly used bacterial inhibitors are alkyl-substituted amine and quatenary ammonium compounds. These chemical compounds are toxic to many bacteria, including *Desulfovibrio desulfuricans,* a bacterial species of major importance in corrosion processes.

Postlude

The application of technology to the control of microbial growth has great economic consequences. The metabolic activities of various microorganisms support several major industries, such as the manufacturing of pharmaceuticals. Biodeterioration, on the other hand, results in large-scale economic losses because of uncontrolled microbial growth. The production of biocides, aimed at preventing microbial degradation of various products, is an industry in itself. The emphasis in much of industrial microbiology is on harnessing microbial activities for beneficial purposes.

The success of using microorganisms in industrial processes, in the final analysis, is judged solely on economic grounds. The microbial fermentation industry is based on using the most efficient strain, least expensive substrate, and optimal environment in a fermentor to produce the highest possible concentrations of a substance that can be efficiently recovered and purified. Changes in the costs of raw materials and energy continuously shift the economic balance between using fermentation and chemical synthesis as the mode of production of a particular product. In some cases using microorganisms is the only effective means of making a product, such as in the production of vaccines. In other cases microbial activities and chemical syntheses are both used for making the final product, such as in the production of modified penicillin derivatives. In still other instances chemical syntheses, at the present time, have replaced microbial fermentations such as in the production of organic solvents.

There is a general maxim that microorganisms can be found or created to carry out any possible chemical transformation and that microorganisms can carry out chemical modifications more efficiently than can be accomplished by using synthetic chemical methods. Based on this philosophy, industrial microbiologists have developed screening procedures to seek, and where necessary, to modify—by purposefully increasing rates of mutation—specific strains of microorganisms capable of producing high yields of desired substances. Many antibiotics routinely used in medicine today were discovered and the microbial strains producing these compounds isolated and developed by using such screening procedures. The promise of recombinant DNA technology to permit the incorporation of foreign genetic information into microorganisms supports the belief that microorganisms will become increasingly important in the production of many goods not hitherto considered possible to produce by microbial fermentation. The full benefits of genetic engineering have yet to be realized. We are already driving cars fueled in part by the products of microbial metabolic activities and eating foods produced with the aid of microorganisms, and in the future our use of such microbial products is certain to increase.

1. How does the industrial use of the term fermentation differ from its use to describe microbial metabolism?

2. How are antibiotic-producing microorganisms found? How is it determined if they are producing substances of industrial importance?

3. Discuss the role of mutation and selection in the history of penicillin production. Why would a similar approach not be suitable for finding an insulin-producing strain of *E. coli*?

4. Why are microorganisms especially important in the production of steroids?

5. What is an immobilized enzyme? Discuss their role in industrial microbiology.

6. What is a fermentor? How are aeration and pH controlled in fermentors?

7. What are the essential properties of a substrate for an industrial fermentation?

8. How was the microbial production of butanol critical in determining the outcome of World War I?

9. How are microorganisms involved in the corrosion of metals?

10. What is bioleaching? How is this process used for the recovery of uranium?

11. Discuss several ways that microorganisms can help meet the current fuel shortage.

12. What is biodeterioration? How do we control biodeterioration of wooden ships? of paints? of paper?

Suggested Supplementary Readings

Aharonowitz, Y., and G. Cohen. 1981. The microbiological production of pharmaceuticals. *Scientific American* 245(3): 140–152.

Atlas, R. M. (ed.). 1984. *Petroleum Microbiology*. Macmillan Publishing Co., New York.

Brierley, C. L. 1982. Microbiological mining. *Scientific American* 247(2): 44–53.

Casida, L. E. 1968. *Industrial Microbiology*. John Wiley & Sons, New York.

Davis, J. B. 1967. *Petroleum Microbiology*. Elsevier Publishing Co., Amsterdam.

Demain, A. L. 1981. Industrial microbiology. *Science* 214: 987–995.

Demain, A. L., and N. A. Solomon. 1981. Industrial microbiology. *Scientific American* 245(3): 66–76.

Eveleigh, D. E. 1981. The microbiological production of industrial chemicals. *Scientific American* 245(3): 154–178.

Hopwood, D. A. 1981. The genetic programming of industrial microorganisms. *Scientific American* 245(3): 91–102.

Miller, B. M., and W. Litsky. 1976. *Industrial Microbiology*. McGraw-Hill Book Co., New York.

Peppler, H. J., and D. Perlman (eds.). 1979. *Microbial Technology* (2 volumes). Academic Press, New York.

Phaff, H. 1981. Industrial microorganisms. *Scientific American* 245(3): 77–89.

Rehm, H. and G. Reed (eds.). 1981–. *Biotechnology: A Comprehensive Treatise* (8 volumes). Verlag Chemie International, Deerfield Beach, Florida.

Riviere, J. *Industrial Applictions of Microbiology.* 1978. Halsted Press, New York.

Rose, A. H. 1977–. *Economic Microbiology* (a multivolume treatise). Academic Press, New York.

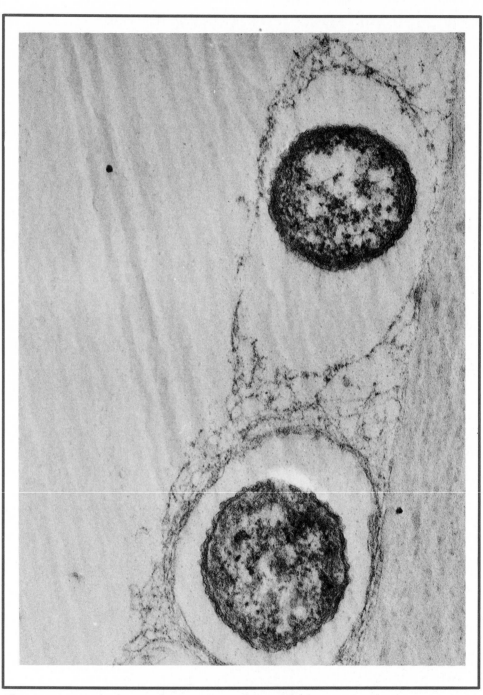

Electron micrograph of rumen bacteria on a cellulose fragment.
(Courtesy William Costerton, University of Calgary.)

Environmental microbiology

SECTION NINE

Microbial ecology

21

The study of **microbial ecology** has received considerable impetus through increased public interest in environmental quality and the proper management of our limited natural resources. Microbial ecology is concerned with examining the interactions of microorganisms and their environment, including both their living and abiotic surroundings. The interactions between member populations within a biological community are dynamic. Microorganisms both respond to and alter their surroundings, with the positive and negative feedback interactions between populations acting to maintain an ecological balance. In an **ecosystem**, the basic self-sustaining and self-regulating ecological structural and functional unit, each population within the community occupies a particular ecological **niche** defined by the role played by that organism within that ecosystem. Ecosystems, although stable, are not static, and as environmental conditions change, the member populations of the community also often change. In many cases, after the introduction of organisms into a newly exposed **habitat**, there is an orderly successional change of populations from an initial unstable pioneer community to a stable climax community, where all of the niches are filled by

diverse populations that are well adapted to survival and functioning within the given ecosystem. A perturbation to the environmental conditions of a habitat can cause population shifts and an instability within the community, but generally, once the perturbation is removed, the succession of populations reestablishes the stable community.

Microorganisms perform essential functions within ecosystems, the integrated systems of biological populations and abiotic factors that comprise the global **biosphere** (that portion of the earth and its atmosphere in which living organisms occur). The activities of microorganisms within the biosphere have a direct impact on the quality of human life. Without the essential biogeochemical cycling activities of microorganisms, higher forms of life, including humans, could not exist. In this chapter we will consider some of the basic aspects of microbial ecology, including population interactions and biogeochemical cycling reactions essential for supporting life on earth. In later chapters we will consider the importance of microorganisms and microbial activities in agricultural practices and in maintaining environmental quality.

Biogeochemical cycling

Biogeochemical cycling describes the movement of materials via biochemical reactions through the global biosphere. The exchange of elements between the biotic and abiotic portions of the bio-

sphere occurs through characteristic cyclic pathways. Just as transformations of compounds occur via central metabolic pathways within cells, there are central pathways through which materials flow within ecosystems. All living organisms carry out chemical transformations that influence their surrounding environment. Changes in the chemical forms of various elements can lead to the physical translocations of materials, sometimes mediating transfers between the atmosphere (air), hydrosphere (water), and lithosphere (land). Because of their ubiquitous distribution and diverse enzymatic activities, microorganisms play a major role in biogeochemical cycling. The major elements of living biomass, carbon, hydrogen, oxygen, nitrogen, phosphorus, and sulfur, are most intensively cycled by microorganisms; other elements are generally cycled to a lesser extent. The biogeochemical cycling activities of microorganisms are essential for the survival of plant and animal populations and determine in large part the potential productivity level of a given habitat. Human activities, such as the release of pollutants, can have a major influence on the rates of microbial cycling activities, leading to significant changes in the biochemical characteristics of a given habitat.

The carbon cycle

Carbon is actively cycled between inorganic carbon dioxide and the variety of organic compounds that compose living organisms. The **biogeochemical cycling of carbon** primarily involves the transfer of carbon dioxide and organic carbon between the atmosphere, where carbon principally occurs as inorganic CO_2, and the hydrosphere and lithosphere, which contain varying concentrations of organic and inorganic carbon compounds (Figure 21.1). The autotrophic metabolism of photosynthetic and chemolithotrophic organisms is responsible for **primary production**, the conversion of inorganic carbon dioxide to organic carbon. Once carbon is fixed (reduced) into organic compounds, it can be transferred from population to population within the biological community, supporting the growth of a wide variety of heterotrophic organisms. The respiratory and fermentative metabolism of heterotrophic organisms returns inorganic carbon dioxide to the atmosphere, completing the carbon cycle. The production of methane by a specialized group of methanogenic archaebacteria represents a shunt to the normal cycling of carbon because the methane that is produced cannot be used by most heterotrophic organisms and thus is lost from the biological community to the atmosphere.

Trophic relationships

The carbon dioxide converted into organic carbon by the **primary producers** in an ecosystem represents the **gross primary production** of the biological community in a given habitat (Figure 21.2). Part of the gross primary production is converted back to carbon dioxide by the respiration of the primary producers, and only the remaining organic carbon in the form of biomass and soluble metabolites—the **net primary production**—is available for heterotrophic consumers in terrestrial and aquatic habitats. The oxidative metabolism of the biological community removes organic carbon and the energy stored in such compounds from the ecosystem and thus represents a decay of the energy stored within a given habitat. If the net primary production is greater than the community respiration, organic matter

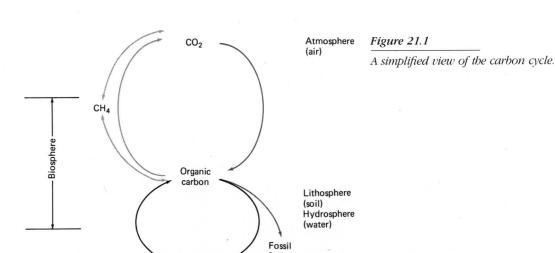

Figure 21.1

A simplified view of the carbon cycle.

Lithosphere
(soil)
Hydrosphere
(water)

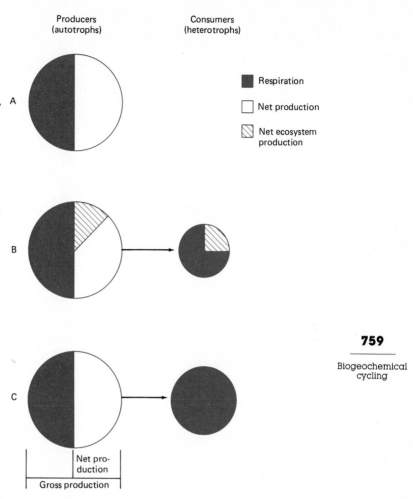

Figure 21.2

Gross and net primary production show the balance of primary productivity, respiration, and community productivity. (A) Gross primary production is represented by the whole circle, and net production, the white area, is what remains after respiration of the primary producers and is available for consumers. (B) Because heterotrophs do not consume all the available net primary production, there is net ecosystem production, which leads to the accumulation of organic matter in the ecosystem. This typically occurs during early successional stages. (C) When consumers use all of the net primary production and total respiration of the community balances total primary productivity, no organic matter accumulates in the system. This situation is typical of climax ecosystems.

accumulates within the ecosystem. If, on the other hand, respiratory activities are greater than the primary production, there must be an input of organic matter from an external source, or the community in that ecosystem will decline. Most ecosystems are dependent on the fixation of carbon dioxide (input of organic matter) by photosynthetic organisms, including plants, algae, and photosynthetic bacteria. The thermal rift areas of the deep ocean regions near the Galapagos Islands represent an interesting exception because the ecosystems associated with these areas are based on the input of organic carbon by sulfur-oxidizing chemolithotrophic bacteria that are able to grow in the warm, hydrogen sulfide-rich waters that enter the ocean through thermal vents (Figure 21.3).

The carbon and energy in organic compounds, formed by primary producers, move through the biological community of an ecosystem. The transfers of energy stored in organic compounds be-

tween the organisms in the community forms a **food web** (Figure 21.4). The feeding relationships between organisms establish the **trophic structure** of an ecosystem, with movement of energy through an ecosystem occurring in steps from one trophic level to another. **Grazers**, which feed upon primary producers, are preyed upon by **predators**, which in turn may be preyed upon by larger predators, establishing a pyramid of biological populations in the food web (Figure 21.5). In phytoplankton-based food webs, algae and cyanobacteria are the primary food source for grazers. In detrital food webs, microbial biomass produced from growth on dead organic matter serves as a primary food source for grazers. Only a small portion of the energy stored in any trophic level is transferred to the next higher trophic level. Normally, 85–90 percent of the energy stored in the organic matter of a trophic level is consumed by respiration during transfer to the next trophic level and enters the decay portion of the food

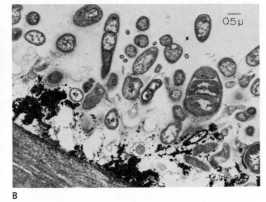

Figure 21.3

Micrographs of bacteria growing in a thermal rift area off the Galapagos Islands. (A) In this scanning electron micrograph, note the dense microbial layer of prosthecate cells, resembling Hyphomicrobium, *on the surface. (B) This transmission micrograph of the bacterial surface shows a loose layer of Gram negative cells and metal deposits. Scale bar = 0.5 μm. These photomicrographs were made as part of a survey of the most commonly observed and morphologically conspicuous microorganisms found attached to natural surfaces or artificial deposits in the area of thermal vents at the Galapagos Rift ocean spreading zone at a depth of 2,550 meters. (Courtesy Holger W. Jannasch, Woods Hole Oceanographic Institution, Woods Hole, Massachusetts, reprinted by permission of the American Society for Microbiology, Washington, D.C., from H. W. Jannasch and C. O. Wirsen, 1981,* Applied and Environmental Microbiology, *41: 528–538.)*

Discovery process

The work of Holger Jannasch of Woods Hole had established that microbial activities occur at very low rates in the deep oceans; deep ocean sediments effectively are biological deserts because of low temperatures, high pressures, and low inputs of organic matter. Bologna sandwiches that were accidentally submerged inside the submersible *Alvin* were not decomposed during several months of exposure to microorganisms of the deep sea. What a surprise it was, then, when investigators found an area of extremely high biological productivity at a depth of 2,550 meters in a region of thermal vents off the Galapagos Islands. The thermal vents warmed the waters, but what was the source of food to support the growth of worms several feet long and clams several feet across? There was no light to support photosynthesis, and transport of organic matter from the surface was unlikely. The most likely explanation was chemolithotrophic primary production based on oxidation of hydrogen sulfide from the vents. Establishing that bacterial chemolithotrophy was the source of organic matter would be difficult; to reach the vents in the *Alvin* would take hours, time on the bottom to carry out experiments would be extremely limited, and working *Alvin's* mechanical arms would be difficult. Nevertheless, this was the task undertaken by Holger Jannasch and Carl Wirsen of the Woods Hole Oceanographic Institution. These investigators found that all surfaces intermittently exposed to H_2S-containing hydrothermal fluid were covered with mats composed of layers of prokaryotic and Gram negative cells interspersed with amorphous Mn-Fe metal deposits; although some of the cells were encased by dense metal deposits, there was little correlation between metal deposition and the occurrence of microbial mats; highly differentiated forms appeared to be analogs of certain cyanobacteria; isolates from massive mats of prosthecate bacterium were identified as *Hyphomicrobium* sp.; intracellular membrane systems similar to those found in methylotrophic and nitrifying bacteria were found in about 25 percent of the cells composing the mats; and thiosulfate enrichments made from mat material resulted in isolations of different types of sulfur-oxidizing bacteria, including the obligately chemolithotrophic genus *Thiomicospira*. These studies established that chemolithotrophic bacteria support the productivity of the thermal rift region.

web and, consequently, the higher the trophic level the smaller its biomass (Figure 21.5).

The decay portions of food webs are dominated by microorganisms (Figure 21.6). Microbial

decomposition of dead plants and animals and partially digested organic matter are largely responsible for the conversion of organic matter to carbon dioxide and the reinjection of inorganic

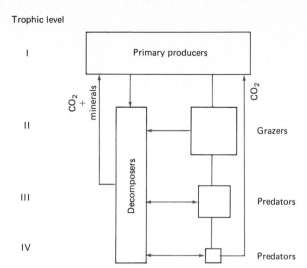

Trophic level

I

II

III

IV

A

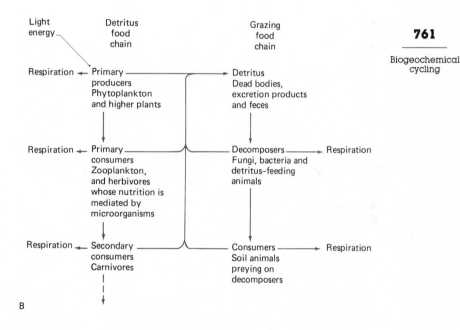

B

Figure 21.4

*Two food webs. (A) This idealized food web shows the transfers between trophic levels. Organic carbon
formed by primary producers is transferred to grazers and predators. CO_2 is returned to primary producers
by decomposers and by respiration of grazers and predators. The population sizes that can be supported
decline at higher trophic levels. (B) The grazing and detritus food chains are interlinked.*

CO_2 into the atmosphere. The rates of organic
matter mineralization depend on various factors,
including environmental conditions—such as pH,
temperature, and oxygen concentration—and the
chemical nature of the organic matter. Some natural organic compounds, such as lignin, cellulose,
and humic acids, are relatively resistant to attack
and decay only slowly. Various synthetic com-

pounds, such as DDT, may be completely resistant to enzymatic degradation (recalcitrant). We
depend on the activities of microorganisms to decompose organic wastes, and when microbial decomposition is ineffective, organic compounds
accumulate. This is evidenced by the environmental accumulation of plastic materials that are
recalcitrant to microbial attack. Many modern

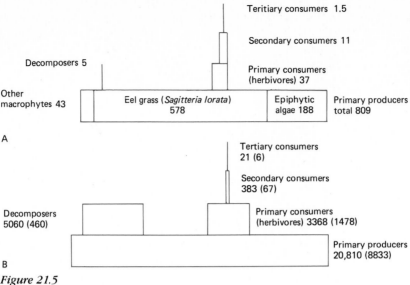

Teritiary consumers 1.5

Secondary consumers 11

Decomposers 5

Primary consumers
(herbivores) 37

Other
macrophytes 43

| Eel grass (*Sagitteria lorata*) 578 | Epiphytic algae 188 |

Primary producers
total 809

A

Tertiary consumers
21 (6)

Secondary consumers
383 (67)

Decomposers
5060 (460)

Primary consumers
(herbivores) 3368 (1478)

Primary producers
20,810 (8833)

B

Figure 21.5

Food webs are structured in the shape of a pyramid. These pyramids represent Silver Springs, Florida. (A) Biomass pyramid: the figures are grams dry weight/m^2. The decomposers have a small biomass but high metabolic activity. (After H. T. Odum, 1957, Ecological Monographs, 27(1): 55–112.) (B) Energy pyramid: the figures are the gross primary production or consumption in kcal/ m^2/yr. The portion of the total energy flow through the group of organisms that is actually fixed as organic biomass and that is potentially available as food for other populations at the next trophic level is indicated in brackets. Note the importance of the decomposers when their activity is expressed in energy terms rather than in biomass as in A. (After E. P. Odum, 1959, Fundamentals of Ecology, W. B. Saunders, Philadelphia.)

problems relating to the accumulation of environmental pollutants reflect the inability of microorganisms to degrade rapidly enough the concentrated wastes of industrialized societies.

Microbial mobilization and immobilization of carbon within the biosphere

The activities of microorganisms affect the accessibility of the biological community to the carbon and energy stored in organic compounds. Some microbial transformations of organic carbon, such as the production of humic acids in soil, tend to reduce the rate of cycling or to immobilize that portion of carbon and energy within the biosphere. **Humic substances** are defined as that portion of the soil that has undergone sufficient transformation to render the parent material unrecognizable. These are complex substances that are not readily metabolized. The formation of humic acids is a two-stage process involving the microbial degradation of organic polymers to their monomeric constituents, followed by chemical polymerization by a series of oxidation reactions, some of which are catalyzed by microbial exoenzymes, to produce the complex soil humic material. Microbial transformations that alter the physical state of carbon-containing compounds, such as the production of gaseous compounds from water-soluble solid substances, also have a major effect on the availability of carbon to the biological community within a particular habitat. The production of methane from low-molecular-weight fatty acids by some bacterial species in anoxic habitats is an interesting microbial process that tends to remove carbon from the biological community. Relatively few organisms can oxidize methane and reintroduce it into the actively cycled carbon pool. Other transformations, such as the anaerobic degradation of lignocellulose, mobilize stored organic carbon by producing simpler biochemicals that can be utilized by a wide variety of organisms. In some cases the growth of microorganisms effectively concentrates energy and nutrients within microbial biomass, making it possible for higher organisms to use the organic matter contained within the ecosystem on an energetically favorable basis; as examples, coprophagous

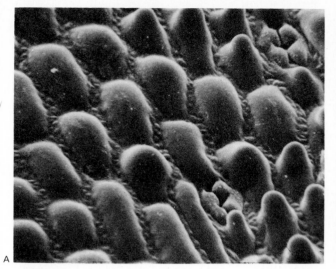

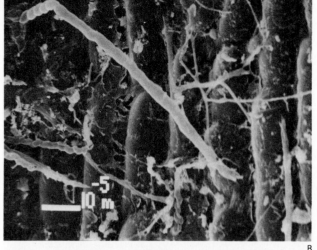

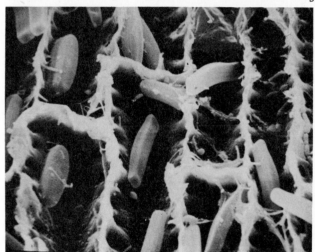

Figure 21.6

The decomposition of dead plant matter is seen in these scanning electron micrographs. (A) The uncolonized ventral surface of Carex *from Toolik Lake, Alaska. (B) Colonized and partially decomposed ventral surface of* Carex *litter exposed to sediment surface microbiota of the littoral zone of Toolik Lake for 13 days. Note the bacteria and fungal filaments enmeshed in a slime matrix. (C) A highly decomposed ventral surface of* Carex *litter that was exposed to the sediment surface microbiota for one year. Note the hollowed cells with only the plant cell walls remaining. The plant surface is now colonized by diatoms. (J. R. Vestal, University of Cincinnati.)*

animals ingest fungal biomass growing on fecal matter as a major source of nutrition, and various shellfish are filter feeders, removing microorganisms suspended in water as their food source.

The hydrogen and oxygen cycles

The cycling of hydrogen and oxygen through the biosphere is closely tied to the carbon cycle (Figure 21.7). Hydrogen is an essential component of all organic compounds, and oxygen occurs in all organic compounds except hydrocarbons. Water represents a major reservoir of both hydrogen and oxygen. When water acts as an electron donor during photosynthesis, molecular oxygen is liberated and hydrogen and electrons are transferred to form reduced coenzymes that are used in biosynthesis. During respiration both oxygen and hydrogen are recycled with the formation of water and carbon dioxide. Oxidation–reduction reactions of organic compounds, including the reduction of carbon dioxide to form organic matter, use hydrogen from reduced coenzymes for biosynthetic metabolism. The oxygen liberated by **photolysis**, the splitting of water that occurs during photosynthesis, is the major source of molecular oxygen upon which aerobic respiration depends.

Oxygen and microbial activities

The concentrations of **molecular oxygen** in a given

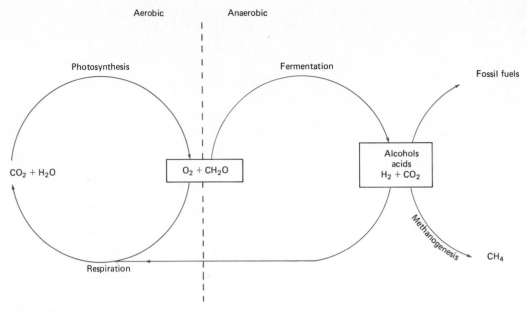

Figure 21.7

This diagram of integrated carbon, hydrogen, and oxygen cycles shows the involvement of oxygen and hydrogen in the aerobic and anaerobic oxidation of organic carbon and in the reduction of CO_2.

habitat are critical in determining the members of the biological community and the potential metabolic activities of the microorganisms within that community. Obligate aerobes depend on the presence of oxygen for active metabolism, but obligate anaerobes are killed by exposure to free oxygen. Numerous habitats, soils for example, are characterized by discontinuities in the concentrations of oxygen, the effect of which is to create **microhabitats**, with aerobic and anoxic zones occurring in relatively close proximity to each other (Figure 21.8). Many microorganisms, including facultative anaerobes, are quite versatile in their relationship to oxygen, altering their metabolism to meet most efficiently their energy needs within the constraints imposed by the availability of oxygen. When there is a lack of free oxygen, some microorganisms carry out anaerobic respiration—using other electron acceptors such as nitrate and sulfate—or fermentative metabolism, which does not require external electron acceptors. When oxygen is depleted from habitats, as occurs in eutrophic lakes when there is rapid degradation of organic compounds, microbial populations persist, but higher animals, which are obligate aerobes that depend on the availability of molecular oxygen for survival, may be killed.

The distribution of aerobic and anaerobic microorganisms is primarily determined by the oxidation-reduction (O/R) or redox (E_h) potential of the environment. The redox potential indicates the relative excess or deficit of electrons available, respectively, for reduction and oxidation reactions. Normally, organisms tolerate only a specific limited range of redox potentials. The specific distribution of microorganisms within the oral cavity, for example, is a reflection of the redox potentials of the various microhabitats found there. Similarly, the specific distributions of microorganisms in aquatic and soil habitats relates to the redox potentials. For example, when soils are flooded, there is limited available oxygen and anaerobic metabolism is favored. Not only does the redox potential influence where organisms reside but also their physiological activities. For example, at low redox potentials, diatoms can change from autotrophic to heterotrophic metabolism. Microbial activities also influence the redox potential of the habitat or microhabitat; some microbial activities introduce oxygen into the habitat and raise the redox potential, others scavenge oxygen and lower the redox potential. The redox potential of aquatic habitat is almost always lower than would be expected for water fully equilibrated with the atmosphere; the lower redox potential values reflect the heterotrophic activities of microorganisms. In polluted habitats, the extensive growth of microorganisms leads to very low redox potentials. Overall, the interrelationships between microbial distribution, substrate utilization, and oxygen concentration form a complex interactive network.

pH and microbial activities

The concentration of hydrogen ions, which determines **pH**, is also extremely important in establishing· the distribution of microorganisms within a given habitat (Figure 21.9). The pH has a marked effect on enzymatic activities and therefore on the other biogeochemical cycling reactions that can occur in that ecosystem. Microorganisms exhibit different tolerance ranges of pH values (Table 21.1). Microbial metabolic activities, particularly those involving oxidation-reduction reactions, influence the pH of the surrounding habitat. For example, the reduction of nitrate to molecular nitrogen and the conversion of sulfate to hydrogen sulfide will cause the pH to become

more alkaline; in contrast, the production of carbon dioxide during respiration and various end products during fermentation causes the pH to become more acidic. The acidic compounds produced by some microorganisms can eliminate other organisms from the biological community. In the extreme case of acid mine drainage, acidophilic *Thiobacillus* species drastically reduce the pH to levels where other organisms cannot survive. In addition to the direct effect of pH on microorganisms and microbial enzymes, the pH also indirectly influences microbial activities by determining the degree of dissociation of many molecules. Of particular importance are the effects of pH on (1) the solubilities of gases, such as carbon dioxide, which is required for primary productivity; (2) the availability of dissolved nutrients, such as phosphorus, which often limits productivity within ecosystems; and (3) the concentrations of toxic substances, such as heavy metals, which may inhibit microbial activities within a given habitat.

Figure 21.8

(A) Microhabitats within soil particles as seen in a section of a soil crumb; this drawing also shows the uneven distribution of bacterial microcolonies and the occurrence of water and air within pore spaces. (B) Anoxic environments develop around plant roots and residues when voids between soil particles are sealed.

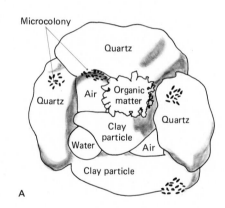

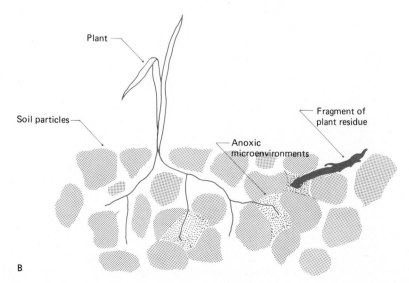

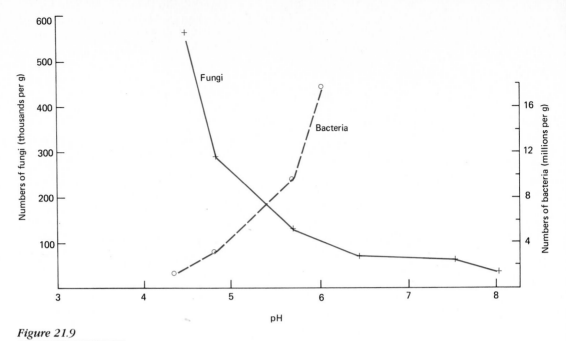

Figure 21.9

This graph shows the relationship between soil pH, the numbers of bacteria, and the numbers of fungal propagules.

The nitrogen cycle

Nitrogen fixation

Whereas carbon, hydrogen, and oxygen are actively cycled by microorganisms, plants, and animals, the **biogeochemical cycling of nitrogen** is largely dependent on the metabolic activities of microorganisms alone (Figure 21.10). Aside from the chemical fixation of molecular nitrogen by human beings to form nitrogen fertilizers, the ability to fix atmospheric nitrogen (the conversion of N_2 to ammonia or organic nitrogen) that can be assimilated into biomass is restricted, almost exclusively, to a limited number of bacterial species. Most microorganisms and all plants and animals are unable to use atmospheric nitrogen directly and depend on the availability of fixed forms of nitrogen for incorporation into their cellular biomass. Other than one exceptional case where a green alga has been shown to fix atmospheric nitrogen, this process is carried out strictly by bacteria. The productivity of many ecosystems is limited by the supply of fixed forms of nitrogen. Ammonia is the first detectable product of nitrogen fixation. It is assimilated into amino acids and subsequently synthesized into proteins and nucleic acids. Proteins, amino acids, and inorganic ammonium ions are used as a source of nitrogen by many organisms that are unable to assimilate atmospheric nitrogen directly.

It is estimated that microorganisms convert approximately 200 million metric tons of nitrogen to fixed forms of nitrogen per year compared to about 30 million metric tons produced by industrial production of nitrogen fertilizers. The fixation of atmospheric nitrogen depends on the **nitrogenase** enzyme system (Figure 21.11). Nitrogenase is very sensitive to oxygen and nitrogen

table 21.1

Lower pH limits for different microorganisms

Organism	Lower limit of pH for growth
Cyanidium caldarium	0
Thiobacillus thiooxidans	0.8
Sulfolobus acidocaldarius	0.8
Chlamydomononas acidophila	1.0
Bacillus spp.	2.0
Synechococcus spp.	4.0
Lactobacillus acidophilus	4.0
Escherichia coli	4.4
Proteus vulgaris	4.4
Enterobacter aerogenes	4.4
Pseudomonas aeruginosa	5.6
Nitrobacter spp.	6.6
Nitrosomonas spp.	7.0

fixation and, therefore, nitrogen fixation often is restricted to habitats with appropriately low levels of free oxygen; the nitrogenase enzyme is protected in some systems by leghaemoglobin, which supplies oxygen to the organisms for respiration without denaturing the nitrogenase. ATP is required to drive the reactions catalyzed by the nitrogenase enzyme system. The fixation of atmospheric nitrogen requires a high energy input (approximately 30 ATP/N_2 fixed) and in terrestrial ecosystems is largely dependent on the availability of relatively high concentrations of organic matter for use in generating ATP.

Nitrogen fixation in soil In terrestrial habitats, the microbial fixation of atmospheric nitrogen is carried out by free-living bacteria and by bacteria living in symbiotic association with plants. The relationships between microorganisms and plants involved in nitrogen fixation will be discussed later in this chapter. **Symbiotic nitrogen fixation** by *Rhizobium* is most important in agricultural fields where this bacterium lives in association with various crop plants. In forests, other symbiotic nitrogen-fixing bacteria, including actinomycetes, live in association with various trees and make significant contributions to the soil nitrogen needed to support the growth of forests. When growing in association with plants or animals,

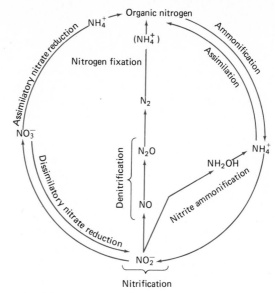

Figure 21.10

The nitrogen cycle: the chemical forms and key processes in the biogeochemical cycling of nitrogen. The critical steps of nitrogen fixation, nitrification, and denitrification are all mediated by bacteria.

Figure 21.11

A generalized scheme for the action of the nitrogenase enzyme.

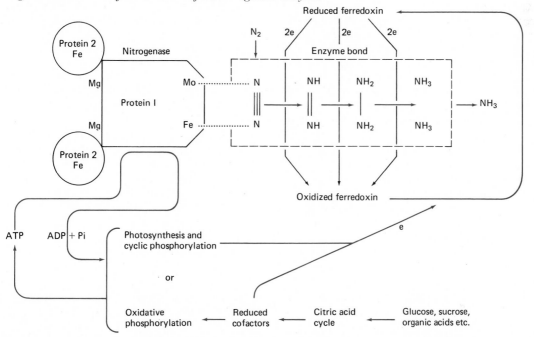

symbiotic nitrogen-fixing bacteria, such as *Rhizobium* species, generally exhibit rates of nitrogen fixation that are two to three orders of magnitude higher than are accomplished by free-living nitrogen-fixing soil bacteria. *Rhizobium* species associated with alfalfa, for example, can account for an input of 250 kg of nitrogen fixed per hectare per year, as compared to 2.5 kg of nitrogen fixed per hectare per year for free-living nitrogen-fixing *Azotobacter* species.

Through using the **acetylene reduction assay**, a method for detecting nitrogen fixation based on the fact that the nitrogenase enzyme system also catalyzes the reduction of acetylene to ethylene, readily detected by using a gas chromatograph, many free-living bacteria have now been shown to be capable of fixing atmospheric nitrogen. Most of these free-living nitrogen-fixing bacteria exhibit nitrogen-fixing activities only at oxygen levels below 0.2 atmosphere. Such conditions frequently occur in subsoil and sediment environments. Although the amount of nitrogen fixed per hectare by free-living soil bacteria is considerably lower than the amount fixed by symbiotic nitrogen-fixing species, the widespread distribution of the free-living bacteria in soil means that they make a significant contribution to the input of nitrogen to terrestrial habitats.

Nitrogen fixation in aquatic habitats   In aquatic habitats cyanobacteria, such as *Anabaena* and *Nostoc,* are very important in determining the rates of conversion of atmospheric nitrogen to fixed forms of nitrogen. Cyanobacteria, capable of nitrogen fixation, are distributed in both marine and freshwater habitats. These cyanobacteria are able to couple the ability to generate ATP through the conversion of light energy and organic matter through the reduction of carbon dioxide, with the ability to fix atmospheric nitrogen to efficiently form nitrogen-containing organic compounds. In low-nutrient aquatic environments the use of light energy for generating ATP is critical to supplying sufficient ATP to drive the nitrogen fixation reactions. Rates of nitrogen fixation by cyanobacteria are typically 10 times higher than shown by free-living bacteria and, as such, cyanobacteria form a very important component of aquatic food webs.

Ammonification

Microorganisms also perform important transformations of organic and inorganic fixed forms of nitrogen that alter the availability of needed nitrogen resources within aquatic and terrestrial ecosystems. Nitrogen in organic matter occurs predominantly in the reduced amino form, such

as occurs in amino acids. Many microorganisms as well as plants and animals are capable of converting organic amino nitrogen to ammonia (ammonification). Deaminase enzymes play an important role in this process of ammonification, which transfers nitrogen from organic to inorganic forms of nitrogen. The microbial decomposition of urea, for example, results in the release of ammonia, which may be returned to the atmosphere or in neutral aqueous environments may occur as ammonium ions. Ammonium ions can be assimilated by a number of organisms, continuing the transfer of nitrogen within the nitrogen cycle.

Nitrification

Although many organisms are capable of ammonification, relatively few are capable of **nitrification**, the process in which ammonium ions are initially oxidized to nitrite ions and subsequently to nitrate ions. Both the oxidation of ammonia to nitrite and the oxidation of nitrite to nitrate, the two steps of nitrification, are energy-yielding processes from which chemolithotrophic bacteria are able to derive needed energy. Relatively low amounts of ATP, however, are generated by the oxidation of inorganic nitrogen compounds. Therefore, large amounts of inorganic nitrogen compounds must be transformed in order to generate sufficient ATP to support the growth of these chemolithotrophic bacteria. The oxidation of approximately 35 moles of ammonia is required to support the fixation of 1 mole of carbon dioxide and approximately 100 moles of nitrite must be oxidized to support the fixation of 1 mole of carbon dioxide. As a consequence of the high amounts of nitrogen that must be transformed to support the growth of chemolithotrophic bacterial populations, the magnitude of the nitrification process is typically very high, whereas the growth rates of nitrifiers are generally relatively low compared to other bacteria.

The two steps of nitrification, the formation of nitrite from ammonium and the formation of nitrate from nitrite, are carried out by different microbial populations (Table 21.2). For the most part, the oxidative transformations of inorganic nitrogen compounds in the nitrification process are restricted to several species of autotrophic bacteria. In addition to the chemolithotrophic nitrifying bacteria, some heterotrophic bacteria and fungi are capable of oxidizing inorganic nitrogen compounds, but the rates of heterotrophic nitrification are normally four orders of magnitude lower than rates of autotrophic nitrification. In soils *Nitrosomonas* is the dominant bacterial ge-

table 21.2

Genera of nitrifying bacteria

Genus	Converts	Habitat
Nitrosomonas	Ammonia to nitrite	Soils, freshwater, marine
Nitrosospira	Ammonia to nitrite	Soils
Nitrosococcus	Ammonia to nitrite	Soils, freshwater, marine
Nitrosolobus	Ammonia to nitrite	Soils
Nitrobacter	Nitrite to nitrate	Soils, freshwater, marine
Nitrospina	Nitrite to nitrate	Marine
Nitrococcus	Nitrite to nitrate	Marine

nus involved in the oxidation of ammonia to nitrite, and *Nitrobacter* is the dominant genus involved in the oxidation of nitrite to nitrate. Several other autotrophic bacteria, including ammonia-oxidizing members of the genera *Nitrosospira, Nitrosococcus,* and *Nitrosolobus,* and nitrite-oxidizing members of the general *Nitrospira* and *Nitrococcus*, are also important nitrifiers in different ecosystems. Many of the nitrifying bacteria contain extensive internal membrane networks that are probably the sites of nitrogen oxidation (Figure 21.12).

Because relatively few microbial genera make significant contributions to the rates of nitrification, it is not surprising that this process is particularly sensitive to environmental stress. Toxic chemicals can block the nitrification process. Nitrification is an obligately aerobic process, and under anaerobic conditions, such as may exist when high concentrations of organic matter are added to soil or aquatic ecosystems, the nitrification process may cease. The process of nitrification is very important in soil habitats because the transformation of ammonium ions to nitrite and nitrate ions results in a change from a cation to an anion. Positively charged cations are bound

Figure 21.12

Micrograph of the internal membranes of the nitrifying bacteria Nitrosomonas europaea *(83,000 ×). (Courtesy Stan W. Watson, Woods Hole Oceanographic Institution, Woods Hole, Massachusetts, reprinted by permission of the American Society for Microbiology, from S. W. Watson and M. Mandel, 1971,* Journal of Bacteriology, *107: 563–569.)*

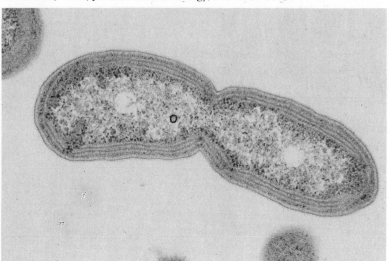

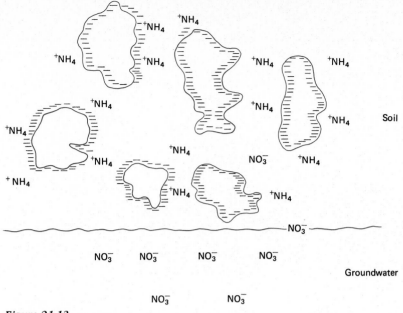

Figure 21.13

Drawing showing the relationship of ammonium and nitrate ions to soil particles; the positively charged ammonium ions are retained and the negatively charged nitrate ions are leached.

by negatively charged soil clay particles and thus are retained in soils, but negatively charged anions, such as nitrate, are not absorbed by soil particles and are readily leached from the soil (Figure 21.13). Nitrification, therefore, represents a mobilization process in soils that results in the transfer of inorganic fixed forms of nitrogen from surface soils to subsurface groundwater reservoirs. In agriculture, inhibitors of nitrification, such as nitrapyrin, sometimes are intentionally added to soils to prevent the transformation of ammonium to nitrate, ensuring better fertilization of crops.

The transfer of nitrate and nitrite ions from surface soil to groundwater supplies is critical for two reasons: (1) it represents an important loss of nitrogen from the soil, where it is needed to support the growth of higher plants; and (2) high concentrations of nitrate and nitrite in drinking water supplies pose a serious human health hazard. Nitrite is toxic to humans because it can combine with blood hemoglobin to block the normal gas exchange with oxygen. Additionally, nitrites can react with amino compounds to form highly carcinogenic nitrosamines. Further, nitrate, although not highly toxic itself, can be microbially reduced in the gastrointestinal tracts of human infants to form nitrite, causing "blue baby syn-

drome"; this reduction of nitrate does not occur in adults because of the low pH of the normal adult gastrointestinal tract. Nitrate and nitrite in groundwater is a particular problem in agricultural areas, such as the "cornbelt" of the Midwestern United States, where high concentrations of nitrogen fertilizers are applied to soil. The use of nitrification inhibitors in combination with the application of ammonium nitrogen fertilizers can minimize the nitrate leaching problem, while at the same time supporting better soil fertility and increased plant productivity.

Denitrification

Denitrification, the conversion of fixed forms of nitrogen to molecular nitrogen, is another important process in the biogeochemical cycling of nitrogen that is mediated by microorganisms. A limited number of bacteria, including some species of *Pseudomonas, Moraxella, Spirillum, Thiobacillus,* and *Bacillus,* are capable of denitrification. Denitrification occurs when nitrate ions serve as terminal electron acceptors in anaerobic respiration, the reduction of nitrate leading to the production of gaseous forms of nitrogen. The reduction of nitrate ions can produce various gaseous nitrogenous compounds, including nitric oxide, nitrous oxide, and molecular nitrogen. Nitrous oxide formation occurs preferentially in habitats with high nitrate concentrations and/or low pH values. Formation of molecular nitrogen is favored when there is an adequate supply of organic matter to supply energy. Dissimilatory nitrate reductase, the enzyme involved in the initiation of the denitrification process, is inhibited by oxygen and denitrification generally occurs under anaerobic conditions. The return of nitrogen to the atmosphere by the denitrification process completes the nitrogen cycle (see Figure 21.10).

The sulfur cycle

Sulfur can exist in a variety of oxidation states within organic and inorganic compounds, and oxidation–reduction reactions—mediated by microorganisms—change the oxidation states of sulfur within various compounds, establishing the **sulfur cycle** (Figure 21.14). Microorganisms are capable of removing sulfur from organic compounds. Under aerobic conditions the removal of sulfur (**desulfurization**) of organic compounds results in the formation of sulfate, whereas under anaerobic conditions **hydrogen sulfide** is normally produced from the mineralization of organic sulfur compounds (Figure 21.15). Hydrogen sulfide may also be formed by sulfate-reducing bacteria that utilize sulfate as the terminal electron acceptor during anaerobic respiration. Hydrogen sulfide can accumulate in toxic concentrations in areas of rapid protein decomposition, is highly reactive, and is very toxic to most biological systems. The predominant source of hydrogen sulfide in different habitats varies. In organically rich soils most of the hydrogen sulfide is generated from the decomposition of organic sulfur-containing compounds. In anaerobic sulfate-rich marine sediments, most of the hydrogen sulfide is generated from the dissimilatory reduction of sulfate by sulfate-reducing bacteria, such as members of the genus *Desulfovibrio*. As discussed in Chapter 20, anaerobic sulfate reduction is important in corrosion processes. This process also is important in the biogeochemical cycling of sulfur.

Use of hydrogen sulfide by autotrophic microorganisms

Although hydrogen sulfide is toxic to many microorganisms, the photosynthetic sulfur bacteria use hydrogen sulfide as an electron donor for generating reduced coenzymes during their

Figure 21.14

The sulfur cycle, showing various transformations of organic and inorganic compounds.

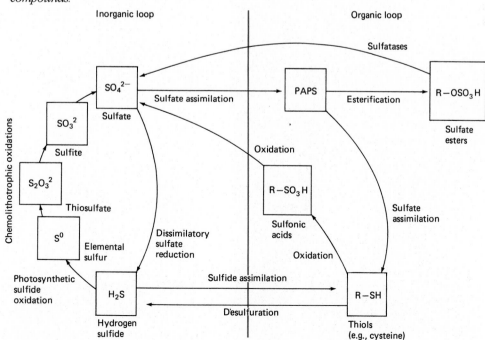

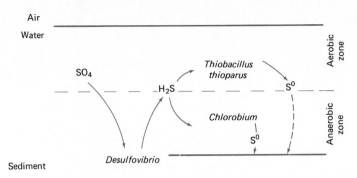

Air

Water

SO$_4$

H_2S

Thiobacillus thioparus

Chlorobium

S^0

S^0

Aerobic zone

Anaerobic zone

Sediment

Desulfovibrio

Figure 21.15

The products of sulfur transformations in aerobic and anaerobic habitats.

photosynthetic metabolism. The anaerobic photosynthetic bacteria often occur on the surface of sediments where there is available light to support their photosynthetic activities and a supply of hydrogen sulfide from dissimilatory sulfate reduction and anaerobic degradation of organic sulfur-containing compounds. Some photosynthetic bacteria deposit elemental sulfur as an oxidation product, whereas others form sulfate.

Some bacteria, including members of the genera *Beggiatoa* and *Thiothrix,* are capable of generating ATP by oxidizing hydrogen sulfide. These bacteria deposit elemental sulfur globules within the cell, which in the absence of hydrogen sulfide can be further oxidized to sulfate (Figure 21.16). *Beggiatoa* and *Thiothrix* are not chemolithotrophs, and although energy is apparently derived from the oxidation of hydrogen sulfide, these organisms require organic carbon for growth. Chemolithotrophic members of the genus *Thiobacillus* oxidize sulfur as their source of energy. The use of *Thiobacillus* species in bioleaching has been discussed in Chapter 19. Some *Thiobacillus* species are acidophilic and grow well at pH 2–3, and the growth of such *Thiobacillus* species can produce sulfate from the oxidation of elemental sulfur, leading to the environmental accumulation of sulfuric acid.

Acid mine drainage

Acid mine drainage is a consequence of the metabolism of sulfur and iron-oxidizing bacteria. Coal, in geological deposits, is often associated with pyrite (FeS$_2$), and when coal mining activities expose pyrite ores to atmospheric oxygen, the combination of autoxidation and microbial sulfur and iron oxidation produces large amounts of sulfuric acid. Any time pyrites are mined as part of an ore recovery operation, oxidation may produce large amounts of acid. The acid draining from mines kills aquatic life and renders the water it contaminates unsuitable as a drinking water supply or for recreational uses. At present, approximately 10,000 miles of U.S. waterways are affected in this manner, predominantly in the states of Pennsylvania, Virginia, Ohio, Kentucky, and Indiana. Strip mining is a particular problem with respect to acid mine drainage because this method of coal recovery removes the overburden, leaving a porous rubble of tailings exposed to oxygen and percolating water. Oxidation of the reduced iron and sulfur in the tailings produces acidic products, causing the pH to drop rapidly and pre-

772

Microbial ecology

Figure 21.16

Micrograph of a Beggiatoa *sp., showing the occurrence of intracellular sulfur granules (1000 ×). (From BPS—Paul W. Johnson, University of Rhode Island.)*

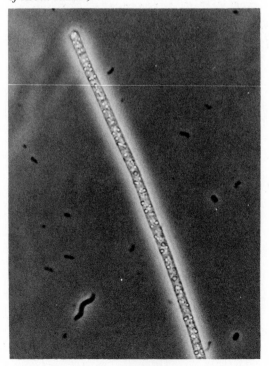

venting the reestablishment of vegetation and a soil cover that would seal the rubble from oxygen. A strip-mined piece of land continues to give rise to acid mine drainage until most of the sulfide is oxidized and leached out; recovery of this land may take 50–150 years.

The overall reaction for the oxidation of pyrite can be summarized as

$$2FeS_2 + 7\tfrac{1}{2}O_2 + 7H_2O \rightarrow 2Fe(OH)_3 + 4H_2SO_4$$

The sulfuric acid produced accounts for the high acidity, and the precipitated ferric hydroxide for the deep brown color of the effluent. The mechanism of pyrite oxidation in acid mine drainage is quite complex. At neutral pH, oxidation by atmospheric oxygen occurs rapidly and spontaneously, but below pH 4.5, autoxidation is slowed down drastically. In the pH range of 4.5 to 3.5, the stalked iron bacterium *Metallogenium* catalyzes the oxidation of iron. As the pH drops below 3.5, the acidophilic bacteria of the genus *Thiobacillus* oxidize the reduced iron sulfide in the pyrite. The rate of the microbial oxidation of FeS

is several hundred times higher than the rate of spontaneous oxidation, and although pyrite oxidation starts spontaneously, microbial oxidation of sulfur and iron is responsible for the continued production of high levels of acid mine drainage.

Other element cycles

Phosphorus
Phosphorus normally occurs as phosphates in both inorganic and organic compounds. Microorganisms assimilate inorganic phosphate and mineralize organic phosphorus compounds; microbial activities are involved in the solubilization or mobilization of phosphate compounds (Figure 21.17). Unlike the other elements discussed, microorganisms normally do not oxidize or reduce phosphorus. In many habitats phosphates are combined with calcium, rendering them insoluble and unavailable to most organisms. Various heterotrophic microorganisms are capable of solubilizing phosphates primarily through the production of organic acids. These actions of

Figure 21.17

The phosphorus cycle, showing various transfers, none of which alters the oxidation state of the phosphate.

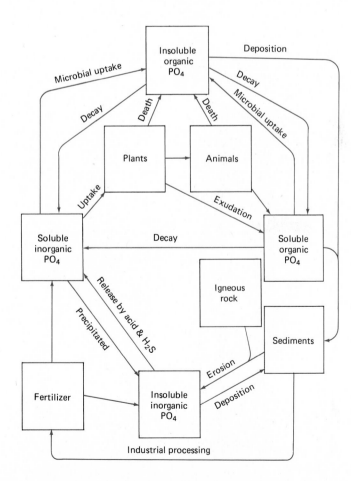

Figure 21.18

An algal bloom in a small lake resulting from eutrophication. (Courtesy Ronald F. Lewis, USEPA, Cincinnati.)

microorganisms mobilize phosphate, and activities of other microorganisms act to immobilize phosphorus. For example, microorganisms compete with plants for available phosphate resources because the assimilation of phosphates by microorganisms removes phosphates from the available nutrient pool required by plants.

In many habitats productivity is limited by the availability of phosphate. When excess concentrations of phosphate enter aquatic habitats, such as when phosphate detergents are added to lakes, there can be a sudden increase in productivity, a process called **eutrophication** (Figure 21.18). The blooms of algae and cyanobacteria associated with eutrophication can greatly increase the concentrations of organic matter in bodies of water. During the subsequent decomposition of this organic matter, oxygen can be severely depleted from the water column, causing major fish kills. The introduction of high concentrations of phosphate from phosphate laundry detergents created such eutrophication problems in many water bodies that some municipalities banned the use of such detergents.

Although phosphate is not normally reduced by microorganisms, it appears that some microorganisms are capable of using phosphate as a terminal electron acceptor in anaerobic respiration pathways under appropriate environmental conditions. Phosphate can serve as a terminal electron acceptor only in the absence of sulfate, nitrate, and oxygen. The product of phosphate reduction is phosphine (PH_3), which is volatile and spontaneously ignites upon contact with oxygen, producing a green glow. The production of phosphine is sometimes observed near burial sites and swamps where there is extensive decomposition of organic matter, giving rise to "ghostly" phenomena.

Iron

The cycling of **iron compounds** has a marked effect on the availability of this essential element for other organisms. Iron is transformed between ferrous (Fe^{2+}) and ferric (Fe^{3+}) oxidation states by microorganisms (Figure 21.19). Ferric and fer-

Figure 21.19

The iron cycle, showing interconversion of ferrous and ferric iron.

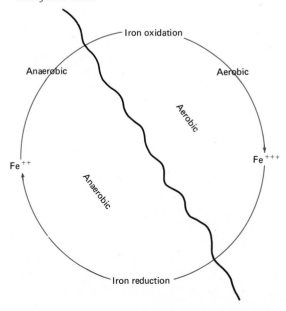

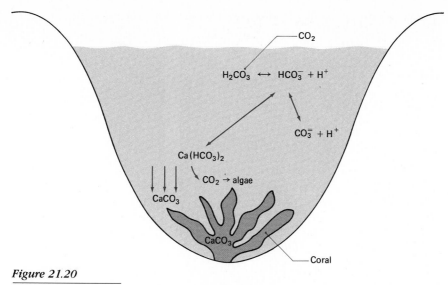

Figure 21.20

The reactions of calcium in seawater lead to the formation of coral reefs.

rous ions have very different solubility properties, ferric compounds tending to be less soluble than ferrous compounds. Various bacteria, including members of the genera *Thiobacillus, Gallionella,* and *Leptothrix*, are capable of oxidizing iron compounds. Some of these bacteria deposit ferric hydroxide in an extracellular sheath. The activities of iron-oxidizing bacteria can lead to the formation of substantial iron deposits. As discussed earlier, these bacteria are also important in corrosion processes and the formation of acid mine drainage.

Calcium

Calcium also exhibits biogeochemical cycling between soluble and insoluble forms. Calcium bicarbonate is soluble, but calcium carbonate is much less soluble. The microbial production of acidic compounds solubilizes precipitated and immobilized calcium compounds. There is an interesting cycling of calcium in marine habitats where dissolved carbon dioxide reacts with available calcium, forming calcium bicarbonate and calcium carbonate (Figure 21.20). During the formation of coral, calcium carbonate precipitates when carbon dioxide held in solution as calcium bicarbonate is removed by algal cells of the coral. This process results in the deposition of calcium carbonate and the formation of coral reefs. Calcium carbonate is also precipitated by various algae to form an outer frustule. Accumulation of algal frustules can lead to the formation of major limestone deposits, such as the famous White Cliffs of Dover on the English Channel coast of Britain (Figure 21.21).

Figure 21.21

The White Cliffs of Dover represent a massive accumulation of diatoms. (From Photoresearchers, Inc., photo by Paolo Koch.)

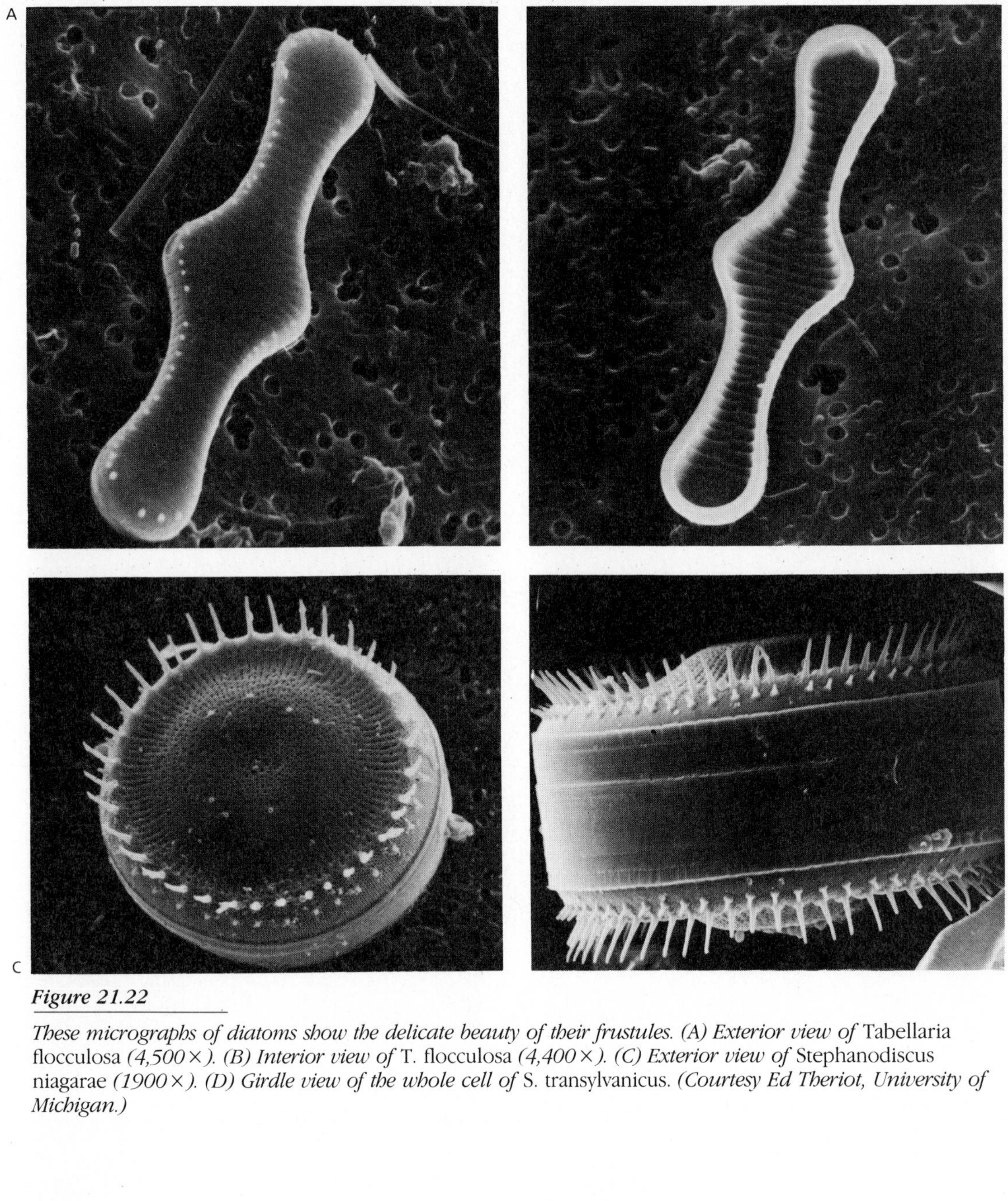

Figure 21.22

These micrographs of diatoms show the delicate beauty of their frustules. (A) Exterior view of Tabellaria
flocculosa *(4,500×). (B) Interior view of* T. flocculosa *(4,400×). (C) Exterior view of* Stephanodiscus
niagarae *(1900×). (D) Girdle view of the whole cell of* S. transylvanicus. *(Courtesy Ed Theriot, University of
Michigan.)*

Silicon

Various algae, most notably the diatoms, form sil-
icon-impregnated structures (Figure 21.22). These
algae precipitate silicon dioxide to build their
delicate and decorative shells. As much as 10 billion metric tons of silicon dioxide are precipitated by microorganisms in the oceans each year.
The shells of these dead microorganisms accumulate and form silicon-rich oozes that later develop into deposits of diatomaceous or Fuller's
earth.

Manganese

Manganese exists as a water-soluble divalent manganous ion and as a relatively insoluble tetravalent manganic ion. The microbial oxidation of
manganous ions forms manganese oxides, which
produce characteristic **manganese nodules** (Figure 21.23). The manganese for the nodules originates in anaerobic sediments; when the manganese enters aerobic habitats it is oxidized and
precipitates to form the nodules. The farming of
manganese nodules in deep ocean sediments is

Figure 21.23

A manganese nodule showing associated bacteria and algae. (A) Scanning electron micrograph showing microcolonies of bacteria. (B) Scanning electron micrograph of a region of the nodule covered with debris of cocolith algae. (Courtesy Paul LaRock, Florida State University.)

considered a possible source for obtaining manganese for industrial usage.

Heavy metals

Mercury, arsenic, and other heavy metals are also subject to microbial biogeochemical cycling. These transformations are important because they alter both the mobility and toxicity of the metals. For example, mercury is released into the environment largely as a consequence of its widespread use in industry and the burning of fossil fuels, although some mercury also leaches from rocks.

The methylation of mercury causes increased toxicity and **biomagnification** (Figure 21.24). Mercury salts, though fairly toxic, are excreted efficiently, and therefore, their release into the environment was not originally viewed with much concern, but in anaerobic sediments, some microorganisms are capable of methylating mercury. The product methylmercury is lipophilic and is readily concentrated in filter-feeding shellfish. Unlike inorganic and phenylmercury compounds, methylmercury is excreted by humans only very slowly, having a half-life of 70 days, and

Figure 21.24

The mercury cycle, showing the biological magnification of methylmercury in the aquatic food chain.

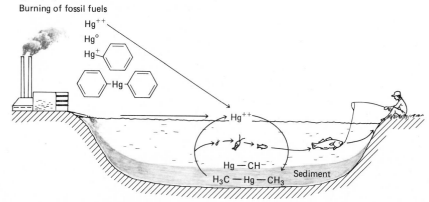

it is highly neurotoxic. In Japan in the 1950s, the ingestion of shellfish containing methylmercury led to outbreaks of **Minamata disease**, a severe disturbance of the central nervous system associated with mercury poisoning. The buildup of methylmercury compounds in Scandinavian freshwater lakes and the North American Great Lakes forced large areas to be condemned for fishing.

Population interactions

After considering the role of microorganisms in modifying their biochemical surroundings, we now turn to another aspect of microbial ecology: **the interactions between diverse microbial populations.** Interactions between two different biological populations can be classified according to whether both populations are unaffected by the interaction, one or both populations benefit from the interaction, or one or both populations are adversely affected by the interaction (Table 21.3). Within a biological community various types of interactions can occur between diverse microbial populations, between microbial and plant populations, and between plant and animal populations. The interactions between populations within a community are dependent on the environmental conditions of the habitat and under different environmental conditions the same populations can exhibit different interpopulation relationships.

The **positive interactions** between biological populations enhance the abilities of the interacting populations to survive within the community of a particular habitat, sometimes permitting populations to exist in a habitat where the individual populations cannot exist alone. Development of positive interactions permits microorganisms to use available resources more efficiently in concert than can be accomplished by an individual microbial population growing alone. Mutualistic relationships between microbial populations create essentially new organisms that are capable of occupying niches that cannot be occupied by either organism alone. For example, lichens, which represent a symbiotic assemblage of microorganisms, grow on rock surfaces where the individual populations cannot.

Negative interactions between populations act as feedback mechanisms that limit population densities. In some cases negative interactions may result in the elimination of a population that is not well adapted for continued existence within the community of a given habitat. Within stable communities negative interactions are themselves adaptive and ensure the maintenance of a balance between populations within the biological community. The negative feedback interactions limit population densities and provide a self-regulation mechanism beneficial for the overall population in the long term because they prevent overpopulation and destruction of the habitat's resources. Negative interactions also tend to preclude the

table 21.3

Classification of population interactions

	Effect of interaction	
Name of interaction	Population A	Population B
Neutralism	0	0
Commensalism	0	+
Synergism (protocooperation)	+	+
Mutualism (symbiosis)	+	+
Competition	−	−
Amensalism	0 or +	−
Predation	+	−
Parasitism	+	−

0 = no effect
+ = positive effect
− = negative effect

invasion of an established community composed of **autochthonous** (indigenous) populations by **allochthonous** (foreign) populations and thus act to maintain community stability. For example, we have already considered the contribution of the natural microbiota of humans to the invasion of that ecosystem by allochthonous pathogens.

Neutralism

The first type of possible interactions that we will discuss, **neutralism**, actually represents a lack of interaction between two populations. Neutralism is rare but can occur between populations that are physically removed from each other. A lack of interaction is more likely at low population densities where organisms are not likely to physically contact each other than when population densities are high. Dormant resting stages of microorganisms are more likely to exhibit relationships of neutralism with other microbial populations than are actively growing vegetative cells. Low rates of metabolic activity, which characterize the resting stages of microorganisms, favor a lack of interaction. Low population densities and the formation of resting stages by some microbial populations allow for temporal and spatial niches within a habitat and permit many populations to coexist within a habitat. Under these conditions organisms coexist without competing for the same available resources in the habitat. Microorganisms do not occupy the same niche within a community at the same time because competition between the microbial populations for the same resources in such cases leads to elimination of the least fit population and retention within the community of the most adaptive population.

Physical separation does not ensure a lack of interaction. Modifications of habitat caused by one microbial population may positively or negatively affect another population. For example, hydrogen sulfide produced by microorganisms can diffuse through the soil, killing plant roots and adversely affecting plant populations. The formation of resting stages likewise does not ensure neutralism with other microbial populations. Some microbial populations produce enzymes that are able to degrade the spores and other resting stages of other microbial populations (Figure 21.25). Many such resting stages, though, are resistant to negative interactions with other microorganisms. The biochemically complex outer layers of endospores, for example, are resistant to enzymatic attack by most other microorganisms.

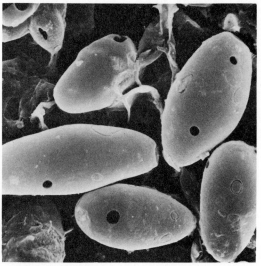

Figure 21.25

The conidia of the fungus Cochliobolus sativus *are being attacked by a giant soil ameba,* Arachuula impatiens; *note the resulting perforations and annular depressions in their spore walls. (Courtesy Kenneth M. Old, CSIRO, Canberra, Australia.)*

Commensalism

In a commensal relationship between two populations one population benefits and the other population is unaffected. By definition **commensalism** is a unidirectional relationship between populations. Commensal relationships between diverse microbial populations are quite common, often occurring when the unaffected population modifies the habitat in such a way that a second population benefits. For example, the removal of oxygen from a habitat, as a result of the metabolic activities of a population of facultative anaerobes, creates an environment that is favorable for the growth of obligately anaerobic populations. The lowered oxygen tension favors the anaerobic bacteria, and assuming a lack of competition for the same available substrates, the obligate anaerobes do not effect the existence of the facultative organisms. Various other chemical modifications of the environment of a given habitat by a given microbial population likewise may benefit other populations without any negative or positive feedback interactions.

In some cases a microbial population can physically alter a habitat, permitting a second population to exist. Production of polysaccharides by *Streptococcus sanguis* on a tooth surface permits the adhesion and successful colonization of the tooth by other bacterial species, including *S. mutans.* Similarly, a primary infection caused by one microbial species may allow opportunistic pathogens to establish secondary infections, with the opportunistic pathogens benefitting from their ability to invade and multiply within the host organism without adversely affecting the primary pathogen population.

Many commensal relationships between microbial populations are based on the production of **growth factors.** Some bacterial populations produce and excrete growth factors, such as vitamins, that can be utilized by other populations. As long as the growth factors are produced in excess and excreted from an organism, a commensal interaction can occur. For example, fastidious microorganisms often depend on growth factors released from other organisms. Some marine bacteria, growing within the water column, depend on specific amino acids and/or vitamins produced by surface algae. Often, it is difficult to culture such bacteria on defined media because of a lack of understanding of the organism's growth factor requirements.

Cometabolism, which occurs when an organism growing on a particular substrate gratuitously oxidizes a second substrate that it is unable to assimilate, forms the basis for many commensal relationships. Although the organism responsible for the transformation does not benefit, other populations may use the oxidation products. For example, *Mycobacterium vaccae,* growing on propane as a source of carbon and energy, will gratuitously oxidize cyclohexane to cyclohexanone (Figure 21.26). The cyclohexanone can be used by other microorganisms, such as populations of *Pseudomonas* species. In such a case the *Pseudomonas* benefit because they are unable to metabolize cyclohexane; the *Mycobacterium* is unaffected because it does not assimilate cyclohexanone. In a similar sense the waste products of one organism may be a favorable substrate for the growth of another organism. Coprophagous fungi, for example, live on the fecal material of animal populations. The fungi benefit from the animal's deposition of fecal material, and the members of that animal population are unaffected by the relationship.

Still another basis for commensal relationships is the **removal or neutralization of toxic materials.** The oxidation of hydrogen sulfide by a microbial population, for example, can lower the concentrations of this toxic material to levels at which other populations can exist. The existence of other microbial populations may not positively or negatively affect the hydrogen sulfide-oxidizing microorganisms.

A commensal relationship may also be estab-

Figure 21.26

The cometabolism of cyclohexane by Mycobacterium vaccae *in the presence of propane is an example of commensalism based on cometabolism, which allows the growth of* Pseudomonas *on the cyclohexane.*

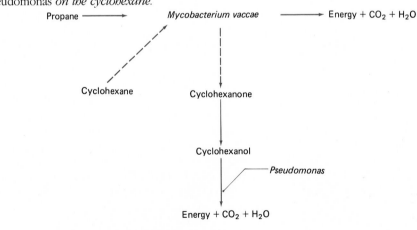

lished when one organism grows on the surface of another organism. For example, bacterial populations growing on the skin surface generally exhibit a commensal relationship with human beings. The bacterial populations benefit from being able to grow on a surface and from the nutrients and water provided in human sweat. Humans are not adversely affected, nor do they necessarily directly benefit from the growth of various microbial populations on the skin surface. Many animals have naturally occurring epiphytic bacterial populations (Figure 21.27). Analogously, **epiphytic bacteria** grow as commensal populations on many plant surfaces. Epiphytic bacteria colonize the surfaces of algae, benefiting from the metabolic activities of the algae (Figure 21.28).

Synergism

Synergism or **proto-cooperation** between two populations indicates that both populations benefit from the relationship but that the association is not obligatory. Both populations are capable of surviving independently, although they both gain advantage from the synergistic relationship. Synergistic relationships are loose in that one member population can readily be replaced by an-

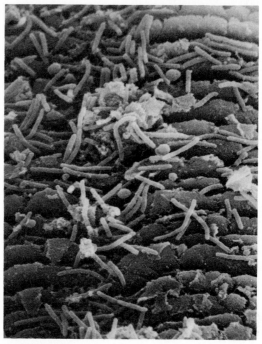

Figure 21.27

Scanning electron micrograph showing epiphytic bacteria growing on a tapeworm removed from the gut of a shark (1700×). Bacteria can be found growing on the surfaces of most animals. (Courtesy Robert Apkarian, University of Louisville.)

Figure 21.28

Algae and plant surfaces are colonized by various epiphytic bacteria. (A) The epiphytic bacteria and diatoms, Licmorphora *sp., on the filamentous surface of the alga* Pylaiella littoralis. *This micrograph was made with a Nomarski differential interference contrast microscope (480×). (B) Filaments of the bacterium* Leucothrix mucor *growing on the surface of the red alga* Polysiphonia *(40×). (From BPS—Paul W. Johnson, University of Rhode Island.)*

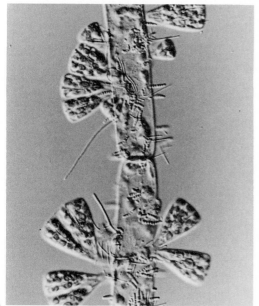

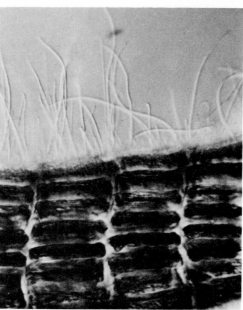

A

B

other population. In some cases it is difficult to distinguish between commensalism and synergism. One form of synergism, **syntrophism**, occurs as a result of **crossfeeding**, in which the two populations supply each other's nutritional needs. Synergistic activities of two microbial populations may allow completion of a metabolic pathway that neither organism is capable of carrying out alone. As a result of such cooperative metabolism, both organisms can derive carbon and energy from a substrate. In a theoretical example of crossfeeding (Figure 21.29), population 1 is able to metabolize compound A, forming compound B, but is unable to carry out the next step in a metabolic pathway; population 2 is unable to use compound A but can use compound B, forming compound C. Both populations 1 and 2 are able to carry out the subsequent metabolic steps of the pathway, producing needed energy and metabolic products, and by acting jointly the two populations are able to complete the pathway.

In a specific example of syntrophism, *Streptococcus faecalis* and *Escherichia coli* are able to convert arginine to putrescine together, although neither organism can carry out the transformation alone (Figure 21.30). *S. faecalis* is able to convert arginine to ornithine that can then be used by a population of *E. coli* to produce putrescine; *E. coli* growing alone can transform arginine to produce agmatine but cannot convert arginine to putrescine. *Lactobacillus arabinosus* and *S. faecalis* similarly can establish a synergistic relationship based on the mutual exchange of required growth factors (Figure 21.31). *L. arabinosus* requires

phenylalanine for growth, which is produced by *S. faecalis*. *S. faecalis* requires folic acid, which is produced by *L. arabinosus*. In a minimal medium both populations can grow together, but neither population can grow alone.

Synergistic interactions between plants and microorganisms are important in providing the nutritional requirements of both the plants and the associated microorganisms. Within the **rhizosphere**, the soil region in close contact with plant roots, the plant roots exert a direct influence on the soil bacteria, known as the **rhizosphere** effect. Likewise, the microbial populations in the rhizosphere have an important influence on the growth of the plant. As a consequence of these interactions, microbial populations reach much higher densities in the rhizosphere than in the free soil, and plants exhibit enhanced growth characteristics. The interactions of plant roots and rhizosphere microorganisms are based largely on interactive modification of the soil chemical environment by processes such as: water uptake by the plant system; release of organic chemicals to the soil by the plant root; microbial production of plant growth factors; and microbially mediated availability of mineral nutrients. Within the rhizosphere, the bacteria obtain required growth factors, such as amino acids, from the plant roots. In turn, the microorganisms in the rhizosphere have a marked influence on the growth of plants. In the absence of appropriate rhizosphere microbial populations, plant growth may be impaired. Microbial populations in the rhizosphere may benefit the plant by (1) removing hydrogen sul-

Figure 21.29

As shown in this theoretical example, cross feeding establishes the basis for synergistic relationships.

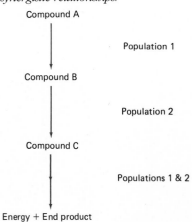

Figure 21.30

The synergistic relationship between E. coli *and* S. faecalis *allows the production of putrescine from arginine.*

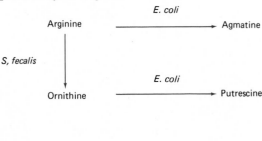

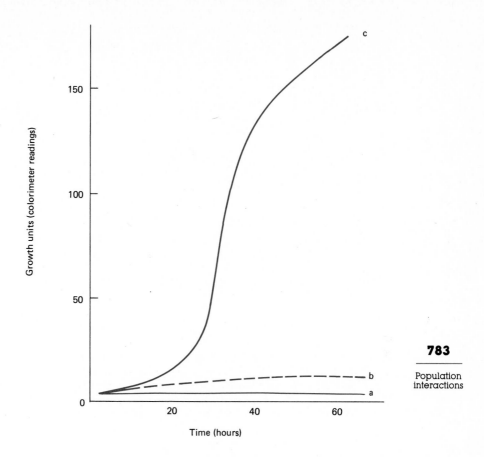

Figure 21.31

The synergistic relationship between S. faecalis *and* L. arabinosus *in a medium lacking phenylalanine and folic acid: (a)* S. faecalis, *which requires folic acid; (b)* L. arabinosus, *which requires phenylalanine; and (c) combined culture.*

fide toxic to the roots; (2) increasing solubilization of mineral nutrients; (3) synthesizing vitamins, amino acids, auxins, and gibberellins that stimulate plant growth; and (4) antagonizing potential plant pathogens through competition and developing amensal relationships based on the production of antibiotics. Allelopathic substances (substances formed by an organism that inhibit another organism) produced by microorganisms in the rhizosphere may allow plants to enter amensal relationships with other plant populations. Allelopathic substances surrounding some plants prevent invasions of that habitat by other plants, and such amensal relationships between plant populations may actually be based on synergistic relationships between plant and microbe.

Mutualism

Mutualism or **symbiosis** is an obligatory interrelationship between two populations that benefits both populations. Because all relationships can be viewed as beneficial in the overall sense of maintaining ecological balance, symbiosis was originally used to denote any intimate relationship between two populations, whether or not the species benefited. The present use of the term symbiosis, however, is restricted to situations where both organisms benefit and where the relationship is obligatory. Mutualism is an extension of synergism, allowing populations to unite and establish essentially a single unit population that can occupy habitats not favorable for the existence of either population alone. Mutualistic relationships may lead to the evolution of new organisms, and various theories of evolution point to the structural similarities between mitochondria, chloroplasts, and prokaryotic cells to indicate that the development of eukaryotic organisms was based on establishing endosymbiotic relationships with bacteria. In a more specific example, the protozoan *Mixotricha paradoxa* is propelled by rows of attached spirochetes rather than by conventional cilia (Figure 21.32).

Microbe–microbe interactions

The relationships between some heterotrophic fungi and their photosynthetic algal or cyanobacterial partners in the formation of **lichens** is an

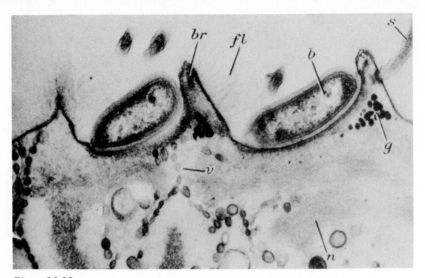

Figue 21.32

A section through the cell surface of the flagellate Mixotricha paradoxa *(31,000 ×). Two bacteria (b) and parts of three brackets (br) are shown, together with numerous spirochetes. Note the long filaments (fl) on the posterior surfaces of the brackets, the dense granules (g) in their bases, the strands of the fibrous network (n), and the chains of small vacuoles (v). (Reprinted by permission of the Royal Society of London, from L. R. Cleveland and A. V. Grimstone, 1964,* Proceedings of the Royal Society of London (series B), *159: 683.)*

Discovery process

Grimstone and Cleveland's careful microscopic studies showed that *Mixotricha paradoxa,* isolated from the intestines of termites, is a bizarre composite form; the host organism is a protozoan, a polymastigote flagellate with one forward and three trailing typical eukaryotic flagella. It moves in a weird straight motion propelled by spirochetes, which are symbionts regularly arranged on the cortex of the host. The four normal complex eukaryote flagella seem to act as rudders, causing changes of direction of the host movement. These cortical spirochetes are attached at specific sites, each connected to a raised portion of the host and simultaneously to a second extracellular symbiotic bacterium. The spirochete, host cortex, and associated cortical bacterium are patterned so specifically that the symbiosis must be the result of many years of evolution. This organism, which is normally subjected to the low oxygen levels of the termite gut, lacks mitochondria. Another symbiont, an endosymbiotic bacterium, was observed surrounded by host endoplasmic reticulum and in a particular spatial arrangement with the cortical spirochetes. Grimstone and Cleveland suggested that this might be the functional replacement of the mitochondria and the source of ATP. Thus, *Mixotricha* normally harbors three independent microbial prokaryotic symbionts: cortical spirochetes and their associated bacteria that provide the anomalous movement; endosymbiotic bacteria that perhaps replace mitochondria; the *Mixotricha* itself is symbiotic in the intestine of its termite host. The discovery of the nature of *Mixotricha* lends support to those who hypothesize that major evolutionary steps resulted from endosymbiotic relationships which eventually led to the formation of new organisms in which the symbiotic partners had lost their individual identities.

excellent example of a mutualistic intermicrobial population relationship that results in the formation of an essentially new organism. Lichens are composed of a primary producer, the **phycobiont**, and a consumer, the **mycobiont** (Figure 21.33). The lichen has totally different physiological properties than either of the individual species of which it is composed. Lichens can grow in habitats, such as on rock surfaces, where neither algae nor fungi can exist alone. Most lichens are resistant to extremes of temperatures and desiccation, a particular advantage on exposed surface habitats. The lichen is very much a self-sufficient organism. The phycobiont utilizes light

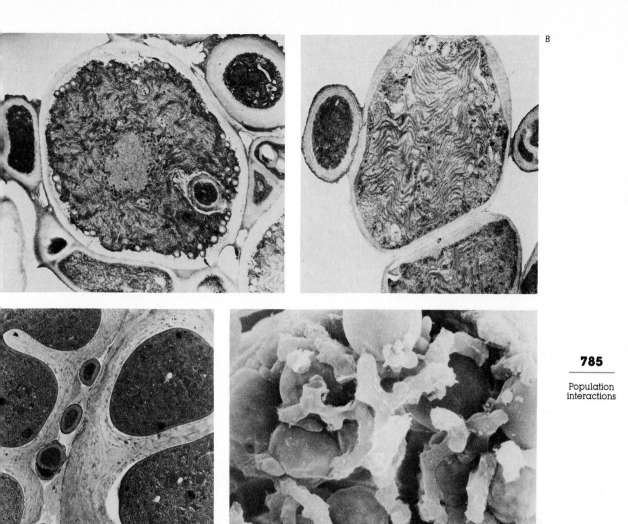

B

D

Figure 21.33

The ultrastructure of a lichen, showing the fungal and algal or cyanobacterial components. (A) Cladonia cristatella, *an algal cell* (Trebouxia erici) *with a large chloroplast that fills most of the cell and a central pyrenoid around which are lipid-containing pyrenoglobuli, small, dark, or black bodies. Note the haustorium, the penetrating hyphae, within the algal cell and the appressoia, the hyphal cells, against the outer algal wall. In lichens with green algal symbionts, most of the carbon fixed photosynthetically by the alga is excreted as some type of sugar alcohol and made available to the fungus. (B) Stereocaulon myriocarpum, an algal cell* (Pseudochlorella *sp.) with attached fungal appressorium.* Pseudochlorella, *which is a green alga, is the primary symbiont of this lichen. (C)* Stereocaulon myriocarpum *section through a cephalodium, a gall-like structure on the lichen thallus, showing cells of the cyanobacterial symbiont* (Scytonema *sp.) that are surrounded by thick sheaths. Note the four hyphal cells of the fungal symbiont between the sheaths of the cyanobacterial symbiont. Most of the cyanobionts in lichens fix atmospheric nitrogen, much of which is released and made available to the fungal symbiont. The released nitrogen is in the form of ammonia or low-molecular-weight peptides. According to some researchers, the mycobiont of lichens containing cyanobacteria exerts some influence on the metabolism of the cyanobiont by specific enzyme-inhibiting compounds. They found that in* Peltigera canina *the activity of the enzyme glutamine synthetase was much lower in the lichenized* Nostoc *symbiont, in contrast to its high activity in the alga growing separately from the fungus. Thus, when the activity of glutamine synthetase is reduced in the lichen thallus, although the bactobiont fixes nitrogen rapidly, the resulting* NH_4 *cannot be assimilated and is released to the mycobiont. (D) In this scanning electron micrograph the fungal hyphae can be seen growing into and around their photosynthetic partner (2200×). (A courtesy J. B. Jacobs and V. Ahmadjian; B, C courtesy K. Withrow and V. Ahmadjian, Clark University; D courtesy Robert Apkarian, University of Louisville.)*

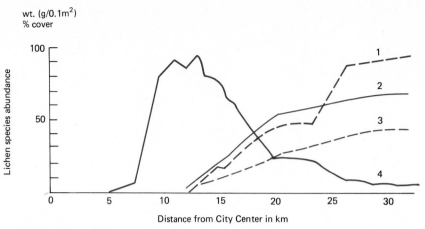

Figure 21.34

The disappearance of lichens from urban areas is illustrated in this graph. By examining lichens on trees it is possible to determine the extent of air pollution and sometimes even the direction of the source of air pollutants.

energy and atmospheric CO_2 to produce the organic matter consumed by the mycobiont. In some lichens the cyanobacterial partner is also capable of fixing atmospheric nitrogen. The mycobiont provides physical protection for the lichen and also produces organic acids that can solubilize rock minerals, making essential nutrients available to the lichen.

Although the mutualistic relationships of algal and fungal populations in lichens are normally stable, they can be disrupted by environmental perturbations. Lichens are extremely sensitive to air pollution, and sulfur dioxide in the atmosphere is particularly inhibitory to lichens (Figure 21.34). Exposure to sulfur dioxide reduces the efficiency of the photosynthetic activity by the phycobiont, allowing the mycobiont to overgrow it and leading to the elimination of the symbiotic relationship. Once the careful metabolic balance is interrupted, the lichen and its member algal and fungal populations disappear from the habitat.

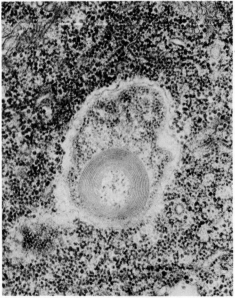

Figure 21.35

Kappa particles of Paramecium aurelia. *(A)* Paramecium aurelia, *showing presence of endosymbiotic bacteria occurring as kappa particles. (B) An isolated kappa particle. (A courtesy of John R. Preer, Jr., Indiana University, reprinted by permission of Cambridge University Press, New York, from J. R. Preer, Jr., and A. Jurand, 1968,* Genetic Research, *12: 331–340; B courtesy Eugene W. McArdle, Northeastern Illinois University.)*

A

B

Another interesting mutualistic relationship occurs between the protozoan *Paramecium aurelia* and various bacterial species. The obligately **endosymbiotic bacteria** live within the protozoan, where they appear as structures such as **kappa particles** (Figure 21.35). There are two classes of populations of *Paramecium aurelia*: killer strains that contain kappa particles and sensitive strains that lack the bacterial endosymbionts. The presence of kappa particles gives an important advantage to killer strains when they are in competition for available resources with sensitive strains. The endosymbiotic bacteria probably derive nutritional benefits from the protozoan.

Plant–microbe interactions

Nitrogen-fixing symbiosis Microorganisms also establish important symbiotic relationships with plant populations. The symbiotic relationship between members of the bacterial genus *Rhizobium* and leguminous plants is of extreme importance for the maintenance of soil fertility. (*Rhizobium* species are able to invade the roots of suitable host plants, leading to the formation of nodules, within which the *Rhizobium* bacteria are able to fix atmospheric nitrogen) (Figure 21.36). The establishment of a symbiotic association between a *Rhizobium* species and a plant is very specific, and there is a mutual recognition between the bacteria and binding sites on the surfaces of the plant roots. The interaction of a *Rhizobium* species and a leguminous plant involves (1) the attraction of the bacteria to the plant roots by amino acids secreted by the plant; (2) the binding of the bacteria to receptors (lectins) on the plant root; (3) the bacterial oxidation of the plant growth substance indoleacetic acid (IAA), leading to curling and branching of the rootlets; (4) entry of the bacteria into the root; (5) development of an infection thread; (6) transformation of the plant cells to form a tumorous growth; (7) transformation of the *Rhizobium* into bacteroid (pleomorphic) forms; and (8) multiplication of *Rhizobium* within the nodule. Within the nodule metabolites are transferred between the bacteria and the plant. Within the infected plant tissue the *Rhizobium* cells multiply, forming unusually shaped pleomorphic cells called **bacteroids**, which are no longer capable of independent reproduction (Figure 21.37). The bacteroid cells contain active nitrogenase enzymes, not found in free-living *Rhizobium* cells, that allow them to fix molecular nitrogen and provide their symbiotic plant partner with an available source of fixed nitrogen for growth. The plants provide organic compounds for the generation of required ATP by the *Rhi-*

zobium. The biochemicals of the nodule supply oxygen to the bacteroids for their respiratory metabolism but maintain a sufficently low concentration of free oxygen so as not to inactivate the nitrogenase enzymes; in particular, leghaemoglobin produced by the plant supplies oxygen to the bacteroids for production of ATP. The control of oxygen is critical because oxygen is both required and inhibitory for the nitrogen fixation process.

In addition to the symbiotic relationship between *Rhizobium* and leguminous plants, various other bacterial species, including cyanobacteria and actinomycetes, are able to enter into similar mutualistic relationships with a restricted number of other types of plants, resulting in the formation of nodules and the ability to fix atmospheric nitrogen. An equivalent sequence of events is involved in the formation of nodules, resulting from the mutualistic relationships between bacteria and nonleguminous plants, although leghaemoglobin is not the specific oxygen carrier in such relationships. *Rhizobium,* for example, can fix nitro-

Figure 21.36

Nodules occur as tumorous growth on the roots of plants infected with nitrogen-fixing bacteria. It is within these tumor-like growths that Rhizobium *fixes atmospheric nitrogen. Only a few types of plants can enter into a symbiotic relationship with nitrogen-fixing bacteria. Shown here is the extensive nodulation of the root system of a soybean plant. (Courtesy Nitragin Co., Milwaukee, Wisconsin.)*

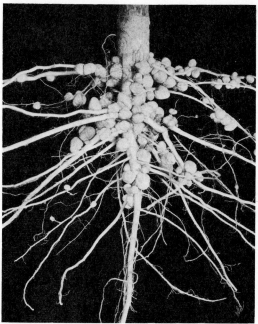

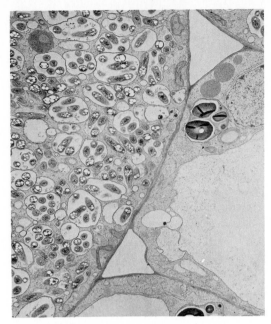

Figure 21.37

Electron micrograph of Rhizobium japonicum *bacteroids in an infected cell (right) of a root nodule of soybean. The left cell is uninfected and shows the appearance of the cell without bacteroids (13,000×). (From BPS—E. H. Newcomb, University of Wisconsin.)*

Figure 21.38

This micrograph shows a section through a root nodule of Alnus glutinosa; *note the vesicular structures and hyphae of the nitrogen-fixing actinomycete* Frankia *(1750×). (Courtesy Jan J. Becking, Institute for Atomic Sciences in Agriculture, Wageningen, The Netherlands.)*

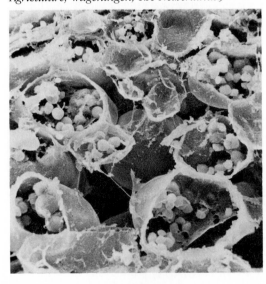

gen in association with *Trema,* a tree found in tropical and subtropical regions. Likewise, actinomycetes establish specific mutualistic relationships that permit the bacteria to fix nitrogen, such as occurs when *Frankia alni* infects the roots of the alder tree, leading to the formation of nodules (Figure 21.38). Such a actinomycete nitrogen-fixing symbiosis is especially important with angiosperms.

Mycorrhizae   The formation of **mycorrhizae** by some fungal species that enter into a mutualistic relationship with plant roots is another important symbiotic relationship between microorganisms and plants. The fungus derives nutritional benefits from the plant roots and contributes to the plant's nutrition. The establishment of mycorrhizal associations involves the integration of plant roots and fungal mycelia into a unified morphological unit. Some plants with mycorrhizal fungi are able to occupy habitats they otherwise could not inhabit. The importance of this microbe–plant interaction is attested to by the fact that 95 percent of all plants have mychorrhizae.

There are several types of mycorrhizal associations differentiated based on the degree of integration of the fungus into the root structure (Figure 21.39). Ectomycorrhizae are characterized by the formation of an external fungal sheath around the root and fungal penetration of the intercellular regions of the root. Such ectomycorrhizal associations occur in most oak, beech, birch, and coniferous trees. Endomycorrhizal associations involve fungal penetration of root cells. The **vesicular-arbuscular** (VA) type of mycorrhizal association in which the root cortex contains specialized inclusions, called **vesicles** and **arbuscules,** is the most common form of mycorrhizae. This association frequently goes unnoticed because it does not have a macroscopic effect on root morphology. Most major agricultural crop plants, including wheat, maize, potatoes, beans, soybeans, tomatoes, strawberries, apples, oranges, grapes, cotton, tobacco, tea, coffee, sugar cane, sugar maple, and others, form VA endomycorrhizal associations. It is apparent from this list that this type of mutualistic association is very common and widely distributed.

Animal–microbe interactions

Insects   There are some particularly interesting mutualistic relationships between microorganisms and animal populations. Some plant-eating insect populations, for example, actually cultivate microorganisms on plant tissues (Figure 21.40). The microorganisms degrade cellulosic plant reslues, providing a digestible source of nutrition for the insects, which lack cellulase enzymes and could not derive any nutritional benefit from simply eating plant material. The microorganisms are provided by the insects with a habitat in which they can proliferate, while the insects process the plant material, preparing a suitable medium for microbial growth and, in some cases, secreting substances that protect the growing microorganisms from invasion by other microbial species. The **fungal gardens** of myrmicine ants, the attini, are an excellent example of an insect population growing fungi in pure culture. The ants macerate leaf material, mix it with saliva and fecal matter, and inoculate the prepared substrate with a pure fungal culture. After growth of the fungus, the ants harvest a portion of the fungal biomass and byproducts that they ingest. Various wood-inhabiting insects, including ambrosia beetles and some termites, maintain similar mutualistic relationships with microbial populations. In these cases the animals rely on the cellulolytic enzymes of microbial populations to convert plant residues into nutritional sources that they can use. The insect provides the microorganism with an optimal habitat for growth.

Ruminant ecosystem Ruminant animals, such as cows, llamas, and camels, establish similar mutualistic relationships with microbial populations. Although plants are the main food sources for these animals, ruminant animals do not produce cellulase enzymes themselves but depend on microbial populations for the degradation of the cellulosic materials they consume. The **rumen**, the large first chamber within the stomach of these animals, provides a stable, constant-temperature, anaerobic environment for establishing mutualistic associations with microbial populations. The plant material ingested by the animal provides a continuous source of nutrients for the microorganisms within the rumen, very much like what occurs in a continuous fermentor.

The overall equation for the fermentation that occurs in the rumen can be stated as

$$57.5 \ (C_6H_{12}O_6) \rightarrow 65 \ \text{acetate} + 20 \ \text{propionate} + 15 \ \text{butyrate} + 60 \ CO_2 + 35 \ CH_4 + 25 \ H_2O$$

Microbial populations within the rumen convert cellulose, starch, and other polysaccharides to carbon dioxide, hydrogen gas, methane, and low-molecular-weight organic acids (Figure 21.41). A portion of the low-molecular-weight fatty acids, carbon dioxide, and molecular hydrogen produced by various fermentative bacteria, such as *Ruminococcus,* are converted by methanogenic

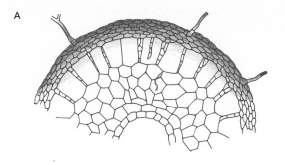

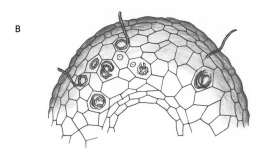

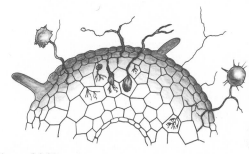

Figure 21.39

The structures of the different types of mycorrhizae. (A) Ectotrophic mycorrhizal rootlet, showing fungal sheath and intercellular penetration. (B) Endotrophic mycorrhizal rootlet showing penetration of hyphae. (C) Rootlet with vesicular-arbuscular mycorrhiza showing penetration of hyphae and "tree-like" and "vesicle-like" hyphal structures.

bacteria to methane. Although hydrogen is produced by many of the fermentative bacteria in the rumen, it never accumulates because of its rapid utilization by methanogenic bacteria. Cows burp considerable amounts of the methane generated by these bacteria within the rumen. The organic acids produced by the microbial populations are absorbed into the bloodstream of the animal where they are oxidized aerobically to produce the ATP needed to meet the animal's energy require-

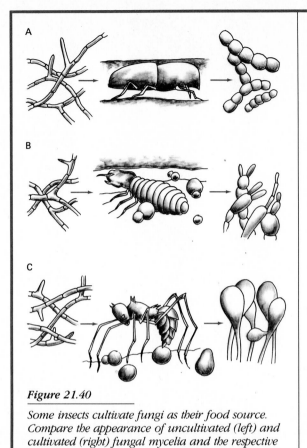

Figure 21.40

*Some insects cultivate fungi as their food source.
Compare the appearance of uncultivated (left) and
cultivated (right) fungal mycelia and the respective
gardening insect (middle): (A) ambrosia beetle; (B)
termite; and (C) attini ant. (Based on Batra and
Batra, 1967, The fungus gardens of insects,*
Scientific American *217(5): 112–120.)*

Discovery process

It has been discovered that insects are unique
among animals in having developed
mutualistic relations with fungi. This may have
come about because so many of them share
the same small habitats. In addition, most
insects are equipped to carry living spores of
fungi either in their guts, in the folds between
their joints that secrete waxy substances, or
among their bristles. Elucidating the
biochemical basis for the fungal–insect
symbiotic relationships has involved various
lines of investigation concerning the
physiological ecology of both the insect and
the fungi. Fungus-gardening insects are
apparently able to produce and spread
antibiotic substances that prevent the growth
of alien fungi; substances secreted by the
insect also transform the mutualistic fungi,
causing either ambrosia, spherules, or
bromatia (particles consisting of swollen tips
of fungal filaments) to appear, rather than
sexual fruiting bodies. Two distinct kinds of
mutualistic associations between insects and
fungi have been found. In the gardens of
wood wasps, ambrosia beetles, termites, and
ants the fungus extracts nourishment from a
substrate, and the insect feeds either on the
fungus, the substrate predigested by the
fungus, or both. The fungus is prevented from
producing sexual fruiting structures but is
supplied by its insect partner with a suitable
habitat and a means of dispersal. In the
colonies of the fungus *Septobasidium* and
scale insects, the situation is reversed; the
insect feeds on the fungus, and the fungus
provides shelter for its castrated insect
partner.

ments. Because the rumen is anaerobic, most of
the caloric content of the ingested plants is maintained in the fatty acids transferred to the bloodstream of the animal.

There are diverse bacterial and protozoan populations within the rumen. Some of these microbial populations are found only within the specialized environment of the rumen, and others
also occur in analogous environments, making this
a borderline case between synergism and mutualism. Clearly, though both animal and microbial populations benefit from this relationship,
there is a delicate balance among the individual
populations within the complex microbial community in the rumen, with each population contributing metabolically to the conversion of substrate to fermentation products.

Bioluminescence The mutualistic relationshi
between some **luminescent bacteria** and marin
invertebrates and fish is particularly interestin¡
Some fish have specific organs in which the
maintain populations of luminescent bacteria, i
cluding members of the genera *Photobacteriu*
and *Beneckea* (Figure 21.42). The bacteria no
mally continuously emit light, but the fish are abl
to manipulate the organs containing the lumine:
cent bacteria so as to emit flashes of light. Th
fish supply the bacteria with nutrients and pr¢
tection from competing microorganisms. The ligl
emitted by the bacteria is used in various wa¡
by different fish. In some cases the pattern of ligl
emission is used in sexual mating rituals. In dee
sea and nocturnal fish, such as the flashlight fisl
Photoblepharon, the light emitted by the bacter

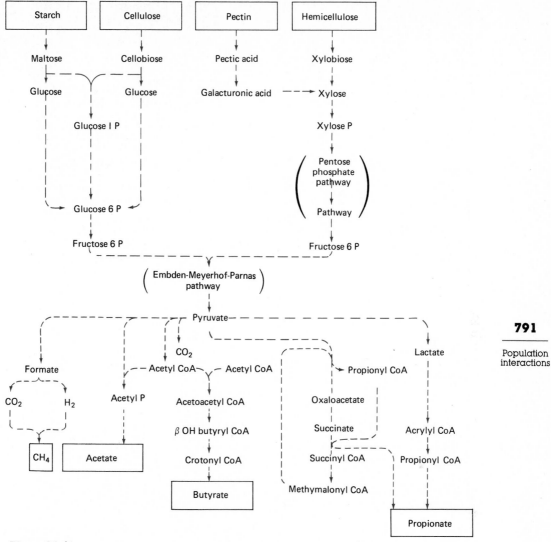

Figure 21.41

Many metabolic transformations occur within the rumen. This diagram shows the bacterial species and pathways involved in the rumen fermentations of the major insoluble carbohydrates present in plants. Compounds in boxes are major substrates or fermentation products.

aids the fish in finding food sources and warding off predators.

Competition

Competition occurs when two populations are striving for the same resource. Often, competition occurs for a nutrient present in limited concentrations, but competition also may occur for other resources, including light and space. As a result of the competition, both populations achieve lower densities than would have been achieved by the individual populations in the absence of competition. Competitive interactions tend to bring about ecological separation of closely related populations, a fact known as the competitive exclusion principle. Competitive exclusion precludes two populations from occupying the same ecological niche. When two populations attempt to occupy

A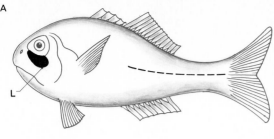

L

B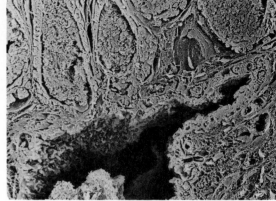

Figure 21.42

Various deep-sea fish have specialized organs containing luminous bacteria. (A) The flashlight fish, Anomalops katoptron, *has such luminous organs situated below the eye. (B) Electron micrograph of a section through the light organelle of the Japanese pinecone fish* Monocentris japonicus *showing large numbers of luminescent bacteria. This fish has a pair of small light organelles under the jaw. (58,000 ×). (Courtesy Margo Haywood and Ken Nealson, Scripps Institution of Oceanography.)*

Figure 21.43

Competition between organisms eliminates the slower-growing species. Here two protozoans that have simiar niches are competing. In mixed culture the faster-growing Paramecium aurelia *replaces* P. caudatum.

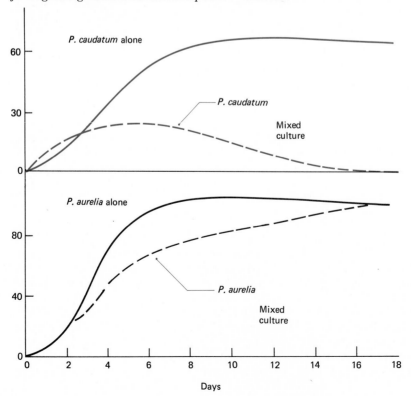

the same niche, one will win the competition and the other will be excluded (Figure 21.43). Chemostats, which by definition have a growth-limiting substrate in the growth medium, are used frequently to study competition between populations under controlled conditions. As a rule, the population with the higher growth rate under the given set of environmental conditions in the habitat will win the competition over the population with the lower growth rate. Fluctuations in environmental conditions can lead to shifts in competitive balances, leading to population oscillations within the microbial community. Spatial separation allows microorganisms to escape competitive pressures, permitting coexistence of competitive populations.

Amensalism

Amensalism, or **antagonism**, occurs when one population produces a substance inhibitory to another population. The first population gains a competitive edge as a result of its ability to inhibit the growth of competitive populations. The production of antibiotics, for example, can give the antibiotic-producing population an advantage over a sensitive strain when competing for the same nutrient resources. The production of lactic acid by *Streptococcus* and *Lactobacillus* species similarly eliminates competitors. The preemptive colonization of food products by lactic acid bacteria precludes the invasion of that food by other bacterial species. This fact is utilized in the production and preservation of food products by the addition of lactic acid as a preservative. Various other chemicals produced by microbial populations, including inorganic compounds such as oxygen, ammonia, mineral acids, and hydrogen sulfide, and organic compounds, such as fatty acids, alcohols, and antibiotics, permit establishment of commensal relationships between microbial populations.

Parasitism

In a relationship of **parasitism**, the parasite population is benefited and the host population is harmed. As a rule, relationships of parasitism are characterized by a relatively long period of contact and the parasite is smaller in size than the host. The parasite normally derives its nutritional requirements from the host cell, and in the process the host cell is damaged. The host–parasite rela-

tionship is typically quite specific. Some microorganisms are obligate parasites, their existence depending on the successful establishment of a parasitic relationship with a host organism. For example, viruses are obligate intracellular parasites, able to multiply only within suitable host cells. Similarly, rickettsia are obligately parasitic bacteria and sporozoans are obligately parasitic protozoa. We have already considered a number of the human diseases that result from infections with microbial pathogens in the medical microbiology section, and some additional diseases of plants and animals will be considered in Chapter 22 as part of the discussion of agricultural microbiology. Such host–parasite relationships that cause disease syndromes clearly exert a negative influence on the susceptible host population and benefit the parasite population.

Parasitic relationships also exist between diverse microbial populations. Bacteriophage invade and multiply within bacterial cells, causing lysis and death of the bacteria; viruses invade fungi, algae, and protozoa. Some bacteria are parasites of other bacteria. For example, *Bdellovibrio* is parasitic on other bacterial populations and is able to invade and multiply by binary fission within cells of *Escherichia coli* (Figure 21.44). As a result of such parasitic interactions, populations of host cells generally decline. Parasitism as such acts as a mechanism for controlling population densities, which in an overall sense is beneficial for maintaining ecological stability.

Predation

Predation involves the consumption of a prey species by a predatory population. Normally, predator–prey interactions are of short duration and the predator is larger than the prey, but this is not always the case. The predatory populations derive nutrition from the prey species, and clearly, the predator population exerts a negative influence on the consumed prey population. Some microbial species are predators and others are prey. Many protozoa prey upon bacterial species (Figure 21.45), and the nondiscriminatory consumption of bacterial populations by protozoan predators is sometimes referred to as **grazing**. Similarly, protozoa and invertebrate animal populations graze on algal primary producers. Predation is an important process in establishing food webs to support the growth of higher organisms. Various filter-feeding animals are able to remove microorganisms from suspension. This grazing activity is important in transferring biomass from

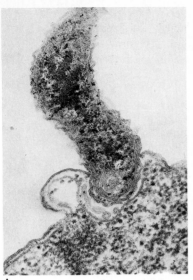

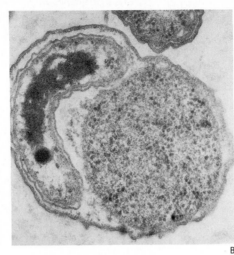

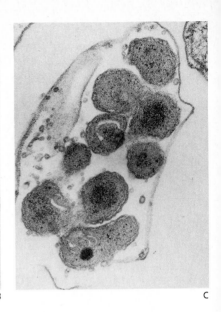

A

B

C

Figure 21.44

(A) Bdellovibrio bacteriovorus *entering an* Escherichia coli *host cell. (B, C) B.* bacteriovorus *contained within an* E. coli *host cell. (Courtesy Jeffrey C. Burnham, Medical College of Ohio, and Sam F. Conti, University of Massachusetts, reprinted by permission of the American Society for Microbiology, from J. C. Burnham, and S. F. Conti, 1968,* Journal of Bacteriology, *96: 1374.)*

Figure 21.45

This micrograph shows a partially lysed bacteria within the vacuole of an ameba that grazes on bacteria. The protozoan derives its nutrition from ingestion of bacteria (12,000×). (Courtesy O. Roger Anderson, Columbia University.)

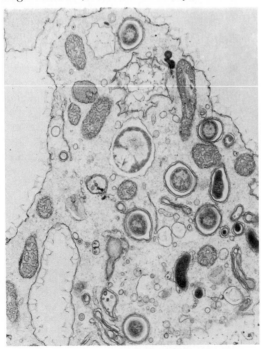

microorganisms to higher trophic levels in aquatic food webs.

Although the predator is normally larger than the prey, there are some interesting cases where a small microbial predator consumes a larger organism. For example, the protozoan *Didinium* can engulf and consume the larger protozoan *Paramecium* (Figure 21.46). Some fungi are able to trap and consume much larger nematodes (Figure 21.47). There are several mechanisms by which these fungi capture nematode prey, including the production of networks of adhesive branches, stalk adhesive knobs, and adhesive or constrictive rings. When a nematode attempts to move past such predatious fungi, the fungus traps the nematode. Even violent movements by the nematode to escape the grasp of the fungus generally fail. The fungal hyphae penetrate and digest the nematode, consuming the animal.

Theoretically, interactions of predator and prey species lead to regular cyclic fluctuations in the populations of the two species (Figure 21.48). As the size of the prey population increases, it can support a larger predator population. The decline in the population size of the prey species means that fewer predators can be supported, and therefore, the population of the predator also declines. If either the predator or the prey were completely eliminated, the population of the other would be

Figure 21.46

Didinium nasutum *is seen here engulfing* Paramecium multimicronucleatum *(1500×). (From BPS—H. S. Wessenberg and G. A. Antipa, 1970,* Journal of Protozoology *17:250–270.)*

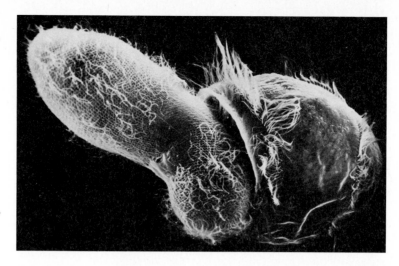

Figure 21.47

Micrographs showing nematode-trapping fungi. Various types of nematode traps are produced by different fungal species. Once trapped the nematode thrashes about, but the fungus does not let go. (A) Nomarski interference micrograph of Dactylaria candida *showing nematode captured by several adhesive knobs. (Courtesy S. Olson and B. Norbring-Hertz, University of Lund.) (B) Scanning electron micrograph of sticky knobs of* D. candida. *(C) Scanning electron micrograph of adhesive network of* Arthrobotrys oligospora. *(B, C courtesy B. Norbring-Hertz, University of Lund; A–C reprinted from* Forum Mikrobiologie *by permission of G-I-T Verlag Ernst Giebler.) (D) Phase contrast micrograph of open and closed constricting ring traps formed by the fungus* Arthrobotrys dactyloides. *(E) Micrograph showing nematode trapped in a constricted ring trap of* A. dactyloides. *(D, E courtesy David Pramer, Rutgers, The State University.)*

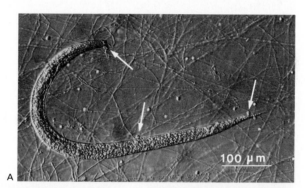

A

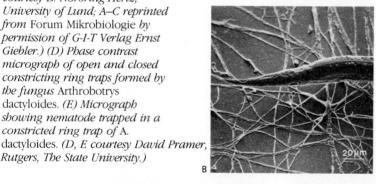

B

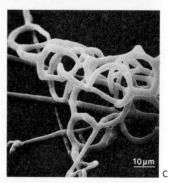

C

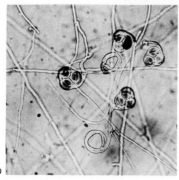

D

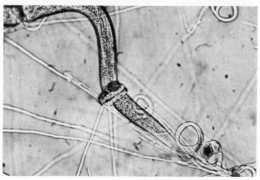

E

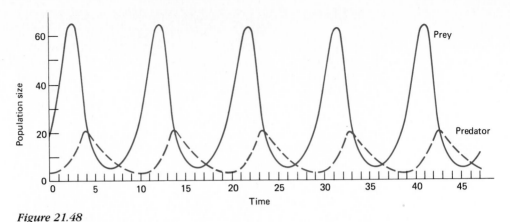

Figure 21.48

Theoretical population oscillations resulting from the interaction of predator and prey.

deleteriously affected. Without the prey as an available food resource, the predator would be eliminated, and without the population control exerted by the predator, the prey species could reach too high a population size, leading to the complete consumption of the available nutrient resources of the habitat.

Microorganisms in their natural habitats

After considering the activities and interrelationships of microbial populations, we can now examine the distribution and functional role of microorganisms in the envirnoment. Microorganisms exist in numerous **habitats** characterized by vastly different environmental conditions. Microbial populations and activities vary greatly between habitats having vastly differing abiotic parameters. The physical and chemical characteristics of a habitat influence the growth, activities, interactions, and survival of the microbial populations found there. Some microorganisms possess unique features that make them "fitter" for survival in a particular ecosystem. The diversity of populations in a given habitat contributes to the ability to change and maintain stability within the community of organisms living in a given habitat.

Some microorganisms are indigenous, that is, **autochthonous**, to a particular habitat. These microorganisms are capable of survival, growth, and metabolic activity within that habitat. The autochthonous microorganisms occupy the niches of that ecosystem. In contrast to the indigenous members of the microbial community, some microorganisms in a given habitat may be foreign, or **allochthonous**. Allochthonous microorganisms typically have grown elsewhere and have been transported into a foreign ecosystem. Such foreign microorganisms do not occupy the niches of that ecosystem and are transient members of the microbial community. Generally, autochtho-

table 21.4

*Microorganisms found in the atmosphere**

Type of organism	Percent-age
Bacteria	
Gram positive pleomorphic rods (e.g., *Corynebacterium*)	20
Gram negative rods (e.g., *Achromobacter, Flavobacterium*)	5
Endospore formers (e.g., *Bacillus*)	35
Gram positive cocci (e.g., *Micrococcus*)	40
Fungi	
Cladosporium	80
Alternaria	5
Penicillium	2
Other (e.g., *Aspergillus, Chaetomium, Dematium, Fumago, Fusarium, Helminthosporium, Sclerotinia, Stachybotrys, Trichoderma, Verticillium*)	13

*In troposphere over North America

nous microorganisms must exhibit adaptive features that make them physiologically compatible with the physical and chemical environment of the habitat. They must be functionally and competitively compatible with the other organisms living in that habitat. Thus, microorganisms occupying extreme environments, such as hot springs, desert soils, and the ocean trenches, must possess physiological adaptations that permit them to survive and function under conditions that normally preclude biological activity.

Atmosphere

The **atmosphere** is characterized by high light intensities, extreme temperature variations, low concentrations of organic matter, and a scarcity of available water, making the atmosphere a hostile environment for microorganisms and a generally unsuitable habitat for microbial growth. Nevertheless, substantial numbers of microorganisms are found in the lower regions of the atmosphere (Table 21.4). These microorganisms do not grow within the atmosphere but rather represent allochthonous populations transported into the atmosphere from aquatic and terrestrial habitats. Transport through the atmosphere is important in the dispersal of many microorganisms, ensuring the continued survival of many microorganisms by permitting **propagules** of microbial populations to reach more favorable habitats (Figure 21.49). Many plant pathogens are transported through the air from one field to the next, and the spread of various fungal diseases of agricultural crops can be predicted by measuring concentrations of airborne fungal propagules. The spread of pathogenic bacteria in droplets is important in the transport of these microorganisms from the infected to a susceptible host.

Although some microorganisms become airborne as growing vegetative cells, more commonly microorganisms enter the atmosphere as spores. Metabolically, dormant spores are better adapted to survival in the atmosphere than actively growing vegetative cells. Some spores have pigments that protect them against exposure to damaging ultraviolet radiation in the atmosphere. Other spores have very thick walls that protect

Figure 21.49

Many fungi depend on spore dispersal through the atmosphere to ensure survival of the species. This drawing shows the discharge of spores into the air from a cup fungus (ascespores) and an agaric (basidiospores). Laminar air flow above the ground (horizontal dashes), and turbulent air flow above this (dashes in spirals) are illustrated. (A) The spores of the cup fungus have just been discharged as a puff into the turbulent air zone. (B) The agaric spores are steadily dropping into the turbulent air. (After C. T. Ingold, 1971, Fungal Spores: Their Liberation and Dispersal, *Oxford University Press.)*

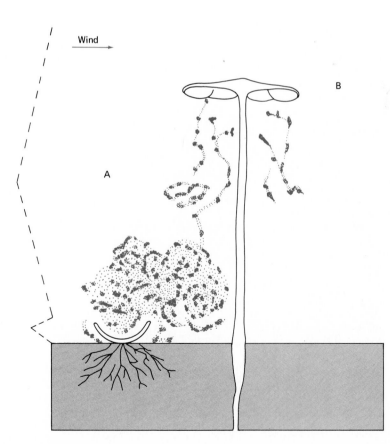

Wind

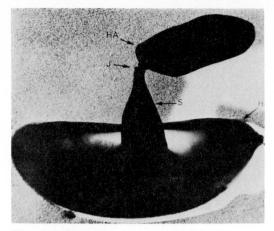

Figure 21.50

An electron micrograph of the ballistospore of
Sporobolomyces roseus *(11,000×).*

Figure 21.51

Scanning electron micrograph of
Hyphomicrobium *showing accumulation of a*
large number of budding cells. The large surface
area of a Hyphomicrobium *cell permits this*
bacterium to obtain its required nutrients from
dilute aqueous solutions and thus to grow in
aquatic habitats with very low nutrient
concentrations. (Courtesy Richard L. Moore,
University of Calgary.)

them against desiccation during aerial transport.
As a rule, spores tend to be light and have aero-
dynamically favorable shapes for extended lateral
travel through the atmosphere. Many microorganisms produce large numbers of spores and have
developed adaptations for discharging their spores
into the atmosphere (Figure 21.50). In some cases
spores are forcibly ejected into the atmosphere
to facilitate aerial dissemination. Aerosols are also
important in the aerial transfer of various microbial populations that do not possess special adaptations for prolonged survival in the atmosphere.

Hydrosphere (aquatic habitats)

Freshwater habitats

Freshwater habitats include lakes, ponds, swamps,
springs, streams, and rivers. There are a number
of important chemical and physical parameters
that make freshwater ecosystems more or less
suitable habitats for microorganisms. Temperature, pH, nutrient concentrations, and oxygen levels are important parameters influencing the existence of particular microbial populations in
freshwater habitats. Oligotrophic lakes have low
concentrations of nutrients and typically have low
rates of primary productivity. Autochthonous heterotrophic microorganisms in such low-nutrient
habitats should be able to grow at low nutrient
concentrations, and many freshwater bacteria exhibit unusual shapes that increase their surface
area-to-volume ratio to permit more efficient uptake of nutrients. Stalked bacteria, such as *Caulobacter* and *Hyphomicrobium* species, are frequently found in lake water (Figure 21.51). In
contrast to oligotrophic lakes, eutrophic lakes have
high nutrient concentrations. These lakes generally have high rates of primary productivity, and
oxygen concentrations are generally low as a result of extensive microbial decomposition of organic nutrients.

Within aquatic habitats there is a zonation in
the distribution of microbial populations based
on the physicochemical parameters of the habitat
(Figure 21.52). Lakes, for example, are divided
into a shallow **littoral zone** where light penetrates
to the bottom **benthos**, a **limnetic zone** where
light penetration is sufficient so that primary productivity exceeds heterotrophic metabolism, and
a **profundal zone** where consumption exceeds
primary productivity (Figure 21.53). The **compensation depth**, separating the limnetic and profundal zones, is the depth of effective light penetration where photosynthesis activity balances
respiratory activity. Cyanobacteria and algae are

Figure 21.52

The vertical distribution of microorganisms in a lake. Cyanobacteria are abundant in the epilimnion; sulfate reducers are abundant in the lower hypolimnion; and maximal concentrations of heterotrophs occur just below the zone of maximal photosynthetic production and at the water–sediment interface. (After Rheinheimer, 1974, Aquatic Microbiology, *John Wiley & Sons and VEB Gustav Fischer Verlag.)*

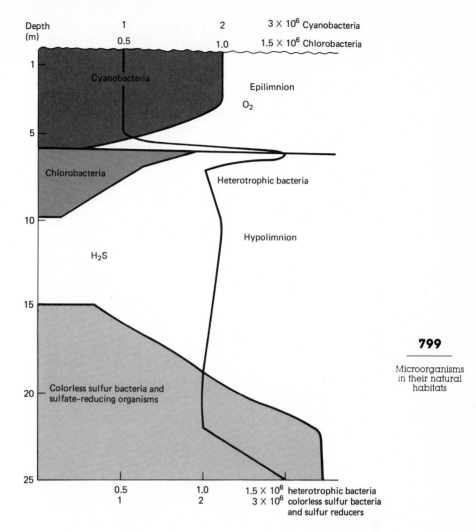

Figure 21.53

The divisions of a lake into zones in relation to light penetration.

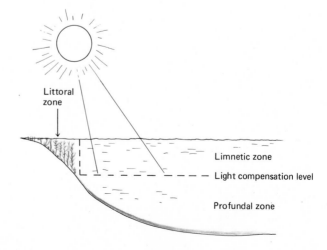

important populations in the community of the limnetic and littoral zones but decline in importance with depth in the profundal zone. The population levels of heterotrophic consumers depend on the concentrations of available nutrients. Normally, much higher populations of bacteria are found in sediment than in the water column because nutrients settle to the bottom of a water body and accumulate in the benthic sediment.

Temperature has a major influence on the distribution of microbial populations in freshwater habitats. Temperature influences the density of water, causing cycling of water within a lake. The highest-density water occurs at 4°C. Many temperate lakes exhibit **thermal stratification**, which undergoes seasonal changes (Figure 21.54). During the summer the upper layer, the **epilimnion**, is warm and oxygen rich, while the lower layer,

Figure 21.54

The divisions of a lake in relation to temperature and oxygen and carbon dioxide concentrations with depth at different times of the year under different systems of stratification. Winter stratification does not occur in some lakes or in most seas, in which case there is a continuous period of mixing from the autumn overturn to the onset of summer stratification.

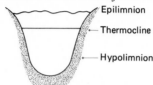

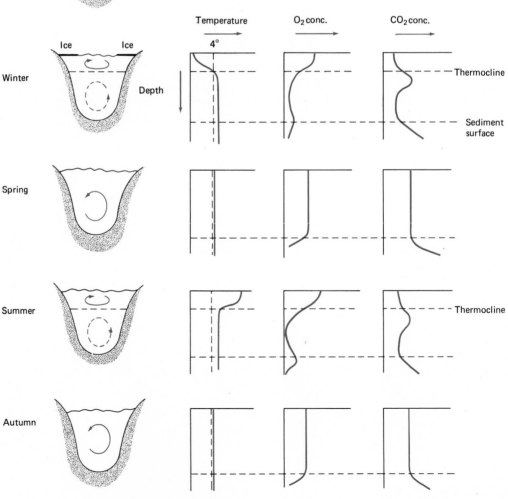

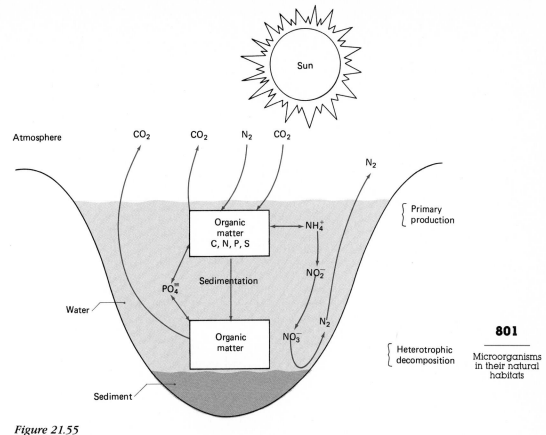

Figure 21.55

Simplified diagram of nutrient cycling within a lake habitat.

the **hypolimnion**, is characterized by low temperatures and low oxygen concentrations. The epilimnion is separated from the hypolimnion by the **thermocline**, a zone of rapid temperature change. Poor light penetration to the hypolimnion restricts oxygen-generating photosynthesis and respiration depletes the existing oxygen. Mineral nutrients are generally depleted in the epilimnion as a result of the extensive growth of photosynthetic and heterotrophic microbial populations.

The principal ecological functions of microorganisms in freshwater habitats include (1) the input of organic matter from primary production; (2) the decomposition of dead organic matter, liberating mineral nutrients for primary production; (3) the assimilation and reintroduction into the food web of dissolved organic compounds; and (4) the biogeochemical cycling of elements (Figure 21.55). Autotrophic bacteria are autochthonous members of the microbiota of lakes and generally play a very important role in the nu-

trient cycling of the lake. The metabolic activities of chemolithotrophic bacteria are important in the cycling of nitrogen, sulfur, and iron within lakes. Members of the genera *Nitrosomonas, Nitrobacter,* and *Thiobacillus* are especially important in the cycling of nitrogen and sulfur within freshwater bodies.

Microorganisms in the sediments of freshwater habitats are generally different than those in overlying waters. In shallow ponds anaerobic photosynthetic bacteria occur at the surface of the sediment, often conferring characteristic colors, such as purple, on such water bodies. Fungi and heterotrophic bacteria grow well on the debris that accumulates at the sediment–water interface. Within the sediment, anaerobic microorganisms become increasingly important. Denitrifying *Pseudomonas* species, hydrogen sulfide-producing *Desulfovibrio* species, and methanogens occur in sediments and are very important in the biogeochemical cycling of carbon, nitrogen, and sulfur within freshwater habitats. The metabolic activities of these bacterial populations produce gaseous compounds that can bubble up through the water column and transfer from the hydrosphere to the atmosphere.

Bacteria are able to use very low concentrations of organic compounds. This is very important in oligotrophic lakes because the metabolic activities of bacteria permit the concentration of dissolved nutrients within bacterial biomass. This ability of bacteria to concentrate organic matter permits the introduction of organic compounds into the food web of the lake that otherwise would be lost to the biological community. The growth of bacteria at such very low concentrations of nutrients is a form of secondary productivity.

Some freshwater microorganisms have developed interesting adaptive features that permit them to function well in freshwater ecosystems. Species of the dinoflagellate *Ceratium,* for example, exhibit seasonal changes in cell shape. During summer when water is warm, the algae form elongated appendages that increase their buoyancy, and during winter when the water is cold and denser, they conserve energy by forming much shorter appendages. Some cyanobacteria possess gas vacuoles that allow them to adjust their buoyancy. By adjusting their height in the water column, depending on the intensity of light, these bacteria can thus maximize their use of light energy for photosynthesis (Figure 21.56).

Marine habitats

The **oceans** cover approximately 71 percent of the earth's surface, providing a sizable habitat for microbial populations. The oceans contain almost every naturally occurring chemical element, but most elements are present in extremely low concentrations. Environmental conditions in the open ocean are relatively constant; normally, the salinity is in the range of 33–37 parts per thousand, the pH is near 8.3, and temperatures below 100 meter depth are between −2 and 5°C. As with freshwater habitats, there is a surface euphotic zone where light penetrates to the compensation depth (Figure 21.57). Algae and cyanobacteria are important primary producers within this euphotic

Figure 21.56

Gas vacuoles allow cyanobacteria to adjust their height in a water column so that they can locate themselves at a level of light intensity that will permit optimal photosynthetic activity. (A) Phase contrast micrograph of Anaebaena *containing gas vacuoles. (B) Electron micrograph of freeze-etched preparation of* Nostoc muscorum, *showing cylindrical structure of the gas vacuoles. (From BPS—J. Robert Waaland, University of Washington.)*

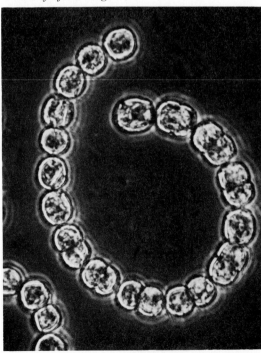

A

B

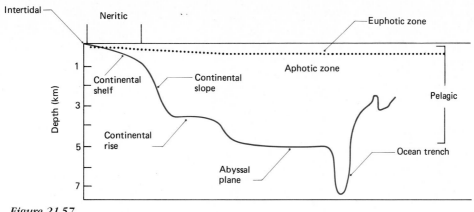

Figure 21.57

The divisions of the ocean, showing horizontal zonation.

zone. The pelagic zone extends from the surface to a depth of 6,000 meters. Within the pelagic zone, recycling of mineral nutrients is extremely slow, and dead organic matter slowly sinks through the pelagic zone to the benthos. In the very deep ocean trenches, rates of organic decomposition are very slow. In fact, lunches that accidently sank with the deep-diving submarine *Alvin* remained intact until the submarine was recovered 10 months later.

Mineral nutrients and dead organic matter accumulate at the benthos. The highest biomass of microorganisms in marine habitats normally occurs near the surface and decreases with depth. Microbial numbers are relatively high in nearshore estuarine waters but are very low, 1–100 per ml, in pelagic waters. Within estuaries, where freshwater and marine habitats interface, num-

bers of microorganisms vary with the tides. Upwelling zones, where this nutrient-rich water moves upward along the continental shelf, are particularly productive (Figure 21.58). Typical productivity values for open ocean, coastal, and upwelling areas of the ocean are 50, 100, and 300 grams of organic carbon per square meter per year, respectively.

Primary production in **pelagic** (open ocean) marine habitats is extremely limited by a lack of mineral nutrients. Because of the expanse and depth of the oceans, planktonic microorganisms have an almost exclusive role in primary production, providing the main source of organic compounds for heterotrophic marine organisms. Higher plants and benthic macroalgae contribute significantly to primary production only in estuaries and littoral, nearshore, areas. Many algal

Figure 21.58

Upwelling of deep ocean water occurs along the continental slope and replaces surface waters driven offshore by the wind.

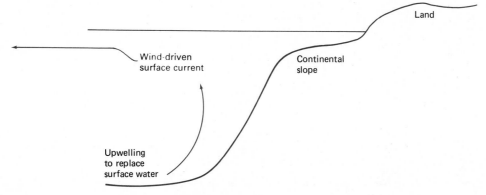

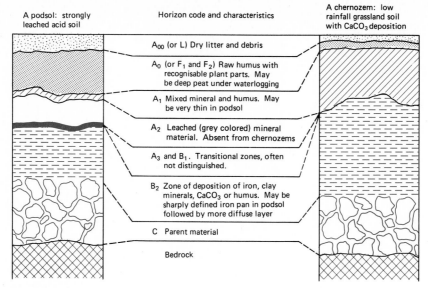

Figure 21.59

Soil profiles of two soil types: a podsol, which forms under conditions of acid leaching; and a chernozem, which is characteristic of much drier conditions, such as grasslands.

species are exclusively marine, where the input of carbon to the marine ecosystem by algae is essential.

Autochthonous marine microorganisms should be capable of growth at salinities between 20 and 40 parts per thousand, and true marine bacteria generally have an optimal salt concentration of 33–35 parts per thousand. Marine bacteria require the ions in seawater to maintain proper membrane functions. Some marine bacteria have multiple membranes surrounding their cells, and exposure to freshwater disrupts these membrane layers. Marine bacteria should be capable of growth below nutrient concentrations found in the oceans. Because most of the ocean has a temperature below 5°C, psychrophilic and psychrotrophic bac-

terial populations are quite common. In the deep ocean trenches barophilic and barotolerant bacteria are important indigenous members of the microbial community. Most marine bacteria are Gram negative and motile. *Pseudomonas* and *Vibrio* species often predominate in marine waters. These marine bacteria are generally nutritionally versatile, can metabolize a wide variety of organic substrates, and are capable of using almost any carbon source as a potential substrate.

Lithosphere (soil habitats)

Soil constitutes the major habitat of **terrestrial microorganisms**. Soil is a favorable habitat for the

table 21.5

The effect of distance from a plant root on numbers of microorganisms

Distance from root (mm)	Microorganisms per gram dry weight of soil		
	Nonfilamentous bacteria	Streptomycetes	Fungi
	($\times 10^7$)	($\times 10^7$)	($\times 10^6$)
0	16.0	4.7	3.6
0–3	5.0	1.6	1.8
3–6	3.8	1.1	1.7
9–12	3.7	1.1	1.3
15–18	3.4	1.0	1.2

Figure 21.60

A soil texture triangle. To determine the soil texture the proportions of sand, silt, and clay are determined. The intersection of lines drawn perpendicular to the sides of the soil triangle indicates the soil texture.

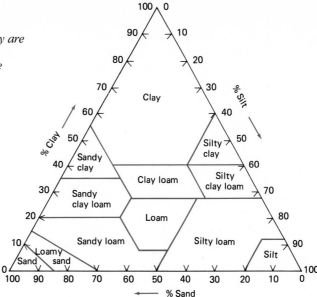

proliferation of microorganisms, and the numbers of microorganisms in soil habitats, typically 10^6–10^9 bacteria per gram, are usually much higher than those found in freshwater or marine habitats. Microorganisms are not evenly distributed through the soil column. Higher numbers of microorganisms occur in the organically rich surface layers than in the underlying mineral soils (Figure 21.59). Particularly high numbers of microorganisms occur in the rhizosphere in association with plant roots (Table 21.5).

There are major chemical and physical differences between soils in different climatic zones. Soils contain varying proportions of clay, silt, and sand particles (Figure 21.60). The soil texture is important because it determines the surface area available for microbial growth. The proportion of various clay minerals also affects the binding properties and chemistry of the soil. The physicochemical differences in temperature (varying from hot soils near the equator to cold polar soils), moisture (varying from dry desert soils to wet partially submerged soils), and pH (varying from extremely acid to strongly alkaline soils) greatly influence which microorganisms occur within the indigenous microbial communities of different soil habitats.

It is difficult to ascribe generalized adaptive features to most indigenous soil microorganisms. Soils have many different microhabitats with differing physicochemical properties, and even at a particular location there may be several different microenvironmental situations favoring the growth of different indigenous microbial populations. A higher proportion of Gram positive bacterial genera are found in soil than in freshwater and marine habitats. Many different bacterial genera are commonly found in soil (Table 21.6). Actinomycetes can comprise a significant proportion of the bacterial community in soil habitats, sometimes accounting for 10–33 percent of the total bacterial community. These organisms are relatively resistant to desiccation and can grow slowly at low nutrient concentrations. At least some soil bacteria are adapted to periods of starvation; *Arthrobacter* species can survive over 80 days without N and C by reducing endogenous rates of metabolism and relying on reserve glycogen and PHB. Soil *Bacillus* and *Clostridium* species sporulate under adverse conditions.

table 21.6

Relative proportions of aerobic and facultative bacteria commonly found in soil

Genus	Percentage
Arthrobacter	5–60
Bacillus	7–67
Actinomycetes	10–33
Pseudomonas	3–15
Agrobacterium	1–20
Alcaligenes	2–12
Flavobacterium	2–10
Corynebacterium	<5
Micrococcus	<5
Staphylococcus	<5
Xanthomonas	<5
Mycobacterium	<5

Fungi constitute the major porportion of the microbial biomass in soils. Most types of fungi can be found in soil habitats, existing either as free-living organisms or in mycorrhizal associations with plant roots. The fungi most frequently isolated from soils are members of the Fungi Imperfecti, but numerous ascomycetes and basidiomycetes also occur in soil habitats. Most soil fungi are opportunistic, growing and carrying out active metabolism when conditions are favorable, having adequate moisture, aeration, and relatively high concentrations of usable substrates. Many soil fungi are able to metabolize polysaccharides, such as cellulose, that are the readily available substrates in soil. Numerous soil fungi also produce resting stages that permit them to remain dormant, even for decades, when conditions are not favorable for growth.

Organic matter in soil is largely composed of humus. Most microorganisms are not capable of using this very complex substance. Some soil microorganisms, designated as autochthonous soil bacteria by Winogradsky, are able to grow slowly in soil, utilizing the refractory humic compounds. In contrast to these slow-growing organisms, most of which are Gram negative rods and actinomycetes, zymogenous or opportunistic soil microorganisms are not able to utilize humic compounds but exhibit high levels of activity and rapid growth on readily metabolizable substrates available in the form of plant litter, animal droppings, and carcasses. Zymogenous organisms are intermittently active when such organic substrates enter soil.

As in aquatic habitats, microorganisms in soil are responsible for biodegradation and mineral cycling. Important plant polymers, such as cellulose and lignin, are degraded almost exclusively by microorganisms. Because soil is a nutrient-rich environment, the numbers and diversity of heterotrophic microorganisms, especially of bacteria and fungi, are characteristically high. Microorganisms in soils, however, play a relatively minor role in the input of organic carbon. Primary productivity in terrestrial habitats is dominated by higher plants. The important role of soil microorganisms in maintaining soil fertility will be discussed further in Chapter 22 when considering agricultural microbiology.

Extreme habitats

The environmental conditions of **extreme environments** greatly restrict the range of microbial species that can grow in such habitats. The extremes of environmental conditions that microorganisms may have to tolerate include: high temperatures approaching boiling water, low temperatures approaching freezing, low acidic pH values, high alkaline pH values, high salt concentrations, low water availability, high irradiation levels, low concentrations of usable nutrients, and high concentrations of toxic compounds. Many microorganisms that inhabit extreme environments, including hot springs, salt lakes, and Antarctic desert soils, possess specialized adaptive physiological features that permit them to survive and function within the physicochemical constraints of these ecosystems. The membranes and enzymes of microorganisms inhabiting extreme environments often have distinct modifications that permit them to function under conditions that would inhibit active transport and metabolic activities in organisms lacking these adaptive features.

Hot springs

The most extreme and extensive high-temperature habitats are found in areas of volcanic activity. Steam vents in such areas may have temperatures of 500°C, and many hot springs have temperatures near 100°C. **Hot springs** occur throughout the world, including within Yellowstone National Park in the United States (Figure 21.61). Microorganisms living in hot springs obviously must be adapted to function at high temperatures. The growth of microorganisms in hot springs is also often limited by low concentrations of organic matter, oxygen, and depending on the particular hot spring, either acid or alkaline pH values. Despite these extreme conditions, several microorganisms do possess the adaptive features necessary to live in hot spring habitats. As the water overflows the hot spring, it flows down channels, establishing a temperature gradient, with a clear zonation of microorganisms occupying habitats of differing maximal temperatures along this temperature gradient, bacteria being the most tolerant of the elevated temperatures. At temperatures above 75°C, only a few bacterial species, including members of the genera *Thermus* and *Sulfolobus,* appear to grow (Figure 21.62). *Bacillus stearothermophilus* is often the dominant bacterial species in hot springs in temperature zones of 55–70°C, but many other microorganisms, including cyanobacteria and algae, also occur in such hot spring habitats. Cyanobacteria occur as layers of growth within specific zones of thermal ponds. The cyanobacteria grow in higher temperature zones than algae, which are restricted to growth below 55°C. Indeed, prokar-

Figure 21.61

Mammoth hot spring in Wyoming's Yellowstone National Park. The terraces are made of travertine, a form of calcium carbonate that has been dissolved from the limestone beneath the ground and carried to the surface by hot water. On the surface the calcium carbonate is no longer able to remain in solution and is deposited as travertine. The travertine itself is white but takes on the colors of the algae and bacteria that grow on it. (Courtesy Wyoming Travel Commission.)

Figure 21.62

Micrograph of Thermus aquaticus *showing neoplasm (n) with dense thin DNA fibrils surrounded by cytoplasm containing numerous ribosomes (ri). Cell envelope comprised of plasma membrane (pm) and wall exhibiting outer dense layers (ow), middle light zone (mw), and inner dense layer (iw). Note cell division by furrowing (f). Where two cells are in contact, the external wall (ow) has separated from the inner wall, which remains adherent to the plasma membrane. (Reprinted by permission of the American Society for Microbiology, Washington, D. C., from T. D. Brock, and M. R. Edwards, 1970,* Journal of Bacteriology *104: 509–517.)*

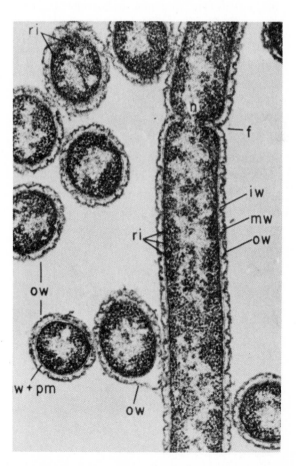

yotes are often more tolerant of extreme environments than are eukaryotes.

Many of the organisms living in hot spring habitats are **obligate thermophiles** and are restricted to growth at high temperatures. These thermophilic microorganisms have adaptive features that allow them to carry out active metabolism at temperatures over 60°C. Many thermophilic microorganisms produce enzymes that are not readily denatured at high temperatures. Sometimes, unusual amino acid sequences occur within the proteins of thermophiles, stabilizing these proteins at elevated temperatures. The membranes of thermophilic microorganisms possess a major proportion of high-molecular-weight and branched fatty acids that permit them to maintain their semipermeable properties at high temperatures. Thermophiles have relatively high proportions of guanine and cytosine in their DNA that raise the melting point and add stability to the DNA molecules of these organisms.

Figure 21.63

Micrograph of Halobacterium *from a salt pond on the shore of the Red Sea. This particular species is most unusual in that it has square cells. (Courtesy Walther Stoeckenius, University of California.)*

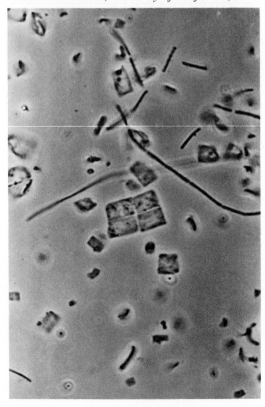

Salt lakes

Salt lakes occur in arid regions where evaporation exceeds freshwater inflow or where a lake is fed by a salt spring. High concentrations of salt dehydrate cells and denature enzymes. Relatively few organisms can grow in highly saline waters, and often the biota of salt lakes is restricted to a **few halophilic and salt-tolerant bacterial species.** Halophilic microorganisms have high internal concentrations of potassium chloride, and the enzymes of such microorganisms must have a greater tolerance to salt than enzymes from microorganisms that are not salt-tolerant. In many cases high concentrations of salt are required by halophiles to maintain their enzymatic activities. Many halophiles have unusual membranes, such as the purple bilayer membranes of *Halobacterium* (Figure 21.63). The role of the halobacterial membrane, which contains bacteriorhodopsin, in phototrophic generation of ATP was discussed earlier. The cell wall of *Halobacterium* lacks murein and appears to be stabilized by sodium ions. The ribosomes of *Halobacterium* require high concentrations of potassium for stability. It is these types of adaptive features that permit halophiles to live in the saturated brine environments of salt lakes.

Desert soils

Desert soils, by definition, receive less than 25 cm of rainfall per year. Many deserts are extremely dry and are subject to extreme diurnal variations in temperature because there is a lack of moisture in the atmosphere overlying these areas to moderate the loss of heat. Microorganisms living in dry desert soils must be able to tolerate long periods of desiccation. In the dry valleys of Antarctica, microorganisms must also tolerate very low temperatures, and during part of the year, high irradiation levels. In such environments many microorganisms have developed adaptations that allow them to survive in a dormant state during unfavorable conditions and to grow actively only during those brief periods of time when conditions are favorable, such as after a rainstorm. Many of the bacteria and fungi living in desert soils form spores that allow them to persist, if necessary, for decades between growth periods. When there is adequate moisture, the spores germinate and for a brief period the organisms can actively grow and reproduce. The lichen symbiosis discussed earlier in this chapter is an adaptive association between microorganisms that permits growth under conditions of severe desiccation. Lichens are important in the dry habitats of Antarctica where they can grow slowly during the relatively warm summer months. The slow growth rates and the ability to retain water permit lichens to exist in such extreme dry habitats.

Extraterrestrial habitats

The planets in our solar system, other than Earth, are hostile habitats for living organisms, lacking water and organic carbon and having toxic concentrations of various gases in their atmospheres to make life as we know it impossible. However, Mars contains some water and therefore experimental life detection systems were sent to Mars as part of the United States National Aeronautics and Space Administration's *Viking* mission. Martian microorganisms would have had to develop adaptations that would permit their existence under these harsh conditions; that is, a microbial Martian would have to be a hardy microbe. **The life detection systems of the *Viking* Mars lander were aimed at detecting microbial life.** More specifically, these systems were capable of detecting the increased turbidity that would be associated with microorganisms growing in solution and the exchange of gases between microorganisms and the overlying atmosphere, including the production of volatile products from the degradation of organic matter and the fixation of carbon dioxide into organic compounds.

Soils from the dry Antarctic valleys were used to test these life detection systems because conditions in these terrestrial Earth habitats are most similar to the dry cold soils of Mars. The results of the *Viking* mission were initially confusing, showing apparent positive test results indicating the presence of living microorganisms, but the results appear to have been due to strictly chemical reactions; the *Viking* exploration project scientists concluded that there are no living organisms in the Martian soils examined.

Postlude

Microorganisms, because of their diversity and ubiquitous distribution, are extremely important in ecological processes. Critical steps of the major global biogeochemical cycles are mediated by microorganisms; for example, virtually all steps in the biogeochemical cycling of nitrogen involve microorganisms. The growth of higher organisms in various habitats relies on the biogeochemical cycling activities of microorganisms; in many cases microorganisms form crucial links in the food webs of an ecosystem that permit the flow of energy and organic compounds to reach higher trophic levels. In some cases animal populations, such as ants, actually cultivate microorganisms in order to derive nutrition from plant sources they cannot digest themselves; cows and other ruminants depend on the degradation of cellulose and other plant polymers by microorganisms within the rumen for their nutrition.

Although many microbes are single-celled organisms, there are a variety of interpopulation interactions involving microorganisms. These interactions establish stability within the biological community of a given habitat, ensuring conservation of the available resources and ecological balance between cohabiting populations. There is a lack of unnecessary duplication of function within stable biological communities; the most fit species fill the available niches, and other populations are competitively excluded. In some cases an interpopulation relationship is beneficial to both interacting species, sometimes giving rise to an essentially new unified "superorganism" with greater functional capacity than possessed by the individual organisms. Even the negative interactions between populations are beneficial to the balance of the overall community, preventing population imbalances and invasion of the indigenous autochthonous community by foreign allochthonous populations.

Microbial populations have developed various adaptations for functioning within diverse habitats. Microorganisms living in extreme environments possess adaptive features that permit them to survive and grow under relatively hostile environmental conditions. In some cases the adaptive features of microorganisms are aimed at survival until conditions are more favorable, such as the formation of spores that may be aerially transported. In other cases microorganisms have developed modified structural proteins and membranes, such as the thermophilic and halophilic bacteria, or strategies for growth, such as lichens, that permit them to grow and carry out active metabolism under normally adverse conditions. The dynamic interactions between microbial populations and their abiotic surroundings contribute to the maintenance of integrated ecosystems in various habitats, the ecological activities of microorganisms being essential for supporting productivity and maintaining environmental quality within a habitat.

1. What is a food web? What are the trophic levels of a food web?

2. What functions do microorganisms play in the cycling of organic matter through food webs?

3. Discuss the role of microorganisms in the biogeochemical cycling of nitrogen; include the different processes and the different microbial populations involved in the global nitrogen cycle. Discuss the differences in nitrogen cycling in aquatic and soil habitats.

4. What are the problems associated with nitrification after fertilizer addition to agricultural soils?

5. How are microorganisms involved in the formation of acid mine drainage?

6. Why is neutralism favored at low population densities?

7. What is cometabolism? Why can this process be important in the degradation of complex organic pollutants?

8. How does commensalism differ from synergism?

9. What are the differences between predation and parasitism?

10. Discuss a mutualistic relationship between:
    a. two microbial populations
    b. microorganisms and a plant
    c. microorganisms and an animal population

12. What is the difference between an autochthonous and an allochthonous microorganism in a particular habitat?

13. What properties are needed for the survival of microorganisms in the atmosphere?

14. What is a true marine bacterium?

15. Discuss the adaptation and zonal separation of microorganisms in a hot spring habitat.

16. Why were the Antarctic dry valleys used as models for the design of the Mars *Viking* lander?

Alexander, M. 1971. *Microbial Ecology.* John Wiley & Sons, New York.

Alexander, M. 1977. *Introduction to Soil Microbiology.* John Wiley & Sons, New York.

Atlas, R. M., and R. Bartha. 1981. *Microbial Ecology: Fundamentals and Applications.* Addison-Wesley Publishing Co., Reading, Massachusetts.

Blakeman, J. P. (ed.). 1981. *Microbial Ecology of the Phylloplane.* Academic Press, New York.

Brock, T. 1966. *Principles of Microbial Ecology.* Prentice-Hall, Englewood Cliffs, New Jersey.

Plate 9

Microorganisms live in diverse and sometimes extreme habitats. (A) Microorganisms growing in the outflows of hot springs in Yellowstone National Park form zones based on temperature tolerances that are clearly visible by their colors. (B) Yellow lichens growing on rocks in the dry Peruvean–Chilean desert. (C) Endolithic lichens growing within rocks in the dry valleys of Antarctica; the black, white, and green zones represent differentiated parts of the organized lichen thallus. (D) The growth of halophilic bacteria and algae in saline lakes, such as Great Salt Lake, Utah, is clearly visible because of the pink color imparted to the water. (A from BPS—Richard Humbert; B from BPS—J. Ordenaal; C courtesy E. I. Friedman, reprinted by permission of AAAS from Science *215: 1045; D courtesy National Geographic Society, Paul Zahl.)*

A

B

C

D

Plate 10

Animals establish various symbiotic relationships with microorganisms. (A) Prochloron symbionts of a small colonial tunicate from Heron Island, Great Barrier Reef, Australia, impart a green color to the animal. (B) This anemone (Antopleura) has a green color because of zoochlorellae (yellow–green algae). (C) Flashlight fish (Photoblepharon palebratus) from Comorro Island, Indian Ocean, have luminescent bacteria, growing in specialized organelles located near the eyes, that provide light in the dark depths where this fish lives. (D) Fungal garden of attine ants; the ants cut leaves and grow fungi as a source of food; in this photograph the ants, a dead yellow leaf, and the white fungal mycelia are clearly visible. (A,B from BPS—J. Robert Waaland; C from BPS—David Powell; D courtesy Smithsonian Institution, Curator Insect Zoo.)

A

B

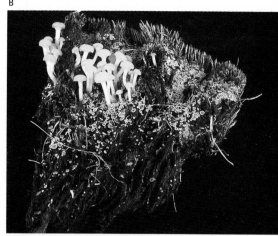

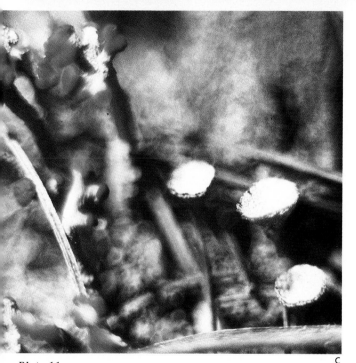

C

D

Plate 11

Lichens are formed by the symbiotic association of fungi and either an algal or cyanobacterial photosynthetic partner. (A) Peltigera aphthosa contains a nitrogen-fixing cyanobacterium and provides an important source of fixed forms of nitrogen in Arctic tundra ecosystems. (B) Most lichens contain ascomycetous fungi, but a few contain basidomycetous fungi, sometimes forming mushroom-like fruiting bodies. (C) The British soldier lichen is so named because of its distinctive coloration. (D) Lichens often grow only on one side of a tree, depending on the prevalent wind direction; the symbiotic lichen association is very sensitive to air pollutants, and lichens have disappeared from urban areas due to poor air quality. (C,D courtesy V. E. Wiedeman.)

Plate 12

The growth of some microorganisms causes plant diseases. (A) Downy mildew of grapes. (B) Apple scab on McIntosh apple, caused by Venturia inequalis. *(C) Cedar apple rust gall on cedar tree. (D) Cedar apple rust on apple leaf. (E) Fire blight on apple tree, caused by* Erwinia amylovora. *(F) Crown gall on willow tree, showing typical tumorous growth. (A,B,F courtesy W. Merrill—from the teaching collection of the Department of Plant Pathology, Pennsylvania State University; C,D,E from United States Department of Agriculture Office of Communications.)*

Brock, T. 1978. *Thermophilic Microorganisms and Life at High Temperatures.* Springer-Verlag, New York.

Burns, R. G., and J. H. Slater (eds.). 1982. *Experimental Microbial Ecology.* Blackwell Scientific Publications, Oxford, England.

Campbell. R. E. 1977. *Microbial Ecology.* Blackwell Scientific Publications, Oxford, England.

Colwell, R. R., and R. Y. Morita (eds.). 1974. *Effect of the Ocean Environment on Microbial Activities.* University Park Press, Baltimore.

Doetsch, R. N., and T. M. Cook. 1973. *Introduction to Bacteria and Their Ecobiology.* University Park Press, Baltimore.

Gould, G. W., and J. E. L. Corry (eds.) 1980. *Microbial Growth and Survival in Extremes of Environment.* Academic Press, New York.

Gray, T. R. G., and D. Parkinson (eds.). 1968. *The Ecology of Soil Bacteria.* University of Toronto Press, Toronto.

Gregory, P. H. 1973. *The Microbiology of the Atmosphere.* John Wiley & Sons, New York.

Laskin, A., and H. Lechevalier (eds.). 1974. *Microbial Ecology.* CRC Press, Boca Raton, Florida.

Lynch, J. M., and N. J. Poole (eds.). 1979. *Microbial Ecology: A Conceptual Approach.* Blackwell Scientific Publications, Oxford, England.

Margulis, L. 1981. *Symbiosis in Cell Evolution: Life and Its Environment on the Early Earth.* W. H. Freeman and Co., San Francisco.

Mitchell, R. 1974. *Introduction to Environmental Microbiology.* Prentice-Hall, Englewood Cliffs, New Jersey.

Rheinheimer, G. 1981. *Aquatic Microbiology.* John Wiley & Sons, New York.

Sieburth, J. McN. 1979. *Sea Microbes.* Oxford University Press, New York.

Wood, E. J. F. 1965. *Marine Microbial Ecology.* Chapman & Hall, London.

811

Suggested
supplementary
readings

LINACRE COLLEGE LIBRARY
UNIVERSITY OF OXFORD

Agricultural microbiology

22

Microbial biogeochemical cycling activities and interpopulation relationships have a major influence on agricultural practices. As in other areas of economic microbiology, some microbial activities are beneficial and others detrimental, and it is the controlled balance of microbial activities that is important in determining agricultural success. Fertilizers and pesticides have become an integral part of modern agricultural practice because crop yields depend on maintaining high levels of soil fertility and limiting destruction due to infectious diseases and pest populations. The environmental fate and usefulness of these chemical fertilizers and pesticides are greatly influenced by the metabolic activities of microorganisms in soil. In some cases the microbial populations themselves provide essential inputs of fixed forms of nitrogen to soil and to effective control of pest populations. In other cases, microorganisms are responsible for the loss of valuable fertilizers and cause numerous harmful diseases of agricultural crops and farm animals.

Soil fertility and management of agricultural soils

Influence of available nitrogen on soil fertility

Microbial biogeochemical cycling activities are extremely important for the maintenance of **soil fertility**. The nutrient in most limited supply normally is **nitrogen**, and thus, the concentration of fixed forms of nitrogen in soil usually determines the potential productivity of an agricultural field. The natural availability of fixed forms of nitrogen in agricultural soils is determined by the relative balance between the rates of microbial nitrogen fixation and denitrification. **Nitrogen-rich fertilizers** are widely applied to soils to support increased crop yields, but proper application of nitrogen fertilizers must take into consideration the solubility and leaching characteristics of the particular chemical form of the fertilizer and the rates of microbial biogeochemical cycling activities. In order to avoid the losses caused by leaching and denitrification, nitrogen fertilizer is commonly applied as an ammonium salt, free ammonia, or urea. When nitrification proceeds too quickly, as it does in some agricultural soils, wasteful losses of nitrogen fertilizer and groundwater contamination with nitrate occur. Nitrification of ammonium compounds also yields acidic products that may have to be neutralized by liming. To prevent the undesirable microbial transformation of nitrogen fertilizers, nitrification inhibitors, such as nitrapyrin, are often applied together with the nitrogen fertilizer. The use of nitrification inhibitors can increase crop yields by 10–15 percent for the same amount of nitrogen fertilizer applied. In addition, by decreasing the rate of nitrification, the problem of groundwater pollution by nitrate is prevented.

Crop rotation

Crop rotation is traditionally used to prevent the exhaustion of soil nitrogen and reduce the cost of nitrogen fertilizer applications. Leguminous crops, planted in rotation with other crops because of their symbiotic association with nitrogen-fixing bacteria, are able to reduce the soil's requirement for expensive nitrogen fertilizer. Leguminous plants produce more fixed nitrogen than they require, and the excess ammonium nitrogen is released to the soil (Table 22.1). Soybeans and corn are often rotated every few years in the Midwestern United States because corn takes up nitrogen from the soil, substantially decreasing the concentration of soil nitrogen, but during the seasons when soybeans are grown, the levels of fixed nitrogen in the soil increase. In some cases nitrogen fixation can be enhanced by inoculation of legume seeds with appropriate *Rhizobium* strains, which increases the extent of nodule formation. In molybdenum-deficient soils, a dramatic improvement in the rate of nitrogen fixation can be achieved by the application of small amounts of molybdenum because this element is a constituent of the nitrogenase enzyme complex, required for nitrogen-fixing activities. It is important that maximal rates of nitrogen fixation be achieved by rotation with leguminous crops to successfully replenish soil nitrogen.

Genetic engineering to create new nitrogen-fixing organisms

Through genetic engineering it is hoped that the ability to fix atmospheric nitrogen can be extended to other plant species, including corn and wheat (Figure 22.1). Much higher crop yields and significant economic savings would be realized if these plants could be grown without the need for added nitrogen fertilizer. The elimination of massive fertilizer applications to agricultural soils would also reduce problems associated with nitrification and groundwater contamination. The creation of nitrogen-fixing plants is probably one the major benefits that could be accomplished through the use of recombinant DNA technology. The advantages of creating new nitrogen-fixing plants, though, is tempered by the fact that photosynthetic generation of ATP will still limit plant productivity, particularly when the energy requirements of nitrogen fixation are considered unless the efficiency of photosynthesis is also increased greatly.

table 22.1

Nitrogen gains in soils in the United States obtained by planting leguminous crops

Crop	Soil nitrogen increase (kg nitrogen fixed/hectare/year)
Alfalfa	100–280
Red clover	75–175
Pea	75–130
Soybean	60–100
Cowpea	60–120
Vetch	80–140

Soil management practices

Soil organic matter, **humus**, is important as well in determining soil fertility. Humus acts as a nutrient reserve, increases ion exchange capacity, and loosens the structure of the soil, which are all important properties of the soil's ability to support the growth of plants. When virgin lands are put to agricultural use, their humus content decreases regularly over the next 40 to 50 years, eventually reaching a stable concentration at a much lower value. The probable causes for this phenomenon are increased aeration of the soil through tilling and the removal of most of the produced organic matter when the crop is harvested. The reduction in soil organic matter lowers the fertility of soil.

Soil management practices vary according to crop and soil characteristics but generally share several common features. With the exception of rice grown in paddies, all major crops require aerobic soil conditions. The aeration status of soils is controlled principally by the soil moisture levels. It is of prime importance to provide adequate drainage for agricultural land to prevent waterlogging of soils that can lead to the formation of anaerobic soil conditions. In addition to the directly injurious effects of prolonged exposure to anaerobic soil conditions on plant roots, oxygen deprivation is indirectly deleterious to plants because of the microbial use of secondary electron acceptors, including nitrate, sulfate, and ferric iron, which results in loss of vital nitrogen due to increased rates of denitrification, production of toxic H_2S, and deposition of sticky greenish ferrous iron in waterlogged soils.

The management of soil pH is often critical in agricultural production. The soil pH regulates solubility of plant nutrients, availability of potentially toxic heavy metals, and biodegradation of plant material. Some mineral cycling activities, such as

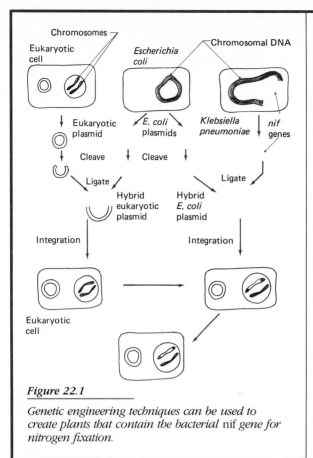

Chromosomes

Eukaryotic cell

Escherichia coli

Chromosomal DNA

↓ Eukaryotic plasmid

↙ *E. coli* plasmids

Klebsiella pneumoniae

↓ *nif* genes

↓ Cleave   ↓ Cleave   ↓

↘ Ligate ↙   Ligate ↘

Hybrid eukaryotic plasmid

Hybrid *E. coli* plasmid

Integration   Integration

Eukaryotic cell

Genetic engineering techniques can be used to create plants that contain the bacterial nif *gene for nitrogen fixation.*

Discovery process

One of the greatest benefits that may be realized through genetic engineering is the introduction of the capacity to fix nitrogen into plants, such as wheat and rice, that are not able to utilize atmospheric nitrogen. For years scientists have been exploring the relationships between *Rhizobium* and the plants with which this nitrogen-fixing bacterium can establish symbiotic relationships. Winston Brill and colleagues at the University of Wisconsin have tried to apply the screening procedures of the pharmaceutical industry in an effort to find especially effective nitrogen-fixing strains of *Rhizobium* that could increase crop yields. Using mutagens and screening procedures, they isolated strains of *Rhizobium* that were capable of very high rates of nitrogen fixation. However, field tests with these efficient *Rhizobium* strains did not increase crop yields; the superior nitrogen-fixing strains could not compete successfully with indigenous strains. Because of the inefficiency and lack of success of the mutation-screening approach, microbiologists are studying the genetics and biochemistry of infection by *Rhizobium* with the aim of employing recombinant DNA techniques to genetically engineer plants containing the bacterial genes for nitrogen fixation. Many research groups are carrying out these investigations. In one series of studies carried out by Aladar Szalay and coworkers at Cornell University, the genes for nitrogen fixation are first inserted into the genome of a yeast; plasmids from *E. coli* and from a yeast cell are cleaved and then fused to form a single hybrid plasmid, which can be recognized by the yeast cell and integrated into its chromosomal DNA. In the next step, the genes that will be introduced into the yeast are isolated from the chromosome of *Klebsiella pneumoniae*, a nitrogen fixer. The genes, collectively designated *nif,* code for some 17 proteins. Another *E. coli* plasmid is cleaved, and the isolated *nif* genes are introduced to form a second hybrid plasmid. Because of the bacterial DNA already inserted into one of the yeast chromosomes, the yeast cell recognizes the hybrid *E. coli* plasmid. The plasmid is then integrated into the yeast chromosome. Although the insertion of the prokaryotic *nif* genes into the eukaryotic yeast cell demonstrates that genetic material can be transferred between different biological systems, the nitrogen-fixing proteins are not expressed in the yeast. More studies are needed to elucidate the factors controlling expression of the *nif* genes before success is obtained. It is increasingly apparent that the ability to engineer organisms, such as eukaryotic plant cells that can fix atmospheric nitrogen, depends on developing a thorough understanding of the molecular biology of gene expression; once we understand the mechanisms of gene regulation in eukaryotes, we will be able to apply this knowledge through genetic engineering to create organisms with novel properties.

sulfur oxidation and nitrification, result in acid production, and in some soils it is necessary to maintain the pH close to neutrality by liming.

Fields that are cleared of crop cover are subject to erosion of the topsoil, especially where the land slopes extensively. "No till farming" is a relatively novel agricultural practice that provides a solution to the combined problems of humus loss, topsoil erosion, and soil compaction. In this type of soil management, ploughing is eliminated or reduced to a minimum, planting is performed with drill-seeding machines, weeds are controlled with herbicides rather than by cultivation, and crop residues are left on the field as a mulch cover. In areas that are especially vulnerable to erosion, grass or other cover is planted for the winter and is killed with herbicides just before the next season's crop is planted. This practice builds up humus, loosens and conserves the topsoil, and helps the soil retain moisture. Besides its soil conservation benefits, "no till" farming in some situations is economically competitive with, or even superior to, conventional farming methods.

Microbial diseases of crops

There are tens of thousands of diseases of cultivated plants, and each agricultural crop is generally subject to over 100 different diseases. **Diseases of agricultural crops** cause serious economic losses, with annual worldwide crop losses due to diseases, insects, and weeds running about $200 billion (Table 22.2). Crop losses resulting from infectious diseases in the United States alone amount to over $4 billion per year, resulting in significant economic hardship to farmers and seriously reducing the world's supply of food. There are major regional differences in the value of crop losses due to diseases, insects, and weeds. The percentage losses differ considerably with continent and are much greater in underdeveloped areas than they are in more developed regions (Table 22.3) because of the lack of economic resources available for pesticides, herbicides, fungicides, and fertilizers for maintaining healthy agricultural crops.

Outbreaks of plant diseases can cause immediate and long-lasting agricultural damage. The chestnut blight disease destroyed the native North American chestnut trees that had provided an important cash crop, especially in the Appalachian area. A leaf blight of maize in 1970 caused the destruction of more than 10 million acres of corn crops in the United States during that one year. Sometimes, plant diseases even have far-reaching historical effects on masses of people. The potato blight in Ireland in 1845 resulted in mass starvation and widespread emigration from Ireland to North America.

Symptoms of plant diseases and mechanisms of microbial pathogenicity

Microbial pathogens are a major cause of plant diseases. The mechanisms by which microbial pathogens cause plant diseases and the symptoms of different plant diseases vary with the causal agent and sometimes with the plant (Table 22.4). A plant is healthy when it can carry out its basic physiological functions, including (1) normal cell division, differentiation, and development; (2) absorption of water and minerals from the soil and translocation of these nutrients throughout the plant; (3) photosynthesis and translocation of the photosynthetic products to areas of utilization or storage; (4) metabolism of synthesized compounds; (5) reproduction; and (6) storage of food supplies for overwintering or reproduction. Plants become diseased whenever any of these functions are significantly disrupted.

Normally, plant pathogens weaken or destroy cells and tissues, reducing or eliminating the ability of these cells and tissues to perform their normal physiological functions and resulting in the onset of disease symptoms and reduced plant growth or death. The kinds of cells and tissues

table 22.2

Agricultural crop losses due to plant diseases

Crop	Loss (millions of tons)
Cereals	135
Potatoes	89
Sugar beets & sugar cane	232
Other vegetables	31
Fruits	33
Coffee, cocoa, & tobacco	3
Vegetable oil crops	14
Fiber crops & natural rubber	3

table 22.3

Regional crop losses due to plant diseases, insects, and weeds

Region	Loss (% total crop)
Europe	25
Oceania	28
North & Central America	29
U.S.S.R. & China	30
South America	33
Africa	42
Asia	43

table 22.4

Some symptoms of microbial diseases of plants

Symptom	Description
Necrosis (rots)	Death of plant cells; may appear as spots in localized areas
Canker	Localized necrosis resulting in lesion, usually on stem
Wilt	Droopiness due to loss of turgor
Blight	Loss of foliage
Chlorosis	Loss of photosynthetic capability due to bleaching of chlorophyll
Hypoplasia	Stunted growth
Hyperplasia	Excessive growth
Gall	Tumorous growth
Scab	Localized lesions usually slightly raised or sunken

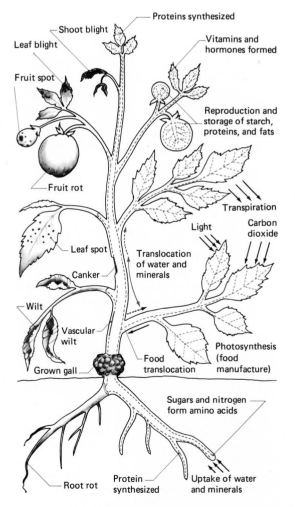

Figure 22.2

Basic plant functions can be interfered with by common types of plant diseases.

that become infected determine which physiological functions of the plant are initially impaired, as shown when (1) infection of the root, as occurs in root rots, interferes with absorption of water and nutrients from the soil; (2) infection of the xylem vessels, as occurs in vascular wilts and certain cankers, interferes with translocation of water and minerals to the crown of the plant; (3) infection of the foilage, as occurs in leaf spots, blights, and mosaics, interferes with photosynthesis; (4) infection of the cortex, as occurs in cortical canker and viral infections of phloem, interferes with the downward translocation of photosynthetic products; (5) infections of reproductive structures, as occur in bacterial and fungal blights and microbial infections of flowers, interfere with reproduction; and (6) infections of fruit, as occur in fruit rots, interfere with reproduction and/or storage of reserve foods for the new plant (Figure 22.2).

Infected plants can develop a variety of morphological abnormalities as a result of infection by a microbial pathogen. Invasion of plant cells by pathogenic microorganisms sometimes results in the rapid death of the plant. In other cases the plants undergo slower changes. Pathogens that penetrate plants directly often elicit a morphological response from the plant that may result in the formation of abscission or gum layers (Figure 22.3). Formation of papillae may be an attempt by the plant's innate defense system to block the spread of the pathogens. The cell walls of infected plant tissues are often modified, resulting in swelling or other distortions of the cell. Cell wall modifications may be due to the production of enzymes by the pathogenic microorganisms that degrade cell wall components. Invasion by plant pathogens may disrupt cell permeability, leading to leakage and death of the plant cells. Changes in permeability may be caused by pectinase enzymes or by toxins produced by the plant pathogens.

Blockage of water transport in a plant can lead to desiccation and symptomatic **wilt**. *Fusarium*, which causes wilt in tomatoes, causes a reduction of water flow through the xylem. Stomatal dysfunction due to invading pathogens alters transpiration and water transport in the plant, and various bacterial pathogens cause wilting by blocking the stomata (Figure 22.4). *Erwinia stewartii* causes wilt of corn, *Erwinia tracheiphila* wilt of cucumbers, *Pseudomonas solanacearum* wilt of tobacco and tomato, and *Corynebacterium insidiosum* wilt of alfalfa. These wilt diseases represent a serious cause of plant loss due to bacterial diseases.

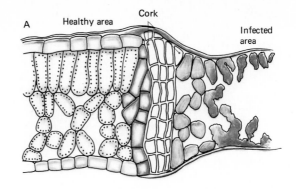

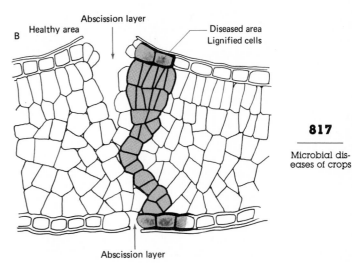

Figure 22.3

Cork (A) or abscission (B) layers are formed by leaves in response to microbial infections.

Plant pathogens may cause altered metabolic activities of the plant, and diseased plants sometimes show decreased growth as a result of changes in respiratory activity and rates of carbon dioxide fixation. Foliar pathogens sometimes produce **chlorosis**, bleaching of chlorophyll, which prevents the plant from carrying out photophosphorylation and producing the ATP needed for carbon dioxide fixation. Plant pathogens may also cause changes in protein synthesis. **Overgrowths** and **gall formation** involve alterations in the nucleic acid function controlling protein synthesis.

Microbial pathogens can disrupt normal plant functions by the production of degradative enzymes, toxins, and growth regulators. Because plant cells are held together by pectin substances, plant pathogens that produce pectinases cause the weakening of plant structures. For example, polygalacturonase enzymes degrade pectic sub-

Figure 22.4

A comparison of a healthy (right) and wilting (left) tomato plants. The wilt is due to infection with Pseudomonas solareacearun. *(Courtesy Thomas Barksdale, USDA, Beltsville, Maryland.)*

stances, producing soft rots and other lesions of plant tissues. Other plant pathogens produce cellulase and/or hemicellulase enzymes that degrade the primary cell wall components of plant cells. Some pathogenic microorganisms of plants produce toxins that interfere with the normal metabolic activities of the plant; for example, *Pseudomonas tabaci* produces an exotoxin, β-hydroxydiaminopimelic acid, which interferes with the metabolism of methionine and causes tobacco wildfire disease.

In some cases plant diseases result from plant cells dividing too rapidly, causing **hyperplasia**, or becoming excessively enlarged, causing **hypertrophy**. Such hyperplastic or hypertrophied cells can result in the development of abnormally large nonfunctional organs, an abnormal proliferation of organs, and/or the production of amorphous overgrowths on normal-looking organs. Overstimulated cells and tissues not only divert much needed carbon and energy away from the normal tissues, but also, by their explosive growth, frequently crush adjacent normal tissues and interfere with the physiological functions of the plant. **Dwarfism** can result when plant pathogens degrade or inactivate plant growth substances. Auxin and indoleacetic acid production by some microorganisms are implicated in disease processes, such as production of galls. Gibberellins and cytokinins produced by some fungal plant pathogens result in excessive elongation of plant stems. Pro-

duction of ethylene by some plant pathogens triggers metabolic changes in the plants, which lead to damage of plant tissues.

Transmission of plant pathogens

The development of plant diseases, which is initiated by the contact of pathogenic microorganisms with a plant surface, involves entry of the pathogen into the plant, growth of the infecting microorganisms, and finally the appearance of disease symptoms. Pathogenic microorganisms may come into contact with the plant via its roots or aerial surfaces. Most fungal plant pathogens are dispersed through the air as spores, often coming to rest on the leaves or stems of plants. Most viral diseases of plants are transmitted by insect vectors, which bring these pathogens into contact with the plant's aerial surfaces. Some bacterial and fungal pathogens are also carried by insect vectors. Nematodes and other animals in soil can transmit pathogens to the plant root system. Motile pathogens in the soil, including plant pathogenic bacteria, such as *Pseudomonas* species, and fungi, such as *Oidium*, are attracted to plant roots through chemotaxis.

Typically, pathogenic microorganisms enter plants through natural openings and wounds caused by animals or injury from farm implements and equipment (Figure 22.5). Viruses nor-

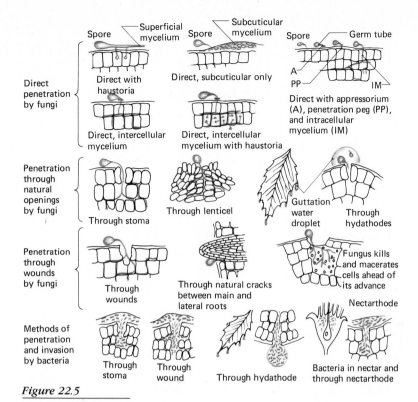

Direct penetration by fungi

Spore — Superficial mycelium
Direct with haustoria

Spore — Subcuticular mycelium
Direct, subcuticular only

Spore — Germ tube
A
PP — IM
Direct with appressorium (A), penetration peg (PP), and intracellular mycelium (IM)

Direct, intercellular mycelium

Direct, intercellular mycelium with haustoria

Penetration through natural openings by fungi

Through stoma

Through lenticel

Guttation water droplet

Through hydathodes

Penetration through wounds by fungi

Through wounds

Through natural cracks between main and lateral roots

Fungus kills and macerates cells ahead of its advance

Methods of penetration and invasion by bacteria

Nectarthode

Through stoma

Through wound

Through hydathode

Bacteria in nectar and through nectarthode

Figure 22.5

As shown in this diagram, microbial pathogens can enter plants in several ways. Fungal pathogens can penetrate plants directly through natural openings or through wounds. Bacterial pathogens can penetrate plants through stomata and wounds.

mally enter plants through wounds caused by the vector carrying the virus. Some viruses enter plants through the roots with the groundwater taken up by the plant. Some plant pathogens, however, are able to penetrate the plant directly. Such penetration of the plant normally involves attachment of the pathogen to the plant surface, followed by formation of a penetration peg passing through the cuticle and the cell wall, accomplished mainly by physical forces. The penetration of the plant pathogen can also be enzymatic when plant tissues are subjected to enzymatic attack by the pathogen that softens the plant tissue in the vicinity of penetration. In some cases the pathogen can penetrate the cuticle, such as occurs with powdery mildews caused by *Erysiphe* species and soft rots caused by *Botrytis cinerea*.

Viral diseases

A necessary attribute of most plant pathogens, especially the viral pathogens of plants, is the ability to survive outside host cells until susceptible plant hosts can be located. Plant pathogens must move from crop to crop and field to field for their continued existence, and they must also successfully overwinter if their crop hosts are seasonal. As obligate intracellular parasites, plant pathogenic viruses of necessity must find suitable plant cells for their replication. If free in soil, viruses would be subject to inactivation by soil microbial enzymes. Persistence of viruses within vectors is one means by which viruses may survive in soil. The distribution of viral plant diseases often follows the spatial distribution pattern of the vectors. The patterns of dissemination of pathogenic viruses are fixed to a large extent by environmental factors that affect the survival and movement of the vector organisms, such as soil texture and moisture.

Many plant pathogenic viruses are transported to susceptible host plants by vectors that acquire the pathogenic viruses from soil or diseased plant tissues. Insects such as aphids, leaf hoppers, mealy bugs, and nematodes often act as the vectors for

table 22.5

Some plant pathogenic viruses

DNA viruses
Caulimovirus group: cauliflower mosaic virus

RNA viruses

Rod-shaped:

Tobavirus group: tobacco rattle virus, pea early browning virus
Tobamovirus group: tobacco mosaic virus

Viruses with flexuosus or filamentous particles:

Potexvirus group: potato virus X
Carlavirus group: carnation latent virus
Potyvirus group: potato virus
Beet yellows virus
Festuca necrosis virus
Citrus tristerza virus

Isometric viruses:

Cucumovirus group: cucumber mosaic virus
Tymovirus group: turnip yellow mosaic virus
Comovirus group: cowpea mosaic virus
Nepovirus group: tobacco ringspot virus
Bromovirus group: brome mosaic virus
Tombushvirus group: tomato bushy stunt
Alfalfa mosaic virus
Pea enation mosaic virus
Tobacco necrosis virus
Wound tumor virus group: wound tumor virus, rice dwarf virus
Tomato spotted wilt virus

Rhabdoviruses:

Lettuce necrotic yellows virus

A   B

Figure 22.6

(A) Viral ring spot on a tobacco leaf. Note the peculiar line patterns, often spreading out on either side of the veins, which make identification of this disease easy. (B) This tobacco leaf shows the characteristic symptoms of mosaic disease. (Courtesy United States Department of Agriculture.)

viral diseases of plants. Even other microbes can serve as vectors for viral pathogens. *Olpidium brassicae*, a chytrid fungus, is the vector of tobacco necrosis virus and probably several other plant pathogenic viruses. Pollen and plant seeds are also involved in the transmission of plant viruses. For example, tobacco rattle virus is detectable on the pollen of infected petunia plants and is disseminated through the air together with the pollen to susceptible plants. The spread of viruses on plant structures involved in the reproduction activities of the plant, such as pollen and seeds, ensures that viruses are maintained with susceptible host plant populations and that viral diseases are endemic to plant populations.

Plant pathogenic viruses are named according to their ability to cause specific diseases (Table 22.5). Often, the only symptom of a viral plant infection is reduced growth rate that results in some degree of dwarfing. In the case of systemic viral diseases of plants, the most common symptoms are mosaics and ringspots (Figure 22.6). **Mosaics** are characterized by the formation of light green, yellow, or white spots intermingled with the normal green aerial plant structures. **Ringspots** are characterized by the appearance of chlorotic or necrotic rings on the leaves. These primary symptoms may be accompanied by a variety of other symptoms in specific viral plant diseases (Figure 22.7).

Recently, **viroids** have been identified as the causative agents of several plant diseases (Figure 22.8). Viroids, which are simply naked RNA molecules, have been implicated as the cause of potato spindle disease, chrysanthemum stunt, and citrus exocortis disease, as well as several other

Figure 22.7

Symptoms of viral diseases of plants.

Tobacco mosaic

Cucumber mosaic on pepper

Cucumber mosaic on squash

Cucumber mosaic on cucumber

Apple mosaic

Bean mosaic

Tomato spotted wilt

Tomato ringspot on grape

Blackberry sterility

Tomato aspermy

Plum pox on apricot — Seed

Potato yellow dwarf

Apple flat limb

Stem necrosis

Pear rough bark

Elm zonate canker

Cherry black canker

Graft brown line

Stunting

Citrus tristeza

Banana bunchy top

Cocoa swollen shoot

Stem pitting

Pear ring pattern mosaic

Prunus necrotic ringspot

Elm ringspot

Tobacco ringspot

Maize dwarf mosaic

Tulip breaking

Clover wound tumor

Citrus woody gall

Apple scar skin

Cucumber mosaic on gladiolus bulb

Pear stony pit

Apple russet ring

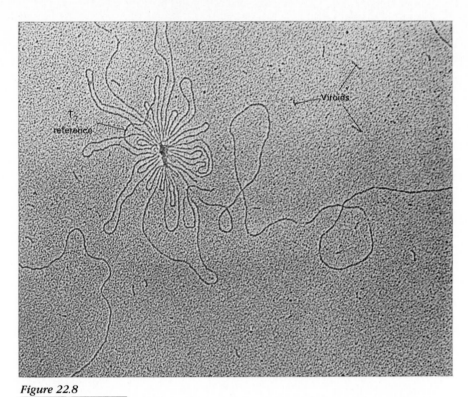

Figure 22.8

Photomicrograph of the potato spindle tuber viroid (23,000×). Note size difference between the reference bacteriophage T7 DNA and the nucleic acid of the viroids. (From T. O. Diener, USDA, and T. Koller and J. S. Sago, Swiss Federal Institute of Technology, Zurich.)

diseases. It has not yet been demonstrated how viroids are transmitted or how they code for their own replication. It is quite interesting, however, that nucleic acid molecules as contrasted with more highly organized microbial pathogens can be the source of disease.

Bacterial diseases

Plant pathogenic bacterial species occur in the genera *Mycoplasma, Spiroplasma, Corynebacterium, Agrobacterium, Pseudomonas, Xanthomanas, Streptomyces,* and *Erwinia.* These bacteria are widely distributed and cause a large number of plant diseases, including hypertrophy, wilts, rots, blights, and galls (Table 22.6). Plant pathogenic bacteria cause many different disease symptoms, and most symptoms of plant disease can be caused by several different bacterial species (Figure 22.9).

Crown gall, which is caused by *Agrobacterium tumefaciens*, is a particularly interesting plant disease (Figure 22.10). Crown gall may occur on fruit trees, sugar beets, or other broad-leaved plants.

The disease process is initiated when viable cells of *A. tumefaciens* enter wounded surfaces of susceptible dicotyledonous plants, usually at the soil–plant stem interface either through the root or a wound. *A. tumefaciens* is able to transform host plant cells into tumorous cells, and the disease is manifested by the formation of a tumor growth, the crown gall. Once the disease is established, the tumor continues to grow even if viable *Agrobacterium* are eliminated. The tumor maintenance principle has been identified as a fragment of a large, **tumor-inducing (Ti) plasmid.** A fragment of this bacterial plasmid is transferred to the plant where it is maintained in the tumor tissue. The Ti plasmid has great potential in recombinant DNA technology to introduce desired genetic information into a wide range of plants of agricultural significance (Figure 22.11). Thus, although *Agrobacterium* possessing the Ti plasmid cause great damage to crops, the Ti plasmic represents the vehicle for the creation of improved crops and is capable of disease resistance and increased yields with decreased management.

The relationship between bacterial plant path-

table 22.6

Some bacterial diseases of plants

Genus	Species	Disease
Pseudomonas	tabaci	Wildfire of tobacco
	angulata	Leaf spot of tobacco
	phaseolicola	Halo blight of beans
	pisi	Blight of peas
	glycinea	Blight of soybeans
	syringae	Blight of lilac
	solanacearum	Moko of banana
	caryophylli	Wilt of carnation
	cepacia	Sour skin of onion
	marginalis	Slippery skin of onion
	savastanoi	Olive knot disease
	marginata	Scab of gladiolus
Xanthomonas	phaseoli	Blight of beans
	oryzae	Blight of rice
	pruni	Leaf spot of fruits
	juglandis	Blight of walnut
	citri	Canker of citrus
	campestris	Black rot of crucifers
	vascularum	Gumming of sugar cane
Erwinia	amylovora	Fire blight of pears and apples
	tracheiphila	Wilt of cucurbits
	stewartii	Wilt of corn
	carotovora	Soft rot of fruit, black leg of potato, blight of chrysanthemum
Corynebacterium	insidiosum	Wilt of alfalfa
	michiganese	Wilt of tomato
	fascians	Leafy gall of ornamentals
Streptomyces	scabies	Scab of potato
	ipomoeae	Pox of sweet potato
Agrobacterium	tumefaciens	Crown gall of various plants
	rubi	Cane gall of raspberries
	rhizogenes	Hairy root of apple
Mycoplasma	sp.	Aster yellows
	sp.	Peach X disease
	sp.	Peach yellows
	sp.	Elm phloem necrosis
Spiroplasma	sp.	Citrus stubborn disease
	sp.	Bermuda grass witches' broom
	sp.	Corn stunt

ogens and their host plant is greatly affected by the stationary nature of the plant, the periodicity of plant growth, and the protective surfaces of the plant. Bacterial plant pathogens must possess an independent mode of dispersal in order to reach new host plants and must also have some mechanism for entering the plant. Because most plant pathogenic bacteria do not form resting stages, they must remain within the confines of the plant tissues. Even during times of plant dormancy, many plant pathogenic bacteria can remain viable on plant seeds and other plant tissues. Some bacterial populations exhibit no significant soil phase.

For example, *Erwinia amylovora*, which causes fire blight in fruit trees, remains within infected tissues and is disseminated in plant exudates by insects or raindrops (Figure 22.12). During the winter *Erwinia amylovora* does not grow but remains dormant within infected tissues of the stems and branches of trees, and in the spring the bacteria are distributed to susceptible plants.

Many bacterially caused plant diseases are seedborne. In these cases the pathogenic bacteria must survive on the seeds for a transient period in the soil. Bacteria may be carried on seeds as a surface contaminant or within the seed. For ex-

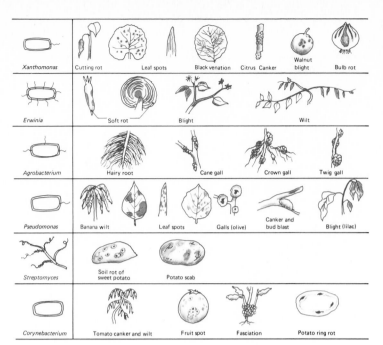

Figure 22.9

Symptoms of bacterial diseases
of plants.

Xanthomonas	Cutting rot	Leaf spots	Black venation	Citrus Canker	Walnut blight	Bulb rot	
Erwinia		Soft rot	Blight		Wilt		
Agrobacterium	Hairy root	Cane gall	Crown gall	Twig gall			
Pseudomonas	Banana wilt	Leaf spots	Galls (olive)	Canker and bud blast	Blight (lilac)		
Streptomyces	Soil rot of sweet potato	Potato scab					
Corynebacterium	Tomato canker and wilt	Fruit spot	Fasciation	Potato ring rot			

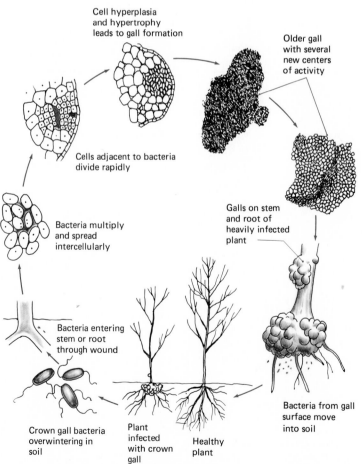

Cell hyperplasia
and hypertrophy
leads to gall formation

Older gall
with several
new centers
of activity

Cells adjacent to bacteria
divide rapidly

Bacteria multiply
and spread
intercellularly

Galls on stem
and root of
heavily infected
plant

Bacteria entering
stem or root
through wound

Crown gall bacteria
overwintering in
soil

Plant
infected
with crown
gall

Healthy
plant

Bacteria from gall
surface move
into soil

Figure 22.10

Disease cycle of crown gall
caused by Agrobacterium
tumefaciens.

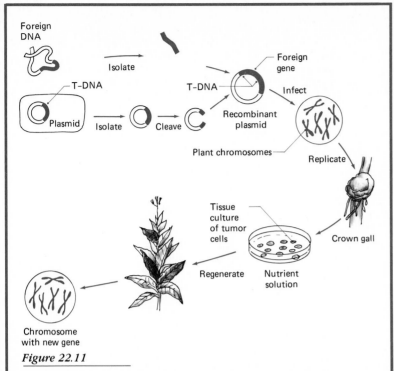

Foreign DNA

Isolate

T-DNA

Plasmid

Isolate

Cleave

T-DNA

Foreign gene

Infect

Recombinant plasmid

Plant chromosomes

Replicate

Crown gall

Tissue culture of tumor cells

Nutrient solution

Regenerate

Chromosome with new gene

Figure 22.11

The Ti plasmid has many potential uses in genetic engineering.

Discovery process

The elucidation of how *Agrobacterium tumefaciens* causes crown gall has raised the possibility of exploiting the natural infection process to introduce foreign DNA into plant cells. *A. tumefaciens,* which causes crown gall tumors in most dicotyledonous plants, carries a plasmid that induces the infected plant cells to synthesize nitrogen compounds called opines. During the normal infectious process, a section of the plasmid, called T-DNA, combines with chromosomal DNA in the nucleus of the plant cell. It may therefore be possible to employ the Ti plasmid as a vector for inserting foreign DNA into plant cells. To do so the plasmid would be cut open at a site within the T-DNA, and the foreign gene would be spliced into it. When the tumor cells are grown in tissue culture, they continue to carry and replicate T-DNA during the normal divisional process. Investigators at the University of Leiden have been able to infect tobacco cells in culture with *A. tumefaciens,* with the finding that tobacco plants regenerated from cultured tumor cells retain T-DNA and continue to make opine synthetase. Moreover, scientists at the Max Planck Institute for Plant Breeding in Cologne demonstrated that the gene carried by T-DNA that codes for the enzyme opine synthetase is passed on through the seed to succeeding generations as if it were an ordinary dominant gene. If foreign genes inserted into T-DNA are also transmitted to plant progeny, new plant strains could be genetically engineered.

ample, *Pseudomonas phaseolicola* is carried in the micropyle and causes the halo blight of beans. *Xanthomonas malvacearum,* the causative organism of cotton blight, can be found on the external cotyledon margins during germination of the seed. Some bacterial diseases are caused by pathogens that have a permanent soil phase, for example, some fluorescent *Pseudomonas* species that cause soft rots in plants. These organisms are abundant as saprophytes in the rhizosphere and

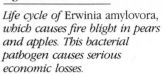

Figure 22.12

Life cycle of Erwinia amylovora, *which causes fire blight in pears and apples. This bacterial pathogen causes serious economic losses.*

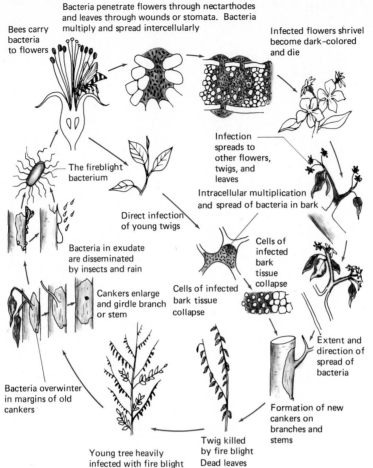

Bees carry bacteria to flowers

Bacteria penetrate flowers through nectarthodes and leaves through wounds or stomata. Bacteria multiply and spread intercellularly

Infected flowers shrivel become dark-colored and die

The fireblight bacterium

Infection spreads to other flowers, twigs, and leaves

Intracellular multiplication and spread of bacteria in bark

Direct infection of young twigs

Bacteria in exudate are disseminated by insects and rain

Cells of infected bark tissue collapse

Cells of infected bark tissue collapse

Cankers enlarge and girdle branch or stem

Extent and direction of spread of bacteria

Bacteria overwinter in margins of old cankers

Formation of new cankers on branches and stems

Young tree heavily infected with fire blight

Twig killed by fire blight Dead leaves cling to twig

infect the plants through the roots, but most plant pathogenic bacteria are unable to survive for long periods of time in the soil.

Fungal diseases

Most plant diseases are caused by pathogenic fungi (Table 22.7). In addition to the large numbers of fungal plant pathogenic species, there are a wide variety of disease symptoms that characterize fungal infections of plants (Figure 22.13). Many fungi are well adapted to act as plant pathogens. Fungal spore production permits aerial transmission between plants and allows plant pathogenic fungi to remain viable outside host plants. Survival and infectivity of most plant pathogenic fungi depend on prevailing environmental temperature and moisture conditions. Spores germinate and mycelia grow when the temperature is between -5 and $45°C$, and there is an adequate supply of

moisture. Spores, though, can retain viability for long periods of time during environmental conditions that do not allow for germination. Plant pathogenic fungi generally exhibit a complex life cycle, spent in part in host plant infection and in part outside host plants in the soil or on plant debris in the soil. As an example, the life cycle of *Rhizoctonia solani*, a fungus that causes a variety of diseases, is illustrated in Figure 22.14. Similarly, the fungus *Monilinia fructicola*, the causative agent of brown rot of stone fruits, overwinters as mycelia or conidia on plant materials in the ground; in the spring new conidia are produced, and the mycelia on mummified fruit in the ground produce ascospores; the ascospores are dispersed by wind, rain, and insects, reaching newly forming fruits and initating the infection of the newly formed fruits that results in brown rot (Figure 22.15).

Fungal plant pathogens usually grow most rapidly and cause the most severe diseases of plants

table 22.7

Some diseases of plants caused by fungi

Slime molds

Plasmodiophora	Clubfoot of crucifers
Polymyxa	Root disease of cereals
Spongospora	Powdery scab of potato
Plasmopara	Downy mildew of grapes

Oomycetes

Albugo	White rust of crucifers
Phytophthora	Late blight of potato
Pythium	Seed decay, root rots

Chitridiomycetes

Olpidium	Root disease of various plants
Physodermai	Brown spot of corn
Synchytrium	Black wart of potato
Urophlyctis	Crown wart of alfalfa

Zygomycetes

Rhizopus	Soft rot of fruits

Ascomycetes

Ceratocystis	Dutch elm disease
Claviceps	Ergot of rye
Diaporthe	Bean pod blight
Dibotryon	Black knot of cherries
Diplocarpon	Black spot of roses
Endothia	Chestnut blight
Erysiphe	Powdery mildew of grasses
Lophodermium	Pine needle blight
Microsphaera	Powdery mildew of lilac
Mycosphaerella	Leaf spots of trees
Ophiobolus	Take all of wheat
Podosphaera	Powdery mildew of apple
Sclerotinia	Soft rot of vegetables
Taphrina	Peach leaf curl
Venturia	Apple scab

Basidiomycetes

Armillaria	Root rots of trees
Cronartium	Pine blister rust
Exobasidium	Stem galls of ornamentals
Fomes	Heart rot of trees
Marasmius	Fairy ring of turf grasses
Polyporus	Stem rot of trees
Puccinia	Rust of cereals
Sphacelotheca	Loose smut of sorghum
Tilletia	Stinking smut of wheat
Typhulai	Blight of turf grasses
Urocystis	Smut of onion
Uromyces	Rust of beans
Ustilago	Smut of corn, wheat, and others.

Deuteromycetes

Alternaria	Leaf spots and blight of various plants
Aspergillus	Rots of seeds
Botrytis	Blights of various plants
Cladosporium	Leaf mold of tomato
Colletotrichum	Anthracnose of crops
Cylindrosporium	Leaf spots of various plants
Fusarium	Root rot of many plants
Helminthosporium	Blight of cereals
Penicillium	Blue mold rot of fruits
Phoma	Black leg of crucifers
Rhizoctonia	Root rot of various plants
Thielaviopsis	Black root rot of tobacco
Verticillium	Wilt of various plants

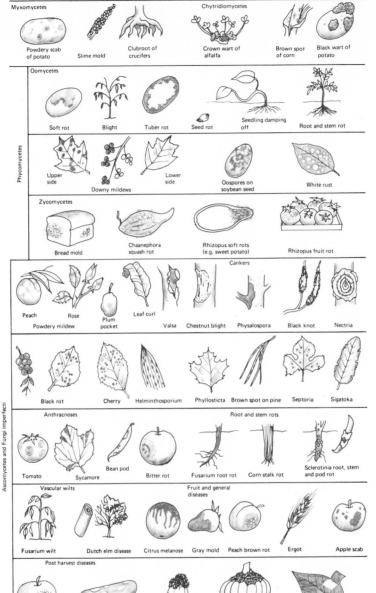

Figure 22.13

Symptoms of fungal diseases of plants.

during the warmer months of the year because during winter many such plant pathogens are inactive. Some fungi, however, such as *Typhula* and *Fusarium*, that cause snow mold of cereals and turf grasses, thrive only in cool seasons or regions. In some cases the optimum temperature for disease development is different than the optimal growth temperature of either the pathogen or the host. For example, in the case of black root rot of tobacco, caused by the fungus *Thielaviopsis*

basicola, the optimum temperature for establishing the disease is lower than the optimal growth temperature of the pathogen, and the host is less able to resist the pathogen at temperatures of 17–23°C. In other cases, such as root rots of wheat and corn caused by the fungus *Gibberella zeae*, the optimum temperature for development of disease is higher than the optimal growth temperature of both the pathogen and the wheat. The soil pH also has a marked effect on the infectivity

Figure 22.14

Life cycle of Rhizoctonia solani, *which causes various plant diseases.*

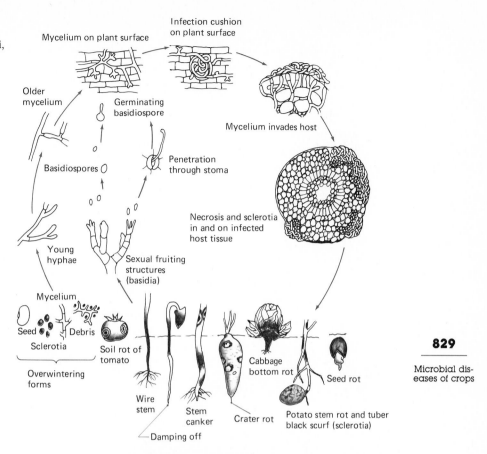

Mycelium on plant surface

Infection cushion on plant surface

Older mycelium

Germinating basidiospore

Mycelium invades host

Basidiospores

Penetration through stoma

Young hyphae

Necrosis and sclerotia in and on infected host tissue

Mycelium

Sexual fruiting structures (basidia)

Seed   Debris

Sclerotia

Soil rot of tomato

Overwintering forms

Wire stem

Stem canker

Damping off

Crater rot

Cabbage bottom rot

Seed rot

Potato stem rot and tuber black scurf (sclerotia)

Figure 22.15

Peaches infected with brown rot. This disease also affects plums, prunes, and cherries. (Courtesy United States Department of Agriculture.)

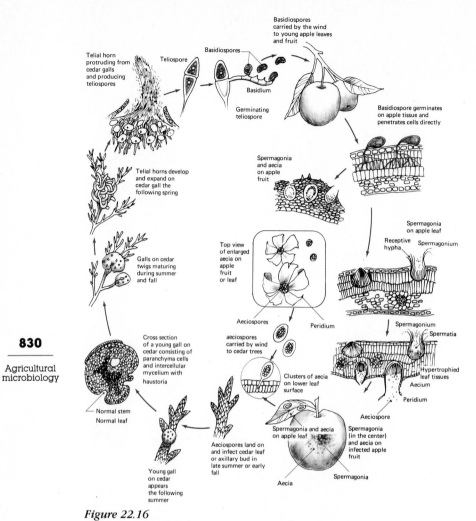

Figure 22.16

Life cycle of Gymnosporangium juniperi-virginianae, *the fungus that causes
cedar-apple rust.*

of soil-borne plant pathogens. For example, *Plasmodiophora brassicae* causes the club root of crucifers at pH values of approximately 5.7, but the disease is completely checked at pH 7.8.

Moisture, like temperature and pH, influences the initiation and development of infectious plant diseases. Some fungal pathogens are dispersed by rain droplets, initiating contact between the pathogen and the susceptible plant. The distribution of rainfall in some regions is closely correlated with the occurrence of plant diseases. For example, downy mildew of grapes and fire blight of pears are more severe when there is high rainfall or high relative humidity. Most fungal pathogens are dependent on high relative humidity for germination of spores in the phylloplane. Some

diseases that affect the plant through the root system are most severe when soil moisture is near the saturation point. The increased moisture allows more rapid multiplication and dispersion of the pathogenic zoospores of *Pythium*.

Perhaps the most important economic fungal diseases of plants are caused by the **rusts** and **smuts**, which are Basidiomycota. There are over 20,000 species of rust fungi, and over 1000 species of smut fungi. The rust fungi require two unrelated hosts for the completion of their complex life cycle. Important plant diseases caused by rusts include black stem rusts of cereals, white pine blister rust, coffee rust, cedar-apple rust, and asparagus rust. An example of the complex life cycle of the cedar-apple rust fungus is illustrated in Figure. 22.16. The smuts are so-named because they produce black dusty spore masses, resembling soot or smut. Important diseases caused by smut fungi include loose smut of oats, corn smut, bunt or stinking smut of wheat, onion smut, etc. Smut and rust fungi cause millions of dollars in annual crop damage.

Control of crop diseases

Since microorganisms that are obligate plant pathogens can remain viable for only a limited period of time outside host plant tissues, appropriate management procedures can be used to control plant pathogens of agricultural crops. The most important of these procedures are the development and planting of **resistant crop varieties** and the use of **crop rotation practices**. Plant pathogens are normally specific for particular host plants; sowing new plant varieties removes the plants that are susceptible to the diseases caused by the particular pathogenic microorganism established in that agricultural field, and rotating crops is an effective way of temporarily removing those plant varieties that are suitable hosts for specific microbial pathogens. Thus, the high populations of infectious plant pathogens produced as a result of the infection of susceptible crops are greatly reduced by crop rotation. The susceptible plant crop can be successfully reestablished when it is rotated back into that field in later years.

Agricultural management practices are frequently directed at avoiding plant disease through modification of host populations. Selective breeding methods have been extensively used in agriculture to develop plant populations that are genetically resistant to disease-causing pathogens, and many new plant strains developed in this way are also successfully producing high crop yields. Pest populations, however, are subject to evolution, selection, and geographical spreading, making it only a matter of time before newly developed plant varieties are subject to serious plant diseases. Thus, in agriculture there is a need for a continuous breeding program for new strains of resistant plants to replace strains of plants that are, or are becoming, susceptible to diseases caused by pathogens and pests. For example, since the 1930s strains of resistant wheat have been continuously developed to resist infections with rust fungi and other pathogenic microorganisms.

Resistance to disease is an inherent feature of a plant designed to restrict the entry and/or subsequent deleterious effects of a pathogen. Plants with thicker cuticles or cork layers are more resistant to plant diseases. Insect vectors carrying plant pathogens may be unable to penetrate these thickened layers. Resistance of some species of barberry to penetration by basidiospores of *Puccinia graminis* has been attributed to the thickness of cuticle in the epidermis of the leaves. The ability to close the stomata when conditions are favorable for infection is an adaptive feature of some plants that renders them relatively resistant to plant pathogens disseminated through the air. Plant-breeding programs are often aimed at developing disease-resistant varieties with such anatomical adaptations.

There are also several adaptive physiological features that make some plant varieties relatively resistant to strains of pathogenic microorganisms that can be selected for in plant breeding programs. Altered metabolic pathways in plants may also render them unsusceptible to the enzymes or toxins produced by pathogenic microorganisms. Such changes in the plant's metabolism may effectively control plant diseases by making plant cells inhospitable habitats for the growth of pathogenic microorganisms. The production of inhibitory substances by the plant can prevent the establishment of an infection. Plants produce a variety of polycyclic and polyaromatic compounds, phytoalexins, that have antimicrobial activities. Plants that have previously been exposed to invading microorganisms and have survived the infection retain such chemicals for a period of time, during which they have an increased resistance to infection by pathogenic microorganisms. Plants can be exposed to attenuated strains of pathogenic microorganisms to induce the formation of such inhibitory biochemicals, but plants do not exhibit the same extensive immune response observed in vertebrate animal populations.

Because many pathogenic microorganisms are transmitted from one infected plant to another, increased spacing between individual plants decreases the likelihood that a pathogenic microorganism will be successfully transmitted from infected to uninfected plants. Dense plantings of hemlock, for example, result in increased losses due to twig rust because the pathogen that causes this disease is spread from plant to plant. The density of host plant populations can also result in environmental modifications that affect development of plant pathogens. Dense planting of tomatoes, for example, results in the retention of humidity that favors development of blight due to the fungus *Botrytis*.

Unfavorable environmental conditions during planting can increase the susceptibility of agricultural crops to microbial infections. At suboptimal temperatures, plants often produce increased amounts of exudates, containing amino acids and carbohydrates, which favors the rapid development of microbial populations in the soil surrounding the plant. Other suboptimal abiotic factors, such as lack of oxygen and excessive moisture content, also result in increased production of such exudates. The subsequent development of pathogenic *Pythium* and *Rhizoctonia* populations in the vicinity of seeds often results in fungal disease of the developing seedlings.

The soil pH may be lowered deliberately for the control of soil-borne plant pathogens, such as *Streptomyces scabies* (Figure 22.17), which cause potato scab disease. The occurrence of this disease is greatly reduced when the soil pH is acidic, and because potatoes grow well in acid soil, it is a simple matter to control this plant disease by acidifying the soil. As it would be dangerous and uneconomical to apply acid directly, soil acidification is conveniently accomplished by applying powdered sulfur to the field. The activity of *Thiobacillus thiooxidans* converts sulfur to H_2SO_4, producing the desired acidification.

Various control measures can be used to eliminate or diminish the reservoirs of pathogenic microorganisms present at the time of planting. In horticultural practices, for example, soils are sometimes fumigated with fungicides or heat sterilized to remove plant pathogens. It is not practical, though, to sterilize soils in large agricultural fields. However, many plant pathogens contaminate plant seeds, and **disinfection of seeds** by chemical or hot water treatments can greatly reduce the incidence of disease within a plant population. Such seed disinfection practices are widely used for many crops, including corn.

The physical removal of infected plant tissues is also an effective method for controlling the spread of disease. The removal of an infected crop from an agricultural field, which is an inherent consequence of crop rotation, removes infected plant tissues harboring microbial pathogens. The specificity of host–pathogen interactions precludes the establishment of infection by that pathogen in the other crops used in a system of ro-

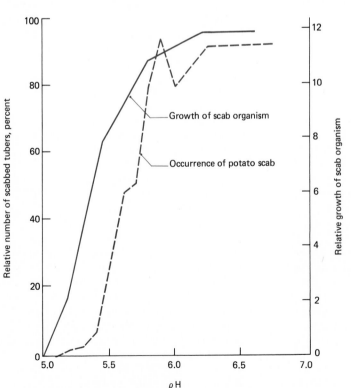

Figure 22.17

This graph illustrates the relationship between pH and Streptomyces scabies, *the causative organism of potato scab disease. From the results shown here, it is apparent that soil acidification is an effective means of controlling this plant pathogen.*

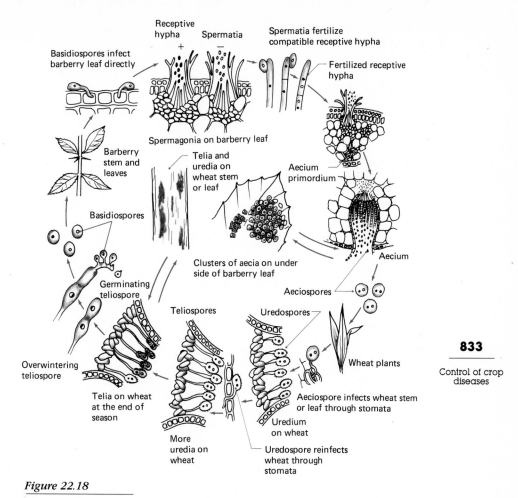

Receptive hypha
+
Spermatia
–
Spermatia fertilize compatible receptive hypha

Basidiospores infect barberry leaf directly

Fertilized receptive hypha

Spermagonia on barberry leaf

Barberry stem and leaves

Telia and uredia on wheat stem or leaf

Aecium primordium

Basidiospores

Aecium

Clusters of aecia on under side of barberry leaf

Aeciospores

Germinating teliospore

Teliospores

Uredospores

Wheat plants

Overwintering teliospore

Telia on wheat at the end of season

Aeciospore infects wheat stem or leaf through stomata

Uredium on wheat

More uredia on wheat

Uredospore reinfects wheat through stomata

Figure 22.18

Life cycle of stem rust of wheat, which is caused by Puccinia graminis. *As with other rusts,* Puccinia *requires alternate hosts.*

tation. Given sufficient time, the infected tissues are eliminated from the soil, and the pathogen population is greatly reduced. The reintroduction of the original plant species can then be accomplished without undue loss, and the practice of periodic cycling of crops can be perpetuated as an effective farm management practice.

In some plant diseases, control can be achieved by limiting the populations of plants that act as alternate hosts for the pathogens. For example, the rust fungus, *Puccinia graminis*, which infects wheat, uses the barberry shrub, *Berberis vulgaris*, as an alternate host (Figure 22.18). By eliminating or reducing the population size of this alternate host, the transmission of disease caused by this rust fungus has been curtailed. Control of the insects and nematodes, which act as vectors for microbial plant pathogens, can be used to control plant diseases. Aphid control programs are espe-

cially important because these insects often act as vectors for plant diseases.

Microbial amensalism and parasitism can be used to control populations of pathogenic microorganisms. Negative interpopulation relationships protect many plants from infection with disease-causing microorganisms. The phenomenon of soil fungistasis, which is believed by some to be due to microbial activities, is widespread in soil. Some bacterial species produce antifungal substances, and the addition of *Bacillus* and *Streptomyces* species to soil has been shown to control damping-off disease in cucumber, peas, and lettuce and several other diseases caused by the fungus *Rhizoctonia solani*. Adding organic compounds to soil can increase soil fungistasis, presumably by stimulating antibiotic-producing streptomycetes. It has also been suggested that cellulase- and chitinase-producing strains of microorganisms contribute to the control of populations of fungal pathogens that have cell walls containing cellulose or chitin. The addition of cellulase-producing myxobacteria to the rhizosphere of young seedlings has been found to control diseases caused by pathogenic fungi such as *Pythium, Rhizoctonia*, and *Fusarium*, which enter the plant through the soil. The treatment of soil with crab and lobster shells, which contain chitin, has been demonstrated to reduce markedly the

severity of root-rot diseases caused by *Fusarium* species, and it has been suggested that the decreased disease incidence reflects higher levels of chitinase-producing microorganisms. However, the role of chitinase and cellulase in controlling populations of plant pathogens has yet to be conclusively shown and has not been developed into a commercially important method for controlling plant diseases.

Pesticide microbiology

The large-scale production of crops requires massive plantings of a single-plant species within an agricultural field. Such monocultures of plants are inherently unstable, and as a result much agricultural production effort and expense goes toward the control of competing weeds, destructive insects, and microbial pathogens. The rising cost of manual labor and the increasingly large-scale agricultural operations in developed countries both have promoted the increased use of chemical pest control. In 1948 there were only three insecticides in agricultural usage, but in 1975, 1170 pesticides, 425 herbicides, 410 fungicides, and 335 insecticides were registered for use in the United

States, and 725 million kilograms of synthetic organic pesticides were produced. In terms of effectiveness, economy, and quality control, pesticides have been an unqualified success for increasing crop production, but there are environmental dangers inherent in the widespread use of agricultural pesticides. There have been many emotionally charged debates over the potential dangers of unwise pesticide use, with sweeping endorsements and condemnations of pesticides made by factions supporting conflicting interests and philosophies. While the total abandonment of all pesticide use would unquestionably cause an international economic and public

table 22.8

Environmental persistence times of some pesticides.

Common name	Chemical structure	Persistence time
Aldrin	1,2,3,4,10,10-1,2,4α,4,8,8α-Hexahydro-endo-1,2-exo-5,8-dimethanonaphthalene	>15 years
Chlordane	1,2,3,4,6,7,8,8-Octachloro-2,3,3α,4,7,7α-hexahydro-4,7-methanoindene	>15 years
DDT	1,1,1-Trichloro-2,2-bis(*p*-chlorophenyl)-ethane	>15 years
Dicamba	3,6-Dichloro-*o*-anisic acid	4 years
Diuron	3-(3,4-Dichlorophenyl)-1,1-dimethylurea	>15 months
2-(2,4-DP)	2-(2,4-Dichlorophenoxy)-propionic acid	>103 days
Endrin	1,2,3,4,10,10-Hexachloro-6,7-epoxy-1,4,4α, 5,6,7,8,8α-octahydro-endo-1,4-endo-5,8-dimethanonaphthalene	>14 years
Fenac	2,3,6-Trichlorophenylacetic acid	>18 months
Fluometuron	*N*'-(3-Trifluoromethylphenyl)-N,N-dimethylurea	195 days
Heptachlor	1,4,5,6,7,8,8-Heptachloro-3α,4,7,7α-tetrahydro-4,7-endomethanoindene	>14 years
Lindane	1,2,3,4,5,6-Hexachlorocyclohexane	>15 years
Monuron	3-(*p*-Chlorophenyl)-1,1-dimethylurea	3 years
Parathion	0,0-Diethyl O-*p*-nitrophenyl phosphorothioate	>16 years
PCP	Pentachlorophenol	>5 years
Picloram	4-Amino-3,5,6-trichloropicolinic acid	>5 years
Propazine	2-Chloro-4,6-bis-(isopropylamino)-*s*-triazine	2-3 years
Simazine	2-Chloro-4,6-bis-(ethylamino)-*s*-triazine	2 years
2,4,5-T	2,4,5-Trichlorophenoxyacetic acid	>190 days
2,3,6-TBA	2,3,6-Trichlorobenzoic acid	2 years
Toxaphene	Chlorinated camphene	>14 years

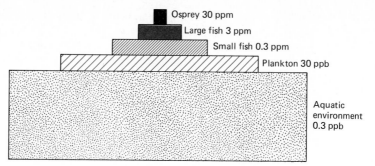

Figure 22.19

The biomagnification of DDT. Very small concentrations of dissolved DDT can move by physical partitioning into plankton. Further concentration of the pollutant takes place through the shrinkage of biomass between successive tropic levels.

Osprey 30 ppm
Large fish 3 ppm
Small fish 0.3 ppm
Plankton 30 ppb
Aquatic environment 0.3 ppb

health crisis of major proportions, continuing uncontrolled and indiscriminant pesticide use would eventually cause irreversible environmental damage. A responsible pesticide policy clearly demands a careful risk–benefit analysis for each compound and its various uses, with particular concern given to the biodegradability and effects on nontarget organisms of each pesticide formulation.

It is essential that pesticides used in agriculture not unduly affect nontarget organisms. Part of the licensing procedure for new pesticides involves toxicity testing to demonstrate that wildlife, fish, and other nontarget organisms are not likely to be exposed to lethal concentrations of the pesticide when applied at recommended levels to agricultural fields. It also is important to show that a pesticide does not disrupt microbial biogeochemical cycling activities that could reduce soil fertility. As discussed earlier in this chapter, the metabolic activities of soil microorganisms, particularly those involved in carbon and nitrogen transformations, are essential for achieving maximal crop yields. A pesticide, aimed at eliminating an agricultural pest that is diminishing crop yield, must not itself reduce plant productivity by interfering with the microbial maintenance of soil fertility.

Biomagnification

In addition to not adversely affecting nontarget organisms, pesticides should be biodegradable in order to minimize possible adverse ecological side effects of their use in agriculture. Different pesticides exhibit vastly different residence times in the environment, with some persisting indefinitely (Table 22.8.). What happens when a persistent organic compound, one that is not subject to biodegradation, is introduced into the biosphere? Chlorinated hydrocarbon insecticides have been detected in remote Antarctic regions, thousands of miles removed from the nearest possible application site. As a result we have learned that it is insufficient to assess the effect of a persistent organic chemical solely on the environment to which it is to be directly applied. It may be dispersed and may elicit unexpectedly detrimental effects on more fragile ecosystems far removed from the original application site.

Even though distribution tends to dilute organochlorines to the low parts per billion (ppb) range, these chemicals still cause concern. They do so because of a phenomenon called "**biological magnification**" or "**biomagnification**" which occurs when an environmental pollutant is both persistent and lipophilic. Because of their lipophilic character, such compounds are partitioned from the surrounding water into the lipids of both prokaryotic and eukaryotic microorganisms, and concentrations in microbial cells may be one to three orders of magnitude higher than the surrounding environment (Figure 22.19). When microorganisms are ingested by members of the next higher trophic level in the food web, the persistent lipophilic pollutant is neither degraded nor excreted to any significant extent and so is preserved practically without loss in the smaller biomass of the higher trophic level. Consequently, its concentration is increased by almost an order of magnitude. The top trophic level organisms, such as birds of prey, carnivores, and large predatory fish, may carry a body burden of the environmental pollutant that exceeds the environmental concentration by a factor of 10^4–10^6. A pesticide may cause the death or serious debilitation of animals at the top of the food web. DDT and other chlorinated hydrocarbons were implicated in the death or reproductive failure of various birds of prey. Because human beings derive their food from various trophic levels, we are in a position of less exposure to biomagnified pesticides than a top-level carnivore. Nevertheless, at the time of unrestricted DDT use, the average American, with no occupational exposure, carried a body burden of 4–6 ppm DDT and its derivatives. Although this amount was not considered dangerous, the trend for increasing contamination of the higher tropic levels of the biosphere became sufficiently clear and led to the ban on the use of DDT in the United States and several other countries for all but emergency situations.

Biodegradation

The majority of the currently used organic pesticides are subject to extensive **biodegradation** within the course of a single growing season. Synthetic pesticides show a bewildering variety of chemical structures, but most contain relatively

Figure 22.20

Degradation pathways for aliphatic (A) and aromatic (B) hydrocarbon pesticide moities.

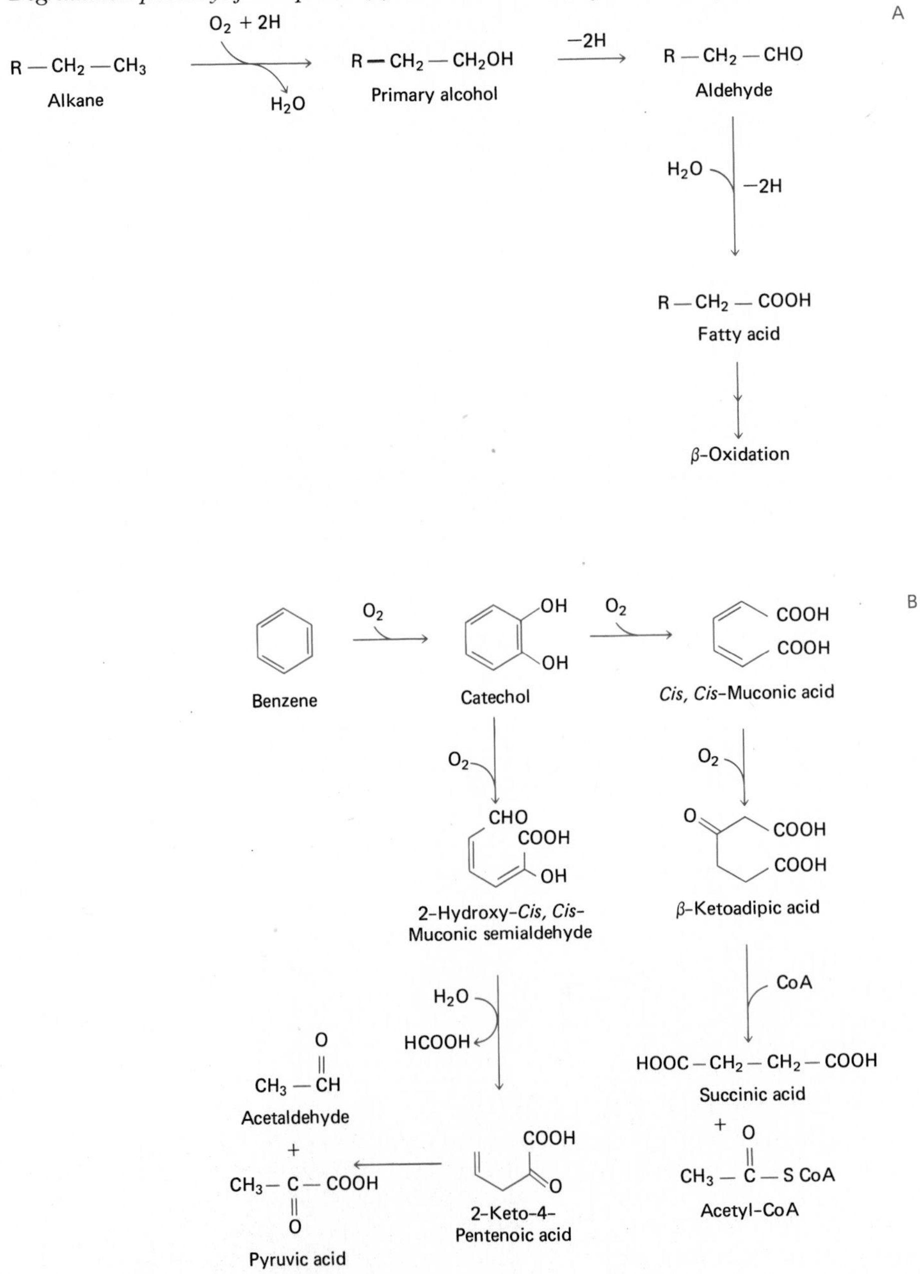

simple hydrocarbon skeletons with a variety of substituents, such as halogens, amino, nitro, hydroxyl, and other functional groups. Aliphatic hydrocarbons are oxidized to fatty acids, which are then degraded via the β-oxidation sequence (Figure 22.20). The resulting C$_2$ fragments are further metabolized via the tricarboxylic acid cycle. Aromatic ring structures are metabolized by dihydroxylation and ring cleavage mechanisms. Prior to these transformations, substituents on the aromatic ring may be completely or partially removed. Substituents uncommon in natural compounds, such as halogens, nitro, and sulfonate

groups, if situated so as to impede oxygenation, will frequently cause recalcitrance.

Often, **a simple change in the substituents of a pesticide may make the difference between recalcitrance and biodegradability.** The chemical structures of some biodegradable and some recalcitrant pesticides are compared in Figure 22.21. The herbicide 2,4-D is biodegraded within days, but 2,4,5-T, which differs only by an additional chlorine substitution in the meta-position, persists for many months. The additional substitution interferes with the hydroxylation and cleavage of the aromatic ring. Propham is cleaved by micro-

Figure 22.21

The structures of some biodegradable and recalcitrant pesticides. The pairs of compounds were selected for overall structural similarity in order to show the molecular features that render the compounds recalcitrant. These features are the 5-chloro substitution in 2,4,5-T, the N-alkyl substitution in propachlor, the multiple chloro substitutions in aldrin, and the two p-chloro substitutions in DDT.

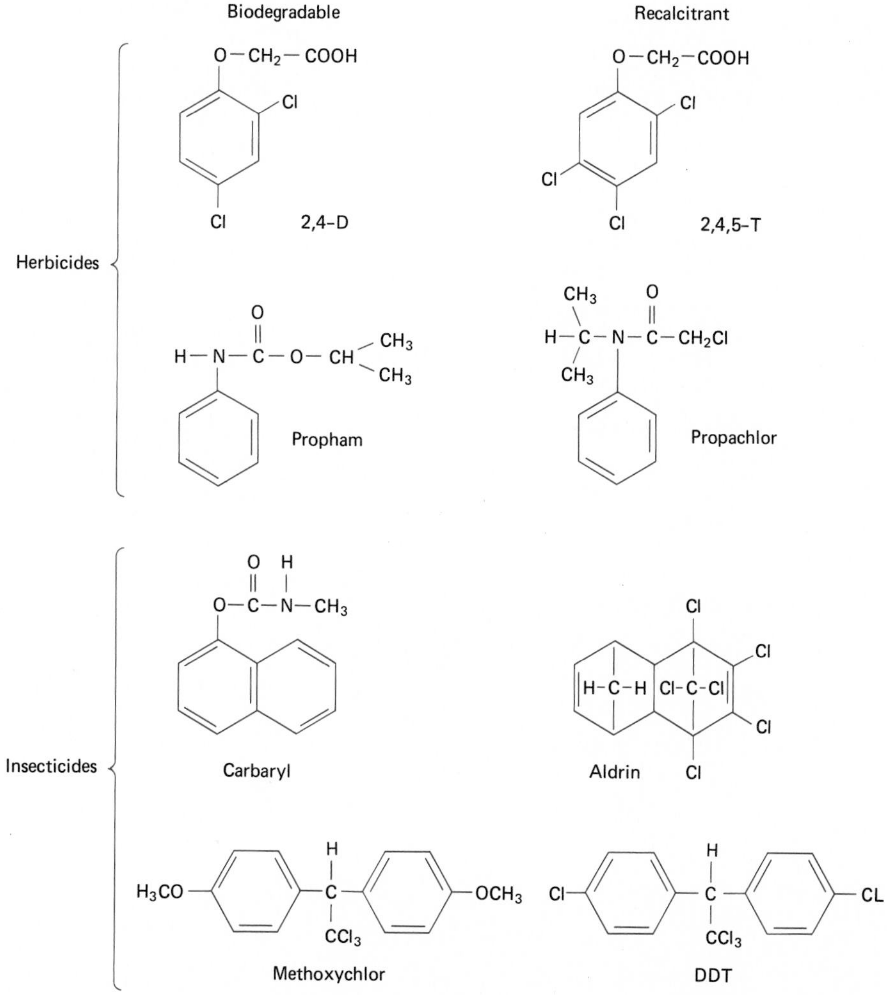

bial amidases so rapidly that for some applications the addition of amidase inhibitors becomes necessary, but propachlor, which has a tertiary amine group, is not subject to attack by such amidases and persists considerably longer. Methoxychlor is less persistent than DDT because the *p*-methoxy groups are subject to dealkylation, and the *p*-chloro substitution endows DDT with great biological and chemical stability.

As the organic molecules that enter the environment and must be degraded become more complex, and as they consist of aliphatic as well as aromatic, alicyclic, or heterocyclic portions, few generalizations can be made about their degradation patterns. If the moieties of the molecule are connected by ester, amide, or ether bonds that can be cleaved by microbial enzymes, the initial attack usually takes this form, and the resulting compounds are subsequently metabolized, as outlined before. If such an attack cannot occur, degradation will commonly be initiated at the aliphatic end of the molecule. However, if degradation at this end is blocked by extensive branching or by other substituents, the attack may start from the aromatic end. The site and mode of the initial attack is determined not only by mo-

lecular structure but also by the enzymatic capabilities of the microorganisms involved as well as by the prevailing environmental conditions, including the availability of oxygen and soil pH. These factors will not only modify the rate but also the pathway and the ultimate products of the degradation.

In some cases one portion of the pesticide molecule is susceptible to degradation, and another is recalcitrant. Some acylanilide herbicides are cleaved by microbial amidases, and the aliphatic moiety of the molecule is mineralized. The aromatic moiety, stabilized by chlorine substitutions, resists mineralization, but the reactive primary amine group may participate in various biochemical and chemical reactions, leading to polymers and complexes that render the fate of such herbicide residues extremely complex. Figure 22.22 shows some of the transformations of the acylanilide herbicide propanil. Microbial acylamidases cleave the propianate moiety, which is subsequently mineralized. A portion of the released 3,4-dichloroaniline (DCA) is acted upon by microbial oxidases and peroxidases, with the result that they dimerize and polymerize to highly stable residues, such as 3,3′,4,4′-tetrachloroazo-

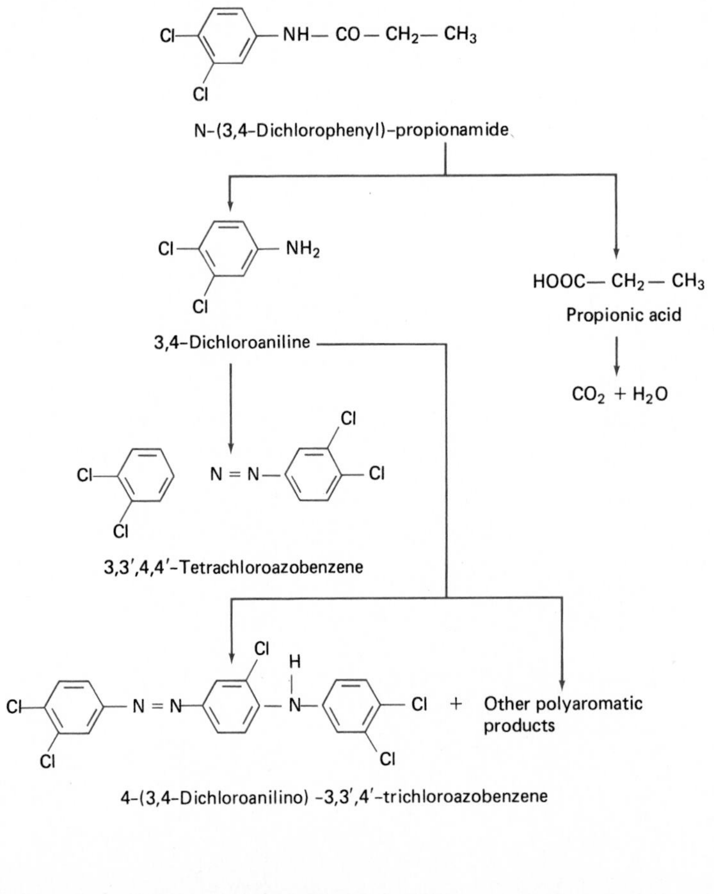

Figure 22.22

The biodegradation pathway for the herbicide propanil, N-(3,4-dichlorophenyl)-propionamide. The aliphatic portion of the molecule is degraded, but the aromatic portion is dimerized and polymerized to persistent residues. The transformation involves microbial synergism.

benzene (TCAB) and related azo compounds. The reasons for such transformations are still somewhat obscure. They may occur by chance when a microbial enzyme that has another metabolic function recognizes and acts upon the man-made residue. In some cases the reaction seems to detoxify the residue from the microbe's point of view, but the overall persistence and environmental impact of the pesticide is increased by such synthetic transformations.

Biological control

Biological control using microbial pathogens offers a method that can augment the use of chemical pesticides in controlling pest populations. Microbial populations can be used directly for controlling plant and animal pest populations. Populations of pathogenic or predatory microorganisms that are antagonistic toward a particular pest population provide a natural means of achieving control of pest populations, and preparations of such antagonistic microbial populations can be called **microbial pesticides**. The effective use of pathogenic microorganisms as pesticides depends on the ability to establish a disease epidemic among susceptible pest populations.

To be effective as microbial pesticides, the prospective microbial pathogen must be virulent and cause disease in the pest population when properly applied at the recommended concentration. The pathogen must not be sensitive to expected environmental variations and after application should survive until the infection within the pest population has been established. The pathogen should be rather specific for the pest population and must not cause disease in nontarget populations. The pathogen should rapidly establish disease in the pest populations so as to minimize destruction caused by the pest populations. Microbial control methods have been developed for the suppression of arthropod pests, especially insects, and several commercial microbial insecticides are available for agricultural use. Microbial suppression of animal populations is aimed at populations that cause crop and other plant damage and at animal populations that act as vectors of disease-causing microorganisms. Controlling insect damage to crops through the use of chemical or microbial pesticides is critical for achieving high crop yields.

Viral pesticides

Pathogenic viruses possess the potential for use as pesticidal agents. Viruses cause diseases in insects and other arthropods. The specificity of the virus–host relationship makes viruses ideal candidates for use to control specific pest insect populations with little or no deleterious effects on humans and other animals. Insect pathogenic viruses frequently cause natural disease epidemics, known as **epizootics**, in insect populations. Viruses that are pathogenic for insects are found in the families Baculoviridae, Poxviridae, Reoviridae, Iridoviridae, Parvoviridae, Picornaviridae, and Rhabdoviridae. Viruses have been used in attempts to control outbreaks of a variety of pests including gypsy moths, Douglas fir tussock moths, pine processionary caterpillars, red-banded leaf rollers (a pest of apples), spruce bud worms, codling moths (a pest of apples, walnuts, and other deciduous fruits), Great Basin tent caterpillars, alfalfa caterpillars, cabbage white butterflies, cabbage loopers, cotton bollworms, corn earworms, tobacco budworms, tomato worms, army worms, and wattle bagworms, among other insect pests. Some of these viruses are **nuclear polyhedrosis viruses (NPV)**, **cytoplasmic polyhedrosis viruses (CPV)**, or **granulosis viruses (GV)**. The nuclear polyhedrosis viruses develop in the host cell nuclei, the virions are occluded singly or in groups in polyhedral inclusion bodies. Cytoplasmic polyhedrosis viruses develop only in the cytoplasm of host midgut epithelial cells; the virions are occluded singly in polyhedral inclusion bodies. Granulosis viruses develop in either the nucleus or cytoplasm of host fat, tracheal, or epidermal cells; the virions are occluded singly or rarely in pairs in small occlusion bodies called capsules.

Baculoviruses are perhaps the most studied insect viruses. Baculoviruses include nuclear polyhedrosis and granulosis viruses. Pathogenic baculoviruses have been principally found for Lepidoptera, Hymenoptera, and Diptera. The infection is often transmitted by ingestion of contaminated food, with cell invasion probably beginning in the midgut. Several nuclear polyhedrosis viruses are produced in the United States on a large scale for control of insect pests. Inoculation of leaves with polyhedrosis viruses can initiate epidemic outbreaks (epizootics) in Lepidoptera and Hymenoptera larvae that feed on plant leaves, causing significant population reductions in these insects.

Nuclear polyhedrosis viruses cause disease in sawflies. When the accidental introduction of the European sawfly into North America in the twentieth century threatened the spruce forests of North America, introduction of nuclear polyhedrosis viruses into the European sawfly population caused a spectacular epizootic that reduced the sawfly populations and saved many spruce forests. Similarly, the European pine sawfly, which was causing serious damage to pines in New Jersey, Ohio, and Michigan, has been controlled by introducing insects containing nuclear polyhedrosis viruses into the population, resulting in epizootics.

An interesting example of the problems associated with long-term use of viral pesticides is found in the attempt to control the rabbit populations of Australia with myxoma virus. Rabbits were introduced into Australia from Europe in 1859, and because there were no natural enemies for the rabbit in Australia, their reproduction was unchecked. Myxoma virus, found naturally among South American rabbits, was found to be a virulent pathogen of the European rabbit, and in an effort to achieve control of the rabbit populations in Australia the myxoma virus was introduced to the Australian rabbit populations in 1960. The virus rapidly spread among the rabbit populations, causing high seasonal morbidity and mortality. Initially 99 percent of the infected rabbits died, but after only a few years the virulence of the myxoma virus for the surviving rabbits declined. The survivors of the initial epidemics in which many rabbits were killed had been selected for their resistance to the virus. The resistance of the rabbits is innate and not due to an immunological defense system. Thus, within only a few years an equillibrium was achieved between the virus and the Australian rabbits. Myxomatosis was effective in lowering the Australian rabbit population to about 20 percent of the level before the introduction of the virus, and now that the virus is firmly established within the rabbit population, there is a pathogen that will maintain control of the rabbit population level.

table 22.9

Some registered uses for B. thuringiensis *products in the United States*

Pest	Crop
Vegetable and field crops	
Alfalfa caterpillar, *Colias eurytheme*	Alfalfa
Artichoke plume moth, *Platyptila carduidactyla*	Artichokes
Bollworm, *Heliothis zea*	Cotton
Cabbage looper, *Trichoplusia ni*	Beans, broccoli, cabbage, cauliflower, celery, collards, cotton, cucumbers, kale, lettuce, melons, potatoes, spinach, tobacco
Diamondback moth, *Plutella maculipennis*	Cabbage
European corn borer, *Ostrinia nubilalis*	Sweet corn
Imported cabbageworm, *Pieris rapae*	Broccoli, cabbage, cauliflower, collards, kale
Tobacco budworm, *Heliothis virescens*	Tobacco
Tobacco hornworm, *Manduca sexta*	Tobacco
Tomato hornworm, *Manduca quinquemaculata*	Tomatoes
Fruit crops	
Fruit tree leaf roller, *Archips argyrospilus*	Oranges
Orange dog, *Papilio cresphontes*	Oranges
Grape leaf folder, *Desmia funeralis*	Grapes
Shade trees, ornamentals	
California oakworm, *Phryganidia californica*	
Fall webworm, *Hyphantria cunea*	
Fall cankerworm, *Alsophila pometaria*	
Great Basin tent caterpillar, *Malacosoma fragile*	
Gypsy moth, *Lymantria* (*Porthetria*) *dispar*	
Linden looper, *Erannis tiliaria*	
Salt marsh caterpillar, *Estigeme acrea*	
Spring cankerworm, *Paleacrita vernata*	
Winter moth, *Operophtera brumata*	

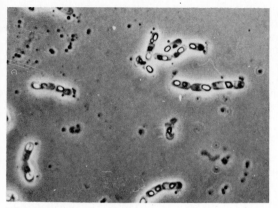

Figure 22.23

(A) Micrograph of Bacillus thuringensis tolworthi *under phase contrast microscopy. Note the spores and parasporal inclusions within the cells. (Courtesy B. N. Herbert, Shell Research Ltd., Kent, England.) (B) Photograph showing commercial products containing* B. thuringensis. *(Courtesy Sandoz, Inc., San Diego, California.) (C) Scanning electron micrograph showing crystals of delta-endotoxin from Thuricide (24,000×). (Courtesy Dr. Guggenheim, University of Basel.)*

Bacterial pesticides

There are several bacterial pathogens of insects that currently are used as, or have potential for use as, insecticides. They include endospore-forming *Bacillus* and *Clostridium* species, and nonendospore-forming species of *Pseudomonas, Enterobacter, Proteus, Serratia,* and *Xenorhabdus.* Of the potential bacterial pesticides, **Bacillus thuringiensis** has been most extensively exploited in the bacterial control of pest insect populations. Commercial preparations of *Bacillus thuringiensis* are registered by at least twelve manufacturers in five countries for use on numerous agricultural crops, forest trees, and ornamentals for control of various insect pests (Table 22.9). *B. thuringiensis* has been tested successfully against more than 140 insect species, including members of the Lepidoptera, Hymenoptera, Diptera, and Coleoptera. *B. thuringiensis* is a crystalliferous bacterium because in addition to endospores, it produces discrete parasporal

bodies within its cell (Figure 22.23). It is the proteinaceous parasporal crystal that is the toxic factor. Four separate toxic substances are produced by *Bacillus thuringiensis.* Use of *B. thuringiensis* brings about commercially acceptable levels of suppression of cabbage worms, cabbage loopers, and many other pests of vegetable crops. *B. thuringiensis* also readily suppresses populations of tent caterpillars, bagworms, and canker worms, pests of forest trees. Gypsy moth and spruce bud worm can be suppressed by *B. thuringiensis,* but only when high application rates are used and uniform foliage coverage is attained.

Of particular interest is the potential use of *B. thuringiensis israelensis* (BTI) to control populations of the mosquito vectors of malaria. In fact, effective use of *B. thuringiensis* now offers the best hope for controlling malaria because unlike DDT it is environmentally safe and mosquitos show no sign of developing resistance. Testing of BTI for control of malaria-carrying mosquitos has begun only recently, but tests in Africa against blackfly,

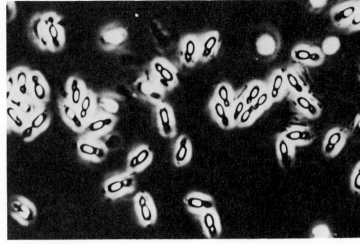

Figure 22.24

(A) A can of Doom, a biological pesticide that is effective in controlling Japanese beetles. (Courtesy Fairfax Laboratories, Clinton Corners, New York.) (B) Micrograph of Bacillus popilliae, one of the two active bacterial components of Doom. (Courtesy Michael Klein, USDA, Wooster, Ohio.)

the carrier of widespread river blindness, are excellent, according to the United Nations World Health Organization.

Other *Bacillus* species cause milky disease of Japanese beetles. In Japan, where the beetle encounters natural antagonists, it is a relatively minor pest. However, in the United States the beetle does not have any natural associated pathogens or other antagonists. The Japanese beetle feeds voraciously on some 300 species of plants and has been responsible for large economic losses. The greatest success in suppressing pest populations of Japanese beetles has been obtained by using bacteria that produce **milky disease**. For many years a mixture of *Bacillus popilliae* and *B. lentimorbus* has been marketed under the trade name Doom (Figure 22.24). *B. lentimorbus*, which infects mainly first- and second-instar grubs, does not produce parasporal crystals; *B. popilliae* produces parasporal bodies and infects a high proportion of third-instar grubs. The use of these *Bacillus* species to produce milky disease in grubs of Japanese beetles is probably responsible for control of pest beetle populations, and although in the past there have been major infestations of Japanese beetles in the United States, today there are relatively few major outbreaks.

Fungal pesticides

Entomogenous fungi are potentially important in the control of pest populations. Pathogenic fungal infections, or mycoses, can cause epizootics. Insect mycoses are caused by Phycomyces, Ascomyces, Basidiomyces, and Deuteromyces. Most studies on entomogenous fungi have been concerned with members of the fungal genera *Beauveria, Metarrhizium, Entomophthora*, and *Coelomomyces. Beauveria bassiana* has been used extensively in the Soviet Union to control the eastward spread of Colorado beetle and against the codling moth; it also infects the corn ear worm. Fungi of the genus *Aschersonia* have been used to control pests of citrus trees in the Soviet Union near the Black Sea. In Florida, *Hirsutella* has been used against the citrus rust mite. The fungus *Metarrhizium* is used in Brazil to control populations of leaf hoppers and frog hoppers.

Predaceous **nematode-trapping fungi** have been suggested for the control of pest nematode populations in soil. Interestingly, nematodes are also considered as possible biological control agents for some soil-borne diseases. Some plants are subject to infection with nematodes. Nematode-trapping fungi have been used in attempts to control root knot disease of pineapples in Hawaii. Successful control of pest nematodes was achieved but was dependent on supplementing soil with organic matter. It is not clear whether control of the nematodes was due to the predaceous fungi or was due to an alteration in food web relationships caused by addition of organic supplements. Subsequent attempts to demonstrate that nematode-trapping fungi, rather than soil supplements, were primarily responsible for controlling nematode populations in soil had ambiguous results. Under appropriate conditions nematode-trapping fungi with supplemental organic matter appear to have the potential for diminishing the effects of nematodes on susceptible plant populations. Other possibilities for controlling pest nematodes include infection with parasitic fungi, protozoa, bacteria, and viruses. Additionally, control of pests may be achieved by using microorganisms to alter the ecological balance involved in disease incidence, besides the direct use of microbial pathogens to eliminate the specific pest population.

General considerations

The development and use of microorganisms for the control of pest populations require considerations that go well beyond finding a pathogen for a particular weed or animal pest population. Questions of economics, production, quality control, application, side effects, and safety must all be satisfactorily answered. In many cases the economics of the development, production, and application of microbial pesticides does not permit commercial development of an otherwise promising biological control system. The effectiveness of the microbial pathogens must be carefully evaluated. Industry must be able to produce and to stockpile large, stable populations of potential **"microbial pesticides"**. Quality control must ensure standardized batches of the "microbial pesticides" that have the same virulence so that recommended application rates may be used with

confidence in the field. Application methods for "microbial pesticides" must permit the microorganisms to remain viable long enough to contact the pest population and to establish widespread disease. Resistant stages, such as spores or cysts, are best for this use. With some candidate microbial pathogens it is difficult to achieve repeatable results. Environmental conditions must be taken into account, as these affect the virulence of "microbial pesticides." Persistence of "microbial pesticides" must also be considered. Some candidate microorganisms have very short survival times. Persistence is required for long-term control of pest populations.

Ideally, "microbial pesticides" should exhibit a high degree of host specificity. They should not adversely affect nontarget populations. The host specificity, though, should not be so narrow as to preclude its effectiveness against a simple genetic variance within the pest population. It is often difficult to predict whether a "microbial pesticide" could establish disease in nontarget populations because it is impossible to test the infectivity of "microbial pesticides" against all possible nontarget populations. Obviously, any "microbial pesticide" used should be harmless to humans and other valued plant and animal populations. The use of "microbial pesticides" is probably best when employed in an integrated program of management practices for agricultural crops and domestic animals that minimizes opportunities for infection or interaction, along with limited applications of appropriate chemical pesticides carefully timed for maximum effect.

Diseases of farm animals

Farm animals are subject to various infectious diseases that can spread through a herd, causing sizable economic losses to animal breeders and farmers. Diseases of animals other than humans are termed **zoonoses**. Some of these diseases can also be transmitted to people and represent an occupational hazard to those working with animals. Cows, horses, swine, and fowl are among the barnyard animals susceptible to serious infectious diseases. As in the case of human diseases, the common portals of entry for the microbial pathogens that cause zoonoses are the respiratory, gastrointestinal, and genitourinary tracts of these animals. Various pathogenic viruses, bacteria, fungi, and protozoa are responsible for disease outbreaks in populations of farm animals.

Because farm animals are normally herded tightly together, these infectious diseases can easily spread through the entire herd, rapidly producing an adverse effect on agricultural economy. Some of the common infectious diseases of farm animals are described in Table 22.10.

Control of animal diseases

Methods of controlling infectious animal diseases are similar to those employed for preventing and treating human diseases. Prevention is the best way of controlling diseases of farm animals. Vaccines have been developed for establishing immunity against many diseases, and an extensive

table 22.10

Some infectious diseases of farm animals

Disease	Causative agent	Description
Anaplasmosis	*Anaplasma marginale*	A protozoan disease of cattle. The protozoa are transmitted through tick, fly, and mosquito vectors. The disease is characterized by severe anemia caused by reproduction of the protozoa within the animal's red blood cells.
Anthrax	*Bacillus anthracis*	Anthrax is a major livestock problem in many parts of the world, affecting cattle, sheep, horses, and swine. It is widespread in Mexico, South America, and Asia. *B. anthracis* occurs in soils and is transmitted to farm animals in contaminated feed and water. Infection results in septicemia, and there is a high mortality rate.
Avian tuberculosis	*Mycobacterium avium*	Chronic pulmonary infection occurring in domestic fowl.
Blackleg	*Clostridium chauvoei*	An acute infection of cattle, sheep, and other ruminant animals. The disease is characterized by severe swelling, and the infected subcutaneous tissues are permeated with blood and gas; the underlying muscles appear dark brown-black, which accounts for the name of this disease. Animals acquire this infection via the gastrointestinal tract by the consumption of contaminated feed or water.
Brucellosis	*Brucella* sp.	A serious infection occurring in cattle, swine, and goats. In cattle *B. abotus* causes abortion and infection of the udders.
Coccidiosis	*Eimeria* sp.	A protozoan disease of cattle, characterized by severe gastroenteritis and bloody diarrhea. Cysts are shed with fecal matter and tranmission is via the gastrointestinal route from ingestion of contaminated water or feed.
Contagious metritis	*Haemophilus* sp.	A sexually transmitted bacterial disease of horses occurring in the United States, France, and Argentina. The disease results in abortion in mares. Quarantining of breeding stock is practiced when outbreaks of this disease occur.
East Coast fever	*Thieleria parva*	A protozoan disease prevalent in cattle along the east coast of Africa. The protozoa are transmitted via a tick vector and reproduce within the red blood cells of infected cattle.
Erysipeloid	*Erysipelothrix rhusiopathiae*	Second only to swine influenza in importance as an infectious disease of swine in the United States. Most common in North America during the summer. Transmission is via the gastrointestinal tract associated with feed or water contaminated with fecal matter or urine containing the causative bacterium. The acute form of this disease can occur as septicemia, chronic endocarditis, and "diamond skin disease."
Foot-and-mouth	Rhinovirus	A highly contagious disease, especially among cattle, that also affects sheep, swine, and goats. The virus is transmitted through direct contact between infected animals and through ingestion of food and water containing the virus. The

Disease	Causative agent	Description
Foot-and-mouth (cont.)	Rhinovirus	disease is characterized by formation of vesicular lesions in the mouth, on the muzzle, and on the udders and teats. The lesions are painful and result in lameness and a reluctance of the animal to eat or move.
Fowl cholera	*Pasteurella multocida*	A highly communicable and often fatal disease of fowl.
Fowl typhoid	*Salmonella gallinarum*	A disease affecting chickens and turkeys that is transmitted from infected to susceptible fowl through fecal contamination of feed and water.
Glanders	*Pseudomonas mallei*	An infectious disease of horses and other equines. Transmitted primarily through the watering trough, feed, and direct contact. Glanders is characterized by formation of nodules and ulcerations of the mouth, respiratory tract, internal organs, and skin. Acute forms of the disease often are fatal.
Hog cholera	RNA virus (Togaviridae)	A highly communicable disease prevalent in the Midwestern United States. The disease is characterized by a high fever, loss of appetite, lack of coordination, purulent conjunctivitis, and diarrhea. In the United States the mortality rate from hog cholera is approximately 90 percent.
Infectious coryza	*Haemophilus gillinarum*	A serious disease of fowl characterized by bloody and mucoid exudates of the nostrils and eyes.
Limberneck	*Clostridium botulinum*	Botulism of fowl resulting in paralysis of the neck muscles; other forms of poisoning that produce symptomatic paralysis of the neck are also classified as limberneck.
Newcastle disease	Newcastle virus (paramyxovirus)	Also known as avian pneumoencephalitis, this disease causes extensive losses to poultry farms. The virus is spread through chicken coops via airborne droplets. The symptoms include coughing, sneezing, and gasping. A sharp decline in egg production occurs in egg-laying flocks. Nervous symptoms include characteristic paralysis of the legs or wings and twisting of the neck.
Piraplasmosis	*Piroplasma bigemina*	Also known as tick fever, this protozoan disease affects cattle, horses, and sheep. Transmission is through tick bites.
Pullorum disease	*Salmonella pullorum*	An economically important disease of chickens that is transmitted congenitally from hen to egg.
Swine influenza	Influenza virus	The disease is transmitted via the respiratory tract and can result in 100 percent infection of susceptible animals in a herd. Swine flu is prevalent in the fall and winter. The symptoms appear suddenly and include fever, anorexia, prostration, cough, and abnormal breathing. Mortality rates are generally less than 2 percent.
Trypanosomiasis	*Trypanosoma brucei* *T. congolense* *T. Vivax*	This protozoan disease affects large numbers of African cattle. It is transmitted by the tsetse fly.

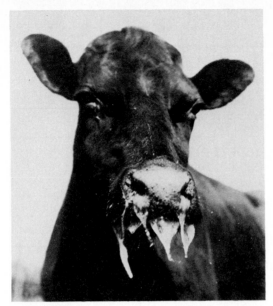

Figure 22.25

Symptoms of foot-and-mouth disease in cattle. This steer has excess saliva flowing from its mouth. The animal was experimentally infected 24 hours previously by inoculating the tongue with the virus of this disease. (Courtesy United States Department of Agriculture.)

immunization program is carried out on most farms. Infectious diseases of farm animals, such as foot-and-mouth disease (Figure 22.25), can be prevented by vaccination, and the use of vaccines reduces farm losses due to many such diseases. Research in genetic engineering has been employed to produce a foot-and-mouth vaccine, and recombinant DNA technology will undoubtedly be used for producing other vaccines to protect farm animals against disease. Antitoxins can be used to produce passive immunity. Antibiotics have been employed as a prophylactic measure to preclude the spread of infectious disease through animal populations; in fact, penicillin and tetracycline antibiotics are sometimes included in bird feed to prevent the spread of bacterial diseases. Anitbiotics can also be used in conjunction with artificial infections to produce immunity. For example, oxytetracycline-controlled infections with *Thieleria parva* establish immunity in cattle against East Coast fever. Sanitary measures in rearing animals are particularly effective in reducing the reservoirs of infectious microorganisms. Proper ventilation in chicken coops is important for preventing the aerial spread of bacterial diseases. The development of resistant animal breeds has reduced the incidence of some diseases, such as

Figure 22.26

(A) Micrograph showing an infected cell and uninfected cell of the tick vector of East Coast fever. An infected cell may contain up to 50,000 sporozoites of Thieleria parva. *The protozoan is transmitted to cattle through tick bites. The* Thieleria *bind to lymphocytes and enter the host white blood cells within about five minutes. (B–F) These micrographs show the uptake of a sporozoite by a lymphocyte. The protozoa later infect red blood cells to complete their life cycle. (Courtesy Don Fawcett, International Laboratory for Research in Animal Diseases, Nairobi, Kenya, reprinted by permission of American Association for the Advancement of Science, from Portraits of a parasite, 1982,* Science 216:504.)*

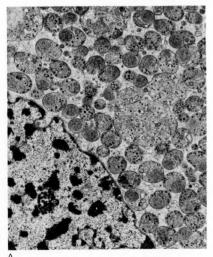

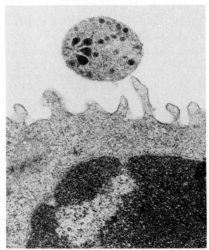

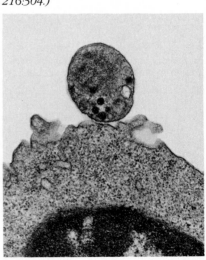

A   B   C

tick-bite fever. Control or elimination of insect vectors is an important method of breaking disease-transmission cycles. Elimination of vector populations is an efficient means of controlling disease because one vector can transmit the disease to many animals. For example, in the case of East Coast fever, one infected tick can contain enough parasites to kill 100 cattle (Figure 22.26).

When prevention fails and an infectious disease becomes established in a herd, it is important to treat or eliminate the infected animals. When the disease is caused by a bacterial species, antibiotics can often be used to cure the infected animals. It is essential that infected animals be removed from the herd, and in some cases it may be necessary to quarantine or even destroy an entire herd to eliminate the source of the infectious pathogens. Treatment methods must be initiated promptly enough to prevent the spread of an infection through an entire herd and from one herd to another. It is critical that effective remedial measures be taken to control the transmission of disease among farm animals to reduce economic losses.

Postlude

The central ecological role of microorganisms in the biosphere is clear when considering agricultural practices. The success of agricultural production of both crops and livestock depends on an integrated program of controlling microbial activities. The large-scale production of agricultural crops, in monocultures, creates an ecologically unstable ecosystem that requires major inputs of energy and intensive management to maintain soil fertility and prevent destruction of the crop by microbial and pest populations. Plants and farm animals, like human beings, are subject to many diseases, and limiting the spread of pathogenic microorganisms through crops and herds is a major task for farmers. Many agricultural practices, including breeding programs to develop heartier plant and animal varieties, are aimed at reducing losses due to microbial diseases. Harvests can be greatly reduced as a

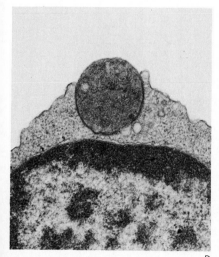

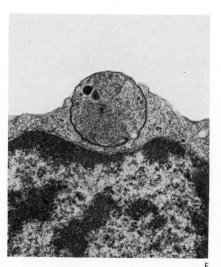

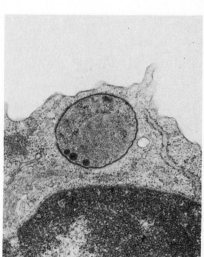

D   E   F

result of microbial plant diseases, causing great economic losses and sometimes creating serious food shortages.

Fertilizers are added to soils in order to overcome nitrogen limitations and increase crop production. The application of nitrogen fertilizers to soil must take into account the microbial nitrification process, lest the fertilizer be wastefully squandered and groundwater problems result. Some farm practices, most notably crop rotation, are aimed at reaping the harvest of microbial transformations of nitrogen that increase soil fertility and eliminating the need to add costly nitrogen fertilizers to the field to achieve high crop yields. The incorporation of the bacterial genes controlling nitrogen fixation into staple crops, such as wheat and corn, through the use of genetic engineering, will go a long way toward ensuring that world food needs can be met when the proper technology is developed.

Modern agriculture has become almost totally dependent on the use of chemical pesticides to control pest populations. Herbicides are applied to fields to prevent competition between the crop plant and weeds; fungicides are used to prevent fungal diseases, the most frequent infectious diseases of plants; and insecticides are used to control damage by insect pests. Biological control, through the controlled establishment of microbial disease in pest populations, represents a useful alternative to conventional chemical pesticides. Here is a case where disease is considered beneficial to humanity. The environmentally safe use of chemical pesticides depends on microbial biodegradation to prevent persistence and biomagnification of these toxic compounds. When microorganisms fail to degrade the pesticide, there may be serious environmental consequences far removed from the site of application, as seen in the case of DDT. The recognition that some compounds are recalcitrant to microbial attack violates the cardinal rule of microbiology—that microorganisms are infallible. As will be discussed in the next chapter, we rely so heavily on microorganisms to remove our wastes that it is difficult, even for microbiologists, to accept the truth that some organic compounds are resistant to biodegradation by microorganisms.

1. How are agricultural soils managed to conserve soil nitrogen?

2. Why is the nitrogen fixation symbiotic relationship so important to soil fertility? How can genetic engineering extend the range of plants that can establish mutualistic relationships with nitrogen-fixing bacteria?

3. What are some of the symptoms of plant disease?

4. Discuss the transmission of plant diseases.

5. Discuss agricultural management practices aimed at controlling plant diseases.

6. What is biomagnification? What properties of a pesticide are important in determining its biomagnification through a food web?

7. Why must pesticides be biodegradable for their safe use in agriculture?

8. What is biological control? How are insect viruses used in the control of plant diseases?

9. What is a zoonosis? Discuss three diseases of farm animals that can be transmitted to humans.

10. Discuss the similarities and differences between the measures used to control disease transmission among farm animals and those used to curb the transmission of human diseases.

Agrios, G. N. 1978. *Plant Pathology*. Academic Press, New York.

Alexander, M. 1965. Biodegradation: problems of molecular recalcitrance and microbial fallibility. *Advances in Applied Microbiology* 7: 35–80.

Alexander, M. 1977. *Introduction to Soil Microbiology*. John Wiley & Sons, New York.

Alexander, M. 1981. Biodegradation of chemicals of environmental concern. *Science* 211: 132–138.

Brill, W. J. 1981. Agricultural microbiology. *Scientific American* 245(3): 198–215.

Burges, H. D. (ed.). 1981. *Microbial Control of Pests and Plant Diseases, 1970–1980*. Academic Press, New York.

Coppel, H. C., and J. W. Mertins. 1977. *Biological Insect Pest Suppression*. Springer-Verlag, Berlin.

Delwiche, C. C. 1970. The nitrogen cycle. *Scientific American* 223(5): 137–146.

Diener, T. O. 1979. *Viroids and Viroid Diseases*. Wiley-Interscience, New York.

Gillespie, J. H., and J. F. Timoney. 1981. *Hagan and Bruner's Infectious Diseases of Domestic Animals*. Cornell University Press, Ithaca, New York.

Gray, T. R. G., and S. T. Williams (eds.). 1975. *Soil Bacteria*. Longman, New York.

Harris, K., and K. Maramorosch. 1981. *Pathogens, Vectors, and Plant Diseases*. Academic Press, New York.

Hill, I. R., and S. J. L. Wright (eds.). 1978. *Pesticide Microbiology: Microbiological Aspects of Pesticide Behavior in the Environment*. Academic Press, London.

Huffaker, C. B., and P. S. Messenger (eds.). 1976. *Theory and Practice of Biological Control*. Academic Press, New York.

Lynch, J. M. 1976. Products of soil microorganisms in relation to plant growth. *CRC Critical Reviews in Microbiology* 5: 67–107.

Merchant, I. A., and R. A. Packer. 1967. *Veterinary Bacteriology and Virology*. The Iowa State University Press, Ames, Iowa.

Mohanty, S. B., and S. K. Dutta. 1981. *Veterinary Virology*. Lea & Febiger, Philadelphia.

Parker, W. H. 1980. *Health and Disease in Farm Animals: An Introduction to Farm Animal Medicine*. Pergamon Press, New York.

Siegmund, O. H. (ed.). 1979. *The Merck Veterinary Manual*. Merck & Company, Rahway, New Jersey.

Stevens, R. B. 1974. *Plant Disease*. John Wiley & Sons, New York.

Walker, N. 1975. *Soil Microbiology*. John Wiley & Sons, New York.

Wheeler, H. 1975. *Plant Pathogenesis*. Springer-Verlag, New York.

Environmental quality: biodegradation of wastes and pollutants

23

850

Environmental
quality: biode-
gradation of
wastes and
pollutants

Human activities create vast amounts of various wastes and **pollutants**. The release of these materials into the environment sometimes causes serious health problems and may preclude desirable usage of our land and water resources. The use of rivers, for example, as a receptacle of sewage, as a habitat for fish, and as a source of irrigation and drinking water depends on the careful management of the amounts of wastes entering the system and the levels of pathogenic microorganisms associated with the release of those wastes. There are limits to the natural capacity of an ecosystem to cope with inputs of wastes and pollutants; that is, the microorganisms in the ecosystem have a limited capacity to biodegrade these unwanted materials without serious deterioration of environmental quality or increased incidence of disease. The level of wastes produced by dense human and domestic animal populations often exceeds the local ecosystem's biodegradative capacity, resulting in serious environmental pollution and epidemic outbreaks of disease. Each year more than 500 million people are afflicted with water-borne diseases, and more than 10 million of these individuals die. Most of these disease outbreaks stem from fecal contamination of drinking water supplies. Proper treatment of wastes, employing microbial biodegradation, and disinfecting potable water supplies in order to kill contaminating pathogenic microorganisms can greatly improve the safety and quality of water supplies and the status of human health. In this chapter we will examine some of the methods involving microbiology that are employed to maintain environmental quality and safely dispose of waste and pollutant materials.

Solid waste disposal

Urban solid waste production in the United States amounts to roughly 150 million tons per year. Much of this material is inert, composed of glass, metal, and plastic, but the rest is decomposable organic waste, such as kitchen scraps, paper, and other household and industrial garbage. Sewage sludge derived from treatment of liquid wastes, animal waste from cattle feed lots, and large-scale poultry and swine farms are also major sources of solid organic waste. In traditional small, family farm operations, most organic solid waste is composted and recycled into the land as fertilizer. In highly populated urban centers and areas of large-scale agricultural production, the disposal of massive amounts of organic waste becomes a difficult and expensive problem.

There are several options for dealing with **solid waste** problems. Today, many of the inert components of solid waste, such as aluminum and glass, are recycled. Even paper, which is relatively resistant to microbial degradation, can be recovered from solid waste, and many books and newspapers are printed on recycled paper. The remaining bulk of the solid waste may be incinerated, creating potential air pollution problems, or the organic components can be subjected to microbial biodegradation in aquatic or terrestrial ecosystems. In many cases the solid waste is dumped at sea or discarded on land, allowing biodegradation to occur naturally without any special treatment, but excessive dumping of organic wastes into terrestrial and marine ecosystems can cause untoward problems unless the operation is carefully managed and monitored.

Sanitary landfills

The simplest and least expensive way to dispose of solid waste is to place the material in **landfills** and to allow it to decompose. In landfill procedures both organic and inorganic solid wastes are deposited together in low-lying land that has minimal real estate value. Because exposed waste can cause aesthetic and odor problems, attract insects and rodents, and pose a fire hazard, each day's waste deposit is covered over with a layer of soil, creating a sanitary landfill (Figure 23.1). When the landfill is full, the site can be used for recreational purposes and may eventually provide a foundation for construction. During the 30–50 years after the establishment of a landfill, the organic content of the solid waste undergoes slow, anaerobic microbial decomposition. The products of anaerobic microbial metabolism include carbon dioxide, water, methane, various low-molecular-weight alcohols, and acids, which diffuse into the surrounding water and air, causing the landfill to settle slowly. Extensive amounts of methane are produced during this decomposition process, potentially providing a source of needed natural gas. Eventually, the decomposition slows greatly, signaling completion of the biodegradation of the solid waste, subsidence ceases, and the land is stabilized and suitable as a site for construction.

Although the use of sanitary landfills is simple and inexpensive, there are several problems associated with this waste disposal method. Pre-

Figure 23.1

Sanitary landfills are used for the inexpensive disposal of solid waste. Several different methods, as shown in these drawings, can be employed. (A) The area method—a bulldozer spreads and compacts solid wastes. Cover material is hauled in and spread at the end of the day's operations. Note that a fence to catch any blowing debris is present in all landfill methods. (B) The trench method—the waste collection truck deposits its load into a trench where it is spread and compacted. At the end of the day, soil is excavated from a future trench and used as the daily cover material. (C) Ramp variation—solid wastes are spread and compacted on a slope. The daily deposit is covered with earth from the base of the ramp. This variation is used with either the area or trench method.

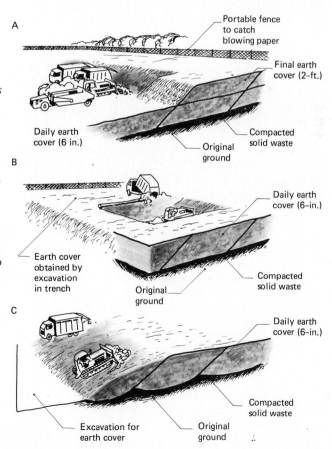

A

B

C

852

Environmental
quality: biode-
gradation of
wastes and
pollutants

Figure 23.2

*These photographs show compost heaps
used for decomposition of solid wastes.
(A) Steam is rising from this compost heap
because of the heat build-up that results
from the rapid decomposition of organic
matter caused by thermophilic
microorganisms. (B) In this heap a straight
row is formed by a machine, which is also
used to turn the heap periodically to
maintain good aeration. (Courtesy Atal E.
Eralp, USEPA, Cincinnati.) (C) This is a
commercial compost facility owned by
Paygro, Inc., in South Charlestown, Ohio.
The T-shaped poles have thermocouples
attached to them to continuously monitor
the temperature at various depths within
the pile. Air is forced through the pile from
underneath by fans to provide aeration.
(Courtesy V. L. McKinley and J. R. Vestal,
University of Cincinnati.)*

mature construction on a still biologically active
landfill site may result in structural damage to the
buildings because of movement of the land base,
and an explosion hazard may exist due to methane seeping into basements and other below-
ground structures. Above-ground plantings may
also be damaged because of methane seepage.
The number of suitable disposal sites available in
urban areas is very limited, often necessitating
long and expensive hauling of the solid waste to
available sites. The possible seepage of anaerobic
decomposition products, heavy metals, and a variety of recalcitrant hazardous pollutants from the
landfill site into underground aquifers, which are
used in many urban areas as a water sources, has

caused many municipalities to place severe restrictions on the location and operation of landfills and to seek alternatives to the landfill technique for disposing of solid waste.

Composting

The organic portion of solid waste can be biodegraded by **composting**, the process by which solid
heterogeneous organic matter is degraded by
aerobic, mesophilic, and thermophilic microorganisms. Composting, like incineration, requires
sorting the solid waste into its organic and inorganic components. This can be accomplished either

at the source, by the separate collections of garbage (organic waste) and trash (inorganic waste), or at the receiving facility, by using magnetic separators to remove ferrous metals and mechanical separators to remove glass, aluminum, and plastic materials. The remaining largely organic waste is ground up, mixed with sewage sludge and/or bulking agents, such as shredded newspaper or wood chips, and then composted. The addition of dehydrated sewage sludge to domestic garbage improves the carbon–nitrogen balance because sewage sludge is high in nitrogen and therefore enhances microbial biodegradation activities as well as provides a means of disposing of some sewage sludge waste. The addition of 10 percent by weight sewage sludge to the material being composted improves the porosity, an important factor because 30 percent air space is needed to optimize the availability of oxygen for microbial respiration in the aerobic compost process and because water must drain out of the composting material to prevent waterlogging and the development of anaerobic conditions.

Composting is a microbial process that converts noxious organic waste materials into a stable, sanitary, humus-like product. Reduced in bulk, it can be used for soil improvement. The different composting methods are differentiated by the physical arrangement of the solid waste; that is, composting can be accomplished in **windrows**, **aerated piles**, or **continuous feed reactors** (Figure 23.2). The windrow method is a simple but relatively slow process, typically requiring several months to achieve biodegradation of the metabolizable components and stabilization of the waste material. Odor and insect problems are controlled in this process by covering the windrows with a layer of soil or finished compost. Unless the decomposing material is turned several times during the process, the quality of the finished compost product is uneven. Because the process is so slow, large amounts of land must be used, causing the same problems as sanitary landfills in densely populated urban areas.

Composting rates can be enhanced in the aerated pile method by increasing the level of aeration. Perforated pipes are buried inside the compost pile and air is pumped through the pile, both oxygenating and cooling the pile. The heat generated in the aerated pile process is used to evaporate water for the final drying of the product. The continuous feed composting process uses a reactor that permits control of the environmental parameters (Figure 23.3). The reactor is analogous to an industrial fermentor and permits the production of a relatively uniform product. Compared to the other compost methods, the continuous feed process requires a high initial financial investment. By optimizing conditions, composting in the reactor is accomplished in just two to

Figure 23.3

The design of this continuous process compost plant permits the efficient and economical use of composting for solid waste disposal.

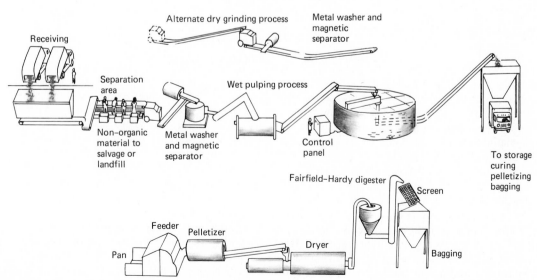

Compost Plant Flow Diagram

four days, although the product requires additional curing for about a month prior to packaging and shipment.

In a compost of domestic garbage and sludge, numerous microbial species that come from soil, water, and human fecal matter are present. The relatively high moisture content of the compost material favors the development of bacterial rather than fungal populations. In the composting of solid organic wastes, the process is initiated by mesophilic heterotrophs, which, as the temperature rises, are replaced by thermophilic microorganisms. The initial temperature increase is probably due to the growth of mesophilic bacteria in the interior portions of the composting material. Thermophilic microorganisms prominent in the composting process include the bacteria *Bacillus stearothermophilus, Thermomonospora* sp., *Thermoactinomyces* sp., and *Clostridium thermocellum* and the fungi *Geotrichum candidum, Aspergillus fumigatus, Mucor pusillus, Chaetomium thermophile, Thermoascus auranticus*, and *Torula thermophila*. In the continuous reactor-type composting process, the reactor is maintained continuously at thermophilic temperatures by using the heat produced within the reactor by the biodegradation of the organic matter.

The control of several conditions is critical for achieving optimal composting. Temperatures needed to achieve maximal rates of organic matter decomposition are in the range of 50–60°C. Self-heating typically raises the temperature inside a static compost pile to 55–60°C or above in two to three days under favorable conditions, but after a few days at optimal temperature, the temperature gradually declines unless the pile is turned to resupply oxygen and ensure that the thermophilic process occurs throughout the pile, instead of only at the core. Moisture must be adequate; 50–60 percent water content is optimal, but excess moisture, 70 percent or above, interferes with aeration and lowers self-heating because of water's large heat capacity. The carbon-to-nitrogen ratio must not be greater than 40:1; a lower nitrogen content would preclude the formation of a sufficient microbial biomass, and a greater nitrogen concentration, such as C:N = 25:1, would lead to volatilization of ammonia, causing odor problems and lowering the usefulness of the compost product as a fertilizer.

Although compost is a good soil conditioner and supplies some plant nutrients, it cannot compete with synthetic fertilizers for use in agricultural production. The sale of compost effectively reduces the cost of the waste disposal operation but generally does not render the waste disposal operation self-supporting. When sewage sludge is used as a major component of the original compost mixture, however, the finished product may contain relatively high concentrations of potentially toxic heavy metals, such as cadmium, chromium, and thallium. Because little is known about the behavior of these metals in agricultural soils, the use of sewage sludge-derived compost in agriculture is not widely practiced, but compost does find extensive applications in parks and gardens for ornamental plants, in land reclamation, particularly after strip mining, and as part of highway beautification projects. Even though landfill operations are less expensive than composting, the long-range environmental costs in terms of groundwater contamination favor the composting process.

Treatment of liquid wastes

Agricultural and industrial operations—along with everyday human activities—produce **liquid wastes**, including **domestic sewage**. These liquid waste discharges flow through natural drainage patterns or sewers, eventually entering natural bodies of water, such as groundwater, rivers, lakes, and oceans. In theory the liquid wastes disappear when they are flushed into such water bodies, according to the adage "the solution to pollution is dilution." Bodies of water into which sewage flows must also serve local communities as the source of water for drinking, household use, industry, irrigation, fish and shellfish production, swimming, boating, and other recreational purposes, making the maintenance of the acceptable high quality of these natural waters essential. Fortunately, **self-purification** is an inherent capability of natural waters, based on the biogeochemical cycling activities and interpopulation relationships of the indigenous microbial populations. Organic nutrients in the water are metabolized and mineralized by autochthonous heterotrophic aquatic microorganisms. Ammonia is nitrified and, along with other inorganic nutrients, used and immobilized by algae and higher aquatic plants. Allochthonous populations of enteric and other pathogens that enter aquatic ecosystems are maintained at low levels and/or eliminated by the pressures of competition and predation of the autochthonous aquatic populations. Consequently, reasonably low amounts of raw sewage can be accepted by natural waters without any significant decline in the level of water quality.

Despite this fact, human demographic patterns of densely populated areas, large-scale agricultural operations, and major industrial activities result in the production of liquid wastes on a scale that routinely overwhelms the self-purification capacity of aquatic ecosystems, causing an unacceptable deterioration of water quality. The relative changes in some environmental parameters and populations in a river receiving sewage are illustrated in Figure 23.4. A prominent feature of river water receiving sewage effluents is the presence of the filamentous aerobic bacterium

Sphaerotilus natans, known as the "sewage fungus" (Figure 23.5). A heterogeneous microbial community also develops amid the filaments of this bacterium below a sewage outfall. *S. natans* and the associated microbial community are efficient degraders of organic matter, consuming oxygen at a rate of two grams O_2 per hour per square meter. The bloom of the sewage fungus exemplifies the aesthetically displeasing results of excessive addition of organic matter to natural water bodies. Depending on the rate of sewage discharge, flow rate, water temperature, and other environmental factors, water quality may reestablish an acceptable quality level at some distance downstream from the sewage outfall, typically within 24 to 60 kilometers. The maintenance of satisfactory water quality means that natural waters should not be overloaded with organic or inorganic nutrients, toxic, noxious, or aesthetically

Figure 23.4

The effects of organic effluent on a stream and the changes that occur downstream of the outfall on materials, plants, and animals are represented in this graph. Water quality studies have found that near the outfall there is a high abundance of Sphaerotilus natans, *the sewage fungus, high ammonium and nitrate concentration, very low oxygen concentration, and a specialized macrofauna not found in clean water. Depending on the rate of sewage discharge, water flow rate, temperature, and other environmental factors, the water quality may return to close to its original state at a typical distance of from a few to several dozen kilometers downstream from the sewage outfall.*

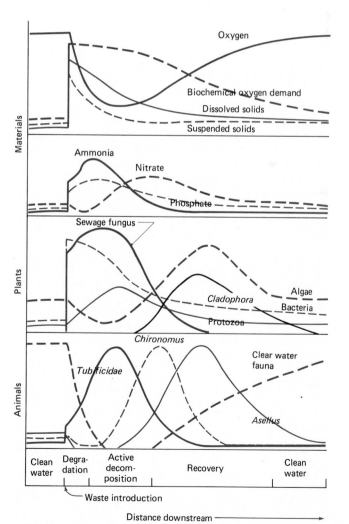

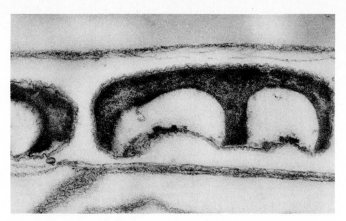

Figure 23.5

Electron micrograph of a thin section of Sphaerotilus natans *showing the sheath surrounding the cells. Cell width is approximately 500 nm. (From BPS— T. J. Beveridge, University of Guelph.)*

856

Environmental quality: biodegradation of wastes and pollutants

unacceptable substances; that their oxygen, temperature, salinity, turbidity, or pH levels should not be altered so significantly that they lose their ability to support fish production and recreational usage of the water body; and that they should not be allowed to become vehicles of disease transmission due to fecal contamination.

Biological oxygen demand (BOD)

We have developed several measures of water quality that help us manage aquatic ecosystems by indicating how much waste can safely be allowed to enter rivers and lakes without serious deterioration of water quality. One widely used measure of water quality, the **biological oxygen demand (BOD)**, represents the amount of oxygen required for the microbial decomposition of the organic matter in the water. The polluting power of different sources of wastes is reflected in the BOD of the material (Table 23.1). The BOD can be easily determined in the laboratory by incubating a water sample and measuring the amount of oxygen consumed during a five-day period (Figure 23.6). A high BOD generally indicates the presence of excessive amounts of organic carbon. The dissolved oxygen in natural waters seldom

exceeds 8 mg/l because of its low solubility, and it is often considerably lower because of heterotrophic microbial activity, making oxygen depletion a likely consequence of adding wastes with high BOD values to aquatic ecosystems. Reaeration, replenishment of dissolved oxygen from the atmosphere and/or from photosynthetic O_2 evolution, is often much slower than the rate of oxygen utilization by heterotrophic microorganisms when abundant organic substrates are available.

Exhaustion of the **dissolved oxygen (D.O.)** content is the principal impact of a sewage overload on natural waters. When the oxygen supply is depleted, the self-purification processes is drastically slowed down. Oxygen deprivation kills obligately aerobic organisms, including some microorganisms, fish, and invertebrates, and the decomposition of dead organisms within the water body creates an additional oxygen demand. Fermentation products and the reduction of the secondary electron sinks of nitrate and sulfate give rise to noxious odors, tastes, and colors, making the water putrid and septic. Turbidity and the presence of toxic metabolic products, such as H_2S, also interfere with photosynthetic oxygen regeneration, further delaying the recovery process.

Sewage treatment

Modern methods of liquid waste treatment are aimed at reducing the BOD of the waste before it is discharged into a water body in order to maintain water quality. There are several different approaches to reducing the BOD, employing combinations of physical, chemical, and microbiological methods. Most communities in developed countries have facilities for treating **sewage**, which is the used water supply containing do-

table 23.1

BOD values for different types of wastes

Type of waste	BOD (mg/liter)
Domestic sewage	200–600
Slaughter house wastes	1,000–4,000
Piggery effluents	25,000
Cattle shed effluents	20,000
Vegetable processing	200–5,000

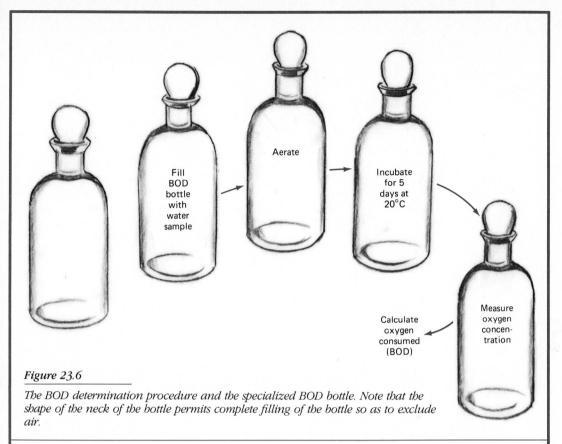

Figure 23.6

The BOD determination procedure and the specialized BOD bottle. Note that the shape of the neck of the bottle permits complete filling of the bottle so as to exclude air.

Discovery process

One consequence of urbanization is the need to remove sewage and other organic wastes away from concentrated population centers. Waterways that are normally used for waste removal under the premise that "the solution to pollution is dilution" can be overwhelmed by such concentrated inputs of organic matter. The BOD procedure, which is used extensively in monitoring water quality and biodegradation of waste materials, is designed to determine how much oxygen is consumed by microorganisms during oxidation of the organic matter present in the sample. The procedure recognizes that the natural biodegradation of organic matter in water consumes the oxygen needed by fish and other higher organisms that live in aquatic ecosystems; excessively high BOD values due to the input of organic wastes are undesirable because of the deterioration of water quality. The amount of oxygen consumed is determined by measuring residual oxygen levels with an oxygen electrode. Incubation at 20°C for five days is commonly used because the test was originally developed in Great Britain, where the average water temperatures are near 20°C and where it takes a maximum of five days for anything entering a local river to reach the ocean. Once the organic matter reaches the oceans, it is no longer considered a threat to water quality. In the United States and other large countries, different incubation conditions are needed for proper use of the BOD; in such countries it is necessary to modify the incubation period used in the standard five-day BOD procedure to account for the extended residence time of organic matter in the waterways receiving organic pollutants. Development of appropriate modifications to the original procedure has been slow, in part because of a lack of understanding of the original assumptions used in establishing the standard five-day incubation procedure. Appropriate modifications to the standard BOD procedure, based on actual residence times in inland waterways and desirable multiple uses of water, are presently being incorporated into water quality standards.

mestic waste, together with human excrement and wash water; industrial waste, including acids, greases, oils, animal matter, and vegetable matter; and storm waters. The use of household garbage disposal units also increases the organic content of domestic sewage. The treatment of liquid wastes is aimed at removing organic matter, human pathogens, and toxic chemicals. The treatment of domestic sewage reduces the biological oxygen demand due to suspended or dissolved organics and the numbers of enteric pathogens so that the discharged sewage effluent will not cause unacceptable deterioration of environmental quality.

Sewage is subjected to different treatments, depending on the quality of the effluent deemed necessary to be achieved to permit the maintenance of acceptable water quality (Figure 23.7). **Primary treatments** rely on physical separation procedures to lower the BOD; **secondary treatments** rely on microbial biodegradation to further reduce the concentration of organic compounds in the effluent; and **tertiary treatments** use chemical methods to remove inorganic compounds and pathogenic microorganisms. Municipal sewage treatment facilities are designed to handle organic wastes but are normally incapable of dealing with industrial wastes containing toxic chemicals, such as heavy metals. Industrial facilities frequently must operate their own treatment plants for dealing with waste materials.

Primary treatment

Primary sewage treatment removes suspended solids in settling tanks or basins (Figure 23.8). The solids are drawn off from the bottom of the tank and may be subjected to anaerobic digestion and/or composting prior to final deposition in landfills or as soil conditioner. Only a low percentage of the suspended or dissolved organic material is actually mineralized during liquid waste treatment; most of it is removed by settling, and as a result the disposal problem is merely "displaced" to the solid waste area rather than being solved. Nevertheless, this displacement is essential because of the detrimental effects of discharging effluents with high BOD into aquatic ecosystems with naturally low dissolved oxygen contents. The liquid portion of the sewage, which contains dissolved organic matter, can be subjected to further treatment or discharged after primary treatment alone. Because liquid wastes vary in com-

Figure 23.7

Flow chart of the stages of sewage treatment. Primary treatment is principally physical, secondary principally biological, and tertiary principally chemical.

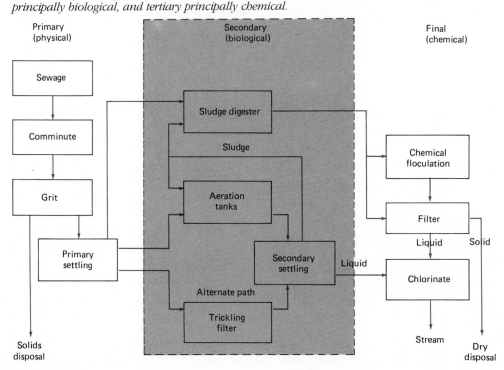

Figure 23.8

Diagram of a settling tank for primary sewage treatment.

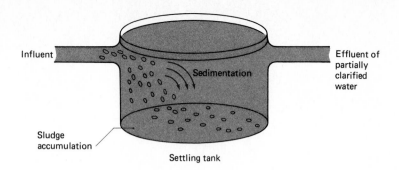

position and may contain mainly solids and little dissolved organic matter, primary treatment may remove 70–80 percent of the BOD and be sufficient. For typical domestic sewage (Table 23.2), however, primary treatment normally removes only 30–40 percent of the BOD.

Secondary treatment

To achieve an acceptable reduction in the BOD, **secondary treatment** by a variety of means is necessary (Table 23.3). In secondary sewage treatment a small portion of the dissolved organic matter is mineralized, and the larger portion is converted to removable solids. The combination of primary and secondary treatment reduces the original sewage BOD by 80–90 percent. The secondary sewage treatment step relies on microbial activity, may be aerobic or anaerobic, and is conducted in a large variety of devices. A well-designed and efficiently operated secondary treatment unit should produce effluents with BOD and/ or suspended solids less that 20 mg/l. Because the secondary treatment of sewage is a microbial process, it is extremely sensitive to the introduction of toxic chemicals that may be contained in industrial waste effluents or that may accidentally contaminate the sewerage system. The accidental introduction of the organic chemicals octachlorocyclopentene and hexachlorocyclopentadiene into the municipal sewage system of Louisville, Kentucky, poisoned the microorganisms in the sewage treatment facility, forcing the dumping of

table 23.2

Characteristics of typical municipal waste water

Component	Concentration (mg/liter)
Solids	
Total	700
Dissolved	500
fixed	300
volatile	200
Suspended	200
fixed	50
volatile	150
BOD (biochemical oxygen demand)	300
TOC (total organic carbon)	200
COD (chemical oxygen demand)	400
Nitrogen (as N)	
Total	40
Organic	15
Ammonia	25
Nitrate	0
Nitrite	0
Phosphorus (as P)	
Total	10
Organic	3
Inorganic	7
Grease	100

7 billion gallons of untreated sewage into the Ohio River during a three-month period before the toxic chemicals could be removed from the system. The accidental introduction of hexanes into the same sewerage system several years later caused a mas-

table 23.3

Efficiency of various types of sewage treatment

Treatment	BOD (% reduced)	Suspended solids (% removed)	Bacteria (% reduced)
Sedimentation	30–75	40–95	40–75
Septic tank	25–65	40–75	40–75
Trickling filter	60–90	0–80	70–85
Activated sludge	70–96	70–97	95–99

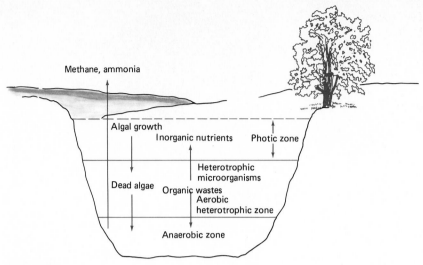

Figure 23.9

This diagram of an oxidation pond illustrates the biological process involved in this form of sewage treatment. The photic zone is highly productive. Dead algae fall into an aerobic heterotrophic zone where they are partially degraded and then fall with dead bacteria into the anaerobic bottom layer of the pond.

sive explosion and the disruption of normal sewage treatment for an extended period of time.

Oxidation ponds **Oxidation ponds**, also known as stabilization ponds and lagoons, are used for

Figure 23.10

In a trickling filter the clarified sewage is sprayed over rocks. The bacterial film on the rocks aerobially decomposes the dissolved organic matter.

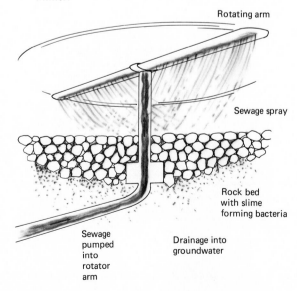

Rotating arm

Sewage spray

Rock bed with slime forming bacteria

Sewage pumped into rotator arm

Drainage into groundwater

the simple secondary treatment of sewage effluents in rural communities and some industrial facilities (Figure 23.9). Heterotrophic bacteria degrade sewage organic matter within the ponds, producing cellular material and mineral products that support the growth of algae. The proliferation of algal populations in these lagoons produces oxygen that replenishes the oxygen depleted by the heterotrophic bacteria, permitting continued organic matter decomposition. Typically, oxidation ponds are shallow, less than 10 feet deep, which maximizes the euphotic zone for algal growth. The bacterial and algal cells formed during the decomposition of the sewage settle to the bottom, eventually filling in the pond. The degradation of organic matter in these ponds is relatively slow, and the residence time for the treatment of domestic sewage may be as long as a week. The effluents containing oxidized products are periodically removed from the ponds, which are then refilled with raw sewage.

Trickling filter The **trickling filter system** is a simple and relatively inexpensive film-flow type of aerobic sewage treatment method (Figure 23.10). The sewage is distributed by a revolving sprinkler suspended over a bed of porous material. The sewage slowly percolates through this porous bed, and the effluent is collected at the bottom. The porous material of the filter bed becomes coated by a dense, slimy bacterial growth, principally composed of *Zooglea ramigera* and similar slime-forming bacteria (Figure 23.11). The slime matrix thus generated accommodates a heterogeneous microbial community, including bacteria, fungi, protozoa, nematodes, and rotifers. The most frequently found bacteria are *Beggiatoa alba, Sphaerotilus natans, Achromobacter* sp., *Flavobacterium* sp., *Pseudomonas* sp., and *Zooglea* sp. This microbial community absorbs and mineralizes dissolved organic nutrients in the sewage, reducing the BOD of the effluent (Figure 23.12). Aeration occurs passively as a result of the movement of air through the porous material of the bed. The sewage may be passed through two or more trickling filters or can be recirculated several times through the same filter to reduce the BOD to acceptable levels. The effluent from the trickling filters may be clarified by allowing sloughed-off biomass to settle prior to discharge. A drawback of this otherwise simple and inexpensive treatment system is that a nutrient overload produces excess microbial slime, which reduces aeration and percolation rates, periodically necessitating renewal of the trickling filter bed. Also, cold winter temperatures strongly reduce the effectiveness of such outdoor treatment facilities.

Biodisk system The **rotating biological contactor or biodisk system** is a more advanced type of aerobic film-flow treatment system. This system requires less space than trickling filters and is more efficient and stable in operation but needs a higher initial financial investment. The biodisk system is used in some communities for the treatment of both domestic and industrial sewage effluents. In the biodisk system, closely spaced disks—usually made of plastic, are rotated in a trough containing the sewage effluent (Figure 23.13). The disks are partially submerged and become coated by a microbial slime, similar to what develops in trickling filters. Continuous rotation of the disks keeps the slime well aerated and in contact with the sewage. The thickness of the microbial slime layer in all film-flow processes is governed by the diffusion of nutrients through the film. Microbial growth on the surface of the disks is sloughed off gradually and is removed by subsequent settling. When the film becomes so thick that oxygen and nutrients fail to reach the inner portions of the film, most of the innermost microorganisms die, causing detachment of the slime layer and temporary disruption of the process.

Activated sludge The **activated sludge** process is a very widely used aerobic suspension type of

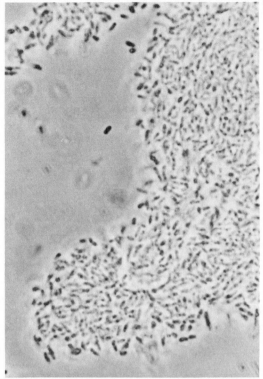

Figure 23.11

Photomicrographs of slime-forming bacteria on a trickling filter. (A) Zooglea ramigera surrounded by extensive slime layer. (Courtesy Patrick Dugan, Ohio State University.) (B) Scanning electron micrograph of microbial community on the surface of a rock from a trickling filter; note the abundance of filamentous algae and fungi that develop on the surface film (500×). (Courtesy Robert Apkarian, University of Louisville.)

Figure 23.12

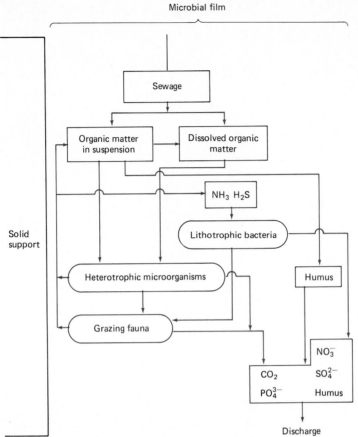

Microbial film

Sewage

Organic matter in suspension

Dissolved organic matter

NH_3 H_2S

Lithotrophic bacteria

Solid support

Heterotrophic microorganisms

Humus

Grazing fauna

NO_3^-

CO_2 SO_4^{2-}

PO_4^{3-} Humus

Discharge

The mineral cycling of organic matter by the microbial film in a trickling filter system.

Figure 23.13

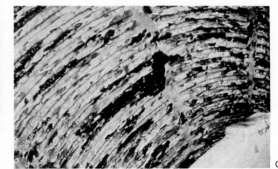

(A) A biodisk assembly; the disks rotate through the sewage treatment and microbes growing on the surfaces of the polyethylene cups degrade the organic matter in the sewage. (B) Close-up view of the normal appearance of the surface of the biodisk with a film of aerobic microorganisms. (C) When the system is overloaded, a heavy film of anaerobic microorganisms develops. Note the black area of anaerobic biomass and the occurrence of gelatinous, filamentous sulfur-oxidizing growths on the outer layer. (Courtesy Autotrol Corp., Milwaukee, Wisconsin.)

A

B

C

Figure 23.14

An activated sludge tank in a
sewage treatment plant. Note
the bubbling due to forced
aeration. (From BPS—Carl
May, Moss Beach, California.)

liquid waste treatment system (Figure 23.14). After primary settling, the sewage, containing dissolved organic compounds, is introduced into an aeration tank. Air injection and/or mechanical stirring provides the aeration. The rapid development of microorganisms is also stimulated by reintroduction of most of the settled sludge from a previous run (Figure 23.15); the process derives its name from this inoculation with "activated sludge."

During the holding period in the aeration tank, vigorous development of heterotrophic microorganisms has taken place. The heterogeneous nature of the organic substrates in sewage allows the development of diverse heterotrophic bacterial populations, including Gram negative rods, predominately *Escherichia, Enterobacter, Pseudomonas, Achromobacter, Flavobacterium*, and *Zooglea* species; other bacteria, including *Micrococcus, Arthrobacter*, various coryneforms and mycobacteria, *Sphaerotilus*, and other large filamentous bacteria; and low numbers of filamentous fungi, yeasts, and protozoa, mainly ciliates.

The protozoa are important predators of the bacteria, along with rotifers. The bacteria in the activated sludge tank occur both in free suspension and as aggregates or flocs. The flocs are composed of microbial biomass held together by bacterial slimes, produced by *Zooglea ramigera* and similar organisms. Most of the ciliate protozoa, such as *Vorticella*, are of the attached filter-feeding type and adhere to the flocs, while feeding predominantly on the suspended bacteria. The floc is too large to be ingested by the ciliates and rotifers and acts a defense mechanism against predation. In the raw sewage, suspended bacteria predominate, but during the holding time in the aeration tank, their numbers decrease, and at the same time those bacteria associated with flocs greatly increase in numbers (Table 23.4).

As a consequence of extensive microbial metabolism of the organic compounds in sewage, a significant proportion of the dissolved organic substrates is mineralized, and another portion is converted to microbial biomass. In the advanced stage of aeration, most of the microbial biomass

Figure 23.15

The flow of materials through an activated sludge secondary sewage treatment system. A portion of the sludge is recycled as inoculum for the incoming sewage.

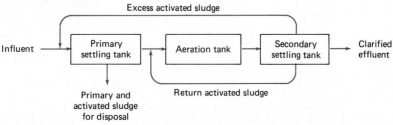

table 23.4

Numbers of bacteria at different stages of sewage treatment

Treatment	Total bacteria (number/ml)	Viable bacteria (number/ml)
Settled sewage	7×10^8	1×10^7
Activated sludge mixed liquor	7×10^9	6×10^7
Filter slimes	6×10^{10}	2×10^9
Secondary effluents	5×10^7	6×10^5
Tertiary effluents	3×10^7	4×10^4

becomes associated with flocs that can be removed from suspension by settling. The settling characteristic of sewage sludge flocs is critical to their efficient removal. Poor settling produces "bulking" of sewage sludge, caused by proliferation of such filamentous bacteria as *Sphaerotilus, Beggiatoa, Thiothrix*, and *Bacillus*, and such filamentous fungi as *Geotrichum, Cephalosporium, Cladosporium*, and *Penicillium*. The causes of bulking are not always understood, but it is frequently associated with high C:N and C:P ratios and/or low dissolved oxygen concentrations. A portion of the settled sewage sludge is recycled for use as the inoculum for the incoming raw sewage, but the rest of the sludge requires additional treatment by composting or anaerobic digestion.

Combined with primary settling, the activated sludge process tends to reduce the BOD of the effluent to 10–15 percent of that of the raw sewage. The treatment also drastically reduces the numbers of intestinal pathogens in the sewage. This reduction is the result of the combined effects of competition, adsorption, predation, and settling. Predation by ciliates, rotifers, and *Bdellovibrio* is probably indiscriminate and affects pathogens as well as nonpathogenic heterotrophs. Also, pathogens tend to grow poorly or not at all under the environmental conditions of an aeration tank, and nonpathogenic heterotrophs proliferate vigorously. Therefore, whereas

nonpathogenic heterotrophs reproduce to compensate for their predatory removal, the pathogens are continuously decimated (Table 23.5). Settling of the flocs removes additional pathogens and numbers of *Salmonella, Shigella*, and *Escherichia coli* typically are 90–99 percent lower in the effluent of the activated sludge treatment process than in the incoming raw sewage. Enteroviruses are removed to a similar degree, and the main removal mechanism appears to be adsorption of the virus particles onto the settling sewage sludge floc.

Septic tank The simplest anaerobic treatment system, the **septic tank**, is used extensively in rural areas that lack sewerage systems (Figure 23.16). Many rural and suburban single-family dwellings use septic tanks. A septic tank acts largely as a settling tank, within which the organic components of the waste water undergo limited anaerobic digestion. The accumulated sludge is maintained under anaerobic conditions and is degraded by anaerobic microorganisms to organic acids and hydrogen sulfide. Residual solids settle to the bottom of the septic tank, and the clarified effluent is allowed to percolate into the soil where the dissolved organic compounds in the effluent undergo biodegradation. These products are distributed into the soil along with the clarified sewage effluent. Septic tank treatment does not reliably destroy intestinal pathogens, and it is

table 23.5

Percentage reductions in the numbers of indicator organisms in different types of sewage treatment processes

Treatment	*Escherichia coli*	Coliforms	Fecal streptococci	Viruses
Sedimentation	3–72	13–86	44–60	—
Activated sludge	61–100	13–83	84–93	79–100
Trickling filter	73–97	15–100	64–97	40–82
Lagoons	80–100	86–100	85–99	95

Cross-section
of septic tank

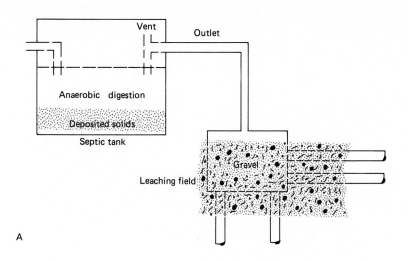

A

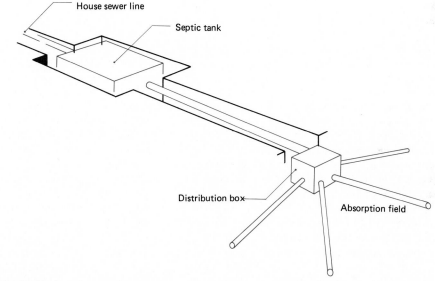

B

Figure 23.16

(A) A typical septic tank used for the disposal of sewage in a rural home is diagrammed. The sewage is degraded by anaerobic digestion. The treated effluent is disposed into a leaching field of permeable soils, with the effluent pipes usually set in gravel. (B) Diagram of the installation of a septic tank for sewage disposal from a home. The sewage effluent is distributed into the surrounding soil.

important that the soils receiving the clarified effluents not be in close proximity to drinking wells to prevent contamination of drinking water with enteric pathogens.

Anaerobic digesters Large-scale **anaerobic digesters** are used for further processing the sew-age sludge produced by primary and secondary treatments (Figure 23.17). Although anaerobic digesters could be used for direct treatment of sewage, economic considerations favor aerobic processes for relatively dilute wastes, and the use of anaerobic digesters is restricted to treating concentrated organic wastes. Therefore, in prac-

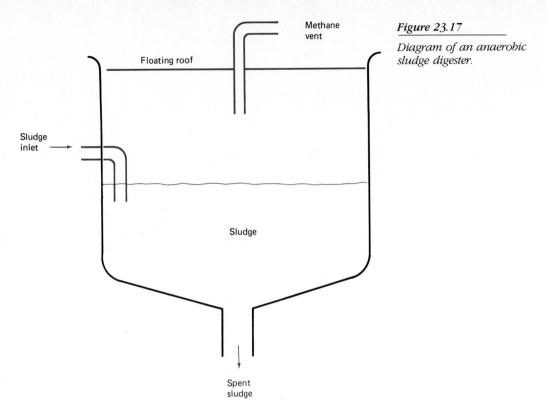

Figure 23.17

Diagram of an anaerobic
sludge digester.

Methane
vent

Floating roof

Sludge
inlet →

Sludge

Spent
sludge

Environmental
quality: biode-
gradation of
wastes and
pollutants

tice large-scale anaerobic digesters are used only
for processing settled sewage sludge and the
treatment of very-high-BOD industrial effluents.
Anaerobic digesters are large fermentation tanks
designed for continuous operation under anaerobic conditions. Provisions for mechanical mixing,
heating, gas collection, sludge addition, and draw-
off of stabilized sludge are incorporated into the
design of a large-scale anaerobic digester to permit effective operation. Anaerobic digesters contain high amounts of suspended organic matter;
between 20 and 100 g/l is considered favorable.
Much of this suspended material is bacterial biomass, and viable counts can be as high as 10^9–10^{10}
bacteria per ml. Fungi and protozoa are present
in very low numbers and do not play a significant
role in anaerobic digestion. A complex bacterial
community is involved in the degradation of organic matter within an anaerobic digester, with
numbers of anaerobic microorganisms typically
two to three orders of magnitude higher than
numbers of aerobes.

The anaerobic digestion of wastes is a two-step
process in which a large variety of nonmethano-
genic, obligately, or facultatively anaerobic bacteria participate. First, complex organic materials,
including microbial biomass, are depolymerized
and converted to fatty acids, CO_2, and H_2 (Figure

23.18). In the next step, methane is generated
either by the direct reduction of methyl groups
to methane or by the reduction of CO_2 to methane by molecular hydrogen or other reduced fermentation products, such as fatty acids. The final
products obtained in an anaerobic digester are a
gas mixture, approximately 70 percent methane
and 30 percent carbon dioxide, microbial biomass, and a nonbiodegradable residue.

The optimal operation of anaerobic digesters
requires good control of several parameters, such
as retention time, temperature, pH, and C:N and
C:P ratios. The optimal performance temperature
is in the range 35–37°C. The pH must remain in
the range 6.0–8.0, with pH 7.0 optimal. Variations
in pH and the inclusion of heavy metals or other
toxic materials in the sludge can easily upset the
operation of the anaerobic digester. In a "stuck"
or "sour" digester the methane production is interrupted, fatty acids and other fermentation
products accumulate, and it is difficult to restore
normal operation. It is usually necessary to clean
out the reactor and charge it with large volumes
of anaerobic sludge from an operational unit, a
costly and time-consuming exercise.

A properly operating anaerobic digester yields
a greatly reduced volume of sludge compared to
the starting material. The product obtained by us-

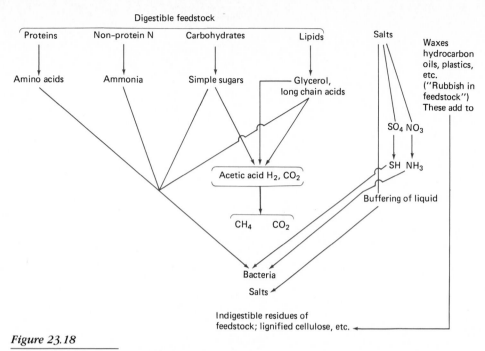

Digestible feedstock

Proteins   Non-protein N   Carbohydrates   Lipids   Salts

Waxes hydrocarbon oils, plastics, etc. ("Rubbish in feedstock") These add to

Amino acids   Ammonia   Simple sugars   Glycerol, long chain acids

SO_4 NO_3

Acetic acid H_2, CO_2

 SH NH_3

Buffering of liquid

CH_4 CO_2

Bacteria

Salts

Indigestible residues of feedstock; lignified cellulose, etc.

Figure 23.18

The metabolic reactions that take place within an anaerobic digestor. Digestible feedstock polymers are hydrolyzed. The hydrolyzed material is converted to acid, hydrogen, and carbon dioxide, which then undergo methanogenesis. These reactions contribute monomers and energy for the formation of bacterial cells, which with indigestible residues of the feedstock form "digested sludge."

ing an anerobic sludge digester, however, still causes odor and water pollution problems and can be disposed of at only a few restricted landfill sites. Aerobic composting can be used to further consolidate the sludge, rendering it suitable now for disposal in any landfill site or for use as a soil conditioner. Several gases are produced as a result of the anaerobic biodegradation of sludge, primarily methane and CO_2. The gas can be used within the treatment plant to drive the pumps and/or to provide heat for maintaining the temperature of the digester, or after purification it may be sold through natural gas distribution systems. Thus, in addition to its primary function in removing wastes, anaerobic digesters can produce needed fuel resources.

Tertiary treatment

The aerobic and anaerobic biological liquid waste treatment processes, just discussed, are designed to reduce the BOD of biodegradable organic substrates and oxidizable inorganic compounds and all represent secondary treatment processes. **Tertiary treatment**, defined as any practice beyond a secondary one, is designed to remove nonbiodegradable organic pollutants and mineral nu-

trients, especially nitrogen and phosphorus salts. The removal of toxic nonbiodegradable organic pollutants, such as chlorophenols, polychlorinated biphenyls, and other synthetic pollutants, is necessary to reduce the toxicity of the sewage effluent to acceptable levels. Activated carbon filters are normally used in the removal of these materials from secondary-treated industrial effluents. Secondary treatment is still required so as to avoid overloading this expensive treatment stage with biodegradable materials that could have been removed in more economical ways.

The release of sewage effluents containing phosphates and fixed forms of nitrogen can cause serious eutrophication in aquatic ecosystems. Sudden nutrient enrichment by sewage discharge or agricultural runoff triggers explosive algal blooms. Because of a variety of causes—some unknown but including mutual shading, exhaustion of micronutrients, presence of toxic products, and/or antagonistic populations—the algal population usually "crashes," and the subsequent decomposition of the dead algal biomass by heterotrophic microorganisms exhausts the dissolved oxygen in the water, precipitating extensive fish kills and septic conditions. Even if the eutrophication does

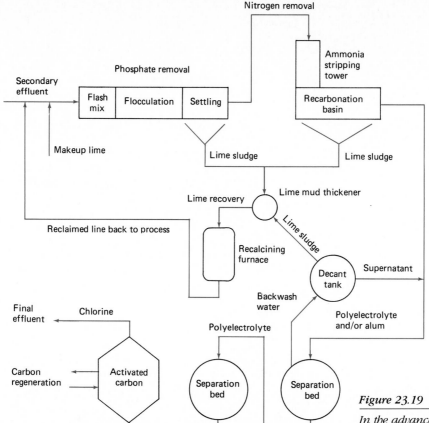

Nitrogen removal

Ammonia stripping tower

Phosphate removal

Secondary effluent

Flash mix | Flocculation | Settling

Recarbonation basin

Makeup lime

Lime sludge   Lime sludge

Lime recovery   Lime mud thickener

Reclaimed line back to process

Recalcining furnace

Lime sludge

Decant tank   Supernatant

Backwash water

Final effluent   Chlorine

Polyelectrolyte   Polyelectrolyte and/or alum

Carbon regeneration   Activated carbon   Separation bed   Separation bed

Organic removal

Additional phosphate removal and clarification

A

Environmental quality: biodegradation of wastes and pollutants

B

Figure 23.19

In the advanced sewage treatment system of South Tahoe, California, phosphate, nitrogen, and recalcitrant organic compounds are removed. (A) The extensive treatment process illustrated in this diagram is operational at relatively few locations where sewage effluents must cause minimal environmental perturbation. (B) Aerial photograph of the South Tahoe sewage treatment facility. Existing plant processes: (a) grit building; (b) primary treatment units; (c) flow equalization basins; (d) aeration basins; (e) secondary clarifiers; (f) chemical addition; (g) chemical clarifiers; (h) ammonia stripping tower (now dismantled); (i) ammonia stripping ponds; (j) filter backwash storage; (k) filters and carbon columns; (l) chlorine storage building; (m) lime mud thickener; (n) backwash decant tank; (o) sludge storage; (p) solids processing building.

not proceed to the above extreme, algal mats, turbidity, discoloration, and shifts in the fish population from valuable species to more tolerant but less valued forms represent undesirable changes due to eutrophication.

To prevent **eutrophication**, phosphate is commonly removed from sewage by precipitation as calcium or iron phosphate. This can be accomplished as an integral part of primary or secondary settling or in a separate facility where the precipitating agent can be recycled. Nitrogen, present mainly as ammonia, can be removed by volatilization as NH_3 at a high pH. Some ammonia eliminated from the sewage in this manner, however, may return to the watershed in the form of precipitation and cause further eutrophication problems. **Breakpoint chlorination** is an alternative procedure for removing ammonia. The addition of hypochlorous acid (HOCl), in 1:1 molar ratio, results in the formation of monochloramine (NH_2Cl), and further addition of HOCl to an approximate molar ratio of 2:1 results in nearly complete oxidation of the ammonia to molecular nitrogen. As chlorination of the sewage effluent is commonly practiced for disinfection purposes, chlorination to this "breakpoint" can be accomplished in the same process. The removal of ammonium nitrogen also lowers the BOD of the effluent because the ammonia would undergo nitrification in waters receiving the sewage effluent, which would consume oxygen dissolved in the receiving water.

A highly advanced tertiary water treatment system is currently under operation in South Tahoe, California, that integrates several of these tertiary treatment processes (Figure 23.19). This advanced treatment system was installed to prevent the eutrophic deterioration of scenic Lake Tahoe, an oligotrophic crater lake. In this system, after conventional primary and secondary treatments, phosphate is precipitated by liming, and ammonia is removed by stripping the high-pH effluent in a stripping tower at elevated temperature with vigorous aeration. After ammonia stripping, the pH is adjusted to neutrality. After additional settling, aided by polyelectrolyte addition and further clarification, nonbiodegradable organics are removed by filtration through activated carbon. The result of this extensive treatment procedure is a very high-quality effluent that can be released into Lake Tahoe without causing eutrophication (Table 23.6).

Disinfection

The final step in the sewage treatment process is **disinfection**, designed to kill enteropathogenic bacteria and viruses that were not eliminated during the previous stages of sewage treatment. Disinfection is commonly accomplished by **chlorination**, using either chlorine gas (Cl_2) or hypochlorite [($Ca(OCl)_2$ or NaOCl]. Chlorine gas reacts with water to yield hypochlorous and hydrochloric acids, the actual disinfectants. Hypochlorite is a strong oxidant, which is the basis of its antibacterial action. As an oxidant, it also reacts with residual dissolved or suspended organic matter, ammonia, reduced iron, manganese, and sulfur compounds. The oxidation of these compounds competes for available HOCl, reducing its disinfecting power. Amounts of hypochlorite sufficient to satisfy these reactions and to allow excess-free residual chlorine to remain in solution for disinfection would result in high salt concentrations in the effluent. Therefore, it is desirable to remove nitrogen and other contaminants by alternate means and use chlorination for disinfection only. A disadvantage of disinfection by chlorination is that the more resistant organic molecules, such as some lipids and hydrocarbons, are not completely oxidized but instead become partially chlorinated. Chlorinated hydrocarbons tend to be toxic and difficult to mineralize. Because alternative means of disinfection, such as ozonation, are more expensive, chlorination remains the principal means of sewage disinfection.

table 23.6

Quality of water after tertiary treatment at the South Tahoe, California, sewage treatment facility

Parameter	Wastewater (mg/l)	Reclaimed water (mg/l)
BOD	200–400	1
COD	400–600	3.25
TOC	—	1.0–7.25
Phosphate	25–30	0.2–1.0
Organic nitrogen	10–15	0.3–2.0
Ammonia nitrogen	25–35	0.3–1.5
Nitrate and nitrite	0	0

Treatment and safety of water supplies

Providing safe water for drinking and other uses, free of pathogens and toxic substances, is closely related to the problem of safe disposal of liquid wastes. Fecal contamination of potable water supplies from untreated or inadequately treated sewage effluents entering lakes, rivers, or groundwaters that serve as municipal water supplies creates conditions for rapid dissemination of pathogens. The primary route of infection is direct ingestion of the pathogens in the drinking water, but additional infection opportunities arise when fruits, vegetables, and eating utensils are washed with contaminated water. The obvious remedy to this situation is to disrupt the transmission of enteropathogenic fecal organisms to water supplies. The pathogenic population is reduced considerably during sewage treatment, but this does not assure adequate treatment of all

Figure 23.20

A flow diagram of the water purification process used by most municipalities.

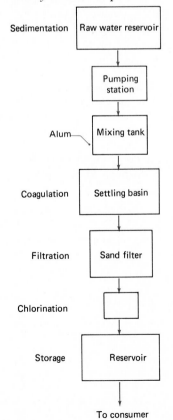

sewage discharge. Therefore, major sanitation efforts are required to treat and safely distribute public water supplies. Such sanitation practices have led to the virtual elimination of water-borne infections in the developed countries, but such infections continue to be major causes of sickness and death in underdeveloped regions.

Disinfection of potable water supplies

Most potable water supplies come from rivers and underground wells and springs. Water from underground sources is partially purified because water is filtered as it passes through the soil column, removing particulate matter and microorganisms. This does not, however, preclude the possibility of bacterial or viral contamination of the water supply, particularly if the source of the water is near a sewage effluent. In some rural areas water is boiled or treated with antimicrobial chemicals to ensure its safety. In densely populated areas municipal water treatment facilities are designed to ensure the safety of the drinking water supply. The principal processes of a water treatment facility are sedimentation, filtration, and disinfection (Figure 23.20). **Sedimentation** is carried out in large reservoirs where the water is held for a sufficient period of time to permit large particulate matter to settle out of the water. Sedimentation rates can be increased by the addition of aluminum sulfate, alum, which forms a floc that precipitates and carries with it microorganisms and suspended organic matter to the bottom of a settling basin. The water is then filtered by passage through sand filter beds that removes up to 99 percent of the bacteria. The water may also be filtered through activated charcoal to remove potentially toxic organic compounds and organic compounds that impart undesirable color and/or taste to the water. The water is then treated with a disinfecting agent to ensure that it does not contain any pathogens that could be a source of water-borne disease. The water may also be treated with fluoride to reduce dental caries in the community, and minerals may be removed to soften the water for washing.

Chlorination is the usual method employed for disinfecting municipal water supplies. This treatment method is relatively inexpensive, and the free residual chlorine content of the treated water represents a built-in safety factor against pathogens surviving the actual treatment period and causing recontamination. The disadvantage of chlorination is the incidental production of trace amounts of organochlorine compounds, some of which are suspected carcinogens. There is no firm evidence at this time that these trace amounts of chlorinated compounds actually represent a serious health hazard, but if this should prove to be the case, there are viable, though more expensive, alternatives.

Ozone (O_3), for example, has been used for water disinfection with good results both in Europe and the United States. **Ozone treatment** kills pathogens reliably and does not result in the synthesis of any undesirable trace organochlorine contaminants. However, because ozone is an unstable gas, water treated with ozone does not have any residual antimicrobial activity and is more prone to chance recontamination than chlorinated water. Ozone has to be generated from air on site in ozone reactors, using electrical corona discharge. Only about 10 percent of the electricity is actually generating ozone; the rest is lost as heat, making disinfection by ozone considerably less cost-effective than chlorination.

Bacterial indicators of water safety

The importance to public health of clean drinking water requires objective test methods to establish high standards of water safety and to evaluate the effectiveness of treatment procedures. To routinely monitor water for the detection of actual enteropathogens, such as *Salmonella* and *Shigella*, would be a difficult and uncertain undertaking. Instead, bacteriological tests of drinking water establish the degree of fecal contamination of a water sample by demonstrating the presence of **indicator organisms**. The ideal indicator organism should (1) be present whenever the pathogens concerned are present; (2) be present only when there is a real danger of pathogens being present; (3) occur in greater numbers than the pathogens to provide a safety margin; (4) survive in the environment as long as the potential pathogens; and (5) be easy to detect with a high degree of reliability of correctly identifying the indicator organism, regardless of what other organisms are present in the sample.

Coliform counts for assessing water safety

The most frequently used indicator organism is the normally nonpathogenic coliform bacterium *Escherichia coli*. Positive tests for *E. coli* do not prove the presence of enteropathogenic organisms but establish the possibility of the presence of such pathogens. Because *E. coli* is more numerous and easier to grow than the enteropathogens, the test has a built-in safety factor for detecting potentially dangerous fecal contamination. *E. coli* meets many of the criteria for an ideal indicator organism, but there are limitations to its use as such, and various other species have been proposed as additional or replacement indicators of water safety.

For *E. coli* to be a useful indicator organism of fecal pollution, it must be differentiated readily from nonfecal bacteria. The classic separation of *Escherichia coli* and *Enterobacter aerogenes*, which indicates the difference between fecal and nonfecal contamination, is based on the **IMViC** (indole, methyl red, Voges Proskauer, citrate) procedure. This test procedure uses a limited number of tests to differentiate coliform from noncoliform bacteria and can easily and inexpensively be performed in water quality testing laboratories (Table 23.7). Isolated bacteria can also be identified by using more detailed identification procedures,

table 23.7

*IMViC patterns for some bacterial species**

Organism	I	M	Vi	C	44°C
Escherichia coli	+	+	−	−	+
Citrobacter freundii	−	+	−	+	−
Klebsiella pneumoniae	−	−	+	+	−
Enterobacter cloacae	+	−	+	+	−
Enterobacter aerogenes	±	−	+	+	−

I = production of indole from tryptophan
M = methyl-red test, which detects acidic products of mixed acid fermentation pathway
Vi = Voges Proskauer test, which detects acetoin production from butanediol fermentation pathway
C = citrate utilization as the sole carbon source
44°C = growth at 44°C
*Variations of these normal patterns can occur and thus identification of genera such as *Klebsiella* cannot be made reliably on these tests alone.

Presumptive test Inoculate lactose or lauryl tryptose broth, incubate 24 h

Gas produced
presumptive test +ve

No gas produced

Incubate 24 h

Gas produced　　No gas produced
　　　　　　　　completed test −ve

Confirmed test　　Inoculate brilliant-green, lactose bile broth, and/or inoculate Endo or EMB plates,

Incubate 48 h　　　　　Incubate 24 h

No gas produced　　Gas produced　　Typical coliform colonies　　Atypical coliform　　No colonies
　　　　　　　　　　　　　　　　confirmed test +ve　　colonies　　confirmed test −ve

Completed test　　Inoculate lactose broth, incubate 48 h
　　　　　　　　and agar slant, incubate 24 h

Gas produced in broth　　　　　No gas produced
Gram-stain smear from agar slant　　presumptive test −ve

Gm −ve, non-sporing rods　　Spores or Gm +ve rods present
completed test +ve　　inoculate formate-ricinoleate broth, incubate 48 h

Gas produced coliforms　　No gas produced
possibly present　　completed test −ve
repeat tests

Figure 23.21

An outline of the water quality testing procedures for determining numbers of coliforms. Such tests are used for determining the safety of drinking and recreational waters; +ve = positive, −ve = negative.

872

Environmental
quality: biode-
gradation of
wastes and
pollutants

including those diagnostic systems discussed in Chapter 16, but these procedures are relatively expensive.

The conventional test for the detection of fecal contamination involves a three-stage test procedure (Figure 23.21). In the first stage, lactose broth

Figure 23.22

The membrane filtration system and colonies of coliform bacteria growing on the surface of a membrane filter. (From BPS—Millipore Corp.)

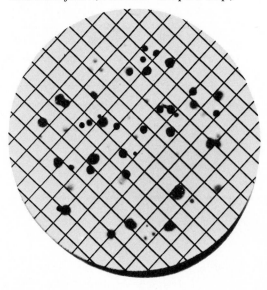

tubes are inoculated with undiluted or appropriately diluted water samples. The tubes showing gas formation are recorded as positive and are used to calculate the most probable number of coliform bacteria in the sample. Gas formation, detected in small inverted test tubes called Durham tubes, gives positive presumptive evidence of contamination by fecal coliforms (called a **presumptive** test). Gas formation in lactose broth at 37°C is not only characteristic of fecal *Salmonella*, *Shigella*, and *E. coli* strains but also of the nonfecal coliform *Enterobacter aerogenes* and some *Klebsiella* species. Therefore, in the second test stage of this procedure, the presence of enteric bacteria is confirmed by streaking samples from the positive lactose broth cultures onto a medium, such as eosin methylene blue agar (EMB). Fecal coliform colonies on this medium acquire a characteristically greenish metallic sheen, *Enterobacter* species form reddish colonies, and nonlactose fermenters form colorless colonies, respectively (called a **confirmed test**). Alternatively, the confirmed test can be accomplished by using brilliant green lactose-bile broth (BGLB). If BGLB is used it is then subcultured onto EMB. Subculturing colonies showing a green metallic sheen on EMB into lactose broth incubated at 35°C should produce gas formation, completing a positive test for fecal coliforms (called a **completed test**).

This three-stage test can be simplified. In the Eijkman test, suitable dilutions are incubated in lactose broth at 44.5°C, a temperature at which fecal coliforms still grow, but nonfecal coliforms are inhibited. Gas formation constitutes a one-step positive test, but precise temperature control is mandatory because temperatures only a few degrees higher inhibit or kill the fecal coliforms. It also is possible to filter known volumes of diluted or undiluted water samples through 0.45-μm pore size bacteriological filters and incubate the filters directly on m-Endo (EMB) or other suitable media. Colonies of fecal coliforms appear with a characteristic metallic sheen and can be counted (Figure 23.22).

Standards for tolerable levels of fecal contamination

The standards for **tolerable limits of fecal contamination** (Table 23.8) vary with the intended type of water use and are somewhat arbitrary with large built-in safety margins, the usefulness of which has been borne out by long experience. The most stringent standards are imposed on the municipal water supplies to be used by many people. Somewhat higher coliform counts are sometimes tolerated in private wells used by only one family because such wells would not become a source of a widespread epidemic. Maintenance of a high drinking water standard does not absolutely exclude the possibility of ingesting enteropathogens with the water but helps keep this possibility to a statistically tolerable minimum. The built-in safety factors are twofold, in that enteropathogens are very likely to be present in much lower numbers than fecal coliforms, and in addition, a few infective bacteria are unlikely to be able to overcome natural body defenses. A minimum infectious dose of several hundred to several thousand bacteria is usually necessary for an actual infection to be established. Drinking water supplies meeting the 1/100 ml coliform standard have never been demonstrated to be the source of a water-borne bacterial infection.

Fecal coliform counts are also used to establish the safety of water in shellfish harvesting and recreational areas. Because shellfish tend to concentrate bacteria and other particles acquired through their filter-feeding activity and are sometimes eaten raw, they can become a source of infection by water-borne pathogens. Therefore, there are relatively stringent standards for waters used for shellfishing. Clinical evidence for infection by en-

table 23.8

United States water standards for coliform contamination

Water use	Maximal permissible coliform count (number/100 ml)
Municipal drinking water	4
Waters used for shellfishing	70
Recreational waters	200

teropathogenic coliforms through recreational use of waters for bathing, wading, and swimming is unconvincing, but as a precaution beaches are usually closed when fecal coliform counts exceed the recreational standard of 1,000/100 ml. Some regional standards require that disinfected sewage discharges not exceed the same limit.

Water quality standards based on fecal coliforms do not account for the possible transmission of viruses associated with fecal matter through municipal water supplies. There is ample evidence, of course, for destructive epidemics by enteroviruses caused by untreated drinking water in various underdeveloped countries. Enteroviruses are somewhat more resistant to disinfection by chlorine or ozone than bacteria and, occasionally, active virus particles are recovered from treated water that meets fecal coliform standards. Thus, the possibility exists that water that meets accepted quality standards may still occasionally be a source of a viral infection. As many as 100 different viral types can be shed in human feces, but practical concern has been mainly with the viruses that cause infectious hepatitis, poliomyelitis, and viral gastroenteritis. Infectious hepatitis is sometimes spread by water supplies, though the more prevalent mode of infection is by the consumption of raw shellfish from fecally contaminated waters. Spread of polio infection through water supplies and/or recreational use of beaches has been suspected in many cases. The situation with viral gastroenteritis is similar. At this point, we can only say that the possibility of an occasional sporadic viral infection through drinking water adequately treated by bacteriological standards cannot be excluded, but there is no hard evidence for any epidemics caused by such water.

Biodegradation of environmental pollutants

Human exploitation of fossil fuel reserves and the production of many novel synthetic compounds, **xenobiotics**, in the twentieth century have introduced into the environment many compounds that microorganisms normally do not encounter and thus are not prepared to biodegrade. Many of these compounds are toxic to living systems, and their presence in aquatic and terrestrial habitats often has serious ecological consequences, including major kills of indigenous biota. The disposal and accidental spillage of these compounds has created serious modern environmental pollution problems, particularly when microbial biodegradation activities fail to remove these pollutants quickly enough to prevent environmental damage. Sewage treatment and water purification systems are usually incapable of removing these substances if they enter municipal water supplies, where they pose a potential human health hazard.

The problems associated with pesticides and their biomagnification through food webs have been discussed in Chapter 22. Here, we will discuss the problems associated with laundry detergents and oil spills. These are representative examples of the growing problem of toxic waste disposal and environmental pollution. Today, there are many sites littered with decaying drums filled with toxic chemicals that are slowly seeping into the environment. The fate and untoward effects of these chemicals remains to determined in the never-ending battle between microbes and human wastes and pollutants.

Alkyl benzyl sulfonates

Alkyl benzyl sulfonates (ABS) are the major components of anionic laundry detergents. Cleaning occurs when ABS molecules form a monolayer around lipophilic droplets or particles, forming an emulsion that can be rinsed out of the fabric with water. The ABS molecule is a surface active molecule, having a polar sulfonate and a nonpolar alkyl end. During laundering ABS molecules orient their nonpolar ends toward lipophilic substances and their sulfonate ends toward the surrounding water. The alkyl portion of the ABS molecule may be linear or branched (Figure 23.23). **Nonlinear ABS** is easier to manufacture and has slightly superior detergent properties than conventional soaps, but nonlinear ABS has proved to be resistant to biodegradation, causing extensive

foaming of rivers receiving ABS-containing wastes. Some communities banned the use of anionic detergents because of their persistence in groundwater supplies used as sources of potable water. It is the methyl branching of the alkyl chain that interferes with biodegradation because the tertiary carbon atoms block the normal β-oxidation sequence. By changing the design of this synthetic molecule to a linear ABS, the blockage can be removed. The detergent industry has switched to linear ABS, free from this blockage and consequently more easily biodegraded. The ABS story is particularly significant because it was one of the first instances when a synthetic molecule was specifically redesigned to remove obstacles to biodegradation while preserving the useful characteristics of the compound. **Biodegradable polymers**, for example, can be synthesized to replace or augment various plastics, including polyethylene, polystyrene, and polyvinylchloride, which are recalcitrant to microbial attack and therefore have been accumulating in the environment. By understanding the role of microbial biodegradation in maintaining environmental quality, human ingenuity has the potential to produce economically profitable synthetic compounds that are biodegradable and that can be disposed of in an environmentally safe manner.

Oil pollution

Over 10 million metric tons of **oil pollutants** enter the marine environment each year as a result of accidental spillages and disposal of oily wastes. Periodically pictures of dying birds floundering in a sea of oil after a major oil spillage face us on the front page of the daily newspaper, evoking images of impending ecological doom. Actually, only a small proportion of all marine oil pollutants comes from major oil spills, and most of the oil pollution problem originates from minor spillages associated with routine operations.

Petroleum is a complicated mixture composed primarily of aliphatic, alicyclic, and aromatic hydrocarbons (Figure 23.24). There are hundreds of individual compounds in every crude oil, the composition of each crude oil varying with its origin. As a result, the fate of petroleum pollutants in the environment is complex. The challenge for microorganisms to degrade all the components of a petroleum mixture is immense. Nevertheless,

Nonlinear alkyl benzyl sulfonate

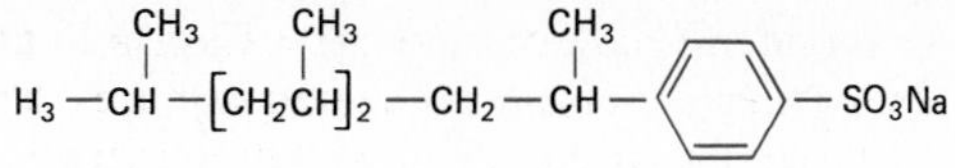

$$CH_3 \quad\quad CH_3 \quad\quad\quad CH_3$$
$$H_3-CH-[CH_2CH]_2-CH_2-CH-\langle\bigcirc\rangle-SO_3Na$$

Linear alkyl benzyl sulfonate

$$H_3C-CH_2-[CH_2]_8-CH_2-CH_2-\langle\bigcirc\rangle-SO_3Na$$

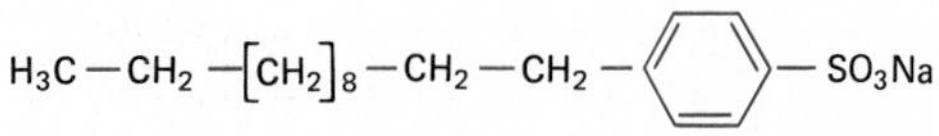

Figure 23.23

Linear and nonlinear alkyl benzyl sulfonate molecules. The methyl branches in the former group of compounds interfere with rapid biodegradation and cause foaming problems in contaminated waters. Linear ABS compounds do not share this problem.

Figure 23.24

Hydrocarbon structures found in petroleum. Homologs, isomers, and combinations result in hundreds of individual hydrocarbon compounds in crude oil samples.

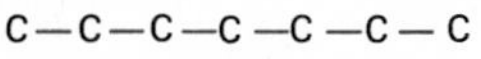

$$C-C-C-C-C-C-C$$
n-Paraffin

Straight chain alkane

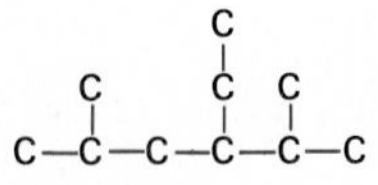

iso-Paraffin

Branched alkane

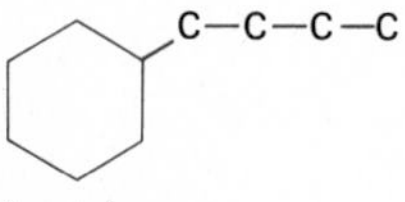

Alkylcyclohexane 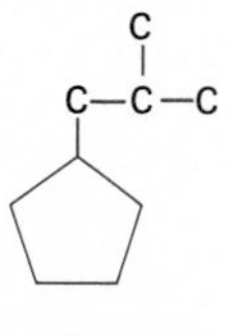   Alkylcyclopentane

Cycloparaffins

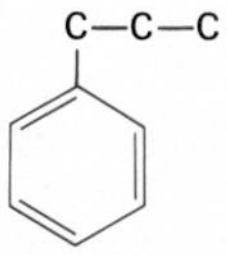

Alkylbenzene 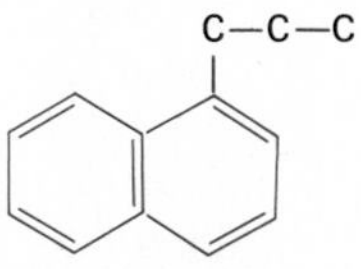   Alkylnaphthalene

Aromatics

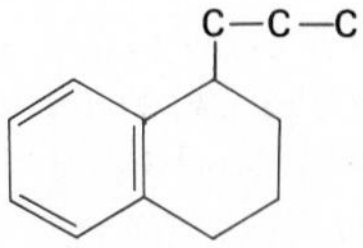

Alkyltetralin 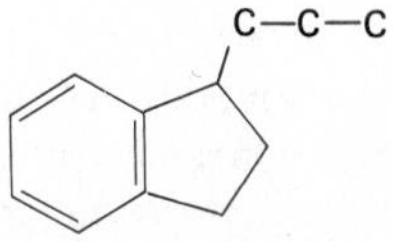   Alkylhydrindene

Cycloparaffinic aromatics

microbial biodegradation of petroleum pollutants is a major process and the reason that the oceans today are not covered with oil. As an example of the ability of microbes to degrade petroleum pollutants, measurements indicate that after the 1978 wreck of the supertanker *Amoco Cadiz* off the coast of France, microorganisms biodegraded 10 tons of oil per day in the impacted area, microbial biodegradation representing the major process responsible for the ecological recovery of the oiled coastal region.

The susceptibility of **petroleum hydrocarbons** to biodegradation is determined by the structure and molecular weight of the hydrocarbon molecule. *n*-Alkanes of intermediate chain length (C_{10}–C_{24}) are degraded most rapidly. Short-chain alkanes (less than C_9) are toxic to many microorganisms, but they generally evaporate rapidly from oil slicks. As alkane chain length increases, so does resistance to biodegradation. Branching, in general, reduces the rate of biodegradation because tertiary and quaternary carbon atoms interfere with the degradation mechanisms or can block degradation altogether. Aromatic compounds, especially of the condensed polynuclear type, are degraded more slowly than alkanes. Alicyclic compounds are frequently unable to serve as sole carbon sources for microbial growth unless they have a sufficiently long aliphatic side chain, but they can be degraded via cometabolism by two or more cooperating microbial strains with complementary metabolic capabilities.

Petroleum has always entered the biosphere by natural seepage but at rates much slower than the forced recovery of petroleum by drilling, which is now estimated to be about 2 billion metric tons per year. The production, transportation, refining, and ultimately the disposal of petroleum and petroleum products results in inevitable environmental pollution. Because the bulk of this load is, of course, heavily centered around off-shore production sites, major shipping routes, and refineries, its input frequently exceeds the self-purification capacity of the receiving waters. Petroleum pollutants in the environment are destructive to birds and marine life and, when driven ashore, cause heavy economic losses due to aesthetic damage to recreational beaches. In addition to killing birds, fish, shellfish, and other invertebrates, oil pollution can also have more subtle effects on marine life. Even at a low ppb concentration, dissolved aromatic components of petroleum can disrupt the chemoreception of some marine organisms. Because feeding and mating responses largely depend on chemoreception, such disruption can lead to elimination of many species from a polluted area even when the pollutant concentration is far below the lethal level. Another disturbing problem is the possibility that condensed polynuclear components of petroleum, some of which are carcinogenic and relatively resistant to biodegradation, may move up marine food chains and taint fish or shellfish. Polynuclear aromatic compounds are among the most resistant components of crude oil to microbial biodegradation and become a major component of the tarry residues left in the sea when oil biodegradative activities slow to a halt.

The successful biodegradative removal of petroleum hydrocarbons from the sea depends on the enzymatic capacities of microorganisms and various abiotic factors. Microbial hydrocarbon biodegradation requires suitable growth temperatures and available supplies of fixed forms of nitrogen, phosphate, and molecular oxygen. In the oceans temperature and nutrient concentrations often limit the rates of petroleum biodegradation. The low concentrations of nitrate and phosphate in seawater are particularly limiting to hydrocarbon biodegradation because petroleum is primarily composed of hydrogen and carbon. For example, after the IXTOC I well blowout, which in 1980 created the largest oil pollution incident, little biodegradation of the oil–water emulsion (mousse) occurred in the surface waters of the Gulf of Mexico because of severe nutrient limitations.

Although many microorganisms can metabolize petroleum hydrocarbons, no single microorganism possesses the enzymatic capability to degrade all, or even most, of the compounds in a petroleum mixture. More rapid rates of degradation occur when there is a mixed microbial community than can be accomplished by a single species. Apparently, the genetic information in more than one organism is required to produce the enzymes needed for extensive petroleum biodegradation. Recombinant DNA technology, however, permits the incorporation of the diverse genetic information extracted from several organisms into a single organism. By using genetic engineering a "superbug" has been created that is capable of degrading many different hydrocarbon structures and that is potentially useful in oil pollution abatement programs. This hydrocarbon-degrading microorganism is the first organism for which a patent has been granted in the United States. The ruling by the Supreme Court that genetic engineering could in essence invent microorganisms has far-reaching consequences for the future use of recombinant DNA technology in the United States to develop microorganisms of economic significance. Microorganisms created by microbiologists have the potential of helping cleanse the environment of man-made pollutants. None of the genetically engineered microbes has been put to the true test of expressing their potential activities in natural ecosystems. Despite the ability to create "superbugs," the usefulness of such organisms in pollution abatement depends on the compatibility of the microbes with their environment. In many cases, environmental factors, rather than the genetic capability of a microorganism, limit the biodegradation of pollutants. Thus, whereas genetically engineered organisms are a useful addition to the arsenal of antipollution measures, there is no panacea for solving human pollution problems.

Postlude

We rely on microorganisms to biodegrade our waste materials and have come to assume that anything thrown out into the environment will disappear, and to an incredibly large extent, this is true. Microorganisms have a vast capacity for rapidly degrading organic materials and thus can be relied upon to act as biological incinerators. The microorganisms associated with waste materials that are released into terrestrial and aquatic ecosystems, however, include pathogens that can cause serious outbreaks of human disease. The multiple usage of water resources as a source of drinking and irrigation water and for recreational purposes mandates careful management of the release of wastes and pollutants into these aquatic ecosystems.

We have a limited freshwater supply, and it is essential that we use that water supply effectively for multiple purposes. It is particularly critical that drinking water supplies be free of human pathogens and that our use of natural water bodies for waste disposals not jeopardize human health and environmental quality. The adequate treatment of sewage to remove pathogens, reduce nutrient levels, and eliminate toxic compounds and the disinfection of potable water supplies—most frequently by using chlorination—have become very important for lowering rates of water-borne infectious diseases. Unfortunately, much of the world's drinking water supplies are contaminated with enteric pathogens due to inadequate sewage treatment and water purification facilities. Even in developed countries over 85 percent of the population in 1970 did not have reasonable access to uncontaminated water supplies. In many regions of the world, there are no sewage treatment nor water purification facilities and human fecal contamination of water supplies is a routine matter. In developed countries, such as the United States, there are established standards of quality based on indicator organisms to ensure that the water is free of potential disease-causing organisms. These water quality standards are subject to review as better standards are revealed to assess more efficiently the safety of water supplies for different purposes. These standards need to be extended to other developing nations to improve worldwide water quality.

Major worldwide efforts are needed to reduce significantly the problem of water pollution. Pollution results when substances enter environments and undesirable reactions occur, such as the depletion of oxygen in the water column of an aquatic ecosystem receiving sewage effluent. The installation of advanced sewage treatment facilities permits the reduction of the BOD of sewage effluents and is essential in the abatement of eutrophication problems in aquatic ecosystems as well as the reduction in the dissemination of human pathogens. These treatment facilities permit aerobic and/or anaerobic biodegradation of organic wastes by a complex assemblage of microorganisms within a controlled environment, where maximal rates of biodegradation can be achieved and environmental damage can be minimized.

When chemicals are released directly into the environment as a result of careless or accidental spillages, we must generally rely on the metabolic activities of microorganisms to remove the polluting substances. When xenobiotics are recalcitrant and microorganisms fail in their role as "biological incinerators," environmental pollutants accumulate. The proper chemical design of synthetic compounds, making them susceptible to microbial attack, is essential for maintaining environmental qualtity. It is possible to create microorganisms, using recombinant DNA technology, capable of degrading complex mixtures of toxic environmental pollutants, and the development of such microorganisms may prove useful in treating the myriad of chemicals in industrial wastes. The human race depends on microorganisms to make the world a better place in which to live and, in the end, we remain dependent on microbial biodegradation for recycling our wastes and maintaining the environmental quality of the biosphere.

1. What are the differences between composting and sanitary landfill operations?

2. Discuss how we treat sewage. What is primary, secondary, and tertiary sewage treatment?

3. What is an indicator organism? Why are coliform counts used to assess the safety of potable water supplies?

4. What is disinfection of a water supply? How is this disinfection of municipal water supplies normally achieved?

5. What is BOD? Why is it important to reduce the BOD of liquid wastes before they are discharged into rivers or lakes?

6. Compare activated sludge and anaerobic digestors for treating sewage. What roles do each play in an integrated liquid waste removal system?

7. How are microorganisms involved in removing oil pollutants? What factors influence the rates at which microorganisms degrade petroleum hydrocarbons?

8. How does its chemical structure influence the rates of biodegradation of a compound? Compare the biodegradability of linear and nonlinear alkyl benzyl sulfonates.

9. How can we use the same waterways as sources of drinking water and for the disposal of liquid wastes?

10. How are the activities of microorganisms essential for the maintenance of environmental quality?

Atlas, R. M. 1981. Microbial degradation of petroleum hydrocarbons: an environmental perspective. *Microbiological Reviews* 45: 180–209.

Atlas, R. M. (ed.). 1984. *Petroleum Microbiology*. Macmillan Publishing Co., New York.

Bonde, G. J. 1977. Bacterial indicators of water pollution. *Advances in Aquatic Microbiology* 1: 273–364.

Brenner, T. E. 1969. Biodegradable detergents and water pollution. *Advances in Environmental Science and Technology* 1: 147–196.

Chakrabarty, A. M. (ed.). 1982. *Biodegradation and Detoxification of Environmental Pollutants*. CRC Press, Boca Raton, Florida.

Curds, C. R., and H. A. Hawkes (eds.). 1975. *Ecological Aspects of Used Water Treatment*. Academic Press, New York.

Dart, R. K., and R. J. Stretton. 1980. *Microbiological Aspects of Pollution Control*. Elsevier Publishing Co., Amsterdam.

Finstein, M. S. 1975. Microbiology of municipal solid waste composting. *Advances in Applied Microbiology* 19: 113–151.

Gaudy, A., and E. Gaudy. 1980. *Microbiology for Environment Science Engineers*. McGraw-Hill Book Co., New York.

Greenberg, A. (ed.). 1980. *Standard Methods for the Examination of Water and Wastewater*. American Public Health Association, Washington, D.C.

Higgins, I. J., and R. G. Burns. 1975. *The Chemistry and Microbiology of Pollution*. Academic Press, London.

LaRiviere, J. W. M. 1977. Microbial ecology of liquid waste treatment. *Advances in Microbial Ecology* 1: 215–259.

Leisinger, T., et al. (eds.). 1982. *Microbial Degradation of Xenobiotics and Recalcitrant Compounds*. Academic Press, New York.

Mitchell, R. (ed.). 1972. *Water Pollution Microbiology*. John Wiley & Sons, New York.

Mitchell, R. 1974. *Introduction to Environmental Microbiology*. Prentice-Hall, Englewood Cliffs, New Jersey.

Skinner, F. A., and G. Sykes (eds.). 1971. *Microbial Aspects of Pollution*. Academic Press, New York.

Taber, W. A. 1976. Wastewater microbiology. *Annual Reviews of Microbiology* 30: 263–277.

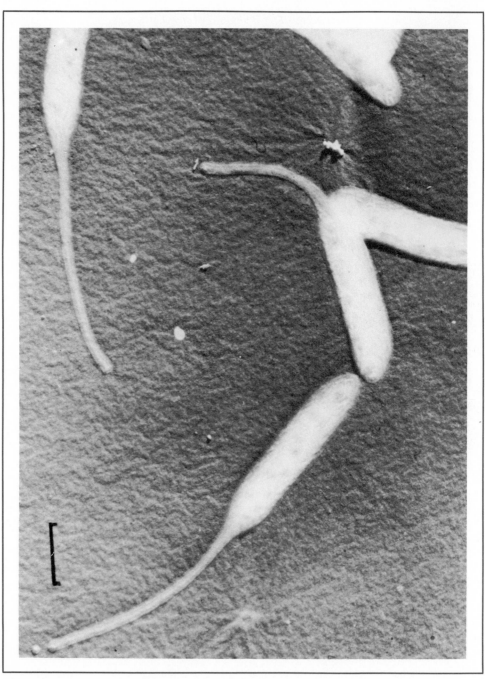

Electron micrograph of Caulobacter.
(Courtesy Jeanne Poindexter, Public Health Research Institute of New York.)

Appendices

SECTION TEN

appendix I
The metric system and some useful conversion factors

appendix II
Glossary

appendix III
Descriptions of bacterial genera

Appendix I. The metric system and some useful conversion factors

Metric system prefixes

pico (p) = 10^{-12}
nano (n) = 10^{-9}
micro (μ) = 10^{-6}
milli (m) = 10^{-3}
centi (c) = 10^{-2}
deci (d) = 10^{-1}
deka (da) = 10
hecto (h) = 10^2
kilo (k) = 10^3
mega (M) = 10^6

Length

1 kilometer (km) = 0.62 mile (mi)
1 mi = 1.609 km
1 meter (m) = 3.28 feet (ft) = 1.09 yard (yd)
1 ft = 0.305 m
1 yd = 0.914 m
1 centimeter (cm) = 0.394 inch (in)
1 in = 2.54 cm
1 ft = 30.5 cm
1 millimeter (mm) = 0.039 in
1 angstrom (Å) = 10^{-10} m

Area

1 hectare (ha) = 10,000 m^2 = 2.471 acres
1 acre = 0.4047 ha
1 sq. km (km^2) = 100 hectares = 0.3861 sq. mile
1 sq. mi = 2.590 km^2
1 sq. m (m^2) = 10,000 cm^2 = 1.1960 sq. yd = 10.764 sq. ft
1 sq. yd = 0.8361 m^2
1 sq. ft = 0.0929 m^2
1 sq. cm (cm^2) = 100 sq. mm = 0.155 sq. in
1 sq. in = 6.4516 cm^2

Mass

1 metric ton (t) =
    1,000 kilograms = 1.10 short tons
1 short ton = 0.91 t
1 kilogram (kg) = 1,000 grams = 2.205 pounds
1 pound = 453.60 g
1 gram (g) = 1,000 milligrams = 0.0353 ounce
1 ounce = 28.35 g
1 milligram (mg) = 10^{-3} g = 0.02 grain
1 microgram (μg) = 10^{-6} g
1 nanogram (ng) = 10^{-9} g
1 picogram (pg) = 10^{-12} g

A–5

The metric system
and some useful
conversion factors

Volume (solids)

1 cubic m (m^3) = 1,000,000 cm^3 = 35.315 cu. ft
    = 1.3080 cu. yd
1 cu. ft = 0.0283 m^3
1 cu. yd = 0.7646 m^3
1 cubic cm (cm^3) = 10^3 mm^3 = 0.0610 cu. in
1 cu. in = 16.387 cm^3

Volume (liquids)

1 liter (l) = 10^3 milliliters = 1.06 quarts (qt)
1 gallon (gal) = 3.785 l
1 kiloliter (kl) = 1,000 liters = 264.17 gallons
1 qt = 0.94 l
1 milliliter (ml) = 10^{-3} liter = 0.034 fluid oz
1 fl. oz. = 29.57 ml.
1 microliter (μl) = 10^{-6} liter

Temperature

degrees Fahrenheit (°F) = 9/5(°C + 32)
degrees centigrade (°C) = 5/9(°F − 32)
0°C = 32°F (freezing point of water)
100°C = 212°F (boiling point of water)

Appendix II. Glossary

α-hemolysis partial hemolysis of red blood cells as evidenced by the formation of a zone of partial clearing (greening) around certain bacterial colonies growing on blood agar.

abiotic referring to the absence of living organisms.

abrasion an area denuded of skin, mucous membrane, or superficial epithelium formed by rubbing or scraping.

abscess a localized accumulation of pus.

absorption the uptake, drinking in, or imbibing of a substance; the movement of substances into a cell; the transfer of substances from one medium to another, e.g., the dissolution of a gas in a liquid; the transfer of energy from electromagnetic waves to chemical bond and/or kinetic energy, e.g., the transfer of light energy to chlorophyll.

accessory pigments pigments that harvest light energy and transfer it to the primary photosynthetic reaction centers.

acellular lacking cellular organization; not having a delimiting cytoplasmic membrane.

acetyl a two-carbon organic radical containing a methyl group and a carbonyl group.

acetyl CoA acetylcoenzyme A; a condensation product of coenzyme A and acetic acid; an intermediate in the transfer of two-carbon fragments, notably in its entrance into the tricarboxylic acid cycle.

achromatic lens an objective lens in which chromatic aberration has been corrected for two colors and spherical aberration for one color.

acid fast the property of those bacteria that retain their initial stain and do not decolorize after washing with dilute acid-alcohol.

acid foods foods with a pH value less than 4.5.

acidic a compound that releases hydrogen (H^+) ions when dissolved in water; a compound that yields positive ions upon dissolution; a solution with a pH value less than 7.0.

acidic stains a stain with a positively charged chromophore (colored portion of the dye) that is attracted to negatively charged cells.

acidophiles microorganisms showing a preference for growth at low pH, e.g., bacteria that grow only at very low pH values, ca. 2.0.

acne an inflammatory disease involving the oil glands and hair follicles of the skin, found chiefly in adolescents and marked by papules or pustules, especially about the face.

acrasin substance secreted by slime molds that initiates aggregation to form a fruiting body; 3′5′-cyclic AMP.

actinomycetes members of an order of bacteria in which species are characterized by the formation of branching filaments and/or true filaments.

activated sludge process an aerobic secondary sewage treatment process using sewage sludge containing active complex populations of aerobic microorganisms to break down organic matter in sewage.

activation energy the energy in excess of ground state that must be added to a molecular system to allow a chemical reaction to start.

active immunity immunity acquired as a result of the individual's own reactions to pathogenic microorganisms or their antigens; attributable to the presence of antibody or immune lymphoid cells formed in response to an antigenic stimulus.

active transport the movement of materials across cell membranes from regions of lower to higher concentration requiring the expenditure of metabolic energy.

acute referring to a disease of rapid onset, short duration, and pronounced symptoms.

adaptive enzymes enzymes produced by an organism in response to the presence of a substrate or a related substance; also called an inducible enzyme.

adenine a purine base component of nucleotides, nucleosides, and nucleic acids.

adenosine a mononucleoside consisting of adenine and D-ribose.

adenosine diphosphate (ADP) a high-energy derivative of adenosine containing two phosphate groups, one phosphate group less than ATP; formed upon hydrolysis of ATP.

adenosine monophosphate (AMP) a compound composed of adenosine and one phosphate group formed upon hydrolysis of ADP.

adenosine triphosphatase (ATPase) an enzyme that catalyzes the reversible hydrolysis of ATP; the membrane-bound form of this enzyme is important in catalyzing the formation of ATP from ADP and inorganic phosphate.

adenosine triphosphate (ATP) a major carrier of phosphate and energy in biological systems, composed of adenosine and three phosphate groups; the free energy released from the hydrolysis of ATP is used to drive many energy-requiring reactions in biological systems.

adhesion factors substances involved in the attachment of microorganisms to solid surfaces; factors that increase adsorption.

adjuvants substances that increase the immunological response to a vaccine; can be added to vaccines to slow down absorption and increase effectiveness; substances that enhance the action of a drug or antigen.

ADP see adenosine diphosphate.

adsorption a surface phenomenon involving the retention of solid, liquid, or gaseous molecules at an interface.

aer- combining form meaning air or atmosphere.

aerial mycelia a mass of hyphae occurring above the surface of a substrate.

aerobes microorganisms whose growth requires the presence of air or free oxygen.

aerobic having molecular oxygen present; growing in the presence of air.

aerobic bacteria bacteria requiring oxygen for growth.

aerosol a fine suspension of particles or liquid droplets sprayed into the air.

aflatoxin a carcinogenic poison produced by some strains of the fungus *Aspergillus flavus*.

agar a dried polysaccharide extract of red algae used as a solidifying agent in various microbiological media.

agglutin an antibody capable of causing the clumping or agglutination of bacteria or other cells.

agglutination the visible clumping or aggregation of cells or particles due to the reaction of surface-bound antigens with homologous antibodies.

AIDS acquired immune deficiency syndrome; an infectious disease characterized by loss of normal immune response system.

akinetes thick-walled resting spores of cyanobacteria and green algae.

alcohols organic compounds characterized by one or more —OH (hydroxyl) groups.

aldehydes a class of substances derived by oxidation from primary alcohols and characterized by the presence of a —CHO group.

algae a heterogeneous group of eukaryotic, photosynthetic, unicellular, and multicellular organisms lacking true tissue differentiation.

alkaline a condition in which hydroxyl (OH^-) ions are in abundance; solutions with a pH of greater than 7.0 are alkaline or basic.

allele any one or more alternative forms of a given gene concerned with the same trait or characteristic; one of a pair or multiple forms of a gene located at the same locus of homologous chromosomes.

allelopathic a substance produced by one organism that adversely affects another organism.

allergen an antigen that induces an allergic response, i.e., a hypersensitivity reaction.

allergy an immunological hypersensitivity reaction; a type of antigen–antibody reaction marked by an exaggerated physiological response to a substance in sensitive individuals.

allochthonous an organism or substance foreign to a given ecosystem.

allosteric effector a substance that can bind to the regulatory site of an allosteric enzyme, resulting in the alteration of the rate of activity of that enzyme.

allosteric enzymes enzymes with a binding and catalytic site for the substrate and a different site where a modulator (allosteric effector) acts.

allosteric inhibitor an allosteric effector that results in reduced rates of activity of an allosteric enzyme.

allotypes antigenically different forms of a given type of immunoglobulin that occur in different individuals of the same species.

amastigotes rounded protozoan cells lacking flagella; a form assumed by some species of Trypanosomatidae, e.g., *Plasmodium*, during a particular stage of development.

amebic dysentery an inflammation of the colon caused by *Entamoeba histolytica*, also known as amebiasis.

amensalism an interactive association between two populations where the interaction is detrimental to one, while not adversely affecting the other.

amino an —NH_2 group.

amino acids a class of organic compounds containing an amino (NH_2) group and a carboxyl (COOH) group.

aminoacyl site site on a ribosome where a transfer RNA molecule attached to a single amino acid initially binds during translation.

amino end the end of a peptide chain or protein with a free amino group, i.e., an alpha amino group not involved in forming the peptide bond.

aminoglycoside antibiotics broad-spectrum antibiotics containing an aminosugar, an amino- or guanido-inositol ring, and residues of other sugars that inhibit protein synthesis, e.g., kanamycin, neomycin, and streptomycin.

ammonification the release of ammonia from nitrogenous organic matter by microbial action.

ammonium ion the cation NH_4^+.

AMP see adenosine monophosphate.

amphibolic pathway a metabolic pathway that has both catabolic and anabolic functions.

amylases enzymes that hydrolyze starch.

anabolism the process of synthesis of cell constituents from simpler molecules, usually requiring the expenditure of energy.

anaerobes organisms that grow in the absence of air or oxygen; organisms that do not use molecular oxygen in respiration.

anaerobic digester a secondary sewage treatment facility used for the degradation of sludge and solid waste.

anaerobic respiration the use of inorganic electron acceptors other than oxygen as terminal electron acceptors for energy-yielding oxidative metabolism.

anamnestic response a heightened immunological response in persons or animals to the second or subsequent administration of a particular antigen given some time after the initial administration; a secondary or memory immune response, i.e., the rapid reappearance of antibody in the blood after the administration of an antibody to which the subject had previously developed a primary immune response.

anaphylactic hypersensitivity an exaggerated immune response reaction of an organism to foreign protein or other substances involving degranulation of mast cells and the release of histamine.

anaphylactic shock a physiological condition of shock resulting from an anaphylactic hypersensitivity reaction, e.g., from a hypersensitive reaction to penicillin or bee bites, which in severe cases death can result within minutes.

anemia a condition characterized by having less than the normal amount of hemoglobin, a condition reflecting a depressed number of circulating red blood cells.

anions negatively charged ions.

anisogametes gametes differing in shape, size, and/or behavior.

annulus a ring-shaped structure; a transverse groove in the cellular envelope of dinoflagellates.

anode the positive terminal of an electrolytic cell.

anorexia absence of appetite.

anoxic absence of oxygen; anaerobic.

anoxygenic photosynthesis photosynthesis that takes place in the absence of oxygen and during which oxygen is not produced; photosynthesis that does not split water and evolve oxygen.

anoxyphotobacteria bacteria that can carry out only anoxygenic photosynthesis.

antagonism the inhibition, injury, or killing of one species of microorganism by another.

anthrax an infectious disease of animals, including humans, cattle, sheep, and pigs, caused by *Bacillus anthracis.*

anti- combining form meaning opposing in effect or activity.

antibacterial agents agents that kill or inhibit the growth of bacteria, e.g., antibiotics, antiseptics, and disinfectants.

antibiotics substances of microbial origin that in very small amounts have antimicrobial activity; current usage of the term extends to synthetic and semi-synthetic substances that are closely related to naturally occurring antibiotics and that have antimicrobial activity.

antibodies glycoprotein molecules produced in the body in response to the introduction of an antigen or hapten that can specifically react with that antigen; also known as immunoglobulins, which are part of the serum fraction of the blood formed in response to antigenic stimulation and which react with antigens with great specificity.

antibody-mediated immunity immunity produced by the activation of the B-lymphocyte population, leading to the production of several classes of immunoglobulins.

anticodon a sequence of three nucleotides in a transfer RNA molecule that is complementary to the codon triplet in messenger RNA.

antifungal agents agents that kill or inhibit the growth and reproduction of fungi; may be fungicidal or fungistatic.

antigen any agent that initiates antibody formation and/or induces a state of active immunological hypersensitivity and that can react with the immunoglobulins that are formed.

antigen–antibody referring to the specific reaction between antigen and complementary antibody molecules.

antihistamines compounds used for treating allergic reactions and cold symptoms by inactivating the histamine that is released as part of the immune response.

antimicrobial agents chemical or biological agents that kill or inhibit the growth of microorganisms.

antiseptics chemical agents used to treat human or animal tissues, usually skin, in order to kill or inactivate those microorganisms capable of causing infection; they are not considered safe for internal consumption.

antisera blood sera that contains antibodies.

antitoxin antibody to a toxin capable of reacting with that poison and neutralizing the specific toxin.

antiviral agents substances capable of destroying or inhibiting the reproduction of viruses.

aplanospores nonmotile, sexual spores.

appendicitis inflammation of the appendix.

apochromatic lens an objective microscope lens in which chromatic aberration has been corrected for three colors and spherical aberration for two colors.

aquatic growing or living in, or frequenting water; a habitat composed primarily of water.

aqueous of, relating to, or resembling water; made from, with, or by water; solutions in which water is the solvent.

archaebacteria prokaryotes with cell walls that lack murein, having ether bonds in their membrane phospholipids; analysis of ribosomal RNA indicates that the archaebacteria represent a primary biological kingdom related to both eubacteria and eukaryotes; considered to be a primitive group of organisms that were among the earliest living forms on earth.

arthropods animals of the invertebrate phylum Arthropoda, many of which are capable of acting as vectors of infectious diseases.

arthrospores spores formed by the fragmentation of hyphae of certain fungi, algae, and cyanobacteria.

ascomycetes members of a class of fungi distinguished by the presence of an ascus, a sac-like structure containing sexually produced ascospores.

ascospores sexual spores characteristic of ascomycetes, produced in the ascus after the union of two nuclei.

ascus the sporangium or spore case of fungi consisting of a single terminal cell.

-ase suffix denoting an enzyme.

asepsis state in which potentially harmful microorganisms are absent; free of pathogens.

aseptic techniques precautionary measures taken in microbiological work and clinical practice to pre-

vent the contamination of cultures, sterile media, etc., and/or infection of persons, animals, or plants by extraneous microorganisms.

asexual lacking sex or functional sexual organs.

asexual reproduction reproduction without union of gametes; formation of new individuals from a single individual.

assay analysis to determine presence, absence, or quantity of one or more components.

assimilation the incorporation of nutrients into the biomass of an organism.

ataxia the inability to coordinate muscular action.

atmosphere the whole mass of air surrounding the earth; a unit of pressure approximating 1×10^6 dynes/cm^2.

ATP see adenosine triphosphate.

ATPase see adenosine triphosphatase.

attenuation any procedure in which the pathogenicity of a given organism is reduced or abolished; reduction in the virulence of a pathogen; control of protein synthesis involving the translation process.

atypical pneumonia bronchopneumonia of unknown etiology, not secondary to any other acute infectious disease; pneumonia that does not respond to penicillin.

autochthonous microorganisms and/or substances indigenous to a given ecosystem; the true inhabitants of an ecosystem; often used to refer to the common microbiota of the body or those species of soil microorganisms that tend to remain constant despite fluctuations in the quantity of fermentable organic matter in the soil.

autoclave apparatus in which objects or materials may be sterilized by air-free saturated steam under pressure at temperatures in excess of 100°C.

autoimmunity immunity or hypersensitivity to some constituent in one's own body; immune reactions with "self" antigens.

autolysins endogenous enzymes involved in the breakdown of certain structural components of the cell during particular phases of cellular growth and development.

autolysis the breakdown of the components of a cell or tissues by endogenous enzymes, usually after the death of the cell or tissue.

autospores sexually formed, nonmotile spores, resembling the parent cell morphologically.

autotrophs organisms whose growth and reproduction are independent of external sources of organic compounds, the requirement for cellular carbon being accomplished by the reduction of CO_2 and the need for cellular energy being met by the conversion of light energy to ATP or the oxidation of inorganic compounds to supply the free energy for the formation of ATP.

autoxidation   the oxidation of a substance upon its exposure to air.

avirulent   lacking virulence; a microorganism lacking the properties that normally promote the ability to cause disease.

axopodia   semipermanent pseudopodia, e.g., the pseudopodia that emanate radially from the spherical cells of heliozoans and some radiolarian species.

β-galactosidase   an enzyme catalyzing the hydrolysis of β-linked galactose within dimers or polymers.

β-hemolysis   complete lysis of red blood cells as shown by the presence of a sharply defined zone of clearing surrounding certain bacterial colonies growing on blood agar.

β-lactamase   an enzyme that attacks a β-lactam ring, such as a penicillinase that attacks the lactam ring in the penicillin class antimicrobials, inactivating such antibiotics.

β-oxidation   metabolic pathway for the oxidation of fatty acids resulting in the formation of acetate and a new fatty acid two carbon atoms shorter than the parent fatty acid.

B cell   a differentiated lymphocyte involved in antibody-mediated immunity.

B memory cells   specifically stimulated B-lymphocytes not actively multiplying but capable of multiplication and production of plasma cells on subsequent antigenic stimuli.

bacteremia   condition in which viable bacteria are present in the blood.

bacteria   members of a group of diverse and ubiquitous prokaryotic, single-celled organisms.

bactericidal   any physical or chemical agent able to kill some types of bacteria.

bacteriological filter   a filter with pores small enough to trap bacteria, about 0.45 μm or smaller, used to sterilize solutions by removing microorganisms during filtration.

bacteriology   the science dealing with bacteria, including their relation to medicine, industry, and agriculture.

bacteriophage   a virus whose host is a bacterium.

bacteriostatic   an agent that inhibits the growth and reproduction of some types of bacteria but need not kill the bacteria.

bacteroids   irregularly shaped forms that some bacteria can assume under certain conditions, e.g., *Rhizobium* in root nodules.

baeocytes   small coccoid reproductive cells produced by some cyanobacteria.

barophiles   organisms that grow best and/or only under conditions of high pressure, e.g., in the ocean's depths.

barotolerant   organisms that can grow under conditions of high pressure but do not exhibit a preference for growth under such conditions.

bartonellosis   Carrion's disease; a bacterial infection of humans endemic to the Andes caused by *Bartonella bacilliformis,* which attacks red blood cells.

basic stain   a dye whose active staining part consists of a cationic, negatively charged group that may be combined with an acid, usually inorganic, that has affinity for nucleic acids.

basidiocarps   the fruiting bodies of basidiomycetes.

basidiomycetes   a group of fungi distinguished by the formation of sexual basidiospores on a basidium.

basidiospores   sexual spores formed on basidiocarps by basidiomycetes.

basidium   club-like structure of basidiomycetes on which basidiospores are borne.

batch process   common simple form of culture in which a fixed volume of liquid medium is inoculated and incubated for an appropriate period of time; cells grown this way are exposed to a continually changing environment; when used in industrial processes, the culture and products are harvested as a "batch" at appropriate times.

benthos   the bottom region of aquatic habitats; collective term for the organisms living along the bottom of oceans and lakes.

Bergey's Manual   reference book describing the established status of bacterial taxonomy; describes bacterial taxa and provides keys and tables for their identification.

binary fission   a process in which two similarly sized and shaped cells are formed by the division of one cell.

binomial nomenclature   the scientific method of naming plants, animals, and microorganisms composed of two names consisting of the species and genus.

bioassay   the use of a living organism to determine the amount of a substance based on the extent of growth or activity of the test organism under controlled conditions.

biochemicals   substances produced by and/or involved in the metabolic reactions of living organisms.

biodegradation the process of chemical breakdown of a substance to smaller products caused by microorganisms or their enzymes.

biodeterioration the chemical or physical alteration of a product that decreases the usefulness of that product for its intended purpose.

biodisk system a secondary sewage treatment system employing a film of active microorganisms rotated on a disk through sewage.

bioenergetics the transfer of energy through living systems; energy transformations in living systems.

biogeochemical cycling the biologically mediated transformations of elements that result in their global cycling including transfer between the atmosphere, hydrosphere, and lithosphere.

bioleaching the use of microorganisms to transform elements so that the elements can be extracted from a material when water is filtered through the material.

biological control the deliberate use of one species of organism to control or eliminate populations of other organisms; used in the control of pest populations.

biological oxygen demand the amount of dissolved oxygen required by aerobic and facultative microorganisms to stabilize organic matter in sewage or water; also known as biochemical oxygen demand.

bioluminescence the generation of light by certain microorganisms.

biomagnification the phenomena of the increase in the concentration of a chemical substance, such as a pesticide, as the substance is passed to higher members of a food chain.

biomass the dry-weight, volume, or other quantitative estimation of organisms; the total mass of living organisms in an ecosystem.

biosphere the part of the earth in which life can exist; all living things together with their environment.

biosynthesis the production (synthesis) of chemical substances by the metabolic activities of living organisms.

biotic of or relating to living organisms; caused by living things.

biotype a variant form of a given species or serotype, distinguishable by biochemical or other means.

biphasic growth curve growth curve reflecting diauxie, i.e., the preferential utilization of one substrate at a given rate before another substrate is metabolized at a different rate.

blastomycosis a chronic mycosis caused by *Blastomyces* in which lesions develop, e.g., in the lungs, bones, and skin.

blight any plant disease or injury that results in general withering and death of the plant without rotting.

blood plasma the fluid portion of the blood minus all blood corpuscles.

blood serum the fluid expressed from clotted blood or clotted blood plasma.

bloodstream the flowing blood in a circulatory system.

blood type an immunologically distinct, genetically determined set of antigens on the surfaces of erythrocytes (red blood cells), defined as A, B, AB, and O.

bloom a visible abundance of microorganisms, generally referring to the excessive growth of algae or cyanobacteria at the surface of a body of water.

BOD see biological oxygen demand.

botulism food intoxication or poisoning that is severe and often fatal, caused by *Clostridium botulinum*.

bradykinin a peptide formed by enzymatic modification of an α-globulin (kininogen) acting as a vasodilator and playing a role in neural recognition of noxious stimuli.

bronchitis inflammation occurring in the mucous membranes of the bronchi often caused by *Streptococcus pneumoniae, Haemophilus influenzae,* and certain viruses.

brucellosis a remittent febrile disease caused by infection with bacteria of the genus *Brucella.*

budding a form of asexual reproduction in which a daughter cell develops from a small outgrowth or protrusion of the parent cell; the daughter cell is smaller than the parental cell.

buffer a solution that tends to resist the change in pH when acid or alkali is added.

Calvin cycle the primary pathway for carbon dioxide fixation in photoautotrophs and chemolithotrophs.

canning method for the preservation of foodstuffs where suitably prepared foods are placed in metal containers that are heated, exhausted, and hermetically sealed.

capillary one of a network of tiny hair-like blood vessels connecting the arteries to the veins.

capsid a protein coat of a virus enclosing the naked nucleic acid.

capsomere the individual protein units that form the capsid of a virus.

capsule a mucoid envelope composed of polypeptides and/or carbohydrates surrounding certain microorganisms; a gelatinous or slimy layer external to the bacterial cell wall.

carbohydrates a class of organic compounds consisting of many hydroxyl (—OH) groups and containing either a ketone or aldehyde.

carbolic acid phenol.

carboxyl end the terminus of a polypeptide chain with a free alpha-carboxyl group not involved in forming a peptide linkage; also known as the C-terminal end.

carboxylic acid an organic chemical having a — COOH functional group.

carcinogen cancer-causing agent.

carditis inflammation of the heart.

caries bone or tooth decay with formation of ulceration; also known as dental caries.

carotenoid pigments a class of pigments usually yellow, orange, red, or purple, that are widely distributed among microorganisms.

carpogonia the basal bodies bearing female gametes in some red algae.

carpospore red algal spore arising from fertilization.

catabolism reactions involving the enzymatic degradation of organic compounds to simpler organic or inorganic compounds with the release of free energy.

catabolite repression repression of the transcription of genes coding for certain inducible enzyme systems by glucose or other readily utilizable carbon sources.

catalases enzymes that catalyze the decomposition of hydrogen peroxide (H_2O_2) into water and oxygen and the oxidation of alcohols to aldehydes by hydrogen peroxide.

catalyst any substance that accelerates a chemical reaction but itself remains unaltered in form and amount.

catalyze to subject to modification, especially an increase in the rate of chemical reaction.

catheterization insertion of a hollow tubular device (a catheter) into a cavity, duct, or vessel to permit injection or withdrawal of fluids.

cathode the electrode at which reduction takes place in an electrolytic cell; a negatively charged electrode.

cations positively charged ions.

cell the functional and structural subunit of living organisms separated from its surroundings by a delimiting membrane.

cell-mediated immunity specific acquired immunity involving T cells, primarily responsible for resistance to infectious diseases caused by certain bacteria and viruses that reproduce within host cells.

cellulase an extracellular enzyme that hydrolyzes cellulose.

cellulose a linear polysaccharide of β-D-glucose.

cell wall structure outside of and protecting the cell membrane, generally containing murein in prokaryotes and composed chiefly of various other polymeric substances, e.g., cellulose or chitin, in eukaryotic microorganisms.

centrifuge an apparatus used to separate by sedimentation particulate matter suspended in a liquid by a centrifugal force.

cephalosporins a heterogeneous group of natural and semisynthetic antibiotics active against a range of Gram positive and negative bacteria, by inhibiting the formation of cross-links in peptidoglycan.

cerebrospinal fluid the fluid contained within the four ventricles of the brain, the subarachnoid space, and the central canal of the spinal chord.

chancre the lesion formed at the site of primary inoculation by an infecting microorganism, usually an ulcer.

chancroid a lesion produced by an infection with *Haemophilus ducreyi* involving the genitalia; a sexually transmitted disease caused by *H. ducreyi*.

chemical oxygen demand the amount of oxygen required to completely oxidize the organic matter in a water sample.

chemical preservative those chemical substances specifically added to prevent the spoilage of a food or the biodeterioration of any substance, by inhibiting microbial growth and/or activity.

chemiosmotic hypothesis the theory that the living cell establishes a proton and electrical gradient across a membrane and that by controlled reentry of protons into the region contained by that membrane, the energy to carry out several different types of endergonic processes may be obtained, including the ability to drive the formation of ATP.

chemoautotrophs microorganisms that derive their energy source from the oxidation of inorganic compounds; organisms that obtain energy through chemical oxidation and use inorganic compounds as electron donors, also known as chemolithotrophs.

chemolithotrophs see chemoautotrophs.

chemostat an apparatus used for continuous-flow culture to maintain bacterial cultures in the log phase of growth based on maintaining a continuous supply of a solution containing a nutrient in limiting quantities that controls the growth rate of the culture.

chemotaxis a type of locomotive response in which the stimulus is a chemical concentration gradient; movement of microorganisms toward or away from a chemical stimulus.

chemotherapy the use of chemical agents for the treatment of disease, including the use of antibiotics to eliminate infecting agents.

chitin a polysaccharide composed of repeating *N*-acetylglucosamine residues, abundant in nature in arthropod exoskeletons and fungal cell walls.

chlamydospores thick-walled, typically spherical or ovoid resting spores produced asexually by certain types of fungi from cells of the somatic hyphae.

chlor- combining form meaning having chlorine as a substitute for hydrogen.

chlorination the process of treating with chlorine as for disinfecting drinking water or sewage.

chlorophyll the green pigment responsible for photosynthesis in plants; the primary photosynthetic pigment of algae and cyanobacteria.

chloroplasts membrane-bound organelles of photosynthetic eukaryotes where the biochemical conversion of light energy to ATP occurs; the sites of photosynthesis in eukaryotic organisms.

chlorosis the yellowing of leaves and/or plant components due to bleaching of chlorophyll, often symptomatic of microbial disease.

cholera an acute infectious disease caused by *Vibrio cholerae* characterized by diarrhea, delirium, stupor, and coma.

chromatic aberration an optical lens defect causing distortion of the image because light of differing wavelengths is focused at differing points instead of at a single focal point.

chromatids fibrils formed from a eukaryotic chromosome when it replicates prior to meiosis or mitosis.

chromatin the deoxyribonucleic acid-protein complex that constitutes a chromosome; the readily stainable protoplasmic substance in the nuclei of cells.

chromosomes structures that contain the nuclear DNA of a cell.

chytrids members of the Chytridiales, which are mainly aquatic fungi that produce zoospores with a single posterior flagellum.

-cide suffix signifying a killer or destroyer.

cilia thread-like appendages, having a 9 + 2 arrangement of microtubules occurring as projections from certain cells, which beat rhythmically, causing locomotion or propelling fluid over surfaces.

ciliophora members of one subphylum of protozoa that possess simple to compound ciliary organelles in at least one stage of their life history: protozoa motile by means of cilia.

circadian rhythm daily cyclical changes that occur in an organism even when it is isolated from the natural daily fluctuations of the environment.

circulatory system the vessels and organs comprising the cardiovascular and lymphatic systems of animals.

cistron the functional unit of genetic inheritance; a segment of genetic nucleic acid that codes for a specific polypeptide chain; synonym for gene.

citric acid cycle see Krebs cycle.

clamp cell connections hyphal structures in many basidiomycetes formed during cell division by dikaryotic hyphal cells, i.e., formed by hyphal cells containing two nuclei of different mating types.

classification the systematic arrangement of organisms into groups or categories according to established criteria.

clone a population of cells derived asexually from a single cell, often assumed to be genetically homologous; a population of genetically identical individuals.

coagglutination an enhanced agglutination reaction based on using antibody molecules whose Fc fragments are attached to cells so that a larger matrix is formed when the Fab portion reacts with other cells for which the antibody is specific.

coagulase an enzyme produced by pathogenic staphylococci, causing coagulation of blood plasma.

cocci spherical or near-spherical bacterial cells, varying in size and sometimes occurring singly, in pairs, in regular groups of four or more, in chains, or irregular clusters.

coccidioidomycosis a disease of humans and domestic animals caused by *Coccidioides immitis,* usually occurring via the respiratory tract.

codon a triplet of adjacent bases in a polynucleotide chain of a messenger RNA molecule that codes for a specific amino acid; the basic unit of the genetic code specifying an amino acid for incorporation into a polypeptide chain.

coenocytic referring to any multinucleate cell, structure, or organism formed by the division of an existing multinucleate entity or formed when nuclear divisions are not accompanied by the formation of dividing walls or septa; multinucleate hyphae.

coenzymes the nonprotein portions of enzymes; biochemicals that act as acceptors or donors of electrons or functional groups during enzymatic reactions.

cofactors see coenzymes.

colicinogenic plasmids plasmids that code for colicins.

colicins proteins produced by some bacteria that inhibit closely related bacteria.

coliforms Gram negative, lactose-fermenting, enteric rods.

colinear two related linear information sequences arranged so that the unit may be taken from the other without rearrangement.

coliphage a virus that infects *E. coli.*

colonization the establishment of a site of microbial reproduction on a material, animal, or person without necessarily resulting in tissue invasion or damage.

colony the macroscopically visible growth of microorganisms on a solid culture media.

Colorado tick fever the only recognized tick-borne viral disease in the United States, an acute febrile disease characterized by sudden onset of fever, headache, and severe muscle pain.

coma a state of unconsciousness.

cometabolism the gratuitous metabolic transformation of a substance by a microorganism growing on another substrate; the cometabolized substance is not incorporated into an organism's biomass, and the organism does not derive energy from the transformation of that substance.

commensalism an interactive association between two populations of different species living together in which one population benefits from the association, while the other is not affected.

common cold an acute self-limiting inflammation of the upper respiratory tract due to a viral infection.

competition an interactive association between two species both of which need some limited environmental factor for growth and thus grow at suboptimal rates because they must share the growth-limiting resource.

competitive inhibition the inhibition of enzyme activity caused by the competition of an inhibitor with a substrate for the active (catalytic) site on the enzyme; impairment of function of an enzyme due to its reaction with a substance chemically related to its normal substrate.

complement group of proteins normally present in plasma and tissue fluids that participates in antigen–antibody reactions, allowing reactions such as cell lysis to occur.

complement fixation the binding of complement to an antigen–antibody complex so that the complement is unavailable for subsequent reactions.

composting the decomposition of organic matter in a heap by microorganisms; a method of solid waste disposal.

condenser lenses the lenses on a microscope used for focusing or directing light from the light source onto the object.

conidia thin-walled asexually derived spores, borne singly or in groups or clusters in specialized hyphae.

conidiophores branches of mycelia-bearing conidia.

conjugation the process in which genetic material is transferred from one microorganism to another involving a physical connection or union between the two cells; a parasexual form of reproduction sometimes referred to synonymously as mating.

conjunctivitis inflammation of the mucous membranes covering the eye, the conjunctiva.

consortium an interactive association between microorganisms generally resulting in combined metabolic activities.

constipation a condition in which the bowels are evacuated at long intervals or with difficulty; the passage of hard, dry stools.

constitutive enzymes enzymes whose synthesis is not altered in response to changes in the environment but rather are continuously synthesized.

contagion the process by which disease spreads from one individual to another.

contagious disease an infectious disease that is communicable to healthy susceptible individuals by physical contact with one suffering from that disease, contact with bodily discharges from that individual, or contact with inanimate objects contaminated by that individual.

contamination the process of allowing uncontrolled addition of microorganisms to an area or substance.

continuous process a process for growing microorganisms without interruption by continual addition of substrates and recovery of products.

contractile vacuoles pulsating vacuoles found in certain protozoa used for the excretion of wastes and the exclusion of water for the maintenance of proper osmotic balance.

controlled atmosphere storage method for preserving foods by altering the concentrations of various gases from normal atmospheric concentrations, usually by increasing the CO_2 concentration and decreasing the O_2 concentration to prevent the spoilage of fruit.

copraphagous capable of growth on fecal matter; feeding on dung or excrement.

cornsteep liquor the concentrated water extract byproduct resulting from the steeping of corn during production of cornstarch, used as a medium adjunct to supply nitrogen and vitamins in industrial fermentations.

corrosion the eating away of a metal resulting from changes in oxidative state.

cortex a layer of a bacterial endospore important in conferring heat resistance to that structure.

coryza an inflammation of the mucous membranes of the nose, usually marked by sneezing and the discharge of watery mucus.

countercurrent immunoelectrophoresis (CIE) a technique based on immunological detection of substances relying on the movement of antibody and antigen toward each other in an electric field resulting in the rapid formation of a detectable antigen–antibody precipitate.

covalent bond a strong chemical bond formed by the sharing of electrons.

cowpox a mild, self-limiting disease caused by a vaccinia virus involving the formation of vesicular lesions in the hands and arms of humans and the udders of cows.

CPE see cytopathic effect.

critical point drying a method for removal of liquids from a microbiological specimen by adjusting temperature and pressure so that the liquid and gas phases of the liquid are in equilibrium with each other; used to minimize disruption of biological structures for viewing by scanning electron microscopy.

cross-feeding the phenomena in which the growth of an organism is dependent on the provision of one or more metabolic factors or nutrients by another organism growing in the vicinity; also termed syntrophism.

crossing over the process in which a break occurs in effect in each of the two adjacent DNA strands, and the exposed 5′-OH and 3′-OH ends unite with the exposed 5′-OH and 3′-OH ends of the adjacent strands so that there is an exchange of homologous regions of DNA.

culture to encourage the growth of particular microorganisms under controlled conditions; a growth of particular types of microorganisms on or within a medium as a result of inoculation and incubation.

curvature of field distortion of a microscopic field of view in which specimens in the center of the field are in clear focus while the peripheral region of the field of view is out of focus.

cutaneous pertaining to the skin.

cyanobacteria prokaryotic, photosynthetic organisms containing chlorophyll *a*, capable of evolving oxygen by the splitting of water; formerly known as blue-green algae.

cyclic photophosphorylation a metabolic pathway involved in the conversion of light energy to chemical energy with the generation of ATP that does not produce the reduced coenzyme, NADPH.

cyst a dormant form assumed by some microorganisms during specific stages in their life cycles, or assumed as a response to particular environmental conditions in which the organism becomes enclosed in a thin- or thick-walled membranous structure, the function of which is either protective or reproductive; a normal or pathological sac with a distinct wall containing fluid.

cystitis an inflammation of the urinary bladder.

cytochromes reversible oxidation–reduction carriers in respiration.

cytokinesis the division of cytoplasm following nuclear division.

cytolysis the dissolution or disintegration of a cell.

cytopathic effects generalized degenerative changes or abnormalities in the cells of a monolayer tissue culture due to infection by a virus.

cytoplasm the living substance of a cell, exclusive of the nucleus.

cytoplasmic membrane the selectively permeable membrane that forms the outer limit of the protoplast, bordered externally by the cell wall in most bacteria.

cytosine a pyrimidine base found in nucleic acid.

cytosis the movement of materials into or out of a cell involving engulfment and formation of a membrane-bound structure rather than by passage through a membrane.

cytoskeleton protein fibers composing the structural support framework of a eukaryotic cell.

cytostomes mouth-like openings of some protozoa, particularly ciliates.

cytotoxic T cells specialized class of T lymphocytes that are able to kill cells as part of the cell-mediated immune response.

cytotoxins substances capable of injuring certain cells without lysis.

D value see decimal reduction time.

dark-field microscope a microscope where the only light seen in the field of view is reflected from the object under examination, resulting in a light object on a dark background.

deaminase an enzyme involved in the removal of an amino group from a molecule, liberating ammonia.

deamination the removal of an amino group from a molecule, especially an amino acid.

decarboxylase an enzyme that liberates carbon dioxide from the carboxyl group of a molecule by hydrolysis.

decarboxylation the splitting off of one or more molecules of carbon dioxide from organic acids, especially amino acids.

decimal reduction time the time required at a given temperature to heat-inactivate or kill 90 percent of a given population of cells or spores.

decomposers organisms, often bacteria or fungi, in a community that convert dead organic matter into inorganic nutrients.

defined medium the material supporting microbial growth in which all the constituents, including trace substances, are quantitatively known; a mixture of known composition used for culturing microorganisms.

dehydrogenase an enzyme that catalyzes the oxidation of a substrate by removal of hydrogen.

denaturation the alteration in the characteristics of an organic substance, especially a protein, by physical or chemical action; the loss of enzymatic activity due to modification of the tertiary protein structure.

dendrograms graphic representations of taxonomic analyses showing the relationships between the organisms examined.

dengue fever a human disease caused by a togavirus and transmitted by mosquitos, characterized by fever, rash, and severe pain in joints and muscles.

denitrification the formation of gaseous nitrogen or gaseous nitrogen oxides from nitrate or nitrite.

dental caries tooth decay.

deoxyribonucleic acid the carrier of genetic information; a type of nucleic acid occurring in cells, containing adenine, guanine, cytosine, and thymine, and D-2-deoxyribose linked by phosphodiester bonds.

deoxyribose a five-carbon sugar having one oxygen less than the parent sugar ribose; a component of DNA.

derepress the regulation of transcription by reversibly inactivating a repressor protein.

dermatitis an inflammation of the skin.

dermatophytes fungi characterized by their ability to metabolize keratin and capable of growing on the skin surface, causing disease.

desert a region of low rainfall; a dry region; a region of low biological productivity.

detergent a synthetic cleaning agent containing surface active agents that do not precipitate in hard water; a surface active agent having a hydrophilic and hydrophobic portion.

detrital food chain a food chain based on the biomass of decomposers rather than on primary producers.

detritus waste matter and biomass produced from decompositional processes.

deuteromycetes fungi with no known sexual stage; also known as Fungi Imperfecti.

diagnostic table a table of distinguishing features used as an aid in the identification of unknown organisms.

diarrhea a common symptom of gastrointestinal disease, characterized by increased frequency and fluid consistency of stools.

diatomaceous earth a siliceous material composed largely of fossil diatoms, used in microbiological filters and industrial processes.

diatoms unicellular algae having a cell wall composed of silica, the skeleton of which persists after the death of the organism.

diauxie the phenomena in which, given two carbon sources, an organism preferentially metabolizes one completely before utilizing the other.

dichotomous key a key for the identification of organisms, using steps with opposing choices until a final identification is achieved.

differential blood count procedure for determining the ratios of various types of blood cells, used to determine the relative numbers of white blood cells as a diagnostic indication of an infectious process.

differential media bacteriological media on which growth of specific types of organisms leads to readily visible changes in the appearance of the media so that the presence of specific types of microorganisms can be determined.

diffraction the breaking up of a beam of light into bands of differing wavelength due to interference.

dikaryotes cells with two different nuclei as a result of the fusion of two cells.

dimorphism the property of existing in two distinct structural forms, e.g., fungi that occur in filamentous and yeast-like forms under different conditions.

dinoflagellates algae of the class Pyrrhophyta, primarily unicellular marine organisms, possessing flagella.

diphtheria an acute, communicable human disease caused by *Corynebacterium diphtheriae*.

diplococci cocci occurring in pairs.

diploid having double the haploid number of chromosomes; having a duplicity of genes.

dipole moment the polarity resulting from the separation of electronic charges in chemical bonds resulting from an unequal distribution of electrons.

disaccharides carbohydrates formed by the condensation of two monosaccharide sugars.

disinfectants chemical agents used for disinfection.

disinfection the destruction, inactivation, or removal of those microorganisms likely to cause infection or give rise to other undesirable effects.

dispersal breaking up and spreading in various directions, e.g., the spread of microorganisms from one place to another.

dissemination the scattering or dispersion of microorganisms or disease, e.g., the spread of disease associated with the dispersal of pathogens.

dissociation separation of a molecule into two or more stable fragments; a change in colony form often occurring in a new environment, associated with modified growth or virulence.

DNA see deoxyribonucleic acid.

DNA homology the degree of similarity of base sequences in DNA from different organisms.

DNA polymerases enzymes that catalyze the phosphodiester bonds in the formation of DNA.

dolipore septa the thick internal transverse openings between cell walls of basidiomycetes.

donor any cell that contributes genetic information to another.

dormant an organism or spore that exhibits minimal physical and chemical change over an extended period of time but that remains alive.

drugs substances used in medicine for the treatment of disease.

dysfunctional immunity an immune response that produces an undesirable physiological state, e.g., an allergic reaction, or the lack of an immune response resulting in a failure to protect the body against infectious or toxic agents.

ecology the study of the interrelationships between organisms and their environments.

ecosystem a functional self-supporting system that includes the organisms in a natural community and their environment.

ectomycorrhizae a stable symbiotic association between a fungus and the root of a plant where the fungal hyphae occur outside the root and between the cortical cells of the root.

effluent the liquid discharge from sewage treatment and industrial plants.

electron a negatively charged subatomic particle that orbits the positively charged nucleus of an atom.

electron acceptors substances that accept electrons during oxidation–reduction reactions.

electron donors substances that give up electrons during oxidation–reduction reactions.

electron transport chain a series of oxidation–reduction reactions in which electrons are transported from a substrate through a series of intermediate electron carriers to a final acceptor establishing an electrochemical gradient across a membrane that results in the formation of ATP.

electrophoresis the movement of charged particles suspended in a liquid under the influence of an applied electron field.

ELISA see enzyme-linked immunosorbent assay.

EMB agar see eosin methylene blue agar.

Embden–Meyerhof pathway a specific glycolytic pathway; a sequence of reactions in which glucose is broken down to pyruvate.

embryonated eggs hen or duck eggs containing live embryos used for culturing viruses and preparing tissue cultures.

encephalitis an inflammation of the brain.

end- combining form indicating within.

endemic peculiar to a certain region, e.g., a disease that occurs regularly in an area.

endergonic a chemical reaction with a positive ΔG; a chemical reaction requiring input of free energy.

endocardium the membrane lining the interior of the heart.

endocytosis the movement of materials into a cell by cytosis.

endogenous produced within; due to internal causes; pertaining to the metabolism of internal reserve materials.

endonuclease an enzyme that catalyzes the cleavage of DNA, normally cutting the DNA at specific sites.

endophytic a photosynthetic organism living within another organism.

endoplasmic reticulum the extensive array of internal membranes in a eukaryotic cell involved in coordinating protein synthesis.

endospores thick-walled spores formed within a parent cell; in bacteria, heat-resistant spores.

endosymbiotic a symbiotic association in which one organism penetrates and lives within the cells or tissues of another organism.

endothelial a single layer of thin cells lining internal body cavities; the inner layer of the seed coat of some plants.

endothermic a chemical reaction in which energy is consumed; a chemical reaction requiring an input of heat energy.

endotoxin toxic substances found as part of some bacterial cells; the lipopolysaccharide component of the cell envelope of Gram negative bacteria, also known as LPS.

end product the chemical compound that is the final product in a particular metabolic pathway.

enrichment culture any form of culture in a liquid medium that results in an increase in the numbers of a given type of organism while minimizing the growth of any other organism present.

enter- combining form meaning the intestine.

enteric of or pertaining to the intestines.

enterotoxins toxins specific for cells of the intestine, causing intestinal inflammation and producing the symptoms of food poisoning.

enthalpy the total heat of a system; ΔH.

entomogenous fungi fungi living on insects; fungal pathogens of insects.

entropy that portion of the energy of a system that cannot be converted to work; ΔS.

enzymatic reactions chemical reactions catalyzed by enzymes.

enzyme-linked immunosorbent assay (ELISA) a technique used for detecting and quantifying specific serum antibodies.

enzymes proteins that function as efficient biological catalysts, increasing the rate of a reaction without altering the equilibrium constant by lowering the energy of activation.

eosin methylene blue agar a medium used for the detection of coliform bacteria; the growth of Gram

positive bacteria is inhibited on this medium, and lactose fermenters produce colonies with a green-metallic sheen.

eosinophil a white blood cell having an affinity for eosin or any acid stain.

epi- prefix meaning upon, beside, among, above, or outside.

epidemic an outbreak of infectious disease among a human population in which for a limited time a high proportion of the population exhibits overt disease symptoms.

epidemiology the study of the factors and mechanisms that govern the spread of disease within a population, including the interrelationships between a given pathogenic organism, the environment, and populations of relevant hosts.

epifluorescence microscopy form of microscopy employing stains that fluoresce when excited by light of a given wavelength, emitting light of a different wavelength; exciter filters are used to produce the proper excitation wavelength, and barrier filters are used so that only fluorescing specimens are visible.

epigenetic direct products derived from an organism's genome, e.g., ribosomal RNA.

epilimnion the warm upper surface layer of an aquatic environment.

epiphytes organisms growing on the surface of another organism, e.g., bacteria growing on the surface of an algal cell.

episomes segments of DNA capable of existing in two alternate forms, one replicating autonomously in the cytoplasm, the other replicating as part of the bacterial chromosome.

epitheca the larger of the two parts of the cell wall (frustule) of a diatom.

epizootic an epidemic outbreak of infectious disease among animals other than humans.

equilibrium a state of balance, a condition in which opposing forces equalize one another so that no movement occurs; in a chemical reaction, the condition where forward and reverse reactions occur at equal rates so that no net change occurs; when a reaction is at equilibrium the amounts of reactants and products remain constant.

equilibrium constant describes the relationship among concentrations of the substances within an equilibrium system regardless of how the equilibrium condition is achieved.

ergotism the condition of intoxication that results from the ingestion of grain contaminated by ergot alkaloids produced by *Claviceps purpurea*.

erythrocytes red blood cells.

erythromycin an antibiotic produced by a strain of *Streptomyces* that inhibits protein synthesis.

estuary a water passage where the ocean tide meets a river current; an arm of the sea at the lower end of a river.

ethanolic fermentation a type of fermentation where glucose is converted to ethanol and carbon dioxide.

etiologic agent causative organism; organism causing disease.

etiology the study of the causation of disease.

eubacteria prokaryotes other than archaebacteria.

eukaryotes cellular organisms having a membrane-bound nucleus within which the genome of the cell is stored as chromosomes composed of DNA; eukaryotic organisms include algae, fungi, protozoa, plants, and animals.

euphotic the top layer of water through which sufficient light penetrates to support the growth of photosynthetic organisms.

eutrophication the enrichment of natural waters with inorganic materials, especially nitrogen and phosphorus compounds, that support the excessive growth of photosynthetic organisms.

evolution the directional process of change of organisms by which descendants become distinct in form and/or function from their ancestors.

exergonic reactions reactions accompanied by a liberation of free energy.

exo- prefix indicating outside, outside layer, or out of.

exoenzymes enzymes that occur either attached to the outer surface of the cell membrane or in the periplasmic space; enzymes released into the medium surrounding a cell, including enzymes that attack extracellular polymers by sequentially removing units from one end of a polymer chain.

exogenous due to an external cause; not arising from within the organism.

exon the region of a eukaryotic genome that is known to encode the information for protein or RNA macromolecules or the regulation of gene expression; a segment of eukaryotic DNA that codes for a region of RNA that is not excised during post-transcriptional processing.

exonucleases enzymes that progressively remove the terminal nucleotides of a polynucleotide chain.

exothermic a chemical reaction that evolves heat.

exotoxins protein toxins secreted by living microorganisms into the surrounding medium.

exponential phase that period during the growth cycle of a population when growth is maximal and constant and there is a logarithmic increase in population size.

extracellular external to the cells of an organism.

extraterrestrial originating or existing outside the earth or its atmosphere.

exudate viscous fluid containing blood cells and debris that accumulates at the site of an inflammation or lesion.

F plasmid fertility plasmid coding for donor strain, includes genes for the formation of the F-pilus.

F value the number of minutes required to heat inactivate or kill an entire population of cells or spores in an aqueous solution at 121°C.

Fab fragment either of two identical fragments produced when an immunoglobulin is cleaved by papain; the antigen-binding portion of an antibody, including the hypervariable region.

FAD flavin adenine dinucleotide.

$FADH_2$ reduced flavin adenine dinucleotide.

Fc fragment the remainder of the molecule when an immunoglobulin is cleaved and the Fab fragment separated; the crystallizable portion of an immunoglobulin molecule containing the constant region; the end of an immunoglobulin that binds with complement.

facultative anaerobes microorganisms capable of growth under either aerobic or anaerobic conditions; bacteria capable of both fermentative and respiratory metabolism.

family a taxonomic group; the principal division of an order; the classification group above a genus.

fastidious an organism difficult to isolate or culture on ordinary media because of its need for special nutritional factors; an organism having stringent physiological requirements for growth and survival.

fatty acids straight chains of carbon atoms with a COOH at one end in which most of the carbons are attached to hydrogen atoms.

feedback inhibition a cellular control mechanism by which the end product of a series of metabolic reactions inhibits the activity of an earlier enzyme in the sequence of metabolic transformations, and thus, when the end product accumulates, its further production ceases.

fermentation a mode of energy-yielding metabolism that involves a sequence of oxidation–reduction reactions in which an organic substrate and the organic compounds derived from that substrate serve as the primary electron donor and the terminal electron acceptor; in contrast to respiration, there is no requirement for an external electron acceptor to terminate the metabolic sequence.

fermenter an organism that carries out fermentation.

fermentor a reaction chamber in which a fermentation reaction is carried out; a reaction chamber for growing microorganisms used in industry for a batch process.

fertility fruitfulness; the reproductive rate of a population; the ability to support life; the ability to reproduce.

fever the elevation of the body temperature above normal.

fibrin the insoluble protein formed from fibrogen by the proteolytic action of thrombin during normal blood clotting.

fibrinogen a protein in human plasma synthesized in the liver that is the precursor of fibrin, used to increase the coagulability of blood.

fibrolysin a proteolytic enzyme capable of dissolving or preventing the formation of a fibrin clot.

filament any elongated thread-like bacterial cell.

filterable virus an obsolete term used to describe infectious agents that were able to pass through bacteriological filters.

filtration the separation of microorganisms from the medium in which they are suspended by passage of a fluid through a filter with pores small enough to trap the microbes.

fission a type of asexual reproduction in which a cell divides to form two or more daughter cells.

flagella flexible, relatively long appendages on cells used for locomotion.

flagellates organisms having flagella; one of the major divisions of protozoans, characterized by the presence of flagella.

flagellin soluble, globular proteins constituting the subunits of bacterial flagella.

flat-field objective a microscope lens that provides an image in which all parts of the field are simultaneously in focus; an objective lens with minimal curvature of field.

floc a mass of microorganisms caught together in a slime produced by certain bacteria, usually found in waste treatment plants.

fluorescence the emission of light by certain substances upon absorption of an exciting radiation; the emitted light is of a different wavelength than the wavelength of the excitation radiation.

fluorescence microscope a microscope in which the microorganisms are stained with some form of fluorescent dye and observed by illumination with short-wavelength light, e.g., ultraviolet light.

fome any inanimate object that can act as a carrier of infectious agents.

fomites objects and materials that have been associated with infected persons or animals and that potentially harbor pathogenic microorganisms.

food additive a substance or mixture of substances other than the basic food stuff, which is intentionally present in food as a result of any aspect of production, processing, storage, or packaging.

food infection disease resulting from the ingestion of food or water containing viable pathogens that can establish an infectious disease, e.g., gastroenteritis from ingestion of food containing *Salmonella*.

food intoxication disease resulting from the ingestion of toxins produced by microorganisms that have grown in a food.

food poisoning a general term applied to all stomach or intestinal disorders due to food contaminated with certain microorganisms, their toxins, chemicals, or poisonous plant materials; disease resulting from the ingestion of toxins produced by microorganisms that have grown in a food.

food preservation the prevention or delay of microbial decomposition, or self-decomposition of food and prevention of damage because of insects, animals, mechanical causes, etc.; the delay or prevention of food spoilage.

food spoilage the deterioration of a food that lessens its nutritional value or desirability, often due to the growth of microorganisms that alter the taste, smell, or appearance of the food, or the safety of ingesting that food.

food web an interrelationship among organisms by which energy is transferred from one organism to another, where each organism consumes the preceding one and in turn is eaten by the following member of the sequence.

formalin a 40 percent solution of formaldehyde, a pungent-smelling, colorless gas used for fixation and preservation of biological specimens and as a disinfectant.

Forssman antigen a heat-stable glycolipid; a heterophile antigen; an immunologically related antigen found in unrelated species.

frame shift mutation a type of mutation that causes a change in the three-base sequences read as codons, i.e., a change in the phase of transcription arising from the addition or deletion of nucleotides in numbers other than three or multiples of three.

free energy the energy available to do work, particularly in causing chemical reactions; ΔG.

freeze etching a technique used to examine the topography of a surface exposed by fracturing or cutting a deep-frozen cell, making a replica and removing the biological material; used in transmission electron microscopy.

fruiting body a specialized fungal structure that bears sexually or asexually derived spores.

frustules the silicaceous cell walls of a diatom.

fungi a group of diverse and widespread unicellular and multicellular eukaryotic organisms, lacking chlorophyll, usually bearing spores, and often filamentous.

fungicides agents that kill fungi.

Fungi Imperfecti fungi with septate hyphae that reproduce only by means of conidia, lacking a known sexual stage; Deuteromycetes.

fungistasis the active prevention or hindrance of fungal growth by a chemical or physical agent.

galls abnormal plant structures formed in response to parasitic attack by certain insects or microorganisms; tumor-like growths of plants in response to an infection.

gametangium a structure that gives rise to gametes or that in its entirety functions as a gamete.

gametes haploid reproductive cells or nuclei, the fusion of which during fertilization leads to formation of a zygote.

gammaglobulin any of the serum proteins with antibody activity.

gas gangrene a disease condition involving tissue death that develops when certain species of toxin-producing bacteria grow in anaerobic wounds or necrotic tissues.

gasohol a mixture of gasoline and ethanol used as a fuel.

gastroenteritis an inflammation of the stomach and intestine.

gastroenterocolitis an inflammation of the gastrointestinal tract accompanied by the formation of pus and blood in the stools.

gastrointestinal syndrome gastroenteritis associated with nausea, vomiting, and/or diarrhea.

gastrointestinal tract the stomach, intestines, and accessory organs.

gas vacuoles membrane-limited, gas-filled vacuoles that occur commonly in groups in the cells of

a number of cyanobacteria and certain other bacteria.

gelatin a protein obtained from skin, hair, bones, tendons, etc., used in culture media for the determination of a specific proteolytic activity of microorganisms.

gelatinase a hydrolytic enzyme capable of liquefying gelatin.

gene a sequence of nucleotides that specifies a particular polypeptide chain.

generation time the time required for the cell population or biomass to double.

genetic engineering the deliberate modification of the genetic properties of an organism either through the selection of desirable traits, the introduction of new information on DNA, or both; the application of recombinant DNA technology.

genetics the science dealing with inheritance.

genital herpes a sexually transmitted disease caused by a herpes virus; an infection by herpes simplex virus marked by the eruption of groups of vesicles.

genitourinary tract the combined urinary and genital systems; the combined reproductive system and urine excretion system, including the kidneys, ureters, urinary bladder, urethra, penis, prostate, testes, vagina, fallopian tubes, and uterus.

genome the complete set of genetic information, as contained in a haploid set of chromosomes.

genotype the genetic information contained in the entire complement of alleles.

genus a taxonomic group, next above species and forming the principal subdivisions of the family.

geometric isomers the formation of nonequivalent structures based on the particular positions at which ligands are attached to a central atom.

germ any microorganism, especially any of the pathogenic bacteria.

German measles rubella; an acute systemic infectious disease of humans caused by rubella viruses invading via the mouth or nose, characterized by a rash.

germ-free animal an animal with no normal microbiota; all its surfaces and tissues are sterile, and it is maintained in that condition by being housed and fed in a sterile environment.

germicide a microcidal disinfectant.

germination a degradative process by which an activated spore becomes metabolically active, involving hydrolysis and depolymerization.

giardiasis an infection with *Giardia* protozoa in the human intestine.

gliding motility movement by some bacteria involving contact with solid surfaces.

globular protein the general name for a group of water-soluble proteins.

glomerulonephritis an inflammation of the filtration region of the kidneys.

gluconeogenesis the biosynthesis of glucose from noncarbohydrate substrates.

glucose the monosaccharide sugar $C_6H_{12}O_6$.

glycogen a nonreducing polysaccharide of glucose found in many tissues and stored in the liver where it is converted when needed into sugar.

glycolysis an anaerobic process of glucose dissemination by a sequence of enzyme-catalyzed reactions to a pyruvic acid.

glycoproteins a group of conjugated proteins that upon decomposition yield a protein and a carbohydrate.

glycosidic bonds bonds in disaccharides and polysaccharides formed by the elimination of water.

glyoxylate cycle a metabolic shunt within the tricarboxylic acid cycle involving the intermediate glyoxylate.

Golgi apparatus a membranous organelle of eukaryotic organisms involved with the formation of secretory vesicles and the synthesis of complex polysaccharides.

gonorrhea a sexually transmitted disease caused by *Neisseria gonorrhoeae;* specific infectious inflammation of the mucous membrane of the urethra and adjacent cavities caused by *N. gonorrhoeae.*

Gram stain differential staining procedure by which bacteria are classified as Gram negative or positive depending on whether they retain or lose the primary stain when subject to treatment with a decolorizing agent, the staining procedure reflects underlying structural differences in the cell walls of Gram negative and Gram positive bacteria.

grana a membranous unit formed by stacks of thylakoids.

granules small intracellular particles, usually staining selectively.

granuloma inguinale the chronic destructive ulceration of external genitalia due to *Donovania granulomatis.*

grazers organisms that prey upon primary producers; protozoan predators that consume bacteria nondiscriminately; filter-feeding zooplankton.

groundwater all subsurface water.

growth any increase in the amount of actively metabolic protoplasm accompanied by an increase in cell number, cell size, or both.

growth curve a curve obtained by plotting increase in size or number of microorganisms against elapsed time.

growth factors any compound, other than the carbon and energy source, that an organism requires and cannot synthesize.

growth rate the increase in the number of microorganisms per unit time.

guanine a purine base that occurs naturally as a fundamental component of nucleic acids.

Guillain-Barré syndrome acute febrile polyneuritis; a diffuse neuron paresis.

H antigen a type of flagella antigen found in certain bacteria.

habitat a location where living organisms occur.

halophiles organisms requiring NaCl for growth; extreme halophiles growing in concentrated brines.

haploid a single set of homologous chromosomes; having half the normal diploid number of chromosomes.

hapten a substance that elicits antibody formation only when combined with other molecules or particles but that can react with preformed antibodies.

helix a spiral structure.

hemagglutination the agglutination or clumping of red blood cells.

heme an iron-containing porphyrin ring occurring in hemoglobin.

hemocytometer a counting chamber used for estimating the number of blood cells.

hemoglobin the iron-containing, oxygen-carrying molecule of red blood cells containing four polypeptides in a heme group.

hemolysin a substance that lyses erythrocytes.

hemolysis the lytic destruction of red blood cells and the resultant escape of hemoglobin.

hemorrhagic showing evidence of bleeding, the tissue becomes reddened by the accumulation of blood that has escaped from capillaries into the tissue.

hepatitis inflammation of the liver.

herbicides chemicals used to kill weeds.

heritable any characteristic that is genetically transmissible.

herpes simplex infections localized blistery skin rash caused by herpes simplex virus, usually on the lips or the genitalia.

hetero- combining form meaning other, other than usual, different.

heterocysts cells that occur in the trichomes of some filamentous cyanobacteria that are the sites of nitrogen fixation.

heteroduplex an intermediate form of DNA occurring during homologous recombination.

heterogamy the conjugation of unlike gametes.

heterogeneous composed of different substances; not homologous.

heterogeneous RNA high-molecular-weight RNA formed by direct transcription in eukaryotes that is then processed enzymatically to form messenger RNA.

heterokaryon a cell containing genetically different nuclei, occurring in some fungal hyphal cells.

heterolactic fermentation fermentation of glucose that produces lactic acid, acetic acid and/or ethanol, and carbon dioxide, carried out by *Leuconostoc* and some *Lactobacillus* spp.

heterophile antibody antibody that reacts with heterophile antigens, commonly found in sera of individuals with infectious mononucleosis.

heterophile antigens immunologically related antigens found in unrelated species.

heterotrophs organisms requiring organic compounds for growth and reproduction, the organic compounds serving as sources of carbon and energy.

Hfr see high-frequency recombinant.

high-frequency recombinant a bacterial strain that exhibits a high rate of gene transfer and recombination during mating; the F plasmid is integrated into the bacterial chromosome.

histamine a physiologically active amine that plays a role in the inflammatory response.

histocompatibility antigens genetically determined isoantigens present on the lipoprotein membranes of nucleated cells of most tissues that incite an immune response when grafted onto a genetically disparate individual and thus determine the compatibility of tissues in transplantation.

histones basic proteins rich in arginine and lysine that occur in close association with the nuclear DNA of eukaryotic organisms.

histoplasmosis a disease of humans and animals caused by the fungus *Histoplasma capsulatum,* characterized by fever, anemia, leukopenia, and emaciation, primarily involving the reticuloendothelial system.

hnRNA see heterogeneous RNA.

holdfast a structure that allows certain algae and bacteria to remain attached to the substratum.

homo- combining form denoting like, common, or same.

homolactic fermentation the fermentation of glucose that produces lactic acid as the sole fermentation product, carried out by certain species of *Lactobacillus.*

homologous pertaining to the structural relation between parts of different organisms due to evolutionary development of the same or corresponding part; a substance of identical form or function.

homologous recombination recombination of regions of DNA containing alleles of the same genes.

homology genetic relatedness.

host a cell or organism that acts as the habitat for the growth of another organism; the cell or organism upon or in which parasitic organisms live.

hot springs thermal springs with water above 98°C.

HTST process high-temperature, short-time pasteurization process at a temperature of at least 71.5°C for at least 15 sec., the most widely used commercial pasteurization.

humic acids any of the various organic acids obtained from humus, the soil matter the origin of which no longer is identifiable; complex polynuclear aromatic compounds comprising the soil organic matter.

humoral referring to the body fluids.

humoral immune defense system see antibody-mediated immunity

humus the organic portion of the soil remaining after microbial decomposition.

hyaluronidase enzymes that catalyze the breakdown of hyaluronic acid.

hybridomas cells formed by the fusion of lymphocytes (antibody precursors) with myeloma (tumor) cells, which produces rapidly growing cells that secrete monoclonal antibodies.

hydr- combining form meaning water.

hydrocarbons compounds composed only of hydrogen and carbon.

hydrogen bond a weak attraction between an atom that has a strong attraction for electrons and a hydrogen atom that is covalently bonded to another atom that attracts the electron of the hydrogen atom.

hydrolysis the chemical process of decomposition involving the splitting of a bond and the addition of the elements of water.

hydrophilic a substance having an affinity for water.

hydrophobia fear of water, one of the symptoms of rabies.

hydrophobic a substance lacking an affinity for water, not soluble in water.

hydrosphere the aqueous envelope of the earth, including bodies of water and aqueous vapor in the atmosphere.

hydrostatic pressure pressure exerted by the weight of a water column, increases approximately 1 atmosphere for every 10 meters depth.

hyperchromatic shift the change in absorption of light exhibited by DNA when it is melted, forming two strands from the double helix.

hyperplasia the abnormal proliferation of tissue cells resulting in the formation of a tumor or gall.

hypersensitivity the state of an exaggerated immunological response upon reexposure to a specific antigen.

hypertonic a solution whose osmotic pressure is greater than that of a standard solution.

hypertrophy an increase in the size of an organ, independent of natural growth, due to enlargement or multiplication of its constituent cells.

hypervariable region a region of immunoglobulins that accounts for the specificity of antigen–antibody reactions; genetically specified terminal regions of the Fab fragments.

hyphae branched or unbranched filaments that constitute the vegetative form of an organism, occurring in filamentous fungi, algae, and bacteria.

hypolimnion the deeper, colder layer of an aquatic environment; the zone below the thermocline.

hypotheca the smaller of the two parts of the cell wall of a diatom.

hypotonic a solution whose osmotic pressure is less than that of a standard solution.

icosahedral virus a virus having cubical symmetry and a complex 20-sided capsid structure.

icosahedron a solid figure contained by 20 plane faces.

identification the process of determining the greatest affinity of an unknown organism to a group that has already been defined.

identification key a series of questions that leads to the unambiguous identification of an organism.

idiophase that phase of metabolism in batch culture in which secondary metabolism is dominant over primary growth-directed metabolism; the phase of antibiotic or other secondary product accumulation.

idiotype individually specific immunoglobulin molecules with distinct variable regions determining the specificity of the antigen–antibody reaction.

IgA, IgD, IgE, IgG, IgM five classes of immunoglobulins.

immobilization the binding of a substance so that it is no longer reactive or no longer circulates freely.

immobilized enzyme an enzyme bound to a solid support.

immune the condition following initial contact with a given antigen in which antibodies specific for that antigen are present in the body; the innate or acquired resistance to disease.

immunity the relative unsusceptibility of a person or animal to active infection by pathogenic microorganisms or the harmful effects of certain toxins; the condition of a living organism whereby it resists disease.

immunization any procedure in which an antigen is introduced into the body in order to produce a specific immune response.

immunodeficiency the lack of an adequate immune response due to inadequate B or T cell recognition and/or response to foreign antigens; a lack of antibody production.

immunoelectrophoresis a two-stage procedure used for the analysis of materials containing mixtures of distinguishable proteins, e.g., serum using electrophoretic separation and immune proteins are analyzed.

immunofluorescence any of a variety of techniques used to detect a specific antigen or antibody by means of homologous antibodies or antigens that have been conjugated with a fluorescent dye.

immunogenicity the ability of a substance to elicit an immune response.

immunoglobulins a varied class of proteins found in plasma and other body fluids, including all known antibodies; the antibody fraction of serum.

immunological referring to the immune response.

immunology the study of immunity.

immunosuppressant a drug that depresses the immune response.

impetigo an acute inflammatory skin disease caused by bacteria, characterized by small blisters, weeping fluid, and crusts.

IMViC tests a group of tests (indole, methyl red, Voges Proskauer, citrate) used in the identification of bacteria of the Enterobacteriaceae family.

incubation the maintenance of controlled conditions to achieve the optimal growth of microorganisms; the period of time between the establishment of an infection and the onset of disease symptoms.

indicator organism an organism used to indicate a particular condition, commonly applied to coliform bacteria, e.g., *E. coli* or *S. fecalis,* when their presence is used to indicate the degree of water pollution due to fecal contamination.

indigenous native to a particular habitat.

inducers substances responsible for activating certain genes, resulting in the synthesis of new proteins.

inducible enzymes enzymes that are synthesized only in response to a particular substance in the environment.

induction an increase in the rate of synthesis of an enzyme; the turning on of enzyme synthesis in response to environmental conditions.

infection a condition in which pathogenic microorganisms have become established in the tissues of a host organism.

infectious a disease that can be transmitted from one person, animal, or plant to another.

infectious dose the number of pathogens that are needed to overwhelm host defense mechanisms and establish an infection.

infectious mononucleosis glandular fever, an acute infectious disease that primarily affects the lymphoid tissues, caused by Epstein-Barr virus, which enters the body via the respiratory tract.

inflammation the reaction of tissues to injury involving localized heat, swelling, redness, and pain.

inflammatory response a nonspecific immune response to injury characterized by redness, heat, swelling, and pain in the affected area.

influenza an acute, highly communicable disease tending to occur in epidemic form, caused by a myxovirus, and characterized by malaise, headache, and fever.

inhibition the prevention of growth or multiplication of microorganisms; the reduction in the rate of enzymatic activity; the repression of chemical or physical activity.

inhibitors substances that repress or stop a chemical action.

initiation in protein synthesis, the stage at which the translating complex of messenger RNA, ribosome, and transfer RNA first assemble.

inoculate to deposit material, an inoculum, onto medium to initiate a culture, carried out with an aseptic technique; to introduce microorganisms into an environment that will support their growth.

inoculum the material containing viable microorganisms used to inoculate a medium.

insecticides substances destructive to insects; chemicals used to control insect populations.

insertion a type of mutation in which a nucleotide or two or more contiguous nucleotides are added to DNA.

in situ in the natural location or environment.

interference microscope type of microscope that relies on destructive and/or additive interference of light waves to achieve contrast.

interferons glycoproteins produced by animal cells that act to prevent the replication of a wide range of viruses by inducing resistance.

intermediary metabolism intermediate steps in the cellular synthesis and breakdown of substances.

intoxication poisoning as by drug, serum, alcohol, or any poison.

intracellular within a cell.

intradermal within or into the skin.

intramuscular within or into the substance of a muscle.

intravenous within or into the veins.

intron an intervening region of the DNA of eukaryotes not coding for a known protein nor regulatory function.

invasiveness the ability of a pathogen to spread through a host's tissues.

in vitro in glass; a process or reaction carried out in a culture dish or test tube.

in vivo within the living organism.

ion an atom that has lost or gained one or more orbital electrons and is thus capable of conducting electricity.

ionic bond a chemical bond resulting from the transfer of electrons between metal and nonmetal atoms; positive and negative ions are formed and held together by electrostatic attraction.

ionization the process that produces ions.

iso- combining form meaning for or from different individuals of the same species.

isogamete a reproductive cell similar in form and size to the cell with which it unites; found in certain protozoans, fungi, and algae.

isogamy fertilization in which the gametes are similar in appearance and behavior.

isomer one of two or more compounds having the same chemical composition but differing in the relative positions of the atoms within the molecules.

isotope an element that has the same atomic number as another but a different atomic weight.

-itis suffix denoting a disease, specifically an inflammatory disease of a specified part.

Jaccard coefficient a measure of similarity used in cluster analysis to show the relationship between individuals that does not consider negative matches.

jaundice yellowness of the skin, mucous membranes, and secretions resulting from liver malfunction.

kappa particles bacterial particles that occur in the cytoplasm of certain strains of *Paramecium aurelia*; such strains have a competitive advantage with other strains of *Paramecium* and are known as killer strains.

karyogamy the fusion of nuclei, as of gametes in fertilization.

keratin a highly insoluble protein that occurs in hair, wool, horn, and skin.

ketone an organic compound derived by oxidation from a secondary alcohol, containing a characteristic group in which a carbon atom forms a double bond with an oxygen atom and two separate single bonds with other carbon atoms.

kingdom a major taxonomic category consisting of several phyla or divisions; the primary divisions of living organisms.

K_m the Michaelis constant; describes the affinity of an enzyme for a substrate; the substrate concentration at half the maximal velocity of an enzyme.

Koch's postulates a process for elucidating the etiologic agent of an infectious disease.

Koplik's spots small red spots surrounded by white areas occurring on the mucous membranes of the mouth during the early stages of measles.

Krebs cycle the tricarboxylic acid cycle; the citric acid cycle; the metabolic pathway in which acetate derived from pyruvic acid is converted to carbon dioxide and reduced coenzymes are produced.

Kupffer cells macrophages lining the sinusoids of the liver.

-labile unstable, readily changed by physical, chemical, or biological processes.

Lac **operon** inducible enzyme system of *E. coli* for the utilization of lactose.

lactam an organic compound that contains an —NHCO— group in ring form.

lactamase an enzyme that breaks a lactam ring.

lactic acid fermentation fermentation that produces lactic acid as the primary product.

lactose a disaccharide in milk; when hydrolyzed it yields glucose and galactose.

lag phase a period following inoculation of a medium during which numbers of microorganisms do not increase.

laminar flow the flow of air currents in which streams do not intermingle; the air moves along parallel flow lines; used in laminar flow hood to provide air free of microbes over a work area.

latent potential; not manifest; present but not visible or active.

latent period the period of time following infection of a cell by a virus before new viruses are assembled.

leach to wash or extract soluble constituents from insoluble materials.

leavening substance used to produce fermentation in dough or liquid; the production of CO_2 that results in the rising of dough.

lecithinases extracellular phospholipid-splitting enzymes.

Legionnaire's disease a form of pneumonia caused by *Legionella pneumophilia*.

leishmaniasis disease caused by protozoa of the genus *Leishmania*.

leprosy Hansen's disease, a chronic contagious disease affecting humans and armadillos, caused by *Mycobacterium leprae*.

leptospirosis disease of humans or animals caused by *Leptospira*.

lesion a region of tissue mechanically damaged or altered by any pathological process.

lethal dose the amount of a toxin that results in the death of an organism.

leukocidin an extracellular bacterial product that can kill leukocytes.

leukocyte a type of white blood cell, characterized by a beaded, elongated nucleus.

leukocytosis an increase above the normal upper limits of the leukocyte count.

leukopenia a decrease below the normal number of leukocytes in the blood.

lichens a large group of composite organisms, each consisting of a fungus in symbiotic association with an alga or cyanobacterium.

life a state that characterizes living systems, encompassing the complex series of physicochemical processes essential for maintaining the organization of the system and the ability to reproduce that organization.

ligases a group of enzymes that catalyze reactions in which a bond is formed between two substrate molecules using energy obtained from the cleavage of a pyrophosphate bond.

lignins a class of complex polymers in the woody material of higher plants.

lipases fat-splitting enzymes.

lipids fats or fat-like substances that are insoluble in water and are soluble in nonpolar solvents.

lipophilic preferentially soluble in lipids or non-polar solvents.

lipopolysaccharides molecules consisting of co-valently linked lipids and polysaccharides.

lipopolysaccharide toxin an endotoxin.

liter a metric unit of volume equal to 1000 milli-liters.

lithosphere the solid part of the earth.

lithotrophs microorganisms that utilize and obtain energy from the oxidation of inorganic matter; au-totrophs.

litmus plant extract dye used as an indicator of pH and oxidation or reduction.

littoral situated or growing on or near the shore; the region between the high and low tide marks.

living system a system separated from its sur-roundings by a semipermeable barrier, composed of macromolecules, including proteins and nucleic acids, having lower entropy than its surroundings, and thus requiring inputs of energy to maintain the high degree of organization, capable of self-repli-cation, and normally based on cells as the primary functional and structural units.

locus the point on a chromosome occupied by a gene.

logarithmic phase see exponential phase.

low acid food food with pH above 4.5.

LPS see lipopolysaccharides.

LTH process low-temperature, hold pasteurization process, e.g., 63°C for 30 minutes.

luminescence the emission of light without pro-duction of heat sufficient to cause incandescence, produced by physiological processes, friction, chemical, or electrical action.

ly-, lys-, lyt- combining forms meaning loosen or dissolve.

lymph a plasma filtrate that circulates through the body.

lymph nodes an aggregation of lymphoid tissues surrounded by a fibrous capsule found along the course of the lymphatic system.

lymphocytes lymph cells.

lymphogranuloma venereum a sexually transmit-ted disease caused by a *Chlamydia* characterized by an initial lesion, usually on the genitalia, followed by regional lymph node enlargement and systemic involvement.

lymphokines a varied group of biologically active extracellular proteins formed by activated T lym-phocytes involved in cell-mediated immunity.

lyophilization the process of rapidly freezing a substance at low temperature, then dehydrating the frozen mass in a high vacuum.

lysins antibodies or other entities that under ap-propriate conditions are capable of causing the lysis of cells.

lysis the rupture of cells.

lysogeny the nondisruptive infection of a bacte-rium by a bacteriophage.

lysosomes an organelle containing hydrolytic en-zymes involved in autolytic and digestive processes.

lysozymes enzymes that hydrolyze peptidoglycan, acting as bactericidal agents when they degrade the bacterial cell walls.

lytic of or relating to lysis or a lysin; viruses that cause lysis of cells within which they reproduce.

MacConkey's agar a solid medium used for the growth of enteric bacteria.

macro- combining form meaning long or large.

macromolecules very large molecules having polymeric chain structures, as in proteins, polysac-charides, and other natural and synthetic polymers.

macrophages mononuclear phagocytes; large ac-tively phagocytic cells found in spleen, liver, lymph nodes, and blood, important factors in nonspecific immunity.

macroscopic of a size visible to the naked eye.

magnetotaxis motility directed by a geomagnetic field.

magnification the extent to which the image of an object is larger than the object itself.

major histocompatibility complex the genetic re-gion in human beings that controls not only tissue compatibility but also the development and activa-tion of part of the immune response.

malaise a general feeling of illness, accompanied by restlessness and discomfort.

malaria an infectious chronic disease caused by the *Plasmodium* protozoa transmitted by mosquitos, characterized by intermittent fever, anemia, chills, and sweating.

maltase an enzyme that converts maltose to glu-cose.

maltose a disaccharide formed upon hydrolysis of starch or glycogen and metabolized by a wide range of fungi and bacteria.

marine of or relating to the oceans.

mash the crushed malt or grain meal steeped and stirred in hot water with amylases to produce wort as a substrate for microorganisms.

mast cells cells that contain granules of histamine, serotonin, and heparin, especially in connective tissues involved in hypersensitivity reactions.

mastigophora a subclass of protozoans characterized by the presence of flagella.

mastitis inflammation of the breast.

mating the meeting of individuals for sexual reproduction.

measles an acute, contagious systemic human disease caused by a paramyxovirus that enters via the oral and nasal route, characterized by the presence of Koplik's spots.

medium the material that supports the growth/reproduction of microorganisms.

meiosis cell division that results in a reduction of the state of ploidy, normally from diploid to haploid during the formation of germ cells.

membrane filter a cellulose-ester membrane used for microbiological filtrations.

meninges the membranes covering the brain and spinal cord.

meningitis inflammation of the membranes of the brain or spinal cord.

mesophiles organisms whose optimum growth is in the temperature range of 20–45°C.

mesosomes intracellular membranous structures found in the infoldings of bacterial cell membranes; their function is as yet unknown.

messenger RNA the RNA that carries information from DNA to ribosomes where it specifies the amino acid sequence for a particular polypeptide chain.

metabolic pathway a sequence of biochemical reactions that transforms a substrate into a useful product for carbon assimilation or energy transfer.

metabolism the sum total of all chemical reactions by which energy is provided for the vital processes and new cell substances are assimilated.

metabolites chemicals participating in metabolism; nutrients.

metabolize to transform by means of metabolism.

metachromatic granules cytoplasmic granules of polyphosphate occurring in the cells of certain bacteria that stain intensively with basic dyes but appear a different color.

methanogens methane-producing prokaryotes; a group of archaebacteria capable of reducing carbon dioxide or low-molecular-weight fatty acids to methane.

methylation the process of substituting a methyl group for a hydrogen atom.

MIC see minimum inhibitory concentration.

micro- combining form meaning small.

microaerophiles aerobic organisms that grow best in an environment with less than atmospheric oxygen levels.

microbes microscopic organisms; microorganisms.

microbiocidal any agent capable of destroying microbes.

microbiology the study of microorganisms and their interactions with other organisms and the environment.

microbiota the totality of microorganisms associated with a given environment.

microfibrils thread-like structures found in the cell walls of filamentous fungi, consisting of chitin.

microfilament an elongated structure composed of protein subunits.

micrometer one millionth (10^{-6}) part of a meter; 10^{-3} of a millimeter.

microorganisms microscopic organisms, including algae, bacteria, fungi, protozoa, and viruses.

microscope an optical or electronic instrument for viewing objects too small to be visible to the naked eye.

microtome an instrument used for cutting thin sheets or sections of tissues or individual cells for examination by light or electron microscopy.

microtubules cylindrical protein tubes that occur within all eukaryotic organisms and aid in maintaining cell shape; they comprise the structure of organelles of cilia and flagella, and serve as spindle fibers in mitosis.

mildew any of a variety of plant diseases in which the mycelium of the parasitic fungus is visible on the affected plant; biodeterioration of a fabric due to fungal growth.

mineralization the microbial breakdown of organic materials into inorganic materials brought about mainly by microorganisms.

minimum inhibitory concentration the concentration of a particular antimicrobial drug necessary to inhibit the growth of a particular strain of microorganism.

mitochondria a semiautonomous organelle found

in eukaryotic cells, the site of respiration and other cellular processes, consisting of an outer membrane and an inner one that is convoluted.

mitosis the sequence of events resulting in the division of the nucleus into two genetically identical cells during asexual cell division, each of the daughter nuclei having the same number of chromosomes as the parent cell.

mixed acid fermentation a type of fermentation carried out by members of the Enterobacteriaceae, converting glucose to acetic, lactic, succinic, and formic acids.

mixotrophs organisms capable of utilizing both autotrophic and heterotrophic metabolic processes, e.g., the concomitant use of organic compounds as sources of carbon and light as a source of energy.

modification the methylation of nucleotide residues in DNA; modification of newly synthesized DNA by specific enzymes in a manner characteristic of the particular bacterial strain.

moiety a part of a molecule having a characteristic chemical property.

mold a type of fungus having a filamentous structure.

mole% G + C the proportion of guanine and cytosine in a DNA macromolecule.

mollicutes a class of prokaryotic organisms that do not form cell walls, e.g., *Mycoplasma*.

Monera prokaryotic protists with a unicellular and simple colonial organization.

mono- combining form meaning single, one, or alone.

monoclonal antibody an antibody produced from a clone of cells making only that specific antibody.

monocytes ameboid agranular phagocytic white blood cells derived from the bone marrow.

mononuclear having only one nucleus.

mononuclear phagocyte system the macrophage system of the body, including all phagocytic white blood cells except granular white blood cells; the reticuloendothelial system.

monosaccharide any carbohydrate whose molecule cannot be split into simpler carbohydrates; a simple sugar.

morbidity the state of being diseased; the ratio of the number of sick individuals to the total population of the community; the conditions inducing disease.

mordant a substance that fixes the dyes used in staining tissues or bacteria; a substance that increases the affinity of a stain for a biological specimen.

morphogenesis morphological changes, including growth and differentiation of cells and tissues during development; the transformations involved in the growth and differentiation of cells and tissues.

morphology the study of the shape and structure of microorganisms.

mortality death; the proportion of deaths in a population.

mortality rate death rate; number of deaths per unit population per unit time.

most probable number (MPN) the statistical estimate of a bacterial population through the use of dilution and multiple tube inoculations.

motility the capacity for independent locomotion.

mRNA see messenger RNA.

mucosa a mucous membrane, the lining of body cavities that communicate to the exterior.

mucous membrane the type of membrane lining cavities and canals that have communication with air.

mucus a viscid fluid secreted by mucous glands consisting of mucin, water, inorganic salts, epithelial cells, and leukocytes.

mumps an acute infectious disease caused by a virus, characterized by swelling of the salivary glands.

murein peptidoglycan; the repeating polysaccharide unit comprising the backbone of the cell walls of eubacteria.

must the fluid extracted from crushed grapes; the ingredients, e.g., fruit pulp or juice, used as substrate for fermentation in wine making.

mutagen any chemical or physical agent that promotes the occurrence of mutation; a substance that increases the rate of mutation above the spontaneous rate.

mutant any organism that differs from the naturally occurring type because its base DNA has been modified, often resulting in an altered protein that gives the cell different properties than its parent.

mutation a stable, heritable change in the nucleotide sequence of the genetic nucleic acid, resulting in an alteration in the products coded by the gene.

mutualism a stable condition in which two organisms of different species live in close physical association, each organism deriving some benefit from the association; symbiosis.

myc- combining form meaning fungus.

mycelia the interwoven mass of discrete fungal hyphae.

mycetoma a chronic infection usually involving the foot, characterized by the presence of pussy nodules and caused by a wide variety of fungi or bacteria; also known as madura foot.

mycobiont   the fungal partner in a lichen.

mycolic acid   fatty acids found in the cell walls of *Mycobacterium* and several other bacteria related to the actinomycetes.

mycology   the study of fungi.

mycorrhizae   a stable, symbiotic association between a fungus and the root of a plant; the term also refers to the root-fungus structure itself.

mycosis   any disease in which the causal agent is a fungus.

mycotoxins   toxic substances produced by fungi, including aflatoxin, amatoxin, and ergot alkaloids.

mycovirus   a virus that infects fungi.

myocardium   the muscular tissue of the heart wall.

myx-   combining form meaning mucus.

myxamoebae   nonflagellated amoeboid cells that occur in the life cycle of the slime molds and members of the Plasmodiophorales.

myxospores   resting cells in the fruiting bodies of members of the Myxobacteriales.

NAD   see nicotinamide adenine dinucleotide.

NADH   reduced nicotinamide adenine dinucleotide.

NADP   see nicotinamide adenine dinucleotide phosphate.

NADPH   reduced nicotinamide adenine dinucleotide phosphate.

nasopharynx   the upper part of the pharynx continuous with the nasal passages.

necrosis   the pathologic death of a cell or group of cells in contact with living cells.

negative stain   the treatment of cells with dye so that the background, rather than the cell itself, is made opaque; used to demonstrate bacterial capsules or the presence of parasitic cysts in fecal samples; a stain with a positively charged chromophore.

Negri bodies   acidophilic, intracytoplasmic inclusion bodies that develop in cells of the central nervous system in cases of rabies.

nematodes   worms of the class Nematoda.

neoplasm   the result of the abnormal and excessive proliferation of the cells of a tissue; if the progeny cells remain localized, the resulting mass is called a tumor.

neurotoxin   a toxin capable of destroying nerve tissue or interfering with neural transmission.

neutralism   the relationship between two different microbial populations where there is a lack of recognizable interaction.

neutropenia   a decrease below the normal standard in the number of neutrophils in the peripheral blood.

neutrophil   a large granular leukocyte with a highly variable nucleus consisting of 3–5 lobes and cytoplasmic granules that stain with neutral dyes and eosin.

niche   the functional role of an organism within an ecosystem; the combined description of the physical habitat, functional role, and interactions of the microorganisms occurring at a given location.

nicotinamide adenine dinucleotide (NAD)   a coenzyme used as an electron acceptor in oxidation–reduction reactions.

nicotinamide adenine dinucleotide phosphate   the phosphorylated form of NAD formed when NADPH serves as an electron donor in oxidation–reduction reactions.

nitrate   a salt of nitric acid; NO_3^-.

nitrate reduction   the reduction of nitrate to reduced forms, e.g., when under anaerobic and microaerophilic conditions, bacteria use nitrate as a terminal electron acceptor for respiratory metabolism.

nitrification   the process in which ammonia is oxidized to nitrite and nitrite to nitrate; a process primarily carried out by the strictly aerobic, chemolithotrophic bacteria of the family Nitrobacteraceae.

nitrifying bacteria   Nitrobacteraceae; Gram negative, obligately aerobic, chemolithotrophic bacteria occurring in fresh and marine waters and in soil that oxidize ammonia to nitrite or nitrite to nitrate.

nitrite   a salt of nitrous acid; NO_2^-; nitrites of sodium and potassium are used as food additives and preservatives.

nitrogenase   the enzyme that catalyzes biological nitrogen fixation.

nitrogen fixation   the reduction of gaseous nitrogen to ammonia, carried out by certain prokaryotes.

nitrogenous   containing nitrogen.

nodules   tumor-like growths formed by plants in response to infections with specific bacteria within which the infecting bacteria fix atmospheric nitrogen.

nomenclature   the naming of organisms, a function of taxonomy governed by codes, rules, and priorities laid down by committees.

noncompetitive inhibition   inhibition of enzyme activity by a substance that does not compete with the normal substrate for the active site and thus cannot be reduced by increasing the substrate concentration.

noncyclic photophosphorylation a metabolic pathway involved in the conversion of light energy for the generation of ATP in which an electron is transferred from an electron donor, normally water, through a series of electron carriers with the eventual formation of a reduced coenzyme, normally $NADPH_2$.

nonhomologous recombination recombination where there is little or no homology between the donor DNA and the region of the DNA in the recipient where insertion occurs.

nonsense codon a codon that does not specify an amino acid but acts as a punctuator of messenger RNA.

nonsense mutation a mutation in which a codon specifying an amino acid is altered to a nonsense codon.

nosocomial infection an infection acquired while in the hospital.

nuclease an enzyme capable of splitting nucleic acids to nucleotides, nucleosides, or their components.

nucleic acid a large acidic chain-like macromolecule containing phosphoric acid, sugar, and purine and pyrimidine bases; the nucleotide polymers RNA and DNA.

nucleoid region the region of a prokaryotic cell in which the genome occurs.

nucleolus an RNA-rich intranuclear body not bounded by a limiting membrane that is the site of ribosomal RNA synthesis in eukaryotes.

nucleoprotein a conjugated protein closely associated with nucleic acid.

nucleosome the fundamental structural unit of DNA in eukaryotes having approximately 190 base pairs folded and held together by histones.

nucleotide the combination of a purine or pyrimidine base with a sugar and phosphoric acid; the basic structural unit of nucleic acid.

nucleus an organelle of eukaryotes in which the cell's genome occurs; the differentiated protoplasm of a cell surrounded by a membrane that is rich in nucleic acids.

nutrient a growth-supporting substance.

O antigens lipopolysaccharide-protein antigens occurring in the cells of Gram negative bacteria.

objective lens the microscope lens nearest the object.

obligate anaerobes organisms that grow only under anaerobic conditions, i.e., in the absence of air or oxygen; organisms that cannot carry out respiratory metabolism.

occluded closed or shut up.

ocular lens the eyepiece of a microscope.

-oid combining form meaning resembling.

oil immersion lens a high-power microscope objective designed to work with the space between the objective and the specimen filled with oil to enhance resolution.

Okazaki fragments the short segments of newly synthesized DNA along the trailing or discontinuous strand that are then linked by a ligase to form the completed DNA.

oligotrophic lakes and other bodies of water poor in those nutrients that support the growth of aerobic photosynthetic organisms; microorganisms that grow at very low nutrient concentrations.

oogamy a form of fertilization that involves either a motile male gamete and a relatively large, non-motile female gamete, or gametangial contact in which the gametangia are morphologically different.

oospores thick-walled resting spores of fungi.

operator region a section of an operon involved in the control of the synthesis of the gene products encoded within that region of DNA; a regulatory gene that binds with a regulatory protein to turn on and off transcription of a specified region of DNA.

operon a group or cluster of structural genes whose coordinated expression is controlled by a regulator gene.

opportunistic pathogens organisms that exist as part of the normal body microbiota but that may become pathogenic under certain conditions, e.g., when the normal antimicrobial body defense mechanisms have been impaired; organisms not normally considered pathogens but that cause disease under some conditions.

opsonization the process by which a cell becomes susceptible to phagocytosis and lytic digestion by combination of a surface antigen with an antibody and/or other serum component.

optical isomers compounds having the same number and kind of atoms and grouping of atoms but differing in their configurations or arrangements in space; specifically their structures are not superimposable.

orchitis inflammation of the testes.

organelle a membrane-bound structure that forms part of a microorganism and that performs a specialized function.

-ose   combining form denoting a sugar.

-osis   combining form meaning disease of.

osmophiles   organisms that grow best or only in or on media of relatively high osmotic pressure.

osmosis   the passage of a solvent through a membrane from a dilute solution into a more concentrated one.

osmotic pressure   the force resulting from differences in solute concentrations on opposite sides of a semipermeable membrane.

osmotic shock   any disturbance or disruption in a cell or subcellular organelle that occurs when it is transferred to a significantly hypertonic or hypotonic medium, with lysis of cells resulting from osmotic pressure.

otitis media   inflammation of the inner ear.

oxidase   an oxidoreductase that catalyzes a reaction in which electrons removed from a substrate are donated directly to molecular oxygen.

oxidation   an increase in the positive valence or decrease in the negative valence of an element resulting from the loss of electrons that are transferred to some other element.

oxidation pond   a method of aerobic waste disposal employing biodegradation by aerobic and facultative microorganisms growing in a standing water body.

oxidation–reduction potential   a measure of the tendency of a given oxidation–reduction system to donate electrons, i.e., to behave as a reducing agent, or to accept electrons, i.e., to act as an oxidizing agent; determined by measuring the electrical potential difference between the given system and a standard system.

oxidative phosphorylation   a metabolic sequence of reactions occurring within a membrane in which an electron is transferred from a reduced coenzyme through a series of electron carriers establishing an electrochemical gradient across the membrane that drives the formation of ATP from ADP and inorganic phosphate.

oxidize   to produce an increase in positive valence through the loss of electrons.

oxyphotobacteria   bacteria capable of evolving oxygen during photosynthesis.

ozonation   the killing of microorganisms by exposure to ozone.

palindrome   a word reading the same backward and forward; a base sequence the complement of which has the same sequence; a nucleotide sequence that reads the same when read in the antiparallel direction.

pandemic   an outbreak of disease that affects large numbers of people in a major geographical region or that has reached epidemic proportions simultaneously in different parts of the world.

parasites   organisms that live on or within the tissues of another living organism, the host, from which it derives its nutrients.

parasitism   an interactive relationship between two organisms or populations in which one is harmed and the other benefits; generally, the population benefiting, the parasite, is smaller than the population that is harmed.

parfocal   pertaining to microscopical oculars and objectives that are so constructed or so mounted that in changing from one to another the image will remain in focus.

parotitis   inflammation of the parotid gland, as in mumps.

passive agglutination   a procedure in which the combination of antibody with a soluble antigen is made readily detectable by the prior adsorption of the antigen to erythrocytes or to minute particles of organic or inorganic materials.

passive immunity   short-term immunity brought about by the transfer of preformed antibody from an immune subject to a nonimmune subject.

Pasteur effect   the slower rate of glucose utilization by a microorganism growing aerobically by respiratory metabolism than by the same organism growing anaerobically, reflecting feedback inhibition; in those organisms capable of both fermentative and respiratory modes of metabolism, the inhibition of glucose utilization in anaerobically grown cells upon exposure to oxygen.

pasteurization   the reduction in numbers of microorganisms by exposure to elevated temperatures but not necessarily the killing of all microorganisms in a sample; a form of heat treatment that is lethal for the causal agents of a number of milk-transferable diseases as well as for a proportion of normal milk microbiota, which also inactivates certain bacterial enzymes that may cause deterioration in milk.

pathogens   organisms capable of causing disease in animals, plants, or microorganisms.

pathology   the study of the nature of disease through the study of its causes, processes, and effects, along with the associated alterations of structure and function.

pelagic zone   that region of an aquatic environment that comprises the entire body of water, excluding the mud or sand that forms the bed or bottom of that environment.

pellicle   a thin protective membrane occurring

around some protozoa, also known as a periplast; a continuous or fragmentary film that sometimes forms at the surface of a liquid culture, consisting entirely of cells or may be largely extracellular products of the cultured organisms.

penicillins a group of natural and semisynthetic antibiotics having a β-lactam ring that are active against Gram positive bacteria inhibiting the formation of cross-links in the peptidoglycan of growing bacteria.

pentose a type of carbohydrate containing five atoms of carbon.

pentose phosphate pathway a metabolic pathway that involves the oxidative decarboxylation of glucose 6-phosphate to ribulose 5-phosphate, followed by a series of reversible, nonoxidative sugar interconversions.

pepsin a proteolytic enzyme.

peptidase an enzyme that splits peptides to amino acids.

peptide bond a bond in which the carboxyl group of one amino acid is condensed with the amino group of another amino acid.

peptides compounds of two or more amino acids containing one or more peptide bonds.

peptidoglycan the rigid component of the cell wall in most bacteria; murein.

peptidyl site the site on the ribosome where the growing peptide chain is moved during protein synthesis.

peptones a water-soluble mixture of proteoses and amino acids produced by the hydrolysis of natural proteins either by an enzyme or an acid.

perfringens food poisoning food poisoning by ingestion of *Clostridium perfringens* type A, a self-limiting condition characterized by abdominal pain and diarrhea.

periodontal disease disease of the tissues surrounding the teeth.

periodontitis inflammation of the periodontium, the tissues surrounding a tooth.

periplasm the region between the cytoplasmic membrane and the cell wall.

periplast see pellicle.

peritrichous referring to the arrangement of a cell's flagella in which the flagella are more or less uniformly distributed over the surface of the cell.

permeability the property of cell membranes that permits transit of molecules and ions in solution across the membrane.

permease an enzyme that increases the rate of transport of a substance across a membrane.

peroxidase an oxidoreductase that catalyzes a reaction in which electrons removed from a substrate are donated to hydrogen peroxide.

peroxide the anion O_2^{2-} or HO_2^-, or a compound containing one of these anions.

peroxisomes microbodies that contain D-amino acid oxidase, α-hydroxy acid oxidase, catalase, and other enzymes, found in yeasts and certain protozoans.

pest a population that is an annoyance for economic, health, or aesthetic reasons.

pesticides substances destructive to pests, especially insects.

petri dish a round, shallow, flat-bottomed dish with a vertical edge together with a similar, slightly larger structure that forms a loosely fitting lid, made of glass or plastic, widely used as receptacles for various types of solid media.

pH the symbol used to express hydrogen ion concentration, signifying the logarithm to the base 10 of the reciprocal of the hydrogen ion concentration.

phage see bacteriophage.

phagocytes any of a variety of cells that ingest and break down certain categories of particulate matter.

phagocytosis the process in which particulate matter is ingested by a cell, involving the engulfment of that matter by the cell's membrane.

phagotrophic referring to the ingestion of nutrients in particulate form by phagocytosis.

pharmaceutical a drug used in the treatment of disease.

pharyngitis inflammation of the pharynx.

phase contrast microscope a microscope that achieves enhanced contrast of the specimen by altering the phase of light that passes through the specimen relative to light that passes through the background, eliminating the need for staining in order to view microorganisms and making the viewing of live specimens possible.

phenetic pertaining to the physical characteristic of an individual without consideration for its genetic makeup; in taxonomy, a classification system that does not take evolutionary relationships into consideration.

phenol coefficient a number that expresses the antibacterial power of a substance relative to the disinfectant phenol.

phenotype the totality of observable structural and functional characteristics of an individual organism, determined jointly by its genotype and environment.

-phile combining form meaning like, or having an affinity for.

-phobic combining form meaning having an aversion to or lacking affinity for.

phosphatases enzymes that hydrolyze esters of phosphoric acid.

phosphodiester bond the bonding of two moieties through a phosphate group; each moiety is held to the phosphate by an ester linkage.

phosphofructokinase an enzyme that mediates the addition of a phosphate group to glucose 6-phosphate with the formation of glucose 1,6-diphosphate, a key step during glycolysis.

phospholipase an enzyme that catalyzes hydrolysis of a phospholipid.

phospholipid a type of lipid compound which is an ester of phosphoric acid and also contains one or two molecules of fatty acid, an alcohol, and sometimes a nitrogenous base.

phosphorylation the esterification of compounds with phosphoric acid; the conversion of an organic compound into an organic phosphate.

photo- combining form meaning light.

photoautotrophs organisms whose source of energy is light and whose source of carbon is carbon dioxide, characteristic of algae and some prokaryotes.

photoheterotrophs organisms that obtain energy from light but require exogenous organic compounds for growth.

photophosphorylation a metabolic sequence by which light energy is trapped and converted to chemical energy with the formation of ATP.

photoreactivation a mechanism whereby the effects of ultraviolet radiation on DNA may be reversed by exposure to radiation of wavelengths in the range 320–500 nm; an enzymatic repair mechanism of DNA present in many microorganisms.

photosynthesis the process in which radiant energy is absorbed by specialized pigments of a cell and is subsequently converted to chemical energy; the ATP formed in the light reactions is used to drive the fixation of carbon dioxide with the production of organic matter.

photosystem I see cyclic photophosphorylation.

photosystem II see noncyclic photophosphorylation.

phototrophs organisms whose sole or principal primary source of energy is light; organisms capable of photophosphorylation.

phycobilisomes granules found in cyanobacteria and some algae on the surface of their thylakoids.

phycobiont the algal partner of a lichen.

phycology the study of algae.

phycomycete a group of true fungi lacking regularly spaced septae in the actively growing portions of the fungus and producing sporangiospores by cleavage as the primary method of asexual reproduction.

phycovirus any virus whose host is a cyanobacterium or alga.

phylogenetic referring to the evolution of a species from the simpler forms; in taxonomy a classification based on evolutionary relationships.

phylum a taxonomic group composed of groups of related classes.

physiology the science that deals with the study of the functions of living organisms and their physicochemical parts and metabolic reactions.

phytoalexin polyaromatic antimicrobial substances produced by higher plants in response to a microbial infection.

phytoplankton passively floating or weakly motile photosynthetic aquatic organisms, primarily cyanobacteria and algae.

phytoplankton food chain a food chain in aquatic habitats based on grazing of primary producers.

pili filamentous appendages that project from the cell surface of certain Gram negative bacteria apparently involved in adsorption phenomena.

pilin a chain of proteins, the subunits of pili.

pitching the inoculation of mash with yeast during production of alcoholic beverages; the inoculation of a substrate with microorganisms.

plague a contagious disease often occurring as an epidemic; an acute infectious disease of humans and other animals, especially rodents, caused by *Yersinia pestis,* that is transmitted by fleas.

planapochromatic lens a flat-field apochromatic objective microscope lens.

plankton collectively, all microorganisms that passively drift in the pelagic zone of lakes and other bodies of water, chiefly microalgae and protozoans.

plasma membrane see cytoplasmic membrane.

plasmids extrachromosomal genetic structures that can replicate independently within a bacterial cell.

plasmodium malaria-causing protozoans; the life stage of acellular slime molds characterized by a motile multinucleate body.

plasmogamy fusion of cells without nuclear fusion to form a multinucleate mass.

plastids a class of membrane-bound organelles found within cells of higher plants and algae, containing pigments and/or certain products of the cell, e.g., chloroplasts.

plate counts method of estimating numbers of microorganisms by diluting samples, culturing on solid media, and counting the colonies that develop to estimate viable microorganisms in the sample.

pleomorphism the variation in size and form between individual cells in a clone or a pure culture.

ploidy the number of complete sets of chromosomes of a eukaryotic nucleus or cell.

pneumonia inflammation of the lungs.

polar flagellum flagella emanating from one or both polar ends of a cell.

polarized light light vibrating in a defined pattern.

poliomyelitis inflammation of the gray matter of the spinal cord, caused by a picornavirus.

poly-β-hydroxybutyric acid a polymeric storage product formed by some bacteria.

polycistronic coding for multiple cistrons; mRNA molecules that code for the synthesis of several proteins; often the proteins are functionally related and under the control of a specific operon.

polymerase an enzyme that catalyzes the formation of a polymer.

polymers the products of the combination of two or more molecules of the same substance.

polymorph a leukocyte having granules in the cytoplasm, also known as a polymorphonuclear leukocyte or PMN.

polymorphonuclear having a nucleus that resembles lobes, connected by thin strands of nuclear substance.

polypeptide a chain of amino acids linked together by peptide bonds but of lower molecular weight than a protein.

polysaccharides carbohydrates formed by the condensation of monosaccharides, e.g., starch and cellulose, having multiple monosaccharide subunits.

polysomes complexes of ribosomes bound together by a single messenger RNA molecule, also known as polyribosomes.

positive stain a stain with a positively charged chromophore.

potable fit to drink.

pour plate a method of culture in which the inoculum is dispersed uniformly in molten agar or other medium in a petri dish; the medium is allowed to set and is then incubated.

precipitin reaction a serological test in which the interaction of antibodies with soluble antigens is detected by the formation of a precipitate.

predation a mode of life in which food is primarily obtained by killing and consuming animals; an interaction between organisms in which one benefits and one is harmed based on the ingestion of the smaller-sized organism, the prey, by the larger organism, the predator.

preemptive colonization phenomenon that occurs when pioneer organisms alter environmental conditions in such a way that discourages further succession.

prey an animal taken by a predator for food.

Pribnow sequence a sequence of nucleotide bases within the DNA that determines the site of transcription initiation.

primary atypical pneumonia pneumonia caused by *Mycoplasma pneumoniae*.

primary immune response the first immune response to a particular antigen that has a characteristically long lag period and relatively low titer of antibody production.

primary producers those organisms capable of converting carbon dioxide to organic carbon, including photoautotrophs and chemoautotrophs.

primary sewage treatment the removal of suspended solids from sewage by physical settling in tanks or basins.

prions infectious proteins; substances that are infectious and reproduce within living systems but appear to be proteinaceous based on degradation by proteases and to lack nucleic acids based on resistance to digestion by nucleases.

progeny offspring.

projector lens the lens of an electron microscope that focuses the beam onto the film or viewing screen.

prokaryotes cells whose genomes are not contained within a nucleus; the bacteria.

promastigote an elongated and flagellated form assumed by many species of the Trypanosomatidae during a particular stage of development.

promoter specific initiation site of DNA where the RNA polymerase enzyme binds for transcription of the DNA.

prophage the integrated phage genome formed when the phage genome becomes integrated with the host's chromosome and is replicated as part of the bacterial chromosome during subsequent cell division.

prophylaxis   the measures taken to prevent the occurrence of disease.

prosthecae   a cell wall-limited appendage forming a narrow extension of a prokaryotic cell.

proteinase   one of the subgroups of proteases or proteolytic enzymes that act directly on native proteins in the first step of their conversion to simpler substances.

proteins   a class of high-molecular-weight polymers composed of amino acids joined by peptide linkages.

protein toxins   proteins secreted by bacteria that act as poisons.

proteolytic enzymes   an enzyme that breaks down proteins.

protista   in one proposed classification system, a kingdom of organisms lacking true tissue differention, i.e., the microbes; in another classification system, a kingdom that includes many of the algae and protozoa.

proto-cooperation   synergism; a nonobligatory relationship between two microbial populations where both populations benefit.

protoplasm   the viscid material constituting the essential substance of living cells upon which all the vital functions of nutrition, secretion, growth, reproduction, irritability, and locomotion depend.

protoplasts   spherical, osmotically sensitive structures formed when cells are suspended in an isotonic medium and their cell walls are completely removed, a bacterial protoplast consists of an intact cell membrane and the cytoplasm contained within.

prototrophs   parental strains of microorganisms that give rise to nutritional mutants known as auxotrophs.

protozoa   diverse, eukaryotic, typically unicellular nonphotosynthetic microorganisms generally lacking a rigid cell wall.

protozoology   the study of protozoa.

pseudopodia   false feet formed by protoplasmic streaming in protozoa and used for locomotion and the capture of food.

psittacosis   an infectious disease of parrots, other birds, and humans, caused by *Chlamydia psittaci*.

psychro-   combining form meaning cold.

psychrophile   an organism that has an optimum growth temperature below 20°C.

psychrotroph   a mesophile that can grow at low temperatures.

puerperal fever   an acute febrile condition following childbirth caused by infection of the uterus and/or adjacent regions by streptococci.

pure culture   a culture that contains cells of one kind; the progeny of a single cell.

purine $C_5H_4N_4$, a cyclic nitrogenous compound, the parent of several nucleic acid bases.

pus   a semifluid, creamy yellow, or greenish yellow product of inflammation composed mainly of leukocytes and serum.

putrefaction   the microbial breakdown of protein under anaerobic conditions.

pyelonephritis   inflammation of the kidneys.

pyknosis   the condition of having a contracted nucleus.

pyoderma   a pus-producing skin lesion.

pyogenic   pus-producing.

pyrimidine   a six-membered cyclic compound containing four carbon and two nitrogen atoms in a ring, the parent compound of several nucleotide bases.

pyrite   a common mineral containing iron disulfite.

pyrogenic   fever-producing.

Q fever   an acute disease in humans characterized by sudden onset of headache, malaise, fever, and muscular pain, caused by *Coxiella burnetii*; the reservoirs of infection are cattle, sheep, and ticks.

quarantine   the isolation of persons or animals suffering from an infectious disease in order to prevent transmission of the disease to others.

rabies   an acute and usually fatal disease of humans, dogs, cats, bats, and other animals, caused by the rabies virus and commonly transmitted in saliva by the bite of a rabid animal.

radappertization   the reduction in the number of microorganisms by exposure to ionizing radiation.

radioimmunoassay   a highly sensitive serological technique used for assaying specific antibodies or antigens employing a radioactive label.

radioisotopes   radioactive isotopes.

radurization   sterilization by exposure to ionizing radiation.

rancid   having the characteristic odor of decomposing fat, chiefly due to the liberation of butyric and other volatile fatty acids.

raphe   a slit or pore in the cell wall of a diatom.

rDNA   recombinant DNA.

reagins   a group of antibodies in serum that react with the allergens responsible for the specific manifestations of human hypersensitivity; a heterophile antibody formed during syphilis infections.

recalcitrant   a chemical that is totally resistant to microbial attack.

recipient strain any strain that receives genetic information from another.

reciprocal recombination occurs as a result of crossing-over in which a symmetrical exchange of genetic material takes place; i.e., the genes lost by one chromosome are gained by the other and vice versa.

recombinant any organism whose genotype has arisen as a result of recombination; also any nucleic acid that has arisen as a result of recombination.

recombinant DNA technology see genetic engineering.

recombination the exchange and incorporation of genetic information into a single genome, resulting in the formation of new combinations of alleles.

reducing power the capacity to bring about reduction.

reduction an increase in the negative valence or a decrease in the positive valence of an element resulting from the gain of electrons.

reductive amination the reaction of an α-carboxylic acid with ammonia to produce an amino acid.

refraction the deviation of a ray of light from a straight line in passing obliquely from one transparent medium to another of different density.

refractive index an index of the change in velocity of light when it passes through a substance causing a deviation in the path of the light.

relapsing fever a human disease characterized by recurrent fever, caused by a *Borrelia* sp. and transmitted by ticks and lice.

replica plating a technique by which various types of mutants can be isolated from a population of bacteria grown under nonselective conditions, based on plating cells from each colony onto multiple plates and noting the positions of inoculation.

replication multiplication of a microorganism; duplication of a nucleic acid from a template; the formation of a mold for viewing by electron microscopy.

replication fork the Y-shaped region of a chromosome that is the growing point during replication of DNA.

replicon a nucleic acid molecule that possesses an origin and is therefore capable of initiating its own replication.

repression the blockage of gene expression.

repressor protein a protein that binds to the operator and inhibits the transcription of structural genes.

reproduction a fundamental property of living systems by which organisms give rise to other organisms of the same kind.

resolution the fineness of detail observable in the image of a specimen.

resolving power a quantitative measure of the closest distance between two points that can still be seen as distinct points when viewed in a microscope field; depends largely on the characteristics of its objective lens and the optimal illumination of the specimen.

respiration a mode of energy-yielding metabolism requiring a terminal electron acceptor for substrate oxidation, with oxygen frequently used as the terminal electron acceptor.

respiratory tract the structures and passages involved with the intake of oxygen and expulsion of carbon dioxide in animals.

reticuloendothelial system see mononuclear phagocyte system.

Reye's syndrome a neurological disease that sometimes occurs after a viral infection.

rheumatic fever a febrile disease characterized by painful migratory arthritis and a predilection to heart damage leading to chronic valvular disease; the cause is unknown.

rhizopods a root-like pseudopodium of protozoa.

rhizosphere an ecological niche that comprises the surfaces of plant roots, and that region of the surrounding soil in which the microbial populations are affected by the presence of the roots.

ribonucleic acid a linear polymer of ribonucleotides in which the ribose residues are linked by $3',5'$-phosphodiester bridges; the nitrogenous bases attached to each ribose residue may be adenine, guanine, uracil, or cytosine.

ribosomal RNA RNA of various sizes that makes up part of the ribosomes, constituting up to 90 percent of the total RNA of a cell; single-stranded RNA but with helical regions formed by base-pairing between complementary regions within the strand.

ribosomes intracellular structures composed of ribosomal RNA and protein, the sites where protein synthesis occurs.

rickettsialpox a disease caused by *Rickettsia akari*, characterized by enlargement of the lymph nodes, fever, chills, headache, secondary rash, and leukopenia following an initial papule at the locale of the bite of a mite.

ringworm any mycosis of the skin, hair, or nails in humans or other animals in which the causal agent is a dermatophyte, tinea.

ripen to bring to completeness or perfection; to age or cure as in cheese; to develop a characteristic flavor, odor, texture, and color.

RNA see ribonucleic acid.

RNA polymerase enzyme that catalyzes the formation of RNA macromolecules.

Rocky Mountain spotted fever a tick-borne human rickettsial disease which occurs in parts of North America, caused by *Rickettsia rickettsii*.

rot any of a variety of unrelated plant diseases characterized by primary decay and disintegration of host tissue.

R plasmid a plasmid encoding for antibiotic resistance.

rRNA see ribosomal ribonucleic acid.

rubella see German measles.

rumen one of the four compartments that form the stomach of a ruminant animal where anaerobic microbial degradation of plant residues occurs, producing nutrients that can be metabolized by the animal.

rusts plant diseases caused by fungi of the order Uredinales, so-called because of the rust-colored spores formed by many of the causal agents on the surfaces of the infected plants; fungi in the order Uredinales.

salinity the concentration of salts dissolved in a solution.

salmonellosis any disease of humans or animals in which the causal agent is a species of *Salmonella*, including typhoid and paratyphoid fever, but most frequently referring to a gastroenteritis.

salt lake an inland water body with a high salt concentration normally approaching saturation.

sanitary landfill a method for disposal of solid waste in low-lying areas, with wastes covered with a layer of soil each day.

sanitize to make sanitary as by cleaning or sterilizing.

saprophytes organisms, e.g., bacteria and fungi, whose nutrients are obtained from dead and decaying plant or animal matter in the form of organic compounds in solution.

Sarcodina a major taxonomic group of protozoans characterized by the formation of pseudopodia.

scanning electron microscope (SEM) a form of electron microscopy in which the image is formed by a beam of electrons that has been reflected from the surface of a specimen; a type of electron microscope in which a beam of electrons systematically sweeps over the specimen, and the intensity of secondary electrons generated at the specimen's surface where the beam impacts is measured and the resulting signal is used to determine the intensity of a signal viewed on a cathode-ray tube that is scanned in synchrony with the scanning of the specimen.

schizogamy a form of asexual reproduction characteristic of certain groups of sporozoan protozoa; coincident with cell growth, nuclear division occurs several or numerous times, producing a schiziont that then further segments into other cells.

SCP see single-cell protein.

secondary immune response the response of an individual to the second or subsequent contact with a specific antigen characterized by a short lag period and the production of a high antibody titer.

secondary sewage treatment the treatment of the liquid portion of sewage containing dissolved organic matter, using microorganisms to degrade the organic matter that is mineralized or converted to removable solids.

secretory pertaining to the act of exporting a fluid from a cell or organism.

sedimentation the process of settling, commonly of solid particles from a liquid.

selective medium an inhibitory medium or one designed to encourage the growth of certain types of microorganisms in preference to any others that may be present.

semiconservative replication the production of double-stranded DNA containing one new strand and one parental strand.

septate separated by cross walls.

septicemia a condition where an infectious agent is distributed through the body in the bloodstream; blood poisoning, the condition attended by severe symptoms in which the blood contains large numbers of bacteria.

septic tank a simple anaerobic treatment system for waste water where residual solids settle to the bottom of the tank and the clarified effluent is distributed over a leaching field.

septum in bacteria the partition or cross wall formed during cell division that divides the parent cell into two daughter cells; in filamentous organisms, e.g., fungi, one of a number of internal transverse cross walls that occur at intervals of length within each hyphae.

serology the *in vitro* study of antigens and antibodies and their interactions.

serotypes the antigenically distinguishable members of a single species.

serum the fluid fraction of coagulated blood.

serum hepatitis a form of viral hepatitis transmitted by the parenteral injection of human blood or blood products contaminated by the causal agent.

serum killing power the antimicrobial activity of the serum of a patient receiving antibiotics; an *in vivo* measure of antibiotic activity.

sewage the refuse liquids or waste matter carried by sewers.

sewage treatment the treatment of sewage to reduce its biological oxygen demand and to inactivate the pathogenic microorganisms present.

sexually transmitted diseases a disease whose transmission occurs primarily or exclusively by direct contact during sexual intercourse.

sexual reproduction reproduction involving the union of gametes from two individuals.

sexual spore a spore resulting from the conjugation of gametes or nuclei from individuals of different mating type or sex.

sheath a tubular structure formed around a filament or around a bundle of filaments, occurring in some bacteria.

shigellosis bacillary dysentary caused by bacteria of the genus *Shigella*.

shunt a diversion from the normal path as an alternative pathway in metabolism.

sigma unit a subunit of RNA polymerase that helps recognize the promoter site.

signal sequence a region of nucleotides at the beginning of a messenger RNA molecule and the corresponding sequence of amino acids in the synthesized protein that indicates that the protein is an exoprotein and is responsible for initiating the export of that protein across the cytoplasmic membrane.

simple matching coefficient a similarity measure used in taxonomic analysis that includes both negative and positive matches in its calculation.

single-cell protein (SCP) protein produced by microorganisms and primarily composed of microbial cells; sources of single-cell protein include bacteria, fungi, and algae.

site-specific recombination see nonhomologous recombination.

slime layer a capsular layer surrounding microbial cells composed of diffuse secretions that adhere loosely to the cell surface.

sludge the solid portion of sewage.

smallpox an extinct disease caused by the variola virus that in humans caused an acute, highly communicable disease characterized by cutaneous lesions on the face and limbs.

smuts plant diseases caused by fungi of the order Ustilaginales and that typically involve the formation of masses of dark-colored teliospores on or within the tissues of the host plant.

sneeze a sudden, noisy, spasmodic expiration through the nose, caused by the irritation of nasal nerves.

sodium a metallic metal of the alkali group.

somatic antigens antigens that form part of the main body of a cell, usually at the cell surface and distinguishable from antigens that occur on the flagella or capsule.

somatic cells any cell of the body of an organism except the specialized reproductive germ cells.

species a taxonomic category ranking just below a genus and including individuals that display a high degree of mutual similarity and that actually or potentially inbreed.

specificity the restrictiveness of interaction; of an antibody, refers to the range of antigens with which an antibody may combine; of an enzyme, refers to the substrate that is acted upon by that enzyme; of a pathogen or parasite, refers to the range of hosts.

spectrophotometer an instrument that measures the transmission of light as a function of wavelength, allowing quantitative measure of intensity of two sources or wavelengths.

spectrum a range, e.g., of frequencies within which radiation has some specified characteristic, such as the visible light spectrum.

spermatia in certain ascomycetes and basidiomycetes, nonmotile, male reproductive cells.

spherical aberration a form of distortion of a microscope lens based on the differential refraction of light passing through the thick central portion of a convex-convex lens and the light passing through the thin peripheral regions of the lens.

spheroplasts spherical structures formed from bacteria, yeasts, and other cells by weakening or partially removing the rigid component of the cell wall.

split genes genes coded for by noncontiguous segments of the DNA so that the messenger RNA and the DNA for the protein product of that gene are not colinear.

sporangium a sac-like structure within which numbers of motile or nonmotile asexually derived spores are formed.

spore an asexual reproductive or resting body that is resistant to unfavorable environmental conditions capable of generating viable vegetative cells when conditions are favorable; resistant and/or disseminative forms produced asexually by certain types of bacteria in a process that involves differentiation of vegetative cells or structures, characteristically formed in response to adverse environmental conditions.

Sporozoa a subphylum of parasitic protozoa in which mature organisms lack cilia and flagella, characterized by the formation of spores.

sporozoite the cells produced by the division of the zygote of a sporozoan.

sporulation the process of spore formation.

sputum the material discharged from the surface of the air passages, throat, or mouth, consisting of saliva, mucus, pus, microorganisms, fibrin, and/or blood.

stab cell an immature lymphocyte.

stain a substance use to treat cells or tissues to enhance contrast so that specimens and their details may be detected by microscopy.

staphylococcal food poisoning an acute, nonfebrile condition caused by the enterotoxins of certain strains of *Staphylococcus*.

stationary phase a growth phase during which the death rate equals the rate of reproduction, resulting in zero growth rate in batch cultures.

statospore a resting spore of some algae consisting of two pieces.

stem cell a formative cell; a blood cell capable of giving rise to various differentiated types of blood cells.

stock culture a culture that is maintained as a source for authentic subcultures; a culture whose purity is ensured and from which working cultures are derived.

strain a cell or population of cells that has the general characteristics of a given type of organism, e.g., bacterium or fungus, or of a particular genus, species, and serotype.

streptomycin an aminoglycoside antibiotic produced by *Streptomyces griseus*, affecting protein synthesis by inhibiting polypeptide chain initiation.

structural gene a gene whose product is an enzyme, structural protein, tRNA, or rRNA, as opposed to a regulator gene whose product regulates the transcription of structural genes.

subcutaneous beneath the skin.

substrate a substance upon which an enzyme acts.

succession the replacement of populations by other populations better adapted to fill the ecological niche.

sulfide a compound of sulfur with an element or basic radical.

superoxide dismutase an enzyme that catalyzes the reaction between superoxide anions and protons, the products being hydrogen peroxide and oxygen.

superoxide radical a toxic oxygen free radical.

suppressor mutation a mutation that alleviates the effects of an earlier mutation at a different locus.

surfactant a surface-active agent.

susceptibility a characteristic reflecting the likelihood that an individual will acquire a disease if exposed to the causative agent.

Svedberg unit the unit in which the sedimentation coefficient of a particle is commonly quoted; when values are given in seconds the basic unit is 10^{-13} seconds.

symbiosis an obligatory interactive association between two populations producing a stable condition in which the two different organisms live together in close physical association to their mutual advantage.

symptom a physiological disorder that results in a detectable deviation from the normal healthy state and is normally reflected by complaints from a patient.

symptomology the symptoms of disease taken together.

synergism in antibiotic action, when two or more antibiotics are acting together, the production of inhibitory effects on a given organism that are greater than the additive effects of those antibiotics acting independently; an interactive but nonobligatory association between two populations in which each population benefits.

syngamy the union of gametes to form a zygote.

syntrophism a phenomenon in which the growth or improved growth of an organism is dependent on the provision of one or more metabolic factors or nutrients by another organism growing in the vicinity.

syphilis a chronic, communicable, sexually transmitted disease peculiar to humans, caused by *Treponema pallidum* and characterized by a variety of lesions.

systematics a synonym of taxonomy; the range of theoretical and practical studies involved in the classification of organisms.

systemic infections infections that are disseminated throughout the body via the circulatory system.

T cells lymphocyte cells that are differentiated in the thymus and are important in cell-mediated immunity, as well as modulation of antibody-mediated immunity.

T helper cells   a class of T cells that enhance the activities of B cells in antibody-mediated immunity.

T suppressor cells   a class of T cells that depress the activities of B cells in antibody-mediated immunity.

taxis   a directional locomotive response to a given stimulus exhibited by certain motile organisms or cells.

taxon   a taxonomic group, e.g., genus, family , order, etc.; a species.

taxonomy   the science of biological classification, the grouping of organisms according to their mutual affinities or similarities.

teichoic acids   polymers of ribitol or glycerol phosphate found in the cell walls of some bacteria.

temperate phage   bacteriophage with the ability to form a stable, nondisruptive relationship within a bacterium, a prophage.

template   a pattern that acts as a guide for directing the synthesis of new macromolecules.

tertiary sewage treatment   a sewage treatment process beyond secondary ones, aimed at the removal of nonbiodegradable organic pollutants and mineral nutrients.

tetanus   lockjaw, a disease of humans and other animals in which the symptoms are due to a powerful neurotoxin formed by the causal agent, *Clostridium tetani*, present in an anaerobic wound or other lesion, characterized by sustained involuntary contraction of the muscles of the jaw and neck.

tetracyclines   a group of natural and semisynthetic antibiotics that have in common a modified naphthalene ring, bacteriostatic with a broad spectrum of activity.

theca   a layer of flattened membranous vesicles beneath the external membrane of a dinoflagellate; an open or perforated shell-like structure that houses part or all of a cell.

thermal death time   the time required at a given temperature for the thermal inactivation or killing of a specified number of microorganisms.

thermocline   zone of water characterized by a rapid decrease in temperature, with little mixing of water across it.

thermodynamics   the basic relationships between properties of matter, especially those that are affected by changes in temperature, and a description of the conversion of energy from one form to another.

thermophiles   organisms having an optimum growth temperature above 45°C.

thylakoids   flattened, membranous vesicles that occur in the photosynthetic apparati of cyanobacteria and algae; the thylakoid membrane contains chlorophylls, accessory pigments, and electron carriers and is the site of light reaction in photosynthesis.

thymine   a pyrimidine component of DNA.

tinea   the lesions of dermatophytosis, ringworm.

tinsel flagellum   flagellum of eukaryotic organisms that bear fine filamentous appendages along their lengths.

tissue culture   the maintenance or culture of isolated tissues and plant or animal cell lines *in vitro*.

titer   the concentration in a solution of a dissolved substance.

tonsilitis   inflammation of the tonsils, commonly caused by *Streptococcus pyogenes*.

toxicity   the quality of being toxic; the kind and quantity of a poison produced by a microorganism or possessed by a nonbiological chemical.

toxic shock syndrome   a disease caused by the release of toxins from *Staphylococcus*, resulting in a physiological state of shock; major outbreaks of the disease have been associated with the use of tampons during menstruation.

toxin   any organic microbial product or substance that is harmful or lethal for cells, tissue cultures, or organisms; a poison.

toxinogenicity   the ability to produce toxins.

toxoid   a modified protein exotoxin that has lost its toxcity but has retained its specific antigenicity.

toxoplasmosis   an acute or chronic disease of humans and other animals, caused by the intracellular pathogen *Toxoplasma gondii*; transmission is by ingestion of insufficiently cooked meats containing tissue cysts.

trachoma   a communicable disease of the eye, caused by *Chlamydia trachomatis*.

transamination   the transfer of one or more amino groups from one compound to another; the formation of a new amino acid by transfer of the amino group from another amino acid.

transcription   the synthesis of mRNA, rRNA, and tRNA from a DNA template.

transduction   the transfer of bacterial genes from one bacterium to another by bacteriophage.

transfer RNA   a type of RNA involved in carrying amino acids to the ribosomes during translation; for each amino acid there is one or more corresponding tRNA that can bind it specifically.

transformation   a mode of genetic transfer in which a naked DNA fragment derived from one bacterial cell is taken up by another and subsequently undergoes recombination with the recipient's chromosome; in tissue culture the conversion of normal cells to cells that exhibit some or all of the properties typical of tumor cells; morphological and other changes that occur in both B and T lymphocytes on exposure to antigens to which they are specifically reactive.

transition a type of point mutation in which one purine is replaced by another or one pyrimidine is replaced by another.

translation the assembly of polypeptide chains with mRNA serving as the template, a process that occurs at the ribosomes.

translocation nonhomologous recombination.

transmission electron microscope (TEM) a type of electron microscope in which the specimen transmits an electron beam focused on it; image contrasts are formed by the scattering of electrons out of the beam, and various magnetic lenses perform functions analogous to those of ordinary lenses in a light microscope.

transposons translocatable genetic elements; genetic elements that move from one locus to another by nonhomologous recombination allowing them to move around a genome.

transversion a type of point mutation in which a purine is replaced by a pyrimidine or a pyrimidine by a purine.

tricarboxylic acid cycle see Krebs cycle.

trichome a chain or filament of cells that may or may not include one or more resting spores.

trickling filter a simple, film-flow-type aerobic sewage treatment system; the sewage is distributed over a porous bed coated with bacterial growth that mineralizes the dissolved organic nutrients.

tRNA see transfer RNA.

-troph combining form indicating relation to nutrition or to nourishment.

trophic level steps in the transfer of energy stored in organic compounds from one organism to another.

trophophase of fungal metabolism, during batch culture that phase in which growth-directed metabolism is dominant over secondary metabolism.

trophozoite a vegetative or feeding stage in the life cycle of certain protozoans.

trypanosomiasis any of a number of human and animal diseases in which the causal organism is a member of the genus Trypanosoma.

tuberculosis an infectious disease of humans and other animals, caused in humans by *Mycobacterium*

tuberculosis and may affect any organ or tissue of the body, most usually the lungs.

tularemia an acute or chronic systemic disease characterized by malaise, fever, and an ulcerative granuloma at the site of infection, caused by *Francisella tularensis*.

turbidity cloudiness, opacity of a solution.

tyndallization a sterilization process aimed at the elimination of endospore formers in which the material is heated to 80–100°C for several minutes on each of three successive days and incubated at 37°C for the intervening periods.

typhoid fever an acute, infectious disease of humans caused by *Salmonella typhi* that invades via the oral route; the symptoms include fever and skin, intestinal, and lymphoid lesions.

typhus fever an acute infectious disease of humans characterized by a rash, high fever, and marked nervous symptoms, transmitted by the body louse and rat flea infected with *Rickettsia prowazekii*.

ultracentrifuge a high-speed centrifuge that will produce centrifugal fields up to several hundred thousand times the force of gravity; used for the study of proteins and viruses, for the sedimentation of macromolecules, and for the determination of molecular weights.

ultraviolet light short-wavelength electromagnetic radiation in the range 4–400 nm.

unicellular having the form and characteristics of a single cell.

uracil a pyrimidine base, a component of nucleic acids.

urea $CO(NH_2)_2$, a product of protein degradation.

ureases enzymes that split urea into carbon dioxide and ammonia.

urethra the canal through which urine is discharged.

urethritis inflammation of the urethra.

urinary tract the system that functions in the elaboration and excretion of urine.

urkaryote the proposed progenetor of prokaryotic and eukaryotic cells; the primordial living cell.

uv see ultraviolet light.

V_{max} the maximal velocity of an enzymatic reaction occurring when the enzyme is saturated with substrate.

VA mycorrhizae see vesicular-arbuscular mycorrhizae.

vaccine any antigenic preparation administered with the object of stimulating the recipient's specific defense mechanisms to pathogens or toxic agents.

vacuoles a membrane-bound cavity within a cell that may function in digestion, secretion, storage, or excretion.

vaginal tract a region of the female genital tract, the canal that leads from the uterus to the external orifice of the genital canal.

vaginitis inflammation of the vagina.

variant a strain that differs in some way from a particular, named organism.

vectors organisms that act as carriers of pathogens and are involved in the spread of disease from one individual to another.

vegetative cells cells that are engaged in nutrition and growth, not acting as specialized reproductive or dormant forms.

vegetative growth production of a new organism from a portion of an existing organism exclusive of sexual reproduction.

vesicular arbuscular mycorrhizae a common type of mycorrhizae characterized by the formation of vesicles and arbuscules.

venereal disease a sexually transmitted disease.

viability the ability to grow and reproduce.

vinegar a condiment prepared by the microbial oxidation of ethanol to acetic acid.

viral of or pertaining to a virus.

virion a single, structurally complete, mature virus.

viroids the causal agents of certain diseases, resembling viruses in many ways but different in the apparent lack of a virus-like structural organization and their resistance to a wide variety of treatments to which viruses are sensitive; naked infective RNA.

virology the study of viruses and viral diseases.

virulence the capacity of a pathogen to cause disease, broadly defined in terms of the severity of the disease in the host.

virulent pathogen an organism with specialized properties that enhance its ability to cause disease.

virus a noncellular entity that consists minimally of protein and nucleic acid and that can replicate only after entry into specific types of living cells; it has no intrinsic metabolism and replication is dependent on the direction of cellular metabolism by the viral genome; within the host cell viral components are synthesized separately and are assembled intracellularly to form mature, infectious viruses.

vitamins a group of unrelated organic compounds some or all of which are necessary in small quantities for the normal metabolism and growth of microorganisms.

volutin see metachromatic granules.

volva a cup-shaped remnant of the universal veil that surrounds the base of the stalk in mature fruiting bodies of certain fungi.

vomiting the forcible ejection of the contents of the stomach through the mouth.

vulvovaginitis inflammation of the vulva and vagina, usually caused by *Candida albicans*, herpes viruses, *Trichomonas vaginalis*, or *Neisseria gonorrhoeae*.

warts small benign tumors of the skin, caused in humans by the human papilloma virus.

water activity (A_w) a measure of the amount of reactive water available, equivalent to relative humidity, the percent water saturation of the atmosphere.

whiplash flagellum smooth flagella of algae and fungi.

whooping cough pertussis, an acute, respiratory-tract disease mainly in children, caused by *Bordetella pertussis* and characterized by sudden episodes of coughing that usually end in loud whooping inspirations.

wilts plant diseases characterized by a reduction in host tissue turgidity, commonly affecting the vascular system; common causal agents are species of the fungi *Fusarium* and *Verticillium* and the bacteria *Erwinia* and *Pseudomonas*.

windrow a slow composting process that requires turning and covering with soil or compost.

wobble hypothesis hypothesis that accounts for the observed pattern of degeneracy in the third base of a codon and says that the third base of the codon can undergo unusual base-pairing with the corresponding first base in the anticodon.

wort in brewing, the liquor that results from the mixture of mash and water held at 40–65°C for one to two hours, during which the starch is broken down by amylases to glucose, maltose, and dextrins, and proteins are degraded to amino acids and polypeptides.

xanthophyll a pigment containing oxygen and derived from carotenes.

xenobiotic a synthetic product not formed by natural biosynthetic processes; a foreign substance or poison.

xerotolerant able to withstand dryness; an organism capable of growth at low water activity.

yeasts a category of fungi defined in terms of morphological and physiological criteria, typically a unicellular, saprophytic organism that characteristically ferments a range of carbohydrates and in which asexual reproduction occurs by budding.

yellow fever an acute, systemic disease that affects humans and other primates, caused by a togavirus, transmitted to humans by mosquitos.

Z pathway the combination of the cyclic and non-cyclic photophosphorylation pathways in oxygenic photosynthetic organisms describing the metabolic reactions accounting for the trapping of light energy, and the generation of ATP, oxygen, and NADPH during photosynthesis.

z **value** the degrees fahrenheit required to reduce the thermal death time 10-fold.

zoonoses diseases of lower animals.

zoospores motile, flagellated spores.

zygospore thick-walled resting spores formed subsequent to gametangial fusion by members of the zygomycetes.

zygote a single, diploid cell formed from two haploid parental cells during fertilization.

zymogenous term used to describe soil microorganisms that grow rapidly on exogenous substrates.

Appendix III. Descriptions of bacterial genera

Acetivibrio Straight to slightly curved rods, 0.5–0.8 by 4–10 μm; chains of cells up to 40 μm may also occur. Gram negative. Monotrichous cells that exhibit tumbling motility; the flagellum is attached to the cell about one-third of the distance from pole to pole. Chemoorganotrophic: fermentative. Cellulose and cellobiose serve as growth substrates. Obligate anaerobes. Optimal growth 35°C. G + C 38 mole%.

Acetobacter Cells ellipsoidal to rod-shaped, straight or slightly curved, 0.6–0.8 by 1.0–3.0 μm, occurring singly, in pairs or in chains. Pleomorphism frequent in some species. Motile by peritrichous flagella or nonmotile. Endospores not formed. Young cells Gram negative, in older cultures some strains become Gram variable. Chemoorganotrophic: metabolism respiratory, never fermentative. Oxidize ethanol to acetic acid at neutral and acid reactions (pH 4.5). Acetate and lactate are oxidized to CO_2. Hexoses and glycerol are used as carbon sources by most strains; mannitol and glutamate are used poorly or not at all; lactose, dextrin, and starch are not hydrolyzed. Generally no pigments produced. Strict aerobes. Optimal growth ca. 30°C. G + C 55–64 mole%.

Acetobacterium Oval-shaped short rods, 1.0 by 2.0 μm, frequently occurring as pairs. Motile by means of one or two subterminal flagella. Endospores not produced. Obligate anaerobes. Facultatively chemolithotrophic: oxidize hydrogen and reduce carbon dioxide. Fermentation of fructose, glucose, lactate, and glycerate yields acetate. Optimal growth 30°C. G + C 39 mole%.

Acholeplasma Cells spherical, with a minimum diameter ca. 125–220 nm, filaments, usually about 2–5 μm in length. Cells bounded by a triple-layered membrane. Lack a true cell wall. Nonmotile. Gram negative. Colonies on solid medium usually have a fried egg appearance, relatively large, up to 3 mm in diameter. Chemoorganotrophic: metabolism fermentative. Serum or cholesterol not required for growth. Arginine and urea not hydrolyzed. Film or spots not produced on horse serum agar or egg yolk medium; no clearing. Facultative anaerobes. Growth occurs at 20–40°C. G + C 30–33 mole%.

Achromatium Cells spherical to ovoid or cylindrical with hemispherical ends. Division by constriction in the middle. Movement, if any, of a slow, rolling, jerky-type, gliding motility. Photosynthetic pigments absent. Cells may contain sulfur droplets and large spherules of calcium carbonate. No resting stages known. Microaerophilic: apparently require sulfides.

Achromobacter Cells straight rods, 0.8–1.2 by 2.5–3.0 μm, with rounded ends. Endospores not produced. Gram negative. Motile with one to nine peritrichous flagella. Chemoorganotrophic: respiratory, never fermentative. Oxidative acid production from various carbohydrates. Strictly aerobic. Catalase positive; oxidase positive. Citrate utilized. Nitrate reduced to nitrite. Urea not hydrolyzed. Gelatin not hydrolyzed. G + C 65–70 mole%.

Acidaminococcus Cocci, 0.6–1.0 μm in diameter, often as oval or kidney-shaped diplococci. Gram negative. Flagella, motility, and spores not demonstrated. Limited metabolism of carbohydrates; only 60 percent of strains ferment glucose. Lactate, fumarate, malate, succinate, citrate, and pyruvate are not used as energy sources; pyruvate suppresses growth completely. Amino acids can serve as sole energy source for growth. Nutritional requirements multiple. Ammonia produced. Anaerobes. No growth on agar surface incubated in air. Optimal growth 30–37°C. G + C 55–57 mole%.

Acidiphilium Cells rod-shaped, 0.3–0.4 by 0.6–0.8 μm, but size varies with growth conditions. Gram negative. Motile or nonmotile; motility by one polar flagellum and two lateral flagella. Catalase positive; oxidase negative or weak. Acidophilic: grows at pH 1.9–5.9, but not 6.1. Chemoorganotrophic: grows best at low concentrations of organic matter. Optimal growth 35–41°C. G + C 68–70 mole%.

Acinetobacter Rods, usually very short and plump, typically 1.0–1.5 by 1.5–2.5 μm in logarithmic phase, approaching coccus shape in stationary phase, predominantly in pairs and short chains. No spores formed. Flagella not present. Gram negative. Chemoorganotrophic: oxidative metabolism. Oxidase negative, catalase positive. Acetoin, indole, and H_2S not produced. Strict aerobes. Optimal growth 30–32°C. G + C 40–47 mole%.

Actinobacillus Cells spherical, oval or rod-shaped, 0.3–0.5 by 0.6–1.4 μm, mostly rods interspersed with coccal elements often occurring at the pole of a rod, giving a characteristic Morse code (dot–dash) form; longer forms up to 6 μm common on media containing glucose or maltose. Cells arranged singly, in pairs, or more rarely in chains. Nonmotile. Do not form endospores. Gram negative. Irregular staining. Cultures very sticky, especially on primary isolation; colonies may be difficult to remove completely from the agar surface. Chemoorganotrophic: metabolism fermentative. Acid but no gas within 24 hours from glucose, fructose, and xylose. Dulcitol, inositol, and inulin are not fermented. Other carbohydrates may be fermented with acid but no gas. β-galactosidase positive. Methyl red negative. Hydrogen sulfide produced by most strains. Nitrates reduced to nitrites. Indole not produced. Urease positive. Facultative anaerobes. Optimal growth 20–42°C. G + C 40.6–42.0 mole%.

Actinomadura Produce true mycelium. Substrate mycelium does not fragment into rod or coccoid elements. Form sort chains of conidia on the aerial mycelium. Chains of arthrospores produced from aerial mycelium; spores occur within a thin fibrous sheath with a verticillate arrangement. Spores are nonmotile. Cell wall contains meso-diaminopimelic acid. G + C 77 mole%.

Actinomyces Gram positive, irregularly staining bacteria; nonacid-fast, nonspore-forming and nonmotile. Filaments with true branching may predominate and are particularly evident in 18–48 hour microcolonies. Diphtheroid cells or branched rods are common; V, Y, and T forms occur. Filaments, 1 μm or less in diameter, varying in length and degree of branching occur in most strains. Chemoorganotrophic: carbohydrates are fermented with the production of acid but no gas. End products from glucose fermentation may include acetic, formic, lactic, and succinic acids but not propionic acid. Proteolytic activity rare, and, if present, is weak. Indole not produced; urease negative. Some species may show greening or complete lysis of rabbit red blood cells. Cell wall peptidoglycan contains alanine, glutamic acid, and lysine with either aspartic acid or ornithine; diaminopimelic acid does not occur in any currently recognized species. Cell wall carbohydrates may include glucose, galactose, rhamnose, and other deoxy sugars; arabinose does not occur. Organic nitrogen is required for growth. Facultative anaerobes; most are preferentially anaerobic, and one species grows well aerobically. CO_2 is required for maximum growth.

Actinoplanes Produce true mycelia with spores borne inside sporangia. Sporangia, 3–20 by 6–30 μm, spherical, subspherical, cylindrical with rounded ends, to very irregular. Produce motile spores. Spores, globular to subglobular, 1–1.5 μm in diameter, in coils, nearly straight chains, or irregularly arranged in sporangia; motile by a tuft of polar flagella 2–6 μm in length. Hyphae, 0.2–2.6 μm in diameter, branched, irregularly coiled, twisted or straight with few septa. Vertical pallisade hyphae formed on certain agars; aerial mycelium generally scant. Most species brilliantly colored on peptone Czapek and certain other agars; colors: orange, red, yellow, violet, and purple. Some form diffusible pigments that color the agar blue, red, yellow, brownish, or greenish. Chemoorganotrophic: metabolism respiratory; oxygen is universal electron acceptor. No organic growth factors are required. H_2S is produced by some species. Strict aerobes. Optimal growth 18–35°C. G + C 72.1–72.6 mole%.

Actinopolyspora Produce true mycelium. Fragmentation occurs in older substrate mycelium. Subtrate growth buff-colored. Spores not produced from substrate mycelium. Aerial mycelium white. Long chains of spores (greater than 20) produced from aerial mycelium. Gelatin liquefied. Starch not hydrolyzed. Extreme halophile. G + C 64.2 mole%.

Actinopycnidium Produce true mycelium. Aerial mycelium with chains of nonmotile conidia. Similar to *Streptomyces* but also form pycnidia-like structures on aerial mycelium. Long chains of spores produced on aerial mycelium.

Actinosporangium Produce true mycelium. Aerial mycelium with chains of nonmotile conidia. Similar to *Streptomyces*, but spores accumulate in drops. Long chains of spores produced on aerial mycelium.

Actinosynnema Produce true mycelium. Substrate hyphae are about 0.5 μm in diameter and coalesce to form a mycelium bearing synnemata (up to 180 μm high) and dome-like bodies on which septate aerial hyphae are formed. The aerial hyphae mature into chains of spores that become flagellated with 5–15 peritrichous flagella. Aerial hyphae fragments into flagellated elements. Gram positive. Aerobic. Gelatin hydrolyzed. Starch hydrolyzed. Acid produced from various carbohydrates. Growth 10–35°C.

Aegyptianella Obligately parasitic, found within erythrocytes. Infect only ruminants, transmitted by arthropod vectors. With Romanowsky-type stains, purple-staining intracytoplasmic inclusions containing bacteria occur in erythrocytes. Inclusions round 0.3–1.0 μm diameter. Thin sections of inclusion bodies examined by electron microscopy show up to 26 bacteria or initial bodies enclosed in a double membrane and with internal morphology similar to *Anaplasma*. Within the inclusion, initial bodies reproduce by binary fission. No appendages to inclusions.

Aerococcus Spheres, 1.0–2.0 μm in diameter, with a strong tendency toward tetrad formation when grown in liquid media. Nonmotile. Gram positive. Growth on solid media is generally sparse and beaded (small discrete colonies). Chemoorganotrophic: acid but no gas is produced from glucose, fructose, galactose, mannose, maltose, and sucrose; acid is usually produced from lactose, trehalose, and mannitol. Microaerophilic. Catalase activity absent or weak. Hydrogen peroxide is produced during aerobic growth. G + C 36–40 mole%.

Aeromonas Cells straight, rod-shaped with rounded ends to coccoid 1.0–4.4 μm, occasionally forming filaments up to 8 μm long; occur singly, in pairs or chains; motile by polar flagella, generally monotrichous; some species nonmotile. No resting stages known. Gram negative. Chemoorganotrophic: metabolism both respiratory and fermentative; carbohydrates broken down to acid or acid and gas. Some species show a butanediol fermentation. Some species grow on a mineral medium with ammonia as the sole source of nitrogen and one of the following as the sole source of carbon: glucose, arginine, asparagine, or histidine; all species grow in a mixture of L-arginine, asparagine, leucine, and methionine. Glucose, fructose, maltose, and trehalose are broken down; adonitol, dulcitol, inositol, inulin, melezitose, sorbose, and xylose are not feremented. Starch, dextrin, and glycerol hydrolyzed. Casein hydrolyzed; gelatin liquefied; deoxyribonuclease, arginine dehydrogenase, and phosphatase produced; glutamic acid not decarboxylated; urea not broken down. Nitrate reduced to nitrite. Oxidase positive; catalase positive. Facultative anaerobes. Optimal growth 38–41°C. G + C 57–63 mole%.

Agitococcus Cocci, 1.0–2.0 μm diameter, which distend with sudanophilic granules. Gram negative. Twitching motility. No flagella. Nonencapsulated. Aerobic. Chemoorganotrophic. Oxidase positive; catalase negative. Nitrate not reduced. G + C 41.2–43.2 mole%.

Agrobacterium Rods 0.8 by 1.5–3.0 μm. Motile by one to four peritrichous flagella; if only one flagellum, lateral attachment is more common than polar. Spores not produced. Gram negative. Growth on carbohydrate-containing media usually accompanied by production of copious extracellular, polysaccharide slime. Colonies nonpigmented; usually smooth, but rough colonies are produced by many strains. Chemoorganotrophic: metabolism respiratory. Molecular oxygen is the terminal electron acceptor. Gelatin slowly liquefied in several weeks or not at all. Casein not hydrolyzed. Utilize a wide range of simple carbohydrates, organic acids, and amino acids as energy sources but not cellulose, starch,

agar, or chitin. Ammonium salts, nitrates, and some amino acids serve as sole nitrogen sources for some species; others require amino acids with additional growth factors. Normally oxidase positive. Aerobic but able to grow under reduced oxygen tensions in plant tissue. Optimal growth 25–30°C. G + C 59.6–62.8 mole%.

Agromyces Initial growth on agar as microcolonies composed of branched, filamentous elements that subsequently undergo septation and fragmentation to yield coccoid and diphtheroid cells. Gram positive, but segments of the mycelium become Gram negative during fragmentation. Aerial mycelium and spores not produced. Not acid-fast. Nonmotile. Microaerophilic to aerobic. Catalase negative; oxidase negative. Weakly proteolytic. Sugars oxidized without gas production. Optimal growth 30°C. G + C ca. 71.4 mole%.

Alcaligenes Cells rods, coccal rods, or cocci, 0.5–1.2 μm by 0.5–2.6 μm, usually occurring singly. Motile with one to four (occasionally up to eight) peritrichous or degenerate peritrichous flagella. No resting stages known. Gram negative. Chemoorganotrophic: metabolism respiratory, never fermentative. Molecular oxygen is the terminal electron acceptor. Strict aerobes. Some strains capable of anaerobic respiration in the presence of nitrate or nitrite. Do not fix gaseous nitrogen. Not actively proteolytic in casein or gelatin media. Cellulose, chitin, and agar not hydrolyzed. Oxidase positive. Optimal growth 20–37°C. G + C 57.9–70 mole%.

Alteromonas Rods, straight or curved, motile with polar flagella. Chemoorganotrophic: metabolism respiratory, not fermentative. Pigments not produced. G + C 40–50 mole%.

Alysiella Cells, 0.6 by 2–3 μm, arranged side by side in pairs to form filaments which are flat, not cylindrical; terminal cells are not rounded. Gliding motility on some surfaces. No spores. No capsules. Gram negative. Chemoorganotrophic: metabolism fermentative; optimal growth in presence of blood or serum. Strict aerobes. Optimal growth 37°C.

Amoebobacter Spherical to ovoid bacteria, multiplying by binary fission. Nonmotile. Gram negative. Contain bacteriochlorophyll *a* and carotenoids, both pigments being located in internal membranes of vesicular type. Contain gas vacuoles in the central part of the cell. Anaerobic. Capable of photosynthesis in the presence of hydrogen sulfide, during which they produce and store, as an intermediate oxidation product, elemental sulfur in the form of globules in the gas vacuole-free peripheral part of the cells. Molecular hydrogen may be used as electron donor. Molecular oxygen is not evolved during photosynthesis. Cell suspensions appear in various shades of pink to purple due to the photosynthetic pigments. Hydrogenase and catalase present. G + C 64.3–65.3 mole%.

Amorphosporangium Produce true mycelium. Spores borne inside sporangia. Sporangia very irregular in shape, multilobed, 6–25 μm wide by 8–15 μm high, width usually greater than height. Spores rod-shaped, 0.5–0.7 by 1.0–1.5 μm; motile by two to three polar flagella. Mycellium brilliant shades of orange. No organic growth factors required.

Ampullariella Produce true mycelium. Spores borne inside sporangia. Sporangia bottle or flask-shaped, finger-like or lobate; 5–20 by 8–30 μm. Spores arranged in parallel chains within the sporangium, rod-shaped, 0.5–1 by 2–4 μm, motile by polytrichous, polar flagella of 3.5–6 μm in length. Optimal growth 18–35°C. G + C 72.3 mole%.

Anaebaena Cyanobacterium. Photoautotrophic: can utilize water as electron donor with evolution of oxygen. Contains chlorophyll *a* and phycobiliproteins. Chlorophyll and carotenoids located in a system of thylakoids internal to the cytoplasmic membrane. Filamentous. A trichome (chain of cells) forms by intercalary division in one plane only. In the absence of combined nitrogen, heterocysts are produced; heterocysts are associated with the ability to fix atmospheric nitrogen. Akinetes also may be formed. Reproduction by random trichome breakage, by formation of hormogonia, and sometimes by germination of akinetes. Hormogonia are distinguishble from mature trichomes by the absence of heterocysts and by one or more of the following characteristics: rapid gliding motility, smaller cell size, cell shape, and gas vacuolation. Heterocysts are intercalary or terminal; position of akinetes (if produced) is variable. Vegetative cells are spherical, ovoid, or cylindrical. G + C 38–44 mole%.

Anaerobiospirillum Cells spiral-shaped. Motile with bipolar tufts of flagella. Gram negative. Spores not produced. No axial filament. Chemoorganotrophic: utilize carbohydrates. Produce mainly acetic and succinic acids from glucose utilization. G + C 44 mole%.

Anaeroplasma Rod-shaped cells. Lack true cell wall. Colonies have fried egg appearance. Chemoorganotrophic: fermentative. Starch fermented. Urea not hydrolyzed. Growth at 30–47°C. Sterols required for growth. Anaerobic.

Anaerovibrio Cells curved or straight rods. Motile. Gram negative. Chemoorganotrophic: fermentative. Propionate produced from fructose. Nitrate not reduced. Anaerobic.

Anaplasma Obligate parasites of vertebrates; transmitted by arthropod vectors. This organism appears within erythrocytes as dense, homogeneous, round structures, 0.3–1.0 μm in diameter. Electron microscopy reveals that these structures are inclusions separated from the cytoplasm of the erythrocyte by a limiting membrane. Each inclusion contains from one to eight subunits or initial bodies, which are the actual parasitic bacteria, each 0.3–0.4 μm in diameter that are dense aggregates of fine granular material embedded in an electron-lucid plasma and all enclosed in a double membrane. Within the vacuole the initial body multiplies by binary fusion and forms an inclusion. Metabolism apparently aerobic, produces catalase. Do not produce pigments. Incorporate amino acids from plasma *in vitro*.

Ancalochloris Multiple appendages with numerous prosthecae. Divides by budding. New buds differentiate 2–4 appendages on the cell surface. Prosthecae do not serve reproductive function. Contains gas vacuoles; otherwise similar to *Prothecochloris*. Gram negative. Nonmotile. Obligate anaerobes. Photosynthetic: in the presence of sulfide, globules of elemental sulfur are deposited outside the cell; assimilate carbon dioxide using hydrogen sulfide as electron donor. Bacteriochlorphyll *a* and *c* or *d* and carotenoids. Green color due to photosynthetic pigments.

Ancalomicrobium Unicellular, Gram negative bacteria having from two to eight prosthecae extending from the cell. These prosthecae attain a length of about 3 μm at maturity, about three times the diameter of the cell. Occasionally, the prosthecae are divided into two parts; the position of the division along the appendage is variable. Cells reproduce by budding. Buds are formed directly from one position on the mother cell, never from the prosthecae. Two to four prosthecae differentiate from the developing bud. Division occurs transversely when the bud has attained approximately the same size as the mother cell. Cells are nonmotile and have no holdfasts. Gas vacuoles may be present in some cells. Chemoorganotrophic. Facultative anaerobes.

Aquaspirillum Helical, curved, or straight rods. Gram negative. Motile with bipolar tufts of flagella; some species have 1–2 flagella at one pole; some species have one flagellum at both poles. Chemoorganotrophic: respiratory; one species facultatively chemolithotrophic capable of hydrogen oxidation. Aerobic. G + C 50–65 mole%.

Arachnia Branched diphtheroid rods 0.2–0.3 μm by 3–5 μm, and branched filaments 5–20 μm or longer. Cells occasionally of uneven diameter, frequently with swollen or clubbed ends. Swollen coccoid cells and/or spheroplasts, of diameter to 5 μm, formed during stationary growth phase. Gram positive, nonacid-fast, nonspore-forming, and nonmotile. Microcolonies definitely filamentous, composed of branched nonseptate and septate filamentous elements often originating from a common center. No aerial filaments. Chemoorganotrophic: metabolism fermentative, carbohydrates being the fermentable substrates; fermentation of glucose in absence of air yields carbon dioxide, acetic acid, propionic acid, and traces of DL-lactic and succinic acids; glucose fermented in air to carbon dioxide and acetic acid, but fermentative products are not oxidized beyond acetate. Does not ferment lactate or pyruvate. Catalase negative. Facultative anaerobes. CO_2 is not required for aerobic or anaerobic growth. Cell wall contains diaminopimelic acid.

Archangium Vegetative cells slender, tapered, flexible rods. Gliding motility. Under appropriate conditions cells aggregate to form fruiting bodies. Sporangia lacking; fruiting bodies irregular in form, swollen or brain-like, consisting internally of twisted-intertwined masses. There is no definite slime wall. Produce rod-shaped microsts (myxospores) not in sporangia. Microcysts are very short rods, ellipsoids or spheres, refractile or optically dense. Vegetative colonies do not etch or erode agar media. Chemoorganotrophic: metabolism respiratory, never fermentative. Strictly aerobic. G + C 67–71 mole%.

Arthrobacter Cells that in complex media undergo a marked change in form during the growth cycle. Older cultures (generally 2–7 days) are composed entirely or largely of coccoid cells. In some strains the coccoid cells are uniform in size and spherical and resemble micrococci; in others they are spherical to ovoid or slightly elongate. In some cultures larger coccoid cells some two to four times the size of the remainder may occur; these may predominate under some cultural conditions. On transfer to fresh complex medium growth occurs by enlargement (swelling) of the coccoid cells followed by elongation from one or occasionally two parts of the cell to give rods which usually have a diameter less than that of the enlarged coccoid cell. In the larger coccoid cells outgrowth may occur at two, three, or rarely four parts of the cell. In both cases subsequent growth and division give rise to irregular rods that vary considerably in size and shape and include straight, bent, and curved, wedge-shaped and club-shaped forms. A proportion of the rods are arranged at an angle to each other giving V formations but more complex angular arrangements often occur. Postfission outgrowths, usually from the proximal ends of one or both cells of a pair of rods, and bud-like outgrowths from segments of septate rods, especially in richer media, may give the appearance of rudimentary branching, but true mycelia are not formed. As the exponential phase proceeds, the rods become shorter and are eventually replaced by the coccoid cells characteristic of stationary phase cultures. The coccoid cells are formed either by a gradual shortening of the rods at each successive division or, especially in richer media, by multiple fragmentation of larger rods. The rods are nonmotile or motile by one subpolar or a few lateral flagella. Do not form endospores. Gram positive; however, the rods may be readily decolorized and may show only Gram positive granules in otherwise Gram negative cells. Coccoid cells are Gram positive but may be weakly so. Not acid-fast. Chemoorganotrophic: metabolism respiratory, never fermentative. Molecular oxygen is the terminal electron acceptor. Little or no acid is produced from glucose in peptone medium. Do not attack cellulose. Catalase positive. Strict aerobes. Optimal growth 20–30°C. G + C 60–72 mole%.

Asticcacaulis Cells rod-shaped (0.5–0.7 μm by 1–3 μm). Typically with an appendage ca. 0.15 μm in diameter, varying in length among isolates and with environmental conditions, arising eccentrically from one pole, or laterally along the cell. In the latter case there may be two appendages per cell. Although structurally similar to stalks of *Caulobacter*, these appendages are called pseudostalks because they lack adhesive material and cannot function in attachment. Each cell typically possesses a small mass of adhesive material at a subpolar (rarely polar) position on the cell. Occur singly; in dense populations, cells may adhere to each other in rosettes, the poles of the cells associated with a common mass of adhesive material. Cell division by transverse, binary, disparate, asymmetrical fission of pseudostalked cells. At the time of separation, one cell is longer and possesses one or two pseudostalks; the shorter sibling has a single subpolar flagellum. The flagellated cell develops a pseudostalk and enters the nonmotile vegetative phase. Resting stages not known. Gram negative. Colonies circular, convex, glistening; colorless. Chemoorganotrophic: strictly respiratory. The principal pathway of intermediary carbon metabolism is the Entner-Doudoroff pathway. Carbon is stored as poly-β-hydroxybutyric acid. Strict aerobes. Optimal growth 15–35°C. G + C 55–60 mole%.

Azomonas Large ovoid cells 2 μm or more in diameter, of varying length down to coccoid morphology. Occur singly, in pairs, or in irregular clumps; show marked pleomorphism. Do not produce endospores or cysts. May produce copious amounts of capsular slime. Motile with polar or peritrichous flagella. Gram negative or variable. Strains that elaborate a water-soluble fluorescent pigment appear white under ultraviolet light. Generally fix at least 10 mg of atmospheric nitrogen nonsymbiotically/g of carbohydrate consumed. Not proteolytic. Catalase positive. Grow well aerobically but can also grow under reduced oxygen tension. Optimal growth between 20 and 30°C. G + C 53–59 mole%.

Azomonotrichon Cells, ellipsoidal and/or rod-shaped, 1.6–2.1 by 2.5–3.5 μm, occurring singly, in pairs, and occasionally in short chains. Gram negative. Nonacid-fast. Motile with polar flagella or nonmotile. Endospores not produced. Chemoorganotrophic: utilize limited number of organic substrates. Aerobic. Catalase positive; oxidase generally negative. Nitrate not reduced. Fix atmospheric nitrogen. Growth range 9–32°C. G + C 58.2–58.6 mole%.

Azorhizophilus Cells rod-shaped, occurring singly and in pairs. Dimorphic as long rods, 1.3–1.7 by 7.0–10.9 μm, or as short rods, 1.6–1.9 by 3.2–4.2 μm. Gram negative. Nonacid-fast. Motile with peritrichous flagella. Chemoorganotrophic: utilize a limited number of substrates. Aerobic. Catalase positive; oxidase negative. Nitrate not reduced. Fix atmospheric nitrogen. Growth range 14–27°C. G + C 63.2–64.6 mole%.

Azospirillum Cells curved rods. Gram negative. Motile with single polar flagellum when grown in liquid media; produce numerous lateral flagella when growing on solid media. Chemoorganotrophic: mainly respiratory, but some strains are weakly fermentative. Aerobic-microaerophilic. Oxidase positive; catalase usually positive. Capable of fixing atmospheric nitrogen under microaerophilic conditions. Accumulate poly-β-hydroxybutyrate. G + C 69–71 mole%.

Azotobacter Large ovoid cells, 2 μm or more in diameter, of varying length down to coccoid morphology. Occur singly, in pairs, or irregular clumps, and rarely in chains of more than four cells. Marked pleomorphism. Do not produce endospores but form thick-walled cysts. May produce copious amounts of capsular slime. Motile with peritrichous flagella or nonmotile. Gram negative, with marked variability. Some strains elaborate a water-soluble fluorescent pigment that appears green under ultraviolet light. Fix atmospheric nitrogen; generally fix at least 10 mg of atmospheric nitrogen nonsymbiotically/g of carbohydrate consumed. Growth media contain carbohydrates (usually glucose), alcohols, or organic acids. Molybdenum required for nitrogen fixation; may be replaced by vanadium. Not proteolytic but can utilize nitrate, ammonia, and amino acids as sources of nitrogen. Catalase positive. Grow well

aerobically but can also grow under reduced oxygen tension. Optimal growth between 20 and 30°C. G + C 63–66 mole%.

Bacillus Cells rod-shaped, straight or nearly so, 0.3–2.2 by 1.2–7.0 μm. Majority motile; flagella typically lateral. Heat-resistant endospores formed; not more than one in a sporangial cell. Sporulation not repressed by exposure to air. Gram reaction: positive, or postiive only in early stages of growth. Chemoorganotrophic: metabolism strictly respiratory, fermentative, or both respiratory and fermentative, using various substrates. The terminal electron acceptor in respiratory metabolism is molecular oxygen, replaceable in some species by nitrate. Catalase formed by most species. Strict aerobes or facultative anaerobes. G + C 32–62 mole%.

Bacterionema Cells are pleomorphic, comprising nonseptate and septate filaments 1–1.5 μm by 20–200 μm and bacilli 1.5–2.5 μm by 3–10 μm. Characteristic morphology is a bacillus attached to a filament ("whip-handle"). Reproduction by filament septation and fragmentation to form rods that germinate to yield one to four filaments. Branching is frequent with aerobic and/or acid conditions. Gram positive, nonacid-fast, no endospores, and nonmotile. Metachromatic granules formed. Chemoorganotrophic: carbohydrates are usually fermented to yield acid and some gas. Products from glucose-grown cultures include acetylmethylcarbinol; with aerobiosis, CO_2 and propionic acid are major, and formic, acetic, and lactic acids are minor and/or variable; with reduced oxygen, CO_2 and lactic acid are major, and formic, acetic, propionic, and succinic acids are minor. Facultative anaerobes.

Bacteroides Gram negative, nonspore-forming rods. Nonmotile with peritrichous flagella. Chemoorganotrophic: metabolize carbohydrates or peptone. Fermentation products of sugar-utilizing species include combinations of succinic, lactic, acetic, formic, or propionic acid, sometimes with short-chained alcohols; butyric acid is usually not a major product. When n-butyric acid is produced, isobutyric and isovaleric acids are also present. From peptone non-sugar-utilizing species produce either trace to moderate amounts of succinic, formic, acetic, and lactic acids or major amounts of acetic and butyric acids (sometimes also succinic acid) with moderate amounts of alcohols and isovaleric, propionic, and isobutyric acids. Carbon dioxide utilized or required by some strains. Obligately anaerobic. Catalase usually not produced, or only in trace amounts. Hippurate usually not hydrolyzed. Lecithinase and lipase usually not produced on egg yolk agar. Optimal growth 37°C. G + C 40–55 mole%.

Bactoderma Branching filaments, but not true mycelium, may be produced. Cells rod or diphtheroid-shaped. Acid-fast. No spores produced. Chemoorganotrophic: assimilate fulvic acids. Do not oxidize ammonia to nitrites or nitrates.

Bartonella In stained blood films the organisms appear as rounded or ellipsoidal forms or as slender, straight, curved, or bent rods occurring either singly or in groups within erythrocytes. Characteristically occur in chains of several segmenting organisms, sometimes swollen at one or both ends and frequently beaded. Within tissues they are situated within the cytoplasm of endothelial cells as isolated elements or are grouped in rounded masses. Reproduce by binary fission. In culture possess a tuft of 1–10 unipolar flagella; flagella have not been demonstrated in tissues. Gram negative. Not acid-fast. Stain poorly or not at all with many aniline dyes but satisfactorily with Romanowsky's and Giemsa's stains.

Bdellovibrio Cells are single, small, curved, motile rods in the parasitic state. Motility by means of a polar, sheathed flagellum, usually monotrichous. Cells elongate to helical (spiral) forms in the nonparasitic stage of their life cycle. Gram negative. May or may not be obligately parasitic on bacteria. Parasitic strains attach to and penetrate into the host. Chemoorganotrophic: metabolism respiratory, not fermentative. Parasitic strains require the presence of suitable host bacteria for intracellular multiplication; only facultative strains or host-independent I mutants derived from parasitic strains are able to grow in complex media. Aerobic. Optimal growth 30°C. G + C 45.5–51.3 mole%.

Beggiatoa Cells colorless, in unattached filaments, 1–30 by 4–20 μm. Motile by gliding. Resting stages not known. Gram negative. Mixotrophs: metabolism respiratory, with molecular oxygen as terminal electron acceptor. Cells contain granules of sulfur when grown in the presence of hydrogen sulfide. Intracellular granules of poly-β-hydroxybutyric acid or volutin may be present. Aerobic to microaerophilic. G + C 37 mole%.

Beijerinckia Cells single, straight, slightly curved, or pear-shaped rods, 0.5–1.5 by 1.7–4.5 μm with distinctly rounded ends. Sometimes large misshapen cells, 3 by 5–6 μm, that are occasionally branched or forked. Characteristic large, highly refractile, lipoid (poly-β-hydroxybutyrate) bodies occur at each end of the cell. Under certain conditions some strains show rounded, coccoid cells without the terminal lipoid globules. Motile or nonmotile. Flagella peritrichous. Cysts (enclosing one cell) and capsules (enclosing several cells) found in some species. Gram negative. On agar medium copious slime is produced and giant colonies develop with a smooth, folded, or plicated surface. The slime is extremely tough, tenacious, or elastic and makes it difficult to remove part of a colony with a loop. Able to fix atmospheric nitrogen. Molybdenum is required for nitrogen fixation, but, unlike *Azotobacter*, this cannot be replaced by vanadium. Also in contrast to *Azotobacter*, calcium is not required for growth and nitrogen fixation. Glucose, fructose, and sucrose utilized by all strains; a wide variability exists in the ability to attack other carbon substrates. Oxidize organic substrates to CO_2 and a small amount of acetic acid so that the pH of medium is lowered during growth. Cells highly acid-tolerant, growth between pH 3 and 9.5 or 10.0. Strict aerobes but also grow and fix nitrogen under reduced oxygen pressure. Optimal growth 20–35°C. G + C 54.7–60.7 mole%.

Beneckea Cells curved or straight rods. Gram negative. Motile with sheathed polar flagella; some strains produce sheathed peritrichous flagella in addition to the polar flagella. Chemoorganotrophic: fermentative and respiratory. Produce acid but no gas from glucose. Oxidase positive. Facultative anaerobes. G + C 45–54 mole%.

Bifidobacterium Rods highly variable in appearance; uniform or branched, bifurcated Y and V forms and club or spatulate forms are generally characteristic of freshly isolated strains. Morphology may be influenced by nutritional conditons and on subculture straight or curved rods, of varying widths within a rod, may possess breaks resembling branching. Gram positive. Nonacid-fast. Nonspore-forming. Nonmotile. Cells often stain irregularly, occasionally two or more granules may stain with methylene blue while the remainder of the cell may be unstained. Metabolize carbohydrates; gas is not produced. Glucose is feremented primarily to acetic and lactic acid. Small amounts of formic and succinic acids produced; CO_2, butyric, and propionic acids are not produced. Catalase negative. Indole negative. Nitrates not reduced. Anaerobic, although there may be infrequent oxygen tolerance in the presence of CO_2. Optimal growth 36–38°C. G + C 57.2–64.5 mole%.

Blastobacter Mature cells rod-shaped, wedge-shaped or club-shaped, often slightly curved, 0.7–1.0 (greatest width) by 2.0–4.5 μm. Several cells attach by their narrow poles directly to a common base to form a rosette in the center of which there is usually a glistening corpuscle. Stalks absent. Multiplication by single buds appearing on the free cell pole. Buds spherical to oblong, 0.3 μm wide and nonmotile.

Blastococcus Divides by binary fission to produce a multicellular structure comprised of numerous (less than 0.5 μm) coccoid subunits. Oval to vibrioid buds formed on surface of some cells. Buds become uni-flagellate swarm cells with lateral flagella. After septation from the mother microcolony they enlarge, become nonmotile rods, and divide by multiple fission to produce the coccoid aggregate. Gram positive. Chemoorganotrophic.

Blattabacterium Straight or slightly curved rods, 0.8–0.9 by 1.5–8 μm, occurring singly, or occasionally in pairs in tandem in a host insect. Characteristically occur as intracellular symbionts in mycetocytes in abdominal fat body tissue and in ovaries and eggs of cockroaches. Microorganisms in mycetocytes are encapsulated by a host-provided membrane which is continuous with host cell organelles. Nonmotile. No resting stages known. Gram positive, becoming indeterminate (speckled, barred, or negative) in malnourished hosts. Cell wall 5–10 nm composed of several alternating electron-opaque and transparent layers. Muramic acid and glucosamine present. Complex fibrillar mesosomes near site of formation of cross walls. Biochemical, nutritional, and physiological characters unknown. Growth in the insect host inhibited by penicillin, streptomycin, chloramphenicol, chlortetracycline, oxytetracycline, sulfathiazole, and egg white lysozyme.

Bordetella Minute coccobacilli, 0.2–0.3 μm by 0.5–1.0 μm, arranged singly or in pairs, more rarely in short chains. Nonmotile or motile by lateral polytrichous flagella. Gram negative, bipolar. Colonies on potato-glycerol-blood agar medium, smooth, convex, pearly, glistening, nearly transparent, surrounded by zone of hemolysis without definite periphery. Chemoorganotrophic: metabolism respiratory, never fermentative. Catalase positive. Indole negative. Gelatin not liquefied. Strict aerobes. Optimal growth 35–37°C.

Borrelia Cells 0.2–0.5 by 3–20 μm, helical, with 3–10 or more coarse, uneven, often irregular coils, some of which may form obtuse angles. Cells up to 1 μm wide and 25 μm long may occur. Electron microscopy shows a foamy, elastic envelope and a cytoplasmic membrane, between which are 15–20 parallel fibrils coiled around the cell body. The fibrils constitute the locomotory apparatus, the coiled fibrils usually rotating in one direction, the cell body in the opposite direction but at the same rate. Motion is in forward and backward waves, laterally by bending and looping and corkscrew-like. Gram negative. Fermentative metabolism. Strict anaerobes. Optimal growth 28–30°C.

Bradyrhizobium Cells short rods, 0.5–0.9 by 1.2–3.0 μm, commonly pleomorphic. Do not produce spores. Gram negative. Motile by one polar or subpolar flagellum. Chemoorganotrophic: metabolism respiratory; do not produce acids. Slow-growing. Stimulate nodule formation; fix atmospheric nitrogen when in symbiotic state within nodules. Aerobic but grows at reduced oxygen tensions. Optimal growth 25–30°C. G + C 62–66 mole%.

Branhamella Cocci commonly arranged in pairs with adjacent sides flattened; division in two planes at right angles to each other. Do not form endospores. Nonmotile. Gram negative with a tendency to resist decoloration. Chemoorganotrophic. Acid not produced from carbohydrates. No xanthophyll pigment. Aerobic. Optimal growth 37°C. Catalase and cytochrome oxidase positive. Nitrates usually reduced. G + C 40–45 mole%.

Brevibacterium Short, nonbranching rods. Exhibits a marked change in form during growth cycle. Older cultures composed largely of coccoid cells that, on transfer to complex medium, give rise to slender irregular rods. Angular division common. As growth continues rods become shorter, giving rise to coccoid cells. Gram positive. Usually nonmotile. Spores not produced. Pigmented: yellow, orange, or red. Nonacid-fast. Chemoorganotrophic. Obligate aerobes.

Brochothrix Cells rod-shaped showing characteristic changes in cellular morphology during growth. In exponential growth phase show regular, unbranched rods that occur singly in short chains and in long, kinked, filamentous chains, which bend and loop to give characteristic knotted masses. In older cultures the rods give rise to coccoid forms. Gram positive. Endospores not produced. Chemoorganotrophic: respiratory and fermentative. Glucose fermented primarily to lactic acid. Facultative anaerobes. Catalase positive; oxidase positive. Optimal growth 20–22°C. G + C 36 mole%.

Brucella Coccobacilli or short rods, 0.5–0.7 by 0.6–1.5 μm, arranged singly, more rarely in short chains. Mammalian parasites and pathogens; facultatively intracellular with a relatively wide host range. No capsules. Nonmotile. Do not form endospores. Gram negative. Do not show bipolar staining. Chemoorganotrophic: metabolism respiratory. Vitamins required for growth. Catalase positive; oxidase usually positive. Nitrates reduced to nitrites (except for *B. ovis*). Citrate not utilized; indole not produced; methyl red and Voges-Proskauer tests negative. Strict aerobes; some require 5–10 percent added CO_2 for growth, especially on initial isolation. Optimal growth 20–40°C. G + C 56–58 mole%.

Butyrivibrio Curved rods, single, or in chains or filaments that may or may not be helical. Motile with monotrichous, polar, or subpolar flagella. No resting stages known. Gram negative. Chemoorganotrophic: metabolism fermetative, carbohydrates being the main fermentable substrates. Glucose is fermented with production of butyrate as one of the important products. Some strains in some media produce much lactic acid and little butyric acid. Strict anaerobes.

Calothrix Cyanobacterium. Photoautotrophic: can utilize water as electron donor with evolution of oxygen. Contains chlorophyll *a* and phycobiliproteins. Chlorophyll and carotenoids located in a system of thylakoids internal to the cytoplasmic membrane. Filamentous. A trichome (chain of cells) forms by intercalary division in one plane only. In the absence of combined nitrogen, heterocysts are produced; heterocysts are associated with the ability to fix atmospheric nitrogen. Akinetes also may be formed. Reproduction by random trichome breakage, by formation of hormogonia, and sometimes by germination of akinetes. Hormogonia are distinguishable from mature trichomes by the absence of heterocysts and by one or more of the following characteristics: rapid gliding motility, smaller cell size, cell shape, and gas vacuolation. Hormogonia give rise to young filaments that bear a terminal heterocyst at only one end of the cellular chain. Mature trichome tapers from base, which bears a terminal heterocyst, to apex. Vegetative cells are disk-shaped, isodiametric, or cylindrical. G + C 40–44 mole%.

Calymmatobacterium Pleomorphic rods, 1–2 μm in length, with rounded ends, occurring singly and in clusters. Exhibit single or bipolar condensation of chromatin, the latter giving rise to the characteristic "safety pin" forms. Usually encapsulated and readily demonstrated by Wright's stain as blue bacillary bodies surrounded by well-defined, dense pinkish capsules. Nonmotile. Gram negative. Facultative anaerobes.

Campylobacter Slender, nonspore-forming, spirally curved rods, 0.2–0.8 μm wide and 0.5–5 μm long. The rods may have one or more spirals. They also appear S-shaped and gull-winged when two cells form short chains. Cells in old cultures form spherical or coccoid bodies. Motile, with a characteristic corkscrew-like motion and a single polar flagellum at one or both ends of the cell. Flagella may be two to three times the length of the cells. Gram negative. Chemoorganotrophic: respiratory; carbohydrates are neither fermented nor oxidized; no acid or neutral end products produced. Energy from amino acids or tricarboxylic acid cycle intermediates, not carbohydrates. Gelatin and urea not hydrolyzed. Methyl red and Voges-Proskauer negative. No lipase activity. Oxidase positive. Pigments are not produced. Microaerophilic to anaerobic. Some species are pathogenic to humans and animals. Found in the reproductive organs, intestinal tract, and oral cavity of animals and humans. G + C 30–35 mole%.

Capsularis Cells rod-shaped, 0.5–0.6 by 1.4 μm, with rounded ends. Encapsulated. Spores not produced. Chemoorganotrophic: fermentation products from carbohydrates or peptones include butyric acid. Acid and gas produced from glucose. Indole not produced. Acid from maltose and mannitol. Optimal growth 37°C.

Cardiobacterium Rods, 0.5–0.75 by 1–3 μm, arranged singly, in pairs, short chains, and clusters. Pleomorphic. No flagella, endospores, or capsules demonstrated. Gram negative with possible retention of stain in parts of the cells. Chemoorganotrophic: fementative. Acid but no gas produced from fructose, glucose, mannose, sorbitol, and sucrose. Glucose fermented largely to lactate and smaller amounts of pyruvate, formate, and propionate. Cytochrome oxidase and small amounts of indole produced. Catalase not produced. Nitrates not reduced. Gelatin not liquefied; does not produce acetylmethylcarbinol, urease, ornithine, or lysine decarboxylase, arginine dihydrolase or phenylalanine deaminase. Facultative anaerobes. Optimal growth 30–37°C.

Caryophanon Large rods or filaments up to 3 μm in diameter, divided by closely spaced cross walls into numerous disk-shaped cells that are less than 1 μm long when growth is active. In living organisms cross walls appear as dark lines, partly complete, partly developing by ingrowth from the external wall. Motile by lateral flagella. Spores not produced. Gram positive. Chemoorganotrophic: respiratory, using oxygen as the terminal electron acceptor. Strict aerobes.

Caseobacter Regular rods, 0.2–0.8 by 2.0–4.0 μm. In older cultures rods become coccoid; after inoculation into fresh medium the coccoid forms germinate and transform into rods. Cells occur singly, in pairs, and in clusters. Nonmotile. Gram positive to Gram variable. Oxidase negative; catalase positive. Chemoorganotrophic. Aerobic. G + C 60–67 mole%.

Caulobacter Cells rod-shaped, fusiform, or vibrioid, 0.4–0.5 μm by 1–2 μm. Typically with a stalk, ca. 0.15 μm diameter, varying in length among isolates and with environmental conditions, extending from one pole as a continuation of the long axis of the cell. The stalk consists of the cell wall layers, the cell membrane, and a core of twisted membranes. A small mass of adhesive material is present at the distal end of the stalk. Occur singly; in dense populations, cells may adhere to each other in rosettes, the distal ends of their stalks embedded in a common mass of adhesive material. Cell division by transverse, binary, or asymmetrical fission of stalked cells. At the time of separation one cell possesses a stalk, the other a single polar flagellum. Each appendage occurs at the cell pole opposite the one formed during fission. The flagellated cell secretes adhesive material at the base of the flagellum, develops a stalk at this site, and enters the immotile vegetative phase. Resting stages not known. Gram negative. Colonies circular, convex, glistening; colorless, pink, red, orange, or yellow. Chemoorganotrophic: strictly respiratory. Most strains can store carbon as poly-β-hydroxybutyric acid. Pigmentation common; pigmented isolates contain carotenoids. Strict aerobes. Optimal growth 20–25°C. G + C 64–67 mole%.

Cedecea Cells rod-shaped. Gram negative. Motile with peritrichous flagella. Chemoorganotrophic: fermentative and respiratory. Acid and gas produced from glucose. Lipase positive. Growth on citrate. Voges-Proskauer positive. Indole not produced. Hydrogen sulfide not produced. Urea not hydrolyzed. Facultative anaerobes. G + C 48–52 mole%.

Cellulomonas In young cultures irregular rods are about 0.5 μm in diameter to 0.7–2.0 μm or more in length, which may be straight, angular, or slightly curved and occasionally club-shaped or beaded; a proportion of the rods are arranged at an angle to each other giving V formations. Occasionally, cells may show rudimentary branching but true mycelia are not formed. As cultures age the rods become shorter, but only a small proportion of coccoid cells is seen in cultures a week or more old. Motile by one (usually polar or subpolar) or a few lateral flagella, or nonmotile. Do not form endospores. Gram positive, but the cells are readily decolorized and often a mixture of Gram positive and Gram negative rods is seen. Not acid-fast. Chemoorganotrophic: metabolism primarily respiratory, using molecular oxygen as terminal electron acceptor; most strains produce acid from glucose under both aerobic and anaerobic conditions. Catalase positive. Cellulose is attacked by all strains. Aerobic; most strains also capable of anaerobic growth. Optimal growth at about 30°C. G + C 71.7–72.7 mole%.

Chainia Form true mycelium. Similar to *Streptomyces*; form aerial mycelium and spore chains. Sclerotia originate by extensive branching and cell division of certain hyphal regions that contain numerous amorphous inclusions. G + C 71 mole%.

Chamaesiphon Cyanobacterium. Photoautotrophic: can utilize water as electron donor with evolution of oxygen. Contains chlorophyll *a* and phycobiliproteins. Chlorophyll *a* and carotenoids located in a system of thylakoids internal to the cytoplasmic membrane. Cells ovoid, occurring singly or forming colonial aggregates, held together by additional outer wall layers. Reproduction by repeated budding at the apical end. Division in one plane only. G + C 47 mole%.

Chitinophaga Flexible rods with rounded ends, 0.5–0.8 by 40 μm. Gliding and flexing motility. Produce a resting stage (myxospore). Myxospores formed by transformation of filaments to spherical bodies without production of a fruiting body. Produce yellow pigment. Chemoorganotrophic: oxidative and fermentative. Chitin is hydrolyzed. Cellulose not hydrolyzed. Catalase positive; oxidase negative. Growth at 12–37°C. G + C 42.9–45.6 mole%.

Chlamydia Nonmotile, spheroidal organisms, 0.2–1.5 μm in diameter depending upon their stage of development in a unique, obligately intracellular growth cycle. The principal developmental stages are (1) the elementary body, which is a small, electron-dense spherule, 0.2-0.4 μm in diameter, containing a nucleus and numerous ribosomes surrounded by a multilaminated wall; (2) the initial body, which is a large, thin-walled, reticulated spheroid, 0.8–1.5 μm in diameter, containing nuclear fibrils and ribosomal elements; (3) an intermediate body representative of a transitional stage between the initial body and elementary body. The elementary body is the infectious form of the organism. The initial body is the vegetative form that divides by fission intracellularly but is apparently noninfectious when separated from the host cells. Gram negative. When separated from host cells, chlamydiae have few detectable metabolic activities, and these are dependent upon an exogenous source of organic and inorganic cofactors including high-energy phosphates (e.g., ATP); nevertheless, when extracellular chlamydiae are provided with cofactors, they can catabolize glucose, pyruvic acid, or glutamic acid and produce carbon dioxide. Because they are unable to synthesize their own high-energy compounds, they have been described as "energy parasites." Optimal growth 33–41°C. G + C 39–45 mole%.

Chlorobium Rod-shaped, ovoid, or vibrio-shaped bacteria, multiplying by binary fission, nonmotile. Gram negative. Contain either bacteriochlorophyll *c* or *d* as the major bacteriochlorophyll component as well as carotenoids of group 5, all pigments being located in elongated, ovoid vesicles ("chlorobium vesicles") underlying and attached to the cytoplasmic membrane. Do not contain gas vacuoles. Anaerobic. Capable of photosynthesis in the presence of hydrogen sulfide, during which they produce and deposit as an intermediate oxidation product, elemental sulfur in the form of globules outside the cells in the medium. Molecular hydrogen may be used as electron donor. Molecular oxygen is not evolved during photosynthesis. Cell suspensions appear in various shades of yellowish green and chocolate brown because of the photosynthetic pigments. Storage material: polyphosphate, not polysaccharides or poly-β-hydoxybutyrate. G + C 49.0–58.1 mole%.

Chloroflexus Narrow, 0.5–1.2 μm, filament of indefinite length. Orange to greenish. Capable of gliding motility. Anaerobic photoautotrophs: under anaerobic conditions, produces bacteriochlorophylls *a* and *c* and carotenes and are capable of photosynthesis in the presence of simple organic compounds that may be photoassimilated and also serve as a source of reducing power for carbon dioxide fixation; in addition carry out oxidative metabolism in the dark under microaerophilic to aerobic conditions. Optimal growth 45–70°C. G + C 53–54 mole%.

Chlorogloeopsis Cyanobacterium. Photoautotrophic: can utilize water as electron donor with evolution of oxygen. Contains chlorophyll *a* and phycobiliproteins. Chlorophyll and carotenoids located in a system of thylakoids internal to the cytoplasmic membrane. Filamentous. A trichome (chain of cells) forms by intercalary division in more than one plane. In the absence of combined nitrogen, heterocysts are produced; heterocysts are associated with the ability to fix atmospheric nitrogen. Akinetes may also be formed. Reproduction by random trichome breakage, by formation of hormogonia, and sometimes by germination of akinetes. Hormogonia are distinguishble from mature trichomes by the absence of heterocysts and by one or more of the following characteristics: rapid gliding motility, smaller cell size, cell shape, and gas vacuolation. Hormogonia composed of small cylindrical cells that enlarge and become spherical. Heterocysts develop in terminal and intercalary positions. Cells in mature trichomes divide in more than one plane; associated detachment of cells leads to irregular *Gloeocapsa*-like aggregates containing terminal heterocysts; hormogonia are produced within such aggregates. G + C 42–43 mole%.

Chloronema Filamentous, 2 μm wide, of indefinite length enclosed in a sheath. Yellow-green. Capable of gliding motility. Possesses gas vacuoles. Anaerobic photoautotrophs: under anaerobic conditions, produces bacteriochlorophylls *a* and *d* and carotenes and are capable of photosynthesis in the presence of simple organic compounds that may be photoassimilated and also serve as a source of reducing power for carbon dioxide fixation; in addition, carry out oxidative metabolism in the dark under microaerophilic to aerobic conditions. G + C 53–54 mole%.

Chondromyces Unicellular rods, capable of aggregation to form fruiting bodies. Gliding motility. Vegetative cells cylindrical, untapered rods, with blunt rounded ends. Sproangia borne singly or in clusters on simple or branched stalks. Myxospores lack capsules and resemble vegetative rods. Vegetative swarms etch, erode, and penetrate agar media. Vegetative slime does not adsorb congo red dye. Gram negative. Aerobic. Optimal growth 28–30°C. G + C 69–70 mole%.

Chromatium Ovoid, bean- or rod-shaped bacteria. Multiplication by binary fission. Motile by means of polar flagella. Under certain unfavorable conditions and in many natural environments cells may be nonmotile and grow in more or less regular aggregates surrounded by slime. Unfavorable mineral salt concentrations cause pronounced swellings, club-, and spindle-shaped forms. Gram negative. Contain bacteriochlorophyll *a* and carotenoids, both pigments being located at internal membranes of vesicular type. Do not contain gas vacuoles. Anaerobic. Capable of photosynthesis in the presence of hydrogen sulfide, during which they produce and store, as an intermediate oxidation product, elemental sulfur in the form of globules inside the cells. Molecular hydrogen may be used as electron donor. Molecular oxygen is not evolved during photosynthesis. Cell suspensions appear in various shades of orange-brown and pink to violet and purple-red because of the photosynthetic pigments. Storage materials: polysaccharides, poly-β-hydroxybutyrate, polyphosphates. G + C 48.0–70.4 mole%.

Chromobacterium Gram negative rods, 0.6–0.9 by 1.5–3 μm. Produce purple-violet pigments (violacein). Motile with polar and lateral flagella. Chemoorganotrophic: oxidative and fermentative metabolism. Citrate utilized as sole carbon source. Oxidase positive. Aerobic to facultative anaerobes. Nitrate reduced to nitrite. Optimal growth 30–35°C. G + C 65.2–67.6 mole%.

Chroococcidiopsis Cyanobacterium. Photoautotrophic: can utilize water as electron donor with evolution of oxygen. Contains chlorophyll *a* and phycobiliproteins. Chlorophyll *a* and carotenoids located in a system of thylakoids internal to the cytoplasmic membrane. Cells occurring singly or forming colonial aggregates held together by additional outer wall layers. Growth and binary fission lead to cubical cell aggregates. Reproduction by multiple fission leads to formation of nonmotile baeocytes with a thin, fibrous outer cell wall layer at the moment of their release. G + C 40–46 mole%.

Citrobacter Motile peritrichously flagellated rods, not encapsulated. Can use citrate as sole C source. Glucose and other carbohydrates fermented with production of acid and gas ($CO_2:H_2$ in the ratio 1:1). Trimethyleneglycol formed from glycerol. Growth not inhibited by KCN.

Clostridium Rods, usually motile by means of peritrichous flagella; occasionally nonmotile. Form ovoid to spherical endospores that usually distend the bacilli. Gram positive but may appear Gram negative in the late stages of growth. Chemoorganotrophic: fermentative. Ferment sugars, polyalcohols, amino acids, organic acids, purines, and other organic compounds. Some species fix nitrogen. Do not reduce sulfate. Most strains are strictly anaerobic, although some may grow in the presence of air at atmospheric pressure. Catalase is usually not produced, but when it is, in small amounts. G + C 23–43 mole%.

Coprococcus Cells cocci, occurring as diplococci and as short to long chains. Gram positive. Chemoorganotrophic: fermentative. Ferment carbohydrates; do not use peptones nor amino acids as primary source of energy and nitrogen. Produce volatile acids with more than three carbons. Anaerobic. G + C 39–42 mole%.

Corynebacterium Straight to slightly curved rods with irregularly stained segments and sometimes granules. Frequently show club-shaped swellings. Snapping division produces angular and palisade (picket fence) arrangements of cells. Generally nonmotile. Gram positive, although some species (e.g., *C. diphtheriae*) decolorize easily, especially in old cultures, and others are more tenacious; granules, however, are strongly Gram positive. Not acid-fast. Chemoorganotrophic: carbohydrate metabolism fermentative and respiratory. Do not produce soluble hemolysins, but on solid media containing blood, lysis of the red cells in contact with the colonies may occur. Aerobic and facultatively anaerobic; grow best aerobically. Catalase positive. G + C 52–68 mole%.

Cowdria Small, pleomorphic, coccoid, or ellipsoidal, occasionally rod-shaped, organisms occurring intracytoplasmically but not intranuclearly, and characteristically localized in clusters inside vacuoles in the cytoplasm of vascular endothelial cells of ruminants. Not passed transovarially in tick vectors. Nonmotile. Gram negative. Have not been cultivated in cell-free media.

Coxiella Short rods, usually 0.2–0.4 by 0.4–1.0 μm, resembling organisms of the genus *Rickettsia* in staining properties, dependence on host cells for growth and close natural association with arthropod and vertebrate hosts. Grows preferentially in the vacuoles of the host cell rather than in cytoplasm or nucleus as do the species of *Rickettsia*. Resists drying and relatively elevated temperatures that generally destroy the viability of *Rickettsia*.

Crenothrix Filaments up to 1 cm long, attached to a firm substrate and may be swollen at the free end. Unbranched but may show what appears to be false

branching. The very thin sheaths surrounding the filaments may be colorless at the tip or encrusted with iron (or manganese) oxides at the base. Cells cylindrical to disk-shaped, dividing by cross-septation in normal filaments and by cross- and longitudinal septation at the tips of enlarged filament ends. Cells in enlarged filament ends are smaller and may round up; sometimes there is only one row of these present. Larger rod-shaped cells may also slip out of the sheath and form new filaments. Gram reaction has not been recorded. Not motile. Has not been grown on artificial media in pure culture.

Cristispira Cells single, flexuous, 0.5–3.0 μm in diameter and helically coiled or planar wave-shaped. Helix spans more than 30 and up to 150 μm, and has 2–10 complete turns. Ends of cells either rounded, or tapered, or pointed with a rigid spicule. Divide by transverse fission. The cells may contain 30–80 seriately aligned ovoid inclusions of unknown composition, which displace the nuclear structure so that it appears as cross-striations in stained preparations. The normal movement is a forward and backward locomotion by means of an irrotational traveling helical wave. The helix may also rotate on its long axis or flex. There are over 100 axial fibrils intertwined with the protoplasmic cyclinder.

Curtobacterium Cells rod-shaped, but older cultures show many coccoid cells. Does not show rod-coccus life cycle characteristic of *Arthrobacter*. Rods motile, with a few lateral flagella, or nonmotile. Cell wall contains ornithine. Chemoorganotrophic: acids formed slowly and weakly from some carbohydrates. Strictly aerobic. Optimal growth ca. 25°C. G + C 66–71 mole%.

Cylindrospermum Cyanobacterium. Photoautotrophic: can utilize water as electron donor with evolution of oxygen. Contains chlorophyll *a* and phycobiliproteins. Chlorophyll and carotenoids located in a system of thylakoids internal to the cytoplasmic membrane. Filamentous. A trichome (chain of cells) forms by intercalary division in one plane only. In the absence of combined nitrogen, heterocysts are produced; heterocysts are associated with the ability to fix atmospheric nitrogen. Akinetes may also be formed. Reproduction by random trichome breakage, by formation of hormogonia, and sometimes by germination of akinetes. Hormogonia are distinguishble from mature trichomes by the absence of heterocysts and by one or more of the following characteristics: rapid gliding motility, smaller cell size, cell shape, and gas vacuolation. Heterocysts are exclusively terminal and are formed at both ends of the trichome. Akinetes are always adjacent to the heterocysts. Vegetative cells are isodiametric or cylindrical. G + C 42–47 mole%.

Cystobacter Unicellular rods that can aggregate to form a fruiting body. Gliding motility. Vegetative cells slender, tapered, flexible rods. Sporangia sessile, occurring singly or in groups; rounded, elongate, or coiled and surrounded by a definite slime envelope or wall; either free or embedded in a second slimy layer. Microcysts rod-shaped, phase-dense or refractile, rigid. Vegetative colonies do not etch or erode agar media; congo red is absorbed. Cellulose is not digested.

Cytophaga Single, short, or elongate flexible rods or filaments, not branched, sheathed, or helical. Cells 0.3–0.7 by 5–50 μm with round or tapered ends. Resting stages not known. Motile by gliding. Gram negative. Chemoorganotrophic: metabolism is respiratory, using molecular oxygen as electron acceptor, nitrate as an alternate electron acceptor, or respiratory and fermentative with acetate, propionate, and succinate among the end products. One or more polysaccharides such as agar, cellulose, or chitin may be decomposed, or carboxymethylcellulose depolymerized. Contain yellow, orange, or red carotenoid pigments. Strict aerobes or facultative anaerobes. Temperature optimum 20–30°C. G + C 29–42 mole%.

Dactylosporangium Produce true mycelium. Substrate mycelium 0.5–1.0 μm in diameter; hyphae long, irregularly branched, twisted, and occasionally coiled. Septation occurs but is not frequent. Do not fragment. Produce tough leathery growth on surface of agar. Gram positive. Not acid-fast. Aerial mycelium not formed or is very rudimentary. Sporangia, 1.0–1.2 by 4.0–6.0 μm, formed in clusters on surface of solid media, emerging from vegetative mycelium on short sporangiophores. Usually straight but in some strains long, wavy, and branched sporangia occur. Each sporangia contains a single row of three to five spores. Large globose spores formed on the mycelium embedded in the agar. Spores motile (zoospores), usually by polytrichous polar flagella. Cell wall contains meso-diaminopimelic acid, pentoses, and hexoses. Utilize arabinose, dextrin, fructose, galactose, glucose, enuline, lactose, maltose, manitol, mannose, rafnose, rhamnose, starch, sucrose, and xylose. Do not utilize dulcitol, glycerol, inositol, or sorbitol. Casien hydrolyzed. H$_2$S produced. Aerobic. Optimal growth 37°C; no growth at 45°C.

Deinococcus Cells spherical, 0.5–3.5 μm in diameter, characteristically dividing in more than one plane to form tetrads or tablets. Nonmotile. Resting stages not produced. Gram positive. Chemoorganotrophic: metabolism respiratory. Glucose may be metabolized, but acid is produced from only a limited number of substrates. Usually pigmented, pink-red. Aerobic. Optimal growth 25–35°C. Resistant to high doses of gamma and ultraviolet radiation. Fatty acid palmitoleate accounts for at least 25 percent of the total fatty acid composition. G + C 70 mole%.

Dermatophilus Produce mycelium of narrow tapering filaments with lateral branching at right angles; septa formed in transversant and horizontal and vertical longitudinal planes. Aerial mycelium ordinarily absent. Substrate mycelium gives rise to up to eight parallel rows of coccoid cells (spores), each of which becomes motile by a tuft of flagella. Gram positive. Chemoorganotrophic: nonfermentative but acid is produced from certain carbohydrates. Catalase positive. Not acid-fast. Aerobic and facultatively anaerobic. Optimal growth at about 37°C.

Dermocarpa Cyanobacterium. Photoautotrophic: can utilize water as electron donor with evolution of oxygen. Contains chlorophyll *a* and phycobiliproteins. Chlorophyll *a* and carotenoids located in a system of thylakoids internal to the cytoplasmic membrane. Cells rounded, occurring singly or forming colonial aggregates held together by additional outer wall layers. Reproduction by multiple fission leads to formation of motile baeocytes that lack a fibrous outer cell wall layer at the moment of their release. G + C 38–44 mole%.

Dermocarpella Cyanobacterium. Photoautotrophic: can utilize water as electron donor with evolution of oxygen. Contains chlorophyll *a* and phycobiliproteins. Chlorophyll *a* and carotenoids located in a system of thylakoids internal to the cytoplasmic membrane. Cells occurring singly or forming colonial aggregates held together by additional outer wall layers. Growth and binary fission lead to pear-shaped structures composed of one or two basal cells and one apical cell. Reproduction by multiple fission of the apical cell leads to formation of motile baeocytes with a thin fibrous outer cell wall layer at the moment of their release. G + C 45 mole%.

Derxia Cells rod-shaped with rounded ends, 1.0–1.2 by 3.0–6.0 μm, occurring singly or in short chains. Cells are rather pleomorphic, some cells occasionally becoming very large. Young cells have homogenous cytoplasm; older cells show typical large refractive bodies. Motile by a short polar flagellum. Resting stages not known. Gram negative. A wide range of sugars, alcohols, and organic acids are oxidized mostly to CO$_2$ and small amount of acid. Atmospheric nitrogen fixed in a nitrogen-deficient medium. Molybdenum required for nitrogen fixation; vanadium cannot replace molybdenum in this process. Strict aerobes, also growing in fixing nitrogen under reduced oxygen pressure. Catalase negative. Optimal growth 25–35°C. G + C content 70.4 mole%.

Desulfobacter Cells rod-shaped, 1–1.5 by 1.7–3.5 μm. Nonmotile or motile with single polar flagellum. Gram negative. Reduce sulfate. Chemoorganotrophic: can use acetate as carbon source and as electron donor for sulfate reduction. Complete oxidation of the organic substrate. Anaerobic. Optimal growth 32°C. G + C 45.9 mole%.

Desulfobulbus Cells ovoid/lemon-shaped, 1–1.3 by 1.8–2 μm, or curved rod/spiral-shaped, 0.5–1.5 by 3–6 μm. Nonmotile or motile with polar flagella. Gram negative. Reduce sulfate; can use fatty acids with 3–18 carbon atoms as carbon source and as electron donor for sulfate reduction. Chemoorganotrophic: organic substrate is incompletely oxidized. Anaerobic. Optimal growth 30–39°C. G + C 52.7–59.9 mole%.

Desulfococcus Cells spherical, 1.5–2.2 μm diameter. Nonmotile. Gram negative. Reduce sulfate; can use benzoate and fatty acids with 1–14 carbon atoms as carbon source and as electron donor for sulfate reduction. Chemoorganotrophic: complete oxidation of organic substrate. Anaerobic. Optimal growth 35°C. G + C 57.4 mole%.

Desulfomonas Cells rod-shaped, 0.8–1.0 by 2.5–10 μm. Nonmotile. Endospores not produced. Gram negative. Reduce sulfate; utilize sulfate and lactate as electron donors. Chemoorganotrophic: utilize sulfate as terminal electron acceptor in anaerobic respiration. Strict anaerobes. Optimal growth 35–40°C. G + C 66–67 mole%.

Desulfonema Filamentous, 2.5–7 μm by several mm. Gliding motility. Gram positive. Reduce sulfate; can use fatty acids with 1–12 carbon atoms and sometimes benzoate as carbon source and as electron donor for sulfate reduction. Chemoorganotrophic: complete oxidation of organic substrate; also can grow autotrophically, using hydrogen as electron donor for sulfate reduction and carbon dioxide as carbon source for growth. Anaerobic. Optimal growth 29–32°C. G + C 34.5–41.6 mole%.

Desulfosarcina Cells ovoid/rod-shaped, 1–1.5 by 1.5–2.5 μm, occurring singly, in pairs, tetrads, and packets of eight. Nonmotile. Gram negative. Reduce sulfate; can use benzoate and fatty acids with 1–14 carbon atoms as carbon source and as electron donor for sulfate reduction. Chemoorganotrophic: complete oxidation of organic substrate; also can grow autotrophically, using hydrogen as electron donor for sulfate reduction and carbon dioxide as carbon source for growth.. Anaerobic. Optimal growth 33°C. G + C 37.5 mole%.

Desulfotomaculum Straight or curved rods, 0.3–1.5 by 3–6 μm, with rounded ends, usually single but sometimes in chains. Motile with peritrichous flagella. Spores oval to round, terminal to subterminal, causing slight swelling of the ends. Gram negative. Produce black colonies in lactate-sulfate agar containing ferrous salt. Chemoorganotrophic: metabolism respiratory. Sulfates, sulfites, and reducible sulfur compounds act as electron acceptors and are reduced to H_2S. Lactate and pyruvate are universal electron donors; limited range of substrates utilized, rarely utilize carbohydrates; acetate not oxidized. Oxidation of organic substrates incomplete leading to formation of acetate or homologue and CO_2. Cells contain a cytochrome of the protoheme class. Strict anaerobes. Optimal growth 35–55°C. G + C 41.7–45.5 mole%.

Desulfovibrio Curved rods, sometimes sigmoid or spirilloid. Morphology influenced by age and environment. Motile by means of polar flagella. Do not form endospores. Gram negative. Chemoorganotrophic: obtain energy by anaerobic respiration reducing sulfates or other reducible sulfur compounds to H_2S. Lactate, pyruvate, and usually malate are oxidized to acetate and CO_2. Carbohydrates rarely utilized. Gas never formed from carbohydrates. Cells contain C_3 cytochromes and desulfoviridin; the latter is responsible for the characteristic red fluorescences of cells when viewed in light of 365 nm following addition of NaOH. Hydrogenase usually present. Gelatin not liquefied, nitrates not reduced. Nitrogen sometimes fixed. Strict anaerobes. Optimal growth 25–30°C.

Desulfuromonas Straight to curved rods, 0.4–1.2 by 1–4 μm. Motile with lateral, subpolar, or polar flagella. Utilize sulfur as electron acceptor with production of hydrogen sulfide; dissimilatory sulfur metabolism essential for growth. Acetate acts as electron donor. Pink to ochre pigments. Obligate anaerobes. G + C 50–63 mole%.

Ectothiorhodospira Spiral to slightly bent rod-shaped bacteria, multiplying by binary fission. Motile by means of polar flagella. Gram negative. Contain bacteriochlorophyll *a* and carotenoids, both pigments being located at lamellar membrane stacks that are continuous with the cytoplasmic membrane. Do not contain gas vacuoles. Anaerobic. Capable of photosynthesis in the presence of hydrogen sulfide, during which they produce and deposit, as an intermediate oxidation product, elemental sulfur in the form of globules outside the cells in the medium. Molecular oxygen is not evolved during photosynthesis. G + C 62.2–69.9 mole%.

Edwardsiella Motile peritrichously flagellated rods; not encapsulated. Citrate and malonate cannot be used as a sole carbon source. H_2S is produced abundantly from TSI agar. Indole is produced. Lysine and ornithine decarboxylases are present. Gram negative. Motile by peritrichous flagella or nonmotile. Not spore-forming. Not acid-fast. Aerobic and facultatively anaerobic. Chemoorganotrophic: metabolism respiratory and fermentative. Acid is produced from the fermentation of glucose. Catalase positive, oxidase negative. Nitrates are reduced to nitrites. Optimal growth 37°C. G + C 50–53 mole%.

Ehrlichia Minute, rickettsia-like organisms pathogenic for certain mammals but not infectious for humans. Cells grow in the cytoplasm but not in the nucleus, usually as compact, single to multiple colonies. Less often seen as single cells (initial bodies). Nonmotile. Gram negative. Small, often pleomorphic, coccoid to ellipsoidal organisms occurring intracytoplasmically, either singly or in compact colonies (morulae) in certain circulating leukocytes of susceptible mammalian hosts.

Eikenella Rods to coccobacilli. Gram negative. Do not require factors in blood (X and V) for growth. Do not show hemolysis on blood agar. Ornithine decarboxylase positive; lysine decarboxylase positive. Catalase negative. Chemoorganotrophic. Acid and gas not produced from carbohydrates. Nitrate reduced. Facultative anaerobes. G + C 56 mole%.

Elytrosporangium Produce true mycelium: slender hyphae, 0.5–2.0 μm diameter. The aerial mycelium at maturity forms chains of three to many spores. Vegetative hyphae develop branched mycelium that does not readily fragment. Mycelium sparingly septate. Form pod-like structures (merosporangia) on the primary mycelium. Gram positive. Chemoorganotrophic: highly oxidative.

Ensifer Rods, 0.7–1.1 by 1.0–1.9 μm, occurring singly or in pairs. Gram negative. Motile with tuft of 3–5 subterminal flagella. Reproduction by budding, with bud elongating to produce asymmetric polar growth; separation of cells occurs by binary fission. Chemoorganotrophic: oxidative metabolism, uses variety of carbon sources. Growth occurs at very low nutrient concentrations. Bacterial predator; attaches to various Gram positive and Gram negative host bacteria and causes lysis of host cells. Aerobic. Weakly catalase positive. Nitrate and nitrite reduced. Optimal growth 27°C. G + C 63–67 mole%.

Enterobacter Motile rods, peritrichously flagellated. Some strains encapsulated. Gram negative. Citrate and acetate can be used as sole carbon source. At 37°C glucose is fermented with production of acid and gas; at 44.5°C gas is not produced from glucose fermentation. Generally Voges-Proskauer positive. Methyl red negative. Dulcitol usually not attacked. Alginate not utilized. Sorbitol is utilized. H_2S not produced on TSI agar. G + C 52–59 mole%.

Eperythrozoon Obligate parasites in the blood of various vertebrate species including some rodents, ruminants, and pigs. Stain blue, pink, or violet with Giemsa's and Romanowsky-type stains. In such preparations cells appear as rings or coccoids (rarely rods), 0.4–1.5 μm in diameter, characteristically round with numerous annular or disk-shaped elements. Occur on erythrocytes and free in plasma with equal frequency. Swarm-like clusters of ring-shaped bacteria cells may occur on the surface of erythrocytes. Rod-shaped forms may occur partly or entirely circling an erythrocyte. Not cultivated in cell-free media.

Erwinia Cells predominantly single, straight rods, 0.5–1.0 by 1.0–3.0 μm. Motile by peritrichous flagella. Gram negative. Produce acid from fructose, glucose, galactose, β-methylglucoside, and sucrose, usually from mannose and ribose but rarely from adonitol, dulcitol, or melezitose. Utilize acetate, fumarate, gluconate, malate, and succinate but not benzoate, oxalate, or propionate, as carbon and energy-yielding sources. Gas production comparatively weak or absent. Fermentation end products from glucose are CO_2 and different combinations of succinate, lactate, formate, and acetate; some strains form butanediol and some ethanol. Do not hydrolyze starch beyond dextrins. Do not decarboxylate glutamic acid. Decarboxylases for arginine, lysine, or ornithine in only a few strains. Rarely produce urease or lipases. Facultative anaerobes. Oxidase negative; catalase positive. Optimal growth 27–30°C. G + C 50–58 mole%.

Erysipelothrix Rod-shaped organisms with a tendency to form long filaments; the filaments may thicken and show characteristic granules. Nonmotile. Nonencapsulated. No endospores produced. Gram positive, but older cultures have a tendency to become Gram negative. Acid but no gas from glucose and from certain other carbohydrates. Catalase negative. Esculin not hydrolyzed. α-Hemolysis but no β-hemolysis on blood agar. Aerobic; grows better in an atmosphere with reduced oxygen and containing 5–10% CO_2. Optimal growth 33–37°C.

Escherichia Straight rods, 1.1–1.5 by 2.0–6.0 μm, occurring singly or in pairs. Gram negative. Motile by peritrichous flagella or nonmotile. Acetate can be used, but citrate cannot be used as sole carbon source. Glucose and other carbohydrates are fermented with production of lactic, acetic, and formic acids; the formic acid is split into equal amounts of CO_2 and H_2. Lactose is fermented by most strains. Facultative anaerobes. Indole positive. No H_2S produced from TSI agar. Methyl red positive. Voges-Proskauer negative. G + C 50–51 mole%.

Eubacterium Gram positive, obligately anaerobic, nonspore-forming rods, uniform to pleomorphic, nonmotile or motile. Chemoorganotrophic: usually produce mixtures of organic acid from carbohydrates or peptone, often including large amounts of butyric, acetic, or formic acids. Do not produce: propionic acid as a major product; lactic acid as the sole product; succinic and lactic acids with small amounts of acetic or formic acids; more acetic acid than lactic acid. Catalase usually not produced. Hippurate usually not hydrolyzed. Optimal growth 37°C.

Excellospora Form single and short chains of conidia on both aerial and substrate mycelia. Chains of spores may form coils. Spores blue-green. Substrate mycelium yellow, orange, red, brown, or black. Optimal growth 45–55°C.

Fischerella Cyanobacterium. Photoautotrophic: can utilize water as electron donor with evolution of oxygen. Contains chlorophyll *a* and phycobiliproteins. Chlorophyll and carotenoids located in a system of thylakoids internal to the cytoplasmic membrane. Filamentous. A trichome (chain of cells) forms by intercalary division in more than one plane. In the absence of combined nitrogen, heterocysts are produced; heterocysts are associated with the ability to fix atmospheric nitrogen. Akinetes also may be formed. Reproduction by random trichome breakage, by formation of hormogonia, and sometimes by germination of akinetes. Hormogonia are distinguishable from mature trichomes by the absence of heterocysts and by one or more of the following characteristics: rapid gliding motility, smaller cell size, cell shape, and gas vacuolation. Hormogonia composed of small cylindrical cells that enlarge and become rounded. Heterocysts develop almost exclusively in an intercalary position. Cells in mature trichome divide in more than one plane to produce a partly multiseriate trichome with lateral branches. Heterocysts in the primary trichome are predominantly terminal or lateral. Hormogonia are produced from the ends of trichomes or from lateral branches. G + C 42–46 mole%.

Flavobacterium Cells vary from coccobacilli to slender rods. Motile with peritrichous flagella or nonmotile; the latter do not show gliding movement or swarming growth on nutrient agar. Endospores not produced. Gram negative. Growth on solid media is pigmented yellow, orange, red, or brown (color varies with media and temperature). Pigments not soluble in media. Chemoorganotrophic: respiratory; fermentation not conspicuous and neither acid nor gas production from carbohydrates is common. Cultures often difficult to maintain following primary isolation. Optimal growth at temperatures below 30°C.

Two distinct ranges of DNA nucleotide base ratios: G + C 30–42 and G + C 63–70 mole%.

Flectobacillus Straight to curved rods, 0.6–1.0 by 1.5–5.0 μm; the degree of curvature varying among individual cells within a culture; the most abundantly occurring are cells in the shape of the letter C. Long sinuous filaments up to 50 μm long are present. Rings 5 to 10 μm in outer diameter are formed by overlapping of the ends of a cell. Coils or helical spirals are infrequently formed. Gram negative. Nonmotile. Not flexible. Resting stages not known. Produce pale pink or rose-colored, nonwater-soluble pigment. Chemoorganotrophic: metabolism respiratory; acids are produced aerobically from a variety of carbohydrates. Strictly aerobic. G + C 39.5–40.3 mole%.

Flexibacter Flexible rods or filaments, 0.5 by 5–100 μm. Resting stages not known. Motile by gliding. In young cultures long (20–30 μm), extremely agile thread cells; in old cultures short, nonmotile rods or coccobacilli. Gram negative. Cell masses are pink, red, orange, or yellow because of carotenoid pigments. Chemoorganotrophic: metabolism is usually respiratory, utilizing molecular oxygen as terminal electron acceptor, but one species is both respiratory and fermentative. Polymers such as agar, alginic acid, cellulose, and chitin are not attacked. Usually strict aerobes. G + C 31.2–42.9 mole%.

Flexithrix Cells, 0.3–0.5 by 5–15 μm, usually as sheathed filaments 0.5 μm in diameter and up to 500 μm long; may show false branching. Resting stages are not known. Unsheathed cells 5–15 μm long, motile by gliding. Gram negative. Chemoorganotrophic: metabolism is respiratory using molecular oxygen as terminal electron acceptor. Strict aerobes. G + C 37.2 mole%.

Fluoribacter Gram negative rods, 0.5–0.7 by 4–5 μm, resembling *Legionella* in cultural characteristics. Fluoresce brightly under ultraviolet light. Green colonies formed on media containing bromcresol purple. Catalase positive; oxidase negative. Chemoorganotrophic. Starch hydrolyzed. Gelatin hydrolyzed. G + C 35.7–41.1 mole%.

Francisella Very small coccoid to ellipsoidal pleomorphic rods, which often show bipolar staining. Nonmotile. Gram negative. Acid but no gas produced in media containing carbohydrates. Catalase negative. H₂S produced. Strictly aerobic. Optimal growth 37°C.

Frankia Hyphae usually 0.3–0.5 μm in diameter. Symbiotic, filamentous, mycelium-forming bacteria which induce and live in root nodules on a wide variety of nonleguminous plants. The mycelium is septate and branched, but branching is not necessarily correlated with cross-wall formation. The nodules are capable of fixing molecular nitrogen. These bacteria also have a free stage in the soil. In active, nitrogen-fixing nodules, the center of the host cell is filled by a hyphal mass; near the periphery, spherical or club-shaped terminal swellings are formed on radially arranged hyphae close to the plant cell wall. These spherical bodies are often called vesicles. The vesicles and club-shaped structures are probably associated with nitrogen fixation. In host cells that have not survived invasion by *Frankia* species, vesicles or club-shaped structures are not found; instead, the mycelium fragment into small particles or bacteria-like cells. Molybdenum and cobalt required for nitrogen fixation. Hyphae, vesicles, and club-shaped structures are Gram variable. The rod-shaped structures or bacteria-like cells are Gram positive. Probably microaerophilic.

Frateuria Gram negative, straight rods, 0.5–0.7 by 0.7–3.5 μm, occurring singly or in pairs, rarely in filaments or as irregular cells and never in chains. Motile by means of polar flagella. Produce brown water-soluble pigment. Chemoorganotrophic. Acid produced from various carbohydrates. Strictly aerobic. Produce hydrogen sulfide. Oxidase negative. Nitrate not reduced. Growth occurs at pH 3.6. G + C 62–64 mole%.

Fusobacterium Gram negative rods. Nonmotile or motile with peritrichous flagella. Spores not produced. Chemoorganotrophic: metabolize carbohydrates or peptone; major products from carbohydrate or peptone include butyric acid, often with acetic and lactic acids and lesser amounts of propionic, succinic, and formic acids and short-chained alcohols. Catalase usually not produced. Obligately anaerobic. Optimal growth 37°C. G + C 26–34 mole%.

Gallionella Cells kidney-shaped or rounded, on the ends of long stalks with the long axis of the cell transverse to that of the stalk. Stalks are formed of bundles of fibrils twisted around one another. Stalks may be covered with iron hydroxide; manganese compounds are not deposited. Multiplication is by fission of cells; the daughter cells remain at first at the end of the stalk; later they may be liberated as swarmer cells that are motile by one or two polar or subpolar flagella. Gram negative. Probably chemolithotrophic: oxidize ferrous to ferric iron and assimilate CO_2. Iron hydroxide may make up to 90 percent of the dry weight of the cell mass. Microaerophilic.

Gardnerella Gram negative to Gram variable rods, 0.5 by 1.5 μm. Nonfilamentous. Nonencapsulated.

Nonmotile. Chemoorganotrophic: acetic acid produced as major product of fermentation. Found in human genitourinary tract where it causes vaginitis. Facultatively anaerobic. Fastidious growth requirements. Catalase negative; oxidase negative. Optimal growth 35–37 °C. G + C 42–44 mole%.

Gemella Cocci occurring singly or in pairs with adjacent sides flattened. Gram positive. Endospores not formed. Nonmotile. Show β-hemolysis when grown on blood agar. Pigments not produced. Chemoorganotrophic: fermentative, acid is produced from several carbohydrates. Nitrates not reduced; nitrites reduced by some strains. Aerobic or facultatively anaerobic. Optimal growth 37°C. G + C 31.4–34.6 mole%.

Gemmiger Cocci, occurring as diplococci and in chains, giving the appearance of budding. Gram negative, but sometimes stain reactions appear Gram positive. Obligate anaerobes. Chemoorganotrophic: requires fermentable carbohydrates. Butyric acid produced from carbohydrate fermentation. Produce volatile acids with more than three carbons. G + C 59 mole%.

Geodermatophilus Mycelium rudimentary; aerial mycelium not produced. Mycelial filaments divide transversely in at least two longitudinal planes to form masses of motile, coccoid elements. Produce a muriform, tuber-shaped, nonencapsulated, holocarpic, multilocular thallus. Some cells develop into elliptical to lanceolate zoospores with a terminal tuft of long flagella. Gram positive. Aerobic.

Gloeobacter Cyanobacterium. Photoautotrophic: can utilize water as electron donor with evolution of oxygen. Contains chlorophyll *a* and phycobiliproteins. Chlorophyll and carotenoids not located in a system of thylakoids internal to the cytoplasmic membrane. Cells rod-shaped, occurring singly or forming colonial aggregates held together by additional outer wall layers. Reproduction by binary fission. Division in one plane only. Sheath present. G + C 64 mole%.

Gloeocapsa Cyanobacterium. Photoautotrophic: can utilize water as electron donor with evolution of oxygen. Contains chlorophyll *a* and phycobiliproteins. Chlorophyll and carotenoids located in a system of thylakoids internal to the cytoplasmic membrane. Cells coccoid, occurring singly or forming colonial aggregates held together by additional outer wall layers. Reproduction by binary fission. Division in two or three planes. Sheath present. G + C 40–46 mole%.

Gloeothece Cyanobacterium. Photoautotrophic: can utilize water as electron donor with evolution of oxygen. Contains chlorophyll *a* and phycobiliproteins. Chlorophyll and carotenoids located in a system of thylakoids internal to the cytoplasmic membrane. Cells rod-shaped, occurring singly or forming colonial aggregates held together by additional outer wall layers. Reproduction by binary fission. Division in one plane only. Sheath present. G + C 40–43 mole%.

Gluconobacter Cells ellipsoidal to rod-shaped, 0.6–0.8 by 1.5–2.0 μm, occurring singly, in pairs or in chains. Motile with 3–8 polar flagella, rarely one flagellum or nonmotile. Endospores not produced. Gram negative but weakly Gram positive in older cultures. Chemoorganotrophic: metabolism respiratory, never fermentative; oxygen is terminal electron acceptor. Oxidize ethanol to acetic acid, sometimes weakly at neutral and acid reactions (pH 4.5). Do not oxidize acetate or lactate to CO_2. Usually pronounced ketogenesis and acid formation from sugars. Catalase positive. Strict aerobes. Optimal growth 25–30°C. Convert glucose to gluconate; produce 5-ketogluconate. G + C 60–64 mole%.

Haemobartonella Obligate parasite on or within erythrocytes of many vertebrate species. Organism coccoid or rod-shaped. Occurs singly, in pairs, or in groups in shallow, or in deep indentations on the erythrocyte surface, sometimes in vacuoles within the erythrocytes, rarely in the plasma. Possess single or double limiting membrane; lack a cell wall. Gram negative. Not acid-fast.

Haemophilus Minute to medium-size, coccobacillary to rod-shaped cells that sometimes form threads and filaments; may show marked pleomorphism. Nonmotile. Gram negative. Strict parasites, requiring growth factors present in blood. Aerobic to facultatively anaerobic. Optimal growth 37°C. G + C 38–42 mole%.

Hafnia Motile, peritrichously flagellated rods. Not encapsulated. Citrate and acetate can be used as sole carbon source. Glucose is fermented with production of acid and gas. Acid is not produced from citrate, tartrate, and mucate. Methyl red usually negative; Voges-Proskauer usually positive. Alginate not utilized. H_2S not produced from TSI agar. Catalase positive; oxidase negative. G + C 52–57 mole%.

Haliscomenobacter Cells rod-shaped, 0.35–0.45 by 3–5 μm, occurring within thin hyaline sheaths. Single cells nonmotile. Do not accumulate iron hydroxide in sheath. Manganese not oxidized. Gram negative. Chemoorganotrophic: metabolism respiratory. Strict aerobes. Optimal growth 25–27°C. G + C 48.3–49.7 mole%.

Halobacterium Cells rod-shaped, 0.6–1.0 by 1–6 μm occurring singly. Motile by a tuft of polar flagella or nonmotile. Resting stages not known. Gram nega-

tive. Require high concentrations (greater than 2 M) of sodium chloride for growth. Cell wall does not contain muramic acid. Cell membrane contains glycerol–ether linkages. Chemoorganotrophic: metabolism respiratory, never fermentative. Carbohydrates used only slightly, if at all. Energy obtained from amino acids. Contain carotenoid pigments. Oxidase positive, catalase positive. Gelatin usually liquefied. H_2S produced from peptone. Indole produced. Urease negative. Strict aerobes. Optimal growth 30–50°C. The DNA shows a major and minor component: the major component G + C 66–68 mole%; the minor component G + C 57–60 mole%.

Halococcus Cocci, 0.6–1.5 μm in diameter, occurring in pairs, tetrads, and irregular clusters. Nonmotile. Gram negative. Grow only in the presence of 2.5 M or higher concentrations of NaCl. Chemoorganotrophic: metabolism respiratory, never fermentative. Aerobic. Cell wall lacks murein. Produce pink-red pigments. G + C 67 mole% for the major component and 59 mole% for the minor component.

Halomonas Cells rod-shaped, occurring singly or in pairs during logarithmic growth phase. Elongated, flexuous rods are produced during stationary growth phase. Gram negative. Motile; lateral or polar flagella. Facultatively anaerobic. Nitrate reduced to nitrite. Catalase positive; oxidase positive. Chemoorganotrophic. Glucose fermented without gas production. Optimal growth at 3–8 percent salt. G + C 60–61 mole%.

Halospora Obligately intracellular symbionts of ciliate protozoa. Life cycle includes two alternate phases, a vegetative phase and an infectious stage. Vegetative forms are short, straight, spindle-shaped rods, 0.5–1.0 by 1–3 μm, which divide by transverse fission. Infectious cells (spores), which are produced when vegetative cells grow without division, are several times longer than the vegetative cells. Infectious cells have a homogeneous area and a cytoplasm-membrane area. Gram negative. Nonmotile. Not flexible.

Herpetosiphon Unbranched, flexible, sheathed rods or filaments, 0.7–1.5 by 5–150 μm or more (to several mm). Resting stages not known. Unsheathed segments motile by gliding. Gram negative. Chemoorganotrophic: metabolism is respiratory, using molecular oxygen as terminal electron acceptor. Agar, alginic acid, and chitin not attacked. Cellulose may be degraded; starch may be hydrolyzed. Gelatin liquefied. Indole not produced. Hydrogen sulfide not produced. Nitrates not reduced to nitrites. Cells possess yellow or orange carotenoid pigments. Strict aerobes. G + C 44.9–53.1 mole%.

Hyphomicrobium   Cells 0.5–1.0 by 1–3 μm; rod-shaped with pointed ends, oval, egg- or bean-shaped forms; produce mono- or bipolar filamentous outgrowths (hyphae) of varying length 0.3–0.4 μm in diameter. The hyphae are not septate but may show true branching. Multiplication by budding at tips of hyphae; mature buds become motile, break off and often attach themselves to surfaces. Motility is lost after attachment. Chemoorganotrophic. Carbon dioxide is required for growth. Oligocarbophilic, i.e., growth can occur in mineral salts medium without added carbon sources. Ammonia not oxidized to nitrate. Aerobic but can grow anaerobically in the presence of nitrate. Optimal growth 25–30°C. G + C 59.2–66.8 mole% except one species with G + C 40 mole%.

Intrasporangium   Mycelium usually rudimentary. Vegetative hyphae bear terminal and subterminal vesicles. No fragmentation of hyphae. No aerial mycelium. No spores. Gram positive. Nonmotile.

Janthinobacterium   Cells rod-shaped, 0.8–1.2 by 2.5–6 μm. Produce violet-purple pigments. Chemoorganotrophic: acids produced from carbohydrates. Strictly aerobic. Optimal growth 25°C. G + C 65.1–66.1 mole%.

Kineosporia   Small, orange-colored colonies are formed by a fine, branched mycelium. Numerous small sporangia borne at the distal ends of the hyphae; no aerial mycelium. Each sporangium contains a single planospore. Gram positive. Not acid-fast. Optimal growth 20–30°C.

Kingella   Gram negative rods. Chemoorganotrophic: oxidative metabolism. Produce acid from glucose. Oxidase positive; catalase negative. Aerobic; weak growth anerobically. Reduce nitrites. G + C 44.5–55 mole%.

Kitasatoa   Produce true mycelium. Sporangia club-shaped, 2–2.5 by 5 μm. Spores diplococcus-like, spherical, ellipsoidal, or cylindrical, arranged in a single chain within the sporangium and motile by a single polar flagellum. Aerial hyphae abundant and of two types: one type, 1.0–1.2 μm wide, produces long chains of cylindrical conidiospores; the other 1.2–1.5 μm wide produces either terminal sporangia or rounded vesicular bodies. Vegetative mycelium produce club-shaped sporangia at the hyphal tips; vegetative hyphae 0.8–1.2 μm wide, highly branched with few septa. Small spores are uniflagellate. Sporangia are finger-like or club-shaped; two to several spores arranged linearly within each sporangium. Optimal growth 10–37°C.

Klebsiella   Nonmotile, encapsulated rods, 0.3–1.5 μm by 0.6–6.0 μm, arranged singly, in pairs, or in short chains. Can use citrate and glucose as sole carbon source. Glucose is fermented with production of acid and gas. Most strains produce butanediol as a major end product of fermentation of glucose. Voges-Proskauer positive; methyl red negative. H$_2$S not produced from TSI. Catalase positive; oxidase negative. Optimal growth 35–37°C. G + C 52–56 mole%.

Kluyvera   Cells rod-shaped, occurring singly. Motile with peritrichous flagella. Gram negative. Chemoorganotrophic: acid and gas produced from glucose fermentation. Catalase positive; oxidase negative. Facultatively anaerobic. Indole produced. Methyl red positive. Voges-Proskauer negative. Citrate utilized. Hydrogen sulfide not produced. Urea negative. G + C 56–57 mole%.

Kurthia   In young cultures (18–24 hours): regular, unbranched rods with rounded ends occurring in chains that are often parallel; the rods are ca. 0.8 μm in diameter and vary in length according to the stage of growth. Older cultures (3–7 days) are usually composed of coccoid cells formed by fragmentation of the rods. The rods are motile by peritrichous flagella. Endospores not produced. Gram positive. Not acid-fast. Chemoorganotrophic: metabolism respiratory, never fermentative, using molecular oxygen as terminal electron acceptor. Catalase positive; oxidase negative. Do not reduce nitrate. Do not produce acid from carbohydrates. Strict aerobes. Optimal growth 25–30°C.

Lachnospira   Curved rods, motile with lateral to subterminal monotrichous flagella. Gram negative but may appear weakly Gram positive. Ferment glucose producing large amounts of ethanol, lactic, formic, and acetic acids, and CO$_2$; small amounts of hydrogen also produced. Not proteolytic. Anaerobic. Found in rumen.

Lactobacillus   Rods, varying from long and slender to short coccobacilli. Chain formation common. Motility unusual; when present by peritrichous flagella. Do not produce spores. Gram positive but becoming gram negative with increasing age. Some strains exhibit bipolar bodies, internal granulations, or a barred appearance with staining. Metabolism fermentative even though growth generally occurs in the presence of air; some are strict anaerobes. Characteristically utilize sugars. Glucose fermented: at least half of the end product is lactate; additional products may be acetate, formate, succinate, CO$_2$, or ethanol. Lactate is not fermented. Volatile acids of more than two carbon atoms are not produced. Nitrate reduction unusual. Gelatin not liquefied. Casein not digested. Catalase negative; oxidase negative. Pigment production rare; if present yellow or orange to rust or brick red. Optimal growth 30–40°C. G + C 33.3–53.9 mole%.

Lamprocystis Spherical to ovoid bacteria, before cell division typically diplococcus-shaped, multiplying by binary fission. Motile by means of polar flagella. Depending on conditions, cells may remain as individuals or attach to tetrads that aggregate into areas of considerable size, the whole being embedded in slime. The aggregates may break up into small clusters in more or less spherical colonies that become motile by the flagella of the composing individual cells. Gram negative. Contain bacteriochlorophyll *a* and carotenoids, both pigments being located in internal membranes of vesicular type. Contain gas vacuoles in the central part of the cell. Anaerobic. Capable of photosynthesis in the presence of hydrogen sulfide, during which they produce and store elemental sulfur in the form of globules in the gas vacuole-free, peripheral part of the cell. Molecular hydrogen may be used as an electron donor. Molecular oxygen is not evolved. G + C 63.8 mole%.

Lampropedia Cells rounded or almost cubical when packed together, 1.0–1.5 by 1.0–2.5 μm, occurring in pairs, tetrads, and regular and/or squared tablets of cells. Division takes place alternately in two planes. The cells forming a tablet are enclosed within a complex structured envelope; each cell is enclosed by a conventional wall of the Gram negative type. Nonmotile and nonflagelated but exhibiting a flickering movement of groups of cells in rapidly growing cultures. No resting stages known. Chemoorganotrophic: respiratory. Energy sources limited to Krebs cycle intermediates; carbohydrates, alcohols, fatty acids, and many related compounds not utilized. Carotenoid or photosynthetic pigments not formed. Prominent inclusions of poly-β-hydroxybutyrate. Do not contain gas vacuoles. Obligate aerobes. Optimal growth 10–35°C. G + C 61 mole%.

Legionella Cells rod-shaped, 0.3–0.4 by 2–3μm, or filamentous. Gram negative. Weakly oxidase positive; catalase positive. Nonmotile. Fastidious, with narrow optimal temperature and pH ranges. Chemoorganotrophic. Do not utilize carbohydrates. Urea not utilized. Nitrate not reduced. Aerobic, will not grow anaerobically. G + C 39 mole%.

Leptospira Cells single, flexuous, helical, 6–20 μm or more by 0.1 μm in diameter; coils 0.2–0.3 μm in overall diameter, pitch 0.3–0.5 μm. One or both ends may be bent more or less at right angles to the long axis, or hooked. Motile by means of axial filaments. Movements are of three basic types: shunting in either direction of the long axis, rapid rotation about the long axis, and flexion. Chemoorganotrophic: metabolism respiratory. Energy-yielding substrates include fatty acids as the sole carbon source. Strict aerobes: microaerophilic. Optimal growth 28–30°C. G + C 35–41 mole%.

Leptothrix Straight rods, 0.6–1.5 by 3–12 μm, occurring in chains within a sheath or free-swimming as single cells, in pairs, and in some species as motile short chains containing up to eight cells. Sheaths have a pronounced tendency to become impregnated or covered with hydrated ferric or manganic oxides. Free cells motile by means of one polar flagellum; one species has a subpolar tuft of several flagella. Resting stages not known. Gram negative. Chemoorganotrophic: metabolism respiratory, never fermentative; molecular oxygen is the universal electron acceptor. Strict aerobes. Optimal growth 20–25°C. G + C 69–70 mole%.

Leptotrichia Straight or slightly curved rods, 1–1.5 μm wide by 5–15 μm long, with one or both ends rounded or pointed; many cells with pointed ends are generally found. Two or more cells arranged in septate filaments of varying length. There is no club formation or branching. Cells nonmotile. No resting stages known. Gram negative. Granules, which may appear Gram positive, are distributed along the long axis. Acid but no gas production from glucose; the major fermentation product is lactic acid. Butyric acid is not produced. Hydrogen sulfide not produced. Indole negative. Catalase negative. Voges-Proskauer negative. Nitrate generally not reduced. Gelatin not liquefied. Anaerobic; 5% CO_2 is essential for optimal growth. Optimal growth 35–37°C. G + C 31.5–34 mole%.

Leuconostoc Cells spherical but often lenticular on agar, usually in pairs and chains. Nonmotile. Gram positive. Spores not produced. Chemoorganotrophic: fermentative. Complex growth factor an amino acid requirement. Growth is dependent on the presence of a fermentable carbohydrate; glucose fermented with production of lactic acid, ethanol, and CO_2. Mannitol is formed from fructose. Catalase negative; oxidase negative. Arginine not hydrolyzed. Nonproteolytic. Nonhemolytic. Indole not formed. Nitrates not reduced. Facultative anaerobes. Optimal growth 20–30°C. G + C 43–44 mole%.

Leucothrix Long filaments composed of short cylindrical or ovoid cells, cross walls clearly in evidence, colorless, unbranched; filaments usually uniform in diameter throughout length although may taper from base to apex under some conditions. Filaments usually attach to solid substrates by means of inconspicuous holdfasts; stalks absent. Filaments do not glide. Rosette formation frequently occurs in culture. Chemoorganotrophic or chemolithotrophic. Strictly aerobic. Sulfur granules are not formed.

Levinea Cells rod-shaped. Gram negative. Similar if not identical to *Citrobacter* except: hydrogen sulfide not produced; indole produced. G + C 54–59 mole%.

Listeria Small, coccoid, Gram positive rods with a tendency to produce chains of 3–5 or more cells and to produce elongated to filamentous forms. Do not produce spores or capsules. Not acid-fast. Gram positive but in older cultures may appear Gram negative. Motile by peritrichous flagella. Acid but no gas from glucose and several other carbohydrates. Esculin is hydrolyzed. Indole not produced. Urea not hydrolyzed. Gelatin not hydrolyzed, usually catalase positive. Aerobic to microaerophilic. Optimal growth 20–30°C. G + C 38 mole%, except one species with G + C 56 mole%.

Lucibacterium Cells rods, axis straight or slightly curved, 0.5–1.0 by 1.2–2.5 μm. Motile by peritrichous flagella. Spores not produced. Capsules not produced. Gram negative. Chemoorganotrophic: metabolism both respiratory and fermentative. Carbohydrate metabolism is fermentative without gas production. Nitrites produced from nitrates. Indole produced. Methyl red positive. Voges-Proskauer negative; oxidase positive. Catalase positive. Starch hydrolyzed. Gelatin liquefied. Growth requires added NaCl. Facultatively anaerobic. Optimal growth 25–30°C. G + C 45–46 mole%.

Lyngbya Cyanobacterium. Photoautotrophic: can utilize water as electron donor with evolution of oxygen. Contains chlorophyll *a* and phycobiliproteins. Chlorophyll and carotenoids located in a system of thylakoids internal to the cytoplasmic membrane. Filamentous. Growth by intercalary cell division in only one plane results in formation of a trichome. Trichome straight, nonmotile, and enclosed in a sheath. G + C 42–67 mole%.

Lysobacter Flexible rods, 0.2–0.5 by 2–70 μm. Gliding motility. Gram negative. Fruiting bodies not produced. White, cream, yellow, pink, or brown pigments; many strains produce a brown water-soluble pigment. Chemoorganotrophic: usually respiratory. Proteolytic. Cellulose not hydrolyzed. Chitin hydrolyzed. Catalase positive; oxidase positive. Indole negative. Methyl red negative. Voges-Proskauer negative. Aerobic. G + C 65.4–70.1 mole%.

Macromonas Cylindrical to bean-shaped cells, motile by a polar flagellum. Cells not embedded in a gelatinous mass. Chemolithotrophic; metabolize sulfur and sulfur compounds. Inclusions of calcium carbonate sometimes accompanied by sulfur globules occur. No resting stages known. Microaerophilic.

Megasphaera Relatively large cocci, 2 μm or more in diameter, in pairs; occasionally the diplococci are arranged in chains. Nonmotile. Spores not produced. Gram negative. Chemoorganotrophic: fermentative. Lactate and glucose are fermented with production of fatty acids, CO_2 and some hydrogen.

Succinate, fumarate, and malate are not used. Complex nutritional requirments. Obligately anaerobic.

Melittangium Vegetative cells, tapered rods, which under appropriate conditions aggregate to form a fruiting body. Sporangia borne singly on a stalk. Microcysts rod-shaped. Chemoorganotrophic: strictly aerobic, energy-yielding metabolism respiratory. G + C 67–71 mole%.

Meniscus Curved or straight rods, 0.7–1.0 by 2.0–3.0 μm. Gas vacuoles arranged randomly within cells. Nonmotile. Resting stages not known. Encapsulated. Gram negative. Chemoorganotrophic: fermentative metabolism with no gas production; does not carry out respiration. Catalase negative; oxidase negative. Facultative anaerobes. Optimal growth 30°C. G + C 44.9 mole%.

Methanobacterium Curved, crooked to straight rods, long and filamentous to coccoid, about 0.5–1.0 μm in width. Gram positive to Gram negative. Nonmotile or motile with monotrichous polar flagella. Spores not produced. Chemolithotrophic: energy for growth obtained by reduction of CO_2 to methane utilizing hydrogen and sometimes formate as the electron donor. Very strict anaerobes. Do not contain murein in cell wall; cell wall contains pseudomurein. G + C 32.7–52 mole%.

Methanobrevibacter Lancet-shaped cocci or short rods, often occurring in chains. Motile or nonmotile. Gram positive. Do not contain murein in cell wall; cell wall contains pseudomurein. Chemolithotrophic: energy for growth obtained by reduction of CO_2 to methane utilizing hydrogen and sometimes formate as the electron donor. Strict anaerobes. Optimal growth 37–39°C. G + C 27–32 mole%.

Methanococcus Spherical cells, occurring singly, in pairs, or in masses. Motile or nonmotile. Gram negative. Chemolithotrophic: energy for growth obtained by reduction of CO_2 to methane utilizing hydrogen and sometimes formate as the electron donor. Strict anaerobes. Do not contain murein in cell wall; cell wall contains protein. Optimal growth 36–40°C. G + C 30.7–31.1 mole%.

Methanogenium Irregular small cocci, 1–3 μm diameter. Motile. Gram negative. Chemolithotrophic: energy for growth obtained by reduction of CO_2 to methane utilizing hydrogen and sometimes formate as the electron donor. Require added NaCl for growth. Strict anaerobes. Do not contain murein in cell wall; cell wall contains protein. Optimal growth 20–25°C. G + C 51.6–61.2 mole%.

Methanomicrobium Short rods, 0.7 by 1.5–2.0 μm, occurring singly. Motile with single polar flagellum. Gram negative. Cell wall lacks murein; cell wall contains protein. Chemolithotrophic: energy for growth obtained by reduction of CO_2 to methane utilizing hydrogen and sometimes formate as the electron donor. Anaerobic. Requires organic growth factors. Optimal growth 40°C. G + C 48.8 mole%.

Methanosarcina Large spherical cells, 1.5–2.5 μm in diameter, generally occurring in regular packets. Nonmotile. Gram positive to Gram variable. Cell wall lacks murein; cell wall contains heteropolysaccharide. The energy metabolism involves formation of methane from acetate and sometimes from methanol, carbon monoxide, and possibly butyrate. Hydrogen may also be utilized in reduction of methanol. Strict anaerobes. Optimal growth 30–37°C. G + C 38.3–51 mole%.

Methanospirillum Curved rods or long, wavy filaments, 0.4 μm wide by up to several hundred μm long. Gram negative. Motile by means of polar flagella. Produce methane from $H_2 + CO_2$ and formate. Cell wall lacks murein; cell wall composed of protein. Anaerobic. Optimal growth 30–40°C. G + C 45–46.5 mole%.

Methylobacillus Short Gram negative rods. Obligate methylotrophs that grow on one-carbon compounds other than methanol and that cannot grow on methane and dimethyl ether. Methanol and methylamine are the only carbon compounds capable of supporting growth. Metabolism respiratory. Catalase positive; oxidase positive. Strict aerobes. Optimal growth 20–30°C. G + C 54.1 mole%.

Methylobacterium Gram negative rods. Grow on one-carbon compounds, methane, methanol, methyl formate, and dimethyl carbonate. Although capable of methane oxidation, not restricted to growth on one-carbon compounds; can utilize substrates such as glucose for growth in addition to methane.

Methylococcus Cells spherical, usually occurring in pairs. Nonmotile. Resting stages not known. Gram negative. Chemoorganotrophic: respiratory, using molecular oxygen as terminal electron acceptor. Methane and methanol can be used as sources of carbon and energy; growth restricted to use of one-carbon substrates. Catalase positive; oxidase positive. Strictly aerobic. Optimal growth 37°C. G + C 62.5 mole%.

Microbacterium Small diphtheroid rods with rounded ends; angular and palisade arrangements of cell masses are typical. In older cultures rods are shorter than in young cultures, but rod–coccus cycle of *Arthrobacter* does not occur. Nonmotile. Gram positive. Not acid-fast. Chemoorganotrophic. Catalase positive. Aerobic. G + C 63–70 mole%.

Microbispora True mycelium produced. Mycelial fragments remain intact. Spores in characteristic longitudinal pairs are formed on aerial mycelium; as a rule no spores formed on substrate mycelium. In most species aerial mycelium is pink. Gram positive. Aerobic to facultatively anaerobic.

Micrococcus Cells spherical, 0.5–3.5 μm in diameter, occurring singly or in pairs and characteristically dividing in more than one plane to form regular clusters, tetrads, or cubical packets. Usually nonmotile. No resting stages known. Gram positive. Chemoorganotrophic: metabolism strictly respiratory. Oxygen is the universal electron acceptor. Glucose oxidized mainly to acetate or carbon dioxide and water. Indole not produced. Catalase positive. Aerobes. Optimal growth 25–30°C. G + C 66–75 mole%.

Microcyclus Curved rods, 0.5–1.0 by 1.0–3.0 μm. Cells infrequently curve to form rings with outer diameter 0.7-3.0 μm. Filaments and coils not produced. Nonpigmented. Chemoorganotrophic: metabolism respiratory. Acid produced from several carbohydrates. Obligate aerobes. G + C 66.3–68.4 mole%.

Microellobosporia Produce true mycelium. Slender hyphae, about 1 μm in diameter, developing a substrate mycelium that grows into and forms a compact layer on the top of the medium and in aerial mycelium. Both substrate and aerial mycelia bear sporangia on short sporangiophores; the sporangia contain a single longitudinal row of nonmotile sporangiospores, usually one to five. Arthrospores not produced. Gelatin hydrolyzed. Starch hydrolyzed. Carbohydrates utilized. Aerobic. Optimal growth 30°C.

Micromonospora Well-developed, branched, septate mycelium averaging 0.5 μm in diameter. Spores borne singly, sessile, or on short or long sporophores that often occur in branched clusters. Aerial mycelium absent. Gram positive. Not acid-fast. Most species aerobic, but two are anaerobic. Optimal growth 20–40°C.

Micropolyspora Produce substrate and aerial mycelium, both bearing short chains of 1–20 spores. Spores 1.2–1.5 μm in diameter. Gram positive.

Microscilla Flexible rods or filaments, 0.5 by 5–100 μm. Do not exhibit life cycle of changing morphological forms. Gram negative. Resting stages not known. Gliding motility. Do not decompose cellulose. Appear to be marine forms related to *Flexibacter*.

Microtetraspora Branched, usually stable substrate mycelium; substrate mycelium does not fragment. Spores not borne on substrate mycelium. Short, sparsely branched aerial mycelium bearing short chains of spores; usually four spores are borne on sporangiophores, but one species produces six spores. Chemoorganotrophic. Optimal growth 28–37°C.

Moraxella Rods, usually very short and plump (coccobacilli), typically 1.0–1.5 by 1.5–2.5 μm, often approaching coccus shape, predominantly in pairs and short chains. Chemoorganotrophic: oxidative metabolism. A limited number of organic acids, alcohols, and amino acids serve as carbon and energy sources. Carbohydrates not utilized. Oxidase positive; catalase usually positive. Indole not produced. Acetoin not produced. Hydrogen sulfide not produced. Strict aerobes. Optimal growth 32–35°C. G + C 40–46 mole%.

Morganella Usually straight rods, 0.4–0.6 by 1.0–3.0 μm. Motile by peritrichous flagella or nonmotile. Chemoorganotrophic: acid and gas produced from fermentation of glucose. Indole produced. Gelatin not liquefied. Urease positive. Oxidatively deaminate a wide range of amino acids. G + C ca. 50 mole%.

Morococcus Colorless, spherical organisms, less than 1 μm in diameter, bound firmly together in tightly packed, mulberry-like aggregates of 10 to 20 cells, with adjacent sides often flattened. Gram negative. Nonmotile. Endospores not formed. Aerobic. Nitrate is reduced. Catalase and cytochrome oxidase are produced. H_2S is produced. Acid is produced from carbohydrates. Hemolytic. Complex growth factors not required. Growth occurs between 23 and 42°C and between pH 5.5 and 9.0. G + C 51.7–52.9 mole%.

Mycobacterium Slightly curved or straight rods 0.2–0.6 by 1.0–10 μm, sometimes branching; filamentous or mycelium-like growth may occur, but on slight disturbance usually becomes fragmented into rods or coccoid elements. Acid–alcohol-fast at some stage of growth. Gram positive, but not readily stained by Gram's method. Nonmotile. No endospores, conidia, or capsules. Does not produce grossly visible aerial hyphae. Lipid content of cells and especially cell walls high. Diffusible pigments rare. Growth: slow to very slow. Optimal growth at about 40°C. G + C 62–70 mole%.

Mycoplana Branching filaments breaking into irregular rods, 0.5–1 by 1.2–4.5 μm, that are motile with subpolar tufts of flagella. Gram negative. Not acid-fast. Chemoorganotrophic: respiratory. No acid from carbohydrates. Nitrate not reduced. Cell wall contains meso-diaminopimelic acid. Aerobic. G + C 64–69 mole%.

Mycoplasma Cells highly pleomorphic, varying in shape from spherical or slightly ovoid, about 125–250 nm in diameter, to slender branched filaments of uniform diameter ranging in length from a few to

150 μm. Cells lack a true cell wall and are bounded by a single triple layered membrane. Usually nonmotile, but gliding motility has been described for some species. No resting stages known. Gram negative. The typical colony is biphasic, with a fried egg appearance. Most species utilize either glucose or arginine as the major carbon source. Chemoorganotrophic: metabolism mainly fermentative, although a few species have oxidative metabolism. Urea not hydrolyzed. All species require cholesterol, or certain other sterols, for growth. Optimal growth 36–37°C. G + C 23–40 mole%.

Myxosarcina Cyanobacterium. Photoautotrophic: can utilize water as electron donor with evolution of oxygen. Contains chlorophyll *a* and phycobiliproteins. Chlorophyll and carotenoids located in a system of thylakoids internal to the cytoplasmic membrane. Cells occurring singly or forming colonial aggregates held together by additional outer wall layers. Growth and binary fission lead to cubical cell aggregates. Reproduction by multiple fission leads to formation of motile baeocytes that lack a thin fibrous outer cell wall layer at the moment of their release. G + C 43–44 mole%.

Nannocystis Vegetative cells short, blunt-ended rods or cocci. Aggregation to form fruiting body occurs. Myxospores resemble vegetative cells and lack a slime capsule. Cells become coccoid. Sporangia sessile. Colonies etch, erode, and penetrate agar media. Congo red is not absorbed.

Neisseria Cocci, 0.6–1.0 μm in diameter, occurring singly but often in pairs with adjacent sides flattened. Endospores not produced. Nonmotile. Capsules may be present. Gram negative. Complex growth requirements. Chemoorganotrophic: few carbohydrates utilized. Aerobic or facultatively anaerobic. Catalase positive; oxidase positive. Optimal growth ca. 37°C. G + C 47.0–52.0 mole%.

Neorickettsia Small, coccoid, often pleomorphic, intracytoplasmic (but not intranuclear) organisms that occur primarily in reticular cells of lymphoid tissues. Infectious cycle includes transovarial transmission in the vector. Gram negative. Nonmotile. Not cultivatable in cell-free media.

Nevskia Rod-shaped cells, 0.7–2.0 by 2.4–12 μm, often slightly bent. Under certain conditions cells are surrounded with slime; more slime is produced on one side, forming acellular, hyaline stalks perpendicular to the long axis of the organism. The stalks branch dichotomously as a result of division of mature cells. Cells generally are filled with refractile globules; the globules are not composed of sulfur nor poly-β-hydroxybutyrate. Reproduction only by binary fission. Gram negative. Aerobic.

Nitrobacter Cells short rods, often wedge or pear-shaped. Reproduction by budding. Cells possess a polar cap of cytomembranes. No resting stages known. Usually nonmotile. Gram negative. Cells rich in cytochromes imparting a yellowish color to cell suspensions; void of other pigments. Chemolithotrophic: some strains are obligate chemolithotrophs that oxidize nitrite to nitrate and fix CO_2; some strains can be grown heterotrophically, but the growth rate is slower than when the cells are grown autotrophically. Strictly aerobic, using oxygen as terminal electron acceptor. Temperature range for growth 5–40°C. G + C 60.7–61.7 mole%.

Nitrococcus Cells spherical, 1.5 μm or larger. Gram negative. Motile by means of one or two subterminally inserted flagella. Cells rich in cytochromes imparting a yellowish to reddish color; void of other pigments. Obligately chemolithotrophic: oxidize nitrite to nitrate and fix CO_2. Strictly aerobic, using oxygen as terminal electron acceptor. Temperature range for growth 15–30°C. G + C 61.2 mole%.

Nitrosococcus Cells spherical, motile or nonmotile. Gram negative. Grow singly, in pairs or in tetrads. Obligately chemolithotrophic: oxidize ammonia to nitrite and fix CO_2. No organic growth factors required. Strictly aerobic, using oxygen as terminal electron acceptor. Temperature range for growth 2–30°C. G + C 50.5–51.0 mole%.

Nitrosolobus Cells pleomorphic and lobate, 1.0–1.5 μm in diameter. Division by constriction. Gram negative. Cells partially compartmentalized by the invagination of the cytoplasmic membrane. Motile by means of peritrichous flagella. Cells are rich in cytochromes that impart a yellowish to reddish color; void of other pigments. Obligately chemolithotrophic: oxidize ammonia to nitrite and fix carbon dioxide. No organic growth factors are required. Strictly aerobic, using oxygen as terminal electron acceptor. Temperature range for growth 15–30°C. G + C 53.6–55.1 mole%.

Nitrosomonas Cells ellipsoidal or short rods, motile or nonmotile, occurring singly, in pairs, or short chains. Gram negative. Possess cytomembranes that occur in flattened vesicles in the peripheral regions of the cytoplasm. Cells rich in cytochromes that impart a yellowish to reddish color; void of other pigments. Obligately chemolithotrophic: oxidize ammonia to nitrite and fix CO_2. No organic growth factors required. Strictly aerobic, using oxygen as terminal electron acceptor. Temperature range for growth 5–30°C. G + C 47.4–51 mole%.

Nitrosospira Cells spiral-shaped. Gram negative. Cells lack cytomembranes. Nonmotile or motile by means of peritrichous flagella. Obligately chemo-

lithotrophic: oxidize ammonia to nitrite and fix CO_2. No organic growth factors required. Strictly aerobic, using oxygen as terminal electron acceptor. Temperature range for growth 15–30°C. G + C 54.1 mole%.

Nitrospina Cells are straight, slender rods; spherical forms are found in senescent cultures. There is no extensive cytomembrane system. Gram negative. Nonmotile. Cells have cytochromes but no other pigments. Obligately chemolithotrophic: oxidize nitrite to nitrate and fix CO_2. No organic growth factors required. Strictly aerobic, using oxygen as terminal electron acceptor. Temperature range for growth 20–30°C. Optimal growth in 70–100% seawater; no growth in distilled water–salts medium even if NaCl added. G + C 57.7 mole%.

Nocardia Produce true mycelium, but mycelium production may be rudimentary. Members of this genus may be divided into three morphological groups according to the degree of mycelial development: group 1 members have limited mycelial development; group 2 members exhibit extensively branched mycelium during early growth phase; group 3 members develop abundant mycelia, including aerial hyphae. Coccoid cells that have been termed microcysts or chlamydospores are produced by some members of groups 1 and 2. Reproductive bodies are mycelial fragments formed irregularly in substrate or aerial hyphae. Spores are not produced on differentiated hyphae. Gram positive. Some species acid-fast to partially acid-fast. Obligate aerobes. Nonmotile. Pigments are produced by several species. G + C 60–72 mole%.

Nocardioides Both substrate and aerial mycelia fragment into rod-shaped and coccoid elements. Aerial mycelium when formed is thin with sparse irregular branching. Gram positive. Not acid-fast. Catalase positive. Chemoorganotrophic. Hydrolyze starch. Produce hydrogen sulfide. Aerobic. Optimal growth 28°C.

Nocardiopsis Sustrate mycelium undergoes fragmentation. Spores not borne on substrate mycelium. Long chains of spores produced on aerial mycelium. Nonmotile. Contains meso-diaminopimelic acid in cell wall but lacks madurose.

Nodularia Cyanobacterium. Photoautotrophic: can utilize water as electron donor with evolution of oxygen. Contains chlorophyll *a* and phycobiliproteins. Chlorophyll and carotenoids located in a system of thylakoids internal to the cytoplasmic membrane. Filamentous. A trichome (chain of cells) forms by intercalary division in one plane only. In the absence of combined nitrogen, heterocysts are produced; heterocysts are associated with the ability to fix atmospheric nitrogen. Akinetes also may be formed. Reproduction by random trichome breakage, by formation of hormogonia, and sometimes by germination of akinetes. Hormogonia are distinguishable from mature trichomes by the absence of heterocysts and by one or more of the following characteristics: rapid gliding motility, smaller cell size, cell shape, and gas vacuolation. Heterocysts are intercalary or terminal; position of akinetes (if produced) is variable. Vegetative cells are disk-shaped. G + C 41 mole%.

Nostoc Cyanobacterium. Photoautotrophic: can utilize water as electron donor with evolution of oxygen. Contains chlorophyll *a* and phycobiliproteins. Chlorophyll and carotenoids located in a system of thylakoids internal to the cytoplasmic membrane. Filamentous. A trichome (chain of cells) forms by intercalary division in one plane only. In the absence of combined nitrogen, heterocysts are produced; heterocysts are associated with the ability to fix atmospheric nitrogen. Akinetes also may be formed. Reproduction by random trichome breakage, by formation of hormogonia, and sometimes by germination of akinetes. Hormogonia are distinguishable from mature trichomes by the absence of heterocysts and by one or more of the following characteristics: rapid gliding motility, smaller cell size, cell shape, and gas vacuolation. Hormogonia give rise to young filaments that bear a terminal heterocyst at both ends of the cellular chain. Vegetative cells are spherical, ovoid, or cylindrical. Akinetes (if produced) are not initiated adjacent to heterocysts and are often formed in chains. G + C 39–46 mole%.

Obesumbacterium Gram negative short rods. Motile with peritrichous flagella or nonmotile. No gliding motility. Pigments not soluble in media. Chemoorganotrophic: acid and gas produced from glucose. G + C 30 mole%.

Oceanospirillum Helical rods, 0.3 by 1.2 μm. Gram negative. Motile, usually with bipolar tufts of flagella. Chemoorganotrophic: strictly respiratory. Carbohydrates not utilized. Oxidase positive; catalase variable. Nitrate not reduced. G + C 42–48 mole%.

Oerskovia Extensively branched substrate mycelium that breaks up into motile, rod-shaped elements. Motile elements have monotrichous flagella when small and peritrichous flagella when long. No aerial mycelium is formed. Gram positive. Not acidfast. Facultative anaerobes. Chemoorganotrophic: respiratory and fermentative metabolism. Catalase positive; oxidase negative. Starch hydrolyzed. Gelatin hydrolyzed. Nitrate reduced to nitrite. G + C 70.5–75 mole%.

Oscillospira Large rods or filaments 3–6 μm in diameter, divided by closely spaced cross walls into numerous disk-shaped cells. Reproduction by transverse fission. Motile by means of numerous lateral flagella. Endospores may be produced. Gram negative.

Oscillatoria Cyanobacterium. Photoautotrophic: can utilize water as electron donor with evolution of oxygen. Contains chlorophyll *a* and phycobiliproteins. Chlorophyll and carotenoids located in a system of thylakoids internal to the cytoplasmic membrane. Filamentous. Growth by intercalary cell division in only one plane results in formation of a trichome. Trichome straight. Cells that comprise the trichome are disk-shaped and not separated by deep constrictions. Trichome motile and sheath formation is not pronounced. G + C 40–50 mole%.

Paracoccus Cells spherical or nearly spherical, occurring singly or in pairs or aggregates. Short, rod-shaped cells may occur in young cultures. Nonmotile. Accumulate poly-β-hydroxybutyrate. No resting stages are known. Gram negative. Chemoorganotrophic: metabolism respiratory never fermentative. One species is facultatively chemolithotrophic: able to use oxidation of hydrogen as energy source for autotrophic growth. Molecular oxygen or nitrate can serve as electron acceptors. Nitrate is reduced to nitrous oxide and molecular nitrogen under anaerobic conditions. Aerobic, but anaerobic respiration using nitrate can occur. Oxidase positive; catalase positive. G + C 64–67 mole%.

Pasteurella Cells ovoid or rod-shaped, 0.1–1.8 μm, occurring singly or less frequently in pairs or short chains. Nonmotile. Do not produce endospores. Gram negative, but bipolar staining is common. Chemoorganotrophic: metabolism fermentative. Small amounts of acid but not gas produced from glucose and other fermentable substrates. Catalase positive. Almost always oxidase positive. Nitrate reduced. Gelatin not liquefied. Methyl red negative. Voges-Proskauer negative. Aerobic to facultatively anaerobic. Optimal growth 37°C. G + C 36.5–43.0 mole%.

Pasteuria Cells, 1–5 by 3–6 μm, nearly spherical to pear-shaped with one end broad and rounded and the other end pointed and attached. Occurs singly or in mass aggregations resembling grape-like clusters. Multiplication is by budding from the top or side of the free rounded end. Nonmotile. Nonpigmented. Do not produce prosthecae. Aerobic.

Pectinatus Slightly curved rods, with rounded ends, occurring singly, in pairs, and only in relatively short chains. Shorter, younger cells do not show helical shape, but the elongated older cells tend to form helices. Motile by means of flagella that emanate from only one side of the cell; attachment of flagella not limited to central portion of concave side of the cell. Chemoorganotrophic: fermentative. Mixed acids produced from carbohydrate fermentation. Nitrate not reduced. Catalase negative. Strictly anaerobic. Optimal growth ca. 32°C. G + C 39.8 mole%.

Pectobacterium Cells predominantly single, straight rods. Motile with peritrichous flagella. Gram negative. Generally considered as the Cartovora group of *Erwinia*. Chemoorganotrophic: produce acid from carbohydrate fermentation. Produce pectinases; actively cause soft rots in plants. Optimal growth 27–30°C. G + C 50–57 mole%.

Pediococcus Cocci occurring in pairs or tetrads. Nonmotile. Do not produce endospores. Gram positive. Chemoorganotrophic: metabolism fermentative. The fermentation is homolactic, producing lactic acid. Starch not fermented. Gelatin not liquefied. Nitrate not reduced. Nutritional requirements complex. Microaerophilic. Generally catalase negative. G + C 34–44 mole%.

Pedomicrobium Cells spherical, oval, rod-shaped, pear- or bean-shaped, 0.4–2.0 μm wide, unevenly staining. Multiplication is primarily by budding at the tips of the cellular extensions (hyphae) of constant diameter (0.15–0.3 μm). Daughter cells may remain attached to their mother hyphae or they may separate, forming uniflagellated swarmers. After increasing in size, mature swarmer cells grow one to numerous hyphae from several sites on their cell surface. Gram negative. Microaerophilic. Ferric and/or manganese depositions occur either on mother cells or on the hyphae.

Pelodictyon Rod-shaped to ovoid bacteria, multiplying by binary fission. Nonmotile. Branching may occur as a result of ternary fission and net-like three-dimensional aggregates or more or less spherical colonies may be formed. Gram negative. Contain either bacteriochlorophyll *c* or *d* and carotenoids, both pigments being located in elongated-ovoid vesicles underlying and attached to the cell membrane. Contain gas vacuoles. Anaerobic. Capable of photosynthesis in the presence of hydrogen sulfide during which they produce and deposit elemental sulfur in the form of globules outside the cells. Molecular hydrogen may be used as electron donor. Molecular oxygen is not evolved. Cell suspension appears yellow-green. G + C 48.5–58.1 mole%.

Peptococcus Spherical bacteria, generally 0.5–1.0 μm in diameter. Occur singly, in pairs, tetrads, or irregular masses. Nonmotile. No spores produced. Gram positive. Chemoorganotrophic: can use protein decomposition products as sole energy source; various combinations of low-molecular-weight volatile fatty acids, CO_2, H_2, and ammonia tend to be the major products from amino acid metabolism. Ability to utilize carbohydrates often limited; gen-

erally, lactate is not a major product of glucose metabolism when glucose is utilized. Weak or variable catalase. Coagulase negative. Anaerobic. Optimal growth 35–37°C.

Peptostreptococcus Spherical to ovoid cells, usually 0.7–1.0 μm in diameter, occurring in pairs, short, or long chains. Nonmotile. Spores not produced. Gram positive. Chemoorganotrophic: with one exception carbohydrates fermented with production of acid, gas, or both. Gas production in carbohydrate-free peptone media by some species. Pyruvate frequently fermented producing acid and/or gas. Malate, citrate, and tartrate not fermented. Acetic products of fermentation may be combinations of acetic, formic, propionic, butyric, isobutyric, valeric, isovaleric, isocaproic, caproic, and succinic acids. Lactate not produced. Ammonia produced from peptone. Catalase negative. Nitrates not reduced. Anaerobic. Optimal growth 35–37°C.

Photobacterium Cells coccobacilli or occasionally rods 1.0–2.5 by 0.4–1.0 μm, axis straight, occurring singly and occasionally in pairs. Do not produce spores. Motile by one or more polar flagella, or occasionally nonmotile. Gram negative. Capsules not produced. Chemoorganotrophic: metabolism both respiratory and fermentative. Carbohydrate metabolism is fermentative and gas is usually produced. Nitrites produced from nitrates. Ammonia produced from peptone. Indole not produced. Methyl red positive. Voges-Proskauer positive. Starch not hydrolyzed. Facultatively anaerobic. Optimal growth 20–30°C. G + C 39–42 mole%.

Pilimelia Hyphae up to 2.2 μm wide and up to 300 μm long, irregularly septate. Sporangia large, globose, or cylindrical. Rod-shaped spores produced end to end in parallel chains, approximately 1000 per sporangia; motile by means of flagella. Require complex organic growth factors. Slight or no penetration of mycelium into agar. Obligate aerobes.

Planctomyces Cells spherical to oblong or pear-shaped, 0.3–1.7 μm in diameter with long and slender stalks. Stalks are 0.3–0.5 μm wide and up to 11 μm long; they often attach to a common holdfast and form rosettes up to 50 cells. Multiplication is by terminal or lateral budding of mature cells.

Planobispora Well-developed, highly branched substrate mycelium. Gram positive. Not acid-fast. Aerial mycelium sparsely branched; hyphae usually parallel to agar surface. Sporangia containing a longitudinal pair of large spores formed only on aerial mycelia. Spores motile by peritrichous flagella. Glucose supports growth. Cellulose not metabolized. Starch hydrolyzed. Nitrates reduced to nitrites. Aerobic. Optimal growth 28–37°C.

Planococcus   Spheres, 1.0–1.2 μm in diameter, occurring singly, in pairs, in threes, or in tetrads. Motile, each cell possessing one or two flagella, occasionally three or four. Do not produce spores. Gram positive. Chemoorganotrophic: metabolism respiratory never fermentative. Do not produce acid or gas from various carbohydrates. Voges-Proskauer negative. Coagulase negative. Nitrate not reduced to nitrite. Catalase positive. Strict aerobes. Optimal growth 20–37°C. G + C 48–52 mole%.

Planomonospora   Substrate and aerial mycelium formed, the former growing profusely into the medium and forming a compact layer on the surface of agar. Sporangia formed on aerial mycelium only, each containing a single large spore motile by peritrichous flagella. Not acid-fast. Gram positive. Starch hydrolyzed. Nitrates reduced to nitrites. Aerobic. Grows on a variety of substrates. Pigments may be produced giving characteristic colors to the aerial and/or vegetative mycelia. Optimal growth 28–37°C.

Pleisomonas   Cells round-ended, straight, rod-shaped, 0.8–1.0 by 3.0 μm, growing singly, in pairs, or short chains. Motile by polar flagella, generally lophotrichous. Resting stages not known. Gram negative. Chemoorganotrophic: metabolism both respiratory and fermentative. Acid but no gas produced from carbohydrate metabolism. Oxidase positive; catalase positive. Facultative anaerobes. Optimal growth 30–37°C. G + C 51 mole%.

Pleurocarpsa   Cyanobacterium. Photoautotrophic: can utilize water as electron donor with evolution of oxygen. Contains chlorophyll *a* and phycobiliproteins. Chlorophyll and carotenoids located in a system of thylakoids internal to the cytoplasmic membrane. Cells occurring singly or forming colonial aggregates held together by additional outer wall layers. Growth and binary fission lead to irregular pseudofilamentous cell aggregates. Reproduction by multiple fission leads to formation of motile baeocytes that lack a thin fibrous outer cell wall layer at the moment of their release. G + C 39–47 mole%.

Polyangium   Under appropriate conditions cells aggregate to form a fruiting body. Vegetative cells cylindrical, of uniform diameter with blunt, rounded ends. Sporangia sessile, solitary or in groups (sori) often bounded by a common envelope or layer of slime. Myxospores resemble vegetative cells, not refractile or phase-dense, lacking a slime capsule. Colonies etch, erode, and penetrate agar. Cellulose may or may not be digested. Congo red not adsorbed.

Prochloron   Cells spherical, 6–25 μm, occurring singly. Photoautotrophic. Contain both chlorophylls *a* and *b*; lack phycobilins. G + C 39–41 mole%.

Promicromonospora   Mycelium, 0.5–1 μm in diameter. Substrate mycelium fragments into nonmotile pleomorphic elements of variable length. Little or no aerial mycelium. No spores (originally described as producing single spores on substrate mycelium, but this observation has not been repeated). Nonmotile. Gram positive. Nocardomycolates not present in cell wall; cell walls do not contain galactose. Chemoorganotrophic: oxidative metabolism only. No growth under anaerobic conditions. Starch hydrolyzed. Nitrate not reduced. G + C 73 mole%.

Propionibacterium   Gram positive, nonspore-forming, nonmotile rods. Usually pleomorphic, diphtheroid, or club-shaped with one end rounded and the other end tapered or pointed. Chemoorganotrophic: metabolize carbohydrates, peptone, pyruvate, or lactate. Fermentation products include combinations of propionic and acetic acids and frequently lesser amounts of isovaleric, formic, succinic, or lactic acid and carbon dioxide. All species produce acid from glucose. Anaerobic to aerotolerant. Optimal growth 30–37°C. G + C 59–66 mole%.

Prosthecobacter   Cells straight or curved rods, 0.5 by 3–10 μm, with a single polar prostheca. Cells attach to surfaces by means of a holdfast located at the tip of the appendage; rosette formation rare. Unlike *Caulobacter,* life cycle is not dimorphic. Nonmotile, swarmer cell; at the time of septation, the daughter cell is usually fully differentiated so that it is a mirror image of the mother cell. Chemoorganotrophic: utilize various carbohydrates. Obligate aerobes. G + C 54.6–60.1 mole%.

Prosthecochloris   Spherical to ovoid bacteria, forming prosthecae, dividing by binary fission in many directions. Nonmotile. Gram negative. Contain bacteriochlorophyll *c* and carotenoids, all pigments being located in elongated, ovoid vesicles underlying and attached to the cytoplasmic membrane. Do not contain gas vacuoles. Anaerobic. Capable of photosynthesis in the presence of hydrogen sulfide during which they produce and deposit elemental sulfur in the form of globules outside the cell. Molecular oxygen is not evolved during photosynthesis. Cell suspensions appear green because of photosynthetic pigments. G + C 50.0–56.1 mole%.

Prosthecomicrobium   Unicellular, Gram negative bacteria with many prosthecae extending in all directions from the cell. Prosthecae are normally less than 2 μm long and conical in shape, tapering distally from the cell toward a blunt tip. Cells divide by binary transverse fission. Motile or nonmotile; motile strains exhibit a tumbling, circular motility. Chemoorganotrophic: nonfermentative. Aerobic. G + C 65.8–69.9 mole%.

Proteus Usually straight rods 0.4–0.6 by 1.0–3.0 μm; coccoid and irregular forms occur under certain conditions. May occur in pairs or chains. Not encapsulated. Motile by peritrichous flagella. Nonpigmented. Acid produced from glucose. Methyl red test usually positive. Nitrates reduce to nitrites. Phenylalanine deaminated to phenylpyruvic acid. Generally indole positive. Growth not inhibited by potassium cyanide. Gram negative. Optimal growth 37°C. G + C 38–42 mole%.

Providencia Straight, short rods. Motile with peritrichous flagella or nonmotile. Gram negative. Do not exhibit swarming on solid surfaces. Chemoorganotrophic: fermentative and oxidative. Oxidatively deaminate a wide range of amino acids. Mannose fermented. Indole produced. Hydrogen sulfide not produced. Urease variable. G + C 40–42 mole%.

Pseudanabaena Cyanobacterium. Photoautotrophic: can utilize water as electron donor with evolution of oxygen. Contains chlorophyll *a* and phycobiliproteins. Chlorophyll and carotenoids located in a system of thylakoids internal to the cytoplasmic membrane. Filamentous. Growth by intercalary division in one plane results in formation of a trichome. Trichome straight. Cells that compose trichome contain gas vacuoles, are isodiametric or cylindrical, and are separated by deep constrictions. The degree of constriction between adjacent cells varies. Reproduction occurs by transcellular or intercellular trichome breakage. Trichome is motile, not ensheathed. G + C 44–52 mole%.

Pseudomonas Cells single, straight, or curved rods but not helical, generally 0.5–1 by 1.5–4 μm. Motile by polar flagella; monotrichous or multitrichous. Do not produce sheaths or prosthecae. No resting stages known. Gram negative. Chemoorganotrophic: metabolism respiratory, never fermentative; some are facultative chemolithotrophs. Molecular oxygen is the universal electron acceptor; some can denitrify, using nitrate as an alternate acceptor. Strict aerobes, except for those species that can use denitrification in anaerobic respiration. Catalase positive. G + C 58–70 mole%.

Pseudonocardia Produce substrate hyphae, 0.4–2.6 μm in diameter, and aerial hyphae, 0.4–1.8 μm in diameter. Growth of the hyphae characteristically shows budding; a constriction is produced just behind the tip of the terminal segment, the tip elongates to form a new segment. Spores are produced on differentiated hyphae; spores, which are borne in chains, may be formed on aerial or substrate mycelium. Gram positive. Not acid-fast. Colonies colorless to yellow or orange. Aerial mycelium white, powdery or forming a thick cover. Starch not hydrolyzed. Grows well on a variety or organic substrates. Aerobic.

Rahmella Short rods. Gram negative. Motile with peritrichous flagella at temperatures below 36°C; nonmotile at 37°C. Chemoorganotrophic: various carbohydrates fermented with production of acid and gas. Methyl red positive. Voges-Proskauer positive. Decarboxylase negative. G + C 51–56 mole%.

Renibacterium Short rods, 0.3–1.0 by 1.0–1.5 μm, often occurring in pairs. Gram positive. Not acid-fast. Nonencapsulated. Nonmotile. Endospores are absent. The cell wall composition is characterized by lysine as the diamino acid of the peptidoglycan. The principal cell wall sugar is glucose. Aerobic. Optimal growth 15–18°C; slow growth occurs at 5 and 22°C; no growth occurs at 37°C. Catalase positive. G + C 53.0 mole%.

Rhizobium Rods, 0.5–0.9 by 1.2–3.0 μm, commonly pleomorphic under adverse growth conditions. Motile by two to six peritrichous flagella or by a polar or subpolar flagellum. Do not produce spores. Gram negative. Growth on carbohydrate media usually accompanied by copius production of extracellular polysaccharide slime. Chemoorganotrophic: metabolism respiratory. Molecular oxygen is the terminal electron acceptor. 3-Ketoglycosides are not produced. Gelatin not liquefied. Agar not hydrolyzed. Aerobic but can grow at reduced oxygen tension. Characteristically able to invade root hairs of leguminous plants and initiate production of root nodules. Within nodules bacteria are pleomorphic (bacteroids). Nodule bacteroids characteristically involved in fixing molecular nitrogen. Optimal growth 25–30°C. G + C 59.1–65.5 mole%.

Rhodococcus Pleomorphic Gram positive actinomycetes generally forming a primary mycelium that fragments into bacillary and coccoid elements. Aerial mycelium absent or very limited. Usually pigmented buff, orange, pink, or red. Chemoorganotrophic: utilize a wide variety of carbon sources for energy and growth. Aerobic. Contain mycolic acids in the cell wall. G + C 59–69 mole%.

Rhodocyclus Cells are half-ring-shaped to ring-shaped before cell division; they are 0.6–0.7 μm wide, and the length of the half-circle shaped cells is about 2.7 μm; the diameter of the circles is 2.0–3.0 μm. Under certain conditions open or compact spirals of cells are formed. Gram negative. Nonmotile. Multiply by binary fission. Photoorganotrophic: grow anaerobically in the light or aerobically in the dark. Bacteriochlorophyll *a* and carotenoids are produced. Color of anaerobic cultures growing photoautotrophically is red-purple. The photosynthetic apparatus consists of the cytoplasmic membrane and only a few invaginations into the cells. Optimal growth 30°C. G + C 65.3 mole%.

Rhodomicrobium Ovoid to elongate-ovoid bacteria, multiplying by budding. Daughter cells originate as spherical buds at the end of filaments from one to several times the length of the mother cell. Mature buds may separate from the filament; they are motile by means of peritrichous flagella. Gram negative. Contain bacteriochlorophyll *a* and carotenoids, both pigments being located at interal membranes of lamellar type. Do not contain gas vacuoles. Anaerobic phototrophs; in addition, most strains are able to grow and carry out oxidative metabolism in the dark under microaerophilic to aerobic conditions. Capable of photosynthesis in the presence of simple organic compounds that can be directly photoassimilated and also serve as a source of reducing power for carbon dioxide fixation. Molecular oxygen is not evolved during photosynthesis. Cell suspensions appear in various shades of pink to reddish brown because of photosynthetic pigments. G + C 61.8–63.8 mole%.

Rhodopseudomonas Rod-shaped and ovoid to spherical bacteria, multiplying either by binary fission or asymmetrical division (budding without stalk formation). Motile by means of polar flagella. Gram negative. Contain bacteriochlorophyll *a* or *b* and carotenoids, both types of pigments being located at internal membrane systems of either vesicular, tubular, or lamellar type. Do not contain gas vacuoles. Anaerobic phototrophs; in addition, some species able to carry out oxidative metabolism in the dark under microaerophilic or aerobic conditions. Capable of photosynthesis in the presence of simple organic compounds that can be directly photoassimilated and also serve as a source of reducing power for carbon dioxide fixation. Not able to utilize elemental sulfur as photosynthetic electron donor. Molecular oxygen not evolved during photosynthesis. Cell suspensions are various shades of yellow-green to brown and red because of photosynthetic pigments. G + C 62.2–72.4 mole%.

Rhodospirillum Spherical-shaped bacteria, multiplying by binary fission. Motile by means of polar flagella. Gram negative. Contain bacteriochlorophyll *a* and carotenoids, both pigments being located at internal membrane systems that originate from and may be continuous with the cytoplasmic membrane. Do not contain gas vacuoles. Anaerobic phototrophs; in addition most species able to carry out oxidative metabolism in the dark under microaerophilic to aerobic conditions. Capable of photosynthesis in the presence of simple organic carbon compounds that can be directly photoassimilated and also serve as source of reducing power for carbon dioxide fixation. Molecular oxygen not evolved during photosynthesis. Cell suspensions are various shades of red to brown because of photosynthetic pigments. G + C 61.7–65.8 mole%.

Rickettsia Parasitic bacteria occurring intracellularly or intimately in association with tissue cells other than erythrocytes or with certain organs in arthropods. Growth generally occurs in the cytoplasm of host cells. Transmitted by arthropod vectors. Short rods, 0.3–0.6 by 0.8–2.0 μm. No flagella. Capsules not produced, but an outer layer of amorphous material may occur. Gram negative. Have not been cultivated in the absence of host cells. G + C 30–32.5 mole%.

Rickettsiella Gram negative rods, usually smaller than those in the genus *Rickettsia*. Pathogenic for their insect larval hosts and other invertebrate hosts but of limited virulence for vertebrates. Have not been cultivated in host-free media.

Rochalimaea Resemble *Rickettsia* but usually occur in an extracellular environment in the arthropod host and can be cultivated in host cell-free media.

Rothia Coccoid, diphtheroid, or filamentous cells. Filamentous forms branched, usually 1 μm in diameter. Gram positive. Nonacid-fast. Nonspore-forming. Nonmotile. No aerial hyphae produced. Catalase positive. Chemoorganotrophic: ferments carbohydrates with production of acid but no gas. Major product of glucose fermentation is lactic acid; other acetic products may be produced but not propionic acid. Cell wall constituents include alanine, glutamic acid, lysine, and galactose but no arabinose or diaminopimelic acid. Aerobic, but some strains able to grow at reduced oxygen tension. Optimal growth 35–37°C.

Ruminococcus Spherical to elongated coccoid cells. Gram positive but may be decolorized easily. Cellulose usually digested; cellobiose always fermented. Major products of cellobiose fermentation are acetic and formic acids; ethanol and/or succinic acids may also be major products, but when ethanol is a major product there is more hydrogen and little or no succinic acid; when succinic acid is a major product there is little hydrogen and little or no ethanol. Chemoorganotrophic. Complex growth factors required. Ammonia is essential as the main nitrogen scource. Catalase negative. Indole not produced. Hydrogen sulfide not produced. Strictly anaerobic. G + C 39.8–45.4 mole%.

Runella Straight to curved rods, 0.5–0.9 by 2.0–4.5 μm, the degree of curvature varying among cells in a culture. The ends of the cell may overlap, producing a ring-shaped structure with a diameter of 2.0–3.0 μm. Filaments of up to 14 μm long may occur. Gram negative. Nonmotile. Not flexible. Resting stages not known. Produce pink pigment. Chemoorganotrophic: respiratory. Acids produced aerobically from various carbohydrates. Strict aerobe. Urease not produced. Gelatin not liquefied. G + C 49–50 mole%.

Saccharomonospora Substrate mycelium may fragment. Single spores are produced on aerial hyphae; spores are not borne on substrate mycelium. The spores are either sessile on the branched hyphae or borne on short lateral sporophores. Spores are not heat-resistant. Produce green pigment. Cell wall contains meso-diaminopimelic acid, arabinose, and galactose.

Saccharopolyspora Substrate mycelium fragments. No spores produced on substrate mycelium. Bead-like long chains of spores produced on aerial mycelium. Cell wall contains arabinose, galactose, and meso-inositol.

Salmonella Rods, usually motile by peritrichous flagella. Can utilize citrate as carbon source. Gram negative. Indole negative. Hydrogen sulfide produced on TSI. Methyl red positive; Voges-Proskauer negative. Catalase positive; oxidase negative. Nitrate reduced. Optimal growth 37°C. G + C 50–53 mole%.

Saprospira Helical filaments, 0.8–1.5 by 10–500 μm; unbranched and unsheathed. Motile by gliding. Resting stages not known. Gram negative. Chemoorganotrophic: metabolism is respiratory, using molecular oxygen as terminal electron acceptor. Agar, cellulose, and chitin not attacked. Cell masses are colored, presumably because of carotenoid pigments. Indole not produced. Nitrates not reduced. Catalase negative. Strict aerobes. G + C 35.4–48.3 mole%.

Sarcina Nearly spherical cells, 1.8–3 μm in diameter, occurring in packets of eight or more. Division occurs in three perpendicular planes. Nonmotile. Gram positive. Chemoorganotrophic: strictly fermentative metabolism; carbohydrates are the fermentable substrates. The main products of glucose fermentation are carbon dioxide, hydrogen acetic acid, as well as ethanol for one species and butyric acid for another. Not pigmented. Catalase negative. Require complex growth factors. Strict anaerobes. G + C 28–31 mole%.

Scytonema Cyanobacterium. Photoautotrophic: can utilize water as electron donor with evolution of oxygen. Contains chlorophyll *a* and phycobiliproteins. Chlorophyll and carotenoids located in a system of thylakoids internal to the cytoplasmic membrane. Filamentous. A trichome (chain of cells) forms by intercalary division in one plane only. In the absence of combined nitrogen, heterocysts are produced; heterocysts are associated with the ability to fix atmospheric nitrogen. Akinetes may also be formed. Reproduction by random trichome breakage, by formation of hormogonia, and sometimes by germination of akinetes. Hormogonia are distinguishable from mature trichomes by the absence of heterocysts and by one or more of the following

characteristics: rapid gliding motility, smaller cell size, cell shape, and gas vacuolation. Hormogonia give rise to young filaments that bear a terminal heterocyst at only one end of the cellular chain. Mature trichome is composed of cells of even width. Heterocysts are predominantly intercalary. Vegetative cells are disk-shaped, isodiametric, or cylindrical. G + C 44 mole%.

Selenomonas Curved to helical rods, ends usually tapered and rounded to give kidney- to crescent-shaped short cells. Long cells and chains of cells are helical. Motile with active tumbling; up to 16 flagella present as a tuft, often near the center of the concave side. Capsules not formed. Resting stages not known. Gram negative. Chemoorganotrophic: metabolism fermentative, carbohydrates and sometimes certain amino acids and lactic acids are the fermentable substrates. Fermentation of glucose yields chiefly acetate, propionate, carbon dioxide, and/or lactate. Require organic growth factors. Strict anaerobes. Catalase negative. Optimal growth 35–40°C. G + C 53–61 mole%.

Seliberia Rod-shaped, spirally twisted cells, 0.5–0.7 by 1–12 μm, forming star-shaped figures or rosettes. Multiplication by transverse fission and budding. Resting stages not known. Chemoorganotrophic. Facultatively anaerobic. Optimal growth 24–25°C.

Serpens Cells rod-shaped, 0.3–0.4 by 8–12 μm, occurring singly. Spores not produced. Gram negative. Stationary growth phase cells are longer (16–25 μm), often with spherical protuberances or blebs. Extremely flexible and motile; rapid serpentine-like movement common on agar surface. Flagella occur in bipolar tufts of 4–10 flagella and, in an intermediate number, over the surface of the cell. Chemoorganotrophic: metabolism respiratory, oxygen serving as terminal electron acceptor. Carbon and energy sources mainly restricted to lactate. Aerobic-microaerophilic. Catalase positive; oxidase positive. Optimal growth 28–33°C. G + C 65.8 mole%.

Serratia Motile, peritrichously flagellated rods. Some strains are encapsulated. Citrate and acetate can be used as sole carbon source. Many strains produce pink, red, or magenta pigments. Glucose is fermented with or without production of a small volume of gas. Methyl red negative. Voges-Proskauer positive. DNase is produced. Gram negative. Facultatively anaerobic. Catalase positive; oxidase negative. G + C 53–59 mole%.

Shigella Nonmotile rods. Not encapsulated. Cannot use citrate or malonate as sole carbon source. Growth inhibited by potassium cyanide. Hydrogen sulfide is not produced. Glucose and other carbohydrates are fermented with the production of acid but not gas.

Gram negative. Generally catalase positive; oxidase negative. Nitrate reduced.

Simonsiella Cells 0.4–0.7 by 2–4 μm, closely apposed to form filaments with free faces of terminal cells rounded. Filaments are flat, not cylindrical. Gliding motility. Spores not produced. No capsules produced. Gram negative. Chemoorganotrophic. Require complex growth factors. Strict aerobes. Optimal growth 30–37°C.

Sphaerotilus Straight rods, 0.7–2.4 by 3–10 μm, occurring in chains, with a sheath of uniform width, which may be attached by means of a holdfast. Sheaths usually thin without incrustation by ferric and manganic oxides. Single cells motile by means of a bundle of subpolar flagella. Resting stages not known. Gram negative. Chemoorganotrophic: metabolism respiratory, never fermentative. Molecular oxygen is the universal electron acceptor. Aerobic but can grow at reduced oxygen tensions. Optimal growth 25–30°C. G + C 65.5–70.5 mole%.

Spirillospora Produce true mycelia. Hyphae 0.2–1 μm in diameter. Sporangia, 5–24 μm in diameter, spherical to vermiform. Spores develop from one or more coils of the sporangium, short to long rods to spiral in shape. Motile by 1–3 subpolar flagella. Optimal growth 25°C. G + C 72.9 mole%.

Spirillum Rigid, helical cells 0.25–1.7 μm in diameter, with less than one turn to many turns. Intracellular granules of poly-β-hydroxybutyrate. Motile by means of polar multitrichous flagella; generally exhibit bipolar flagellation. Gram negative. Chemoorganotrophic: strictly respiratory metabolism, with oxygen as terminal electron acceptor; a few species grow anaerobically, using nitrate as terminal electron acceptor. Oxidase positive; catalase positive. Many species produce yellowish-green, fluorescent-water-soluble pigments. Aerobic-microaerophilic. Optimal growth 30°C. G + C 38–65 mole%.

Spirochaeta Helical cells, 0.20–0.75 by 5–500 μm, axial fibrils. Motile, probably by means of the axial fibrils. Chemoorganotrophic: ferment carbohydrates. Obligate and facultative anaerobes. G + C 50–66 mole%.

Spiroplasma Cells pleomorphic, varying in shape from spherical to slightly ovoid, with a minimal diameter of about 100–250 μm, to helical branched nonhelical filaments. Cells lack a true cell wall and are bounded by a single triple layered membrane. An additional outer layer of short projections is present on the surface of the bounding membrane. Helical filaments are motile, showing two types of motility; a rapid rotary motion and a slow undulation. Gram positive. Colonies on solid medium show typical biphasic fried egg appearance. Chemoorgan-

otrophic: acid produced from glucose and mannose. Cholesterol and perhaps other sterols required for growth. Facultative anaerobes, maximal growth occurs at 5% CO_2. Optimal growth 32°C. G + C 25.0–26.35 mole%.

Spirosoma Rods, 0.5–1.0 by 1.5–50 μm, with various degrees of curvature, often resulting in rings, coils, and undulating filaments. Outer diameter of ring 1.5–3.0 μm. Gram negative. Resting stages not known. Produce pale yellow pigment. Chemoorganotrophic: respiratory. Acids produced from carbohydrates. G + C 51.0–52.9 mole%.

Spirulina Cyanobacterium. Photoautotrophic: can utilize water as electron donor with evolution of oxygen. Contains chlorophyll *a* and phycobiliproteins. Chlorophyll and carotenoids located in a system of thylakoids internal to the cytoplasmic membrane. Filamentous. Growth by intercalary division in one plane results in formation of a trichome. Trichome helical. The shape of the cells that compose the multiple trichome is not fixed. There is little or no constriction between cells. Reproduction occurs probably by transcellular trichome breakage. Sheath formation not pronounced. G + C 44–53 mole%.

Sporichthya Hyphae, 0.5–1.2 μm in diameter, short (10–25 μm), branched; form only aerial mycelia attached to the surface of solid media by means of holdfasts which originate from the wall of the hyphal base. The aerial hyphae divide into smooth-walled spores. Spores become motile when exposed to water. The cell wall contains diaminopimelic acid. Gram variable. Heterotrophic: acids generally not detectable. Starch hydrolyzed. Casein hydrolyzed. Gelatin not liquefied. Facultative anaerobes. Optimal growth 28–37°C.

Sporocytophaga Flexible rods with rounded ends, 0.3–0.5 by 5–8 μm, occurring singly. A resting stage, the microcyst, is formed. Motile by gliding. Gram negative. Chemoorganotrophic: metabolism is respiratory, using molecular oxygen as terminal electron acceptor. Cellobiose, cellulose, glucose, and—for some strains—mannose are the only known sources of carbon and energy. Catalase positive. Strict aerobes. Optimal growth 30°C. G + C 36 mole%.

Sporolactobacillus Cells straight rods, 0.7–0.8 by 3–5 μm, occurring singly, in pairs, and rarely in short chains. Motile by means of a number of long peritrichous flagella. Endospores formed. Gram positive. Chemoorganotrophic: metabolism fermentative. Carbohydrates metabolized by homolactic fermentation. Catalase negative; lack cytochromes. Microaerophilic. Optimal growth 30–35°C.

Sporosarcina Cells spherical, occurring in regular tetrads or packets. Motility variable. Endospores formed. Gram positive. Chemoorganotrophic: energy-yielding metabolism strictly respiratory, oxygen acting as terminal electron acceptor. Strict aerobes.

Staphylococcus Cells spherical, 0.5–1.5 μm in diameter, occurring singly, in pairs, and characteristically dividing in more than one plane to form irregular clusters. Nonmotile. No resting stages known. Gram positive. Cell walls contain peptidoglycan and associated teichoic acids. Chemoorganotrophic: metabolism respiratory and fermentative. Catalase positive. A wide range of carbohydrates may be utilized with the production of acid. Facultative anaerobes. Optimal growth 35–40°C. G + C 30–40 mole%.

Stibiobacter Cells, rod-shaped, 0.5 by 0.6–1.8 μm. Gram positive. Motile with a single flagellum. Facultatively chemolithotrophic: oxidize antimony in form of SbO_3 to its highest state of valency; also can grow on certain organic substrates. Obligate aerobes. Optimal growth 28–30°C.

Stigmatella Vegetative cells, rods with tapered ends. Form fruiting bodies. Sporangia borne singly or in clusters of stalked fruiting bodies. Myxospores short, rigid, phase-dense rods surrounded by a slime capsule. Vegetative colonies do not etch, erode, or penetrate agar medium. Congo red is adsorbed. Urea usually hydrolyzed. Aerobic. Optimal growth 30°C. G + C 68.5–78.7 mole%.

Stomatococcus Cells spherical, 0.9–1.3 μm, occurring in clusters and less frequently as pairs. Gram positive. Encapsulated. No spores produced. Nonmotile. Chemoorganotrophic: various sugars used as carbon sources. Facultative anaerobes. Catalase variable. Optimal growth 30–37°C. G + C 56–60.4 mole%.

Streptobacillus Rods, 0.3–0.7 by 1–5 μm long, with rounded or pointed ends, frequently in chains and filaments, 0.5–0.9 by 10–150 μm long. Rods may show central thickening; filaments often have a series of oval to elongated bulbous swellings. True branching does not occur. Nonencapsulated. Nonmotile. No resting stages known. Gram negative. Not acidfast. Chemoorganotrophic: metabolism fermentative. Facultative anaerobes. Optimal growth 35–37°C. G + C 24–26 mole%.

Streptococcus Cells spherical to ovoid, less than 2 μm in diameter, occurring in pairs or chains. Gram positive. Chemoorganotrophic: metabolism fermentative; predominant end product of glucose fermentation is lactic acid. Catalase negative. Facultative anaerobes. Minimal nutritional requirements generally complex. Optimal growth ca. 37°C. G + C 33–42 mole%.

Streptomyces Produce true mycelia; slender hyphae, 0.5–2.0 μm diameter. The aerial mycelium at maturity forms chains of three to many spores, 0.5–2.0 μm in diameter. Vegetative hyphae develop branched mycelium that does not fragment readily. Mycelium sparingly septate. Reproduction by germination of the aerial spores, sometimes by growth of fragments of the vegetative mycelium. Gram positive. Produce a wide variety of pigments. Heterotrophs: highly oxidative, acids not generally detectable as fermentation products. Generally hydrolyze gelatin, casein, and/or starch. Nitrates generally reduced. Aerobes. Optimal growth 25–35°C. G + C 69–73 mole%.

Streptosporangium Hyphae highly branched, sparingly septate. Sporangia spherical to ovoid, 7–48 μm. Sporangiospores arranged in a single coil within the sporangium, spherical to ovoid, nonmotile. Strict aerobes. G + C 69.5–70.6 mole%.

Streptoverticillium Substrate mycelium, 1–2 μm in diameter, exhibits branching. The aerial mycelium characteristically has 3–6 short branches that are produced in a whorl (verticil) at more or less regular intervals along the aerial mycelium, giving a barbed wire appearance. These branch in turn to produce 2–12 or more secondary branches; chains of rounded to oblong spores form at the ends of the secondary branches. Gram positive. Produce a variety of pigments. Aerobic. Optimal growth ca. 25–35°C. G + C 69–73 mole%.

Succinimonas Straight rods with rounded ends, usually short to coccoid. Motile, with monotrichous polar flagella. No spores produced. Gram negative. Chemoorganotrophic: metabolism fermentative, carbohydrates are the fermentable substrates. Large amounts of succinate and some acetate are the major products of glucose fermentation. No butyrate or gases formed. Strict anaerobes. Catalase negative.

Succinivibrio Curved rods with pointed ends. Cells are helical with less than one to three or more coils per cell and may become straight to slightly curved with pointed ends. Rapid progressive vibrating motility with monotrichous and polar flagellation. No resting stages known. Gram negative. Chemoorganotrophic: metabolism fermentative; carbohydrates are the fermentable substrates. Major products from glucose fermentation are succinate, acetate, formate, and sometimes lactate. Butyrate or hydrogen are not formed. Strict anaerobes.

Sulfolobus Spherical cells with lobes, 0.8–1 μm. Nonmotile; no flagella produced. Endospores not produced. Gram negative. Cell wall does not contain peptidoglycan. Facultatively chemolithotrophic: use elemental sulfur as energy source; can also use glutamate, ribose, and yeast extract as source of carbon and energy. Aerobic. Optimal growth 70–75°C. pH optimum 2–3. G + C 60–68 mole%.

Symbiotes Pleomorphic *Rickettsia*-like cells, living chiefly intracellularly in specialized paired organs, called mycetomes, or bedbugs. Intracellular forms coccoid or diplococcoid. Extracellular bacillary or thread forms, migrate to other tissues in maturing hosts. Biochemical, nutritional, and physiological characteristics unknown.

Synechococcus Cyanobacterium. Photoautotrophic: can utilize water as electron donor with evolution of oxygen. Contains chlorophyll *a* and phycobiliproteins. Chlorophyll and carotenoids located in a system of thylakoids internal to the cytoplasmic membrane. Cells rod-shaped, occurring singly or forming colonial aggregates held together by additional outer wall layers. Reproduction by binary fission. Division in one plane only. Sheath absent. Group I G + C 39–43 mole%; group II G + C 47–56 mole%; group III G + C 66–71 mole%.

Synechocystis Cyanobacterium. Photoautotrophic: can utilize water as electron donor with evolution of oxygen. Contains chlorophyll *a* and phycobiliproteins. Chlorophyll and carotenoids located in a system of thylakoids internal to the cytoplasmic membrane. Cells coccoid, occurring singly or forming colonial aggregates held together by additional outer wall layers. Reproduction by binary fission. Division in two or three planes. Sheath absent. Group I G + C 35–37 mole%; group II G + C 42–48 mole%.

Tatlockia Gram negative rods resembling *Legionella* in cultural characteristics. Colony blue-gray on medium containing bromcresol purple. Colony does not fluoresce. Catalase positive; oxidase weakly positive. Starch not hydrolyzed. Gelatin not hydrolyzed. G + C 43.6–45.0 mole%.

Thermoactinomyces Formation of single spores on both the aerial and substrate mycelium. Occasionally longitudinal pairs of spores are formed on the aerial mycelium. Growth occurs between 45 and 60°C. Gram variable. Aerobic.

Thermomicrobium Pleomorphic rods, often appearing as pairs. Gram negative. Produce pink pigment. Obligate aerobes. Chemoorganotrophic: grow well on sucrose or glycerol with glutamate. Restricted carbon metabolism compared to *Thermus*. Optimal growth 70–75°C. G + C 64.3 mole%.

Thermomonospora Single spores form only on aerial mycelium. Vegetative hyphae flexuous and nonseptate. Aerial mycelium white. Single spores are usually formed at the tip of the sporophores. Gram variable. Not acid-fast. Facultative aerobes. Thermophilic; optimal growth 45–55°C.

Thermoplasma Cells pleomorphic, varying in shape from spherical (0.3–2 μm) to filamentous structures. Reproduction apparently by budding. Lack a true cell wall; bounded by a single triple-layered membrane. No resting stages known. Colonies have fried egg appearance. Saprophytic: grows on basal salts medium containing yeast extract. Does not require sterols for growth. Strict aerobes. Optimal growth 59°C. Optimum pH 1–2. Contains a histone-like protein in association with its DNA. G + C ca. 46 mole%.

Thermothrix Cells, 0.5–1.0 by 3–20 μm, forming chains of cells up to 1 cm in length. The filaments do not possess a sheath. Gram negative. Sulfur is deposited extracellularly. Facultatively chemolithotrophic: oxidizes reduced sulfur compounds using nitrate as terminal electron acceptor; also grows heterotrophically, using carbohydrates both aerobically and anaerobically. Optimal growth 70–73°C. Facultative anaerobes.

Thermus Rods and filaments, 0.5–0.8 by 5–200 μm. Large spheres, 10–20 μm in diameter, form in older cultures. Gram negative. Nonmotile; no flagella. Endospores not produced. Generally produce yellow-orange pigments. Chemoorganotrophic: grows on amino acids, sugars, organic acids. Obligately aerobic. Optimal growth 70–72°C. G + C 64–67 mole%.

Thiobacillus Small rod-shaped cells. Motile by means of a single polar flagellum, but two species are nonmotile. No resting stages known. Gram negative. Chemolithotrophic: energy derived from the oxidation of one or more reduced or partially reduced sulfur compounds; final oxidation product is sulfate. Obligate aerobes. Optimal growth 28–30°C. G + C 50–68 mole%.

Thiocapsa Spherical to slightly ovoid, multiplying by binary fission. Nonmotile. Tetrads may be formed. Under certain conditions may grow in more or less regular aggregates surrounded by slime. Gram negative. Contain bacteriochlorophyll *a* or *b* and carotenoids, both types of pigments being located on internal membranes of vesicular or tubular type. Do not contain gas vacuoles. Anaerobic phototrophs. Capable of photosynthesis in the presence of hydrogen sulfide, during which they produce and store elemental sulfur in the form of globules inside the cells. Molecular oxygen is not involved during photosynthesis. Cell suspensions appear various shades of orange-brown to pink and purple-red because of the photosynthetic pigments. G + C 63.3–69.9 mole%.

Thiocystis Spherical to ovoid bacteria, before cell division typically diplococcus-shaped, multiplying by binary fission, motile by means of polar flagella. Under certain conditions cells may be nonmotile and grow in more or less regular aggregates surrounded by slime. Gram negative. Contain bacteriochlorophyll *a* and carotenoids, both pigments being located at internal membranes of vesicular type. Do not contain gas vacuoles. Anaerobic. Capable of photosynthesis in the presence of hydrogen sulfide, during which they produce and store elemental sulfur in the form of globules inside the cells. Molecular oxygen is not evolved during photosynthesis. Cell suspensions appear various shades of brown to purple-red and violet because of the photosynthetic pigments. Catalase positive. G + C 61.3–67.9 mole%.

Thiodictyon   Cells rod-shaped with rounded ends, sometimes appearing spindle-shaped, multiplying by binary fission, nonmotile under all conditions. May form aggregates in which the cells become arranged end to end in an irregular net-like structure. Gram negative. Contain bacteriochlorophyll *a* and carotenoids, both pigments being located at internal membrane of vesicular type. Contain gas vacuoles in the central part of the cell. Anaerobic. Capable of photosynthesis in the presence of hydrogen sulfide, during which they produce and store elemental sulfur in the form of globules in the gas vacuole-free peripheral part of the cells. Molecular oxygen is not evolved during photosynthesis. Cell suspensions appear in various shades of purple-violet due to the photosynthetic pigments. G + C 65.3–66.3 mole%.

Thiomicrospira   Cells curved rods or spirals, 0.2–0.3 by 2–6 μm. Obligately chemolithotrophic: oxidize reduced sulfur compounds as source of energy; some strains exhibit denitrification, using nitrate as terminal electron acceptor. G + C 36–44 mole%.

Thiopedia   Cells spherical to ovoid, multiplying by binary fission. Cells arranged in flat sheets with typical tetrads as the structural units. Nonmotile. Gram negative. Contain bacteriochlorophyll *a* and carotenoids, both pigments being located at internal membranes of vesicular or stack-lamellar type. Contain gas vacuoles in the central part of the cell. Anaerobic. Capable of photosynthesis in the presence of hydrogen sulfide, during which they produce and store elemental sulfur in the form of globules in the gas vacuole-free peripheral part of the cells. Molecular oxygen is not evolved during photosynthesis. Cell suspensions appear pink to purple due to photosynthetic pigments.

Thioploca   Flexible filaments made up of numerous segments, generally with sulfur inclusions; occur in parallel or braided fascicles, enclosed by a common sheath. Individual filaments show independent gliding motion.

Thiospira   Cells spiral-shaped, usually with pointed ends, with sulfur inclusions. Motile by mono- or polytrichous polar flagella. No resting stages known. Not cultivated in pure culture.

Thiospirillum   Cells spiral-shaped, dividing by binary fission. Motile by means of polar flagella. Under certain conditions cells may be nonmotile and grow in regular aggregates surrounded by slime. Gram negative. Contain bacteriochlorophyll *a* and carotenoids, both pigments being located at internal membranes of vesicular type. Do not contain gas vacuoles. Anaerobic. Capable of photosynthesis in the presence of hydrogen sulfide, during which they produce and store elemental sulfur in the form of globules inside the cells. Cell suspensions appear various shades of orange-brown to purple-red because of photosynthetic pigments. Molecular oxygen not evolved during photosynthesis. G + C 45.5 mole%.

Thiothrix   Long filaments composed of short cylindrical or ovoid cells; filaments colorless and unbranched. Filaments usually attached to solid substrates by means of an inconspicuous holdfast; stalks absent. Filaments do not glide. Deposit sulfur granules inside the cells. Cells form rosettes. Strictly aerobic.

Thiovulum   Cells round to ovoid, 5–25 μm in diameter. Cytoplasm normally contains sulfur inclusions. Motile by peritrichous flagella, forward movement accompanied by rotation around the long axis. Cells characterized by presence of one polar fibrillar organelle. No resting stages known. Multiplication by constriction followed by fission. Gram negative. Catalase negative. Microaerophilic. Chemolithotrophic: obtain energy by oxidation of reduced sulfur compounds.

Toxothrix   Cells cylindrical, colorless, 0.5–0.75 by 3–6 μm, in filaments up to 400 μm long. Filaments often U-shaped and rotate while slowly moving forward. A mucoid substance, excreted from several sites on the trailing ends, is deposited as a double track. Oxidized iron may be deposited onto the mucoid threads, rendering them yellowish brown. Not grown in pure culture.

Treponema   Unicellular, helical rods 5–20 μm long, with tight regular or irregular spirals. Cells have one or more axial fibrils inserted at each end of the protoplasmic cylinder. Motile. Gram negative. Chemoorganotrophic: metabolism fermentative, using amino acids and/or carbohydrates. Catalase negative. Oxidase negative. Urease negative. Voges-Proskauer negative. Nitrate not reduced. Strict anaerobes. G + C 32–50 mole%.

Ureaplasma   Cells coccoid. 0.3 μm diameter, occurring singly and in pairs. Lack a true cell wall; cells bounded by a triple-layered membrane. Colonies exhibit cauliflower head or fried egg appearance. Urea hydrolyzed with simultaneous production of carbon dioxide and ammonia. Catalase negative. Glucose fermented aerobically and anaerobically. Optimal growth 35–37°C. G + C 26.9–29.8 mole%.

Vampirovibrio Cells are small curved or straight rods. Gram negative. Similar to *Bdellovibrio* but parasitic for *Chlorella*.

Veillonella Small cocci, 0.3–0.5 μm in diameter. Nonmotile. Spores not produced. Gram negative. Carbohydrates and polyalcohols not fermented. Produce acetate, propionate, CO_2, and H_2 from lactate. Oxidase negative. Chemoorganotrophic: complex nutritional requirements. Carbon dioxide required. Anaerobic. Optimal growth 30–37°C. G + C 40–44 mole%.

Vibrio Short rods, axis curved or straight, 0.5 by 1.5–3.0 μm, occurring singly or occasionally united into spiral shapes. Motile by a single polar flagellum, or in some species two or more flagella in one polar tuft. Gram negative. Not acid-fast. Endospores not produced. Nonencapsulated. Chemoorganotrophic: metabolism both respiratory and fermentative. Fermentation of carbohydrates produces mixed products but no CO_2 or H_2. Oxidase positive. Nonpigmented or yellow. Frequently Voges-Proskauer positive. Nitrates reduced to nitrites. Urease negative. Facultative anaerobes. G + C 40–50 mole%.

Vitreoscilla Colorless filaments, 1.2–2 by 3–70 μm, composed of cylindrical or barrel-shaped cells. Reproduction by fragmentation. Resting stages not known. Motile by gliding. Gram negative. Chemoorganotrophic: metabolism presumably respiratory. Complex nutritional requirements. Intracellular sulfur granules not formed from hydrogen sulfide. G + C 43.6 mole%.

Wolbachia Heterogeneous group of small coccoid forms, rods, and spheres which resemble *Rickettsia*. Gram negative. Associated with anthropods, seldom pathogenic for their hosts and seldom develop within mycetomata.

Wolinella Helical, curved, or straight unbranched cells 0.5–1 by 2–6 μm, having rounded or tapered ends. Gram negative. Rapid, darting motility by means of a single polar flagellum. Endospores not produced. Carbohydrates are not fermented and do not support growth. Growth stimulated by formate and fumarate. Hydrogen and formate are utilized as energy sources. Nitrates but not sulfates reduced. Anaerobic, but some strains grow under microaerophilic conditions. Oxidase negative; catalase negative. G + C 42–46 mole%.

Xanthobacter Small, rod-shaped cells, 0.4–1.0 by 0.8–6.0 μm, producing pleomorphic and coccoid cells under some conditions. Cells accumulate poly-β-hydroxybutyrate. Nonmotile. Resting stages not known. Gram negative but may appear Gram variable. Produce yellow pigment. Fix atmospheric nitrogen. Facultatively chemolithotrophic: oxidize hydrogen and fix carbon dioxide; also grow on various alcohols and amino acids, carbohydrate utilization limited. Obligate aerobes. Catalase positive. Optimal growth 25–30°C. G + C 69–70 mole%.

Xanthomonas Cells single, straight rods, 0.2–0.8 by 0.6–2.0 μm. Motile by means of a polar flagellum. Do not produce sheaths. No resting stages known. Gram negative. Generally produce yellow pigments. Chemoorganotrophic: metabolism respiratory, never fermentative. Oxidase negative or weakly positive. Catalase positive. Nitrates not reduced. Complex growth factor requirements. Strict aerobes. Optimal growth 25–27°C. G + C 63.5–69.2 mole%.

Xenococcus Cyanobacterium. Photoautotrophic: can utilize water as electron donor with evolution of oxygen. Contains chlorophyll *a* and phycobiliproteins. Chlorophyll and carotenoids located in a system of thylakoids internal to the cytoplasmic membrane. Cells rounded, occurring singly or forming colonial aggregates held together by additional outer wall layers. Reproduction by multiple fission that leads to the formation of nonmotile baeocytes with a fibrous cell wall layer at the moment of their release. Groups G + C 44 mole%.

Xenorhabdus Cells large rods, 0.8–2.0 by 4.0–10.0 μm. Motile with peritrichous flagella. Endospores not produced. Gram negative. Luminescent. Associated with terrestrial invertebrates. Pigmented. Chemoorganotrophic: respiratory and fermentative. Acid with no gas produced from glucose. Oxidase negative. Methyl red negative. Voges-Proskauer negative. Chitinase negative. Growth inhibited by 3% NaCl. G + C 43.0–44.0 mole%.

Yersinia Cells ovoid or rods, 0.5–1.0 by 1–2 μm. Nonmotile or motile with peritrichous flagella. Nonencapsulated. Various carbohydrates fermented without production of gas. Methyl red positive. Voges-Proskauer negative. Gelatin not hydrolzyed. Nitrates reduced, except by one species. Optimal growth 30–37°C. G + C 45.8–46.8 mole%.

Zoogloea Cells rod-shaped, 0.5–1.0 by 1–3 μm, becoming pleomorphic with age. No spores or cysts produced. Young cells actively motile with polar monotrichous flagella. Naturally aggregate in macroscopic flocs. Pleomorphic with age. Old cells contain poly-β-hydroxybutyrate. Gram negative. Chemoorganotrophic: metabolism respiratory, never fermentative. Molecular oxygen is terminal electron acceptor. Not proteolytic. Strict aerobes. Optimal growth 28–30°C.

Zymomonas   Plump rod-shaped cells with rounded ends, occurring singly or in pairs. Motile by means of lophotrichous flagella. No resting stages known. Gram negative. Chemoorganotrophic: fermentative. Produce ethanol and CO_2 by Entner-Doudoroff pathway. Catalase positive. Produce hydrogen sulfide. Gelatin not liquefied. Indole not produced. Nitrate not reduced. Methyl red negative. Anaerobic but able to tolerate some oxygen. Optimal growth 30°C. G + C 47–48 mole%.

Index

I-7

Index

LINACRE COLLEGE LIBRARY
UNIVERSITY OF OXFORD

I-28

Index

The Runaway Woman

Also by Josephine Cox